BUILDING TECHNOLOGY

BUILDING TECHNOLOGY
Mechanical and Electrical Systems
Second Edition

BENJAMIN STEIN

Consulting Engineer

JOHN WILEY & SONS, INC. New York • Chichester • Brisbane • Toronto • Singapore • Weinheim

DISCLAIMER

The information in this book has been derived and extracted from a multitude of sources, including building codes, fire codes, industry codes and standards, manufacturer's literature, engineering reference works and personal professional experience. It is presented in good faith, but, although the author and the publisher have made every reasonable effort to make the information presented accurate and authoritative, they do not warrant, and assume no liability for, its accuracy or completeness or its fitness for any specific purpose. It is primarily intended as a learning and teaching aid and not as a source of information for the actual final design of building systems. It is the responsibility of users to apply their professional knowledge in the use of the information presented in this book and to consult original sources for current detailed information as needed, in actual design situations.—*Benjamin Stein P.E.*

This text is printed on acid-free paper.

Copyright © 1997 by John Wiley & Sons, Inc.

All rights reserved. Published simultaneously in Canada.

Reproduction or translation of any part of this work beyond that permitted by Section 107 or 108 of the 1976 United States Copyright Act without the permission of the copyright owner is unlawful. Requests for permission or further information should be addressed to the Permissions Department, John Wiley & Sons, Inc., 605 Third Avenue, New York, NY 10158-0012.

This publication is designed to provide accurate and authoritative information in regard to the subject matter covered. It is sold with the understanding that the publisher is not engaged in rendering legal, accounting, or other professional services. If legal advice or other expert assistance is required, the services of a competent professional person should be sought.

Library of Congress Cataloging in Publication Data:
Stein, Benjamin.
    Building technology : mechanical & electrical systems / Benjamin Stein.—2nd ed.
       p. cm.
    Rev. ed. of: Building technology / William J. McGuinness, Benjamin Stein. c1977.
    Includes bibliographical references and index.
    ISBN 0-471-59319-2 (cloth : alk. paper)
    1. Buildings—Mechanical equipment. 2. Buildings—Electric equipment. 3. Buildings—Environmental engineering.
    I. McGuinness, William J. Building technology. II. Title.
TH6010.M24   1996
696—dc20   96-6186

Printed in the United States of America

10 9 8 7 6 5 4 3 2 1

Preface

This book was written to provide the knowledge required for the job position of technologist in the building industry. This job position also carries the title of draftsman, junior designer, field service engineer, inspector and technician, depending on the company, the project and the type of work involved.

The building process generally can be divided into three phases: the planning phase, the construction phase, and the maintenance and field service phase after beneficial occupancy of the facility. The technologist will find opportunities to apply his or her knowledge in all three phases.

The planning phase of a building project involves the architect and the engineering consultants. The function of the architect is programming and space planning. The architect confers with the various project engineers with respect to mechanical and electrical systems in order to determine the type of systems that will best suit a facility. At this stage of planning, the technologist generally is not involved. Once the overall system planning is completed, the project mechanical (HVAC), electrical and plumbing engineers begin detailed design. It is at this stage that the technologist is called upon.

The usual arrangement for a project at the design stage, in an engineering office, involves the following technical positions:

Project engineer.
Design engineers (designers).
Technologists (junior designers, draftsmen).

In addition, specialists such as lighting designers, computer personnel and energy specialists may be involved, depending on the type and size of the job. The project engineer coordinates with the architect and the various design specialties. The design engineer(s) perform the actual system design and calculations, with the assistance of one or more technologists. The technologist is responsible for producing the working drawings; that is, he or she is primarily concerned with the "nuts and bolts" of the job. The working drawings are contract documents that, along with the job specifications, tell the contractor, in detail, what the job requirements are. Their preparation involves a thorough knowledge of basic principles, design procedures, equipment and hardware, installation procedures and the techniques of drawing preparation

Working drawings were, until recently, generally produced by a draftsman or junior designer who sat at a drafting table and physically drew them, usually with pencil on paper but also sometimes with ink on vellum or cloth. The advent of computers, CAD (computer aided design) programs and plotters has, in large measure, replaced hand drawing, although manual preparation of working drawings is still found occasionally in design offices and frequently in field construction offices. For the most part, however, the job title of draftsman is no longer applicable; technologist or junior designer is more appropriate. Although this person may now "draw" at a computer console, the knowledge required for the job is the same as previously required.

Once the planning is complete and contracts let, the construction phase of building begins. At this stage, technologists in the engineer's office handle design changes (of which there are always some), shop drawings (manufacturer's detailed equipment drawings), some material inspection and a considerable portion of the field construction inspection work, including solving coordination and space allocation problems. Technologists in the contractor's office prepare large-scale, detailed, coordinated installation drawings, for the use of the trade construction personnel. Much of this work is still hand drawn. They also handle equipment shop drawings and sometimes are involved with material acquisition. Technologists on the contractor's staff may work in the contractor's field office on large jobs, doing many types of field work and coordination work with equipment suppliers.

Technologists working for equipment manufacturers prepare detailed equipment drawings and provide the necessary coordination with the engineer and contractor. When construction is complete, technologists from the engineer's and contractor's offices will assist in the very extensive job of field testing, adjusting and balancing of mechanical and electrical systems.

After a building is turned over to the owner for "beneficial occupancy," it becomes the owner's responsibility to maintain the building's systems. With today's complex M/E systems, this is not a simple task. Often the owner hires a service company to perform this function, which, in turn, hires field service engineers and technicians who have a construction and/or manufacturing background. Here again, the M/E technologist can find constructive use for his or her knowledge and capabilities.

From the foregoing it should be evident that the building industry has a continuing need for competent mechanical and electrical technologists. It is to provide the necessary technical background for such persons that this book is intended.

Since the publication of the first edition of this book 20 years ago, the building industry has undergone many changes. Design must meet the ever more demanding requirements of energy codes, laws delineating the needs of handicapped persons, environmental impact standards and improved human comfort criteria. Also, modern buildings not only must meet the complex technological needs of its occupants but also must be designed for extreme flexibility in order to meet the rapid technical changes that typify our society.

In the construction phase, new materials and installation techniques have been developed that are less labor-intensive yet provide better serviceability and longer life. No field has developed more rapidly than that of controls and automation, which make life easier for the user but much more complicated for the engineering staff.

All of these developments have been considered in the preparation of this completely revised second edition. In addition, the underlying (and unchanging) principles upon which all technical design is based have been restated, refreshed, and re-presented in what I trust is an interesting and informative fashion.

Benjamin Stein

Acknowledgments

No modern book is ever the exclusive product of one person's labor. I am deeply indebted to all those who have given aid and advice during the years of preparing this text. First and foremost, however, I wish to acknowledge and thank my wife Lila Stein, who strongly encouraged me to prepare this revision and who took upon herself the demanding task of corresponding with manufacturers whose products are illustrated, plus the burden of all phases of manuscript typing and preparation. Without her diligent and excellent work, this revision would not have been possible.

Dr. Hamilton A. Chase provided me with some of the source material for the HVAC and plumbing sections of this book, plus astute comments on their use. Mr. Shlomo Chen of Tel Aviv, Israel, provided valuable suggestions at the outset of this project. Mr. Arthur Fox's excellent review comments were much appreciated. All of the manufacturers whose products I have used to illustrate the various sections are credited in the illustrations. Organizations whose data appear in this volume are similarly credited.

Every author is indebted to the publisher's staff who help throughout the lengthy preparation process. The original suggestion for a second edition came from Mr. Everett Smethurst of John Wiley's staff, who encouraged me to undertake rewriting the first edition, despite the fact that my coauthor and friend, Professor William J. McGuinness, had died some years previously. Amanda Miller, who replaced Everett Smethurst, has offered assistance throughout, for which I am indebted. I am also indebted to the staff at John Wiley for their competent, and diligent work: to MaryAlice Yates in the administrative area, to Dean Gonzalez for his experienced and highly professional handling of more than 1000 illustrations, and to Donna Conte and Diana Cisek for their editorial and production work.

Contents

Preface v
Acknowledgments vii

1 Heat and Human Comfort   1

1.1   Human Comfort Indoors   2
1.2   Body Temperature Control   2
1.3   Heat and Temperature   3
1.4   Sensible Heat and Latent Heat   5
1.5   Properties of Atmospheric Air   6
1.6   Body Heat Balance   8
1.7   Basic Rules of Thermodynamics   8
1.8   Mechanics of Sensible Heat Transfer   9
1.9   Thermal Comfort Criteria   15
1.10 Measurements   15
1.11 Units and Conversions   16
      Key Terms   18
      Supplementary Reading   18
      Problems   19

2 Thermal Balance of Buildings   21

HEAT TRANSFER   22

2.1   Heat Transfer in Buildings   22
2.2   Conduction   22
2.3   Convection   28
2.4   Radiation   29
2.5   Heat Flow Through Air Spaces   31
2.6   Heat Flow Through Built-up Sections   32
2.7   Heat Loss Through Surfaces on and Below Grade   34
2.8   Heat Flow Through Windows and Doors   34

HEATING   41

2.9   Heat Loss from Air Infiltration and Ventilation   41

2.10	Heat Loss to Adjacent Unheated Spaces	42
2.11	Summary of Building Heat Losses	42
2.12	Building Heat Loss Calculation Procedure	43
2.13	Outside Design Conditions	50
2.14	Evolution of Modern Climate Control Systems	51
2.15	Heating Systems Fuels	52
2.16	Heating Systems	54
	COOLING	**54**
2.17	Building Heat Gain Components	54
2.18	Residential Heat Gain (Cooling Load) Calculations	58
2.19	Nonresidential Heat Gain Calculations	63
2.20	Cooling Load Calculation Forms	64
	PSYCHROMETRICS	**64**
2.21	The Psychrometric Chart	64
2.22	Components of the Psychrometric Chart	70
2.23	Basic HVAC Processes on the Psychrometric Chart	73
2.24	Air Equations for HVAC Processes	78
	Key Terms	79
	Supplementary Reading	79
	Problems	79

3 Hydronic Heating 83

3.1	Hydronic Systems	84
3.2	Components of a Hydronic Heating System	84
3.3	Hot Water Boilers	86
3.4	Definitions and Basic Equipment Functions	92
3.5	Expansion Tank	95
3.6	Circulator	98
3.7	Terminal Units	100
3.8	Piping Arrangements	114
3.9	Hydronic Heating System Control	120
3.10	Piping Design Factors	125
3.11	Design Procedure	127
3.12	Basic Residential Hydronic Heating System Design	128
3.13	Preliminary Design Considerations	138
3.14	Hydronic Heating Design of a Large Residence	142
3.15	Hydronic Heating Design of an Industrial Building	151
3.16	Miscellaneous Design Considerations	164
	Key Terms	165
	Supplementary Reading	166
	Problems	166

4 Electric Heating 167

4.1	Electric Resistance Heating	168
4.2	Resistance Heaters	168
4.3	Resistance Heating Equipment	169
4.4	Residential Electric Heating Design	179
4.5	Nonresidential Electric Heating Design	184
	Key Terms	195
	Problems	195

5 Air Systems, Heating and Cooling, Part I 197

	ALL-AIR SYSTEMS	**198**
5.1	Air System Characteristics	198
5.2	Components of All-Air Systems	198
5.3	Heat-Carrying Capacity of Ducted Air Flow	200
5.4	Air Pressure in Ducts	201
	WARM AIR FURNACES	**203**
5.5	Furnace Components and Arrangements	203
5.6	Furnace Energy and Efficiency Considerations	204
5.7	Furnace Types	208
5.8	Warm Air Furnace Characteristics	222
5.9	Noise Considerations	224
5.10	Warm Air Furnace Blowers	224
5.11	Furnace Accessories	226
	AIR DISTRIBUTION SYSTEM	**227**
5.12	Ducts	227
5.13	Duct Fittings and Air Control Devices	229
5.14	Duct Insulation	229
5.15	Air Distribution Outlets	235
5.16	Outlet Characteristics	253
5.17	Outlet Selection, Location and Application	258
5.18	Outlet Air Patterns	262

ALL-AIR SYSTEMS 265
- 5.19 System Types 266
- 5.20 Single-Zone System Duct Arrangements 270

AIR FRICTION IN DUCT SYSTEMS 274
- 5.21 Air Friction in Straight Duct Sections 275
- 5.22 Noncircular Ducts 282
- 5.23 Air Friction in Duct Fittings 288
- 5.24 Sources of Duct System Pressure Loss 294

DUCT SIZING METHODS 294
- 5.25 Equal Friction Method 295
- 5.26 Modified Equal Friction Method 298
- 5.27 Extended Plenum Method 299
- 5.28 Semi-Extended Plenum (Reducing Plenum) Method 303

SYSTEM DESIGN 307
- 5.29 Design Procedure 307
- 5.30 Designer's Checklist 309
- Key Terms 311
- Supplementary Reading 312
- Problems 312

6 Air Systems, Heating and Cooling, Part II 315

REFRIGERATION 315
- 6.1 Unit of Mechanical Refrigeration 316
- 6.2 Cooling by Evaporation 316
- 6.3 Refrigeration Using a Closed Vapor Compression System 317

HEAT PUMPS 321
- 6.4 Basic Heat Pump Theory 321
- 6.5 Heat Pump Performance (Heating Mode) 323
- 6.6 Equipment Sizing Considerations 326
- 6.7 Heat Pump Configurations 328
- 6.8 Physical Arrangement of Equipment, Small Systems 329
- 6.9 Physical Arrangements of Equipment, Large Systems 353

REFRIGERATION EQUIPMENT 354
- 6.10 Condensers 354
- 6.11 Evaporators 355
- 6.12 Compressors 357
- 6.13 Air-Handling Equipment 357

DESIGN 357
- 6.14 Advanced Design Considerations 357
- 6.15 Design Example—The Basic House 359
- 6.16 Design Example—Mogensen House 362
- 6.17 Design Example—Light Industry Building 369
- Key Terms 373
- Supplementary Reading 373
- Problems 374

7 Testing, Adjusting and Balancing (TAB) 377

INSTRUMENTATION 378
- 7.1 Temperature Measurement 378
- 7.2 Pressure Measurement 382
- 7.3 Air Velocity Measurements 382

MEASUREMENTS 390
- 7.4 Velocity Measurement Techniques 390
- 7.5 Air Flow Measurement 392

BALANCING PROCEDURES 395
- 7.6 Preparation for Balancing an Air System 395
- 7.7 Balancing an Air System 396
- 7.8 Preparation for Balancing a Hydronic System 397
- 7.9 Balancing a Hydronic System 397
- Key Terms 399
- Supplementary Reading 399
- Problems 399

8 Principles of Plumbing 401
- 8.1 Introduction 401
- 8.2 Basic Principles of Plumbing 403
- 8.3 Plumbing System Design Constraints 404
- 8.4 Minimum Plumbing Facilities 404
- 8.5 Hydraulics 406
- 8.6 Static Pressure 406
- 8.7 Pressure Measurement 409
- 8.8 Liquid Flow 410
- 8.9 Flow Measurement 413
- 8.10 Dimensions and Conversions 413
- 8.11 Plumbing Materials 413
- 8.12 Piping Materials and Standard Fittings 414
- 8.13 Piping Installation 436

xii / CONTENTS

8.14	Valves	440
8.15	Fixtures	450
8.16	Drawing Presentation	480
	Key Terms	494
	Supplementary Reading	495
	Problems	495

9 Water Supply, Distribution and Fire Suppression 497

9.1	Design Procedure	498
9.2	Water Pressure	498
9.3	Water Supply System	500
9.4	Water Service Sizing	505
9.5	Friction Head	508
9.6	Water Pipe Sizing by Friction Head Loss	514
9.7	Water Pipe Sizing by Velocity Limitation	518
	DOMESTIC HOT WATER	519
9.8	General Considerations	519
9.9	Instantaneous Water Heaters	522
9.10	Storage-Type Hot Water Heaters	524
9.11	Hot Water Circulation Systems	526
9.12	Sizing of Hot Water Heaters	528
	WATER SUPPLY DESIGN	531
9.13	Valving	531
9.14	Backflow Prevention	534
9.15	Water Hammer Shock Suppression	537
9.16	Residential Water Service Design	538
9.17	Nonresidential Water Supply Design	545
	WATER SUPPLY FOR FIRE SUPPRESSION	547
9.18	Standpipes	548
9.19	Sprinklers	548
	Key Terms	554
	Supplementary Reading	555
	Problems	555

10 Drainage and Wastewater Disposal 557

	SANITARY DRAINAGE SYSTEMS	558
10.1	Sanitary Drainage—General Principles	558
10.2	Sanitary Drainage Piping	562
10.3	Hydraulics of Gravity Flow	562
10.4	Drainage Piping Sizing	568
10.5	Drainage Accessories	572
10.6	Traps	575
	VENTING	577
10.7	Principles of Venting	577
10.8	Types of Vents	580
10.9	Drainage and Vent Piping Design	583
10.10	Residential Drainage Design	589
10.11	Nonresidential Drainage Design	594
	STORM DRAINAGE	602
10.12	Storm Drainage—General Principles	602
10.13	Roof Drainage	604
10.14	Storm Drainage Pipe Sizing	605
10.15	Controlled Roof Drainage	615
	PRIVATE SEWAGE TREATMENT SYSTEMS	616
10.16	Private Sewage Treatment	616
	Key Terms	624
	Supplementary Reading	625
	Problems	626

11 Introduction to Electricity 627

11.1	Electrical Energy	628
	SOURCES OF ELECTRICITY	628
11.2	Batteries	628
11.3	Electrical Power Generation	631
	CIRCUIT BASICS	631
11.4	Voltage	631
11.5	Current	633
11.6	Resistance	633
11.7	Ohm's Law	634
	CIRCUIT ARRANGEMENTS	634
11.8	Series Circuits	634
11.9	Parallel Circuits	636
	ALTERNATING CURRENT (A-C)	638
11.10	General	638
11.11	A-C Fundamentals	638
11.12	Voltage Levels and Transformation	639
11.13	Voltage Systems	641
11.14	Single-Phase and Three-Phase	641
	CIRCUIT CHARACTERISTICS	641
11.15	Power and Energy	643
11.16	Energy Calculation	647
11.17	Circuit Voltage and Voltage Drop	649

11.18	Ampacity	649	13.6	Computer Use in Electrical Design	753
11.19	Electrical Power Demand and Control	651	13.7	The Architectural-Electrical Plan	754
11.20	Energy Management	652	13.8	Residential Electrical Criteria	754
	MEASUREMENT IN ELECTRICITY	652		CIRCUITRY	755
11.21	Ammeters and Voltmeters	652	13.9	Equipment and Device Layout	755
11.22	Power and Energy Measurement	653	13.10	Circuitry Guidelines	755
11.23	Specialty Meters	661	13.11	Drawing Circuitry	763
	Key Terms	661	13.12	Circuitry of The Basic House	776
	Supplementary Reading	662	13.13	The Basic House; Electric Heat	777
	Problems	662		Key Terms	779
				Supplementary Reading	780
				Problems	780

12 Branch Circuits and Outlets 665

12.1	National Electrical Code	666
12.2	Drawing Presentation	666
	BRANCH CIRCUITS	667
12.3	Branch Circuits	667
12.4	Branch Circuit Wiring Methods	670
	CONDUCTORS	673
12.5	Wire and Cable	673
12.6	Connectors	680
	RACEWAYS	684
12.7	Conduit	684
12.8	Flexible Metal Conduit	690
12.9	Surface Raceways	692
12.10	Floor Raceways	695
12.11	Ceiling Raceways	702
12.12	Other Branch Circuit Wiring Methods	706
	OUTLETS	710
12.13	Outlets	710
12.14	Receptacles and Other Wiring Devices	719
	Key Terms	727
	Supplementary Reading	727
	Problems	728

14 How Light Behaves; Lighting Fundamentals 783

14.1	Reflection of Light	784
14.2	Light Transmission	785
14.3	Light and Vision	787
14.4	Quantity of Light	787
14.5	Illumination Level; Illuminance	787
14.6	Luminance and Luminance Ratios	789
14.7	Contrast	790
14.8	Glare	791
14.9	Diffuseness	793
14.10	Color	793
14.11	Illuminance Measurement	795
14.12	Reflectance Measurement	795
	HOW LIGHT IS PRODUCED: LIGHT SOURCES	796
14.13	Incandescent Lamps	797
14.14	Quartz Lamps	801
14.15	Fluorescent Lamps—General Characteristics	801
14.16	Fluorescent Lamp Ballasts	807
14.17	Fluorescent Lamp Types	810
14.18	Special Fluorescent Lamp Types	814
14.19	HID (High Intensity Discharge) Lamps	815
14.20	Mercury Vapor Lamps	817
14.21	The Metal Halide (MH) Lamp	820
14.22	High Pressure Sodium (HPS) Lamps	821
14.23	Induction Lamps	823
	HOW LIGHT IS USED: LIGHTING FIXTURES	825
14.24	Lampholders	826
14.25	Reflectors and Shields	828
14.26	Diffusers	830

13 Building Electric Circuits 729

	OVERCURRENT PROTECTION	730
13.1	Circuit Protection	730
13.2	Fuses and Circuit Breakers	734
	SYSTEM GROUNDING	737
13.3	Grounding and Ground Fault Protection	737
13.4	Panelboards	745
	ELECTRICAL PLANNING	751
13.5	Procedure in Wiring Planning	751

14.27	Lighting Fixture Construction, Installation and Appraisal	835
	HOW LIGHT IS USED: LIGHTING DESIGN	**841**
14.28	Lighting Systems	841
14.29	Lighting Methods	844
14.30	Fixture Efficiency and Coefficient of Utilization	845
14.31	Illuminance Calculations by the Lumen Method	846
14.32	Lighting Uniformity	851
14.33	Determining Coefficient of Utilization by Approximation, Using the Zonal Cavity Method	853
14.34	Lighting Calculation Estimates	857
14.35	Conclusion	857
	Key Terms	858
	Supplementary Reading	859
	Problems	859

15 Residential Electrical Work 861

15.1	General	862
15.2	Social, Meeting and Family-Functions Areas	863
15.3	Study; Work Room; Home Office	877
15.4	Kitchen and Dining Areas	878
15.5	Sleeping and Related Areas	881
15.6	Circulation, Storage, Utility and Washing Areas	886
15.7	Exterior Electrical Work	890
15.8	Circuitry	891
15.9	Load Calculation	891
15.10	Climate Control System	893
15.11	Electrical Service Equipment	895
15.12	Electrical Service—General	896
15.13	Electric Service—Overhead	896
15.14	Electric Service—Underground	900
15.15	Electric Service—Metering	902
15.16	Signal and Communication Equipment	903
	Key Terms	918
	Supplementary Reading	919
	Problems	919

16 Nonresidential Electrical Work 921

16.1	General	922
16.2	Electrical Service	923
16.3	Emergency Electrical Service	926
16.4	Main Electrical Service Equipment	928
16.5	Electrical Power Distribution	928
16.6	Motors and Motor Control	941
16.7	Motor Control Centers	951
16.8	Guidelines for Layout and Circuitry	951
16.9	Classrooms; Electrical Facilities	956
16.10	Typical Commercial Building	960
16.11	Forms, Schedules and Details	968
16.12	Conclusion	969
	Key Terms	969
	Supplementary Reading	970
	Problems	970

APPENDIXES 975

A Metrication; SI Units; Conversions 975

A.1	General Comments on Metrication	975
A.2	SI Nomenclature, Symbols	975
A.3	Common Usage	976
A.4	Conversion Factors	977

B Equivalent Duct Lengths 981

C Loss Coefficients for Duct Fittings 991

C.1	Loss Coefficients for Elbows	992
C.2	Loss Coefficients for Offsets	995
C.3	Loss Coefficients for Tees and Wyes, Diverging Flow	997
C.4	Loss Coefficients, Diverging Junctions (Tees, Wyes)	998
C.5	Loss Coefficients, Converging Junctions (Tees, Wyes)	1003
C.6	Loss Coefficients, Transitions (Diverging Flow)	1005
C.7	Loss Coefficients, Transitions (Converging Flow)	1007
C.8	Loss Coefficients, Entries	1008
C.9	Loss Coefficients, Exits	1009
C.10	Loss Coefficients, Screens	1011
C.11	Loss Coefficients, Obstructions (Constant Velocities)	1012

D HVAC Field Test Report Forms 1015

E Symbols and Abbreviations 1013

Glossary	1033
Index	1039

Tables

Table 1.1	Human Metabolic Activity	3
Table 1.2	Latent Heat of Vaporization of Water	6
Table 1.3	Heat Generated by Typical Physical Activities	8
Table 1.4	Factors in Human Heat Balance	15
Table 2.1	Typical Thermal Properties of Common Building and Insulating Materials-Design Values	25
Table 2.2	Summary of Heat Transfer Terms and Symbols	29
Table 2.3	Emittance Values of Various Surfaces and Effective Emittances of Airspaces	32
Table 2.4	Surface conductances, C_s (Btu/h/ft^2-F), and Resistances, R, for Air	32
Table 2.5	Thermal Resistances of Plane Air Spaces, ^{0}F-ft^2-h/Btu	33
Table 2.6	Perimeter Heat Loss Factor F2; Concrete Slab on Grade	40
Table 2.7	Representative U and R Factors for Vertical Windows	41
Table 2.8	R-Factors for Typical Door Constructions	41

Table 2.9
Winter ACH as a Function of Construction Air Tightness ... 42

Table 2.10
CLTD Values for Single-Family Detached Residences ... 60

Table 2.11
CLTD Values for Multifamily Residences ... 60

Table 2.12
Window Glass Load Factors (GLF) for Single-Family Detached Residences ... 61

Table 2.13
Window Glass Load Factors GLF for Multifamily Residences ... 62

Table 2.14
Shade Line Factors (SLF) ... 63

Table 2.15
Summer Air Change Rates as Function of Outdoor Design Temperature ... 63

Table 3.1
Typical I=B=R Output Ratings for Residential Baseboard Radiation, Btuh/ft ... 102

Table 3.2
Output Derating Factors for Baseboard Radiation Operated at Low Water Temperatures ... 102

Table 3.3
Correction Factors for Commercial Finned-Tube Radiation Operating at Temperatures Other Than 215°F (102°C) ... 103

Table 3.4
Derating Factors for Commercial Finned-Tube Radiation Operating at Water Flow Rates Below 3 fps ... 103

Table 3.5
Correction Factors for Commercial Finned-Tube Radiation Installed at Other Than Recommended Installed Height ... 106

Table 3.6
Typical I=B=R Heat Output Ratings for Commercial Finned-Tube Radiation ... 106

Table 3.7
Heat Emission Rates for Cast Iron Radiators ... 115

Table 3.8
Water Velocity For Varying Flow Rates in Type L Copper Tubing ... 126

Table 3.9
Flow Rates for Varying Water Velocities in Type L Copper Tubing ... 127

Table 3.10
Water velocity For Varying Flow Rates in Type M Copper Tubing ... 127

Table 3.11
Flow Rates for Varying Water Velocities in Type M Copper Tubing ... 127

Table 3.12
Baseboard Radiation Calculation Based on Average Temperature ... 132

Table 3.13
Baseboard Radiation Based on Actual Water Temperatures ... 133

Table 3.14
Pressure Loss Due to Friction in Type M Copper Tube ... 135

Table 3.15
Allowance for Friction Loss in Valves and Fittings Expressed as Equivalent Length of Tube ... 136

Table 3.16
Zonal Heating Loads—Mogensen House ... 145

Table 4.1
Summary of Design, Electric Heating, Basic Plan (Example 4.3) ... 180

Table 4.2
Summary of Design, Electric Heating, Mogensen House (Example 4.4) ... 183

Table 4.3
Design Summary; Electric Heating of a Light Industry Building (Example 4.5, Figure 4.13) ... 194

Table 5.1
Recommended NC Criteria for Various Occupancies and Maximum Supply Outlet Face Velocity ... 261

Table 5.2
Application of Air Supply Outlets ... 266

Table	Title	Page
Table 5.3	Duct Material Roughness Factors Based on Good Workmanship	280
Table 5.4	Equivalent Rectangular Duct Dimension	284
Table 5.5	Size of Equivalent Oval Duct to Circular Duct	287
Table 5.6	Recommended Air Velocities (fpm) for Noise Limitations in Residences	297
Table 5.7	Maximum Velocities for Low Velocity Systems (fpm)	297
Table 6.1	Design Data for Heating/Cooling Loads, The Basic House	359
Table 7.1	Anemometers	390
Table 8.1	Minimum Number of Plumbing Fixtures	405
Table 8.2	Physical Properties of Ferrous Pipe	415
Table 8.3	Physical Characteristics of Copper Pipe	431
Table 8.4	Linear Thermal Expansion of Common Piping Materials	438
Table 8.5	Nominal Valve Dimensions	442
Table 8.6	Plumbing Fixture Usage	451
Table 9.1	Minimum Pressure Required by Typical Plumbing Fixtures	499
Table 9.2	Recommended Flow Rates for Typical Plumbing Fixtures	499
Table 9.3	Water Supply Fixture Units and Fixture Branch Sizes	506
Table 9.4	Table for Estimating Demand	507
Table 9.5	Demands at Individual Water Outlets	508
Table 9.6a	Flow (gpm), Velocity (V, fps) and Friction Head Loss (H ft of water) for Schedule 40 Thermoplastic Pipe Per 100 ft. of Equivalent Length	512
Table 9.6b	Flow (gpm), Velocity (V, fps) and Friction Head Loss (H ft of water) for Schedule 80 Thermoplastic Pipe Per 100 ft. of Equivalent Length	513
Table 9.7	Equivalent Length of Plastic Pipe (ft) for Standard Plastic Fittings	514
Table 9.8	Equivalent Length of Metal Pipe (ft) for Standard Metal Fittings and Valves	515
Table 9.9	Sizing Tables Based on Velocity Limitation	520
Table 9.10	Estimated Hot Water Demand	530
Table 9.11	Minimum Residential Water Heater Capacities	532
Table 9.12	Runout Pipe Size According to Number and Size of Outlets Supplied	545
Table 10.1	Approximate Discharge Rates and Velocities in Sloping Drains Flowing Half Full	566
Table 10.2	Drainage Fixture Unit Values for Various Plumbing Fixtures	569
Table 10.3	Minimum Size of Non-Integral Traps	569
Table 10.4	Horizontal Fixture Branches & Stacks	570
Table 10.5	Building Drains and Sewers	570
Table 10.6	Maximum Length of Trap Arm	577
Table 10.7	Size and Length of Vents	584

Table 10.8 Size of Roof Gutters — 611

Table 10.9 Size of Vertical Conductors and Leaders — 611

Table 10.10 Size of Horizontal Storm Drains — 612

Table 10.11 Capacity of Septic Tanks — 621

Table 10.12 Sewage Flows According to Type of Establishment — 622

Table 10.13 Required Absorption Area in Seepage Pits for Each 100 gal of Sewage per Day — 623

Table 10.14 Tile Lengths for Each 100 gal of Sewage per Day — 623

Table 12.1 Physical Properties of Bare Conductors — 675

Table 12.2 Allowable Ampacities of Insulated Copper Conductors, Rated 0-2000 v, 60°-90° C (140°-194°F). Not More Than Three Conductors in Raceway or Cable or Direct Buried. Based on Ambient Temperature of 30°C (86°F) — 676

Table 12.3 Conductor Application and Insulation — 677

Table 12.4 Dimensions of Rubber-Covered and Thermoplastic-Covered Conductors — 678

Table 12.5 Comparative Dimensions and Weights of Metallic Conduit — 686

Table 12.6 Dimensions of Conduit Nipples and Bushings — 690

Table 12.7 Rigid Steel Conduit Spacing at Cabinets — 691

Table 12.8 Dimensions of Rigid Steel Conduit Elbows — 692

Table 12.9 Characteristics of Selected WIREMOLD® Metallic Surface Raceways — 698

Table 13.1 Wiring Capacities of Outlet Boxes — 772

Table 14.1 Illuminance Categories & Illuminance Values for Generic Types of Activities in Interiors — 789

Table 14.2 Typical Incandescent Lamp Data — 799

Table 14.3 Miniature Mirrored-Reflector Tungsten Halogen Lamp (M-16) — 802

Table 14.4 Typical Fluorescent Lamp Ballast Data — 810

Table 14.5 Typical Fluorescent Lamp Data: Standard Lamps, 60 Hz, Conventional Core and Coil Ballasts — 811

Table 14.6 Fluorescent Lamps Interchangeability: Standard Lamps & Ballasts Only — 813

Table 14.7 Typical Characteristics of Linear Triphosphor Lamps — 814

Table 14.8 Typical Data for Mercury Vapor Lamps — 818

Table 14.9 Typical Metal-Halide Lamp Data — 821

Table 14.10 Comparative Characteristics; Standard Mercury Vapor, Metal Halide and High Pressure Sodium Lamps, with Magnetic Ballasts — 822

Table 14.11 Typical Data for High Pressure Sodium Lamps — 822

Table 14.12 Efficacy of Various Light Sources — 825

Table 14.13 Effect of Illuminant on Object Colors — 826

Table 14.14 Typical Coefficient of Utilization for a Single-Lamp, Open Reflector Fluorescent Luminaire — 856

Table 15.1
Load, Circuit and Receptacle Chart for
Residential Electrical Equipment 879

Table 16.1
Nominal Service Size in Amperes 930

Table 16.2
Current and Volt-Amperage Relationships 930

Table 16.3
System and Utilization Voltages 931

Table 16.4
Rating and Approximate Dimensions of a-c
Full-Voltage Conventional Single-Speed
Motor Controllers, Three-Phase
Combination Circuit Breaker Type 950

Table 16.5
Control Equipment Enclosures 950

Table A.1
Prefixes That May Be Applied to All SI
Units 976

Table A.2
Typical Abbreviations: All Systems of Units 977

Table A.3
Lighting Units—Conversion Factors 977

Table A.4
Acoustic Units and Conversions 978

Table A.5
Common Approximations 978

Table A.6
Useful Conversion Factors: Alphabetized 978

BUILDING TECHNOLOGY

1. Heat and Human Comfort

The first seven chapters of this book are devoted to a study of indoor heating, ventilating and air conditioning, universally known by the abbreviation HVAC. Study of this first chapter will enable you to:

1. Appreciate the problems that an HVAC system is designed to solve.
2. Learn the difference between heat and temperature.
3. Convert temperatures between Fahrenheit, Celsius, Kelvin and Rankine scales.
4. Understand the basic temperature control mechanisms of the human body, the concept of human metabolism, its relation to activity level and the effects of thermal stress on the body.
5. Understand the basic physics of heat, including units, specific heat, sensible heat, latent heat, enthalpy and the essential laws of thermodynamics.
6. Understand the properties of moist (atmospheric) air including dry and wet bulb temperatures, and absolute and relative humidity.
7. Be familiar with important HVAC instruments, including the sling psychrometer, a thermal comfort meter and an indoor climate analyzer.
8. Understand the three fundamental steady-state heat transfer mechanisms of conduction, convection and radiation.
9. Understand the operation of evaporative cooling as a body temperature control mechanism.
10. Learn about human thermal comfort criteria.

1.1 Human Comfort Indoors

The basic purpose of HVAC design is to provide a comfortable, usable indoor "climate" throughout the year. Since outdoor temperatures and other climate factors vary widely with the season, it is apparent that the HVAC system must be dynamic to compensate for these changes. Similarly, the indoor comfort needs of occupants vary greatly, depending on the type of space occupancy. A gym, a residential bedroom and a chemistry laboratory have widely different requirements, which the HVAC system must satisfy to provide the conditions that will permit the space to be used as intended.

The first question that must be answered before any design can begin is how to define *human thermal comfort*. The best definition of human thermal comfort is a negative one; that is, thermal comfort is the absence of thermal discomfort. A thermally comfortable person feels nothing at all; he or she is simply unaware of the thermal environment. This means that the space occupant is neither too warm nor too chilly, is not uncomfortable due to stuffiness or drafts, is not conscious of perspiration (a feeling of body wetness) and is not disturbed by strong odors. Ventilation is included in the concept of thermal comfort used here. To achieve this desirable thermal environment, HVAC designers rely on a flexible HVAC system and on the automatic temperature control system of the human body. The HVAC system could be called upon to do any or all of the following:

- Maintain a uniformly warm indoor temperature in cold weather.
- Maintain a uniformly cool indoor temperature during hot weather.
- Add humidity to the indoor air in winter and reduce it in summer.
- Assure that interior wall surfaces in winter will not have a chilling effect on nearby occupants.
- Recirculate interior air and filter out air-borne dust.
- Control the velocity of the recirculated air; it should be fast enough to provide freshness and slow enough to avoid drafts.
- Exhaust odor-laden air from rooms such as kitchens and laboratories.
- Introduce appropriately tempered (warm or cool) makeup fresh air to reduce indoor stuffiness and to replace the air exhausted from rooms.

In densely occupied commercial or institutional buildings, all the foregoing functions may be provided. In structures of lesser demand, such as residences, it is not always considered essential to include all these items, which constitute a complete air conditioning system. The most basic function of any HVAC system is heating, which is required throughout the United States for at least some part of the year. Cooling, and perhaps more important, dehumidification, was once considered a luxury. Today it is recognized to be almost as important as heating, particularly in the hot and humid summer climates of coastal regions. The HVAC indoor design requirements are chosen to meet human body needs under the specific occupancy conditions of a space. In order to understand the basis of these conditions, it is first necessary to understand the fundamentals of body temperature control and the physical principles of heat transfer.

1.2 Body Temperature Control

Human beings are constant-temperature (warm-blooded) creatures, with a normal deep body temperature of 98.6°F (37°C). We emphasize that this is an internal temperature, because the external (skin) temperature can vary from a low of about 40°F (4.4°C) to a high of about 106°F (41.1°C). These extremes can be maintained for a limited time without physiological damage. Indeed, wide variation in skin temperature is one of the techniques used by the body's highly sophisticated automatic temperature control system to regulate heat transfer to the environment.

The amount of heat generated by the body depends on the person's activity. Table 1.1 shows roughly the amount of energy generated by the body during different activities. This energy is produced by metabolizing ("burning") the food we eat and is, therefore, referred to as the body's metabolic rate. The entire process is known as *metabolism*. The body is only about 20% efficient in converting food to muscular energy; the other 80% is converted to heat that must be disposed of continuously, to avoid overheating the body. (Food not required to sustain bodily functions and body activity is stored as fat on the body.) The body disposes of heat by one or more of the four physical processes for heat transfer and exchange: conduction, convection, radiation and evaporation. In or-

Table 1.1 Human Metabolic Activity[a]

Activity	Btuh[b]	Watts[c]	Metabolic Units, Met[d]	Kcal/h	Horsepower Power, hp
Running up stairs	3600	1055	10.0	907	1.41
Running uphill	3250	952	9.0	819	1.28
Wrestling	2890	846	8.0	728	1.13
Basketball	2525	740	7.0	636	0.99
Running	2165	634	6.0	546	0.85
Very heavy work	2000	586	5.5	504	0.79
Walking uphill	1800	527	5.0	454	0.71
Heavy work	1600	469	4.4	403	0.63
Rapid walk	1445	423	4.0	364	0.57
Medium machine work	1200	351	3.3	302	0.47
Ballroom dancing	1085	318	3.0	273	0.43
Light bench work	800	234	2.2	202	0.31
Light sendentary work	720	211	2.0	181	0.28
Sitting at rest	400	117	1.1	101	0.16
Sleeping	361	106	1.0	91	0.14

[a] Chart of approximate rate of human energy expenditure for various activities. The values are for an average-size adult male. See Table 1.3 for more exact figures and their breakdown into sensible and latent heat.

[b] Btuh is the common abbreviation for Btu *per* hour. It is also written Btu/h, which is mathematically more accurate. 1 Btuh = 0.2929 w.

[c] 1 w = 3.412 Btuh.

[d] 1 met = 18.43 Btuh/ft^2. Since an average adult male is assumed to have 19.6 ft^2 of body surface area, an activity rate of 1 met is equal to an energy rate of 361 Btuh:

$$19.6 \text{ ft}^2 \times 18.43 \text{ Btuh/ft}^2 = 361 \text{ Btuh}$$

der to understand these concepts and processes, it is first necessary to familiarize oneself with the basic concepts of thermal engineering, including temperature, sensible and latent heat, thermodynamic laws and units of measurement.

1.3 Heat and Temperature

Heat is a form of energy. *Temperature* is simply an arbitrary scale invented in order to indicate the amount of heat energy contained in an object. In other words, temperature is a measure of the density of heat energy in an object; the higher the heat content of an object (including air), the higher its temperature. In the United States, the temperature scale most commonly employed is the Fahrenheit scale, on which water freezes at 32°F and boils at 212°F. In recent years the more logical Celsius scale has made some inroads into HVAC work, although Fahrenheit is still the most commonly used. On the Celsius scale, water freezes at 0°C and boils at 100°C (hence the name formerly used—"centigrade"—that is, 100 degrees). Conversion between the two is quite simple using the relations

$$°C = \frac{5}{9}(°F - 32) \tag{1.1}$$

and

$$°F = \frac{9}{5}°C + 32 \tag{1.2}$$

A simple conversion chart is given in Figure 1.1. The technologist may also come across two other temperature scales in advanced HVAC work—Rankine and Kelvin. Both scales are absolute temperature scales, starting at 0°, which represents a complete absence of heat. The Rankine scale is related to the Fahrenheit scale by the relation

$$°R = °F + 460 \tag{1.3}$$

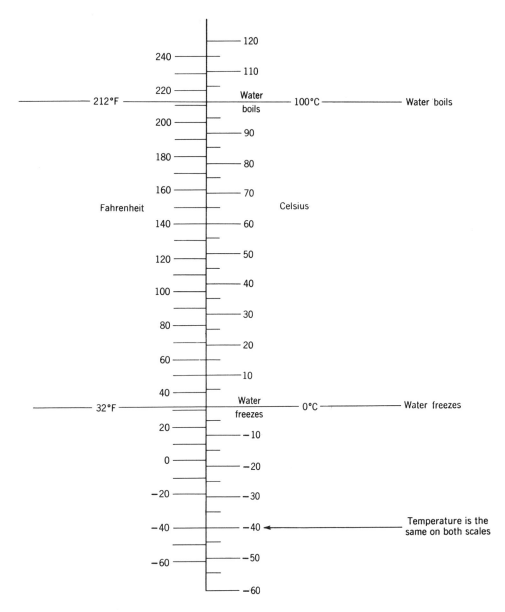

Figure 1.1 Conversion chart between Fahrenheit and Celsius temperature scales.

and has the same degree size as the Fahrenheit scale. The Kelvin scale is related to the Celsius scale by the relation

$$°K = °C + 273 \qquad (1.4)$$

and has the same degree size as the Celsius scale.

The unit of heat energy in the English system of units (which is the system used in the United States in HVAC work) is the *British thermal unit,* or Btu. It is defined as the amount of heat required to raise 1 lb of water by 1 F°. Put another way, the addition of 1 Btu of heat energy to 1 lb of water can be measured as a 1°F increase in the water temperature. The international system of units (SI) does not have an equivalent heat energy unit. The older CGS (centimeter-gram-second) system used the calorie, which is defined as the amount of heat required to raise the temperature of 1 g of water by 1 C°. Since this unit is very small, the *kilocalorie* is used more often. It is obviously the amount of heat required to raise the temperature of 1 kg of water (1 liter) by 1 C°. Conversion factors for Btu, calories, kilocalories and joules (the SI unit of energy) can be found in Appendix A.

The amount of heat that must be added (or removed) from a unit mass of a substance in order to change its temperature by one degree is known as the *specific heat* of that substance. In the English system of units used in this book, specific heat is expressed in Btu/lb/°F. By definition, the specific heat of water is 1.0. The specific heat of most other substances, both liquid and solid, are less than 1.00. For instance, the specific heat of copper is 0.0918, that of iron is 0.1075, that of silver is 0.0558, that of gasoline is about 0.5, that of alcohol varies between 0.5 and 0.7, depending on the type, that of concrete is 0.15 and that of wood varies between 0.4 and 0.7. These lower values indicate immediately the usefulness of water to transfer (and store) heat, as in hot water heating systems.

1.4 Sensible Heat and Latent Heat

Sensible heat is that which causes a change in temperature when it is added or removed. *Latent heat* is that which causes a change of state in the substance, as for instance, from solid to liquid or liquid to gas while the temperature remains constant.

To illustrate the meanings of these terms, we can use water as an example. Refer to Figure 1.2. If we begin adding heat to a container holding 1 lb of water, we will be able to measure a temperature increase of 1 F° for every Btu added. Thus, if we start with water at room temperature (70°F), the water will come to a boil after adding 142 Btu:

Total added sensible heat
$$= \text{Specific heat} \times \text{Temperature change}$$
$$= 1.0 \frac{\text{Btu}}{\text{°F(lb)}} \times (212°F - 70°F) \times 1 \text{ lb}$$
$$= 142 \text{ Btu}$$

(This relation assumes an ideal heating process in which no heat is lost, that is, 100% efficiency.) This amount of heat energy—142 Btu—is sensible heat, because all of it caused a corresponding temperature change. The sensible heating process is shown on Figure 1.2 as a straight line whose slope is 1°F/Btu.

Water at atmospheric pressure cannot get hotter

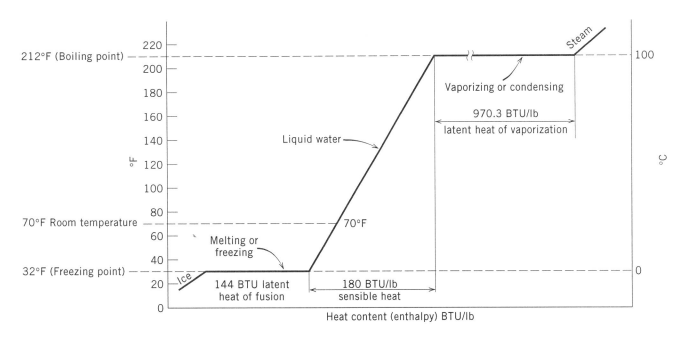

Figure 1.2 Heat content (enthalpy) chart of water in Btu per pound. At the 32°F (0°C) lower change-of-state point, freezing water loses 144 Btu/lb and melting ice gains 144 Btu/lb of latent heat. At the 212°F (100°C) upper change-of-state point, vaporizing water absorbs (gains) 970.3 Btu/lb and condensing steam releases (loses) 970.3 Btu/lb of latent heat. In the liquid state between 32 and 212°F, water gains or loses 1 Btu of sensible heat per pound, for each 1 F° of temperature change, that is, 180 Btu/lb for the 180 F° temperature change between 30 and 212°F.

than 212°F. Therefore, if we continue to add heat, the water will turn to steam at a temperature of 212°F (100°C). All the heat added after the water temperature reaches 212°F is used to vaporize the water, that is, to change its state from liquid (water) to gas (steam). This heat is called latent heat. Quantitatively, it takes 970 Btu to vaporize 1 lb of water at 212°F. This quantity is called the *latent heat of vaporization*. The total heat change in 1 lb of water, beginning at 70°F and ending with the vaporization of the entire 1 lb of water is, therefore, the heat required to raise the water temperature to 212°F (sensible heat) plus the heat required to vaporize the water at 212°F (latent heat); that is,

Heat change = Sensible heat + Latent heat
= 142 Btu + 970 Btu
= 1112 Btu

Notice that water will vaporize at temperatures all the way down to freezing (32°F, 0°C). This effect is the familiar phenomenon of evaporation, in which water seems to disappear spontaneously into the air. In actual fact, this vaporization, or evaporation, requires the addition of heat in the quantities listed in Table 1.2. The heat comes from the water that is evaporating, causing it to cool. This is the physical reason for the cooling effect of evaporation. This subject is discussed in detail in Chapter 6. The speed at which water evaporates depends on a number of factors, including temperature, humidity and air movement.

The latent heat of vaporization of water varies inversely with the water temperature. It is highest at low temperatures and lowest at the boiling point. Table 1.2 shows this effect.

Table 1.2 Latent Heat of Vaporization of Water

Temperature		Heat of Vaporization (Evaporation), Btu/lb
°F	°C	
32	0	1075.5
40	4.4	1071
60	15.6	1059.7
70	21.1	1054.0
75	23.9	1051.2
80	26.7	1048.4
90	32.2	1042.7
100	37.8	1037.1
120	48.9	1025.6
150	65.6	1008.2
180	82.2	990.2
212	100	970.3

Returning to Figure 1.2, and again beginning at 70°F room temperature, 1 lb of water will release 1 Btu of sensible heat for each degree of temperature drop. At 32°F (0°C), a change of state from liquid to solid (or vice versa) occurs. Removing 144 Btu/lb from liquid water will cause it to freeze into ice. This quantity is known as the *latent (change-of-state) heat of fusion*. Conversely, adding 144 Btu will convert 1 lb of ice to water at 32°F (0°C).

1.5 Properties of Atmospheric Air

In order to understand human comfort conditions, we must first understand the basic properties of atmospheric air, which surrounds the body at all times. The study of the physical properties and thermal processes of atmospheric air is called *psychrometrics*. We shall have more to say on this subject in our discussions on warm air heating, humidification, dehumidification and cooling processes. At this point, we will limit the discussion to definitions and their explanations.

a. Dry Bulb (DB) Temperature

This is the air temperature that is registered by a thermometer with a dry bulb, that is, a bulb that has not been deliberately made wet. See Figure 1.3. The dry bulb temperature is that which is meant, in normal speech, when referring to the air temperature. It is the temperature stated by the weather bureau in its reports and that which shows on an ordinary household thermometer.

b. Humidity

This term refers to the amount of water vapor in the atmosphere. The higher the (dry bulb) temperature of the air, the more water vapor the air is capable of carrying. The actual amount of water in the air is expressed in two ways: *absolute humidity* in pounds of water per cubic foot of dry air or *specific humidity* in terms of weight of water (grams or pounds) per pound of dry air. (Specific humidity is also referred to as *humidity ratio*.) None of these terms is as important to the technologist as relative humidity. *Relative humidity (RH)* is the ratio of the actual amount of water vapor in the air to the maximum amount of water vapor that the air can absorb at that dry bulb temperature, expressed as a percentage. It is, therefore, a measure of the "wetness" of the air at that temperature.

When air has absorbed as much water as it can hold at that dry bulb temperature, it is saturated. Any additional moisture in the air can exist only as water droplets that are visible to the eye (as in a steam room) and not as water vapor, which is invisible. When moist air is cooled, it will eventually reach saturation, since, as stated previously, the amount of water vapor that air can hold is proportional to the dry bulb temperature. The temperature at which the air becomes saturated (during cooling) is known as the *dew-point (temperature)*, since at that point droplets of water (dew) begin to form on surfaces. If moist air is cooled rapidly, the water vapor that condenses does not have time to settle on surfaces (as dew droplets). Instead, the droplets simply appear suspended in the air, as fog. In humid climates, fog normally forms just before dawn, when temperatures are lowest. The fog is then "burned off" by the sun. What actually happens, of course, is that the sun causes a rise in air temperature, in turn, causing the water droplets in the fog to re-evaporate into the air as water vapor.

c. Wet Bulb (WB) Temperature

Wet bulb temperature measures the cooling that results from evaporation. As we have seen from Figure 1.2 and Table 1.2, when water evaporates, it absorbs heat from the surrounding air and from the surface on which the water rests. This is the reason that we feel chilled when stepping out of a shower, even in a very warm but relatively dry bathroom. The exposed skin from which water is evaporating is actually giving up large amounts of sensible heat, which is being absorbed by the evaporating water as latent heat, thus rapidly cooling the body. This action is the basis of evaporative cooling. However, this is not the case in a steam room where the air is completely saturated with water vapor. In that environment, the water on the body cannot evaporate, and, therefore, no cooling is felt. Consequently, we see that the cooling effect of water evaporation can be used as a measure of the relative humidity of the surrounding air. The higher the air humidity, the slower water evaporates and the lower its cooling effect on the surface from which it evaporates.

This effect is used to advantage in a device called a *sling psychrometer*, which is illustrated in Figure 1.3. The device is simply a piece of wood or other rigid material onto which are affixed two ordinary thermometers. One thermometer simply reads the air temperature, that is, the dry bulb temperature.

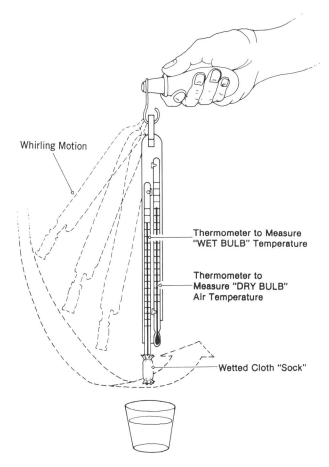

Figure 1.3 Sling psychrometer, so called because the device is whirled around like a sling, until the temperature on the thermometer with the wet sock stabilizes. This is the wet bulb temperature. Given the dry bulb and wet bulb temperatures, the relative humidity of the air (and other pertinent data) can be determined from tables or from a psychrometric chart. From Bradshaw, *Building Control Systems*, 2nd ed., 1993, reprinted by permission of John Wiley & Sons.)

On the bulb of the other thermometer, a piece of absorbent cloth, called a sock, is placed and the sock is then wet by immersion in water. Water evaporating from the sock will cool the bulb on which it is placed. The amount of cooling is inversely proportional to the relative humidity; the drier the air, the greater the cooling by evaporation and the lower the temperature reading of this wet bulb thermometer. The drop in temperature of the wet bulb is called, logically, the *wet bulb depression* and its reading is called the wet bulb (WB) temperature. By comparing the DB and WB readings using a table or a psychrometric chart, the relative humidity and all the other relevant characteristics of the air can be read off directly. The entire device is called a sling psychrometer, because it is whirled in the air like a sling via its rotating handle (see Figure 1.3) until the wet bulb thermometer reading stabilizes. The purpose of the whirling is to prevent formation of a layer of wet (saturated) air around the wet sock that would prevent evaporation. This same air motion effect is used in a hot air blower-type hand dryer, such as is found in public rest rooms. The heat from this device increases the DB temperature of the air surrounding the hands, thus permitting rapid water evaporation. The air movement prevents accumulation of a wet air layer around the hands that would slow or halt evaporation. The evaporative cooling effect is so strong that, despite the high temperature of the air being blown onto the hands, the skin feels cool until all the water evaporates, at which point the heat of the blower is quickly felt.

1.6 Body Heat Balance

As stated in Section 1.2, the human body produces heat continuously, in quantities that depend on the body's physical activity. Table 1.3 shows the total heat generated for typical activities plus the division between sensible heat and latent heat. Note that the latent portion is constant, because it represents the water vapor that we exhale. Human breath is essentially saturated air (100% RH) at body temperature of about 98°F. This is the reason that we can "see our breath" in winter; the saturated exhaled air is immediately cooled to the dew point, producing a mini-fog, as was explained previously. The remaining heat produced by the body (in the form of sensible heat) must be removed to maintain the body's heat balance.

The thermal processes by which the body interacts with its environment are conduction, convection, radiation and evaporation. The net heat gain or loss between the body and the environment must be such that the body's heat balance is maintained within body temperature limits. These relations are shown schematically in Figure 1.4 and will be discussed individually. Expressed mathematically, the body's sensible heat thermal equation is

$$M \pm CD \pm CO \pm R - E = HS \quad (1.5)$$

where
- M is the body metabolic heat production,
- CD is the conductive heat gain or loss,
- CO is the convective heat gain or loss,
- R is the radiation heat gain or loss
- E is the evaporate heat loss, and
- HS is the body heat storage gain or loss.

Notice that the body can gain or lose heat by conduction, convection or radiation. Evaporation, however, always cools the body and results in a heat loss. The net result of these processes is the heat storage factor, which may be positive, negative or zero. If it is positive, the body overheats and is subjected to heat stress; if negative, the body cools and is subjected to the unpleasant and even dangerous effects of chilling. If it is zero, we are thermally balanced, although not necessarily comfortable.

Table 1.3 Heat Generated by Typical Physical Activities[a]

| | Heat, Btuh | | |
Activity	Total	Sensible[b]	Latent[c]
Seated, reading	400	295	105
Seated, desk work	450	345	105
Office work, general	475	370	105
Standing light work	550	445	105
Moderate work, walking	600	495	105
Light factory work	800	695	105
Medium factory work	1200	1095	105
Heavy factory work	1500	1395	105
Active athletics	2000	1800	200[d]

[a] Figures are average for an average size adult male.
[b] Between 20 and 60% of the sensible heat is radiated, depending on air velocity, assuming a 75°F dry bulb air temperature.
[c] Latent heat loss is constant with normal breathing rate.
[d] Latent heat loss increases with the rapid breathing associated with active athletics.

1.7 Basic Rules of Thermodynamics

Throughout the following discussion, and indeed in all HVAC work, it is important to remember several basic rules of thermodynamics.

a. Energy Can Be Neither Created Nor Destroyed

This rule is variously known as the law of conservation of energy or as the first law of thermodynamics. To the HVAC technologist, it simply means that, when heat is transferred, a gain in one place is balanced by an equal and opposite loss in another

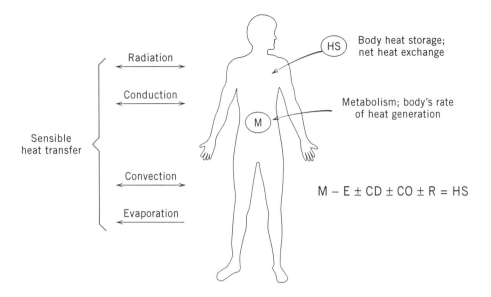

Figure 1.4 Body heat balance is maintained by transferring heat with the environment. The transfer processes are conduction, convection, radiation and evaporation. The net heat gain or loss appears as heat storage gain or loss in the body. Conduction, convection and radiation are all two-way processes that can result in body heat gain or loss; evaporation always results in body heat loss.

place. Sensible heat can be converted to latent heat by evaporation, and latent heat can be converted to sensible heat by condensation. Stated another way, we can say that the net change of energy in a system is the difference between the energy added and the energy removed. In terms of body heat balance, this means that the difference between the (heat) energy gained and the heat energy lost by the body must equal the change in heat storage; that is,

Body heat storage change = Heat gained − Heat lost

If the gain exceeds the loss, the body overheats. If the loss exceeds the gain, the body overcools.

b. Heat Flows "Downhill"

Heat flows from an area or object of higher temperature to an area or object of lower temperature. Indeed, heat is sometimes defined as the form of energy that is transferred by temperature difference. In general, the larger the temperature difference, the more rapid is the heat transfer. In order to reverse this flow, energy must be added to the system. Thus, to keep the inside of a refrigerator cold in a warm kitchen, energy must be removed by the refrigerator's cooling system. Conversely, to keep a body warm in a cold climate, energy must be added by the body's internal heating system, that is, the metabolic system.

c. There Is No 100% Efficient Thermal Process

This is the essential meaning of the second law of thermodynamics. Every heat transfer involves losses in the form of wasted heat. Losses can be minimized, however. Two of the factors that increase losses are large temperature differences and friction. At this point, we will return to the discussion of the processes involved in the transfer of sensible heat between the body and its environment.

1.8 Mechanics of Sensible Heat Transfer

a. Conduction

Heat is transferred by conduction when two items at different temperatures physically touch each other. The rate of heat transfer depends on the temperature difference and on the conductivity of the item at lower temperature. When we place our hands on a metal surface, it feels cool to the touch, because the metal is an excellent conductor of heat. It therefore conducts heat rapidly away from our warm hands, giving us the sensation of coolness. (This excellent heat conduction that is characteristic of metals accounts for the very hot handle of a frying pan and the extreme rapidity with which a metal teaspoon heats up when placed in a cup of boiling hot tea or coffee.) However, when we touch a piece of wood or cloth at the same temperature as the metal, we feel warmth, because wood and cloth are good insulators and do not conduct heat easily.

Air is an excellent thermal insulator. (Indeed, the trapped air pockets in thermal insulation provide the material's insulating property.) Since only a small portion of the body is directly exposed to the air (hands and head) and the remainder through a layer of insulation (clothing), the body's heat loss by conduction to the surrounding air is very small.

b. Convection

Cool air immediately adjacent to the body is heated by conduction. Since warm air is lighter than cool air, it rises, and cooler air takes its place. When this cool air, in turn, is heated by contact with warm skin, it too rises. This constant air movement is called a *convective air current*. It is "fueled" by the temperature difference between the body and the air surrounding it. See Figure 1.5. The larger the temperature difference between the skin and the surrounding air, the faster the convective air current will move, and the more heat it will transfer. If air motion is increased by a fan or other device, heat transfer is also obviously increased and convection becomes more effective as a heat transfer process.

Convection is also useful as a heating process when the air temperature is above that of the body. In that situation, warm air contacting the body is cooled when it transfers heat to the skin. It becomes heavier and falls, to be replaced by lighter, warmer air. Heating convectors operate on this principle by establishing a convection loop within a room. See Figure 1.6. Convectors will be discussed at length in the heating sections of this book. Notice in Figure 1.4 that convection (and radiation) are most effective in heat loss transfer when the ambient temperature surrounding the body is low, that is, when the temperature difference driving the convective air currents is large. The convective factor *CO* in the heat balance equa-

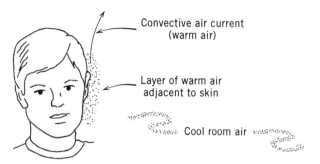

(a) Skin temperature higher than air temperature

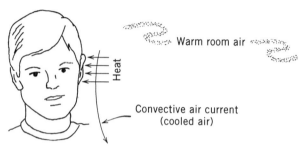

(b) Ambient air temperature higher than skin temperature

Figure 1.5 A convective air current is set up when air is warmed *(a)* or cooled *(b)* by contact with the skin. In *(a)*, the effect is cooling; in *(b)*, it is heating (see Figure 1.6).

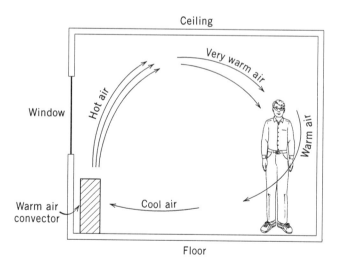

Figure 1.6 Warm air convectors are best placed under windows to temper the cold air dropping from the window cooled by conduction; falling due to cooling; see Figure 1.5*b*. As the warm air rises, it is cooled by the window and other cool surfaces in the room, including occupants. As the air cools, it drops and returns to the bottom of the convector, to be reheated.

tion will be negative if the body loses heat to the surrounding air and positive if it gains heat convectively.

c. Radiation

Radiation is the transfer of heat through space from a mass at a higher temperature to a mass at a lower temperature. The classic example of radiation occurs between the sun and the earth, through a vast expanse of vacuum. The air is not necessary; the atmosphere actually interferes somewhat with radiant heat transfer by absorbing and dispersing some of it. Radiation of heat occurs between all bodies that exist in line-of-sight to each other. Thus, in a kitchen with a stove at 160°F, a convector at 150°F, a person at about 80°F, walls and furnishing at 70°F and a window at 50°F, the stove radiates to everything; the convector, to the occupant, furnishings, walls and window; and everything, to the window. See Figure 1.7. We emphasize line-of-sight, because radiant heat cannot go around corners; it is blocked by any solid object. For this reason, standing in the shade of a tree on a sunny day is effective in making a person feel cooler. All the sun's powerful radiant energy is blocked. As a result, the body can regulate its temperature more easily.

The amount of radiant energy transferred from one mass to another is proportional to the temperature differential, the thermal absorption characteristic of the receiver (the mass at lower temperature) and the angle of exposure of the cooler mass to the warmer one. This angle is inversely proportional to the distance between them, as will be explained in detail later. The exact calculation of these factors is complex and beyond our scope here. Suffice it to say that, in still air (where convection and evaporation are minimized), radiation is the principal form of heat exchange between the human body and its environment. This accounts for the fact that even in cold winter air, with no wind, a person exposed to strong sunlight actually feels uncomfortably warm even when very lightly dressed. This person would immediately feel cold if the sun were blocked by a tree, because he or she would then be radiating more heat to the cold surfaces surrounding the body than the body receives from the sun.

We mentioned previously, in passing, that one of the factors involved in radiant heat transfer is the heat absorption characteristic of the receiver—that is, the mass at lower temperature. This factor is extremely important in building insulation but much less so for interior surfaces. We all know that a black matte surface exposed to the sun gets hotter than a white surface or a reflective aluminum surface. This effect is related to a material's surface absorption characteristic. Since interior room surfaces are normally finished with a paint that is neither specifically absorptive nor reflective, we can assume that radiative heat transfer is proportional to temperature difference only and ignore the type of interior surface finish. (Surface absorption characteristics are very important in thermal insulation and are discussed at length in Section 2.4.)

In order to quantify the radiant heating effect of the environment on the human body in an enclosed space, a concept called *mean radiant temperature (MRT)* was developed. This is simply the weighted arithmetic average of the surface temperatures in the room. See Figure 1.8, which shows the basis of an MRT calculation for the room of Figure 1.7. Strictly speaking, the solid angle between a room occupant and each room element should be used in the calculation. However, in spaces that do not have radiant floors or heated ceilings, an acceptable approximation of the space's MRT can be made using the angles in a two-dimensional plan view, such as an architectural plan. This assumes that walls, floors and ceilings are at approximately 70–75°F and, therefore, do not appreciably affect a

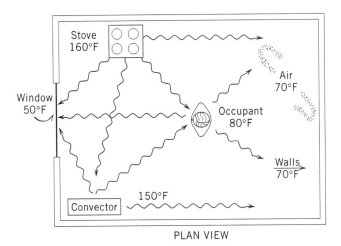

Figure 1.7 Arrows indicate the radiant heat interchange in this space. Each body radiates heat to every other body at a lower temperature. The net radiant heat gain or loss is different for each mass in the space.

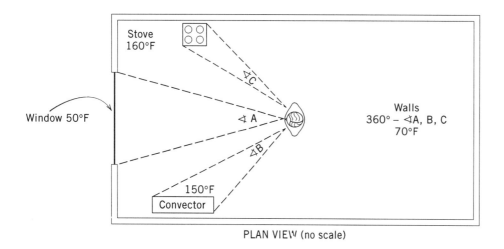

$$\text{MRT} = \frac{30(150) + 22(150) + 15(160) + [360 - (30 + 22 + 15)](70)}{360} = \frac{28510}{360} = 79°F\ (26°C)$$

Figure 1.8 Approximate calculation of the MRT of this room with respect to an occupant at the location shown. An occupant at another location would have a different MRT. The calculated MRT of 79°F (26°C) is so close to body surface temperature that net radiation heat loss of the occupant is approximately zero. This will make the occupant feel uncomfortably warm unless humidity is low and air motion is sufficient to permit adequate heat loss by convection and evaporation.

lightly clothed occupant whose average surface temperature is 78–80°F. Further approximations in this calculation assume that all hot or cold surfaces are the same height and that occupants can be approximated by narrow vertical cylinders. In a room with a hot or cold floor, wall or ceiling, these approximations are not applicable.

The calculation of the approximate MRT for the room occupant in Figure 1.8 is performed as follows:

$$\text{MRT} = \frac{\angle A \times t_A + \angle B \times t_B + \angle C \times t_C + \cdots + \angle N \times t_N}{360°} \quad (1.6)$$

where $\angle A$ is the subtended (exposure) angle between the occupant and surface A,

t_A is the average temperature of surface A, and so on, and

$$\angle A + \angle B + \cdots + \angle N = 360°.$$

The following rules of thumb are useful in designing a space where the previously stated approximations are reasonable and in evaluating the results of an MRT calculation:

(1) Design the space so that wall temperatures are not more than ±5 F° different from the air temperature and the ceiling not more than ±10 F° different from the air temperature.
(2) The relative humidity of the space should be in the range of 35–50%, assuming occupants are lightly clothed.
(3) Air velocity in the room should not exceed 40 ft/min (fpm).
(4) If the calculated MRT is 10°F or more hotter or colder than the design air temperature, the occupant will feel uncomfortable, given the humidity and air velocity conditions in rules 2 and 3.
(5) An MRT above average body temperature [about 80°F (27°C)] indicates a net radiant heat gain in the heat balance equation; an MRT below 80°F (27°C) indicates a net radiant heat loss.
(6) A calculated MRT of 70°F and an air temperature of 70°F will be satisfactory to most people. Every degree of MRT above or below 70°F must be compensated by a 1.5°F change in air tem-

perature in the opposite direction, in order to maintain the 70/70 comfort condition.
(7) The preceding calculation procedure is not applicable to space using radiant ceilings or floors for heating.

d. Evaporation

As stated before, the human body can either gain or lose heat by conduction, convection and radiation. By contrast, evaporation is a one-way thermal process, causing bodily heat loss only. The reason should be evident from what has already been discussed. Moisture on the skin (perspiration) evaporates into the air to become air-borne water vapor. In so doing, it absorbs sensible heat from the skin and changes it to the latent heat of the water vapor. Total heat energy remains the same, according to the first law of thermodynamics. (As we will learn later on, evaporative coolers—so-called desert coolers—operate on the same principle; they remove sensible heat from the air by converting it to latent heat. In so doing, they both lower the air temperature and increase the RH to more comfortable levels.)

As mentioned previously, the average overall surface temperature of a lightly clothed person indoors is about 80°F (27°C). When the ambient temperature is below this figure, the body can easily rid itself of the heat it generates by convection and radiation, particularly if there is air movement to help convection. When the ambient temperature approaches 80°F (27°C) and the room surfaces approach this temperature, radiation and convection drop sharply. This is because both processes depend on a temperature difference to drive them, and this differential has disappeared. The body's internal heat-regulating mechanism reacts by pumping blood into the skin and by activating the perspiration process. The increased blood flow increases the skin temperature and by so doing re-energizes the radiation and convection loss process. More important, however, it causes the perspiration on the skin to evaporate rapidly, thus removing large amounts of sensible heat from the body. From Table 1.2, we see that, at 80°F skin temperature, the body can lose about 1050 Btu/lb of water (perspiration) by evaporation.

The effectiveness of this evaporative cooling process is increased by air motion, which serves to continually remove the saturated layer of air adjacent to the skin. See Figure 1.9. This is the reason that an electric fan makes us feel cooler even

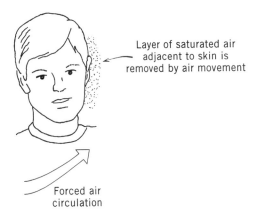

Figure 1.9. The evaporative cooling process is assisted by air motion. The saturated layer of air immediately adjacent to the skin is blown away by air motion. This permits dryer air to contact the skin, resulting in further evaporation and resultant cooling.

though its motor is adding heat to the room. In the absence of air motion, a saturated layer of air remains adjacent to the skin, since convective air motion is absent due to the high ambient air temperature. (See Figure 1.5.) This results in a feeling of wetness, causing considerable thermal discomfort.

Conversely, the effectiveness of the evaporative cooling process is reduced by high ambient relative humidity. Without going into the technicalities of partial vapor pressures, the reaction is easy to understand. When the air already contains a large amount of water vapor, the addition of more water vapor by evaporation is slow. It comes to a complete halt at 100% RH, that is, saturated air. On the other hand, dry air with a low RH readily absorbs additional moisture. This is why we are not conscious of perspiring in hot dry air, such as a desert climate, despite the fact that we can lose as much as a quart of water per hour through perspiration! In a desert-like situation, the body gains heat by radiation from hot surfaces and can keep cool only by constant and intensive evaporative cooling.

e. Summary of Heat Transfer Processes

The interaction of the four heat transfer processes with the body's metabolic rate (heat production) is complex. The variables in the equation are DB air temperature, RH, amount and type of clothing, MRT of the space, air motion, activity level of the person involved and that person's physical position in a room. A typical situation is graphed in Figure 1.10 for a lightly clothed seated person at rest (reading) and producing about 400 Btuh. (Btu per hour is normally written Btuh, although more properly it should be Bth/h.) See Table 1.3.

Note that the metabolic rate of 400 Btuh (about 130 w) for a person at rest is constant, since metabolic rate is governed only by a person's activity level. The graph is drawn for a room at 45% RH and an air velocity between 50 and 100 fpm. At low ambient temperatures the major portion of the body's heat loss is by radiation to the cool room surfaces with a small convection loss and an even smaller evaporation loss. As the ambient temperature rises, all the room surfaces are warmed to about air temperature except for the windows. At about 80°F, the radiation/convection and evaporation components are equal. At higher room temperatures, skin body temperature rises as does the evaporative cooling component of the heat transfer equation. When the room temperature reaches over 100°F, the radiation/convection component goes to zero and the body relies completely on evaporation for cooling. At this point, air motion becomes absolutely necessary to avoid body overheating. Table 1.4 summarizes the factors involved in body heat transfer processes with the environment.

f. Thermal Stress

The question that immediately arises when looking back at the body heat balance equation (1.5) is: What happens when the heat storage factor is not zero? The answer is that very serious physical

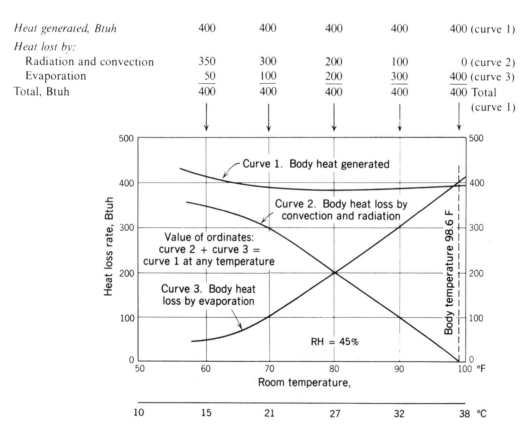

Figure 1.10 The total body heat generated remains substantially constant regardless of room temperature, depending only on the body's activity level. The methods that the body uses to rid itself of this heat vary with ambient temperature, relative humidity and air motion. For a fixed RH of 45%, the relation between convection/radiation loss and evaporation loss is shown by the curves, as a function of room temperature. (From Stein and Reynolds, *Mechanical and Electrical Equipment for Buildings*, 8th ed., 1992, reprinted by permission of John Wiley & Sons.)

Table 1.4 Factors in Human Heat Balance

Factor	Affected by
Metabolism (heat production)	Physical activity
Conduction	Temperature and conductivity of contact surfaces (including air)
Convection	DB air temperature
	Air motion
	Amount and type of clothing
Radiation	Room MRT
	Finish of room surfaces
	Overall body temperature
Evaporation	DB temperature
	RH
	Air motion
	Amount of exposed skin surface

effects occur. As the body gains heat, skin temperature rises and sweating begins. If heating continues, the deep body temperature will rise above normal (as with a fever). If this rise continues, it will result in nausea, exhaustion, fainting (heat stroke or prostration), eventual brain damage at about 107°F and finally death.

The same finality results from overcooling. As the body is chilled, shivering and "goosebumps" appear as a heating mechanism. Further chilling results in the loss of the power of speech, body rigidity, loss of consciousness and finally death. These extremes are, of course, not relevant to HVAC work. They are mentioned simply to show the extreme sensitivity of the human body to even small body temperature changes and the importance of designing a flexible thermal comfort control system.

1.9 Thermal Comfort Criteria

The HVAC standards universally accepted in the United States are those published by the American Society of Heating, Refrigerating and Air Conditioning Engineers, Inc. (ASHRAE, 1791 Tullie Circle, N.E., Atlanta, GA 30329). ASHRAE Standard 55-1992, Thermal Environmental Conditions for Human Occupancy, defines the indoor conditions that 80% (and 90%) of the population will find comfortable, for summer and winter. It describes the interactions between DB temperature, MRT, RH, air speed, weight of clothing and activity level. All these conditions are interrelated, as we have shown in the preceding discussion. The standard gives approximately the following comfort criteria:

(a) For summer comfort, lightly dressed people (short sleeves) doing light office work (450 Btuh) will be comfortable at DB temperatures between 73 and 79°F (23 and 29°C), RH of 40% (range of 25–60%) and an air speed not exceeding 50 fpm. The lower the temperature in the 73–79°F range, the lower should be the air speed. (Elderly people are particularly disturbed by high air speeds.) Higher temperatures require lower RH.

(b) For winter comfort, people dressed in indoor winter clothing (heavy suit, dress, sweater) will be comfortable in a DB temperature range of 68–74°F (20–23°C), RH of about 40% and an air speed not exceeding 40 fpm.

In general, comfort is maximal when the MRT is about the same as the DB air temperature and air speed is about 20–40 fpm. Below 20 fpm, the room feels stuffy; above 50 fpm, the space feels drafty. RH below 20–25% will cause static electricity problems; RH above 60% will cause wetness, mildew and condensation on single glazing in winter months.

The HVAC technologist is not usually responsible for establishing indoor design criteria, except for small projects in which comfort criteria are not critical. He or she should consult the referenced ASHRAE standard and experienced HVAC design engineers before deciding on design comfort criteria for any project.

1.10 Measurements

As we have seen, thermal comfort depends on six factors—DB temperature, RH, MRT, air velocity, metabolic rate (activity) and insulation (clothing). In testing an existing installation or testing and balancing a new one, the first two factors can be measured with a sling psychrometer (Figure 1.3); air velocity can be measured with an anemometer; and MRT can be calculated by measuring surface temperature with a pyrometer. Metabolic rate and the effect of clothing are input data when using a comfort chart. Sophisticated electronic instruments exist that not only perform the required measurements of ambient conditions but also predict the indoor comfort acceptability.

The device shown in Figure 1.11 is called a Thermal Comfort Data Logger by its manufacturer. Since it is made by a foreign manufacturer, it uses

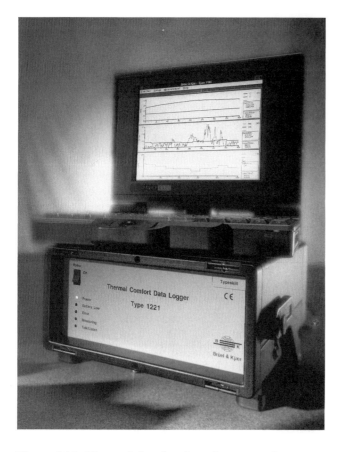

Figure 1.11 Thermal Comfort Data Logger with attached screen for data display. The unit accepts temperature, humidity and air motion data from sensors (Figure 1.12) and calculates thermal comfort indices via an on-board computer. (Photo courtesy of Brüel & Kjaer.)

Figure 1.12 Tripod stand with sensor/transducers that provide the required input data to the Thermal Comfort Data Logger. (Photo courtesy of Brüel & Kjaer.)

a comfort equation somewhat different from that of the ASHRAE standard. The sensors shown in Figure 1.12 provide input data on temperature, humidity and air velocity. The temperature data include not only dry bulb air temperature but also radiation from surrounding surfaces and air convection. These data are processed by the on-board computer and thermal comfort indices calculated. The data are displayed on the device screen, as seen in Figure 1.11.

1.11 Units and Conversions

With this chapter, we began our study of HVAC, plumbing and electrical systems. The design of these systems requires expressing quantities of energy, flow, length, velocity, diameter, resistance, time and so on. In the United States, these quantities have been expressed historically in units known as English units, or units in the English system. The other major system of units in use almost every place else in the world today is the metric system, abbreviated SI for Systeme Internationale.

All measurement systems are based on three basic units—length, weight (mass) and time. The English unit uses foot, pound and second for these three units. The SI system uses meter, kilogram and second. For this reason, the SI system used to be called the MKS system. The English system uses odd subdivisions and divides fractionally. Thus, we have 12 inches to a foot, 16 ounces to a pound and ¼, ⅛, and ¹⁄₃₂ parts of an inch. This makes the system extremely unwieldly and makes calculation time-consuming. In contrast, the SI system divides and multiples decimally. Thus, there are 1000 millimeters, 100 centimeters and 10 decimeters (a unit rarely used) to the meter. Unfortunately, as of this writing, the movement to change over to SI units in the United States in general is slow, and in the building trades it is barely perceptible.

Many authors (including this writer) have dutifully supplied dual sets of units in their books, at least in part. In the field, however, the English system continues to hold sway almost exclusively. As a result, we have decided to avoid the clumsiness of dual units and have supplied only English units in this book. The two exceptions to this have been the use of degrees Celcius (°C) for temperatures and meters per second for air velocities, since these units are beginning to be used in the United States. Even so, dual units are only shown where we feel it will be helpful, rather than throughout. Appendix A provides an extensive listing of conversion factors that will enable you to convert English units to SI units (or the reverse) easily.

In all types of technical design work, it is constantly necessary to convert from one unit to another, not between systems, but within one system. In HVAC work, it is necessary to convert air changes per hour to cubic feet per minute (cfm) or gallons per minute (gpm) to cubic feet per second (cfs) and so on. In plumbing work, we make conversion of flow from cubic feet per hour (cfh) to gpm, of volume from cubic feet (cf) to gallons, of pressure from the weight of cubic feet of water to pounds per square inch (psi) and so on. Of course, you can always do these conversions in a single step by simply consulting a table of conversion factors, such as the one in Appendix A. However, sometimes such a table is not handy, or it is simply inconvenient. On such occasions, you can convert simply and accurately by going through a step-by-step conversion of units, with cancellation at each step. In order to do this, of course, it is necessary to know, from memory, a few basic conversion equivalences in either their exact or approximate forms. Some of these conversions are between standard English units and SI (metric), but most are within the English system. They include the following:

Unit	Exact	Approximate
Length	1 meter = 3.28 feet	1 m = 3.3 ft
	1 foot = 12 inches	1 ft = 12 in.
	1 inch = 2.54 centimeters	1 in. = 2.5 cm
	1 inch = 25.4 millimeters	1 in. = 25 mm
Volume	1 cubic foot = 7.481 gallons	1 cf = 7.5 gal
	1 cubic foot = 28.32 liters	1 cf = 28 l
Weight	1 cubic foot of water weighs 62.41 pounds	1 cf = 62.4 lb
	1 kilogram = 2.204 pounds	1 kg = 2.2 lb
Pressure (head)	1 foot of water = 0.433 pounds per square inch	1 ft water = 0.43 psi
Time	1 minute = 60 seconds	1 min = 60 sec
Power	1 watt = 3.412 Btuh	1 w = 3.4 Btuh
	1 horsepower = 746 watts	1 hp = ¾ kw

With these few equivalences, a technologist can handle the vast majority of unit conversions that must be done. The technique consists of simply multiplying the original quantity by a series of conversions, each of which is equal to one and, therefore, does not change the original quantity but does change the units. At every step, units are cancelled. This method is foolproof, unlike use of conversion factors that can easily be misapplied. A series of examples of increasing complexity follow, to show both the technique and its simplicity.

(a) Convert 14 feet to inches

$$14 \text{ ft} \times \frac{12 \text{ in.}}{1 \text{ ft}} = 168 \text{ in.}$$

(b) Convert 148 square inches to square feet

$$148 \text{ in.}^2 \times \frac{1 \text{ ft}}{12 \text{ in}} \times \frac{1 \text{ ft}}{12 \text{ in}} = \frac{148}{12 \times 12} \text{ ft}^2 = 1.028 \text{ ft}^2$$

(c) Convert 8.1 cubic feet to cubic inches

$$8.1 \text{ ft}^3 \times \frac{12 \text{ in.}}{\text{ft}} \times \frac{12 \text{ in.}}{\text{ft}} \times \frac{12 \text{ in.}}{\text{ft}}$$
$$= 8.1 \times 12 \times 12 \times 12 \text{ in.}^3$$
$$= 13{,}996.8 \text{ in.}^3$$

(d) Convert 8.6 cubic foot per second to gallons per minute

$$\frac{8.6 \text{ ft}^3}{\text{sec}} \times \frac{60 \text{ sec}}{\text{min}} \times \frac{7.5 \text{ gal}}{\text{ft}^3} = 8.6 \times 60 \times 7.5 \text{ gpm}$$
$$= 3870 \text{ gpm}$$

(e) Convert a pressure of 0.31 kilograms per square millimeter to pounds per square inch

$$\frac{0.31 \text{ kg}}{\text{mm}^2} \times \frac{25.4 \text{ mm}}{\text{in.}} \times \frac{25.4 \text{ mm}}{\text{in.}} \times \frac{2.2 \text{ lb}}{\text{kg}} = 440 \frac{\text{lb}}{\text{in.}^2}$$
$$= 440 \text{ psi}$$

(f) Convert a pressure of 6 inches of water to pounds per square inch

Since we want pressure in pounds per square inch (psi), we will calculate the weight of a 6-in. column of water, 1 in.^{2} in cross section. Its volume is, obviously, 6 in.^{3}. Therefore,

$$6 \text{ in.}^3 \times \frac{62.4 \text{ lb}}{\text{ft}^3} \times \frac{\text{ft}}{12 \text{ in.}} \times \frac{\text{ft}}{12 \text{ in.}} \times \frac{\text{ft}}{12 \text{ in.}} = 0.2167 \text{ lb}$$

This weight, exerted on 1 in.^{2} of area, give a pressure of 0.2167 psi. Alternatively, remembering that the pressure exerted by a column of water 1 ft high is 0.43 psi (see previous list),

$$6 \text{ in. water} = 0.43/2 = 0.215 \text{ psi},$$

which is close enough for most requirements. For a detailed explanation of water pressure calculation, see Section 8.6 and Figures 8.1 and 8.2.

(g) Calculate ventilation rate

What ventilation rate in cubic feet per minute is required to give 6 air changes per hours (ACH) to a room 3 m by 4 m by 2.8 m high?

Although the calculation can be done in a single step, it will be clearer in two steps. A ventilation rate of six air changes simply means that the entire volume of air in the room is changed six times per hour. Therefore,

$$\text{Room volume} = 3 \times 4 \times 2.8 = 33.6 \text{ m}^3$$
$$6 \text{ ACH} = 6 \times 33.6 \text{ m}^3 = 201.6 \text{ m}^3/\text{h}$$

Converting this to cubic feet per minute is now a simple procedure:

$$201.6 \frac{\text{m}^3}{\text{h}} \times \frac{\text{h}}{60 \text{ min}} \times \frac{(3.28)^3 \text{ ft}^3}{\text{m}^3} = 118.8 \frac{\text{ft}^3}{\text{min}} \text{ (cfm)}$$

Using the approximation of 3.3 ft/m, and doing the entire calculation in one step, we have

$$3 \times 4 \times 2.8 \text{ m}^3 \times (3.3)^3 \frac{\text{ft}^3}{\text{m}^3} \times \frac{6 \text{ ACH}}{\text{h}} \times \frac{\text{h}}{60 \text{ min}}$$
$$= 120.7 \text{ cfm}$$

again, sufficiently accurate.

(h) Calculate heat generated

How much heat in Btuh is obtained from a 7.5-kw auxiliary electric heater in a heat pump?

$$7.5 \text{ kw} \times \frac{1000 \text{ w}}{\text{kw}} \times \frac{3.4 \text{ Btuh}}{\text{watt}} = 25,500 \text{ Btuh}$$

Key Terms

Having completed the study of this chapter, you should be familiar with the following key terms. If any appear unfamiliar or not entirely clear, you should review the section in which these terms appear. All key terms are listed in the index to assist you in locating the relevant text.

Absolute humidity
Anemometer
Btu (British thermal unit)
Conduction
Convection
Degree Celsius
Degree Fahrenheit
Degree Kelvin
Degree Rankine
Dew point
Dry bulb (DB) temperature
Enthalpy
Evaporative cooling
Humidity ratio
Latent heat
Latent heat of fusion

Latent heat of vaporization
Mean radiant temperature (MRT)
Met
Metabolic rate
Metabolism
Psychrometrics
Pyrometer
Radiation
Relative humidity (RH)
Sensible heat
Sling psychrometer
Specific heat
Specific humidity
Thermal comfort criteria
Wet Bulb depression
Wet bulb (WB) temperature

Supplementary Reading

ASHRAE Handbook of Fundamentals, 1993, Chapters 1 and 2.

Pita, E. G. *Air Conditioning Principles and Systems*, Wiley, New York, 1981, Chapters 1 and 2.

Stein, B., and Reynolds, J. *Mechanical and Electrical Equipment for Buildings*, 8th ed., Wiley, 1992, Chapters 2 and 4.

McQuiston, F. C., and Parker, J. D. *Heating, Ventilating and Air Conditioning*, 3rd ed., Wiley, New York, 1988, Chapters 1 and 4.

Problems

1. Express the boiling point and freezing point of water in °F, °C, °K, °Rankine.
2. A block of ice, ½ ft³ in volume, is taken from a freezer, where it was stored at 32°F (0°C). How many Btu of heat will be required to convert the ice to water at 75°F?
3. What conclusions about room humidity can you draw from the following results of measurement with a sling psychrometer?
   (a) DB >> WB
   (b) DB = WB
   (c) DB < WB
4. A silver spoon, a marble ashtray and a newspaper have been lying on a wooden table in a room for several hours. List the items in order of coolness to the touch of a normally warm hand. Why is this so?
5. Why does a person standing in the shade of a tree when the surrounding air is at 90°F and 75% RH with a 1-mph breeze feel warmer than with the same conditions but with 30% RH?
6. Explain why a person wearing dark-colored clothing feels warmer standing in the sun than a person wearing light-colored clothing of the same weight. Is this true regardless of air temperature and RH?

2. Thermal Balance of Buildings

This chapter will concentrate on the four basic areas of study that form the foundation of all HVAC work. They are
- Elements of heat transfer theory.
- Building heat loss transfer theory.
- Building heat gain calculations.
- Elements of psychrometrics.

Study of this chapter will enable you to:

1. Understand the fundamental heat transfer processes of conduction, convection and radiation as they apply to the building "envelope."
2. Calculate conductive heat loss through building envelope components.
3. Calculate heat loss through air spaces.
4. Calculate heat loss through built-up wall sections.
5. Calculate heat loss through surfaces on and below grade.
6. Calculate heat loss through glass doors and windows.
7. Calculate heat loss of air infiltration and mechanical ventilation.
8. Calculate heat gain (cooling load) from all the sources listed in items 2–7.
9. Select outside design conditions and understand degree-day calculations.
10. Understand the construction and use of the psychrometric chart in analyzing HVAC processes.

Heat Transfer

2.1 Heat Transfer in Buildings

We learned in Chapter 1 how the human body maintains its thermal balance. The same principles can be applied to the thermal balance of a building, if we make the necessary analogies. The interior design temperature for which we will design the HVAC system is the equivalent of the deep body temperature that was discussed in Chapter 1. The heat transfer mechanisms of a building are similar to that of the body except for the absence of evaporative skin cooling. That is, a building loses (or gains) heat by conduction, convection and radiation. The amount of the heat transferred by each of these mechanisms depends on the construction of the building envelope (skin), that is, how the walls, roof and ground level are built. If we use internal heat gain as the analogy of body metabolism, then the heat storage *(HS)* term in Equation (1.5) would be the amount of heating required in the winter. Summer cooling is more complex because of the energy required to remove latent heat, where dehumidification is necessary. For the moment, however, we will confine our discussion to heating.

The building steady-state heat balance equation is

$$\text{Heating} + M = CD + CO + R \qquad (2.1)$$

where CD, CO and R are the conduction, convection and radiation heat losses and
M is the internal heat gain.

Note that the E term (evaporative cooling) of Equation (1.5) is absent. As we shall see, the CD and CO terms are always a heat loss in winter. The radiation term R is usually a loss, although buildings with large glass areas in their outside walls can show a radiative heat gain from strong sunlight, even in winter. The term M in equation (2.1) is always a positive number because it represents the total internal heat gain from occupants, lighting and machinery. The heating energy required is, therefore, the total of the building envelope's net heat loss less the internal heat gain. Although the three heat transfer mechanisms (conduction, convection and radiation) are the same as previously studied, the factors affecting them in a building are somewhat different from those of the body. It is these factors that we will discuss in the following sections.

2.2 Conduction

Conductive heat transfer takes place through a solid material any time there is a temperature difference between the two sides of the material. See Figure 2.1(a). This temperature difference can be thought of as the driving force that causes the heat transfer. Therefore, the rate at which heat is

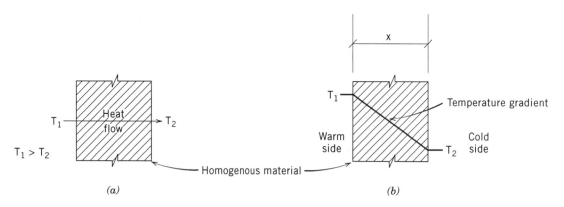

Figure 2.1 *(a)* Heat flow by conduction through any homogeneous solid material occurs from the side at higher temperature T_1 to the side at lower temperature T_2. The flow of heat is continuous, uniform and uninterrupted as long as the temperature difference exists. *(b)* The temperature gradient through a homogeneous material x inches thick is a straight line whose slope is $(T_1-T_2)/x$ degrees per inch. The temperature at any point inside the material is, therefore, $T_1-(T_1-T_2)/x. \times$ (Depth in inches from the outside surface).

transferred is directly proportional to temperature difference: the larger the difference, the faster the heat flow. As we have learned, the direction of heat flow is from the higher temperature to the lower one. The heat transfer will continue, without interruption, as long as the temperature difference remains. If the material is homogeneous throughout, then the temperature gradient between the two sides can be represented as a straight sloping line. See Figure 2.1 *(b)*. This means that we can easily determine the temperature at any point within the material. The importance of this item will become clear in Section 2.6.

As stated, the rate of heat transfer is proportional to the temperature difference. It is obviously also proportional to the area of the piece of material; the larger it is, the more area is exposed to the temperature difference and, therefore, the more heat is transferred. The heat transfer rate is also proportional to the thermal conductivity of the specific (homogeneous) material involved. Different materials conduct heat at different rates. Metals are very good conductors. This is the reason that they feel cool to the touch; they are simply conducting away body heat rapidly, leaving the skin cool. Hence, the metal feels cool, although what we are actually feeling is our cool skin. In point of fact, the metal is no cooler than anything else in its vicinity. Other materials such as wood, fibers, cloth and cork are poor thermal conductors. They feel warm to the touch because they do not conduct away body heat rapidly; consequently, the skin at the contact point remains warm. Such materials are called *thermal insulators*. (As we will learn, the best of these insulating materials owe their insulating properties to entrapped air.)

Thermal conductivity is usually represented by the lowercase letter k. In the English system its unit is Btu per hour, per square foot of surface area (exposed to the temperature difference), per inch of material thickness, per °F of temperature difference.

Obviously, the thicker a material, the slower it will conduct heat because the heat has that much farther to go to get to the other side. Since most building materials are made in specific thicknesses (e.g., 4-in. brick, 8-in. block, ⅝-in. plywood), it is usually more convenient to speak of the overall *thermal conductance C* of a specific building material rather than its conductivity per inch of thickness. Obviously then,

$$C = \frac{k}{\text{Thickness in inches}} \quad (2.2)$$

Overall conductance C is smaller than k for materials thicker than 1 in. and larger than k for materials thinner than 1 in. In units,

$$C = \frac{\text{Btuh}}{\text{ft}^2\text{-°F}} \quad \text{or} \quad \frac{\text{Btu}}{\text{h-ft}^2\text{-°F}}$$

that is, Btu per hour, per square foot of area, per °F of temperature difference between the two sides.

The *rate of heat transfer Q* through any homogeneous material of given thickness, is

$$Q = C \times A \, (T_2 - T_1) \quad (2.3)$$

where

Q is the overall rate of heat transfer in Btuh,
C is the material's conductance in Btuh/ft²-°F or Btu/h-ft²-°F,
A is the material surface area in square feet and $(T_2 - T_1)$ is the temperature differential in °F.

Checking units we have

$$Q = \frac{\text{Btuh}}{\text{ft}^2\text{-°F}} \times \text{ft}^2 \times \text{°F} = \text{Btuh}$$

To avoid confusion, remember that k (conductivity) is the thermal heat transfer characteristic of a type of material—stone, wood, plaster, and so on—per inch of thickness. The symbol C (conductance) refers to a specific material of specific thickness. Both of these factors have largely fallen into disuse in recent years because of the emphasis on energy conservation that began more than two decades ago.

Prior to the Arab oil embargo of 1973, fuel was cheaper than insulation. Building design then took little if any notice of energy use. Designers dealt with thermal conductance figures and another factor called overall thermal transmittance U, which we will explain later. When the price of fuel skyrocketed in 1973, it rapidly became apparent that previous design techniques were producing buildings that were too expensive to operate because of energy costs. The solution to this problem was to redesign building envelopes to reduce heat transfer drastically. A very important part of this redesign involved the use of materials with low conductivity plus the addition of insulation, particularly in retrofit work. Because the goal of the design was to *resist* heat loss (and gain), the concept of thermal resistance became easier to use than thermal conductance. *Thermal resistance R* is simply the reciprocal of conductance, that is,

$$R = \frac{1}{C} \quad (2.4)$$

where R is measured in °F-ft²-h/Btu.

It can be thought of as the number of hours it takes for 1 Btu to penetrate 1 ft² of the material, for each F° temperature difference. Thermal resistance is a more logical quantity to use than conductance, because the thicker the material the higher its resistance. Furthermore, in built-up ceiling or wall sections consisting of different materials, the overall thermal resistance is simply the arithmetic sum of the thermal resistances of the components; that is,

$$R_T = R_1 + R_2 + \cdots + R_N \quad (2.5)$$

(See Figure 2.5 for typical resistance calculations of built-up wall sections.) Calculation of overall conductance is much more laborious. To use an electrical analogy (see Figure 11.9, page 636), thermal resistances of built-up sections are similar to electrical resistances connected in series. The use of thermal resistance R has become so common, that today insulation is described as R-19, R-40 and so forth.

Although it is not often used, thermal resistivity r is the resistance of a material per inch of thickness; that is,

$$r = \frac{1}{k} \quad (2.6)$$

and its units are h-ft²-°F-in./Btu.

Typical values of k, C, R and r are given in Table 2.1 for some common building materials. For more information, refer to the second reference in the supplementary reading list at the end of this chapter. Figure 2.2 shows conductivity k, conductance C, resistivity r and resistance R calculations for a

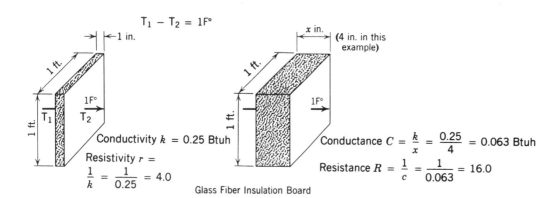

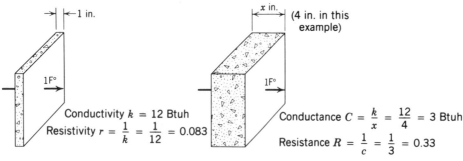

Figure 2.2 Heat transfer coefficients (k,r,C,R) for two very common construction materials, per unit area of 1 ft² and unit temperature difference between the two sides of 1 F°. Glass fiber board is a good heat insulation material. A 4-in. thickness has a resistance of 16 and is, therefore, called R-16 insulation. Ordinary concrete has poor thermal resistance. A 4-in. thickness has a resistance of only 0.33. It is, therefore, a good heat conductor. (From Stein and Reynolds, *Mechanical and Electrical Equipment for Buildings*, 8th ed., John Wiley 1992, reprinted by permission of John Wiley & Sons.)

Table 2.1 Typical Thermal Properties of Common Building Materials—Design Value[a]

Description	Density, lb/ft³	Conductivity[b] (k), Btu-in./ h-ft²-°F	Conductance (C), Btu/ h-ft²-°F	Resistance[c] (R) Per Inch Thickness (1/k), °F-ft²-h/ Btu-in.	Resistance[c] (R) For Thickness Listed (1/C), °F-ft²-h/ Btu
BUILDING BOARD					
Asbestos-cement board	120	4.0	—	0.25	—
Asbestos-cement board, 0.25 in.	120	—	16.50	—	0.06
Gypsum or plaster board, 0.5 in.	50	—	2.22	—	0.45
Plywood (Douglas fir)	34	0.80	—	1.25	—
Plywood (Douglas fir), 0.375 in.	34	—	2.13	—	0.47
Plywood (Douglas fir), 0.5 in.	34	—	1.60	—	0.62
Plywood (Douglas fir), 0.625 in.	34	—	1.29	—	0.77
Vegetable fiber board					
Sheathing, regular density, 0.5 in.	18	—	0.76	—	1.32
Shingle backer, 0.375 in.	18	—	1.06	—	0.94
Sound deadening board, 0.5 in.	15	—	0.74	—	1.35
Tile & lay-in panels, plain or acoustic	18	0.40	—	2.50	—
0.5 in.	18	—	0.80	—	1.25
0.75 in.	18	—	0.53	—	1.89
Hardboard[e]					
Medium density	50	0.73	—	1.37	—
High-density, service-tempered grade & service grade	55	0.82	—	1.22	—
Particleboard[e]					
Medium density	50	0.94	—	1.06	—
High density	62.5	1.18	—	0.85	—
Underlayment, 0.625 in.	40	—	1.22	—	0.82
Wood subfloor, 0.75 in.	—	—	1.06	—	0.94
BUILDING MEMBRANE					
Vapor—permeable felt	—	—	16.70	—	0.06
Vapor—seal, 2 layers of mopped 15-lb felt	—	—	8.35	—	0.12
Vapor—seal, plastic film	—	—	—	—	Negl.
FINISH FLOORING MATERIALS					
Carpet & fibrous pad	—	—	0.48	—	2.08
Carpet & rubber pad	—	—	0.81	—	1.23
Cork tile, 0.125 in.	—	—	3.60	—	0.28
Terrazzo, 1 in.	—	—	12.50	—	0.08
Tile-asphalt, linoleum, vinyl, rubber	—	—	20.00	—	0.05
Wood, hardwood finish, 0.75 in.	—	—	1.47	—	0.68
INSULATING MATERIALS					
Blanket and batt[e]					
Mineral fiber, fibrous form processed from rock, slag, or glass					
approx. 3–4 in.	0.4–2.0	—	0.091	—	11
approx. 3.5 in.	1.2–1.6	—	0.067	—	15
approx. 5.5–6.5 in.	0.4–2.0	—	0.053	—	19
approx. 6–7.6 in.	0.4–2.0	—	0.045	—	22
approx. 8.25–10 in.	0.4–2.0	—	0.033	—	30
approx. 10–13 in.	0.4–2.0	—	0.026	—	38
Board and Slabs					
Cellular glass	8.0	0.33	—	3.03	—
Glass fiber, organic bonded	4.0–9.0	0.25	—	4.00	—

Table 2.1 *(Continued)*

Description	Density, lb/ft^3	Conductivity[b] (k), Btu-in./ h-ft^2-°F	Conductance (C), Btu/ h-ft^2-°F	Resistance[c] (R) Per Inch Thickness (1/k), °F-ft^2-h/ Btu-in.	For Thickness Listed (1/C), °F-ft^2-h/ Btu
Expanded perlite, organic bonded	1.0	0.36	—	2.78	—
Expanded polystyrene, molded beads	1.0	0.26	—	3.85	—
	1.5	0.24	—	4.17	—
	2.0	0.23	—	4.35	—
Mineral fiber with resin binder	15.0	0.29	—	3.45	—
Mineral fiberboard, wet felted					
Core or roof insulation	16–17	0.34	—	2.94	—
Acoustical tile	18.0	0.35	—	2.86	—
Acoustical tile	21.0	0.37	—	2.70	—
Interior finish (plank, tile)	15.0	0.35	—	2.86	—
Cement fiber slabs (shredded wood with Portland cement binder)	25.0–27.0	0.50–0.53	—	2.0–1.89	—
Loose Fill					
Cellulosic insulation (milled paper or wood pulp)	2.3–3.2	0.27–0.32	—	3.70–3.13	—
Perlite, expanded	2.0–4.1	0.27–0.31	—	3.7–3.3	—
	4.1–7.4	0.31–0.36	—	3.3–2.8	—
	7.4–11.0	0.36–0.42	—	2.8–2.4	—
Mineral fiber (rock, slag or glass)[e]					
approx. 3.75–5 in.	0.6–2.0	—	—	—	11.0
approx. 6.5–8.75 in.	0.6–2.0	—	—	—	19.0
approx. 10.25–13.75	0.6–2.0	—	—	—	22.0
Spray Applied				—	30.0
Polyurethane foam	1.5–2.5	0.16–0.18	—	6.25–5.56	—
Cellulosic fiber	3.5–6.0	0.29–0.34	—	3.45–2.94	—
Glass fiber	3.5–4.5	0.26–0.27	—	3.85–3.70	—
ROOFING					
Asbestos-cement shingles	120	—	4.76	—	0.21
Asphalt roll roofing	70	—	6.50	—	0.15
Asphalt shingles	70	—	2.27	—	0.44
Built-up roofing, 0.374 in.	70	—	3.00	—	0.33
Slate, 0.5 in.	—	—	20.00	—	0.05
Wood shingles, plain & plastic film faced	—	—	1.06	—	0.94
PLASTERING MATERIALS					
Cement plaster, sand aggregate	116	5.0	—	0.20	—
Gypsum plaster:					
Lightweight aggregate, 0.5 in.	45	—	3.12	—	0.32
Lightweight aggregate on metal lath, 0.75 in.	—	—	2.13	—	0.47
Perlite aggregate	45	1.5	—	0.67	—
Sand aggregate	105	5.6	—	0.18	—
Sand aggregate on metal lath, 0.75 in.	—	—	7.70	—	0.13
Vermiculite aggregate	45	1.7	—	0.59	—
MASONRY MATERIALS					
Masonry Units					
Brick, fired clay	150	8.4–10.2	—	0.12–0.10	—
	100	4.2–5.1	—	0.24–0.20	—
				0.40–0.33	—

Table 2.1 *(Continued)*

Description	Density, lb/ft^3	Conductivity^b (k), Btu-in./ h-ft^2-°F	Conductance (C), Btu/ h-ft^2-°F	Resistance^c (R) Per Inch Thickness (1/k), °F-ft^2-h/ Btu-in.	Resistance^c (R) For Thickness Listed (1/C), °F-ft^2-h/ Btu
Concrete blocks					
Normal weight aggregate (sand & gravel)					
8 in., 33–36 lb, 126–136 lb/ft^3 concrete,					
2 or 3 cores	—	—	0.90–1.03	—	1.11–0.97
Same with verm.-filled cores	—	—	0.52–0.73	—	1.92–1.37
12 in., 50 lb, 125 lb/ft^3 concrete,					
2 cores	—	—	0.81	—	1.23
Lightweight aggregate (expanded shale, clay, slate or slag pumice)					
6 in., 16–17 lb, 85–87 lb/ft^3 concrete,					
2 or 3 cores	—	—	0.52–0.61	—	1.93–1.65
Same with perlite-filled cores	—	—	0.24	—	4.2
Same with verm.-filled cores	—	—	0.33	—	3.0
8 in., 19–22 lb, 72–86 lb/ft^3					
concrete	—	—	0.32–0.54	—	3.2–1.90
Same with perlite-filled cores	—	—	0.15–0.23	—	6.8–4.4
Same with verm.-filled cores	—	—	0.19–0.26	—	5.3–3.9
12 in., 32–36 lb, 80–90 lb/ft^3 concrete,					
2 or 3 cores	—	—	0.38–0.44	—	2.6–2.3
Same with verm.-filled cores	—	—	0.17	—	5.8
Gypsum partition tile					
3 by 12 by 30 in., solid	—	—	0.79	—	1.26
Concretes					
Sand & gravel or stone aggregate concretes	150	10.0–20.0	—	0.10–0.05	—
Lightweight aggregate concretes					—
Expanded shale, clay, or slate;					—
expanded slags; cinders; pumice					—
(with density up to 100 lb/ft^3)	100	4.7–6.2	—	0.21–0.16	—
	80	3.3–4.1	—	0.30–0.24	—
	40	1.3	—	0.78	—
Perlite, vermiculite, and polystyrene beads	50	1.8–1.9	—	0.55–0.53	—
	30	1.1	—	0.91	—
Foam concretes	120	5.4	—	0.19	–
	80	3.0	—	0.33	—
Foam concretes and cellular concretes	40	1.4	—	0.71	—
SIDING MATERIALS (on flat surface)					
Shingles					
Asbestos-cement	120	—	4.75	—	0.21
Wood, plus insulation backer board, 0.3125 in.	—	—	0.71	—	1.40
Siding					
Asbestos-cement, 0.25 in., lapped	—	—	4.76	—	0.21
Asphalt insulating siding (0.5 in. bed)	—	—	0.69	—	1.46
Wood, bevel, 0.5 × 8 in. lapped	—	—	1.23	—	0.81
Aluminum or Steel^g over sheathing					
Insulating-board backed nominal 0.375 in.	—	—	0.55	—	1.82

Table 2.1 *(Continued)*

				Resistance[c] (R)	
Description	Density, lb/ft[3]	Conductivity[b] (k), Btu-in./ h-ft[2]-°F	Conductance (C), Btu/ h-ft[2]-°F	Per Inch Thickness (1/k), °F-ft[2]-h/ Btu-in.	For Thickness Listed (1/C), °F-ft[2]-h/ Btu
Insulating-board backed nominal 0.375 in., foil backed	—	—	0.34	—	2.96
Architectural (soda-lime float) glass	158	6.9	—	—	—
WOODS (12% moisture content)					
Hardwoods					
Oak	41.2–46.8	1.12–1.25	—	0.89–0.80	—
Birch	42.6–45.4	1.16–1.22	—	0.87–0.82	—
Maple	39.8–44.0	1.09–1.19	—	0.92–0.84	—
Ash	38.4–41.9	1.06–1.14	—	0.94–0.88	—
Softwoods					
Southern pine	35.6–41.2	1.00–1.12	—	1.00–0.89	—
Douglas fir–larch	33.5–36.3	0.95–1.01	—	1.06–0.99	—
Southern cypress	31.4–32.1	0.90–0.92	—	1.11–1.09	—
Hem–fir, spruce–pine–fir	24.5–31.4	0.74–0.90	—	1.35–1.11	—
West Coast woods, cedars	21.7–31.4	0.68–0.90	—	1.48–1.11	—
California redwood	24.5–28.0	0.74–0.82	—	1.35–1.22	—

[a] Values are for a mean temperature of 75°F (24°C). Representative values for dry materials are intended as design (not specification) values for materials in normal use. For properties of a particular product, use the value supplied by the manufacturer or by unbiased tests.

[b] To obtain thermal conductivities in Btuh-ft-°F, divide the K-factor by 12 in./ft.

[c] Resistance values are the reciprocals of C before rounding off C to two decimal places.

[d] Does not include paper backing and facing, if any. Where insulation forms a boundary (reflective or otherwise) of an air space, see Tables 2.3 and 2.6 for insulating value of an air space with the appropriate effective emittance and temperature conditions of the space.

[e] Conductivity varies with fiber diameter. Batt, blanket, and loose-fill mineral fiber insulations are manufactured to achieve specified R-values, the most common of which are listed here.

[f] Insulating values of acoustical tile vary, depending on density of the board and on type, size, and depth of perforations.

[g] Values for metal siding applied over flat surfaces vary widely.

Source. Data reprinted by permission of the American Society of Heating, Refrigerating and Air-Conditioning Engineers, Atlanta, Georgia, from the 1993 *ASHRAE Handbook—Fundamentals*.

common insulating material and a common construction material. See also Table 2.2 for a summary of the terms that relate to heat transfer by conduction.

2.3 Convection

As described in Section 1.8, convection is a heat transfer mechanism that relies on fluid flow to carry the heat from one place to another. Although convective currents can occur in any fluid, convective heat transfer in building work is caused by air movement. The basic natural, or free convective flow is illustrated in Figure 1.5. (Forced convection, as illustrated in Figure 1.6, is not under discussion here.) This same natural convection occurs at a cold window in a heated room. See Figure 2.3 *(a)*. The cold outside air reduces the temperature of the inside surface of the window (by conduction). This in turn cools the layer of air immediately adjacent to the window, making it heavier than the warm room air. As a result, it drops towards the floor and is replaced by warm room air. This air in turn is cooled and falls, thus creating a convective air current, as shown. (For this reason, heaters are placed below windows—to reheat the cold air dropping from them. This practice prevents what can be a quite strong and unpleasant cold "draft" on a cold day from a single-glazed window.)

Table 2.2 Summary of Heat Transfer Terms and Symbols

Symbol	Term	Units	Definition
k	Conductivity	Btuh/ft²-in.-°F	The rate of heat flow through a homogeneous material, per inch of thickness, per ft², per h, per °F
c	Conductance	Btuh/ft²-°F	The rate of heat flow through a given thickness of homogeneous material, per ft², per h, per °F
r	Resistivity	ft²-°F-in./Btuh	The characteristic of a specific homogeneous material that defines its resistance to the passage of heat by conductance. Numerically, the reciprocal of k
R	Resistance	ft²-°F-h/Btu	The resistance of an homogeneous material of given thickness to the passage of heat, by conduction; the reciprocal of C
U	Overall conductance	Btuh/ft²-°F	The overall conductance of a number of materials combined into a single construction assembly; the reciprocal of total resistance.
R_T	Overall resistance	ft²-°F/Btuh	The overall thermal resistance of a number of materials combined into a single construction assembly; the arithmetic sum of all the individual resistances; the reciprocal of U

When a window is double-glazed, the heat transfer is more complex. The warm inside air loses heat through the inside pane, cools and becomes heavier, falls along the window and sets up an inside-the-room convective current. Because the inside glass pane is warmer than the outside pane, a circulating current, which transfers heat across the space, is set up in the air between the two panes. Finally, a convective air current is set up on the outside of the outside pane. Thus heat is transferred through a double-pane window by three convective air currents and direct conduction through the two panes of glass. See Figure 2.3 *(b)*.

The best thermal (and acoustic) insulation would be achieved if the space between the two panes of a double-glazed window were evacuated. The reason that this is not done is simply that air pressure would easily crack the glass. (A recent patent for an evacuated double-glazed window uses transparent glass supports between the panes to prevent this breakage. Also, sealed double-glazed windows that contain inert gases are available and show excellent insulation characteristics.)

Convective currents are easily established when the barrier between the high and low temperatures is vertical, as with a window. When the barrier is horizontal, the establishment of a convective air flow depends on the direction of heat flow, that is, whether the heat flow is up or down. In Figure 2.4 *(a)*, which represents a winter condition, with heat flow upward, convective air currents are set up in the attic. Their strength depends on the type and location of insulation, which governs the amount of conductive heat transfer. In Figure 2.4 *(b)*, which represents a summer condition, with heat flow downward, there is no convective air current in the attic because the hottest and, therefore, lightest air is trapped against the attic ceiling. This is also true inside the room below, where a hot blanket of stagnant air forms at the ceiling and stays there. This example demonstrates that convective air currents in horizontal air spaces will flow only if the heat flow direction is upward. (It also demonstrates the need for an air outlet near the peak of a sloped roof, which will permit hot air to escape, thus establishing a convective air current. An opening for air to enter at the attic eaves is also required.)

2.4 Radiation

The basic mechanism of radiation was explained in Section 1.8. Radiant heat energy is an electromagnetic wave phenomenon in the infrared range. When such a wave strikes a barrier, part of the energy is reflected, part is absorbed and, sometimes, part is transmitted, depending on the mate-

30 / THERMAL BALANCE OF BUILDINGS

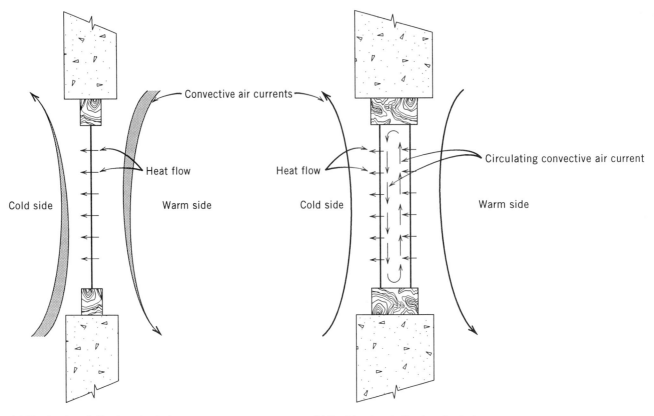

(a) Single-glazed, fixed-sash window *(b)* Double-glazed, fixed-sash window

Figure 2.3 *(a)* Since heat flow through a single pane of glass is large, strong convective currents will exist on both sides of the glass (assuming still air). *(b)* With a double-glazed, fixed-sash window, overall heat transmission is reduced. Consequently, the inside and outside convective currents are also smaller than for single glazing. The air in the space between the panes of glass transfers heat from the warmer inside surface of the inside pane to the cooler inside surface of the outside pane by means of a circulating convective current. This current, shown by arrows on the diagram, is driven by the temperature difference between the two inside surfaces of the glass panes.

rial of which the barrier is made. Transparent materials such as glass transmit much of the energy, particularly if the energy is in the short wavelength portion of the infrared range such as is the energy from the sun. Long wavelength infrared energy, such as is typical of heat radiated from low temperature objects, is blocked by glass. This causes the familiar greenhouse effect. Heat from the sun enters through the glass in the space's envelope and heats up the objects in the space. They then reradiate heat in the long wavelength range because of their low temperature. This heat is largely blocked by the same glass that transmits the short wavelength infrared energy of the sun, causing the temperature in the space to rise rapidly.

Shiny surfaces such as aluminum foil, polished metalized paper or plastic, and the like reflect most of the heat striking them, absorb very little and transmit even less. They are essentially opaque to heat, just as similar surfaces are opaque to light. These materials, unlike dark-colored, dull-finish materials, cannot radiate heat well at all. Since they cannot radiate, they cannot absorb, and since they do not transmit, they must necessarily reflect the heat striking them. As a result, these materials become, in effect, excellent thermal barriers. This inability to radiate heat is measured by a factor

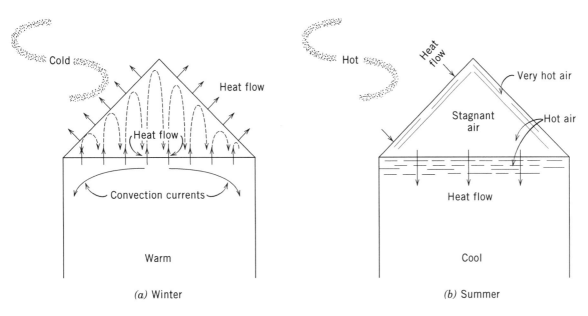

Figure 2.4 *(a)* In winter, convective air currents in the attic continuously circulate because heat flow is upward causing the warm light air to rise and cooler heavier air to drop. *(b)* In summer, heat flow is down. The lighter warmer air stays at the top of a sealed, unventilated attic and at the ceiling of the rooms in the house. Convective currents are negligible.

called *emittance*, which is the ratio of the radiation of the material in question to that of an ideal "black-body" radiator, at a specific temperature. The emittance of bright aluminum foil is 0.05, whereas that of wood is about 0.9. Table 2.3 lists the emittances of common construction materials both alone and when used on one or both sides of closed air spaces. It is important to appreciate the difference in action between a thermal barrier (such as bright foil) and thermal insulation. Thermal barriers act to reflect *radiant* heat; insulation acts to slow down the transmission of *conducted* heat. The two actions are frequently combined in such materials as foil-backed insulation batts.

2.5 Heat Flow Through Air Spaces

Air is a thermal insulator. When there is little or no air motion, the thin layer (film) of air immediately adjacent to a surface shows considerable thermal resistance, depending on the position and emittance of the surface on which the air rests. Refer to Table 2.4. Notice there that the resistance of an air film in still air on a vertical highly reflective surface is 1.70. This is the equivalent of ¾-in. fiberboard sheathing or of 1½-in. plywood. On a nonreflective horizontal surface with heat direction down (a plaster ceiling in the summer, for instance), this film has a resistance of 0.92, which is the equivalent of ¾-in. plywood or an 8-in. concrete block. When the surface air layer is dispersed by air movement, its thermal resistance value rapidly becomes negligible.

When air is confined between two surfaces so that it is effectively still air, the heat transfer across the space is conductive, convective and radiative. Conduction varies with the width of the space and the temperature difference between both sides. Convection varies with the space position (horizontal, vertical, sloped), the direction of heat flow (up, down, diagonal), the width of the space and the temperature difference. Radiative heat transfer depends on the effective emittance of the space (see Table 1.6), and the temperature between the sides. (Mean temperature of the space slightly affects heat transfer by conduction and convection.) Table 2.5 gives the thermal resistance of air spaces of

Table 2.3 Emittance Values of Various Surfaces and Effective Emittances of Air Spaces[a]

Surface	Average Emittance e	Effective Emittance E of Air Space	
		One Surface Emittance e; the Other 0.90	Both Surfaces Emittances e
Aluminum foil, bright	0.05	0.05	0.03
Aluminum foil, with condensate just visible (>0.7 gr/ft^2)	0.30	0.29	—
Aluminum foil, with condensate clearly visible (>2.9 gr/ft^2)	0.70	0.65	—
Aluminum sheet	0.12	0.12	0.06
Aluminum-coated paper, polished	0.20	0.20	0.11
Steel, galvanized, bright	0.25	0.24	0.15
Aluminum paint	0.50	0.47	0.35
Building materials: wood, paper, masonry, nonmetallic paints	0.90	0.82	0.82
Regular glass	0.84	0.77	0.72

[a] These values apply in the 4- to 40-μm range of the electromagnetic spectrum.

Source. Data extracted and reprinted by permission of the American Society of Heating, Refrigerating and Air-Conditioning Engineers, Atlanta, Georgia, from the 1993 *ASHRAE Handbook—Fundamentals.*

Table 2.4 Surface Conductances C_s (Btuh-ft^2-°F) and Resistances $R \; \dfrac{°F \cdot ft^2 \cdot h}{Btu}$ for Air[a]

Position of Surface	Direction of Heat Flow	Surface Emittance					
		Non-reflective, e=0.90		e=0.20		e=0.05	
		C_s	R	C_s	R	C_s	R
Still Air							
Horizontal	Upward	1.63	0.61	0.91	1.10	0.76	1.32
Sloping (45°)	Upward	1.60	0.62	0.88	1.14	0.73	1.37
Vertical	Horizontal	1.46	0.68	0.74	1.35	0.59	1.70
Sloping (45°)	Downward	1.32	0.76	0.60	1.67	0.45	2.22
Horizontal	Downward	1.08	0.92	0.37	2.70	0.22	4.55
Moving Air (any position)							
15-mph wind (for winter)	Any	6.00	0.17				
7.5-mph wind (for summer)	Any	4.00	0.25				

Note: A surface cannot take credit for both an air space resistance value and a surface resistance value. No credit for an air space value can be taken for any surface facing an air space of less than 0.5 in.

[a] Conductances are for surfaces of the stated emittance facing virtual black-body surroundings at the same temperature as ambient air. Values are based on a surface-air temperature difference of 10°F and for surface temperature of 70°F.

Source. Reprinted by permission of the American Society of Heating, Refrigerating and Air-Conditioning Engineers, Atlanta, Georgia, from the 1993 *ASHRAE Handbook—Fundamentals.*

different thicknesses, taking all these factors into account. These resistance values are particularly useful when calculating the overall thermal resistance of built-up wall and roof sections, as we shall see in subsequent sections. As a rough rule of thumb, an air space with nonreflective sides (wood, insulation, plaster, etc.) has a thermal resistance of about 1, that is, R-1; a narrow space with reflective sides is R-2; and a wide space is R-4. For accurate calculations, use the figures in Table 2.5.

The most effective way to take advantage of the insulating properties of air is to isolate the air into many small pockets. Doing so prevents the formation of convective air currents and utilizes film resistance. This is exactly the situation in common insulating materials such as mineral fiber, glass fiber and expanded plastic of various types. (See Table 2.1.) In these materials, air is trapped between fibers and in closed microcells, forming millions of small dead-air pockets, which gives the material its thermal insulating properties. As a rule of thumb, these materials have an R-4 rating per inch of material thickness. Thus, a 4-in. batt is about R-16, and an 8-in. batt is approximately R-32. Here, too, accurate calculations require use of accurate figures from Table 2.1.

2.6 Heat Flow Through Built-up Sections

In the preceding sections, you have studied the three processes of heat transfer through materials, including air. All three usually operate simultaneously. Furthermore, practical building construction almost always consists of wall, floor, ceiling and roof sections that are built-up of layers of different materials. To demonstrate overall steady-state heat transfer calculations for practical building sections, we will now analyze a few such com-

Table 2.5 Thermal Resistances of Plane Air Spaces, °F-ft²-h/Btu[a,b]

Position of Air Space	Direction of Heat Flow	Air Space[b] Mean Temperature[c], °F	Temperature Difference[c], °F	Effective Emittance (E)[c] 0.5-in. Air Spaces[b]					Effective Emittance (E)[c] 0.75-in. Air Space[b]				
				0.03	0.05	0.2	0.5	0.82	0.03	0.05	0.2	0.5	0.82
Horizontal	Up ↑	90	10	2.13	2.03	1.51	0.99	0.73	2.34	2.22	1.61	1.04	0.75
		50	30	1.62	1.57	1.29	0.96	0.75	1.71	1.66	1.35	0.99	0.77
		50	10	2.13	2.05	1.60	1.11	0.84	2.30	2.21	1.70	1.16	0.87
		0	20	1.73	1.70	1.45	1.12	0.91	1.83	1.79	1.52	1.16	0.93
		0	10	2.10	2.04	1.70	1.27	1.00	2.23	2.16	1.78	1.31	1.02
45° Slope	Up ↗	90	10	2.44	2.31	1.65	1.06	0.76	2.96	2.78	1.88	1.15	0.81
		50	30	2.06	1.98	1.56	1.10	0.83	1.99	1.92	1.52	1.08	0.82
		50	10	2.55	2.44	1.83	1.22	0.90	2.90	2.75	2.00	1.29	0.94
		0	20	2.20	2.14	1.76	1.30	1.02	2.13	2.07	1.72	1.28	1.00
		0	10	2.63	2.54	2.03	1.44	1.10	2.72	2.62	2.08	1.47	1.12
Vertical	Horizontal →	90	10	2.47	2.34	1.67	1.06	0.77	3.50	3.24	2.08	1.22	0.84
		50	30	2.57	2.46	1.84	1.23	0.90	2.91	2.77	2.01	1.30	0.94
		50	10	2.66	2.54	1.88	1.24	0.91	3.70	3.46	2.35	1.43	1.01
		0	20	2.82	2.72	2.14	1.50	1.13	3.14	3.02	2.32	1.58	1.18
		0	10	2.93	2.82	2.20	1.53	1.15	3.77	3.59	2.64	1.73	1.26
45° Slope	Down ↘	90	10	2.48	2.34	1.67	1.06	0.77	3.53	3.27	2.10	1.22	0.84
		50	30	2.64	2.52	1.87	1.24	0.91	3.43	3.23	2.24	1.39	0.99
		50	10	2.67	2.55	1.89	1.25	0.92	3.81	3.57	2.40	1.45	1.02
		0	20	2.91	2.80	2.19	1.52	1.15	3.75	3.57	2.63	1.72	1.26
		0	10	2.94	2.83	2.21	1.53	1.15	4.12	3.91	2.81	1.80	1.30
Horizontal	Down ↓	90	10	2.48	2.34	1.67	1.06	0.77	3.55	3.29	2.10	1.22	0.85
		50	30	2.66	2.54	1.88	1.24	0.91	3.77	3.52	2.38	1.44	1.02
		50	10	2.67	2.55	1.89	1.25	0.92	3.84	3.59	2.41	1.45	1.02
		0	20	2.94	2.83	2.20	1.53	1.15	4.18	3.96	2.83	1.81	1.30
		0	10	2.96	2.85	2.22	1.53	1.16	4.25	4.02	2.87	1.82	1.31
				1.5-in. Air Space[b]					3.5-in. Air Space[b]				
Horizontal	Up ↑	90	10	2.55	2.41	1.71	1.08	0.77	2.84	2.66	1.83	1.13	0.80
		50	30	1.87	1.81	1.45	1.04	0.80	2.09	2.01	1.58	1.10	0.84
		50	10	2.50	2.40	1.81	1.21	0.89	2.80	2.66	1.95	1.28	0.93
		0	20	2.01	1.95	1.63	1.23	0.97	2.25	2.18	1.79	1.32	1.03
		0	10	2.43	2.35	1.90	1.38	1.06	2.71	2.62	2.07	1.47	1.12
45° Slope	Up ↗	90	10	2.92	2.73	1.86	1.14	0.80	3.18	2.96	1.97	1.18	0.82
		50	30	2.14	2.06	1.61	1.12	0.84	2.26	2.17	1.67	1.15	0.86
		50	10	2.88	2.74	1.99	1.29	0.94	3.12	2.95	2.10	1.34	0.96
		0	20	2.30	2.23	1.82	1.34	1.04	2.42	2.35	1.90	1.38	1.06
		0	10	2.79	2.69	2.12	1.49	1.13	2.98	2.87	2.23	1.54	1.16
Vertical	Horizontal →	90	10	3.99	3.66	2.25	1.27	0.87	3.69	3.40	2.15	1.24	0.85
		50	30	2.58	2.46	1.84	1.23	0.90	2.67	2.55	1.89	1.25	0.91
		50	10	3.79	3.55	2.39	1.45	1.02	3.63	3.40	2.32	1.42	1.01
		0	20	2.76	2.66	2.10	1.48	1.12	2.88	2.78	2.17	1.51	1.14
		0	10	3.51	3.35	2.51	1.67	1.23	3.49	3.33	2.50	1.67	1.23
45° Slope	Down ↘	90	10	5.07	4.55	2.56	1.36	0.91	4.81	4.33	2.49	1.34	0.90
		50	30	3.58	3.36	2.31	1.42	1.00	3.51	3.30	2.28	1.40	1.00
		50	10	5.10	4.66	2.85	1.60	1.09	4.74	4.36	2.73	1.57	1.08
		0	20	3.85	3.66	2.68	1.74	1.27	3.81	3.63	2.66	1.74	1.27
		0	10	4.92	4.62	3.16	1.94	1.37	4.59	4.32	3.02	1.88	1.34
Horizontal	Down ↓	90	10	6.09	5.35	2.79	1.43	0.94	10.07	8.19	3.41	1.57	1.00
		50	30	6.27	5.63	3.18	1.70	1.14	9.60	8.17	3.86	1.88	1.22
		50	10	6.61	5.90	3.28	1.73	1.15	11.15	9.27	4.09	1.93	1.24
		0	20	7.03	6.43	3.91	2.19	1.49	10.90	9.52	4.87	2.47	1.62
		0	10	7.31	6.66	4.00	2.22	1.51	11.97	10.32	5.08	2.52	1.64

[a] Values apply for ideal conditions, that is, air spaces of uniform thickness bounded by plane, smooth, parallel surfaces with no air leakage to or from the space.

[b] A single resistance value cannot account for multiple air spaces; each space requires a separate resistance calculation that applies only for the established boundary conditions. Resistances of horizontal spaces with heat flow downward are substantially independent of temperature difference.

[c] Interpolation is permissible for other values of mean temperature, temperature difference and effective emittance E. Interpolation and moderate extrapolation for air spaces greater than 3.5 in. are also permissible.

Source. Data extracted and reprinted by permission of the American Society of Heating, Refrigerating and Air-Conditioning Engineers, Atlanta, Georgia, from the 1993 ASHRAE *Handbook—Fundamentals*.

pound constructions. (Note that all the discussion thus far has related to steady-state heat transfer, that is, heat transfer after all the transient phenomena have passed. The most important of these transient phenomena—thermal lag—will be discussed later on.)

Figure 2.5 *(a–d)* shows how the resistance of individual components in a wall assembly add arithmetically to make the overall thermal resistance R_T. Such wall assemblies have traditionally been identified by an overall *coefficient of thermal transmission*, called U, whose units are Btuh/ft²-°F. The relation between overall thermal resistance and U is simply

$$U = \frac{1}{R_T} \quad (2.7)$$

Despite the fact that the modern approach is to use thermal resistance R and not conductance, most tables still list the transmission coefficient U of built-up constructions assemblies. Calculation of the overall resistance of these sections is quite simple using Equation 2.7. Table 2.2 gives a summary of heat transfer terms and symbols.

Note that the addition of an uninsulated nonreflective air space to the simple block construction of Figure 2.5a increases the R of the wall by about 16%. Making the air space reflective [Figure 2.5 *(c)*], increases the original resistance by almost 50%. Adding only 3½ in. of insulation, with or without a reflective layer, more than triples the original resistance and more than doubles that of the air space construction. In modern construction, 3½ in. of insulation is considered insufficient in all but the mildest climate; 4–6 in. is much more commonly used. This much insulation will increase the wall section to about R-30.

In actual construction, with the possible exception of hot dry climates, a vapor barrier would be installed on the warm side (inside) of built-up assemblies of the type shown in Figure 2.5. This barrier, which is usually no more than a sheet of polyethylene plastic, serves to prevent moisture from inside the building from "migrating" through the wall. If such air-borne water vapor, in winter, is permitted to pass into the wall, it will reach its dew point someplace inside the wall and condense, forming droplets of water. This condensation seriously depreciates the R value of insulation and the effectiveness of reflective layers. Figure 2.6 shows the technique of calculating the temperature gradient through a built-up wall assembly, such as in Figure 2.6 *(d)*. In the absence of a vapor barrier, and assuming an inside RH of 50%, and inside and outside temperature of 70 and 20°F, respectively, the dew point of the air is at 50°F. It will be reached in the center of the insulation, as shown in Figure 2.6. Figure 2.7 shows how the total resistance factor R_T and transmittance factor U are calculated for two typical roof constructions.

2.7 Heat Loss Through Surfaces on and Below Grade

It has been found through testing that most of the loss from a slab on grade is through its perimeter and not through the slab itself into the earth. See Figure 2.8. The formula for use in this calculation is

$$q = F_2 (P) (t_i - t_o) \quad (2.8)$$

where

q = heat loss in Btuh of the slab
F_2 = perimeter factor (see Table 2.6)
P = length of slab perimeter in feet
t_i = inside temperature, °F
t_o = outside temperature, °F

The perimeter heat loss factor depends not only on the type and thickness of insulation used (if any) but also on the severity of the area's winter climate, the type of wall construction above the slab, the use of vapor barriers and the exact method of insulation installation. For this reason, a range of values is given in Table 2.6. For specific designs, consult the project architect, engineer and local insulation supplier for accurate data, based on local experience and testing.

Calculation of heat loss through below-grade walls and below-grade floor is a complex procedure because the earth temperature changes with depth. The procedure involves calculation of the loss of 1-foot-high strips of below-grade walls and summing the total. For a detailed description of the method, refer to the second supplementary reading reference at the end of this chapter.

2.8 Heat Flow Through Windows and Doors

Windows are usually the source of the largest winter heat loss (and summer heat gain) in a building. For this reason, the technologist/designer performing the heat transfer calculation must take particular care in determining an accurate R or U

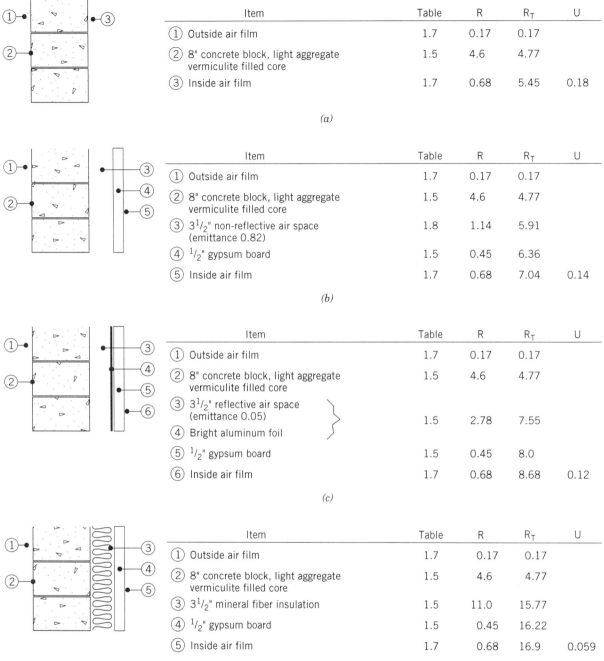

Figure 2.5 Calculation of total thermal resistance R_T and overall transmittance U for built-up wall section assemblies. *(a)* Simple concrete block wall; *(b)* block wall with a nonreflective air space; *(c)* block wall with a reflective air space; *(d)* block wall with an insulated, reflective cavity. In practical calculations, the overall wall resistance would be reduced by thermal "bridges" at studs and possibly by the ceiling construction.

36 / THERMAL BALANCE OF BUILDINGS

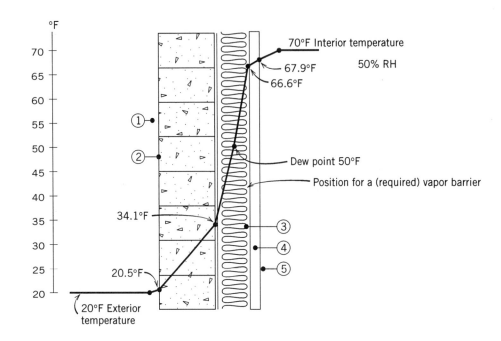

Item	Component Temperature rise	Temperatures Cold side	Warm side
① Air film	$\frac{0.17}{16.9} \times 50°F = 0.5°$	20°F	20.5°F
② 8" block	$\frac{4.6}{16.9} \times 50°F = 13.6°$	20.5°F	34.1°F
③ 3½" insulation	$\frac{11}{16.9} \times 50°F = 32.5°$	34.1°F	66.6°F
④ gypsum board	$\frac{0.45}{16.9} \times 50°F = 1.3°$	66.6°F	67.9°F
⑤ Inside air film	$\frac{0.68}{16.9} \times 50°F = 2.1°$	67.9°F	70°F

Figure 2.6 Calculation of the temperature gradient through the wall section of Figure 2.5(d). Refer to Figure 2.5(d) for a description of components and their thermal resistances. Note that moisture from the inside air at RH 50% will condense at 50°F. This temperature occurs exactly in the middle of the insulation blanket. This moisture will drastically reduce the thermal resistance of the mineral fiber insulation. A vapor barrier between the gypsum board and the insulation is required to block the movement of water vapor into the insulation.

Coefficients of Transmission, U (Btu/h ft^2°F), of Flat Masonry Roofs with Built-Up Roofing, with and Without Suspended Ceilings: Winter Conditions, Upward Flow

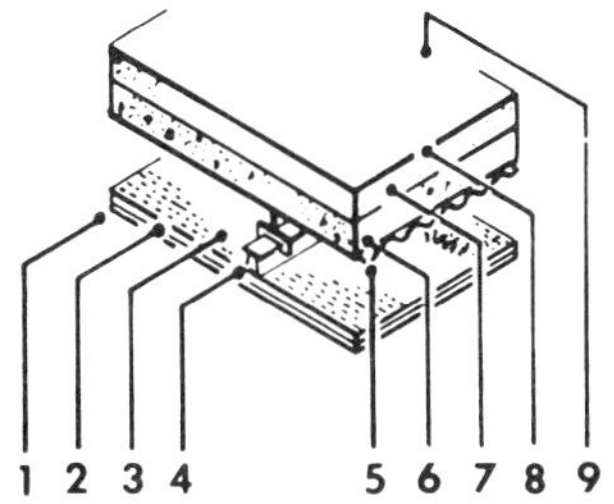

	Base Case Resistance R	Construction (Heat Flow Up)	Resistance R[c]: New Item 7 ("Construction")
	0.61	1. Inside surface (still air)	0.61
	0.47	2. Metal lath and light-weight aggregate plaster, 0.75 in.	0.47
	0.93[a]	3. Nonreflective air space, greater than 3.5 in. (50 F mean; 10 F° temperature difference)	0.93[a]
	0[b]	4. Metal ceiling suspension system with metal hanger rods	0[b]
	0	5. Corrugated metal deck	0
	2.22	6. Concrete slab, lightweight aggregate, 2 in. (30 lb/ft^3)	2.22
	—	7. Rigid roof deck insulation (none)	4.17
	0.33	8. Built-up roofing, 0.375 in.	0.33
	0.17	9. Outside surface (15-mph wind)	0.17
	4.73	Total Thermal Resistance, R	8.90

Base Case
$$U = \frac{1}{4.73} = 0.211$$

New Item 7
$$U = \frac{1}{8.90} = 0.112$$

[a] Use largest air space (3.5 in.) value shown in Table 2.5.
[b] Area of hanger rods is negligible in relation to ceiling area.
[c] When rigid roof deck insulation added, $C = 0.24$ ($R = 1/C = 4.17$).

Figure 2.7 *(a)* Coefficients of transmission, U (Btu/h-ft^2-°F), of flat masonry roofs with built-up roofing, with and without suspended ceilings; winter conditions, upward heat flow. *(b)* Coefficients of transmission, U (Btu/h-ft^2-°F), of 45° pitched roofs. (Reprinted by permission from: Copyright © by ASHRAE, Atlanta, GA. *1981 Handbook of Fundamentals.*)

Coefficients of Transmission, U (Btu/h ft² °F), of 45° Pitched Roofs

Part A. Reflective Air Space

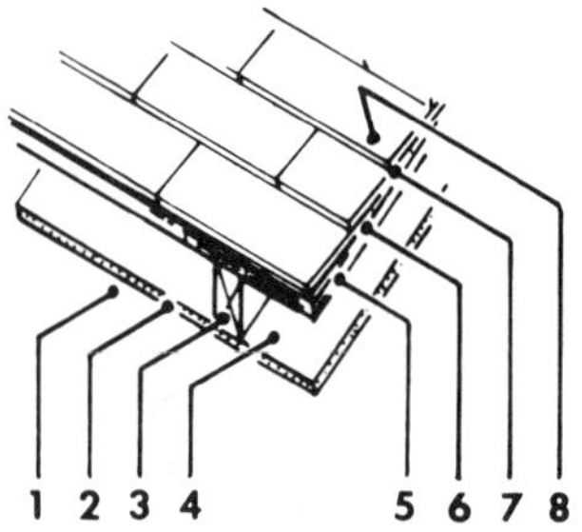

	Resistance for Heat Flow Up: Winter Conditions			Resistance for Heat Flow Down: Summer Conditions	
	Between Rafters, R_i	At Rafters, R_s	Construction	Between Rafters, R_i	At Rafters, R_s
	0.62	0.62	1. Inside surface (still air)	0.76	0.76
	0.45	0.45	2. Gypsum wallboard 0.5 in., foil backed	0.45	0.45
	—	4.35	3. Nominal 2-in. × 4-in. ceiling rafter	—	4.35
	2.17	—	4. 45° slope reflective air space, 3.5 in. (50 F mean, 30 F° temperature difference), $E = 0.05$	4.33[a]	—
	0.77	0.77	5. Plywood sheathing, 0.625 in.	0.77	0.77
	0.06	0.06	6. Permeable felt building membrane	0.06	0.06
	0.44	0.44	7. Asphalt shingle roofing	0.44	0.44
	0.17	0.17	8. Outside surface (15-mph wind)	0.25[b]	0.25[b]
	4.68	6.86	Total Thermal Resistance, R	7.06	7.08

Heat Flow Up
$$U_i = \frac{1}{4.68} = 0.213$$

Heat Flow Down
$$U_i = \frac{1}{7.06} = 0.141$$

To adjust U values for the effect of framing: With 10% framing (typical of 2-in. rafters at 16-in. o.c.), these adjusted U_{av} values are, respectively:

$U_{av} = 0.206$ (3% less heat loss) $\qquad\qquad U_{av} = 0.141$ (unchanged)

Figure 2.7 *(b)*

Part B. Nonreflective Air Space

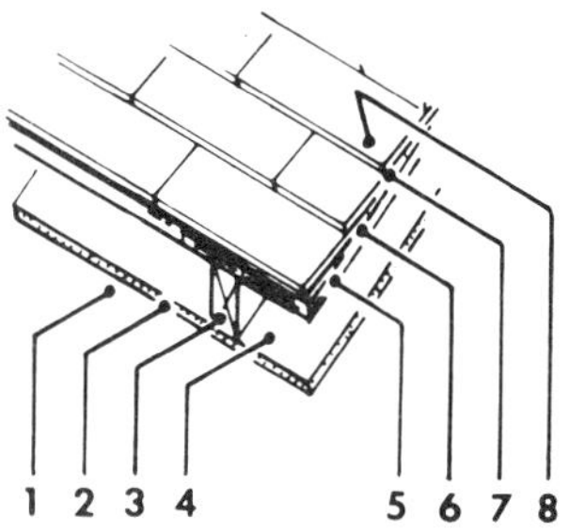

	Resistance for Heat Flow Up: Winter Conditions			Resistance for Heat Flow Down: Summer Conditions	
	Between Rafters, R_i	At Rafters, R_s	Construction	Between Rafters, R_i	At Rafters, R_s
	0.62	0.62	1. Inside surface (still air)	0.76	0.76
	0.45	0.45	2. Gypsum wallboard, 0.5 in.	0.45	0.45
	—	4.35	3. Nominal 2-in. × 4-in. ceiling rafter	—	4.35
	0.06	—	4. 45° slope, nonreflective air space, 3.5 in. (50 F mean, 10 F° temperature difference)	0.90[a]	—
	0.77	0.77	5. Plywood sheathing, 0.625 in.	0.77	0.77
	0.06	0.06	6. Permeable felt building membrane	0.06	0.06
	0.44	0.44	7. Asphalt shingle roofing	0.44	0.44
	0.17	0.17	8. Outside surface	0.25[b]	0.25[b]
	3.47	6.86	*Total Thermal Resistance, R*	3.63	7.08

Heat Flow Up

$U_i = \dfrac{1}{3.47} = 0.287$

Adjusted for 10% framing, as above:

$U_{av} = 0.273$ (5% less heat loss)

Heat Flow Down

$U_i = \dfrac{1}{3.63} = 0.275$

$U_{av} = 0.262$ (5% less heat loss)

[a] Air space value of 90 F mean, 10F° temperature difference.
[b] Outside wind velocity 7.5 mph.

Figure 2.7(b) *(Continued)*

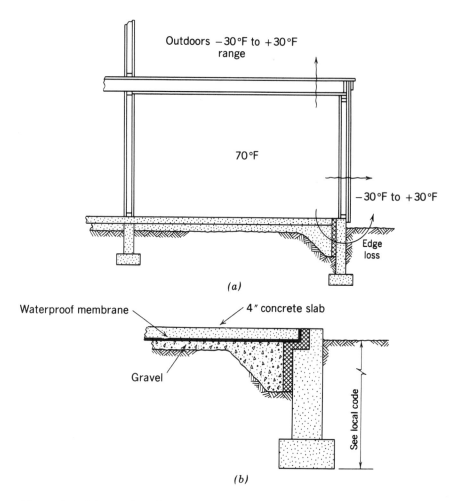

Figure 2.8 (a) Heat loss from a slab on grade is proportional to its perimeter, because it is primarily edge loss. The loss increases with lower exterior temperatures. Edge loss is also higher with metal-stud wall construction than with masonry walls. See Table 2.6. (b) Insulation can be installed in a number of ways.

Table 2.6 Perimeter Heat Loss Factor F_2; Concrete Slab on Grade

Insulation	F_2 Range
R-0	0.62–0.9
R-5.4	0.45–0.58
R-11	0.27–0.36

value (overall transmission coefficient) for them. Among the factors affecting R and U are thickness and number of glazing panes, size of air spaces between panes, gas fill type if used (argon or krypton), operation (movable or fixed), type of operation movement (slide, swing, hinged), type of sash construction (aluminum, wood, metal with thermal break), coatings on glass, aspect (vertical, horizontal, sloped), heat flow direction and proportion of glass to sash and mullion area. Because of the large number of variables, accurate data must be obtained from manufacturers for actual designs. In the absence of such data, tables in various publications can be used for preliminary calculations. (See the first two references on the supplementary reading list at the end of this chapter.) Table 2.7 lists some approximate U values for representative window types. You can readily see that, because of the extremely low R values of even the best windows (R-3), rooms with appreciable glass area will lose most of their heat through the windows. (As we

Table 2.7 Representative U and R Values for Vertical Windows

	U, $Btuh/ft^2$-°F (R, ft^2-°F/Btuh)	
	U(R)	U(R)
Type	Metal Frame	Wood Frame
Single-glazed	1.3 (0.77)	0.9 (1.1)
Double-glazed, air-filled	0.85 (1.1)	0.5 (2.0)
Double-glazed, inert gas-filled	0.80 (1.25)	0.40 (2.5)
Triple-glazed air-filled	0.77 (1.3)	0.40 (2.5)
Triple-glazed, inert gas-filled	0.72 (1.4)	0.35 (2.9)

will see, the situation is even more critical in cooling load calculation.)

Doors can also be a considerable heat loss source not only because of infiltration around the door but also because of the poor R value of the door itself. Table 2.8 gives some typical R figures for common door constructions. Here, too, the technologist should seek reliable information based on tests of specific products. In the absence of such data, tabular data from this or other authoritative sources can be used for preliminary calculations.

Table 2.8 R-Factors for Typical Door Constructions

Description	Door Alone	Door with Storm Door
Wood doors		
1⅜", 1¾" hollow-core flush door	R-2.1	R-3.3
1⅜", 1¾" solid-core flush door	R-2.6	R-3.8
2¼" solid-core flush door	R-3.7	R-5.0
1¾" Steel doors		
Polyurethane foam core, no thermal break	R-3.5	—
Polyurethane foam core, with thermal break	R-5.0	—
Solid urethane foam core, no thermal break	R-2.5	—
Solid urethane foam core, with thermal break	R-5.0	—

Heating

2.9 Heat Loss from Air Infiltration and Ventilation

Every building requires ventilation to rid itself of odors, carbon dioxide and moisture that accumulates as a result of normal occupancy. In almost all residential and in many commercial buildings, this ventilation occurs naturally. Outside air enters through cracks around doors, windows, walls, crawl spaces, roof and attic doors, fireplaces, cable and piping entries and the like. Infiltration occurs on the windward side of the building, and exfiltration (loss of air) occurs on the leeward side through similar crack-type openings. With the emphasis on energy conservation came the realization that in most buildings the heat loss due to cold air infiltration in winter amounted to 25–50% of the building's heat loss. This caused a revision in construction techniques that resulted in much "tighter" buildings. These tight buildings have much lower infiltration loss but, in many cases, have resulted in what has come to be known as the *Sick Building Syndrome (SBS)*, caused by insufficient fresh air. The solution to both problems—excessive infiltration heat loss and SBS—has been to design and construct buildings for controlled infiltration and deliberate ventilation. There are two methods for calculating infiltration heat loss: the *air change method* and the *effective leakage area* method. They are analyzed separately in the discussion that follows.

The air change method assumes a number of total *air changes per hour (ACH)* based on the tightness of construction and then calculates the heat loss using the heat storage capacity of air. The number of air changes per hour for residential buildings, in winter, is given in Table 2.9. Constuction tightness definitions, on which Table 2.9 is based, follow:

(a) *Tight construction*—close fitting doors, windows and framing, no fireplace, and use of a vapor barrier. Small new houses (less than 1500 square feet) are frequently in this category.
(b) *Medium tighness*—older large houses with average maintenance. Fireplaces must have a damper or glass enclosure; windows and doors of standard commercial construction.
(c) *Loose construction*—old houses, poorly fitted

42 / THERMAL BALANCE OF BUILDINGS

Table 2.9 Winter ACH as a Function of Construction Air Tightness[a]

	\multicolumn{10}{c}{Outdoor Design Temperatures, °F}
	50	40	30	20	10	0	−10	−20	−30	−40
Tight	0.41	0.43	0.45	0.47	0.49	0.51	0.53	0.55	0.57	0.59
Medium	0.69	0.73	0.77	0.81	0.85	0.89	0.93	0.97	1.00	1.05
Loose	1.11	1.15	1.20	1.23	1.27	1.30	1.35	1.40	1.43	1.47

[a] Values for 15-mph wind and indoor temperature of 68°F.

Source. Reprinted by permission of the American Society of Heating, Refrigerating and Air-Conditioning Engineers, Atlanta, Georgia, from the 1993 *ASHRAE Handbook—Fundamentals.*

windows and doors, unenclosed fireplace. Poorly constructed houses fit into this category, as do most mobile homes.

To calculate infiltration heat loss Q in Btuh/°F, we use the formula

$$Q = V \times ACH \times 0.018 \quad (2.9)$$

where
  V = volume of the space in cubic feet
  ACH = air changes per hour
  0.018* = heat capacity of air in $\frac{Btuh}{°F\text{-}ft^3/h}$

The effective leakage area method is more accurate because it is more detailed. It requires accurate knowledge of window and door construction, overall building construction, wind velocity and site data. The method then uses tables of leakage area for walls, windows, doors, floor and the like (see, for instance, Supplementary Reading, Reference 2 [Table 3, Chapter 23] in a formula that combines leakage area, wind velocity, stack effect and temperature difference to arrive at an infiltration flow rate in cubic feet per minute (cfm). This figure is then used in the following formula to arrive at infiltration heat loss:

$$Q = 1.08 \, (cfm)(t_i - t_o) \quad (2.10)$$

where
  Q = the heat loss in Btuh
  1.08 = the heat content of air in Btuh/cfm-°F
  t_i = inside air temperature
  t_o = outside air temperature

*The factor 0.018 is derived by multiplying the density of air by its specific heat; that is,

$$\frac{0.075 \, lb}{ft^3} \times \frac{0.240 \, Btu}{lb\text{-°}F} = \frac{0.018 \, Btu}{ft^3\text{-°}F} \times \frac{hour}{hour} = \frac{0.018 \, Btuh}{°F\text{-}ft^3/h}$$

The factor 1.08 is derived by using the previously calculated factor of .018 Btuh/°F × ft³/h and converting air flow to ft³/min (cfm):

$$\frac{0.018 \, Btu}{°F \times ft^3/h} \times \frac{60 \, min}{hour} = \frac{1.08 \, Btu}{ft^3/min \times °F} = 1.08 \frac{Btu}{cfm\text{-°}F}$$

This same formula can be used to calculate the heat loss whenever the air flow in cubic feet per minute is known, including forced ventilation and stack effect ventilation.

2.10 Heat Loss to Adjacent Unheated Spaces

In many buildings, heated spaces adjoin unheated ones such as garages, basements, utility rooms and enclosed stairwells. These unheated spaces are always at temperatures somewhere between the outside and inside temperature. Although detailed calculations are possible most designers will use outside temperature for all such areas except basements. This is because basements are surrounded by earth, which is at a higher temperature than outside (winter) air, and because most basements have a heat gain from building heating plants. As a result, the ambient temperature in a basement is estimated as one-half to two-thirds of the way between outside and inside temperatures depending on heat gain, windows and depth.

2.11 Summary of Building Heat Losses

We are now able to summarize the preceding discussion on heat loss calculation. The total heat loss of a structure is the sum of the following individual components:

a. Heat Loss Through Walls

$$\Sigma Q_w = \frac{A_w \times \Delta t}{R_w} \text{ or } A_w \times U_w \times \Delta t \quad (2.11)$$

where
 Q_w is the wall heat loss in Btuh,
 A_w is the wall area in square feet,
 R_w is the thermal resistance of the wall (or U_w, the overall thermal transmittance) and
 Δt is the temperature difference between the inside and outside wall surfaces.

Note that for all walls, and in particular insulated walls with metal stud construction, the R factor of the wall must be reduced at thermal bridging points caused by ceiling beams, wall studs, columns and the like. See for instance Supplemental Reading, Reference 1 (Table 4.8).

b. Heat Loss Through Ceilings and Roofs

Heat loss occurs only between spaces with a temperature difference between them. Therefore, in multistory construction ceiling/roof loss occurs only on the top floor. Calculation is similar to (a), using the appropriate R factor for upward heat flow.

c. Heat Flow Through Floors

This calculation is applicable only to floors above unheated areas (basements) and those on grade. With the latter, the heat loss calculation is a perimeter calculation as explained in Section 2.7.

d. Heat Flow Through Windows and Doors

Particular care must be taken here with window calculations because of the large losses involved. Include glass doors as a special case of windows. The direction of heat flow is especially important with nonvertical windows such as skylights.

e. Infiltration (and Ventilation) Losses

For residential work, the air change method of infiltration calculation is normally adequate. For commercial buildings that must adhere to increasingly restrictive governmental agency energy guidelines, the more complex crack-length method should be used. Ventilation losses are straightforward calculations using Equation 2.10.

2.12 Building Heat Loss Calculation Procedure

As should be apparent from Section 2.11, a detailed heat loss calculation for even a small residence is a time-consuming affair. Furthermore, if the data are not well organized, the calculations can easily become confused. Today, most such calculations are performed on a computer with menu-driven programs. When they are done manually, as for instance in small engineering offices or by entry-level technologists who are learning the procedures, a calculation form is invariably used. Unfortunately, there are about as many styles of such forms as there are offices. Several are shown in Figure 2.9.

The form shown in Figure 2.9a is the simplest type requiring a separate sheet for each space. U and R factors are taken from tables similar to Tables 2.1–2.4. Summaries of the total structure heat loss must be made on a separate form. Figure 2.9b is a similar type of calculation form with space for calculation of several rooms on a single sheet. Here, too, U factors are taken from tables similar to Tables 2.1–2.4. Notice that infiltration data refer only to door and windows as these are the principal sources of infiltration air. The 1.1 factor is an approximation of the more accurate value 1.08 [see Equation (2.10)].

Figure 2.9 *(c-1)* is a form recommended by the Air Conditioning Contractors of America (ACCA, 1513 16th Street, N.W., Washington, D.C. 20036). This form is used primarily for residential construction but can be used for small commercial buildings as well. Notice that this procedure uses HTM factors (Heat Transfer Multipliers), which are nothing more than the U factors multiplied by the design temperature difference ($U \times \Delta t$). This saves one calculation step since all that is then required to obtain the heat loss Q is to multiply the HTM factor by area, since

$$Q = A \times U \times \Delta T$$

where

$$U \times \Delta T = \text{HTM}$$

therefore,

$$Q = A \times \text{HTM}$$

A sample page from the extensive HTM tables used by this method is shown in Figure 2.9 *(c-2)*. Design conditions are recorded on a separate form; they are shown here at the bottom of Figure 2.9 *(c-1)*.

Another contractor-type form is shown in Figure

44 / THERMAL BALANCE OF BUILDINGS

```
PROJECT _____     PREPARED BY _____
LOCATION _____     DATE _____
```

HEATING LOAD CALCULATIONS

INSIDE DESIGN TEMP _____
OUTSIDE DESIGN TEMP _____
DIFFERENCE _____

ITEM	QUANTITY × U-VALUE × ΔT = BTUH (W)
GLASS	FT² (M²) °F (°C)
NET WALL (ABOVE GRADE)	FT² (M²) °F (°C)
NET WALL (BELOW GRADE)	FT² (M²) °F (°C)
ROOF	FT² (M²) °F (°C)
OUTSIDE AIR	CFM (L/S) 1.08 (1.20) °F (°C)

TOTAL _____

(a)

Figure 2.9 Typical building heat loss calculation forms. For explanation see text. (a) From Bradshaw, *Building Control Systems,* 2nd ed., Wiley, 1993, reprinted by permission of John Wiley & Sons. (b) From Pita, *Air Conditioning Principles and Systems,* Wiley, 1981, reprinted by permission of Prentice Hall, Inc., Upper Saddle River, NJ. (c) Reprinted with permission from *ACCA Manual J.* (d) Reprinted with permission from Hydronics Institute, Publication H-22.

Heating Load Calculations Form

| HEATING LOAD CALCULATIONS | Project _____ Location _____ | Outdoor _____ °F. Calc. by _____ Engr. _____ Indoor _____ °F. Check by _____ |

Room																	
Plan size																	
Heat transfer	U	A	TD	BTU/hr	U	A	TD	BTU/hr	U	A	TD	BTU/hr	U	A	TD	BTU/hr	
Wall																	BTU/hr
Window																	
Door																	BTU/hr
Roof/ceiling																	
Floor																	
Partition																	
Subtotal																	
Infiltration	CFM/ft	L, A			CFM/ft	L, A			CFM/ft	L, A			CFM/ft	L, A			
Window 1.1 ×																	
Door 1.1 ×																	
Total heat loss																	

Room																	= = = =
Plan size																	
Heat transfer	U	A	TD	BTU/hr	U	A	TD	BTU/hr	U	A	TD	BTU/hr	U	A	TD	BTU/hr	Building heat transfer / Infiltration heat loss / Ventilation heat loss / Total building heat loss
Wall																	
Window																	
Door																	
Roof/ceiling																	
Floor																	
Partition																	
Subtotal																	
Infiltration	CFM/ft	L, A			CFM/ft	L, A			CFM/ft	L, A			CFM/ft	L, A			
Window 1.1 ×																	
Door 1.1 ×																	
Total heat loss																	

NOTES:

(b)

Figure 2.9 *(Continued)*

46 / THERMAL BALANCE OF BUILDINGS

HEAT LOSS CALCULATION
(DO NOT WRITE IN SHADED BLOCKS)

					Entire House		1 Living		2 Dining		3 Laundry		4 Kitchen		5 Bath-1	
1	Name of Room															
2	Running Ft. Exposed Wall				160		21		25		18		11		9	
3	Room Dimensions Ft.				51 x 29		21 x 14		7 x 18		7 x 11		11 x 11		9 x 11	
4	Ceiling Ht. Ft. Directions Room Faces				8		8 West		8 North		8		8 East		8 East	
	TYPE OF EXPOSURE	Const No.	HTM Htg.	HTM Clg.	Area or Length	Btuh Htg.	Area or Length	Btuh Htg.	Area or Length	Btuh Htg.	Area or Length	Btuh Htg.	Area or Length	Btuh Htg.	Area or Length	Btuh Htg.
5	Gross	a 12-d			1280		168		200		144		88		72	
	Exposed	b 14-b			480											
	Walls &	c 15-b			800											
	Partitions	d														
6	Windows	a 3-A	41.3		60	2478	40	1652	20	826						
	& Glass	b 2-C	48.8		20	976										
	Doors Htg.	c 2-A	35.6		105	3738							11	392	8	285
		d														
7	Windows	North														
	& Glass	E&W														
	Doors Clg.	South														
8	Other Doors	11-E	14.3		37	529					17	243				
9	Net	a 12-d	6.0		1078	6468	128	768	180	1080	127	762	77	462	64	384
	Exposed	b 14-b	10.8		460	4968										
	Walls &	c 15-b	5.5		800	4400										
	Partitions	d														
10	Ceilings	a 16-d	4.0		1479	5916	294	1176	126	504	77	308	121	484	99	396
		b														
11	Floors	a 21-a	1.8		1479	2662										
		b														
12	Infiltration HTM		70.6		222	15673	40	2824	20	1412	17	1200	11	777	8	565
13	Sub Total Btuh Loss = 6+8+9+10+11+12					47808		6420		3822		2513		2115		1630
14	Duct Btuh Loss		0%		—		—		—		—		—		—	
15	Total Btuh Loss = 13 + 14					47808		6420		3822		2513		2115		1630
16	People @ 300 & Appliances 1200															
17	Sensible Btuh Gain = 7+8+9+10+11+12+16															
18	Duct Btuh Gain		%													
19	Total Sensible Gain = 17 + 18															

ASSUMED DESIGN CONDITIONS AND CONSTRUCTION (Heating):

From Table 2

		Const. No.	HTM
A.	Determing Outside Design Temperature -5° db-Table 1		
B.	Select Inside Design Temperature 70°db		
C.	Design Temperature Difference: 75 Degrees		
D.	Windows: Living Room & Dining Room - Clear Fixed Glass, Double Glazed - Wood Frame - Table 2	3A	41.3
	Basement - Clear Glass Metal Casement Windows, with Storm - Table 2	2C	48.8
	Others - Double Hung, Clear, Single Glass and Storm, Wood Frame - Table 2	2A	35.6
E.	Doors: Metal, Urethane Core, no Storm - Table 2	11E	14.3
F.	First Floor Walls: Basic Frame Construction with Insulation (R-11) ½" Board - Table 2	12d	6.0
	Basement wall: 8" Concrete Block - Table 2		
	Above Grade Height: 3 ft (R = 5)	14b	10.8
	Below Grade Height: 5 ft (R = 5)	15b	5.5
G.	Ceiling: Basic Construction Under Vented Attic with Insulation (R-19) - Table 2	16d	4.0
H.	Floor: Basement Floor, 4" Concrete - Table 2	21a	1.8
I.	All moveable windows and doors have certified leakage of 0.5 CFM per running foot of crack (without storm), envelope has plastic vapor barrier and major cracks and penetrations have been sealed with caulking material, no fireplace, all exhausts and vents are dampered, all ducts taped.		

(c-1)

Figure 2.9 (Continued)

Table 2 (Continued)

No. 14 Masonry Walls, Block or Brick, Finished or Unfinished - Above Grade

	_____ Winter Temperature Difference _____																
	20	25	30	35	40	45	50	55	60	65	70	75	80	85	90	95	U
	HTM (Btuh per sq. ft.)																
A. 8" or 12" Block, No Insul., Unfin.	10.2	12.8	15.3	17.8	20.4	22.9	25.5	28.0	30.6	33.1	35.7	38.2	40.8	43.3	45.9	48.4	.510
B. 8" or 12" Block + R-5	2.9	3.6	4.3	5.0	5.8	6.5	7.2	7.9	8.6	9.4	10.1	10.8	11.5	12.2	13.0	13.7	.144
C. 8" or 12" Block + R-11	1.5	1.9	2.3	2.7	3.1	3.5	3.8	4.2	4.6	5.0	5.4	5.8	6.2	6.5	6.9	7.3	.077
D. 8" or 12" Block + R-19	1.0	1.2	1.4	1.7	1.9	2.2	2.4	2.6	2.9	3.1	3.4	3.6	3.8	4.1	4.3	4.6	.048
E. 4" Brick + 8" Block, No. Insul.	8.0	10.0	12.0	14.0	16.0	18.0	20.0	22.0	24.0	26.0	28.0	30.0	32.0	34.0	36.0	38.0	.400
F. 4" Brick + 8" Block + R-5	2.7	3.3	4.0	4.7	5.3	6.0	6.6	7.3	8.0	8.6	9.3	10.0	10.6	11.3	12.0	12.6	.133
G. 4" Brick + 8" Block + R-11	1.5	1.9	2.2	2.6	3.0	3.3	3.7	4.1	4.4	4.8	5.2	5.5	5.9	6.3	6.7	7.0	.074
H. 4" Brick + 8" Block + R-19	.9	1.2	1.4	1.6	1.9	2.1	2.3	2.6	2.8	3.1	3.3	3.5	3.8	4.0	4.2	4.5	.047

No. 15 Masonry Walls, Block or Brick, Finished or Unfinished - Below Grade*

	_____ Winter Temperature Difference _____																
	20	25	30	35	40	45	50	55	60	65	70	75	80	85	90	95	U
	HTM (Btuh per sq. ft.)																
Walls Extend 2'-5' Below Grade																	
A. 8" or 12" Block + No Insul.	2.5	3.1	3.7	4.4	5.0	5.6	6.2	6.9	7.5	8.1	8.7	9.4	10.0	10.6	11.2	11.9	.125
B. 8" or 12" Block + R-5	1.5	1.8	2.2	2.6	3.0	3.3	3.7	4.1	4.4	4.8	5.2	5.5	5.9	6.3	6.7	7.0	.074
C. 8" or 12" Block + R-11	1.0	1.3	1.5	1.8	2.0	2.3	2.6	2.8	3.1	3.3	3.6	3.8	4.1	4.3	4.6	4.8	.051
D. 8" or 12" Block + R-19	.7	.9	1.0	1.2	1.4	1.5	1.7	1.9	2.0	2.2	2.4	2.6	2.7	2.9	3.1	3.2	.034
Walls Extend More Than 5' Below Grade																	
E. 8" or 12" Block + No Insul.	1.7	2.2	2.6	3.0	3.5	3.9	4.3	4.8	5.2	5.6	6.1	6.5	6.9	7.4	7.8	8.2	.087
F. 8" or 12" Block + R-5	1.2	1.5	1.8	2.1	2.3	2.6	2.9	3.2	3.5	3.8	4.1	4.4	4.7	5.0	5.3	5.6	.059
G. 8" or 12" Block + R-11	.9	1.1	1.3	1.5	1.7	2.0	2.2	2.4	2.6	2.8	3.0	3.3	3.5	3.7	3.9	4.1	.043
H. 8" or 12" Block + R-19	.6	.8	.9	1.1	1.2	1.4	1.5	1.7	1.8	2.0	2.1	2.3	2.4	2.6	2.8	2.9	.031

No. 16 Ceilings Under a Ventilated Attic Space or Unheated Room

	_____ Winter Temperature Difference _____																
	20	25	30	35	40	45	50	55	60	65	70	75	80	85	90	95	U
	HTM (Btuh per sq. ft.)																
A. No Insulation	12.0	15.0	18.0	21.0	24.0	27.0	29.9	32.9	35.9	38.9	41.9	44.9	47.9	50.9	53.9	56.9	.599
B. R-7 Insulation	2.4	3.0	3.6	4.2	4.8	5.4	6.0	6.6	7.2	7.8	8.4	9.0	9.6	10.2	10.8	11.4	.120
C. R-11 Insulation	1.8	2.2	2.6	3.1	3.5	4.0	4.4	4.8	5.3	5.7	6.2	6.6	7.0	7.5	7.9	8.4	.088
D. R-19 Insulation	1.1	1.3	1.6	1.9	2.1	2.4	2.6	2.9	3.2	3.4	3.7	4.0	4.2	4.5	4.8	5.0	.053
E. R-22 Insulation	1.0	1.2	1.4	1.7	1.9	2.2	2.4	2.6	2.9	3.1	3.4	3.6	3.8	4.1	4.3	4.6	.048
F. R-26 Insulation	.8	1.0	1.1	1.3	1.5	1.7	1.9	2.1	2.3	2.5	2.7	2.8	3.0	3.2	3.4	3.6	.038
G. R-30 Insulation	.7	.8	1.0	1.2	1.3	1.5	1.6	1.8	2.0	2.1	2.3	2.5	2.6	2.8	3.0	3.1	.033
H. R-38 Insulation	.5	.7	.8	.9	1.0	1.2	1.3	1.4	1.6	1.7	1.8	2.0	2.1	2.2	2.3	2.5	.026
I. R-44 Insulation	.5	.6	.7	.8	.9	1.0	1.1	1.3	1.4	1.5	1.6	1.7	1.8	2.0	2.1	2.2	.023
J. R-57 Insulation	.3	.4	.5	.6	.7	.8	.8	.9	1.0	1.1	1.2	1.3	1.4	1.4	1.5	1.6	.017
K. Wood Decking, No Insulation	5.7	7.2	8.6	10.0	11.5	12.9	14.3	15.7	17.2	18.6	20.0	21.5	22.9	24.3	25.8	27.2	.287

No. 17 Roof on Exposed Beams or Rafters

	_____ Winter Temperature Difference _____																
	20	25	30	35	40	45	50	55	60	65	70	75	80	85	90	95	U
	HTM (Btuh per sq. ft.)																
A. 1½" Wood Decking, No Insul.	6.3	7.9	9.4	11.0	12.6	14.2	15.8	17.3	18.9	20.5	22.0	23.6	25.2	26.8	28.3	29.9	.315
B. 1½" Wood Decking + R-4	2.9	3.6	4.3	5.0	5.8	6.5	7.2	7.9	8.6	9.4	10.1	10.8	11.5	12.2	13.0	13.7	.144
C. 1½" Wood Decking + R-5	2.4	3.1	3.7	4.3	4.9	5.5	6.1	6.7	7.3	7.9	8.5	9.1	9.8	10.4	11.0	11.6	.122
D. 1½" Wood Decking + R-6	2.2	2.7	3.3	3.8	4.4	4.9	5.4	6.0	6.5	7.1	7.6	8.2	8.7	9.3	9.8	10.4	.109
E. 1½" Wood Decking + R-8	1.8	2.2	2.7	3.1	3.6	4.0	4.4	4.9	5.3	5.8	6.2	6.7	7.1	7.6	8.0	8.5	.089
F. 2" Shredded Wood Planks	4.3	5.4	6.5	7.6	8.7	9.8	10.8	11.9	13.0	14.1	15.2	16.3	17.4	18.4	19.5	20.6	.217
G. 3" Shredded Wood Plank	3.2	4.0	4.8	5.6	6.4	7.2	7.9	8.7	9.5	10.3	11.1	11.9	12.7	13.5	14.3	15.1	.159
H. 1½" Fiber Board Insulation	3.5	4.4	5.3	6.1	7.0	7.9	8.8	9.6	10.5	11.4	12.3	13.1	14.0	14.9	15.8	16.6	.175
I. 2" Fiber Board Insulation	2.8	3.5	4.2	4.9	5.6	6.3	7.0	7.7	8.4	9.1	9.8	10.5	11.2	11.9	12.6	13.3	.140
J. 3" Fiber Board Insulation	2.0	2.5	3.0	3.5	4.0	4.5	4.9	5.4	5.9	6.4	6.9	7.4	7.9	8.4	8.9	9.4	.099
K. 1½" Wood Decking + R-13	1.2	1.5	1.8	2.1	2.4	2.7	3.0	3.3	3.6	3.9	4.2	4.5	4.8	5.1	5.4	5.7	.060
L. 1½" Wood Decking + R-19	.8	1.0	1.2	1.4	1.6	1.8	2.0	2.3	2.5	2.7	2.9	3.1	3.3	3.5	3.7	3.9	.041

Footnotes are found on page 71.

(c-2)

Figure 2.9 *(Continued)*

Figure 2.9 *(Continued)*

Table 2   **HEAT LOSS FACTORS (HLF)**   Transmission

EXTERIOR DOORS
With or without glass, treated the same as Windows.

WINDOWS (GLASS)

No. 1. Windows
- (a) Single (no storm sash) 1.13
- (b) With storm sash .... .56
- (c) Double glazed with ¼" air space .... .65
- (d) Triple glazed with two ½" air spaces .... .36

EXPOSED WALLS
The factor for lath and plaster is the same as for ½" dry wall (gypsum board).

No. 2. Frame, Not Insulated
- (a) Clapboards or wood siding, studs, ½" dry wall (gypsum board) (no sheathing) .... .33*
- (b) Asbestos-cement siding over wood siding, paper, studs, ½" dry wall (gypsum board) (no sheathing) .... .30
- (c) Wood siding, paper, wood sheathing, studs, ½" dry wall (gypsum board) .... .25
- (d) Asbestos-cement siding over wood siding, paper, wood sheathing, studs, ½" dry wall (gypsum board) .... .23
- (e) Asbestos-cement shingles, paper, wood sheathing, studs, ½" dry wall (gypsum board) .... .29

No. 3 Frame, Insulated
- (a) Wood siding, paper, wood sheathing, studs, ½" insulating board, plaster .... .19
- (b) Wood siding, $^{25}/_{32}$" insulating board, studs, ½" dry wall (gypsum board) .... .22
- (c) Wood siding, paper, wood sheathing, ½" flexible insulation in contact with sheathing, studs, ½" dry wall (gypsum board) .... .18
- (d) Wood siding, paper, wood sheathing, ½" flexible insulation with an air space on both sides of insulation, studs, ½" dry wall (gypsum board) .... .15
- (e) Wood siding, paper, wood sheathing, 3⅝" rockwool or equivalent, studs, ½" dry wall (gypsum board) .... .07
- (f) Wood siding, paper, wood sheathing, 2" rockwool or equivalent, studs, ½" dry wall (gypsum board) .... .10
- (g) Wood siding, 1" styrofoam board sheathing, 3⅝" rockwool insulation or equivalent, studs, ½" dry wall (gypsum board) .... .06
- (h) ¾" x 10" wood siding, wood sheathing, 2" x 6" studs on 24" centers, 5½" rockwool or equivalent insulation, vapor seal, ½" dry wall (gypsum board) .... .05
- (i) ¾" x 10" wood siding, 1" styrofoam sheathing, 2" x 6" studs on 24" centers, 5½" rockwool or equivalent insulation, vapor seal, ½" dry wall (gypsum board) .... .04
- (j) Wood foundation above grade, ⅝" treated plywood, 2" x 6" studs on 24" centers, 5½" rockwool or equivalent insulation, ½" dry wall (gypsum board) .... .06
- (k) Wood foundation below grade, ⅝" treated plywood, 2" x 6" studs on 24" centers, 5½" rockwool or equivalent insulation, ½" dry wall (gypsum board) .... .03

No. 4. Brick, Not Insulated
- (a) 8" brick, ½" plaster one side .... .47
- (b) 8" brick, furred, lath and plaster one side .... .31
- (c) 12" brick, ½" plaster one side .... .33
- (d) 12" brick, furred, lath and plaster one side .... .25
- (e) 4" brick, 8" hollow tile, ½" plaster one side .... .31
- (f) 4" brick, 8" hollow tile, furred, ½" dry wall (gypsum board) .... .23
- (g) 4" brick, paper, wood sheathing, studs, ½" dry wall (gypsum board) .... .29
- (h) 4" brick, 4" light weight aggregate block, furred, ½" dry wall (gypsum board) .... .25

No. 5. Brick, Insulated
- (a) 8" brick, furred, ½" insulating board, ½" plaster one side .... .22
- (b) 12" brick, furred, ½" insulating board, ½" plaster one side .... .20
- (c) 4" brick, 8" hollow tile, ½" insulating board, ½" plaster one side .... .18
- (d) 4" brick, 4" light weight aggregate block, ½" insulating board, ½" plaster one side ... .19
- (e) 4" brick, paper, wood sheathing, studs, ½" insulating board, ½" plaster .... .22
- (f) 4" brick, $^{25}/_{32}$" insulating board, studs, ½" dry wall (gypsum board) .... .23
- (g) 4" brick, paper, wood sheathing, 3⅝" rockwool or equivalent, studs, ½" dry wall (gypsum board) .... .08
- (h) 4" brick, paper, wood sheathing, 2" rockwool or equivalent, studs, ½" dry wall (gypsum board) .... .09

*For this type of construction, use Item No. 22 for Infiltration Factor.

NOTES: (For new types of construction or materials, use this space to add new factors.)

(d-2)

Figure 2.9 *(Continued)*

2.9 *(d-1)*. This form is published by The Hydronics Institute (35 Russo Place, P.O. Box 218, Berkeley Heights, N.J. 07922). It is widely known as the I=B=R form because it was originally developed by the Institute of Boiler and Radiator Manufacturers (I=B=R) whose facilities are now operated by the Hydronics Institute. This form requires the use of tables of heat loss factors (see column 2 of the form). These factors are identical with U factors in ASHRAE tables. A sample page of these factors as published by The Hydronics Institute is shown in Figure 2.9 *(d-2)*.

Some forms involve simplifications of the detailed ASHRAE calculation methods. Other simplifications and rules of thumb are used by experienced designers. For the technologist, however, our recommendation is that detailed calculations be performed under the guidance of experienced HVAC engineers. When sufficient experience is accumulated, the technologist will be capable of using approximations and rules of thumb where they are applicable.

2.13 Outside Design Conditions

Notice that throughout the preceding discussion no mention was made of how to select the outside design temperature. This figure is needed to determine the temperature difference between inside and outside temperatures (Δt), in order to be able to calculate heat loss. Many tables exist that list outside design temperatures for both summer and winter for almost every area in the United States. They are all based on weather records. Some tables list all-time records or 25 year highs and lows; others list percentages, along with design temperatures. Most designers today use the 97.5% temperature figure for winter conditions. This means that the stated temperature will be exceeded 97.5% of the time. Conversely, the outdoor temperature will drop below the design temperature 2.5% of the time. Since most heating systems have a safety factor built in, these very low temperatures should not present a problem. Other designers take a more conservative approach, and use a 99% temperature figure. This means that weather bureau records indicate lower winter temperatures only 1% of the time. The difference between these two approaches can be significant. Designing for the 99% will increase the size of the heating system by about 10%. The same statistical considerations hold for summer design conditions, which include both dry bulb and wet bulb temperatures (or DB and RH). Actually the summer design conditions are more critical than the winter figures because both cooling and dehumidification are usually required. For these reasons, selection of outdoor design conditions should be made by experienced design engineers rather than inexperienced technologists. Refer to Supplementary Reading Reference 1 (Appendix A) for an extensive table of outside design conditions for the United States and Canada, plus an explanation of its use.

The HVAC technologist should also be familiar with another heating season concept—that of degree days. This concept is based on the idea that heating will be required any time the exterior temperature falls below 65°F. (The concept was obviously developed before the era of energy conservation and heavy thermal insulation.) The number of degree days in a winter day is computed as follows:

$$\text{DD} = 65 - \frac{t_{\max} + t_{\min}}{2} \quad (2.12)$$

where
  DD = number of degree days
  $t_{\max}$ = day's maximum dry bulb temperature
  $t_{\min}$ = day's minimum dry bulb temperature

Thus, for a day with a maximum temperature of 70°F and a minimum of 50°F, its degree day count would be

$$\text{DD} = 65 - \frac{70 + 50}{2} = 65 - 60 = 5 \text{ (degree days)}$$

(A negative number of degree days counts as zero in the season total.) With today's construction and insulation standards and night thermostat setback, such a temperature variation would not require heating. However, in an uninsulated structure without night setback, such as a pre-1973 house, the heating plant would probably have operated for a period of time.

The concept of degree days is very useful in predicting fuel consumption and energy requirements. However, because of the changes in construction and comfort criteria, many modern structures use a 60°F base temperature rather than 65°F. This gives a more accurate and realistic energy use figure. Figure 2.10 shows a degree-day map of the United States based on 65°F. This can be useful for estimating when accurate data are unavailable.

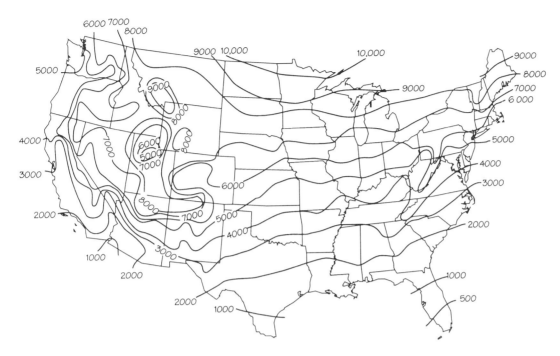

Figure 2.10 Map of the United States showing heating degree-day lines based on 65°F. (From Bobenhausen, *Simplified Design of HVAC Systems*, Wiley, New York, 1994, reprinted by permission of John Wiley & Sons.)

2.14 Evolution of Modern Climate Control Systems

Having considered the subjects of human comfort and heat loss from enclosed spaces, we will now proceed to a detailed study of heating systems. When cooling is to be provided, cooling systems must be considered as separate entities or combined in heating/cooling installations. It would, however, be a mistake after learning the details of presently popular climate control methods to assume that they will necessarily be applicable after a number of years of technical practice. Something always occurs to change the aspect of any technical discipline. Accordingly, let us review briefly the history of what has happened, study well what is now current and prepare to accept something different.

At the present time, energy conservation and environmental problems predominate. Although fossil fuels are not in short supply, their use causes pollution and degradation of the environment. As a result, recent years have seen the increasing use of renewable, nonpolluting energy sources, with particular emphasis on solar energy. Also, comfort standards have been made more realistic without affecting thermal comfort to any appreciable degree. A brief history of American heating systems shows the following developments, practices and trends.

Circa 1870 Petroleum was known but not widely used. An eight room house was heated by eight open coal fires, a dirty, labor-intensive method that provided only fair results. See Figure 2.11.

1900 Coal-fired boilers supplied steam to one-pipe systems or gravity-circulated hot water to cast-iron radiators. Alternatively, coal-fired furnaces warmed air that circulated by gravity. These systems were very bulky, very dirty and a lot of work.

1902 Dr. Willis Carrier developed the compressive refrigeration cycle for cooling, but several decades would pass before it was much used.

1935 Boilers fired by oil or gas were used to produce steam. Hot water systems similarly fired now used forced circulation. Warm air systems had blowers to circulate the air. These methods were more compact, automatically controlled and efficient.

Figure 2.11 Brooklyn, New York, in the 1800s. Party wall of a demolished brownstone residence. The residents had to climb three flights up with the coal and three down with the ashes. Eight chimney flues signaled the early beginning of air pollution. There was a small circle of warmth around each fire while the flues carried away most of the warmed air. To fill this void, outside cold air was thus drawn in around un-weather-stripped window sash edges, accelerating the infiltration and chilling the backs of occupants facing the fires. It took a hundred years, but designers found several better methods. Sociologists may find interest in the small fireplaces in the servants' sitting room, ground floor front, and in the servants' bedroom, top floor front. Adequate cooking space appears, however, in the large kitchen hearth, ground floor rear.

1945   The demise of the steam/cast-iron radiator heating system occurred. Hot water radiant heating made its appearance, and cooling started to become popular.

1955   Hot water radiant heating declined in favor because it was expensive and hard to balance. Electric heating and air systems were now much in use. Boilers became smaller and more efficient. Individual room control and multizone operation gained prominence.

1976   Figure 2.12 shows three commonly chosen comfort systems that are still in use. Other systems still in use are discussed elsewhere in this book.

1996   Steam heating is rarely used. Buildings are superinsulated to very high R values. Solar-assisted heating systems with heat recovery devices are common. Hot water heating, radiant heating and warm air heating systems are all in common use, depending on climate, type of fuel and cooling requirements. Passive solar heating for at least part of the year is increasingly used in mild climate areas. Cooling and dehumidification are considered to be almost as important as heating in all types of buildings.

2.15 Heating Systems Fuels

The three principal fuels in use today for both residential and small commercial buildings are oil, gas and electricity. Large commerical buildings sometimes use commercially available steam or steam generated on the premises. Since this book deals only minimally with such large buildings, we will discount steam as a heating system fuel for our purposes. Figure 2.12 shows schematically a hot water heating installation for each of these fuels and points out the major differences among the systems. A hot water heating system was chosen arbitrarily, for the purpose of illustration. In point of fact, hot water (hydronic) heating for residences is popular only in the Northeast. It is unusual in the Midwest and West, where all-air or combination air/water systems with all-year-round cooling are used for residential work. On the other hand, hydronic heating systems for commercial buildings are found throughout the United States, probably because of architectural and space considerations.

The choice of fuel type for a heating system is primarily one of cost and availability. Gas, both natural and propane, is available at competitive prices throughout the United States. As a result, it is the most popular countrywide residential building fuel. Another advantage of gas is the reliability of supply. Oil is popular only in the northeastern states, partly by tradition. Reliability is somewhat poorer than either gas or electricity because of the reliance on truck delivery. The use of electricity as a heating fuel is basically a function of price, which

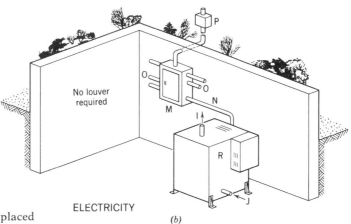

LEGEND

A Louver admits air to support combustion
B Oil tank, 275 gal. Larger tanks are usually placed outdoors and below ground.
C Oil gauge
D Vent pipe relieves tank air
E Oil fill pipe
G Oil supply to oil burner
H Air to support combustion
I Hot water to heating system
J Hot water return to boiler
K Smoke flue
L Chimney flue, terra cotta lined
M Electric panelboard
N Electric circuit to boiler
O Other electric circuits
P Electric meter on exterior wall
Q Oil-fired hot water boiler
R Electric hot water boiler
S Gas-fired hot water boiler
U Gas to other equipment
V Gas service entry (below ground) and master shutoff
W Gas to the burning unit
Z Gas meter

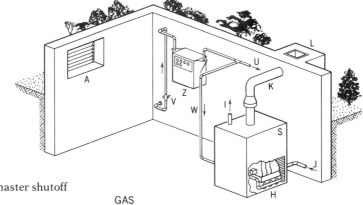

Figure 2.12 Most commonly used heating system fuels are (a) oil, (b) electricity, and (c) gas. Gas eliminates fuel storage; electricity also eliminates the flue and chimney and the need for combustion air.

varies widely throughout the United States. In areas where electricity is cheap such as parts of the Northwest and Southeast, it is a widely used heating fuel. In areas such as the Northeast where prices in excess of 15¢/kwh are common, the use of electricity as a heating fuel is unusual except in superinsulated, all-electric, specially designed buildings. Among these are the so-called "smart houses" where the easy control and rapid response of electric heating units make their use reasonable despite a high fuel cost. Another advantage of electricity as a fuel is low initial cost when the heating system is decentralized. Such systems use individual room heaters and controls instead of a centralized electric boiler.

2.16 Heating Systems

The selection of the type of heating system to be used in a building is not the responsibility of the technologist. However, it is important that he or she understand the factors that affect the decision. The systems themselves will be discussed in detail in the following chapters. The type of heating system used depends in large measure on the climate. In very cold areas, residential buildings are superinsulated. Since this may cause an *internal air quality* (IAQ) problem, mechanical ventilation with heat recovery is often used. Due to the large heating load, high efficiency furnaces or boilers are common. Systems are generally ducted warm air or radiant heat. Heat pumps are not used due to their low efficiency in cold climates. In milder climates, systems are either all-air for heating and cooling or hybrid systems with hot water coils in an air delivery system. Heat pumps for both heating and cooling are common as are condensing air conditioning systems. In warm climates all-air cooling systems are the rule, with baseboard or electric duct heaters supplying the small amount of heating required.

Cooling

2.17 Building Heat Gain Components

Calculation of a building's total cooling load is required to be able to design an adequate and efficient cooling system. This calculation, which involves determining all of the building's heat gain components, is much more complex than the heat loss calculations that were studied in the previous sections. This is so for two reasons:

- There are many more factors involved in heat gain than in heat loss.
- Heat gain varies sharply during the day. This makes it necessary often, in nonresidential work, to do hour-by-hour calculations.

The total cooling load of a structure involves

- Sensible heat gain through windows.
- Sensible heat gain through walls, floors and ceiling/roof.
- Sensible and latent heat gain from infiltration and ventilation.
- Sensible and latent heat gain due to occupancy.

In order to calculate these four components, separate data are required for each structural block being studied. This is because solar loads and internal loads are so large that different areas in the same building may have vastly differing cooling needs. The data required for these calculations include

(a) Time of day.
(b) Orientation.
(c) Latitude.
(d) Heat gain through glass.
(e) Type of construction (thermal lag).
(f) Shading, external and internal.
(g) Internal sensible heat loads.
(h) Internal latent heat loads.
(i) Daily temperature range.
(j) Acceptable internal temperature swing.

The complexity of these calculations, whether they are done manually or by computer program, requires that an experienced air conditioning designer perform them. A technologist can perform these calculations, beginning with less complex buildings, under the direct supervision of a designer. For this reason, we will discuss the foregoing factors in the following sections. As will become clear, detailed calculations require extensive data tables. The methods that we will explain are those recommended by ASHRAE, as they appear in the 1993 *ASHRAE Handbook—Fundamentals*. The required data tables appear in that volume, which is a necessary part of every engineering office library.

a. Time of Day

In heat loss calculations, time of day is not considered since we assume that the heating system will be designed to heat adequately whenever the exterior design temperature occurs (usually at night). With heat gain calculations, the effect of direct solar radiation is so strong that the time of day must be known to perform the necessary hour-by-hour calculations. Eastern exposures have their maximum solar load in the morning; western exposures, in the afternoon; and roofs, at noon. In these calculations, solar time, and not clock time, is used.

b. Orientation

Orientation of the building determines which side of the building is exposed to the sun at a particular hour. With long narrow buildings, the solar heat gain can be drastically reduced by orienting the building's narrow ends east and west, so as to present minimum area to the sun. In the northern hemisphere, another solar control technique is to locate glass on the south side of the building and to have it shaded properly. Proper shading reduces summer sun load and permits direct solar insolation during the winter, for "free" warming.

c. Latitude

The position of the sun during the day depends on the time of day and the latitude (0° at the equator, 90° N at the north pole, 90° S at the south pole). The sun travels from east at sunrise to west at sunset. Its altitude above the horizon varies from minimum (0°) at sunrise and sunset to maximum at solar noon. Knowledge of the altitude of the sun is required to determine insolation (heat from direct solar radiation) and to design shading.

d. Heat Gain Through Glass

This component of the building's heat gain consists of two parts: direct solar radiation gain through unshaded glass exposed to the sun and conductive heat gain through shaded glass and glass not exposed to the sun. Shaded glass is treated the same as north-facing glass, that is like glass not exposed to the sun. The glass heat gain factor is frequently the largest heat gain of a building and must, therefore, be accurately calculated, including all shading influences. Shading can be external, internal or both.

e. Type of Construction (Thermal Lag)

In our discussion on heat loss, we explained that the calculations were steady-state. That means that the heat loss is continuous. This is appropriate for long winter nights and dull overcast days where for many hours the same difference between indoor and outdoor temperature exists. We did not take into account any transient phenomena; instead, we assumed that the inside-outside temperature difference exists throughout the day. Actually this is not true, but for heating calculation that are intended for sizing the heating plant, the transient phenomena can be neglected. For cooling calculations, however, we want to avoid oversizing the refrigeration equipment for both economic and comfort reasons. One of the common errors in early cooling design practice was to oversize cooling units. The result was blasts of cooling with long periods of shutdown, which permitted humidity to accumulate to the discomfort of occupants. In recent practice cooling units, especially in residences, are slightly undersized. This results in almost continuous operation and, consequently, adequate dehumidification.

The phenomenon of thermal lag simply means that it takes time for heat to get through a structural element. The more massive the element and the higher its specific heat, the longer it takes. For instance, the morning sun strikes the east wall of a building. If the structure is of light construction (frame), the heat will penetrate in 1–2 hours, while the sun is still shining on the wall, thus causing a large heat gain. If the structure is heavy (masonry), it might take 4–8 hours for the heat to penetrate the wall. By that time the sun has moved to the south so that the net effect on the building heat gain is lower. The effect is even more pronounced with roofs that are exposed to direct sunlight throughout the day.

It is very important to understand the difference between the action of insulation and that of structural mass (thermal lag). Insulation reduces the rate at which heat can get through a structural item (wall, roof, ceiling, floor) and establishes a fixed thermal gradient. That gradient remains as long as the temperature difference between inside and outside remains. If a wall is very light, consisting only of plywood and insulation for instance, it would take only a very short time for the heat to get through and to cause a (limited) continuous heat gain. On the other hand, the amount of heat getting through an uninsulated massive wall would be greater, but the time required for the heat to get through and establish a steady-state heat gain would be much longer. The time delay is related only to mass. It would be lengthened only slightly if the same massive wall were insulated, since most thermal insulation (rock-wool, fiberglass) is very light.

These actions are illustrated in Figure 2.13. Wall (a) is composed of concrete blocks and insulation. Assuming that the wall has cooled all night to temperature t_i and that the wall faces southeast, the situation a short time after sunrise would be as shown in Figure 2.13*(a-1)*. A thin layer of concrete has been warmed by the sun, which causes a temperature of t_o on the outside of the block. Beyond this layer the temperature is still t_i, as indicated by the sharp drop in the temperature gradient. After another period of time, the block has warmed to about two-thirds of its depth, as shown in Figure 2.13*(a-2)*. As soon as the heat penetrates the block, it relatively rapidly penetrates the insulation to establish the steady-state condition shown in Figure 2.13*(a-3)*. If the wall had no insulation, then the situation at time T and $2T$ would be the same as previously, as in the gradients in Figure 2.13*(b-1)* and *(b-2)*. However, as soon as the heat penetrates the block, the lack of insulation permits a large flow of heat into the space, raising the inside temperature t_{i-2} to a much higher level than in Figure 2.13*(a-3)*, provided the exterior solar load remains the same.

Absence of any substantial mass to cause thermal lag is shown in Figure 2.13*(c)*. Here the lack of mass causes the steady-state situation to be established relatively rapidly Figure 2.13*(c-2)*. This situation continues as long as the *sol-air temperature* remains the same; see Figure 2.13*(c-2)* and *(c-3)*. Note that inside temperature t_{i-3} is lower than t_{i-2} because of the insulation but higher than t_{i-1} because the wall lacks the R value of the concrete blocks. (The sol-air temperature is a theoretical outdoor air temperature that in the absence of solar radiation, would give the same rate of heat entry to a surface as the actual combination of solar radiation and convective heat exchange from the outdoor air.)

Notice the advantage of mass in a structure as shown in Figure 2.13*(b)*. Here, if the wall were an eastern exposure, the outside sol-air temperature would drop after a few hours because the sun would move south and away from an east wall. The result would be a final, steady-state temperature lower than t_{i-2} and, therefore, a lower cooling load, even without any insulation in the wall. This is the principle behind the massive construction of native desert architecture. The massive (uninsulated) walls delay the transmission of the sun's heat until after sundown. Then, the low night desert temperature is tempered by the reradiation of heat from the walls into the interior spaces (and also outward into the night air).

f. Shading, External and Internal

Since solar heat gain through the building envelope glass can constitute an extremely large cooling load, engineers and architects have developed three different approaches to solve this problem. The first is to treat the glazing itself either with heat-reflecting coatings or by coloring in the glass. This procedure reduces the transmitted solar load, and its effect is considered in a factor called the *Shading Coefficient* (SC), which is discussed later. It has nothing to do with either external or internal shading and is only mentioned here because of the (misleading) name of this coefficient.

External shading, the second approach, is part of the architectural design of the building. It consists of overhangs, baffles, shields and other devices designed to shade glass from the outside and to shade other parts of the building "envelope." Shaded glass is treated as north-facing glass in calculations. In the tropics where the sun is directly overhead for most of the day, roofs are especially vulnerable. There, it is not uncommon to provide a "roof shade" (a second roof or "sun-intercepter") some distance above the principal roof with a fully ventilated air space between. External shading is much more effective than internal shading because solar heat never enters the building.

Internal shading, the third approach, is accomplished with devices such as drapes, window shades and venetian blinds. These act to reduce cooling load by adding insulation (drapes and shades) and by positioning thermal barriers (venetian blinds and some types of shades).

g. Internal Sensible Heat Loads

These loads include lighting and other electrical loads, heat loads plus a load for each space occupant. The latter is usually taken at about 270 Btuh for people in sedentary activities (sitting, desk work, household activities) and more if the activity warrants. See Table 1.1. For residential calcula-

BUILDING HEAT GAIN COMPONENTS / 57

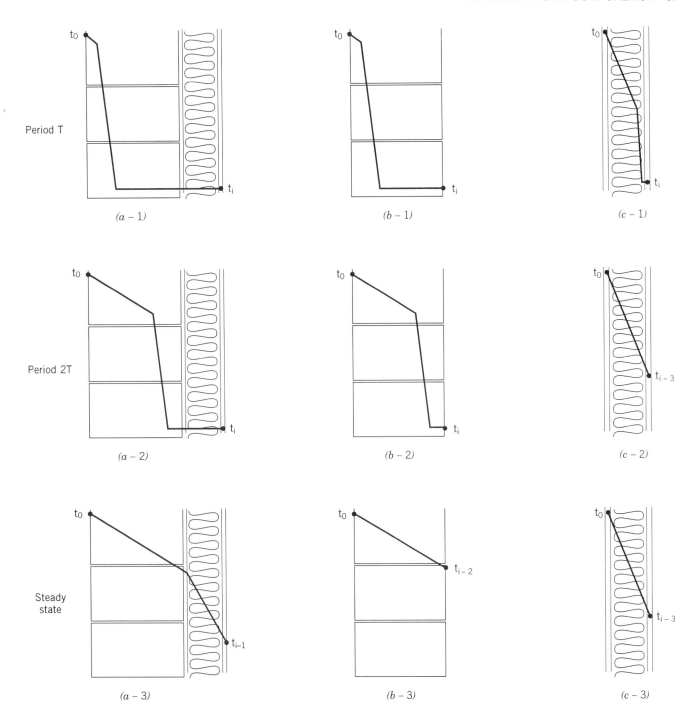

Figure 2.13 Effect of thermal lag on heat transmission through *(a)* a massive insulated wall, *(b)* a similar uninsulated wall, and *(c)* an insulated wall with little mass. (The diagrams and gradients are simplified for clarity. Additional factors that would affect outdoor and indoor temperatures t_o and t_i are neglected.)

tions, sedentary activities are assumed. Electrical loads include motors, computers, sound equipment and office equipment. For nonresidential buildings, loads are obtained from the electrical designer or technologist. Additional heat loads come from kitchen appliances, hobby equipment and any other electrical or gas-powered equipment in the space under consideration.

h. Internal Latent Heat Loads

Very often, a third or more of the refrigeration energy supplied to air-condition a space is required to condense the moisture in the air in order to reduce the relative humidity to a comfortable level. The source of this air-borne moisture is occupants, showers and baths, floor washing, laundries, infiltration of moist outside air and mechanical ventilation. Calculation procedures for these latent heat loads differ between residential and nonresidential buildings, as will be explained later.

i. Daily Temperature Range

The difference between the 24-hour high and low dry bulb temperatures is called the daily temperature range. Mountainous and arid areas have a high range of up to 30 F°. Coastal areas and areas of high humidity have a daily range as low as 10 F°. Areas with a low daily range require a larger cooling capacity than high range areas because the structure does not cool off at night.

j. Acceptable Interior Temperature Swing

Recommended practice in residences is to design the cooling system for continuous operation in hot weather by slight undersizing. This is the cause of an indoor temperature rise that is known as swing. It occurs for an hour or so in the hottest part of the afternoon. It is considered acceptable because the continuous cooling action has kept the walls and other room surfaces cool. (They have a low mean radiant temperature). Thus, occupants have a feeling of surrounding coolness and, therefore, do not mind the slight temporary increase in indoor air temperature. ASHRAE recommends that in a residence, this swing not exceed 3°F on a design (temperature/humidity) day, with the thermostat set at 75°F.

2.18 Residential Heat Gain (Cooling Load) Calculations

In theory, there should be no difference between the heat gain calculations for residential and nonresidential buildings. For any structure, the total heat gain is the sum of the sensible heat gains through windows, walls, roofs, ceilings and floors plus the sensible and latent heat gain from occupancy, infiltration and ventilation. In practice, however, it was found that using standard procedures for calculation of the foregoing loads always resulted in too large a cooling load for residential buildings. That is, after the residential building was completed, measurements indicated a smaller cooling load than calculated. Since oversized cooling equipment will not properly dehumidify, as noted previously, a more realistic and accurate procedure for calculation of residential building cooling loads was developed by ASHRAE. It differs from the nonresidential calculation in its treatment of heat gain through windows; its simplified calculation of wall, ceiling and roof gain; and its calculation of latent heat gain from occupancy and infiltration. The calculation procedure is detailed in the material that follows.

Refer to Figure 2.14 while studying the procedure description. Note that the calculation must be performed on a room-by-room basis. This technique enables the designer to distribute the cooling air properly (assuming an all-air system). Note also that heat gain from above-grade exterior walls only is calculated. Basement walls are not considered, since they represent a heat loss and not a heat gain. Basements are naturally cool and do not normally require additional cooling.

a. Above-Grade Outside Walls

The heat gain through an outside wall is a combination of direct solar insolation, conduction/convection and thermal lag effects. The combined effect is expressed in a factor called *Cooling Load Temperature Difference* or simply CLTD. Table 2.10 for single-family residences gives CLTD factors as a function of design temperature, orientation of the wall and daily temperature range—low, medium and high.

Table 2.11 for multifamily residences has as an additional variable: the type of wall construction, that is, lightweight (LW), medium weight (MW)

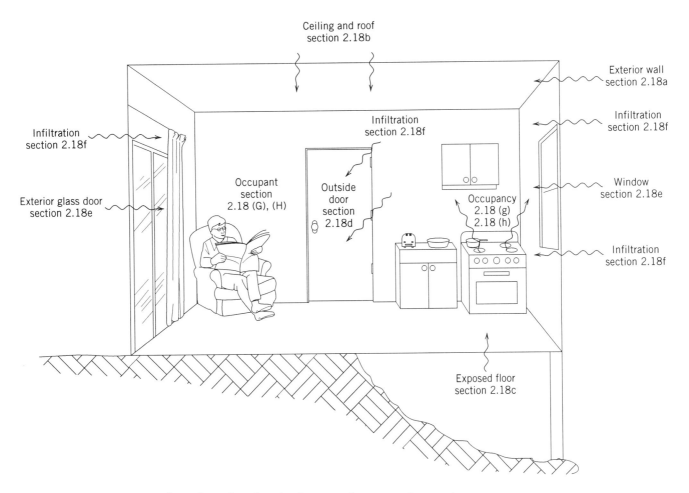

Figure 2.14 Components of residential cooling load. Text reference is shown adjacent to each component.

and heavyweight (HW). This factor is not used in Table 2.10 for detached single-family residences. There, an average weight is used to determine the CLTD given in the table. Multifamily residence construction, however, varies from lightweight frame construction in many garden apartments to heavyweight concrete and masonry in high-rise residences, necessitating the use of this additional factor. In both tables, a dark wall color is assumed. For light colors, the CLTD can be reduced slightly.

Once the CLTD is determined, the wall heat gain is calculated as

$$Q_w = U_w A \,(\text{CLTD}) \qquad (2.13)$$

or

$$Q_w = A(\text{CLTD})/R_w \qquad (2.14)$$

where
  Q_w is the wall heat gain in Btuh
  A is the wall area in ft^2
  CLTD is the cooling load temperature difference
  U_w is the wall thermal transmittance in Btuh/ft^2-°F or
  R_w is the wall thermal resistance in ft^2-°F/Btuh

b. Ceilings and Roofs

This calculation is identical to that for exterior walls, using Tables 2.10 or 2.11.

$$\begin{aligned}Q_{\text{roof}} &= U_r(A)\,(\text{CLTD}) \\ &= A(\text{CLTD})/R_r\end{aligned} \qquad (2.15)$$

Table 2.10 CLTD Values for Single-Family Detached Residences[a]

	Design Temperature °F									
	85		90			95			100	
Daily Temp. Range[b]	L	M	L	M	H	L	M	H	M	H
All walls and doors										
North	8	3	13	8	3	18	13	8	18	13
NE and NW	14	9	19	14	9	24	19	14	24	19
East and West	18	13	23	18	13	28	23	18	28	23
SE and SW	16	11	21	16	11	26	21	16	26	21
South	11	6	16	11	6	21	16	11	21	16
Roofs and ceilings										
Attic or flat built-up	42	37	47	42	37	51	47	42	51	47
Floors and ceilings										
Under conditioned space, over unconditioned room, over crawl space	9	4	12	9	4	14	12	9	14	12
Partitions										
Inside or shaded	9	4	12	9	4	14	12	9	14	12

[a]Cooling load temperature differences (CLTDs) for single-family detached houses, duplexes, or multifamily, with both east and west exposed walls or only north and south exposed walls, °F.

[b]L denotes low daily range, less than 16°F; M denotes medium daily range, 16 to 25°F; and H denotes high daily range, greater than 25°F.

Source. Extracted and reprinted by permission of the American Society of Heating, Refrigerating and Air-Conditioning Engineers, Atlanta, Georgia, from the 1993 *ASHRAE Handbook—Fundamentals.*

Table 2.11 CLTD Values for Multifamily Residences[a]

		Design Temperature, °F								
		85		90			95			100	
Daily Temp. Range[b]		L	M	L	M	H	L	M	H	M	H
Walls and doors											
	LW	14	11	19	16	12	24	21	17	26	22
N	MW	13	10	18	15	11	23	20	16	25	21
	HW	9	6	15	11	7	20	16	12	21	17
	LW	23	17	28	22	17	33	27	22	32	26
NE	MW	20	15	25	20	16	30	25	21	29	25
	HW	16	12	21	17	13	26	22	18	26	22
	LW	32	27	37	32	27	43	38	32	42	37
E	MW	30	24	34	29	24	40	34	29	39	33
	HW	23	18	28	23	18	34	29	23	33	28
	LW	31	27	35	31	26	41	37	31	42	37
SE	MW	28	22	32	27	22	37	32	27	37	33
	HW	21	16	26	22	17	32	27	22	31	27
	LW	25	22	29	26	22	35	31	26	36	32
S	MW	22	18	26	22	18	31	26	22	31	27
	HW	16	11	20	16	12	26	21	17	26	21
	LW	39	36	44	40	35	50	46	40	51	47
SW	MW	33	29	37	34	29	44	40	35	45	40
	HW	23	18	28	24	19	36	31	25	35	30
	LW	44	41	48	45	40	54	51	46	56	52
W	MW	37	33	41	38	33	46	42	38	48	43
	HW	26	22	31	27	23	37	32	27	37	32
	LW	33	30	37	34	30	43	39	34	44	40
NW	MW	28	25	32	29	24	37	33	29	39	35
	HW	20	16	25	20	16	31	26	21	31	26
Roof and ceiling											
Attic or N,S,W		58	53	65	60	55	70	65	60	70	65
Built-up East		21	18	23	21	18	25	23	21	25	23
Floors and ceiling											
Under or over unconditioned space, crawl space		9	4	12	9	4	14	12	9	14	12
Partitions											
Inside or shaded		9	4	12	9	4	14	12	9	14	12

[a]Cooling load temperature differences (CLTDs) for multifamily low-rise or single-family detached if zoned with separate temperature control for each zone, °F.

[b]L denotes low daily range, less than 16°F; M denotes medium daily range, 16 to 25°F; and H denotes high daily range, greater than 25°F.

Source. Extracted and reprinted by permission of the American Society of Heating, Refrigerating and Air-Conditioning Engineers, Atlanta, Georgia, from the 1993 *ASHRAE Handbook—Fundamentals.*

where the U or R factors are the applicable ones for the specific type of construction, with heat direction downward.

c. Exposed Floors

Buildings built on columns or with portions of the building cantilevered over open air have a heat gain through the underside of the exposed floor. Since this surface is always shaded, the CLTD is quite low. (See Tables 2.10 and 2.11.) Use the same heat gain formula

$$Q_f = U_f A(\text{CLTD}) \text{ or } A(\text{CLTD})/R_f \quad (2.16)$$

where U and R are the floor construction's thermal transmittance or resistance, respectively

d. Doors

Doors are treated exactly as exterior walls, with the appropriate U or R factor in Equation 2.12.

e. Windows and Glass Doors

Tables 2.12 and 2.13 give the *glass load factors (GLF)* to be used in calculating the glass heat gain from the following:

$$Q_{glass} = A(\text{GLF}) \quad (2.17)$$

where
Q_{glass} is the glass heat gain in Btuh and
A is the area of the specific glass in ft².

The variables in Table 2.12 and 2.13 are orientation of the window or glass door, type of interior shading, type of glazing and outdoor design temperature. The GLF includes both solar radiation heat gain and transmission (conduction/convection) heat gain.

Shaded glass is treated as north-facing glass. The shading effect of a simple overhang can be calculated by using the *shade line factor (SLF)* given in Table 2.14. Shading due to other exterior shading devices must be calculated individually.

f. Infiltration Heat Load

In the previous section we noted that the heat load of infiltrated air is both sensible and latent. The sensible heat load can be calculated as

$$Q_{in\text{-}s} = 1.1(\text{cfm})(\Delta t) \quad (2.18)$$

where
$Q_{in\text{-}s}$ is the sensible heat load of infiltrated air,
cfm is the infiltration air flow rate in ft³/min and
Δt is the difference in dry bulb temperatures between outside and inside

Note that this formula is almost identical to that of Equation 2.10 except for the factor 1.1 instead of 1.08. The 1.1 factor is correct for cooling calculations due to the higher specific heat of the more

Table 2.12 Window Glass Load Factors (GLF) for Single-Family Detached Residences[a]

Design Temp., °F	Regular Single Glass				Regular Double Glass				Heat-Absorbing Double Glass			
	85	90	95	100	85	90	95	100	85	90	95	100
No inside shading												
North	34	36	41	47	30	30	34	37	20	20	23	25
NE and NW	63	65	70	75	55	56	59	62	36	37	39	42
E and W	88	90	95	100	77	78	81	84	51	51	54	56
SE and SW[b]	79	81	86	91	69	70	73	76	45	46	49	51
South[b]	53	55	60	65	46	47	50	53	31	31	34	36
Horizontal	156	156	161	166	137	138	140	143	90	91	93	95
Draperies, venetian blinds, translucent roller shade fully drawn												
North	18	19	23	27	16	16	19	22	13	14	16	18
NE and NW	32	33	38	42	29	30	32	35	24	24	27	29
E and W	45	46	50	54	40	41	44	46	33	33	36	38
SE and SW[b]	40	41	46	49	36	37	39	42	29	30	32	34
South[b]	27	28	33	37	24	25	28	31	20	21	23	25
Horizontal	78	79	83	86	71	71	74	76	58	59	61	63
Opaque roller shades, fully drawn												
North	14	15	20	23	13	14	17	19	12	12	15	17
NE and NW	25	26	31	34	23	24	27	30	21	22	24	26
E and W	34	36	40	44	32	33	36	38	29	30	32	34
SE and SW[b]	31	32	36	40	29	30	33	35	26	27	29	31
South[b]	21	22	27	30	20	20	23	26	18	19	21	23
Horizontal	60	61	64	68	57	57	60	62	52	52	55	57

[a] Glass load factors (GLFs) for single-family detached houses, duplexes, or multi-family with both east and west exposed walls or only north and south exposed walls, Btu/h·ft².

[b] Correct by +30% for latitude of 48° and by −30% for latitude of 32°. Use linear interpolation for latitude from 40 to 48° and from 40 to 32°.

Source. Extracted and reprinted by permission of the American Society of Heating, Refrigerating and Air-Conditioning Engineers, Atlanta, Georgia, from the 1993 *ASHRAE Handbook—Fundamentals*.

Table 2.13 Window Glass Load Factors (GLF) for Multifamily Residences[a]

Design Temp., °F	Regular Single Glass				Regular Double Glass				Heat-Absorbing Double Glass			
	85	90	95	100	85	90	95	100	85	90	95	100
No inside shading												
North	40	44	49	54	34	36	39	42	23	24	26	29
NE	88	89	91	95	78	79	80	83	52	52	53	55
East	136	137	139	142	120	121	122	125	79	79	81	83
SE	129	130	134	139	109	113	116	119	72	75	77	79
South[b]	88	91	96	101	76	78	81	84	50	52	54	56
SW	154	159	164	169	134	137	140	143	89	91	93	95
West	174	178	183	188	151	154	157	160	100	102	104	106
NW	123	127	132	137	107	109	112	115	71	72	75	77
Horizontal	249	252	256	261	218	220	223	226	144	146	148	150
Draperies, venetian blinds, translucent roller shades, fully drawn												
North	21	25	29	33	18	21	23	26	15	17	19	21
NE	43	44	46	50	39	40	41	44	33	33	34	36
East	67	68	70	74	61	62	63	65	50	50	51	54
SE	64	65	69	73	58	59	61	63	48	48	50	52
South[b]	45	48	52	56	40	42	44	47	33	34	36	39
SW	79	83	87	91	70	72	75	78	57	59	62	64
West	89	92	96	100	79	81	84	86	65	66	69	71
NW	63	66	70	74	56	58	61	63	46	48	50	52
Horizontal	126	128	132	135	113	115	117	120	93	94	96	98
Opaque roller shades, fully drawn												
North	17	21	25	29	15	17	20	23	14	15	18	20
NE	33	34	35	39	31	32	33	36	29	28	30	32
East	51	52	53	57	48	49	50	53	45	45	46	48
SE	49	50	53	57	46	47	49	52	42	43	45	47
South[b]	35	38	42	46	32	34	37	40	29	31	33	35
SW	61	65	69	73	57	59	62	65	52	54	56	58
West	68	71	75	80	64	66	68	71	58	60	62	64
NW	49	52	56	60	45	47	50	53	41	43	45	47
Horizontal	97	99	102	106	91	93	95	97	83	85	87	89

[a] Glass Load factors (GLFs) for multi-family low-rise or single-family detached if zoned with separate temperature control for each zone, Btu/h·ft².

[b] Correct by +30% for latitude of 48° and by −30% for latitude of 32°. Use linear interpolation for latitude from 40 to 48° and from 40 to 32°.

Source. Extracted and reprinted by permission of the American Society of Heating, Refrigerating and Air-Conditioning Engineers, Atlanta, Georgia, from the 1993 *ASHRAE Handbook—Fundamentals.*

humid summer air. Some designers use 1.08 for both calculations, and others use 1.1 for both. The difference between the two is less than 2%, which is within engineering accuracy.

Since the cfm air flow for infiltration is not usually known, whereas the air changes per hour (ACH) are known, the formula can be rewritten as

$$Q_{\text{in-s}} = \frac{1.1 \, (\text{ACH}) \, (V) \, (\Delta t)}{60} = 0.0183 \, (V) \, (\text{ACH}) \, (\Delta t)$$

(2.19)

where

V is the room volume in ft³,
ACH is the air changes per hour and
Δt is the temperature difference, as before.

Note that this formula is essentially identical with the heating formula, Equation (2.9). The number of air changes per hour as a function of the outdoor design temperature is given in Table 2.15. The latent heat load of infiltrated air will be discussed later.

Table 2.14 Shade Line Factors (SLF)

Direction Window Faces	Latitude, Degrees N						
	24	32	36	40	44	48	52
East	0.8	0.8	0.8	0.8	0.8	0.8	0.8
SE	1.8	1.6	1.4	1.3	1.1	1.0	0.9
South	9.2	5.0	3.4	2.6	2.1	1.8	1.5
SW	1.8	1.6	1.4	1.3	1.1	1.0	0.9
West	0.8	0.8	0.8	0.8	0.8	0.8	0.8

Shadow length below the overhang equals the shade line factor times the overhang width. Values are averages for the 5 hours of greatest solar intensity on August 1.

Source. Reprinted by permission of the American Society of Heating, Refrigerating and Air-Conditioning Engineers, Atlanta, Georgia, from the 1993 *ASHRAE Handbook—Fundamentals.*

Table 2.15 Summer Air Change Rates as Function of Outdoor Design Temperatures

Class	Outdoor Design Temperature, °F					
	85	90	95	100	105	110
Tight	0.33	0.34	0.35	0.36	0.37	0.38
Medium	0.46	0.48	0.50	0.52	0.54	0.56
Loose	0.68	0.70	0.72	0.74	0.76	0.78

Values for 7.5 mph wind and indoor temperature of 75°F.

Source. Reprinted by permission of the American Society of Heating, Refrigerating and Air-Conditioning Engineers, Atlanta, Georgia, from the 1993 *ASHRAE Handbook—Fundamentals.*

g. Building Occupancy Loads

As explained previously, the heat loads due to occupancy consist of lighting, appliances and people. These loads may have a latent component in addition to the sensible component. The sensible component is calculated by direct conversion of electrical appliance wattage to Btuh (1 w = 3.412 Btuh) and use of 270 Btuh per sedentary occupant. The heat output of gas appliances such as stoves can also be converted to Btuh using manufacturer's data on fuel consumption. The same is true for oil-burning devices, although these are unusual in residences.

In the absence of more accurate data, approximate sensible heat loads of typical residential appliances can be found in Supplementary Reading, References 1 and 2. Since the spaces (rooms) in residences are individually calculated in order to be able to size cooling system terminals, an estimate must be made of people occupancy in the various rooms. It is customary to spread the occupancy throughout the house rather than concentrate all the occupants into one room. The latent component of the occupancy load is discussed next.

h. Cooling Load for Latent Heat

The sources of latent heat for which cooling must be provided are occupants (approximately 130 Btuh per sedentary occupant), infiltrated air and mechanical ventilation air, plus water vapor resulting from such normal residential activities as cooking, bathing, laundry and cleaning. However, because of the use of exhaust fans and appliance vents, most of the actual latent load is due to infiltration. Experience has shown that the ratio of total cooling load (sensible plus latent) to the sensible load rarely exceeds 1.3 and varies with construction tightness and design humidity ratio. See Figure 2.15. Construction tightness was discussed in Section 2.9. Humidity ratio, also known as specific humidity, was discussed briefly in Section 1.5b and is explained in greater detail in the following psychrometry section. The total cooling load, for residential buildings only, is the sum of all the sensible loads (items a–g) times the latent load factor selected from Figure 2.15.

2.19 Nonresidential Heat Gain Calculations

As stated at the beginning of Section 2.18, the theoretical basis of all heat gain calculations is the same for any type of building. The cooling load consists of the total sensible and latent heat gain of the building envelope plus the sensible and latent heat gain due to occupancy. The ASHRAE calculation techniques, however, are different for nonresidential calculations in that they are more exact and more detailed. Thus, for instance, CLTD factors for walls and roofs as in Table 2.10 are replaced by much more detailed data, depending on the specific type of wall or ceiling construction, including the surface color. The same is true for calculation of solar load through glass, occupancy cooling loads and infiltration cooling load. Furthermore, many of the load calculations are hourly rather than the daily maximum used in residential work. Of course, these detailed calculations are not

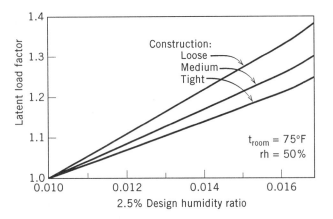

Figure 2.15 Latent load factor (LF) as a function of construction tightness and outside air humidity ratio, in residential buildings, with a 2.5% design humidity ratio. The factor is the ratio of total cooling load to sensible load, that is,

$$LF = \frac{\text{Total load}}{\text{Sensible load}}$$

(Reprinted by permission of the American Society of Heating, Regrigeration and Air-Conditioning Engineers, Atlanta, Ga., from the 1993 *ASHRAE Handbook—Fundamentals*.)

required for small, relatively simple nonresidential buildings. The choice of method and permissible approximations requires the judgement of an experienced designer. For our purpose, we will restrict our study to residential and small commercial buildings. Large complex structures are beyond the scope of this text.

2.20 Cooling Load Calculation Forms

As with heating load calculation forms, so too with cooling load calculation forms, there are many types in circulation, each with its own characteristics. The form shown in Figure 2.16(a) is a general one applicable to both residential and nonresidential use. Note that there are blanks for wall and roof color, *solar heat gain factor (SHGF)*, peak load, date and time, and full latent load. This form is intended primarily for calculation of the peak load of a nonresidential building. It can be adapted to a residential building by using the tables and factors explained in Section 2.18. Figure 2.16(b) is a somewhat older cooling load calculation form for a single room of a residential building. The ETD (equivalent temperature difference) and CLF (cooling load factor) factors in the form have been replaced with the more up-to-date CLTD and GLF factors of Tables 2.10, 2.11, 2.12 and 2.13. The total room sensible heat gain (RSHG) can be modified by the latent load factor (Figure 2.15) as explained at the end of Section 2.18 to arrive at the total cooling load.

The form shown in Figure 2.16(c) is published by the ACCA [see Figure 2.9(c)] as part of their *Manual J: Load Calculation for Residential Winter and Summer Air Conditioning*. The form is used with tables of heat transfer multipliers (HTM) for cooling loads, parts of which are shown in Figure 2.16d and e. This form calculates each room of the residence individually, as is necessary. Design conditions and construction details are filled in at the bottom of the form. Ignore the numbers filled in on the form; they are part of an illustrative example in *Manual J*.)

Note that on the form in Figure 2.16(a) the humidity ratio of the outside air is required, and on the form in Figure 2.16(c) the specific humidity (grains of water) is required. These numbers and terms are part of the science of *psychrometrics*, which is the study of the characteristics of moist air. A knowledge of the fundamentals of psychrometrics is absolutely essential to the understanding of air conditioning processes, both heating and cooling. An explanation of the elements of psychrometrics is presented next.

Psychrometrics

2.21 The Psychrometric Chart

The psychrometric chart is simply a graphic representation of the seven interrelated characteristics

Figure 2.16 Cooling load (heat gain) calculation forms. *(a)* From Bradshaw, *Building Control Systems*, 2nd ed., 1993, reprinted by permission of John Wiley & Sons. *(b)* From Pita, *Air Conditioning Principles and Systems*, 1981, reprinted by permission of Prentice-Hall, Inc., Upper Saddle River, NJ. *(c–e)* Reprinted with permission from *ACCA Manual J*.

PROJECT ______ PREPARED BY ______
LOCATION ______ DATE ______

COOLING LOAD CALCULATIONS

SPACE USE ______
FLOOR AREA ______ VOLUME ______
PEAK LOAD DATE ______ TIME ______ HRS/DAY OF OPERATION ______
GLAZING ______ SHADING ______
WALL COLOR: LT ☐ DK ☐ ROOF COLOR: LT ☐ DK ☐ LATITUDE ______

CONDITIONS	DB	WB	% RH	DP	HUM RATIO GR/LB (KG/KG)
OUTDOOR					
ROOM					
DIFFERENCE					

SENSIBLE LOADS

SOLAR

EXPOSURE	AREA	X	SHGF	X	SC	X	TLF	=	BTUH (W)	SUBTOTAL	TOTALS

TRANSMISSION

ITEM	EXP	AREA	X	ΔT	X	U-VALUE	X	TLF	=	BTUH (W)	SUBTOTAL
GLASS											
NET WALL											
ROOF											

O.A.

INFILTRATION
______ CFM (L/S) X 1.08 (1.20) X ______ °F (°C) = ______

VENTILATION
______ CFM (L/S) X 1.08 (1.20) X ______ °F (°C) = ______

INTERNAL HEAT

BTUH (W) SUBTOTAL

PEOPLE: ______ X ______ X ______ X ______ = ______
PEOPLE BTUH(W) LA BSF DIVERSITY

LIGHTS: ______ X 3.4* X ______ X ______ X ______ = ______
WATTS BALLAST BSF DIVERSITY

EQUIPT: ______ X 3.4* X ______ X ______ = ______
WATTS BSF DIVERSITY

APPLIANCES: ______ X 3.4* X ______ X ______ = ______
WATTS BSF DIVERSITY

OTHER: ______ = ______ ______

* Omit for SI Units

TOTAL SENSIBLE LOAD ______

LATENT LOADS

BTUH (W)

PEOPLE: ______ X ______ X ______ = ______
PEOPLE BTUH(W)/PERSON DIVERSITY

APPLIANCES: ______ ______

OUTSIDE AIR: ______ X ______ X 0.68 (2.808) = ______
CFM (L/S) GR/LB (KG/KG)

TOTAL LATENT LOAD ______

TOTAL LOAD (SENSIBLE & LATENT) ______

(a)

Residential Cooling Load Calculations Form

RESIDENTIAL COOLING LOAD CALCULATIONS Project ______ Location ______ Outdoor ______ F. Inside ______ F. Outdoor Daily Range ______ F. Engr. ______ Calc. ______ Check ______

Room Name																
Plan Size																
	U	A	ETD	BTU/hr	U	A	ETD	BTU/hr	U	A	ETD	BTU/hr	U	A	ETD	BTU/hr
Wall gross/net																
Roof/ceiling																
Floor																
Partition																
Door																
Windows	Dir.		CLF		Dir.		CLF		Dir.		CLF		Dir.		CLF	
Infiltration																
People																
Appliances																
RSHG																

Room Name																
Plan Size																
	U	A	ETD	BTU/hr	U	A	ETD	BTU/hr	U	A	ETD	BTU/hr	U	A	ETD	BTU/hr
Wall gross/net																
Roof/ceiling																
Floor																
Partition																
Door																
Windows	Dir.		CLF		Dir.		CLF		Dir.		CLF		Dir.		CLF	
Infiltration																
People																
Appliances																
RSHG																

Building Total

Calc. RSHG = ______
Duct gain % = ______
Duct leak % = ______
Total RSHG = ______
Cool. Load = X RSHG = ______

NOTES:

(b)

Figure 2.16 *(Continued)*

of moist air, which are, in conventional (English) units:

(a) Dry bulb (DB) temperature, °F.
(b) Wet bulb (WB) temperature, °F.
(c) Moisture content of air, which is is expressed in one of two ways:
 (1) grains of water per pound of dry air. This is called *specific humidity*.
 (2) pounds of water per pound of dry air, as a percentage. This is called *humidity ratio*.

 As a matter of interest, a pound of dry air can absorb 0.001–0.025 pounds of water (7–175 grains), depending on the DB temperature. There are 7005 grains in a pound. One pound of air occupies about 14 ft^3.
(d) Relative humidity (RH), %.
(e) Dew point, °F. (The *dew point* is that temperature at which the moisture in a given mixture of dry air and water vapor will condense. As explained previously, it is called dew point because this is the process in nature by which dew is formed.)
(f) Heat content, Btu/lb of dry air. (As mentioned previously, the total heat content of air is known as its enthalpy. It consists of the sum of

HEAT GAIN CALCULATION
DO NOT WRITE IN SHADED BLOCKS

				Entire House			1 Living			2 Dining			3 Laundry			4 Kitchen			5 Bath-1		
1	Name of Room			Entire House			1	Living		2	Dining		3	Laundry		4	Kitchen		5	Bath-1	
2	Running Ft. Exposed Wall			160			21			25			18			11			9		
3	Room Dimensions Ft.			51 x 29			21 x 14			7 x 18			7 x 11			11 x 11			9 x 11		
4	Ceiling Ht. Ft. Directions Room Faces			8			8	West		8	North		8			8	East		8	East	
	TYPE OF EXPOSURE	Const. No.	HTM Htg. / Clg.	Area or Length	Btuh Htg.	Btuh Clg.	Area or Length	Btuh Htg.	Btuh Clg.	Area or Length	Btuh Htg.	Btuh Clg.	Area or Length	Btuh Htg.	Btuh Clg.	Area or Length	Btuh Htg.	Btuh Clg.	Area or Length	Btuh Htg.	Btuh Clg.
5 Gross	a	12-d		1280			168			200			144			88			72		
Exposed	b	14-b		480																	
Walls &	c	15-b		800																	
Partitions	d	13N		232																	
6 Windows & Glass Doors Htg.	a b c d																				
7 Windows		North	14	20		280				20		280									
& Glass		E & W	44	115		5060	40		1760							11		484	8		352
Doors Clg.		South	23	30		690															
		Basement	70/36	8/8		848															
8 Other Doors		10-e	3.5	37		130							17		60						
9 Net	a	12-d	1.5	1078	1617		128	192		180	270		127	191		77	116		64	96	
Exposed	b	14-b	1.6	233	373																
Walls &	c	15-b	0																		
Partitions	d	13-n	0																		
10 Ceilings	a	16-d	2.1	1479		3106	294		617	126		265	77		162	121		254	99		208
	b																				
11 Floors	a	21-a	0																		
	b	19-f	0																		
12 Infiltration HTM			7.18	218		1565	40		287	20		144	17		122	11		79	8		57
13 Sub Total Btuh Loss = 6+8+9+10+11+12																					
14 Duct Btuh Loss			%																		
15 Total Btuh Loss = 13 + 14																					
16 People @ 300 & Appliances 1200						3000	3		900	3		900			—			1200			—
17 Sensible Btuh Gain = 7+8+9+10+11+12+16						16669			3756			1859			535			2133			713
18 Duct Btuh Gain			%			—			—			—			—			—			—
19 Total Sensible Gain = 17 + 18						16669			3756			1859			535			2133			713

NOTE: USE CALCULATION PROCEDURE D TO CALCULATE THE EQUIPMENT COOLING LOADS
*Answer for "Entire House" may not equal the sum of the room loads if hall or closet areas are ignored or if heat flows from one room to another room.

From Tables 4 & ??

ASSUMED DESIGN CONDITIONS AND CONSTRUCTION (Cooling) — Const. No. / HTM

A. Outside Design Temperature: Dry Bulb 88 Rounded to 90 db   38 grains - Table 1
B. Daily Temperature Range: Medium - Table 1
C. Inside Design Conditions: 75F, 55% RH   Design Temperature Difference = ... (90-75 = 15)
D. Types of Shading: Venetian Blinds on All First Floor Windows - No Shading, Basement
E. Windows: All Clear Double Glass on First Floor - Table 3A
   North 14
   East or West 44
   South 23
   All Clear Single Glass (plus storm) in Basement - Table 3A Use Double Glass
   East 70
   South 36
F. Doors: Metal, Urethane Core, No Storm, 0.50 CFM/ft. .... 11e / 3.5
G. First Floor Walls: Basic Frame Construction with Insulation (R-11) x ½" board - Table 4 .... 12d / 1.5
   Basement Wall: 8" Concrete Block, Above Grade: 3 ft (R-5) - Table 4 .... 14b / 1.6
   8" Concrete Block Below Grade: 5 ft (R-5) - Table 4 .... 15b / 0
H. Partition: 8" Concrete Block Furred, with Insulation (R-5), Δ T approx. 0°F - Table 4 .... 13n / 0
I. Ceiling: Basic Construction Under Vented Attic with Insulation (R-19), Dark Roof - Table 4 .... 16d / 2.1
J. Occupants: 6 (Figured 2 per Bedroom, But Distributed 3 in Living, 3 in Dining)
K. Appliances: Add 1200 Btuh to Kitchen
L. Ducts: Located in Conditioned Space - Table 7B
M. Wood & Carpet Floor Over Unconditioned Basement, Δ T approx. 0°F .... 19 / 0
N. The Envelope was Evaluated as Having Average tightness - (Refer to the Construction details at the Bottom of Figure 3-3)
O. Equipment to be Selected From Manufacturers Performance Data.

(c)

Figure 2.16 *(Continued)*

Table 3C
Glass Heat Transfer Multipliers (Cooling)
External Shade Screen, Shading Coefficient = .25

Clear Glass

Design Temperature Difference	Single Pane						Double Pane Single Pane & Low e Coating						Triple Pane Double Pane & Low e Coating					
	10	15	20	25	30	35	10	15	20	25	30	35	10	15	20	25	30	35
DIRECTION WINDOW FACES	NO INTERNAL SHADING																	
N	23	27	31	35	39	43	19	21	23	25	27	29	17	18	19	20	21	22
NE and NW	31	35	39	43	47	51	26	28	30	32	34	36	24	25	26	27	28	29
E and W	38	42	46	50	54	58	31	33	35	37	39	41	28	29	30	31	32	33
SE and SW	35	39	43	47	51	55	29	31	33	35	37	39	26	27	28	29	30	31
S	27	31	35	39	43	47	23	25	27	29	31	33	20	21	22	23	24	25
	DRAPERIES OR VENETIAN BLINDS																	
N	14	18	22	26	30	34	12	14	16	18	20	22	10	11	12	13	14	15
NE and NW	19	23	27	31	35	39	16	18	20	22	24	26	14	15	16	17	18	19
E and W	23	27	31	35	39	43	20	22	24	26	28	30	17	18	19	20	21	22
SE and SW	21	25	29	33	37	41	18	20	22	24	26	28	16	17	18	19	20	21
S	17	21	25	29	33	37	14	16	18	20	22	24	12	13	14	15	16	17
	ROLLER SHADES — HALF DRAWN																	
N	17	17	17	17	17	17	16	18	20	22	24	26	14	15	16	17	18	19
NE and NW	23	27	31	35	39	43	22	24	26	28	30	32	19	20	21	22	23	24
E and W	28	32	36	40	44	48	26	28	30	32	34	36	23	24	25	26	27	28
SE and SW	26	30	34	38	42	46	24	26	28	30	32	34	21	22	23	24	25	26
S	20	24	28	32	36	40	19	21	23	25	27	29	17	18	19	20	21	22

Tinted (Heat Absorbing) Glass

Design Temperature Difference	Single Pane						Double Pane Single Pane & Low e Coating						Triple Pane Double Pane & Low e Coating					
	10	15	20	25	30	35	10	15	20	25	30	35	10	15	20	25	30	35
DIRECTION WINDOW FACES	NO INTERNAL SHADING																	
N	16	20	24	28	32	36	12	14	16	18	20	22	9	0	11	12	13	14
NE and NW	22	26	30	34	38	42	16	18	20	22	24	26	12	13	14	15	16	17
E and W	26	30	34	38	42	46	20	22	24	26	28	30	15	16	17	18	19	20
SE and SW	24	28	32	36	40	44	18	20	22	24	26	28	14	15	16	17	18	19
S	19	23	27	31	35	39	14	16	18	20	22	24	11	12	13	14	15	16
	DRAPERIES OR VENETIAN BLINDS																	
N	12	16	20	24	28	32	9	11	13	15	17	19	6	7	8	9	10	11
NE and NW	17	21	25	29	33	37	12	14	16	18	20	22	8	9	10	11	12	13
E and W	20	24	28	32	36	40	15	17	19	21	23	25	10	11	12	13	14	15
SE and SW	18	22	26	30	34	38	14	16	18	20	22	24	9	10	11	12	13	14
S	14	18	22	26	30	34	11	13	15	17	19	21	7	8	9	10	11	12
	ROLLER SHADES — HALF DRAWN																	
N	14	18	22	26	30	34	10	12	14	6	18	20	7	8	9	10	11	12
NE and NW	19	23	27	31	35	39	14	16	18	20	22	24	10	11	12	13	14	15
E and W	23	27	31	35	39	43	17	19	21	23	25	27	12	13	14	15	16	17
SE and SW	21	25	29	33	37	41	15	17	19	21	23	25	11	12	13	14	15	16
S	17	21	25	29	33	37	12	14	16	18	20	22	9	10	11	12	13	14

Reflective Coated Glass

Design Temperature Difference	Single Pane						Double Pane Single Pane & Low e Coating						Triple Pane Double Pane & Low e Coating					
	10	15	20	25	30	35	10	15	20	25	30	35	10	15	20	25	30	35
DIRECTION WINDOW FACES	NO INTERNAL SHADING																	
N	14	18	22	26	30	34	10	12	14	16	18	20	6	7	8	9	10	11
NE and NW	19	23	27	31	35	39	14	16	18	20	22	24	8	9	10	11	12	13
E and W	23	27	31	35	39	43	16	18	20	22	24	26	10	11	12	13	14	15
SE and SW	21	25	29	33	37	41	15	17	19	21	23	25	9	10	11	12	13	14
S	17	21	25	29	33	37	12	14	16	18	20	22	7	8	9	10	11	12
	DRAPERIES OR VENETIAN BLINDS																	
N	11	15	19	23	27	31	8	10	12	14	16	18	5	6	7	8	9	10
NE and NW	15	19	23	27	31	35	11	13	15	17	19	21	7	8	9	10	11	12
E and W	18	22	26	30	34	38	14	16	18	20	22	24	8	9	10	11	12	13
SE and SW	17	21	25	29	33	37	13	15	17	19	21	23	8	9	10	11	12	13
S	13	17	21	25	29	33	10	12	14	16	18	20	6	7	8	9	10	11
	ROLLER SHADES — HALF DRAWN																	
N	12	16	20	24	28	32	9	11	13	15	17	19	5	6	7	8	9	10
NE and NW	17	21	25	29	33	37	12	14	16	18	20	22	7	8	9	10	11	12
E and W	20	24	28	32	36	40	15	17	19	21	23	25	8	9	10	11	12	13
SE and SW	19	23	27	31	35	39	14	16	18	20	22	24	8	9	10	11	12	13
S	15	19	23	27	31	35	11	13	15	17	19	21	6	7	8	9	10	11

(d)

Figure 2.16 (Continued)

Table 4
Heat Transfer Multipliers (Cooling)

No. 1 through 9 - Windows and Glass Doors - Refer to Table No. 3 for Summer HTM Values.

No. 10 - Wood Doors	\multicolumn{11}{c	}{Summer Temperature Difference and Daily Temperature Range}
	10		15			20			25	30	35	U
	L	M	L	M	H	L	M	H	M	H	H	H	
	\multicolumn{12}{c	}{HTM (Btuh per sq. ft.)}
A. Hollow Core	9.9	7.6	12.7	10.4	7.6	15.5	13.2	10.4	16.0	13.2	16.0	18.8	.560
B. Hollow Core & Wood Storm	5.8	4.5	7.5	6.1	4.5	9.1	7.8	6.1	9.4	7.8	9.4	11.1	.330
C. Hollow Core & Metal Storm	6.3	4.9	8.1	6.7	4.9	9.9	8.5	6.7	10.3	8.5	10.3	12.1	.360
D. Solid Core	8.1	6.3	10.4	8.6	6.3	12.7	10.9	8.6	13.2	10.9	13.2	15.5	.460
E. Solid Core & Wood Storm	5.1	3.9	6.6	5.4	3.9	8.0	6.8	5.4	8.3	6.8	8.3	9.7	.290
F. Solid Core & Metal Storm	5.6	4.4	7.2	6.0	4.4	8.8	7.6	6.0	9.2	7.6	9.2	10.8	.320
G. Panel	11.8	9.1	15.1	12.5	9.1	18.5	15.8	12.5	19.2	15.8	19.2	22.5	.670
H. Panel & Wood Storm	6.3	4.9	8.1	6.7	4.9	9.9	8.5	6.7	10.3	8.5	10.3	12.1	.360
I. Panel & Metal Storm	7.2	5.6	9.3	7.6	5.6	11.3	9.7	7.6	11.7	9.7	11.7	13.8	.410

No. 11 - Metal Doors	10		15			20			25	30	35	U
	L	M	L	M	H	L	M	H	M	H	H	H	
	\multicolumn{12}{c	}{HTM (Btuh per sq. ft.)}
A. Fiberglass Core	10.4	8.0	13.3	11.0	8.0	16.3	13.9	11.0	16.9	13.9	16.9	19.8	.590
B. Fiberglass Core & Storm	6.5	5.0	8.3	6.8	5.0	10.1	8.7	6.8	10.5	8.7	10.5	12.3	.367
C. Polystyrene Core	8.3	6.4	10.6	8.7	6.4	13.0	11.1	8.7	13.4	11.1	13.4	15.8	.470
D. Polystyrene Core & Storm	5.6	4.3	7.2	5.9	4.3	8.7	7.5	5.9	9.1	7.5	9.1	10.7	.317
E. Urethane Core	3.3	2.6	4.3	3.5	2.6	5.2	4.5	3.5	5.4	4.5	5.4	6.4	.190
F. Urethane Core & Storm	3.0	2.3	3.8	3.2	2.3	4.7	4.0	3.2	4.9	4.0	4.9	5.7	.170

No. 12 Wood Frame Exterior Walls With Sheathing and Siding or Brick Veneer or Other Exterior Finish.	10		15			20			25	30	35	U
	L	M	L	M	H	L	M	H	M	H	H	H	
	\multicolumn{12}{c	}{HTM (Btuh per sq. ft.)}
A. None ½" Gypsum Board (R-0.5)	4.8	3.7	6.1	5.0	3.7	7.5	6.4	5.0	7.8	6.4	7.8	9.1	.271
B. None ½" Asphalt Board (R-1.3)	3.8	3.0	4.9	4.0	3.0	6.0	5.1	4.0	6.2	5.1	6.2	7.3	.217
C. R-11 ½" Gypsum Board (R-0.5)	1.6	1.2	2.0	1.7	1.2	2.5	2.1	1.7	2.6	2.1	2.6	3.0	.090
D. R-11 ½" Asphalt Board (R-1.3) R-11 ½" Bead Brd. (R-1.8) R-13 ½" Gypsum Brd. (R-0.5)	1.4	1.1	1.8	1.5	1.1	2.2	1.9	1.5	2.3	1.9	2.3	2.7	.080
E. R-11 ½" Extr Poly Brd. (R-2.5) R-11 ¾" Bead Brd. (R-2.7) R-13 ½" Asphalt Brd. (R-1.3) R-13 ½" Bead Brd. (R-1.8)	1.3	1.0	1.7	1.4	1.0	2.1	1.8	1.4	2.1	1.8	2.1	2.5	.075
F. R-11 1" Bead Brd. (R-3.6) R-11 ¾" Extr Poly Brd. (R-3.8) R-13 ½" Extr Poly Brd (R-2.5) R-13 ¾" Bead Brd. (R-2.7)	1.2	1.0	1.6	1.3	1.0	1.9	1.7	1.3	2.0	1.7	2.0	2.4	.070
G. R-13 ¾" Extr Poly Brd. (R-3.8) R-13 1" Bead Brd (R-3.6)	1.1	.9	1.5	1.2	.9	1.8	1.5	1.2	1.9	1.5	1.9	2.2	.065
H. R-11 1" Extr Brd. (R-5.0) R-13 1" Extr Poly Brd. (R-5.0) R-19 ½" Gypsum Brd. (R-0.5)	1.1	.8	1.4	1.1	.8	1.7	1.4	1.1	1.7	1.4	1.7	2.0	.060
I. R-19 ½" Asphalt Brd. (R-1.3) R-19 ½" Bead Brd. (R-1.8)	1.0	.7	1.2	1.0	.7	1.5	1.3	1.0	1.6	1.3	1.6	1.8	.055
J. R-11 R-8 Sheathing R-13 R-8 Sheathing R-19 ½" or ¾" Extr Poly R-19 ¾" or 1" Bead Brd.	.9	.7	1.1	.9	.7	1.4	1.2	.9	1.4	1.2	1.4	1.7	.050
K. R-19 1" Extr Poly Brd (R-5.0)	.8	.6	1.0	.8	.6	1.2	1.1	.8	1.3	1.1	1.3	1.5	.045
L. R-19 R-8 Sheathing	.7	.5	.9	.7	.5	1.1	.9	.7	1.1	.9	1.1	1.3	.040
M. R-27 Wall	.7	.5	.8	.7	.5	1.0	.9	.7	1.1	.9	1.1	1.2	.037
N. R-30 Wall	.6	.4	.7	.6	.4	.9	.8	.6	.9	.8	.9	1.1	.033
O. R-33 Wall	.5	.4	.7	.6	.4	.8	.7	.6	.9	.7	.9	1.0	.030

Footnotes to Table 4 are found on page 84.

(e)

Figure 2.16 (Continued)

the sensible heat content of the dry air plus the latent heat content of the moisture in the air.)
(g) Density of the air, ft³/lb of dry air.

As we shall see, knowledge of any two of the seven factors will immediately give us the other five by inspection of the psychrometric chart. (An exception to this rule occurs with dew point and moisture content, which are essentially the same piece of information, as will become clear shortly.) The chart describes all possible conditions of moist air. Since the purpose of air conditioning is to "condition" the air (so that it is comfortable), it follows that all HVAC processes can be shown on the psychrometric chart. It is this HVAC process plotting that makes the chart so useful.

In our discussion we will demonstrate only the basic processes as they are shown on the chart. Use of the chart to plot complex processes, select coils based on sensible heat ratio and perform other advanced pyschrometric design functions is beyond our scope here. However, once the psychrometric principles involved are grasped, these advanced techniques can be learned fairly readily.

2.22 Components of the Psychrometric Chart

Refer to Figure 2.17. Do not be put off by the apparent complexity of the chart. Actually, the chart in Figure 2.17 is a simplified version of the full chart, which can be found in Figure 2.19. For our purposes, however, this simplified version contains all the data necessary to learn our way around the chart.

a. Dry Bulb (DB) Temperature Lines

The vertical lines on the chart represent dry bulb temperatures. We have drawn a heavy line on 75°F, DB. This means that every condition of air that has as one of its characteristics a 75°F DB will fall somewhere along this line.

b. Specific Humidity Lines

The horizontal lines on the chart represent specific humidity. We have drawn a heavy line at a specific humidity of 80 grains (per pound) of air. Notice that this corresponds to 0.011 lb of water on the second vertical scale. That this is correct can easily be checked:

$$80 \text{ grains} \times \frac{\text{lb}}{7005 \text{ grains}} = 0.011 \text{ lb}$$

We stated previously that setting any two of the seven characteristics of air, establishes the other five. The intersection point of the 75°F DB line and the 80 grain specific humidity line fixes the other aspects of the air. We will call this intersection point A and so label it on the chart.

c. Wet Bulb (WB) Temperature Lines

The lines that are drawn diagonally from upper left to lower right at an angle of 30° from the horizontal are the WB lines. The lines extend between the curved left edge of the diagram to the horizontal and vertical axes of the diagram. The curved left edge is labelled on its extension "saturation line—100% RH," and along its center "wet bulb temperature, °F." If we now return to the intersection of the DB 75°F line and the 80 grain line, we find that the WB temperature at this point is 66°F (by visual interpolation). We have drawn in this line for explanation only.

d. Relative Humidity Lines

These are the curved lines that extend from lower left to upper right. They are labeled in percent, from 10 to 100%. The 100% RH is the left boundary of the chart. Referring again to intersection point A, we see by visual interpolation that it corresponds to an RH of 61%. Here again, we have drawn in this RH line for explanation only.

e. Enthalpy Lines

The heat content (enthalpy) lines correspond almost exactly to the slope of the WB temperature lines and are, therefore, not drawn in separately. Their values are shown on the diagonal scale at the left of the chart. Therefore, by extending the 66°F DB line to this scale, we find that the enthalpy of the air corresponding to point A is 30.6 Btu/lb of air. This value is the sum of the sensible and latent heat in the air whose mixture is indicated by point A.

f. Density Lines

These lines extend steeply from the upper left to the lower right and are labelled 12.5–15 ft³/lb of dry air. We have drawn in the density line of the air that we are studying and find it to be 13.7 ft³/lb of dry air.

COMPONENTS OF THE PSYCHROMETRIC CHART / 71

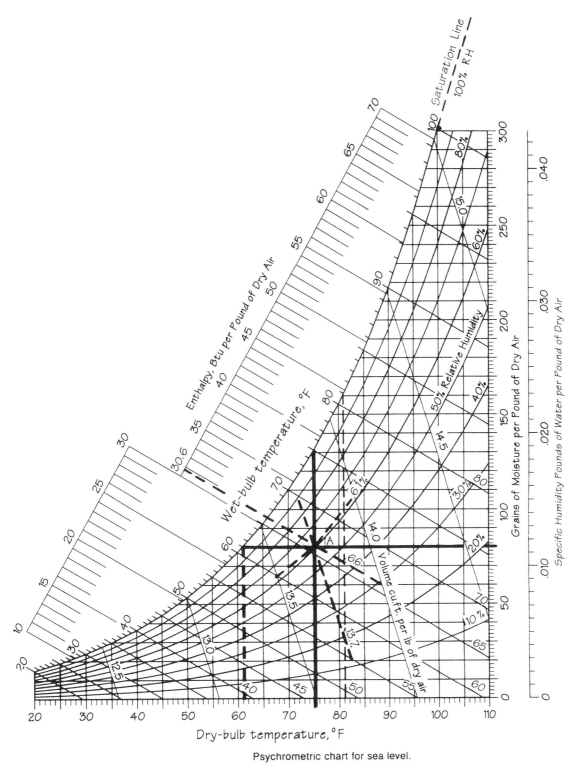

Psychrometric chart for sea level.

Figure 2.17 Psychrometric chart in conventional units. See text for an explanation of the lines drawn on the chart. From Bobenhausen, *Simplified Design of HVAC Systems*, 1994, reprinted by permission of John Wiley & Sons.)

g. Dew Point

As we explained previously, the dew point of an air mixture is reached when its dry bulb temperature is lowered until the water in the air condenses. At that point, the air is completely saturated. This, of course, corresponds to 100% relative humidity since the air at that point cannot absorb any additional water. To find the dew point on the chart, we simply extend the specific humidity line to the left of point A, until it strikes the 100% RH line. That is, we graphically take air containing a given amount of water and cool it (left extension) until the RH is 100%. At that point the WB temperature is 61°F. If we then drop a vertical line to the DB line, we find that it too is 61°F. That is, at the dew point the DB and WB temperatures are equal. A moment's thought will confirm that this must be so always. When we measure the WB temperature of air with a sling psychrometer, it is usually lower than DB because of the evaporative cooling effect of the wet sock on the WB thermometer. However, at 100% RH, no water can evaporate from the wet sock, and the WB temperature does not drop; that is, it remains the same as the DB reading. Therefore, we have established that at the dew point and the DB and WB temperatures are always equal.

At this point we have shown that knowing two facts about any air/water vapor mixture determines all seven characteristics. We began with DB temperature and specific humidity. From the intersection of these two lines, we determined WB temperature, RH, density, enthalpy and dew point. Notice, however, that the dew point can be determined by specific humidity alone, because both are essentially the same fact. Simply extend the specific humidity line to the left until it strikes the 100% RH line to determine the dew point. Thus, air with 50 grains of water has a 48°F DB/WB dew point; 100 grains condenses at 67°F DB/WB, 150 grains at 79°F DB/WB and so forth. Therefore, knowledge of specific humidity and dew point constitutes only one piece of information about the air, and another is needed to establish a particular air condition. This is what we meant when we stated at the beginning of this section that knowing any two facts about air establishes all the others, except moisture content (specific humidity) and dew point.

It is very important that you not confuse specific humidity and relative humidity. *Specific humidity* is exactly what it says—a specific amount (weight) of water per pound of air. For this reason we strongly advise against using the alternate "humidity ratio" formulation since it can lead to confusion. Specific humidity tells us simply how much water there is in the air. On the other hand, *relative humidity* is a ratio between the amount of water vapor in the air and the maximum water-carrying capacity of the air. For this reason, it is expressed as a percentage. You must understand that for a given specific humidity, a whole range of RH is possible, and vice versa. Using our example on the chart of Figure 2.17, for a specific humidity of 80 grains, relative humidity can vary between 20 and 100% as the DB temperature varies between 110°F and 61°F. That is, 80 grains of water per pound of air represents only 20% of the air's water-carrying capacity at 110°F DB, and all (100%) of the air's water carring capacity at 61°F DB.

Example 2.1

(a) At what DB temperature will a specific humidity of 80 grains of water represent 50% RH?
(b) What is the maximum specific humidity and dew point at that temperature?

Solution:

(a) Follow the 80 grain line on the chart to the point where it intersects the 50% RH line. At that point, drop down vertically to the DB scale and read 81.5°F DB.
(b) If 80 grains is the specific humidity (SH) at 50% RH, then by definition

$$\frac{80 \text{ grains}}{\text{Maximum capacity}} = 50\%$$

$$\text{Maximum capacity} = \frac{80}{0.5} = 160 \text{ grains}$$

Graphically, maximum SH occurs at 100% RH. Therefore, extending the 81.5°F DB line vertically to the saturation line (100% RH), we find, as we expect, 81.5°F WB. Extending horizontally across, we find 160 grains of water, also as expected. Of course, since we know that at the dew point DB=WB, we could have gone directly to 81.5% WB on the 100% RH saturation line, without going through the additional step of extending a line vertically from the DB scale.

As stated, for a given RH, a whole range of specific humidities is possible. Taking 50% RH as an example, the SH varies from 207 grains at 110°F DB down to 10 grains at 25°F DB. The 50% RH means that at each DB temperature, the air is capable of carrying (absorbing) twice the amount of water, at which point it will be saturated.

Example 2.2

(a) What is the range of specific humidities shown on the chart of Figure 2.17 for a 40% RH?
(b) What specific humidity corresponds to a 40% RH at 50°F DB and 90°F DB?
(c) How much water must be removed from a pound of air at 80% RH and 90°F DB to "condition" it to 50% RH when it is cooled to 78°F?

Solution:

(a) By inspection, the 40% RH line runs between 165 grains of water at 110°F DB to 6 grains at 20°F DB.
(b) By inspection of the chart,
   (1) at 40% RH and 50°F DB, SH = 22 grains.
   (2) at 40% RH and 90°F DB, SH = 86 grains.
(c) By inspection of the chart,
   (1) at 80% RH and 90°F DB, SH = 176 grains/lb.
   (2) at 50% RH and 78°F DB, SH = 72 grains/lb.

   Amount of water to be removed is 104 grains/lb of air.

A simple and very practical example of the application of the psychrometric chart is to find the dew point of a specific air-water vapor condition.

Example 2.3 Cold water piping at 40°F DB runs along the ceiling of a restaurant kitchen. During cooking hours the air in the kitchen reaches 95°F DB and 70% RH. Does the piping require insulation to keep water from condensing on it and dripping?

Solution: Refer to Figure 2.19(b). Follow the 95°F DB line vertically up to 70°F RH and then horizontally across to the 100% RH saturation line. Read 83.40°F WB. If we drop vertically, we will, of course, also read 83.40°F DB since at the dew point, WB = DB. Since the pipe wall temperature is only slightly above 40°F (the metal pipe has very little insulation value), water will certainly condense on the pipe. The pipe, therefore, requires enough insulation so that its outside surface is above 83.4°F. The insulation should also have an outside vapor barrier to prevent condensation inside the insulation.

2.23 Basic HVAC Processes on the Psychrometric Chart

Now that you understand the data presented in the pychrometric chart, we can begin to use it to show basic HVAC processes. There are four fundamental actions involved in any air-conditioning process:

- Sensible heating.
- Sensible cooling.
- Latent heating (humidification).
- Latent cooling (dehumidification).

Any HVAC process can be described (and plotted) as some combination of these four actions, simply because there are no others. A process is plotted from one point on the chart representing a specific condition of air to a second point, representing another condition of air.

a. Sensible Heating Only

Any horizontal line drawn from left to right represents a process of sensible heating only. Since the line is horizontal, the water content (grains) is constant (i.e., there is no change in latent heat). Because the line is drawn from a lower to higher DB temperature (left to right), it must represent a sensible heating process. Referring to Figure 2.18, line O-SH represents a sensible heating process. A closed space containing an electric resistance heater or a fan coil unit with a hot water coil would be two examples of a space containing equipment that produces such an HVAC process.

b. Sensible Cooling Only

It follows from the preceding explanation that a horizontal line drawn from right to left, such as line O-SC on Figure 2.18 would represents a process of sensible cooling only. A space containing an air conditioner with a dry cooling coil would represent such a sensible cooling process. (A wet coil dehumidifies also.)

c. Latent Heating Only

This is simply a process of humidification. You must become accustomed to the fact that the addition of water vapor to a space simply means that

74 / THERMAL BALANCE OF BUILDINGS

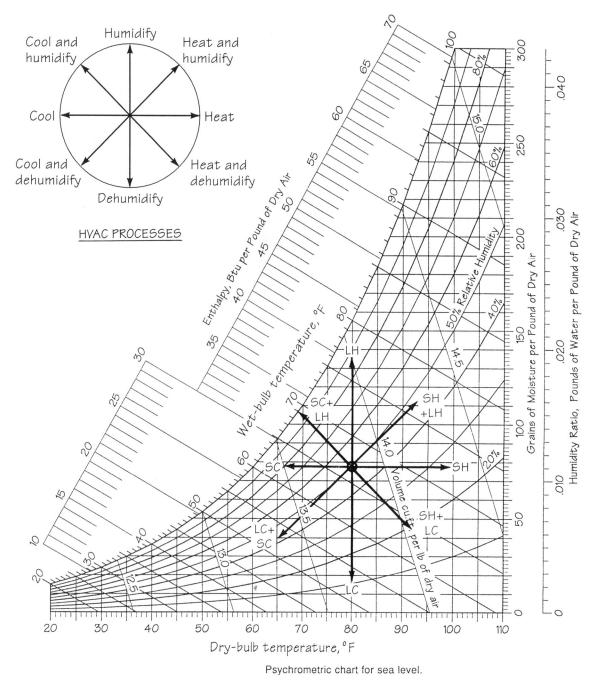

Psychrometric chart for sea level.

Figure 2.18 Psychrometric chart with HVAC processes plotted and analyzed. Processes in the upper right quadrant represented by line O-SH+LH indicate sensible heating (SH) and latent heating (humidification). Lines proceeding down and to the right indicate sensible heating and latent cooling (dehumidification). Lines drawn from upper right to lower left indicate sensible and latent cooling (SC+LC). Finally, lines drawn up and left indicate sensible cooling and latent heating. A summary of these process lines is shown by the process "star" at the upper left of the illustration. (The psychrometric chart is from Bobenhausen, *Simplified Design of HVAC Systems*, 1994, reprinted by permission of John Wiley & Sons.)

latent heat has been added to the space. The amount of latent heat is the (latent) heat of vaporization of that quantity of water. In Figure 2.18, line O-LH is a latent heating line because the amount of water has changed while the DB temperature remains the same. Any vertical process line that extends upward such as line O-LH is a latent-heating-only line. In more conventional language, such a line represents humidification only. The change in latent heat can be read off the enthalpy scale. Remember that

Enthalpy = Sensible heat + Latent heat

Since there is no change in sensible heat, the entire enthalpy change represents latent heat change. A humidifier with its own source of heat that introduces water vapor into a space at room temperature would entail such a process. A steamer-type humidifier adds sensible heat also. A cold-spray humidifier removes sensible heat from a space by evaporative cooling, as will be explained later.

d. Latent Cooling Only

It should be apparent that a process line extending vertically downward, such as line O-LC represents a latent cooling process. In other words, such a line shows dehumidification only; that is, the dry bulb temperature remains unchanged while the specific humidity decreases. Dry dehumidifiers and liquid absorption equipment are representative equipment used in this type of process.

Most HVAC processes, however, are not pure sensible or pure latent heat actions but rather a combination of two of the preceding processes. They are also easily shown on the psychrometric chart.

e. Heat and Humidify

In more technical terms, this process adds sensible heat and latent heat (humidification). It is represented graphically by line O = SH + LH, extending from the index point O up to the right. You should understand that any line in the upper right quadrant represents heating plus humidification. We have drawn it at a 45° angle, but it can be at any angle between 0° and 90°. The vertical component of such a sloped line represents added latent heat (humidification) and the horizontal component represents added sensible heat. A steamer-type humidifier, as already mentioned, would be one device that performs this HVAC process. The line for such a device's HVAC process would be at a steep angle of 70° to 85° from the horizontal, since most of its energy is expended in humidification (vertical component) and only a little incidental energy in sensible heating (horizontal component). Another device that provides heating and humidification is a common residential hot air furnace with a humidifier. The line for this device would be shallow, probably less than 30°. This is because the vast majority of its energy is used to provide sensible heating and only a bit for humidification.

f. Cool and Humidify

Proceeding counterclockwise, line O = SC + LH represents sensible cooling plus latent heating. Again, any line pointing at an angle upward and left, between 90° and 180° from the 0° axis (shown as the SH line), represents this process. This process is best illustrated by evaporative cooling devices. These "desert coolers" take hot dry air and simply add water by blowing air through a wetting device of some sort. Water, picked up by the air, vaporizes. In order to do so, it needs to absorb its heat of vaporization. This it does by absorbing sensible heat from the air in the air stream and in the room into which it is pumped. In so doing, it cools the air. The result is a lower DB temperature (sensible cooling) and a higher humidity (latent heating), at a constant WB temperature.

g. Cool and Dehumidify

Continuing around the "star" drawn on Figure 2.18, we reach the lower left quadrant. Any line drawn from upper right to lower left represents sensible and latent cooling (cool and dehumidify). It is shown on the diagram as line O = LC + SC. This is the action of the classic wet coil air conditioner, which cools by pumping cold air into a space and dehumidifies by condensing room moisture on a cold coil.

h. Heat and Dehumidify

The last combination action is represented by lines extending from upper left to lower right, that is, in the lower right quadrant. We have drawn on such a line O = SH + LC in Figure 2.18. This process, heating and dehumifying, is not a common one. It can be represented by absoption dehumification equipment.

76 / THERMAL BALANCE OF BUILDINGS

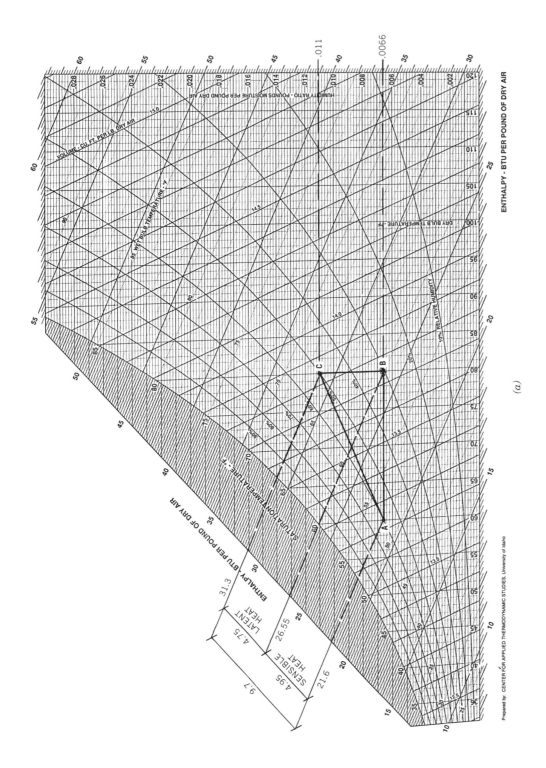

Figure 2.19(a)

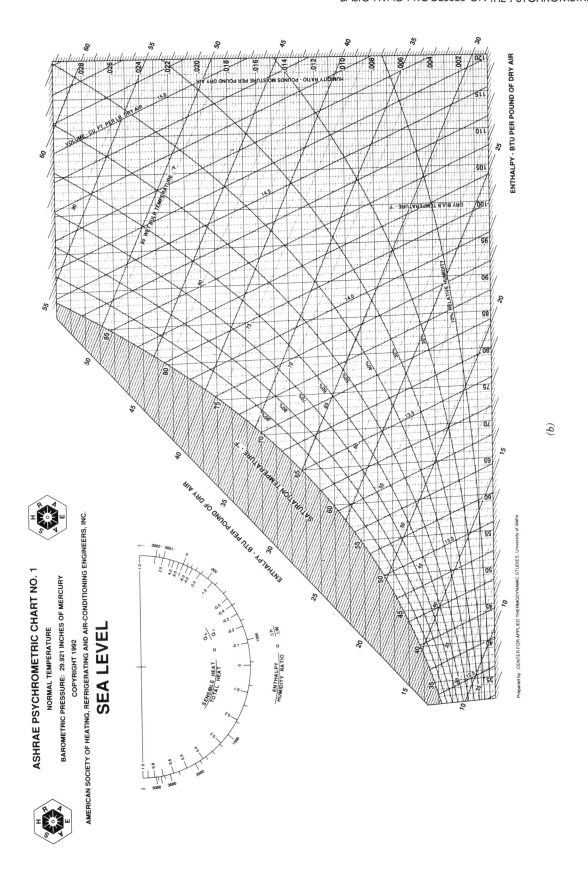

Figure 2.19(b).

2.24 Air Equations for HVAC Processes

We are now in a position to use the air equations and a psychrometric chart plot to solve actual HVAC problems. Remember that we calculated the sensible heat gain of infiltrated air by using Equation 2.18 (Section 2.18):

$$\text{Sensible load} = 1.1(\text{cfm})(\Delta t) \text{ in Btuh} \quad (2.18)$$

Example 2.4 What size electric convector is required to heat a space from 60°F DB to 80°F DB, assuming a 1000-cfm blower? Assuming an initial 60% RH, what is the final RH?

Solution: Using Equation 2.18, we have

$$\text{Sensible load} = 1.1 \, (\text{cfm}) \, (\Delta t)$$
$$= 1.1 \, (1000) \, (80-60) = 22{,}000 \text{ Btuh}$$
$$22{,}000 \text{ Btuh} \times \frac{\text{kw}}{3{,}412 \text{ Btuh}} = 6.5 \text{ kw}$$

For a graphic solution, draw the process line on the accurate psychrometric chart of Figure 2.19(a) from point A to point B. Since it is purely a sensible heating process, the line is horizontal, as shown on the chart. We can read off the chart that the new RH is 30%. The change in enthalpy is from 21.6 Btu/lb at point A to 26.55 Btu/lb at point B. The heat load/enthalpy equation is

$$\text{Heat load} = 4.45 \, (\text{cfm}) \, (\Delta \text{ enthalpy, Btuh}) \quad (2.20)$$

Using this formula, we have

$$\text{Heat load} = 4.45 \, (1000) \, (26.55 - 21.6) = 22{,}000 \text{ Btuh}$$

which corresponds to the preceding solution.

Example 2.5 We wish to add a humidifier to the air stream heated by the convector of Example 2.4 that will raise the relative humidity of the heated air to 50%. The humidifier should have a built-in heater just large enough to supply the latent heat required for the vaporization. The humidifier has a 100-cfm blower. What must be the rating of the humidifier heating element?

Solution: The chart shows the specific humidity in pounds of water per pound of air. We need to convert this to grains. The process is purely latent heating (humidification) and is shown on the chart in Figure 2.19(a) as vertical line BC. Point C is at the desired air condition of 80°F DB and 50% RH. The SH of point B is

$$\text{Specific humidity} = 0.0066 \times \frac{7000 \text{ grains}}{\text{lb}}$$
$$= 46.2 \text{ grains/lb}$$

The SH of point C is

$$0.011 \times \frac{7000 \text{ grains}}{\text{lb}} = 77.2 \text{ grains/lb}$$

The formula for latent load is

$$\text{Latent load} = 0.68 \, (\text{cfm}) \, (\Delta \text{ grains}) \quad (2.21)$$

Therefore, the latent load $= 0.68 \, (100) \, (77.2 - 46.2) = 2108$ Btuh.

$$2108 \text{ Btuh} \times \frac{1000 \text{ w}}{3412 \text{ Btuh}} = 618 \text{w}$$

The humidifier, therefore, requires a 600-w heater. Checking this result with the enthalpy equation (2.20), we have

$$\text{Heat load} = 4.45 \, (\text{cfm}) \, (\Delta \text{enthalpy})$$
$$= 4.45 \, (100) \, (31.3 - 26.55) = 2113 \text{ Btuh}$$

This is well within engineering accuracy and confirms the previous calculation.

Note that if the two processes, that is, heating and humidification, can be combined into a single process using a heater with a built-in humidifier, the process would then be plotted from point A at 60°F DB and 60% RH directly to point C at 80°F DB and 50% RH. This process line could then be broken down into its sensible component AB and its latent component BC, with exactly the same results as calculated previously. You can find detailed explanations of advanced graphic techniques for solution of complex HVAC process problems with the psycrometric chart in the supplementary reading references listed at the end of the chapter. They are not presented here because they are beyond the scope of this text.

Key Terms

Having completed the study of this chapter, you should be familiar with the following key terms. If any appear unfamiliar or not entirely clear, you should review the section in which these terms appear. All key terms are listed in the index to assist you in locating the relevant text.

Air changes per hour (ACH)
Air equations
Btuh
Coefficient of thermal transmission (U)
Cooling load temperature difference (CLTD)
Conduction
Convection
Degree days (DD)
Dew point
Effective leakage area
Emittance *(e)*
Enthalpy
Evaporative cooling
Glass load factor (GLF)
Grains of water
Heat transfer rate *(Q, q)*
Humidity ratio
Infiltration
Interior temperature swing
Latent cooling
Latent heating
Natural (free) convection
Outside design conditions

Perimeter heat loss factor
Psychrometrics, psychrometric chart
Radiation
Radiation barrier
Radiative heat transfer
Relative humidity
Sensible cooling
Sensible heating
Shade line factor (SLF)
Shading
Shading coefficient (SC)
Sick Building Syndrome (SBS)
Sol-air temperature
Solar heat gain factor (SHGF)
Specific humidity
Temperature gradient
Thermal conductance *(C)*
Thermal conductivity *(k)*
Thermal insulation
Thermal lag
Thermal resistance *(R)*
Thermal resistivity *(r)*

Supplementary Reading

Stein, B., and Reynolds, J. *Mechanical and Electrical Equipment for Building,* 8th ed., Wiley, New York, 1992, Chapter 4.

ASHRAE Handbook—Fundamentals, 1993, Chapters 3, 6, 20, 21, 22, 25, and 27.

McQuiston, F. C., and Parker, J. D. *Heating, Ventilating and Air Conditioning,* 3rd ed., Wiley, New York, 1988, Chapters 3, 5, 6, 7, 8.

Pita, E. G. *Air Conditioning Principles and Systems,* Wiley, New York, 1981, Chapters 3, 6, and 7.

Problems

1. Calculate the hourly heat loss of a small one-story, slab-on-grade industrial building in Boston, Massachusetts, with the following conditions:

    a. Plan dimensions: 50 × 100 ft
    b. Orientation: long dimension facing south
    c. Height: 12 ft to the underside of the roof construction

d. Walls: use the wall shown in Figure 2.5 *(d)*
e. Roof: use the roof detail in Figure 2.7 *(a)*, base case
f. Windows: North wall, 200 ft², single-glazed metal frame
South wall, 360 ft² glass area, single-glazed metal frame
East and West walls, 120 ft², double-glazed, air-filled space, metal frame (use Table 2.7)
g. Doors: 1¾-in. steel, polyurethane foam core, no thermal break; total area 260 ft²
h. Edge insulation: R-5.4
i. Infiltration: Assume ACH = 3
j. Outside winter 97.5% design condition: 9°F DB
Make any assumptions necessary to perform the calculations, and justify the assumptions.

2. Recalculate the hourly heat loss of Problem 1 using wall insulation of 6-in. glass fiber instead of the original 3½ in. of insulation. What is the percentage reduction in overall heating load?

3. Recalculate the hourly heat loss of Problem 1 using
   a. reflective air space with $e = 0.05$ or
   b. roof deck insulation as in "new item 7" of Figure 2.7*(a)*
   Which method is more effective in winter? Which method is more effective in summer?

4. Draw the temperature gradient through the roof section of Problem 1, in winter, with the
   a. Original roof.
   b. Reflective air space roof.
   c. Insulated roof.

5. Inspect and sketch the heating plant in your classroom building, dormitory or home. Discuss the following:
   a. Fuel and fuel storage.
   b. Provision for combustion air.
   c. Flue and chimney.
   d. Space required for the plant.
   Do not include controls, ducts, piping or heating elements in the occupied space.

6. Refer to Figure 3.38 for The Basic House plan. Using the construction data given there plus the following data, calculate the heat losses for each room in Btuh. Use the calculation form shown in Figure 2.9*(b)* amended as required to use the ACH method of air infiltration heat loss. Assume medium tight construction.

The windows are double hung, wood frame, with sizes as follows:

Living Room: 2 @ 4 ft × 3 ft
Kitchen: 2 @ 2 ft 6 in. × 2 ft 2 in.
Dining Room: 2 @ 3 ft × 3 ft 6 in.
Bath: 1 @ 1 ft 6 in. × 2 ft 6 in.
BR #1 and BR #2: 2 @ 3 ft × 3 ft in.

There are two doors:

Front: 1¾-in. solid, flush, weather stripped, with storm sash
Back: 1⅜-in., hollow core, with glass panes, no storm sash

Make any necessary assumptions and justify them briefly. Indicate the source of all data used.

7. The Basic House is to be constructed in a cold area of the United States where insulation standards call for the following insulation values:

Ceilings below unheated, uninsulated attic: R-33
Floors over unheated basement or crawl space: R-22

The windows are double-glazed, with a storm sash, and the air space between window and storm sash is 3½ in.

Recalculate all the room losses with this revised data. Again, make all necessary assumptions and indicate the source of any data used.

8. Calculate the maximum cooling load (heat gain) of each room of The Basic House plan in Figure 3.38. Use the following design data:

Outside design temperature (2.5%): 87°F DB
Orientation: due South
Daily temperature range: medium
Windows: no exterior shading devices; interior venetian blinds

Assume four occupants and normal household activities. Show all calculations. Select any cooling load calculation form, or make one of your own. As in all previous calculations, justify all assumptions, and indicate the source of all data.

9. *Psychrometrics.* Use the chart in Figure 2.19.
   a. What is the specific humidity, in grains, of air whose dew point is 65°F?
   b. What is its relative humidity at 65°F? 75°F? 85°F?
   c. Air at 85°F DB and 80% RH is cooled and dehumidified to 75°F DB and 50% RH.

What is the sensible cooling in Btu/lb? What is the latent cooling in Btu/lb? What is the total cooling in Btu/lb?

Show the process line on the psychrometric chart.

10. Use the data in Table 2.7 for metal frame windows. Calculate the maximum outdoor temperature at which condensation will form on the inside of the windows, if the room air is held at 75°F DB and 50% RH, for

   a. single-glazed.
   b. double-glazed, air-filled.
   c. triple-glazed, inert gas-filled.

   (*Hint:* Use the technique shown in Figure 2.6 to calculate the inside glass temperature. The inside and outside air films are the other thermal resistance components.)

11. Based on the data given in Table 2.7, prepare two tables—one for wood frame windows and one for metal frame windows—showing the maximum indoor humidity for no window condensation, as a function of outdoor temperature, for

   a. single-glazed.
   b. double-glazed, air-filled.
   c. double-glazed, inert gas-filled.
   d. triple-glazed, air-filled.

   Use the following outdoor temperatures: 40°F, 30°F, 25°F, 15°F, 10°F, 5°F, 0°F, −5°F, −10°F, −20°F. The indoor air is at 75°F DB. The table format follows:

Maximum RH (%) for No Condensation

Outdoor Temperature, °F	Window (a)[a]	Window (b)[b]	Window (c)[c]	Window (d)[d]

[a] Window (a) is single-glazed.
[b] Window (b) is double-glazed, air-filled.
[c] Window (c) is double-glazed, inert gas-filled.
[d] Window (d) is triple-glazed, air-filled.

3. Hydronic Heating

Active heating and cooling systems for buildings use either air or water as the heat transfer medium. Systems that use water are properly referred to as *hydronic systems*, and their design is called *hydronics*. Some HVAC engineers reserve this term for water-based heating systems only, referring to them as "hydronic heating systems," while using the term "chilled-water systems" for water-based cooling arrangements. In this chapter, we will study hydronic heating systems intensively. Hydronic cooling systems, which are used only in commercial buildings, including multiple residences, will only be touched on in this chapter. A fuller discussion will be found in Chapter 6.

Careful study of this chapter will enable you to:

1. Identify and understand the functioning of all components of a hot water (hydronic) heating system.
2. Understand the different control arrangements used in hydronic heating systems.
3. Design and lay out single and multiple circuit series loop hydronic heating systems.
4. Calculate temperatures and water flow in series system piping loops.
5. Design and lay out single and multiple circuit one-pipe hydronic heating systems.
6. Select residential baseboard radiation units based on the system temperatures and heat losses.
7. Establish heating zones in large residential and nonresidential buildings.
8. Calculate friction head in all types of heating system piping arrangements.
9. Select commercial finned-tube radiation units based on the system factors and space heat loss.
10. Design a two-pipe reverse return hot water heating system, including pipe sizing.

3.1 Hydronic Systems

Water is a very efficient means of transmitting heat to different areas of a building. Because of its very high specific heat (1 Btu/lb-°F), a relatively small flow of water can carry a large amount of heat. For instance, a flow of only 1 gallon per minute (gpm) will deliver 500 Btu/hour/°F. This is calculated as follows:

$$\frac{1 \text{ gal}}{\text{min}} \times \frac{60 \text{ min}}{\text{h}} \times \frac{8.33 \text{ lb}}{\text{gal}} \times \frac{1 \text{ Btu}}{\text{lb-°F}} = \frac{500 \text{ Btu}}{\text{h-°F}}$$

For a system that operates at a temperature drop of 20°F between outgoing and return water (a common design figure), this flow of 1 gpm will deliver 10,000 Btuh:

$$\frac{500 \text{ Btu}}{\text{h-°F}} \times 20 \text{ F°} = 10,000 \text{ Btuh}$$

You should remember these two important pieces of data because they are used repeatedly in the design of water systems:

1 gpm carries 500 Btuh per °F and
1 gpm will deliver 10,000 Btuh for a 20°F drop in temperature.

Hot water carries almost 3500 times more heat than the same volume of air. This gives the hydronic system one of its principal advantages—compactness. Whereas air systems require large bulky ducts that have a major architectural impact, hydronic systems use small pipes that can be run unobtrusively to any part of a building, usually without causing any space coordination problems. Furthermore, because the piping in a modern hydronic heating system is a closed loop around which water is pumped (forced circulation) rather than a system that relies on gravity as in the older hydronic systems, the piping can be installed without a slope. This removes another installation constraint, making the piping installation very simple. (Water is drained from the lines by using small drain fittings at system low points.)

Hydronic heating systems have the additional advantages of quiet operation, long life, very high efficiency, low maintenance, ease of zoning and control, choice of fuel (oil, gas, electricity) and great flexibility. The disadvantages of hydronic heating systems are the need for a separate building ventilation system, the requirement for separate humidification devices and the difficulty and expense of incorporating cooling into the system. Because most small residential air conditioning systems in today's construction must provide both heating and cooling, preference is given to air systems, since cooling is easily incorporated there. However, in multiple residences and commercial buildings hydronic heating/cooling systems are frequently used.

The diagrams of Figure 3.1 show, in a purely schematic fashion, the equipment arrangement for an hydronic heating-only system [Figure 3.1(a)], a system that will provide either heating or cooling [Figure 3.1(b)] and a system that will provide heating and cooling simultaneously [Figure 3.1(c)]. The system shown in Figure 3.1(c) is actually two separate and distinct systems: a heating system and a cooling system, operating simultaneously. This arrangement is also called a four-pipe system for the obvious reason. Because the large amount of piping makes this system expensive, a cheaper version uses a three-way inlet valve to each fan coil unit and a single return pipe for all units. The three-way inlet valve selects either hot or cold water as needed. The disadvantage of this arrangement is that the single return water pipe mixes hot water from some terminal units with cold water from others. This practice wastes a large amount of energy. For this reason, three-pipe systems have fallen into disuse despite their lower first cost. The terminal units in system of Figure 3.1(a) are suitable for heating only. The terminal units of the systems in Figure 3.1(b) and (c) are suitable for either heating or cooling. Fan coil units are shown because they are the most common dual-use terminal units. (See Figure 3.19.) The remainder of this chapter will deal with hydronic heating systems only.

3.2 Components of a Hydronic Heating System

Although systems vary in design, all hydronic heating systems contain certain basic elements. These elements are

(a) A hot water boiler and its controls and safety devices.
(b) An expansion tank, also referred to as a compression tank.
(c) A water pump, also called a circulator.
(d) Terminal devices that transfer heat from the circulating water to the various building spaces.
(e) Piping, including valves and fittings.

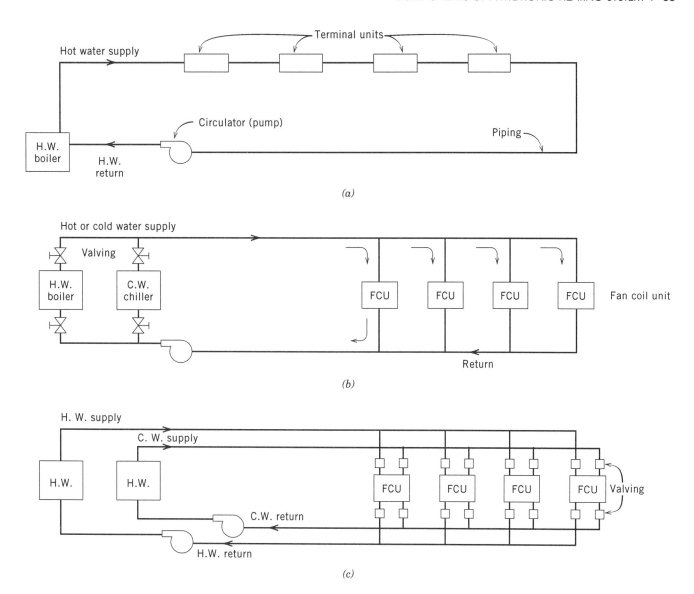

Figure 3.1 Schematic diagrams of typical hydronic systems. *(a)* Series hydronic heating loop. Controls are not shown for the sake of clarity. *(b)* Hydronic heating and cooling system with single-coil, fan coil units. All units are supplied either hot or cold water. Either the hot water boiler or the cold water chiller operates, not both. *(c)* Four-pipe heating/cooling hydronic system. Both the boiler and the chiller operate. Each double-coil, four-pipe fan coil unit operates independently to supply heating or cooling as required for a specific space.

(f) Controls that regulate the operation of the boiler burner, the system circulator(s), and any zone and flow valves.

These items will be discussed individually.

3.3 Hot Water Boilers

Boiler is the term applied to a device that produces either hot water or steam. By contrast, *furnace* is the term normally applied in HVAC work to a device that heats air. We are restricting our study to hot water systems because steam is no longer widely used except in high-rise buildings, and they are beyond the scope of this book. Therefore, when we use the term *boiler*, we are specifically referring to a hot water boiler. Refer to Figure 3.2, which shows the principal parts of a typical medium-size cast-iron boiler, suitable for multiple residences and commercial buildings. Figure 3.3 shows the hydronic and electrical controls usually found on a hydronic heating boiler. Figure 3.4 shows two common piping arrangements for a boiler with

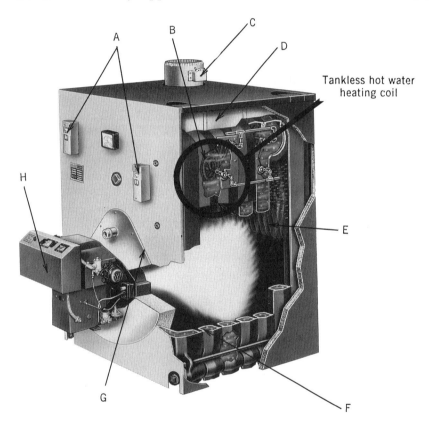

A. Front-mounted boiler controls
B. Tankless domestic hot water heater
C. Adjustable lock-type damper
D. Steel flue canopy
E. Cost iron vertical flue
F. Wet base
G. Burner mounting plate and observation port
H. Gas, oil- or dual-fuel burner

(a)

Figure 3.2 *(a)* Cutaway photo of a typical medium-size cast-iron boiler suitable for multiple residence or commercial use. Standard equipment not specifically labeled includes pressure/temperature safety relief valve(s), pressure and temperature gauges, burner controls and safety devices. (Photo courtesy of Burnham Corp.)

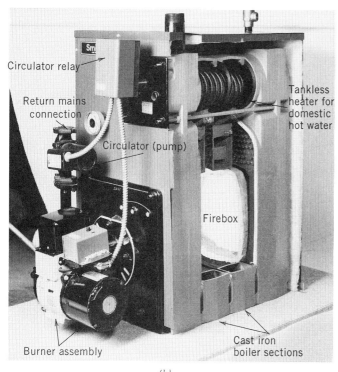

Figure 3.2 *(b)* Cast-iron packaged boiler, fired with light oil. Side panel has been removed to expose the firebox, the water channels and the tankless domestic hot water heater. DOE capacity ratings for this design range from 91 to 250 MBH. All units are 18 in. wide and 35 in. high. Length varies from 33 in. overall (including the burner assembly) for the smallest unit (91 MBH) to 47 in. long for the largest unit (250 MBH). Standard working pressure is 30 psi. (Courtesy of Smith Cast Iron Boilers.)

either vertical or horizontal supply piping and the location of a *tankless coil* for domestic hot water.

Boilers are built in accordance with the Boiler and Pressure Vessel Code, Section IV, published by the American Society of Mechanical Engineers (ASME). The usual hydronic heating boiler is rated for a 30-psig (pounds per square inch, gauge) working pressure and a maximum pressure of 160 psig. (See Section 8.6 for an explanation of static pressure measurement, gauge pressure, and absolute pressure.) Boilers can produce hot water up to a temperature of 250°F, although the usual residential and commercial setting is somewhere between 160°F and 200°F. The temperature setting depends on the type of terminal units employed, the length of the piping circuits and whether the boiler is used to heat domestic hot water (in which case the higher temperatures are used). Temperatures lower than 160°F are employed when boilers supply floor or ceiling panels in radiant panel system arrangements. Boilers designed to supply water at temperatures below 240°F are referred to as low water temperature (LWT) units.

Boilers, baseboard radiation and finned-tube radiation are tested and rated by The Hydronics Institute [formerly the Institute of Boiler and Radiation Manufacturers (I = B = R) and the Steel Boiler Institute (SBI)]. (*Radiation* is the term employed to describe the terminal units that perform the actual space heating, because in early systems these units were cast-iron radiators.) The Hydronics Institute, whose members are manufacturers of these items, publishes lists of equipment that meet its efficiency and other operational standards. Such units carry an I = B = R emblem. All I = B = R-rated boilers are manufactured in accordance with the requirements of Section IV of the ASME Boiler and Pressure Vessel Code. This is the universally recognized code (in the United States) covering the manufacture of heating boilers. Boilers with capacity rat-

88 / HYDRONIC HEATING

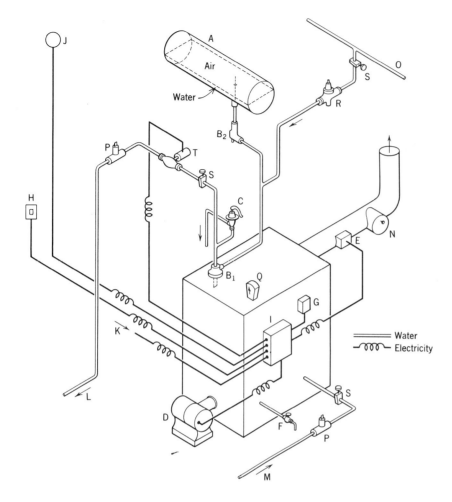

A *Compression Tank.* Accommodates the expansion of the water in the system.
B *Air Control Fittings.* Vent out unwanted air in the boiler and maintain the level in the compression tank.
C *Pressure Relief Valve.* Usually set for 30 psi. Initial cold pressure about 12 psi. Relieves excessive system pressure.
D *Oil Burner.* Responds to aquastat or thermostat.
E *Stack Temperature Control.* Senses stack temperature and stops oil injection if ignition has not occurred.
F *Drain Valve.* At low point in the water system.
G *Aquastat.* Maintains temperature of boiler water by starting the oil burner when temperature of water drops below the aquastat's setting. Set between 140 and 200°F.
H *Remote Switch.* At a safe distance from the boiler so that the plant can be turned off in case of trouble, without approaching the boiler.
I *Electrical Control Center.*
J *Thermostat.* When the room temperature drops below its setting, it turns on either the burner or the circulating pump or both, depending on the control arrangement.

K *Electrical Power Source.* Operates from an individual circuit at the electric panel.
L *Hot Water Supply.* Copper tubing to convectors or baseboards (radiation).
M *Hot Water Return.* Copper tubing from convectors or baseboards.
N *Draft Adjuster.* Regulates the draft (combustion air) over the flame.
O *House Cold Water Main.* (Make-up water line) from which water is fed automatically into boiler..
P *Flow Control Valves.* Prevent casual flow of water by gravity when the circulator is not running.
Q *Temperature/Pressure Gauge.* Indicates water temperature and pressure. Sometimes supplemented by immersion thermometers in supply and return mains.
R *Pressure Reducing Valve.* Admits water into the system when the pressure there drops below about 12 psi. Has a built-in check valve to prevent backflow of boiler water into the water main.
S *Shutoff Valves.* Normally open. Can be closed to isolate the system and permit servicing of components.
T *Circulator.* Centrifugal circulating pump that moves the water through the tubing and heating elements.

Figure 3.3 Isometric schematic of a typical oil-fired boiler, showing the usual electrical and hydronic controls and indicating devices.

HOT WATER BOILERS / 89

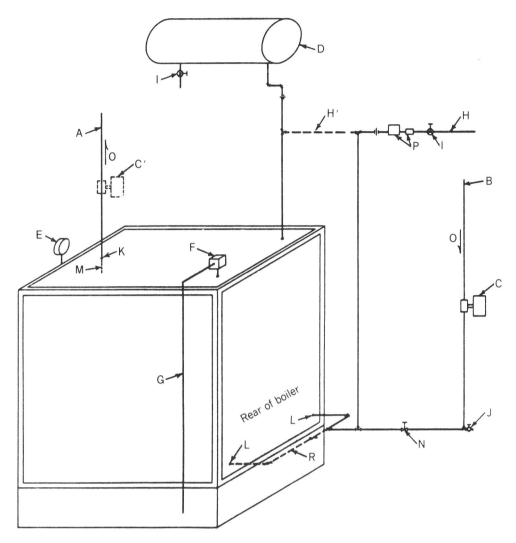

Boiler with Vertical Supply Tapping

A Supply pipe
B Return pipe
C Pump. Location may be either in a horizontal or vertical position, depending on instructions supplied with unit
C′ Pump, alternate location
D Air cushion tank
E Altitude gage
F Safety relief valve
G Overflow from safety relief valve
H Make-up water line (house cold water supply line). Always install between pump and air cushion tank
H′ Make-up water line, alternate location
I Globe valve
J Drain cock
K Vertical supply tapping
L Boiler return tapping
M Dip tube
N Purge valve (optional)
O Direction of flow water
P Pressure reducing valve and check valve
R Yoke return into both return tappings, if recommended by manufacturer

(a)

Figure 3.4 Schematic piping for boilers with *(a)* vertical supply tapping *(b)* and horizontal supply tapping. Typical valving and the location of a tankless coil for domestic hot water is shown in *(c)*. (Courtesy of The Hydronics Institute.)

90 / HYDRONIC HEATING

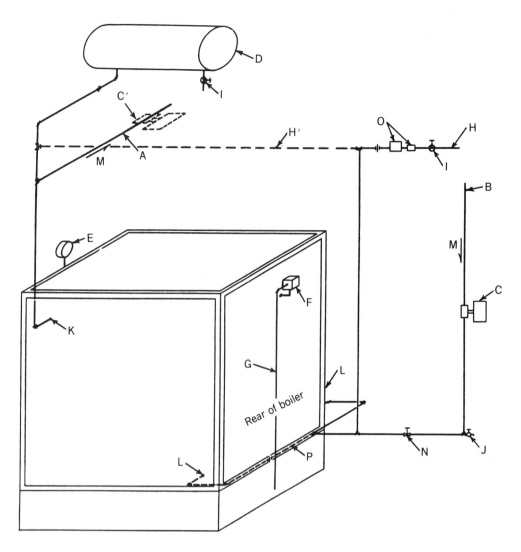

Boiler with Horizontal Supply Tapping

- A Supply pipe
- B Return pipe
- C Pump. Location may be either in a horizontal or vertical position, depending on instructions supplied with unit
- C′ Pump, alternate location
- D Air cushion tank
- E Altitude gage
- F Safety relief valve
- G Overflow from safety relief valve
- H Make-up water line (house cold water supply line). Always install between pump and air cushion tank
- H′ Make-up water line, alternate location
- I Globe valve
- J Drain cock
- K Horizontal supply tapping
- L Boiler return tapping
- M Direction of flow of water
- N Purge valve (optional)
- O Pressure reducing valve and check valve
- P Yoke return into both return tappings, if recommended by manufacturer

(b)

Figure 3.4 *(Continued)*

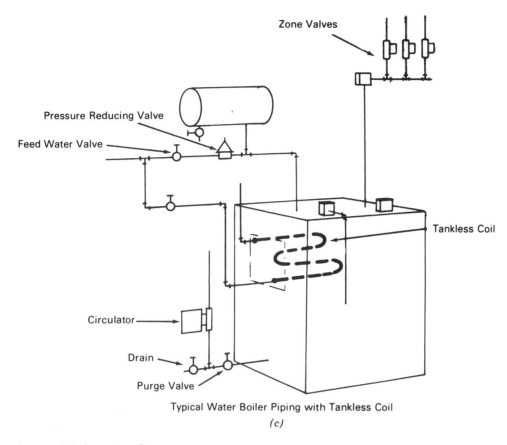

Typical Water Boiler Piping with Tankless Coil
(c)

Figure 3.4 *(Continued)*

ings of up to 300,000 Btuh (usually written 300 MBH) are referred to as *residential boilers.* Boilers with higher ratings are known as *commercial boilers.* I = B = R standards now include the requirements of the U.S. Department of Energy (DOE) and the National Appliance Energy Conservation Act (NAECA) of 1987.

When selecting a boiler for a particular installation, the designer frequently applies a safety factor to the maximum calculated heating load, depending on the type of building. Commercial buildings, which cool down on weekends and holidays when not in use, require a quick heat pickup. To provide this rapid pickup, their boilers are oversized up to 25%. On the other hand, residential units operate continuously during the heating season (even with night setback) and, therefore, require minimum oversizing, if any at all. ASHRAE Code 90.1/1989, Energy Efficient Design of New Buildings Except Low Rise Residential Buildings, calls for a maximum of 25% system oversizing in the interest of energy conservation. Boilers are most efficient when operated at or near their full rating.

All residential boilers and commercial boilers up to about 2000 MBH (2,000,000 Btuh) are made of cast iron. Cast iron is highly corrosion-resistant, making such boilers relatively maintenance-free and giving their users long years of service. Cast-iron boilers are also highly efficient, easy to cast and relatively simple to field assemble where necessary. Larger commercial boilers are made of steel. Residential boilers and small commercial units are usually supplied as preassembled package units that include the boiler body, burner with controls and safety devices, circulating pump, wiring, operating controls and safety devices.

In recent years, small, wall-mounted, gas-fired package boilers intended for individual apartments have been manufactured. These units enable apartment occupants to control their own heating rather than rely on a central heating system. Further, these units eliminate any need for heat-use billing, which has become common recently. The units have sealed combustion chambers and are vented directly through an outside wall. They also draw combustion air through the outside wall on which they are mounted.

3.4 Definitions and Basic Equipment Functions

Because many of the items that appear in Figures 3.3 and 3.4 and in subsequent figures may not be familiar to you, the following list of terms is provided for easy reference. Major items of equipment are discussed in individual sections and are so indicated in the list.

Air Cushion Tank See *Expansion Tank*, Section 3.5.

Air Vent A device intended to release air from the closed hydronic heating system. The devices, which may be manual or automatic, are placed at high points in the system where air accumulates and on terminal units. See Figure 3.5.

Altitude Gage See *Pressure Gage*.

Balancing Valve See *Valves*.

Baseboard See *Terminal Units*, Section 3.7.

Branch Piping The piping that connects a terminal unit into a hot water circulation loop. The pipe feeding the terminal unit is called the *supply branch*, and the pipe returning water to the main loop is called the *return branch*. (See Figures 3.24, 3.25, and 3.43.)

Circuit Also called loop. A water circuit is a complete, closed piping loop originating at the boiler and returning to the boiler. Hot water heating systems can be single circuit or multiple circuit. See Section 3.8 and Figure 3.23.

Circulator See *Pump*.

Compression Tank See *Expansion Tank*, Section 3.5.

Convector See *Terminal Units*, Section 3.7.

Diaphragm Tank See *Expansion Tank*, Section 3.5.

Dip Tube A piping arrangement or special fitting at the boiler that acts to release air accumulation in a closed, forced hot water heating system. The fitting shown in Figure 3.6 discharges air that rises into the top of the boiler through its side connection, into the expansion tank located above the boiler. The lower tube of the fitting is an extension of the vertical water supply, inside the boiler.

Drain Valve Valves installed at the base of the boiler and at low points of the piping circuits, for the purpose of draining the entire system.

Expansion Tank Also referred to as the *air cushion tank*, the *compression tank* and the *diaphragm tank*. Its purpose is to absorb the additional water volume in the closed system, due to water heating. This item is discussed separately in Section 3.5.

Finned-Tube Radiation See *Terminal Units*, Section 3.7.

Flow Control Valve See *Valves*.

Heat Distribution Units See *Terminal Units*, Section 3.7.

Main, Main Pipe A single pipe coming from the boiler that supplies all branch piping and multiple circuits. Also, the pipe that carries return water to the boiler from all the branches and separate piping circuits. (See Figure 3.27.)

Make-up Water Pipe The connection to the boiler that is used for filling the system and adding water as it becomes necessary. (See Figure 3.4 for location.)

Mono-Flow Fitting See *One-Pipe System*.

Multiple Circuit System A hot water heating (or cooling) system comprising more than one piping loop. (See Figures 3.23 and 3.27.)

One-Pipe System A system of piping in which the terminal units are connected to a single closed pipe loop with water diverter fittings, also called MONO-FLO fittings. (See Section 3.8.)

Pressure Gage Also known as an *altitude gage*, this meter indicates the boiler water pressure. Calibration is usually in psi and/or feet of water. Occasionally the meter is a combination type, including a boiler water thermometer. Most often the temperature gage is a separate instrument.

Pressure Reducing Valve See *Valves*.

Safety Valve See *Valves*.

Pump Also known as a *circulator*. (See Section 3.6.)

Purge Valve See *Valves*.

Radiation, Radiator See *Terminal Units*.

Return Branch See *Branch Piping*.

Supply Branch See *Branch Piping*.

Terminal Units The devices that deliver heat into the building spaces by radiation and convection. Typical terminal units include baseboards, finned-pipe radiation, cast-iron radiators, radiant panels, unit heaters, convectors and fan coil units. (See Section 3.7.)

Valves

a. Balancing Valves

Valves used in multiple loop (circuit) systems to control the amount of hot water in each circuit.

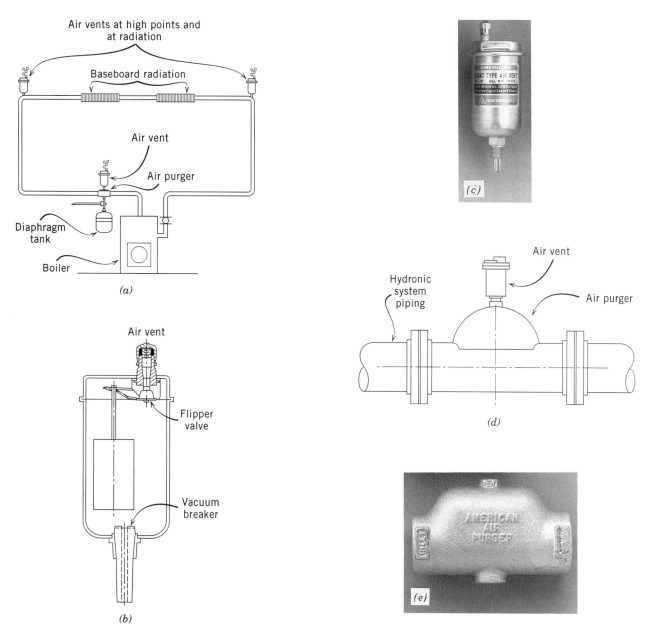

Figure 3.5 *(a)* Air vents should be installed at high points of a hydronic heating system, as shown schematically. They are typically placed at heaters, baseboards and convectors and at the high point of the return piping. *(b)* Section through a float-type air vent. The valve assembly automatically varies the vent opening (port) in accordance with the amount of air to be vented. *(c)* A typical float vent is 3–4 in. high and about 2 in. in diameter. *(d)* Air is removed from the water in a hydronic system by using an air purger in the water line and a float vent. The purger separates the air from the circulating water, and the air vent releases it to the atmosphere. *(e)* Typical air purger for a 1½-in. pipe measures 8 in. long by 5 in. high and 3½ in. deep. (Photos and drawing courtesy of Amtrol.)

Figure 3.6 Boiler air control fitting or "dip-tube." (See Figure 3.3, item B1.) Air collecting at the top of the boiler rises into the compression tank through the side connection of this fitting. See Figure 3.10 (b) for piping connections. (Courtesy of ITT Fluid Handling Sales.)

Figure 3.7 Flow control valve. (See Figure 3.3, item P.) Usually placed near the boiler on both the supply and the return mains. The valve opens when the circulator (pump) is activated and closes when the circulator stops, to prevent gravity flow. (Courtesy of the ITT Fluid Handling Sales.)

These valves are usually placed in the return piping of each loop. They serve to balance the heat distribution of the entire system. Additional valves can be placed in the return lines of terminal units, branches, risers and headers. Each valve is provided with a metering connection to permit accurate flow measurement during system operation. Valves are sized for flow rather than pipe size.

b. Flow Control Valve

A check valve that prevents reverse flow due to gravity, when the pump is not operating. Used in systems containing indirect water heaters and frequently in zoned systems. Installed in the supply piping. See Figure 3.7.

c. Pressure-Reducing Valve

A valve installed in the make-up water line, for the purpose of reducing the pressure of the city water supply to that of the boiler. The valve pressure setting is field adjustable. (See Figure 8.26, page. 449.)

d. Purge Valve

A valve installed in piping between the pressure-reducing valve on the make-up water line and the drain valve. Its purpose is to purge the system of trapped air by forcing water through the system at high velocity.

e. Safety Valve

A pressure relief valve that operates when boiler pressure exceeds the preset operating pressure. It will reset automatically when boiler pressure falls below the valve setting. The valve outlet is piped to a drain so that hot water (and steam) released when the valve opens will not cause injury or damage. (See Figure 8.25, page 448.) Some safety valves are combined high pressure, high temperature valves and will open when subjected to excessive pressure or excessive temperature.

f. Zone Control Valve

A thermostatic valve that controls the flow of hot water in a single zone of a multizone heating system. See Figure 3.8.

3.5 Expansion Tank

The expansion tank is a closed vessel normally located just above the boiler. See Figures 3.3 and 3.4. Its purpose is to compensate for the increase in volume of the water in the closed hydronic system, when the water is heated. Water volume increases approximately 1% for every 40 F° of temperature increase. If a hydronic heating system were not provided with some means of absorbing this expansion in volume, it would simply leak at some point in the piping due to the high pressure caused by the volume increase. The traditional approach to this problem was to install a tank (as in Figures 3.3 and 3.4) in such a fashion that when the system is initially filled, air is trapped in the upper portion of this tank. Then, when the heated water expands, the air cushion in the tank simply compresses, thus allowing space for the water to expand. This action gave rise to the various names—expansion tank, air cushion tank and compression tank.

The tank is designed and sized to absorb the full water volume expansion over the entire range of boiler operating temperatures, without the pressure relief valve operating. A typical tank of this design is shown in Figure 3.9. The water level in the tank can be controlled by using a tank air-control fitting, as illustrated in Figure 3.10. This arrangement, however, has a disadvantage. The direct contact between hot water and pressured air in the tank can cause air to be entrained in the water. This in turn can cause air blockages in the piping, accelerated pipe and fitting corrosion, noisy water flow and even total water logging of the tank. Trapped air is released by installing manual or automatic air vents at all high points in the piping system and at all terminal units. The latter are particularly susceptible to trapped air, which can considerably reduce their effectiveness. (Figure 3.5 illustrates air vents.)

To avoid the problem of air/water contact in the expansion tank, modern systems use a diaphragm tank. See Figure 3.11. These tanks have a flexible inert diaphragm that physically separates the air

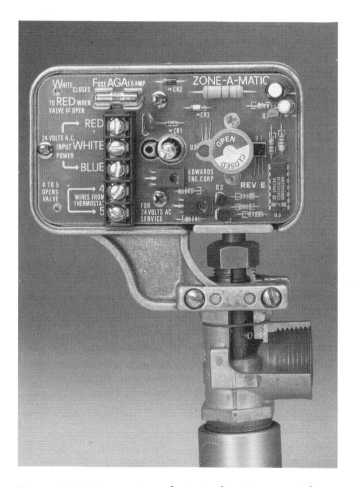

Figure 3.8 Cutaway view of a typical motor-operated zone control valve. The valve motor is actuated by the zone thermostat, and it acts either to admit or cut off hot water to the zone piping circuit. The valve contains auxiliary contacts that can be used for burner control. The valve assembly is small enough to fit inside the end space of a normal baseboard heater. (Courtesy of Edwards Engineering.)

Figure 3.9 Typical expansion tank. Bottom tappings are for connection to the boiler. See Figure 3.10 (c) for piping connections to air vent/air passage fitting and to the boiler. The tappings on the end of the tank are intended for gauge glasses. (Courtesy of ITT Fluid handling Sales.)

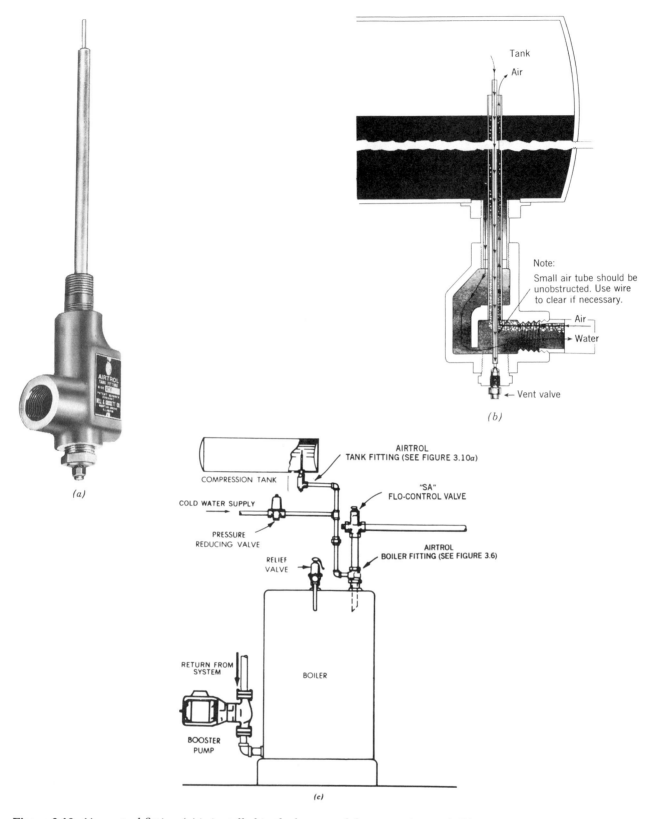

Figure 3.10 Air control fitting *(a)* is installed in the bottom of the expansion tank *(b)*. By venting air through the vent valve at the bottom, the water level in the tank can be controlled. *(c)* Piping of the tank, tank air fitting and boiler air fitting (Figure 3.6). (Courtesy of ITT Fluid Handling Sales.)

EXPANSION TANK / 97

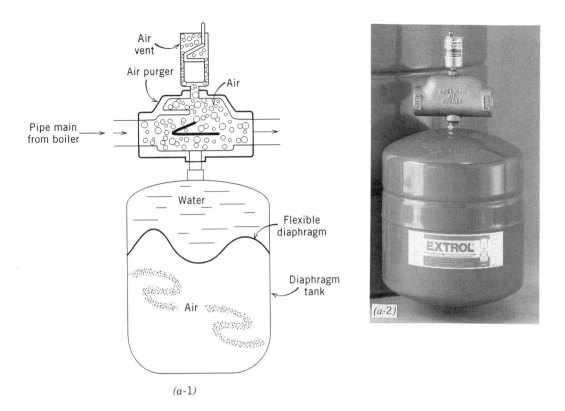

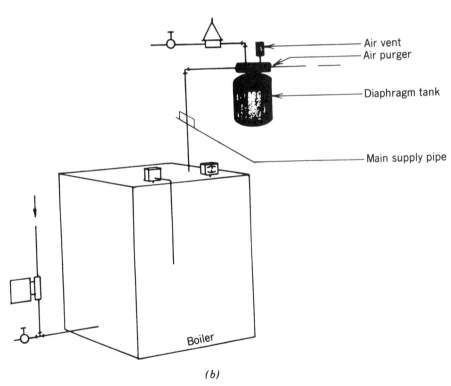

Figure 3.11 *(a)* Diaphragm tank mounted directly on the main piping of a small hydronic heating system. Trapped air is separated by the purger and expelled to atmosphere by the automatic air vent. (Courtesy of Amtrol.) *(b)* The combination of air purger, air vent and diaphragm tank is mounted directly on the main supply piping, immediately above the boiler. (Courtesy of The Hydronics Institute.)

in the top of the tank from the water in the bottom. As shown in Figure 3.11, diaphragm tanks for small hydronic heating systems, such as in one-family residences, are frequently installed directly in the piping, in conjunction with an air purger and an automatic air vent. The purger extracts and isolates any air trapped in the water, which is then vented to the atmosphere by the air vent devices. Air venting is a continuing process since air is always entering the system via make-up water, which contains dissolved air. Air also enters during maintenance and repair procedures.

Expansion tanks are frequently furnished as part of the overall boiler package and are, therefore, sized by the supplier. In small systems, the tank size is based on the system Btuh capacity and to a lesser extent on the piping arrangement. A good rule of thumb for tanks calls for 1 gal of capacity per 5000 Btuh of system capacity. In large systems, the tank size depends on the volume of water in the system, system temperature and system pressure. These parameters are not the responsibility of a technologist. Since the expansion tank effectively determines the system operating pressure, only one tank may be used in any system.

3.6 Circulator

At this point, we strongly advise that you study Sections 8.5–8.8 and Sections 9.5 and 9.6 or, having already studied them, review the material. These sections deal with basic hydraulics, static and friction head and the use of friction head charts. A thorough understanding of these subjects is necessary to understand the material that follows.

A *circulator* is simply a small pump whose purpose is to circulate the water in the hydronic heating system. Refer to Figure 3.12(a), which shows a sample pump curve. This curve is typical of small centrifugal pumps, which are the type normally used as circulators. The curve shows that the pump can lift 20 gpm to a height of 10.5 ft. That is, the pump has a capacity of 20 gpm at 10.5 ft of head. The curve also shows that the same pump can lift 10 gpm to 16 ft above the pump inlet. Put another way, it can pump 10 gpm while overcoming 16 ft of head. At 0 gpm, it can lift water 17.5 ft. This simply means that the impeller of the pump (see Figure 3.13) will churn away inside its casing to sustain a column of water 17.5 ft high. Therefore, if this pump were connected to a vertical pipe 20 ft high (and open at the top), it would run continuously, raising the water to 17.5 ft inside the pipe, but no higher. (This assumes zero suction head.) In such a case, the pump is operating only against static head, with no friction, because there is no flow.

In an open system, such as the one illustrated in Figure 3.12(b), the pump operates against static head and friction. The static head is the vertical lift, and the friction is the resistance of the piping to the water flow. In the diagram of Figure 3.12(b), the pump delivers 10 gpm against a total head of 16 ft; this total head can be any combination of lift and friction that totals 16 ft of water head.

Modern hydronic heating systems are closed, as compared, for instance, to domestic water systems, which are open to the atmosphere. In a closed system, there is no static head for the pump to overcome since the weight of water in riser piping is exactly counterbalanced by an equal weight of water in the return piping. The circulator, therefore, must overcome only the friction head in the piping, which rarely exceeds 15 ft of head, and is most often about one half that figure. See Figure 3.12(c) and (d). Hydronic heating systems are commonly designed for friction heads between 3 and 15 ft. The circulator is, therefore, usually driven by a small, fractional-horsepower electric motor. Figure 3.13 illustrates a typical hydronic system circulator.

Circulator pumps are normally low head, low flow units; they are low head, as explained previously, because of the absence of static head, and low flow, as was explained in Section 3.1, because of the high heat-carrying capacity of water. They either are installed in-line when the water flow from pump inlet to outlet is in the same direction or are base mounted where a right angle change in water direction from inlet to outlet is desired. Most circulators are installed in-line. The circulator can be installed either in the main supply line, outgoing from the boiler (item T, Figure 3.3), or in the main return line coming back to the boiler [item C, Figures 3.4 (a,b)]. In systems with high head (10 ft or more), it is normal practice to place the pump so that it pumps away from the expansion tank and boiler.

Figure 3.14(a) shows a typical graphical design solution. The pump curve is obtained from manufacturer's literature and is selected to supply the head and flow design requirements. The system design curve is simply a plot of system friction versus flow. Once the piping system has been laid out, lengths are measured on the drawings, losses in fittings are either estimated or calculated (de-

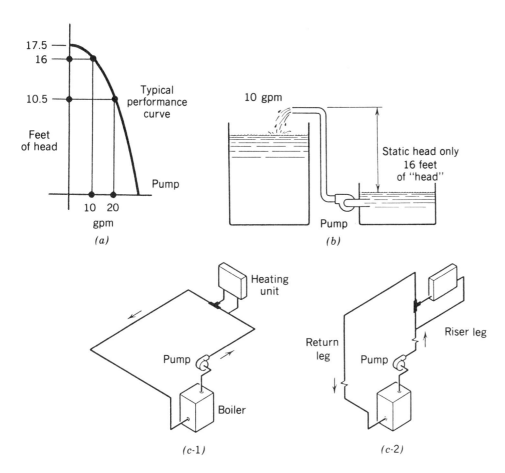

Figure 3.12 *(a)* Typical performance curve of a centrifugal pump. This is the type used as a circulator in hydronic heating systems. *(b)* When the pump is used in an open system, the total developed head is the sum of the static (lifting) head and the friction head caused by flow in a pipe. In the illustration, the pump will deliver 10 gpm at a static head plus friction head of 16 ft of water. *(c)* In a closed system, the pump has to overcome only friction head because the water being lifted in the riser leg is exactly counterbalanced by the weight of water in the return leg. Therefore, the pump in a one-story system *(c-1)* does exactly the same amount of work as the pump in a two-story system *(c-2)* provided the piping is the same length. In both systems, the pump has to overcome only the friction head (pipe friction) in the system to circulate the water. No lifting (static head) is involved because both systems are completely closed.

pending on the accuracy required) and total system friction is calculated for different flow rates, using pipe friction charts of the type shown in Figures 3.34 and 9.6. The system design curve is then drawn. The point A where it intersects the pump curve will be the operating point. This point must correspond to the head and flow requirements. Note that the system design curve starts at zero head for zero flow. This shows that the system is closed and has no static head. In an open system, the pump must develop the full head before any flow begins. In open systems, the design curve begins somewhere along the vertical axis, at a head corresponding to the system's static head.

The lower system curve, labelled "installed system," represents the actual field condition more closely because it shows lower friction head. In design calculation, most designers are generous with friction calculations for fittings. Actual friction is often much lower. The result will be a lower

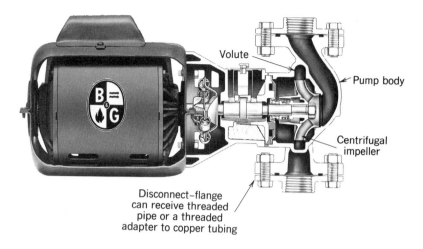

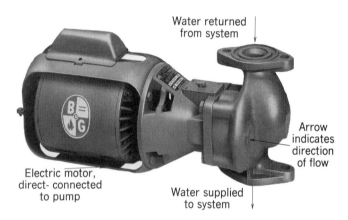

Figure 3.13 Typical circulating pump for a small hot water heating system. Motors are usually fractional horsepower. The sectional view shows the pumping action. Water is drawn in through the upper opening and flows through the pump body to the receiving opening of the centrifugal impeller element. Fast rotation throws the water into the volute under pressure and from there it goes to the lower opening. (See Figure 3.3, item T.) Flow may be horizontal or vertical. (Courtesy of ITT Fluid Handling Sales.)

operating point corresponding to higher flow at lower (friction) head. Typical actual pump curves are shown in Figure 3.14(b).

3.7 Terminal Units

The term *terminal units* is normally used to describe the various devices that deliver heat into the spaces in which they are placed. They are referred to as terminal units since they are the final stage, or terminal, of the hydronic heating system. The term used in the past was *radiation*. It described the most common (and unsightly) terminal unit of steam heating systems—the ribbed cast-iron radiator. (See Figure 3.21.) Although cast-iron radiators are still in use in both steam and hydronic heating systems, they have largely been replaced in new construction hydronic systems by finned-tube radiation and convectors of various designs. In any case, the term *radiation* is misleading since most of the heat from these units, at least in hydronic systems, is from convection and not radiation.

TERMINAL UNITS / 101

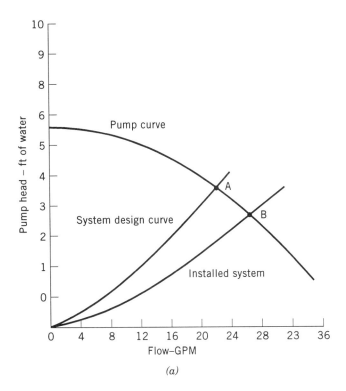

Figure 3.14 *(a)* Graph of typical circulator (pump) curve superimposed on a system design curve. The point of intersection A is the system operating point. Since the installed system friction head is generally less than the design-calculated head, the actual system operating point will be at B.

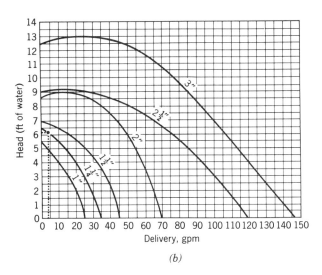

Figure 3.14 *(b)* Typical performance curves for curculating pumps commonly used in hydronic heating systems. The sizes marked on each curve represent the pipe size that the circulator is constructed to accommodate.

The terminal units used in modern hydronic heating systems are

- Baseboard radiation.
- Finned tube (commercial) radiation.
- Convectors.
- Unit heaters.
- Unit ventilators.
- Fan coil units.
- Radiant panels.
- Cast-iron radiators.
- Steel panel radiators.

a. Baseboard Radiation

These units, illustrated in Figure 3.15, are the most common terminal units in use today in hydronic heating systems. They are called baseboards because they are installed at the base of a wall, in the same position that a baseboard molding is placed. A typical baseboard unit consists of a ½-, ¾- or

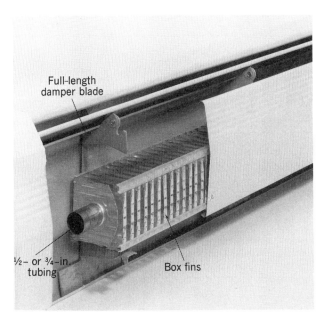

Figure 3.15 Cutaway of typical residential-type baseboard terminal unit for hot water heating systems. The illustrated unit is 8⅛ in. tall and 2⅝ in. deep overall. It is mounted at floor level for maximum effectiveness. Output varies from a minimum of 450 Btuh/ft for a flow of 1 gpm of 160°F water, to a maximum of 820 Btuh/ft for a flow of 4 gpm of 210°F water, for ¾-in. tubing. (Courtesy of Edwards Engineering.)

1-in. copper tube on which are mounted square aluminum or copper fins. This finned-tube assembly is then installed in a metal enclosure designed to encourage the establishment of a convective air current over the finned pipe. Hot water passing through the copper pipe heats the pipe and, in turn, the fins. Cool air enters the baseboard enclosure at the bottom (floor level), is heated as it passes over the hot finned pipe assembly and exits the top of the enclosure. The heated air then rises into the room, is cooled by the room heat loss, drops to the floor and reenters the baseboard to be reheated, thus completing the heating cycle in the room. This room air circulation pattern is similar to that shown in Figure 1.6.

Residential baseboard (and commercial finned-tube radiation) rated by the Hydronics Institute carries an I=B=R emblem. These ratings are based on the active length of baseboard, that is, the length of finned pipe rather than the total baseboard length. The latter includes space at both ends for valves and piping connections. Ratings of baseboards are based on a flow of 1 gpm (500 lb/h). Many manufacturers also publish a heat rating for a flow of 4 gpm (2000 lb of water/h). As can be seen from Table 3.1, the heat output of a typical baseboard increases only slightly with this increased flow. This is because heat output is determined by the temperature of the finned pipe and the quantity of air passing through the unit. Neither of these is appreciably affected by the quantity of water flow through the unit above the 1 gpm minimum. I=B=R ratings begin at a water temperature of 150°F (65°C) and increase to 220°F (104°C). For water temperatures lower than 150°F, output reduction factors are listed in Table 3.2.

b. Finned-Tube (Commercial) Radiation

This terminal unit is essentially the same item as the residential baseboard unit described and illustrated in the preceding section, except for construction details. Finned-tube commercial units can use multiple finned tubes in a single enclosure, either side by side (not more than two) or in vertical tiers (not more than three). See Figure 3.16. The units can be placed at the base of a wall or in a trench or mounted on a wall as desired and required architecturally. Like baseboard radiation, commercial finned-tube radiation is rated by the Hydronics Institute and given an I=B=R heat output rating. The basis of the rating is, however, quite different than residential baseboard ratings. Important aspects of commercial finned-tube I=B=R ratings follow.

1. The heat output rating in Btuh/ft is based on hot water flowing through the tube at 3 fps (ft per

Table 3.1 Typical I=B=R Output Ratings for Residential Baseboard Radiation, Btuh/ft

Water Temperature,		Water Flow	
°F	°C	1 gpm (500 lb/hr)	4 gpm (2000 lb/hr)
160	71	450	480
170	77	520	550
180	82	580	610
185	85	610	640
190	88	650	690
195	91	680	720
200	93	710	750
210	99	780	820
220	104	840	890

Notes:
1. Figures are for 3/4-in. copper pipe, unpainted aluminum fins 2 3/8-in. square, 55 per foot.
2. Ratings include 15% increase due to low level installation, as approved by I=B=R.

Source. Ratings extracted, with permission, from published literature of Edwards Engineering Corp., Model B-34B.

Table 3.2 Output Derating Factors for Baseboard Radiation Operated at Low Water Temperatures[a]

Water Temperature		Copper-Aluminum Construction
°F	°C	
150	66	1.0
140	60	0.85
130	54	0.69
120	49	0.55
110	43	0.41
100	38	0.28
90	32	0.17

[a] For 1 gpm flow and 65°F (18.3°C) ambient air temperature.

Source. Reproduced, with permission, from I=B=R Ratings of The Hydronics Institute.

second), at 215°F (102°C). Multiplying factors for temperatures other than 215°F are given in Table 3.3; multiplying factors for water velocities other than 3 fps are given in Table 3.4.

Note from Table 3.4 that heat output drops only 10% at 25% water flow rate. This corresponds to what we saw in Table 3.1 for heat output of residential baseboard, for flows of 1 and 4 gpm. The conclusion is clearly that heat output of finned-tube-type radiation is almost independent of water flow. This is a particularly useful characteristic of these terminal units because actual water flow is often quite different from the design figure due to inaccurate assumptions in the design process.

Table 3.3 Correction Factors for Commercial Finned-Tube Radiation Operating at Temperatures Other Than 215°F (102°C)

°F	°C	Factor
100	38	0.15
110	43	0.21
120	49	0.26
130	54	0.33
140	60	0.40
150	66	0.45
155	68	0.49
160	71	0.53
165	74	0.57
170	77	0.61
175	79	0.65
180	82	0.69
185	85	0.73
190	88	0.78
195	91	0.82
200	93	0.86
205	96	0.91
210	99	0.95
215	102	1.00
220	104	1.05
225	107	1.09
230	110	1.14
235	113	1.20
240	116	1.25

(Average Radiator Temperature)

Source. Extracted, with permission, from I=B=R Ratings of The Hydronics Institute.

Table 3.4 Derating Factors for Commercial Finned-Tube Radiation Operating at Water Flow Rates Below 3 fps

Flow Rate	Factor
3.0	1.00
2.75	0.996
2.5	0.992
2.25	0.988
2.0	0.984
1.75	0.979
1.5	0.973
1.25	0.966
1.0	0.957
0.75	0.946
0.5	0.931
0.25	0.905

Source. Extracted, with permission, from I=B=R Ratings of The Hydronics Institute.

2. The height above floor level at which the radiation is installed is important since it seriously affects the convective air flow through the enclosure. I=B=R output ratings are based on the manufacturers recommended installed height (also referred to as mounting height). Installation at other than the recommended height requires adjusting the rated heat output by the factors in Table 3.5 as follows:

Installed heat rating = (I=B=R rating)
$$\times \frac{\text{Installed height factor}}{\text{Recommended height factor}}$$

3. Commercial finned tubing may be ¾, 1 or 1¼ in. Output ratings vary only slightly with the different pipe sizes.

An example will demonstrate the use of Tables 3.3–3.5.

Example 3.1 A commercial finned-tube heater is rated at 1320 Btuh/ft, with a recommended installed height of 16 in. The unit is actually installed at 26 in. Water flow is 2 fps, and water temperature is 180°F. Determine the actual heat rating.

Solution: From Table 3.3 we obtain a temperature factor of 0.69. From Table 3.4 we obtain a flow

104 / HYDRONIC HEATING

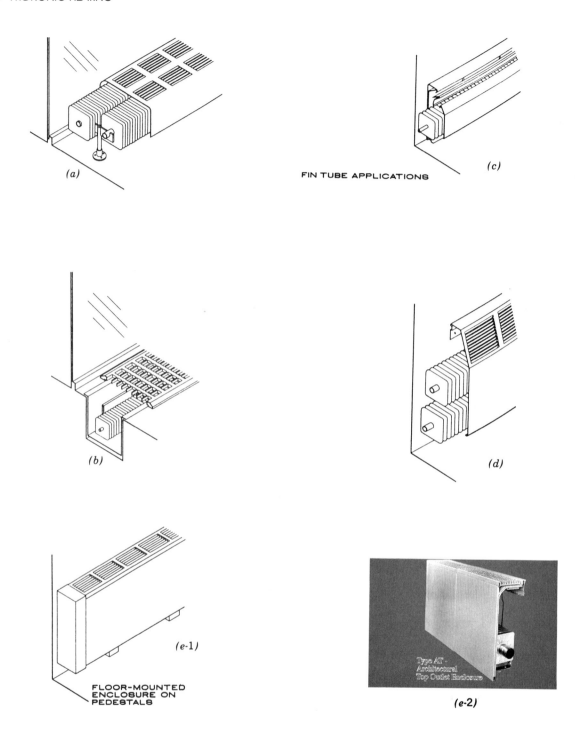

Figure 3.16 Finned-type radiation. Units can be installed adjacent to glass walls *(a)*, recessed into the floor at a glass wall *(b)* or installed in single tier *(c)* or double tier *(d)* at the base of a wall. Commercial units are available in a wide variety of architectural enclosures including top outlet enclosures *e-1, e-2)*, front outlet enclosures *(f-1, f-2)*, low enclosures *(c, h)*, and single- and double-slope enclosures *(g-1, g-2)*. (Drawings from *Architectural Graphic Standards*, 8th ed., 1988. Reprinted by permission of John Wiley & Sons., Photos *e-2, f-2,* and *h* courtesy of Dunham-Bush, Inc. Photo *g-1* courtesy of Slant Fin Corporation.)

TERMINAL UNITS / 105

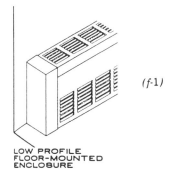
(f-1)
LOW PROFILE FLOOR-MOUNTED ENCLOSURE

Type F - Flat Top Front Outlet Enclosure
(f-2)

(g-1)

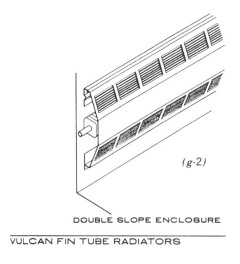
(g-2)
DOUBLE SLOPE ENCLOSURE

VULCAN FIN TUBE RADIATORS

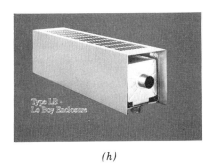
Type LB - Lo Boy Enclosure
(h)

Figure 3.16 *(Continued)*

Table 3.5 Corrections Factors for Commercial Finned-Tube Radiation Installed at Other Than Recommended Installed Height

Installed Height, in.	Factor
36 or more	1.00
34	1.01
32	1.02
30	1.03
29	1.04
28	1.05
27	1.06
26	1.07
25	1.08
24	1.09
23	1.10
22	1.11
21	1.12
20	1.13
19	1.14
18 or less	1.15

Note:

$$\text{New rating} = \text{IBR rating} \times \frac{\text{Factor for actual installed height}}{\text{Factor for recommended installed height}}$$

Source. Extracted, with permission, from I=B=R ratings of The Hydronics Institute.

Table 3.6 Typical I=B=R Heat Output Ratings for Commercial Finned-Tube Radiation

Enclosure Height, in.	Recommended Installed Height, in.	Rating, Btuh/ft
9	9$\frac{1}{16}$	1380
12$\frac{1}{2}$	12$\frac{5}{8}$	1650
18$\frac{1}{2}$	18$\frac{3}{4}$	1880
24$\frac{1}{2}$	24$\frac{1}{2}$	1950

Notes:
1. Pipe is ¾-in. copper tube.
2. Fins are 2⅜-in. square aluminum, 0.010 in. thick, mounted 55 per foot of pipe.
3. Each unit contains two finned-pipe elements, side by side.

Source. Extracted, with permission, from data of Edwards Engineering Corp.

factor of 0.984. From Table 3.5 we obtain a height factor ratio of 1.07/1.15. Combining all these factors, we have

$$\text{New rating} = 1320 \times 0.69 \times 0.984 \times \frac{1.07}{1.15}$$

$$= 834 \text{ Btuh/ft}$$

Note that the most important derating factor is that of water temperature. Table 3.6 gives ratings for typical commercial finned-pipe radiation.

c. Convectors

These units, which are also called cabinet heaters, consist of one or more vertical tiers of finned-tube radiation. See Figure 3.17. They differ from the commercial finned-tube radiation described in the preceding section only in that convectors have larger capacity. Convectors are normally floor mounted and are either fully exposed, semi-recessed or fully recessed. Units installed on outside walls should have rear insulation to prevent a large heat loss through the wall. This is especially important for recessed units since the R value of the wall is usually reduced by the recess for the convector. Heat output ratings are given in manufacturers' literature for individual convector designs.

d. Unit Heaters

All the previously discussed radiation depend on natural convection for their proper operation. For this reason, mounting according to the manufacturer's recommendation is very important. Indeed, improper mounting height can reduce the natural convective airflow severely and, as a result, reduce the heat output by as much as 50%. A unit heater does not have this limitation because it contains a propeller fan or blower, which draws air from outside the unit and blows it over the heating element. The heated air then exits the unit at fairly high velocity to warm the adjacent space. Because a unit heater does not depend on natural convective currents, it can be mounted anywhere in a space. When mounted at floor level, it is frequently referred to as a cabinet heater. When mounted at ceiling level, it is often called a space heater. The heating element in unit heaters is either finned-tube or a hot water heating coil. Unit heaters are most often used in large spaces and where ambient air currents are fairly strong so that natural convection cannot be relied upon. Among such areas are storage areas, corridors and vestibules. See Figure 3.18.

TERMINAL UNITS / 107

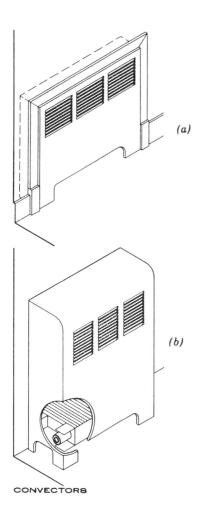

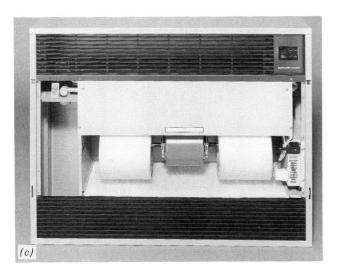

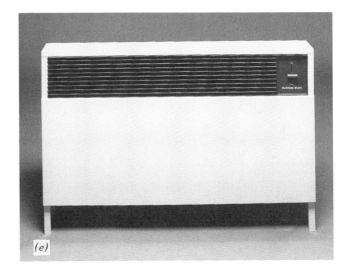

Figure 3.17 Convectors (cabinet heaters). These units use one to three tiers of finned-tube radiation in recessed (a), semi-recessed, or completely exposed (b) enclosures. Units can be floor mount (c, d) or pedestal (e) and can use a forced draft fan (c, d) or natural convection (e). Cabinets can be sloped (d) or flat (c, e). (Drawings from *Architectural Graphic Standards*, 8th ed., 1988, reprinted by permission of John Wiley & Sons. Photos courtesy of Dunham-Bush Ltd.)

108 / HYDRONIC HEATING

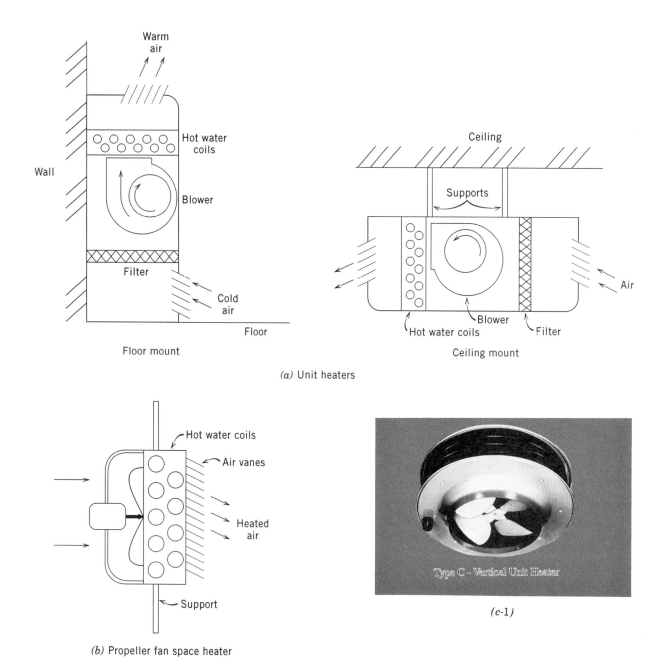

Figure 3.18 (a) Unit heaters can be mounted anywhere since they do not depend on heat-driven convective air currents. Instead they use a blower as shown to force air over the hot water (or steam) coils or finned-tube radiation. (b) Propeller fan unit heaters are usually known as space heaters. Units are of horizontal blow design (illustrated) or similar vertical (blow down) type (c-1). Piping is shown pictorially (c-2) and schematically (c-3). Because of noise, application is limited to industrial spaces. (Photos c-1 and illustration c-2 courtesy of Dunham-Bush Ltd.)

TERMINAL UNITS / 109

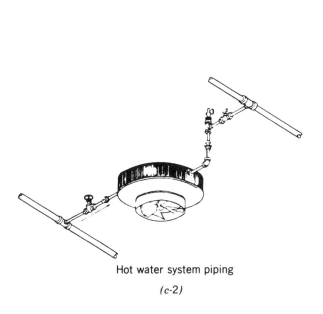

Hot water system piping
(c-2)

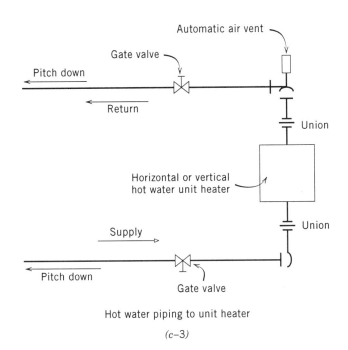

Hot water piping to unit heater
(c-3)

Figure 3.18 *(Continued)*

e. Unit Ventilators

Unit ventilators are similar to unit heaters except that they can also provide fresh air via ducted connection to the outside of the building. The units generally have large capacities, using blowers rather than the propeller fans found in some unit heaters. Heating is provided, as with unit heaters and space heaters, by blowing room air, or a mixture of room air and outside air, over a hot water coil. (Large capacity units also use steam as the heating medium.)

f. Fan Coil Units

See Figure 3.19. As is shown in Figure 3.1, fan coil units are used to provide heating, cooling or both, depending upon piping connections and controls. Essentially, a fan coil unit is a cabinet containing a hydronic coil assembly and a motor-driven blower. Hot and/or cold water from a central source is piped to the unit. Control of the blower is often given to the user in residences and office buildings. In public spaces and other areas where there is no fixed occupancy or where it is not desirable that the blower speed be altered, user controls are omitted. Fan coil units normally do not provide humidification. Dehumidification is provided automatically by cooling room air below its dew point as it passes over the cooling coil. (Note the condensate drain pan and drain line in Figure 3.19.) Recirculated room air is filtered as it passes through the return air filter at the base of a typical unit.

If it is desired to provide fresh air through a fan coil unit, it must be mounted on an outside wall and provided with a manually operated damper or the through-the-wall connection. Because control of this damper can be difficult and because of the problems of noise and dirt entering through this outside air connection, particular care must be taken in the design and control of fresh air connections. Single-coil fan coil units provide either heating or cooling, depending on the season of the year. This can be problematic in large buildings (where fan coil unit use is very common) in "swing" seasons such as spring and fall. Then, parts of the building may require heating while other parts require cooling. To solve this problem, dual coil units are used in high-quality construction, and the central plant supplies both hot and cold water to each unit. Local controls then operate the unit's valves, to permit circulation of hot or cold water, as required. Single-coil units may be thermostatically controlled, or, in cheaper installations, the degree of heating or cooling is regulated manually by user control of the blower speed. Where this is

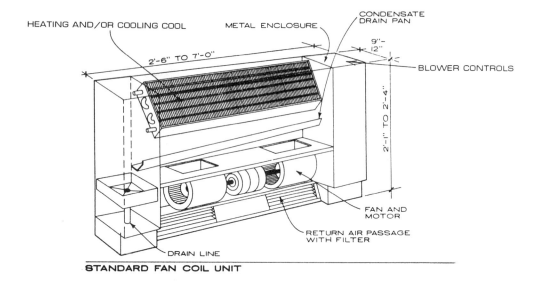

Figure 3.19 A standard fan coil unit contains one or two hydronic coil assemblies, a multispeed blower and various controls, enclosed in a metal cabinet. Recirculated room air is taken in at the bottom of the unit through a filter, forced over the hydronic coil by the blower and expelled at the top. Humidification is not provided. (Illustration from *Architectural Graphic Standards*, 8th ed., 1988, reprinted by permission of John Wiley & Sons.)

not desirable, the blower speed is set by the building's maintenance personnel and is not changed. In buildings where dual-coil fan coil units are not used, an electric heating coil is sometimes provided to furnish a limited amount of heating to spaces requiring it, when chilled water is circulated.

g. Radiant Panels

As explained in Section 1.8, a person can be quite comfortable thermally, even at low air temperatures, if the mean radiant temperature (MRT) is sufficiently high. This fact is applied in spaces that use radiant heating panels. In hydronic systems, the only room surfaces that can be used for radiant panel piping are the floor and ceiling, with the ceiling taking preference because people frequently put (insulating) rugs and carpets on the floor. The principal reason that radiant heating is not often used is economic; it is simply too expensive to design and install in small installations. However, in large tract housing, where hundreds of identical units are constructed, radiant floor panels can prove economical both in first cost and operating costs. The mass of concrete in which the coils are buried has the great advantage of thermal mass, which acts to even out temperature changes and keep a space uniformly heated and comfortable. An often stated disadvantage of radiant heat is that there is a considerable time lag, because of this mass, between turning on the heat and feeling its effect. In commercial buildings that are shut down evenings and weekends, this is true; in residences that are continuously heated, this thermal lag characteristic is a definite advantage as explained previously. Heating units with fans or blowers that operate by heating the air in a space such as those discussed previously (fan coil units, unit heaters), have very little carry-over, since the specific heat of air is negligible and the hot water in the unit will cool very rapidly. As a result, in such spaces, a chill is felt almost immediately when the units are shut off. The same is true of hot air heating systems, as will be discussed later. Cast-iron radiation, which is described later, has some carry-over due to the mass of the radiator and the relatively large quantity of hot water that it contains.

Radiant panel heating has the additional advantage of operating at a low water temperature; usually about 120°F (49°C) for floor panels and 140°F (60°C) for ceiling panels. An uncarpeted floor operating with ¾-in. tubing on 12-in. centers, carrying 120°F water, will deliver about 50 Btuh/ft² of floor. When covered with carpets and rugs, this figure is reduced considerably. A panel-type radiant ceiling with 3/8-in. tubing on 6-in. centers carrying 140°F water will deliver about 60 Btuh/ft². Radiant floors induce convective air currents in the room because of the natural tendency of heated air to rise. This adds to the comfort of the occupants. In radiant ceiling installations, a blanket of heated air forms at the ceiling and remains there since it cannot rise. This leads to a condition of stagnant air in a closed room, which can become uncomfortable. Further, in such rooms, there is a considerable vertical temperature gradient between the cold floor and the warm ceiling, which can also cause discomfort to occupants.

As stated previously, hydronic radiant heating is expensive to use because of the large amount of piping and valving involved. In recent years, plastic pipe capable of continuously carrying 120°F (49°C) water has become readily available. As a result radiant floor heating is being increasingly used because of its inherent advantages. Typical construction details of hydronic radiant floor installations are shown in Figure 3.20.

h. Cast-Iron Radiators

The classic rib-type cast-iron radiator was originally developed for use in steam systems and later adapted to hydronic hot water systems. Typical units are shown in Figure 3.21. Each unit is connected to a supply and return branch and is normally equipped with an air vent, either manual or automatic. Hot water passing through the unit heats the cast-iron ribs, which then heats the room by a combination of radiation and convection. The thermal mass of the radiator itself plus the contained water cause a thermal lag that acts to smooth out rapid temperature variations.

Cast-iron radiators are rated in square feet steam. The rating does not indicate the surface area of the radiator. It is based on a radiation rate of 240 Btuh/ft² of surface, using steam at 215°F (102°C) as the heat transfer medium. Cast-iron ra-

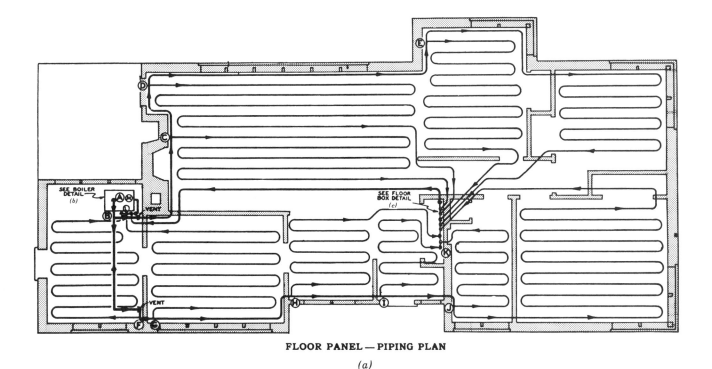

FLOOR PANEL — PIPING PLAN

(a)

Figure 3.20 (a) (Partial) floor plan of floor slab piping of a radiant floor hydronic heating system. Shaded areas indicate exterior walls and interior partitions. The circled letters indicate connections between individual piping loops and the supply and return mains. (Drawing courtesy of The Hydronics Institute.)

112 / HYDRONIC HEATING

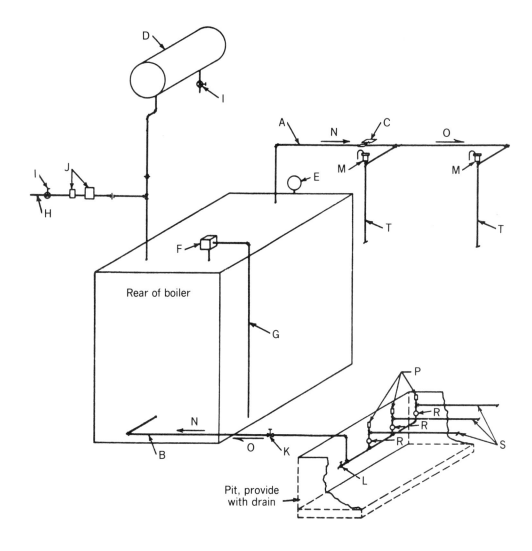

Piping for Radiant Floor Panel System

A   Supply pipe
B   Return pipe
C   Pump
D   Air cushion tank
E   Altitude gage
F   Safety relief valve
G   Overflow from safety relief valve
H   Make-up water line. Always install between pump and air cushion tank
I   Globe valve
J   Pressure reducing valve and check valve
K   Purge valve
L   Drain cock
M   Automatic air vents
N   Pitch upward in direction of arrow
O   Direction of flow water
P   Radiator vent valves
R   Balancing valves
S   Return from coils
T   Supply to coils

(b)

Figure 3.20 *(b)* Schematic of piping at the boiler supplying the radiant floor system. Balancing valves in the three main return pipes and in each of the individual loop returns (see Figure 3.20 *c*) permit accurate balancing of water flow in each loop and in mains. Balancing is performed after the installation is complete. (Drawing courtesy of The Hydronics Institute.)

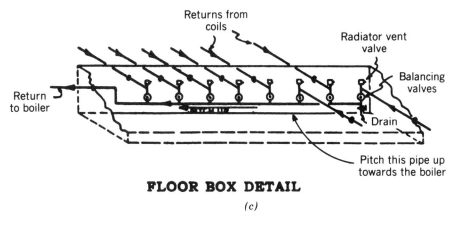

FLOOR BOX DETAIL
(c)

Figure 3.20 *(c)* Multiple-loop radiant floor hydronic installations require an installation and access point for the loop-balancing valves. A floor box is frequently installed in a closet, as in this installation. (Drawing courtesy of The Hydronics Institute.)

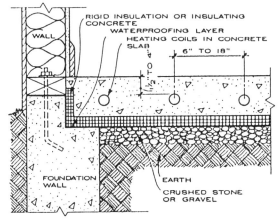

HOT WATER HEATING COILS IN FLOOR SLAB ON GRADE

Plastic, ferrous, or nonferrous heating pipes are used in floor slabs that rest on grade. It is recommended that perimeter insulation be used to reduce thermal losses at the edges. Coils should be embedded completely in the concrete slab and should not rest on an interface. Supports used to position the coils while pouring the slab should be nonabsorbent and inorganic. A layer of waterproofing should be placed above grade to protect insulation and piping.

(d)

Figure 3.20 *(d)* Typical detail showing the method of hot water coil installation in the floor slab; in this case, on grade. Pipe lateral spacing depends on the total length of pipe required to supply the heat loss for a particular space. (Drawing from *Architectural Graphic Standards*, 8th ed., 1988, reprinted by permission of John Wiley & Sons.)

diators used in hydronic systems have lower heat emission rates in proportion to the system supply water temperature. Table 3.7 lists the heat emission rates for various water temperatures, per square foot (of radiator rating). An example should make this clear.

Example 3.2 A particular cast-iron radiator has a 5-ft^2 rating. What is its Btuh-rated emission when supplied with 180°F water?

Solution: From Table 3.7, the heat emission of cast-iron radiation using 180°F water is 170 Btuh/ft^2. Therefore,

$$\text{Heat rating} = 5(170) = 850 \text{ Btuh}$$

Figure 3.21*b* illustrates a baseboard type of cast-iron radiation. This type uses a finned tube, which was described previously, mounted in a heavy cast-iron enclosure. The unit combines the thermal mass advantage of cast-iron radiation, with the architecturally pleasing shape. The extended shape, assists in providing uniform room heating.

i. Steel Panel Radiators

The unsightliness of the classic ribbed radiator and the physical hazard to children posed by the sharp edge of the fins (Figure 3.21) were two of the factors that led to the development of flat steel radiators.

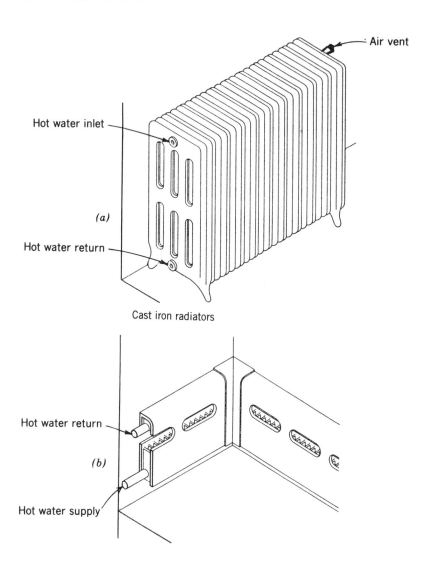

Figure 3.21 Cast-iron radiation. *(a)* Column-type ribbed radiator. Radiators are connected to the hot water piping via supply and return branches, through shutoff valves. Valves permit removal of an individual radiator without extended shutdown of the entire system. *(b)* Baseboard-type cast-iron radiation. (Reproduced from *Architectural Graphic Standards*, 8th ed., 1988, reprinted by permission of John Wiley & Sons.)

Several designs are illustrated in Figure 3.22. The principle of operation is essentially the same as described for finned-tube radiation and cast iron radiation. The units have smaller thermal mass than cast iron radiation.

3.8 Piping Arrangements

As was pointed out in the beginning of this chapter, the components of a hydronic heating system are a boiler, a circulator, an expansion tank, terminal units and various valves, meters and protective devices, plus the piping that connects all the component parts. In the preceding sections, we discussed the major components in some detail. At this point, we will examine the piping arrangements by which the terminal units are supplied with hot water. After that, we will discuss the remaining item of importance in hydronic heating systems, that of system controls.

Four common piping arrangements are used to connect terminal units in hydronic heating systems. They are

Table 3.7 Heat Emission Rates for Cast-Iron Radiators

Design or Average Water Temperature		Heat Emission Rates, Btuh/ft²
°F	°C	
170	77	150
175	79	160
180	82	170
185	85	180
190	88	190
195	91	200
200	93	210
205	96	220
210	99	230
215	102	240

Source. Reprinted with the permission of The Hydronics Institute.

- Series loop system.
- One-pipe system.
- Two-pipe direct-return system.
- Two-pipe reverse-return system.

(Note that we are not including radiant floor and ceiling systems, because in these systems each piping loop is itself a terminal unit, since it provides the heating. It is, therefore, not appropriate to discuss the piping system that supplies the terminal units. Radiant panels are essentially piped as multiple zones and subzones of a one-pipe system.)

a. Series Loop System

This system is often also referred to (incorrectly) as a perimeter system, because most small one-story perimeter-heated structures use this piping arrangement. In point of fact, a *perimeter system* simply means a heating system in which the heating units (terminal units) are installed around the perimeter of the building or space. Such perimeter units can be piped in any one of the four methods mentioned previously.

A series loop system is one in which the terminal units are piped in series, with the output of the first unit directly connected to the input of the second unit, and so on, exactly as devices are connected in a series electrical circuit. Figure 3.23(a) shows series loop piping schematically, and Figure 3.23(b) shows the same series system in isometric projection. Figure 3.23(c) shows the equivalent electrical circuit. The advantage of the series system is its simplicity and economy. Only a single pipe is used to connect the terminal units. When they are arranged in a perimeter loop around the building, a minimum of piping and fittings is required. The system also requires a minimum of labor and is,

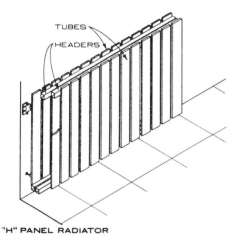
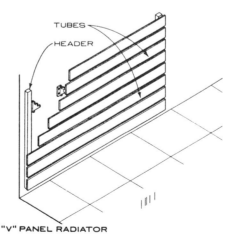

Figure 3.22 Steel panel radiators. The "ribs" of these units are flat steel tubes of rectangular cross section, arranged either vertically or horizontally. Suply and return headers are connected at the top and bottom or at the sides, respectively. (From *Architectural Graphic Standards*, 8th ed., 1988, reprinted by permission of John Wiley & Sons.)

116 / HYDRONIC HEATING

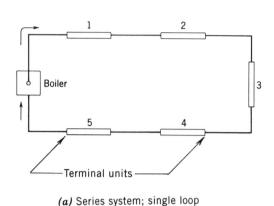

(a) Series system; single loop

Figure 3.23 (a) Single-loop series piping system. All the terminal units are connected in one continuous loop. This arrangement is sometimes referred to as a perimeter system. All auxiliary devices are omitted for clarity.

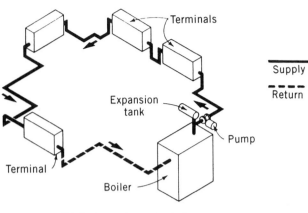

(b) Series system; single loop

Figure 3.23 (b) Isometric view of a single-loop series piping system. (From *Architectural Graphic Standards* 8th ed., 1988, reprinted by permission of John Wiley & Sons.)

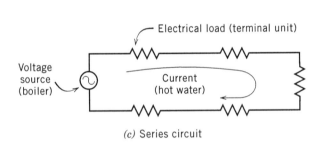

(c) Series circuit

Figure 3.23 (c) A series electrical circuit is the electrical analog of a series piping system.

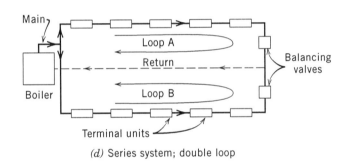

(d) Series system; double loop

Figure 3.23 (d) Schematic of a double-loop series piping system. The balancing valves, normally placed in the return piping, as shown, control the division of hot water between the two loops.

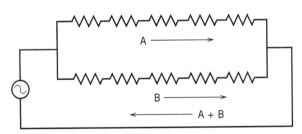

(e) Double series circuit (series-parallel circuit)

Figure 3.23 (e) The electrical equivalent of a two-loop series piping system is a series parallel circuit. The two load sections are in parallel, and the combination is in series with the voltage source. There is no simple electrical analogy to the balancing valves of the piping system shown in Figure 3.23 (d).

therefore, the most economical installation overall.

The principal disadvantage of the system is that water temperature keeps dropping as we proceed around the loop, so that the unit nearest the boiler gets the hottest water and the last unit gets the coldest. This is usually not a major problem if the overall system temperature drop is limited to 20°F or less. Some designers size the radiation for each space based on the average temperature of the system. Others calculate the temperature drop in each terminal unit and size radiation based on the actual water temperature at each unit. Since there is no way to control the amount of water flow in each unit, the only possibility for heat control at each unit is mechanical. For baseboards, this means use of the mechanical damper. (See Figure 3.15.)

A second disadvantage of a series loop is that any problem with any terminal unit requires that the entire system be shut down, since all the hot water flows through each unit. Even repairing or replacing a terminal unit means a shutdown at least long enough to replace the defective unit with a bypass, and a second shutdown to install a new or repaired unit.

In large buildings where the large temperature difference between the first and last terminal units would cause design problems, a dual-circuit (loop) system can be used. This arrangement is shown in Figure 3.23(d), and its electrical equivalent is shown in Figure 3.23(e). The flow into each loop is controlled by a balancing valve. With this system the temperature drop around each of the two loops can be reduced, so that radiation design and balancing is simplified. Do not confuse multiple circuit systems with multiple zone systems. Multiple circuit systems have only a single control arrangement that affects all the loops equally. In multiple zone systems (see Section 3.8.e), each zone has its own control arrangements and acts independently of the other zones.

b. One-Pipe System

This piping system, which is shown schematically in Figure 3.24(a), in isometric in Figure 3.24(b) and in a typical application in Figure 3.24(c), uses a single-pipe loop (hence the system name) to which each terminal unit is connected with a supply branch and a return branch. Water is diverted from the main loop pipe into the individual terminal units by use of special diverter tees. These are usually placed in the return branch but may also be placed in the supply branch, or both, according to manufacturers' recommendations and according to whether the terminal unit is physically above or below the main pipe. See Figures 3.25 and 3.26. These tees are frequently called monoflow fittings after the trade name of one of the major manufacturers of these fittings. The one-pipe system overcomes the disadvantage of the series loop in that, by valving both branch lines, individual terminal units can be throttled and shut off, and even removed from the circuit, without disrupting operation of the entire system.

The major disadvantage of the one-pipe system is that, like the series loop, each convector causes a temperature drop in the circuit. This drop is not as large as in a series circuit because the cooler water returning to the loop pipe mixes with hotter circulating water. Still, the terminal units closest to the boiler receive the hottest water. Another disadvantage of this system is the increased system friction caused by the diverter tees. Despite these disadvantages, the one-pipe system is very widely used because of its economy, simplicity and reliability. Large one-pipe systems can use multiple loops, as already shown with series circuits, or multiple zones as will be discussed later. Note the use of a flow control valve in Figure 3.27. This special check valve is normally placed in the supply main to prevent reverse flow of water by gravity when the pump is not running. It is required when the boiler is physically below the terminal heating units or when using an indirect domestic hot water heater (that requires an external heat exchanger).

c. Two-Pipe Direct-Return System

This system, illustrated in Figure 3.28, uses separate pipes for the supply and return mains, with each terminal unit connected to both pipes. Use of two pipes solves the problem of water temperature variation at the various heating units because the cooler return water is separately piped. However, as you can readily see in Figure 3.28, unit 1 is much closer to the boiler than unit 5. This means that the pipe friction to unit 1 is much lower than that to unit 5. It will, therefore, take a larger flow of water. This, in effect, "short-circuits" unit 5 and the other units as well. The same is true for each unit with respect to the units following it. Attempts to balance the flow by installing orifice plates with restricted openings at each unit have been only partially successful. Because of the difficulty in balancing flow to the heating units in this piping system, it is not commonly used.

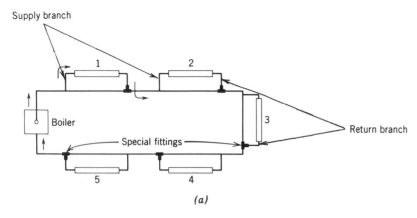

Figure 3.24 *(a)* Schematic diagram of a one-pipe system. In general, the special venturi tee fittings that divert water from the main pipe to the terminal unit are placed in the return branch.

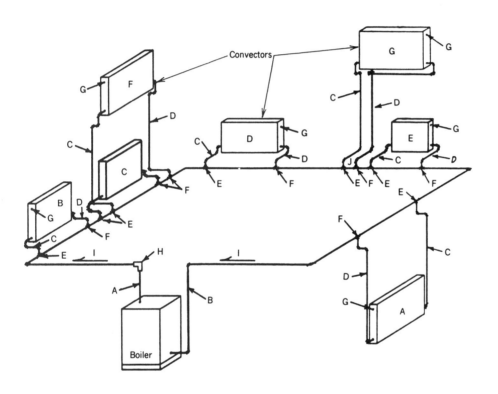

One Pipe Forced Hot Water Heating System—Single Circuit

A Supply pipe
B Return pipe
C Supply branches
D Return branches
E If one pipe fitting is designed for *supply* connection to heat distributing units, install here
F If one pipe fitting is designed for *return* connection from heat distributing units, install here
G Air vent on each unit
H Flow control valve required if an indirect water heater is used and optional if an indirect water heater is not used
I Direction of flow of water
J Not less than 6 inches

(b)

Figure 3.24 *(b)* Isometric drawing of a three-level, single-circuit, one-pipe hydronic heating system. Convector A, situated below the main pipe, should use special diverter tees at both connections to the main pipe. Convectors B–E will use only a single special tee. Convectors F and G at the upper level may require special tees in both branches. See Figure 3.26. See text for an explanation of the function of a flow control valve. (Courtesy of The Hydronics Institute.)

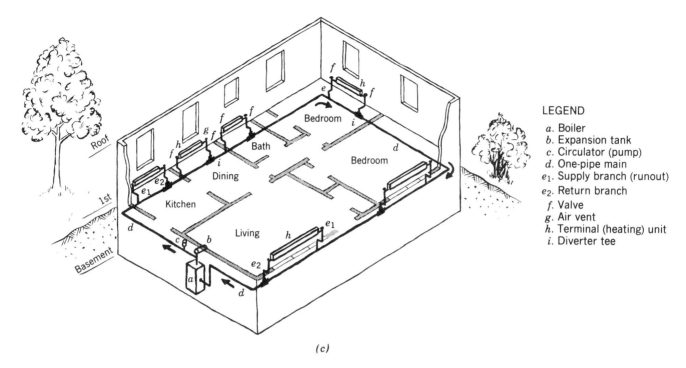

Figure 3.24 *(c)* One-pipe system as applied to a small single-floor residence. Diverter tees are placed in the return branches only.

LEGEND
a. Boiler
b. Expansion tank
c. Circulator (pump)
d. One-pipe main
e_1. Supply branch (runout)
e_2. Return branch
f. Valve
g. Air vent
h. Terminal (heating) unit
i. Diverter tee

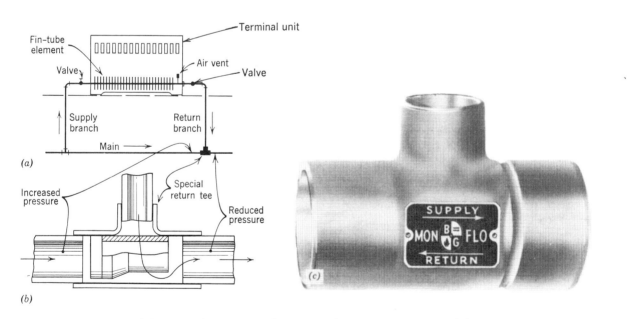

Figure 3.25 Details of the special venturi tee fittings used in one-pipe systems. *(a)* The fitting is usually placed in the return branch from the terminal unit, which can be any of the types discussed in Section 3.7. *(b)* Section through the tee indicates its venturi action. The tee induces flow through the terminal unit by retarding the flow in the main. This forces water into the supply branch and produces a low pressure area behind the tee in the main that draws water back through the return branch and into the main. *(c)* Photo of a typical one-pipe diverter tee. These tees are known as monoflow units, after the trade name of the unit shown. (Photo courtesy of ITT, Bell & Gossett.)

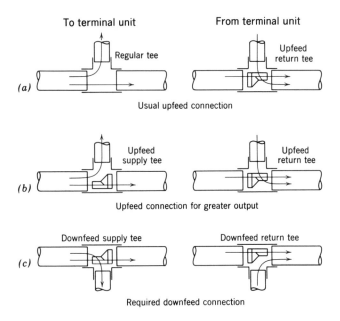

Figure 3.26 Application of special venturi tees in one-pipe systems. *(a)* The usual arrangement uses a common tee in the supply branch to the terminal unit and a venturi tee (monoflow tee) in the return branch. *(b)* Terminal units requiring greater upfeed flow than is obtained with a single diverter tee in the return branch can use an additional special tee in the supply branch. Note that the supply and return branch tees are different in their internal construction. *(c)* Downfeed connections to terminal heating units below the main pipe often use special tees in both branches.

d. Two-Pipe Reverse-Return System

To overcome the balancing problem of the direct-return system, a reverse-return arrangement is used, as illustrated in Figure 3.29. In this system, the unit nearest the boiler is connected to the shortest supply pipe but the longest return pipe. Similarly, the farthest unit is connected to the longest supply main but the shortest return main. In this fashion all the terminal units have equal length piping circuits, and the flow is automatically equalized to each unit. The price that is paid for this automatic balancing is the cost of the additional piping. As with other systems, single or multiple loops can be used, depending on the size of the installation. Because of the relatively high piping cost of this arrangement, most residential installations use the one-pipe system, either single- or multicircuit and single- or multizone.

e. Zoning

As we already noted, it is important that you understand the difference between a multiple zone system and a multiple circuit system. The essential difference is one of controls. A multiple circuit (loop) system is a single system with a single set of controls. It is split into two or more circuits for technical reasons, such as avoiding excessively long piping runs or limiting pipe sizes, and for economic considerations. Multiple zones are really separate systems with separate controls, equipment and functions, all obtaining hot water from a single boiler. As described previously, the type of piping system used is independent of the zoning. Therefore, a heating system can be a multizone one-pipe system [Figure 3.30(*a*)], a multizone, two-pipe reverse-return system [Figure 3.30(*b*)], a multizone, multicircuit series, one-pipe system [Figure 3.30(*c*)], or any other combination that suits the technical requirements of the building being heated.

Multiple zones are used in large buildings, including residences, that have defined areas with different heating requirements. In a residence, for instance, one zone could include sleeping accommodations; a second zone, the living area; a third zone, the basement and garage; and a fourth zone, the recreation room and shop. In larger buildings, zoning can be set up for different orientations and interior/exterior areas or by space usage. Two basic zoning systems are in common use. One uses a single pump for the entire system plus zone control valves that control the water flow in each zone. The second arrangement uses a separate circulator for each zone. In both cases each zone is individually controlled by its own thermostat and other control devices, independently of the other zones. The choice of systems and zoning are tasks not usually assigned to a technologist, although he or she must understand how such systems operate.

3.9 Hydronic Heating System Control

The type of control system used to govern the operation of a residential hydronic heating system depends on a number of factors. These include whether or not the heating system also supplies domestic hot water and, if so, how; the type of terminal units used; whether the heating system is multizoned; whether radiant panels are used; and what outside design temperature is used. As must

HYDRONIC HEATING SYSTEM CONTROL / 121

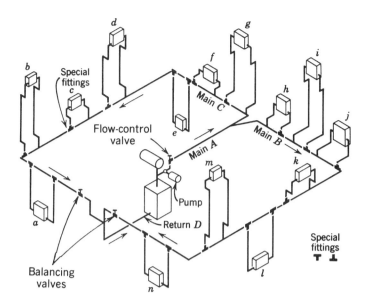

Figure 3.27 Three-level, two-circuit (double-loop), one-pipe hot water heating system. Note that terminal units below the main pipe have special tees in both branch pipes, as shown in Figure 3.26. All the other units use only a single venturi tee in the return. Balancing valves are required in the loop returns to divide the water flow properly between the two piping circuits. (From McGuinnes and Stein, *Mechanical and Electrical Equipment for Buildings*, 6 ed., 1979, reprinted by permission of John Wiley & Sons.)

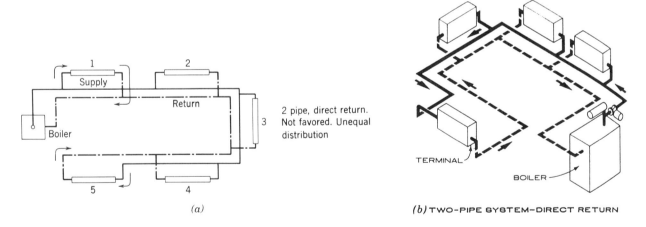

Figure 3.28 Schematic *(a)* and isometric *(b)* drawings of a two-pipe direct return system. Because of unequal piping circuit lengths to each heating unit, this system requires the use of flow-balancing devices (orifice plates) at each unit. This difficulty in flow regulation makes the system undesirable for most applications. (Drawing *(b)* from *Architectural Graphic Standards*, 1988, 8th ed., reprinted by permission of John Wiley & Sons.)

122 / HYDRONIC HEATING

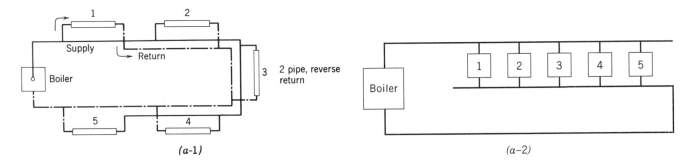

(a-1)

(a-2)

Figure 3.29 Two-pipe, reverse-return piping system. *(a-1)* Drawing shows the physical arrangement of the equipment as a perimeter heating system. *(a-2)* Drawing shows the same piping schematically, demonstrating clearly that the reverse-return pipe results in equal piping circuit lengths to each of the five terminal units.

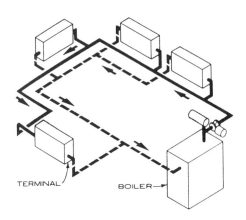

(b) TWO-PIPE SYSTEM—REVERSE RETURN

Figure 3.29 *(b)* Isometric drawing of a two-pipe, reverse-return piping system. This system results in automatic flow balance to each terminal unit but uses more piping than any of the previously described systems. (From *Architectural Graphic Standards*, 8th ed., 1988, reprinted by permission of John Wiley & Sons.)

be obvious, the decision is not a simple one, and it is, therefore, the responsibility of the system designer or engineer. We will describe a few of the principal systems in use, so that the technologist can become familiar with the different types and their principal characteristics.

a. Thermostat/Aquastat Control

This is the basic control system, and it is shown schematically in Figure 3.31*(a)*. The room (house) thermostat controls only the system circulator. The desired water temperature is preset and is controlled by an aquastat (water thermostat), which operates the burner. When the room thermostat calls for heat, the pump starts and continues until the room temperature reaches the thermostat upper limit. Independently, the aquastat will start the burner when water temperature drops to its lower limit and will shut down the burner when the water temperature rises to the aquastat's upper limit. It is understood, of course, that most thermostats and aquastats operate at low voltage (24 v normally). Therefore, in the actual circuitry, they operate relays that, in turn, control the line voltage pump and burner circuits. These intermediate devices are not shown in the schematic control diagrams for the sake of clarity.

This control arrangement has a number of disadvantages. They are

HYDRONIC HEATING SYSTEM CONTROL / 123

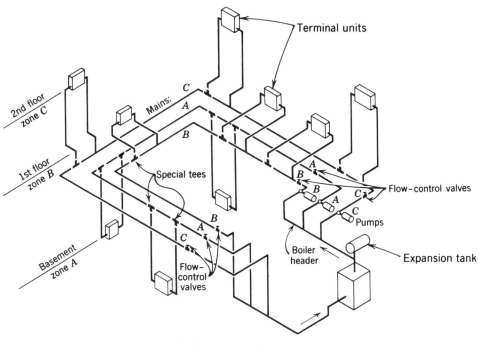

(a) Multizone, one-pipe system

Figure 3.30 (a) Multizone, one-pipe system. A residence is split into three zones—basement, living area and sleeping area. Each zone has its own controls and circulator (pump). Note that two special diverter tees are used to connect branches feeding heating units below the main pipe as shown in Figure 3.26.

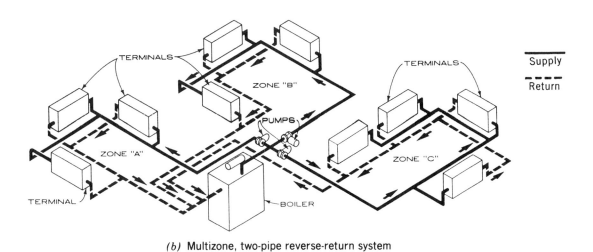

(b) Multizone, two-pipe reverse-return system

Figure 3.30 (b) Multizone, two-pipe reverse-return system. Zones are completely independent except for reliance on the central boiler for hot water supply. (From *Architectural Graphic Standards*, 8th ed., 1988, reprinted by permission of John Wiley & Sons.)

124 / HYDRONIC HEATING

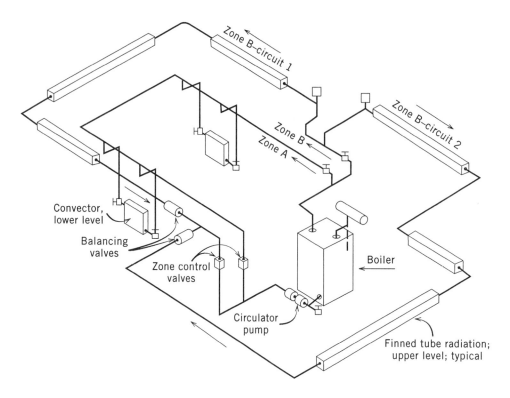

(c) Multiple zone multiple circuit hydronic heating system

Figure 3.30 *(c)* Two-zone hybrid system. Zone A is a straightforward one-pipe loop. Zone B is a two-circuit series system. Flow in each circuit is controlled by a balancing valve.

(1) Frequent cycling of the pump causes excessive wear and excessive temperature cycling.
(2) Maintaining water at an elevated temperature in the boiler causes high energy losses.
(3) The system is not suitable when the boiler also supplies domestic hot water due to the high boiler water temperature required for this function.
(4) When heat is not required, the burner will keep cycling on and off to maintain water temperature, which serves no function except to waste energy.

As a result of this last consideration, designers generally agree that water temperature should vary with outside temperature and should not be fixed. That is, as the outside temperature falls, boiler water temperature should rise and vice versa. Further, when heat is not required, both the circulator (also called pump or booster) and the burner should be shut off. The control schemes described later operate in this fashion and are among those most commonly used.

b. Burner Control Only (Pump Operates Continuously)

This system can be used if the boiler does not supply domestic hot water. See Figure 3.31*b*. When the room thermostat calls for heat, its contacts will turn on the boiler burner, provided water temperature is below the high level cutout. The circulator operates continuously unless shut off manually or by an optional exterior thermostat. This arrangement gives the following results:

(1) When the exterior temperature drops, the room thermostat will operate the burner frequently, thus raising the average water temperature.

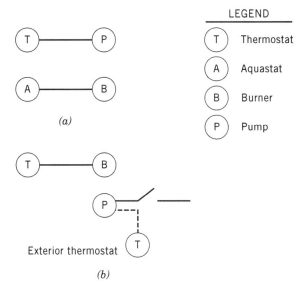

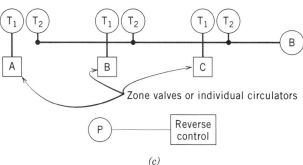

Figure 3.31 Heating system control schemes. *(a)* Thermostat/aquastat control. *(b)* Thermostatic control only, with continuous operation of the system pump. Optional pump cutoff with an external thermostat. *(c)* Two-stage thermostat control of a three-zone system (A, B, C). The pump operates continuously except when shut down by a reverse control that operates when none of the zones require heat. See text for full description of the control operation.

This keeps room surface warmer and raises the space's MRT, making occupants more comfortable. When exterior temperatures rise, the burner will operate infrequently, causing the average water temperature to drop. The result is a self-regulating water temperature system.

(2) The continuous water circulation smooths out temperature variations, increasing the space's comfort level.
(3) Since water temperature varies with demand, boiler and piping heat losses are reduced, and system efficiency is increased. Fuel consumption is considerably reduced.
(4) The additional cost of continuous operation of the fractional horsepower circulator is minimal. If desired, in areas of high electrical energy costs, an exterior thermostat can be installed that will cut out the circulator when exterior temperatures exceed a preset limit. Usually, however, a manual switch is sufficient.

This system is obviously not appropriate to installations requiring domestic hot water from the heating boiler because of the large drop in water temperature on warm winter days. The system is sometimes referred to as *water temperature reset control*.

c. Two-Stage Thermostat

The two-stage thermostat system, which is shown schematically in Figure 3.31(c), is frequently used in zoned systems. Each zone uses a two-stage thermostat, shown in the diagram as T_1 and T_2. The first stage of each zone's thermostat T_1 operates the circulator or zone valve for that zone. When zone valves are used, the main circulator runs continuously and will shut down only if none of the zones calls for heat. When the second stage of the thermostat in any zone (which is always set below the first stage) calls for heat, the burner is activated and remains on as long as any second stage is calling for heat. In zones with lower heat requirements, the first stage of the thermostat will open, shutting off its circulator or zone valve, thus preventing overheating.

In addition to these operating controls, all hydronic systems are equipped with a group of safety devices. Among these devices are a high water temperature burner cutout, an oil burner relay that shuts off the fuel valve with loss of flame or a gas pilot switch that shuts off the gas valve with loss of the pilot flame, and a low water cutoff that cuts off the boiler's burner if boiler water level drops below a specific level.

3.10 Piping Design Factors

Hydronic systems usually use copper pipe and tubing in sizes up to and including 1½ in., schedule 40 steel pipe in larger sizes and heat resistant plastic pipe in all sizes, where permitted by building codes. A detailed description of ferrous and nonferrous piping materials can be found in Section 8.12 (page 414). Dimensional data for ferrous pipe is found in Table 8.2 (page 415) and for nonferrous metallic pipe (copper and brass) in Table 8.3 (page 431). Particularly important in the design of piping for hydronic heating systems are the arrangements made for pipe expansion. Table 8.4 (page 438) gives the coefficients of linear expansion for metallic and nonmetallic (plastic) pipe, plus expansion data for specific temperature rises. Figures 8.17, 8.18 and 8.19 show some typical details of fittings and hangers that are used to accommodate pipe expansion. When using copper tubing, the maximum length of straight run that can be installed without an expansion loop, joint or fitting is roughly as follows:

Average Water Temperature, °F	Maximum Length of Straight Run for Copper Pipe, ft
210	28
200	30
190	33
180	35
170	39
160	42
150	47

The relationship between water velocity and flow (quantity) in pipe depends only on the pipe diameter. Most hydronic heating systems are designed for a water velocity of 3 fps. At this velocity, turbulence, noise and wear are minimal. To translate this figure into flow (gpm), we need to remember only that flow is the product of pipe cross-sectional area and flow velocity. Expressed mathematically,

$$Q = AV$$

This subject is fully developed in Section 8.8, which begins on page 410. Turn to that section now and study it thoroughly. From that section, we can copy the equation

$$Q = 2.45 \, d^2 V \quad (3.1)$$

where

Q is flow in gpm,
d is the pipe inside diameter in inches and
V is the water velocity in fps.

Using type L copper tubing, we find from Table 8.3 (page 431) that the ID of ½-in. pipe is 0.545 in., of ¾-in. pipe is 0.785 in. and of 1-in. pipe is 1.025 in. Using Equation 3.1, we find that for these three sizes the flow is $0.728V$, $1.51V$ and $2.57V$. We can, therefore, easily calculate water velocity for various flow rates and flow rates for various water velocities for these three commonly used sizes of type L copper tubing. The results are tabulated in Tables 3.8 and 3.9. Similar data for type M copper tubing, with inside diameters of 0.569, 0.811 and 1.055 in. for ½-, ¾- and 1-in. tubing, respectively, are given in Tables 3.10 and 3.11.

Note (and remember) that for a design flow of 3 fps, ¾-in. type L copper tubing will carry 4.5 gpm. Refer to Section 3.1 where we developed the very useful facts that 1 gpm carries 500 Btuh/°F and that, therefore, for a design temperature drop in a hydronic system of 20°F, a flow of 1 gpm will deliver 10,000 Btuh. Most systems are targeted for a velocity of 3 fps and use of ¾-in. tubing. This gives, from Table 3.8, a flow of 4.5 gpm. Since 1 gpm can deliver 10,000 Btuh at 20°F water temperature drop, 4.5 gpm will deliver 4.5 × 10,000 Btuh or 45,000 Btuh (this is also written as 45 MBH). These figures were not chosen at random. In residential work, ¾-in. tubing and 3 fps water velocity are in very common use. The significance of the preceding calculation is that if the heat loss calculation for a residence totals more than 45,000 Btuh (and less than 77,000 Btuh) the designer then has three choices:

(a) Increase the tubing size to 1 in. This raises the flow (at 3 fps) to 7.7 gpm and the system capacity to 77,000 Btuh (77 MBH).
(b) Use a multiple-loop system as shown, for instance, in Figures 3.23(d), 3.27 and 3.30(a) and (c). The main supply tubing from the boiler would then be 1 in., and two individual loops would be ¾ in.
(c) Increase the system water velocity to a maximum of 6 fps. Velocities higher than 6 fps are undesirable primarily because of noise. From Table 3.8 we find that, for a ¾-in. pipe, a

Table 3.8 Water Velocity for Varying Flow Rates in Type L Copper Tubing

Tubing Size, in.	Velocity, fpm					
	1 gpm	2 gpm	3 gpm	4 gpm	5 gpm	6 gpm
½	1.4	2.7	4.1	5.5	6.9	8.2
¾	0.66	1.3	2.0	2.7	3.3	4.0
1	0.39	0.78	1.2	1.6	1.9	2.3

Table 3.9 Flow Rates for Varying Water Velocities in Type L Copper Tubing

Tubing Size, in.	Flow, gpm							
	1 fps	1.5 fps	2 fps	2.5 fps	3 fps	3.5 fps	4 fps	5 fps
1/2	0.73	1.1	1.46	1.8	2.2	2.6	2.9	3.6
3/4	1.5	2.3	3.0	3.8	4.5	5.3	6.0	7.6
1	2.6	3.9	5.1	6.4	7.7	9.0	10.3	12.9

Table 3.10 Water Velocity for Varying Flow Rates in Type M Copper Tubing

Tubing Size, in.	Velocity, fps					
	1 gpm	2 gpm	3 gpm	4 gpm	5 gpm	6 gpm
1/2	1.26	2.5	3.8	5.0	6.3	7.6
3/4	0.62	1.2	1.9	2.5	3.1	3.7
1	0.37	0.73	1.1	1.5	1.8	2.2

velocity of 3.5 fps gives a flow of 5.3 gpm (53 MBH at 20°F drop) and 4.0 fps gives a flow of 6.0 gpm (60 MBH at 20°F drop). Larger systems are normally either split into multiple circuits or multiple zones to avoid the difficulty and expense involved in using tubing larger than 1 in.

3.11 Design Procedure

As with the architectural and structural design of a building, so also with the hydronic heating system design, there are many possible solutions, all of which will provide the necessary thermal comfort. Designs and design procedures will vary from one designer to another and even with one designer, from one project to another. The reason is that the large number of variables in a hydronic system lend themselves to a large choice of solutions. All designers start with the same set of data—an architectural plan and the results of the heat loss calculations for the building. In most cases, there is no specific budget, but there is always the requirement to produce a design that is technically satisfactory, at the lowest possible cost. (Whether "cost" is first cost or life-cycle cost, that is, owning and operating cost, depends on whether the construction is speculative or occupant-builder. This subject is not in the province of a technologist's work, but it is very important and should be understood.) Further, there are energy codes that must be complied with and minimum system efficiency standards that have to be met. These also are not usually the responsibility of the technologist, and they are mentioned here only to make a beginning designer aware of some of the system design constraints.

The variables in a design are:

- Type of piping system (arrangement).
- Type of terminal units.
- System water temperature.

Table 3.11 Flow Rates for Varying Water Velocities in Type M Copper Tubing

Tubing Size, in.	Flow, gpm							
	1 fps	1.5 fps	2 fps	2.5 fps	3 fps	3.5 fps	4 fps	5 fps
1/2	0.79	1.2	1.6	2.0	2.4	2.8	3.2	4.0
3/4	1.6	2.4	3.2	4.0	4.8	5.6	6.4	8.1
1	2.7	4.1	5.5	6.8	8.2	9.5	10.9	13.6

- Type and, to an extent, size of piping.
- Characteristic of circulator.

Many of these items are interrelated. Thus, for instance, a larger pipe size means lower friction (head), higher flow and greater system capacity.

Most designers will proceed as follows in residential design.

Step 1. Select the type of terminal units to be used, based on the architectural layout and the quality of construction.

Step 2. Select a piping arrangement based on the building size and layout and on the total building heat loss. At this stage, multiple loops and zoning would be considered.

Step 3. Calculate the size of all terminal units and the system water flow based on the total heating load and an assumed water temperature drop. Also, select the input and return water temperatures.

For residences using baseboard radiation, boiler output temperatures of 160–180°F are common for small houses, and 180–200°F, for large houses. (Exposed piping at these temperatures must be insulated to reduce heat loss and to prevent burns from bodily contact.) Increasing the system temperature drop increases the delivered Btuh/gpm and, therefore, reduces the required flow for a given heat loss. This in turn reduces the required pipe sizes, making the entire system more economical. Limits of temperature drop for the various types of terminal units are:

Baseboard—10–50°F, 20°F most common
Convectors—10–30°F
Cast-iron radiation—maximum of 30°F

In all cases, for specific units, manufacturers' guidelines and recommendations should be consulted. In series systems with low to average overall temperature drop (up to 20°F), an average temperature for all terminal units can be assumed without introducing excessive error, even though the units receive different temperature water, as explained in Section 3.8. With higher system temperature drop or where greater accuracy is desired, individual unit temperatures should be calculated in series piping systems, and terminal units should be sized accordingly.

Step 4. Size the convectors based on room heat loss, average or actual temperature drop, and flow.

Step 5. Having established the flow rate, calculate water velocities in all parts of the system for assumed pipe sizes.

Water velocity should not exceed 6 fps to avoid excessive noise and turbulence. If velocities are unsatisfactory, pipe size can be altered, or the system temperature drops determined in the previous design step can be changed. (The design is often a trial-and-error procedure, where each trial brings the design closer to a satisfactory solution. Seldom are more than two tries necessary, particularly for an experienced designer.) Minimum flow rates for baseboard radiation and commercial finned-tube radiation, for which heat output is at least 90% of rated 3 fps output, are:

$1/2$-in. tubing   0.3 gpm
$3/4$-in. tubing   0.5 gpm
1-in. tubing   0.9 gpm

Step 6. Calculate the system head. (This step is frequently combined with Step 5.)

Step 7. Select a pump that will supply the required flow and head.

Step 8. Select a boiler.

Step 9. Recheck the system temperature drop based on actual flow figures.

At this point, illustrative examples of the design of actual buildings will clarify this procedure.

3.12 Basic Residential Hydronic Heating System Design

For the first illustrative example of hydronic design, we will use a straightforward architectural plan of a small residence that we call The Basic House plan. This plan will be used throughout this book to demonstrate the application of HVAC, plumbing and electrical design. The house, shown in Figure 3.32, is a small, well-insulated two bedroom residence situated in the New York City vicinity. Figures 3.32(a) and (b) show the house plan, and Figure 3.32(c) shows a wall section and gives the house insulation data plus the results of a heat loss calculation.

Example 3.3 Design a hydronic heating system for The Basic House plan of Figure 3.32, using the heat loss data given there.

Solution: We will follow the design procedure outlined in Section 3.11 to the extent possible. The following numbered steps correspond to the numbers of the design steps in Section 3.11.

Step 1. Select an appropriate type of terminal unit.

The ideal place for a heating unit in any installation is at the point of maximum heat loss. In most buildings, this is at the doors and windows. This is also true of The Basic House. The simplest and least objectionable way, architecturally, to do this is to use finned-tube baseboard heaters below the windows in each room. In the kitchen, this is not practical since the window is above the sink countertop. We would, therefore, use the area on one or both sides of the outside door instead. Rather than use specific data from one manufacturer, we will use the typical ratings given in Table 3.1. We encourage you to compare these data with actual catalog data. The differences will be quite small. Furthermore, even though one manufacturer is specified, in actual construction, the HVAC contractor may supply another unit with similar but not identical characteristics. For this reason, we will use typical data throughout this design problem.

Step 2. Select a piping arrangement appropriate to the structure.

This architectural plan is ideal for a perimeter loop around the outside walls of the main level. Only a single circuit is necessary because the total building heat loss (without the basement) is only 34,000 Btuh. As noted several times previously, 1 gpm will deliver 500 Btuh/F° temperature difference. Therefore, a flow of 1 gpm will deliver 10,000 Btuh for a temperature drop of 20°F and 40,000 Btuh for a flow of 4 gpm. Further, the building is quite small so that a series loop piping arrangement can be used. This design avoids the expense of the branch piping, balancing and shutoff valves required in a one-pipe system. If temperature drop is held to below 20 F°, the difference in water temperature between convectors at the beginning and end of the series loop can be ignored, and an average loop temperature can be used for design.

Step 3. Select the supply and return water temperatures and the system temperature drop.

The total main floor calculated heat loss is 34,000 Btuh [see tabulation in Figure 3.32*(c)*]. We arbitrarily select a flow of 4 gpm and a boiler output water temperature of 190°F. (This is somewhat higher than the 180°F recommended. It is used in order to avoid excessively long baseboards in a series loop.) We select a flow of 4 gpm. For these parameters, the overall temperature drop in the perimeter loop around the main level would be

$$\Delta T = \frac{34{,}000 \text{ Btuh}}{[500 \text{ Btuh/gpm/ F°}] \times 4 \text{ gpm}} = 17 \text{ F°}$$

Since this is below 20 F°, the calculation of baseboard lengths can be based on an average temperature of

$$190°F - 17 \text{ F°}/2 = 190°F - 8.5 \text{ F°} = 181.5°F$$

This is the system average (design) water temperature. Referring now to Table 3.1 on page 000 for baseboard radiation output at various temperatures, we will have to interpolate to obtain the value we need.

Temperature, °F	Output at 4 gpm, Btuh
180	610
181.5	?
185	640

$$\begin{aligned}
\text{Output at } 181.50°F &= 610 + \frac{181.5 - 180}{185 - 180} \times (640 - 610) \\
&= 610 + \frac{1.5}{5}(30) \text{ Btuh} \\
&= 610 + 9 \\
&= 619 \text{ Btuh/ft of radiation}
\end{aligned}$$

Step 4. Size the convectors based on room heat loss, average or actual temperature drop and flow.

To demonstrate that the error introduced by using average water temperature is within engineering accuracy, we will size the baseboard radiation by both methods, that is, by average temperature and by actual temperature, and then compare the results.

(a) *Average temperature calculation.* The length of each section of radiation is calculated simply by dividing the loss in that space by the average heat output of 619 Btuh/ft of radiation. For instance, the kitchen that has a calculated heat loss of 4800 Btuh requires

$$\text{Length} = \frac{4800 \text{ Btuh}}{619 \text{ Btuh/ft}} = 7.75 \text{ ft}$$

Unfortunately, there is not sufficient space in the kitchen to place even 7 ft of baseboard. We, therefore, use two two-tier units; a 3 ft, 6-in. unit to the left of the door and a 2-ft unit on the wall containing the sliding (pocket) door to the right of the outside door. The remainder of the units do not present any space problem. The calculation results are shown in Table 3.12. All lengths are calculated for the average loop radiation of 619 Btuh/ft.

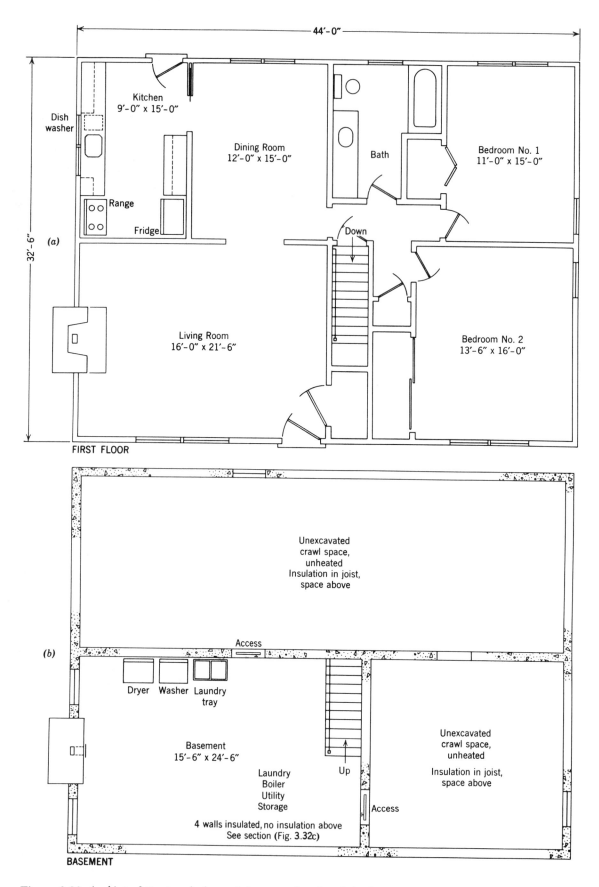

Figure 3.32 *(a, b)* Architectural plans of the street level and the basement level respectively of The Basic House. *(c)* Wall section of The Basic House plus specification and heat loss data.

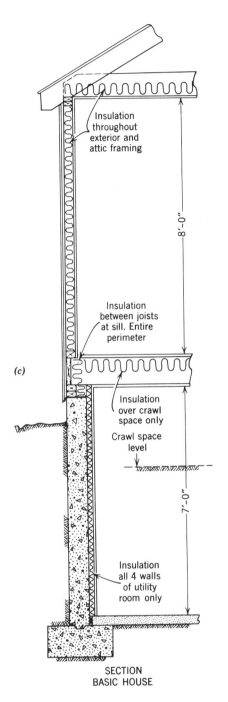

Figure 3.32 (Continued)

Specification, Significant Items Only

Superstructure	Wood frame
Foundation	8-in. poured concrete
Windows	All insulating glass, double, ½-in. air space
Insulation	k value = 0.27 for both batts and rigid insulation

9-in. mineral wool batts   R = 33.3
   In ceiling joists between first floor and attic
   In floor joists over crawl space
   Behind wood sill, entire perimeter
3-in. mineral wool batts   R = 11.1
   In stud space, all exterior frame walls
2-in. rigid insulation   R = 7.4
   Four walls of utility space in basement
NONE In floor joists between living room and utility room below

Vapor barriers	Plastic sheet on *warm* side of *all* insulation
Ventilation	Vents to crawl space and attic Open in summer/closed in winter

Heating Design Data: New York area

Indoor temperature, winter	75°F
Outdoor temperature, winter	0°F

Heat Losses

Living Room	9,300
Kitchen	4,800
Dining Room	4,300
Bath	2,900
Bedroom 1	5,200
Bedroom 2	7,500
Total, main floor	34,000 Btuh
Basement	4,900
Total, building	38,900 Btuh

Table 3.12 Baseboard Radiation Calculation Based on Average Temperature

Space	Heat Loss, °F	Baseboard Length Required, ft	Baseboard Length Used, ft
Kitchen	4800	7.75	see text
Dining room	4300	6.95	7
Bath	2900	4.68	5
BR #1	5200	8.4	8
BR #2	7500	12.1	1 @ 8 1 @ 4
Living room	9300	15	15

The actual lengths of baseboard depend on the specific manufacturer. Thus, the 15-ft length required in the living room might be made up of a 7-ft section and an 8-ft section or some other combination. The actual overall length of baseboard is at least 6–12 in. longer than the active finned-tube section, to allow space for connections, valves, reducers and other fittings.

(b) *Accurate baseboard radiation calculation.* Having established the flow (4 gpm) and the boiler output temperature (190°F), we can calculate the temperature drop in each room's radiation individually. First, however, we must determine the direction of water flow in the loop. Because the kitchen has limited baseboard space, we would want it to receive the hottest water. We would, therefore, establish a loop that starts at the kitchen, proceeds to the dining room and so on, with the living room receiving the coolest water. The calculations must be made in the proper order as we proceed around the loop. The first calculation is for the kitchen.

Heat loss—4800 Btuh
Entering water temperature—190°F

Using the fact that 1 gpm delivers 500 Btuh/F° drop and knowing that we have established a flow of 4 gpm, the water delivers

4 gpm (500 Btuh/gpm/F°) = 2000 Btuh/F°

The temperature drop in the baseboard that will deliver 4800 Btuh is, therefore,

$$\Delta T = \frac{4800 \text{ Btuh}}{2000 \text{ Btuh/F°}} = 2.4\text{F°}$$

Therefore, the average water temperature in the kitchen baseboard is

$$190°\text{F} - \frac{2.4°\text{F}}{2} = 188.8°\text{F}$$

From Table 3.1, the heat output of baseboard at 188.8°F is, by interpolation:

Temperature, °F	Output, Btuh
185	640
188.8	?
190	690

$$\text{Output} = 640 + \frac{188.8 - 185}{190 - 185} \times (690 - 640)$$
$$= 640 + \frac{3.8}{5}(50) = 678 \text{ Btuh}$$

$$\text{Length required} = \frac{4800 \text{ Btuh}}{678 \text{ Btuh/ft}} = 7.08 \text{ ft, say 7 ft}$$

Here again we use two-tiered radiation because of the space limitation. For the purpose of the calculation, however, we will retain the 7-ft figure. (We are treating the two sections of baseboard in the kitchen as one, for the purpose of our study. The outlet water temperature of the kitchen radiation is 190°F − 2.4°F = 187.6°F. This then is the inlet temperature of the dining room baseboard. The results of similar calculations for the other rooms are given in Table 3.13.

Note that the return temperature is 173°F, which is 17 F° below the entering temperature of 190°F. This corresponds to the temperature drop calculated in Step 3, as it should, since it is based on the total building heat loss of 34,000 Btuh. A comparison of the results of the average and accurate calculations follows.

Space	Average T Calculation, ft	Accurate T Calculation, ft
Kitchen	8	7
Dining room	7	7
Bath	5	5
Bedroom #1	8	8
Bedroom #2	12	12
Living Room	15	16

Table 3.13 Baseboard Radiation Based on Actual Water Temperatures

Space	Heat Loss, Btuh	Entering Temperature, °F	ΔT, F°	Average Temperature, °F	Leaving Temperature, °F	Baseboard Output, Btuh/ft	Baseboard length,[a] ft	Actual Space Heating
Kitchen	4800	190.0	2.4	188.8	187.6	678	7	4746
Dining room	4300	187.6	2.15	186.5	185.45	655	6.6/7	4585
Bath	2900	185.45	1.45	184.7	184.0	638	4.5/5	3190
BR #1	5200	184.0	2.6	182.7	181.4	626	8.3/8	5008
BR #2	7500	181.4	3.75	179.5	177.7	607	12.4/12	7284
Living Room	9300	177.7	4.65	175.4	173.1	569	16.4/16	10,104
	34,000							34,917

[a] Calculated length/design length.

Note that the differences occur at the ends of the loop where the difference between actual and average temperature is greatest. Knowing this, a designer using the simpler average temperature method would shorten the baseboard length at the beginning of the series loop and lengthen it at the end of the loop, thus obtaining the same results that are obtained by accurate calculation, with a lot less work.

The last column in Table 3.13 gives the actual baseboard output for each space. It shows that in the kitchen, the two bedrooms and the living room the output is very slightly below calculated heat loss. In the kitchen, the heat output of the refrigerator more than adequately makes up the difference. In the two bedrooms, the 3–4% difference is insignificant. If anything, it is probably desirable since most people like the bedrooms cool. In the living room, the 2% difference is also negligible. Finally, remember that the heat loss calculation is made for a once-in-a-great-while low temperature (see Section 2.13). This means that for 97.5% (or 99%) of the time the heating system will be more than adequate to maintain design temperatures. Finally, if desired, the boiler temperature can be raised slightly several degrees. This will more than compensate for the 2–4% difference between calculated and design heat output.

You may have noticed that no radiation has been supplied in the basement despite the calculated 4900 Btuh heat loss. Experience has shown that basements receive sufficient heat from the boiler losses and exposed piping to adequately heat them, and no additional radiation is required.

Steps 5, 6 and 7. Calculate water velocities and system head for assumed pipe sizes and select circulator.

The first thing that must be done to proceed with this stage of the design process is to show on the building plan the baseboards and the piping connections between them. These is done on Figure 3.33. Note that we have split the 8 ft of baseboard in Bedroom #1 into two 4-ft sections in order to place radiation below both windows. This will give the room a much more even heat distribution than placing all the radiation below the double window, at only a small increase in cost. The basement plan shows the pipe connections between baseboards. They are run directly under the main level floor joists and are insulated with at least ¾-in. thick fiberglass sleeves in the unheated crawl space. This serves to greatly reduce heat loss. In the basement area, the pipes are left uncovered except in places where they might be touched by occupants. The uncovered pipes help keep the basement warm in winter.

At this point we can measure the total length of piping in the series loop, including the ¾-in. pipe of the finned-pipe radiation. The total length as measured comes to 194 ft. This is known as the *developed length* of piping. Since the piping is installed in areas where it is not subject to physical damage, thin wall type M copper tubing can be used. The pressure loss (in psi) due to friction in type M copper tubing is given in Table 3.14. In addition to the straight runs of pipe, there are various fittings in the loop including couplings, elbows and valves. The designer can account for the friction loss in these fittings in one of two ways. He or she can list all the fittings of each type and then, using Table 3.15, can find the equivalent length of straight tubing for the total. This equivalent length is then added to the developed length of piping (the measured length of piping in the

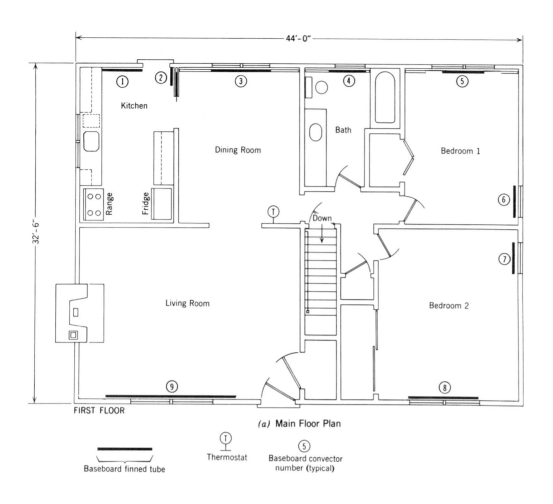

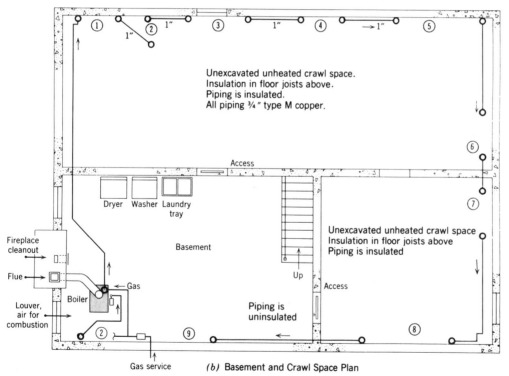

Figure 3.33 Plans of the main (a) and basement levels (b) of The Basic House, showing the designed baseboard radiation and the system piping. (c) Isometric schematic of the piping arrangment.

BASIC RESIDENTIAL HYDRONIC HEATING SYSTEM DESIGN / 135

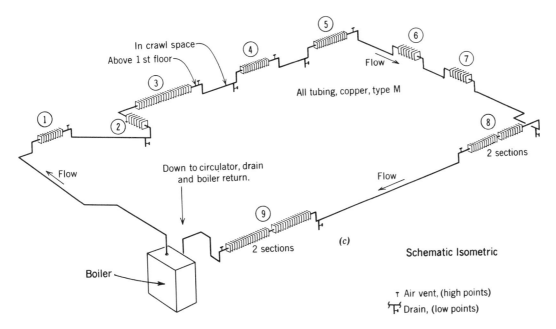

Figure 3.33 *(Continued)*

Table 3.14 Pressure Loss Due to Friction in Type M Copper Tube

Flow, gpm	Pressure Loss per 100 ft of Tube, psi											
	Standard Type M Tube Size, in.											
	3/8	1/2	3/4	1	1 1/4	1 1/2	2	2 1/2	3	4	5	6
1	2.5	0.8	0.2									
2	8.5	2.8	0.5	0.2								
3	17.3	**5.7**	1.0	**0.3**	0.1							
4	28.6	9.4	1.8	**0.5**	0.2							
5	**42.2**	13.8	**2.6**	0.7	0.3	0.1						
10		**46.6**	8.6	2.5	0.9	0.4	0.1					
15			17.6	**5.0**	1.9	0.9	0.2					
20			**29.1**	8.4	3.2	1.4	0.4	0.1				
25				12.3	4.7	2.1	0.6	0.2				
30				17.0	6.5	2.9	0.8	0.3	0.1			
35					8.5	3.8	1.0	0.4	0.2			
40					11.0	4.9	1.3	0.5	0.2			
45					13.6	6.1	1.6	0.6	0.2			
50						7.3	2.0	0.7	0.3			
60						10.2	2.7	1.0	0.4			
70						13.5	3.6	1.2	0.5	0.1		
80							4.6	1.6	0.7	0.2		
90							5.7	2.0	0.9	0.2		
100							7.5	2.7	1.0	0.3	0.1	
200								8.5	3.6	1.0	0.3	0.1
300									8.0	2.0	0.7	0.3
400										3.3	1.2	0.5
500											1.7	0.7
750											3.6	1.5
1000												2.5

Note: Numbers in boldface correspond to flow velocities of just over 10 fps.
Source. Courtesy of Copper Development Association.

Table 3.15 Allowance for Friction Loss in Valves and Fittings Expressed as Equivalent Length of Tube

Fitting Size, in.	Equivalent Length of Tube, ft						
	Standard Ells		90° Tee				
	90°	45°	Side Branch	Straight Run	Coupling	Gate Valve	Globe Valve
3/8	0.5	0.3	0.75	0.15	0.15	0.1	4
1/2	1	0.6	1.5	0.3	0.3	0.2	7.5
3/4	1.25	0.75	2	0.4	0.4	0.25	10
1	1.5	1.0	2.5	0.45	0.45	0.3	12.5
1 1/4	2	1.2	3	0.6	0.6	0.4	18
1 1/2	2.5	1.5	3.5	0.8	0.8	0.5	23
2	3.5	2	5	1	1	0.7	28
2 1/2	4	2.5	6	1.3	1.3	0.8	33
3	5	3	7.5	1.5	1.5	1	40
3 1/2	6	3.5	9	1.8	1.8	1.2	50
4	7	4	10.5	2	2	1.4	63
5	9	5	13	2.5	2.5	1.7	70
6	10	6	15	3	3	2	84

Note: Allowances are for streamlined soldered fittings and recessed threaded fittings. For threaded fittings, double the allowances shown in the table.

Source. Courtesy of Copper Development Association.

circuit), and the total equivalent length (TEL) of the circuit is determined. An alternative method of arriving at the TEL is simply to add a percentage of the developed length, depending on the complexity of the piping. For a relatively simple piping installation with only couplings, elbows and a few valves, 50% additional piping is sufficient. For a complex installation with many fittings and valves, up to 100% should be added. We will use the second method here. Since the series loop in The Basic House is quite simple, a 50% increase in developed length will be more than sufficient to account for fittings. The total equivalent circuit length is, therefore:

TEL = 194 ft × 150% = 291 ft

From Table 3.14 or Figure 3.34 we see that the friction loss for 3/4-in. type M copper tubing, with a flow of 4 gpm, is 1.8 psi per 100 ft. Therefore, for 291 ft of pipe, the total friction head loss is

Friction head = 291 ft × 1.8 psi/100 ft = 5.24 psi

Converting this to feet of water (using the conversion factor of 1 ft of water = 0.433 psi), we have

Head, in feet of water
$= 5.24 \text{ psi} \times \dfrac{\text{ft}}{0.433 \text{ psi}} = 12.1 \text{ ft}$

Thus, the system circulator must provide a flow of 4 gpm at a head of 12 ft of water. The electric motor required to drive such a circulator pump is a small fractional-horsepower unit. If it were desired to use a circulator that develops lower head, as is the case with circulators supplied in some small package boilers, the piping runs in the crawl space and basement could be increased to 1 in. This increase would reduce the friction in the total developed length to 0.5 psi per 100 ft of pipe (see Figure 3.31), although the addition of 1 in. to 3/4-in. reducer fittings at each baseboard would increase the fittings loss. Overall, the loop head would probably not exceed 6–7 ft of water.

As a rule of thumb, friction head in heating system piping should fall between 0.25 and 0.6 in./ft of TEL, or 25–60 in./100 ft. The friction head in our system is 1.8 psi/100 ft. This is equal to

$\dfrac{1.8 \text{ psi}/100 \text{ ft}}{0.433 \text{ psi/ft}} = 4.15 \text{ ft per } 100 \text{ ft} = 50 \text{ in.}/100 \text{ ft}$

This indicates that using 3/4-in. type M tubing throughout results in a reasonable design.

The remaining item in this step of the design is to check water velocity. This we can do using Table 3.10. Note that the water velocity in 3/4-in. pipe is

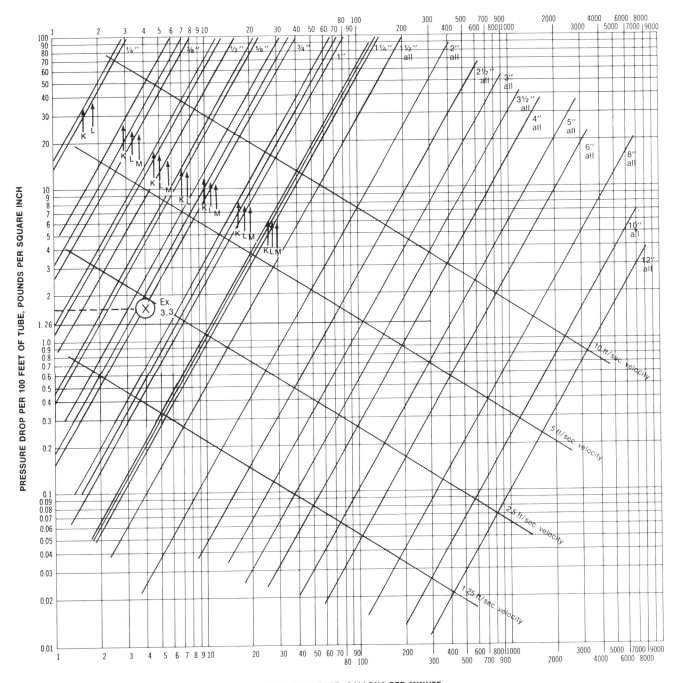

Figure 3.34 Chart showing pressure drop and water velocity for varying flow rates of water in copper pipe, types K, L and M. The horizontal line drawn at 1.26 psi represents a friction drop of 350 millinches (of water) per foot of pipe. See text in Example 3.5. (Courtesy of Copper Development Association.)

2.5 fps, and in 1-in. pipe it is 1.5 fps. Both of these figures are far below the upper limit of 5–6 fps and will, therefore, not produce any appreciable noise, turbulence or rapid corrosion.

Step 8. Select a boiler.

This step is purely mechanical. It involves checking through manufacturers' catalogs to find a boiler unit that will meet the following requirements:

(a) Minimum output 45 MBH (45,000 Btuh). This will provide about a 15% safety factor and will assist in producing a desirable "quick pickup."
(b) Fuel type as required (gas, oil or dual-fuel). We have selected a gas-fired unit and have shown the incoming gas line in the basement in Figure 3.33.
(c) Dimensions of the unit must allow entry into the basement through the house, without the necessity to dismantle the unit or any structural member of the house.

Step 9. Recheck the system temperature drop based on actual flow figures.

Since we have maintained the 4-gpm flow throughout the design, the system temperature drop has remained at 17 F° and no change is required.

This essentially completes the hydronic heating system design for The Basic House. In the next section, we will apply what we have learned to the design of an hydronic heating system to a much larger residence that does not lend itself to a simple single series loop arrangement.

3.13 Preliminary Design Considerations

Having successfully completed the hydronic heating design of a small residence, we will now proceed to a large residence that requires multiple heating zones. The residence shown in Figures 3.35–3.39 presents a typical problem of climate control that must be solved by the architect and the mechanical engineer. The house is an actual building, constructed some years ago. With the permission of the architect, Mr. Budd Mogensen, AIA, the structure will be a clinical framework for our study. As in all well-coordinated projects, the scheme for interior climate control was considered as a basic element of the general design. It was

Figure 3.35 Photograph taken during construction of the Mogensen house, located in Sands Point, Long Island, New York. The photo shows the west elevation of the house. Terraces face Long Island Sound. (Reproduced with permission of B. Mogensen, AIA.)

selected and developed along with the architectural plans. It is not good practice to delay the heating design until after the architectural design is complete.

a. General Architectural Information

Located on the north shore of Long Island and occupying a large plot, this house looks out over the waters of Long Island Sound. All the principal rooms face the view. Nestled into a hill, the house presents a two-story facade to the west. As we can see in Figure 3.37, the (uphill) east elevation resembles that of a one-story house. Two skylight dormers reach up to trap the morning sun, lighting the entry foyer and the master bedroom. Conventional windows provide east light for the living room and master bath. The upstairs guest bath accepts the sun through a plastic roof bubble but is otherwise windowless. On the drawings, the elevations are identified directly as the points of the compass. Actually, they are 45° away from these directions. See the north arrows in Figure 3.38. Thus, the front elevation faces southwest rather than directly west. In our discussion, we will call it west.

b. The Structure

Figure 3.39 shows that the footings and the east wall below grade are to be of poured concrete. This east wall turns the corner at both ends to extend

Figure 3.36 Construction photos, partial views. *(a)* Master bedroom and study. Garages below. *(b)* Kitchen and dining room. Two bedrooms below. *(c)* Living room. Family room below. *(d)* Living room interior, looking south. (Reproduced with permission of B. Mogensen, AIA.)

partially on the south and north elevations. See Figure 3.38. The construction photograph (Figure 3.35), indicates wood frame construction on the west facade. The entire upper story is of wood frame construction. Wood studs, joists and rafters make up the structural frame. A few steel beams carry long spans. Otherwise, the house is wall bearing, using stud walls. Throughout there is heavy thermal insulation. Windows and doors are weather-stripped. All glazing, fixed and movable, is of the double (insulating) type.

c. Form and Geometry

The construction photograph (Figure 3.35) shows clearly that the house is divided into three sections. The divisions are evident in both floor plans (Figure 3.38). Views a, b, and c of Figure 3.36 show the left, center and right-hand sections of this three-part scheme. Independent of this three-section arrangement, the upper and lower floors are each planned for their respective uses. The upper or principal living unit affords access to all rooms from a central foyer. The lower story, intended for family and guests, places all its rooms conveniently around the central hall.

d. Space Study

One sometimes has to search for areas suitable for boilers, air ducts or other equipment. This house has no basement or crawl space. The garage is of conventional width, but it does have a generous 27-ft depth. This, however, would be adequate only

140 / HYDRONIC HEATING

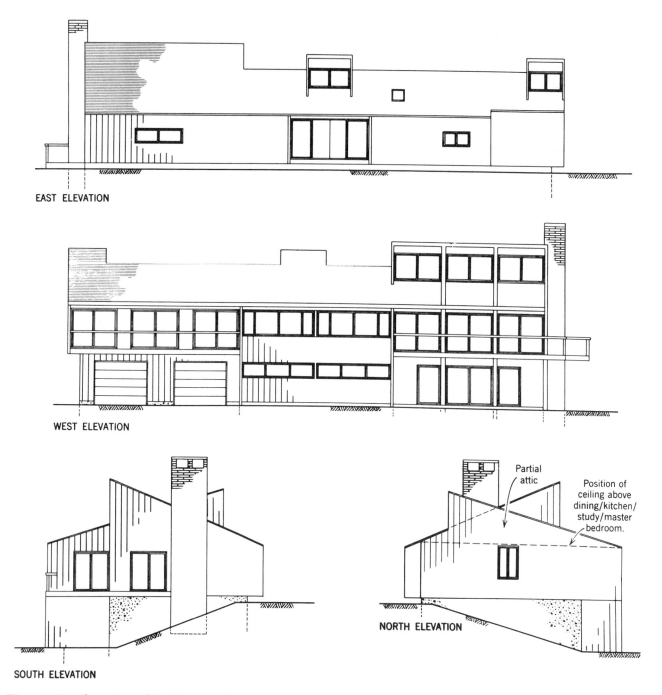

Figure 3.37 Elevations of the Mogensen house. See north arrows on Figure 3.38 for exact orientation of the elevations. (Reproduced with permission of B. Mogensen, AIA.)

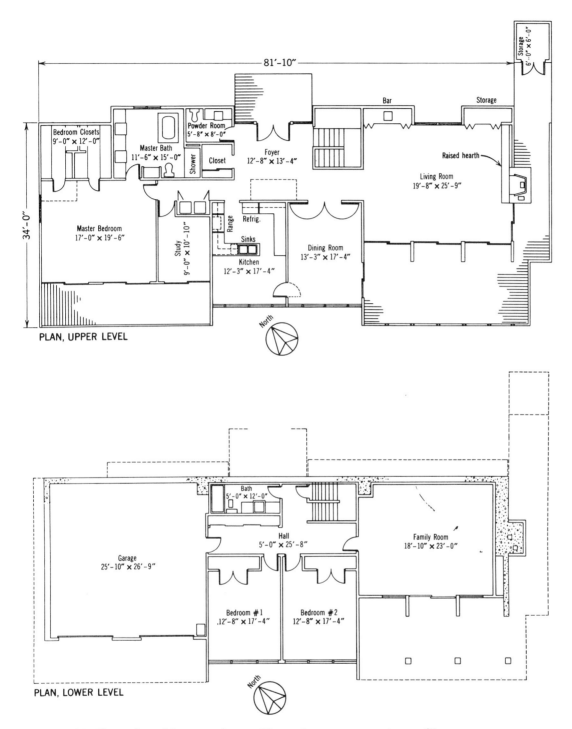

Figure 3.38 Floor plans, Mogensen house. Dimensions are approximate. (Reproduced with permission of B. Mogensen, AIA.)

142 / HYDRONIC HEATING

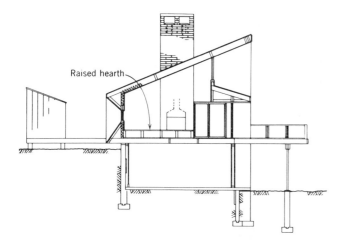

Figure 3.39 Section through the living room and family room, looking south. (Reproduced with permission of B. Mogensen, AIA.)

for long cars, possible boat storage, or a workbench and a few garden tools. The outdoor storage shed is intended for terrace furniture. Therefore, no space is available on the lower level for a boiler room.

Looking upstairs, the section in Figure 3.39 and the interior view (Figure 3.36) tell us that there is no attic above the living room. There is, however, a very small wedge-shaped passage above the glass doors. It might be suitable for tubing or air ducts but not for a boiler. There is, in addition, a somewhat larger attic. See Figure 3.39. It, too, is wedge-shaped and is about 18 ft wide and 7 ft high. It extends over the northern two thirds of the upper story. Heating equipment may, if necessary, be located at high points in a structure. In selecting a location for the boiler room of a hot water heating system, this attic would not be suitable. It would be a poor decision to place a heavy boiler above habitable rooms in this light wood structure. Relatively lightweight air-handling equipment could be placed in such a space, but heavy hydronic equipment could not.

e. The Boiler Room

It is apparent that we need a boiler room in which the boiler can stand on a concrete slab. The architect granted us permission (for study purposes only) to modify his design. His recommendation is that we use the south one third of the family room. This plan involves eliminating the glass on the end of this new room. Because the assigned space is larger than required for a boiler room, a beach shower room is created at the front. Compare the plans in Figure 3.38 with those in Figure 3.40. A flue for the boiler can be provided in the masonry of the chimney. The family room is reduced in size as indicated.

3.14 Hydronic Heating Design of a Large Residence

The design of the hydronic heating system for the Mogensen house follows the same procedure used for The Basic House. Since the house is large, custom-designed and owner-occupied, economy is not the prime consideration. Instead, comfort, convenience and quality are of major importance. This follows the general rule that the quality of the mechanical systems in a building must match the overall quality of the building.

Example 3.4 Design an hydronic heating system for the residence shown in Figure 3.38. The results of the engineer's heat loss calculations for the building are shown in the first two columns of Table 3.16.

Solution: We will follow the design procedure outlined in Section 3.11.

Step 1. Select the terminal units to be used.

Finned-tube radiation is chosen for its efficiency, architecturally pleasing appearance and flexibility. Refer to Figure 3.40, which shows the locations selected for the required radiation. In rooms where the exterior glass extends to the floor level, baseboard radiation cannot be used. Therefore, in the living room, master bedroom, study and foyer on the upper level and in the family room on the lower level, finned-tube radiation recessed into the floor in front of the glass was used. A detail of the installation, plus typical heat output ratings are given in Figure 3.41. The remaining spaces on the upper level (and part of the foyer) plus the remaining spaces on the lower level are all heated by conventional, single-tier baseboard, such as shown in Figure 3.15. Heat output ratings for these baseboards are given in Table 3.1.

Notice that the two master-bedroom closets and the study are provided with electric heaters. The closets do not normally require heat although they will be cool because of the outside walls that are part of both. The electric heaters are intended to

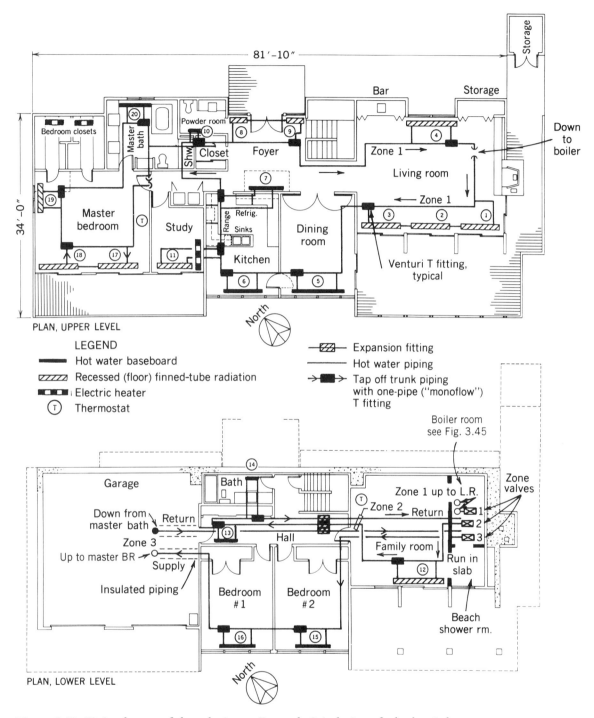

Figure 3.40 Piping layout of the solution to Example 3.4, design of a hydronic heating system for the residence shown in Figure 3.38. See text for a detailed explanation of the solution.

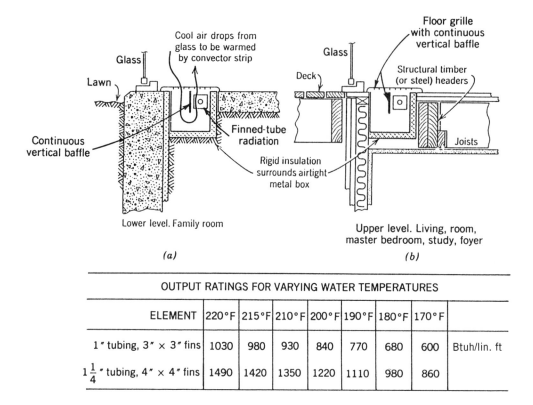

Figure 3.41 Details of recessed finned-tube radiation for the lower level (a) and upper level (b). This design depends on the convective air current set up by cold air dropping off the glass into the trench, as shown in (a). Trench dimensions are 8 in. W × 7 in. D for 1-in. finned-tube and 10 in. W × 9 in. D for 1¼-in. tube. The baffle should be at least 3 in. above the trench bottom for 1-in. tubing and 4 in. for 1¼-in. tubing. The trench should begin no farther then 6 in. from the glass.

warm them on the occasions that it is necessary to spend more than a minute or two in the closet. The electric heater in the study is there for an entirely different reason, which will be explained later in the discussion on zoning.

Step 2. Select the piping arrangement to be used, and establish zones and multiple circuits.

Large residences can easily be divided into two or more zones, each of which has a different heating schedule. In this house, Zone 1 consists of the living room, dining room, kitchen, powder room, foyer and study. These areas will be in use during daytime hours and when entertaining at any hour. Zone 2, consisting of the family room, bath and hall on the lower level, will usually be in use in the evening, when Zone 1 is usually inactive. Zone 3, consisting of the bedrooms on both levels and the master bath is the "sleeping zone," which is normally kept cooler than the remainder of the house.

The study is a special case because its occupancy may not fit into any of these zones. In today's homes, the study often serves as a home office from which business is conducted. As such it may be occupied during any hours of the day (or night), requiring heat when the remainder of the house can be set to lower temperatures. For this reason we have placed the study in Zone 1 but have added a fairly large (1-kw) electric heater. This will provide thermal comfort even when the Zone 1 thermostat is set back to a nighttime setting of 65°F or possibly lower. The rooms in each zone, the heaters in each room and the calculated heat loss for each room and zone are listed in Table 3.16.

The piping arrangement is shown on the architectural plan of Figure 3.40 and schematically in Figure 3.42. We have chosen to use the one-pipe arrangement for all three zones, with separate piping loops for all zones. An alternative arrangement

Table 3.16 Zonal Heating Loads—Mogensen House

Zone	Room	Calculated Heat Loss,[a] Btuh	Heating Element No. and Rating,[c] Btuh	Baseboard Element Length, ft	Recessed Element Length, ft
1	Living room	28,600	1—7770	—	7[d]
			2—7770	—	7[d]
			3—7770	—	7[d]
			4—7770	—	7[d]
	Dining room	8000	5—7800	12	—
	Kitchen	5100	6—5200	8	—
	Foyer	9400	7—3900	6	—
			8—2775	—	2.5[d]
			9—2775	—	2.5[d]
	Powder room	1500	10—1950	3	—
	Study	3900	11—4620	—	6[e]
	Zone total	56,500	60,100		
2	Family room	7900[b]	12—7700	—	10[e]
	Hall	3700	13—3900	6	—
	Bath	1500	14—1950	3	—
	Zone total	13,100	13,550		
3	Bedroom #2	4900	15—5200	8	—
	Bedroom #1	5200	16—5200	8	—
	Master bedroom	15000	17—5390	—	7[e]
			18—5390	—	7[e]
			19—5390	—	7[e]
	Master bath	2800	20—3250	5	—
	Zone total	27,900	29,820		

[a] Heat losses were calculated for a design condition of 0°F outside and 70°F inside.
[b] The heat loss shown is for the shortened family room, as shown in Figure 3.40. The original family room has a calculated heat loss of 12,700 Btuh.
[c] All radiation ratings are based on an *average* water temperature of 190°F. See text discussion.
[d] 1¼-in. finned tube.
[e] 1-in. finned tube.

might use a two-pipe reverse-return system for Zone 1. We have decided not to use it in this study example because the pipe sizing and friction calculations are very complex and would not be done by an HVAC or architectural technologist. We could also have used a single one-pipe loop for all the rooms fed by Zones 2 and 3, with thermostatic control valves in each room. The trade-off in such a plan is the cost of valves and fittings against the cost of additional zone piping. Here again the decision involves an economic study that is not the responsibility of a technologist. For these reasons, we have decided to use three one-pipe zones, which will give a satisfactory, cost-effective, flexible heating system.

Steps 3 and 4. Calculate water flow in all zones using assumed temperature drop and boiler output temperature, and calculate the length of all baseboards and finned-tube radiation.

A system temperature drop of 20°F is commonly used in residential work and will be used here as well. Since the residence is large and the piping runs are fairly long, we will use a boiler output temperature of 200°F. This gives an average loop temperature of

$$T_{AVG} = 200°F - \frac{20°F}{2} = 200°F - 10°F = 190°F$$

We will, therefore, use 190°F for all the flow calculations.

146 / HYDRONIC HEATING

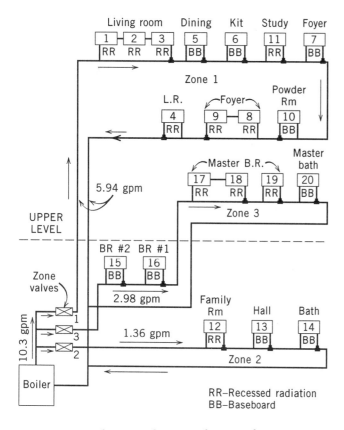

Figure 3.42 Schematic diagram showing the arrangement of terminal units in the three zones of the heating system for Example 3.4.

The one-pipe loop, like the series loop used in The Basic House design, has the disadvantage that the radiation nearest the boiler receives higher water temperature than the units farther into the loop. However, as with the series loop, if system temperature drop is limited to a maximum of 20 F°, use of the average loop temperature in calculation is permissible. The typical connections of finned radiation in a one-pipe system are shown in Figure 3.43. In addition to the shutoff valve on the supply branch, a balancing valve can be used in the return line to limit the water flow. As mentioned previously, a thermostatic valve can also be used if room-by-room control is desired. We will use a balancing valve that can regulate water flow.

Zone 1:

Total calculated zone heating load = 56,5000 Btuh

We can calculate the flow and require radiation length for each of the rooms as follows:

Living Room:

Total load = 28,600 Btuh
Number of baseboard sections = 4 (No. 1, 2, 3, 4)

$$\text{Required Btuh of each} = \frac{28{,}600}{4} = 7150 \text{ Btuh}$$

From Figure 3.41; output per foot of recessed 1¼-in. radiation at 190°F = 1110 Btuh. (Use of 1-in. tubing would result in excessively long units.)

$$\text{Required length of each section} = \frac{7150 \text{ Btuh}}{1110 \text{ per foot}}$$
$$= 6.44 \text{ ft, use 7 ft}$$

We would therefore use four 7-ft sections, giving a total output of 31,080 Btuh. The additional length will ensure quick morning pickup.

These lengths and the average output are entered in Table 3.16. The result of similar calculations for the remaining rooms, using 650 Btuh per foot of baseboard radiation (Table 3.1) and 770 Btuh per foot for 1-in. recessed finned tube in the study (see Figure 3.41) are shown in Table 3.16. Work out the calculations to verify the figures in Table 3.16. The total radiation designed for Zone 1 is 60,100 Btuh as compared to 56,500 Btuh required. This gives the following for the main pipe:

$$\text{Flow} = \frac{60{,}100 \text{ Btuh}}{10{,}000 \text{ Btuh/gpm (for 20 F° drop)}} = 6.0 \text{ gpm}$$

Refer to Figure 3.44, which is a schematic diagram of Zone 1 piping and radiation. We have calculated the flow into each radiation unit and the temperatures all along the loop. They are shown on the diagram. Typical calculations follow.

(a) The flow into units 1, 2 and 3, which are connected as a single extended unit, is

$$\text{Flow} = \frac{23{,}310 \text{ Btuh}}{10{,}000 \text{ Btuh/gpm (for 20 F° drop)}}$$
$$= 2.33 \text{ gpm}$$

The rating we used in our calculation was for a flow of 1 gpm. However, since the rating for 4 gpm is only 5% higher, the output at 2 gpm is only about 2½% above the figure used (1110 Btuh/ft). This is well within engineering tolerance. In any case, adjusting the unit's valve so that the flow is exactly 2.3 gpm is very difficult and the actual flow may be considerably different. The heat output, fortunately, is almost independent of flow. Therefore, we need not readjust the calculation.

(b) The temperature drop between points A and B is calculated as follows:

HYDRONIC HEATING DESIGN OF A LARGE RESIDENCE / 147

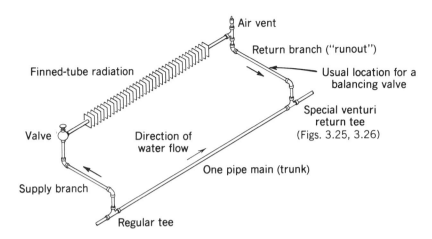

Figure 3.43 Typical piping of finned-tube radiation on a one-pipe loop. See text and Section 3.8(b) for further information.

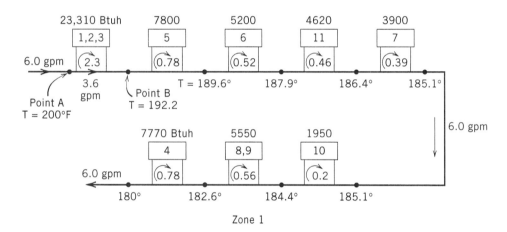

Figure 3.44 Details of heat output, temperatures and hot water flow in Zone 1 of the heating system of Example 3.4.

$$\Delta T_{AB} = \text{Total temperature drop} \times \frac{\text{partial load}}{\text{total load}}$$
$$= 20°F \times \frac{23{,}310 \text{ Btuh}}{60{,}100 \text{ Btuh}}$$
$$= 7.76°F$$

Note the very important fact that this temperature drop depends only on the heat output of the radiation between points A and B. As we know, this heat output is almost independent of flow through the finned tube. If flow is low, temperature drop in the terminal unit will be higher than the specified (20°F), and vice versa. However, if flow is higher, say 3.3 gpm instead of 2.3 gpm in the first group of radiation, then the water returned to the main line is cooler than specified. But there it mixes with less hot water, because it took an extra gpm, and the result is that the temperature at point B remains essentially the same. The same thing happens if the flow is less than designed. As a result, the system is effectively self-balancing and self-regulating, within fairly wide limits.

We can now proceed to the two remaining zones. Both are arranged as one-pipe loops. Again, using a design temperature drop of 20°F and referring to Figure 3.42 for details of the two-zone loops, we have the following flow calculations.

Zone 2:

Total load = 13,100 Btuh
Family room load = 7900 Btuh
Output of recessed 1-in. finned pipe radiation at 190°F average temperature = 770 Btuh/lineal foot (see table in Figure 3.41).

$$\text{Required tubing length} = \frac{7900 \text{ Btuh}}{770 \text{ Btuh/lineal ft}}$$
$$= 10.3, \text{ say } 10 \text{ ft}$$
$$\text{Actual baseboard output} = 10 \text{ ft } (770 \text{ Btuh/lineal ft})$$
$$= 7700 \text{ Btuh}$$

Hall load = 3700 Btuh

Using baseboard with an output of 650 Btuh/lineal ft:

$$\text{Required baseboard length}$$
$$\frac{3700 \text{ Btuh}}{650 \text{ Btuh/lineal ft}} = 5.7 \text{ ft}$$

Here we would use 6 ft, because flow is less than 1 gpm and average water temperature is below 190°F. Actual output is, therefore, 6 ft × 650 Btuh/lineal ft = 3900 Btuh.

$$\text{Bath load} = 1500 \text{ Btuh}$$
$$\text{Required baseboard length} = \frac{1500 \text{ Btuh}}{650 \text{ Btuh/lineal ft}}$$
$$= 2.3 \text{ ft}$$

Here we would use 3 ft due to low flow and temperature.

Actual output = 3 ft × 650 Btuh/lineal ft = 1950 Btuh

The total load for Zone 2 is, therefore,

$$7700 + 3900 + 1950 = 13,550 \text{ Btuh}$$

This gives a total zone flow of

$$\text{Zone 2 flow} = \frac{13,550 \text{ Btuh}}{10,000 \text{ Btuh/gpm (for 20°F drop)}}$$
$$= 1.36 \text{ gpm}$$

Zone 3:

Total calculated heat loss = 27,990 Btuh
Bedroom #2 load = 4900 Btuh

Using baseboard with an output of 650 Btuh/lineal ft, the length required is

$$\text{Length} = \frac{4900 \text{ Btuh}}{650 \text{ Btuh/lineal ft}} = 7.54 \text{ ft, use 8 ft}$$
$$\text{Actual baseboard output} = 8 \text{ ft} \times 650 \text{ Btuh/lineal ft}$$
$$= 5200 \text{ Btuh}$$
$$\text{Bedroom #1 load} = 5200 \text{ Btuh}$$
$$\text{Required baseboard length} = \frac{5200 \text{ Btuh}}{650 \text{ Btuh/lineal ft}}$$
$$= 8 \text{ ft exactly}$$
Actual baseboard output = 5200 Btuh

Master bedroom load is 15,000 Btuh, which is divided into three sections of recessed finned-tube radiation. Using 1-in. tubing with an output of 770 Btuh/ft,

$$\text{Total length required} = \frac{15,000 \text{ Btuh}}{770 \text{ Btuh/lineal ft}} = 19.5 \text{ ft}$$

Since there are three units, each will be 7 ft long, with an individual output of 7 ft × 770 Btuh/lineal ft = 5390 Btuh and a total output of 16,170 Btuh for all three units. The actual output will be less than this, because the actual average water temperature in each unit will be below 190°F.

$$\text{Master bath load} = 2800 \text{ Btuh}$$

Using baseboard,

Required baseboard length =
$$\frac{2800 \text{ Btuh}}{650 \text{ Btuh/lineal ft}} = 4.3 \text{ ft, use 5 ft}$$
$$\text{Actual output} = 5 \text{ ft } (650 \text{ Btuh/lineal ft})$$
$$= 3250 \text{ Btuh}$$

The total load in Zone 3 amounts to 29,820 Btuh as compared to the calculated heat loss of 27,900 Btuh.

The total required supply of 200°F hot water is

Flow =
$$\frac{29,820 \text{ Btuh}}{10,000 \text{ Btuh/gpm (for 20°F drop)}} = 2.98 \text{ gpm}$$

The total flow required for all three zones is

Zone 1   6.0 gpm
Zone 2   1.36 gpm
Zone 3   2.98 gpm
Total   10.34, say 10.3 gpm

Steps 5 and 6. Calculate system head and water velocity in the branches of the system.

We will calculate the requirements of each zone individually.

Zone 1:

By scaling the piping layout of Figure 3.40 we obtain a total length of 168 ft. To this we add 50%

to account for fittings, which gives us 252 ft. The Hydronics Institute (I=B=R) recommends that 12 ft of equivalent pipe be added to the circuit for each one-pipe tee. This is because these special Venturi tees operate by introducing friction into the system. Since in Zone 1 we have eight such tees, we add 96 ft to the total previously calculated of 252 ft, for a grand total of 348 ft. This is the total equivalent length (TEL) of the Zone 1 piping circuit. Referring now to the friction chart (Figure 3.34), we find that, using 1-in. type M copper tubing, the friction head is exactly 1 psi/100 ft for a flow of 6.0 gpm. Since Zone 1 has a TEL of 348 ft, the total friction head is

$$\text{Friction head of Zone 1} = 348 \text{ ft} \times \frac{1 \text{ psi}}{100 \text{ ft}}$$
$$= 3.48 \text{ psi}$$

Converting this to feet of water, we have

$$\text{Friction head of Zone 1} = 3.48 \text{ psi} \times \frac{\text{ft of water}}{0.433 \text{ psi}}$$
$$= 8.0 \text{ ft head}$$

that is, using 1-in. type M tubing, the circuit of Zone 1 has a friction head of 8 ft (of water).

Zone 2:
Following the same procedure for Zone 2, we have

Actual piping length	103 ft
TEL (150% of length)	155 ft
Additional length for three tee fittings	36 ft
Zone 2 total length	191 ft

Using ½-in. pipe with a flow of 1.36 gpm, from Figure 3.34 we obtain a friction head of 1.35 psi per 100 ft. Therefore, for 191 ft of ½-in. pipe the total friction drop is 2.58 psi. Converting this to feet of water, we have

$$\text{Friction head of Zone 2} = \frac{2.58 \text{ psi}}{0.433 \text{ psi/ft of head}}$$
$$= 6 \text{ ft (head)}$$

Zone 3:

Piping length	222 ft
TEL	333 ft
Additional length for five tee fittings	+60 ft
Zone 3 total length	393 ft

Using ¾-in. type M tubing, with a flow of 3 gpm (2.98), we obtain from Figure 3.34 a friction head of 1 psi/100 ft, or 3.93 psi for the entire loop. Converting this to feet of water, we have

$$\text{Friction head of Zone 3} = \frac{3.93 \text{ psi}}{0.433 \text{ psi/ft of head}}$$
$$= 9.1 \text{ ft (head)}$$

Summarizing this step, we have

Zone 1   6 gpm @ 8 ft of head
Zone 2   1.4 gpm @ 6 ft of head
Zone 3   3 gpm @ 9.1 ft of head

In all three zones water velocity is below 2.5 fps.

Figure 3.42 shows three zone valves (assuming one circulator, not shown). This is one possible arrangement. Another would be a separate circulator for Zone 1, and a circulator plus two zone valves for Zones 2 and 3. These decisions would be made by the project's engineer based on the selection of the boiler and its standard equipment, plus economic considerations. In any case, balancing cocks in all three loops are required to adjust water flow.

Each zone valve is controlled by a zone thermostat. Zone 1 thermostat should be in the living room; Zone 2, in the family room; and zone 3, in the master bedroom. They are shown so located in Figure 3.40. Figure 3.45 shows typical details of the boiler room and the oil tank for an oil-fired boiler.

Steps 7 and 8. Selection of pump and boiler.
The selection of a pump or pumps depends on whether a single pump for the entire building or separate zone circulators are used. Actual pump selection is fairly complex, as it requires use of pump characteristic charts. As such, it is somewhat beyond our scope here. Boiler selection is a matter of selecting a unit of sufficient capacity to supply the entire building radiation of 103,470 Btuh (see Table 3.16), which will physically fit into the space allotted. In addition, it must be suitable and have sufficient capacity to heat hot water if this is required, and it must have satisfactory efficiency. These items also are matters generally handled by the project engineer. It is common practice to oversize the boiler by at least 10% but no more than 25%. This helps provide the fast pickup desired after night setback of the space temperatures.

One technical point should be noted before ending our design discussion. The long straight runs of piping in the basement for Zones 2 and 3 require the use of expansion loops and/or fittings. Refer to Section 3.10 for a table of maximum straight run lengths for various sizes of copper pipe. See also Figures 8.17, 8.18 and 8.19 for details of expansion loops fittings and hangers. A typical commercial

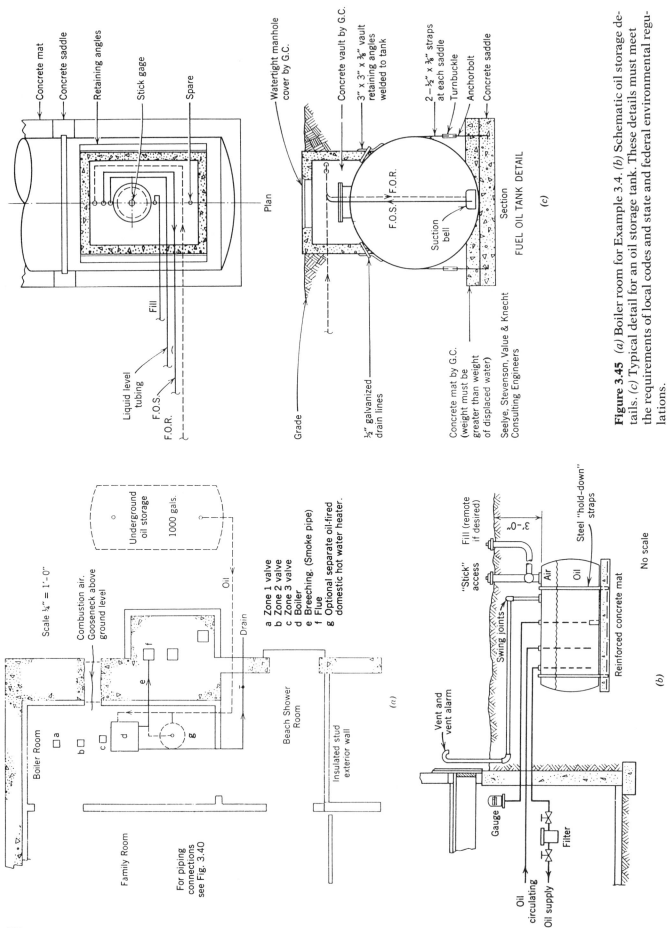

Figure 3.45 (a) Boiler room for Example 3.4. (b) Schematic oil storage details. (c) Typical detail for an oil storage tank. These details must meet the requirements of local codes and state and federal environmental regulations.

Seelye, Stevenson, Value & Knecht Consulting Engineers

expansion fitting is shown in Figure 8.18, page 441.

This completes our solution of the design problem of Example 3.4.

3.15 Hydronic Heating Design of an Industrial Building

The work of a technologist is not limited to residential buildings. In the following sections, we will apply the hydronic heating techniques that we have learned to an industrial building. First, however, you should become somewhat familiar with the building itself.

The building is designed for light manufacturing in the clothing industry. However, with only minimal changes, it can serve other industrial uses just as well. The architect's rendering of the building, the plot plan and the architectural plans essential to our work are shown in Figures 3.46–3.50.

The plan has several principal components:

Administrative wing.
Work area.
Storage area.
Shipping and receiving.
Parking field.

Industrial buildings have headroom in accordance with the space usage. In the administration and work areas the ceiling clearances are 9 ft. In the storage area, the clearance is 12 ft to the underside of the roof trusses. Exterior walls are of 8-in. concrete block, or 4-in. brick backed with 4-in. concrete block, depending on location. A one-level concrete floor slab has been designed for possible use of rolling carriers for materials and merchandise. Windows are of commercial or architectural steel sash single-glazed. Perimeter insulation reduces heat loss at slab edges. Rigid insulation is placed between the concrete plank of the roof and the built-up roofing.

Example 3.5 Design a hydronic heating system for the light industry building shown in Figures 3.46–3.50.

Solution: We will use the same procedure as used in Examples 3.3 and 3.4. Refer to Section 3.11 for a listing and explanation of the design steps involved.

Step 1: Select the terminal units to be used.

Industrial buildings frequently use unit heaters and space heaters (see Figure 3.18) to heat large, open, high-ceiling areas. In this building, the ceiling in the storage area is only 12 ft high, making it somewhat low for high output space heaters. Instead, we will use commercial finned-tube radiation in the storage area, and in the remainder of the building as well. The choice is between unit convectors (see Figure 3.17) and enclosed multi-tier commercial finned-tube radiation (see Figure 3.16). We have chosen the latter because

Figure 3.46 Architect's rendering, building for light industry. North elevation. (Courtesy of Scheiner and Swit, Architects.)

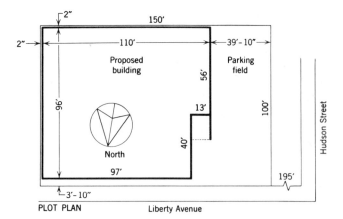

Figure 3.47 Plot plan of the light industry building of Example 3.5. (Courtesy of Scheiner and Swit, Architects.)

(a) High output convectors are narrow (not long) and deep. They protrude from the wall and obstruct traffic. If they are recessed or semi-recessed, they require insulation to prevent high heat loss to the outside.

(b) The short (narrow) convectors do not cover the width of wide industrial windows, thus allowing cold air to reach the floor.

(c) Recessed convectors require additional construction work to form their niches. This is not cheap or desirable when recessing into a simple concrete block wall, as is used in this building.

(d) Convector dimensions vary from one manufacturer to another, making the recessing or semi-recessing that much more complicated.

For all these reasons, and because commercial finned-tube radiation is so flexible in application, we have chosen it for our terminal units.

Step 2: Select the piping arrangement to be used.

In most large buildings, the heating system is zoned according to the usage of the different areas. This building is no exception. It can be divided readily into three zones according to function. Zone 1 is the administrative wing of the building, which includes all the numbered spaces in Figure 3.48. This area is relatively small and centralized. A one-pipe system will serve it adequately and efficiently. Zone 2 covers the large storage area. This area constitutes a separate zone because its temperature requirements and schedule are different from the rest of the building. In general, storage areas are designed for a constant temperature 24 hours a day, 7 days a week. This temperature is normally cooler in the winter (and warmer in the summer) than active areas of the building. Because of the size of the building and the length of piping runs, a single-pipe system is inadvisable. We have, therefore, decided to use a two-pipe reverse-return system. Zone 3, the work area, also has its own work schedule and temperature requirements. Here too because of the length of piping runs, we will use a two-pipe reverse-return piping system. The location of the thermostats for all three systems is shown on Figure 3.53.

Steps 3 and 4: Calculate the size of all terminal units and the water flow, based on calculated heating load and assumed water temperature drop.

The input required for this step is the list of heat losses calculated for the various areas of the building. They are:

HYDRONIC HEATING DESIGN OF AN INDUSTRIAL BUILDING / 153

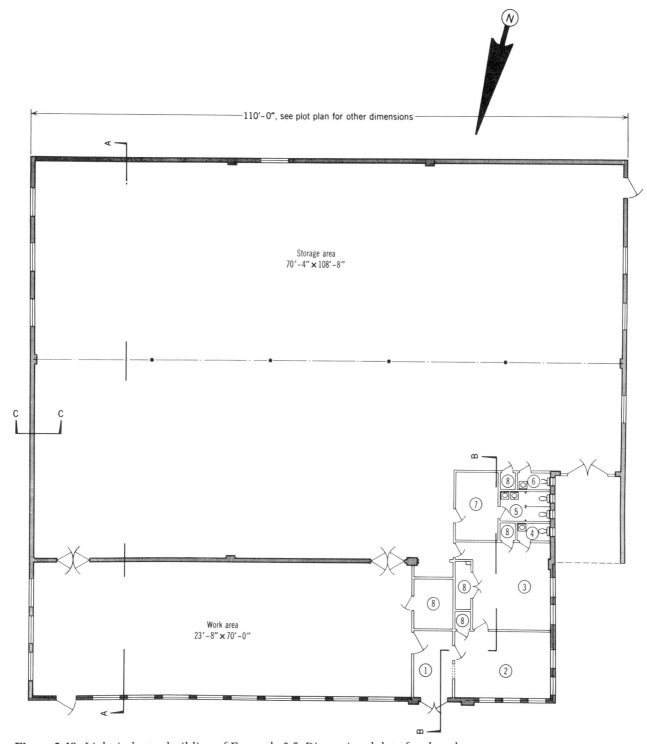

Figure 3.48 Light industry building of Example 3.5. Dimensional data for the administrative wing are:

1. Lobby— 7 ft in. × 12 ft 0 in.
2. Administrative office— 12 ft 0 in. × 18 ft 0 in.
3. Private office— 14 ft 6 in. × 15 ft 0 in.
4. Private toilet— 3 ft 0 in. × 6 ft 8 in.
5. Women's toilet— 6 ft 0 in. × 9 ft 6 in.
6. Men's toilet— 3 ft 0 in. × 6 ft 8 in.
7. Women's rest room— 8 ft 0 in. × 12 ft 0 in.

For sections A-A, B-B and C-C see Figure 3.49. For building elevations see Figure 3.50. (Courtesy of Scheiner and Swit, Architects.)

154 / HYDRONIC HEATING

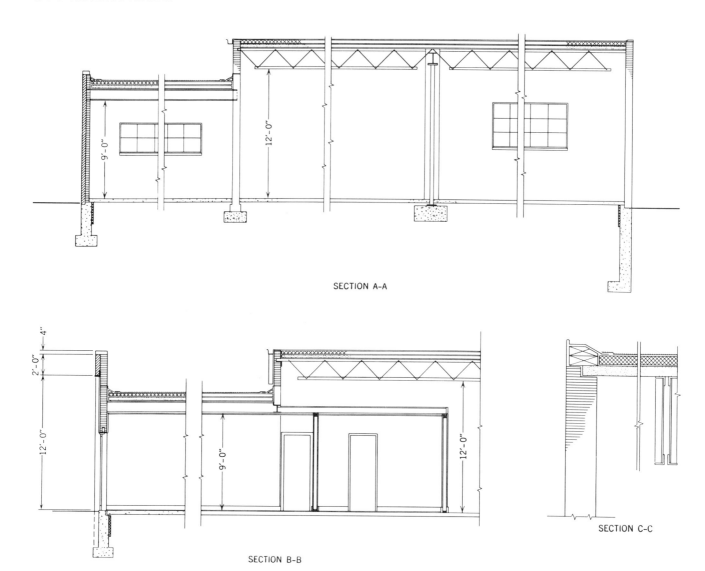

Figure 3.49 Building sections. For location of sections, see building plan in Figure 3.48. (Courtesy of Scheiner and Swit, Architects.)

HYDRONIC HEATING DESIGN OF AN INDUSTRIAL BUILDING / 155

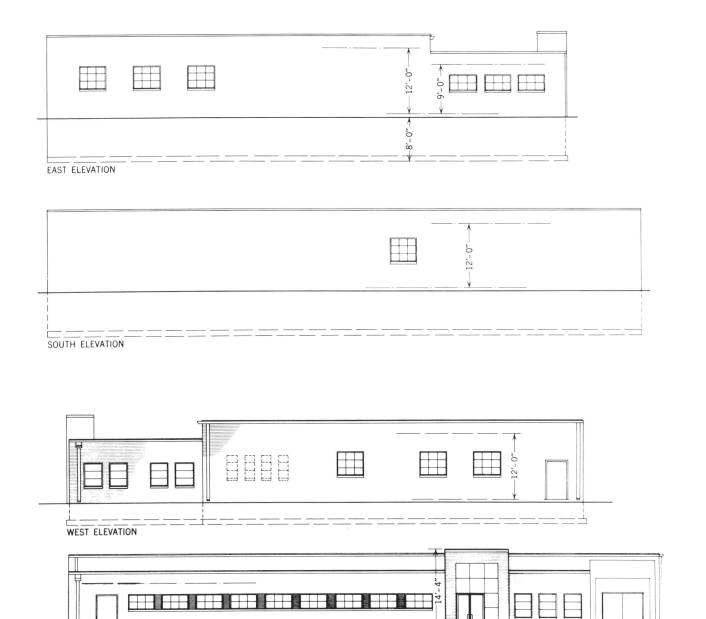

Figure 3.50 Elevations of light industry building. (Courtesy of Scheiner and Swit, Architects.)

Zone 1—Administration Wing
- Administration ..... 14.4 MBH
- Private office ..... 8.5 MBH
- Men's toilet ..... 1.5 MBH
- Women's toilet ..... 3.2 MBH
- Private toilet ..... 1.5 MBH
- Foyer ..... 4.6 MBH
- Women's rest room ..... (Negligible)
- Zone 1 subtotal ..... 33.7 MBH
- Zone 2—Storage Area ..... 196.6 MBH
- Zone 3—Work Area ..... 52.6 MBH
- Building total ..... 282.9 MBH

The number of terminal units in each space is dictated by the room layout. As stated previously, the ideal position for a heating unit is below the room's windows, if any. If none, then a convenient spot is found in the room. Refer to Figure 3.51, which shows the locations selected for the terminal units. The heating unit design selected for use in the administrative wing is illustrated in Figure 3.52, along with a listing of heat output data. For the sake of architectural consistency the same enclosure and mounting height is used in all the rooms in this wing. The variables are the lengths of the units and the number of fins per foot on the ¾-in. hot water tubing. A boiler output temperature of 200°F is selected, with a 20F° drop in water temperature (in each terminal unit and around the entire one-pipe loop). This gives an average temperature of 190°F, which is used to determine the heat output per foot of the finned tube. This is then used to determine the unit's required length.

The results are tabulated next. A typical space calculation will demonstrate the technique.

Administration: See Figure 3.51. Three units required.

$$\text{Btuh per unit} = \frac{\text{Room Heat loss}}{\text{No. of units}}$$
$$= \frac{14{,}400 \text{ Btuh}}{3} = 4800 \text{ Btuh}$$

Because the administration office is half way along the loop, the average water temperature will actually be nearer 180°F than 190°F. We will, therefore, oversize the units based on the 190°F rating. From the table in Figure 3.52(b), we select a 75HC unit. The S420 configuration gives 1410 Btuh/ft. Therefore,

$$\text{Finned-tube length required} = \frac{4800 \text{ Btuh}}{1410 \text{ Btuh/lineal ft}} = 3.4 \text{ ft}$$

We would oversize somewhat and use a 4-ft long unit for the reason given previously. Three such units would give a total actual heat output of 3 units × 4 ft × 1410 Btuh = 16920 Btuh. The unit in the lobby is oversized for the same reason as given previously.

Remembering that 1 gpm will deliver 10,000 Btuh for a 20°F temperature drop, the total supply flow in Zone 1 can easily be calculated.

$$\text{Zone 1 flow} = \frac{38{,}240 \text{ Btuh}}{10{,}000 \text{ Btuh/gpm}} = 3.8 \text{ gpm}$$

We can now similarly select the terminal units and calculate the flow for Zones 2 and 3.

Administrative Area Finned-Tube Terminal Units

| Space | Calculated Heat Loss, Btuh | Terminal Unit[a] | | | No. of Units | Actual Btuh in Space |
		Btuh/ft[b]	Length, ft	Btuh per Unit		
Admin[c]	14400	1410	4	5640	3	16920
Office	8500	1120	4	4480	2	8960
Mens' room	1500	1120	1.5	1680	1	1680
Women's room	3200	1120	3	3360	1	3360
Private	1500	1120	1.5	1680	1	1680
Lobby[c]	4600	1410	4	5640	1	5640
Total	33,700					38,240

[a] All terminal units are Dunham-Bush Valvector®, type S, single-tier finned-tube radiation, 40 fins per foot, sloping top S420 enclosure, ¾-in. tubing, 24 in. installed height. Data extracted with permission from Dunham-Bush publications. See Figure 3.52 and its accompanying data.

[b] Btuh/ft at 190°F average temperature.

[c] Fins for Administration and Lobby units are 2¾ in. × 4 in. All others are 2¾ in. × 3 in.

HYDRONIC HEATING DESIGN OF AN INDUSTRIAL BUILDING / 157

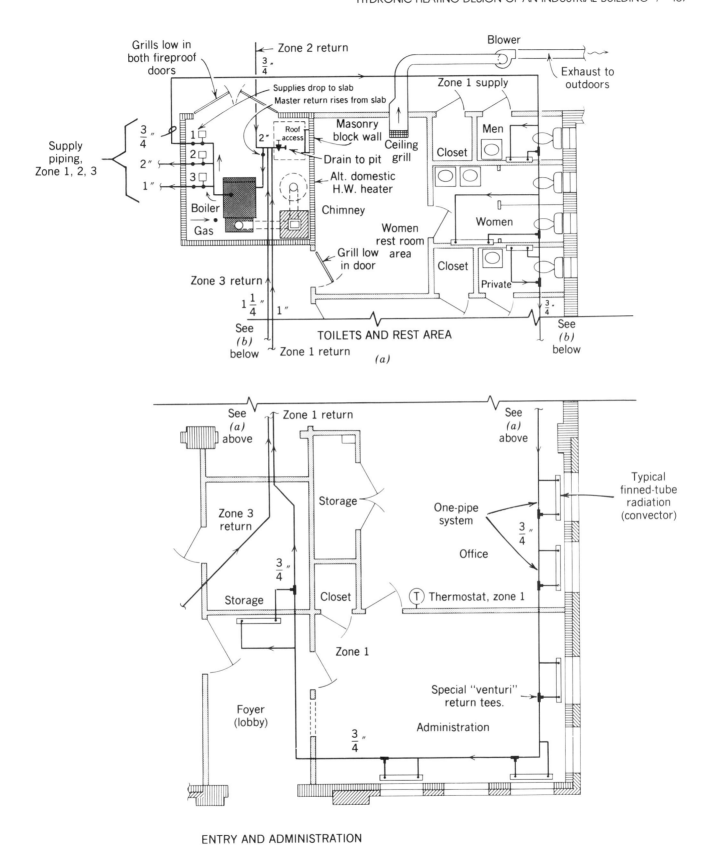

Figure 3.51 Piping in the administrative wing of the industrial building for Example 3.5. See text for a detailed explanation of the design.

158 / HYDRONIC HEATING

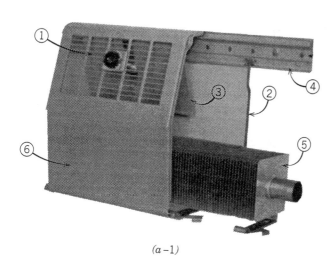
(a-1)

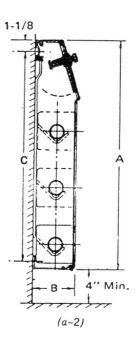
(a-2)

Dimensions

Type	A	B	C
312	12½"	3½"	10⅜"
412	12½"	4½"	10⅜"
512	12½"	5¼"	10⅜"
420	20"	4½"	17⅞"
520	20"	5¼"	17⅞"
424	24"	4½"	21⅞"
524	24"	5¼"	21⅞"

Figure 3.52 *(a)* Cutaway *(a-1)* and section *(a-2)* of typical commercial finned-tube radiation. The labelled items in the cutaway photo are

1. Damper operator handle.
2. Finned-pipe support hanger.
3. Full-length damper for control of the unit's output.
4. Wall-mounted channel support for the enclosure and finned pipe.
5. Finned-pipe heating element.
6. Metal enclosure units in this design are available in one, two or three tiers as seen in the section *(a-2)*

(Courtesy of Dunham-Bush, Inc.)

Type S/SG Commercial Finned-Tube Radiation Capacity with Copper/Aluminum Elements

Tube Dia.	Fin Style	Fin Size	Fins Per Ft.	Tiers	Encl. Type	Average Water Temperature							Installed Height	
						215	210	200	190	180	170	160		
3/4″	75HH	2¾″ × 3″	40	1	S312	1140	1080	980	890	790	700	600	16½	
				1	S412	1180	1120	1020	920	820	720	630	16½	
				1	S420	1430	1360	1230	1120	990	870	760	24	Zone 1
				1	S424	1520	1450	1310	1190	1050	930	810	28	
				2	S420	1720	1630	1480	1340	1180	1050	910	24	
				2	S424	1840	1740	1580	1430	1270	1120	970	28	
3/4″	75HC	2¾″ × 4″	40	1	S412	1500	1420	1290	1170	1030	910	790	16½	
				1	S420	1810	1720	1560	1410	1250	1100	960	24	Zone 1
				1	S424	1930	1830	1660	1500	1330	1180	1020	28	
				2	S420	2170	2060	1870	1690	1500	1320	1150	24	
				2	S424	2320	2200	2000	1810	1600	1420	1230	28	Zone 3
3/4″	75HS	4″ × 4″	40	1	S412	1670	1580	1430	1300	1150	1020	880	16½	
				1	S420	2020	1920	1740	1570	1390	1230	1070	24	
				1	S424	2150	2040	1850	1680	1480	1310	1140	28	
				2	S420	2420	2300	2080	1890	1670	1480	1280	24	
				2	S424	2650	2460	2220	2020	1780	1580	1370	28	
3/4″	75HS	4″ × 4″	48	1	S412	1770	1680	1520	1380	1220	1080	940	16½	
				1	S420	2140	2040	1840	1670	1480	1310	1140	24	
				1	S424	2280	2170	1960	1780	1570	1390	1210	28	
				2	S420	2570	2440	2210	2000	1770	1570	1360	24	
				2	S424	2750	2610	2360	2140	1890	1670	1450	28	Zone 2
1″	1HH	2¾″ × 3″	48	1	S312	1180	1120	1010	920	810	720	620	16½	
				1	S412	1190	1130	1030	930	820	730	630	16½	
				1	S420	1440	1370	1240	1130	1000	880	760	24	
				1	S424	1540	1460	1320	1200	1060	940	810	28	
				2	S420	1730	1640	1490	1350	1190	1050	920	24	
				2	S424	1850	1760	1590	1440	1270	1130	980	28	

Source. Data extracted from Dunham-Bush catalog. Reproduced with permission.

Figure 3.52 *(b)* Tabulation of a few of the configurations available in this design and their heat output at various average water temperatures. The highlighted units were selected for the Zones shown, in Example 3.5. See text. (Courtesy of Dunham-Bush, Inc.)

Zone 2—Storage Area:
The calculated heat loss is 196.6 MBH. From Figure 3.53, we see that a good layout utilizes 12 equally sized terminal units. Therefore, the rating of each unit must be at least

$$\text{Unit rating} = \frac{196.6 \text{ MBH}}{12} = 16.38 \text{ MBH} = 16{,}380 \text{ Btuh}$$

The windows in the storage area are 5 ft wide. The terminal units should, therefore, be at least that long. Since we are using a two-pipe reverse-return system, each terminal unit will receive water at 200°F. Assuming a 20 F° temperature drop, the average temperature in each convector will be 190°F throughout. For reasons that will become clear in the discussion on pipe sizing, we wish to restrict ourselves to a maximum tubing size of ¾

160 / HYDRONIC HEATING

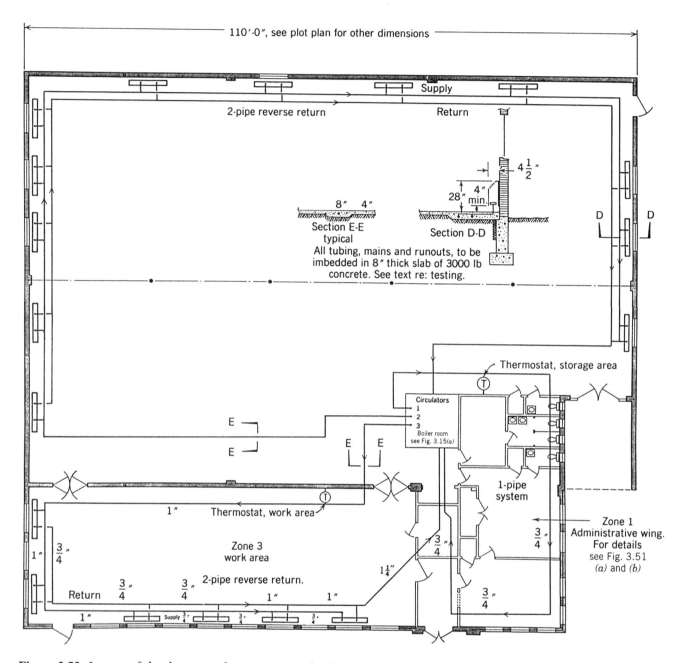

Figure 3.53 Layout of the three-zone heating system for the industrial building of Example 3.5. Enlarged details are shown in Figures 3.51 and 3.54.

in. inside the convector. Referring to the table in Figure 3.52b, we select a 75HS unit that uses two tiers of ¾-in. tubing, with 48 fins, each 4 in. × 4 in. Using a S424 enclosure with mounting height of 28 in., the output of each unit per foot is 2140 Btuh. The required unit length is, therefore,

$$\text{Heating unit length} = \frac{16{,}380 \text{ Btuh}}{2140 \text{ Btuh/lineal ft}}$$
$$= 7.7 \text{ ft, use 8 ft}$$

The actual output per unit is, therefore,

Unit output = 8 ft × 2140 Btuh/lineal ft = 17,120 Btuh

The total output of 12 units is

Total zone 2 output = 12 × 17,120 Btuh
= 205,440 Btuh = 205.4 MBH

This is 4% above the calculated heat loss. Utilizing the fact that a flow of 1 gpm delivers 10,000 Btuh (10 MBH) for a 20 F° drop, we can readily calculate the required water flow

$$\text{Flow} = \frac{205.4 \text{ MBH}}{10. \text{ MBH/gpm (for 20 F° drop)}} = 20.5 \text{ gpm}$$

Zone 3—Work Area:
  Calculated heat loss = 52.67 MBH (52,600 Btuh)
Six units are in the space. See Figure 3.53.

$$\text{Minimum rating of each unit} = \frac{52{,}600 \text{ Btuh}}{6}$$
$$= 8676 \text{ Btuh}$$

Since the windows in the work area are also 5 ft wide, we should select convection units of at least that length. Referring to Figure 3.52 we find that the 75HC unit with two tiers of ¾-in. pipe, carrying 40 fins per foot, each 2¾ in. × 4 in., in an S424 enclosure, with 28 in. mounting height, has a heat output of 1810 Btuh ft at 190°F average water temperature. We have deliberately selected this enclosure and mounting height because it matches the units selected for the storage area. The difference between them is internal—the number of fins per foot and their size. Using this unit, the required length of each unit is

$$\text{Heating unit length} = \frac{\text{Required rating}}{\text{output/ft}}$$
$$= \frac{8676 \text{ Btuh}}{1810 \text{ Btuh/ft}}$$
$$= 4.79 \text{ ft, use 5 ft}$$

The actual output per unit is, therefore,

Unit output = 5 ft × 1810 Btuh/lineal ft
= 9050 Btuh = 9.05 MBH

Six of these units have a total output of

Total Zone 3 output = 6(9.05 MBH) = 54.3 MBH

This is 1.7 MBH above the calculated 52.6 MBH required, or 3% larger than required. This is well within required engineering accuracy. The zone flow would be

$$\text{Zone 3 flow} = \frac{54{,}300 \text{ Btuh}}{10{,}000 \text{ Btuh/gpm (for 20 F° drop)}}$$
$$= 5.43 \text{ gpm}$$

Steps 5 and 6: Calculate system head, pipe sizes and water velocities.

Zone 1—Administrative Wing:
The first item in this step is to measure the length of piping runs and calculate the TEL of the one-pipe loop. Scaling the drawings (see Figure 3.51), we find the following

Developed length	124 ft
50% fitting allowance	62 ft
Friction of nine special venturi tees (at 12 ft equivalent each)	108 ft
TEL	294 ft

Following the same procedure as was demonstrated in the calculation of the Mogensen house piping, we can assume a pipe size and then check total friction head in the circuit using Figure 3.34 (or Figure 9.6, page 510). The criteria for our choice are cost and water velocity. Larger pipe means higher cost but lower water velocity and noise. As previously noted, water velocity should not exceed 6 fps. From Figure 3.34 we find that 1-in. type M copper pipe will give a friction head loss of 0.45 psi/100 ft for the flow of 3.8 gpm in Zone 1. Expressed in feet of water this is

$$0.45 \text{ psi} \times \frac{1 \text{ ft}}{0.433 \text{ psi}}$$
$$= 1.04 \text{ ft of friction head/100 ft of pipe}$$

(This same figure can be read off directly from Figure 9.6, page 510, which shows head loss in feet of water in addition to psi.)

The water velocity for this pipe size would be less than 1.5 fps. The friction head in inches of water per foot of pipe is

$$1.04 \text{ ft} \times \frac{12 \text{ in.}}{\text{ft}} = \frac{12.5 \text{ in.}}{100 \text{ ft}} = 0.125 \text{ in./ft}$$

This is far below the recommended 0.2–0.6 in./ft and indicates that ¾-in. piping can safely be used. (See rule of thumb in Step 7 of Example 3.3, page 133.) Returning to the friction charts of Figure 3.34 (or Figure 9.6), we find that for 3.8 gpm and ¾-in. pipe:

Friction head = 1.6 psi/100 ft = 3.7 ft of water/100 ft
and water velocity = 2.4 fps.

Friction head in inches of water per foot of pipe =
$$\frac{3.7 \times 12}{100} = 0.44 \text{ in./ft}$$

which is in the acceptable range of 0.2–0.6 in./ft. Given a water velocity of 2.4 fps,

Total loop friction head = 294 ft (of pipe)
$$\times \frac{3.7 \text{ ft of head (water)}}{100 \text{ ft (of pipe)}}$$
= 10.9 ft of head

We will, therefore, use ¾-in. type M copper pipe for Zone 1 main piping. Runouts to the finned-pipe radiation and the finned-pipe radiation itself are all also ¾ in. These data are shown on the drawings in Figures 3.51 and 3.53.

If separate circulators are used for each zone, a 1-in. circulator would readily supply the required 3.8 gpm at 10.9 ft of head. Some designers prefer to keep the main pipe at least one pipe size larger than the runout (branch piping) and the radiation tubing. They would, therefore, use a minimum size of 1-in. pipe for the main. In any case, the main pipe size should never be smaller than the branch piping feeding the radiation. Similarly, the branch pipes should not be smaller than the radiation tubing. They can, however, be the same size.

Zone 2—Storage Area:

We mentioned in our discussion of the heating system for the Mogensen house (Example 3.4) that pipe sizing for a two-pipe reverse-return system is complex and is not normally done by technologists. It is presented here as an advanced technique, for use by advanced-level HVAC technologists and for designers.

To make the piping layout for Zone 2 easy to understand, we have drawn it in Figure 3.54 with the terminal units "folded out." Each section of pipe in the supply and return lines is labeled by letters from A to Z. The amount of water in the supply pipe decreases every time it feeds a terminal unit. Similarly, the amount of water in each section of return pipe increases at each convection unit. Each section of pipe has a different flow and, therefore, must be treated separately.

The technique used in this design problem is to size the piping (including fitting losses) for a uniform, fixed friction head loss throughout the system, with the pipe size varying as required. (This same technique is used in sizing piping for water supply systems. It is explained in detail, for that application, in Section 9.6.) Refer now to Figure 3.54. Beginning at point A (at the boiler), the pipe size is largest. As it passes units 1, 2, 3, 4 and so on, the flow drops off, and the pipe size can be reduced to maintain the same friction head loss. The same is true in reverse for the return pipe. Starting at terminal unit 1 flow is at a minimum, and the pipe size can be small. As we proceed around the loop, return water is picked up at each terminal unit. The pipe size must be increased accordingly to maintain uniform friction head.

To determine the uniform friction head loss to be used throughout, we require the TEL of the system. Notice that regardless of the path taken, because of the reverse return, the length of path is constant. The path A-B-N-Z is exactly the same as A-M-Y-Z. Measuring these distances on the plan we obtain a developed length of 320 ft. Adding 50% for fittings, we arrive at a system TEL of 480 ft. Since circulators rarely develop more than 15 ft of head, we would have for our system

Maximum friction head per 100 ft = $\frac{15 \text{ ft (of head)}}{4.8 (100) \text{ (ft)}}$
= 3.125 ft

In terms of pressure in pounds per square inch, this converts to

$$3.125 \text{ ft of water} \times \frac{0.433 \text{ psi}}{\text{ft}} = 1.35 \text{ psi}$$

that is, the maximum uniform friction head drop to be used in design is 1.35 psi (3.1 ft of water) per 100 ft of pipe. These head figures can then be used in the charts of Figure 3.34 or Figure 9.6 (page 510) to determine pipe sizes.

We will now apply the previously described procedure to Zone 2 of our building. Technologists working in the field will come across pipe-sizing tables for hydronic heating systems in which friction head is given in millinches per foot (of pipe). A millinch is simply one-thousandth of an inch. These tables usually list pipe sizes and MBH capacity for a given temperature drop for values of friction head in millinches, ranging from 100 (.1 in.) to 600 (0.6 in.)/ft, or 10–60 in./100 ft. In our example the maximum pressure drop already calculated is 3.1 ft/100 ft. This is equal to 37.2 in./100 ft or 372 millinches/ft of pipe. We will calculate pipe sizes in Zone 2 using Figure 3.34, for the equivalent of 350 millinches friction head, because it is a value found in all readily available tables. This will permit you to check the results using a millinch chart. Since we are using psi friction charts, we will convert all friction values to psi.

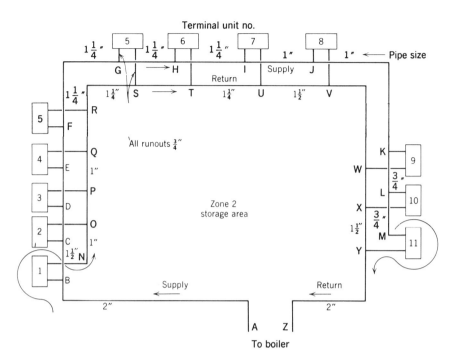

Figure 3.54 Foldout diagram of Zone 2, industrial building, Example 3.5. See text for a detailed description of the calculation technique for arriving at required pipe sizes.

$$\frac{350 \text{ millinches (head)}}{\text{ft (pipe)}} \times \frac{\text{in.}}{1000 \text{ millinches}} \times \frac{\text{ft}}{12 \text{ in.}}$$
$$= 0.02917 \frac{\text{ft of head}}{\text{foot of pipe}} \times \frac{0.433 \text{ psi}}{\text{ft}} = 0.0126 \text{ psi/ft}$$
$$= 1.26 \text{ psi friction head/100 ft of pipe}$$

This gives an overall friction head of

$$\frac{1.26 \text{ psi}}{100 \text{ ft pipe}} \times 4.8 \text{ (100 ft) TEL}$$
$$= 6.048 \text{ psi} = 13.97 \text{ ft of head, say 14 ft}$$

The pipe-sizing calculation procedure is shown in the following tabulation. The pipe (tubing) sections of both supply and return pipes are listed in the first two columns. The MBH and water flow corresponding to each section are then listed. For instance, pipe section AB carries the entire load, which is 205.4 MBH, and the entire flow, which is (at 20F° drop) 20.5 gpm. The same flow is carried by pipe section ZY, except that it is return water, not supply water. For the purpose of pipe sizing, both pipe sections are identical. Pipe section BC carries the entire load, less one terminal unit, that is

$$205.4 \text{ MBH} - 17.1 \text{ MBH} = 188.3 \text{ MBH}$$

the water flow in this section is

$$\text{Flow} = \frac{188.3 \text{ MBH}}{10 \text{ MBH/gpm (at 20 F}° \text{ drop)}} = 18.8 \text{ gpm}$$

This same flow occurs in pipe section YX and so on, until the first four columns of the tabulation are complete.

At this point we take Figure 3.34, and draw a line horizontally at 1.26 psi/100 ft (350 millinches/ft), which we calculated previously. We then select the pipe size appropriate to each value of gpm flow in column four of our tabulation and list it as shown. Maximum water velocity is about 4 fps in the 2-in. pipe, dropping to 2 fps in the ¾-in. pipe. At these low velocities, noise will not be a problem. The pipe sizes selected are shown in Figure 3.54.

Notice that the smallest mains are ¾ in. We did not use radiation piping larger than ¾ in., since as stated previously, branch piping should not be larger than mains piping or smaller than radiation piping. Since the minimum size radiation tubing is ¾ in., the branch piping (runouts) must be at least

Calculation of Zone 2 Pipe Sizes, Example 3.5

Tubing Section		MBH	20 F° Drop Flow, gpm	Pipe Size, in.
Supply	Return			
AB	ZY	205.4	20.5	2
BC	YX	188.3	18.8	1½
CD	XW	171.2	17.1	1½
DE	WV	154.0	15.4	1½
EF	VU	136.9	13.7	1½
FG	UT	119.8	12.0	1¼
GH	TS	102.7	10.3	1¼
HI	SR	85.6	8.6	1¼
IJ	RQ	68.4	6.8	1
JK	QP	51.3	5.1	1
KL	PO	34.2	3.4	¾
LM	ON	17.1	1.7	¾

¾ in., and, therefore, the mains must not be less than ¾ in. as well.

One final point: We strongly advise against using pipe sizing tables and other calculation shortcuts until you thoroughly understand the calculation process. Only then will you be able to use such tables effectively and to spot the errors that frequently occur.

The same procedure as just detailed must be followed to calculate the pipe sizes in Zone 3. This will be left for you to do as an exercise. For this purpose, we have measured the piping in Zone 3 and found the developed length of pipe to be 190 ft. This would make the TEL 285 ft, based on the assumption that 50% additional footage will account for the friction caused by fittings. This is, of course, an estimate. If a designer feels that it is either too high or too low, he or she can calculate the actual fitting friction using Table 3.15. A TEL of 285 ft at a friction head loss of 1.26 psi per 100 ft gives a total head of 3.6 psi, or 8.3 ft. of water.

Step 7: Select pump(s).

Here again, as with the Mogensen house, the choice is between the use of a single pump (circulator) and three zone valves, three circulator pumps, or two pumps, one of which serves two zones. Zone requirements are:

Zone 1   3.8 gpm @ 10.9 ft head
Zone 2   20.5 gpm @ 14.0 ft head
Zone 3   5.4 gpm @ 8.3 ft head

We have shown, for the sake of simplicity, separate circulators for each zone. The decision as to which arrangement to use rests with the project engineer.

Step 8: Select a boiler.

Here too, this task is not usually performed by a technologist because there are architectural, economic and domestic hot water considerations involved. Generally speaking, the boiler selected will be gas- or oil-fueled (we selected gas) of 330-MBH capacity or slightly larger and of a physical size to permit easy installation. The architect will provide the required chimney to provide good draft for discharge of flue gases. Ventilation and combustion air enter the boiler room through grilles in the fire doors of the room. See Figure 3.51. The boiler room itself is of fireproof construction with masonry walls and a fire-rated ceiling.

This completes the design of a hydronic heating system for the industrial building of Figures 3.46 through 3.50.

3.16 Miscellaneous Design Considerations

a. Maintenance of Equipment

Select equipment that will have a long, maintenance-free life. As an example, the radiation enclosures selected for the industrial building in Example 3.5 were wall mounted, with a sloping top. See Figure 3.52. The wall mount permits cleaning and mopping under the units to avoid dirt and trash accumulation. The sloping top prevents objects from being placed on top of the convection unit, which would interfere with the natural convective air currents. Piping is embedded in the concrete floor slab for physical protection and ease of installation.

b. Thermal Expansion of Pipes

Embedding piping in the concrete slab raises the question of thermal expansion of the piping. We have previously given pipe expansion data (see Section 3.10) and have shown techniques for compensating for pipe expansion (see Figure 8.18, page 441). Expansion fittings were used in the long pipe runs shown in Figure 3.40 in the lower level plan of the Mogensen house. The scheme there is to anchor the ends of long runs and take up expansion with a fitting in the middle. Even in short runs, there is a small amount of motion. Vertical branches through the slab pass through oversize metal sleeves. This avoids any strain on the branch or its connection to the main.

In Sections D-D and E-E of Figure 3.53, the

embedment of the tubing in 8 in. of concrete is shown. The concrete prevents any expansion of the tubing. By restraining it, a stress, uniformly distributed, is set up in the metal. It is well within the stress that the metal can take, and no damage will result. It is most important that the bond between concrete and copper be effective. Concrete with an ultimate strength of 3000 psi is used. To prevent leaks, the tubing is tested at increased water pressure before the concrete is poured. If leaks are found, they are sealed, and then the concrete is poured. This careful routine is most important to avoid future leakage of the mains and to prevent motion of the tubing.

c. Ventilation

You may have seen that we provided ventilation of the women's rest room in Figure 3.51. Ventilation is a construction code requirement.

d. Piping Symbols

For a list of piping symbols see Figures 8.41 and 8.42 (pages 490 and 491).

Key Terms

Having completed the study of this chapter, you should be familiar with the following key terms. If any appear unfamiliar or not entirely clear, you should review the section in which these terms appear. All key terms are listed in the index to assist you in locating the relevant text.

Air cushion tank
Air vent
Absolute pressure
Aquastat
Average water temperature
Balancing valve
Baseboard radiation
Booster
Branch piping
Boiler
Cabinet heaters
Cast-iron boilers
Cast-iron radiators
Circulation pump
Circulator
Commercial boilers
Commercial finned-tube radiation
Compression tank
Convector
Developed length
Diaphragm tank
Dip tube
Drain valve
Expansion fitting
Expansion tank
Fan coil units
Finned-tube assembly
Finned-tube radiation
Flow control valve
Friction head
Furnace
Gauge pressure
Hydronic heating systems

Hydronic systems
Hydronics
I = B = R
Indirect domestic hot water heater
Inlet temperature
In-line
Installed height (of radiation)
Low water temperature (LWT) boiler
MBH
Make-up water pipe
Millinch
Monoflow fitting
Mounting height (of radiation)
Multiple circuit system
Multiple loops
One-pipe system
Orifice plates
Perimeter
Pump curve
Purge valve
Radiant panels
Radiation, radiator
Residential boilers
Return branch
Runout
Series loop system
Space heater
Static head
Static pressure
Steel panel radiators
Supply branch
System head
System temperature drop

166 / HYDRONIC HEATING

Tankless coil
Terminal devices
Terminal units
Total equivalent length (TEL)
Two-pipe direct-return system
Two-pipe reverse-return system
Two-stage thermostat

Two-tiered radiation
Unit heaters
Unit ventilators
Water temperature reset control
Zone control valve
Zoning

Supplementary Reading

Stein, B., and Reynolds, J., *Mechanical and Electrical Equipment for Buildings*, 8th ed., Wiley, New York, 1992, Chapters 5–7.

ASHRAE Handbook—Fundamentals, 1993, Chapter 2.

Hydronics Institute:
Publication 200; *Installation Guide; Residential Hydronic; (Hot Water and Steam) Heating*
Publication 224; *Economizer Controls*
Publication 231; *Boiler Sizing and Replacement*
Publication 232; *Residential System Layouts*
Publication 250; *Advanced Installation Guide for Hydronic Heating Systems*

Problems

1. Calculate the static friction head of piping in The Basic House plan using 1-in. piping between baseboards. Use the approximate method of accounting for fittings.
2. Calculate the size of the electric motor required to drive the circulator of Example 3.3. The pump must deliver 4 gpm at 12 ft of head. Assume an overall efficiency of the motor-pump combination of 70%.
3. Verify the tabulation of heating loads and radiation lengths shown in Table 3.16 for Zone 1 of the Mogensen house.
4. Demonstrate that the water temperature at point B of Figure 3.44 will remain constant for flows of 1.5 and 3.0 gpm through the finned-pipe radiation group 1-2-3, instead of the design value of 2 gpm (1.998).
5. A hot water heating system has a developed length of 200 ft of 1-in. type M tubing. In this circuit there are soldered fittings, also 1 in. in size, as follows: 60 standard 90° ells, 10 standard 45° ells and 4 gate valves. Calculate the total equivalent length of the system.
6. The total equivalent length of a water circuit is 300 ft. The total pressure to be lost in friction is 9 psi. The flow is 20 gpm. Select a type M copper tube size.
7. For the Mogensen house of Example 3.4, calculate the water flow in each terminal unit and the water temperatures around the loops for Zones 2 and 3. Prepare sketches similar to Figure 3.44 for each zone.
8. For Zone 3 of Example 3.5, determine all the required pipe sizes. Developed circuit length is 190 ft. Use a TEL of 285 ft. Prepare a foldout sketch similar to that of Figure 3.54 with a tabulation similar to that in the text for Zone 2. Use a uniform friction head of 350 millinches (1.26 psi). (*Hint:* Follow the procedure explained in the text for Zone 2.)

4. Electric Heating

There are two principal methods of electrical heating—resistance heating and heat pumps. There are also two basic arrangements of systems—centralized and distributed. *Centralized systems* use electric boilers or furnaces with water or air to carry the heat around the building. Such systems are essentially no different than fossil fuel installations, except for the fuel used in the boiler or furnace. *Distributed* or *decentralized systems* produce the heat at the point of use. They are independent of other units or of any central heat source or heat distribution system. In this chapter, we will deal primarily with decentralized resistance units such as electric baseboards, wall-mounted convectors and the like. Heat pumps, which are actually electrical refrigeration units run in reverse, are discussed at length in Chapter 6.

Study of this chapter will enable you to:

1. Understand the advantages and disadvantages of electric heating as compared to fossil fuel use.
2. Calculate the effect of overvoltage and undervoltage on the output of resistance heaters.
3. Identify and describe the construction and function of all the major types of electric resistance heaters.
4. Understand and distinguish between convectors, radiant heaters and infrared heaters.
5. Understand the functioning of high temperature safety devices on heaters.
6. Understand the differences in application between natural convection and forced-air heaters.
7. Select resistance cables for use in a radiant floor or ceiling to provide a specific wattage density.
8. Size unit heaters to supply a calculated heating load in a nonresidential building.
9. Select the appropriate electric resistance heater for a specific function and space.

10. Lay out an electric resistance heating system, including controls, for both residential and nonresidential buildings.

4.1 Electric Resistance Heating

Decentralized electric resistance heating systems have a number of distinct advantages over fossil fuel systems. Installation costs are low because there is no need for a boiler or furnace, no requirement for fuel storage and no expensive piping or bulky ductwork. The absence of a boiler or furnace also saves space, making electric heating particularly convenient for small slab-on-grade buildings. The absence of combustion equipment is also environmentally desirable since there are no flue gases and no soot generated at the user's building. (These problems are efficiently handled at the electric company's power plant.) This means that electric resistance heating is clean and quiet.

In addition, electric heating has very definite operating advantages. The system is almost completely maintenance-free since there are no moving parts or piping. Point-of-use control is simple and efficient. Individual controls of each heating unit or in each room are easily arranged and afford great flexibility. Energy conservation can be achieved by the fact that heat may be turned down or off in unused rooms. When turned on again, the response is rapid, and room comfort is speedily restored. Comfort is served because the temperature in each room can be adjusted to the pleasure of the occupant by the use of the room controls.

With all these advantages, why isn't electric heating used more widely? There are a number of very good reasons, the most important of which is operating cost. Electricity in many parts of the country is more expensive (sometimes much more expensive) in dollars per Btu than oil or gas. The decision then as to whether to use electric heat requires an economic study. That study, in turn, depends on many factors. These include whether the building is speculative construction or owner occupied and how the life-cycle costing is calculated. (Cost studies are not an area that normally concerns a technologist.) Another reason for not using direct-resistance heating is that most construction today also includes a cooling system, either hydronic or ducted air. The thought then is that if piping or ductwork is going to be installed in any case for cooling, why not use it for heating as well? This has led, in some air systems, to the use of electric furnaces to supply warm air heating and electric refrigeration for cooling, using the same ductwork. Alternatively, an electric heat pump can supply both heating and cooling via the ductwork. Finally, new high-efficiency oil and gas furnaces have tilted the fuel cost equation even more in favor of fossil fuels. This, too, is part of the required economic study.

Overall, electric systems are gaining in popularity, particularly where ease of submetering is important. That being so, the architectural and HVAC technologist must be familiar with the equipment used in electric heating and the methods of application and use. That is the subject of this chapter.

4.2 Resistance Heaters

Electric resistance heaters all operate on the same principle. Current passes through a wire with high electrical resistance, generating heat in the process. (A review of Sections 11.4–11.7 and 11.15 would be helpful at this point.) The amount of heat generated depends on the wire resistance and the voltage. In the United States, heaters rated up to 1650 w are usually operated at a line voltage of 120 v. Larger heaters use line voltage of 208, 220, or 277 v depending on the electrical system. A heater rated for 277 v can be used at lower voltages, but it will produce less heat. An example should make this clear.

Example 4.1

(a) What is the electrical resistance of a cabinet heater that is rated to produce 6000 w (6 kw) of heat at 277 v.
(b) How much heat will the same heater produce if operated at 220 v?
(c) How many Btuh is produced in each case?

Solution:

(a) From Section 11.15, we know that power in an a-c circuit is

   Power = Voltage × Current × Power factor

   Since the power factor in a resistive circuit is 1.0, this equation becomes

   power (P) = voltage (V) × current (I) or
   $P = VI$

   Using the data given in this example

$$6000 \text{ w} = 277 \text{ v} \times I$$

or

$$I = \frac{6000 \text{ w}}{277 \text{ v}} = 21.67 \text{ amp}$$

Remembering Ohm's Law that

$$V = IR \quad \text{or} \quad R = \frac{V}{I}$$

we can calculate the unit's electrical resistance as

$$R = \frac{V}{I} = \frac{277 \text{ v}}{21.67 \text{ amp}} = 12.79 \text{ ohms}$$

Actually, we could have done this calculation in a single step:

$$P = IV = \left(\frac{V}{R}\right) \times V = \frac{V^2}{R}$$

or

$$R = \frac{V^2}{P}$$

Substituting the data given,

$$R = \frac{V^2}{P} = \frac{(277 \text{ v})^2}{6000 \text{ w}} = 12.79 \text{ ohms}$$

(b) The heat produced at 220 v would be

$$P = \frac{V^2}{R} = \frac{(220 \text{ v})^2}{12.79 \text{ ohms}} = 3785 \text{ w}$$

Notice that using a 277-v heater at 220 v reduces its output by 37%! As you can see, it is extremely important to use heaters at their rated voltage.

(c) The heat produced by this heater in the two instances would be

1. 6000 w × 3.412 Btuh/w = 20472 Btuh
    = 20.472 MBH
2. 3785 w × 3.412 Btuh/w = 12914 Btuh
    = 12.914 MBH

It is particularly important not to use electric heaters at voltages above their rated voltage.

Example 4.2. How much heat would be produced by a heater rated 6000 w at 220 v, if it were connected to 277 v?

Solution:

$$P = \frac{V^2}{R}$$

$$R = \frac{V^2}{P} = \frac{(220 \text{ v})^2}{6000 \text{ w}} = 8.066 \text{ ohms}$$

At 277 v,

$$P = \frac{V^2}{R} = \frac{(277 \text{ v})^2}{8.066 \text{ ohms}} = 9512 \text{ w}$$

This is almost 60% above the heater's rating, and it would probably burn out the heater if the heater's high temperature cutout did not disconnect it.

All electric heaters are rated for a specific voltage or a small range of voltages such as 110/120 v or 220/240 v. They should never be used for voltages outside the rating range. Another important fact that emerges from what we have just learned about resistance heaters is that they are built with a specific resistance. Therefore, electric baseboard, for instance, unlike hot water baseboard, comes with a specific wattage rating for a specific length and voltage. If a different Btuh (wattage) rating is required, an entirely different unit must be specified. This is true for all electric heating units. As we will see, it is particularly important when using electric resistance wire in ceilings, walls and floors.

4.3 Resistance Heating Equipment

The principal types of fixed (nonportable) resistance heating equipment in use today are

- Baseboard convector.
- Wall-mounted heater, gravity or forced air.
- Floor (recessed) convector.
- Radiant panels, wall or ceiling type.
- Embedded resistance cable, ceiling or floor.
- Unit heaters and unit ventilators.

Duct insert heaters are a special case because they are distributed around the structure but depend on a central air supply. These units are placed at the air outlet of a duct in a space and are usually locally controlled. The unheated central air supply is heated by the electric resistance coils in the duct heater as the air enters the room. Most texts classify this type of system as a centralized heating system.

Before proceeding with a detailed description of these heaters, a word about classification is in order. Some heaters are called convectors, others are called radiant heaters and still others are called infrared heaters. The last, *infrared heaters*, are more properly referred to as radiant heaters because most of the energy output is radiated. All heaters without fans or blowers are both radiant and convective. That is, they furnish heat by establishing convective air currents and by radiating heat in the infrared range. If a unit gives most of its heat to convective air currents, it is properly referred to as a *convector*. On the other hand, if most of its heat output is radiant, then it is properly referred to as a *radiant heat source*. The construction difference between the two types of heaters is mainly one of power density and, therefore, the temperature of the heating element. Generally speaking, the higher the element temperature, the more heat is radiated as compared to the convective component. Physical design of a heater also affects its radiating and convecting ability.

A heater intended for convection must be constructed in such a fashion that an air current can enter the unit, pass over the hot element and leave without obstruction. On the other hand, a heater intended for radiation must have its hot element exposed; it must "see" the area in front of it in order to radiate. In some cases, the same unit will act differently, depending on the installation location. A radiant floor has a considerable convective component because warm air rises. A similarly constructed radiant ceiling has almost no convective component because the air heated by contact with the ceiling cannot fall since it is lighter than the cold air below it. Therefore, a warm ceiling at the same temperature as a warm floor is almost completely radiant, whereas the floor is radiant and convective. Radiant wall panels also set up substantial convective air currents and are, therefore, radiant/convective rather than only radiant.

a. Baseboard Convectors

Baseboard convectors are similar in construction and function to the hydronic units described in Chapter 3. See Figure 4.1. The difference is that instead of a finned hot water pipe as the heating element, electrical units use one of several designs of electric resistance heating elements. The design of the heater element governs its operating temperature and, therefore, the division between radiant and convective heat transfer. The enclosure is metallic, usually 10–12 in. high and 2–4 in. deep. Dimensions vary among manufacturers. Units are manufactured in specific lengths for specific wattages. They contain a wiring channel so that they can be installed in continuous runs. Space is also provided for junction wiring and, in some models,

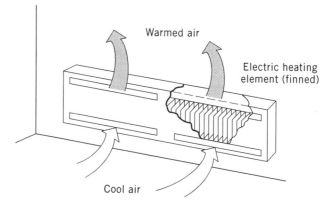

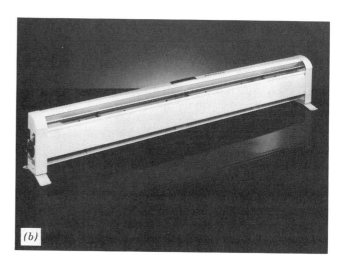

Figure 4.1 (a) Typical electric baseboard heater, also referred to as an electric baseboard convector since it operates by natural convection. Cool air enters the unit at the bottom, is heated by the electric element, rises and exits the top of the unit. It should be installed so that the convective air flow is not blocked by any permanent construction or by furniture. Heating units for wall mounting, recessed or surface, are made with elements of incandescent bare wire, low temperature bare wire or sheathed elements. An inner liner or reflector is usually placed between elements so that part of the heat is distributed by convection and part by radiation. Small units with ratings up to 1650 w operate at 120 v. Higher wattage units are made for 208 or higher voltages and requires heavy-duty receptacles or fixed wiring. Ratings range from 300 to 2000 w. (b) Units are also available in portable format. (Photo *b* courtesy of SLANT/FIN Corporation.)

for an integral line-voltage thermostat. The alternative control schemes are a remote line-voltage thermostat or low voltage control using a relay and a remote low voltage thermostat. Most units are equipped with an overheat safety cutout that will disconnect the unit if air circulation is blocked, as, for instance, by a curtain. Safety cutouts of the thermostatic type will cause the heater to cycle on and off to prevent overheating. Manual reset safety cutouts will shut off the heater entirely if it is blocked. The advantage of this type of safety cutout is that a nonoperating heater will quickly be checked and the blockage discovered and removed.

Electric baseboards, as other heat sources, are best located under windows. Additional locations that are somewhat less desirable are walls adjacent to outside doors and outside walls. Simplicity and economy of installation and control have made electric baseboard convectors the most popular type of electric heating in residential and commercial installations. An additional advantage is ease of relocation and electrical rearrangement that is often necessary when a space is refurbished. Baseboard units, as with all natural convection units, have no moving parts. They are, therefore, essentially maintenance-free and should have a minimum life of 20–30 years.

b. Wall-Mounted Heaters

Wall-mounted heaters fall into three categories

- Gravity flow convectors (Figure 4.2).
- Forced air heaters (Figure 4.3)
- Radiant panel heaters (Figure 4.5).

Cabinet convectors operate on the same principle as the baseboard convectors described previously and differ primarily in shape, mounting and design of the heating element. Recessed units are only semirecessed to allow a vertical path for incoming cool air at the bottom and outgoing warm air at the top. The heating element is designed for a large surface area so that air passing through the unit is heated to the maximum extent. This establishes the convective air currents that can, if the system is properly designed, maintain an even floor-to-ceiling temperature throughout the space. Since the air currents are natural rather than forced, occupants are not aware of them, although the velocity is sufficient to avoid the uncomfortable feeling of stagnant room air. The same is true of a good convective baseboard installation.

Like baseboards, they are completely silent in operation and essentially maintenance-free. Wall convectors are equipped with high temperature cutouts and, in the smaller sizes, with built-in thermostats (optional). Control of the units can be by line voltage thermostat, either local or remote, or by thermostat-controlled relay. This latter arrangement is usual in large spaces with multiple convectors controlled by a single thermostat.

Forced air, wall-mount heaters are used where:

(1) Very quick temperature pickup is required.
(2) The space to be heated is too large to rely on natural convective air currents; that is, the extended throw of a forced air unit is required to heat the entire floor area adequately. Maximum effective throw for a wall heater is about 12–15 ft.
(3) The space to be heated is architecturally unsuitable for the use of natural convection heaters. One such limitation is space. For the same wattage, a natural convection unit is at least twice as wide as a fan-type forced air unit.
(4) High wattage units are required. Fan-type units are available in larger heating capacities than natural convection units.
(5) The space to be heated has air currents that must be overcome by the heating units. This is frequently the case in industrial areas with much movement of material and personnel. It is also true of spaces containing operating machinery.

Forced air heaters are designed with concentrated high temperature heating elements capable of raising the temperature of fast moving air by the required amount. Failure of the fan would result in a very rapid overheating and burnout of the heating element. For this reason all forced air wall heaters contain high temperature cutouts. Although the fans are a noise source, modern units are designed for quiet operation and are usable in all but quiet areas such as libraries, music rooms and the like. Because of the fan, maintenance is higher than with natural convection heaters. However, here also, well built units can be expected to give years of trouble-free service. Switching is generally by relay, controlled by a remote thermostat.

In spaces where neither a natural convection unit nor a forced-air unit will provide satisfactory heating, radiant sources are frequently used. These are discussed in Section 4.3.d.

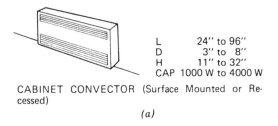

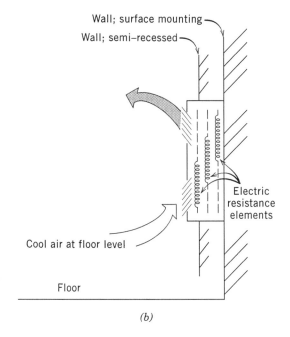

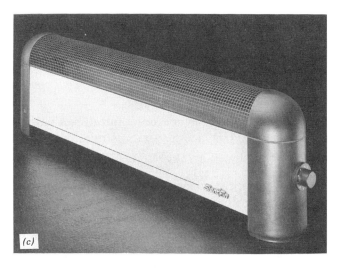

Figure 4.2 *(a)* Typical natural convection electric resistance-type, wall-mounted cabinet heater. These units are most often used in decentralized (distributed) electric heating systems. A decentralized electric system applies heating units to individual rooms or spaces. Often the rooms are combined into zones with automatic temperature controls. *(b)* Units can be surface mounted or semi-recessed. *(c)* Convector heaters are available in portable format for easy transfer between spaces. (Drawing *a* from *Architectural Graphic Standards*, 8th ed., 1988, reprinted by permission of John Wiley & Sons. Photo *c* courtesy of SLANT/FIN Corp.)

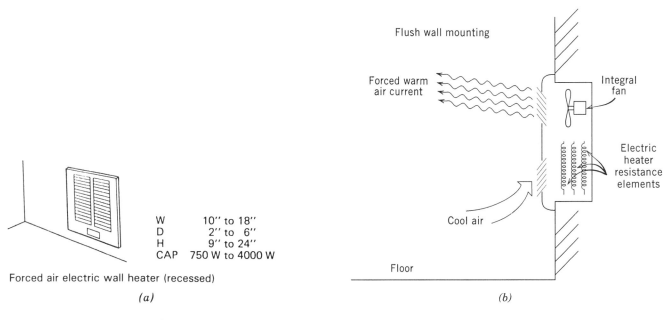

Figure 4.3 Forced-air wall heaters are used where recessing into the wall is required and where the throw provided by the heater fan is necessary. *(a)* Typical dimensional data. *(b)* Section through a unit showing construction. (Drawing *a* from *Architectural Graphic Standards*, 8th ed., 1988, reprinted by permission of John Wiley & Sons.)

c. Recessed Floor Convectors

Recessed floor convectors are used where baseboards are not applicable, such as at windows or glass panels that extend to the floor. See Figure 4.4. See also Figure 3.16 *(b)*. The heating depends on establishing a natural convective air current. Cool air drops from the adjacent glass surface into the rear of the floor convector box and rises from the front. As with other quality convector units, a high temperature cutout is provided to prevent the heating element from overheating when the intake is blocked. Control is always remote.

d. Radiant Heaters

As previously explained, every heater transmits its heat to the space where it is installed, partly by convection and partly by radiation. Heaters designed specifically to give up a large percentage of their heat by radiation are referred to as radiant heaters or infrared heaters. The term *infrared* is actually misleading since all heat is, by definition, radiation in the infrared region of the wave spectrum. In HVAC work, the term *infrared heater* is usually used to describe high temperature sources such as quartz tubes, IR lamps and high temperature electric resistance coils. All these sources operates at so high a temperature that they glow red, giving some light in addition to heat. They are focusable sources and are, therefore, particularly useful in open spaces such as truck docks and loading platforms. In such areas, it is impossible to keep the surrounding air warm because of constant exposure to currents of outside temperature air. High-capacity infrared heaters are generally gas-fired rather than electrical for reasons of operating economy.

Low temperature radiant electric heaters such as the one illustrated in Figure 4.5 are used in bathrooms and other areas requiring rapid direct warming without the use of a fan. These heaters have a large exposed hot surface that radiates in the far infrared range (long wavelength). This type of infrared radiation (heat) is readily absorbed by the room surfaces and reradiated so that eventually all room surfaces are warmed. Initially, however, occupants of the room are immediately conscious of the radiant heat even though the air temperature may be quite low. Low temperature,

174 / ELECTRIC HEATING

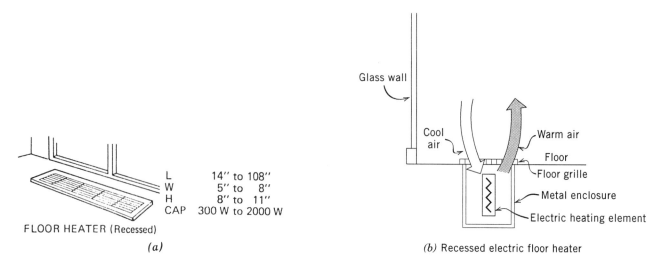

Figure 4.4 (a) Recessed electric floor convectors are used primarily at floor-length glass doors, windows or walls. The heating action depends on establishing a natural convective air current. (b) It is, therefore, important that the space between the glass and the grille (3–8 in.) and the grille itself be completely unobstructed. (Drawing a from *Architectural Graphic Standards*, 8th ed. 1988, reprinted by permission of John Wiley & Sons.)

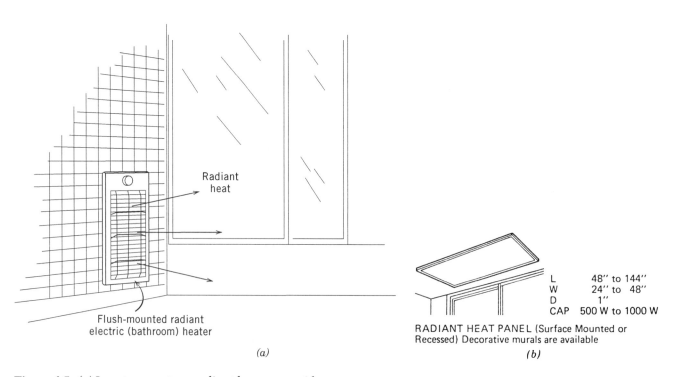

Figure 4.5 (a) Low temperature radiant heaters provide thermal comfort in a bathroom, where use of a fan is undesirable because of evaporative cooling. The long wave infrared output is readily absorbed by the room surfaces, which then increase in temperature. Comfort is achieved by a relativly high mean radiant temperature. See also Figure 4.7 b.

Figure 4.5 (b) Ceiling-mounted radiant heat panels are used where wall units are not applicable due to space limitations or blockage of the space between the heater and the room occupants (Reproduced with permission from *Architectural Graphic Standards*, 8th ed., 1988, reprinted by permission of John Wiley & Sons.)

RESISTANCE HEATING EQUIPMENT / 175

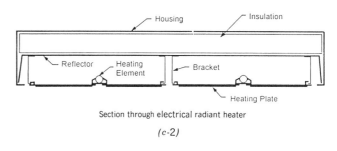

Section through electrical radiant heater

(c-2)

Figure 4.5 *(c-1)* Typical commercial radiant heating panel. *(c-2)* Section through a panel shows its construction. The heating element is a stainless steel, sheathed electric resistance heating rod. It transfers its heat to the adjoining heating plates, which then radiate infrared energy into the room. Units are available from 500 w (38 in. × 6 in. × 2 in.) to 3000 w (66 in. × 17 in. × 2 in.) in a wide range of voltages. *(d)* Typical application in a commercial showroom. (Courtesy of Energotech USA, Inc.)

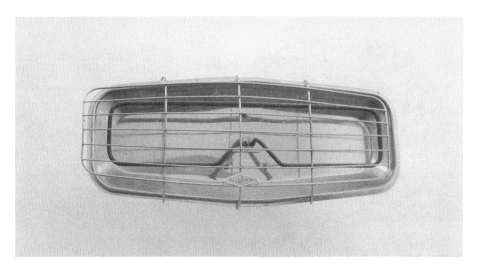

Figure 4.6 Typical bathroom-type auxiliary radiant electric heater. The nichrome wire heating element is encased in a nickel-plated stainless steel tube. Heat generated by the element is reflected by the polished metal backplate. (Photo by Stein.)

large-surface heaters of this type transmit about 60% of their output as radiant heat and 40% as convection when installed on a wall.

Similar type radiant heating panels can be installed on the room's ceiling [Figure 4.5 *(b)*] if wall installation is not desired or possible. In such installations, the radiant component of heat output increases because convection is difficult, as explained previously. With proper design, panels can also be installed under floor covering. Wall and ceiling panels have a wattage density of 50–100 w/ft^2 (170–340 Btu/ft^2), with the higher densities intended for ceiling use. Panels embedded in the ceiling concrete are limited to 33w /ft^2 by the National Electric Code (Article 424-98). Similarly, panels installed under floor coverings are limited by the NEC to 15/ft^2 (about 59 Btu/ft^2). High temperature quartz-tube bathroom heaters (see Figure 4.6) give about 80–90% of their heat in radiation and only 10–20% in convection. The action of convectors, low temperature radiation units and high temperature infrared radiation heaters is shown in Figure 4.7.

e. Embedded Resistance Cables

The **National Electric Code** (Article 424-41) sets upper limits on the wattages and minimum side-to-side spacing of embedded cables as a protection against overheating. Cables installed on plasterboard (dry-wall) ceilings or wet plaster ceilings (Figure 4.8) are limited to a maximum of 2.75 w/ lineal foot of cable, installed no closer than 1.5 in. on centers. This results in a maximum power density of 22 w or 75 Btu/ft^2. This wattage gives a cable temperature of about 150°F and a plaster surface temperature of about 120°F. When using plasterboard, a wider spacing of cables is common. This results in a ceiling temperature of about 100°F. Alternately, a cable rated 2.2 w/lineal foot is used to achieve the 100°F surface temperature. The entire embedded cable installation must meet the requirements of the **National Electrical Code**, Article 424.

Installation of embedded cables in walls is not permitted. Cables embedded in concrete or poured masonry floors can have a power density as high as 198 w/ft^2 (679 Btu/ft^2), although wattage densities that high are rarely used indoors. (The **NEC**, Article 424-44, permits cables in concrete floors to have a maximum wattage rating of 16.5 w/lineal foot and to be installed at a minimum side-to-side spacing of 1 in.) Resistance cable for embedment is made up in specific lengths for given wattages and voltages. The cable cannot be shortened or lengthened as that would change its overall resistance and wattage. In any case, embedment cable is purchased preterminated, with nonresistance leads, for the necessary electrical connections.

A comparison of ceiling and floor cable shows advantages and disadvantages for both. Ceiling cable is cheaper to install and maintain, thermal

RESISTANCE HEATING EQUIPMENT / 177

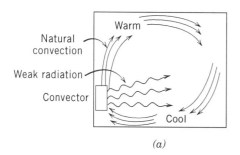

Figure 4.7 (a) Convection heaters establish natural convection air currents in a room, particularly if installed below a window. Direct radiation from these heaters is weak.

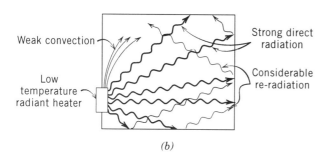

Figure 4.7 (b) Low temperature radiant heaters produce thermal comfort by strong direct radiation and considerable absorption by the room surfaces and reradiation. This raises the overall MRT for an occupant. See also Figure 4.5. Convection effects are weak.

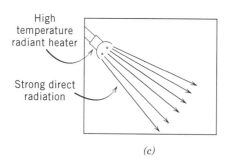

Figure 4.7 (c) High temperature radiant heaters such as quartz-tube or incandescent coil units heat almost entirely by direct radiation. See Figure 4.6. Any blockage of this radiation is immediately apparent. Reradiation and convection currents are weak.

response of a radiant ceiling is quicker than that of a floor and no rugs, carpeting and other coverings that would cause a problem exist. Floor systems have the advantage of greater thermal comfort particularly in rooms with constant cold drafts and in spaces such as playrooms, kindergartens and the like, where children sit on the floor. Also, as explained previously, floor heating causes thermal air currents to flow, which increase the feeling of comfort. Ceiling systems often cause a sensation of stagnant air due to the absence of convection. To overcome this unpleasant feeling and to provide needed fresh air, some means of continuous ventilation is required when using a ceiling system.

f. Unit Heaters and Ventilators

Electric unit heaters are similar to hydronic units (see Figure 3.18) except that the heating element is an electric resistance coil. These heaters are arranged for ceiling or wall suspension and are rated 3–60 kw. Units larger than 10 kw are generally arranged for three-phase wiring. Single-phase units usually take 240 v; three-phase units are wired for 208, 240 or 480 v. A single-phase motor-driven propeller fan provides the required air current. All units are equipped with a high temperature cutout. Control of these heaters is normally via thermostatic control of a contactor (relay). Unit heaters find their greatest application in industrial facilities. A typical unit is illustrated in Figure 4.9.

The function of a unit ventilator is to introduce outside air into a room for ventilation (and cooling) and, if necessary, to temper (heat) the incoming fresh air to room temperature. That is, the heating function is secondary to the fresh air ventilating function. These units are commonly found in schools, offices and mercantile buildings where constant occupancy requires ventilation, plus cooling or heating as necessary. When using electric

178 / ELECTRIC HEATING

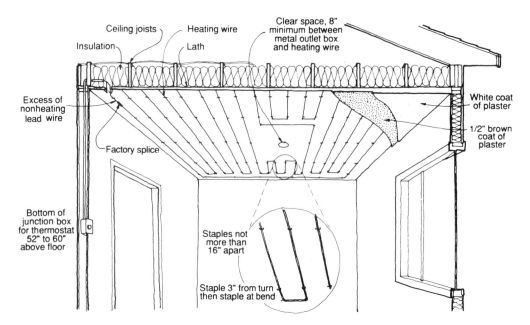

Figure 4.8 Typical detail showing electric resistance heating cable installed in a plaster ceiling. The cable is furnished in factory-assembled lengths from 75 to 1800 ft. Most cables are rated 2.75 w/lineal foot and are available for 120, 208 and 240 v. Cables are preterminated at the factory. (From *Mechanical and Electrical Equipment for Buildings*, 8th ed., 1992, reprinted by permission of John Wiley & Sons.)

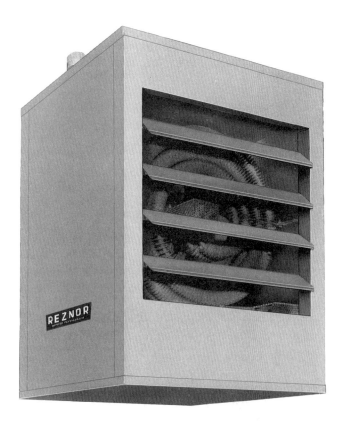

Figure 4.9 Typical electric unit heater. This unit, rated 30 kw, measures 19 in. wide, 24 in. high and 18 in. deep. It is provided with an integral high temperature cutout and a 24v coil contactor for on-off control. Its propeller fan is powered by a single-phase motor controlled by a separate time delay fan switch that prevents the fan from operating (and blowing cold air) until the heating coil is functioning. (Courtesy of REZNOR.)

unit ventilators, the cooling function is usually limited to the use of outside air. These devices are frequently quite large. They are usually installed one to a room, to supply all the ventilation air requirement of the space. This varies from 15 to 50 cfm per occupant, depending on the room use. The electric heating element is thermostatically controlled to heat the incoming air as required by the room temperature and the outside air temperature. Units are best installed under windows, like fan coil units. The outside air inlet is normally a louvered screened wall opening below the window arranged to deflect wind and resist entrance of driving rain. A schematic section through an electric unit ventilator is shown in Figure 4.10.

4.4 Residential Electric Heating Design

One way to minimize fuel cost in an electrically heated facility is to increase the thermal insulation of the structure. This can be accomplished with increased insulation thickness, use of insulating materials with better insulation properties (higher R value per inch), double or triple glazing, effective weather-stripping to reduce infiltration losses and reduction of ventilation to a minimum. All these techniques were applied in the 1970s and 1980s as a result of skyrocketing fuel prices, resulting in what came to be called superinsulated buildings. It soon become apparent however that there is an economic break-over point beyond which the cost of additional insulation and insulating techniques is higher than the fuel savings. This point varies with the type of building, outside design conditions and type of heating system. Today, electrically heated buildings are more heavily insulated than fossil fuel-heated structures, but only marginally, and that only in cold climates. In the electric heating design examples that follow, modern recommendation for insulation thicknesses are followed. For actual design, local and national insulation and energy-use standards should be consulted.

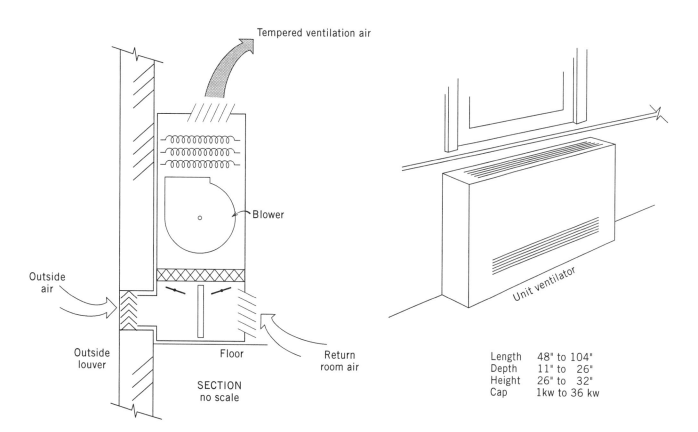

Figure 4.10 Electric unit ventilators supply tempered (warmed) outside air for ventilation of occupied areas. Heaters are thermostatically controlled by inside and outside temperature. Units all provide cooling with outside air. Typical dimensions and ratings are shown.

ASHRAE Standard 90.1 is particularly important and useful in this regard.

An undesirable side effect that resulted from supertight buildings, that is, buildings with very low infiltration, was the accumulation of undesirable odors and gases. This phenomenon is known as the Sick Building Syndrome or simply SBS. To cure SBS, ventilation standards were revised to ensure a minimum number of air changes per hour in all occupied structures. This required ventilation can be accomplished either mechanically or by deliberate infiltration.

Example 4.3. Design an electric heating system for The Basic House plan shown in Figure 3.32 (page 130). For information only, the insulation value for which heat losses were calculated are:

Ceilings and floors over crawl spaces	R33
Opaque walls	R11
Masonry basement walls	R7

All windows are double-glazed. Calculated heat losses are given in Table 4.1.

Solution: Although many different kinds of electrical heating methods could be used, this solution is confined to the use of resistance baseboard heaters and one wall heater (in the kitchen). We have chosen standard power density baseboards that have an output of 250 w/lineal foot. (See the tabulation in Table 4.1.) This is a value common to many manufacturers. The use of 240-v baseboards in preference to 120 v reduces the size of the required electrical conductors. Thermostats for control of each room are shown on Figure 4.11 at appropriate locations. The kitchen thermostat is integral with the wall heater. A summary of the design is shown in Table 4.1 and Figure 4.11. The electrical plan for the building is shown in Figure 13.27 (page 777).

Example 4.4. Design an all-electric heating system for the Mogensen house whose architectural plans are shown in Figure 3.38 (page 141).

Solution: In designing an electric heating system, the hourly heat losses are, of course, expressed in

Table 4.1 Summary of Design, Electric Heating, Basic Plan (Example 4.3)[a]

	Engineer's Calculations		Design Results			
	Heat Loss Rate		Heater No.[b]	Output		Length, in.
Space	Btuh	Watts		Btuh	Watts	
Living room	9300	2720	1	5120	1500	72
			2	5120	1500	72
Dining room	4300	1250	3	4265	1250	60
Bedroom #2	7500	2200	4	5120	1500	72
			5	2560	750	36
Bedroom #1	5200	1520	6	1707	500	30
			7	3415	1000	48
Bath	2900	850	8	3415	1000	48
Basement	4900	1430	9	2560	750	36
			10	2560	750	36
Kitchen[c]	4800	1420	11	5122	1500	H:16 W:13 D: 2
Totals	38,900	11,390	Totals	40,964	12,000	
	Required 11.4 kw			Specified 12.0 kw		

[a] Design temperatures: outside 0°F; inside 75°F.
[b] All heaters except kitchen are 240-v, aluminum finned-tube electric baseboard, 250 w/lineal foot.
[c] Kitchen heater is a 240-v forced air-type wall heater with integral thermostat. (See Figure 4.3.)

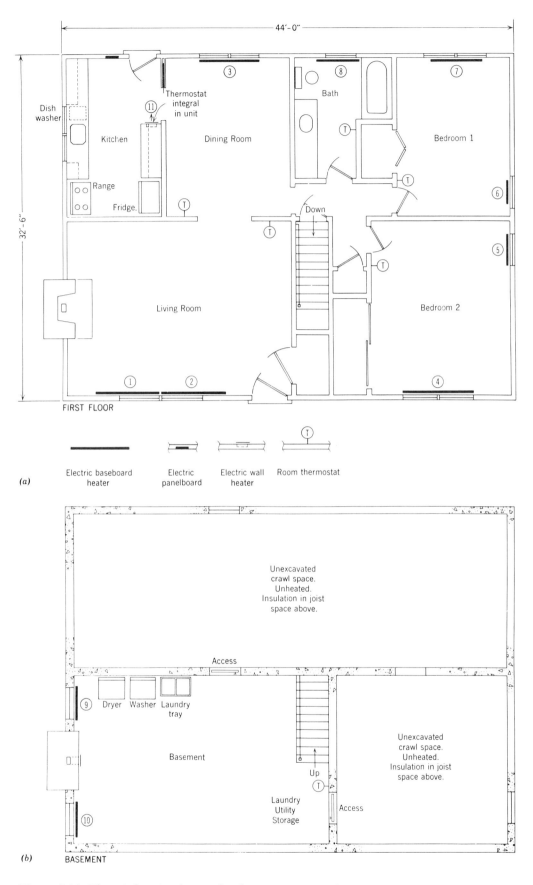

Figure 4.11 Electric heating layout for the Basic House plan, Example 4.3. For details see Table 4.1 and text.

182 / ELECTRIC HEATING

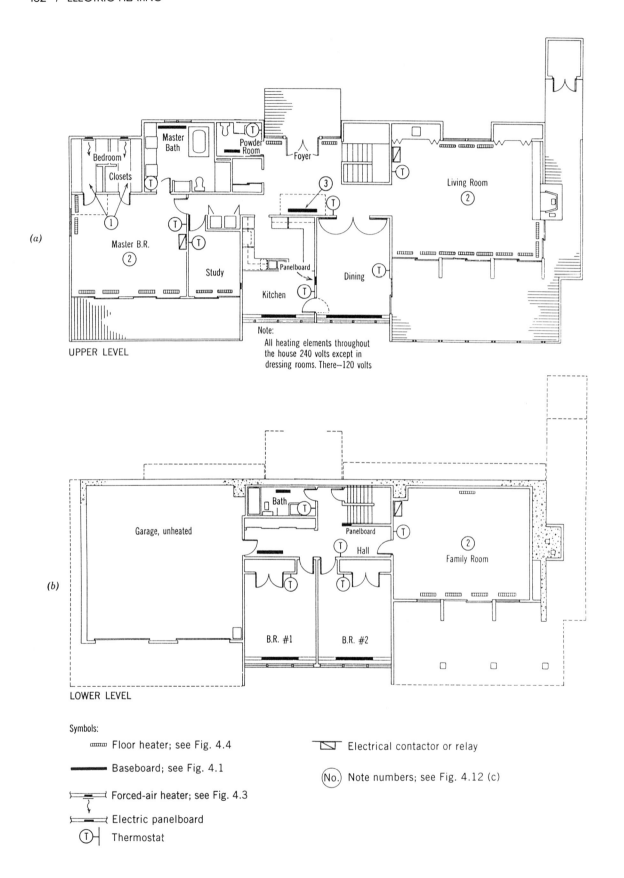

1. Minimal heat loss in the closets. Two 750-w forced air heaters for occasional use.

2. 750 watt recessed floor heaters controlled by a single thermostat and two pole contactor.

3. Foyer; two 750 watt recessed floor heaters plus one 1250 watt baseboard. Controlled via a single thermostat and two pole relay.

4. All spaces are controlled by individual thermostats, as shown. Two pole, line voltage (240 volts) thermostats that can carry up to 2000 watts load are used. Larger loads are switched by relay.

(c)

Figure 4.12 *(a, b)* Electric heating plan for upper and lower levels. See also Table 4.2 for a listing of calculated heat loss and Btuh output of selected heating units. *(c)* Notes to the electric heating layout of Example 4.4.

(1) Minimal heat loss in the closets. Two 750-w forced air heaters for occasional use.

(2) 750-w recessed floor heaters controlled by a single thermostat and two-pole contactor.

(3) Foyer; two 750-w recessed floor heaters plus one 1250-w baseboard heater. Controlled via a single thermostat and two-pole contactor.

(4) All spaces are controlled by individual thermostats, as shown. Two-pole, line voltage (240 v) thermostats that can carry up to 2000-w load are used. Larger loads are switched by contactor.

Table 4.2 Summary of Design, Electric Heating System, Mogensen House (Example 4.4)

Space	Engineer's Heat Loss Calculations, w	Output of Units Selected,[a,b] w
Living room	8400	9000
Dining room	2300	2500
Kitchen	1500	1500
Foyer	2800	2750
Powder room	400	500
Master bath	800	1000
Master bedroom	4400	4500
Study	1100	1500
Family room	3700	3750
Bedroom #1	1600	1500
Bedroom #2	1400	1500
Bath	400	500
Hall	1100	1250
	29,900 w (30 kw)	31,750 w (31.75 kw)
		Upper level 23,250
		Lower level 8500

[a] All baseboards are rated at 250 w/lineal foot.
[b] Recessed floor heaters are rated 750 w at 240 v.

watts instead of in Btuh. The standard conversion is 1 w = 3.412 Btuh. Table 4.2 lists the losses, room by room, in watts. The family room is calculated for a greater heat loss than it was for hot water heating. The reason is that there is no need for a boiler room. The family room is built as shown in the original drawings, one-third larger than in the hot water scheme of Example 3.4.

Refer to Figure 4.12. Recessed-type convectors must be used in the living room, master bedroom, family room and study. See Figure 4.4. The 750-w fan-type heaters in the master bedroom closets are for occasional use when spending more than a minute or two in the closet. All other heaters are electric baseboards with aluminum-finned grid elements, at 250 w/lineal foot. See Figure 4.1. All heaters are 240 v except the closet heaters, which are 120 v.

The use of 240 v for heaters reduces the number of circuits and in some cases the wire sizes. Control of all heaters is by a single thermostat in each space. When the total electrical heating load in a single space exceeds 2000 w a latching relay or contactor is used to connect and disconnect the loads. Below that load a line voltage thermostat with adequate current rating is satisfactory. The results of the design are shown on Figure 4.12.

4.5 Nonresidential Electric Heating Design

When compared to a residential building, industrial buildings present an entirely different set of problems to the technologist laying out the heating system. In residential buildings, appearance, convenience and (usually) first cost are the governing factors. In industrial buildings, ruggedness, low maintenance, long life and low operating costs (efficiency) are most important. Simplicity of operation is also important. We will use the industrial building that we studied in Chapter 3 as our nonresidential design problem.

Example 4.5. Design an electric heating system for the industrial building shown in Figure 3.48 (page 153).

Solution: The calculated heat losses for the various spaces follow:

Space	MBH	Watts
Work area	52.6	15,416
Storage area	196.6	57,620
Administration	14.4	4220
Private office	8.5	2491
Men's toilet	1.5	440
Women's toilet	3.2	938
Private toilet	1.5	440
Lobby	4.6	1348
Women's rest area	negligible	

We will consider the areas individually.

(a) *Work area.* Since there is no machinery or furniture layout given us, the heating units should be installed on the outside walls for two reasons:
- Most of the area's heat loss is through the windows.
- The 9 ft, 0 in. ceiling height is too low to use any ceiling-mounted heaters.

For this area we would choose commercial cabinet convectors of the type shown in Figure 4.2. Six surface-mounted units, each rated 3 kw at 277 v are used, distributed as shown on Figure 4.13. This gives a total load of 18 kw. Control is by thermostat and a three-pole contactor. The units are rated 277 v because most industrial buildings take 277/480-v service. Natural flow convectors rather than forced air fan units are chosen because:

- The area is sufficiently narrow so that natural warm air diffusion will maintain even temperatures throughout the space.
- Work positions are always adjacent to the windows to take advantage of daylight. Forced air units would be partially blocked and would almost certainly cause discomfort to people working close to the windows.

(b) *Storage area.* Here, too, as in the work area, the designer is given an open space to work with. He or she has no knowledge of the location or type of storage equipment that will be utilized. The best approach, therefore, is to place heat sources at the locations of heat losses. At the same time the design should consider the probability that tall storage racks may block air currents. We, therefore, decided on two types of heating units:

(1) Forced-air convectors below each window. We will use surface mounting because the wall construction is concrete block and recessing would require additional construction, plus insulation behind the units. The areas nearest the windows will almost certainly be an aisle so that forced air units can be used without fear of their being blocked. We selected 7 units, each rated 3 kw at 277 v. Each unit is 36 in. long, surface wall mounted, with louvering for top discharge and front inlet. See Figure 4.13 for convector locations and Figure 4.14 for convector data. This totals 21 kw, leaving another 36 kw of heating to be supplied.

(2) For the remaining heat requirement, we have chosen 3-kw ceiling-mounted unit heaters with louver diffusers that will produce an elliptical air diffusion pattern. See Figure 4.13 for unit heater layout and air patterns, and Figure 4.15 for heater data. We choose these units for several reasons. First, a considerable portion of the space's heat loss is through the roof. Therefore, in accordance with the principle of supplying heat at the point of heat loss, we should use ceiling heaters. Second, the 12-ft ceiling height is sufficient to permit use of a wide

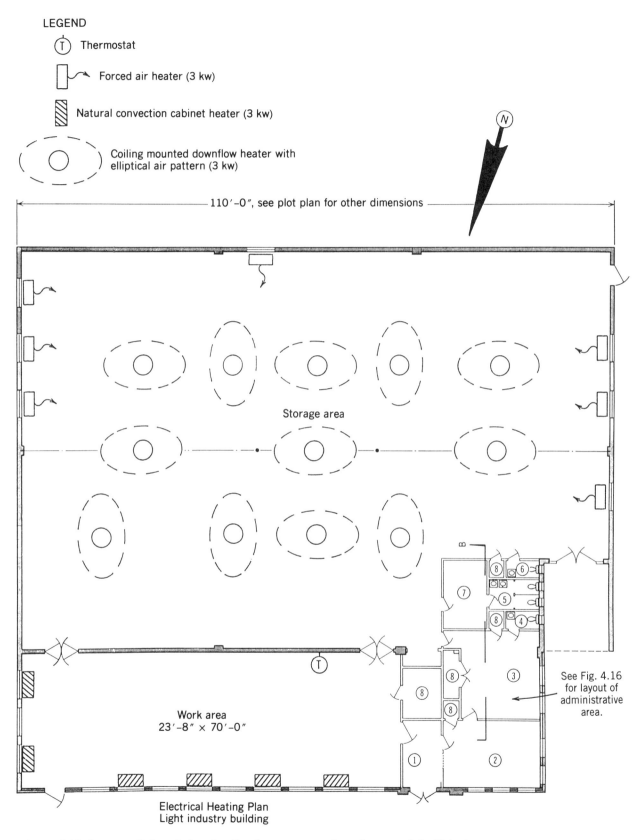

Figure 4.13 Layout of electric heating for the storage and work areas of the light industrial building of Example 4.5.

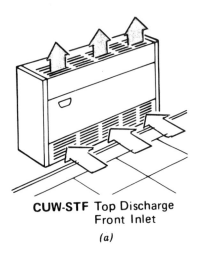

CUW-STF Top Discharge
Front Inlet

(a)

Selection Chart

Cabinet Series	Cabinet Length	Number Fintube Elements	Total Heating Capacity KW	Total Heating Capacity BTU/Hr.	Fan* Speed and Heat Setting	CFM Std. Air	Heat Output (KW)	Final Temp. (°F.)†	208V 1-Phase Amps.	208V 3-Phase Amps.	240V 1-Phase Amps.	240V 3-Phase Amps.	277V 1-Phase Amps.	480V 3 Phase 4-Wire Amps.	480V 3 Phase 3-Wire Amps	Maximum Total Integral Power Relay Holding Coil Volt Amperes Inrush	Sealed
1200		2	2.0	6,800	High / Low	250 / 200	2.0 / 1.0	84 / 75	10	9	9	8	8	4	4		
1300		3	3.0	10,500	High / Low	250 / 200	3.0 / 2.0	96 / 90	15	9	13	8	12	4	4		
1400	36"	4	4.0	13,700	High / Low	250 / 200	4.0 / 2.0	108 / 90	20	14	17	12	15	8	6	80	30
1500		5	5.0	17,100	High / Low	250 / 200	5.0 / 3.0	120 / 100	25	17	22	15	19	8	8		
1600		6	6.0	20,500	High / Low	250 / 200	6.0 / 3.0	132 / 105	30	17	26	15	22	8	8		
2200		2	4.0	13,700	High / Low	500 / 400	4.0 / 2.0	84 / 75	20	17	17	15	15	8	8		
2300		3	6.0	20,500	High / Low	500 / 400	6.0 / 4.0	96 / 90	30	17	26	15	22	8	8		
2400	47"	4	8.0	27,300	High / Low	500 / 400	8.0 / 4.0	108 / 90	39	28	34	24	30	15	12	80	30
2500		5	10.0	31,400	High / Low	500 / 400	10.0 / 6.0	120 / 100	48	34	42	30	37	15	15		
2600		6	12.0	41,000	High / Low	500 / 400	12.0 / 6.0	132 / 105	59 / 2 ckts @ 29.5	34	51 / 2 ckts @ 25.5	30	44	15	15		

(b)

Figure 4.14 (a) Cabinet-type forced-air convector, with front inlet and top discharge. (b) Manufacturer's data showing the electrical characteristics of selected convectors. (c) Dimensional data for the convectors shown in (a) and (b). Convectors are also available with cabinets designed for top inlet, bottom inlet, front discharge, bottom discharge, and semi-recessed mounting, in various combinations. (Courtesy of Q-Mark, a division of Marley Electric Heating.)

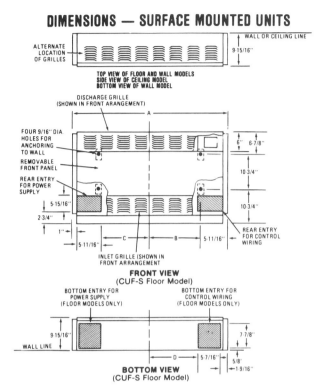

Surface Mounted

Cabinet Series	A	B	C	D
1000	36	12 5/16	11 5/16	11
2000*	47	17 13/16	16 13/16	16 1/2
3000	58	23 5/16	22 5/16	22
4000	69	28 13/16	27 13/16	27 1/2
5000	80	34 5/16	33 5/16	33

Note: (All dimensions in inches)
*Not U.L. Listed for CUI-S only.

(c)

Figure 4.14 *(Continued)*

188 / ELECTRIC HEATING

Diffuser Selection for Vertical (Blow-Down) Mounting

Cat. No.	Description	Used on	Max. M.H.	Dimension	Diffuser Pattern and Area
MLD-S		MUH-03 & MUH-05	9	25(A) 12(B)	
MLD-M	LOUVER DIFFUSER	MUH-07 & MUH-10	14	39(A) 19(B)	
MLD-M	Permits directional (straight line)	MUH-15	18	50(A) 25(B)	
MLD-M	air flow as in air curtain application	MUH-20	20	56(A) 28(B)	
MLD-L	over doorways. Rectangular	MUH-25	23	72(A) 36(B)	
MLD-L	coverage. Louvers can be turned	MUH-30	20		
MLD-L	in either direction	MUH-40	24	88(A) 44(B)	
MLD-L		MUH-50	22	80(A) 40(B)	

(a)

Selection Chart

Catalog No.	Electrical Data					Control Voltage(1)	2 Stage Element Control	Air Delivery Data			Fan Motor Data			Maximum Effective Mounting Height		Horiz. Air Throw	Wire Size	Installed Weight (lbs.) w/ bracket
	Volts	Phase	kw	Btu/hr. (000)	amps(3)			CFM(3)	FPM(3)	Δt(°F)	Volts	RPM(3)	hp	Horiz.	Vert.			
MUH03-81	208	1ø	3.0	10.2	14.5	208	N/A	350	800	27°	208	1600	1/100	8	9	12	AWG 12	27
MUH03-21	208/240	1ø	2.2/3.0	7.5/10.2	11.0/12.5	208/240	N/A	350	800	27°	208/240	1600	1/100	8	9	12	AWG 12	27
MUH03-71	277	1ø	3.0	10.2	11.0	277	N/A	350	800	27°	277	1600	1/100	8	9	12	AWG 14	27
MUH03-41	480	3ø	3.0	10.2	3.6	24	N/A	350	800	27°	480	1600	1/100	8	9	12	AWG 14	27
MUH05-81	208	1-3ø	5.0	17.0	24.0	208	5A	350	800	45°	208	1600	1/100	8	9	12	AWG 10	27
MUH05-21	208/240	1-3ø	3.7/5.0	12.6/17.0	18.0/21.0	208/240	5A	350	800	45°	208/240	1600	1/100	8	9	12	AWG 10	27
MUH05-71	277	1ø	5.0	17.0	18.0	277	N/A	350	800	45°	277	1600	1/100	8	9	12	AWG 10	27
MUH05-41	480	3ø	5.0	17.0	6.0	24	N/A	350	800	45°	480	1600	1/100	8	9	12	AWG 14	27

(b)

Dimensions

Catalog No.	Height	Width	Depth
MUH-03 & 05	16"	14"	7 1/2"
MUH-07 & 10	21 3/4"	19"	7 1/2"
MUH-15 & 20	21 3/4"	19"	12 3/4"
MUH-25 & 30	30"	26 5/8"	11 3/4"
MUH-40 & 50	30"	26 5/8"	17 1/8"

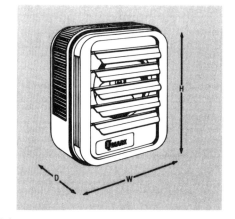

(c)

Figure 4.15 Technical data for an overhead unit heater arranged to blow downwards. (a) The basic unit heater can be equipped with different diffusers. Illustrated is the elliptical distribution diffuser used in Example 4.5. (b) Electrical data for selected unit heaters as published in the manufacturer's catalog. (c) Dimensional data. (Courtesy of Q-Mark, a division of Marley Electric Heating.)

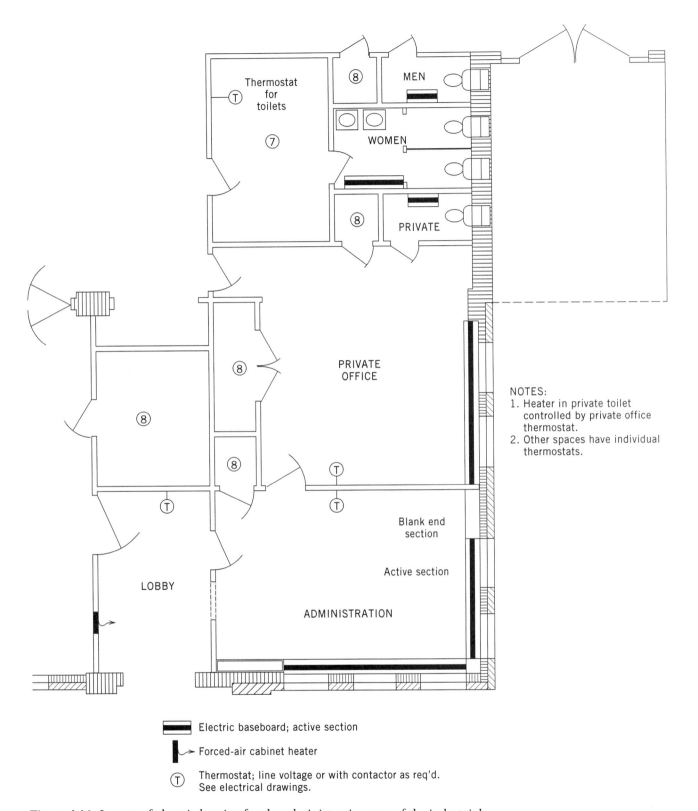

Figure 4.16 Layout of electric heating for the administrative area of the industrial building of Example 4.5.

coverage downflow air pattern. Finally, ceiling heat diffusers can cover the area effectively without our being concerned about blockage of air currents due to storage racks.

The ceiling units are spaced evenly on the ceiling but away from the exterior walls, which are effectively covered by the cabinet heaters. We use twelve 3-kw, 277-v units for a total of 36 kw. The total heating load is, therefore, 21 kw for wall convectors plus 36 kw for ceiling heaters, making a total of 57 kw. This matches the requirement almost exactly.

(c) *Administration.* See Figure 4.16. The total heat loss for this area is 4220 w. An architectural sill-height convector around the outside periphery of the room would be an effective and attractive solution to the heating problem. See Figure 4.17. The outside wall length totals 30 ft, of which 22 ft is window wall. We would, therefore, use 22 ft of active convector and two blank sections at the ends, for appearance. The wattage per foot required is

$$\text{Watts per foot} = \frac{4220 \text{ w}}{22 \text{ ft}} = 192 \text{ w/ft}$$

We, therefore, choose the convector with 188 w/ft, for 277 v. The total electrical load is

$$\text{Load} = 22 \text{ ft } (188 \text{ w/ft}) = 4136 \text{ w}$$

(d) *Private office.* The total load is 2491 w. The total window wall length up to the column is just over 10 ft. We would, therefore, use the same sill-height convector as used in the administration area, with heating elements giving 250 w/ft. See Figure 4.17. This gives a total heat load of 10 ft at 250 w/ft, or 2500 w. This is almost

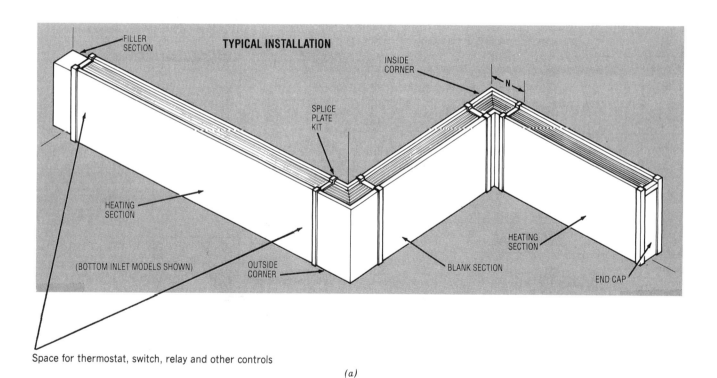

Figure 4.17 Details of architectural sill-height (baseboard-type) convector, as used in Example 4.5. *(a)* Typical installation. *(b)* Dimensional data for sections and accessories, as published in the manufacturer's catalog. *(c)* Electrical data for single-element sections. The same enclosure as in *(b)* is usable for two or three elements per unit. Electrical data for these constructions are also published by the manufacturer. *(d)* Specifications for optional built-in controls. (Courtesy of Q-Mark, a division of Marley Electric Heating.)

Accessories

Catalog No.	Use w/ Ash	Dimensions (Inches)		
		H	D	N*
Left End Caps				
ASH-14-ECL***	14	14-1/4	5	3/16
ASH-14-ECR***	14	14-1/4	5	3/16
ASH-14-FL3				0-3
ASH-14-FL6***				3-6
ASH-14-FL9***	14	12-7/8	4-3/8	6-9
ASH-14-FL12***				9-12
ASH-14-FL18***				15-18
Inside Corners				
ASH-14-IC**	14	14-1/4	5	9
Outside Corners				
ASH-14-OC**	14	14-1/4	5	4
Splice Plate Kits (Left and Right Hand Pair)				
ASH-14-SP	14	14-1/4	5	1/16
Blank Sections (Ash14)				
ASH-14-BL2**				28
ASH-14-BL3**				36
ASH-14-BL4**				48
ASH-14-BL5**	14	14-1/4	5	60
ASH-14-BL6**				72
ASH-14-BL8**				96
ASH-14-BL10**				120

* N is the additional length the accessory adds to the total installation length.

** Add suffix "-1" for bottom inlet, top outlet; add suffix "-2" for front inlet, top outlet.

*** Built-in duplex receptacle available. See page 44.

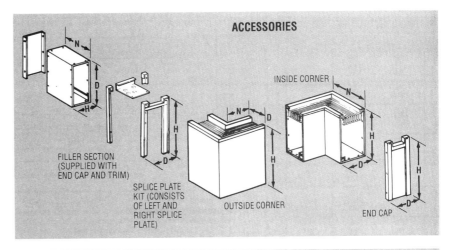

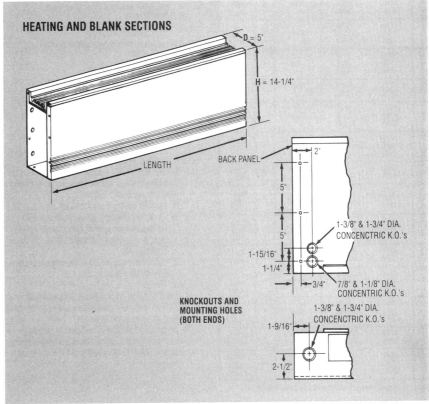

(b)

Figure 4.17 *(Continued)*

Convector Specifications

Length	Amperage				Nominal Watts/Ft.	Total Heating Capacity		No. of Elements	Catalog Number*
	120V	208V	240V	277V		Watts	BTU/Hr.		
28 in.	2.4	1.2	1.0	0.9	125	250	853	One	*{ −2125 −2188 −2250
	3.1	1.8	1.6	1.4	188	375	1280		
	4.2	2.4	2.1	1.8	250	500	1706		
3 ft.	3.1	1.8	1.6	1.4	125	375	1280	One	*{ −3125 −3188 −3250
	4.7	2.7	2.4	2.0	188	564	1925		
	6.2	3.6	3.1	2.7	250	750	2560		
4 ft.	4.2	2.4	2.1	1.8	125	500	1706	One	*{ −4125 −4188 −4250
	6.2	3.6	3.1	2.7	188	750	2560		
	8.3	4.8	4.2	3.6	250	1000	3413		
5 ft.	5.2	3.0	2.6	2.2	125	625	2133	One	*{ −5125 −5188 −5250
	7.8	4.5	3.9	3.4	188	940	3208		
	10.4	6.0	5.2	4.5	250	1250	4266		
6 ft.	6.2	3.6	3.1	2.7	125	750	2560	One	*{ −6125 −6188 −6250
	9.4	5.4	4.7	4.1	188	1125	3840		
	12.5	7.2	6.2	5.4	250	1500	5120		
8 ft.	—	4.8	4.2	3.6	125	1000	3413	One	*{ −8125 −8188 −8250
	—	7.2	6.2	5.4	188	1500	5120		
	—	9.6	8.3	7.2	250	2000	6826		
10 ft.	—	6.0	5.2	4.5	125	1250	4266	One	*{ −10125 −10188 −10250
	—	9.0	7.8	6.7	188	1875	6400		
	—	12.0	10.4	9.0	250	2500	8532		

(c)

Figure 4.17 *(Continued)*

Optional Built-in Control Specifications

Optional Built-in Control (Catalog No. Suffix)	Ratings
1-Pole Thermostat (-T)	Thermostat adjustable through grill; tamper resistant; range 60–120°F; rated 24 amps @ 120-240 VAC and 22 amps @ 277 VAC; Pilot Duty rating of 125 VA @ 24-277 VAC.
2-Pole Thermostat (-2T)	Thermostat adjustable through grill; tamper resistant; range 60–120°F; rated 24 amps @ 120-240 VAC and 22 amps @ 277 VAC; Pilot Duty rating of 125 VA @ 24-277 VAC.
2-Stage Thermostat (-2ST)	Thermostat adjustable through grill; tamper resistant; range 60–120°F; rating (per stage) 24 amps @ 120-240 VAC and 22 amps @ 277 VAC; Pilot Duty (per stage) 125 VA @ 24-277 VAC; 3°F differential between stages.
Disconnect Switch* (-DS)	Disconnect switch energized through grill; tamper resistant; double pole single throw switch rated 20 amps (per pole) @ 120-277 VAC.
Transformer Relay (-TR)	Single pole relay with 24 volt holding coil and built-in transformer; relay contacts rated 24 amp @ 120-240 VAC and 22 amps @ 277 VAC for 07 and 14 units; 22 amps @ 120-240 VAC and 19 amps @ 277 VAC for 05 units. 24 volt control.
Power Relay (-PR)	Single pole magnetic relay rated 18 amps @ 120-277 VAC; available with 24, 120, 208/240, or 277 VAC holding coil.

(continued)

Optional Built-in Control (Catalog No. Suffix)	Ratings
1-Pole Thermostat and Disconnect Switch (-TDS)	Line voltage control, both thermostat and disconnect in power circuit; thermostat adjustable through grill (range 60-120°F); disconnect switch energized through grill; control combination rated 20 amps @ 120-277 VAC.
Disconnect Switch and Transformer Relay (-DSTR)	Line voltage control (requires a remote 24V Pilot Duty thermostat); both disconnect switch and transformer relay in power circuit; disconnect switch energized through grill; control combination rated 20 amps @ 120-240 VAC and 19 amps @ 277 VAC.
Disconnect Switch and Power Relay (-DSPR)	Line voltage control; both disconnect switch and power relay in power circuit; requires a remote control voltage and thermostat for power relay (holding coil voltages available: 24, 120, 208/240, 277 VAC); disconnect switch energized through grill. Control combination rated 18 amps @ 120-277 VAC.
Pilot Duty Thermostat (-PDT)	Thermostat adjustable through grill; tamper resistant; range 60–120°F; thermostat (rated 125 VA @ 24-277 VAC) is wired for Pilot Duty operation of Power Relay (PR) or Transformer Relay (TR). See circuit amperage restrictions with -PR or -TR.
120V Duplex Receptacle (-R)	20 amp duplex receptacle built into left or right end cap or 6, 9, 12 or 18-inch filler section.

Notes: These control options are available on all models as built in components. In cases where the amperage of the heater (or heaters) exceeds the rated limit of the control, multiple controls must be specified.

(d)

Figure 4.17 *(Continued)*

precisely the calculated heating load of 2491 w.

(e) *Lobby.* The calculated load is 1348 w. Since the lobby has an outside door, cold air currents will be common. To overcome them, a 1500-w forced-air wall heater of the type shown in Figure 4.3 is recommended. It is recessed into the wall adjacent to the exterior doors. For consistency a 277-v unit would be chosen.

(f) *Toilet areas.* For these areas we would use commercial-quality baseboards, rated 500 w for the private toilet and men's toilet and 1000 w for the women's toilet. The units would be 120 or 277 v as recommended by the project's electrical designer. The units must be mounted at least 6 in. above the floor to permit the floors to be cleaned without touching the heaters.

This completes the electric heating design of the industrial building. The results are summarized in Table 4.3.

Table 4.3 Design Summary; Electric Heating of a Light Industry Building (Example 4.5, Figure 4.13)

Space	Calculated Heat Loss, w	Heater Type	Heating Load, w
Work area	15,416	Natural convection cabinet heaters	6 @ 3 kw = 18,000
Storage area	57,620	Forced-air convectors	7 @ 3 kw = 21,000
		Ceiling unit heaters	12 @ 3 kw = 36,000
Administration	4220	Sill-height convector	22 ft @ 188 w/ft = 4136
Private office	2491	Sill-height convector	10 ft @ 250 w/ft = 2500
Lobby	1348	Forced-air wall heater	1500
Toilets:			
Men's	440	Baseboard	500
Private	440	Baseboard	500
Women's	938	Baseboard	1000
Total	82.9 kw	Total	85.1 kw

Key Terms

Having completed the study of this chapter, you should be familiar with the following key terms. If any appear unfamiliar or not entirely clear, you should review the section in which these terms appear. All key terms are listed in the index to assist you in locating the relevant text.

Centralized heating systems
Decentralized heating systems
Duct insert heaters
Embedded resistance cable
Forced-air heaters
Gravity flow convectors

Infrared heaters
Overheat cutout
Radiant heaters
Radiant panels
Resistance heating
Unit ventilator

Problems

1. A cabinet heater is rated 3000 w at 277 v, single-phase. It is connected to a 240-v source. What is its heat output in watts? in Btuh?
2. A contractor purchased a factory-packaged resistance wire assembly of a certain length. It was rated 240 v and 200 w. He noted that it was too long for his needs, so he decided to cut the length in half and compensate by using half the voltage—that is 120 v. To his surprise, it did not give the rated heat output. Why not? How much heat did it give? What should he have done in these circumstances?
3. What type of electric heater would you recommend for each of the following, and why?
    a. An open truck dock.
    b. An elementary school classroom.
    c. A high school chemistry lab.
    d. A music practice room.
4. Draw a floor plan of your apartment or house. Show the electric heating equipment that you would recommend with a short explanation. Locate the control thermostats. (Sizing the equipment is not required.)
5. Why is radiant heat preferable to forced-air heating in each of the following?
    a. Open loading areas.
    b. Shower rooms.
    c. High ceiling industrial areas.
6. Rework Example 4.5, the light industry building, using different electrical heat sources than those in the text solution. Select the equipment from the data in the text or from electric heating catalogs. Give details and catalog numbers of all equipment. Explain all your choices. Draw floor plans showing the location of all equipment.

5. Air Systems, Heating and Cooling, Part I

In Chapters 3 and 4, we learned how to design and draw hydronic and electrical heating systems. Both these systems are excellent for their purpose. Not so many years ago, particularly in cold climates, heating was all that was required for a building HVAC system. Adequate ventilation was usually provided by natural infiltration in construction that was deliberately not airtight. In warmer climates, summer heat was relieved, somewhat, by the use of fans. With the advent of economically and mechanically practical refrigeration machines, the demand for interior space cooling, in addition to the usual heating, grew quickly. This led rapidly to the development of ducted air systems, which could provide cool dry air for summer comfort and humidified warm air for winter comfort. In this chapter, we will learn the principles of ducted air systems and their application to the design of warm air heating. In Chapter 6, the use of refrigeration machines to provide cooling (and heating) will be studied, using the information on ducted air distribution that we will learn in this chapter. Study of this chapter will enable you to:

1. Recognize and understand all the components of low pressure, low velocity, all-air heating systems.
2. Calculate the heat-carrying capacity of ducted airflow.
3. Be familiar with the characteristics of air pressure in ducted air flow, including measurement techniques.
4. Calculate duct air pressures including static pressure, velocity pressure and total pressure.
5. Be familiar with the major types of warm air furnaces, including components, accessories, duct arrangements and operating characteristics.
6. Understand the components and construction of duct systems, including ducts and fittings.

7. Be completely familiar with air distribution outlets, including registers, diffusers and grilles. This includes understanding the outlets' operating characteristics and how to select, locate and properly apply them.
8. Understand the various types of all-air systems in use today. This includes understanding the duct arrangements of single-zone systems.
9. Calculate air friction in duct systems using charts, tables and duct slide rule-type calculators. This includes round, rectangular and oval ducts of all materials.
10. Understand all stages of the design procedure for warm air duct systems. This includes duct size calculation by four different methods, as applicable to the duct system.

All-Air Systems

Our discussion in this chapter will be restricted to arrangements that are known as all-air systems: specifically, low pressure, low velocity, all-air systems. These systems use ducts to carry warm or cool air from the central point where it is "made" to the various spaces in a building. At these terminations, the air is distributed within the space by specially designed air outlets. This type of system is completely different from water/air systems, such as those shown schematically in Figure 3.1(b) and (c). There, the heat-carrying medium is water, which transfers its heat (or coolness) to air at the terminal point. Such a system is really hydronic. It uses a heat exchange device such as a fan coil unit to deliver the heating or cooling in the form of warm or cool air at the terminals. Water/air systems are commonly used in large buildings where the distances involved make piping more practical and economical than ductwork. There are many other considerations involved in the selection of a system type for a building. They are, however, not the responsibility of the technologist and will, therefore, not be discussed here.

5.1 Air System Characteristics

The great advantage of all-air systems is their ability to provide year-round comfort air conditioning with a single system. Originally, the term *air conditioning* referred only to cooling, and even today it is used in this sense. However, the HVAC profession tends to use the term in its broadest sense, that is, conditioning of room air to provide year-long comfort. This means control of air temperature and humidity. It also means controlled ventilation and air purification.

It might be helpful at this point to review the human comfort material in Chapter 1. Briefly, most people are comfortable with the following indoor conditions:

Winter: 68–74°F DB, 35–50% RH, maximum air velocity 40 fpm
Summer: 73–79°F DB, 25–60% RH, maximum air velocity 50 fpm
All year: mean radiant temperature (MRT) approximately the same as the air temperature

These are ideal conditions, which, as we will learn, are almost impossible to maintain uniformly throughout a space. Even in well-designed spaces, the temperature may vary vertically as much as 5°F, and from the center of the room to the walls, it varies even more. The extent of these variations depends on the design and placement of the room air outlets. This subject will be discussed in detail in the section dealing with air registers and diffusers.

The principal disadvantage of air as a heat-carrying medium is its required volume. Air weighs approximately 0.075 lb/ft^3 (slight variation with temperature). This means that 1 lb of air occupies 13.3 ft^3. Since the specific heat of air is only 0.24 Btu/ft^3/°F, it takes 13.3/0.24 or 55 ft^3/°F to carry 1 Btu of heat! This accounts for the large cross-sectional area of duct required to supply even a relatively small heating or cooling load, such as in a residential installation. (By way of comparison, 55 ft^3 of water carries 3458 Btu/°F). On the other hand, air is easily humidified (for winter requirements) and almost as easily dehumidified. It is also easily filtered, cleaned and exchanged with fresh air when required. In the United States, the overwhelming majority of residential heating and cooling installations are all-air systems, as are a large portion of commercial installations. It is, therefore, obvious that their advantages outweigh their disadvantages, from both engineering and economic points of view.

5.2 Components of All-Air Systems

Although the details of air systems vary from one installation to another, the essential components

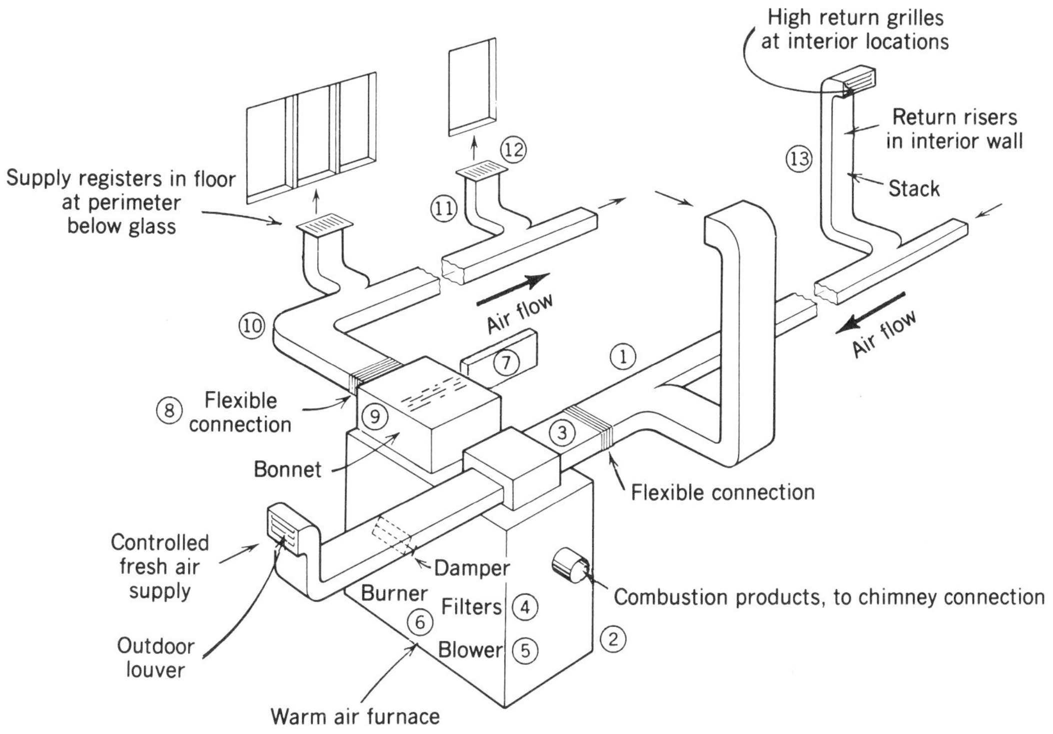

COMPONENTS OF ALL-AIR SYSTEMS / 199

LEGEND

① Return air duct
② Alternate return air location
③ Return air duct connection
④ Air filter
⑤ Blower
⑥ Burner
⑦ Humidifier
⑧ Flexible duct connection
⑨ Evaporator coil location (optional)
⑩ Main supply duct
⑪ Branch supply duct
⑫ Supply air register
⑬ Branch return duct

Figure 5.1 Typical warm air furnace installation showing components and accessories. When cooling is required, a larger bonnet, capable of containing an (A-frame) evaporator coil, is constructed. The arrangement shown has the furnace in the basement of a one-story building. Warm air ducts are arranged in a perimeter system with supply registers in the floor under windows and return grilles high on inside walls. This arrangement is suitable for year-round heating and cooling.

of all forced-air systems are shown in Figure 5.1. Follow the description of the parts with the labels on the illustration. Air from the building spaces is returned to the furnace via a system of return air ducts ①. This return air may enter at the top of the furnace as shown or at the bottom ②. The return air passes through a filter located either at the duct entrance ③ or within the furnace enclosure ④. The filter can be either mechanical or electrostatic. The air then passes through the blower ⑤, which adds static pressure and velocity to the air stream. It proceeds to the burner heat exchange mechanism ⑥, where it is heated and its temperature raised between 45 and 80°F. The air is then humidified by a humidifier ⑦ located at the bonnet or immediately thereafter, in the first supply duct section ⑧. Notice that the supply and return ducts are connected to the furnace with flexible connections (usually treated canvas). This is done to prevent the vibration noise of the furnace enclosure, which is caused by the blower, from being transmitted to the ductwork, and thence throughout the building.

If the system is to provide cooling as well, an evaporator coil ⑨ is placed in an enlarged bonnet (plenum). The air passing over this evaporator coil is cooled and then circulated throughout the building. Obviously, when the cooling system is operating, the heater is shut down, and the humidifier is disconnected. The refrigerant lines that connect the evaporator coil to the remote condensing unit are not shown, for clarity. The treated air—that is, humidified filtered heated air in winter and cool dry air in summer—enters the system of supply ducts ⑩ and is distributed throughout the building. The branch supply ducts ⑪ terminate in registers ⑫ or diffusers in the conditioned space. Air is returned through return air grilles and branch return air ducts ⑬. In this illustration, the supply registers are placed in the floor, and the return grilles high on inside walls. This is one of many possible arrangements that will be discussed in detail in the section on air outlets later in this chapter.

Notice also in Figure 5.1 that there is a controlled fresh air supply that connects into the return air system. This provides make-up air to compensate for air that is exhausted from kitchens and bathrooms. It is customary not to return air from these spaces because of odors and high humidity. The amount of make-up air is easily controlled by a damper in the intake air duct. Combustion air can be taken from the basement or from a separate combustion air intake (not shown). The latter is the preferred method, particularly in cold climates, because it avoids infiltration of cold outside air into the basement, which can appreciably increase the building heating load.

Additional components of air systems that are not shown in Figure 5.1 include air flow (volume) dampers in branch ducts and special duct fittings. In commercial systems, there are many sophisticated air temperature and air volume controls devices such as mixing boxes, variable air volume (VAV) boxes and terminals and the like. These, however, are not normally the responsibility of HVAC technologists except for showing them on the working drawings. All the components of air systems will be discussed in detail later in this chapter. First, however, an understanding of the properties of moving air is required. This, therefore, will be the subject we turn to next.

5.3 Heat-Carrying Capacity of Ducted Air Flow

As was calculated in Section 5.1, it takes about 55 ft³/°F to carry 1 Btu of heat. Since we must be able to calculate the amount of air required to provide a given heating (or cooling) load, we will now derive the equation that relates air quantity to heating (cooling) load, in Btuh. Refer to Figure 5.2. We need to know the amount of heat carried by a specific quantity flow of air in a duct. If we call the flow of air Q as measured in cubic feet per minute (cfm), we can calculate the heat it carries very simply as follows:

- Calculate the weight of air flowing (in lb/min) corresponding to this flow (in cfm), by multiplying Q by air density. This gives us flow (in lb/min). (*Note:* We have deliberately used the letter Q to represent air flow. Many texts use the letter V. However, we have found this to be very confusing to students, since V is always used for volume and velocity. The letter Q normally represents volume flow such as cfm.)
- Multiply the flow (lb/min) by the specific heat to obtain the Btu/min/°F heat flow.
- Convert the heat flow to Btuh/°F.
- Using the design temperature change in the air, calculate the heat flow in Btuh, corresponding to a flow of Q cfm.

The calculation is, therefore, as follows:

(a) We begin with an air flow in a duct of Q cfm.
(b) Multiplying by density, we have

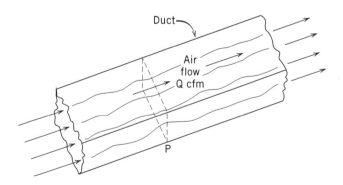

Figure 5.2 The quantity of air flowing in a duct is measured in cubic feet per minute (cfm), as it flows past a cross-sectional plane P.

$$\frac{Q\,\mathrm{ft}^3}{\min} \times \frac{0.075\ \mathrm{lb}}{\mathrm{ft}^3} = 0.075\ Q\ \mathrm{lb/min}$$

as the weight of air flowing in the duct.

(c) Multiplying by specific heat, we have

$$\frac{0.075\ Q\ \mathrm{lb}}{\min} \times \frac{0.24\ \mathrm{Btu}}{\mathrm{lb/°F}} = 0.018\ Q\ \mathrm{Btu/min/°F}$$

as the heat carried by Q cfm/°F.

(d) Multiplying by 60 min/h, we have

$$\frac{0.018\ Q\ \mathrm{Btu}}{\mathrm{min/°F}} \times \frac{60\ \mathrm{min}}{\mathrm{h}} = 1.08\ Q\ \mathrm{Btu/h/°F}$$
$$= 1.08\ Q\ \mathrm{Btuh/°F}$$

as the heat carried per hour by Q cfm, per °F. (Remember that the abbreviation Btuh means Btu per hour, and is written more accurately, mathematically, as Btu/h. We are following industry convention in the use of Btuh.)

(e) Assuming a drop in temperature Δt, the amount of heat H in Btuh, lost by Q cfm is

$$H = 1.08\ Q\ \frac{\mathrm{Btuh}}{°F} \times \Delta t\ (°F) = 1.08\ Q\ \Delta t \quad (5.1)$$

where

H is heat delivered in Btuh

Q is airflow in cubic feet per minute and

Δt is the change (drop) in temperature of the air from supply to return.

(f) Conversely, if we know the amount of heat flow required and want to know how much air is needed to supply the heat, we can use the same equation. Since $H = 1.08\ Q\ \Delta t$

$$Q = \frac{H}{1.08\ \Delta t} \quad (5.2)$$

where the terms are the same.

An example should make the use of this important equation clear.

Example 5.1 A heat loss calculation for a new residence indicates a heat loss of 84,000 Btuh for the entire house. What is the required air output of the furnace, assuming a return air temperature of 68°F and a furnace air supply temperature of 140°F.

Solution: The temperature difference between incoming and outgoing air is 140°F − 68°F = 72°F. Using Equation (5.2), we have

$$Q = \frac{84{,}000\ \mathrm{Btuh}}{1.08\ (72°F)} = 1050\ \mathrm{cfm}$$

The same equation, with one small change, can be used for cooling. Due to the higher density of cold air, the equivalent equation to Equation (5.1), which is used for cooling, is

$$H = 1.1\ Q\ \Delta t \quad (5.3)$$

or in its alternate form:

$$Q = \frac{H}{1.1\ \Delta t} \quad (5.4)$$

where all the terms are as defined previously.

Example 5.2 The house in Example 5.1 has a calculated sensible cooling load of 46,000 Btuh. Return air temperature is 79°F, and supply air temperature is 57°F. What is the blower air output requirement?

Solution: Using Equation (5.4), we have

$$Q = \frac{46{,}000\ \mathrm{Btuh}}{1.1\ (79-57)°F} = 1900\ \mathrm{cfm}$$

Notice that this is larger than the heating air requirement, because Δt for heating is larger than Δt for cooling.

5.4 Air Pressure in Ducts

Refer to Figure 5.1. The furnace blower (also called the system fan) delivers energy to the air in the system. This energy takes the form of air pressure, which causes the air to circulate through the supply and return ducts. This pressure can be expressed as the sum of two quantities: static pressure and velocity pressure (energy). Expressed in an equation, this is written

Total pressure = Static pressure + Velocity pressure

or

$$P_T = P_S + P_V \quad (5.5)$$

Static pressure, also called static head, is the pressure that the air has at rest. It is sometimes called spring pressure because it can be thought of as the pressure that pushes on the sides of the duct. See Figures 5.3 and 5.4. It can also be thought of as the potential energy of the system that is gradually converted to kinetic energy in order to keep the air moving against the system friction. Indeed, it is sometimes defined as the pressure, or static head, required to overcome the system friction. Since that is so, static pressure drops gradually and continuously as we move away from the blower along the supply duct, provided that the duct dimension does not change.

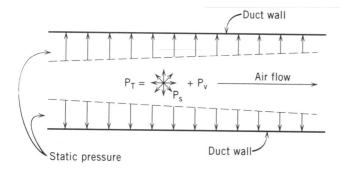

Figure 5.3 The total pressure (P_{TOT}) of air flowing in a duct is the sum of the static pressure P_S and the velocity pressure P_V. The static pressure acts like a spring and pushes against the duct walls. The energy of velocity pressure is felt only in the direction of flow and remains constant as long as the duct size does not change. However, the static pressure drops as we proceed in the direction of flow. It is "used up" by friction.

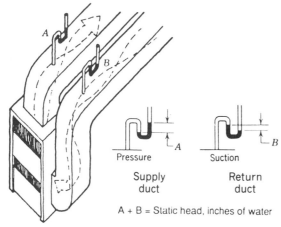

Figure 5.4 The static pressure in the supply duct is positive and can be measured by a manometer as A inches of water column. The pressure in the return duct is negative (suction) and is measured as B inches of suction.

Velocity pressure is not pressure at all; it is the kinetic energy of the moving air stream. It is converted to pressure by a process called *static regain* when the velocity of the moving air is changed. When you extend your hand through the window of a moving automobile, you feel this air pressure. The same is true when a stream of water from a hose strikes the hand. The fluid velocity in both instances drops sharply, and its kinetic energy is converted to pressure. To give you an impression of the magnitude of this pressure, a simple calculation will help. A maximum air velocity of 1000 fpm is common in residential main supply ducts. This corresponds to a speed of 11.4 mph. The velocity pressure is, therefore, quite small. It is, however, important in calculating losses in duct fittings, as we will learn later on. It is also very important in commercial installations where an air velocity of 2000 fpm (22.7 mph) and even higher is common. In residential duct work, velocity pressure is frequently ignored because it is so small. Velocity pressure (or head) can be accurately calculated from Equation (5.6)

$$P_V = (V/4005)^2 \tag{5.6}$$

where

P_V is the velocity pressure in inches water gauge and

V is the air velocity in feet per minute (fpm).

A few accurate calculations will help you to get a feel for the pressures involved.

Example 5.3 A residential duct system is designed with air velocities of 900 fpm in the main and 600 fpm in the branches. What are the velocity pressures in both?

Solution: Using Equation (5.6), we have

(a) in the main duct
$$P_V = (900/4005)^2 = 0.05 \text{ in. w.g.}$$
(b) in the branch duct
$$P_V = (600/4005)^2 = 0.022 \text{ in. w.g.}$$

Since the entire pressure available for the ductwork in a residential system rarely exceeds 0.3 in. w.g., the velocity pressure in the mains can be important in marginal designs, whereas generally it can be ignored in the branch ducts. Exact calculations, as we will learn, should consider velocity pressure. However, residential duct systems are almost always oversized to reduce noise, and air flow is regulated and balanced by dampers. Also, the loss in fittings is usually overestimated. For these reasons, most designers do not calculate velocity pressures in small or medium-size residential design.

The diagrams in Figure 5.5 should help you understand the pressure relationships of air motion in a duct. Because the pressures involved are very small, an inclined tube manometer is used in actual field work. By inclining the tube, a very small pressure can cause a large movement of the liquid

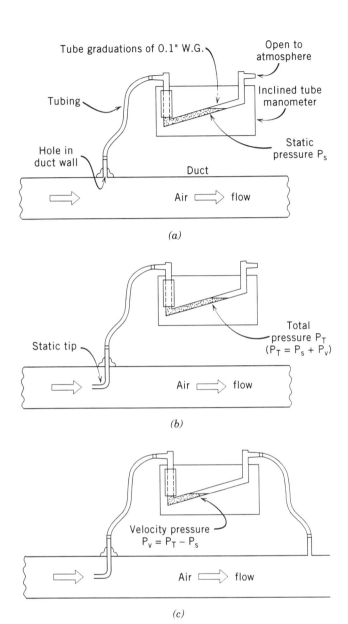

Figure 5.5 (a) Static pressure in a duct can be readily measured using an inclined tube manometer, which is also known as a draft gauge. (b) Placing a static tip into the center of the duct will measure total pressure. (c) Velocity pressure is the difference between total pressure and static pressure. It can be measured by developing a back pressure P_S acting against a total pressure P_T.

(oil) in the tube. The tube is marked (graduated) in tenths of an inch of water. Commercial manometers can be read to an accuracy of 0.03 in. w.g. (water gauge). Some units are marked as inches water column (WC), which is identical to water gauge (w.g.). The two terms are used interchangeably, with w.g. being more common.

In Figure 5.5(a) a hole in the duct is connected by a piece of flexible hose to the inclined tube manometer, which is also known as a *draft gauge*. Since the other end of the inclined tube is open to the atmosphere, the gauge will read the static (spring) pressure that is pushing on the duct walls. In Figure 5.5(b), a simple tube called a static tip is inserted in the hole, facing the air stream. It now measures the static pressure plus the velocity pressure, that is, the total pressure P_V. Imagine the face of the tube as another duct wall. As such, it measures static pressure. To this pressure P_S is added the velocity pressure P_V caused by the drop in air velocity inside the tube to zero, giving a reading of total pressure P_T.

If we now connect the other end of the gauge to another hole slightly downstream, we will develop a back-pressure of P_S pushing against a forward pressure of P_T in the gauge. The result is velocity pressure because

$$P_T = P_V + P_S \qquad (5.5)$$

and, therefore,

$$P_T - P_S = P_V$$

We will have a great deal more to say about duct pressures when we study friction losses in ducts and fittings. At this point, you should remember two facts:

1. Static pressure is required to overcome duct friction.
2. Velocity pressure is usually small and remains constant as long as the cfm and duct area do not change.

Warm Air Furnaces

The heart of a warm air heating system is either a warm air furnace or a heat pump. In this chapter, we will study furnaces; in Chapter 6 heat pumps, which supply both heating and cooling, will be studied. Warm air furnaces are used primarily in residences and small commercial and institutional buildings.

5.5 Furnace Components and Arrangements

The components of a warm air furnace, as we can see in Figure 5.1 are standard. They consist of the heating unit itself (which can be either gas, oil or electric fueled), the blower, humidifier, air filter, supply duct plenum and controls. The supply duct plenum may contain an evaporator coil if cooling is being supplied in addition to heating. In such installations, the refrigeration compressor and condenser are remotely located—usually outside the building. The arrangement of the components and the overall size and shape of the furnace depend on where the furnace is to be installed and, more particularly, where the supply and return ducts are to go. The four basic physical designs of furnaces follow.

(a) *Upflow (high-boy) units.* See Figures 5.6 and 5.7. These units are full-height. The supply ducts exit the top plenum, which has space for an evaporator cooling coil. The return duct runs overhead but drops to enter the furnace housing at the bottom, through a side panel. The blower is normally installed at the bottom of the enclosure, adjacent to the entry point of the return duct. Units of this design can be installed in a full-height basement or in a closet/utility room that can accommodate overhead supply ducts and overhead and bottom return ducts.

(b) *Downflow (counterflow) units.* See Figures 5.8 and 5.9. These units were originally designed specifically for use with perimeter heating-only systems in slab-on-grade or low crawl space houses. A typical unit is installed in a closet or utility room, directly over a small sheet metal plenum. (An enclosed crawl space should not be used as a plenum because of problems with pesticides and insects.) Ducts emerge from the sheet metal plenum to feed perimeter floor outlets or a perimeter loop. (See Figure 5.46, page 271.) The almost universal requirement for summer cooling in new construction has led to the addition of cooling coils in these downflow units. The blower, which is located above the heat exchanger, supplies conditioned air directly into the underfloor plenum. Return air enters the furnace enclosure at the top.

(c) *Low-boy units.* See Figure 5.10. These units are similar to the upflow (high-boy) units. The return duct enters at the top of the furnace enclosure. The blower is located at floor level as in the high-boy unit. The heat exchanger, however, is also at this level, to the side or in front of the blower. This requires a wider casing as shown. A typical application for this design is a low ceiling basement where a full-height unit with plenum cooling coil would exceed the available ceiling height.

(d) *Horizontal units* (also called lateral units). See Figure 5.11. Horizontal units are installed in-line between the supply and return air ducts. The blower is mounted directly behind the heat exchanger. Their low overall height permits installation in a crawl space, attic or garage or at the ceiling level. Cooling coils (if required) are normally installed in a small plenum constructed in the trunk supply duct as shown. The entire unit is designed horizontally rather than vertically to achieve the low profile necessary for cramped-space or overhead installation, as shown. The two principal problems with this type of installation are servicing space and vibration isolation for units attached to wood framing.

5.6 Furnace Energy and Efficiency Considerations

There is a bewildering array of alphabet soup performance and efficiency ratings for furnaces, heat pumps and refrigeration units. These include COP (coefficient of performance), EER (energy efficiency ratio), SEER (seasonal energy efficiency ratio), HSPF (heating season performance factor) and AFUE (annual fuel utilization efficiency). Although these ratings are not of major concern to technologists, they should be understood because of the requirements of The National Appliance Energy Conservation Act of 1987. This federal law sets out minimum efficiency requirements for all sorts of HVAC equipment, including furnaces, heat pumps and refrigeration units. COP and EER are steady-state efficiency ratings normally applied to water-source heat pumps. HSPF and SEER are seasonal efficiency ratings that are usually used for air source heat pumps.

Of all the ratings, the only one applicable to warm air furnaces is AFUE. It is applied to gas- and oil-fired heating equipment and is listed in the *Directory of Certified Furnace and Boiler Efficiency Ratings* published by GAMA (Gas Appliance Manufacturer's Association). The AFUE furnace rating was developed to take into account actual operating conditions rather than only laboratory-style tests. These conditions include flue losses and other variable factors such as on-off cycling. Cycling is related to weather patterns, design conditions selected and deliberate oversizing. The federal law, effective January 1992, mandates a minimum AFUE of 78% for warm air furnaces larger than 45 MBH and smaller than 225 MBH

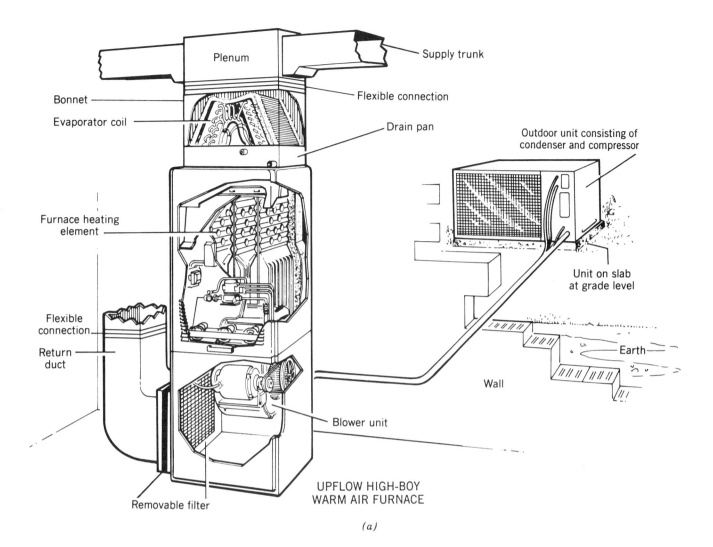

(a)

BLOWER SIZE	MOTOR H.P.	BLOWER SPEED	EXTERNAL STATIC PRESSURE—INCHES WATER COLUMN													
			0.1		0.2		0.3		0.4		0.5		0.6		0.7	
			CFM	TEMP. RISE	CFM	TEMP. RISE	CFM	TEMP. RISE	CFM	TEMP. RISE	CFM	TEMP. RISE	CFM	TEMP. RISE	CFM	TEMP. RISE
11.8x8	½	HIGH	1609	32.4	1550	33.6	1491	35.0	1431	36.4	1365	38.2	1280	40.7	1202	43.4
		MED.	1294	40.3	1268	41.1	1231	42.4	1188	43.9	1131	46.1	1070	48.9	992	52.6
		LOW	1045	49.9	1021	51.1	995	52.4	960	54.3	924	56.5	875	59.6	820	63.6
11.8x8	½	HIGH	1609	40.1	1550	41.6	1491	43.3	1431	45.1	1365	47.3	1280	50.4	1202	53.7
		MED.	1294	49.9	1268	50.9	1231	52.4	1188	54.3	1131	57.1	1070	60.3	992	65.1
		LOW	1045	61.8	1021	63.2	995	64.9	960	67.2	924	69.8	875	73.8	820	78.7
11.8x10.6	¾	HIGH	1642	52.4	1568	54.9	1505	57.2	1441	59.7	1359	63.3	1270	67.8	1170	78.5
		MED. HI	1621	53.1	1557	55.3	1487	57.9	1415	60.8	1357	63.4	1252	68.7	1152	74.7
		MED. LOW	1465	58.7	1399	61.5	1343	64.1	1285	67.0	1197	71.9	1122	76.7	1030	83.6
		LOW	1335	64.5	1295	66.4	1249	68.9	1207	71.3	1152	74.7	1084	79.4	1010	85.2

(b)

Figure 5.6 (a) Typical upflow warm air furnace. Units of this design force heated air into a top plenum to which the main supply duct is connected. The return duct, also usually overhead, drops down at the furnace and connects at the bottom of one side of the furnace enclosure. Flexible canvas connectors at the supply duct exit and the return duct entry reduce transmitted noise and vibration from the furnace. Upflow units can be installed in full-height basements, closets and utility rooms. Refrigerant piping, which connects the cooling coil (evaporator) to a remote refrigeration compressor and condenser, is shown for information. (Drawing reproduced with permission from *ACCA Manual C*, p. 17.) (b) Typical external static pressure data for a residential upflow furnace.

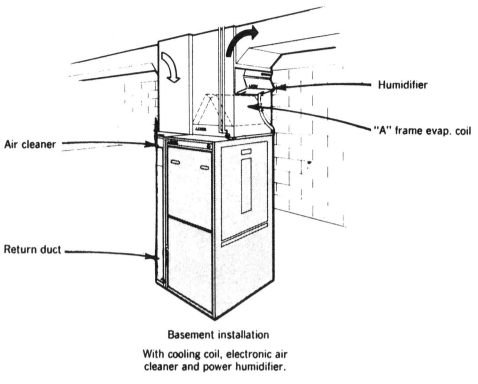

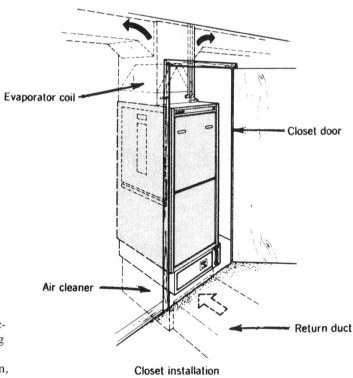

Figure 5.7 Typical upflow furnace installations. *(a)* Standard high-ceiling basement installation, with electronic air cleaner in lieu of a mechanical filter. Cooling coil piping is not shown for clarity. Humidifier is usually mounted on the supply duct. *(b)* Closet installation, with return air entering from below. The air cleaner is mounted at the junction of the return air duct and the base of the furnace. This type of installation is appropriate for a slab-on-grade or a low crawl space house.

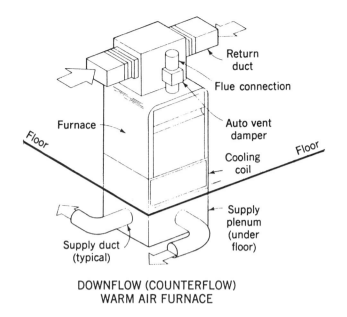

DOWNFLOW (COUNTERFLOW)
WARM AIR FURNACE

Figure 5.8 Typical downflow warm air furnace. These units are designed to serve floor-level perimeter heating (and cooling) outlets. These outlets are supplied by ducts connected to a subfloor plenum under the furnace. The furnace is also called counterflow because it supplies warm air downward into the plenum. This flow is opposite (counter) to the more common upflow design. (From Ramsey and Sleeper, *Architectural Graphic Standards*, 8th ed., 1988, © John Wiley & Sons, reprinted by permission of John Wiley & Sons.)

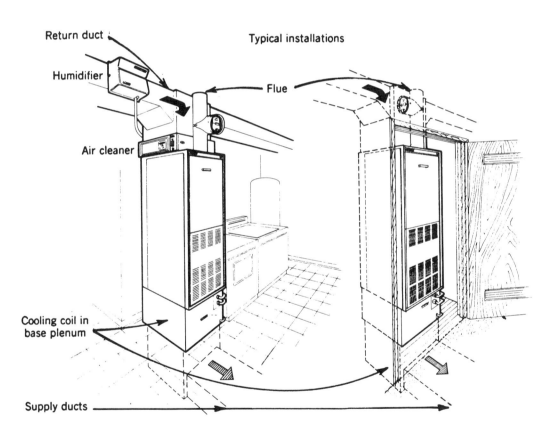

UTILITY ROOM INSTALLATION WITH ELECTRONIC
AIR CLEANER, COOLING COIL AND HUMIDIFIER.

CLOSET INSTALLATION
WITH COOLING COIL

Figure 5.9 Typical installations of downflow units. These furnaces are specifically designed for slab-on-grade or low crawl space houses using perimeter heating (and cooling). The blower forces air down into an underfloor plenum from where it is distributed to perimeter outlets. An optional cooling coil can be installed in a base plenum. Return air enters overhead. Filters, air cleaners and humidifiers are mounted overhead as shown. (Built-in humidifiers are mounted inside the furnace enclosure, in the supply air path.)

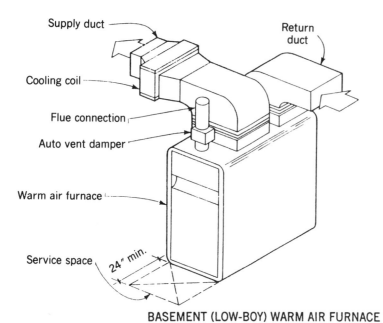

Figure 5.10 Typical low-boy warm air furnace. These units are upflow designs that are wider and shorter than standard high-boy furnaces. They are intended for use where ceiling height is limited such as in a low basement or low ceiling utility room. The return air enters the top of the unit and makes a loop through the blower, heater and filter before exiting through the main supply duct. The humidifier and optional cooling coil are installed in the supply trunk duct. (From Ramsey and Sleeper, *Architectural Graphic Standards*, 8th ed., 1988, © John Wiley & Sons, reprinted by permission of John Wiley & Sons.)

(residential and small commercial usage). All modern warm air furnaces meet and exceed this efficiency requirement provided that the load is at least 25% of the unit's rating. This is because their cycling efficiency is only marginally lower than steady-state efficiency. Therefore, even with a great deal of on-off cycling, AFUE will still exceed the minimum AFUE requirement. For technical data on residential furnace cycling efficiency, refer to Manual S published by Air Conditioning Contractors of America (ACCA).

5.7 Furnace Types

There are three principal types of warm air furnaces being produced today: conventional, condensing and pulse.

a. Conventional Gas or Oil Furnaces

Older furnaces use an atmospheric heat exchanger that takes combustion air from the surrounding space. This produces a slight negative pressure in the space, which increases infiltration in standard construction to make up for the air lost. In cold weather (which is exactly when the furnace is in use), this can add appreciably to a building's heat load. If the furnace is installed in a basement, it may well necessitate a heating outlet there. In very tightly constructed basements, or where the furnace is installed in a closet or utility space, infiltration would be insufficient to supply combustion. In these installations, a combustion air duct from the furnace enclosure to the outside is required. The outside louver should be sized for $\frac{1}{2}$ in.^{2} of free area for every 1000 Btuh of furnace input rating. The input rating is 10–25% greater than the output rating.

Older furnaces do not meet the 78% AFUE requirement; they run somewhere between 50 and 65% efficiency. Modern conventional gas and oil furnaces have a sealed heat exchanger, which takes combustion air from outside, usually using a forced draft or induced draft fan to do so. Many units also

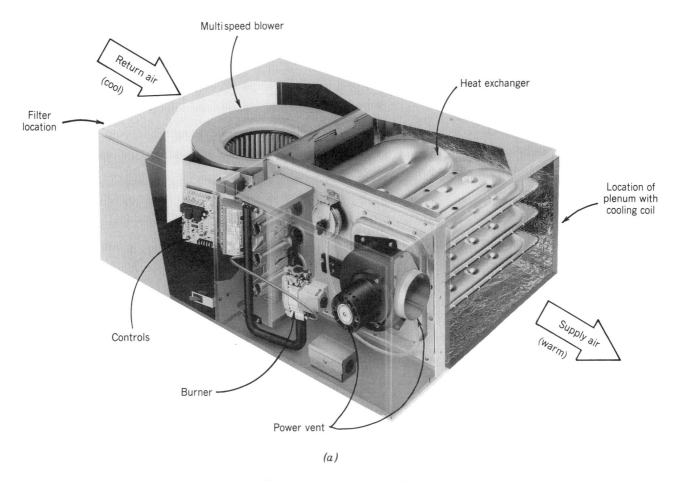

(a)

Figure 5.11 (a) Typical modern horizontal flow warm air furnace. This unit can also be used as an upflow furnace. Furnace characteristics for various models (b) and dimensional data (c). These units are designed to be installed in attics or low pitched roofs or overhead in areas where floor space is at a premium. The furnace is placed in-line between the incoming return air duct and the outgoing supply air duct. Like the low-boy design, the humidifier and the cooling coil (optional) are installed outside the furnace in the supply air duct. (Courtesy of Armstrong Air Conditioning, a Lennox International company.)

Blower Performance

Model	Motor Size (hp)	Blower Size	Temp. Rise	Blower Speed	CFM @ Ext. Static Pressure—in. W.C. with Filter(s) †							
					.20	.30	.40	.50	.60	.70	.80	.90
GHJ050D10	1/4	10 x 6	40–70	Hi	1170‡	1130‡	1110‡	1060‡	1020‡	990‡	910	840
				Med	900	900	870	850	830	790	740	650
				Low	720	710	690	670	650	610	550	460‡
GHJ050D14	1/3	10 x 8	20–50	Hi	1580	1540	1490	1420	1340	1250	1160	1050
				Med	1430	1420	1350	1310	1250	1780	1090	990
				Low	1210	1220	1220	1190	1160	1100	1020	920
GHJ075D09	1/4	10 x 6	50–80	Hi	1050	1000	960	930	870	810	740	650‡
				Med	850	830	810	760	720	680‡	600‡	540‡
				Low	670‡	660‡	650‡	620‡	580‡	550‡	490‡	410‡
GHJ075D14	1/3	10 x 8	40–70	Hi	1480‡	1430‡	1390	1350	1280	1200	1110	1000
				Med	1270	1270	1250	1210	1160	1090	920	830
				Low	1070	1090	1080	1060	1030	980	920‡	820‡
GHJ075D16	1/2	10 x 8	36–65	Hi	1870	1790	1710	1630	1540	1440	1350‡	1260‡
				Med	1500	1450	1410	1370	1300	1240	1160‡	1070‡
				Low	1170	1170	1160	1150	1090	1050	990‡	910‡
GHJ100D14	1/3	10 x 8	45–75	Hi	1450	1430	1370	1350	1290	1230	1160	1070
				Med	1200	1180	1180	1170	1150	1100	1030	960
				Low	980	980	1000	990	970	940	900	830
GHJ100D20	3/4	12 x 9	35–65	Hi	2290‡	2210‡	2130	2060	1980	1900	1820	1740
				Med	1820	1760	1710	1670	1630	1600	1500	1450
				Low	1300	1280	1250	1220	1210	1160	1120‡	1000‡
GHJ125D20	3/4	12 x 9	50–80	Hi	2260‡	2180‡	2100‡	2020‡	1950‡	1870	1780	1680
				Med	1900	1850	1810	1730	1700	1610	1550	1450
				Low	1440	1420	1390	1360	1320	1270	1210	1130‡

Notes: †.50 in. w.c. max. approved ext. static pressure. Airflow rated with AFILT524-1 filter kit.
‡ Not recommended for heating; Temperature rise may be outside acceptable range.

Physical and Electrical

Model	Input (Btuh)	Output (Btuh)	AFUE (ICS)	Nom. Cooling Cap.	Gas Inlet (in.)	Flue Size (in.)	Volts/Ph/hz	Min. Time Delay Breaker or Fuse	Nominal F.L.A.	Trans. (V.A.)	Appr. Weight (lbs.)
GHJ050D10	50,000	40,000	80.7	1.5–2.5	1/2	4	115/1/60	15	5.7	40	120
GHJ050D14	50,000	40,000	81.7	2.5–3.5	1/2	4	115/1/60	15	8.2	40	130
GHJ075D09	75,000	60,000	80.4	1.5–2.5	1/2	4	115/1/60	15	5.7	40	125
GHJ075D14	75,000	60,000	80.4	2.5–3.5	1/2	4	115/1/60	15	8.3	40	135
GHJ075D16	75,000	60,000	80.5	3.0–4.0	1/2	4	115/1/60	15	8.3	40	145
GHJ100D14	100,000	80,000	80.2	2.5–3.5	1/2	4	115/1/60	15	9.1	40	155
GHJ100D20	100,000	80,000	80.6	3.5–5.0	1/2	4	115/1/60	15	12.2	40	165
GHJ125D20	125,000	100,000	80.6	3.5–5.0	1/2	5*	115/1/60	15	12.2	40	170

*Connection to the combustion blower is 4 inch. Vent must be 5 in. for Cat. 1 installation

(b)

Figure 5.11 *(Continued)*

Dimensions (in.)

Model	A	B	C	D
GHJ050D10	14½	13½	13¼	4⅞
GHJ050D14	17½	16½	16¼	6⅜
GHJ075D09	14½	13½	13¼	4⅞
GHJ075D14	17½	16½	16¼	6⅜
GHJ075D16	22	21	20¾	8⅝
GHJ100D14	22	21	20¾	8⅝
GHJ100D20				
GHJ125D20	22	21	20¾	8⅝

Clearances (in.)
Upflow

Model	Left Side	Right Side	Front	Back	Vent	Top
GHJ050D10	3[1]	0	4[2]	0	6[3]	1
GHJ050D14	2[1]					
GHJ075D09	3[1]					
GHJ075D14	2[1]	0	4[2]	0	6[3]	1
GHJ075D16	0					
GHJ100D14	0	0	4[2]	0	6[3]	1
GHJ100D20						
GHJ125D20	0	0	4[2]	0	6[3]	1

[1] 0" if B1 vent is used
[2] 2" if B1 vent is used
[3] 1" if B1 vent is used

Horizontal

Model	Left Side	Right Side	Front	Back	Vent	Air Flow L to R Top	Air Flow L to R Bottom	Air Flow R to L Top	Air Flow R to L Bottom
GHJ050D10	1	1	18	0	6[2]	1	3[1]	3[3]	0
GHJ050D14							2[1]	2[3]	
GHJ075D09							3[1]	3[3]	
GHJ075D14	1	1	18	0	6[2]	1	2[1]	2[3]	0
GHJ075D16							0	1	
GHJ100D14	1	1	18	0	6[2]	1	0	1	0
GHJ100D20									
GHJ125D20	1	1	18	0	6[2]	1	0	1	0

[1] 0" if B1 vent is used
[2] 1" if B1 vent is used
[3] 1" if B1 vent is used

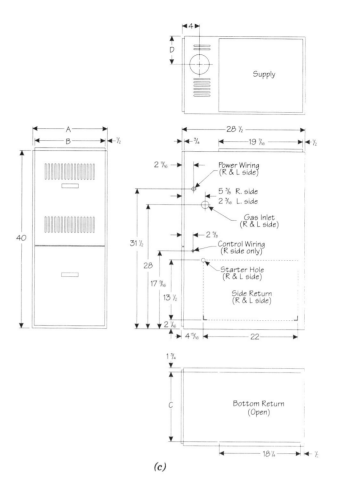

(c)

Figure 5.11 (Continued)

TYPICAL INSTALLATIONS

(a) Electronic air cleaner
(b) Automatic humidifier
(c) Gas-fired air furnace
(d) Cooling coil
(e) Flue

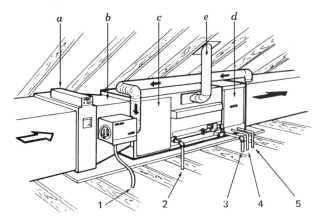

Attic installation with cooling coil, electronic air cleaner and automatic humidifier

1. Tap water *in* to humidifier.
2. Gas *in* to furnace.
3. Drain. Condensate *out*.
4. Liquid refrigerant *in* to evaporator.
5. Expanded (gas) refrigerant *out* to condenser.

(d)

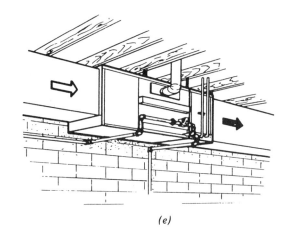

Basement installation with cooling coil

(e)

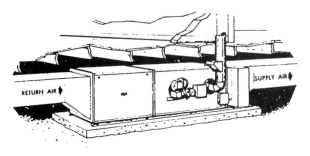

Crawl Space Horizontal Installation

(f)

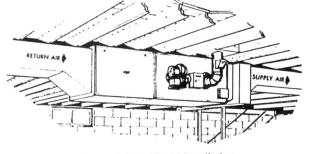

Suspended Horizontal Installation

(g)

Figure 5.11 *(d–g)* Typical installations of horizontal-type warm air furnaces in an attic *(d)*, suspended in a basement *(e, g)*, and in a shallow crawl space *(f)*. Units *(d)* and *(e)* contain cooling coils mounted in the supply trunk duct. Units attached to, or resting on, wood framing as in *(d)*, *(e)* and *(g)*, must be mounted on vibration isolators. Crawl space units *(f)* are installed on concrete pads below the house and, therefore, require less vibration isolation.

have automatic vent dampers or power venting to reduce losses. These modern units have AFUE efficiencies of 80–84%. They are referred to as mid-efficiency units. One such unit and its characteristics is shown in Figure 5.11(a–c).

b. Condensing Furnaces

Condensing furnaces achieve AFUE efficiencies above 90% by recovering heat that normally goes up the chimney as hot flue gas at 400–500°F. Most of this heat is carried in superheated water vapor (steam). By channeling the flue gas through a secondary heat exchanger inside the furnace, it is possible to reduce its temperature to about 150°F. In so doing, the superheated water vapor (steam) condenses into water and gives up 1000 Btu/lb of water. This warm water condensate is then drained away into the sewer system, while the heat recovered is added to the furnace output. Another advantage of this design is that it eliminates the need for an expensive masonry chimney. All that is required is a small diameter (2–3 in.) plastic pipe to exhaust the few remaining flue gases. A typical unit of this design and its characteristics are shown in Figure 5.12. Compare the characteristics of this unit to those of the conventional unit in Figure 5.11, for an appreciation of the high efficiency of condensing furnace.

c. Pulse Combustion Furnace

The pulse combustion furnace burns fuel in a manner very similar to combustion in an automobile engine. A spark plug ignites a mixture of fuel and air in the combustion chamber. The resulting hot gases are forced through the furnace heat exchanger where they give up their heat to the moving air stream. Once the process is started, it is self-igniting. The spark plug is required only to begin the process. The miniexplosions occur 60 to 80 times a minute, creating a fairly loud humming sound, similar to an idling engine. For this reason, pulse furnaces should be carefully isolated acoustically. Efficiencies as high as 96% are attained with pulse furnaces operating under ideal conditions. In the field, AFUE efficiencies of over 90% are common. A warm air furnace that operates on the principle of pulse combustion and also condenses flue water vapor is shown in Figure 5.13.

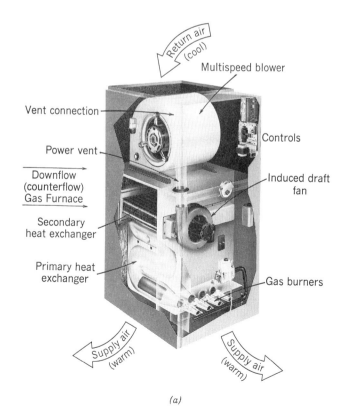

(a)

Figure 5.12 (a) Modern condensing-type downflow (counterflow) warm air furnace. (See Figures 5.8 and 5.9.) Note the use of a secondary heat exchanger that functions to remove most of the heat from the furnace flue gas. The vent connection is a plastic pipe that vents to outside air. (Courtesy of Armstrong Air Conditioning, a Lennox International company.)

Blower Specifications

Model	Motor Size (hp)	Blower Size	Temp. Rise	Blower Speed	CFM @ Ext. Static Pressure—in. W.C. with Filter(s)[1]							
					.20	.30	.40	.50	.60	.70	.80	.90
GCK050D10	1/4	10 x 6	40–70	Hi	1080[2]	1030	1010	960	900	840	770	680
				Med	870	840	820	790	740	700	640	550
				Low	680	670	650	620	590[2]	560[2]	490[2]	430[2]
GCK050D12	1/3	10 x 8	30–60	Hi	1350	1310	1230	1180	1100	1030	940	850
				Med	1260	1230	1170	1120	1050	980	910	830
				Low	1150	1110	1080	1040	990	920	850	830
GCK075D14	1/2	10 x 9	40–70	Hi	1700[2]	1640[2]	1560	1490	1410	1330	1240	1150
				Med	1550	1470	1410	1340	1280	1200	1130	1030
				Low	1330	1280	1240	1190	1120	1070	1010	910
GCK075D20	3/4	12 x 9	40–70	Hi[2]	2240[2]	2160[2]	2070[2]	1980[2]	1890[2]	1800[2]	1700[2]	1610[2]
				Med	1910[2]	1830[2]	1780[2]	1700[2]	1630[2]	1550	1470	1390
				Low	1420	1380	1360	1310	1270	1220	1160	1100
GCK100D14	1/2	10 x 9	55–85	Hi	1670[2]	1580[2]	1500	1420	1340	1230	1110	1000
				Med	1540	1450	1380	1310	1220	1130	1020	890
				Low	1360	1300	1240	1180	1100	1010	910[2]	780[2]
GCK100D20	3/4	12 x 9	45–75	Hi	2260[2]	2160[2]	2090[2]	1990[2]	1910[2]	1810	1700	1590
				Med	1740	1670	1640	1580	1550	1460	1390	1310
				Low	1400	1380	1350	1310	1270	1230	1180	1110
GCK125D20	3/4	12 x 9	45–75	Hi	2190	2130	2030	1960	1850	1790	1690	1600
				Med	1910	1850	1770	1730	1640	1590	1520	1420
				Low	1500	1460	1420	1400	1360[2]	1310[2]	1250[2]	1180[2]

[1] .50 in. w.c. max. approved ext. static pressure. Airflow rated with AFILT525-1 filter kit.
[2] Not recommended for heating; Temperature rise may be outside acceptable range.

Physical and Electrical

Model	Input (Btuh)	Output (Btuh)	AFUE (ICS)	Nom. Cooling Cap.	Gas Inlet (in.)	Volts/Ph/hz	Min. Time Delay Breaker or Fuse	Nominal F.L.A.	Trans. (V.A.)	Appr. Weight (lbs.)
GCK050D10	50,000	45,000	90.0	1.5–2.5	1/2	115/1/60	15	5.5	40	150
GCK050D12	50,000	45,000	90.0	2.5–3.0	1/2	115/1/60	15	7.5	40	150
GCK075D14	75,000	67,500	90.0	2.5–3.5	1/2	115/1/60	15	9.4	40	180
GCK075D20	75,000	67,500	90.0	3.5–5.0	1/2	115/1/60	15	12.1	40	190
GCK100D14	75,000	90,000	90.0	2.5–3.5	1/2	115/1/60	15	9.2	40	190
GCK100D20	100,000	90,000	90.0	3.5–5.0	1/2	115/1/60	15	12.0	40	195
GCK125D20	125,000	112,500	90.0	3.5–5.0	1/2	115/1/60	15	12.0	40	215

(b)

Figure 5.12 (b) Blower specifications and physical and electrical data.

Dimensions (in.)

Model	A	B	C	D
GCK050D10	17½	16½	16	3⅝
GCK050D12	17½	16½	16	3⅝
GCK075D14	22	21	21½	3⅜
GCK075D20	26½	25½	25	5⅝
GCK100D14	22	21	20½	3⅜
GCK100D20	26½	25½	25	5⅝
GCK150D20	26½	25½	25	3⅛

Clearances (in.)

Model	Top	Side	Front	Back	Vent
GCK050D10	1	0	2	0	0
GCK050D12	1	0	2	0	0
GCK075D14	1	0	2	0	0
GCK075D20	1	0	2	0	0
GCK100D14	1	0	2	0	0
GCK100D20	1	0	2	0	0
GCK150D20	1	0	2	0	0

Vent Length Specifications—Maximum

Model	Pipe Size (in.)		
	2	2½	3
GCK050	50 ft.	50 ft.	50 ft.
GCK075	50 ft.	50 ft.	50 ft.
GCK100	50 ft.	50 ft.	50 ft.
GCK125	N/A	50 ft.	50 ft.

Notes: Allowance for vent terminal included in lengths shown.
One 90° elbow equals 5 ft. of pipe.
Min. length 5 ft. and 1 elbow not including the vent terminal.

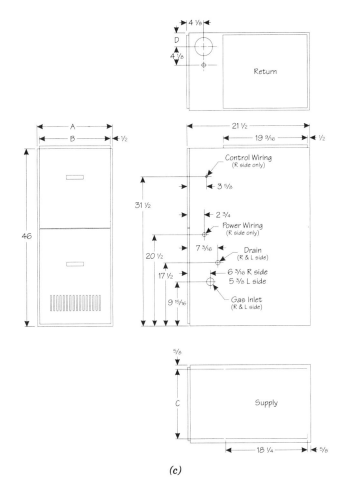

(c)

Figure 5.12 *(c) Dimension data and venting information.*

216 / AIR SYSTEMS, HEATING AND COOLING, PART I

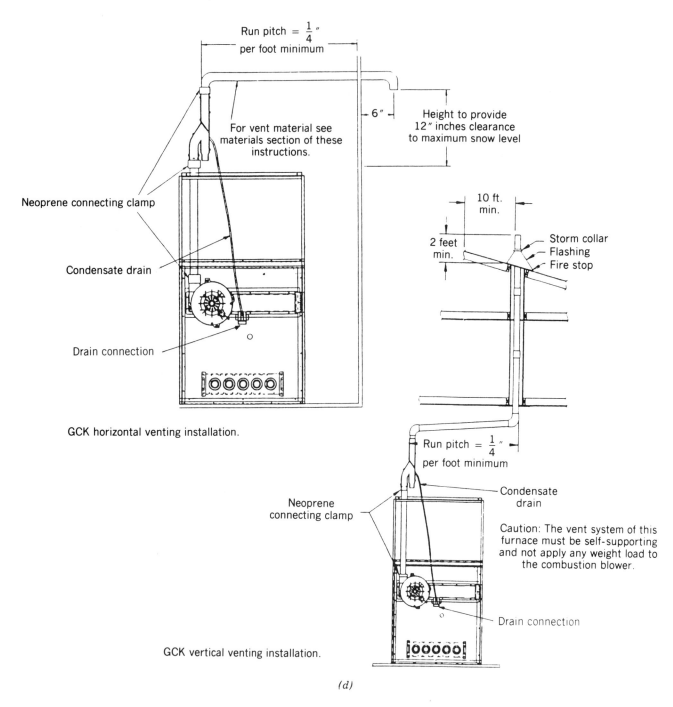

Figure 5.12 *(d)* Venting and condensate drain arrangements.

FURNACE TYPES / 217

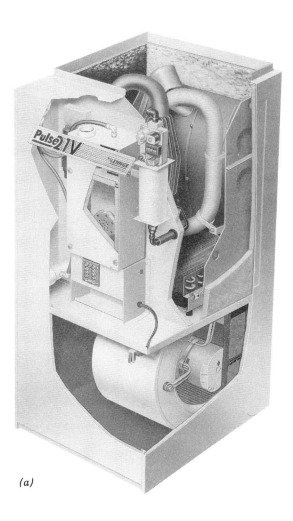

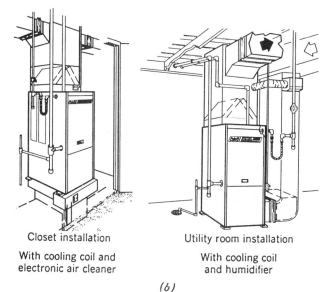

Figure 5.13 *(a)* The illustrated gas-fired warm air furnace operates on the principle of pulse combustion and also condenses the water in the flue gas. As a result, no flue or chimney is required. Combustion air to the sealed combustion unit is drawn in through the same PVC pipe that exhausts the remaining cool flue gases. Burner control is automatic, producing heat output in proportion to the air delivered by the blower. The blower is electronically speed controlled to maintain a specified air flow (cfm) throughout the entire static pressure range. These units are rated 60,000–100,000 Btuh input, with a maximum external static pressure of 0.80 in. w.g., including filter resistance. *(b)* Typical installations. (Photo and data courtesy of Lennox Industries.)

Specifications

Model No.		G21V3-60	G21V3-80	G21V5-80	G21V5-100
Input—Btuh (kW)		60,000 (17.6)	80,000 (23.4)	80,000 (23.4)	100,000 (29.3)
Output—Btuh (kW)		57,000 (16.7)	76,000 (22.3)	75,000 (22.0)	95,000 (27.8)
*A.F.U.E.		94.3%	94.5%	93.4%	94.5%
California Seasonal Efficiency		92.5%	92.4%	90.9%	91.5%
Temperature rise range—°F(°C)		40–70 (22–39)	45–75 (25–41)	35–65 (19–36)	40–70 (22–39)
High static certified by A.G.A/C.G.A.—in wg. (Pa)		.80 (200)	.80 (200)	.80 (200)	.80 (200)
Gas Piping Size I.P.S.—in. (mm)	Natural	1/2 (13)	1/2 (13)	1/2 (13)	1/2 (13)
	**LPG/Propane	1/2 (13)	1/2 (13)	1/2 (13)	1/2 (13)
Vent/Intake air pipe size connection —in. (mm)		2 (51)	2 (51)	2 (51)	2 (51)
Condensate drain connection—in. (mm) SDR11		1/2 (13)	1/2 (13)	1/2 (13)	1/2 (13)
Blower wheel nom. diameter x width —in. (mm)		10 x 8 (254 x 203)	10 x 8 (254 x 203)	11½ x 9 (279 x 229)	11½ x 9 (279 x 229)
Blower motor hp (W)		1/2 (373)	1/2 (373)	1 (746)	1 (746)
Blower motor minimum circuit ampacity		12.0		17.4	
Maximum fuse or circuit breaker size (amps)		15.0		25	
Electrical characteristics		120 volts—60 hertz—1 phase (All models)			
Number and size of filters—in. (mm)		(1) 16 x 25 x 1 (406 x 635 x 25)		(1) 20 x 25 x 1 (508 x 635 x 25)	
Nominal cooling that can be added	Tons	1½, 2, 2½ or 3	2, 2½ or 3	3½, 4 or 5	3½, 4 or 5
	kW	5.3, 7.0, 8.8 or 10.6	7.0, 8.8 or 10.6	12.3, 14.1 or 17.6	12.3, 14.1 or 17.6
Shipping weight—lbs. (kg) 1 package		250 (113)	250 (113)	297 (135)	297 (135)
External Filter Cabinet (furnished) ·Filter size—in. (mm)		(1) 16 x 25 x 1 (406 x 635 x 25)		(1) 20 x 25 x 1 (508 x 635 x 25)	
**LPG/Propane kit		LB-65810B (46J46)	LB-65810B (46J46)	LB-65810B (46J46)	LB-65810C (46J47)

·Filter is not furnished with cabinet. Filter cabinet utilizes existing filter supplied with G21V unit.
*Annual Fuel Utilization Efficiency. Isolated combustion system rating for non-weatherized furnaces.
**LPG/Propane kit must be ordered extra for field changeover.

(c)

Figure 5.13 (c) Specification and blower data.

FURNACE TYPES / 219

Blower Data

G21V3-60-80 BLOWER PERFORMANCE
0 through 0.80 in. w.g. (0 Through 200 Pa) External Static Pressure Range
VSP2-1 Blower Control—Factory Settings

G21V3-60	G21V3-80
Low Speed—1	Low Speed—1
High Speed—4	High Speed—4
Heat Speed—1	Heat Speed—2

	BDC2 Jumper Speed Positions											
	"LOW" Speed (Cool Or Continuous Fan)				"HIGH" Speed (Cool)				"HEAT" Speed			
"ADJUST" Jumper Setting	1		2		3		4		1		2		3		4		1		2		3		4	
	cfm	L/s	cfm	L/s	cfm	L/s	cfm	L/s	cfm	L/s	cfm	L/s	cfm	L/s	cfm	L/s	cfm	L/s	cfm	L/s	cfm	L/s	cfm	L/s
+	540	225	700	330	830	390	1000	470	1150	545	1260	595	1400	660	1410	665	1150	545	1250	590	1350	635	1420	670
NORM	490	230	630	295	740	350	880	415	1040	490	1140	540	1240	585	1265	595	1030	485	1140	540	1220	575	1300	615
−	440	210	560	265	670	315	800	380	940	445	1030	485	1140	540	1160	545	920	435	1020	480	1100	520	1190	560

NOTE—The effect of static pressure and filter resistance is included in the air volumes listed.

G21V5-80-100 BLOWER PERFORMANCE
0 through 0.80 in. w.g. (0 Through 200 Pa) External Static Pressure Range
VSP2-1 Blower Control—Factory Settings

G21V5-80	G21V5-100
Low Speed—1	Low Speed—1
High Speed—4	High Speed—4
Heat Speed—1	Heat Speed—2

	BDC2 Jumper Speed Positions											
	"LOW" Speed (Cool Or Continuous Fan)				"HIGH" Speed (Cool)				"HEAT" Speed			
"ADJUST" Jumper Setting	1		2		3		4		1		2		3		4		1		2		3		4	
	cfm	L/s	cfm	L/s	cfm	L/s	cfm	L/s	cfm	L/s	cfm	L/s	cfm	L/s	cfm	L/s	cfm	L/s	cfm	L/s	cfm	L/s	cfm	L/s
+	800	380	1050	495	1410	665	1620	765	1710	805	2030	960	*2270	*1070	*2270	*1070	1900	895	2140	1010	*2270	*1070	*2270	*0
NORM	720	340	950	450	1280	605	1500	710	1570	740	1850	875	2100	990	2220	1050	1700	800	1940	915	2080	980	2200	0
−	620	295	850	400	1120	530	1310	620	1420	670	1650	780	1860	880	1990	940	1520	715	1730	815	1860	880	1940	9

NOTE—The effect of static pressure and filter resistance is included in the air volumes listed.
*2300 cfm (1085 L/s) at 0.2 in. w.g. (50 Pa).
2100 cfm (990 L/s) at 0.5 in. w.g. (125 Pa).
2000 cfm (990 L/s) at 0.8 in. w.g. (200 Pa).

(c)

Figure 5.13 (c) (Continued)

DIMENSIONS — inches (mm)

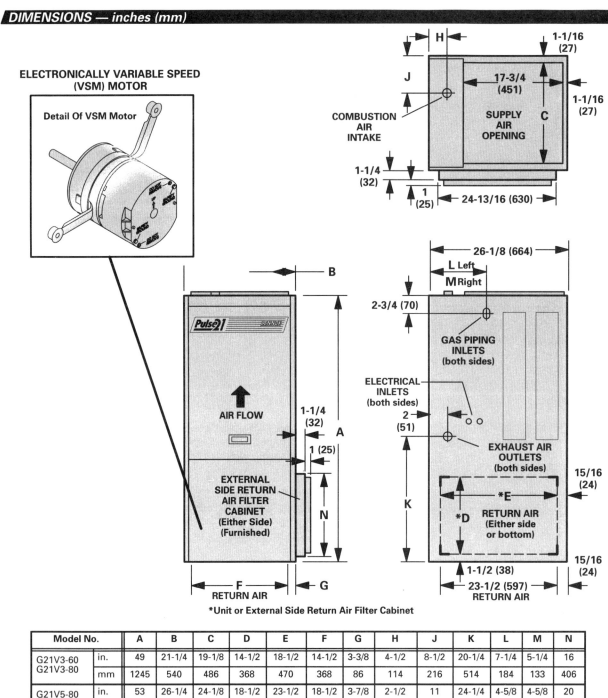

Model No.		A	B	C	D	E	F	G	H	J	K	L	M	N
G21V3-60 G21V3-80	in.	49	21-1/4	19-1/8	14-1/2	18-1/2	14-1/2	3-3/8	4-1/2	8-1/2	20-1/4	7-1/4	5-1/4	16
	mm	1245	540	486	368	470	368	86	114	216	514	184	133	406
G21V5-80 G21V5-100	in.	53	26-1/4	24-1/8	18-1/2	23-1/2	18-1/2	3-7/8	2-1/2	11	24-1/4	4-5/8	4-5/8	20
	mm	1346	667	613	470	597	470	98	64	279	616	117	117	508

INSTALLATION CLEARANCES — ALL MODELS

Sides	0 inch (0 mm)
Rear	0 inch (0 mm)
Top	1 inch (25 mm)
Front	0 inch (0 mm)
Front (service)	36 inches (914 mm)
Floor	Combustible
Exhaust Pipe	0 inches (0 mm)
Exhaust Pipe Side	6 inches (152mm) (service only)

(d)

Figure 5.13 *(d)* Physical dimensions and clearances.

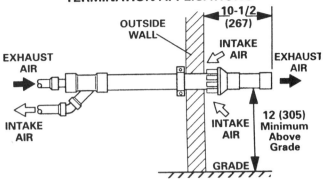

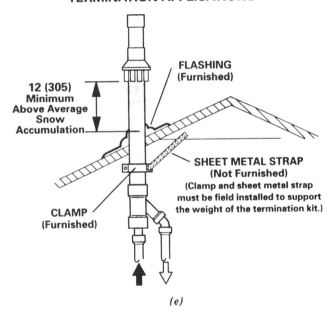

Figure 5.13 *(e)* Flue termination details.

5.8 Warm Air Furnace Characteristics

The data required by a system designer are available in manufacturers' published material. Typical dimensional and technical data are given in Figures 5.11–5.13. Of particular importance to the engineering technologist are the furnace output rating in Btuh and air delivery information. The latter gives the cfm of air delivered and the corresponding air temperature rise as a function of static pressure, for the various blower speeds.

a. Dimensional Data

The data required here are the furnace length and width ("footprint"), overall height including air plenum if any, service accessibility clearance, required clearances to combustible materials, clearance for filter replacement and venting data. Each of these items must be considered individually after a preliminary duct layout has been made, as we will explain.

(1) Overall dimensions of the furnace and the required clearances control its placement in the building. Basements in residences usually have low ceilings that may not permit the use of a full-size high-boy unit with a cooling coil in the plenum. Remember also to include the required overhead clearance to combustible materials. Removable filters are often placed in the return air duct connection. This not only requires a full filter length of clearance, for filter removal, but also sets off the return duct, making the assembly of return duct, filter assembly and furnace very wide.

(2) The minimum dimensions of the stripped furnace unit must be obtained from the manufacturer to ensure accessibility to the proposed location. This is particularly important with replacement units and with furnaces intended for installation in attics, closets, utility rooms and other tight, confined spaces with limited access.

(3) A vent connection to the furnace flue collar is required for every furnace, to exhaust flue gases. The vent pipe size, maximum length and required clearances depend on the type of furnace used. Venting tables are published by GAMA. Specific venting information for each furnace is provided by the manufacturer. Remember that flue gases can be as hot as 600°F with conventional furnaces. This requires not only sufficient clearances to combustible materials but also fire stops where passing through floors, walls and ceilings, as required by applicable fire codes. Insulation may also be required. On the other hand, modern conventional high efficiency furnaces produce flue gases in the 250°F range (near-condensing). Excessively long vent connections can cause water condensation in the flue and/or chimney with resulting damage to the chimney and to the furnace. As noted previously, a condensing furnace requires only a plastic pipe to vent cool flue gas and a drain connection to remove condensate. See Figure 5.12*(d)*.

b. Output Rating

See Figure 5.12. Residential-type furnaces are rated up to 250,000 Btuh input and 200,000 Btuh output. Larger units are classified as commercial furnaces. The rating of residential units are usually written out in full, such as 250,000 Btu/h or 250,000 Btuh. Commercial units usually abbreviate the number of thousand Btu with the letter *M*. Thus, 300,000 Btuh is written 300 MBH. Furnaces, like boilers and *unlike* air conditioners, are normally oversized by 10–25%. The building heat loss is used as the required Btuh base figure. The output of the furnace selected should be at least as large if it is installed in a conditioned area, and at least 10% larger if installed in a cold area, to compensate for uninsulated plenum and ducts.

Oversizing is said to be helpful in delivering very rapid space heating after a period of night temperature setback. However, warm air heating is by nature rapid, so that oversizing should be held to a minimum. Furthermore, oversizing causes frequent heating system cycling. This results in unpleasant temperature swings because air has almost negligible heat retention properties (specific heat). Therefore, as soon as the blower stops, the air temperature in rooms and ducts begins to drop rapidly. This effect can be overcome to an extent by continuous blower operation, combined with burner cycling.

c. Air Delivery Data

Refer to the table of air delivery information in Figure 5.6. Refer also to Figure 5.14, which shows the location of static pressure losses in a typical air system. Figure 5.14 is a schematic drawing showing essentially the same elements as Figure 5.1. If

WARM AIR FURNACE CHARACTERISTICS / 223

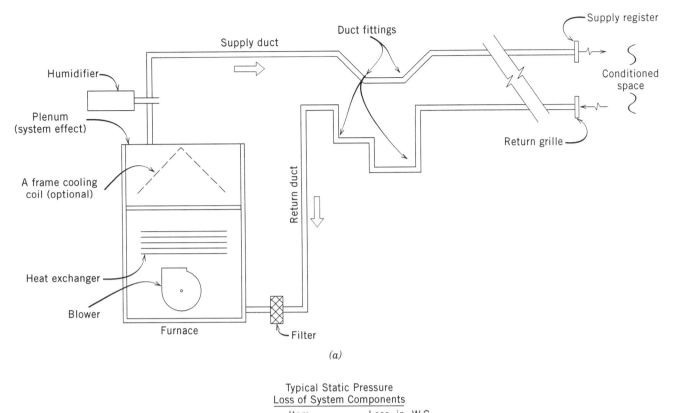

Figure 5.14 (a) Schematic drawing of a warm air heating system with (optional) cooling coil and ductwork. Sources of static friction are indicated. (b) List of typical static pressure losses of system components.

you consult the air delivery table in Figure 5.6, you will notice that the air delivery columns are listed under various values of external static pressure. These figures refer to the static pressure available external to the basic furnace. Since different manufacturers supply different items as part of the furnace, it is essential for the designing technologist to know, in advance, which items are included and which are not.

Specifically, the most common items that may or may not be included are the filter and the humidifier. Refer now to Figure 5.14. Mechanical filters installed in the return duct line are not included as part of the furnace, whereas electrostatic filters usually are. Similarly, plate-type humidifiers installed in the supply duct are not included in the basic furnace, whereas atomizing types frequently are. This means that in an installation such as shown in Figure 5.14, the designer must subtract the static pressure loss of the filter (which is considerable) and the static pressure loss in the humidifier, from the external static pressure given in the table, to obtain the pressure available at the plenum. On the other hand, if the manufacturer's literature indicates that the filter and humidifier are included as part of the furnace, then the pressure figures can be used directly.

The static pressure loss in a cooling coil is large, usually exceeding the static pressure loss in the entire duct system. Some manufacturers publish separate tables of external static pressure with and without a cooling coil. If it is not specifically stated as included, the coil's static loss must be subtracted from the published external static pressure to obtain the net static pressure available when cooling is included in the system. Because of the large pressure drop in the coil, it is important to know at the outset whether cooling will be part of the system, even at some future date. Otherwise, the blower will not be adequate nor will the ducts be adequately sized, as we will learn.

The procedure for handling other system static pressure losses shown on the diagram of Figure 5.14 will be discussed in the following sections that cover friction losses in ducts and fittings. The significance of external static pressure and how it is used will become clear later on. For the moment, however, the technologist must keep in mind the clarifications necessary with regard to the humidifier and filter.

Note further from the table in Figure 5.6 that, as the cfm rises, the temperature rise drops. This is entirely logical. The air is heated as it passes through the heat exchanger. Since the combustion rate is fixed, the more air that passes per minute, the less it is heated, and vice versa. The use of these figures will also become clear later on, in our system design discussion.

5.9 Noise Considerations

All mechanical equipment generates noise. Ducts are excellent carriers of sound. Since they interconnect all the rooms in a building, they comprise a very effective but generally undesirable noise-conducting system. Not only will combustion and blower noise be heard in every room, but sound will also carry from room to room. Treatment of these acoustical problems is not usually the responsibility of the engineering technologist. We will, therefore, mention only a few of the considerations that are of interest and importance to technologists.

The best way to avoid noise problems is not to generate the noise in the first place. As applied to warm air furnaces this means that designers should

- Place the furnace as far as possible from sleeping quarters and other quiet living spaces.
- Make sure that vibration mounts are used, if the furnace is not installed on a concrete floor, such as in a basement, garage or utility room. This is particularly important if a furnace is to be attached to wood frame construction, as for instance in an attic or overhead in a basement.
- Determine that the blower and any forced or induced draft fan are installed on vibration isolators.
- Ensure that trunk connections to the furnace plenum are flexible, sound-isolating connections.
- Make sure that duct friction and duct area calculations take into account acoustical insulation if it is to be installed on the inside of ducts.
- Consider the high static pressure loss of acoustical attenuators in ducts, in pressure calculations. Such attenuators are very effective.
- Use thermal insulation on ducts passing through nonconditioned spaces because it can be very effective in reducing vibration of metal ducts, particularly those with large dimensions. This reduction is most effective when the thermal insulation is glued to the metal duct surface.

5.10 Warm Air Furnace Blowers

The terms *fan* and *blower* are used interchangeably when referring to the furnace fan. In most furnaces, the unit is actually a centrifugal blower of the type shown in Figure 5.15a. The function of the blower is to circulate the air in the duct system. Most blowers are belt driven from a multisheave pulley. This permits changing the belt position on the drives, and thereby the blower speed. In some furnaces, the blower is directly coupled to the motor. There, a speed-controlled motor is used to permit changing the blower speed. As demonstrated previously, speed control of the blower is necessary when changing over from heating to cooling because of the different air quantities required for the two services. A speed change may also be required in the initial system balancing after installation. Typical furnace blower curves are shown in Figure 5.15b. The system friction curve is simply a graph of the ductwork static friction at different values of air flow in the system.

WARM AIR FURNACE BLOWERS / 225

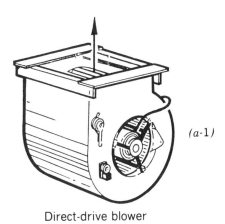

Direct-drive blower

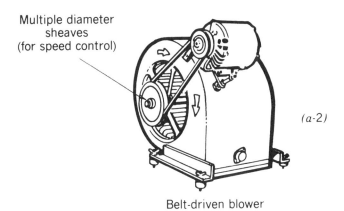

Belt-driven blower

Figure 5.15 *(a-1)* Direct-drive blowers use an electric motor inside the housing, direct coupled to the blower. These units are normally operated through a variable-speed motor controller to achieve the required speed control. *(a-2)* Speed control of belt-driven blowers is usually accomplished by use of multiple sheaves on the motor and blower. A speed change requires physically moving the drive belt from one sheave to another. (Reproduced with permission from *ACCA Manual C*, p. 24.)

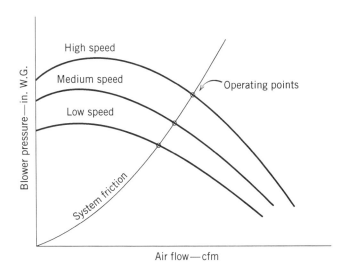

Figure 5.15 *(b)* Typical static pressure/airflow characteristics of a multispeed centrifugal blower. This is the type normally used in residential-type warm air furnaces. The operating points of the duct system are noted at the intersection of the blower curves and the system friction curve.

The intersection of the system curve with the blower curves shows the operating points possible at different blower speeds.

Residential-type furnaces are usually supplied as package units with a blower design that the furnace manufacturer has found to be satisfactory for most applications. The only leeway the designer has with respect to the blower is speed control. In contrast, commercial furnaces can be equipped with blowers that meet the specifications of the system designer. For our purposes—residential and small commercial installations—the package-type furnace and blower will be used. The manufacturer usually presents blower data in tabular form as shown in Figure 5.13. Some manufacturers prefer to present the data in graphic form, similar to that shown in Figure 5.15(b) (without the system curve, of course).

5.11 Furnace Accessories

The two essential warm air furnace accessories are a humidifier and an air filter/cleaner.

a. Humidification

Humidification is almost always required in winter. This is particularly true in residences because the humid air created in the house in areas such as bathrooms, kitchens and laundries is exhausted to the outside. Make-up air from the outside is dry, because cold air carries very little moisture. In small commercial buildings, ventilation achieves the same result by bringing in dry air to replace stale but more humid exhausted air. In both cases, humidification is, therefore, required.

As an example of the dryness of heated make-up air, consider a typical condition. If exterior air at 30°F and 50% RH (typical winter air conditions) is heated in a warm air furnace to 140°F, it will have a relative humidity approaching zero. (See the psychrometric chart in Figure 2.18.) Even after it mixes with room air and its temperature drops to 74°F, this air will have only a 10% RH. This dry air will act to reduce the humidity of all the air in the room. The exact effect will depend on the ventilation rate, occupancy and so on. However, it is clear that humidification is required to compensate for the introduction of very dry air into the building.

Excess humidity is also undesirable. From a purely comfort point of view, an RH of up to 50% is acceptable. High humidity can cause condensation on windows and metal sash. As a rule of thumb, indoor relative humidity should not exceed

- 15 + outdoor temperature (in °F) for single-glazing or
- 25 + outdoor temperature for double-glazing.

Thus for an outdoor temperature of 15°F, indoor RH should not exceed 30% for single-glazed spaces and 40% for double-glazed spaces.

This rule of thumb, like all such rules, is only approximate. For an actual installation, a calculation based on the actual R/U factors for the window being used should be performed. In the absence of manufacturers' data, the data given in Table 2.7 can be used. Two important facts with regard to winter inside humidity should be remembered.

- The upper limit of humidity drops as outside temperature drops, if we wish to avoid condensation on windows.
- Comfort humidity also drops with colder outside temperature. For this reason, winter comfort humidity is usually given as 35–50%. The lower figure applies at low outside temperatures, and the upper figure applies at higher temperatures.

Refer to the results of Problem 2.11 for more data on this important subject.

There are many types of humidifiers available. The simplest types use pans or multiple porous plates from which water is evaporated. Evaporation can be natural or aided by heat and/or forced air. More complex types use atomizers that spray water droplets into the duct air stream. The droplets evaporate into water vapor within a few feet of travel in the duct. Most units are connected to a piped source of water, which is controlled by a water-level device, such as a float switch. Selection of the appropriate type of humidifier requires calculation (or estimation) of the building humidification requirements. This is turn depends on building volume, occupancy, activity and construction quality plus data on outdoor conditions. Once the humidification requirements are established, a particular unit can be selected. Most furnace units are rated for a specific furnace plenum temperature and for continuous furnace operation. Other plenum temperatures and furnace on-off cycling must be considered in sizing a humidifier. Since all humidification calculations can only approximate actual field conditions, a humidistat and/or other control device is recommended to control humidification.

b. Air Cleaning

A mechanical filter is the simplest type of air-cleaning device. It can be of the renewable (washable) or throw-away type. It is usually placed in the return air duct, adjacent to the connection to the furnace. See Figure 5.6(a). These units consist of a frame containing treated coarse fibrous material such as glass wool, plastic thread or a combination of materials. Throw-away units called viscous impingement filters remove dust and dirt from the air, not only by filtering but also by capturing impinging particles on their treated sticky surfaces. Renewable, washable units are less effective, because they operate only as mechanical filters. They are seldom used in modern installations.

So-called electronic air cleaners operate in a number of ways. Some add an electrical charge to dust and dirt particles passing through an ionizing section and then trap the charged particles at collection electrodes. Other types attract dirt particles to electrostatically charged plates. Still others trap dust and dirt particles by using the particle's own electrical charge. All these units become "loaded" with dirt and must be cleaned periodically.

As noted in our discussion on external static pressure in Section 5.8.c, the static pressure loss in filters and humidifiers must be considered. Most furnaces are tested and rated with a clean filter installed. This is true for most, but not all. The static pressure loss in a clean filter varies from 0.1 to 0.35 in. w.g. and is 50% higher in a dirty filter. If we remember that the maximum static pressure developed by a residential-type warm air furnace blower rarely exceeds 0.8 in. w.g., we can readily see how important it is to clarify whether the blower static pressure data includes the filter. Static drop in humidifiers is much smaller, depending on type. It varies between negligible for the atomizing types to about .025 in. w.g. maximum for the multiple wetted plate type. Exact static pressure drop figures should be obtained from the manufacturer, using the system's design air velocity.

Air Distribution System

Having discussed the warm air furnace and its accessories, which is the heart of any warm air heating system, we will now proceed to the air distribution system. This system carries the warm air from the furnace to the various rooms in the building, via a network of supply ducts and fittings. At the duct terminations in the various rooms, the supply air is dispersed by registers specifically designed to distribute the air in optimal fashion. The "used" air is then collected at return air grilles and brought back to the furnace in return ductwork for filtering, reheating, humidification and recirculation. If the building HVAC system is to supply summer cooling as well as winter heating, the duct system may have to be larger than for heating alone. Also, the room air outlets may be different, both in design and placement. These items will become clear as our study progresses.

If we refer again to Figure 5.1, we can see all the essential elements of an air distribution system: ducts, duct fittings, air outlets and air control devices. An understanding of the construction and functioning of each of these system components is essential to an understanding of the system functioning, as a whole.

5.12 Ducts
a. Rectangular Ducts

The most commonly used materials for construction of rectangular ducts in low pressure, low velocity forced-air systems are galvanized steel, aluminum and rigid fibrous glass.

(1) Galvanized steel is probably the most widely used material for supply and return ducts. When used outdoors, painting is recommended even though the zinc galvanizing acts as an effective weatherproofing for at least 5 years. Galvanized sheet steel has high strength and is rust-resistant; nonporous; highly durable; readily cut, drilled and welded; and easily painted. It is also widely available in the United States in a variety of qualities. The most commonly used type is called lock-form quality, which describes the usual method of joint closure.

Minimum metal gauges for steel and aluminum duct are given in SMACNA standards. The principal disadvantages of galvanized steel duct are its weight and its acoustical characteristics. In addition to being an excellent channel for noise transmission, it is also a source of noise from vibration, particularly in large ducts. Addition of thermal insulation on the outside, especially if glued, will dampen the metal vibration but will not attenuate the duct's noise transmission ability. For that purpose, acoustical damping inside the duct is required.

(2) Aluminum is often substituted for galvanized steel because it is lighter and much more corrosion- and weather-resistant. It also has a more attractive appearance, which is a consideration in installations using exposed ductwork. Among the disadvantages of aluminum duct are high cost, low physical strength in the thicknesses used for ducts and a thermal expansion coefficient more than double that of steel. Aluminum is also difficult to weld. The large thermal expansion is not normally a problem in residential work but can be a duct length limiting factor in commercial work. Aluminum is smoother than galvanized steel and, therefore, has a lower static friction drop. Comparative figures for this characteristic are found in Figure 5.54.

(3) Rigid fibrous glass board, normally 1-in. thick, is used frequently to fabricate rectangular duct. The material is a composite of fibrous glass board with a factory-applied facing of plain or reinforced aluminum. This facing acts as a finish and as a vapor barrier. Rigid fibrous glass board has distinct advantages over metal duct including (light) weight, good thermal insulation and acoustical qualities and simplicity of fabrication and installation. The principal disadvantages of this material are relatively high cost, low physical strength, sensitivity to moisture and pressure limitations (2 in. w.g.). In addition, in some areas of the country, it is not acceptable according to local codes. Another disadvantage is its higher static friction loss as compared to metal duct. Although this last item is rarely a deciding factor, it may cause the designer to use larger and, hence, more expensive ducts. Use of fibrous glass duct is limited to locations where the duct is not subject to physical damage. Also, because of its lack of physical strength, risers are limited to about 20 ft.

(4) Other materials including black carbon steel, stainless steel, fiberglass-reinforced plastic (FRP), gypsum board and polyvinyl chloride (PVC) are all used for special-purpose rectangular duct. They are not considered to be general-purpose duct materials and will, therefore, not be discussed here. You can consult publications of the Sheet Metal and Air Conditioning Contractors National Association, Inc. (SMACNA), for further information on these duct materials as well as construction and installation standards for all ductwork.

b. Round Ducts

Round metal ducts are very common in HVAC systems because they are strong, rigid, efficient and economical. Round ducts have the lowest static friction loss of any shape, and the highest ratio of cross-sectional area to perimeter of any shape. That means that for a given cross-sectional area, a round duct will use less material than any other shape. It is, therefore, cheaper than any other shape. The considerations involved in duct shapes are discussed at length in Section 5.22 where duct friction is analyzed.

Round ducts also have the advantage of great rigidity as a result of shape. This makes them ideal for installation in locations where the ducts are subject to physical abuse. This rigidity also minimizes noise and vibration transmission. Another advantage of round ducts is economy of insulation, due to the minimal perimeter of the round shape. The only major disadvantage of round ducts is that they often will not fit into tight locations such as hung ceilings or between joists, because of their shape. Such locations require rectangular ducts. Round ducts are generally metallic, although prefabricated round rigid fibrous glass duct is available.

c. Oval Ducts

Oval ducts were developed to solve the bulky shape problem of round ducts without losing their advantage. They are almost as efficient and rigid as round ducts, and their flatter profile allows their use in tight locations such as between wall studs and between joists in framed houses. Their only disadvantage is, at this writing, their price premium. Table 5.5 (page 287) lists oval duct sizes and their round duct equivalents.

d. Flexible Ducts

Flexible ducts are available in two basic designs: metallic, both insulated and bare, and mesh covered, insulated, metallic helix (see Figure 5.16). All flexible ducts have considerably higher friction than their nonflexible equivalent. This, however, is not usually an important factor since flexible ducts are most often used in short runs. Despite their

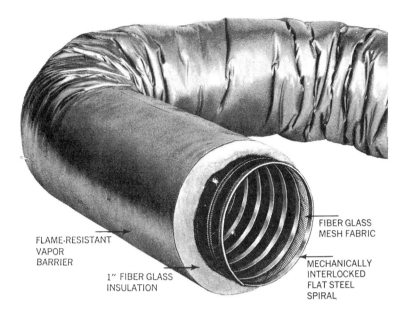

Figure 5.16 Insulated flexible duct, usable in low velocity systems. It consists of a steel helix (spiral) covered with a fiberglass fabric and thermal insulation. These ducts can be spliced, clamped and formed into oval ends. Their principal use is for short, angled runs and sharp turns.

high cost, they are considerably cheaper than the custom-made joints and fittings used with rigid duct.

5.13 Duct Fittings and Air Control Devices

A well-designed duct system has low static friction loss. Any change in duct size or direction can cause turbulence in the air stream. This is turn increases the static head loss in the duct system. (This subject will be treated more full in Section 5.21.) As a result, a good duct system uses duct fittings that are specifically designed to minimize air turbulence. This means that transitions are smooth [Figure 5.17(a)], branch takeoffs are gradual [Figure 5.17(b)], elbows are long and rounded, [Figure 5.17(c)], and so on. Figure 5.18 shows a few of the many types of duct fittings and a typical duct hanging detail.

Figure 5.19 shows some of the more commonly used duct air control devices including turning vanes, splitters and volume dampers. Volume dampers are constructed in single-leaf (e) or multileaf (opposing blade) design (c). The purpose of volume dampers is simply to control the quantity of air in a duct. Every branch duct is equipped with a volume damper, which is capable of being locked in position. Turning vanes (f) are used where there is insufficient space for a long sweep elbow and a sharp turn right angle elbow must be used. Figure 5.20 shows the drawing symbols and abbreviations commonly used on HVAC working drawings.

5.14 Duct Insulation

Duct insulation is generally provided on ducts in accordance with energy codes and ASHRAE standards. In their absence, use the following guidelines:

(a) Do not insulate ducts passing through a conditioned space, or through furred interior spaces.
(b) Do not insulate ducts of "heat-only" systems that pass through unheated basements.
(c) Do insulate ducts of heating/cooling systems passing through unheated basements. This will probably result in a requirement for a heating register in the duct to compensate for the lack of basement heating from duct heat loss, in winter.

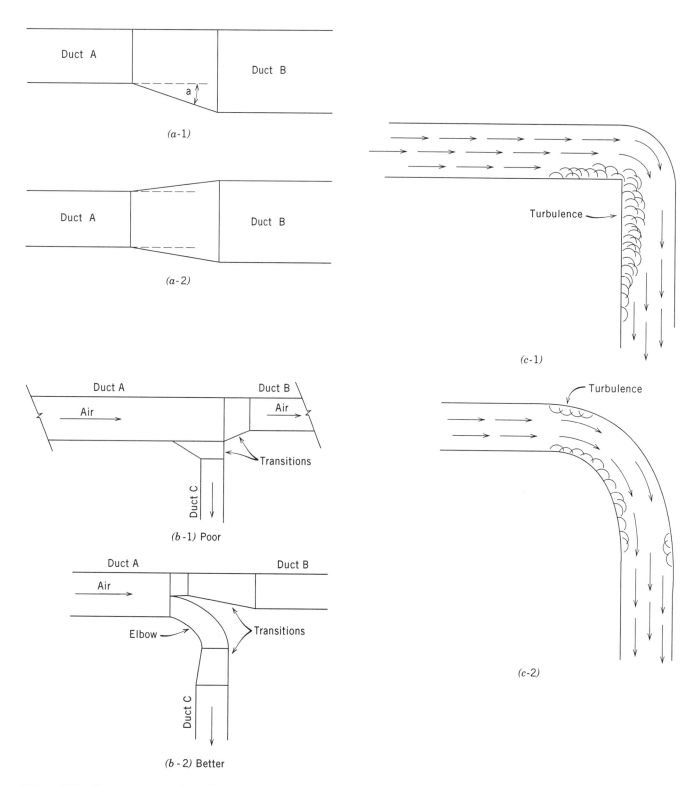

Figure 5.17 Transitions can be either single *(a-1)* or double *(a-2)*. Angle *a* should be as small as possible, to reduce turbulence, and should not exceed 20°. If space is tight, requiring a larger angle, a double transition *(a-2)* can be used. Takeoff connections should not be at right angles *(b-1)* but should use a long elbow *(b-2)* connected in the direction of airflow. Sharp turns *(c-1)* cause severe turbulence and result in high static friction loss. Long gradual elbows *(c-2)* have minimal turbulence and low losses.

Figure 5.18 (a) Air boot fittings are terminations of branch ducts, onto which registers, diffusers and grills are mounted. Of the boot fittings shown, types H, I and J connect to round branch ducts; types A, C and O connect to square branches, and types M and N connect to rectangular branches. Note that transitions are gradual and that right angles are avoided wherever possible. For an explanation of "equivalent length" see Section 5.23. (From Ramsey and Sleeper, *Architectural Graphic Standards*, 8th ed., 1988, © John Wiley & Sons, reprinted by permission of John Wiley & Sons.)

Figure 5.18 (b) Angles, elbows and offsets can be of standard design (B, G, I, E) or fabricated specifically for job conditions (K, L. M). Offsets are fabricated from two elbows and a straight section. Gradual offsets (L, M) have much lower losses than sharp offsets like type K.

Figure 5.18 (c) The application of elbows and angle fittings to trunk ducts is shown. Right angle elbow D uses internal turning vanes to reduce turbulence. See text.

Figure 5.18 (d) Two typical trunk ducts with takeoffs are shown. At the left, the trunk reduces at each takeoff, and the takeoff connects at the transition fitting. Compare this to Figure 5.17b. At the right, the duct size reduction is one-sided, and all takeoffs are with round duct. This design is common in residential work. Both trunk designs are known as reducing trunk or reducing plenum designs.

Figure 5.18 (e) Typical hanger detail for rectangular duct. (From Ramsey and Sleeper, *Architectural Graphic Standards*, 8th ed., p. 628, 1988, © John Wiley & Sons, reprinted by permission of John Wiley & Sons.)

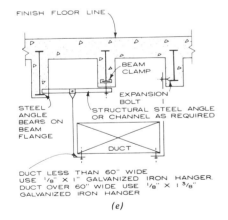

(e)

NOTE
On ducts over 48 in. wide hangers shall turn under and fasten to bottom of duct. When cross-sectional area exceeds 8 sq ft duct will be braced by angles on all four sides.

DUCT SUPPORT DETAIL

William G. Miner, AIA, Architect, Washington, D.C.

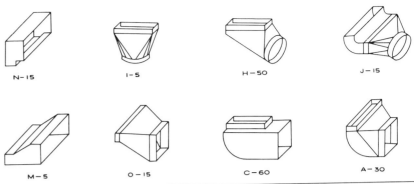

AIR BOOT FITTINGS NOTE: N-15 ← NUMBER = EQUIVALENT LENGTH (FT) LETTER = SHAPE DESIGNATION

(a)

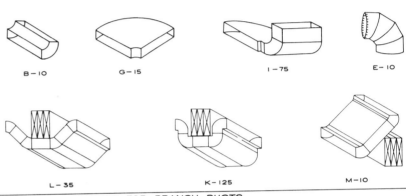

ANGLES AND ELBOWS FOR BRANCH DUCTS

(b)

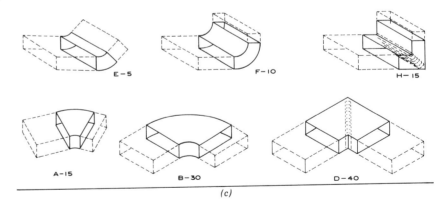

(c)

TRUNK DUCTS AND FITTINGS

(d)

231

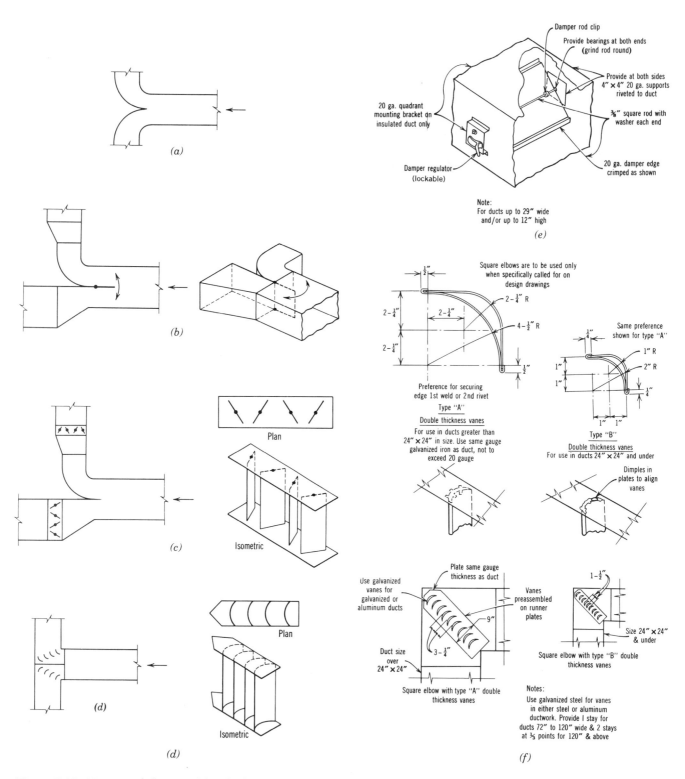

Figure 5.19 Air control devices. *(a)* A double elbow will split the air flow in proportion (inversely) to the friction in each branch. *(b)* A vertical splitter vane is used to control the division of air at a juncture. *(c)* In higher pressure systems (above 0.5 in. w.g.), opposed-leaf volume dampers are used to control the airflow into each branch of a junction. *(d)* At tee joints, turning vanes are used to reduce turbulence and friction. *(e)* Single-blade volume dampers are used in all branch ducts and in some trunk ducts. All volume dampers should be provided with a locking mechanism. *(f)* Details of turning vanes for large ducts in a commercial installation. [Drawing details *(e)* and *(f)* courtesy of Seelye, Stevenson, Value and Knecht, Consulting Engineers.]

SYMBOL MEANING	SYMBOL	SYMBOL MEANING	SYMBOL
POINT OF CHANGE IN DUCT CONSTRUCTION (BY STATIC PRESSURE CLASS)		SUPPLY GRILLE (SG) (U.S. UNITS)	20 × 12 SG / 700 CFM
DUCT (1ST FIGURE, SIDE SHOWN 2ND FIGURE, SIDE NOT SHOWN)	20 × 12	RETURN (RG) OR EXHAUST (EG) GRILLE (NOTE AT FLR OR CLG) (METRIC UNITS)	500 × 300 / 350 ℓ/s
ACOUSTICAL LINING DUCT DIMENSIONS FOR NET FREE AREA		SUPPLY REGISTER (SR) (A GRILLE + INTEGRAL VOL. CONTROL)	20 × 12 SR / 700 CFM
DIRECTION OF FLOW		EXHAUST OR RETURN AIR INLET CEILING (INDICATE TYPE)	20 × 12 GR / 700 CFM
DUCT SECTION (SUPPLY) (U.S. UNITS)	S 30 × 12	SUPPLY OUTLET, CEILING, ROUND (TYPE AS SPECIFIED) INDICATE FLOW DIRECTION	500 mm / 350 ℓ/s
DUCT SECTION (EXHAUST OR RETURN) (METRIC UNITS)	E OR R / 500 × 300	SUPPLY OUTLET, CEILING, SQUARE (TYPE AS SPECIFIED) INDICATE FLOW DIRECTION	12 × 12 / 700 CFM
INCLINED RISE (R) OR DROP (D) ARROW IN DIRECTION OF AIR FLOW	R	TERMINAL UNIT. (GIVE TYPE AND/OR SCHEDULE)	T.U.
TRANSITIONS: GIVE SIZES NOTE F.O.T. FLAT ON TOP OR F.O.B. FLAT ON BOTTOM IF APPLICABLE		COMBINATION DIFFUSER AND LIGHT FIXTURE	
		DOOR GRILLE	DG / 12 × 6
STANDARD BRANCH FOR SUPPLY & RETURN (NO SPLITTER)	S R	SOUND TRAP	ST
SPLITTER DAMPER		FAN & MOTOR WITH BELT GUARD & FLEXIBLE CONNECTIONS	
VOLUME DAMPER MANUAL OPERATION	VD	VENTILATING UNIT (TYPE AS SPECIFIED)	
AUTOMATIC DAMPERS MOTOR OPERATED	SEC / MOD	UNIT HEATER (DOWNBLAST)	
ACCESS DOOR (AD) ACCESS PANEL (AP)	OR / AD	UNIT HEATER (HORIZONTAL)	
FIRE DAMPER: SHOW ◀ VERTICAL (Wall) SHOW ◆ HORIZ. (Floor)	FD / AD	UNIT HEATER (CENTRIFUGAL FAN) PLAN	
DAMPERS: (SD) SMOKE— Ŝ FIRE/SMOKE (F/SD)— ▲	AD / SD	THERMOSTAT	T
HEAT STOP FOR FIRE RATED CLG		POWER OR GRAVITY ROOF VENTILATOR-EXHAUST (ERV)	
TURNING VANES		POWER OR GRAVITY ROOF VENTILATOR-INTAKE (SRV)	
FLEXIBLE DUCT FLEXIBLE CONNECTION		POWER OR GRAVITY ROOF VENTILATOR-LOUVERED	
GOOSENECK HOOD (COWL)		LOUVERS & SCREEN	36 × 24L
BACK DRAFT DAMPER	BDD		

Figure 5.20 *(a)* Symbols commonly used in HVAC work. (Reproduced with permission from *SMACNA HVAC Systems Duct Design Manual*, 1990.)

AHU	Air-handling Unit	FTR	Fin-tube Radiation
C	Condensate (Steam)	GC	General Contractor
CD	Cold Duct	HP	Heat Pump
CDR	Condensing Water Return	HPS	High Pressure Steam
CDS	Condensing Water Supply	HWR	Heating Water Return
CFM	Cubic Feet/Minute of Air	HWS	Heating Water Supply
CHWR	Chilled and Heating Water Return	LPS	Low Pressure Steam
CHWS	Chilled and Heating Water Supply	MB	Mixing Box
CWR	Chilled Water Return	RA	Return Air
CWS	Chilled Water Supply	RAD	Return Air Damper
EF	Exhaust Fan	SA	Supply Air
FA	Fresh Air	UH	Unit Heater
FAD	Fresh Air Damper	UV	Unit Ventilator
FCU	Fan Coil Unit	VD	Volume Damper
FD	Fire Damper		

Figure 5.20 *(b)* List of abbreviations commonly used in HVAC work.

(d) Do not insulate perimeter heating ducts run in the concrete floor slab. These ducts help to heat the slab and contribute materially to comfort in the space.

(e) Do insulate ducts run outside and in attics, garages, crawl spaces and any other space where the heat loss (or gain) is clearly wasted.

(f) If at all possible, avoid installation of ducts in exterior walls and attics. If unavoidable, insulate the ducts according to the formula in the next paragraph.

As a rule, the R value of duct insulation should be

$$R = \frac{\Delta t \, (°F)}{15}$$

where Δt is the temperature difference between the air in the duct and surrounding ambient air temperature.

Thus, for a duct passing through an uninsulated attic and carrying cooling air at 55°F, we would first estimate the summer attic temperature to be 140°F. Then

$$R = \frac{140 - 55}{15} = \frac{85}{15} = 6, \text{ use R-6 insulation.}$$

For the same duct carrying warm air at 140°F, with a winter attic temperature of 20°F:

$$R = \frac{140 - 20}{15} = \frac{120}{15} = 8, \text{ use R-8 insulation}$$

We would, therefore, probably insulate the duct for the larger of the two requirements, that is, R-8. We say probably because it is essentially an economic decision. As an absolute minimum, use R-2 insulation on ducts in insulated enclosed crawl spaces and in uninsulated basements and R-4 everywhere else.

Since blanket insulation is necessarily compressed when installed around a duct, even if glued, most authorities recommend doubling the normal thickness required. That assumes a 50% compression. Using this assumption, typical duct insulation would be:

- R-2: 1 in. of (compressed) glass fiber blanket or ½ in. of rigid glass fiber board or liner
- R-4: 2 in. of (compressed) glass fiber blanket or 1 in. of rigid glass fiber board or liner
- R-6: 3 in. of (compressed) glass fiber blanket or 1½ in. of rigid glass fiber board or liner
- R-8: 2 in. (compressed) glass fiber blanket covered with 1 in. of rigid glass fiber board or liner

In all cases, a vapor barrier on the outside of the insulation should be provided if the duct is carrying cool conditioned air.

As an indication of the importance of duct insulation, we can calculate temperature gain and loss in air carried by an uninsulated duct. An uninsulated duct 50 ft long, carrying 2000 cfm of 55°F air through a 140°F attic at 700 fpm (common conditions for a trunk feeder duct) will gain about 11°F. That means that the cool air supplied at the end will be at 66°F instead of 55°F. Obviously, the desired cooling will not be achieved. Similarly, in winter, the same duct carrying 1500 cfm of warm air at 140°F through the same attic at 20°F will lose almost 20°F. That means that the warm air will be delivered at 120°F instead of 140°F. Since the total temperature rise is about 70°F (from 70°F room temperature to 140°F supply temperature), this 20°F loss represents a loss of almost 30% of the heating capacity of the duct! Obviously then, duct insulation can be critical for ducts passing through unconditioned, uninsulated interior spaces or through exterior areas.

5.15 Air Distribution Outlets

Conditioned air originates at the warm air furnace (or heat pump/refrigeration unit) is distributed by the duct system and terminates at the room outlets. After mixing with room air, the "deconditioned" air is picked up at a return air outlet. It is then carried back to the furnace for reprocessing and recirculation. Supply air outlets are normally registers or diffusers, and occasionally grilles. Return air outlets are almost always grilles, and occasionally registers.

A *grille* is any slotted, louvered or perforated cover that fits onto a duct termination. See Figure 5.21. Louvered grilles have horizontal and/or vertical vanes. These vanes may be fixed or movable. The purpose of the grille is to permit free passage of air while generally obscuring direct vision or access into the duct. Fixed-opening grilles are normally mounted on return ducts and are wall-, ceiling- or floor-mounted. Adjustable vane units are normally used on supply ducts and are wall- or floor-mounted.

a. Definitions

A *register* is a grille that is equipped with some type of volume damper for the control of air flow. See Figure 5.22. It normally has movable vanes for directing the air stream. Registers are almost always used on supply ducts. In very special cases requiring volume damping, registers may also be used on return ducts.

A *diffuser* is a supply air outlet normally designed to distribute the air it supplies in a widespread pattern roughly parallel to the surface in which it is mounted. Most diffusers are intended for ceiling mounting, although wall diffusers are used as well. The majority of diffusers are equipped with devices that permit changing their air distribution patterns. They may or may not be equipped with volume dampers, as needed. Diffusers can be round, square, rectangular or linear as required. See Figure 5.23.

b. Comfort Zone

The purpose of a supply register or diffuser is to introduce conditioned air into a space in such a fashion that comfort is maintained. After all, the whole purpose of supplying conditioned air is to establish comfortable conditions in the room. Let us review very briefly what these comfort conditions are, as regards air temperature and air speed (see Section 5.1).

Winter: 68–74°F DB, 40 fpm maximum
Summer: 73–79°F DB, 50 fpm maximum

It has also been found that the difference in temperature between head and feet should not exceed 4–5°F to maintain comfort. It is important to note that the moving air in the previously stated comfort criteria is assumed to be at the same temperature as the room air. If the air is colder, these acceptable velocities drop because of the feeling that there is an unpleasant "draft" in the room. This condition is most critical with air movement around the head and neck and least critical at ankle level. It has also been found that elderly people are much more "draft" sensitive than young people.

Now consider that the air in the branch ducts strikes the supply register or diffuser at 500–600 fpm, at a winter temperature of 130–150°F or a summer temperature of 53–57°F. It should be obvious that no register or diffuser, however well designed, can instantaneously convert this air to the previously listed comfort conditions. This is one of the reasons that the American Society of Heating, Refrigeration and Air Conditioning Engineers (ASHRAE), in its standard 55, which defines comfort conditions, specifies an "occupied zone" in a room, inside which comfort conditions must be maintained. Another reason is that, in rooms with cold walls or ceilings, there is always a temperature gradient between the comfort conditions in the occupied zone and the area near cold walls. This occupied zone is shown in Figure 5.24. It is simply the area from the floor to a height of 6 ft and the room volume that is 2 ft from any wall. Establishing this occupied zone means that air from a wall, floor or ceiling supply outlet has about 2 ft within which to alter its temperature and

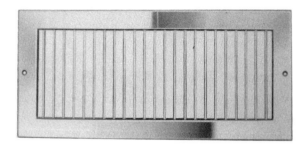

Single deflection
supply grille

(a-1)

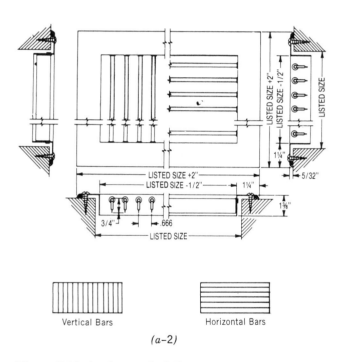

(a-2)

Figure 5.21 *(a-1)* A single deflection supply grille controls the air stream flow in two directions. Units are available with horizontal or vertical bars. *(a-2)* Section through the grille showing construction of both designs. *(a-3)* Performance data of the grille. (Courtesy of Carnes Company.)

PERFORMANCE DATA

SIZE		V=Duct Vel.	300			400			500			600			700					800	900	1,000	1,200
		Blade Set°	0	22½	45	0	22½	45	0	22½	45	0	22½	45	0	22½	45						
		P_t	.010	.012	.027	.014	.021	.050	.023	.034	.082	.034	.051	.120	.047	.071	.165						
6x6		CFM	75			100			125			150			175			12x10 24x5		250	330	420	500	580
8x4		Throw	8			10			13			16			18			14x8 30x4		14	18	23	26	31
		NC	L			L			L			L			L			20x6		L	L	L	22	26
8x5		CFM	85			110			140			165			195			12x12 18x8 36x4		300	400	500	600	700
10x4		Throw	8			11			14			17			19			14x10 24x6		15	20	25	29	34
		NC	L			L			L			L			L			16x8 28x5		L	L	L	24	28
8x6		CFM	100			135			170			200			235			16x10 40x4		330	430	540	650	760
10x5		Throw	9			12			15			18			21			26x6		16	21	26	30	36
12x4		NC	L			L			L			L			L			30x5		L	L	20	25	29
14x4		CFM	115			155			195			235			270			14x14 24x8 40x5		410	540	680	820	950
		Throw	9			12			16			19			23			16x12 32x6 48x4		17	22	29	34	40
		NC	L			L			L			L			L			20x10 34x6		L	21	28	31
8x8 16x4		CFM	130			180			220			260			310			16x14 48x5		470	620	780	930	1090
10x6		Throw	9			13			17			20			24			18x12		18	24	31	37	43
12x5		NC	L			L			L			L			20			36x6		L	L	22	28	32
12x6		CFM	150			200			250			300			350			16x16 26x10		530	710	890	1070	1250
14x5		Throw	10			14			18			21			25			18x14 32x8		20	26	33	40	46
18x14		NC	L			L			L			L			21			22x12 42x6		L	L	23	29	33
10x8 20x4		CFM	170			230			280			340			390			18x16 30x10		600	800	1000	1200	1400
14x6		Throw	11			15			19			22			27			20x14 36x8		21	27	35	42	49
16x5		NC	L			L			L			L			22			24x12 48x6		L	L	24	30	34
10x10 20x5		CFM	210			280			350			400			480			18x18 28x12 40x8		680	900	1120	1350	1580
12x8 24x4		Throw	13			17			21			24			29			20x16 30x12		22	29	37	44	52
16x6		NC	L			L			L			L			24			24x14 36x10		L	L	26	31	35
18x6		CFM	230			310			380			460			530			20x20 28x14 48x8		830	1110	1390	1670	1950
28x4		Throw	14			18			22			25			30			22x18 34x12		24	32	41	49	57
		NC	L			L			L			21			25			24x16 40x10		L	21	27	33	37

Various values for various blade settings and damper openings

SYMBOLS

V = Duct velocity in fpm.
CFM = Quantity of air in cubic ft./min.
NC = Noise criteria (8 db room attenuation). re 10^{-12} watts.
P_t = Total pressure inches H_2O.
T = Throw in feet.
L = NC less than 20

SINGLE DEFLECTION

Notes:
(1) Additions and factors (listed below) have to be applied for varying blade settings and damper openings.
(2) For sizes, CFM, blade settings or damper openings, etc., not listed, interpolate as necessary.

NC — Add the following db to the NC obtained from Table No. 1 for various blade settings.

Duct Velocity	300	400	500	600	700	800	900	1,000	1,200
45°			6	6	5	5	5	5	5

T — Multiply the T listed in Table No. 1 by the following F_1 factor for various blade settings.

Blade Setting	0°	22½°	45°
Factor	1	.89	.60

(a-3)

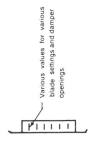

Figure 5.21 (*Continued*)

Fixed blade return air grille
(b-1)

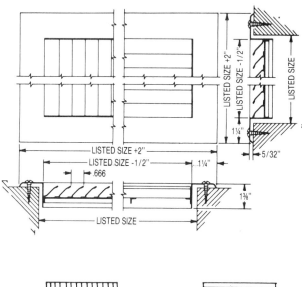

(b-2)

Figure 5.21 *(b-1)* Return air grille with stationary curved blades. The blades are available in vertical or horizontal position. Performance is unaffected by blade orientation. *(b-2)* Section through the grille showing construction of both designs. *(b-3)* Performance data. (Courtesy of Carnes Company.)

Return Air Performance Data

45° Blade Settings

Various values for damper openings.

SYMBOLS:
V = Duct velocity in fpm.
CFM = Quantity of air in cubic ft./min.
NC = Noise criteria (8 db room attenuation). re 10^{-12} watts.
L = NC less than 20
P_t = Total pressure inches H_2O.

Size			V P_t	200 .03	400 .09	600 .16	800 .25	1000 .35
4x4			CFM NC	22 L	44 L	66 L	88 20	110 23
6x6 8x4			CFM NC	50 L	100 L	150 L	200 24	250 28
8x6 10x5	12x4		CFM NC	65 L	130 L	200 L	270 25	340 31
10x6 12x5	16x4 18x4		CFM NC	80 L	160 L	240 21	320 27	400 33
8x8 12x6	14x5		CFM NC	90 L	180 L	260 22	350 28	440 34
10x8 14x6	16x5 20x4		CFM NC	110 L	220 L	330 23	440 30	550 36
10x10 28x4	20x5 18x6	12x8	CFM NC	140 L	280 L	400 25	550 32	690 39
12x10 30x4	24x5 20x6	14x8	CFM NC	160 L	320 L	480 25	640 33	800 40
14x10 36x4	28x5 22x6	16x8	CFM NC	190 L	380 L	570 27	760 35	950 41
12x12 30x5	26x6 18x8	16x10 40x4	CFM NC	200 L	400 L	600 28	800 36	1000 42
14x14 34x6	24x8 20x10	16x12	CFM NC	270 L	540 L	820 30	1090 38	1360 44
16x14 18x12	48x5 36x6		CFM NC	310 L	620 20	930 31	1240 39	1550 46
16x16 18x14	24x10 22x12	30x8	CFM NC	360 L	710 21	1070 33	1420 41	1780 48
18x16 20x14	30x10 24x12	36x8 48x6	CFM NC	400 L	800 22	1200 34	1600 42	2000 49
18x18 20x16	30x12 24x14	36x10 40x8	CFM NC	450 L	900 23	1350 35	1800 43	2200 50
22x20 24x18	30x14 26x16	36x12	CFM NC	600 L	1200 26	1800 37	2400 46	3000 52
24x24 30x18	48x12 36x16		CFM NC	800 L	1600 28	2400 40	3200 49	4000 55
28x24 32x20	48x14 36x18		CFM NC	900 L	1800 29	2700 41	3600 50	4500 56
30x24 48x16	36x20		CFM NC	1000 L	2000 30	3000 42	4000 51	5000 57
36x24 48x18			CFM NC	1200 20	2400 31	3600 43	4800 52	6000 59

(b-3)

AIR DISTRIBUTION OUTLETS / 239

Linear grille/register
(c-1)

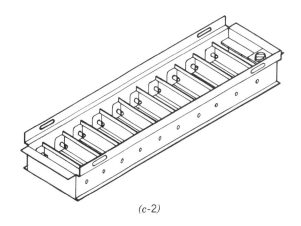

(c-2)

FLOOR APPLICATION MODELS CCGB and CCHB

DIMENSIONAL DATA

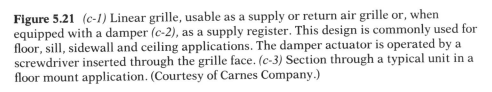

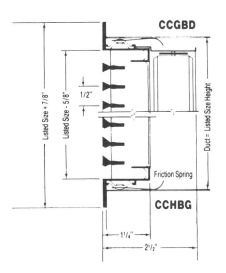

Model	Degree Blade Deflection	Damper	Straightening Vanes	Available Sizes*			
				Listed Size Height		Listed Size Width	
				Min.	Max.	Min.	Max.
CCGBG	0	NO	NO	2	12	6	72
CCHBG	15	NO	NO	2	12	6	72
CCGBD	0	YES	NO	2	12	6	72
CCHBD	15	YES	NO	2	12	6	72
CCGBS	0	NO	YES	2	12	6	72
CCHBS	15	NO	YES	2	12	6	72
CCGBB	0	YES	YES	2	12	6	72
CCHBB	15	YES	YES	2	12	6	72

*Minimum & Maximum single unit sizes. Larger list widths furnished as separate pieces.

(c-3)

Figure 5.21 *(c-1)* Linear grille, usable as a supply or return air grille or, when equipped with a damper *(c-2)*, as a supply register. This design is commonly used for floor, sill, sidewall and ceiling applications. The damper actuator is operated by a screwdriver inserted through the grille face. *(c-3)* Section through a typical unit in a floor mount application. (Courtesy of Carnes Company.)

Sound ratings are based on a 4 foot unit with the damper full open, and 10 db room attenuation. For lengths other than 4 feet, use the table below to determing the increase in noise level.

No. of 4 foot lengths	db to be added
1	0
2	3
3	5
4	6
6	8
10	10

Tests show that drastic dampering at the grille will result in considerable db increase. Dampering at the grille should be reserved for fine balancing. Gross balancing should be provided for by dampers upstream in the supply ductwork.

NC values shown in the performance tables are for the damper in the full open position. Partially closed dampers will increase the NC level as shown in the table below.

Effective Damper Opening %	db to be added
100	0
82	8
71	13
50	21

"L" indicates an NC value less than 20.

The total and static pressure is with damper in the full open position and is given in inches water gage (W.G.)

THROW

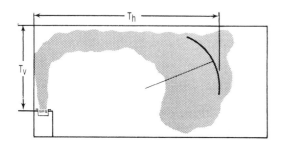

SILL & FLOOR APPLICATION

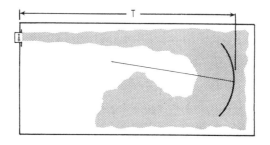

SIDEWALL APPLICATION

Throw values are based on a 4 foot length of grille having 0° or 15° blade deflection and supply air temperature equal to room air temperature. The maximum throw value shown is based on a V_t of 50 FPM and the minimum throw value on 150 FPM. Throw values for sidewall application are based on an 8 to 10 foot mounting height (See sketches above).

Cooler supply air will result in shorter throw vaules.

Warmer supply air will result in longer throw values. Use the multiplication factors in the table below to determine throw values depending on supply air temperature.

V_t FPM	Isothermal	$\Delta_t = -20°$ F	$\Delta_t = +20°$ F
150	1.00	1.00	1.00
50	1.00	.90	1.10

(c-4)

PERFORMANCE DATA – 0° BLADE DEFLECTION

List Size Height	A_k Per Ft. of Length	Duct Velocity - FPM		200	300	400	500	600	700
		Total Pressure P_t		.010	.025	.046	.073	.107	.147
		Static Pressure P_t		.008	.020	.037	.058	.085	.117
2"	.038	CFM/FT.		33	50	67	84	100	117
		NC		L	22	31	38	44	48
		Throw in Ft.	Sidewall	8 - 4	9 - 6	11 - 6	11 - 8	12 - 8	15 - 9
			Sill-Floor	12 - 8	13 - 10	14 - 11	15 - 12	16 - 12	18 - 13
2½"	.063	CFM/FT.		42	62	83	104	125	146
		NC		L	L	22	29	35	39
		Throw in Ft.	Sidewall	8 - 4	10 - 6	12 - 7	12 - 8	13 - 8	15 - 9
			Sill-Floor	13 - 9	14 - 11	15 - 12	16 - 12	17 - 12	18 - 14
3"	.089	CFM/FT.		50	75	100	125	150	175
		NC		L	L	L	22	29	33
		Throw in Ft.	Sidewall	8 - 4	10 - 6	12 - 7	13 - 8	14 - 8	15 - 9
			Sill-Floor	13 - 9	15 - 11	16 - 12	17 - 12	18 - 13	19 - 14
3½"	.114	CFM/FT.		58	88	117	146	175	204
		NC		L	L	L	L	23	28
		Throw in Ft.	Sidewall	8 - 4	11 - 7	13 - 8	14 - 9	15 - 9	16 - 10
			Sill-Floor	14 - 9	15 - 11	17 - 12	18 - 13	19 - 13	20 - 15
4"	.139	CFM/FT.		67	100	133	176	200	233
		NC		L	L	L	L	22	26
		Throw in Ft.	Sidewall	10 - 4	12 - 7	14 - 8	15 - 9	16 - 10	17 - 11
			Sill-Floor	14 - 9	16 - 11	18 - 12	19 - 13	20 - 14	21 - 15
4½"	.164	CFM/FT.		75	113	150	188	225	263
		NC		L	L	L	L	22	27
		Throw in Ft.	Sidewall	11 - 4	13 - 7	15 - 9	16 - 10	17 - 10	18 - 12
			Sill-Floor	15 - 9	16 - 12	19 - 13	20 - 14	21 - 14	22 - 16
5"	.189	CFM/FT.		83	125	167	209	250	292
		NC		L	L	L	L	22	27
		Throw in Ft.	Sidewall	12 - 4	14 - 8	16 - 9	17 - 10	18 - 11	20 - 13
			Sill-Floor	15 - 9	17 - 12	20 - 13	21 - 14	22 - 15	23 - 16
6"	.238	CFM/FT.		100	150	200	250	300	350
		NC		L	L	L	L	23	39
		Throw in Ft.	Sidewall	12 - 4	15 - 8	17 - 10	19 - 11	20 - 12	22 - 13
			Sill-Floor	16 - 9	18 - 12	20 - 13	22 - 15	23 - 16	25 - 17
8"	.322	CFM/FT.		133	200	267	334	400	467
		NC		L	L	L	L	24	28
		Throw in Ft.	Sidewall	13 - 4	16 - 9	18 - 11	21 - 12	23 - 13	25 - 14
			Sill-Floor	17 - 10	19 - 13	21 - 14	23 - 16	25 - 17	27 - 18

Note: A_k is the effective free area, in square feet

(c-5)

Figure 5.21 *(c-5)* Typical performance data for 0° blade deflection. A similar table is available for 15° deflection.

Double-deflection register
(a-1)

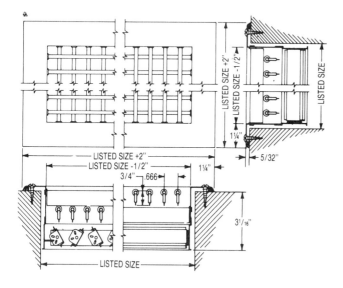

(a-2)

Figure 5.22 *(a-1)* Double-deflection supply register combines a double deflection grille with an opposed blade damper. This combination provides air deflection in one, two, three or four directions, together with positive volume control. Units are available with vertical front bars and horizontal rear bars, or the reverse. *(a-2)* Section through the register showing construction of both designs. (Courtesy of Carnes Company.)

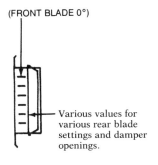

(FRONT BLADE 0°)

Various values for various rear blade settings and damper openings.

DOUBLE DEFLECTION

SYMBOLS:
 V = Duct velocity in fpm
 CFM = Quantity of air in cubic ft./min.
 NC = Noise criteria (8 db room attenuation) re 10^{-12} watts.
 P_t = Total pressure inches H_2O.
 T = Throw in feet.
 L = NC less than 20

Performance Data

Size			V = Duct Vel. Blade Set° P_t	300 0 .011	22½ .015	45 .032	400 0 .020	22½ .026	45 .055	500 0 .030	22½ .040	45 .085	600 0 .043	22½ .058	45 .120	700 0 .058	22½ .079	45 .165
6x8 8x4			CFM Throw NC	75 8 L			100 11 L			125 14 L			150 16 L			175 19 L		
8x5 10x4			CFM Throw NC	85 8 L			110 11 L			140 15 L			165 17 L			195 21 L		
8x6 10x5 12x4			CFM Throw NC	100 9 L			135 12 L			170 16 L			200 19 L			235 23 20		
14x4			CFM Throw NC	115 9 L			155 13 L			195 17 L			235 20 L			270 24 21		
8x8 10x6 12x5	16x4		CFM Throw NC	130 10 L			180 14 L			220 18 L			260 21 L			310 26 22		
12x6 14x5 18x4			CFM Throw NC	150 11 L			200 15 L			250 19 L			300 22 L			350 27 23		
10x8 14x6 16x5	20x4		CFM Throw NC	170 12 L			230 16 L			280 20 L			340 24 L			390 29 24		
10x10 12x8 16x6	20x5 24x4		CFM Throw NC	210 14 L			280 18 L			350 22 L			400 26 20			480 31 25		
18x6 28x4			CFM Throw NC	230 15 L			310 19 L			380 24 L			460 28 21			530 33 26		
12x10 14x8 20x6	24x5 30x4		CFM Throw NC	250 15 L			330 20 L			420 25 L			500 29 22			580 34 27		
12x12 14x10 16x8	18x8 24x6 28x5	36x4	CFM Throw NC	300 16 L			400 21 L			500 27 L			600 31 23			700 36 28		
16x10 26x6 30x5	40x4		CFM Throw NC	330 17 L			430 22 L			540 28 L			650 32 24			760 38 28		
14x14 16x12 20x10	24x8 32x6 34x6	40x5 48x4	CFM Throw NC	410 18 L			540 24 L			680 31 L			820 36 25			950 43 29		
16x14 18x12 36x6	48x5		CFM Throw NC	470 19 L			620 26 L			780 33 20			930 49 26			1090 46 30		
16x16 18x14 22x12	26x10 32x8 42x6		CFM Throw NC	530 21 L			710 28 L			890 35 21			1070 43 27			1250 49 32		
18x16 20x14 24x12	30x10 36x8 48x6		CFM Throw NC	600 22 L			800 29 L			1000 37 22			1200 45 27			1400 52 32		

Notes: (1) Additions and factors (listed below) have to be applied for varying blade settings and damper openings.
(2) For sizes, CFM, blade settings or damper openings, etc., not listed below, interpolate as necessary.

Model RTDA—Register (Front Blade 0°)
NC—Add the following db to the NC obtained from Table for various rear blade settings.

Dual Velocity	300	400	500	600	700	800	900	1,000	1,200
Rear Blade 0°	2	2	2	1	1	1	1	1	1
Rear Blade 45°	12	12	12	11	11	10	10	10	10

NC—Add the following db to the NC obtained above for various damper openings.

Damper Opening	100%	75%	50%
db Add	0	10	22

P_t—Multiply the P_t listed in Table by the following F_2 factor for the wide open damper.

Blade Setting	0°	22½°	45°
Factor	1.70	1.50	1.10

T—Multiply the T listed in Table by the following F_1 factor for various blade settings.

Rear Blade Setting	0°	22½	45°
Factor	1	.89	0.60

(a-3)

Figure 5.22 (a-3) Performance data of the double-deflection register with front blade set at 0° deflection.

Performance Data

Size				V = Duct Vel. Blade Set° P_t	300 0 .016	22½ .019	45 .034	400 0 .028	22½ .033	45 .059	500 0 .043	22½ .051	45 .091	600 0 .062	22½ .074	45 .130	700 0 .084	22½ .099	45 .175
6x6 8x4				CFM Throw NC	75 7 L			100 10 L			125 12 L			150 14 20			175 17 24		
8x5 10x4				CFM Throw NC	85 7 L			110 10 L			140 13 L			165 15 21			195 18 25		
8x6 10x5 12x4				CFM Throw NC	100 8 L			135 11 L			170 14 L			200 17 22			235 20 26		
14x4				CFM Throw NC	115 9 L			155 12 L			195 15 L			235 18 22			270 21 26		
8x8	16x4	10x6	12x5	CFM Throw NC	130 9 L			180 12 L			220 16 L			260 19 23			310 23 27		
12x6	14x5	18x4		CFM Throw NC	150 10 L			200 13 L			250 17 L			300 20 24			350 24 28		
10x8	20x4	14x6	16x5	CFM Throw NC	170 11 L			230 14 L			280 18 20			340 21 24			390 26 28		
10x10	20x5	12x8	24x4 16x6	CFM Throw NC	210 12 L			280 16 L			350 20 20			400 23 25			480 28 29		
18x6	28x4			CFM Throw NC	230 13 L			310 17 L			380 21 21			460 24 26			530 29 30		
12x10	24x5	14x8	30x4 20x6	CFM Throw NC	250 13 L			330 17 L			420 22 21			500 25 28			580 30 30		
12x12	18x8	36x4	14x10 24x6 16x8 28x5	CFM Throw NC	300 14 L			400 19 L			500 24 22			600 28 27			700 32 31		
16x10	40x4	26x6	30x5	CFM Throw NC	330 15 L			430 20 L			540 25 22			650 29 27			760 34 31		
14x14	24x8	40x5	16x12 32x6 48x4 20x10 34x6	CFM Throw NC	410 16 L			540 21 L			680 28 23			820 32 28			950 38 32		
16x14	48x5	18x12	36x6	CFM Throw NC	470 17 L			620 23 L			780 29 23			930 34 28			1090 41 33		
16x16	26x10	18x14	32x8 22x12 42x6	CFM Throw NC	530 19 L			710 25 L			890 31 24			1070 38 29			1250 44 34		
18x16	30x10	20x14	36x8 24x12 48x6	CFM Throw NC	600 20 L			800 26 L			1000 33 24			1200 40 30			1400 47 34		

Notes: (1) Additions and factors (listed below) have to be applied for varying blade settings and damper openings.
(2) For sizes, CFM, blade settings or damper openings, etc., not listed below, interpolate as necessary.

Model RTDA—Register (Front Blade 22½°)
NC—Add the following db to the NC obtained from Table for various rear blade settings.

Duct Velocity	300	400	500	600	700	800	900	1,000	1,200
Rear Blade 0°	2	2	2	1	1	1	1	1	1
Rear Blade 45°	7	7	6	6	6	6	6	5	5

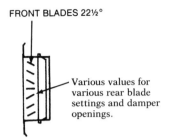

FRONT BLADES 22½°

Various values for various rear blade settings and damper openings.

DOUBLE DEFLECTION

SYMBOLS:
V = Duct velocity in fpm
CFM = Quantity of air in cubic ft./min.
NC = Noise criteria (8 db room attenuation) re 10^{-12} watts.
P_t = Total pressure inches H_2O.
T = Throw in feet.
L = NC less than 20

NC—Add the following db to the NC obtained above for various damper openings.

Damper Opening	100%	75%	50%
db Add	0	10	22

P_t—Multiply the P_t listed in Table by the following F_2 factor for the wide open damper.

Blade Setting	0°	22½°	45°
Factor	1.50	1.25	1.08

T—Multiply the T listed in Table by the following F_1 factor for various blade settings.

Rear Blade Setting	0°	22½°	45°
Factor	1	.89	.60

(a-4)

Figure 5.22 *(a-4)* Performance data of the register with front blade set at 22½°.

AIR DISTRIBUTION OUTLETS / 245

Four-way air supply register
(b-1)

REGISTER DIMENSIONS

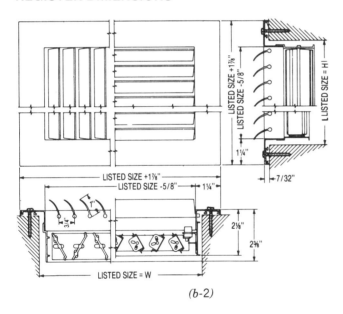

(b-2)

Figure 5.22 *(b-1)* Register with four sets of adjustable deflectors. This unit establishes a four-way air pattern and apportions equal quantities of primary air in four directions. *(b-2)* Section through the register showing construction of blades and volume dampers. (Courtesy of Carnes Co.)

PERFORMANCE DATA

CURVED BLADE

CFM	SIZE W X H (WIDE X HIGH) DIRECTIONAL THROW	10 x 4 / 8 x 5 / 6 x 6				12 x 4 / 10 x 5 / 8 x 6				14 x 5 / 12 x 6 / 8 x 8				24 x 6 / 18 x 8 / 14 x 10 / 12 x 12			
		1	2	3	4	1	2	3	4	1	2	3	4	1	2	3	4
50	Duct Velocity	200				165											
	Total Pressure	.02				.01											
	Throw	4-6	3-5	2-4	2-4	3-5	2-4	2-4	2-4								
	Min. Ceiling Height	8	8	8	8	8	8	8	8								
75	Duct Velocity	300				250											
	Total Pressure	.03				.03											
	Throw	5-8	5-8	4-7	4-7	4-7	3-6	3-5	3-5								
	Min. Ceiling Height	8	8	8	8	8	8	8	8								
100	Duct Velocity	400				330				200							
	Total Pressure	.05				.03				.02							
	Throw	7-11	6-10	5-8	5-8	6-10	5-8	4-6	4-6	5-7	4-6	3-5	3-5				
	Min. Ceiling Height	8	8	8	8	8	8	8	8	8	8	8	8				
125	Duct Velocity	500				415				250							
	Total Pressure	.08				.06				.02							
	Throw	9-14	8-12	7-16	6-10	7-11	6-10	6-10	5-11	5-7	4-6	4-6	3-5				
	Min. Ceiling Height	8	8	8	8	8	8	8	8	8	8	8	8				
150	Duct Velocity	600				450				300				150			
	Total Pressure	.12				.07				.03				.01			
	Throw	12-18	10-16	8-12	7-11	9-14	8-12	7-11	6-10	7-11	5-7	5-7	4-6	4-6	3-5	3-5	3-5
	Min. Ceiling Height	8	8	8	8	8	8	8	8	8	8	8	8	8	8	8	8
200	Duct Velocity	800				670				400				200			
	Total Pressure	.18				.16				.05				.02			
	Throw	14-20	12-18	10-16	9-14	11-16	9-14	8-12	8-12	9-14	7-11	7-11	6-9	5-8	4-6	4-6	4-6
	Min. Ceiling Height	9	8	8	8	9	8	8	8	9	8	8	8	9	8	8	8
300	Duct Velocity									600				300			
	Total Pressure									.12				.03			
	Throw									13-19	12-18	10-16	9-14	8-12	7-11	6-9	5-7
	Min. Ceiling Height									9	8	8	8	9	8	8	8
400	Duct Velocity									800				400			
	Total Pressure									.23				.05			
	Throw									18-27	14-21	13-19	12-18	11-16	9-14	8-12	7-11
	Min. Ceiling Height									10	8	8	8	10	8	8	8
500	Duct Velocity									1000				500			
	Total Pressure									.37				.08			
	Throw									20-30	18-27	16-24	14-21	13-19	12-18	10-16	9-14
	Min. Ceiling Height									10	9	8	8	10	9	8	8
600	Duct Velocity													600			
	Total Pressure													.12			
	Throw													16-24	13-19	12-18	11-17
	Min. Ceiling Height													12	10	8	8
800	Duct Velocity													800			
	Total Pressure													.23			
	Throw													22-33	18-27	15-22	14-21
	Min. Ceiling Height													13	10	9	8
1000	Duct Velocity													1000			
	Total Pressure													.37			
	Throw													27-40	20-30	18-27	16-24
	Min. Ceiling Height													15	12	10	9

(b-3)

Figure 5.22 (b-3) Performance data of the four-way register.

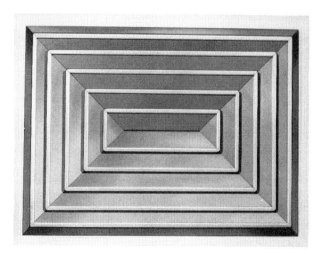

Rectangular four-way blow ceiling diffuser
(a-1)

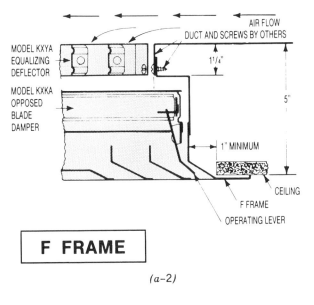

(a-2)

FOUR-WAY BLOW

NECK SIZE IN INCHES Y and Z	VELOCITY	300	400	500	600	700	800
	PRESSURE DROP	.028	.049	.080	.111	.145	.195
6 x 9 Area .375 Sq. Ft.	Total CFM	113	150	190	225	260	300
	NC	L	14	20	25	29	34
	CFM Each Side Y	38	50	65	75	85	100
	CFM Each Side Z	19	25	30	38	45	50
	Throw Each Side Y	2-3	3-4	5-6	6-7	7-8	9-10
	Throw Each Side Z	1-2	2-3	3-4	4-5	5-6	6-7
6 x 12 Area .50 Sq. Ft.	Total CFM	150	200	250	300	350	400
	NC	L	13	20	25	29	33
	CFM Each Side Y	56	75	93	112	131	150
	CFM Each Side Z	19	25	32	38	44	50
	Throw Each Side Y	4-5	5-6	6-7	7-9	9-11	10-12
	Throw Each Side Z	1-2	2-3	3-4	4-5	5-6	6-7
6 x 15 Area .625 Sq. Ft.	Total CFM	188	250	312	375	438	500
	NC	L	15	21	27	31	35
	CFM Each Side Y	75	100	124	150	175	200
	CFM Each Side Z	19	25	32	38	44	50
	Throw Each Side Y	5-7	6-8	8-10	9-11	11-13	12-14
	Throw Each Side Z	1-2	2-3	3-4	4-5	5-6	6-7
6 x 18 Area .75 Sq. Ft.	Total CFM	225	300	375	450	525	600
	NC	L	15	21	27	31	35
	CFM Each Side Y	94	125	156	188	218	250
	CFM Each Side Z	19	25	32	38	44	50
	Throw Each Side Y	6-8	7-9	9-11	10-12	11-13	12-15
	Throw Each Side Z	1-2	2-3	3-4	4-5	5-6	6-7
6 x 21 Area .875 Sq. Ft.	Total CFM	263	350	438	525	612	700
	NC	L	16	23	28	33	37
	CFM Each Side Y	112	150	187	224	262	300
	CFM Each Side Z	19	25	32	38	44	50
	Throw Each Side Y	7-9	8-10	10-12	11-13	12-15	13-16
	Throw Each Side Z	1-2	2-3	3-4	4-5	5-6	6-7
6 x 24 Area 1.00 Sq. Ft.	Total CFM	300	400	500	600	700	800
	NC	L	17	23	28	33	37
	CFM Each Side Y	131	175	218	262	306	350
	CFM Each Side Z	19	25	32	38	44	50
	Throw Each Side Y	8-10	9-11	11-13	12-15	13-16	14-18
	Throw Each Side Z	1-2	2-3	3-4	4-5	5-6	6-7
9 x 12 Area .75 Sq. Ft.	Total CFM	225	300	375	450	525	600
	NC	L	15	21	27	31	35
	CFM Each Side Y	70	94	118	141	165	188
	CFM Each Side Z	42	56	70	84	98	112
	Throw Each Side Y	4-6	5-7	7-9	8-10	10-12	11-14
	Throw Each Side Z	2-3	3-4	5-7	6-8	7-9	8-10
9 x 15 Area .938 Sq. Ft.	Total CFM	282	375	470	563	656	750
	NC	L	16	23	28	33	37
	CFM Each Side Y	99	132	165	198	230	263
	CFM Each Side Z	42	56	70	84	98	112
	Throw Each Side Y	5-7	6-8	8-10	10-13	12-15	13-16
	Throw Each Side Z	2-3	3-4	5-7	6-8	7-9	8-10
9 x 18 Area 1.125 Sq. Ft.	Total CFM	338	450	562	675	788	900
	NC	L	16	24	29	34	37
	CFM Each Side Y	127	169	211	254	296	338
	CFM Each Side Z	42	56	70	84	98	112
	Throw Each Side Y	6-9	7-10	9-12	11-14	12-15	14-18
	Throw Each Side Z	2-3	3-4	5-7	6-8	7-9	8-10
9 x 21 Area 1.31 Sq. Ft.	Total CFM	393	524	655	786	917	1050
	NC	L	17	24	29	34	38
	CFM Each Side Y	155	206	258	309	360	413
	CFM Each Side Z	42	56	70	84	98	112
	Throw Each Side Y	6-8	7-9	10-13	12-15	13-17	15-19
	Throw Each Side Z	2-3	3-4	5-7	6-8	7-9	8-10
9 x 24 Area 1.50 Sq. Ft.	Total CFM	450	600	750	900	1050	1200
	NC	10	18	25	30	35	39
	CFM Each Side Y	183	244	305	366	427	488
	CFM Each Side Z	42	56	70	84	98	112
	Throw Each Side Y	9-12	10-13	12-15	13-16	14-18	16-20
	Throw Each Side Z	2-3	3-4	5-7	6-8	7-9	8-10
12 x 15 Area 1.25 Sq. Ft.	Total CFM	375	500	625	750	875	1000
	NC	10	18	25	30	35	39
	CFM Each Side Y	113	150	188	225	263	300
	CFM Each Side Z	75	100	125	150	175	200
	Throw Each Side Y	5-7	6-8	8-10	10-13	12-15	13-16
	Throw Each Side Z	4-6	5-7	7-9	9-11	10-13	11-14
12 x 18 Area 1.50 Sq. Ft.	Total CFM	450	600	750	900	1050	1200
	NC	10	18	25	30	35	39
	CFM Each Side Y	150	200	250	300	350	400
	CFM Each Side Z	75	100	125	150	175	200
	Throw Each Side Y	5-7	6-8	8-11	12-15	13-17	15-19
	Throw Each Side Z	4-6	5-7	7-9	9-11	10-13	11-14

(a-3)

Figure 5.23 *(a-1)* Typical rectangular four-way blow steel ceiling diffuser. *(a-2)* Mounting detail with an F-frame, in a hung ceiling. Deflector and damper are also shown. *(a-3)* Diffuser performance data as a function of duct air velocity. (Courtesy of Carnes Company.)

248 / AIR SYSTEMS, HEATING AND COOLING, PART I

SOUND RATINGS

Figure 1: Noise rating chart for Model SK Diffusers without Damper

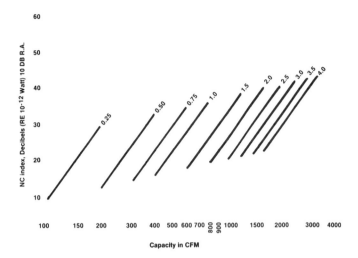

To use Noise Rating Chart, Figure 1, determine the square footage of the diffuser neck area from the chart in Figure 5.23 (a-1). Follow the vertical CFM line on the chart until it intersects the diagonal square footage line. Read the NC Level on the left hand side of the chart.
For the effect of dampering, use chart Figure 2.

Figure 2: Effect of Damper on NC Index

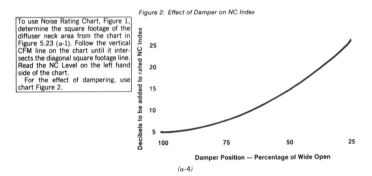

(a-4)

Figure 5.23 *(a-4)* Noise ratings of the four-way ceiling diffuser as a function of neck area.

Round steel fixed pattern ceiling diffuser

(b-1)

Figure 5.23 *(b-1)* Fixed-pattern round steel ceiling diffuser, with four cones. Similar units are available with adjustable patterns. (Courtesy of Carnes Company.)

DIMENSIONS

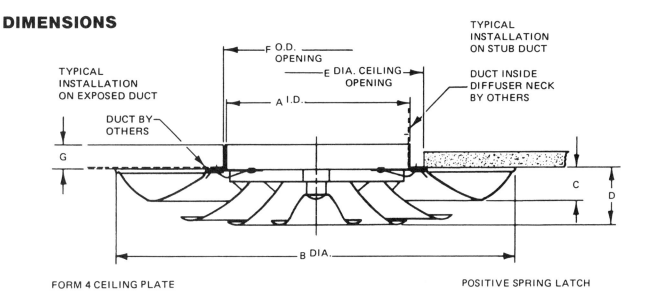

FORM 4 CEILING PLATE

POSITIVE SPRING LATCH

Size	Form 4						
	A	B	C	D	E	F	G
04	4¹/₁₆	14½	1³/₁₆	1¹⁵/₁₆	7¼	4⅛	3/4
05	5¹/₁₆	14½	1³/₁₆	1¹⁵/₁₆	7¼	6⅛	1⅜
06	6¹/₁₆	14½	1³/₁₆	1¹⁵/₁₆	7¼	6⅛	1
08	8¹/₁₆	17⁹/₁₆	1⅜	2⅜	9¼	8⅛	1
10	10¹/₁₆	19¹³/₁₆	1³/₁₆	1⅞	11¼	10⅛	1
12	12¹/₁₆	22¹/₁₆	1⅜	2⁵/₁₆	13¼	12⅛	1
14	14¹/₁₆	24¹/₁₆	1½	2⁵/₁₆	15¼	14⅛	1
16	16¹/₁₆	28¹³/₁₆	2⅛	2¾	17¼	16⅛	1
18	18¹/₁₆	32¹/₁₆	2⅛	3⅛	19¼	18⅛	1
20	20¹/₁₆	33⁹/₁₆	2¼	3⅛	21¼	20⅛	1
24	24¹/₁₆	39⁵/₁₆	2⅝	3¹¹/₁₆	25¼	24⅛	1
28	28¹/₁₆	48⁹/₁₆	2¹¹/₁₆	4½	29¼	28⅛	1
32	32¹/₁₆	60¹³/₁₆	2⅞	4¹⁵/₁₆	33¼	32⅛	1
36	36¹/₁₆	60¹³/₁₆	2⅞	4¹⁵/₁₆	37¼	36⅛	1

Dimensions are given in inches.

(b-2)

Figure 5.23 *(b-2)* Installation detail for diffuser mounting directly on a stub duct. Units can also be mounted on a damper, a splitter/damper, a radial deflector, an equalizing damper or a direction changer as required by the particular installation.

Performance Data—Model SSEA

Neck Size (Inches)		Duct Velocity - FPM				
		200	400	600	800	1000
4	CFM	17	35	52	70	87
	PT	.005	.02	.05	.09	.14
	Minimum Throw	3	3	4	4	4
	Maximum Throw	5	5	5	6	6
	NC	L	L	L	24	30
5	CFM	27	55	82	109	136
	PT	.01	.03	.07	.11	.17
	Minimum Throw	3	3	4	4	4
	Maximum Throw	5	5	6	6	7
	NC	L	L	20	28	34
6	CFM	39	78	115	155	195
	PT	.01	.04	.08	.13	.20
	Minimum Throw	3	3	4	4	5
	Maximum Throw	5	6	6	6	7
	NC	L	L	22	30	36
8	CFM	70	140	210	280	350
	PT	.01	.03	.07	.12	.18
	Minimum Throw	4	4	5	5	6
	Maximum Throw	6	6	8	8	10
	NC	L	L	27	34	40
10	CFM	109	218	325	435	545
	PT	.01	.06	.12	.24	.36
	Minimum Throw	4	5	6	7	7
	Maximum Throw	7	9	11	13	14
	NC	L	L	29	37	42
12	CFM	158	315	475	630	785
	PT	.01	.04	.10	.20	.31
	Minimum Throw	5	6	6	7	10
	Maximum Throw	8	10	12	14	18
	NC	L	21	31	39	45
14	CFM	215	430	645	855	1070
	PT	.02	.05	.11	.20	.31
	Minimum Throw	6	8	11	13	18
	Maximum Throw	9	13	15	19	23
	NC	L	22	32	40	46
16	CFM	280	660	840	1120	1400
	PT	.01	.07	.14	.26	.41
	Minimum Throw	7	10	13	17	19
	Maximum Throw	10	14	17	22	25
	NC	L	22	33	41	47

(b-3)

Figure 5.23 *(b-3)* Typical performance data for the illustrated diffuser.

RECTANGULAR LOUVERED FACE DIFFUSER: Available in 1, 2, 3, or 4-way pattern, steel or aluminum. Flanged overlap frame or inserted in 2 X 2 ft or 2 X 4 ft baked enamel steel panel to fit tile modules of lay-in ceilings. Supply or return.

ROUND LOUVERED FACE DIFFUSER: Normal 360° air pattern with blank-off plate for other air patterns. Surface mounting for all type ceilings. Normally of steel with baked enamel finish. Supply or return.

RECTANGULAR PERFORATED FACE DIFFUSER: Available in 1, 2, 3, or 4-way pattern, steel or aluminum. Flanged overlap frame or 2 X 2 ft and 2 X 4 ft for replacing tile of lay-in ceiling can be used for supply or return air.

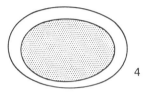

ROUND PERFORATED FACE DIFFUSER: Normal 360° air pattern with blank-off plate for other air patterns. Steel or aluminum. Flanged overlap frame for all type ceilings. Can be used for supply or return air.

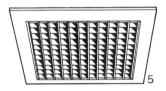

LATTICE TYPE RETURN: All aluminum square grid type return grille for ceiling installation with flanged overlap frame or of correct size to replace tile.

(c)

Figure 5.23 (c) Common air distribution outlets and their principal characteristics. (From Ramsey and Sleeper, *Architectural Graphic Standards*, 8th ed., p. 628, 1988, © John Wiley & Sons, reprinted by permission of John Wiley & Sons.)

252 / AIR SYSTEMS, HEATING AND COOLING, PART I

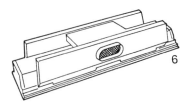

SADDLE TYPE LUMINAIRE AIR BOOT: Provides air supply from both sides of standard size luminaires. Maximum air delivery (total both sides) approximately 150 to 170 cfm for 4 ft long luminaire.

SINGLE SIDE TYPE LUMINAIRE AIR BOOT: Provides air supply from one side of standard size luminaires. Maximum air delivery approximately 75 cfm for 4 ft long luminaire.

LINEAR DIFFUSER: Extruded aluminum, anodized, duranodic, or special finishes, one way or opposite direction or vertical down air pattern. Any length with one to eight slots. Can be used for supply or return and for ceiling, sidewall, or cabinet top application.

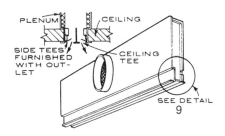

INTEGRATED PLENUM TYPE OUTLET FOR "T" BAR CEILINGS: Slot type outlet, one way or two way opposite direction air pattern. Available in 24, 36, 48, and 60 in. lengths. Replaces or integrates with "T" bar. Approximately 150 to 175 cfm for 4 ft long, two-slot unit.

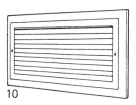

SIDEWALL OR DUCT MOUNTED REGISTER: Steel or aluminum for supply or return. Adjustable horizontal and vertical deflection. Plaster frame available. Suitable for long throw and high air volume.

(c)

Figure 5.23 (c) (Continued)

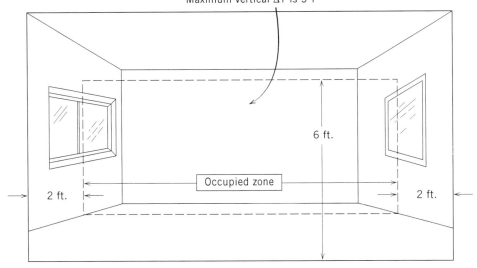

Figure 5.24 The occupied zone of a room is defined as the volume 2 ft from each wall and 6 ft high off the floor. In this space, design comfort conditions must be maintained. A maximum vertical temperature differential of 5 F° between ankle height at 4 in. above finished floor (AFF) and neck height of 67 in. AFF should be maintained year round.

velocity to acceptable limits. Further, within this area it must at least begin to mix with room air, in order to transfer its heat (or coolness) without leaving stagnant spaces.

5.16 Outlet Characteristics

The operating characteristics of a register or diffuser describe the flow of air from the unit and how this supply air mixes with the room air. The following glossary defines the terms that describe these actions in two categories: outlet performance and air mixing.

a. Outlet Performance

(1) *Drop.* When the supply air is colder than the room air, it drops as it travels across the space. The performance criterion known as drop is the vertical distance that the lower edge of the air pattern falls, between the outlet and the end of its throw. See Figure 5.25a. See also the term *Rise*.

(2) *Face velocity* (outlet velocity). The average air velocity coming out of the outlet, measured in the plane of the opening. Since the velocity will vary over the face of the outlet, multiple measurements must be taken and averaged.

(3) *Free area.* The open area of a register or grille through which air can pass, unobstructed. The free area, which varies between 60 and 90% of the gross area, determines the face velocity and pressure drop of the outlet. See Figure 5.26.

(4) *Gross area.* The area of the inside dimensions of the frame, that is, the grille area, not including the device frame. See Figure 5.26.

(5) *Isothermal jet.* An air jet at the same temperature as the room air.

(6) *Noise Criterion (NC).* An indication of the background noise level acceptable in a specific space. The NC rating of an outlet is determined by plotting the decibel values of the noise it creates on a special graph. The outlet

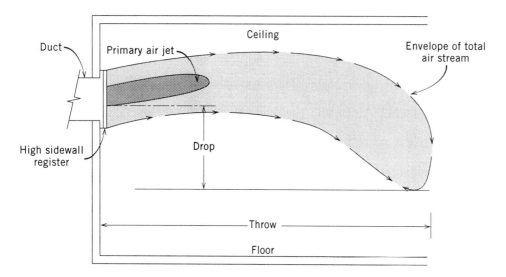

Figure 5.25 *(a)* The drop of a cold air stream is measured between the bottom of the primary air jet and the bottom of the total air stream envelope, at the end of its throw. Drop increases (and throw decreases) as the temperature difference between room air and incoming air increases, simply because the entering colder air is heavier. See also Figure 5.37.

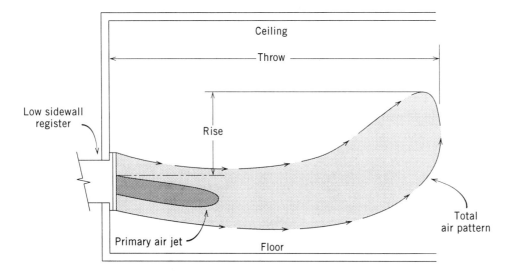

Figure 5.25 *(b)* The rise of a warm air stream is measured between the top of the primary air jet and the top of the total air pattern, at the end of its throw. Rise increases (and throw decreases) as the temperature difference between incoming air and room air increases, because the incoming warmer air is lighter. See also Figure 5.39.

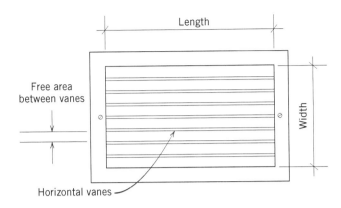

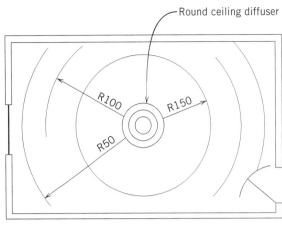

Figure 5.26 The gross area of a supply outlet is the entire area inside the frame. For the rectangular diffuser shown, it is the product of length and width. The free area is the open area between vanes, that is, the gross area less the area of vanes. The illustrated unit shows only horizontal vanes. For units equipped with vertical vanes as well, their area must also be subtracted from gross area to obtain net, or free, area.

Figure 5.27 The radius of diffusion of a round ceiling diffuser is its throw, to a specified terminal velocity. In the diagram, R_{150} is the throw of the air stream to a terminal velocity of 150 fpm. R_{100} and R_{50} are the (radial) throws to terminal velocities of 100 and 50 fpm, respectively.

NC rating is then compared to the maximum permissible NC of that space, to determine whether the outlet is usable in that space.

(7) *Radius of diffusion.* The horizontal distance that an air stream travels after leaving a ceiling outlet before the maximum velocity drops to a specified level, usually between 50 and 200 fpm. This is the term used for the throw of a ceiling diffuser that discharges radially. See Figure 5.27.

(8) *Rise.* When the entering supply air is warmer than the room air, it tends to rise because it is also lighter than the room air. Rise is the vertical distance that the upper edge of the total air pattern rises, between the outlet and the end of its throw. See Figure 5.25(b). See also *Drop*.

(9) *Spread.* A measure of the dispersion of an air stream from a wall or floor outlet caused by the vertical vanes in the face of the outlet. As the spread increases, the throw decreases. When the vanes are set for zero deflection, the air pattern has approximately the same dispersion and throw as a free discharge. See Figure 5.28.

(10) *Surface effect.* See Figure 5.29. When an outlet is located within about 1 ft of a room surface, the motion of the air stream creates a low pressure area between the air stream and the room surface (floor, wall or ceiling). This forces the air stream against the surface, reducing air entrainment, increasing throw, and decreasing drop. This effect is useful in perimeter heating and cooling systems, to blanket walls with conditioned air without causing drafts.

(11) *Temperature differential.* The difference in temperature between supply air and average room DB temperature.

(12) *Terminal velocity.* The maximum air stream velocity at the end of the throw. See Figure 5.30.

(13) *Throw.* See Figure 5.30. The horizontal (or vertical) axial distance an air stream travels from the outlet until the maximum stream velocity drops to a specified minimum, usually between 50 and 200 fpm. Throw data listed for outlets in manufacturers' catalogs are for isothermal air streams discharging into an open, clear, unobstructed space. It, therefore, does not take into account surface effect for high sidewall outlets or floor perimeter outlets, or drop for cold air, or rise for warm air.

(14) *Vane.* A portion of the register face designed to direct the air stream. See Figure 5.26.

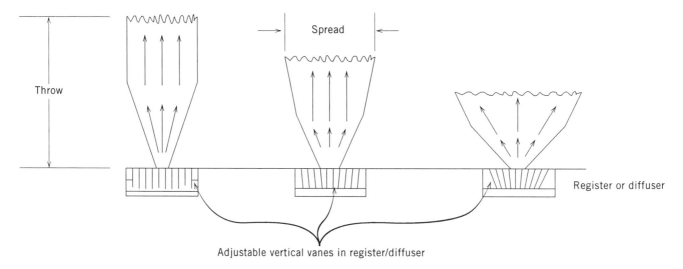

Figure 5.28 Vertical vanes in the register face can be adjusted to vary the width or spread of the air stream pattern. As the stream widens, its length, or throw, shortens.

(15) *Velocity.* Generally refers to the maximum axial velocity of an air stream. Peripheral velocity is considerably lower.

(16) *Vertical temperature gradient.* See Figure 5.24. Unless otherwise specified, the temperature differential between air at 4 in. above the floor (ankle height) and air at 67 in. above the floor (neck height). In spaces where people are normally seated, such as an auditorium, neck height is taken to be 42 in. AFF (above finished floor). These gradients occur when there is incomplete mixing of primary and secondary air and leads to areas of stagnant air.

b. Characteristics of the Air Stream

(1) *Diffusion.* The distribution in a space, of supply air from an outlet, and its mixing with room air.

(2) *Entrainment; entrained air.* The action by which room air moves into, and mixes with, the stream of primary air from the supply outlet; the air so entrained. See also *Induction.* See Figure 5.30.

(3) *Induction.* The process by which room air is drawn to an outlet by aspiration of the primary air stream. The combined air then constitutes the air stream. See Figure 5.30.

(4) *Envelope.* The outer boundary of the moving air stream, where motion is caused by the primary air jet. It does not include air moving from convective air currents.

(5) *Primary air.* The air delivered by the supply duct to the outlet. See Figure 5.30.

(6) *Primary air pattern.* The shape of the air stream from the supply outlet, where the air velocity at the outer edges of the air stream envelope is not less than 150 fpm. The air in the envelope consists of primary air, induced room air and entrained room air. See Figure 5.30.

(7) *Secondary air; room air.* The amount of secondary (room) air in the total air pattern is usually 10–20 times the amount of primary air. Also refers to the room air drawn into the primary air stream.

(8) *Stagnant air; stagnant zone.* Still room air that is substantially unaffected by the primary air stream. The zone or area of a space containing stagnant air. See Figure 5.31. A condition caused by insufficient mixing of primary and secondary air. Air motion in the stagnant zone is caused by convective currents only.

(9) *Stratification.* The formation of areas of stagnant air that are unaffected by the primary air stream. The air in the room is, therefore, stratified, with a moving strata (layer) and a stagnant (still) layer. Air motion in the stagnant layer is caused only by natural convec-

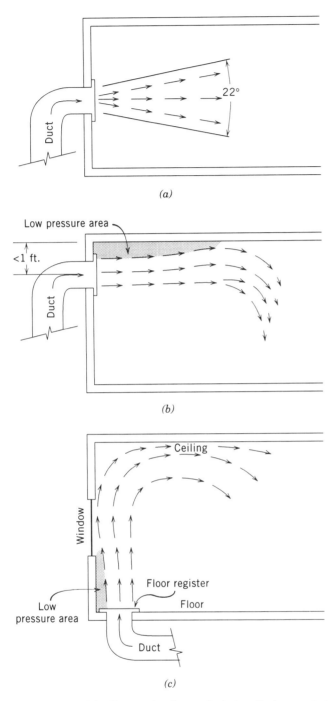

Figure 5.29 *(a)* When an isothermal air jet discharges into an open unobstructed space, it expands in a 22° cone. *(b)* If the air stream is within about 1 ft of a wall or ceiling, a low pressure area is formed between the jet and the boundary. (This area is shown shaded.) This forces the air stream against the boundary (wall, ceiling) and increases the air stream's throw. This action is called the surface effect. The surface effect can be induced in air streams from sidewall outlets somewhat farther from the ceiling by setting the register's horizontal vanes so that the air stream strikes the ceiling at a glancing angle. This will produce the effect illustrated and will lengthen the throw. Too sharp an angle of incidence will cause turbulence that will shorten the throw. *(c)* By placing a floor heating register within 1 ft of a window wall, the warm air jet is forced against the window by the surface effect. There it mixes with the cold air sliding down from the window. The warm mixture rises as shown.

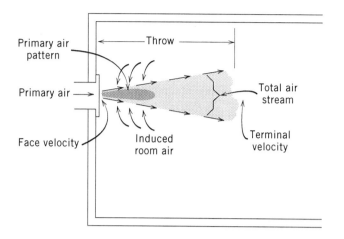

Figure 5.30 Throw length of an unobstructed air stream is measured from the register to the point at which maximum air stream velocity has dropped to a specified level—usually between 75 and 200 fpm. Primary air mixes with room air by a process of induction to form the total air stream. Room air induced into the stream is called entrained air.

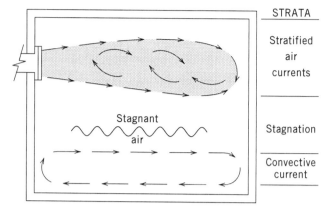

Figure 5.31 Stratification occurs when the moving air currents set up by the primary air jet remain at one elevation, or strata, and do not mix with all, or most of, the room air. In the illustration, a high sidewall register introduces warm air, which remains close to the ceiling because it is lighter than the cooler room air. Just below this level is a layer of stagnant, or dead air. At the floor level, low velocity (below 15 fpm) convective air currents move, powered by the heat transfer into and out of the room.

tion currents and is typically less than 20 fpm. See Figure 5.31.
(10) *Supply air.* See *Primary air.*
(11) *Total air pattern.* The envelope of all the air moving in a space as a result of the supply air stream. It does not include air moving in convective currents.

5.17 Outlet Selection, Location and Application

Outlets are selected and located according to application (heating, cooling or both), duct location (overhead, floor level), type of heating/cooling system (perimeter, radial, extended plenum, etc.) and the characteristics of the supply system including supply air velocity and temperature differentials. The simplest selection is for an all-heating or all-cooling system. Heated air, being lighter than the room air, tends to rise. Therefore, for an all-heating system, warm air should be supplied at or near the floor level to allow it to rise; return air, which has cooled and dropped, can be picked up near the floor level across the room. Conversely, cooled air is heavier than the room air. Therefore, in an all-cooling system, cool air should be introduced at or near the ceiling to allow it to drop and circulate. After warming, the air will rise, and it can be picked up near the ceiling, at a remote location. Supply and return outlets at the same elevation must be widely separated to avoid "short-circuiting" the air flow before it has a chance to mix properly with room air. For this reason, heating returns are sometimes mounted high, and cooling returns, low.

Location of outlets for a combined heating/cooling system is much more difficult. Here the specific characteristic of the supply outlet (spread, throw, adjustability) must be examined in order to make a proper selection. The return outlet should be placed in the stagnant air area so as to increase circulation. (Note, however, that return outlets do not "draw" and, therefore, will not affect the supply air pattern. See Figure 5.32.) When using low perimeter supply outlets, stagnant air will concentrate near the ceiling during cooling [Figure 5.33(b)] and near the floor during heating [Figure 5.33(a)]. Since it is impossible to satisfy both conditions with a single return outlet, two choices remain. Either use a high wall return grille because the stagnant air problem is more severe during cooling or use a high and a low return register on a

OUTLET SELECTION, LOCATION AND APPLICATION / 259

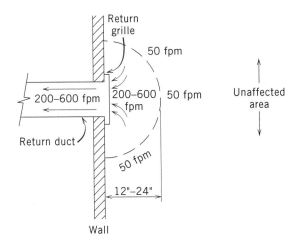

Figure 5.32 The location of a return grille does not affect the air pattern in a space. Within 12–24 in. of the return grille face, air velocity is approximately 50 fpm and static pressure is zero. At the grille, face, velocities range from 200–600 fpm and a slight negative static pressure exists.

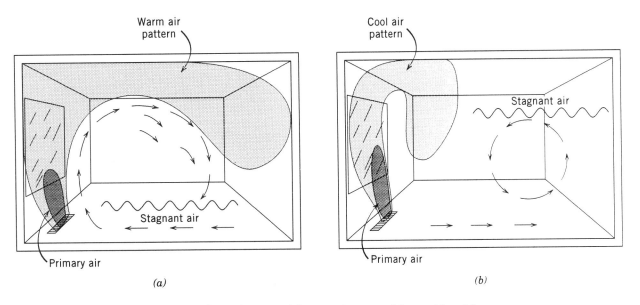

Figure 5.33 (a) Warm primary air from the spread floor register combines with cold air coming off the window and induced room air, to form a strong circular current. The room is free of drafts and uniformly heated. Linear diffusers and low sidewall diffusers will produce the same result. See Figure 5.35. (b) The cooling air pattern falls back on itself near the window, resulting in inadequate mixing of primary air and room air. The higher the throw, the better the room cooling. The throw should at least reach the ceiling to achieve satisfactory room cooling.

single duct, with volume dampers. During cooling, the upper register is opened; during the heating season, the lower register is opened and the upper is closed. Since two registers are considerably more expensive than a single grille, this solution is seldom used. Furthermore, the occupant must be sufficiently knowledgeable to perform this function. In practice, most designers would use a single high outlet. See Figure 5.1.

In combined heating/cooling systems using high sidewall outlets (Figure 5.34), the largest stagnant air pool during heating occurs at floor level. This then would be the location of the return outlet, which is also effective for cooling.

a. Principles of Outlet Selection

The general rules governing the selection of outlets are these:

(1) The supply air stream shape should be selected to mix with room air so that no stratification and stagnant air remains in the occupied zone.
(2) The location of the supply outlet(s) and the air stream shape must be such that drafts are avoided (i.e., air velocity in the occupied zone should not exceed 50 fpm).
(3) During the heating season, perimeter heating should be provided to counteract cold air dropping from windows, outside doors and cold walls.
(4) The position of the return grille(s) should be such that supply air is not "short-circuited" before it can condition the room air.
(5) The throw of supply outlets should be such that high velocity air does not reach room boundaries (walls or ceiling). Such a situation would indicate excessive velocity and insufficient mixing of primary and secondary air. Terminal velocity should occur at the boundary. Therefore,

  (a) The total air pattern of vertical throw outlets should reach the ceiling (Figure 5.33).
  (b) The total air pattern of high sidewall outlets should reach the opposite wall (Figure 5.34).
  (c) The total air pattern of ceiling diffusers should reach the room walls.

(6) Large rooms require multiple supply outlets. These must have air streams whose total air

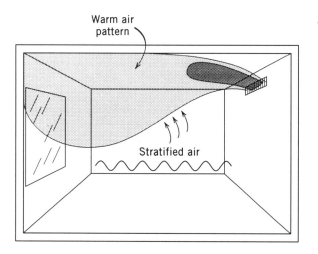
(a)

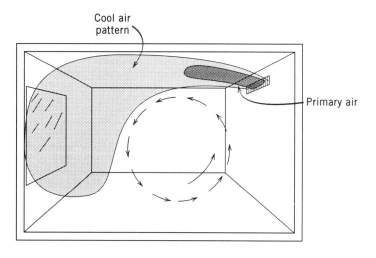
(b)

Figure 5.34 *(a)* The warm air will remain along the ceiling leaving a cool stagnant layer below. With a long enough throw, the warm air will reach the cool window and establish proper circulation. Excessive throw will result in undesirable drafts. *(b)* Cool air will readily entrain rising warm room air, and a good circulation will be established. The HSW inside wall register location is recommended for cooling and ventilation using outside air.

envelopes overlap to provide adequate coverage, without collision of high velocity air streams.

(7) The NC rating of outlets must be no higher than the NC rating of the room in which they are installed.

b. Data Analysis

Application of these principles requires an analysis of three groups of data:

(1) The architectural details of the space, with particular reference to the location of heat loss or gain such as windows, doors, ceiling under a roof and outside walls.
(2) The amount (cfm) of air required and its temperature, as calculated from the heat loss or heat gain required.
(3) The characteristics of supply registers and diffusers, as given in the technical data section of manufacturers' catalogs, with particular reference to velocity, throw, diffusion and NC.

c. Outlet Selection Procedure

Using these general rules, or principles, and the three groups of data, the selection procedure for outlets is as follows:

(1) Decide on the outlet types and locations based on duct locations, window and door positions and type of HVAC system being designed.
(2) Use a manufacturer's catalog that gives complete outlet data including cfm, throw for specific temperature(s), spread and noise. Select outlet(s) that will deliver the maximum cfm required (usually the cool air cfm).
(3) Check the throw for a specific terminal velocity. Throws terminating in the occupied zone should not exceed 50 fpm. If the throw ends at a room boundary (wall or ceiling), it can have a terminal velocity up to 150 fpm, but preferably 50–100 fpm.
(4) Check the outlet pressure drop against the system static pressure drop calculations.
(5) Check the outlet noise level (NC value) at the face velocity being used. Table 5.1 is a short list of recommended NC values for a few common occupancies, plus maximum face velocity for most outlets that correspond to this NC criteria. In actual practice, always check face velocity data for a specific outlet. The velocities in

Table 5.1 Recommended NC Criteria for Various Occupancies and Maximum Supply Outlet Face Velocity

Occupancy	NC Criteria	Maximum Face Velocity
Large auditoriums	20–25	500
Small auditoriums, theaters, houses of worship, conference rooms, executive office	25–30	600
Sleeping room	25–35	700
Private office, libraries, small conference rooms	30–35	700
Living rooms, recreation rooms	35–40	800
Lobbies, drafting rooms, computer areas	40–45	800
Merchandising areas	40–50	900
Light industry, kitchen, equipment rooms	50–60	1000

Table 5.1 are provided only as a preliminary guideline.

You should be aware that the location of the return grille does not affect the air distribution in a room, except in the extremely unusual case that it is so close to the supply duct that the air flow is "short-circuited." (See Figure 5.32.)

As a rule of thumb, plain return outlets should be designed for a face velocity of 300–400 fpm and a maximum of 500 fpm. Return outlets above the occupied zone such as ceiling or high sidewall units may have face velocities up to 600 fpm. Door louvers, wall louvers and undercut doors should be limited to 300 fpm. Filter grilles should not have a face velocity above 300 fpm.

Using these figures, calculation of return grille size is straightforward.

Example 5.4 What size low sidewall plain return grille is required for a room supplied with 120 cfm of air for summer cooling and 85 cfm for winter heating?

Solution: We would size the return grille for the larger of the two airflow requirements. Using Equation (5.7), which is derived in Section 5.23.a (page 288)

$$Q = AV \tag{5.7}$$

where

Q is the airflow in cubic feet per minute,
A is the open area of the grille in square feet and
V is the air velocity in feet per minute

and trying a solution with a face velocity of 400 fpm (see rule in preceding paragraph), we have

$$120 \text{ cfm} = A \times 400 \text{ fpm}$$
$$A = \frac{120 \text{ ft}^3/\text{min}}{400 \text{ ft/min}} = 0.3 \text{ ft}^2$$
$$A = 0.3 \text{ ft}^2 \times \frac{144 \text{ in.}^2}{\text{ft}^2} = 43.2 \text{ in.}^2$$

Assuming a free area of 80% of gross area, we would calculate

$$A_{\text{gross}} = \frac{43.2}{0.8} = 54 \text{ in.}^2$$

Therefore, a 6×9-in. grille (or larger would be satisfactory.

5.18 Outlet Air Patterns

There are four basic air flow patterns in a space:

- Vertical, spreading
- Vertical, nonspreading
- Horizontal, at, or close to, ceiling level
- Horizontal, at, or close to, floor level

These characteristics of each pattern and its application are explained next.

a. Vertical Flow, Spreading Air Pattern

The vertical flow, spreading air pattern is produced by low sidewall diffusers, floor-mounted diffusers, and low wall linear diffusers such as the baseboard type. See Figure 5.35. Units are normally placed under windows and on cold walls to prevent cold air drafts from forming during the heating season. For heating, floor outlets below windows are preferable, particularly as part of a perimeter heating system. The total air pattern for heating appears approximately as in Figure 5.33(a). Throw should reach the ceiling in order to establish a good circular air motion pattern in the room. The return outlet is placed low on the opposite (inside) wall. This type of air pattern is not highly recommended for a combined heating and cooling system because the cooling air pattern will fall back on itself and a thorough mixing of primary and secondary air will not occur. [See Figure 5.33(b).] Careful selection of the outlet characteristic, however, can result in an acceptable cooling system.

b. Vertical Flow, Nonspreading Pattern

The vertical flow, nonspreading distribution is similar to the spreading type except that throw is much longer due to the smaller spread. The outlet types are the same as for the spread distribution. This distribution is usable for combined heating and cooling because the longer throw will substantially eliminate the stagnant air shown in Figure 5.33(b).

c. Horizontal Flow

High Side Wall Outlets. The best position for a high sidewall (HSW) register with respect to heating is on an inside wall opposite the room window, as shown on Figure 5.36. The total air heating and cooling patterns produced are shown in Figure 5.34. As can be seen, the high sidewall register location is ideal for cooling (and ventilation), but much less effective for heating due to the tendency of hot air to rise. It is, therefore, important that the primary air throw be long enough to reach the outside wall and window. There, cool air entrainment will lower the total air temperature, causing the air to drop and establish the desired circulation. Insufficient throw will leave a blanket of hot air on the ceiling and stagnant cold air below.

If a sufficiently long throw is not possible due to the architecture of the space, an alternate (and expensive) solution is to use separate outlets for heating and cooling. A single return is usually possible. Register throw is normally adjustable by using the vertical vanes on the register face. Typical throw patterns are shown in Figure 5.28. The register's horizontal vanes can also be used to increase throw by setting them so that the primary air strikes the ceiling at a glancing angle, within 3–4 ft of the wall. This will induce the surface effect and will markedly increase the air stream throw. See Figure 5.29(b). Too great a horizontal vane angle will cause the primary air stream to collide with the ceiling. This will cause turbulence that will shorten the air stream throw.

Ceiling Diffuser. This type of outlet is used primarily for cooling. If heating is required, a diffuser with a vertical discharge characteristic is required.

OUTLET AIR PATTERNS / 263

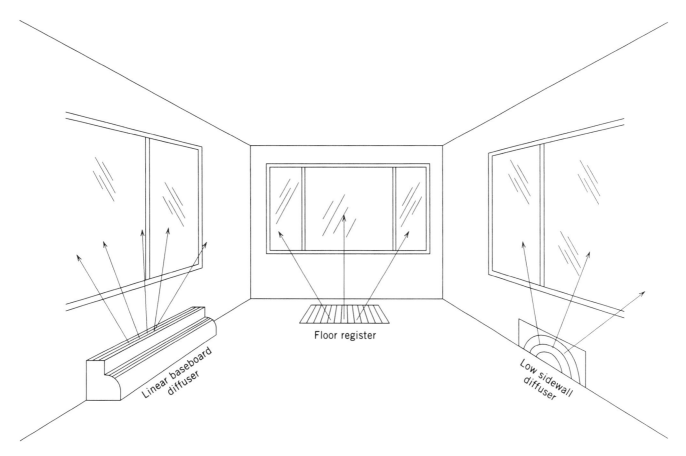

Figure 5.35 Typical vertical air stream spread-type outlets. These outlets are best located below windows. For heating duty, the spread patterns of warm air should blanket the windows. This prevents cold air drafts and increases the mixing of primary air with room air. Throw should reach the ceiling.

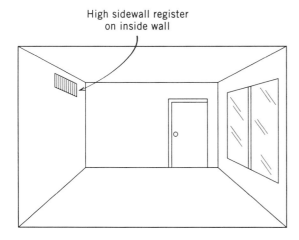

Figure 5.36 Typical horizontal flow HSW register. The HSW position is ideal for cooling and ventilation and marginally acceptable for heating if throw is adequate and the outlet is on an inside wall. See also the total air stream patterns in Figure 5.34.

Horizontal discharge diffusers are available with radial patterns and directional patterns. The former are usually round; the latter are square or rectangular. Horizontal flow ceiling diffusers have the ability to entrain a large amount of room air. As a result, and because the horizontal flow covers a large area, horizontal flow ceiling diffusers can handle large flow rates (high cfm) without producing unpleasant drafts. In cooling use, the diffuser's throw depends heavily on the temperature difference between the incoming air and the room air. See Figure 5.37. As a result, it is important for the design technologist to check the catalog throw data for a particular temperature difference. Typical cooling and heating air flow patterns are shown in Figure 5.38. The throw selected should reach the room walls, or, in large rooms, it should reach the air pattern of the adjacent diffusers. See Figure 5.27.

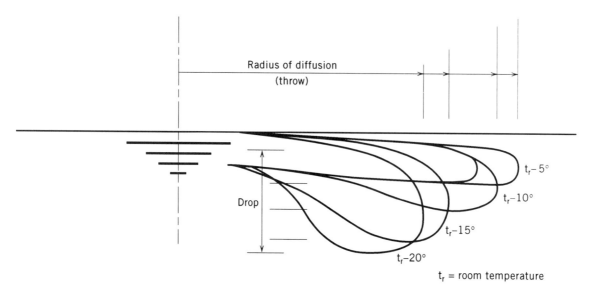

Figure 5.37 The radius of diffusion (radial throw) of a symmetrical ceiling diffuser depends on the temperature of the primary air. As the temperature of the primary air is lowered, the air becomes heavier, throw decreases and drop increases.

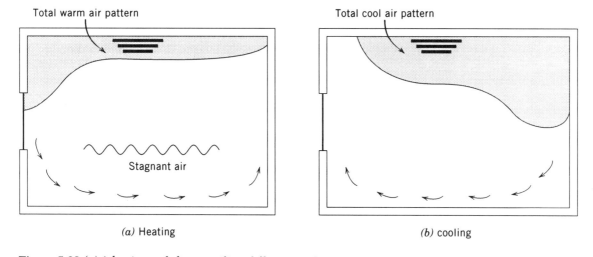

Figure 5.38 (a) A horizontal throw ceiling diffuser produces a layer of warm air that clings to the ceiling and does not mix with room air. This produces stratification, with warm air above and cool air below. The blanket is lower near the window where it mixes with cool air and, as a result, drops. (b) Cooling is effective with this distribution. The cool air drops on the inside wall and rises convectively from heat picked up at the window.

d. Horizontal Flow, Low Sidewall Outlets

This group of outlets includes sidewall registers and linear diffusers designed to direct the primary air stream essentially parallel to the floor (or at a low angle). The outlets are mounted on an inside wall, generally opposite a window. The shape of the primary air stream is a widening beam as it crosses the floor. The exact shape depends, of course, on the specific diffuser. The total air stream envelopes for both heating and cooling are shown in Figure 5.39.

Air entrainment for heating will substantially eliminate stratification and stagnation. However, since the primary air is projected directly into the occupied zone, some discomfort from excessive temperature and drafts may be felt. To eliminate these, it is recommended that supply outlets of this type be placed on inside walls, that face velocity not exceed 300 fpm and that supply air temperature not exceed 115–120°F. The problem will be even more acute in the cooling mode because cold drafts are much more annoying than warm ones.

Also, since the cold supply air does not rise but lays on the floor, entrainment is minimal, and air velocity remains high over the entire room floor. For this reason, low sidewall (LSW) outlets with horizontal throw are not recommended for cooling service. Table 5.2 summarizes the application of the four types of airflow distribution discussed previously.

All-Air Systems

The number of different types of all-air comfort conditioning systems that have been devised runs into the dozens. These systems attempt, often unsuccessfully, to condition the separate spaces in large buildings on an individual basis. The reason that so many systems have been invented is that in large buildings there is often a simultaneous demand for heating, cooling and ventilation. Rooms on the sunny side of a building need cooling, rooms on the shaded side need heating, and

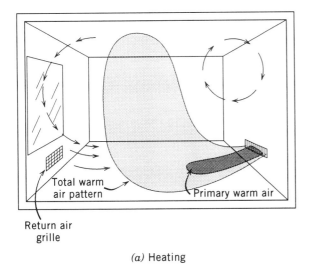

(a) Heating

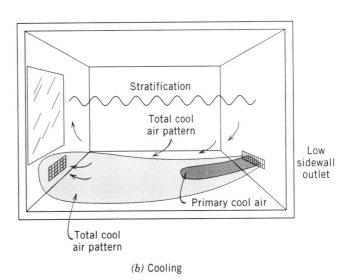

(b) Cooling

Figure 5.39 (a) The total air pattern from a horizontal discharge low sidewall diffuser will rise somewhere in the center of the room as shown. Higher air velocity must be avoided to prevent undesirable drafts. Induction and entraining occur on both sides of the total air pattern, with very little stagnant air remaining. This system, with proper balancing, is adequate for heating. (b) Cool air will simply lie at the floor level with this distribution. Very little mixing with room air occurs, causing stratification of the entire upper section of the occupied area. A large temperature differential develops between the cold floor and the warm upper level. This arrangement is not recommended for cooling.

Table 5.2 Application of Air Supply Outlets

Flow Pattern	Type of Outlet	Preferred Location	Recommended Application
Vertical, spread	LSW diffuser, linear baseboard, spread floor register	Exterior walls, below windows	Preferably heating, also cooling
Vertical, nonspread	LSW diffuser, linear baseboard, nonspread floor register	Exterior walls, below windows	Heating and cooling
Horizontal, high sidewall	HSW register	Inside wall	Cooling; also heating if carefully designed and adjusted
Horizontal, ceiling diffuser	Ceiling diffuser	Ceiling	Cooling
Horizontal, low sidewall	LSW register, linear diffusers	Interior walls, opposite windows	Heating

rooms in the building core need primarily ventilation. These large building systems are beyond the scope of the book as far as design is concerned. However, the engineering technologist will definitely be called upon to work on such systems under the direction of an HVAC engineer. That being so, he or she must have an overall concept of what these systems are and how they operate.

5.19 System Types

Of the many year-round comfort conditioning all-air systems in use, the most common are

- Single zone.
- Multiple zones.
- Single-duct reheat.
- Single-duct variable air volume (VAV).
- Dual duct.

a. Single-Zone System

See Figure 5.40. This is the system most often used for small single-use buildings that operate as a single zone. It is also the system that will be studied most intensively in this book. These systems are always low pressure and low velocity. A single air-handling unit, which supplies fixed quantities of air to the building spaces, is used. If a change in air volume is required for seasonal changeover (heating to cooling and vice versa), the duct volume dampers must be reset. The total air quantity can be changed at the air handler, generally by motor speed control. In modern systems, air quantities can be regulated automatically.

Multiple subzones can be established by thermostatic control of volume dampers in branch ducts feeding different areas, as shown in Figure 5.41. All these zones, however, are in the same mode—heating or cooling. Where this zoning arrangement is not satisfactory because of air temperature and volume requirements, several whole systems can operate in parallel; each supplies a separate zone with its own specific requirements. This system has low first cost, simple maintenance and long life. However, because it is designed to handle a structure as a single zone, this system is often not sufficiently flexible to maintain comfort conditions in all rooms, particularly during periods of load variation. As a result, larger buildings, including large residences, sometimes use a multizone system.

b. Multizone System

See Figure 5.42. There are several variations of this design. Essentially, a multizone system consists of a single air handler plus individual zone duct systems. Each duct system has a heating and cooling source supplied by a central heating device and a central refrigeration device. In some systems, each zone has heating and cooling coils; in other systems, hot and cold air are supplied, to be mixed

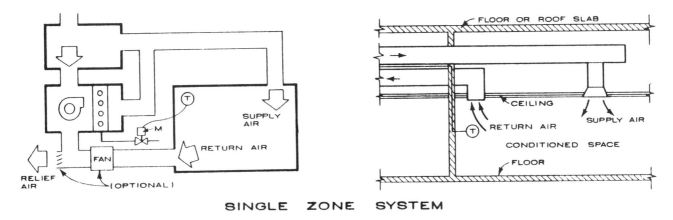

Figure 5.40 Single-zone systems are low pressure, low velocity installations, best applied to small residential and commercial buildings. The entire building is treated as a single control zone, controlled by a single thermostat. See text. (From Ramsey and Sleeper, *Architectural Graphic Standards*, 8th ed., 1988, John Wiley & Sons, reprinted by permission of John Wiley & Sons.)

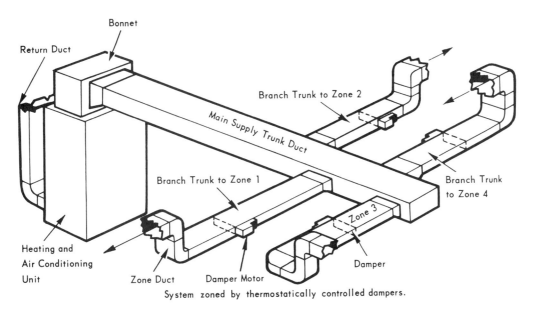

Figure 5.41 System zoned by thermostatically controlled motorized dampers. This system is essentially a single-zone extended-plenum duct system with automatic damper control. The single heating/cooling unit supplies only warm or only cool air, at a temperature suitable to supply the heaviest zone load. The other zones then throttle the air supply to satisfy their load requirements. (Reproduced with permission from *ACCA Manual C*, p. 21).

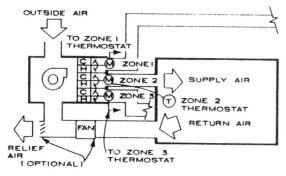
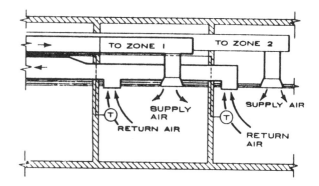

MULTIZONE SYSTEM

Figure 5.42 Multizone systems are used for medium-size buildings where different zones have different conditioned air requirements. Hot and cold air are produced centrally and provided to each zone. Therefore, heating and cooling can be provided simultaneously to different zones. Limiting factors are the number of zones and energy costs. See text. (From Ramsey and Sleeper, *Architectural Graphic Standards*, 8th ed., 1988, © John Wiley & Sons, reprinted by permission of John Wiley & Sons.)

at the entry to each zone's ductwork. Thus, one zone can use heating, another cooling and a third ventilation with outside air. As can be readily imagined, the ductwork for this multiple zone low pressure, low velocity system rapidly becomes enormous. This is the system's principal disadvantage. Other disadvantages are high first cost, difficult control and high energy cost. This latter is due to the fact that the return air from all the zones is mixed, since there is only one air handler. As a result, warm air is mixed with cold air, resulting in a large waste of energy. For small to medium-size buildings with no more than four zones, this system is a reasonable choice.

c. Single Duct with Reheat

See Figure 5.43. This system was developed before the energy crisis of 1973 made HVAC designers (and others) take a long hard look at their comfort conditioning systems. It is not commonly used today because it is notorious for energy waste. However, careful design can make it useful for some climates. The system was designed to solve the problem of massive ductwork in large multiple zone buildings using the multizone system already described. This system uses a single duct that provides air (generally very cold) to the entire building. At each zone, a small reheat coil heats this cold air to the temperature required for that zone. The central system must provide air cold enough to meet the maximum cooling load of the building's warmest zone. All other zones must reheat this cold air. These systems use air as cold as 40°F, which is then reheated as required. The system is applicable to large, multiple zone buildings using low, medium or high pressure systems. This system has high first cost and high energy cost. However, it provides excellent control and constant air volume and does not require a changeover to switch from heating to cooling or the reverse. Today, this system is used for labs, hospitals and other facilities requiring accurate temperature control and constant volume for ventilation requirements.

d. Single-Duct Variable Volume (VAV) System

See Figure 5.44. This system has become the most popular design for medium-size to large buildings because of low first cost, low energy cost and small ductwork. The system, as its name implies, compensates for variable loads by varying the volume of air supplied rather than its temperature. This is the "secret" of its energy economy. The air volume variation is accomplished by a thermostatically controlled variable air-volume box. This box takes main duct air from the single supply duct and modulates the air quantity supplied to a space, to match its load. The central supply will furnish either cold air or warm air to the entire building, depending on outdoor conditions and prevailing

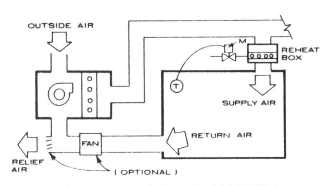
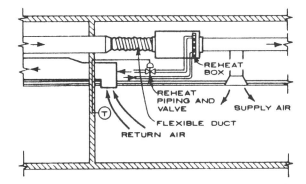

SINGLE DUCT REHEAT SYSTEM

Figure 5.43 The single-duct reheat system circulates constant temperature cold air, which is then reheated at each zone as required. This arrangement occupies little space and provides excellent control. However, it is extremely energy wasteful. As a result, it is seldom used in modern design. See text. (From Ramsey and Sleeper, *Architectural Graphic Standards*, 8th ed., 1988, © John Wiley & Sons, reprinted by permission of John Wiley & Sons.)

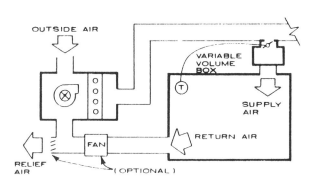
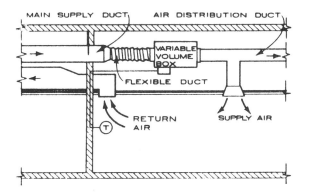

SINGLE DUCT VARIABLE VOLUME SYSTEM

Figure 5.44 The single-duct variable volume system is economical of building space since it runs only a single duct and excels at energy conservation. It operates by supplying fixed temperature air via a single duct throughout the building. At each zone location, a thermostatically controlled variable air volume box takes only the volume of air required to meet its load. The system cannot simultaneously provide both heating and cooling. See text. (From Ramsey and Sleeper, *Architectural Graphic Standards*, 8th ed., 1988, © John Wiley & Sons, reprinted by permission of John Wiley & Sons.)

indoor needs. Obviously then, this system is more suited to buildings that always need cooling (large interior zone) than to buildings with perimeters requiring heating and cooling simultaneously. Also, because the volume of air supplied to a space varies with load, this system cannot be used in buildings requiring constant air changes, such as labs and medical facilities.

e. Dual Duct Systems

See Figure 5.45. This system comes in two designs—variable volume and constant volume. The constant-volume system consists of two complete duct distribution systems—one with hot air and one with cold air. A mixing box at each zone location provides air at the temperature required for

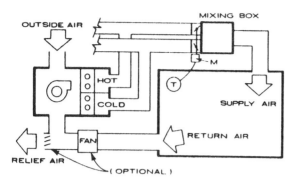
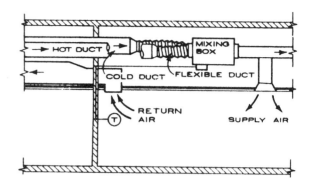

DOUBLE DUCT SYSTEM

Figure 5.45 The dual-duct system supplies hot and cold air in separate ducts, to be mixed as required by the load of each zone. This system is available in a constant volume and a variable volume design. The constant volume design has better control but is less economical than the variable volume design. See text. (From Ramsey and Sleeper, *Architectural Graphic Standards*, 8th ed., 1988, © John Wiley & Sons, reprinted by permission of John Wiley & Sons.)

the load. Control is excellent with this system. Disadvantages are high first cost, high energy cost and a large volume of building space occupied by the two duct systems.

The variable volume arrangement uses one duct to supply primary air in accordance with the major demand (heating or cooling). This air has variable temperature but constant volume. The second air duct has a fixed temperature but variable volume. The two air streams are mixed at each zone. This system uses smaller duct work than the constant volume system, is cheaper to install and uses less energy. Control, however, is not as rapid and accurate as with the constant volume system.

5.20 Single-Zone System Duct Arrangements

As previously noted, single-zone systems are used in buildings where the entire space can be considered as a single zone. This generally includes small to medium-size residences, repair shops, stores, small industrial buildings and the like. The duct arrangements most frequently used are:

- Perimeter loop.
- Radial (perimeter).
- Radial (overhead).
- Extended plenum.
- Reducing plenum.

Construction and application of these arrangements are discussed here. Design of the systems and, more specifically, duct design is covered in detail in Sections 5.25–5.28.

a. Perimeter Loop Duct

Refer to Figure 5.46. Experience has shown that this duct arrangement is ideal for heating slab-on-grade and crawl space structures in cold climates. The perimeter duct, installed directly in the concrete floor, heats the slab, thereby providing a large radiant heat source for the entire structure. The perimeter floor outlets, which should be located under all windows, will temper the cold air sliding down from the windows and prevent cold drafts. See Figures 5.33(a) and 5.35. Additional perimeter floor outlets are installed to supply additional warm air, as indicated by the load calculations. If the system is to be used for cooling as well as heating, nonspread floor registers should be used. See Section 5.18b and Table 5.2. The ducts themselves can be metallic (galvanized sheet metal or steel), concrete, asbestos-cement, ceramic or organic fiber. The installation is shown in Figure 5.46. The usual duct size is 6–8 in. depending on the air quantities being carried. A typical residential perimeter loop system is shown in Figure 5.47. The recommended floor outlet is shown in Figure 5.48.

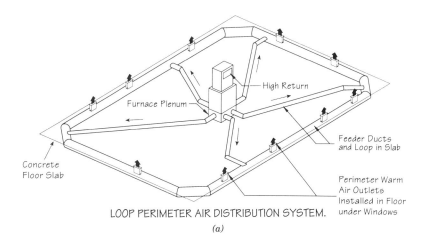

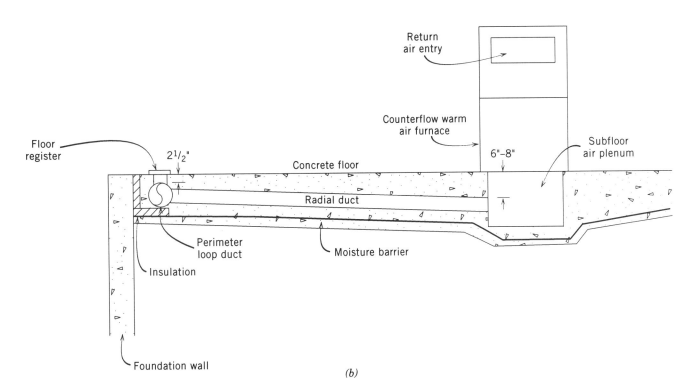

Figure 5.46 *(a)* Perimeter loop air distribution system. A downflow warm air furnace forces air into a subfloor plenum. Radial ducts 6–8 in. in diameter connect the perimeter loop duct to the air plenum. Floor registers around the structure supply warm air into the various rooms. The concrete floor slab is heated by the radial feeder ducts and loop perimeter ducts. *(b)* Installation detail of the loop duct. [*(a)* Bobenhausen, *Simplified Design of HVAC Systems*, 1994, © John Wiley & Sons, reprinted by permission of John Wiley & Sons.]

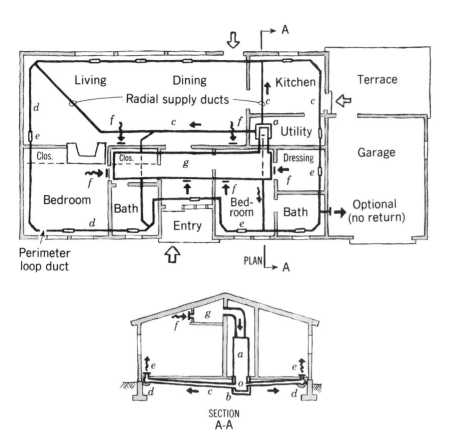

Figure 5.47 Forced, warm air perimeter loop system, adaptable for cooling. No returns from kitchen, baths or garage. *(a)* Downflow air furnace. *(b)* Supply plenum. *(c)* 8-in. (plus) subslab supply ducts (encased in concrete. *(d)* 8-in. perimeter duct (encased in concrete). *(e)* Floor register, adjustable for direction and flow rate (Figure 5.47). *(f)* Return grille. *(g)* Return plenum. (From Stein and Reynolds, *Mechanical and Electrical Equipment for Buildings*, 8th ed., 1992, © John Wiley & Sons, reprinted by permission of John Wiley & Sons.)

b. Radial Ducts with Perimeter Outlets

See Figure 5.49. This duct arrangement is used where floor slab heating is not of primary importance. This might be in buildings with a low ceiling basement or an enclosed crawl space or in a mild climate area. In buildings with a basement, the radial ducts are run uninsulated, under the floor slab. In buildings with a crawl space, the radials can be run in or under the floor slab, as desired. When run under the floor slab in an enclosed crawl space, they are usually uninsulated; in an open crawl space, they are insulated.

As with the perimeter loop system, a downflow furnace supplies air to the radial ducts via an underfloor plenum. A single return is usually adequate as the furnace is centrally located. An outlet is placed at the termination of each radial, normally under each window. Each outlet requires a separate radial. Therefore, an economic breakover point occurs where it becomes more economical to use a perimeter loop with only a few radials. This is one of the first decisions that the project mechanical engineer makes.

c. Radial Duct Arrangement (Overhead)

This system is used where the primary function of the comfort conditioning system is cooling and the air handler (furnace, heat pump, air conditioner) is centrally located so that branch duct lengths to the various building spaces are roughly equal in

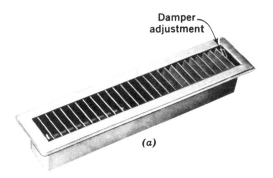

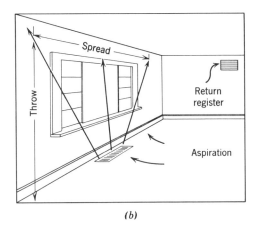

Figure 5.48 Floor register *(a)* and its distribution *(b)*. The register's vanes control the spread and, thereby, also the throw. See Figure 5.28. The very warm air from the register mixes with cold air dropping from the window and warmish room air to set up a total room air circulation. See Figure 5.33. (From Stein and Reynolds, *Mechanical and Electrical Equipment for Buildings*, 8th ed., 1992, © John Wiley & Sons, reprinted by permission of John Wiley & Sons.)

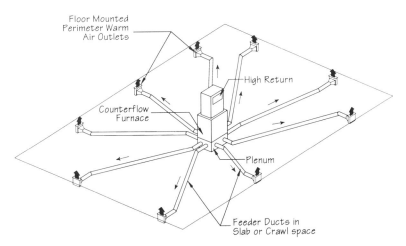

RADIAL PERIMETER AIR DISTRIBUTION SYSTEM.

Figure 5.49 Radial duct system with perimeter outlets. This arrangement is used in structures with basements and with open or enclosed crawl space below the first floor (slab). Ducts are either encased in the slab (open crawl space) or run below the floor (basement or enclosed crawl space). (Bobenhausen, *Simplified Design of HVAC Systems*, 1994, © John Wiley & Sons, reprinted by permission of John Wiley & Sons.)

length. An upflow air handler is used, feeding radial ducts that terminate in ceiling diffusers. The specific design of diffusers depends on whether their use is cooling only or primarily cooling and secondarily heating. (See Figure 5.38.) If heating is important, a diffuser with a downward throw would be utilized.

d. Extended Plenum Air Distribution System

An extended plenum system is simply a trunk duct extending from the supply plenum of the main air-handling unit (warm air furnace or air conditioning unit) with multiple outlets and/or branch ducts connected to it. See Figures 5.41 and 5.50. Because the trunk duct does not change in size from its connection to the supply plenum until the end of its run, it is, in effect, an extended plenum; hence its name. The extended plenum may run in a basement or crawl space as in Figure 5.50, in which case the outlets would be floor or sidewall units. Alternatively, the extended plenum can run overhead in a dropped ceiling or attic, in which case the outlets would be ceiling diffusers or sidewall registers as required. Extended plenums have a number of advantages including:

- Low first cost because of the absence of expensive duct size change fittings.
- Low operating cost because of the absence of energy-using fittings (high static pressure losses).
- Ease of balancing due to low pressure losses and few trunk pressure changes.
- Ease of making changes. Branches can be added, moved and removed without upsetting the system.

The extended plenum arrangement can be used efficiently when the overall trunk length is not more than 50 ft long with the air handler in the center or not more than 30 ft long with the air handler at one end. Longer duct lengths not only become uneconomical but also require the use of a reducing plenum to maintain air velocity. This will become clear in our duct design discussion further on in this chapter.

e. Reducing Plenum Air Distribution System

This arrangement is also known as a semi-extended plenum. See Figure 5.18(d) (page 231) and Figure B.3. When the plenum is more than about 25–30 ft long with the air handler at one end, it becomes necessary to reduce the plenum (trunk duct) size. The number of such reductions depends on the total length, the number of takeoff branch ducts and the air velocities required. It is neither necessary nor advisable to reduce the plenum size after each branch takeoff. As a rule, no size reduction less than 2 in. in the duct width should be made. (As a general rule, not restricted to plenum design, duct size should not be changed less than 2 in. in width or 2 in. in diameter.)

Plenums with a minimum number of size reductions are usually referred to as semi-extended plenums and sometimes as semi-reducing plenums. The two terms mean the same thing. Semi-extended plenums have all the advantages of extended plenums and can be used for duct lengths up to about 50 ft.

There is some confusion between a reducing plenum and a reducing trunk. See Figures B.2 and B.3. Essentially, the trunks are identical; the differences occur at takeoff. Branch takeoffs on reducing trunks are usually at reducing fittings and have relatively low total equivalent length (TEL) and no velocity factor. See Section 5.27 for a detailed explanation of this factor and Appendix B, Figure B.3 for typical values. Reducing plenum takeoffs are right angles fittings on the trunk body. They have generally high TEL to which is added a velocity factor. The velocity factor is an additional TEL for each takeoff, proportional to the number of downstream fittings after the takeoff.

Air Friction in Duct Systems

We have studied the components of all-air systems and the duct arrangements usually used in single-zone systems. The next step in our study is to learn how air flows in duct systems, through straight duct sections and through fittings. Most important, we must learn how to calculate the friction losses of air movement in ducts. Once we have mastered this skill, we will be in position to approach overall duct system design for small to medium-size residential and commercial buildings.

Refer to Figure 5.14 (page 223), which lists all the sources of static pressure loss in a system. Note that there are two sources of static pressure loss—items of equipment and ductwork. The loss in an equipment item—evaporator coil, humidifier, filter, supply register and return grille—can be ob-

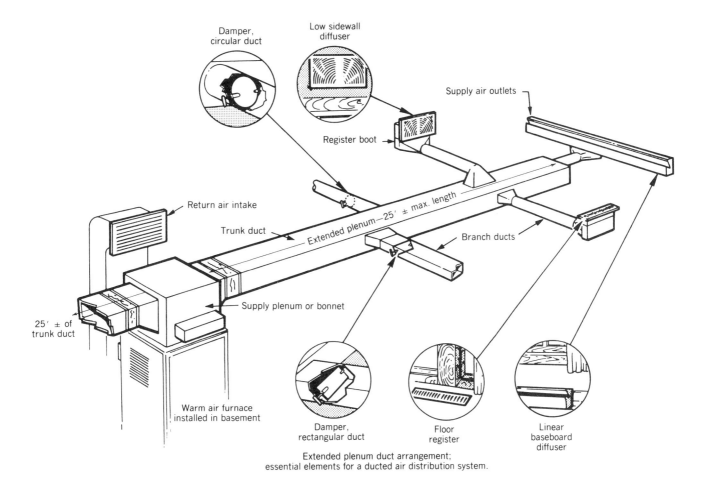

Figure 5.50 Extended plenum duct arrangement, showing the essential elements of a ducted air system. It is understood, of course, that a single system would normally use only one type of branch duct and one type of outlet. A floor-level single return intake is shown only for the sake of simplicity. In practice, the return could be a high sidewall outlet or an entire return duct system. (Reproduced with permission from *ACCA Manual C*, p. 28.)

tained by looking it up in the manufacturer's catalog. The losses in ductwork, which include the supply plenum, must be calculated. We will study that subject next.

You may have noticed that there is one large part of the overall system that is completely ignored as far as static pressure loss is concerned. That part is the section between the supply register output and the return register input, that is, the building spaces or rooms. The reason that these are neglected is that the air velocity in them is so low that the static and velocity pressure drop in them is close to zero and, therefore, negligible. The pressure at the face of the supply register is usually taken to be zero and that at the face of the return grille is assumed to be very slightly negative (suction).

5.21 Air Friction in Straight Duct Sections

At this point, you should review Section 5.4, which introduced the subject of air pressure in ducts. Very briefly, we learned there that:

- The source of all duct pressure is the system air handler, usually the furnace blower.
- Total pressure at any point is the sum of static pressure plus velocity pressure.

- Velocity pressure in low velocity systems is very small, so small, indeed, that it is often neglected in branch duct calculations.
- Static pressure is that required to overcome system friction. It can be thought of as being "used up" by air friction in the ducts, as the air moves around the system.

The total pressure loss in any section of duct is the sum of the static pressure loss and the dynamic pressure loss. Dynamic pressure loss is caused by major air turbulence, which, in turn, is caused by a change in duct size or direction. See, for instance, Figure 5.17(c-1) (page 230). Static pressure loss in straight sections of duct is caused by the friction of the moving air "rubbing" against the walls of the duct. This rubbing results in a loss of energy in the moving air stream, which expresses itself as a drop in static pressure, or head. This friction is proportional to the roughness of the duct walls, to the quantity of air moving (cfm) and to its velocity (fpm). It is also proportional to the ratio of duct perimeter to cross-sectional area. This means that the more surface there is per unit of area, the higher the friction, simply because there is greater air-to-duct surface contact.

It was pointed out in Section 5.12.b that round duct has the highest ratio of area to perimeter of any shape. Conversely, round duct has the lowest perimeter to cross-sectional area ratio. This means that round duct has the lowest friction loss of any shape, for a given air flow and velocity. The reason is that its minimum perimeter means minimum contact between duct wall and moving air to cause friction. Round duct is used as the basis of all friction calculations. If other shapes are used, they are calculated as equivalents to round duct, as we will explain shortly.

Figure 5.51 is the standard duct friction chart for round galvanized steel duct. It is based on air at 70°F and sea level air pressure, weighing 0.075 lb/ft^3, flowing in galvanized steel duct constructed with longitudinal seams and beaded slip couplings on 4-ft centers. This duct has an absolute roughness ϵ of 0.0003. (There are several other round galvanized steel constructions that have the same roughness. The chart in Figure 5.51 applies equally to such a duct.) This duct construction gives a roughness category of "medium smooth." No correction to the chart data is required for air at temperatures from 40 to 100°F, at elevations up to 1500 ft and at duct pressures +20 in. of water relative to the ambient pressure. Note that this chart dates from 1987. Older charts were all based on ducts with 40 joints per 100 ft, a roughness category of "average," and an absolute roughness of $\epsilon = 0.005$. Use of these older charts will give a somewhat higher friction loss than that of modern duct construction.

The duct friction chart relates air flow, air velocity, duct diameter and friction loss per 100 ft of duct, measured in inches w.g. Knowing or selecting any two of these will establish the third and fourth. In practice, we usually know the air quantity and the maximum friction loss. We then select a combination of air velocity and duct diameter from the chart. The shaded area of the chart indicates recommended combinations of parameters. A few examples should make use of the chart clear.

Example 5.5 A branch duct is required to supply 350 cfm. The system is being designed for a static friction drop of 0.2 in. w.g./100 ft of duct. Find the required duct size. What would be the air velocity?

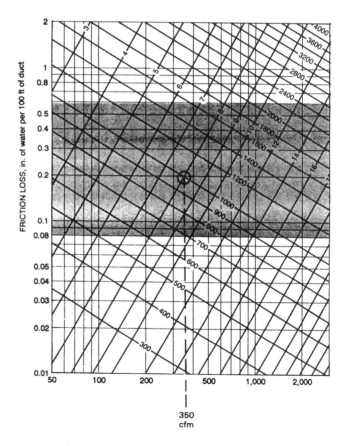

Example 5.5

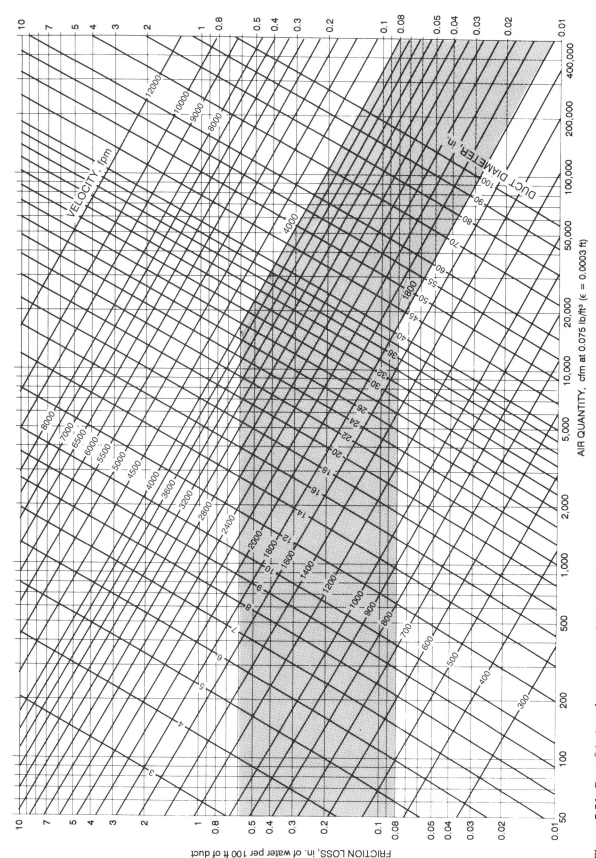

Figure 5.51 Duct friction chart, standard (English) units. Air motion in round galvanized steel duct, medium smooth, roughness coefficient $\epsilon = 0.0003$. Air at 70°F, sea level atmospheric pressure, 0.075 lb/ft³ density, 0.075 lb/ft³ density. (Reproduced by permission of the American Society of Heating, Refrigeration and Air Conditioning Engineers, Atlanta, Georgia, from the 1993 *ASHRAE Handbook—Fundamentals*.)

Solution: Enter the chart at 350 cfm along the bottom of the chart. Draw a line vertically until it intersects the horizontal line representing 0.2 in. w.g. The intersection shows 8-in. duct (line sloping up to the right) and 1000 fpm (line sloping down to the right). The fact that this point is in the shaded part of the chart means that the combination of air quantity, velocity, duct diameter and friction is an acceptable one. (In practical duct design, if this were a residence, we would probably use a larger duct in order to reduce the air velocity, so as to limit duct noise.)

Example 5.6 A trunk duct 14 in. in diameter and 30 ft long carries 1100 cfm. What is its static friction loss? What is the air velocity?

Solution: Enter the chart at 1100 cfm on the horizontal axis. (You will have to estimate the position of 1100 cfm, by eye. Remember that the chart is logarithmic. That means that 1100 is much farther from 1000 than 1900 is from 2000.) Extend a line vertically until it hits the 14-in. duct line. Read off the chart (by approximation) 0.11 in./100 ft friction loss and 1100 fpm. Since we have only 30 ft of duct, the total friction loss is

$$f = \frac{0.11 \text{ in.}}{100 \text{ ft}} \times 30 \text{ ft} = 0.03 \text{ in. w.g.}$$

Remember that the chart gives loss per 100 ft. Loss for any other length has to be calculated, as before. Failure to do this is the most common error of novice designers.

Example 5.7 A branch duct 15 ft long will carry 100 cfm to a room in a residence. The noise criteria recommends a maximum velocity of 500 fpm. Select an appropriate duct size. Friction is not critical since the length of duct is very short.

Solution: Enter the chart at 100 cfm on the horizontal scale. Extend a line vertically. Note that it intersects the following combinations:

6-in. duct   500 fpm   0.08 in./100 ft
5-in. duct   730 fpm   0.2 in./100 ft
4-in. duct   1170 fpm   0.6 in/100 ft

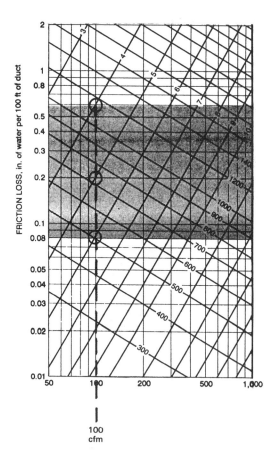

Example 5.6 (Reprinted with permission from 1993 *ASHRAE Handbook—Fundamentals*.)

Example 5.7 (Reprinted with permission from 1993 *ASHRAE Handbook—Fundamentals*.)

We would probably choose 6-in. duct, although if the duct were serving a noisy room such as a kitchen or bath, the 5-in. duct is also a reasonable choice.

As you have surely concluded, the friction chart of Figure 5.51 is not easy or convenient to use because of its logarithmic air quantity scale and because it is necessary to make visual interpolations between values. Recognizing these difficulties, a number of companies and professional organizations have produced slide rule-type calculators that give the same data as the friction chart, plus sizes of equivalent rectangular ducts and other important data. Two of the best known of these calculators are shown in Figures 5.52 and 5.53. The calculator shown in Figure 5.53 gives the friction loss for straight duct sections on one side and duct fitting losses on the other side. This calculator was produced in 1989 and is based on the current duct construction data. The unit shown in Figure 5.52 was produced in 1976 and is based on older data. Users of these and similar calculators should always check to determine which duct roughness data was used to calibrate the calculator.

Table 5.3 gives duct roughness factors for materials other than the galvanized steel used in the friction chart of Figure 5.51. Figure 5.54 is a chart of correction factors to be used with other ducts. Illustrative examples should make their use clear.

Example 5.8 A technologist using a duct calculator based on the old average smoothness duct arrives at the following data.

Duct Section	Q, cfm	Diameter, in.	f/100 ft
A	1000	12	0.20
B	600	10	0.195
C	200	6	0.32

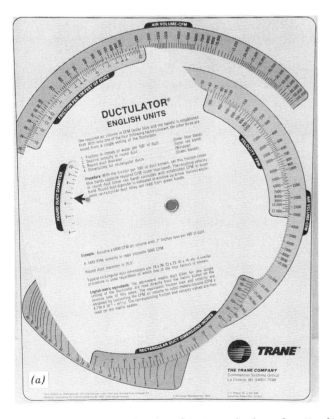
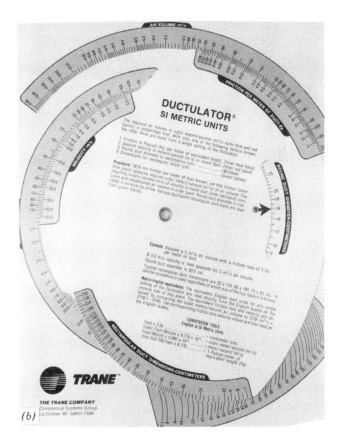

Figure 5.52 A popular duct friction calculator has English units on one side (a) and metric units on the reverse side (b). Scales on the calculator represent air flow Q, air velocity V, friction, round duct size and equivalent rectangular duct size.

Table 5.3 Duct Material Roughness Factors Based on Good Workmanship

Duct Material	Roughness Category	Roughness Index (ϵ_1)
Thermoplastic duct (PVC)	Smooth	0.0001
Uncoated carbon steel duct, clean	Smooth	0.0001
Aluminum sheet duct (12 joints/100 ft)	Smooth	0.0001
Galvanized sheet duct (continuous rolled—12 joints/100 ft)	Smooth	0.0001
Spiral galvanized duct (12 joints/100 ft)	Medium	0.0003
Aluminum sheet duct (40 joints/100 ft)	Smooth	
Galvanized sheet duct (hot dipped—40 joints/100 ft)	Average	0.0005
Fibrous glass duct (rigid)	Medium	0.003
Fibrous glass liners (airside with facing material mechanically fastened)	Rough	0.01
Fibrous glass liners (airside spray coated, mechanically fastened)	Rough	0.01
Flexible duct, metallic (fully extended)	Rough	0.01
Flexible duct, all types of fabric & wire (fully extended)	Rough	0.01
Concrete duct	Rough	0.01

Reproduced with permission from *ACCA Manual Q*, 1990.

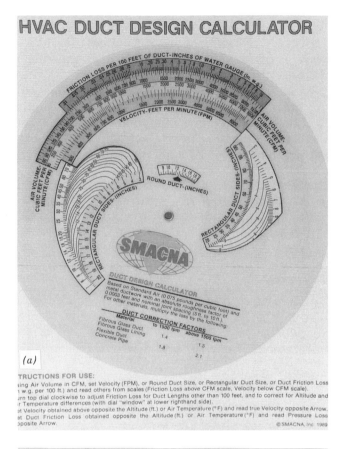

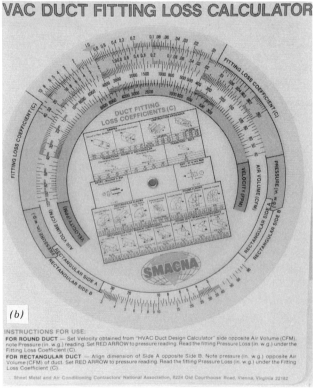

Figure 5.53 This modern duct friction calculator gives friction data for straight duct on one side *(a)* and fitting friction loss data on the other side *(b)*. The fitting losses calculations are based on loss coefficients of common fittings, tabulated on the calculator.

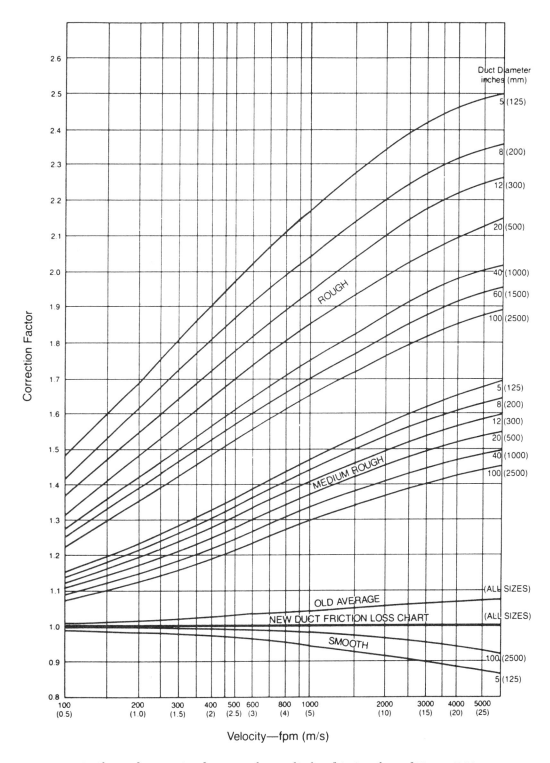

Figure 5.54 Chart of correction factors to be applied to friction data of Figure 5.51 when using ducts of roughness other than medium smooth. See Table 5.3. (Reproduced with permission from *SMACNA HVAC Systems Duct Design Manual,* 1990.)

Using Figure 5.54, calculate the friction/100 ft based on current data for galvanized steel duct. Would the duct sizes need to be changed?

Solution: From Figure 5.54, we note that the correction curve for "old average" duct is the same for all duct sizes for each value of Q. From the chart, we have:

Q, cfm	Factor
1000	1.045
600	1.03
200	1.015

Since the "old average" was for rougher duct (See Table 5.3), we must divide the friction values found with the old calculator by the correction factors to obtain friction of the new smoother duct. We, therefore, have:

Duct section A:
friction/100 ft = 0.2/1.045 = 0.191 in. w.g.
Duct section B:
friction/100 ft = 0.195/1.03 = 0.189 in. w.g.
Duct section C:
friction/100 ft = 0.32/1.015 = 0.315 in. w.g.

We can immediately see that these corrections are minor and, for the sizes indicated, would not affect the choice of duct size.

Example 5.9 A 100 ft long run of round duct is to carry 2000 cfm at a friction rate of 0.10 in. w.g. What size duct is required of the following materials:

(a) galvanized steel, roughness 0.0003
(b) galvanized steel, roughness 0.0005
(c) PVC thermoplastic duct, roughness 0.0001
(d) Fibrous glass-lined duct, roughness 0.01

Solution: Using Figure 5.51 and Table 5.3, we find the following information.

Duct Material	Friction Desired	Duct Size, in.	Correction Factor	Chart Friction	Duct Size, in.
(a)	0.10	18	1.0	0.095	17.9
(b)	0.10		1.06	0.094	18
(c)	0.10		0.91	0.11	17.5
(d)	0.10		1.92	0.052	20.4

There are, of course, no decimal size ducts. Ducts (a), (b) and (c) would all be 18 in. The decimal sizes are shown in case the designer wants to convert to rectangular sizes. The conclusion, however, is clear; only the very rough fibrous glass duct causes a change in duct size.

Additional correction factors for altitudes above sea level and temperatures other than 70°F are given in Figure 5.55. The method of use of these factors is the same as already demonstrated.

5.22 Noncircular Ducts

Despite the efficiency of circular ducts in carrying air with minimum friction, the round shape has a number of disadvantages. The most important of these is its space requirement. In modern construction, space is almost always at a premium, particularly in hung ceilings, pipe chases and mechanical spaces. Therefore, even though rectangular ducts are more expensive than circular ones, they are used almost exclusively in commercial work. In design, the required round duct size is found from charts such as Figure 5.51, and then the rectangular duct that gives the same friction loss per 100 ft is found in tabulations such as that given in Table 5.4. When using a calculator of the type shown in Figures 5.52 and 5.53, this equivalent rectangular duct can be found directly.

Note from Table 5.4 that, for each circular duct size, there are a number of equivalent rectangular duct configurations, each with a different aspect ratio. The aspect ratio of a rectangular duct is simply the ratio of width to height. Thus, the aspect ratio of a square is 1.0, of a duct 16 in. wide by 8 in. high is 2.0 and so on. See Figure 5.56. The higher the aspect ratio is, the more perimeter per area the duct has. This means that as aspect ratio increases, so do cost, friction and vibration noise. It is, therefore, good design to use the lowest possible aspect ratio that fits the construction space conditions.

Note that aspect ratios higher than 4 are not listed in Table 5.4 because such ducts are not recommended for use. They can be built and, in special cases, are used. In large sizes, internal supports must be used to keep the long dimension from sagging and vibrating. This increases costs radically. Also, fittings for high aspect ratio ducts are expensive and very inefficient (high pressure losses). Simply as a matter of interest, relative installed costs of ducts, taking square ducts (1.0 aspect ratio) as 100%, are:

NONCIRCULAR DUCTS / 283

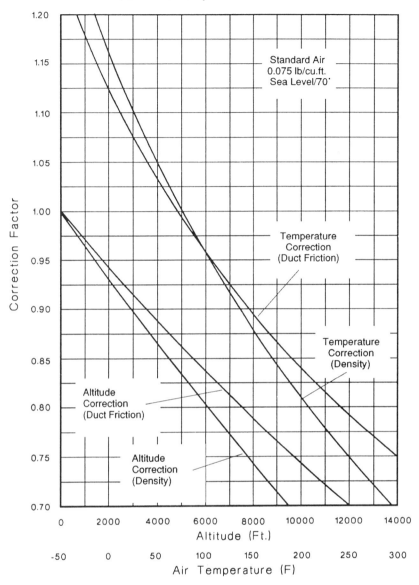

Figure 5.55 Graph of altitude and temperature correction factors to be applied to duct friction data obtained from Figure 5.51. (Reproduced with permission from *ACCA Manual Q*, 1990.)

Table 5.4 Equivalent Rectangular Duct Dimension

Duct Diameter, in.	Rectangular Size, in.	Aspect Ratio										
		1.00	*1.25*	*1.50*	*1.75*	*2.00*	*2.25*	*2.50*	*2.75*	*3.00*	*3.50*	*4.00*
6	Width	—	6									
	Height	—	5									
7	Width	6	8									
	Height	6	6									
8	Width	7	9	9	11							
	Height	7	7	6	6							
9	Width	8	9	11	11	12	14					
	Height	8	7	7	6	6	6					
10	Width	9	10	12	12	14	14	15	17			
	Height	9	8	8	7	7	6	6	6			
11	Width	10	11	12	14	14	16	18	17	18	21	
	Height	10	9	8	8	7	7	7	6	6	6	
12	Width	11	13	14	14	16	16	18	19	21	21	24
	Height	11	10	9	8	8	7	7	7	7	6	6
13	Width	12	14	15	16	18	18	20	19	21	25	24
	Height	12	11	10	9	9	8	8	7	7	7	6
14	Width	13	14	17	18	18	20	20	22	24	25	28
	Height	13	11	11	10	9	9	8	8	8	7	7
15	Width	14	15	17	18	20	20	23	25	24	28	28
	Height	14	12	11	10	10	9	9	9	8	8	7
16	Width	15	16	18	19	20	23	23	25	27	28	32
	Height	15	13	12	11	10	10	9	9	9	8	8
17	Width	16	18	20	21	22	25	25	28	27	32	32
	Height	16	14	13	12	11	11	10	10	9	9	8
18	Width	16	19	21	23	24	25	28	28	30	32	36
	Height	16	15	14	13	12	11	11	10	10	9	9
19	Width	17	20	21	23	24	27	28	30	30	35	36
	Height	17	16	14	13	12	12	11	11	10	10	9
20	Width	18	20	23	25	26	27	30	30	33	35	40
	Height	18	16	15	14	13	12	12	11	11	10	10
21	Width	19	21	24	26	28	29	30	33	33	39	40
	Height	19	17	16	15	14	13	12	12	11	11	10
22	Width	20	23	26	26	28	32	33	36	36	39	44
	Height	20	18	17	15	14	14	13	13	12	11	11
23	Width	21	24	26	28	30	32	35	36	39	42	44
	Height	21	19	17	16	15	14	14	13	13	12	11
24	Width	22	25	27	30	32	34	35	39	39	42	48
	Height	22	20	18	17	16	15	14	14	13	12	12
25	Width	23	25	29	30	32	36	38	39	42	46	48
	Height	23	20	19	17	16	16	15	14	14	13	12
26	Width	24	26	30	32	34	36	38	41	42	46	52
	Height	24	21	20	18	17	16	15	15	14	13	13
27	Width	25	28	30	33	36	38	40	41	45	49	52
	Height	25	22	20	19	18	17	16	15	15	14	13
28	Width	26	29	32	35	36	38	43	44	45	49	56
	Height	26	23	21	20	18	17	17	16	15	14	14
29	Width	27	30	33	35	38	41	43	44	48	53	56
	Height	27	24	22	20	19	18	17	16	16	15	14
30	Width	27	31	35	37	40	43	45	47	48	53	60
	Height	27	25	23	21	20	19	18	17	16	15	15
31	Width	28	31	35	39	40	43	45	50	51	56	60
	Height	28	25	23	22	20	19	18	18	17	16	15
32	Width	29	33	36	39	42	45	48	50	54	56	60
	Height	29	26	24	22	21	20	19	18	18	16	15
33	Width	30	34	38	40	44	47	50	52	54	60	64
	Height	30	27	25	23	22	21	20	19	18	17	16

Table 5.4 (Continued)

Duct Diameter, in.	Rectangular Size, in.	Aspect Ratio										
		1.00	1.25	1.50	1.75	2.00	2.25	2.50	2.75	3.00	3.50	4.00
34	Width	31	35	39	42	44	47	50	52	57	60	64
	Height	31	28	26	24	22	21	20	19	19	17	16
35	Width	32	36	39	42	46	50	53	55	57	63	68
	Height	32	29	26	24	23	22	21	20	19	18	17
36	Width	33	36	41	44	48	50	53	55	60	63	68
	Height	33	29	27	25	24	22	21	20	20	18	17
38	Width	35	39	44	47	50	54	58	61	63	67	72
	Height	35	31	29	27	25	24	23	22	21	19	18
40	Width	37	41	45	49	52	56	60	63	66	70	76
	Height	37	33	30	28	26	25	24	23	22	20	19
42	Width	38	43	48	51	56	59	63	66	69	74	80
	Height	38	34	32	29	28	26	25	24	23	21	20
44	Width	40	45	50	54	58	61	65	69	72	81	84
	Height	40	36	33	31	29	27	26	25	24	23	21
46	Width	42	48	53	56	60	65	68	72	75	84	88
	Height	42	38	35	32	30	29	27	26	25	24	22
48	Width	44	49	54	60	62	68	70	74	78	88	92
	Height	44	39	36	34	31	30	28	27	26	25	23
50	Width	46	51	57	61	66	70	75	77	81	91	96
	Height	46	41	38	35	33	31	30	28	27	26	24
52	Width	48	54	59	63	68	72	78	83	84	95	100
	Height	48	43	39	36	34	32	31	30	28	27	25
54	Width	49	55	62	67	70	77	80	85	90	98	104
	Height	49	44	41	38	35	34	32	31	30	28	26
56	Width	51	58	63	68	74	79	83	88	93	102	108
	Height	51	46	42	39	37	35	33	32	31	29	27
58	Width	53	60	66	70	76	81	85	91	96	105	112
	Height	53	48	44	40	38	36	34	33	32	30	28
60	Width	55	61	68	74	78	83	90	94	99	109	116
	Height	55	49	45	42	39	37	36	34	33	31	29
62	Width	57	64	71	75	82	88	93	96	102	112	120
	Height	57	51	47	43	41	39	37	35	34	32	30
64	Width	59	65	72	79	84	90	95	99	105	116	124
	Height	59	52	48	45	42	40	38	36	35	33	31
66	Width	60	68	75	81	86	92	98	105	108	119	128
	Height	60	54	50	46	43	41	39	38	36	34	32
68	Width	62	70	77	82	90	95	100	107	111	123	132
	Height	62	56	51	47	45	42	40	39	37	35	33
70	Width	64	71	80	86	92	99	105	110	114	126	136
	Height	64	57	53	49	46	44	42	40	38	36	34
72	Width	66	74	81	88	94	101	108	113	117	130	140
	Height	66	59	54	50	47	45	43	41	39	37	35
74	Width	68	76	84	91	98	104	110	116	123	133	144
	Height	68	61	56	52	49	46	44	42	41	38	36
76	Width	70	78	86	93	100	106	113	118	126	137	148
	Height	70	62	57	53	50	47	45	43	42	39	37
78	Width	71	80	89	95	102	110	115	121	129	140	152
	Height	71	64	59	54	51	49	46	44	43	40	38
80	Width	73	83	90	98	104	113	118	124	132	144	156
	Height	73	66	60	56	52	50	47	45	44	41	39

Source. Data extracted and reprinted by permission of the American Society of Heating, Refrigerating and Air Conditioning Engineers, Atlanta, Georgia, from the 1993 *ASHRAE Handbook—Fundamentals*.

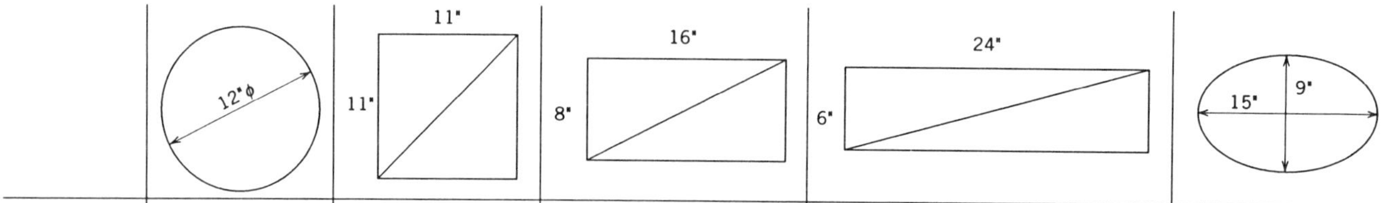

Dimension	12" Diam.	11" x 11"	16" x 8"	24" x 6"	Major axis - 15" Minor axis - 9"
Aspect ratio	—	1.0	2.0	4.0	—
Perimeter (in.)	37.7	44	48	60	38.3
Area (in.²)	113.1	121	128	144	106
Ratio of perimeter: area	0.333	0.364	0.375	0.417	0.361
Air velocity (constant CFM)	100%	93.4%	88.4%	78.5%	106.7%

Figure 5.56 Comparative characteristics of equivalent friction ducts. Duct cost is proportional to the quantity of metal used, that is, the perimeter. Note that if air volume (cfm) is held constant, the air velocity will drop as cross-sectional duct area increases. See Section 5.23. Duct equivalent sizes are taken from Tables 5.4 and 5.5.

Aspect ratio	1	2	3	4	5	5	7
Cost, %	100	115	130	145	165	185	210

In addition, since higher aspect ratio means higher friction, it also means more energy use by the blower and, therefore, higher operating costs.

Oval ducts have recently become fairly popular, particularly in residential work. This is because the clear space in a stud construction wall is only 3⅝ in., which will accept only a 3-in. round duct. However, oval ducts as large as 3 × 15 in. will fit into a stud wall with studs 16 in. on centers. This is equivalent to a 7-in. round duct. Oval equivalents to circular ducts are given in Table 5.5. Figure 5.57 shows the use of circular, rectangular and oval ducts in a typical installation situation.

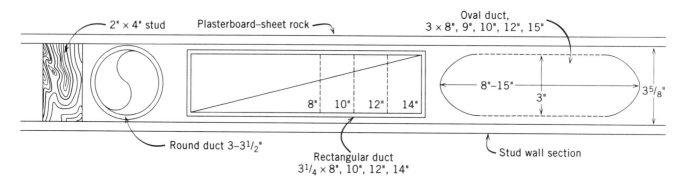

Figure 5.57 The restricted width of a stud wall (3⅝ in.) requires use of rectangular or oval ducts. Rectangular ducts for use as stacks in stud walls are available in the sizes shown.

Table 5.5 Size of Equivalent[a] Oval Duct to Circular Duct

Equivalent Circular Duct Diameter, in.	Minor Axis, in.											
	3	4	5	6	7	8	9	10	11	12	14	16
5	8											
5.5	9	7										
6	11	9										
6.5	12	10	8									
7	15	12	10	8								
7.5	19	13	—	9								
8	22	15	11	—								
8.5		18	13	11	10							
9		20	14	12	—	10						
9.5		21	18	14	12	—						
10			19	15	13	11						
10.5			21	17	15	13	12					
11				19	16	14	—	12				
11.5				20	18	16	14	—				
12				23	20	17	15	13				
12.5				25	21	—	—	15	14			
13				28	23	19	17	16	—	14		
13.5				30	—	21	18	—	16	—		
14				33	—	22	20	18	17	15		
14.5				36	—	24	22	19	—	17		
15						27	23	21	19	18		
16						30	—	24	22	20	17	
17						35	—	27	24	21	19	
18						39	—	30	—	25	22	19
19						46	—	34	—	28	23	21
20						50	—	38	—	31	27	24
21								43	—	34	28	25
22								48	—	37	31	29
23								52	—	42	34	30
24										45	38	33
25										50	41	36
26										56	45	38
27											49	41
28											52	46
29											58	49
30											61	54
31												57
32												60
33												66

[a] Equivalent duct friction.

Source. Data extracted and reprinted by permission of the American Society of Heating, Refrigerating and Air Conditioning Engineers, Atlanta, Georgia, from the 1993 *ASHRAE Handbook—Fundamentals.*

5.23 Air Friction in Duct Fittings

A duct system consists of straight sections and fittings. In Sections 5.21 and 5.22, we learned how to calculate the friction loss in straight sections of ducts of various shapes. We will now learn how to do this calculation for fittings. The term *fittings* applies to every part of a duct system except straight duct sections of unchanging size. Fittings, therefore, include transitions, inlets, outlets, elbows, angles, offsets, wyes, tees, dovetails, branches, exit connections and so on. Indeed, the list is so long that ASHRAE has developed a computer duct fitting data base (1993) to assist designers in duct system calculations.

On the average, pressure losses in fittings comprise at least one-half of the total pressure loss in a system and sometimes as much as 75%. It is, therefore, apparent that these losses must be carefully calculated in preparing any but the smallest and simplest duct system. There are two methods of determining the pressure loss in a fitting—equivalent length and loss coefficient calculation. Equivalent length should be used only when there is no difference in air velocity between the entrance and the exit of the fitting. Loss coefficient calculations are applicable in all situations. To understand why this is so, we must first study the effect of duct (or fitting) cross-sectional area on the velocity of airflow.

a. Air Velocity in Ducts

Refer to Figure 5.58. In the section of duct illustrated there is an air flow of Q cfm at a velocity of V fpm. The volume of air flowing past section P_1 in 1 min, when the air velocity is 1 fpm is a column of air 1 ft long of cross-sectional area A. Numerically, the volume of this column is the area A ft^2 times 1 ft length, or A ft^3. Since this volume flows past in 1 min, the flow Q is A ft^3/min (or cfm).

If the air velocity were 2 fpm, the column would be 2 ft long and its volume would be $2A$ ft^3. Since it too flows by in one minute, the flow Q would then be $2A$ ft^3/min (or cfm). Therefore, if the air velocity is V, the volume of the air column passing section P_1 in 1 min would be VA ft^3, and the flow rate Q would be VA ft^3/min or AV cfm.

We have, therefore, developed the fundamental flow equation:

$$Q = AV \qquad (5.7)$$

where

Q is the volume of air flow in cubic feet per minute,

A is the cross-sectional area of the duct in square feet and

V is the air velocity in feet per minute.

An example will help you understand this extremely useful equation.

Example 5.10 A design technologist has the choice of using a 10-in. round duct or a 12×7-in. rectangular duct to carry 350 cfm of conditioned air. Both give the same friction (see Table 5.4). Calculate the air velocity in each.

Solution: We will use Equation (5.7), remembering to convert duct area into square feet. For a round 10-in. duct:

$$A = \frac{\pi D^2}{4} = 0.7854 D^2 = 0.7854(10)^2 \text{ in.}^2$$

$$A = 78.54 \text{ in.}^2 \times \frac{\text{ft}^2}{144 \text{ in.}^2} = 0.545 \text{ ft}^2$$

$$V = \frac{Q}{A} = \frac{350 \text{ ft}^3/\text{min}}{0.545 \text{ ft}^2} = 642 \text{ ft/min}$$

For a rectangular 12×7-in. duct:

$$A = 12 \text{ in.} \times 7 \text{ in.} = 84 \text{ in.}^2 \times \frac{\text{ft}^2}{144 \text{ in.}^2} = 0.583 \text{ ft}^2$$

$$V = \frac{350 \text{ ft}^3/\text{min}}{0.583 \text{ ft}^2} = 600 \text{ ft/min}$$

The numbers in Example 5.10 were chosen to demonstrate a very important point. When using the friction chart of Figure 5.51, the air velocity shown is that of air in a round duct. In this case,

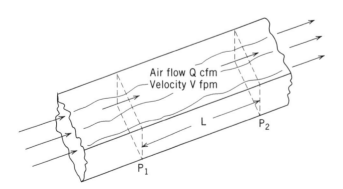

Figure 5.58 Q cfm of air volume traveling through a duct of A ft^2 cross-sectional area, will move at V ft/min, where $Q = AV$. See text for derivation of this relation.

an inexperienced technologist might be inclined to select a larger round duct because, in residential work, it is recommended that air velocities above 600 fpm be avoided, due to noise. He or she would then find the equivalent (oversized) rectangular duct. What should be done is first to check air velocity in the rectangular duct that is equivalent to the original (10-in.) round duct. It is always lower than that in a round duct. In this case, it is down to 600 fpm, which meets the recommendation limit. An equivalent rectangular duct of higher aspect ratio would have an even lower air velocity. (We are assuming, of course, that air flow Q is held constant at the design value.)

b. Equivalent Duct Lengths for Fitting Losses

We stated previously that one of the two methods for figuring fitting static pressure loss is by using an equivalent length of straight duct. We also stated that this method should be used only where fitting input and output velocities are the same. We will now explain why this is so. Refer to Figure 5.59(a), which shows graphically the pressure loss in a straight section of duct. Total pressure at any point, as we learned in Section 5.4, is the sum of static pressure and velocity pressure. This means that at the inlet of the duct

$$P_{\text{inlet}} = P_{S_1} + P_{V_1}$$

and at the outlet

$$P_{\text{outlet}} = P_{S_2} + P_{V_2}$$

where

P_{S_1} is static pressure at point 1,
P_{S_2} is static pressure at point 2,
P_{V_1} is velocity pressure at point 1 and
P_{V_2} is velocity pressure at point 2.

However, since the size of the duct does not change nor does the air volume Q that flows, the outlet air velocity V_2 must equal the inlet velocity V_1, because

$$V = \frac{Q}{A}$$

and both Q and A remain unchanged. Since we know that the velocity pressure is

$$P_V = \left(\frac{V}{4005}\right)^2$$

it follows that velocity pressure is the same at the outlet as at the inlet. Therefore, all the pressure change between inlet and outlet must be static pressure change. This is shown as a continuous drop in the diagram. We can, therefore, say that a certain length L of straight duct causes a specific amount of static pressure loss.

Now refer to Figure 5.59(b), which shows a typical duct offset fitting of the type used to dip under some physical obstruction. (See also Figure 5.18.) Here again, as in Figure 5.59(a), the cross-sectional area of duct and the air flow Q remains constant. Therefore, as before, the velocity pressures at the inlet and outlet are the same ($P_{V_2} = P_{V_1}$), and the drop in total pressure is simply a drop in static pressure. This is obviously also true of any fitting with equal inlet and outlet areas because velocity pressure remains unchanged. However, since a drop in static pressure can be expressed in terms of a specific length of straight duct [see Figure 5.59(a)], it follows that the total pressure loss in any fitting with equal inlet and outlet areas can be expressed as an equivalent length of straight duct.

Let us now look at an even more complicated fitting in Figure 5.59(c). This fitting has a restricted throat portion in which the air velocity does change because the area changes. However, what happens inside this throat does not concern us, because the fitting outlet has the same area as the inlet. Therefore, even for a fitting of this type, inlet and outlet velocity pressures are equal. As a result, we can express the total pressure loss of this fitting too as an equivalent length of straight duct. (For a detailed explanation of what happens to pressure inside this type of fitting, refer to *ACCA Manual Q* or *ASHRAE Handbook—Fundamentals*. Very briefly, in the restricted throat section, air velocity and velocity pressure increase, and static pressure decreases. When the area enlarges, most of the increased velocity pressure is reconverted to static pressure in a process known as *static regain*.)

A numerical example should help to clarify the principle of equivalent length. Assume that the three items discussed previously, that is, the straight section of duct in Figure 5.59(a), the offset fitting in Figure 5.59(b) and the restricted throat fitting of Figure 5.59(c) are all part of one duct system, with the same duct size. Assume also that pressure measurements indicate the following:

(a) Straight section; static pressure loss = 0.1 in. w.g./100 ft
(b) Offset section; static pressure loss = 0.035 in. w.g.
(c) Throat section: static pressure loss = 0.075 in. w.g.

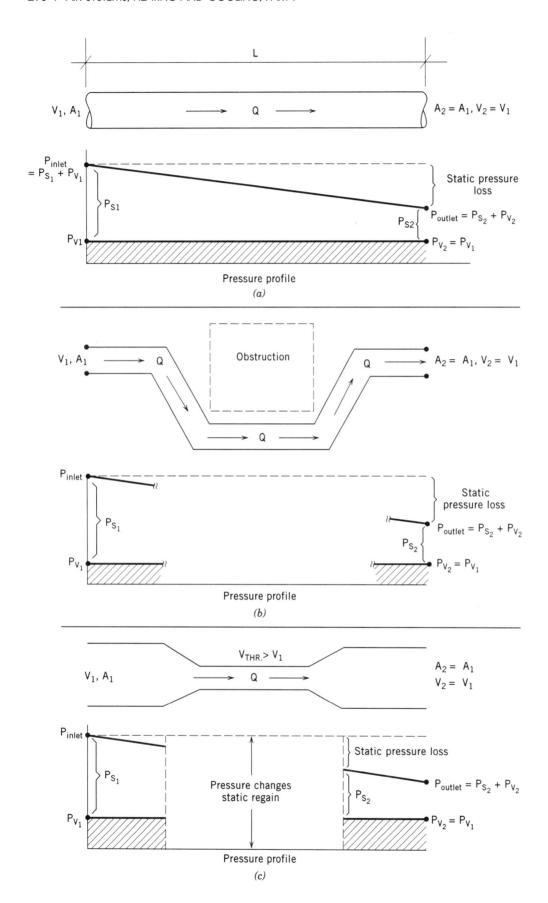

Say that the offset fitting is the equivalent of L_1 ft of straight duct, then:

$$\frac{0.035 \text{ in.}}{L_1 \text{ ft}} = \frac{0.1 \text{ in.}}{100 \text{ ft}}$$
$$L_1 = 35 \text{ ft}$$

Similarly, the throat section is equivalent to L_2 ft of straight duct:

$$\frac{0.075 \text{ in. w.g.}}{L_2} = \frac{0.1 \text{ in. w.g.}}{100 \text{ ft}}$$
$$L_2 = 75 \text{ ft}$$

Now refer to Figure 5.18 (page 231). Adjacent to each fitting sketch is a number. It represents the equivalent length of straight duct that will cause the same static friction loss as the fitting. For fittings with equal inlet and outlet areas, these equivalent lengths are accurate. For fittings with different areas, these equivalent lengths are only an approximation and should only be used in small to medium low velocity, low pressure systems.

A comprehensive listing of equivalent lengths for residential fittings is given in Appendix B. Note that some of these fittings have different inlet and outlet areas. This means that for accurate work, equivalent length should not be used. In residential work, most ducts are sized for noise criteria, which makes them larger than would be required by friction calculations. Also, at the low velocities used in residential systems, velocity pressures are almost negligible. Finally, all ducts have volume dampers that allow balancing the system after installation. As a result, the inaccuracies introduced by using equivalent length for all fittings do not result in an unworkable design for a small-to-medium-size low velocity residential type design.

c. Loss Coefficients

The loss coefficient method of figuring pressure loss in a duct fitting is always applicable because it considers total pressure loss. This is different from the equivalent length technique, which considers only static pressure loss (because velocity pressure remains constant). Consider a common transition fitting such as shown in Figure 5.60. Note that the outlet velocity pressure is higher than the inlet velocity pressure because the air velocity is higher at the outlet than at the inlet. This is so because the outlet area is smaller than the inlet area. Remembering that $Q = AV$ and, therefore, $V = Q/A$, it follows that with Q constant, a drop in area means a corresponding rise in velocity. And, since $P_V = (V/4005)^2$, outlet velocity pressure must be higher than inlet velocity pressure. The pressure loss in this fitting is, therefore, a combination of static pressure loss and velocity pressure gain. For this reason, the equivalent length method, which considers only static pressure drop, is not accurate.

An abbreviated list of loss coefficients for common fittings appears in Appendix C. A complete list of 228 different types is available in electronic form, in a data base, from ASHRAE. In this form, the data can be used directly in any of the major duct design programs. Alternatively, the data can be used for manual calculation. Somewhat shorter lists are printed in *SMACNA Duct Design Manual*, *ACCA Manual Q* and the 1993 *ASHRAE Handbook—Fundamentals*. See the bibliography at the end of Chapter 7 for additional sources.

Pressure loss in a fitting is calculated, using its loss coefficient, as follows:

$$P_{\text{LOSS}} = P_V \times C \qquad (5.8)$$

where
P_{LOSS} is the fitting pressure loss in inches w.g.,
P_V is the velocity pressure at the fitting in inches w.g. and
C is the loss coefficient.

When calculating the pressure loss in a fitting with different inlet and outlet areas, as for instance a

Figure 5.59 Outlet velocity of any fitting or duct section is the same as inlet velocity provided flow is constant, and inlet and outlet areas are equal. *(a)* In a straight section of duct, velocity pressure is constant, and static pressure drops off uniformly due to friction. Total pressure is always the sum of static and velocity pressure. *(b)* Regardless of the fitting shape—in this case an underpass type—outlet velocity will equal inlet velocity if flow is constant and cross-sectional inlet and outlet areas are equal. The pressure profile does not show the pressure changes in the underpass. The net overall result is a loss of static pressure. *(c)* In the throat section of this fitting, velocity is high and so is velocity pressure. However, because outlet area equals inlet area, there is a partial static pressure regain and overall a net static pressure loss. See text for a full explanation.

292 / AIR SYSTEMS, HEATING AND COOLING, PART I

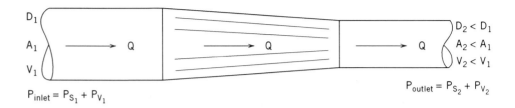

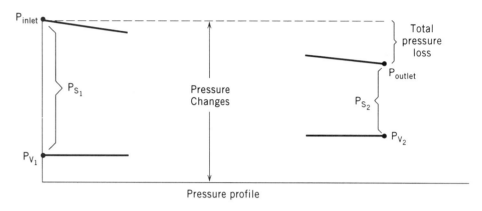

Pressure profile

Figure 5.60 Transition fitting between round ducts of different diameters. Because the outlet diameter is smaller than the inlet diameter its area is also smaller. Therefore, the outlet velocity pressure is higher ($P_{V_2} > P_{V_1}$). As a result, the total fitting pressure loss is not a static pressure loss as in Figure 5.59 but a combination of static pressure loss and velocity pressure gain.

transition fitting, use the velocity pressure at the smaller opening in Equation 5.8. This will be, of course, the higher of the two pressures. Examples will make the calculation method clear.

Example 5.11 A smooth radius 10 in. diameter 90° elbow, with a radius of 15 in., carries 500 cfm. Calculate its pressure loss.

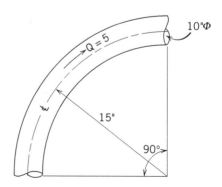

Example 5.11

Solution: We note in the Appendix C data for this fitting that an additional piece of information is required before we can select the loss coefficient; the R/D ratio. In our case, $R = 15$ in. and $D = 10$ in., so

$$\frac{R}{D} = \frac{15}{10} = 1.5$$

We then select from the table, for a 90° elbow with $R/D = 1.5$, a loss coefficient of 0.15. Using Equation 5.8, we have

$$P_{\text{LOSS}} = P_V \times C = 0.15 P_V$$

We know that

$$P_V = (V/4005)^2;$$
$$V = Q/A;$$

and

$$A = \frac{\pi}{4} D^2$$

We, therefore, calculate

$$A = \frac{\pi}{4} D^2 = \frac{\pi}{4} (10)^2 = 78.54 \text{ in.}^2$$

$$V = \frac{Q}{A} = \frac{500 \text{ ft}^3/\text{min}}{78.54 \text{ in.}^2 \times \text{ft}^2/144 \text{ in.}^2} = 917 \text{ ft/min}$$

$P_V = (V/4005)^2 = (917/4005)^2 = 0.05$ in.w.g.
$P_{LOSS} = 0.15 (.05) = 0.008$ in. w.g.

This is a very small loss, as we would expect from a smooth, large radius elbow.

Example 5.12 A round conical transition from 8 to 12-in. diameter duct carries 500 cfm. The cone angle is 45°. Find the fitting pressure loss.

Solution: The pressure loss is calculated using equation (5.8)

$$P_{LOSS} = P_V \times C$$

The solution procedure is, therefore,

Step 1. Using tabular data, find loss coefficient C.
Step 2. Calculate P_V.
Step 3. Calculate fitting pressure loss P_{LOSS}.

We will now perform the calculation using this three-step procedure.
Step 1: In Appendix C, Section 6, we find the loss coefficient data for the fitting. It is reproduced here for ease of reference.

From the illustration we see that the one piece of data still required is the ratio A_1/A where A_1 and A are the upstream and downstream fitting areas, respectively. (The other required data—cfm and cone angle—are given.)

$$\frac{A_1}{A} = \frac{A_{upstream}}{A_{downstream}} = \frac{(\pi/4)(D_1)^2}{(\pi/4)(D)^2} = \frac{12^2}{8^2} = \frac{144}{64} = 2.25$$

Using the three pieces of data:

$$\text{cfm} = 500$$
$$\Phi = 45°$$
$$A_1/A_2 = 2.25$$

we find from the table by interpolation:

A_1/A	Coefficients
2	0.33
2.25	C
4	0.61

Appendix C - Section 6

Loss Coefficients, Transitions (Diverging Flow)

Use the velocity (Vc) in the upstream section to determine the reference velocity pressure (Pv)

Pt = C x Pv (In. Wg.)

A) Transition, Round, Conical (Upstream Pv)

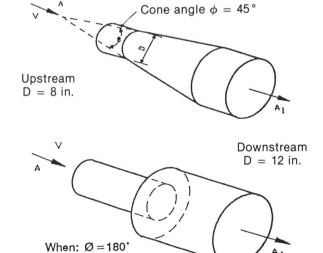

Example 5.12

CFM	A_1/A	Coefficient C ∅							
		16°	20°	30°	45°	60°	90°	120°	180°
<2400	2	0.14	0.19	0.32	0.33	0.33	0.32	0.31	0.30
	4	0.23	0.30	0.46	0.61	0.68	0.64	0.63	0.62
	6	0.27	0.33	0.48	0.66	0.77	0.74	0.73	0.72
	10	0.29	0.38	0.59	0.76	0.80	0.83	0.84	0.83
	>16	0.31	0.38	0.60	0.84	0.88	0.88	0.88	0.88
2400 to 12000	2	0.07	0.12	0.23	0.28	0.27	0.27	0.27	0.26
	4	0.15	0.18	0.36	0.55	0.59	0.59	0.58	0.57
	6	0.19	0.28	0.44	0.90	0.70	0.71	0.71	0.69
	10	0.20	0.24	0.43	0.76	0.80	0.81	0.81	0.81
	>16	0.21	0.28	0.52	0.76	0.87	0.87	0.87	0.87
>12000	2	0.05	0.07	0.12	0.27	0.27	0.27	0.27	0.27
	4	0.17	0.24	0.38	0.51	0.56	0.58	0.58	0.57
	6	0.16	0.29	0.46	0.60	0.69	0.71	0.70	0.70
	10	0.21	0.33	0.52	0.60	0.76	0.83	0.84	0.83
	>16	0.21	0.34	0.56	0.72	0.79	0.85	0.87	0.89

$$C = 0.33 + \frac{2.25 - 2}{4 - 2} \times (0.61 - 0.33)$$
$$C = 0.365$$

Step 2: Calculate P_v.

We know from Equation (5.6) that $P_v = (V/4005)^2$, and $V = Q/A$. As stated previously, the air velocity is calculated at the smaller opening. Therefore,

$$V = \frac{Q}{A} = \frac{500 \text{ cfm}}{(\pi/4)D^2}$$

Since Q is stated in cubic feet per minute, and V is required in feet per minute, we must convert the fitting diameter from inches to feet.

$$8 \text{ in.} \times \frac{1 \text{ ft}}{12 \text{ in.}} = 0.667 \text{ ft}$$

Then

$$V = \frac{Q}{A} = \frac{500 \text{ cfm}}{(\pi/4)D^2} = \frac{500 \text{ cfm}}{0.785(0.667)^2 \text{ ft}^2}$$
$$V = 1432 \text{ fpm}$$

Therefore,

$$P_v = (V/4005)^2 = (1432/4005)^2 = 0.128 \text{ in. w.g.}$$

Step 3: Calculate P_{LOSS}.

$P_{LOSS} = P_v \times C = 0.128 \text{ in. w.g.} \times 0.364$
$P_{LOSS} = 0.047 \text{ in. w.g.}$

This is an appreciable pressure loss.

The preceding calculations serve two purposes. The first is to demonstrate the loss coefficient method of pressure loss calculation for fittings. The second is to give you an appreciation of the value of a fitting data base and a computer program that performs all of these laborious arithmetic calculations in the twinkling of an eye.

5.24 Sources of Duct System Pressure Loss

There are five principal sources of pressure loss in a duct system. They are:

- Straight sections of duct.
- Duct fittings.
- Supply outlets and return inlets.
- Blower inlet and outlet structures.
- Air system devices.

We have studied the first three items in some detail, and the you should be able, at this point, to determine the pressure losses in each category.

The blower (or fan) inlet and outlet structures are the plenums that are used to connect ductwork to the air-handling unit. In residential work, where velocities are low and plenums are simple, losses are very low, rarely exceeding 0.05 in. w.g. In commercial work, losses can be as high as 0.4 in. w.g. for inlets with sharp changes of direction. The inlet and outlet losses are frequently referred to in the literature as the "system effect." Loss coefficients for various inlet and outlet configurations are given in the previously referenced SMACNA, ACCA and ASHRAE publications. They should be consulted for all commercial work and for large residential designs.

The final category of pressure loss sources are known as air side devices. These include:

- Filters of all types, including air washers
- Humidifiers
- Heat exchangers, including energy recovery devices and duct heaters
- Dampers, air-flow controls and smoke control devices
- Louvers and screens
- Sound traps and acoustic linings
- Heating and cooling coils, including DX cooling coils, steam and hot water coils and electrical heating elements
- Air distribution equipment, including mixing boxes of all types and valves
- Monitoring devices and measuring equipment permanently installed in the airflow

Here, too, most of these items are found only in commercial equipment. Since it is assumed that an HVAC technologist will work on many projects, including large commercial ones, he or she should be aware of these pressure loss sources. The actual pressure loss in each is usually given in the manufacturer's catalog along with the other technical data. A useful fact to remember in this connection is that pressure drop varies as the square of air flow (Q). Therefore, if pressure loss is given or known for one flow Q_1, it can be found for another air flow Q_2 by using the relation

$$P_2 = P_1 \times \left(\frac{Q_2}{Q_1}\right)^2 \qquad (5.9)$$

where P_1 and P_2 are in the same units and Q_1 and Q_2 are in the same units.

Duct Sizing Methods

There are two extremely detailed and time-consuming procedures in the design of an HVAC system. The first is the determination of heat losses and gains, as we studied in Chapter 2. The second, when designing an all-air system, is the duct-sizing procedure. Before the advent of computers, forms and schedules were used (and still are) in an effort to systematize and simplify these complex and wearying calculations. The advantage of using a prepared form or schedule is that it forces you to plug in the numbers in the right places, making it difficult (but far from impossible) to make a mistake. A computer program does exactly the same thing, but with the great additional advantage that it does all the calculating. The disadvantage of both, at least for a beginner, is that they make the design procedure mechanical, and this can lead to errors. Since most of you are novice designers, we will avoid extensive design examples that may overwhelm you with numbers in favor of small sectional designs that clearly demonstrate the methods involved.

There are four duct-sizing methods in common use for design of single-zone, low velocity systems. They are:

- Equal friction method
- Modified equal friction method
- Extended plenum method
- Semi-extended plenum (reducing plenum) method

For larger, complex and/or high velocity systems, duct-sizing methods include static regain, constant velocity and the T-method, among others. These methods require considerable experience, are usually done by computer and are beyond the scope of this text. Refer to the bibliography at the end of this chapter for more information.

5.25 Equal Friction Method

The equal friction method is used very frequently in the design of duct systems for small to medium-size residences and commercial structures. The basic idea of this method is to use the same friction rate (friction per 100 ft) to size all the ducts in the system. The friction rate to be used is arrived at by one of three methods:

- Velocity limitation in the first trunk duct section
- Total pressure available divided by the equivalent total length (TEL) of the longest duct run
- Rule of thumb, which states that for such systems the friction rate should be between 0.08 and 0.12 in. w.g. per 100 ft.

An example of the use of the equal friction method should help to make its application clear.

Example 5.13 Use the equal friction method to size the ducts shown in Figure 5.61(a). Justify all assumptions.

Solution: The example could represent the supply duct layout for a medium-size ranch-style residence or a single-level commercial building. We will assume that this is a residential installation. The design steps that precede preparation of the duct diagram will be detailed later in this chapter, in the discussion of overall design procedure.

(1) *Determine friction rate to be used.* Consult Table 5.6, which lists maximum velocities permitted for ductwork and outlets in residential installations. (Table 5.7 gives the same data for other types of buildings.) We will use the maximum trunk velocity permitted, because the duct is insulated and is relatively small. Both of these factors help reduce the noise generated and transmitted by the ductwork. Using either the chart in Figure 5.51 or one of the calculators shown in Figures 5.52 or 5.53 we determine that a flow of 1050 cfm at 900 fpm gives a friction rate of 0.08 in. w.g./100 ft. This friction rate is just at the edge of the shaded recommended design area of Figure 5.51 and complies with the lower value of the rule-of-thumb range (0.08–0.12) listed previously.

As a further check on this proposed friction rate, we should work out the third method listed, that is, divide the pressure available by the TEL of the longest duct run. Referring to Figure 5.61(a), this would be duct run AB-CDEFG. In order to do this accurately, we would need a detailed duct diagram showing all fittings. We would also need to calculate the net external pressure available from the furnace data such as is given in Figure 5.6(b). (The furnace data given in Figure 5.12(b) is not sufficient. The technologist would have to contact the manufacturer for static pressure data corresponding to the cfm figures tabulated.) The net supply duct static pressure available is only a portion of the total external static pressure. That calculation will be explained in the following duct system design procedure.

The TEL is calculated by adding all the straight duct lengths to the sum of the TEL

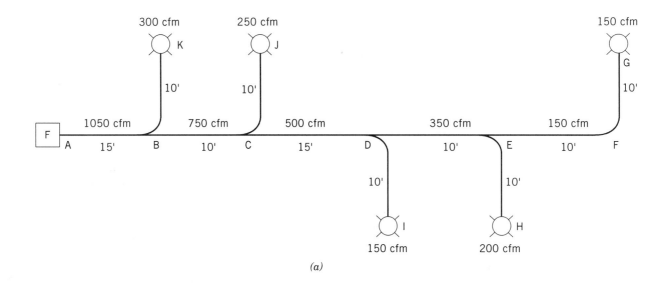
(a)

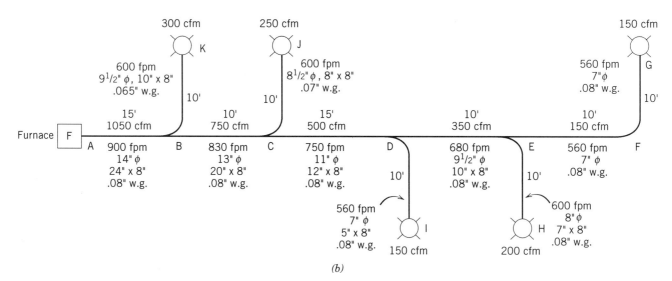
(b)

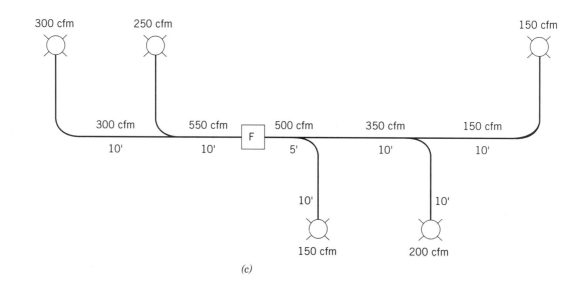
(c)

Table 5.6 Recommended Air Velocities (fpm) for Noise Limitation in Residences

	Supply Side				Return Side			
	Recommended		Maximum		Recommended		Maximum	
Location	Rigid	Flex	Rigid	Flex	Rigid	Flex	Rigid	Flex
Trunk ducts	700	600	900	600	600	600	700	600
Branch ducts	600	600	900	600	400	600	600	600
Supply outlet face velocity	Size for Throw		700		—		—	
Return grille face velocity	—		—		—		500	
Filter grille face velocity	—		—		—		300	

Source. Reprinted with permission from ACCA *Bulletin 116*.

Table 5.7 Maximum Velocities for Low Velocity Systems (fpm)

	Controlling Factor—Noise Generation, Main Ducts	Controlling Factor—Duct Friction			
		Main Ducts		Branch Ducts	
Application		Supply	Return	Supply	Return
Residences	See Table 5.6	1000	800	600	600
Apartments, hotel bedrooms, hospital bedrooms	1000	1500	1300	1200	1000
Private offices, director's rooms, libraries	1200	2000	1500	1600	1200
Theaters, auditoriums	800	1300	1100	1000	800
General offices, high-class restaurants, high-class stores, banks	1500	2000	1500	1600	1200
Average stores, cafeterias	1800	2000	1500	1600	1200
Industrial	2500	3000	1800	2200	1500

Source. Reprinted with permission from *ACCA Manual Q*, 1990.

Figure 5.61 *(a)* Single-line diagram of a duct arrangement, showing duct lengths and required air quantities. *(b)* Duct sizes, air velocities and friction rate for all sections of the system, as calculated by the equal friction rate method. *(c)* Placement of the air handler (furnace) in the center of the duct system makes balancing the system much simpler.

values of all the fittings as obtained from Appendix B. Since, for our example, we have neither the fan pressure nor the system TEL, we will assume that this calculation gives a friction rate close to the 0.08 in. w.g. we have already obtained. If an actual calculation were to show a value much higher, a lower motor speed would be selected to drop the pressure. If the friction rate is too low, a higher motor speed is needed. Too high a friction rate means excessive velocity and noise; too low a friction rate means excessively large ducts.

(2) *Sizing all sections of trunk duct.* Using the friction rate of 0.08 in. w.g./100 ft, we now proceed to size all sections of trunk duct, again using either Figure 5.51 or a duct calculator. The round duct sizes obtained for duct sections AB, BC, CD, DE, and EF are shown on Figure 5.61*b*. Air velocity and the friction rate are also indicated for each trunk section. Depending on the architectural layout, the technologist might choose to use either round ducts or rectangular ducts. The equivalent rectangular sizes are also shown on Figure 5.61*b*. Note that all the ducts are 8 in. deep. This is the depth commonly used in residential work, because this size duct will fit in the space between floor joists. If the duct is run across the joists, any reasonable depth is usable. The entire main duct assembly is constructed as a reducing trunk duct. See Figure 5.18*(d)* and Appendix B, Figure B.2.

(3) *Branch ducts.* Again, using the friction chart or a duct friction calculation, duct sizes for branches can be determined using the same friction rate. Maximum velocity should not exceed 600 fpm to limit noise levels and friction rate adjusted accordingly. Round branch ducts are commonly used in residences, although rectangular ducts may be preferable to fit the architecture. Note that section FG is simply an extension of trunk section EF. For this reason, the velocity in EF was also held below 600 fpm. This essentially completes the design of the duct system.

The equal friction method gives best results when the TELs of all runs are approximately equal. This would be the case, for instance, if the furnace were in the center of the duct run as in Figure 5.61*(c)* and not at the end, as is Figure 5.61*(a)*. A glance at Figure 5.61*(a)* shows why. The total pressure drop between the furnace and outlet G (duct run ABCDEFG) is much higher than the pressure drop to outlet K (duct run ABK). That means that the pressure at outlet K is much higher than that at outlet G. The result will be too much air at K and too little at G, in other words, an unbalanced system. (A good rule of thumb to follow is that the pressures at all outlets in a system should not vary, one from another, by more than 0.05 in. w.g.)

To compensate for this unbalance, it is standard practice to install volume dampers in all branches and runouts. Field adjustment of these dampers adds friction to short runs, permitting the system to be balanced. The problem with this "fix" is that dampers cause noise, which is exactly what we want to avoid. A much more satisfactory solution, from an engineering point of view, is to design the required additional friction into the system. This, indeed, is exactly what is done in the modified equal friction method.

5.26 Modified Equal Friction Method

The modified equal friction method is more accurately called the equal pressure loss method because it attempts to give all runs approximately the same TEL. This makes the system (almost) self-balancing.

The procedure for this method is:

(1) Prepare a detailed duct layout showing all fittings.
(2) Using the required air quantities in each section, calculate the pressure drop in all straight duct sections. Use the velocity limitations imposed by noise criteria.
(3) Find the TEL of all fittings using Appendix B or calculate their pressure drops using Appendix C.
(4) Find the pressure drop of the longest run.
(5) Redesign the friction in other branches and fittings so that the total pressure drop from the furnace to each supply outlet is approximately equal.

This last step is the most difficult. Refer to Figure 5.61*a*. In order to make the pressure drop to outlet K the same as that to outlet G, a large pressure drop must be introduced into runout BK. One way to do this is to reduce the size of the duct from $9\frac{1}{2}$ in. to 4–5 in. This, however, will increase the air velocity enormously, causing noise, vibration and severe drafts in the room being served. A much better technique is to use a high resistance takeoff such as types P or Q in Figure B.2 or types A or F in Figure B.3. Even these, however, are not sufficient in short runs, and balancing dampers will still be required in all branches and runouts.

In more complex systems, with several main branches, the calculation becomes one of trial and error. We begin with an educated guess at friction rates for the various branches. Then TELs are found, and pressure drops are calculated and compared. At that point, friction rates, fitting types, air velocities and duct sizes are all juggled in an attempt to balance the system. True balance is almost always impossible, which is one reason that volume dampers are almost always used. Another important reason is that seasonal changeover between heating and cooling modes always requires changes in air flow. These changes are accomplished, in part, with dampers.

The advantage of the modified equal friction method is that it will give a nearly balanced system that will require only slight field adjustment. The disadvantage of the method is that to design it correctly, for anything but a small system, is tedious and time-consuming. Actually, to perform the calculations accurately, loss coefficients should be used for fittings rather than TEL. This is because almost all the fittings involve velocity changes between inlet and outlet. As explained in Section 5.23, the loss in such fittings cannot be calculated accurately using TEL. And, as we saw in that section, manual calculation of fitting pressure loss, using its loss coefficient, is an involved and time-consuming operation. Fortunately, in modern engineering offices, computers have relieved designers of these burdensome calculations.

5.27 Extended Plenum Method

See Figures 5.50 and B.3. An extended plenum, as explained in Section 5.20d, is simply a relatively short straight section of trunk duct that feeds a number of branch outlets, usually not exceeding six. The trunk cross section does not change throughout its length. It is called an *extended plenum* because, like a plenum, its size is constant, and it extends over a length of 25–30 ft. In small systems, it may represent the entire duct system. In large systems, such trunk ducts are found at the discharge of a fan, VAV box, mixing box and the like. They too are known as extended plenums.

The extended plenum method is not so much a specific calculation method, as it is the characteristics of air flow in this type of trunk duct. The air flow in an extended plenum is governed by principles that we have already learned. This air flow can be summarized as follows.

(a) As we proceed along the trunk duct, air velocity and friction rate will decrease after each takeoff. This is necessarily so, since $Q = AV$, that is, air volume is the product of duct cross-sectional area and air velocity. (See Section 5.23, Equation 5.7, page 288.) Since air volume Q decreases after each takeoff and area A remains constant, velocity V must also decrease proportionately.

(b) Since trunk velocity decreases after each takeoff, it will frequently occur that runout (branch) velocity is higher than trunk velocity. This will cause a small conversion pressure loss. This loss is not appreciable at trunk velocities below 600 fpm.

(c) At takeoffs where trunk velocity is much higher than branch velocity, the pressure loss at the takeoff will be high. This can result in a "starved" takeoff, if the takeoff is not properly designed. This will become clear in the following calculations.

The design criteria usually applied to an extended plenum are:

- The trunk duct is sized for volume and design velocity. This velocity is usually friction-limited rather than noise-limited.
- Branch takeoffs are velocity-limited by noise criteria.
- Maximum trunk length should not exceed 30 ft.
- Takeoffs are usually round duct, but rectangular is also acceptable.
- Takeoff connections for round or rectangular duct should be made with a 45° angle connection. (A 90° connection can have very high pressure drop if trunk velocity is higher than branch velocity.)
- Extractors or scoops at takeoff points should be avoided. They cause turbulence and high pressure losses.
- Since the system is not self-balancing, balancing dampers should be installed in each branch.
- Since the friction rate changes after each takeoff, the friction loss in each section of trunk duct must be calculated individually.

A numerical example should clarify the preceding system characteristics and design principles.

Example 5.14 Refer to the extended plenum shown in Figure 5.62. An extended trunk of constant size feeds five branches with outlets at their ends. (For the sake of simplicity, we have made all the branches identical.) Each branch feeds an outlet with a 400-cfm design air quantity. Design this extended plenum duct system. Do not exceed an air velocity of 1000 fpm in any section. Calculate friction rates and pressure drops throughout the system.

Is the system balanced or nearly so?
What conclusions can be drawn about the balance of an extended plenum?

Solution:

(1) The first step is to size the main trunk. Totaling all the air volumes required, we note them on the drawing:

Each branch	400 cfm
Duct section AB	2000 cfm
Duct section BC	1600 cfm
Duct section CD	1200 cfm
Duct section DE	800 cfm
Duct section EF	400 cfm

Using the specified maximum velocity of 1000 fpm, we find that duct section AB must be 19 in. in diameter. The equivalent rectangular section is 20×16 in., giving a friction rate of 0.075 in. w.g./100 ft for this section. These data are marked on the drawing.

(2) Since we know the air quantity in each section of the plenum, we can calculate the air velocity in each section very simply as follows:

$$Q_{AB} = A \times V_{AB}$$

and

$$Q_{BC} = A \times V_{BC}$$

Therefore,

$$\frac{Q_{AB}}{Q_{BC}} = \frac{V_{AB}}{V_{BC}}$$

since A remains constant.
or

$$V_{BC} = V_{AB} \times \frac{Q_{BC}}{Q_{AB}} = 1000 \text{ fpm} \times \frac{1600 \text{ cfm}}{2000 \text{ cfm}} = 800 \text{ fpm}$$

Similarly,

$$V_{CD} = V_{AB} \times \frac{Q_{CD}}{Q_{AB}} = 1000 \times \frac{1200 \text{ cfm}}{2000 \text{ cfm}} = 600 \text{ fpm}$$

and

$$V_{DE} = 400 \text{ fpm}$$

and

$$V_{EF} = 200 \text{ fpm}$$

These values are marked on the duct diagram.

(3) Since we know the Q, and V and size of each section, we can use the chart to find the friction rate. This too is marked on the diagram for each section. Note that the air velocity in the last section, EF, is so low (200 fpm) that the friction rate and, therefore, also the actual friction, are negligible. The friction rates for sections BC, CD and DE are 0.048, 0.03 and 0.013 in. w.g., respectively.

(4) We select a velocity of 500 fpm for the branches to avoid noise problems. Knowing Q and V for the branches, we find that a 12 in. round or 10×12-in. rectangular duct is required. The friction rate is 0.035 in. w.g. These data apply to all branches, and this is marked on the duct drawing. At this point, we can calculate the pressure losses in all the straight duct sections of the system. These are summarized in Table A.

(Example 5.14) Table A Pressure Loss in Straight Duct Sections

Duct Section	Length, ft	Friction Rate, in. w.g./100 ft	Pressure Loss, in. w.g.
AB	5	0.075	0.00375
BC	5	0.048	0.0024
CD	5	0.03	0.0015
DE	5	0.013	0.00065
EF	5	0	0
BG, CH, DI, EJ, FK	20	0.035	0.007

(5) Next we need to find the pressure drop of the takeoff fitting at each branch connection. Note that we used a 45° takeoff to reduce pressure loss. Figure 5.63 shows why the loss in such a takeoff is less than that in a right angle takeoff. Referring to Appendix B, Figure B.3, we find that a type D fitting has an TEL of 10 ft. To this TEL must be added an additional length to reflect the high takeoff fitting pressure loss due to the difference in air velocities between main and branch. The greater this difference is, the higher the pressure drop will be. This *velocity*

EXTENDED PLENUM METHOD / 301

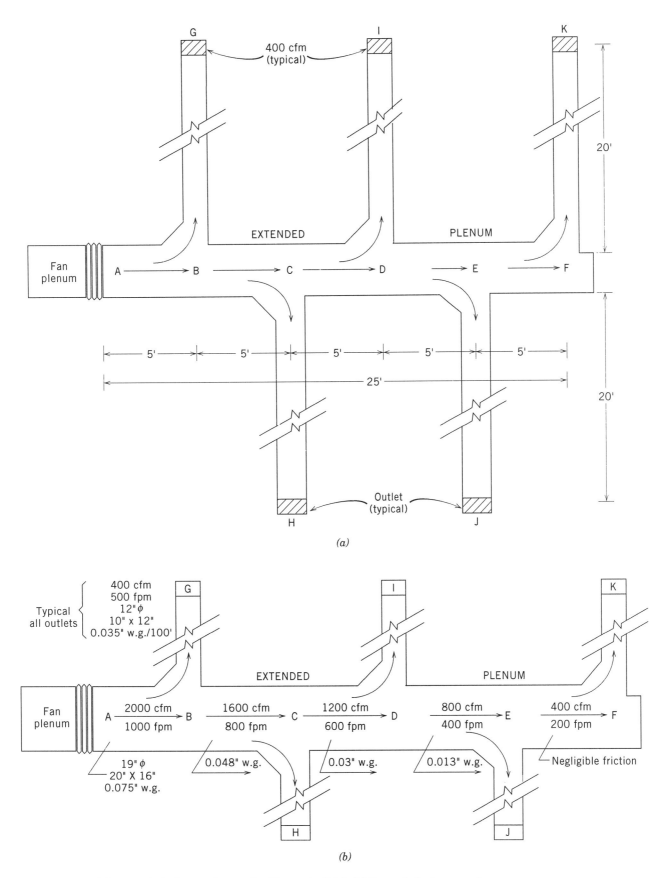

Figure 5.62 (a) Extended plenum layout for Example 5.14. (b) Extended plenum of Example 5.14 with all duct sizes, airflows and sectional friction losses indicated.

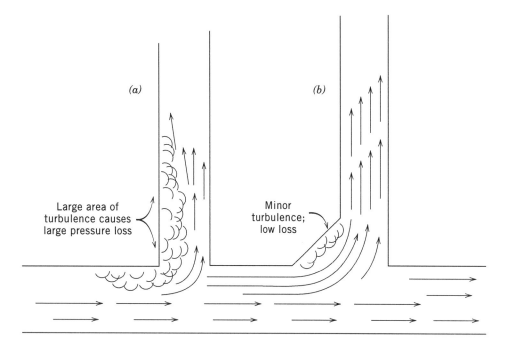

Figure 5.63 (a) Right angle (butt) connection of branch duct into trunk causes a large area of turbulence and large pressure loss. (b) An angle connection permits a smooth turn of the air flow with minimal turbulence, friction and pressure loss.

factor is usually taken as 10 ft of TEL for every downstream branch *after* the takeoff. (See the table at the bottom of Figure B.3.) Therefore, for branch BG, we would add to the 10 ft TEL of the fitting another 40 ft representing the four downstream takeoffs.

The number of downstream takeoffs in an extended plenum is an indication of trunk velocity at the upstream takeoff. It is, therefore, convenient to use this number to determine the velocity factor. Physically, the additional pressure drop is easy to understand. Air in the plenum is rushing by the takeoff. The higher its velocity, the more difficult it is to get a portion of it to turn off into the branch. The turn causes turbulence and, therefore, high pressure loss. The pressure loss in all fittings, as calculated using ETL and velocity factor, is tabulated in Table B. Figure 5.64 shows diagramatically the effect of the velocity factor.

We have repeatedly stated that using a total equivalent length (TEL) for fittings through which a velocity change occurs, is not entirely accurate. The pressure loss in such fittings should be calculated using loss coefficients. To demonstrate this, we have calculated the takeoff fitting losses in our system using loss coefficients. The results are tabulated in Table C. The data were taken from Appendix C, fitting P. Note that the pressure loss calculated using loss coefficients is 1½ to 2 times as large as that calculated by using the TEL method. This is a very significant difference and shows the importance of knowing when to use each method. For purposes of comparison, we have indicated on Figure 5.64 the additional branch length represented by fitting pressure loss as calculated with loss coefficients.

(6) We have summarized all the pressure loss data in Table D, including a comparison of the overall pressure loss to each outlet as calculated using TEL and loss coefficient. Using TEL, the pressure losses from fan to outlets G, H and I are almost identical. The losses to outlets J and K are lower. This would indicate that branches G, H and I are essentially balanced and that J and K need to be throttled slightly, using dampers. Notice, however, that when using the

SEMI-EXTENDED PLENUM (REDUCING PLENUM) METHOD / 303

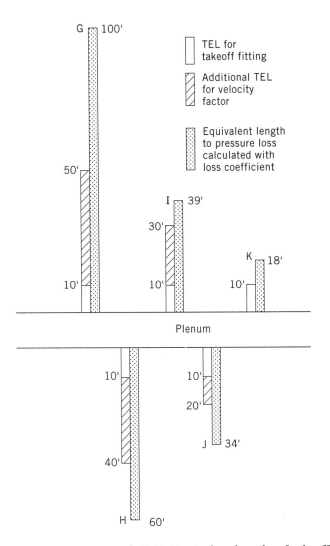

Figure 5.64 Example 5.14. Equivalent lengths of takeoff fittings for branches, as calculated using TEL plus velocity factor and loss coefficients.

more accurate loss coefficient calculation it is outlets I, J and K that are almost balanced, and not G and H.

However, the most important and most surprising result of this calculation is that total pressure drop to an outlet *decreases* as we move away from the fan. That means that the outlets nearest the fan get the least air and those farthest away get the most air. This is exactly opposite to what we would logically expect. It indicates clearly that balancing dampers must be installed in all branches because the system is inherently unbalanced.

5.28 Semi-Extended Plenum (Reducing Plenum) Method

An extended plenum has disadvantages. In addition to inherent imbalance and, therefore, the need for balancing dampers in branches, the duct is far too large beyond approximately its midpoint. This last disadvantage is offset by simplicity and economy of fabrication, since no expensive transitions are involved. However, because the oversized duct causes such a radical imbalance, with the remotest outlets getting the most air, many designers prefer to use one, or at most two, size reductions. This is also done when the plenum extends more than 25–30 ft. No criterion exists for the placement of these transitions. Some designers suggest that a transition be used wherever the trunk velocity falls below the branch velocity.

Example 5.14, Table B Calculation of Fitting Pressure Loss by TEL Method

Location of Fitting	Fitting Loss TEL, ft	Velocity Factor, ft	Total TEL, ft	Friction Rate, in. w.g./100 ft	Total Pressure Loss, in. w.g.
B	10	40	50	0.035	0.0175
C	10	30	40	0.035	0.014
D	10	20	30	0.035	0.0105
E	10	10	20	0.035	0.007
F	10	0	10	0.035	0.0035

Example 5.14, Table C Fitting Pressure Loss Calculation Using Loss Coefficients

Location of Fitting	$\dfrac{Q_{branch}}{Q_{main}}$	$\dfrac{V_{branch}}{V_{main}}$	Loss Coefficient	$P_V = \dfrac{(V_{main})^2}{(4005)^2}$, in. w.g.	P_{LOSS}, in. w.g.
B	0.2	0.5	~0.55	0.0625	0.035
C	0.25	0.625	0.514	0.04	0.021
D	0.33	0.83	0.615	0.0225	0.0138
E	0.5	1.25	1.19	0.01	0.012
F	1.0	~2.5	~2.5	0.0025	0.0063

Example 5.14, Table D Total Pressure Loss from Furnace to Outlets

Outlet Location	Pressure Loss in Straight Ducts[a]		Fitting Loss[a]			Total Pressure Loss from Furnace to Outlet, in. w.g.	
			By TEL		By Loss Coefficient		
	Path	Loss, in. w.g.	TEL, ft	P_{LOSS}, in. w.g.	P_{LOSS}, in. w.g.	By TEL	By Coefficient
G	ABG	0.01075	50	0.0175	0.035	0.0283	0.0458
H	ABCH	0.01315	40	0.014	0.021	0.0272	0.0342
I	ABCDI	0.01465	30	0.0105	0.0138	0.025	0.0285
J	ABCDEJ	0.0153	20	0.007	0.012	0.016	0.0273
K	ABCDEFK	0.0153	10	0.0035	0.0063	0.016	0.0216

[a]The pressure losses in the trunk duct caused by the takeoff fittings have been ignored to avoid complicating the calculation. In actual practice, they should be included.

To illustrate this semi-extended plenum method, we have placed a transition in the duct of Example 5.14 (see Figure 5.62) at point D, beyond the third takeoff. At this point velocity drops to 400 fpm in the original duct, as compared to a branch velocity of 500 fpm. We have resized the plenum for 1000 fpm and recalculated all the losses based on these revised data. See Figure 5.65. Table A1 shows the pressure losses in the straight sections of duct; Table B1 lists the fitting losses, including both takeoff fittings and the duct transition fitting. Note that the loss for takeoff fittings at B, C and D are calculated with loss coefficients. The loss coefficient table does not cover the flow and velocity conditions at points E and F. As a result, we were forced to use the TEL method for these fittings. This is indicated in Tables B1 and C1.

Table D1 summarizes the results. Notice that the total pressure losses to all outlets are much more uniform than for the full extended plenum. Balanc-

Table A1 Pressure Loss in Straight Duct Sections, Semi-extended Plenum

Duct Section	Length, ft	Friction Rate, in. w.g./100 ft	Pressure Loss, in. w.g.
AB	5	0.075	0.00375
BC	5	0.048	0.0025
CD	5	0.03	0.0015
DE	5	0.125	0.0063
EF	5	0.03	0.0015
All branches	20	0.035	0.007

ing dampers would still be needed in all runouts for minor field adjustment. The disadvantage of this method is the additional cost due to the addition of a duct transition(s). This cost is at least partially offset by the cheaper smaller duct section after each transition.

SEMI-EXTENDED PLENUM (REDUCING PLENUM) METHOD / 305

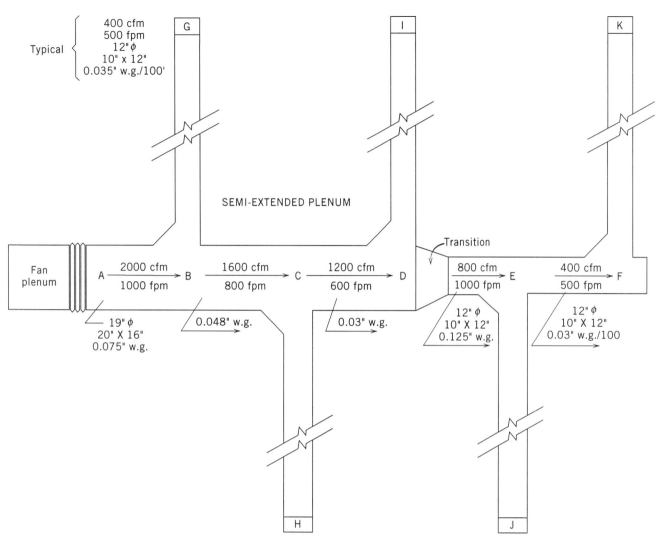

Figure 5.65 Semi-extended plenum layout. This trunk duct arrangement is also called a reducing plenum.

306 / AIR SYSTEMS, HEATING AND COOLING, PART I

Table B1 Calculation of Fitting Pressure Loss by TEL Method; Semi-extended Plenum

Location of Fitting	Fitting Loss TEL, ft	Velocity Factor, ft	Total TEL, ft	Friction Rate, in. w.g./100 ft	Total Pressure Loss, in. w.g.
B[a]	—	—	—	—	—
C[a]	—	—	—	—	—
D[a]	—	—	—	—	—
E	10	10	20	0.035	0.007
F	10	0	10	0.035	0.0035

[a]Calculated by loss coefficient. See Table C1.

Table C1 Fitting Pressure Loss Calculation Using Loss Coefficients; Semi-extended Plenum

Location of Fitting	$\dfrac{Q_{branch}}{Q_{main}}$	$\dfrac{V_{branch}}{V_{main}}$	Loss Coefficient	$P_v = \dfrac{(V_{main})^2}{(4005)^2}$, in. w.g.	P_{LOSS}, in. w.g.
B	0.2	0.5	~0.55	0.0625	0.035
C	0.25	0.625	0.514	0.04	0.021
D	0.33	0.83	0.615	0.0225	0.0138
E[a]	—	—	—	—	—
F[a]	—	—	—	—	—
Duct transition fitting[b]	—	—	—	—	0.00313

[a]Pressure loss calculated by TEL method. See Table B1.
[b]Horizontal and vertical transition angles 19° and 15°, respectively. See Appendix C.

Table D1 Total Pressure Loss from Furnace to Outlets; Semi-extended Plenum

| Outlet Location | Pressure Loss in Straight Ducts[a] | | Fitting Loss[a] | | By Loss Coefficient | Total Pressure Loss from Furnace to Outlet, in. w.g. |
| | Path | Loss, in. w.g. | By TEL | | | |
			TEL, ft	P_{LOSS}, in. w.g.	P_{LOSS}, in. w.g.	
G	ABG	0.01075	—	—	0.035[c]	0.0458
H	ABCH	0.01315	—	—	0.021[c]	0.0342
I	ABCDI	0.01465	—	—	0.0138[c]	0.0285
J	ABCDEJ	0.021	—	0.007[b]	0.00313[d]	0.0311
K	ABCDEFK	0.022	—	0.0035[b]	0.00313[d]	0.0286

[a]The pressure losses in the trunk duct caused by the takeoff fittings have been ignored to avoid complicating the calculation. In actual practice they should be included.
[b]From Table B1.
[c]From Table C1.
[d]Loss in the duct transition.

System Design

Having studied air system components and duct system design, we are now in a position to outline the overall design process. In all probability, a technologist will not be involved in the preliminary stages of design. These include codes and jurisdictions, system type choices and economic decisions. However, since we believe that most technologists will progress to designer status rapidly, it is important to be familiar with the overall design process, from the beginning.

The steps in a typical design follow. Depending on the specific project, the order in which these steps are performed may be different from that listed. Too, some of the steps may be omitted as not being relevant to the project at hand. For instance, in residential design, the heating (and cooling) unit is purchased as a package, with a multispeed blower supplied as part of the package. The technologist then works with the blower characteristics, designing the duct system to use the pressure and air quantity available. In commercial projects, the designer lays out the desired system, calculates the required air volume and pressure and selects a blower to meet these needs. These differences will obviously change the order of certain design steps.

5.29 Design Procedure

Keeping in mind that each project has its own peculiarities that may require variations, the typical warm air system design procedure is described next.

a. Codes and Ordinances

Every construction project is subject to local construction ordinances and codes. Most of these refer to national codes. Some of the larger cities have their own codes in addition to requiring adherence to national codes. Determining which codes apply is the responsibility of the project engineer. Fulfilling code requirements is the designer's responsibility. One or more of the following codes will apply to any duct system:

1. The BOCA Basic Mechanical Code of Building Officials and Code Administrators International, Inc., Homewood, Illinois.
2. The Uniform Mechanical Code of International Conference of Building Officials (ICBO), Whittier, California.
3. The Standard Mechanical Code of Southern Building Code Congress International, Birmingham, Alabama.
4. The National Building Code of American Insurance Association, New York, Chicago and San Francisco.
5. National Fire Protection Association (NFPA), Quincy, Massachusetts.
6. National Building Code (by the National Research Council of Canada), Ottawa, Ontario, Canada.

Some of the most important ordinance requirements relate to fire and smoke control. The subject itself is beyond the scope of this book. Refer to the following NFPA Codes for more information:

NFPA 90A: Installation of Air Conditioning and Ventilating Systems
NFPA 90B: Installation of Warm Air Heating and Air Conditioning Systems
NFPA 92A: Smoke Control Systems

A complete listing of codes and standards appears in *ASHRAE Handbook—Fundamentals*, Chapter 38.

A detailed presentation of fire protection in HVAC systems is given in *ASHRAE Handbook—Applications*, Chapter 47.

b. Load Calculation

In this step, the heating (and cooling) loads for each space in the structure are calculated. They are then summarized to determine the total building load for heating, and for cooling if required. A stripped architectural drawing should be prepared at this stage. This is simply an architectural plan showing walls, doors and windows. Dimensional data (including ceiling heights) should be indicated. On this plan, the calculated loads for each space are indicated. The next stage of the load calculation procedure is to calculate air quantities required for each space, again for heating and cooling, using the relations developed in Section 5.3:

$$\text{cfm} = \frac{\text{Heating load in Btuh}}{1.08 \, \Delta T}$$

and

$$\text{cfm} = \frac{\text{Cooling load in Btuh}}{1.1 \, \Delta T}$$

The temperature difference ΔT is in the range of 45–75 F° for heating and 17–21 F° for cooling. The difference figures are based on a winter room return-heating air temperature of 70°F and heated air entering the room at 115–145°F. It is recommended that the entering air not exceed 135°F if a person can stand next to the supply register, since 145°F air is uncomfortably hot.

Similarly, the cool air temperature differential is based on a return-air temperature of 75°F and cool room air entering at 54–58°F. Here, also, if entering air can strike an occupant, 58°F is a better choice. Air at 54°F is uncomfortably cold, particularly when blown across the skin at velocities up to 300 fpm. The larger of the two air quantities (heating or cooling) is obviously the one that will determine the required duct sizes to all spaces and trunk duct sizes.

c. System Type

At this point, a study is made of the building operation in order to decide whether a multizone system is required. If zoning is required, the type of zoning arrangement is the next decision. See Section 5.19. Once the system is decided upon, a single-line duct diagram can be drawn on the working drawing(s) showing air quantities in all sections. At this stage, if the technologist is working with calculation forms or computer input forms, all the air quantity (and temperature) data can be entered. The dimensional data for all spaces was entered on the load forms at the load calculation stage.

d. Furnace Selection

On the basis of the calculated building load, a furnace of sufficient capacity (MBH) is selected. The furnace rating must include spare capacity as described in Section 5.8.b. Having determined the furnace MBH and the total cfm required for the building, the furnace temperature rise requirement should be calculated:

$$\text{Temperature rise} = \frac{\text{Furnace rating in MBH}}{1.08 \text{ (Total cfm)}}$$

With these three items of data—the furnace capacity, temperature rise and cfm required—a specific unit can be chosen from manufacturers' data tables, similar to those of Figure 5.6b and Figure 5.12.

From these same tables dimensional information can be taken. These data are now used to determine the space requirements for the furnace, its plenum, duct connections and so on. Of course, the decision as to the type of furnace to be used, that is, upflow high-boy, upflow low-boy, downflow or horizontal has already been made, based on the building architecture and the duct plan to be used.

If the furnace is a residential type, the external static friction available for different motor speeds and air flow quantities is also available from the manufacturer's data. If the unit is a commercial furnace, the blower will be selected at a later stage.

e. Supply and Return Outlets

On the basis of air quantities calculated in Section 5.29b, and considering the selection criteria detailed in Sections 5.15–5.18, supply outlets and return outlets can be selected, located and sized. Use manufacturers' data such as that shown in Figures 5.21, 5.22 and 5.23. Show locations and sizes on the working drawings. Record the static pressure drops at each outlet, for future use. Except in unusual installations, the total pressure drop in supply registers or diffusers should not exceed 0.07 in. w.g. and 0.04 in. w.g. in a return grille.

A convenient rule of thumb for determining the number of outlets required in a space is that one outlet is required for every 8000 Btuh of heating load and every 4000 Btuh of cooling load. It is also convenient to translate these figures into cfm. Using a common heating temperature rise of 55 °F (125°F supply) and a cooling temperature drop of 19 °F (56°F supply), we can calculate the cfm per outlet. For heating:

$$Q = \frac{8000}{1.08 \ (55°F)} = 135 \text{ cfm/outlet}$$

For cooling:

$$Q = \frac{4000}{1.1 \ (19°F)} = 190 \text{ cfm/outlet}$$

These figures can vary ±15% depending on air temperatures, outlet locations and outlet face velocity. However, as a guide and as a quick check on calculations, the figures given are reliable.

f. Duct Design

The duct design stage begins with establishing a target friction rate as described in Section 5.25. Assuming that we are dealing with a package unit, the external pressure available can be determined as described in Section 5.8.c. As a rule this pressure will be somewhere between 0.1 and 0.4 in w.g. Pressures above 0.4 in. w.g. are necessary only for very long duct runs or runs with many fittings and turns. Pressures below 0.1 are suitable only for straight trunk runs 25 ft long or less, with short branches. The duct design procedures described in Sections 5.21–5.28 can now be applied. A detailed duct plan is required showing all fittings, turns, junctions and the like to enable accurate calculation of pressure losses in the various system branches.

For low velocities, the use of TEL figures for fitting losses will not introduce large errors. In all cases, loss coefficients will give more accurate pressure loss data. If pressure loss calculations result in an overall duct system pressure loss below 0.1 in. w.g. or above 0.4 in. w.g., the friction rate should be altered (if possible), so that the pressure loss falls in this range. Pressures above 0.4 in. w.g. are available at high blower speeds but are preferably avoided because of high air velocity and the attendant noise problems.

Return-air ducts are included in the preceding calculation. Some designers prefer to divide the available pressure between the supply and return duct systems on an estimated basis. In our opinion, this is justified only if the return system is not ducted but consists of door undercuts, door and wall louvers and the like.

The particular duct design procedure selected depends on the duct layout. In most cases, the modified equal friction procedure as described in Section 5.26 will be adequate. Once the design is complete, the technologist can decide whether to use round ducts or convert to rectangular ducts. Of course, the decision regarding the duct material was made at the beginning of the duct design stage. Once all the ducts are sized, the entire system should be rechecked for velocity and noise problems. When the designer is satisfied that the system is workable, it can be drawn as a two line duct layout on the plans. At this stage, physical problems of installation and coordination with other trades will arise. Such problems will frequently lead to minor (and sometimes major) changes in the duct design.

g. Additional Design Items

As the technologist gains experience, he or she will almost intuitively know where to place fittings, turning vanes, balancing dampers and the like. A few useful rules in duct design are these:

(1) Use the longest possible radii in turn fittings.
(2) Where sharp turns are unavoidable, use elbows with turning vanes. A sharp turn is one where the inside radius of the duct is less than one-third of the duct width.
(3) When calculating pressure drop, use dampers in their open position as a fitting pressure loss.
(4) Do not use register dampers for balancing. They are a source of unacceptable noise. Instead, place a balancing damper in the branch, as far upstream as possible.
(5) In residential work, remember that the smallest stack readily available is $3\frac{1}{4} \times 10$ in. This is equivalent to a 6 in. round duct, which is often too large for the air requirement in a heating system. A damper is, therefore, always required. See Figure 5.1 for location of stacks.
(6) Most residential air-conditioning contractors will use joist spaces for return air. Since this space is 16 in. wide and 8–12 in. deep, it is equivalent to a 11–15 in. duct, even with the additional roughness of the wood. As a result, additional pressure is available in the supply system.

5.30 Designer's Checklist

Most designers use a checklist to ensure that the design includes all the required items, properly applied. Such checklists are developed over years of design experience and vary from one designer to another. In many offices, such checklists are standardized for use by all designers. One such checklist is reprinted here with permission from *ACCA Manual Q—Commercial Low Pressure, Low Velocity Duct System Design*.

- Make sure that all duct velocities are in the correct range.
- Make sure that the velocity through each air-side component (such as filters, coils, louvers and dampers) is in the correct range.
- Branch takeoffs should not be close to the fan.
- A branch runout fitting should not be installed behind an elbow or upstream runout (leave six diameters).
- A branch runout fitting should not be installed behind an upstream extractor.
- The proportions or sizes of "split flow" fittings should be based on the cfm requirements of each resulting branch.
- Turning vanes should be installed with leading and trailing edges that are perpendicular to the air flow.
- Fan inlet and outlet fittings should minimize system effect losses.

- Fans should be selected to operate near the middle of the recommended operating range.
- Economizer cycle requires a return fan or an exhaust fan.
- A return fan is recommended if the pressure drop in the return duct system exceeds 0.10 in. w.g.
- Relief air dampers are required if the outdoor air cfm that is introduced into the space exceeds the cfm that is exhausted from the space and there is no return or exhaust fan (space pressurized).
- Corrections for surface roughness are required. Use the appropriate friction chart or use a sheet metal friction chart and apply the required correction factor.
- Corrections for elevation and temperature are required.
- Allowances (in the load calculations) should be made for duct losses and duct leakage.
- Fan motor should be selected for the largest power requirement that will be experienced during start-up or during normal operation.
- Fan curve and system curves should be checked to ensure that the fan operating point will remain within the recommended operating range during all possible operating conditions.
- Variable pitch (adjustable) pulleys should be specified to provide a way to adjust fan performances at the job site.
- Air distribution outlets and return inlets should be selected and sized according to the manufacturer's recommendations.
- Air outlets should have integral dampers (registers) that can be used for making minor adjustments.
- Splitter dampers should be used as diverters (only), and they should not be used to control air volume.
- A balancing damper should be installed at each branch duct takeoff from the main (supply or return) duct.
- A balancing damper should be installed in each runout or duct drop from a main duct or branch duct to a supply outlet or return inlet.
- Balancing dampers should be installed in each zone duct of a multizone duct system.
- Show all dampers, including fire dampers, in their proper locations on the plans.
- Provide open/closed-type dampers in outside and return air duct entrances.
- Show damper locations at accessible points and, whenever possible, at an acceptable distance from a duct transition or fitting.
- Avoid attaching diffusers, registers or grilles directly onto the bottom or sides of a duct.
- Provide (SMACNA-approved) boots, necks and extractors at all 90° branch duct connections to sidewall registers or ceiling diffusers.
- Short discharge ducts between mixing boxes and supply registers may cause excessive discharge velocities and air noise at face of register.
- Do not allow return air from one space or zone to pass through another space or zone to reach a return air register.
- Door louvers do not provide an acceptable return air path when the return air system operates at low pressure (e.g., ceiling return plenum).
- A perforated static pressure plate downstream from the fan discharge may be required if the fan discharges through a "blow through" coil.
- Screens are required on outdoor air intake and exhaust openings.
- Provide access doors of adequate size within working distance of all coils, volume dampers, fire dampers, pressure-reducing valves, reheat coils, mixing boxes, blenders, constant volume regulators and the like.
- Provide access for making pressure, temperature and tachometer readings at all critical points.
- All duct seams, duct connections, casing and plenum connections should be sealed to minimize leakage.
- Avoid "line of sight" installation of opposing supply outlets or return inlets (unless they serve the same room).
- Sound attenuation should be provided downstream from air-side devices that generate excessive noise.
- Merging air streams should be thoroughly mixed before they enter any type of air-side device or component.
- Provide a change of filters just prior to balancing.
- Make sure that there is no "short-circuiting" of discharge air from cooling towers, condensing units, relief exhausts, roof exhausters and the like to the inlet of any outside air intake.

Key Terms

Having completed the study of this chapter, you should be familiar with the following key terms. If any appear unfamiliar or not entirely clear, you should review the section in which these terms appear. All key terms are listed in the index to assist you in locating the relevant text.

ACCA
Air stream diffusion
Air stream drop
Air stream envelope
Air stream spread
Air stream terminal velocity
Air stream throw
All-air systems
Annual fuel utilization efficiency (AFUE)
Aspect ratio
Back-pressure
Balancing damper
Comfort zone
Condensing furnaces
Counterflow furnace
Diffuser
Downflow furnace
Draft gauge
Dual-duct system
Duct transition
Dynamic pressure loss
Electronic air cleaners
Entrained air
Entrainment
Equal friction method
Equal pressure loss method
Equivalent length
Evaporator coil
Extended plenum
Extended plenum method
External static pressure
Flexible ducts
Grille
Heat exchanger
High-boy furnace
High sidewall outlets
Horizontal furnace
Humidistat
Inclined tube manometer
Induction
Isothermal jet
Lateral furnace
Loss coefficient (fitting)
Low-boy furnace
Low sidewall outlets

MBH
Make-up air
Mixing boxes
Modified equal friction method
Multileaf damper
Net static pressure
Noise criterion (NC)
Occupied zone
Outlet face velocity
Oval ducts
Perimeter loop
Pitot tube
Primary air
Primary air pattern
Pulse combustion furnace
Radius of diffusion
Reducing plenum
Register
Register free area
Register gross area
Roughness Index
Semi-extended plenum
SMACNA
Secondary air
Single-leaf damper
Single-zone system
Spring pressure
Stagnant air
Stagnant zone
Static
Static head
Static pressure
Static regain
Stratification
Supply air
Supply air rise
Supply plenum
Surface effect
System effect
Takeoff fittings
Total air pattern
Upflow furnace
Variable air volume (VAV)
Vane

Velocity factor
Velocity pressure
Vertical temperature gradient
Viscous impingement filter

Volume dampers
Water column
Water gauge

Supplementary Reading

American Society of Heating, Refrigeration and, Air Conditioning Engineers (ASHRAE)
1791 Tullie Circle, N.E.
Atlanta, Ga. 30329 Tel.404-636-8400

Handbook—Fundamentals, 1993

Air Conditioning Contractors of America (ACCA)
1513 16th Street, N.W.
Washington, D.C. 20036

Manual 4—Perimeter Heating and Cooling, 1990
Manual B—Principles of Air Conditioning, 1970
Manual C—What Makes a Good A/C System
Manual D—Duct Design for Residential Winter and Summer Air Conditioning and Equipment Selection, 1984
Manual G—Selection of Distribution Systems
Manual J—Load Calculation, 1986

Manual Q—Commercial Low Pressure Low Velocity Duct System Design, 1990
Manual S—Residential Equipment Selection
Manual T—Air Distribution Basics

Sheet Metal and Air Conditioning Contractors National Association, Inc. (SMACNA)
8224 Old Courthouse Road
Tysons Corner, Vienna, Va. 22180

HVAC System Duct Design, 1990
HVAC Systems Applications, 1987

Air Movement and Control Association, Inc. (AMCA)
30 West University Drive
Arlington Heights, Ill. 60004

Publication 200, Air Systems, 1987

Problems

Use the friction chart of Figure 5.51 or a duct calculator and Tables 5.3, 5.4 and 5.5, for all duct calculations.

1. a. A trunk duct must carry 1500 cfm at a maximum air velocity of 900 fpm. What size round galvanized steel duct is required?
   b. What size rectangular duct for the same air friction would you choose if the duct is to be lined with fibrous glass insulation? Why?
2. a. A friction rate of 0.08 in. w.g. has been established for a duct run. What size rectangular duct would you select to carry 1000 cfm?
   b. For the same conditions as in part a, a velocity limit of 700 fpm has been set to limit noise. Does this affect the choice of duct size? How? What would you do in this situation?
3. A duct 50 ft long has an acoustic trap that causes a total pressure loss of 0.01 in. w.g. at 600 cfm. This duct is part of a duct system that is being designed with a uniform friction rate of 0.08 in. w.g. What size duct is required for an air flow of 600 cfm? Explain.
4. a. A contractor uses a $16 \times 9^{5}/_{8}$ in. space between joists as a return duct to carry 1050 cfm. What is the friction rate?
   b. The return in part a is 40 ft long. It is used with a furnace whose external static head at 1050 cfm is 0.2 in. w.g. How much head is available for the supply system?

5. Rework Problems 1–4, if the ductwork is installed in Denver, Colorado, at an elevation 5000 ft.
6. A duct handling 2000 cfm must be installed in a hung ceiling with 12 in. clearance for ductwork. The duct has 1 in. of rigid insulation all around. What size duct is a good economic choice if maximum air velocity should not exceed 1200 fpm?
7. A building manager complains that a floor in her building is not getting enough warm air. The technologist assigned to check out the complaint sees from the plans that the 16 × 12-in. trunk feeding that floor should be carrying 2400 cfm. He takes manometer readings 50 ft apart on the trunk duct and reads 1.25 in. and 1.35 in. How much air is flowing?
8. A load calculation for a residence indicates a heating load of 78,000 Btuh and a sensible cooling load of 52,000 Btuh. What are the blower air quantities requirements for heating and cooling. State the assumptions made for comfort conditions and the supply and return air temperatures.
9. A duct carries air at 1200 fpm at a static pressure of 1.55 in. w.g. What is the total pressure (including velocity pressure)?
10. A test shows that a furnace blower delivers a specific quantity of air at 0.65 in. w.g. The duct system contains the following air-side devices. Their pressure loss at the blower air output is listed:
    Evaporator coil—0.35-in. pressure loss
    Filter—0.17-in. pressure loss
    Plate-type humidifier—0.03-in. pressure loss
    What is the furnace's external static pressure?
11. The noise criteria (NC) requirement of a particular conference room is NC 30. The total air requirement of the room is 600 cfm. Duct velocity is 600 fpm. Using four-way blow ceiling diffusers of the type shown in Figure 5.23*a* with dampers set at 75% open, select the outlet(s) that will meet the air and NC requirements. *Note:* Two identical diffusers operating at the same time give a noise level 3 db higher than a single diffuser.
12. What grade insulation (*R* value) would you put on a duct passing through an open carport? Outdoor design conditions are 96°F for summer and 15°F for winter. The duct is installed under the uninsulated carport roof and is boxed in with ⅜-in. weatherproof plywood. Make any necessary assumptions. Draw a sketch showing the installation.
13. What type of outlets would you recommend and where should they be placed for the following installations? State any necessary assumptions.
    a. Slab-on-grade house, cold climate, heating only.
    b. Slab-on-grade house, cold climate, heating and cooling (Midwest).
    c. One story office building, crawl space, heating and cooling.
    d. Large residence, slab construction, cooling only.
    e. Residence, full basement, heating and cooling.
    Explain your choice in each case. A sketch would be helpful.
14. What size return grille is required to handle 115 cfm. Make any required assumptions and justify them.
15. A 10-in. round galvanized steel duct carries 400 cfm.
    a. What is an equivalent rectangular duct of 2:1 aspect ratio?
    b. What is an equivalent oval duct with a major/minor axis ratio not exceeding 2.0?
    c. What is the air velocity in each?
16. A 16-in. round duct is connected to a 10-in. round duct with a 90° angle conical transition. Calculate the fitting pressure loss, using a loss coefficient, if the duct is carrying 1000 cfm. Make a sketch of the transition. Show all calculations.
17. What are the equivalent duct lengths for the following fittings:
    a. A round duct diverging wye.
    b. A round duct right angle tee.
    c. A butt takeoff on the side of an extended plenum, with rectangular branch duct.
18. Using the duct arrangement shown in Figure 5.61(*c*), work out all duct sizes in round and rectangular format, air velocities and friction rates using the equal pressure loss method. Indicate all results on a one-line sketch of the system. Make any assumptions necessary and justify them. Compare the results to the layout of Figure 5.61(*b*) as to duct sizes, velocities, noise, pressures and balance.

6. Air Systems, Heating and Cooling, Part II

Refrigeration

In Chapter 5, we learned the design principles of all-air systems and concentrated on warm air heating. In this chapter, we will learn the principles of mechanical refrigeration as applied to comfort cooling systems. Particular emphasis will be placed on the operation of the heat pump, which has found extensive use in providing both cooling and heating in a wide variety of buildings. Finally, we will apply the knowledge gained in this chapter and in Chapter 5 in design exercises for specific buildings. Study of this chapter will enable you to:

1. Understand the operation of the vapor compression refrigeration cycle and all the components in such a system.
2. Understand the operating theory of a heat pump and distinguish between the various types of heat pumps.
3. Compare heat pump performance in the heating mode on the basis of coefficient of performance (COP) and heating season performance factor (HSPF) figures.
4. Evaluate air conditioner and heat pump performance on the basis of energy efficiency ratio (EER) and seasonal energy efficiency ratio (SEER) ratings.
5. Calculate the seasonal operating cost of a heat pump based on its SEER.
6. Find a system's heat pump balance point using a graphical construction method.
7. Determine the proper size DX coil to use based on sensible and latent cooling loads.
8. Understand the application of centralized systems with terminal units and decentralized systems with package, incremental units.
9. Understand the use of packaged terminal air conditioner (PTAC) and packaged terminal

heat pump (PTHP) equipment and the application of split air conditioners and heat pumps.
10. Draw and assist in the design of complete heating/cooling residential and small commercial buildings. This includes selection and location of equipment and ductwork design.

6.1 Unit of Mechanical Refrigeration

A practical mechanical refrigeration cycle was developed in 1902 by Dr. Willis Carrier, although several decades would pass before it began to be practically applied for comfort cooling. Most modern air conditioning units use this same compressive cycle, although in a much improved form. Prior to World War II, comfort cooling was primarily used in theaters and was based not on mechanical refrigeration but rather on ice. Large fans would blow air across blocks of ice, picking up cool damp air. This air, when blown into the theater, cooled the space, although the increased humidity often led to an uncomfortable feeling of clamminess in humid climates. To this day, the unit of refrigeration in common use is based on this original application of ice for cooling.

As we learned in Chapter 1, the latent heat of fusion of water is 144 Btu/lb. That is the amount of heat that must be extracted from a pound of water to turn it into ice. Conversely, when ice melts, each pound of ice absorbs from its surroundings 144 Btu (at 32°F). Therefore, when 1 ton of ice melts, it absorbs

$$2000 \text{ lb} \times 144 \text{ Btu/lb} = 288,000 \text{ Btu}$$

When a ton of ice melts over a 24-hr period, the rate of heat absorption from the surroundings (refrigeration) is

$$\frac{288,000 \text{ Btu}}{24 \text{ hr}} = 12,000 \text{ Btu/h or } 12,000 \text{ Btuh}$$

Therefore, cooling at a rate of 12,000 Btuh is called one ton of cooling.

Example 6.1 What is the cooling capacity in Btuh of the following air conditioning units:

(a) ¾ ton
(b) 2 tons
(c) 5 horsepower

Solution:

(a) ¾ ton × 12,000 Btuh/ton) = ¾ (12,000) = 9000 Btuh
(b) 2 tons (12,000 Btuh/ton) = 24,000 Btuh
(c) 5 horsepower: Insufficient data. Many nontechnically trained people equate compressor horsepower to tons, simply because in small air conditioning units there is a rough correspondence between horsepower and tonnage. This is, however, a mere coincidence, and using it as a rule is incorrect. In large units, more than 1 ton of refrigeration is produced for each horsepower of compressor motor size. In all air conditioners, the motor size depends on the efficiency of the entire assembly, and this varies with each unit. The only accurate way of determining the refrigeration capacity and the electrical requirements of an air conditioning unit is to consult the manufacturer's published data.

6.2 Cooling by Evaporation

The compressive refrigeration cycle is based on the cooling effect of evaporation. In Chapter 1, we learned that when a liquid vaporizes it absorbs heat. The quantity of heat absorbed per unit weight is called the latent heat of vaporization. For water, this quantity is 970 Btu/lb. Note that this is almost seven times as large as the latent heat of fusion of water (144 Btu/lb). This heat-absorbing characteristic is used very effectively by the body for cooling by perspiration evaporation, as was pointed out in Chapter 2.

Vaporization of a liquid will occur in two ways: slowly by evaporation or rapidly by boiling, which is essentially forced evaporation. Since water is the liquid that is most common in our environment, we will use it in our explanation, although the same principles apply to any liquid. Refer to Figure 6.1, which represents a bowl of water exposed to the atmosphere. Some of the water molecules at the surface will escape into the air and become water vapor. This process is called *evaporation*. It takes place constantly at every exposed body of water on earth, including all ponds, rivers, lakes and oceans. It is precisely this evaporation action, caused by the heat of the sun, that drives the earth's weather systems.

The rate at which evaporation occurs depends on

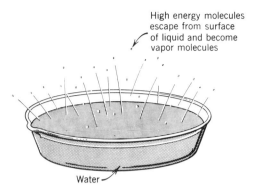

Figure 6.1 A liquid at any temperature will evaporate spontaneously into the atmosphere. The rate of evaporation increases with water temperature and surrounding air temperature. It also increases as the surrounding ambient pressure drops. (From Dossat, *Principles of Refrigeration*, 1961, John Wiley & Sons. Reprinted by permission of Prentice-Hall, Inc., Upper Saddle River, NJ.)

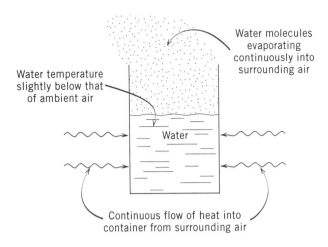

Figure 6.2 As water evaporates, it takes heat from the water in the container, thus lowering its temperature. This heat is replaced by heat flow from the surrounding space.

the water temperature and air temperature and pressure. The evaporation rate increases with increasing water and air temperature and decreasing air pressure. As the water temperature increases, it will eventually boil. The temperature at which this occurs is called the liquid's boiling point, or more accurately its *saturation temperature*. At this temperature, all the water will change to gas (water vapor) and in so doing will absorb 970 Btu/lb from the heat source.

The cooling effect of evaporation is caused by this heat absorption characteristic. Refer to Figure 6.2. Water evaporating from an open container absorbs heat from the water in the container, thus lowering its temperature slightly. This heat is replaced, through the container walls, from the heat in the surrounding air. This process will continue until all the water in the container evaporates. In the case of the evaporation of perspiration, the body supplies the continuous heat being absorbed, thus cooling itself very effectively. Obviously, an open system such as that seen in Figures 6.1 and 6.2 is not commercially practical since the refrigerant (in this case water) is constantly being used up. What is required for a practical mechanical refrigeration system is a closed arrangement that will reuse the refrigerant continuously. That is the essential idea behind the compressive refrigeration cycle.

6.3 Refrigeration Using a Closed Vapor Compression System

Water is not a practical refrigerant because of its high saturation temperature (boiling point). What is required in a closed system is a liquid that will reach its saturation temperature (boil) at a low temperature. Various fluids have been used as refrigerants, including ammonia and a group of fluorinated hydrocarbons generally known by the commercial name Freon. Unfortunately, these materials have been found to be environmentally unsuitable and are now replaced with other, environmentally neutral fluids. The exact chemical composition of modern refrigerants is not important to us. What is important is to understand how refrigerants like Freon, with a boiling point of about −20°F, are used in the refrigeration cycle. A thorough understanding of this compressive cycle is very important to the technologist. For this reason, it is developed in detail here.

a. Refrigerant Vaporization

Refer to Figure 6.3. An insulated space such as a refrigerator box can be rapidly cooled by simply

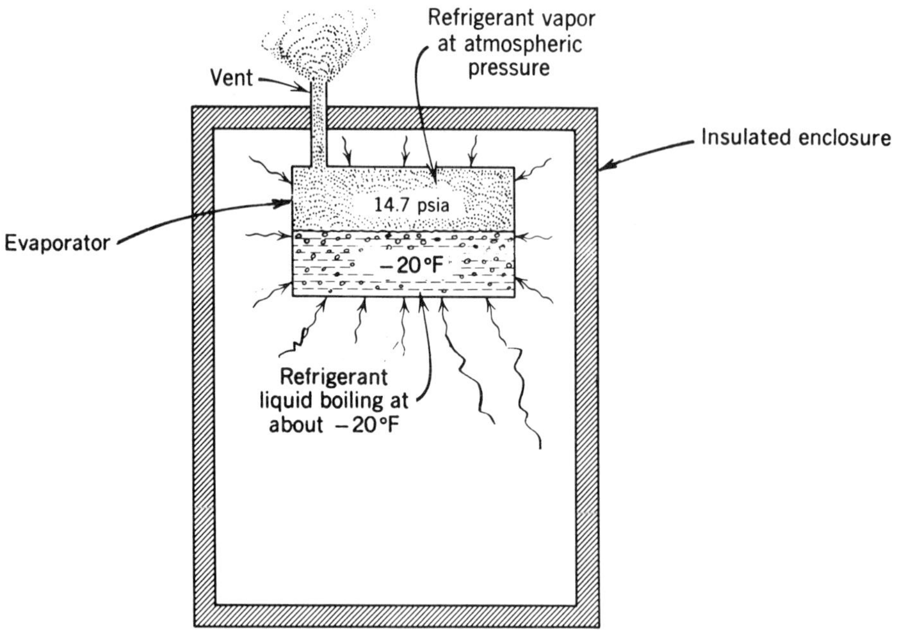

Figure 6.3 The refrigerant in the evaporator boils at about −20°F, at atmospheric pressure. It will, therefore, rapidly cool the inside of the container. Cooling will continue until all the refrigerant is exhausted into the surrounding air. (From Dossat, *Principles of Refrigeration*, 1961, John Wiley & Sons. Reprinted by permission of Prentice-Hall, Inc., Upper Saddle River, NJ.)

allowing a low boiling point refrigerant liquid to flash (boil) into vapor at atmospheric pressure. In so doing, the evaporating vapor absorbs large amounts of heat from the surroundings, thus cooling the interior of the refrigerator box and its contents. Of course, we would quickly lose the refrigerant, and the cooling would be rapid, severe and of short duration. By placing a throttling device in the vent, as in Figure 6.4, we can control the flow of vapor. This, in turn, controls its pressure and therefore also its boiling point. The valve is, therefore, effectively a cooling temperature control. The device containing the boiling refrigerant is called the system *evaporator* because it is there that the refrigerant evaporates.

b. Refrigerant Flow Control

Refer to Figure 6.5. A valve is required at the input of the evaporator. It must control the flow of refrigerant into the evaporator from a storage container, to correspond exactly to the flow out of the evaporator. The specific design of this device is not of concern to us; only its operation is important. It is called the *refrigerant flow control device*, and in most modern systems it is a thermostatically controlled expansion valve.

c. Recycling the Refrigerant

Obviously, the refrigerant should not be exhausted to free air, as is schematically shown in Figures 6.1 and 6.2. It must somehow be recycled. Since it evaporated by absorbing heat, all that is required to return it to its liquid state is to remove the same amount of heat. This will condense the vapor back into a liquid. The device used to perform this function is, therefore, called a *condenser*. See Figure 6.6. The problem that immediately arises is how the condenser will draw off this heat. The condenser is simply a coil through which the saturated vapor passes. It is cooled by blowing ambient air over it or by running it through a heat exchanger that uses cooling water. Since the temperature of the vapor at the evaporator output is somewhere between 30 and 50°F, ambient air at 80–110°F or cooling water at 60–90°F are useless as cooling mediums. (Cooling water comes from municipal lines, cooling ponds or cooling towers.) Heat will not flow "uphill," that is, from the cool refrigerant

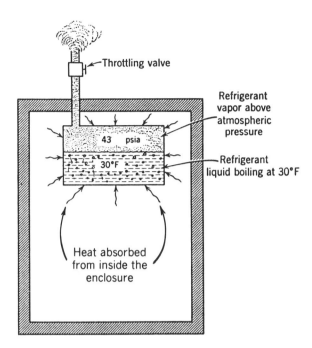

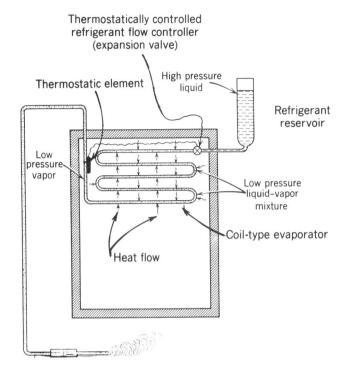

Figure 6.4 A throttling valve in the evaporator outlet line can be used to vary the evaporator pressure and, therefore, its boiling point. Here the valve is almost completely closed. This raises the boiling point from −20°F at atmospheric pressure to 30°F at about 3 atmospheres. Shutting the valve would raise the pressure until the vapor boiling point equalled the container temperature. At that point, all heat transfer stops. (From Dossat, *Principles of Refrigeration*, 1961, John Wiley & Sons. Reprinted by permission of Prentice-Hall, Inc., Upper Saddle River, NJ.)

Figure 6.5 The refrigerant flow control valve supplies refrigerant to the evaporator at exactly the rate at which it is evaporated. The thermostat bulb senses the outlet (boiling) temperature. In passing through the expansion valve, the liquid refrigerant expands, thereby reducing pressure and creating a liquid-vapor mixture at relatively low pressure. (From Dossat, *Principles of Refrigeration*, 1961, John Wiley & Sons. Reprinted by permission of Prentice-Hall, Inc., Upper Saddle River, NJ.)

vapor to the warmer condenser cooling medium (ambient air or cooling water). To overcome this problem, the vapor coming out of the evaporator is compressed, raising its temperature to about 130°F and its pressure to about 8 atmospheres (120 psi). This hot compressed gas then enters the condenser coils where it is cooled by air or water to about 100°F and partially condensed into liquid. Pressure remains at 7–8 atmospheres. This warm, high pressure liquid and gas refrigerant mixture then travels past the refrigerant tank and on to the thermostatically controlled valve at the input of the evaporator, to begin the cycle all over again.

d. A Closed Recycling System

See Figure 6.6, which is a flow diagram of a simple vapor compression closed-cycle system. The principal parts of the system, shown schematically, are:

(1) *Evaporator.* The function of an evaporator is to provide a heat-absorbing surface. It is usually a coil of pipe, inside which the refrigerant is vaporizing and absorbing heat. Air blown over the surface of this pipe is cooled. This cool air is the end product of the refrigeration process. In a common household refrigerator, the cool air is confined to a closed box containing perishables. In a comfort air conditioning system, recirculated room air is blown over the evaporator coil. Since the room air is warmer than the evaporator coil, it will cause a 10–20 °F temperature rise in the vapor temperature. This is shown on Figures 6.7 and 6.8.

(2) *Suction line.* The suction line is the line through which the slightly warmed refrigerant vapor passes on its way to the compressor.

(3) *Compressor.* The function of the compressor is to change the low temperature, low pressure

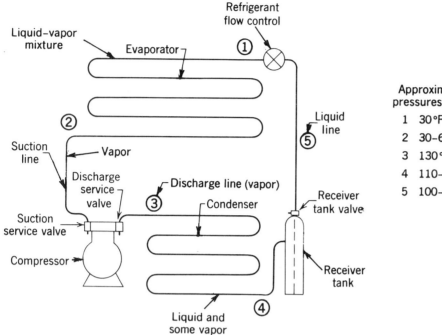

Figure 6.6 Schematic diagram showing the major components of a closed-cycle vapor compression refrigeration system. Typical temperatures and pressures at various points in the system are also shown. (From Dossat, *Principles of Refrigeration*, 1961, John Wiley & Sons. Reprinted by permission of Prentice-Hall, Inc., Upper Saddle River, NJ.)

vapor coming out of the evaporator to a high temperature, high pressure gas.

(4) *Condenser.* The condenser receives hot vapor from the compressor and condenses it to a gas-liquid mixture by cooling. It is usually constructed as a long folded pipe over which cooling air passes when air is used as the cooling medium. Alternatively, cool water and a heat exchanger can be used to perform the required cooling and vapor condensation.

(5) *Receiver tank.* The receiver tank is used to store a quantity of liquid refrigerant.

(6) *Flow control device.* Located at the input of the evaporator, the flow control device reduces the temperature and pressure of the high temperature liquid refrigerant and supplies a low temperature, low pressure liquid-gas mixture to the evaporator.

Figure 6.7 shows the basic closed-cycle vapor compression refrigeration system as applied to comfort cooling. Two blowers (fans) are added to the basic equipment described previously. The indoor blower recirculates room air. The exterior fan cools the condenser with outside air. Remember that the terms *evaporator* and *condenser* refer to actions performed on the refrigerant. Some confusion arises because warm humid air condenses on the evaporator (see Figure 6.7) and is drained off. This condensate has no relation to the system condenser, which condenses refrigerant vapor. Another source of confusion arises because the evaporator is frequently called a cooling coil due to its cooling action. This is particularly common when the evaporator is installed in a warm air furnace and connected to a remote compressor and condenser. See for instance Figures 5.8, 5.9, 5.10, 5.11 and 5.13. Finally, the evaporator is also frequently referred to as a DX (direct expansion) coil in systems such as that shown in Figure 6.7. A DX coil absorbs heat by the direct expansion of the refrigerant liquid to a gas. The desired cooling action occurs when recirculated room air is blown over the coil. When the cold surface of an evaporator is

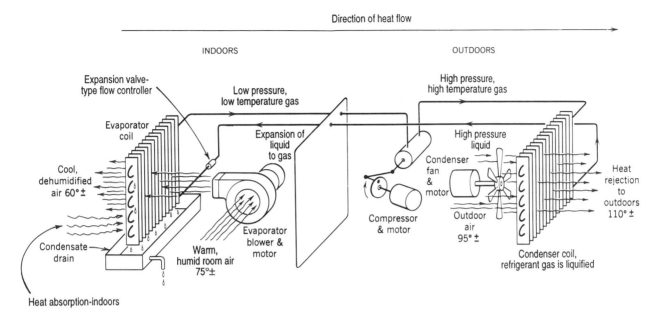

Figure 6.7 Pictorial representation of the components of a vapor compression refrigeration system as applied to comfort cooling. The indoor and outdoor sections can be installed in a single enclosure, as in the common window unit or through-the-wall package unit. Alternatively, the two sections can be separated by up to 100 ft with only insulated refrigerant pipes connecting the two sections. See for instance Figure 5.6(a).

used to cool water that will then be circulated through the building as a cooling medium, the entire assembly is known as a chiller rather than an air conditioner.

Heat Pumps

6.4 Basic Heat Pump Theory

In Section 6.3, we explained the basic theory of the vapor compression cooling cycle. Refer to Figure 6.7. Notice that the entire complex system—evaporator, compressor, condenser, piping and valving—accomplishes only one simple task. That task is to transfer heat from one place to another. The system does not create heat (except for machinery friction); it transfers heat. In the air-to-air system shown in Figure 6.7, the system takes heat from the warm indoor space and dumps it outside. Since the temperature outdoors far exceeds the indoor temperature, the heat is being transferred (pumped) "uphill." We learned in Section 1.7 that, according to the basic laws of thermodynamics, heat naturally flows "downhill," that is, from a point of higher temperature to a point of lower temperature. To reverse this process and pump heat from a lower to a higher temperature, we must add energy. That is what the compressor does—it adds energy to the system, which acts using the system machinery, to pump the heat "uphill." The two air movers shown (condenser fan and evaporator blower) are simply devices that aid in the heat transfer. The actual heat transfer is accomplished by phase changes of the refrigerant (liquid to gas to liquid). To simplify matters somewhat, we can redraw Figure 6.7 as the block diagram of Figure 6.8(a), showing only energy considerations. There we see that heat is pumped from a lower temperature to a higher temperature, using electrical energy input to do so.

Look at Figure 6.8(a), and forget for the moment that it is a block diagram of an air conditioner

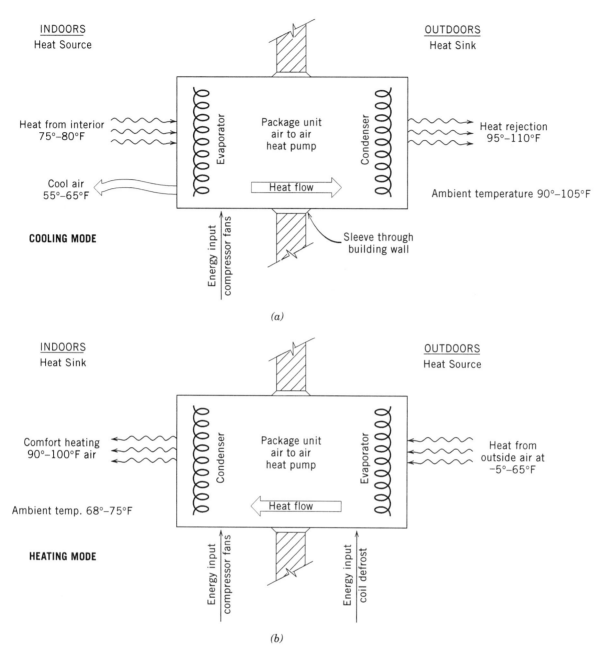

Figure 6.8 Heat flow diagrams for the two modes of operation of an air-to-air heat pump. *(a)* In the cooling mode, operation is identical to that of an air-to-air through-the-wall package-type air conditioner. Some of the heat in warm (humid) inside air is picked up by the indoor evaporator, carried through the unit and rejected outdoors. The exterior condenser is cooled by ambient outside air. Energy for the heat pumping is supplied by the compressor plus the evaporator and condenser fans.
(b) In heating mode, heat is extracted from outside air and is pumped into the building interior. It is rejected inside in the form of warm air for comfort heating. Energy for the heat-pumping operation is supplied by the compressor. Additional energy is taken by fans and by electric defrosting heaters for the exterior evaporator.

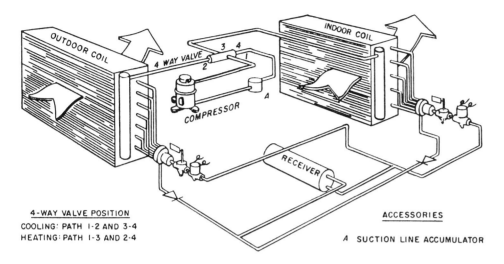

Figure 6.9 Pictorial drawing of an air-to-air heat pump. The four-way reversing valve is controlled by a room thermostat that calls for heating or cooling as required. For the sake of clarity, auxiliary valves and control/safety devices are not shown. (From Ambrose, *Heat Pumps and Electric Heating*, © John Wiley & Sons, reprinted by permission of John Wiley & Sons.)

intended for cooling. It should be obvious that the vapor compression system can also be used for heating by simply turning it around, as is shown schematically in Figure 6.8(b). In actual practice, it is not necessary to turn the unit around physically. All that is needed is a four-way valve that will reverse the flow of refrigerant. Since the evaporator and condenser are both coils, they can act as either one or the other, depending on refrigerant flow. That means that reversing flow into the evaporator changes it into a condenser. Similarly, reversing flow through a condenser changes it into an evaporator. Such a complete assembly, including the reversing valve, is called a *heat pump*. It is capable of heating or cooling, depending on its control valve position. In the cooling mode, it is identical to what is commonly called an air conditioner. In its heating mode, it is always referred to as a heat pump. Figure 6.9 schematically shows the refrigerant flow and flow valve position in both modes. The position of the valve is controlled by the indoor thermostat. When heating is required, it sets the four-way flow valve into heating position; when cooling is required, it resets the valve into the cooling position.

Another attractive characteristic of the heat pump is its very efficient operation in the heating mode. This will be discussed in Section 6.5. The question that arises at this point is something like, If the heat pump can supply both cooling and heating, and the latter very efficiently, why would anyone buy separate heating and cooling systems, as is so often done? The answer to that excellent question has to do with both economics and engineering. It will become clear in the following discussion.

6.5 Heat Pump Performance (Heating Mode)

a. Heat Extraction

Refer to Figure 6.8(b). The question that usually arises when referring to heating mode performance is how usable heat can be extracted from cold outside air. A moment's thought, however, will answer this question. A complete absence of heat occurs at absolute zero temperature, which corresponds to 0° Rankine or −460°F. At any temperature above absolute zero, air contains heat. The specific heat of dry air is 0.24 Btu/lb-°F or °R. That means that at 100°F (560°R) air contains 134 Btu/lb. At a typical winter temperature of 40°F (500°R), air contains 120 Btu/lb or 90% of the heat content at 100°F! Therefore, despite the fact that we tend to think of 40°F air as cold because of our high body temperature, there is a good deal of heat available for use in such air. Of course, the lower the temperature of the heat source is (in this case outside air), the more work that must be done to pump it up to the heat sink temperatures (in our case 68–75°F indoor room temperature). In other words, the lower the outside air temperature, the lower is the efficiency of a heat pump.

A word about terminology: the point from which heat is taken is called the *heat source*. The point to which heat is delivered is called the *heat sink*. Thus, for a heat pump in cooling mode, the heat source is the room being cooled and the heat sink is outside air. In heating mode, the heat source is outside air, and the room is the heat sink.

b. Heat Pump Heating Efficiency

The law of conservation of energy states that energy cannot be created or destroyed. Refer again to Figure 6.8b; the energy (heat) extracted from the outside air plus the energy input to the compressor appear as heat input into the space. In other words, the heat produced by the heat pump is greater than its energy input. The ratio of heat output to energy input is called its coefficient of performance (COP). The COP of a well-designed heat pump varies between 1.5 and 3.0, depending on the outside temperature. It is defined as

$$\text{COP} = \frac{\text{Heat delivered in Btuh}}{\text{Energy supplied in Btuh}} \quad (6.1)$$

The difficulty with heat pumps is that heat output and, therefore, also COP drop as outside temperature drops. That means that as the weather turns colder and the demand for heat increases, the heat pump output decreases, simply because the heat must be pumped over a larger temperature difference. This is shown graphically in Figure 6.10(a).

To make things just a bit more complicated, three other heat pump performance factors are in common use. The informed technologist should understand their meaning.

(1) *Heating Seasonal Performance Factor (HSPF).* This factor is more meaningful than COP because it considers the heat pump's performance over the entire heating season. COP is a measure of what the heat pump is doing at a particular instant and, in turn, depends on indoor and outdoor temperatures. HSPF is an indicator of seasonal efficiency, including supplemen-

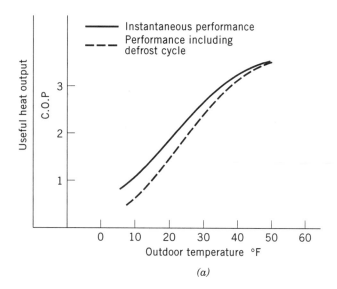

Figure 6.10 *(a)* Heat pump heating cycle performance. Note that useful output and coefficient of performance both drop sharply as the outdoor temperature drops. Outside air is used as the heat source for this air-to-air heat pump characteristic.

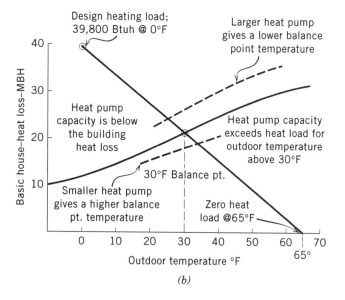

Figure 6.10 *(b)* The intersection of the building heat loss curve (straight line) and the heating characteristic of a proposed heat pump (curved line) determines the system balance point (see text). The balance point temperature can be raised by using a smaller unit or lowered by using a larger unit.

tal heat and on-off cycling, among other variables. Manufacturers publish the HSPF figure for each heat pump model, based on tests according to ARI (Air Conditioning and Refrigeration Institute) Standard 240. The tests use a climate of somewhat less than 6000 degree days.

(2) *Energy Efficiency Ratio (EER)*. This factor is defined as the ratio of cooling capacity to input power (in Btuh per watt) for a specific set of operating conditions. Expressed as a formula,

$$\text{EER} = \frac{\text{Cooling capacity in Btuh}}{\text{Input power in watts}} \quad (6.2)$$

If we compare this to Equation (6.1), which defines COP, we see that the two factors are related by the conversion factor of 3.412 Btuh/w, that is

$$\text{EER} = 3.412 \times \text{COP} \quad (6.3)$$

therefore, a COP of 2.0 is the same as an EER of 6.8 and so on. EER factors are most frequently applied to consumer refrigeration products such as room air conditioners and household refrigerators.

(3) *Seasonal Energy Efficiency Ratio (SEER)*. This factor is an indication of the heat pump's performance over the entire cooling season. It is determined by testing a unit with up to four separate tests, each at varying indoor and outdoor conditions that simulate varying weather conditions. Therefore, it is to cooling what HSPF is to heating—an attempt to simulate efficiency over an entire season.

The EER and SEER ratings of two heat pumps can be compared directly only if their Btuh capacities are identical. If this is not true, the seasonal electrical energy bill for each unit can be calculated and the results compared. The seasonal electrical energy cost is

$$\text{Electrical cost} = \frac{\text{Btuh} \times \text{Hours} \times \text{Rate}}{(\text{EER or SEER})(1000)} \quad (6.4)$$

where
Btuh is the unit's rating,
Hours is the anticipated number of operating hours per season and
Rate is the local electrical rate in dollars per kilowatt-hour.

Example 6.2 Compare the seasonal cost of two heat pump units:

Unit A: 28,000 Btuh, SEER = 7.4
Unit B: 32,000 Btuh, SEER = 8.6

Assume an average electrical cost of 8.5¢/kwh and 1680 hr of operation (4 months at 14 hr per day).

Solution: Seasonal cost of unit A:

$$\text{Cost} = \frac{28{,}000 \text{ Btuh} \times 1680 \times 0.085 \text{ \$/kwh}}{7.4 \, (1000)} = \$540.00$$

Seasonal cost of unit B:

$$\text{Cost} = \frac{32{,}000 \times 1680 \times 0.085 \text{ \$/kwh}}{8.6 \, (1000)} = \$531.00$$

The units are, therefore, equal in energy expense, and the choice between them would be made on some other basis.

Typical performance figures for residential heat pump performance are:

COP   2.0–3.5
HPSF   6.0–8.5
EER   6.8–11.9
SEER* 10–12.5

All air source heat pumps will build up a layer of frost on the outdoor evaporator as the outside temperature drops below 45°F. Since this ice layer will seriously degrade heat pump performance, heat pumps are provided with automatic defrost cycles. This requires energy and serves to reduce overall efficiency further as the outdoor temperature falls. The heat pump capacity rating, when considering the defrost heat "penalty," is referred to as the unit's *integrated capacity*. It too is shown on the graph of Figure 6.10(a).

Most heat pumps use electricity as fuel. Since electricity is generally more expensive than fossil fuels in dollars per Btuh, it is generally not economical to operate a heat pump for comfort heating when the COP falls below about 2.0. This usually occurs at an outdoor temperature between 20 and 30°F for air-to-air units. The subject of the economics of heat pump operation is complex and is not normally the concern of a technologist. It is touched on here to give the technologist an appreciation of the considerations involved in equipment and system selection.

---

*The National Appliance Energy Conservation Act of 1987 prohibits the manufacture of any air conditioner with a SEER less than 10.

6.6 Equipment Sizing Considerations

a. Heat Pump Balance Point

In order to determine the theoretical ideal size for a heat pump in heating mode, a graphical solution method is used. The method consists of plotting the building heating load characteristic and a heat pump curve together and finding their intersection. Refer to Figure 6.10(b). It is assumed that a building's heating load varies linearly with temperature. This simply means that a change in outdoor temperature will cause a proportional change in building heat loss (load). This is substantially true if we consider only the heat loss through the building envelope and ignore infiltration and deliberate ventilation losses. For the purpose of heat pump sizing, when designing residential systems, this assumption can be made without introducing an excessive error.

To draw the heat load characteristic, plot two points: the calculated design heating load at the design temperature and the no-load (no heat) point at 65°F. If we take The Basic House plan from Figure 3.32(c) as an example (page 131) we see that the calculated heating load at 0°F design temperature is 39,800 Btuh. Plotting the two points on Figure 6.11 and connecting them with a straight line gives us the theoretical building heat load at every outside temperature from 0 to 65°F. If we now plot on the same sheet the performance curve of the heat pump being considered, the intersection point of the two lines is called the *system balance point*. (The heat pump performance curve data are available from the equipment manufacturer's published technical material.) At the balance point, the heat pump output exactly matches the building's heat loss. At outdoor temperatures above the balance point, the heat pump has excess capacity. At temperatures below the balance point, the heat pump output is insufficient, and additional heat is required. This is indicated on Figure 6.10(b).

For commercial buildings, the concept of a system balance point is still valid. However, the building load line is not as easily determined as for a residence. In determining the heat load, the ventilation portion of the heat load cannot be ignored. Furthermore, the no-load condition cannot be set simply at 65°F outdoor temperature. When the building is occupied, it has internal heat gain from occupants, machinery and solar load. These will reduce the no-load outside temperature to 45°F or even lower, depending on the building details. Since most commercial building cooling calculations are done today by computer, the design technologist can easily calculate the building heat load at 5–10 F° intervals from the outside design temperature up to no-load. These points can then be used to plot the building heat loss characteristic. In commercial buildings with multiple zones, separate calculations have to be made for each zone. Design considerations for such buildings are beyond the scope of this book.

A balance point at 30°F or lower will ensure excellent heating efficiency from the heat pump for all of the building's needs. A balance point between 30 and 35°F will give good heating efficiency except on the few days every year or so when the outdoor temperature actually drops to the design temperature figure. On those few days, the house may be cooler than the design indoor condition (68–75°F), and some additional heat may be desirable. However, the actual operating balance point is almost always lower than the one calculated because no allowance for internal heat gains and solar gain is taken in residential heat load calculations. As a result, a balance point between 30 and 35°F will seldom require the use of supplemental heating equipment. A balance point above 35°F means either a very cold climate, an undersized heat pump or both. The solution to that situation will require one of the following:

1. Separate supplemental heating, either electric or fossil fuel.
2. A larger heat pump unit.
3. A modified heat pump system that will supply the required additional heat. This may be a hybrid heat pump—fossil fuel system or simply a heat pump with auxiliary electric resistance heating elements. This latter is the cheapest alternative. As a result, most heat pumps intended for residential use are provided with supplementary electric resistance heating elements.

An additional consideration in the sizing a heat pump is its functioning in cooling mode. This is discussed next.

b. Heat Pumps in Cooling Mode and Air Conditioners

As already stated, the heat pump in cooling mode is identical to what is commonly known as an air

EQUIPMENT SIZING CONSIDERATIONS / 327

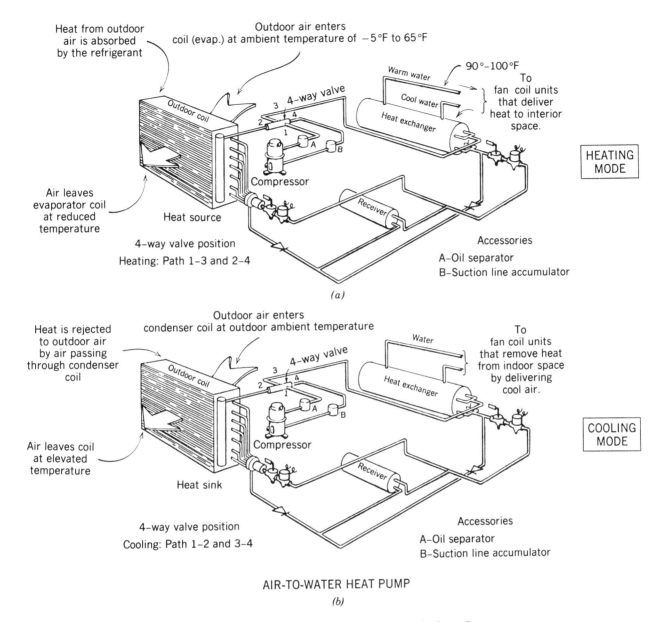

Figure 6.11 Air-to-water heat pump energy flow. *(a)* In the heating mode, heat flows into the refrigerant at the outdoor coil and is rejected to the water in the heat exchanger. The heated water is then circulated to fan coil units to heat the indoor space. *(b)* In the cooling mode, heat is picked up from the indoor space by circulating room air over the (cool) water coils in the fan coil unit. This heat is rejected at the outdoor coil acting as a refrigerant condenser. (From Ambrose, *Heat Pumps*, © John Wiley & Sons, reprinted by permission of John Wiley & Sons.)

conditioner. When constructed as an air-to-air unit as shown in Figures 6.8 and 6.9 it operates as a DX coil air conditioner. One difference between the two is in the area of equipment sizing. When calculating the cooling load of a structure, and particularly for residential work, many designers will calculate only the sensible heat gain. This is particularly common in residential work. To the sensible cooling load will then usually be added an additional one-third of the sensible load, to represent latent load. This gives a sensible heat ratio of 0.77 (1/1.3). This rule of thumb may be very far from accurate for the particular structure because of climate and occupancy. In such cases, the total cooling capacity may be adequate, but the cooling coil selected for an *SHR (sensible heat ratio)* of 0.77 can be far off the mark. Depending on the actual sensible/latent ratio, the result can be:

- A system that is short sensible capacity, resulting in an excessively warm house.
- A system that is short latent capacity, resulting in an uncomfortably humid house.
- A system that attempts to adjust these two problem by changing room air flows. This will upset air balance and may result in drafts or air-starved rooms.

To avoid some of these problems, designers will frequently oversize a unit's cooling capacity as much as 25%. With air conditioning equipment, this will not only unnecessarily increase first cost but will result in excessive cycling, cold drafts and, in humid climates, insufficient dehumidification. As a result, oversizing of air conditioning equipment is discouraged. Indeed, some designers recommend deliberate undersizing, particularly in humid climates. This ensures almost continuous unit operation and adequate dehumidification. The solution to these problems, of course, is to calculate both sensible and latent loads as accurately as possible. A cooling coil matching the calculated requirements can then be selected from manufacturers' published data.

When sizing a heat pump for cooling capacity, oversizing the cooling capacity will result in better heating performance and less need to use supplemental electric or fossil fuel heat. The following suggestions, based on experience, should be helpful in sizing heat pumps for heating and cooling service:

- In warm climates, the cooling load will probably exceed the heating load. Selecting a unit for the design cooling load should provide adequate heating as well.
- In cool climates, the heating load will equal or exceed the cooling load. Heat pumps may be oversized 10–15% in order to satisfy the heating requirement without use of auxiliary electric heat. However, oversizing should be strictly limited in high humidity climates.
- In cold climates, the heating load is much larger than the cooling load. Any attempt to satisfy the heating load without supplemental electric or fossil fuel heating will drastically oversize the cooling capacity and will always result in poor cooling performance. This is especially true in humid areas. As a result, many designers use 6000 degree days as an upper limit when deciding whether to use a heat pump for year-round heating and cooling. In geographic areas with a higher degree day total (cold climates), use of a heat pump is simply not practical because of degraded performance in both heating and cooling modes.

6.7 Heat Pump Configurations

In technical literature, heat pumps are listed in four configurations. They are

- Air to air
- Air to water
- Water to air
- Water to water

The first word refers to the heat source and the second to the heat sink. These names refer to the heat pump in its heating mode.

a. Air-to-Air Heat Pump

See Figures 6.8 and 6.9. In the heating mode, this arrangement extracts heat from outside air at temperatures of −5°–65°F and delivers it to inside spaces at temperatures of 90–100°F. This type of unit is the most common arrangement for residential work, as it requires only an electrical connection and no special piping. In cooling mode [Figure 6.8 *(a)*], the unit operates like a common air conditioner, taking heat from inside air at about 75–80°F and pumping it to an outside heat sink at temperatures of 100–110°F.

b. Air-to-Water Heat Pump

See Figure 6.11. This configuration is not common because the temperature change in the output water is small. That means that, in heating mode, warm water will be delivered to a hydronic heating system, instead of the hot water usually used. This requires the use of large radiation surfaces, which increase the system first cost. In the cooling mode, the output water is cool and not cold, making it suitable for use in fan coil units. As a result, air-to-water arrangements are used only in mild climates with low to moderate heating and cooling loads. The arrangement is useful in zoned systems, multi-family residences and small commercial applications.

c. Water-to-Air Heat Pump

See Figure 6.12. These units extract heat from a water source and deliver it to the indoor supply air when operating in the heating mode. The system is used when an adequately large water source such as a well, pond or lake is available. The source must be large enough so that its temperature is not seriously affected by seasonal temperature changes. In the cooling mode, heat is extracted from inside air and dumped into the (constant temperature) water sink. When an adequate water source is available, this system is highly efficient. First cost, however, can be high due to piping and heat exchanger costs.

d. Water-to-Water Heat Pump

See Figure 6.13. These units use heat exchangers at both ends of the system. Like the water-to-air system, a large reliable water source/sink is also required. As a result, this arrangement is most often used in commercial applications. Efficiency of the system is high because no defrosting cycle is required.

In addition to these four standard arrangements, there are designs using coils buried in the earth, hybrid designs using solar collectors and others. These types are highly specialized and must be designed for a specific application. Technologists interested in these special designs will find additional information in the references listed in the bibliography at the end of this chapter.

6.8 Physical Arrangement of Equipment, Small Systems

We will arbitrarily define a small system as being limited to about 15 tons of cooling (and somewhat more heating) simply because most manufacturers do not make larger heat pumps. In addition to defining the heat source and heat sink of a heat pump as explained in the previous section, heat pumps (and air conditioners) are also classified by the location of the system parts. For small and medium-size units, there are two arrangements: a self-contained packaged unit and a split unit.

a. Self-Contained Packaged Unit

A packaged unit contains the entire system in one enclosure. Window units, through-the-wall units and rooftop package units are of this type. These packages are also referred to as unitary heat pumps or air conditioners because they contain all the equipment necessary for a complete installation, including the air-handling equipment. A packaged unit that supplies air to a number of outlets via ducts is still a unitary package. Packaged units typically contain the evaporator and condenser coils, a compressor, an air mover, plus all the required auxiliaries, piping and controls. The most common packaged unit is the window or through-the-wall air conditioner shown schematically in Figure 6.14. This type of through-the-wall unit is referred to as an incremental unit, a PTAC (packaged terminal air conditioner) or a PTHP (packaged terminal heat pump). The same equipment can be arranged for rooftop mounting with air discharging at the ceiling level of the space to be cooled. Since cold air is heavier than warm air, a rooftop arrangement is ideal for cooling. It is less effective in the heating mode, because the warm air output will simply stay at the ceiling level, and circulation will be poor. See Figure 5.38 (page 000). For this reason, packaged rooftop units are most commonly used for cooling only.

Figure 6.15 shows a typical through-the-wall climate control package consisting of an electric air conditioner and a gas heater in a single enclosure. Warm or cool air, as called for by the interior

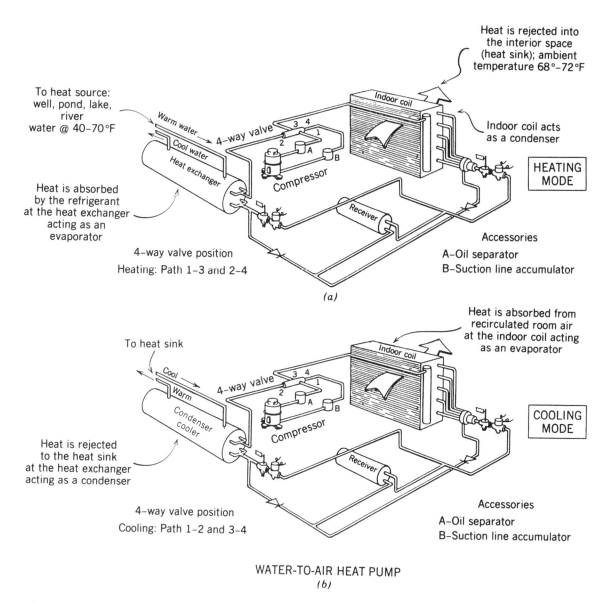

Figure 6.12 Water-to-air heat pump. *(a)* In the heating mode, water pumped from the water source (heat source) gives up heat at the heat exchanger. This heat is transferred by the refrigerant to the indoor coil. There, the indoor coil acting as a condenser rejects the heat into the indoor space (heat sink). *(b)* In the cooling mode, heat is absorbed from the indoor space air by circulation over the indoor coil, which acts as an evaporator. The heat is transferred to the heat exchanger where it is picked up by circulating water from the water source heat sink. (From Ambrose, *Heat Pumps*, © John Wiley & Sons, reprinted by permission of John Wiley & Sons.)

PHYSICAL ARRANGEMENT OF EQUIPMENT, SMALL SYSTEMS / 331

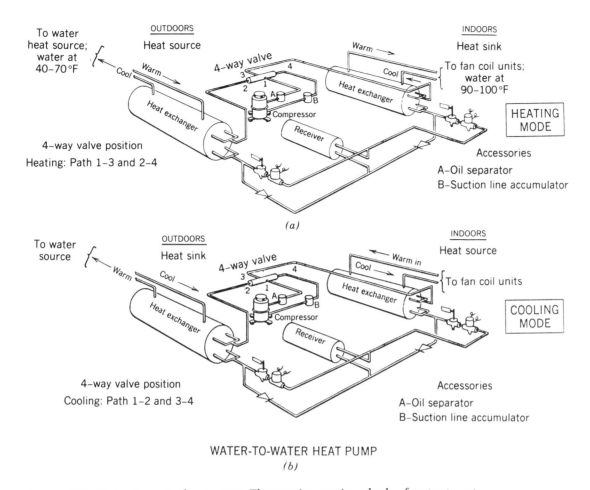

Figure 6.13 Water-to-water heat pump. These units require a body of water to act as both a heat source and heat sink. The indoor section of the heat pump supplies fan coil units or similar terminals that will operate on relatively small temperature differentials in water. *(a)* In the heating mode, source water flows through the heat exchanger acting as an evaporator. Heat is drawn into the refrigerant source and delivered to the indoor heat exchanger, which acts as a condenser. Heat is rejected there, warming the interior space. *(b)* In the cooling mode, heat picked up from the fan coil water circuit by the indoor coil acting as an evaporator is transferred to the outdoor heat exchanger, which acts as a condenser. From the heat exchanger, heat is transferred to the water source, which acts as a heat sink. (From Ambrose, *Heat Pumps*, © John Wiley & Sons, reprinted by permission of John Wiley & Sons.)

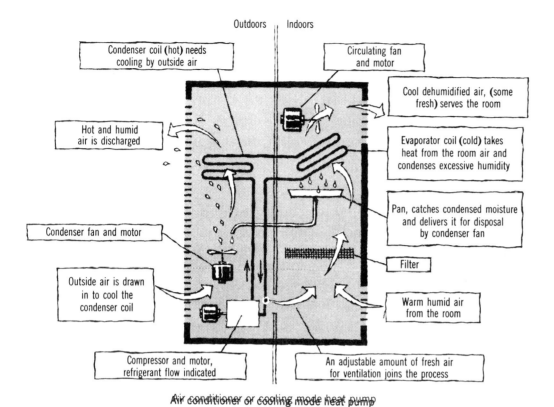

Figure 6.14 Schematic diagram of a package air conditioner (or heat pump operating in cooling mode). This through-the-wall arrangement is very common for small units up to about 2½ tons. See also Figure 6.7, which shows a similar through-the-wall (or window) installation of a package unit. Note that the evaporator and room air blower are always inside and that the compressor, condenser and condenser fan are always outside.

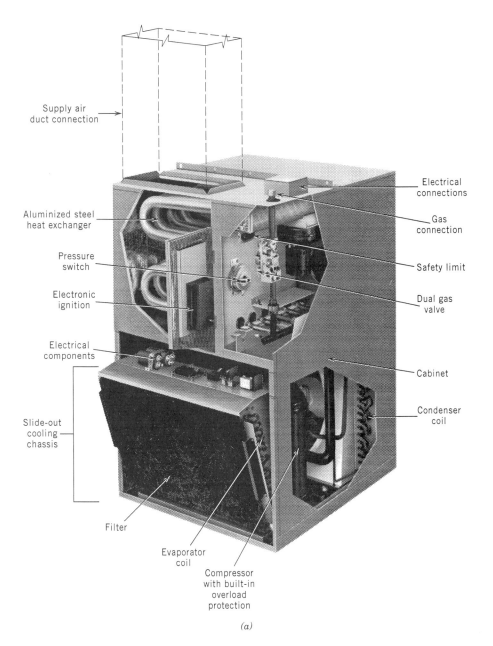

Figure 6.15 *(a)* Combination through-the-wall gas heating and electric cooling (air conditioner) unit. The cool air supply duct is connected to the top of the unit, as shown. Combustion air is drawn in through the exterior exposed surface of the unit (the rear in this photo). A built-in power vent eliminates the need for a chimney. Similar units are available as all-electric heat pumps and as electric heating, electric cooling (air conditioner) units. *(b)* Dimensional and technical data for the units shown in *(a)*. Btuh capacities vary from 27 to 50 MBH in heating and 1½ tons (17,600 Btuh) to 2½ tons (29,200 Btuh) in cooling at a seasonal energy efficiency ratio of 8. (Courtesy of Armstrong Air Conditioning, a Lennox International company.)

HWC Series
Higher Efficiency Cooling Unit Dimensions and Specifications

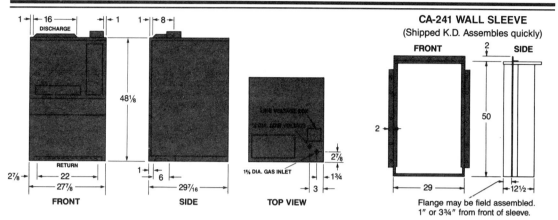

RATINGS AND SPECIFICATIONS

HEATING SECTION										
MODEL NUMBER	36HWC181	48HWC181	60HWC181	66HWC181	36HWC241	48HWC241	60HWC241	66HWC241	60HWC301	66HWC301
Rated Input - Btu/hr.	36,000	48,000	60,000	66,000	36,000	48,000	60,000	66,000	60,000	66,000
Capacity - Btu/hr. ①	27,000	36,000	45,000	50,000	27,000	36,000	45,000	50,000	45,000	50,000
Efficiency - A.F.U.E. ●	70.8	70.8	71.6	71.2	70.8	70.8	71.6	71.2	71.6	71.2
CFM @ Heating Speed	500 Lo Spd	500 Lo Spd	700 Hi Spd	700 Hi Spd	750 Lo Spd	750 Lo Spd	750 Lo Spd	750 Lo Spd	750 Lo Spd	750 Lo Spd

COOLING SECTION	208-230 VOLTS — SINGLE PHASE, 60 Hz, 197 MINIMUM OPERATING VOLTS		
■ Capacity - Btu/hr.	17,600	23,800	29,200
Efficiency - S.E.E.R.	8.00	8.05	8.05
CFM (Hi-Spd.)	650	800	780
Filter Size	13" x 25"	18" x 25"	18" x 25"
Compressor	P.S.C. Hermetic	P.S.C. Hermetic	P.S.C. Hermetic
R.L.A.	8.0	12.8	15.5
L.R.A.	43.3	60.0	70
Min. Circuit Ampacity	13.3	21	23.6
Max. Circuit Fuse Size	20 Amps	30 Amps	35 Amps
Approx. Ship. Weight	400 lb.	415 lb.	430 lb.

Provision for condensate drain must be provided inside structure.
① Capacity ratings are based on D.O.E. standard tests.

■ Certified ratings per A.R.I. Standard 210 and ASHRAE Standard 116. BTUH ratings shown apply to 230-volt operation when tested under standard A.R.I. rating conditions of 95°F outside dry bulb, 80°F inside dry bulb and 67 inside wet bulb temperatures.
● Energy efficiency ratings are based on U.S. Government standard tests.

BLOWER PERFORMANCE DATA

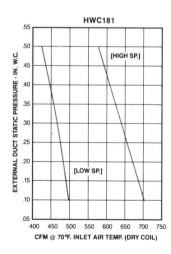
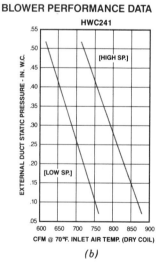
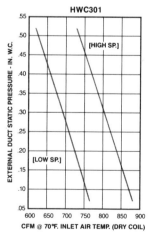

(b)

Figure 6.15 *(Continued)*

thermostat, is supplied via ductwork connected to the package. This type of unit is frequently installed in a niche or closet on an outside wall, in single or multifamily residential buildings. The exterior location is required for both heating (combustion air and venting) and cooling (condenser cooling with ambient air).

Figure 6.16 shows a small packaged heat pump suitable for rooftop mounting, or mounting at the level of the conditioned space, using side entry ducts. This unit would be used in an all-electric installation. Models in this design have cooling ratings from 2 to 3½ tons. Heat output varies with outside temperature and is tabulated in Figure 6.16*(b)*. To compensate for the drop in heat output at low outdoor temperatures, resistance heating packages of 5–20 KW are available. Other heat pump units of this design and slightly larger dimensions are available up to 5 tons cooling. These ratings make the units suitable for small to medium-size residences, stores and offices. Figure 6.16 *(c)* and *(d)* give electrical data, blower performance and dimensional data for models of the design.

Figure 6.17 shows a medium-size, commercial-grade packaged unit that also supplies gas heating and electric cooling. Heating and cooling capacities are given in Figure 6.17*(b)*. This unit is designed for rooftop mounting [Figure 6.17*(c)*] but can be adapted for side entry of ducts as well [Figure 6.17*(d)*]. Units in this design have ratings that range from 77 MBH heating and 35.6 MBH total cooling capacity (3 tons) up to 150 MBH heating and 58 MBH total cooling capacity (5 tons). These ratings make the units suitable for large residences and commercial installations.

For all the illustrated packaged units, technical specifications and dimensional data are provided with the illustration. These should assist you in obtaining a "feel" for the relation between HVAC rating and the physical size of equipment. A very common application of a small package through-the-wall heat pump is shown in Figure 6.18. Here the noise generated by the heat pump can be useful in masking traffic noise and noise from adjoining rooms.

Another type of packaged unit is the PTAC mentioned previously. Although a through-the-wall heat pump is properly referred to as a PTHP, it is frequently (incorrectly) lumped with packaged air conditioners as a PTAC. Most through-the-wall PTAC units in speculative construction use a standard cooling chassis (air conditioner) plus a resistive electrical section (chassis), as shown schematically in Figure 6.19*(a)*. A typical unit of this design is shown in Figure 6.19*(b)*. Other designs combine the air conditioner and the resistive heating elements into one compact chassis.

b. Split Unit

If we look at Figures 6.7 and 6.15, we immediately see that the package air conditioner or heat pump is really two parts connected by a couple of refrigerant pipes. The indoor unit has a coil and blower; the outdoor unit contains the compressor, filter and another blower or fan. There is no good technical reason that would prevent separating the two parts. On the contrary, there are very good reasons, discussed previously, and listed next, to do so.

- *Noise.* As already noted, a package unit is noisy, even when it is installed outside, and air is ducted to the interior. Ducts are excellent noise channels.

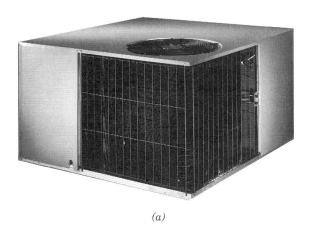

(a)

Figure 6.16 *(a)* Package heat pump, suitable for rooftop mounting, with vertical supply and return, or mounting adjacent to the conditioned space, with horizontal duct connections [see *(d)*]. (Courtesy of Armstrong Air Conditioning, a Lennox International Company.)

Performance Data

	Cooling					Heating				
							47° F.		17° F.	
Model	BTUH	SEER	EER	S/T	HSPF	BTUH	COP	BTUH	COP	CFM
PHP10A24A	23000	10.00	9.20	.72	6.80	22600	3.10	12600	2.00	800
PHP10A30A	29000	10.00	9.10	.72	7.00	28600	3.10	15800	2.00	1000
PHP10A36A	34000	10.00	9.00	.74	7.20	34000	3.15	21000	2.20	1200
PHP10A36B	34000	10.00	9.00	.74	7.20	34000	3.15	21000	2.20	1200
PHP10A42A	40000	10.00	9.00	.74	7.20	40000	3.20	22600	2.20	1400

Cooling Performance—Extended Ratings

		Outdoor Temp.—DB														
	Indoor Temp	65 Deg.			82 Deg.			95 Deg.			105 Deg.			115 Deg.		
Model	DB\WB	BTUH	S/T	KW	BTUH	S/T	KW	BTUH	S/T	KW	BTUH	S/T	KW	BTUH	S/T	KW
PHP10A24A	85/72	26,400	.61	2.03	27,600	.65	2.43	24,000	.66	2.52	22,600	.68	2.72	19,600	.71	2.88
	80/67	25,200	.67	2.00	26,000	.70	2.40	23,000	.72	2.50	21,200	.76	2.68	18,000	.80	2.85
	75/62	23,800	.74	1.98	24,600	.78	2.38	20,000	.81	2.47	29,200	.85	2.65	15,800	.90	2.80
PHP10A30A	85/72	33,000	.60	2.55	33,600	.64	3.03	30,000	.64	3.20	28,200	.67	3.38	26,000	.69	3.58
	80/67	31,600	.67	2.54	32,200	.70	3.01	29,000	.72	3.18	27,000	.76	3.35	23,500	.80	3.49
	75/62	30,200	.74	2.52	30,600	.78	2.98	25,400	.81	3.06	24,000	.85	3.30	20,200	.91	3.37
PHP10A36A	85/72	37,000	.60	3.20	38,800	.65	3.63	36,000	.65	3.99	34,200	.68	4.25	32,000	.70	4.51
	80/67	36,200	.67	3.18	38,000	.71	3.60	34,000	.74	3.77	32,600	.77	4.03	28,200	.79	4.36
	75/62	34,800	.73	3.14	36,200	.78	3.55	31,000	.82	3.75	28,400	.86	3.98	23,800	.92	4.19
PHP10A36B	85/72	37,000	.60	3.20	38,800	.65	3.63	36,000	.65	3.99	34,200	.68	4.25	32,000	.70	4.51
	80/67	36,200	.67	3.18	38,000	.71	3.60	34,000	.74	3.77	32,600	.77	4.03	28,200	.79	4.36
	75/62	34,800	.73	3.14	36,200	.78	3.55	31,000	.82	3.75	28,400	.86	3.98	23,800	.92	4.19
PHP10A42A	85/72	49,600	.61	3.71	46,400	.65	4.20	44,800	.66	4.61	41,000	.70	4.80	36,800	.74	5.07
	80/67	48,200	.67	3.66	45,000	.71	4.16	40,000	.74	4.44	36,800	.78	4.65	32,000	.84	4.85
	75/62	44,600	.76	3.6	42,800	.78	4.08	38,500	.81	4.25	34,400	.86	4.46	30,000	.90	4.62

Heating Performance—Extended Ratings

	Outdoor Temp DB/WB									
	0/0		17/15		35/33		47/43		62/56	
Model	BTUH	KW	BTUH	KW	BTUH	KW	BTUH	KW	BTUH	KW
PHP10A24A	9,400	1.75	12,600	1.85	17,600	2.15	22,600	2.14	27,000	2.20
PHP10A30A	10,600	2.22	15,800	2.31	22,200	2.52	28,600	2.70	35,000	2.85
PHP10A36A	12,400	2.60	21,000	2.80	26,600	3.02	34,000	3.16	41,000	3.33
PHP10A36B	12,400	2.60	21,000	2.80	26,600	3.02	34,000	3.16	41,000	3.33
PHP10A42A	14,000	2.93	22,600	3.01	27,800	3.36	40,000	3.66	48,000	3.91

(b)

Figure 6.16 (b) Performance data, including extended ratings for high and low temperatures. Cooling ratings of models in this configuration vary from 23,000 Btuh (2 tons) to 40,000 Btuh (3½ tons). Auxiliary electric resistance heaters are available in ratings of 5 kw to maintain heating capacity at low outside temperatures.

Physical and Electrical Data

Model	Voltage Hz Phase	Normal Voltage Range	Min. Circuit Ampacity	Max. Fuse/ HACR Brkr.	Compressor Rated Load (amps)	Compressor Locked Rotor (amps)	Dia. (in.)	Outside Fan Nom. RPM	Outside Fan Rated Load (amps)	Rated Watts HP	Indoor Blower Wheel d x w (in.)	Indoor Blower Rated Watts HP/AMP	Refrig. Charge (oz.)	Weight (lbs)
PHP10A24A	208-230/60/1	197–253	15.8	25	9.8	56.0	18	1075	.90	1/8	10 x 6	1/2 2.6	75	260
PHP10A30A	208-230/60/1	197–253	20.6	30	13.7	75.0	18	1075	.90	1/8	10 x 8	1/2 2.6	73	280
PHP10A36A	208-230/60/1	197–253	21.7	30	13.8	78.8	18	1075	1.80	1/4	10 x 8	1/2 2.6	93	300
PHP10A36B	200-230/60/3	187–253	17.3	25	10.3	75.0	18	1075	1.80	1/4	10 x 8	1/2 2.6	93	300
PHP10A42A	208-230/60/1	197–253	26.6	35	17.1	105.0	18	1075	1.80	1/4	10 x 9	1/2 3.4	102	330

Blower Performance Data

Model	Blower Speed	CFM @ Ext. Static Pressure—in. W.C. w/o Filter(s)*						
		0.2	0.3	0.4	0.5	0.6	0.7	0.8
PHP10A24A	Hi	1100	1060	1000	940	880	800	720
	Med	940	890	870	840	800	720	660
	Low	850	800	790	770	750	670	600
PHP10A30A	Hi	1400	1350	1280	1200	1120	1030	920
	Med	1160	1120	1080	1030	980	900	780
	Low	1050	1020	1000	950	910	840	750
PHP10A36A	Hi	1400	1350	1280	1200	1120	1030	920
	Med	1160	1120	1080	1030	980	900	780
	Low	1050	1020	1000	950	910	840	750
PHP10A36B	Hi	1400	1350	1280	1200	1120	1030	920
	Med	1160	1120	1080	1030	980	900	780
	Low	1050	1020	1000	950	910	840	750
PHP10A42A	Hi	1640	1560	1500	1400	1300	1260	1160
	Med	1570	1500	1440	1340	1270	1200	1100
	Low	1480	1430	1360	1290	1230	1170	1050

*Add .10 to duct static for downflow CFM equivalent.

(c)

Figure 6.16 (c) Physical, electrical and blower data.

338 / AIR SYSTEMS, HEATING AND COOLING, PART II

Dimensions

Model	A	B	C
PHP10A24	25 1/4	17 7/16	20 11/16
PHP10A30	25 1/4	17 7/16	20 11/16
PHP10A36	25 1/4	17 7/16	20 11/16
PHP10A42	29 1/4	21 7/16	24 11/16

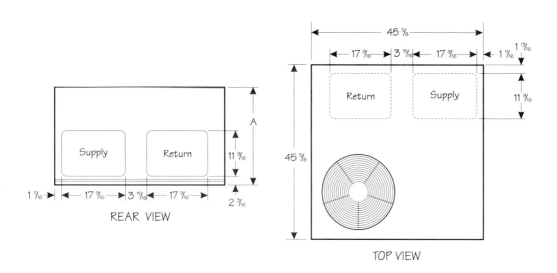

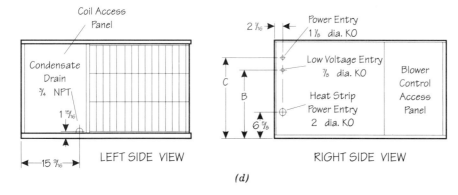

(d)

Figure 6.16 (d) Dimensional data.

PHYSICAL ARRANGEMENT OF EQUIPMENT, SMALL SYSTEMS / 339

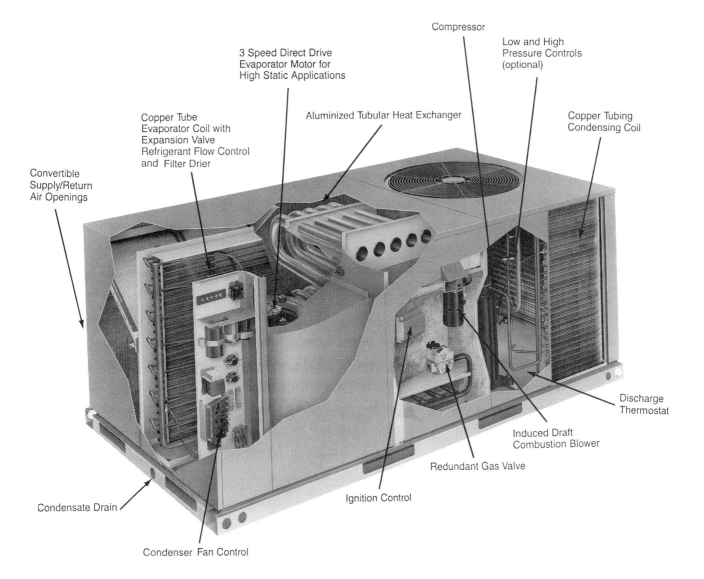

(a)

Figure 6.17 (a) Package electric cooling, gas heating unit for rooftop mounting. The air conditioner section is available in 3–5 ton rating in this configuration. (Courtesy of Armstrong Air Conditioning, a Lennox International company.)

Ratings and Specifications

Model No.	Unit Supply Voltage	Cooling Specifications			Heating Specifications		
		Cooling MBTUH	SEER	Sound Rating BELS	Heating Input MBTUH	AFUE %	Heating Output MBTUH
GRTC36A	208/230-1-60	35.6	9.80	8.4	100-150	78	77-116
GRTC36B	208/230-3-60	36.0	8.50	8.4	100-150	78	77-116
GRTC42A	208/230-1-60	40.0	9.80	8.4	100-150	78	77-116
GRTC42B	208/230-3-60	42.0	8.50	8.4	100-150	78	77-116
GRTC48A	208/230-1-60	46.6	8.50	8.4	100-150	78	77-116
GRTC48B	208/230-3-60	46.6	—	8.4	100-150	78	77-116
GRTC60A	208/230-1-60	58.0	8.50	8.4	100-150	78	77-116
GRTC60B	208/230-3-60	58.0	—	8.4	100-150	78	77-116

Heating efficiency for all models is expressed in AFUE. (Annualized Fuel Utilization Efficiency).
Cooling efficiency for single phase models is expressed in SEER (Seasonal Energy Efficiency Ratio).

Physical Data—Basic Units

		GRTC36	GRTC42	GRTC48	GRTC60
Evaporator Blower	Centrifugal Blower (Dia. x Wd. in.)	11 x 9	11 x 9	11 x 9	11 x 9
	Fan Motor HP	1/2	3/4	3/4	1
	Max. Ext. SP @ Nom. CFM	.85	.85	.85	.85
Evaporator Coil	Rows Deep	3	3	3	3
	Fins Per Inch	11 - 14	11 - 14	11 - 14	11 - 14
	Face Area (Sq. Ft.)	3.6	3.6	4.3	5.1
Condenser Fan	Propeller Dia. (in.)	24	24	24	24
	Fan Motor HP	1/2	1/2	1/2	1/2
	Nom. CFM Total	4050	4150	4650	4650
Condenser Coil	Rows Deep	1	1	1	1
	Fins Per Inch	12 - 16	12 - 16	12 - 16	12 - 16
	Face Area (Sq. Ft.)	10.6	14.0	14.0	16.9
Air Filters (See Note)	Quantity Per Unit (14" x 20" x 1")	1	1	1	1
	Quantity Per Unit (15" x 20" x 1")	1	1	1	1
	Quantity Per Unit (14" x 25" x 1")	1	1	1	1
	Total Face Area (Sq. Ft.)	6.5	6.5	6.5	6.5
Charge	Refrigerant 22 (oz.)	87	83	90	105

Note: Filter racks will accept 1" or 2" thick filters.

Cooling Capacity—MBTUH

Model No.	Total Capacity	Sensible Capacity	Nominal CFM	KW
GRTC36A	35.6	25.2	1300	3.95
GRTC36B	36.0	25.6	1300	4.46
GRTC42A	40.0	27.4	1400	4.37
GRTC42B	42.0	29.0	1400	5.22
GRTC48A	46.6	31.6	1600	5.81
GRTC48B	46.6	31.6	1600	—
GRTC60A	58.0	40.2	2000	6.71
GRTC60B	58.0	40.2	2000	—

Note: Blower motor heat has been deducted from all of the above capacity ratings. The KW ratings include the KW of both the supply air blower motor and the condenser fan motor.

(b)

Figure 6.17 (b) HVAC and electrical data for 3-, 4- and 5-ton units (35.6–58 MBH).

Heating Data

Model	Gas Heating Capacity		AFUE %	Temp. Rise Range	Corresponding CFM Min/Max
	Input MBTUH	Output MBTUH			
150GRTC60A(1PH)/B(3PH)	150	115	78	45-75 F	1450/2400
125GRTC60A(1PH)/B(3PH)	125	96	78	45-75 F	1200/2000
100GRTC60A(1PH)/B(3PH)	100	77	78	35-65 F	1100/2050
150GRTC48A(1PH)/B(3PH)	150	115	78	45-75 F	1450/2400
125GRTC48A(1PH)/B(3PH)	125	96	78	45-75 F	1200/2000
100GRTC48A(1PH)/B(3PH)	100	77	78	35-65 F	1100/2050
150GRTC42A(1PH)/B(3PH)	150	115	78	45-75 F	1450/2400
125GRTC42A(1PH)/B(3PH)	125	96	78	45-75 F	1200/2000
100GRTC42A(1PH)/B(3PH)	100	77	78	35-65 F	1100/2050
150GRTC36A(1PH)/B(3PH)	150	115	78	45-75 F	1450/2400
125GRTC36A(1PH)/B(3PH)	125	96	78	35-65 F	1400/2550
100GRTC36A(1PH)/B(3PH)	100	77	78	35-65 F	1100/2050

The gas-furnaces can be converted to propane at ratings shown above.

Electrical Data

Model	Unit Supply Voltage	Min. Operating Voltage	Min. Circuit Ampacity	Max. Overcurrent Device	Compressor		Condenser Fan Motor		Indoor Blower Motor	
					RLA	LRA	RLA	LRA	RLA	LRA
GRTC36A	208/230-1-60	197	29.1	35	17.9	90.5	2.3	6.5	4.4	9.0
GRTC36B	208/230-3-60	187	22.3	30	12.5	66.0	2.3	6.5	4.4	9.0
GRTC42A	208/230-1-60	197	32.2	40	19.9	107.0	2.3	6.5	5.0	9.0
GRTC42B	208/230-3-60	187	26.4	35	15.3	82.0	2.3	6.5	5.0	9.0
GRTC48A	208/230-1-60	197	37.9	50	24.5	114.0	2.3	6.5	5.0	9.0
GRTC48B	208/230-3-60	187	28.3	35	16.8	84.0	2.3	6.5	5.0	9.0
GRTC60A	208/230-1-60	197	47.0	60	30.5	135.0	2.3	6.5	6.6	14.8
GRTC60B	208/230-3-60	187	33.4	40	19.6	105.0	2.3	6.5	6.6	14.8

Blower CFM vs. Available External Static Pressure (in W.G.) Bottom Air in/out—230–Volt Unit Dry Coil with Air Filter Only

Model	Motor Speed	External Static Pressure - IWG								
		.20	.30	.40	.50	.60	.70	.80	.90	1.0
GRTC36A	Hi	1680	1630	1560	1480	1410	1320	1240	1110	1000
	Med	1440	1400	1360	1300	1220	1160	1060	960	820
	Low	1320	1280	1240	1200	1140	1060	980	860	710
GRTC42A	Hi	1860	1780	1710	1630	1540	1420	1340	1220	1080
	Med	1780	1740	1670	1600	1510	1400	1320	1200	1060
	Low	1700	1640	1580	1500	1410	1340	1300	1120	1020
GRTC48A	Hi	1940	1890	1820	1740	1680	1560	1460	1360	1240
	Med	1840	1760	1700	1640	1560	1460	1360	1240	1100
	Low	1610	1580	1520	1460	1360	1280	1160	1040	860
GRTC60A	Hi	2460	2380	2300	2230	2160	2070	1960	1840	1680
	Med	2440	2340	2280	2200	2140	2050	1950	1800	1640
	Low	2340	2260	2180	2100	2000	1890	1760	1640	1500

(b)

Figure 6.17 (b) (Continued)

342 / AIR SYSTEMS, HEATING AND COOLING, PART II

UNIT DIMENSIONS
ROOF MOUNTING

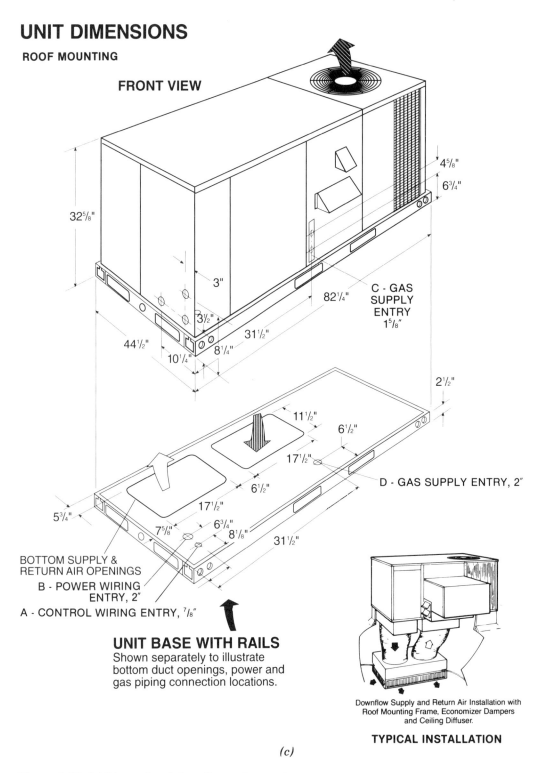

(c)

Figure 6.17 (c) Dimensional data for roof mounting.

PHYSICAL ARRANGEMENT OF EQUIPMENT, SMALL SYSTEMS / 343

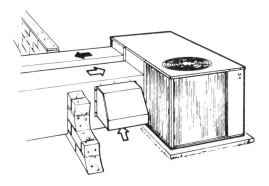

Figure 6.17 (d) Dimensional data for duct side entry.

Figure 6.18 Motel rooms are a common application of through-the-wall heat pumps. The exterior coil is exposed to outside ambient air behind a panel set slightly forward of the outside wall (a). The interior cabinets recirculate room air through a top grille (b). All such units have a dampered opening that permits fresh air to be admitted and stale room air to be exhausted.

344 / AIR SYSTEMS, HEATING AND COOLING, PART II

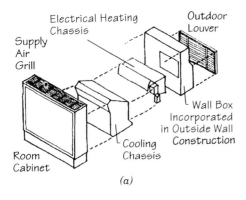

(b-1)

(b-2)

Figure 6.19 *(a)* Through-the-wall nonducted package air conditioner with separate heating chassis. *(b-1)* Inside view of a typical unit. *(b-2)* Outside view of the same unit. [*(a)* From Bobenhausen, *Simplified Design of HVAC Systems*, 1994, © and *(b-1, b-2)* from Bradshaw, *Building Control Systems*, 2nd ed., 1993, © John Wiley & Sons, reprinted by permission of John Wiley & Sons.)

- *Size.* Because all the equipment is contained in a single package, the enclosure is large (and heavy). The size factor is particularly important where space is at a premium.
- *Maintenance.* Manufacturers are interested in making package units as small as possible. These small, densely packed units are often installed in small, out-of-the-way niches and closets. Both of these facts make maintenance difficult and often impossible without removing and dismantling the unit. This leads to time delay and high costs.

A two-piece package unit is called a split air conditioner or heat pump. In private residences, this arrangement for air conditioning has become almost standard. See Figure 5.6*(a)* (page 205) and Figure 5.12*(b)* (page 214). The noisy compressor and the outside coil (condenser) can be placed up to 100 ft from the house. An A-frame evaporator placed in a warm air furnace bonnet is connected to the outside (condenser) unit by two small-diameter pipes. This arrangement reduces noise, improves air flow on the condenser coil and saves space inside the house. A similar arrangement with the condenser unit on the roof is very common in low rise multifamily dwellings.

A schematic drawing showing the two alternative arrangements—a complete roof package and a split unit with roof condenser and indoor air handler—are shown in Figure 6.20*(a)* and *(b)*. The same arrangement with a slab-on-grade condenser is shown in Figure 6.20*(c)*.

The physical arrangement of a split system depends on whether the equipment is an air conditioner providing cooling only or a heat pump providing both cooling and heating.

Split Air Conditioner. The outdoor unit, called the outdoor condenser unit [see Figure 6.21*(a)*], always consists of the compressor and the condenser coil and its fan. It can be installed on a slab at grade level [Figure 6.21*(c)*] or installed on the roof [Figure 6.21*(d)*]. The indoor (evaporator) portion of the system has two configurations, depending on whether the building heating system is an air system or not. With an air system, we can use the heating system blower and ducts by simply placing an A-frame evaporator coil in the warm air furnace bonnet. This is the arrangement most frequently used in residences. It is illustrated in Figures 5.6*(a)* (page 205) and 5.7 (page 206). The A-frame evaporator coil, so called because of its shape, is illustrated in Figure 6.22*(a)*.

If the building uses a hydronic heating system or any other system that does not provide the necessary blower (and ductwork), the indoor portion of the split air conditioner must contain a blower in addition to the evaporator coil. Such a unit is variously called an air handler, a blower evaporator and a DX fan coil unit. Typical units are shown in Figure 6.22 *(b–d)*. A typical installation is shown in Figure 6.22*(e)*. The novice HVAC technologist should be forewarned that these terms do not always means the same thing. An air handler can be only a blower, but it can also contain a cooling coil and electric heaters. A blower evaporator always has a blower and evaporator coil, but it can also be equipped with an electric heater, although this is not common. A fan coil unit is normally a fan unit with hot/cold water coils, as illustrated in Figure 3.19 (page 110). A DX fan coil unit (more properly a DX coil fan unit) is an assembly of blower and DX coil (evaporator). Such units are frequently also equipped with resistance-type electric heaters.

Cooling performance of the two configurations, that is, furnace blower with A-frame in the ductwork and blower evaporator unit is quite similar. This assumes that the furnace blower was selected knowing that an A-frame evaporator coil would be installed. (This knowledge is also necessary to size the building ductwork properly, as was explained in detail in Chapter 5.) Performance data for both arrangements are given in Figure 6.21*(b)*.

Split Heat Pump. In this arrangement the exterior unit is called an outdoor heat pump. It consists of the system compressor plus a coil and fan. The coil acts as a condenser in the cooling mode and as an evaporator during the heating mode. Units are very similar in construction to the outdoor condensing unit seen in Figure 6.21. Since there is obviously no warm air heating system in the building, the heat pump indoor unit, called the heat pump air handler, always consists of a coil and blower, such as those shown in Figure 6.22. The coil acts as an evaporator during cooling and as a condenser during heating.

Split air conditioners and heat pumps up to about 15 tons of cooling, and somewhat higher heating ratings are most often used in residential and small commercial applications, including stores. Because of noise and space considerations, better than 90% of residential systems are of the split design.

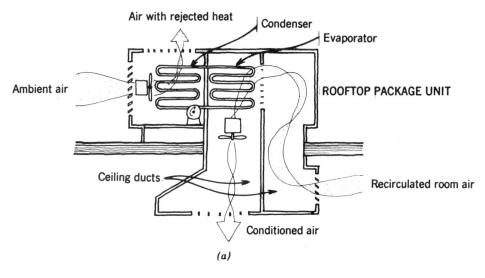

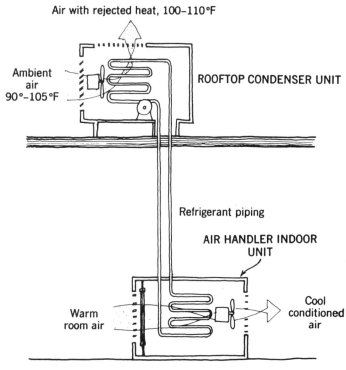

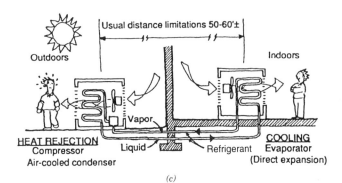

Figure 6.20 An air conditioner or a heat pump can be mounted as a single package on the roof (a), or it can be split into two units. The split unit may be a rooftop condenser and indoor evaporator/air handler (b) or a slab-on-grade outdoor condenser and an indoor evaporator and air handler (c). See the text for a discussion of the advantages and disadvantages of the three configurations.

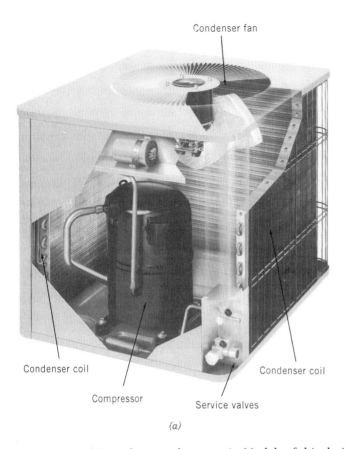

Figure 6.21 *(a)* Outdoor condenser unit. Models of this design have a total cooling capacity of 18,000 Btuh (1½ tons) to 57,000 Btuh (5 tons). Performance data with evaporator coils or blower evaporators is shown in *(b)*, along with electrical data. Dimensional data, plus a schematic of a typical physical arrangement for a slab-on-grade installation, are shown in *(c)*. A typical installation of a roof-mounted condensing unit and ceiling-mounted indoor evaporator unit is shown in *(d)*. Also shown are the piping, wiring and some of the control items. [*(a)–(c)* Courtesy of Armstrong Air Conditioning, a Lennox International company. *(d)* Courtesy of Carrier Corp., a subsidiary of United Technologies.]

Cooling Performance with Evaporator Coils

Basic Cond. Unit Model	Indoor Model	SEER	ARI Capacity	Sensible Capacity	CFM	Air Fric.	Refrigerant Line Size Suction	Refrigerant Line Size Liquid	Indoor Coil Orifice Size*
SCU10A12A-1	CAU,CAC18	11.0	13,000	9,100	400	.10	3/4	3/8	.043
SCU10A18A-1	CAU,CAC24	10.1	17,800	12,700	600	.22	3/4	3/8	.049
	CSH24	10.0	17,800	13,200	600	.15	3/4	3/8	.053
SCU10A24A-1	CAU,CAC24	10.0	23,600	17,300	800	.30	3/4	3/8	.057
	CAU,CAC30	10.1	23,600	17,300	800	.26	3/4	3/8	.059
	CSH24	10.0	23,600	17,300	800	.18	3/4	3/8	.057
SCU10A30A-1	CAU,CAC30	10.1	30,000	22,500	1,000	.27	3/4	3/8	.063
	CSH36	10.0	29,600	22,500	1,000	.22	3/4	3/8	.065
SCU10A36A-1	CAU,CAC36	10.1	35,400	24,800	1,200	.30	7/8	3/8	.065
	CAU,CAC42	10.1	35,400	24,800	1,200	.28	7/8	3/8	.065
SCU10A42A-1	CAU,CAC36	10.1	39,500	28,900	1,300	.30	7/8	3/8	.074
	CAU,CAC42	10.1	39,500	28,900	1,300	.29	7/8	3/8	.074
SCU10A48A-1	CAU49	10.1	47,500	35,200	1,600	.28	7/8	3/8	.084
	MC48	10.1	47,500	35,200	1,600	.28	7/8	3/8	.084
SCU10A60A-1	CSH60	10.0	57,000	42,100	1,800	.24	7/8	3/8	.090

*Required to achieve ARI rating.

Cooling Performance with Blower Evaporators

Basic Cond. Unit Model	Indoor Model	SEER	ARI Capacity	Sensible Capacity	CFM	Suction	Liquid	Indoor Coil Orifice Size*
SCU10A18A-1	AH24	10.10	18,000	13,500	600	3/4	3/8	.049
	MB08/MC24	10.10	17,800	12,700	600	3/4	3/8	.049
SCU10A24A-1	AH24	10.10	23,800	17,600	800	3/4	3/8	.059
	MB08/MC24	10.00	23,600	17,300	800	3/4	3/8	.057
SCU10A30A-1	AH36	10.10	30,000	22,500	1,000	3/4	3/8	.063
	MB12/MC29	10.10	30,000	22,500	1,000	3/4	3/8	.063
SCU10A36A-1	AH36	10.10	35,200	25,700	1,200	7/8	3/8	.068
	MB12/MC36	10.10	35,400	24,800	1,200	7/8	3/8	.065
	AH36	10.00	40,500	29,500	1,300	7/8	3/8	.074
SCU10A42A-1	MB12/MC36	10.00	39,500	28,900	1,300	7/8	3/8	.074
	MB14/MC42	10.05	40,500	30,000	1,400	7/8	3/8	.074
SCU10A48A-1	MB16/MC48	10.10	47,500	35,200	1,600	7/8	3/8	.084
	MB16/MC42	10.00	47,000	34,300	1,600	7/8	3/8	.084
SCU10A60A-1	MB20/MC60	10.00	57,000	42,200	1,800	7/8	3/8	.090

Physical and Electrical

Model	Voltage Hz/Phase	Nominal Voltage Range	Min. Circuit Ampacity	Max. Overcurrent Device (amps)	Compressor Rated Load	Compressor Locked Rotor	Fan Dia. (in.)	Fan Motor Rated HP	Fan Motor Nominal RPM	Fan Motor Full Load (amps)	Refrig. Charge (oz.)	Weight (lbs.)
SCU10A18A-1	208/230-60-1	197–253	11	15	8	45	18	1/5	1075	0.75	72	160
SCU10A24A-1	208/230-60-1	197–253	16	25	12	60	18	1/5	1075	0.75	72	160
SCU10A30A-1	208/230-60-1	197–253	17	25	12.5	76.1	18	1/8	1075	0.9	79	170
SCU10A36A-1	208/230-60-1	197–253	21	30	15	78.8	18	1/3	1075	1.6	89	190
SCU10A42A-1	208/230-60-1	197–253	24	35	17.5	105	18	1/3	1075	1.6	102	220
SCU10A48A-1	208/230-60-1	197–253	30	45	22.5	119	24	1/3	1075	1.6	130	235
SCU10A60A-1	208/230-60-1	197–253	35	50	26.5	141	24	1/3	1075	1.6	133	235
SCU10A12A-1	208/230-60-1	197–253	7	15	5	26.3	18	1/5	1075	0.75	76	130

Figure 6.21 (b) Performance data.

Dimensions (in.)

Model	A	B	C
SCU10A12	28 1/4	22 1/4	23 1/8
SCU10A18	28 1/4	22 1/4	23 1/8
SCU10A24	28 1/4	22 1/4	23 1/8
SCU10A30	28 1/4	22 1/4	23 1/8
SCU10A36	28 1/4	22 1/4	25 1/8
SCU10A42	30 1/4	26 1/4	25 1/8
SCU10A48	38 1/4	34 1/4	29 1/8
SCU10A60	38 1/4	34 1/4	29 1/8

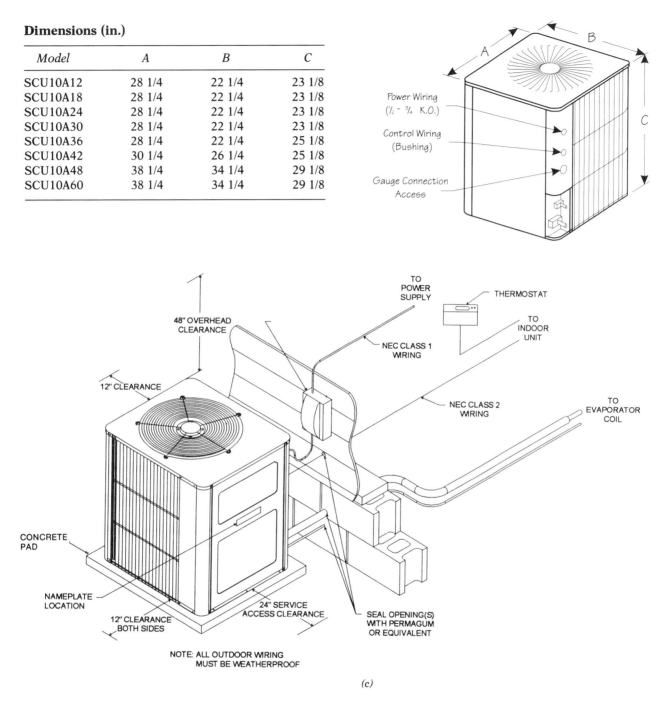

(c)

Figure 6.21 (c) Installation data.

Typical piping and wiring

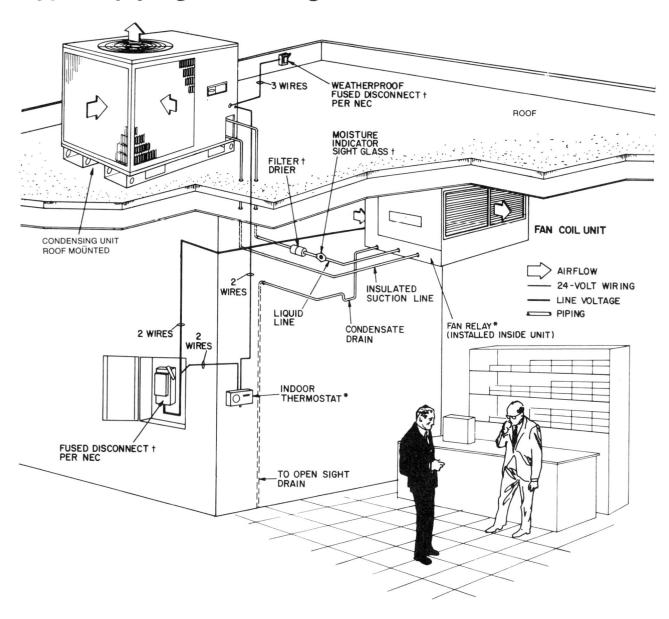

*Accessory item.   †Field supplied.

NOTES:
1. All piping must follow standard refrigerant piping techniques. Refer to Carrier System Design Manual for details.
2. All wiring must comply with the applicable local and national codes.
3. Wiring and piping shown are general points-of-connection guides only and are not intended for or to include all details for a specific installation.

Figure 6.21 *(d)* Typical roof installation.

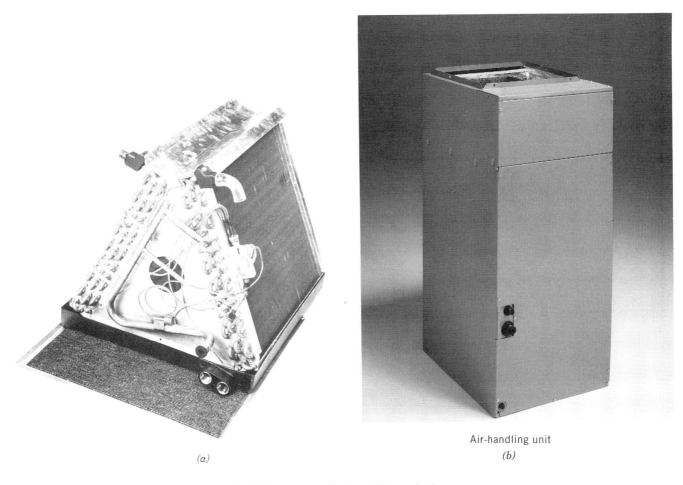

Air-handling unit
(b)

(a)

Figure 6.22 (a) A-frame evaporator coil, which is normally installed in the bonnet ductwork of warm air furnaces and connected by refrigerant lines to an external condenser unit. (b) Air-handling unit contains an evaporator coil plus a blower. (c) DX fan coil unit is identical in function to the air-handling unit shown in (b). This DX coil is a longitudinal design direct expansion evaporator coil. (d) DX fan coil unit, which uses an A-frame DX evaporator coil, is similar in function to (b) and (c). (e) Typical application of a DX fan coil unit (blower evaporator unit). Note the location of the system humidifier and electronic air cleaner. [Illustrations (a, c–e) courtesy of Carrier Corp., a subsidiary of United Technologies. Illustration (b) courtesy of Armstrong Air Conditioning, a Lennox International company.]

352 / AIR SYSTEMS, HEATING AND COOLING, PART II

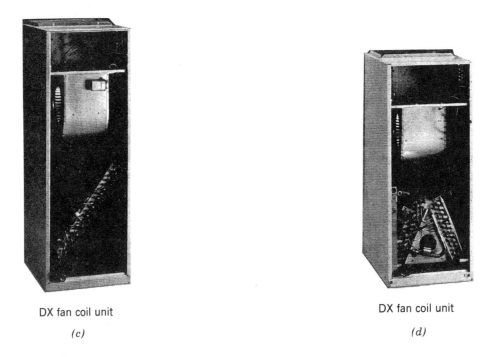

DX fan coil unit
(c)

DX fan coil unit
(d)

Figure 6.22 (Continued)

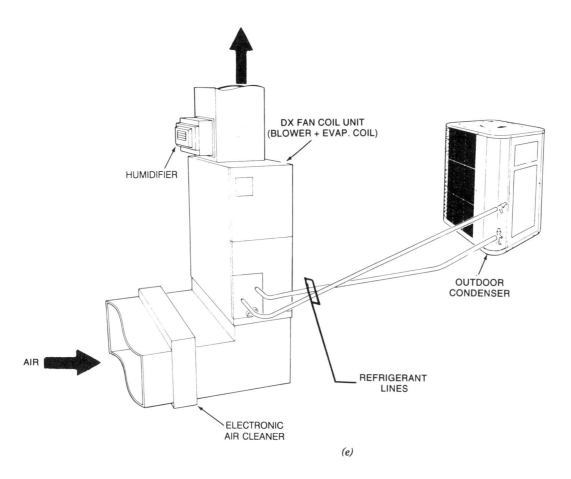

(e)

Figure 6.22 (Continued)

When selecting a split unit, the choice of whether to use a slab-on-grade outdoor unit or a rooftop unit is usually decided on architectural grounds. Some of the other considerations are:

- Slab-on-grade units can be unsightly and a source of annoying noise when placed close to an occupied building.
- Slab-on-grade outdoor units are very easy to maintain, can be screened, are cheap to install and can be placed a considerable distance from the indoor unit, if required.
- Rooftop units can be either part of a split unit or a complete package ducted to the space below. See Figure 6.20.
- Rooftop units require a massive concrete base plus sound traps and isolation to prevent extremely annoying low frequency vibration and noise into the space below.
- Rooftop units are completely out of sight and have short piping and duct runs that reduce installation costs.
- Rooftop units must be carefully installed so that ducts and piping do not cause water leaks into the spaces below.
- Because access is not convenient, maintenance of rooftop equipment is generally poor and is more expensive than that for grade-level equipment.

6.9 Physical Arrangement of Equipment, Large Systems

Cooling (and heating) systems larger than about 15 tons are classed as large systems. Here also two arrangements are common: the unitary or incremental package system and the central system. In the former, which is also called a distributed system, complete package units are used to provide cooling and heating to individual space units. In other words, a large system is broken up into many small systems. Thus, in a large multistory apartment house with, say, 100 apartments, each apartment might have a closet-mounted package like that shown in Figure 6.15(a). If the building were a complex of garden apartments, then roof-mounted packages like that of Figure 6.16 could be used. In a large motel, each room can be satisfactorily handled with a unit like that shown in Figure 6.18. These installations are all called *distributed systems* because the climate control equipment is distributed throughout the building(s).

Alternatively, a *central system* furnishing hot and/or cold water (or air) to each space can be designed. With such an arrangement, the individual spaces have only terminal units such as fan coil units, induction units and the like. See Figure 3.19 (page 110) and Figures 5.42–5.45 (pages 268–270).

The advantages of using distributed package units (rooftops, PTACs or PTHPs) follow:

- Mechanical breakdown of a unit affects only that unit. Other individual package units continue to operate.
- Maintenance is simplified and affects only one machine at a time.
- An air or water distribution system is eliminated. This reduces first cost and maintenance.
- Zoning and individual unit control is simplified.
- Units can be added, subtracted and moved easily. For this reason, systems using package units are called incremental systems; each unit is an increment of the whole installation.
- First cost is lower than that of a ducted central system.
- Installation is simple and cheap.
- Sizing and selection of equipment is much simpler than for a single package. See the design example in Section 6.17.

The disadvantages of using incremental (distributed package) units follow:

- Control of temperature, humidity, fresh air intake and air distribution are relatively crude.
- An economizer cycle (use of fresh air for cooling) is generally not provided.
- Each unit requires access to an outside wall. This restricts use of these units to perimeter zones, or buildings with open spaces, if ducting is to be avoided. Incremental units like those in Figure 6.15 are ducted. Units of the type shown in Figure 6.19 are not ducted.
- Because the compressor and condenser fan are included in a package (incremental) unit, it will be much noisier than a terminal device, such as a fan coil unit, fed from a remote, central chiller. In some installations, this disadvantage can be turned into an advantage by using the compressor and fan noise to mask or blanket unwanted noise, such as from traffic or nearby industry. (This type of application is based on the well-known fact that the noise created by your equipment is much less disturbing than that coming from a neighbor's equipment.)

These are only a few of the many considerations involved. Refer to Section 6.17 for a design example using incremental units.

Refrigeration Equipment

By this point, you should have a firm grasp of the principal components in a vapor compression refrigeration cycle and their function. A rapid survey of the operation of these components plus some of the other equipment often encountered in larger systems is in order.

6.10 Condensers

As we have seen, the function of the condenser is to condense the refrigerant vapor by removing heat from it. There are three principal types of condensers:

- Air-cooled
- Water-cooled
- Evaporative-cooled

a. Air-Cooled Condensers

Air-cooled condensers reject heat to the atmosphere by blowing ambient (outside) air over the condenser coil. Obviously then, the higher the outdoor temperature is, the more air will have to be passed over the coil to reject its heat and condense the refrigerant vapor. Air-cooled condensers, such as those illustrated in Figures 6.17 and 6.21, almost always use a propeller fan because of its high air quantity, low static pressure characteristic. Fan motors are about 0.1 to 0.2 hp per ton of cooling. This type of condenser is low in cost and reliable. However, the large amount of air that must be moved to reject heat limits this condenser to systems up to about 50 tons. Larger systems normally use water-cooled condensers. (Specially designed air-cooled condensers are available for loads up to 100 tons.)

b. Water-Cooled Condensers

Water-cooled condensers reject their heat into a heat exchanger that is cooled by water. The problem then is how to arrange a continuous supply of cooling water. In the early days of air conditioning, city water was used as the cooling agent. The water was passed through the heat exchanger once and then discarded. Because of the wastefulness of this procedure and the load it places on the municipal water and sewer systems, most local authorities prohibit this practice today. If no lake, river, pond or wells are available to supply the required cooling water, a cooling tower must be used.

A *cooling tower* is a device that uses air to cool water, which is then recirculated to the condenser heat exchanger. Although cooling towers are available in many designs, they all operate on the principle of evaporative cooling. Water is pumped to the top of a structure (tower), where it is sprayed inside the tower. As it falls, some of it evaporates. In so doing, it absorbs 1000 Btu/lb of water vapor, as we have already learned. This heat can only come from the falling water. (Actually, the evaporating water cools the air around it, which in turn draws heat from the warm falling water.) Cooled water collects at the bottom of the tower, from which point it is recirculated to the condenser heat exchanger. See Figure 6.23. Since the water that evaporates is lost to the atmosphere, make-up water must be supplied from the city water mains.

There are three principal types of cooling tower designs: natural draft, forced draft and induced draft. See Figure 6.24. In the *natural draft tower* (Figure 6.23(a), air circulation through the tower is by natural convection (stack effect). The sprayed water simply falls inside the tower and is partially evaporated by natural convective air currents. To assist evaporation, most towers contain a "fill" that slows the flow of water, allowing more to evaporate. In *mechanical draft towers* [Figures 6.23(b) and (c)] (forced draft or induced draft), a large fan or blower greatly increases the motion of air through the tower. This serves to increase the amount of water evaporation and, thereby, increases the cooling effect. Total water temperature drop in most towers is about 10–20F°.

Cooling towers have a number of disadvantages. Because they depend on evaporative action for their cooling, their efficiency is affected by humidity in the surrounding air. The higher the humidity (wet bulb temperature), the lower is the tower's efficiency. They require a good deal of make-up water, which in many areas is expensive. They also require frequent cleaning and corrosion protection. The noise created by their fans can be a source of annoyance to neighbors, even when towers are installed on roofs. In winter, air and water flow

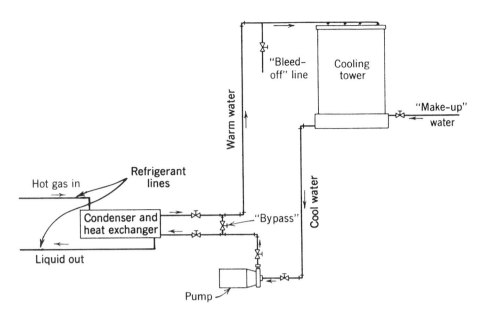

Figure 6.23 Schematic diagram of a recirculating water system between a water-cooled condenser and a cooling tower. Make-up water comes from the municipal water supply. The bypass valve permits removing the cooling tower and its accessories for maintenance. (From Dossat, *Principles of Refrigeration*, 1961, John Wiley & Sons. Reprinted by permission of Prentice-Hall, Inc., Upper Saddle River, NJ.)

must be carefully regulated to prevent freezing. Occasionally, electric heaters must be installed for this purpose. (Large buildings frequently require cooling for inside zones, even in winter.) Finally, towers not in use must be drained, cleaned and repaired.

c. Evaporative Condensers

The evaporative condenser uses the cooling effect of water evaporating directly on the condenser coil. See Figure 6.25. Water from a local tank is sprayed directly on to the hot condenser coils. An induced draft fan above or to the side of the condenser coil increases the draft and, thereby, the evaporation and cooling rate. Evaporative condensers are more efficient than either air-cooled or water tower-cooled condensers. Unlike cooling towers, they must be installed close to the compressor. Depending on the hardness of spray water, scale accumulation on the condenser coils can be a problem.

6.11 Evaporators

It is important to keep in mind that the terms *evaporator* and *condenser* refer to processing of the refrigerant. Otherwise, terms such as *evaporative condenser* (Section 6.10.c) will lead to considerable confusion. The evaporator is the piece of equipment that performs the functional space cooling, by absorbing heat from the conditioned space. The principal type of heat-transfer-to-air evaporator in use today is simply a coil of pipe equipped with fins to aid in heat transfer. See Figure 6.26(c).

Years ago there were two types of such evaporators in common use: flooded [Figure 6.26(a)] and dry expansion [Figure 6.26(b)]. The flooded evaporator, which was always filled with liquid refrigerant, is no longer in common use. The dry expansion (DX) evaporator uses a thermostatically controlled expansion valve that meters the flow of liquid refrigerant into the evaporator coil. The flow rate is controlled such that all the liquid refrigerant is vaporized by the time it reaches the end of the coil. The flow rate varies with the variation of heat load on the evaporator. (Although DX stands for dry expansion most industry people today refer to DX as direct expansion.)

The second principal type of evaporator in common use is the shell-and-tube, water-cooled evaporator. The refrigerant passes through a series of pipes encased in a shell. Water circulating through the shell is chilled by the evaporating refrigerant.

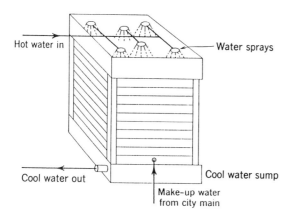

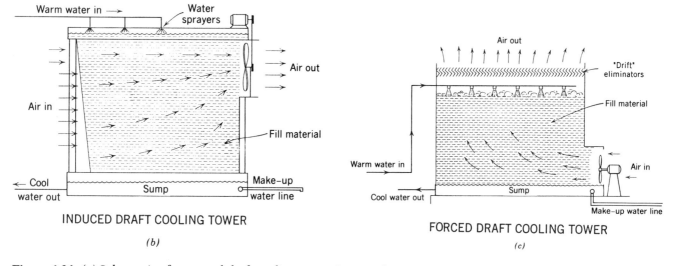

Figure 6.24 *(a)* Schematic of a natural draft cooling tower. In actual construction, these towers have hyperbolic curved sides that improve the reliability and the predictability of their stack effect. *(b)* Induced draft cooling tower. The purpose of the fill material is to expose the falling water droplets to as large a surface area as possible, in order to increase vaporization of the warm water. *(c)* Forced draft cooling tower. The "drift" eliminators reduce the quantity of water that is lost as it "drifts" away. This in turn will reduce the amount of make-up water required. (From Dossat, *Principles of Refrigeration*, 1961, John Wiley & Sons. Reprinted by permission of Prentice-Hall, Inc., Upper Saddle River, NJ.)

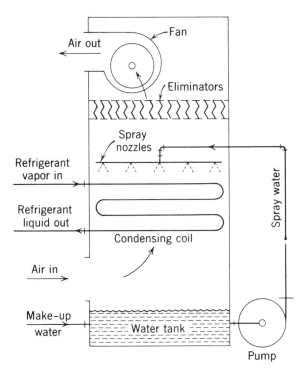

Figure 6.25 The evaporative condenser uses direct water evaporation to provide the required refrigerant cooling. Recirculated water is sprayed directly onto the hot condenser coils. The purpose of the eliminators at the top of the unit is twofold: to prevent water being carried by the air stream from entering (and damaging) the blower and to reduce the amount of water lost by drift. (From Dossat, *Principles of Refrigeration*, 1961, John Wiley & Sons. Reprinted by permission of Prentice-Hall, Inc., Upper Saddle River, NJ.)

The chilled water is then used in a fan coil or similar terminal unit to cool the building interior space. In large versions, shell-and-tube evaporators are referred to as *chillers*.

6.12 Compressors

There are five types of compressors in use for various sizes and types of refrigeration equipment. They are reciprocating (piston type), rotary, helical (screw type), scroll (orbital) and centrifugal. Construction details are not the concern of the technologist or, in general, the HVAC engineer. Most air conditioning equipment is packaged by the manufacturer who selects the type of compressor best suited to the system requirements. The types are listed simply for information purposes.

6.13 Air-Handling Equipment

Air-handling equipment includes fans, blowers and terminal devices. As with compressors, the type of blower or fan supplied with an evaporator in a package unit is selected by the manufacturer. Large systems, using separate air-handling equipment, are beyond the scope of our study. The various types of air distribution systems were covered in Section 5.19 and are shown in Figures 5.40–5.45. A typical fan coil unit, which is the most common hydronic cooling terminal device, is shown in Figure 3.19. Air distribution equipment is covered in Chapter 5.

Design

6.14 Advanced Design Considerations

Equipment-sizing considerations are discussed in some detail in Section 6.6. The material in this section is more technical. It is intended for technologists who have acquired the necessary background knowledge and are engaged in actual design work.

Remember that, to calculate the air quantities required in heating and cooling, we used the following formulas. For heating:

$$\text{cfm} = \frac{\text{(Sensible) heating load}}{1.08 \times \Delta t}$$

For cooling:

$$\text{cfm} = \frac{\text{Sensible cooling load}}{1.1 \times \Delta t}$$

(Rereading Sections 5.3 and 5.29 at this point would be helpful.)

The information that follows should assist you in determining the loads and the temperature differences required for these formulas.

a. Cooling Load Calculation

When using split unit air conditioners or heat pumps, the location of the air-cooled condenser or outdoor heat pump unit is important. When units

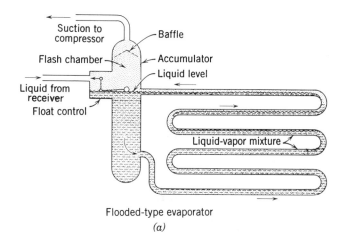

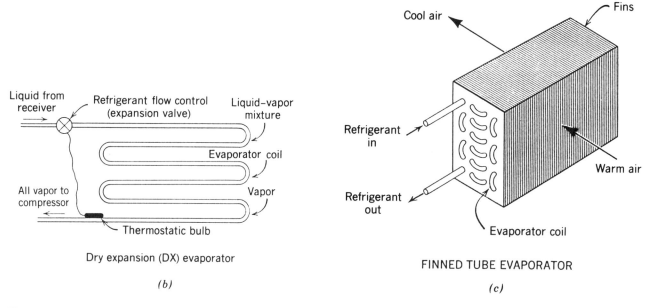

Figure 6.26 Evaporators of the flooded type (a) are no longer widely used. The dry expansion (DX) type (b) uses a thermostatically controlled expansion valve to control the flow of hot refrigerant into the evaporator coil. Fins are added to the evaporator coil (c) to aid in heat transfer. (From Dossat, *Principles of Refrigeration*, 1961, John Wiley & Sons. Reprinted by permission of Prentice-Hall, Inc., Upper Saddle River, NJ.)

are installed at grade level, either exposed or shaded from the sun, summer design conditions available in ASHRAE or other authoritative sources can be used. If the unit is roof-mounted, a penalty of 2–5F° should be added to the outside design temperature to allow for lower efficiency at the condenser (outdoor coil).

b. Cooling Design Conditions, Indoor Coil

As previously stated, design calculation for residential work usually does not include any load for mechanical ventilation. For these buildings, using a return air temperature of 75°F, the entering air conditions for the indoor coil would be 75°F DB and 62°F wet bulb, assuming a 50% RH design condition. For commercial structures with some mechanical ventilation, both dry and wet bulb temperatures are about 2F° higher. This gives indoor coil air conditions of 76–77°F DB and 63–64°F WB.

c. Heating Design Conditions, Heat Pump Indoor Coil

In the heating mode, with no ventilation considered (residential calculation), the return air temperature (entering air, indoor coil) can be taken at 70°F. If ventilation is considered, this temperature drops from 2 to 5F° depending on the outdoor design temperature.

d. Cooling Temperature Differential (Δt)

As stated in Section 6.6, the preferred procedure in cooling design is to calculate the latent as well as the sensible heat load. This permits an accurate calculation of the sensible heat ratio (SHR). A typical small structure (residence or other) might have a sensible heat load of 40,000 Btuh and a latent heat load of 8000 Btuh. This would give a total load of 48,000 Btuh and a sensible heat ratio of

$$\text{SHR} = \frac{40,000 \text{ Btuh}}{48,000 \text{ Btuh}} = 83\%$$

Having determined the SHR, we can determine the temperature difference to use in our air quantity calculations. A high SHR means a small latent heat load, less required dehumidification and, therefore, a warmer coil and a smaller Δt. Similarly, a low SHR means a high latent load, more required de-humidification and, therefore, a colder indoor coil and a high Δt. As a guide, use the following figures:

Calculated SHR	Temperature Difference (Δt), F°
less than 0.8	19.6–21
0.8–0.85	18.1–19.5
above 0.85	17–18

6.15 Design Example—The Basic House

At this point, we have considered all the factors necessary to perform a year-round, all-air climate control design for a residence. We select for our first design example The Basic House. The architectural plan is found in Figure 3.32 (page 130). Design heat loads are also found in that illustration. Cooling loads, both sensible and latent, were calculated. We do not see any value in reproducing the calculation sheets here since the forms we used are proprietary. It is, therefore, very doubtful that you will use this particular form. Instead, the results are tabulated in Table 6.1. The total calculated latent load, assuming no mechanical ventilation, comes to 5522 Btuh.

Therefore,

Total cooling load = Sensible load + Latent load
= 24,487 Btuh + 5,522 Btuh
= 30,009 Btuh = 2½ tons

Table 6.1 Design Data for Heating/Cooling Loads, The Basic House

Space	Heat Loss, Btuh[a,b]	Sensible Heat Gain, Btuh	Cooling, cfm[b]
Living room	9300	6726	330
Dining room	4300	3408	167
BR #1	5200	4108	201
BR #2	7500	5020	246
Bath	4800	Negligible	—
Kitchen	4900	5225	256
Subtotal	34,000	24,487	1199
Basement	4900	Negligible	—
Total	38,900	24,487	1,200

[a] From Figure 3.32.

[b] cfm for heating = $\dfrac{38,900 \text{ Btuh}}{(1.08 \times 55\text{F°})}$ = 655 cfm

The sensible heat ratio is

$$\text{SHR} = \frac{\text{Sensible load}}{\text{Total load}} = \frac{24{,}487 \text{ Btuh}}{30{,}009 \text{ Btuh}} = 0.816$$

From the tabulation in the previous section we can find the temperature differential required, by interpolation:

SHR	Δt
0.8	18.1 F°
0.816	?
0.85	19.5 F°

$$\Delta t = 18.1\text{F°} + \frac{0.016}{0.05} \times (19.5 - 18.1)$$
$$= 18.1\text{F°} + 0.448$$
$$= 18.55\text{F°}$$

Therefore, the cooling (cfm) figures in the third column of Table 6.1 are all calculated by using the expression

$$\text{cfm} = \frac{\text{Sensible load}}{1.1\,(18.55)}$$
$$= \frac{\text{Sensible load in Btuh}}{20.41}$$

The total cfm required for cooling is 1200 (see Table 6.1).

Calculating the air flow required for heating, and using a temperature difference of 55F° (125F° supply air and 70°F return air), we have

$$\text{cfm for heating} = \frac{38{,}900 \text{ Btuh}}{1.08\,(55\text{F°})} = 655 \text{ cfm}$$

Duct sizes will, therefore, be calculated using the cooling air requirements as listed in Table 6.1, since the cooling air requirements are greater than the heating air requirements.

The procedure that we would use for a building of this type and size follows:

Step 1 Select a system type based on the building architecture.
Step 2 Locate and size supply registers and return grilles.
Step 3 Select and locate heating unit, cooling unit and thermostats.
Step 4 Make a duct layout including all sizing.

Following this procedure, we will develop the design, explaining each design decision.

Steps 1 and 2. In these steps, we select a system type and locate and size supply registers and return grilles.

Although a heat pump can be selected to serve this structure, we will go by a more conventional route, using a gas- or oil-fired furnace and a split air conditioner. The partial basement and crawl space are ideally suited to a basement furnace and duct system. The supply air terminals will be floor registers below exterior glass (windows) and adjacent to exterior doors. A single central return duct will be centrally located in this small house. We will equip it with high and low return grilles to receive return cooling and heating air, respectively. All doors will be undercut to permit passage of return air.

Since the return duct is large and close to the furnace blower, it will be furnished with a duct liner that will serve to reduce vibration, sound transmission and heat loss. All ducts in the unexcavated crawl space will be insulated. Although the basement has a calculated heat loss of 4900 Btuh (see Table 6.1), the furnace losses should keep it at a comfortable 65–68°F without additional heating. However, as a "safety net" a small register (8 × 5 in.) will be provided, tapped from one of the two main feeder ducts. Register and grille locations are shown in Figure 6.27.

We have chosen to use heavy-duty floor registers that deliver about 100–175 cfm each and to use multiple registers in each room rather than a single large register. See Figure 5.21 *(c)*. The reason for this is the need for good circulation and coverage, particularly in cooling mode, where floor registers are problematic. See Figure 5.33 *(b)*. In the heating mode, placement of registers below each window is ideal, since it will prevent annoying cold drafts at the floor level. See Figures 5.29 *(c)* and 5.33 *(a)*. The increased cost of multiple registers is, therefore, justified.

Air velocity in all ducts is kept considerably below the noise limits given in Tables 5.6 and 5.7. The cfm figures shown in Figure 6.27 at all registers are the final flow figures after balancing. All branch ducts are equipped with dampers in the duct for this purpose. It is important to remember that great precision in calculation of air quantities is not necessary. A common rule of thumb calls for 450–500 cfm per ton total air for cooling. For the 2½ tons required here (30,000 Btuh total load), this would come to 1125–1250 cfm, which matches closely the 1200 cfm calculated (Table 6.1). Note that this is almost double the 655 cfm required for heating. This indicates the requirement for a multispeed blower motor in the furnace.

Steps 3 and 4. In these steps, we select and locate the heating and cooling equipment and design the duct system.

DESIGN EXAMPLE—THE BASIC HOUSE / 361

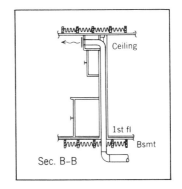

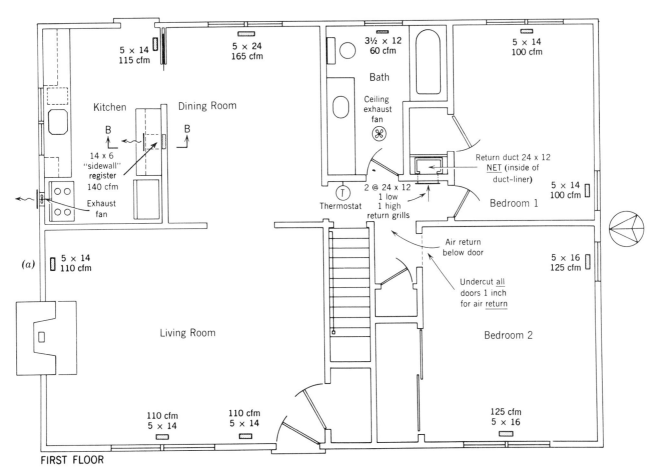

Figure 6.27 Layout of supply air registers and return air grilles for The Basic House plan.

As stated at the beginning of the design solution for The Basic House plan, we have chosen to use a warm air furnace for heating and a split air conditioner for cooling. The configuration used for cooling is a slab-on-grade condenser unit outside the building connected to an A-frame evaporator coil in the furnace plenum. See Figures 5.6 *(a)* and 6.28. The furnace is a gas-fired condensing unit of the type shown in Figures 5.12 and 5.13. Since these units have flue vent temperatures as low as 100–130°F, the vent is a small diameter PVC pipe extended through the roof or basement wall. The distance between the furnace and the vent opening can be as much as 50 ft with up to four 90° bends. We have taken advantage of this flue-less characteristic by placing the furnace in the middle of the building. This has a number of advantages:

- Reduced duct sizes
- Equal pressure drops to remote outlets
- Reduced pressure in the system
- Lower blower speed
- Reduced noise from the furnace

These advantages are particularly important in the cooling mode, which requires higher flow and static pressure than the heating mode. The smallest furnace shown in Figure 5.12 has an output rating of 45,000 Btuh. Since the building load is 40,000 Btuh, this furnace is oversized by 12%. Recall that furnaces may be oversized by no more than 25%. Therefore, this furnace is suitable for our design.

The outdoor condenser unit should be placed far enough away from the house so that its noise is not a nuisance. The unit is similar to the outdoor condensing unit shown in Figure 6.21*(a)*. The A-frame evaporator is similar to that shown schematically in Figure 5.6*(a)*. We see from the data in Figure 6.21*(b)* that the SCU10A 36 A-1 condensing unit with CAU indoor A-frame evaporator coil can supply the load. The duct system has been designed for low velocity, with a static head below 0.1 in. w.g. The duct sizing, pressure and velocity calculations are left as an exercise. See Problem 6.39.

One duct on the suction (return side) of the air system pulls in fresh air. A damper in a convenient place (near an access opening) can partially or entirely close this duct. It can be fully open for ventilation or fully closed for the greatest fuel economy. Balance of air flow between north and south ends of the house may be adjusted by the splitter damper where the main duct divides. Balancing the system is accomplished using dampers in all the branch ducts. Final adjustments can be made at the registers, which are equipped with opposed-blade dampers. Because of the large air flow differences between heating and cooling, damper settings for each season should be marked on the ducts. That will eliminate the time-consuming and highly technical job of system rebalancing.

6.16 Design Example—Mogensen House

We have used this actual structure as a design example previously. The architectural plans are found in Chapter 3 (Figures 3.35–3.39). A piping layout for a hydronic heating system is shown in Figure 3.40. An electric heating layout for upper and lower levels in shown in Figure 4.12. Now we are going to consider heating and cooling by use of a heat pump. First, however, we want to bring to your attention some of the considerations that preceded the HVAC design.

The Mogensen house was planned for economy of construction and operation. There is no basement, and attic space for HVAC equipment is severely limited. At the ridge, the height of the small partial attic is only 5 ft. This means that a central heating/cooling plant must use compact equipment. To reduce summer heat gain, the house is well insulated and equipped with double-pane glass windows. Recognizing that solar heat gain through windows is frequently the largest single heat gain component in a building, the architect has provided shading to reduce this load. Figure 3.35 (page 138) is a photograph taken on a summer afternoon. Notice that most of the western exposure glass is in full shade. These glass areas include the living room, family room, master bedroom and study. The only glass that receives the full impact of the afternoon summer sun is the row of small bedroom windows on the lower level.

It is sometimes a good decision to cool only part of a house. In the house we are now studying, the actual choice was to cool only the upper-level rooms. The lower level has only a very small heat gain. The east wall of that story is below grade against the cool earth. The north and south walls have no glass. The west glass (in the family room) is in shade. Finally, the windows and sliding glass doors can be opened to the cooling breezes from Long Island Sound.

Adding strength to the decision for cooling only the upper level rooms, was the planned occupancy and use of the house. Most of the family living is in the upper level. It was decided to place a heat pump in the attic since, as mentioned, space for

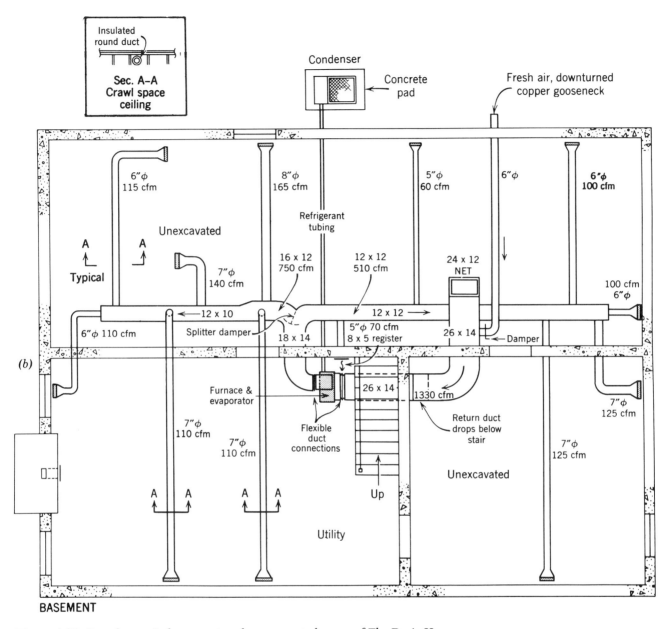

Figure 6.28 Duct layout in basement and unexcavated space of The Basic House.

mechanical equipment is at a premium. This location has the advantage of short duct runs to ceiling registers. Return air can be picked up at a few central locations. The heat pump will heat and cool only the upper level. The lower level will use electric baseboard heating and, as mentioned, will not be cooled. Since this design is an actual structure, we were able to photograph some of the HVAC system during construction. See Figures 6.29 and 6.30.

Based on the preceding design decisions, a suggested design procedure for the Mogensen house climate control system would be:

Figure 6.30 Warm or cool air enters all rooms at the ceiling level. Except in the living room, one-way throw, curved-blade registers deliver a flat layer of air into the room, inducing a secondary flow of room air up across the glass (see Section C of Figure 6.32) As shown, the flexible insulated air duct turns down to a metal adapter. A register will be fitted into this square opening. An opposed-blade damper above the curved blades can be adjusted to regulate the flow of air into the room.

Figure 6.29 Construction photographs, Mogensen house. *(a)* View of living room looking south. *(b)* Close-up of air distribution system. Three two-way throw, curved-blade registers will deliver a warm or cool air blanket in the region of the glass doors. Two double-deflection wall registers will deliver air horizontally to effect good circulation in the room. Flexible insulated ducts will be trimmed flush with wall before installation of the two wall registers. See Figure 6.32, Section B.

- Perform the required calculations to obtain the heat gains and heat losses for the upper level of the house.
- Calculate air quantities for all spaces, based on assumed comfort conditions and the required temperature differentials between supply and return air. Remember that, for heat pumps in heating mode, the temperature differential is much smaller than for a furnace. Typically, warm air is supplied at a temperature between 90–100°F. This results in a temperature difference (Δt) of 15–20 F°, assuming a return temperature of 75°F. Modern design frequently uses a lower room and return temperature, in the interest of economy and energy conservation. A return temperature of 68°F would give a temperature rise of 27 F° for 95°F entering air. This means that air quantities required for heating with heat pumps are much larger than those required by heating furnaces. These latter usually supply air at 125–140°F, giving a temperature rise (over 68°F return air) of 57 F° to 72°F).

- Select heat pump equipment.
- Select air outlet locations and types.
- Make a duct layout.
- Size ducts and registers.
- Consider ventilation requirements.

The following design considerations and procedures for the heat pump climate control of this building were evaluated.

(1) Due to the architecture and exposure of this structure, it was decided to use two zones. The zoning makes possible short duct runs and better control. The heat losses and gains for the rooms in each zone follow.

	Winter Heat Loss, Btuh	Summer Heat Gain, Btuh
Zone 1		
Living room	28,600	10,800
Hall/foyer	9,400	4,900
Dining room	8,000	9,000
Total	46,000	24,700
Zone 2		
Master bedroom	15,000	9,400
Master bath	2,800	1,600
Study	3,900	3,000
Powder room	1,500	300
Dressing rooms	—	600
Kitchen	5,100	9,600
Total	28,300	24,500

(2) For rooms where air quantities are radically different for the two operating modes (heating and cooling), motor-operated splitter/dampers are used. See Figure 5.19(b). (Winter/summer settings will be determined by air flow measurements during the balancing procedure after installation.) These rooms include the living room, dining room, master bedroom and kitchen. Air flow in other rooms can be adjusted by the occupant by using the opposing blade dampers installed at each register. Ducts to all rooms are sized for the larger of the heating/cooling air flow requirements. Duct dampers cannot be used here because all of the duct work is enclosed in the building wall and ceilings, making it inaccessible.

(3) Sizing the heat pumps to supply the heat load would oversize the cooling capacity to the point that it would not dehumidify satisfactorily. As a result, a smaller unit was used, equipped with a resistance heating element to pick up the additional heat load on very cold days.

(4) The indoor air handlers and coils are suspended in the attic. See Figures 6.22(c) and 6.32. They are connected to the two exterior slab-on-grade mounted outdoor units with insulated refrigeration lines. Access to these indoor units, for servicing, is available through removable ceiling panels. Removal of a unit is possible through an access door in the wall of the skylight shaft in the master bedroom.

(5) Registers and return grills are selected and located as shown on Figure 6.31. As already stated, all supply registers and two return grilles are equipped with opposed-blade dampers for seasonal changes and to suit occupant comfort. Thermostats for the two zones are located as shown.

(6) The duct layout is shown on Figure 6.32. All ducts are insulated to reduce vibration and heat loss. Return ducts are lined with a layer of acoustical, sound-absorbing material. This reduces the blower noise that reaches the rooms and doubles as thermal insulation. An overall equipment layout is shown in Figure 6.33.

(7) Thermostats T_1 and T_2 control the operation of the two heat pumps independently. Each unit will operate in heat or cool mode as required by the thermostat. All modern thermostats have a continuous run setting for blower operation. Continuous operation of the system blower, even after the heat pump is shut off, makes for temperature uniformity in the space and very gradual temperature changes over time. Too, continuous air motion, particularly in the cooling season, definitely adds to occupant comfort. The energy cost of continuous blower operation is not high and is generally considered to be well worth the expense.

(8) Figure 6.33 is an overall equipment layout showing both the indoor and outdoor heat pump components and the piping and wiring connections schematically.

(9) Sources of odor include the kitchen range, laundry dryer, laundry room, garage and bathrooms. By exhausting air from these spaces to the outdoors, these odors are reduced, and some humidity is eliminated. See Figure 6.34. The air that is drawn out of the house during seasons of heating or cooling must be replaced by outdoor air drawn in and conditioned by the central equipment. Figure 6.32 shows how this is done. In both zones, fresh air is admitted to the suction side of the blower coil unit. Its rate of flow may be adjusted by volume dampers in the fresh air duct near each unit.

(10) Figure 6.32 is an engineering layout. Before installation, the contractor is required to submit, for approval of the engineer and archi-

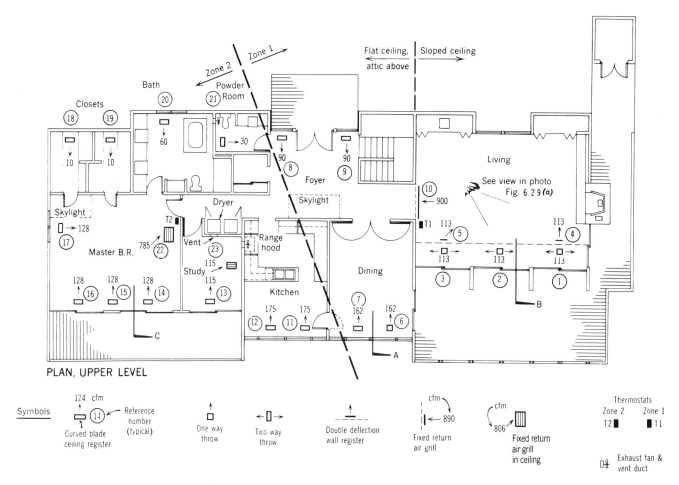

Figure 6.31 Air quantities and the layout of registers and grilles for the upper-level climate control system of the Mogensen house.

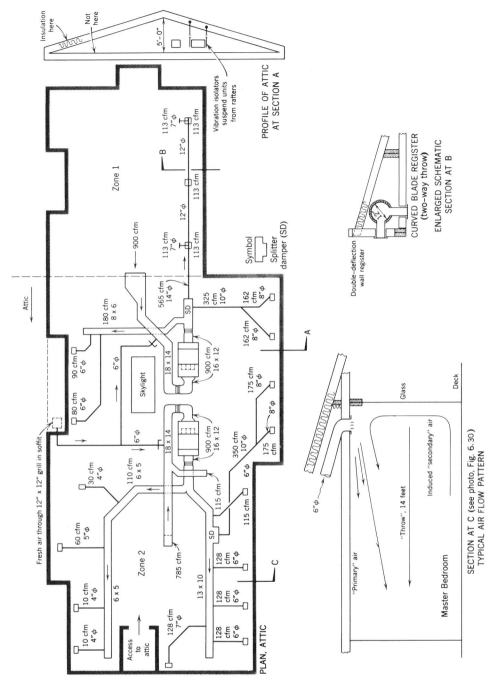

Figure 6.32 Equipment and duct layout, upper level, Mogensen house.

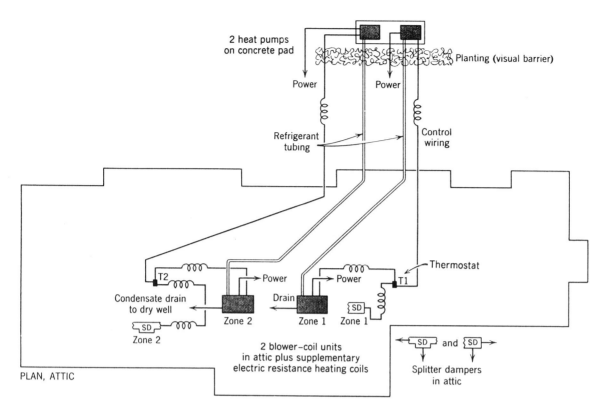

Figure 6.33 Electrical power and control requirement for the heating/cooling system, upper-level Mogensen house.

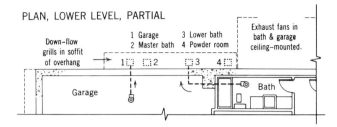

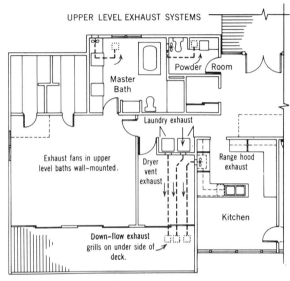

Figure 6.34 Exhaust ventilation. This design avoids units or ducts that would protrude through exterior walls or roofs. Interior ducts 10 × 3¼ in. between studs or joists carry all exhausted air to inconspicuous down-flow soffit grilles. Dryer vent is self-powered by the fan, which is part of the dryer unit.

tect, shop drawings of the duct system. Figure 6.35 shows how ductwork is presented in a well-prepared shop drawing.

6.17 Design Example—Light Industry Building

In the previous two (residential) design examples, we used a furnace/split air conditioner combination and a split heat pump, both with ductwork. In this example, we will use nonducted through-the-wall incremental units of the type shown in Figure 6.19. As explained in Section 6.9, there are two methods of supplying the HVAC requirements of a large space. One method is to use a central proces-

sor and a distribution system. That was the approach used in Examples 3.4 (page 142), 3.5 (page 151) and 5.14 (page 300).

The second method is to use individual package units (PTAC units), with individual local control. The advantages and disadvantages of this approach were listed in detail in Section 6.9 (page 353) and should be reviewed at this time. This design approach is often described as "decentralized" or "incremental." It is decentralized because each PTAC is separately controlled. It is incremental because each PTAC operates as a separate unit, unconnected to the other units in the building yet part of the overall building system. In recent years, designers have used central computer control of individual incremental units. This type of system combines the advantages of central control with the advantages of incremental package units.

Example 6.3 Design a decentralized, incremental heating-cooling system for the work area of the light industry building shown in Figures 3.48–3.50 (page 153).

Solution: Heat loss and heat gain calculations give these results:

Sensible heat gain—66,200 Btuh
Sensible heat loss—52,600 Btuh

Although latent heat gain was calculated, it is not shown because PTAC units do not give the designer any control over dehumidification. Units are selected on the basis of cooling capacity for sensible heat gain. The only control that a designer has with respect to dehumidification is by size of the cooling unit. Undersized units operate continuously and, therefore, dehumidify well. They are therefore chosen for high humidity areas. Oversized units cycle on and off. They dehumidify poorly but pick up load rapidly. They are therefore more applicable to dry (hot) climates.

The following tabulation shows typical data for PTAC units that are physically suitable.

Refer to Figure 3.53, which shows the location of six hydronic heating units for the same space. A designer could alternatively have elected to use PTAC units with hydronic heating coils and electric cooling. However, we have chosen to make this an all-electric design using package units.

The hydronic units shown in Figure 3.53 give good coverage. We will, therefore, use a similar layout for six PTAC units. They are shown in Figures 6.36 and 6.37. Using six units, the required

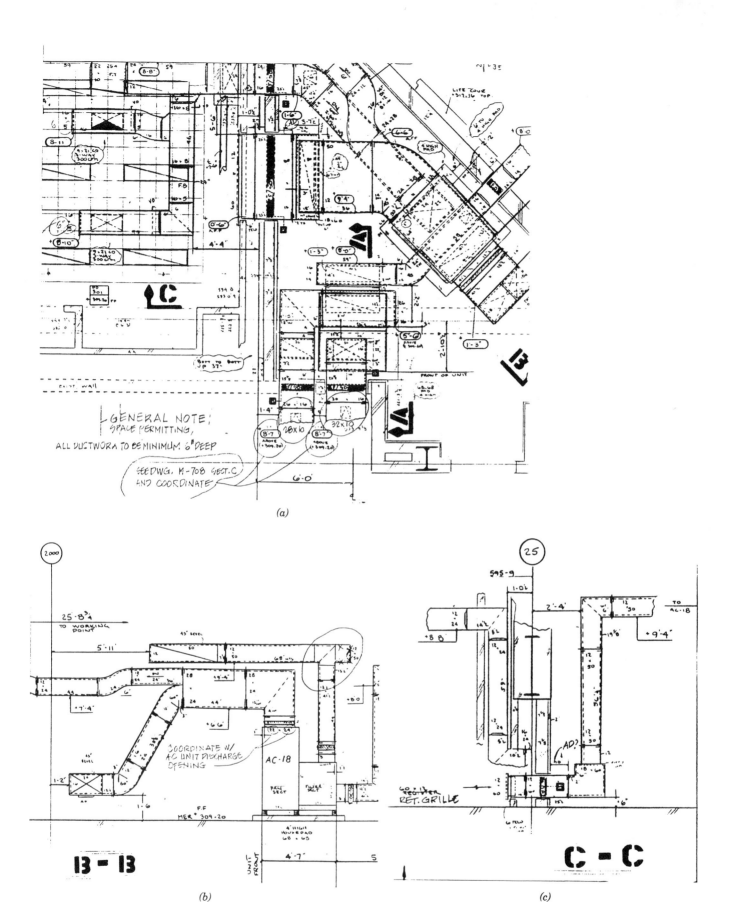

Figure 6.35 (a) Part of a large ductwork shop drawing. Since the ducts will be installed in layers, one above the other, vertical sections (b) and (c) are drawn to show details. (Courtesy of Cool Sheet Metal Co.)

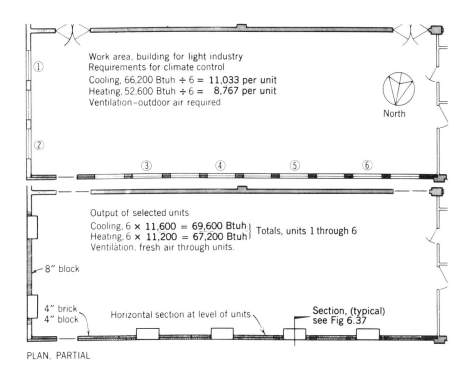

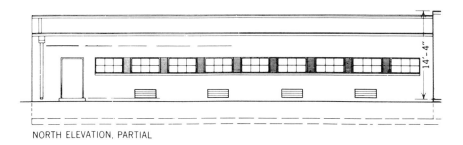

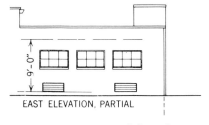

Figure 6.36 Layout of the solution to Example 6.3. The space is heated and cooled using through-the-wall all-electrical PTAC units.

ABC Manufacturing Company

Item	Model 1	Model 2	Model 3	Model 4
Cooling capacity, Btuh	6,600	8,700	11,600	14,200
EER	10.0	10.6	10.2	9.7
Heating capacity, Btuh	6,200	7,900	11,200	13,200
CFM—2-speed blower	250/180	300/220	380/275	420/300
CFM—ventilation	100	100	100	100

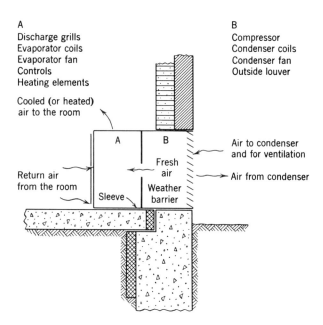

Figure 6.37 Through-the-wall PTAC unit installed.

heating and cooling capacity for each unit are:

Cooling:

$$\frac{\text{Total load}}{6} = \frac{66,200 \text{ Btuh}}{6} = 11,033 \text{ Btuh}$$

Heating:

$$\frac{\text{Total load}}{6} = \frac{52600 \text{ Btuh}}{6} = 8767 \text{ Btuh}$$

PTAC units are chosen rather than PTHP (heat pump) units, because the winter design temperature of 0°F would require large supplemental resistance heaters to compensate for the drop in heat pump output. This would raise the initial cost and the operating cost. An economic analysis performed by the engineer indicated that PTAC units are more economical in this design. (This type of analysis, which includes owning and operating life cycle costs, is not normally performed by technologists. For more information, consult the bibliography for a more technical book by this author.

The Model 3 PTAC unit is chosen from the preceding table. Its ratings are:

11,600 Btuh cooling
11,200 Btuh heating

This gives us these design margins (oversizing).

For cooling:

$$\frac{11,600}{11,033} = 1.05 \text{ or } 5\% \text{ oversize}$$

For heating:

$$\frac{11,200}{8767} = 1.28 \text{ or } 28\% \text{ oversize}$$

The cooling size is excellent, as the unit will run almost continuously and, therefore, provide the required dehumidification. The heating capacity is excessive. However, since the next smaller unit—Model 2—is too small, Model 3 would be used. It is frequently difficult to meet both the heating and cooling requirements without some oversizing or undersizing. In this case, since the majority of PTAC units with resistance heaters have high/low settings in addition to thermostatic control, the heating capacity oversize can be reduced by use of the low heat setting.

Key Terms

Having completed the study of this chapter, you should be familiar with the following key terms. If any appear unfamiliar or not entirely clear, you should review the section in which these terms appear. All key terms are listed in the index to assist you in locating the relevant text.

A-frame evaporator
Air-cooled condensers
Air handler
Air-to-air heat pump
Air-to-water heat pump
Blower evaporator
Central system chiller
Compressive refrigeration cycle
Condenser
Cooling temperature differential
Cooling tower
Coefficient of performance (COP)
Distributed system
Direct expansion (DX) coil
DX fan coil unit
Energy efficiency ratio (EER)
Evaporative condensers
Evaporator
Expansion valve
Fluorinated hydrocarbons
Forced draft
Freon
Heat pump air-handling unit
Heat pump balance point
Heat pump indoor unit
Heat sink
Heat source
Heating mode

Heating seasonal performance factor (HSPF)
Hybrid heat pump
Incremental package system
Incremental units
Induced draft
Integrated capacity
Make-up water
Mechanical draft tower
Natural draft tower
Outdoor heat pump
Packaged terminal air conditioner (PTAC)
Packaged terminal heat pump (PTHP)
Refrigerant flow controller
Rooftop units
Saturation temperature
Seasonal energy efficiency ratio (SEER)
Sensible heat ratio (SHR)
Slab-on-grade units
Split air conditioners
Split heat pump
Splitter/dampers
System balance point
Thermostatic expansion valve
Through-the-wall unit
Unitary heat pump
Water-cooled condensers
Water-to-air heat pump
Water-to-water heat pump

Supplementary Reading

B. Stein and J. Reynolds *Mechanical and Electrical Equipment for Buildings*, 8th ed., John Wiley & Sons, New York, 1992.

ASHRAE, American Society of Heating, Refrigeration and Air Conditioning Engineers
1791 Tullie Circle, N.E.
Atlanta, Ga. 30329 Tel. 404-636-8400

   1993 *Handbook—Fundamentals*

ACCA, Air Conditioning Contractors of America
1513 16th Street, N.W.
Washington, D.C. 20036

   Manual CS—Commercial Applications, Systems and Equipment, 1993
   Manual S—Residential Equipment Selection

SMACNA, Sheet Metal and Air Conditioning Contractors National Association, Inc.
8224 Old Courthouse Road
Tysons Corner, Vienna, Va. 22180

   HVAC Systems Applications, 1987

Problems

1. How many tons of refrigeration are required to cool a space with a sensible heat gain of 65,000 Btuh?
2. Will an undersized or oversized refrigeration unit provide better dehumidification? Why?
3. A gas-fired stowaway furnace in an attic provides full winter and summer air conditioning. It is served by an outdoor compressor-condenser.

    a. Name five connections to the attic unit other than electricity.
    b. Draw a sketch showing the units and their connections.

4. a. Is the fresh (outdoor) air supply duct connected to the return duct or to the supply air duct?
    b. Give the reason for your answer.
5. In a heat pump, the compressor operates whenever heating or cooling is needed.

    a. Where does evaporation of the refrigerant take place in summer: outdoors or indoors?
    b. Where does condensing of the refrigerant take place in winter: outdoors or indoors?

6. Name four locations in a residence from which it is desirable to have exhaust ventilation.
7. In a house with an air heating-cooling system, how is the air that is drawn out of the house by exhaust fans replaced?
8. When a heat pump is in use for heating and the outdoor temperature drops from 50 to 30°F, does the heat pump become less efficient or more efficient? Explain.
9. Explain briefly why COP applies to heat pumps and not to air conditioners.
10. a. Why is calculation of latent cooling load more important for nonresidential buildings than for residential ones?
    b. In the same climate, which requires more cooling per square foot, a residence or a department store? Why?
    c. Which require more heating? Why?
11. Define briefly the following terms:

    a. Compressive refrigeration cycle—condenser, evaporator
    b. Air-to-water heat pump
    c. Heat source
    d. Heat sink
    e. Evaporative cooling
    f. COP
    g. EER, SEER
    h. Balance point
    i. DX, dry expansion, direct expansion
    j. PTAC, PTHP
    k. Cooling tower

12. Does outside humidity affect the performance of a cooling tower? How? Why?
13. What is the function of a four-way flow switch in a heat pump?
14. Is a heat pump more efficient in the heating cycle or the cooling cycle? Explain.
15. Why is a heat pump called that? What is being pumped? Explain.
16. a. What is it that a condenser condenses? How?
    b. What is it that an evaporator evaporates? How?
17. How can a heat pump in its heating cycle deliver more (heat) energy than is taken from the electrical input? Doesn't this contradict the law of conservation of energy?
18. How much heat per pound of dry air is contained in air at 0°F? 32°F? 100°F?
19. In an air-to-air PTHP, what are the heat sources and heat sinks in the heating mode? Cooling mode?
20. What heat sources and sinks can be used with a water-coupled heat pump?
21. What is the EER of a heat pump that operates in heat mode with a COP of 3.2?
22. What limitations does the National Appliance Energy Conservation Act place on air conditioner and heat pump performance?
23. Why are defrosters necessary on heat pumps?
24. A store has a 0°F design condition winter heat load of 44,000 Btuh. Using the heating performance data of heat pump models 10A given in Figure 6.18(b), plot the building and heat pump curve and find the balance point.

    a. Which model heat pump will supply all the heating required (if any)?
    b. Will auxiliary heat be required? When?

25. Use the same heat pump characteristics as plotted in Problem 24. What is the balance point for a residence with a 10°F design condition heating load of 36 MBH. Which heat pump model is best? Explain.
26. List four types of heat pumps, with different heat source and heat sinks. Draw a block diagram of each and label the parts in heating and

cooling modes. Show the heat flow through the evaporator and condenser, with approximate temperatures. Justify all assumptions.

27. List five advantages and five disadvantages of using unitary, incremental units.
28. How does latent heat load affect the choice of an evaporator coil? (*Hint:* Read Section 6.22.)
29. The heat losses and gains for The Basic House design problem are based on a 15°F winter design condition, and 89°F DB, 75°F WB summer conditions.
    a. Assuming that heat loss varies linearly with outside temperature, recalculate the heat loss for all spaces using an winter design temperature of 0°F.
    b. In the interest of energy conservation, we are changing the summer inside design temperature from 75 to 78°F. Outside design conditions remain the same. Recalculate the cooling load for each space. Assume that cooling load, like heating load, is linearly proportional to the required temperature difference. (This is not strictly correct.)
    c. A conventional gas furnace will be used instead of a condensing unit. This requires moving the furnace (and the A-frame evaporator) to a location near the chimney. Redraw the duct layout of Figure 6.28, and recalculate all duct sizes. Be specific about all assumptions. Show pressures and velocities being used for each duct section.

7. Testing, Adjusting and Balancing (TAB)

All HVAC systems, regardless of size, require adjustment after installation. This does not mean that the installation was incorrect. It simply means that, even in a small simple system, all sorts of field adjustments must be made to achieve the design intention. These adjustments include motor speeds, pressure adjustments, valve settings, fuel supply control, liquid-level controls, temperature settings, damper positions and so on. This work is completely separate from the work of a field inspector. An inspector's task is to see to it that the system is installed according to plans and specifications. Once the installation is complete and approved by the inspection team, the work of testing, adjusting and balancing (abbreviated TAB) begins. Strictly speaking, the three portions of TAB are different from each other. *Testing* is the procedure that checks that equipment operates as it is supposed to. Motors turn, pumps deliver liquids, manual and automatic controls perform as required and so on. *Adjusting* is the work of setting and regulating variables such as flow, pressure, speed and temperature. *Balancing* is closely connected to adjusting because it is concerned with the flow quantities of air and water. The individual TAB processes are so interrelated that the whole procedure is simply referred to as *system balancing*.

The work is very technical and highly specialized. It requires a good hands-on working knowledge of all HVAC systems, an equally thorough familiarity with a whole range of TAB instruments plus, of course, an ability to understand HVAC plans. TAB specialists very often start their careers as HVAC technologists. It is for this reason that a chapter on TAB is included here. Study of this chapter will enable you to:

1. Understand the purpose and function of testing and balancing of HVAC systems.
2. Be familiar with the functioning and applica-

tion of instruments used in HVAC testing and balancing work.
3. Perform traverse measurements in ducts using Pitot probes and manometers.
4. Understand how to use various types of anemometers in air velocity measurements.
5. Calculate average air velocities and air flow in ducts.
6. Perform the necessary preparations for balancing an air system.
7. Accomplish the balancing of straightforward limited-size air systems and assist in the balancing of large complex air systems.
8. Make the many necessary preparations for balancing a hydronic system.
9. Balance a residential hydronic heating system and assist in balancing large hydronic heating/cooling systems.
10. Prepare the report forms containing all the balancing data for air and hydronic HVAC systems.

Instrumention

TAB work is possible only with adequate instrumentation. The physical quantities that require measurement include temperature, humidity, pressure, flow velocity and quantity, rotational speeds and electrical power and energy. For each of these physical quantities, instruments are available to suit the range and the physical accessibility of the quantity being measured. Before use, the requirement for calibration of each instrument should be checked. Some instruments maintain their accuracy for long periods of time or do not require calibration at all. Others require frequent recalibration. The manufacturers' instructions on this point should be carefully observed in order to ensure accurate measurements. There are so many instruments on the market today that a comprehensive survey would fill an entire volume. In particular, new electronic instruments appear almost daily. They offer such desirable and time-saving features as auto-ranging, digital readout, memories, and programmability. In the material that follows, we will review the basics of HVAC instrumentation, leaving the details of a specific instrument to the ability and intelligence of the technologist.

7.1 Temperature Measurement

a. Glass Tube Thermometer

See Figure 7.1(a). The simplest and most common type of thermometer is the glass tube design. All such units operate on the same principle. A reservoir at the base of the tube contains a liquid. The liquid expands and contracts according to the temperature of its surroundings—generally air or water—forcing liquid up through a calibrated glass tube. The liquid most frequently used is mercury. Mercury-filled glass tube thermometers have a useful temperature range of −40 to 1000°F, and the tubes are calibrated accordingly. Glass tube thermometers have the advantages of accuracy, indefinite life, no need for calibration and accuracies of up to 0.5% (or one-third of a scale division) depending on the scale.

Their principal disadvantage is that the entire bulb (liquid reservoir) must be immersed in the fluid whose temperature is being measured. If the fluid is a liquid, full immersion can be seen. If the fluid is air, the technician must be careful to shield the bulb from surrounding surfaces at substantially different temperatures. An ambient air temperature reading will be highly inaccurate if an unshielded reading is taken near a boiler or furnace. Since glass tube thermometers take a while to achieve their final reading, several readings should be taken, a few minutes apart, each one lasting several minutes. When the same reading occurs at least twice in succession it can be recorded as the correct temperature.

Most TAB technicians use a range of glass tube thermometers with different scale graduations and different physical size. Each type is useful for a limited range of applications. Some technicians use bulb-type glass tube thermometers to measure the temperature of a pipe by placing the bulb against the pipe and wrapping the two with insulating tape. This procedure should be avoided, because it is inaccurate. The line contact between the thermometer bulb and the pipe is inadequate for proper measurement. Furthermore, the insulated wrapping will prevent heat radiation from the pipe causing an artificially high reading. When surface temperature measurement is required, a special type of thermometer, called a *pyrometer*, should be used. This instrument is discussed in Section 7.1.d.

b. Dial Thermometer with Bimetallic Element

This type of thermometer uses a bimetallic element similar to that in a simple thermostat to measure temperature changes. When two metals that have different coefficients of expansion are joined together, a change in temperature causes the combination to bend or twist, depending on how they are joined. This motion is transmitted to a circular dial by a mechanical linkage. The dial is graduated in degrees of temperature. These thermometers are made in a wide variety of temperature ranges and physical designs. When used to measure liquid temperatures, the bimetallic element is mounted in a hollow metal stem attached to the dial. This stem is then immersed in the liquid whose temperature is to be measured. Domestic baking/meat thermometers are made in this design. When used to measure moving air temperature, the bimetallic element is installed inside the meter case, and so arranged that the air to be checked passes over the element. This type is illustrated in Figure 7.1*(b-1)*. A chart-recording unit of this design is shown in Figure 7.1*(b-2)*. These units have the advantages of ruggedness (unlike the glass tube type) and indefinite life. Although they should not require recalibration, they should be checked against a mercury glass tube unit periodically because the linkage can be damaged. This would result in an incorrect reading. These units have limited accuracy (±5–10%) and are useful for quick checks.

c. Capillary Tube Thermometer

See Figure 7.1*(c)*. One design of this type of thermometer uses a Bourdon tube, which is identical to that found in a pressure gauge. (See Figure 8.4, p. 410.) The tube is connected at one end to a long flexible fluid-filled capillary tube that ends in a relatively large sensing bulb. The other end is connected to a Bourdon gauge that is graduated in temperature degrees. The capillary tube and bulb are filled with a liquid or gas. Changes in temperature cause the fluid to expand or contract. This, is turn, changes the pressure in the Bourdon tube and, by means of a mechanical linkage, moves the dial pointer. This design is highly accurate (±½%), fairly rugged and needs infrequent calibration. Its principal advantage is the ability to read temperatures remotely. The standard length of capillary tubing is 6 ft. Units are available to measure temperature of −60 to 500°F in individual ranges of about 100°F.

Another design of capillary tube thermometer uses the liquid in the tubing to read directly in a graduated glass tube. This type is essentially the same as the glass tube thermometer described previously, except that the liquid reservoir, in the form of a sensing bulb, is connected to the glass tube by a long capillary tube. This permits remote sensing. Remote sensing is very useful when measuring temperatures at locations that are difficult or hazardous to get at. A recording unit of this type is illustrated in Figure 7.1*(c-2)*.

d. Pyrometers

These units are normally used to measure surface temperature of pipes, ducts and equipment. The unit's sensing element contains a bimetallic thermocouple. This thermocouple generates a small voltage, which is proportional to its temperature. This voltage can be measured by a millivoltmeter that is calibrated in degrees. The sensing element is connected to the instrument by wires and can, therefore, be remote at almost any distance. The great advantage of this type of instrument is that a single meter can be used to monitor as many thermocouples as desired by simply switching between the wires. Thermocouples are frequently permanently installed in equipment to permit continuous or periodic temperature checking (and alarm functions). A digital electronic pyrometer that can display temperature in either °F or °C at selectable precision is illustrated in Figure 7.1*(d)*. Pyrometers are highly accurate, require calibration once or twice a year and cover a huge range of temperatures, according to the type of thermocouple used. When used for surface temperature sensing of a pipe or duct, the manufacturer's directions should be carefully followed, because incorrect readings can result from improper probe use.

e. Thermal Anemometers

These devices, described in Section 7.3*(b)*, are primarily intended to measure air velocity. As a secondary function, some of these units also measure temperature. Two such units are illustrated in Figure 7.5*(c)* and *(d)*. They are mentioned here because of their auxilliary temperature measuring capability. However, because their principal function is to measure air velocity, they are discussed in detail in Section 7.3.

380 / TESTING, ADJUSTING AND BALANCING (TAB)

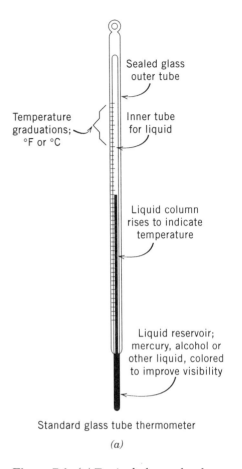

Standard glass tube thermometer
(a)

Figure 7.1 *(a)* Typical glass tube thermometer. They are available in a very wide range of physical sizes and temperature ranges and graduation precision.

(b-1)

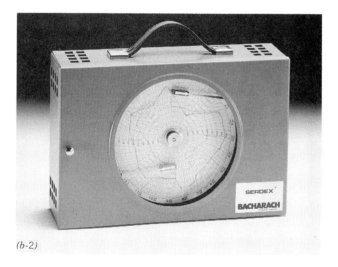

(b-2)

Figure 7.1 *(b-1)* Temperature/humidity indicator. Temperature of air passing through and around the unit is measured by a bimetallic coil-type sensing element within the unit. The dial is calibrated from 0 to 130°F. The meter also measures relative humidity with a membrane diaphragm that responds rapidly to humidity changes. The humidity range is 0–100% RH. *(b-2)* Chart recorder that measures temperature and humidity as described for *(b-1)* and records the measurements on a 6-in. diameter, 24-hr circular chart. (Photos courtesy of Bacharach, Inc.)

TEMPERATURE MEASUREMENT / 381

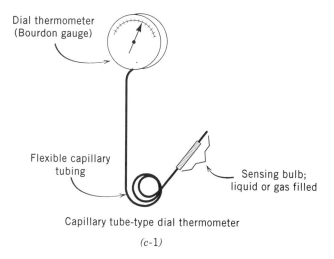

Capillary tube-type dial thermometer

(c-1)

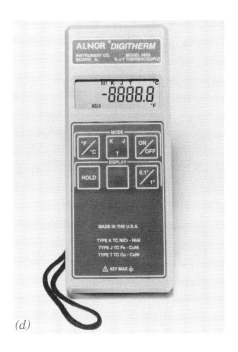

(d)

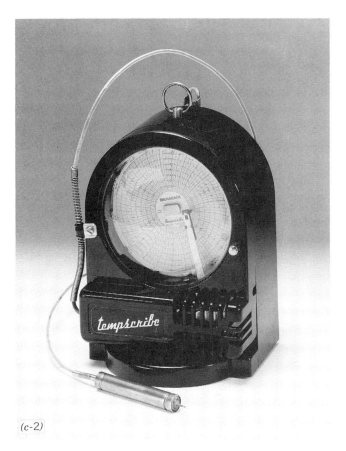

(c-2)

Figure 7.1 *(c-1)* Capillary tube-type dial thermometer. Expansion and contraction of the fluid in the sensing bulb causes a change in pressure in the Bourdon gauge. This pressure change causes movement of the dial pointer on a scale that is graduated in temperature degrees. *(c-2)* Capillary tube-type recording thermometer. The temperature-sensing bulb (in the lower foreground of the photo) senses temperature and transmits the signal through 6 ft of capillary tubing to the 24-hr or 7-day chart recorder. Temperature ranges are available in different models from −30 to 120°F and −35 to +50°C. *(d)* Modern digital electronic hand-held pyrometer, with a range of thermocouple probles and sensors (not shown). Primarily intended for surface temperature measurements, although usable for measuring immersion temperatures of gases as well. The illustrated unit is programmable, will hold and retain readings and will measure in a number of ranges over an extremely wide range with good accuracy. [*(c-2)* Courtesy of Bacharach, Inc. *(d)* Courtesy of Alnor Instruments Company.]

7.2 Pressure Measurement

Air and water are both fluids. Pressure and its measurement in water piping and vessels is discussed in Sections 8.5–8.7 (pp. 406–410). If you have already studied those sections, a review at this point would be useful. Otherwise, after studying this section, we would advise you to read those sections in order to appreciate both the similarities and the differences. In either case, you should review Section 5.4, which explains in detail the concept of air pressure in ducts and the use of manometers of various types to measure duct air pressure and air velocity.

a. Manometers

See Figure 7.2. The action of a manometer in measuring duct air pressure is shown in Figures 5.4 and 5.5 (p. 202). Because these pressures are low, the liquid used in an air pressure manometer is either colored water or oil. Mercury manometers are used to measure the much higher pressures in water lines. The simplest manometer is a U tube. See Figure 7.2(a). Its use is shown in Figure 5.4. Air pressure manometers of all types are frequently made of a transparent block of plastic with the manometer tubing cast directly into the block. This makes the instrument almost indestructible yet capable of excellent accuracy. See Figure 7.2(b).

The units are calibrated directly in inches of water. Since manometers have no moving parts, they maintain their accuracy without recalibration almost indefinitely. These manometers, also referred to as draft gauges, are standard in the industry. The inclined scale type shown in Figure 7.2(b) can be read to an accuracy of 0.03 in. w.g. Most units have an adjusting piston in the liquid tube that permits setting the liquid's meniscus (level) at the zero pressure line. The unit is set level on a stable surface and read directly. Connection to ducts is made with flexible tubing and a probe that is introduced into the duct being tested through a test hole.

b. Magnetic (Differential) Pressure Gauge

This gauge measures the difference in pressure between two sealed compartments. This difference causes a diaphragm between the two compartments to move. The motion is transmitted through a magnetic linkage to a dial pointer. The instrument is highly accurate, can measure air pressure differences down to 0.01 in w.g. and is fairly inexpensive. The usual pressure ranges of this meter are 0–0.5 and 0–1.0 in w.g. It is very useful in reading pressure differences across filters and between two points in a duct. The unit is maintenance-free and requires adjustment only of the pointer zero setting before use. A unit of this type is illustrated in Figure 7.2(c) along with schematic diagrams of typical applications.

7.3 Air Velocity Measurements

a. Pitot Tube and Manometer

The simplest and most common field technique for measuring air flow is by using a Pitot tube and a manometer. The theory behind this measurement is given in Section 5.4 and Figure 5.5 and is repeated briefly here for convenience. The total pressure in a duct, P_T is the sum of the static (spring) pressure P_S and the velocity pressure P_V. That is,

$$P_T = P_T + P_V$$

Since we also know that

$$P_V = (V/4005)^2 \qquad (5.6)$$

it follows that air velocity

$$V = 4005 \ \sqrt{P_V} \qquad (7.1)$$

Therefore, by measuring velocity pressure, we can calculate air velocity very accurately. The method is shown in Figure 5.5(c) using two probes: a duct wall type that measures static pressure and a probe tip that measures total pressure. Connecting them as shown there to measure the differential pressure gives the velocity pressure directly since

$$P_V = P_T - P_S$$

The same differential pressure measurement can be made with a single hole in a duct or pipe, by using a Pitot tube. See Figure 7.3(a). A Pitot tube is simply two concentric tubes. The inner tube has an opening that is placed facing the air stream. It measures total pressure P_T as shown in Figures 5.5(b) and 7.3(b). The outer tube has holes drilled into the circumference of the tube. These holes are, therefore, at right angles to the air stream. They measure static pressure P_S as shown in Figures 5.5(a) and 7.3(b). When the two output points are connected to opposite ends of a manometer, it reads the difference between total and static pressure, that is, velocity pressure P_V. See Figures 5.5(c)

AIR VELOCITY MEASUREMENTS / 383

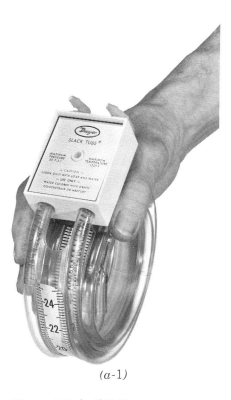

(a-1)

Figure 7.2 *(a-1)* U-Type manometer made with flexible tubing for carrying convenience. The tube can be filled with colored water or mercury. Various models in this design have a pressure measuring range up to 60 in. w.g. (Courtesy of Dwyer Instruments, Inc.)

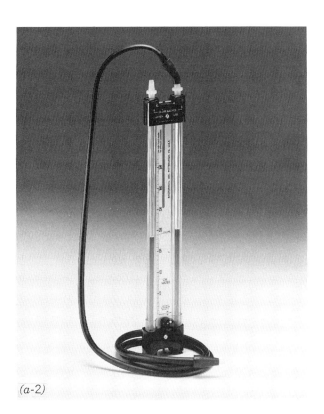

(a-2)

Figure 7.2 *(a-2)* Standard U-tube manometer calibrated in centimeters of water for metric calculations. (Courtesy of Bacharach, Inc.)

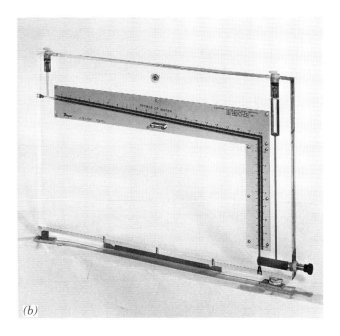

(b)

Figure 7.2 *(b)* Inclined/vertical manometer. The inclined portion is used for pressures up to 2 in. of water and can be read to an accuracy of ± ¼%. The vertical section is graduated from 2 to 10 in. w.g. The entire manometer is cast into a thick block of clear acrylic plastic measuring 16 × 25 in. (Courtesy of Dwyer Instruments, Inc.)

(c-1)

Figure 7.2 *(c-1)* Magnetic differential pressure meter measures the difference in pressure between lines connected to its two outlets. When the low pressure connection is left open as in *(c-2)*, the meter will read gauge pressure, that is, pressure above atmospheric pressure. The inclined manometer shown in *(c-2)* is simply another means of measuring the same gauge pressure. *(c-3)* Velocity pressure is measured with a magnetic gauge by connecting the center tube of a Pitot probe to the high pressure inlet and the outer tube to the low pressure inlet. The difference is velocity pressure. The same measurement technique using an inclined tube manometer is shown for information only. See also Figure 7.4. *(c-4)* A differential pressure gauge can be used to measure directly the pressure drop across a filter, when set up as shown. Any appreciable change in this reading indicates a change in the condition of the filter. *(c-5)* Closure of the duct damper will cause an immediate increase in the upstream pressure and will indicate on the meter. This connection is useful to monitor and check the operation of fire dampers. (Courtesy of Dwyer Instruments, Inc.)

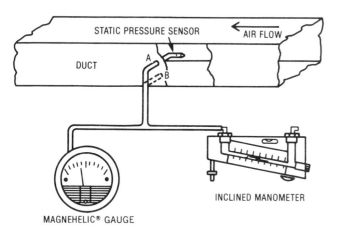

Measuring static pressure in an air duct or plenum.

(c-2)

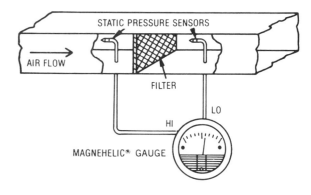

Dwyer differential pressure gauges used to monitor filter condition.

(c-4)

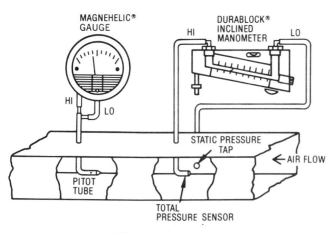

Differential pressure gauge used to measure velocity pressure.

(c-3)

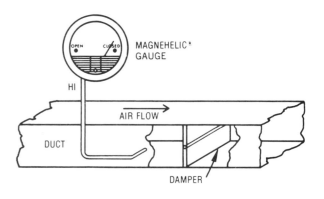

Differential pressure gauge used for pressure sensing.

(c-5)

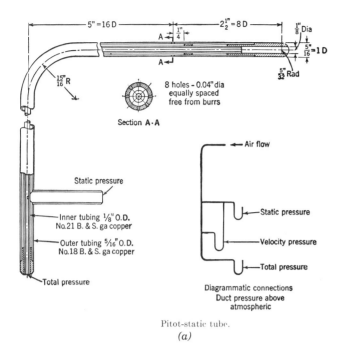

Figure 7.3 *(a)* Construction details of a standard Pitot tube. The tube consists of two concentric tubes. The inner tube terminates in a "nose" opening, which is placed facing into the air stream. This tube, therefore, measures the total air pressure in a duct. The outer tube has eight holes spaced around the circumference at right angles to the inner tube and, therefore, at right angles to the airflow direction. This tube, therefore, measures the flow static pressure. Takeoff points at the opposite end of the tube provide for connection of manometers. (From Severns and Fellows, *Air Conditioning and Refrigeration*, 1962, © John Wiley & Sons, reprinted by permission of John Wiley & Sons.)

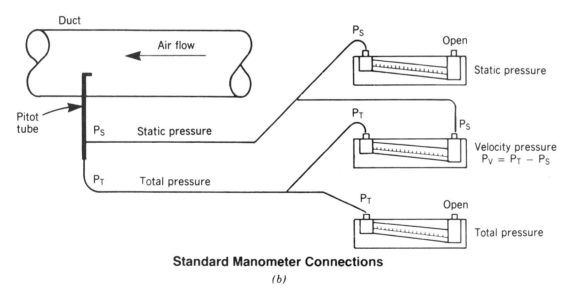

Figure 7.3 *(b)* Standard manometer connections to a Pitot tube probe will give static, velocity and total pressure readings as shown. Use of three manometers simultaneously permits measuring all three pressures without reconnecting. Elimination of the center manometer will still permit measurement of all three pressure quantities, since $P_V = P_T - P_S$. If only velocity pressure is required, a single manometer connected as shown for the center manometer will provide the desired measurement. (Reproduced with permission from SMACNA HVAC Systems, Testing, Adjusting and Balancing, 1983.)

and 7.4. Although only one ordinary manometer is required to take all readings, in practice two are frequently used to save time, since a series of readings must be taken over the cross section of a duct. The third reading is easily obtained from the equation $P_T = P_S + P_V$. A manometer graduated in velocity as well as pressure is shown in Figure 7.4. It is connected as shown for the center manometer in Figure 7.3(b).

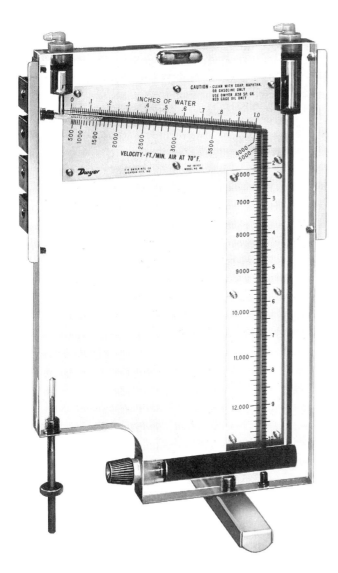

Figure 7.4 Combination inclined/vertical manometer and air velocity meter. The unit, which measures 16½ × 11 in., is encased in a block of clear acrylic plastic, making it ideal for field work. When used with a Pitot tube and connected as shown in Figure 7.3(b), it will measure pressure from 0 to 10 in. w.g. and air velocity from 400 to 12,000 fpm. (Courtesy of Dwyer Instrument, Inc.)

The technique for taking the necessary traverse measurements (over the cross section) is explained in Section 7.4.a.

b. Anemometers

1. Rotating Anemometers. In contrast to the previously detailed pressure measurement technique, from which velocity is calculated, an anemometer measures air current velocity directly. A wind sock and a rotating cup-type wind gauge are both simple anemometers. A rotating vane anemometer operates on the same principle as a child's pinwheel. Air passing through the unit causes the propeller blades to turn at a speed proportional to air velocity. The blades are connected through a gear train to a dial that is calibrated in feet. It measures the length of an imaginary tube of air passing through the blades. The faster the air blows, the faster the blades spin and the longer this imaginary cylinder of air. By timing the flow, for a ½ minute or a minute, the air velocity is easily calculated.

For instance, a dial reading of 600 ft timed in 1 min means an air velocity of 600 ft/min. Similarly, a dial reading of 600 ft in ½ min means a velocity of 1200 fpm, since

$$\text{Velocity} = \frac{\text{Distance}}{\text{Time}}$$

Therefore,

$$V = \frac{600 \text{ ft}}{0.5 \text{ min}} = 1200 \text{ ft/min or } 1200 \text{ fpm}$$

Mechanically linked anemometers are not accurate at air velocities below 200 fpm. Their useful range is 200–2000 fpm. Units are made in 3-, 4- and 6-in. diameters. A modern self-timing unit that reads velocity directly is shown in Figure 7.5(a). When used to find the air velocity over a large surface such as a coil or filter, the unit must be moved (traversed) over the entire surface, because air velocities vary over these surfaces. Although these anemometers do not require recalibration, older units are frequently used with a calibration curve that compensates for gear train drag, particularly at low velocities.

2. Deflecting Vane Anemometer. A deflecting vane anemometer is illustrated in Figure 7.5(b). This unit contains a movable vane that is deflected (pushed) by the air current. The amount of deflection is proportional to the air speed. The vane is connected to a pointer that reads air speed directly on its scale. This device is not highly accurate, but

(a)

Figure 7.5 *(a)* A modern rotating vane anemometer with built-in timer that averages air flow every few seconds. This permits direct reading of average air velocity without external timer and calculator. It also permits rapid sweeping of large area grilles and registers. The rotating vane shown transmits its rotational speed electrically. This maintains accuracy over the entire range and eliminates the need for calibration curves at low air speeds. The illustrated unit has a useful range of 50–6000 fpm. It can also be used to measure volumetric flow rate. (Courtesy of Alnor Instrument Company.)

(b)

Figure 7.5 *(b)* Deflecting vane anemometer. The unit is held directly in the air stream. Air pressure causes the pivoted vane in the meter to deflect, moving the pointer over the meter face. The amount of deflection, which is proportional to air velocity, is read directly on the meter scale, which is calibrated and marked in air velocity. (Courtesy of Alnor Instrument Company.)

388 / TESTING, ADJUSTING AND BALANCING (TAB)

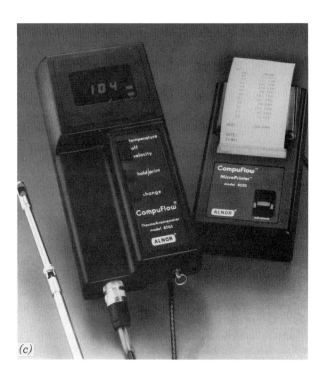

Figure 7.5 *(c)* Modern digital thermal anemometer and associated microprinter. The unit is autoranging with a total velocity range of 20–3000 fpm. Accuracy is ±3% or better on all scales. Automatic averaging of readings permits rapid scanning of grilles, registers and other surfaces with variable air volumes. The unit is also arranged to read air temperature over a range of 0–70°C (32–158°F). (Courtesy of Alnor Instrument Company.)

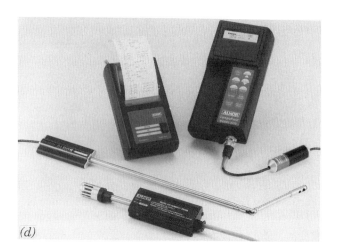

Figure 7.5 *(d)* This thermal anemometer is a multifunction instrument capable of measuring, storing and printing (with the illustrated printer) air velocity, volumetric flow, temperature and relative humidity, by use of different probes. The air velocity probe gives the unit a range of 20–6000 fpm and 0–50°C (32–122°F) in various ranges and accuracies. The RH probe measures relative humidity from 0 to 100% and temperatures from 0 to 60°C (32–140°F). (Courtesy of Alnor Instrument Company.)

it will quickly give an approximate air velocity reading. When used to measure air velocity over a large surface, a "profile" or traverse must be made, and the readings, averaged. It is most useful for measuring velocity over a small (3-in. square) specific area. Meters are available in single and multiple ranges from 0 to 3000 fpm.

3. Thermal Anemometers. A third type of anemometer operates on the principle that the electrical resistance of a hot wire will change with temperature. If air is passed over such a wire, it will be cooled in proportion to air velocity. This instrument is called, logically, a *hot wire anemometer*. The wire resistance is measured in an extremely sensitive electrical bridge circuit, making the instrument highly sensitive although not very accurate (±10%). It is, therefore, particularly useful in detecting low velocity air movement such as leaks or drafts.

Modern units that operate on the principle of measurement of the cooling effect of moving air are known by the more general name of *thermal anemometers*. These units may use the traditional "hot wire," known today as a resistance temperature device (RTD). Alternatively, they can use a thermistor sensor or sensitive thermocouple junction. The principle of operation, however, remains unchanged. Two modern, electronic, digital readout units of this type are shown in Figure 7.5(c) and (d). In addition to their primary function as anemometers, they are also usable as thermometers to measure the air current temperature.

4. Velometer. See Figure 7.6. A velometer is an anemometer that operates on the principle of a swinging vane. Sampled air passes through a Pitot tube-type circular tunnel in which the vane is mounted. The vane motion and the corresponding pointer motion are proportional to air velocity. Unlike many of the instruments discussed, and especially modern digital units, the velometer is a purely mechanical instrument. Nevertheless, it is highly accurate and is very widely used for TAB work. The meter has scales of 0–300, 0–1250, 0–2500, 0–5000 and 0–10,000 fpm. Three probes are

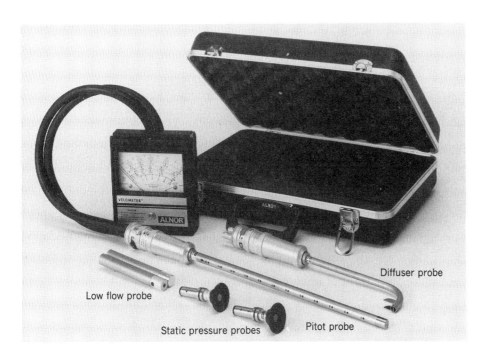

Figure 7.6 Velometer kit includes low flow probe for velocities up to 300 fpm, Pitot probe for measuring air velocities in ducts; diffuser probe for measuring air velocity at diffusers, registers and grilles; and two static pressure probes. These components are sufficient to perform a complete balancing procedure for an all-air system. (Courtesy of Alnor Instrument Company.)

provided with the meter: a low velocity probe useful for measuring in-room terminal velocities, a Pitot tube-type probe for medium air velocities as in ducts and a high velocity probe. The instrument will also measure static pressure when used with a static pressure probe. The velometer should have periodic accuracy checks, although it maintains its calibration for extended periods, depending on usage. Accuracy of readings depends on the scale and is normally better than ±2%. Table 7.1 summarizes the uses and characteristics of anemometers commonly used in HVAC work.

All the instruments that we have discussed for measuring air velocity do so over a very limited area. Pitot tubes and velometers measure velocity at a point. Vane-type anemometers measure velocity over an area equal to their face area. This varies between 3 and 30 in.^{2}. Furthermore, the air velocity in a duct or at a register is not constant. It varies over the cross section of a duct and over the face of a register or grille. Therefore, what is needed is a measuring technique, using the instruments just discussed, that will give up an average air velocity over the entire area of the item being measured. Whether the averaging is done manually by the TAB technique or automatically by the instrument is not important. These measuring procedures are discussed in Section 7.4.

Measurements

7.4 Velocity Measurement Techniques

Air velocity in a duct varies with position in a duct. It is slower near the duct walls because of friction (drag). Therefore, to obtain average air flow velocity, it is necessary to perform a traverse or profile over the cross section of the duct. In order to make the air flow as linear as possible (without turbulence) and to make the measurements as accurate as possible, a good TAB technician will perform the following before taking any readings:

- Insert an egg-crate type of flow straightener (or other type) into the duct at least five duct diameters (or duct diagonals for rectangular duct) upstream of the Pitot tube entry.
- Perform the test in a straight section of duct, as far as possible from elbows, fittings of all types, size changes and the like. Minimum distances should be eight diameters upstream and two diameters downstream from the Pitot tube.

Table 7.1 Anemometers

Type	Use and Characteristics	Calibration	Accuracy
Manometer U, vertical, inclined	Use with probes and Pitot tube to measure total, static and velocity pressures in ducts and across filters, coils, etc. Very rugged.	Zero adjustment only	±1–5%
Rotating vane	Supply and exhaust air velocity and flow measurement. Also useful for terminal device face velocity. Simple and rugged.	Periodic accuracy check	±3–10%
Deflecting vane	Use for measuring face velocity of terminal devices, grilles, registers. Simple to use, rugged.	Frequent accuracy check	±5–10%
Hot-wire	Use to measure low and very low air currents such as room circulation and drafts. Very sensitive. Requires careful use.	Zero adjustment; periodic accuracy check	±2–5%
Velometer	All types of air motion measurements; duct (with probe), face velocity, supply and exhaust air velocity. Requires careful use, in accordance with manufacturer's recommendations.	Periodic accuracy checks are recommended	±2% depending on range

- Duct diameter should be at least 30 times the diameter of the Pitot tube.
- Minimum duct dimension for a rectangular duct should be at least 30 times the diameter of the Pitot tube.

a. Rectangular Duct Traverse

If the minimum duct dimension is not at least 8 in., use Figure 7.7(a). For larger ducts, use Figure 7.7(b). Divide the cross section evenly into rectangular areas, not less than 16 and not more than 64. Minimum dimensions of a single test area should be 3 in. square. (In small ducts such as 10 × 14 in., it will be necessary to use a 2½-in. dimension for one side of a test rectangle in order to have a minimum of 16 readings). Position the Pitot tube carefully at the center of each test rectangle and take a velocity pressure reading. Number each test area and record the readings on a chart or on a numbered tabulation. Do not average the pressure readings. Convert all pressure readings to velocity, and then take the average. This is then the average air velocity in the duct.

If a duct air flow straightener cannot be used or if a fitting is close by, considerable turbulence will be present in the air stream. This may result in negative readings in the traverse. Record these readings as zero velocity, but use the total number of divisions to find the average. An example should make the measurement method clear.

Example 7.1 A traverse of a 10 × 18-in. duct gives the pressure readings shown in Figure 7.8. Find the average air velocity in the duct.

Solution: The pressures are tabulated, and the velocity in each test area is calculated using the expression

$$V = 4005 \sqrt{P_V}$$

where
V is air velocity in feet per minute and
P_V is velocity pressure in inches of water (in. w.g.).

This formula holds true only for dry air at 0.075 lb/ft^3, at 70°F and 29.92 in. of mercury barometric pressure. Correction factors for other conditions, particularly humidity and altitude (atmospheric pressure), can be found in the manufacturer's literature that accompanies the Pitot tube and the manometer. After calculating all velocities, an arithmetic average is taken. This then is the duct air velocity. It is shown on Figure 7.8.

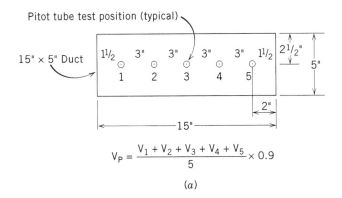

$$V_P = \frac{V_1 + V_2 + V_3 + V_4 + V_5}{5} \times 0.9$$

(a)

Figure 7.7 (a) For narrow ducts, it is sufficient to make a linear traverse with a Pitot tube, along the center line. Each position should represent the same proportion of total duct area. In this case the area is 3 × 5 in. The calculated average velocity must be multiplied by an arbitrary factor of about 0.9, to compensate for lower air velocity at the duct walls.

$$V_P = \frac{V_1 + V_2 + \ldots V_n}{n}$$

(b)

Figure 7.7 (b) Rectangular ducts are divided into 16–64 equal areas for a Pitot tube traverse. The average velocity is the arithmetic average of the individual area velocities. Accuracy increases as the size of individual rectangles decreases. Sides of measurement rectangles should be between 2½ and 6 in. Rectangles should be as nearly square as possible. (See also Appendix D, Form D.7.)

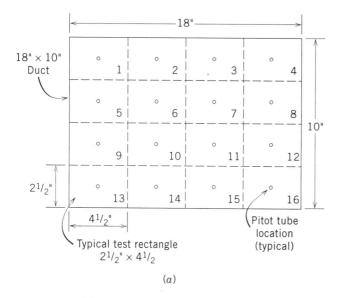

Figure 7.8 *(a)* Solution to Example 7.1. *(a)* The 18 × 10-in. duct is divided into 16 equal area sections, each measuring 2½ × 4½ in. A measurement of velocity pressure is taken with a Pitot tube at the center of each rectangle and recorded in table *(b)*.

Rectangle position	Pressure (Pv), in. w.g.	Velocity, fpm
1	0.027	658
2	0.031	705
3	0.033	728
4	0.027	658
5	0.032	716
6	0.041	811
7	0.043	830
8	0.032	716
9	0.033	728
10	0.043	830
11	0.041	811
12	0.033	728
13	0.027	658
14	0.031	705
15	0.033	728
16	0.027	658
Total		11,558

$$\text{AVG V} = \frac{11558}{16} = 722 \text{ fpm}$$

(b)

Figure 7.8 *(b)* The tabulated pressures are converted to velocity using the relationship given in the text. The 16 velocities are then averaged arithmetically to give the overall average duct air velocity.

Manufacturers also publish tables and distribute slide rules that will perform the required calculation and make any corrections for air at other conditions. One such curve for dry air at 70°F is given in Figure 7.9.

b. Circular Duct Traverse

A traverse in a circular duct is done following the same principles. Readings are taken on two diameters, at right angles to each other. Since we want each reading to represent the same (annular) area, the test points get closer together as they proceed from the center outward. In very small ducts, say 3–4 in., a single reading at the duct center, multiplied by 0.9 to account for low peripheral velocity, will give a usable velocity figure. In ducts from 6 to 9 in., take six readings across. For ducts 10 and 12 in. in diameter, use eight readings. For all larger diameters, use ten readings. The positioning of Pitot tube points for these ducts is shown in Figure 7.10.

c. Face Velocity of a Register

Since the velocity of air exiting from a register is not uniform over the register face area, a traverse of some type must be made, and the readings, averaged. When using a vane type anemometer, it is placed against the face of the register and covers a certain area depending on its size. This procedure should be repeated over the entire face of the register, taking care not to measure the same face area twice. The resultant arithmetic average is usable as the device's face velocity. When using a point-type anemometer such as a velometer, a traverse of the type shown in Figure 7.8 should be made. No less than 12 readings should be taken. For large registers, up to 48 point readings can be taken and averaged. Return grill air flow is generally more uniform over its face area, and a smaller number of measurements is possible.

7.5 Air Flow Measurement

Field measurement of air flow can be accomplished by two methods: direct measurement and calculation.

a. Direct Measurement

See Figure 7.11. The illustrated device is used by placing it over a ceiling or wall register or grille so

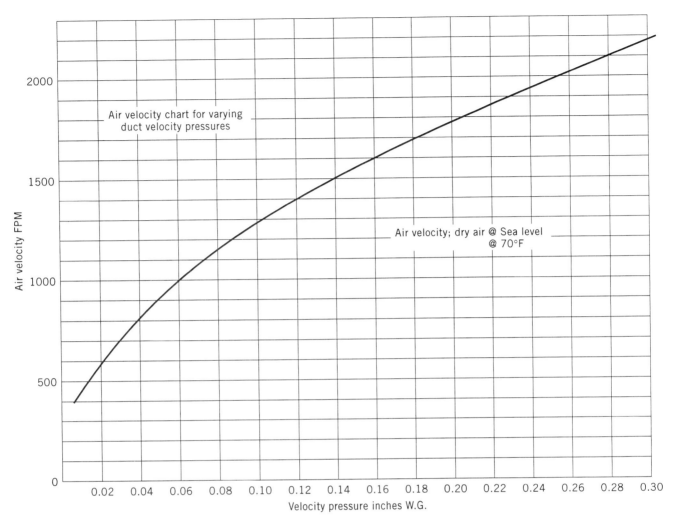

Figure 7.9 Instead of calculating air velocity using the formula given in the text, the velocity can be picked off the chart directly. The values shown are for dry air at sea level and 70°F.

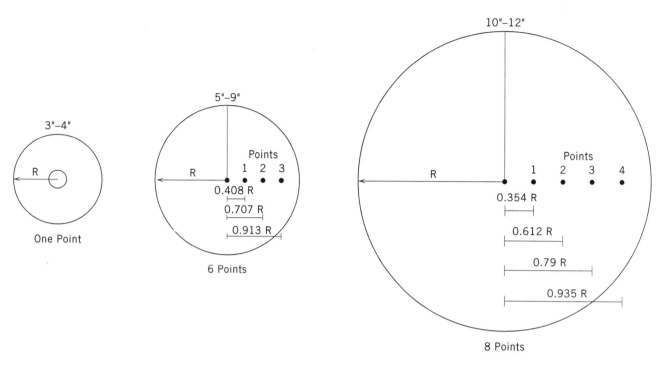

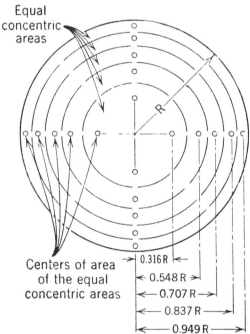

Locations of a Pitot tube for a 10–point duct traverse

Figure 7.10 Traverse points on round ducts. For 3- and 4-in. ducts, a single measurement in the center will give a satisfactory velocity when multiplied by 0.9. For 5- to 9-in. ducts, two 6-point traverses at right angles (one horizontal and one vertical) are required. For 10- and 12-in. diameter ducts, use two 8-point traverses; for larger ducts, use two 10-point traverses on diameters at right angles to each other. The spacing shown for test points will ensure that each test point represents the same percentage of the total cross-sectional area. (Ten-point traverse diagram from Severns and Fellows, *Air Conditioning and Refrigeration*, 1962, © John Wiley & Sons, reprinted by permission of John Wiley & Sons.)

Figure 7.11 Flow-measuring hood. This device consists of a hood that covers the supply or return terminal plus an instrument base that contains a modified anemometer. The anemometer performs an automatic traverse over the air channelled through the base and reads directly in cfm. The illustrated unit is available with a range of hood openings. It measures flow in four ranges, up to 2000 cfm, with an accuracy of ± 3% of full range. (Courtesy of Alnor Instrument Company.)

that the entire supply or return terminal is covered. Air flow is channeled through the base that is instrumented with a modified anemometer. The anemometer samples air velocity at 16 points over its area, determines average velocity and converts this to air flow quantity, which is then indicated on the instrument's meter. In effect, this device performs automatically the traverse that is described in Section 7.4.c. These flow-measuring hoods have several limitations.

1. They lose accuracy as register velocities increase and should not be used for velocities over 2000 fpm.
2. The hood and its instrumented base must be held manually over the register or grille. Care must be taken that the entire air supply device is covered, with no leakage. This is frequently difficult when measuring large area devices because of the bulk and weight of the instrument. This is particularly true when measuring air flow from ceiling diffusers and air-supply lighting troffers.

b. Calculated Flow

This technique relies on the well-known relationship

$$Q = AV$$

where
 Q is flow in cubic feet per minute,
 A is area in square feet and
 V is air velocity in feet per minute.

In Section 7.4, we discussed the instruments for field measurement of air velocity. Once velocity is known, the preceding equation can be used to determine flow. As already explained, the air velocity figure to use in this equation is the average velocity in the duct or over the face area of a register. For area A, one uses the cross-sectional area of a duct or the net face area of a register. Net area is listed as such in the manufacturer's catalog. Alternately, register manufacturers will use a constant K to represent the ratio between net and gross area of a register or grille.

c. Other Methods

A third method of flow measurement for both air and water involves the use of a sharp edge orifice plate or a smooth Venturi tube, placed into a duct or pipe. Pressure measurements made on both sides of either of these devices can be related to flow by a series of calculations and graphic plots. These are complex, advanced techniques that are beyond our scope here. Refer to the bibliography at the end of this chapter for more information.

Balancing Procedures

7.6 Preparation for Balancing an Air System

Before starting a TAB procedure, a number of preliminary steps are advisable. They will help to make the actual TAB work smooth, rapid, accurate and efficient. They are:

(a) Obtain a complete set of as-built HVAC drawings. These will be either contractor prepared field drawings or as-built-corrected contract drawings. In addition, shop drawings for equipment and ductwork must be readily available. If drawings for any part of the system do not show as-built conditions, prepare a simple single-line drawing showing all equipment and outlets.

(b) Mark on these drawings design air velocities and flow rates for each duct and outlet.

(c) At each fan or blower, mark design cfm, rpm, pressure and motor data including running current. Show speed controls and interlocking.

(d) For each filter, show type, cfm, pressure loss, area and air flow.

(e) For coils, show pressure drop, cfm, area, temperatures and capacity.

(f) Show location and type of all dampers.

(g) Record any special equipment information that will be checked during the TAB procedure.

(h) Prepare TAB report forms for recording test data. (A few sample forms are given in Appendix D.)

(i) Select the instruments that will be needed for all tests.

(j) Mark on the drawings where all measurements will be taken. If special access fittings are required, such as those for Pitot tubes, make sure that they are in place.

(k) Check with the field inspector that all systems are operative including all controls. The field inspector should also have the required data on all damper positions. If not, these must be ascertained before any TAB work can begin.

(l) Coordinate the TAB work with the contractor. It is necessary to have a contractor's representative available during TAB work. A TAB technologist is authorized to perform testing and balancing only. Any procedures, work or changes required to accomplish this TAB work must be performed by the HVAC contractor. This includes starting and operating all systems in all the design modes for which they are intended. Actual operation of the equipment by the TAB technologist can create problems of responsibility for malfunctioning. This is because TAB work is almost always performed before the system is turned over to the owner, that is, while the contractor is still fully responsible.

7.7 Balancing an Air System

The actual balancing procedure can be very complex if the system is large. Before going out into the field, the TAB technologist must plan the work precisely. Large systems are always made up of subsystems. Proper TAB procedure would be first to balance subsystems that are independent of the overall system. This demands a complete understanding of, and familiarity with, the design. If anything is unclear, check the design intent with the project engineer. Out on the job site, the TAB technologist is expected to know exactly what he or she wants to do, how to do it and what the results are supposed to be. A brief listing of the TAB procedure in the field follows. In some jobs, several steps can be combined or done in a different order. Remember that client satisfaction depends on an adequate TAB job.

(a) Check that all the preparatory steps listed in the preceding section have been taken.

(b) Turn on all fans. Measure fan speeds and adjust to design values. Check motor running-current. If running current is above or more than 10% below the design value, shut down the fan until the cause is determined.

(c) Measure and record initial cfm at supply fans. A Pitot tube traverse is the preferred method. If this is not possible for some reason, use anemometer readings across coils in the air-handling units. The cfm must be within ±10% of the design value before proceeding with the next step. Adjustments of fan cfm is normally made by adjustment of the drive speed. Rotational speeds are most easily checked with a simple hand-held tachometer.

An air quantity of more than ±10% from the design value indicates one of the following problems. They should be checked in the order listed:

- Incorrect damper positions—probably closed
- Incorrect filter
- Equipment malfunction
- Incorrect installation (This should not be possible if proper inspection of the installation has been performed. The job inspector should be called in, if this seems to be the problem.)
- Incorrect design (Consultation with the design engineer is required.)

(d) Measure the flow (cfm) in major duct branches and adjust to within ±10% of design. Adjustment is normally made with splitter dampers. Dampers should be fixed in position, and the positions marked.

(e) Measure and adjust air flow to all air outlets to ±10% of design requirement. Some TAB technicians start at the last outlet (farthest from the fan), and some start at the first outlet (nearest to the fan). Our recommendation is to use the latter method; it seems to require less readjustment. Use a velometer or vane anemometer to measure outlet air velocity. Take profile (traverse) readings to arrive at average air velocity. Calculate cfm using average velocity and net face areas of outlets. Record all flows on the appropriate TAB form.

- Measure and adjust the cfm in multiple outlet branches before adjusting the flow at each outlet.
- Measure and adjust air flow at all air outlets.

(f) After all terminal outlets are adjusted, repeat the entire procedure. This is necessary because each adjustment affects the pressures and flow in the entire system. As a result, the quantities previously measured in main and branch ducts will have changed. Keep repeating the procedure until flow readings remain the same when remeasured. Record the velocity and flow at each outlet for each round of adjustments. The TAB form should contain space for three sets of entries. It should not be necessary to repeat the sequence of measurements more than twice. In small systems, one repetition is frequently sufficient.

(g) Measure and record performance of all equipment. This includes:
- Static pressure at fans, filters and coils
- Motor currents
- All motor and fan speeds

(h) Measure and record WB and DB temperature at all coils along with the load condition. It may not be possible to operate the equipment at design loads. If this is so, record the operating conditions (partial load). This will enable the project engineer to determine, using the manufacturers' published data, whether the equipment is operating correctly at part load. It should also then be possible to extrapolate, to determine if the equipment will operate satisfactorily at full load.

(i) Perform air velocity and flow checks on the return air system.

The preceding description of TAB procedures for all air systems is brief but covers all the important aspects of the work. In practice, an experienced TAB technologist will make a quick survey after the system is up and running during which he or she will detect any major deviations from the desired operating conditions. These are usually not malfunctions. Instead, simple oversights, such as an open window, door or duct access panel or a blocked return grille or duct, can play havoc with system pressures and cfm quantities. This ability to locate trouble spots quickly comes with experience and a sharp eye for detail. For the novice TAB technologist, a very detailed step-by-step procedure list is the best course to follow.

7.8 Preparation for Balancing a Hydronic System

Before beginning any TAB work, the following preparatory steps should be taken. Thorough preparation always results in time savings in the field.

(a) On a set of as-built drawings, mark pressures, flow rates, temperatures and motor data. Clearly mark the actual field location of valves and other controls. Check with the field inspector on any special conditions that arose during construction. Familiarize yourself with the control system. A single-line diagram of the control schemes will be extremely helpful, particularly if the system has automatic controls and interlocks that are not field adjustable.

(b) Prepare appropriate TAB test forms for field use. (See Appendix D.)

(c) Mark on the drawings all points of measurements and the items to be measured. This will prevent anything from being overlooked. Have orifice plates and/or Venturi tubes installed at points where flow is to be read.

(d) Coordinate the TAB work with the construction contractor. A contractor's representative must be available to perform all hands-on system operation.

7.9 Balancing a Hydronic System

Having accomplished the preparatory work just outlined, proceed with the actual balancing procedure as detailed next.

(a) On a preliminary visit to the site with the field inspector and a contractor's representative, check that all systems, controls and safety devices are functional and that all hydronic systems have been drained, flushed, refilled and vented as required.

(b) Before any testing begins, confirm that manual valves are open, controls are set in their proper operating position and any seasonal controls have been properly set. This may involve overriding some automatic controls, in order, for instance, to test a heating system in the summer or a cooling system in the winter.

(c) With the pumps off, measure (and record) static pressure at each pump outlet.

(d) Start all systems. Immediately check the operating currents of all motor-driven equipment. If motors are drawing excessive currents, shut down the system to determine the cause.

(e) At each pump, perform the following test:

(1) With the pump discharge valve wide open, record the operating characteristics—flow, discharge and suction head, speed and electric motor data.

(2) Gradually close the discharge line valve to shutoff point. At several points during this closing procedure, take full measurements. Use one gauge to measure all pressures; this avoids introducing a metering error. Do not permit the pump to run with the discharge line closed for any length of time, because it may overheat. Using the data recorded, plot a pump characteristic, and compare it to the manufacturer's published data. If there is any significant difference, clarify the reason with the pump manufacturer.

(3) Gradually open the valve, take head and flow readings and check that they fall on the curve just plotted. If not, repeat these steps until an accurate pump curve is obtained. Record the total head and flow in full-open valve position. A total head higher than design means a maximum flow lower than design and vice versa. If flow is greater than design, close down the output valve until flow is about 110% of the design value. At this point, record pressures, flow and motor data. All these readings should be within system tolerances.

(f) Some hydraulic systems use automatic balancing valves. For such systems, manual balancing of flow rates in mains and branches is unnecessary. Where manual balancing is to be done, adjust manual-balancing valves with all systems operating. Read flow rates at orifice plates and/or Venturi tubes that were installed previously. Water flow rates (as with air flow rates) within ±10% of design are considered to be on target. Using balancing valves, adjust flow rates to terminal units to ±10% of design.

Note: Keep in mind that flow rates in hydronic heating are not critical. Terminal units will deliver about 90% of their rated output with 50% flow, because the heat output of a hydronic terminal unit (radiator or baseboard) depends primarily on the difference between ambient air temperature and hot water temperature. Chilled water-cooling systems are not so forgiving with inaccuracies in liquid flow. There a drop in flow will cause a serious drop in cooling effect.

(g) Repeat the balancing process for chillers, large coils and terminal units until the values remain unchanged. This may require two or three repetitions.

(h) Make a final check of pump flow and pressures and of pump electrical data. Record this information. It represents the balanced system data and can be used in the future, if any parts of the system are repaired or replaced.

(i) Mark and record the position of all valves and balancing cocks and the readings of all gauges and thermometers. This, too, is data for future reference.

We have not discussed the TAB work required on condenser water systems, cooling towers, large chillers, heat exchangers and other parts of large systems because they are beyond our scope here. Technologists will begin TAB work on small projects. After gaining experience with the design and field aspects of small systems, many will go on to similar work on large complex systems.

Key Terms

Having completed this chapter, you should be familiar with the following key terms. If any appear unfamiliar or not entirely clear, you should review the section in which these terms appear. All key terms are listed in the index to assist you in locating the relevant text.

Anemometers
Balancing
Bimetallic element
Bourdon gauge
Bourdon tube
Capillary tubing
Capillary tube thermometer
Deflecting vane anemometer
Dial thermometer
Differential pressure
Draft gauges
Flow straightener
Hot-wire anemometer

Magnetic pressure gauge
Manometers
Pitot tube
Pyrometers
Resistance temperature device (RTD)
Rotating vane anemometer
TAB
Thermal anemometers
Thermocouple
Traverse measurements
Velocity pressure
Velometer

Supplementary Reading

B. Stein and J. S. Reynolds, *Mechanical and Electrical Equipment for Buildings*, 8th ed., John Wiley & Sons, New York, 1992. This book covers the same areas of study as the present book, but in greater detail and scope. It is very useful for further study.

American Society of Heating, Refrigerating and Air Conditioning Engineers, Inc. (ASHRAE)
1791 Tullie Circle, N.E.
Atlanta, GA 30329

Handbook—HVAC Applications, Chapter 34, 1991
Sheet Metal and Air Conditioning Contractors National Association, Inc. (SMACNA)
8224 Old Courthouse Road
Tysons Corner, Vienna, VA 22180
HVAC Systems; Testing, Adjusting and Balancing, 1983
E. G. Pita, *Air Conditioning and Systems: An Energy Approach*, Chapter 16, John Wiley & Sons, New York, 1981.

Problems

1. A manometer will be used to test the pressures in an air system. Maximum blower pressure is 0.6 in. w.g. Would you use a water manometer or a mercury manometer? Why?
2. The following temperature measurements must be made in a TAB project. What type of thermometer would you use? Why?

   a. Motor bearing temperature.
   b. Oil reservoir temperature.
   c. Air stream temperature.
   d. Pipe surface temperature.
3. What is a capillary tube? What does it contain when used to connect a temperature sensing bulb to a thermometer dial?
4. a. What is a thermocouple? How is it used to measure temperature?
   b. What is a bimetallic element? How is it used to measure temperature?
5. Is an inclined scale manometer more accurate

than a vertical unit? Is it more precise? Explain.

6. A Pitot tube traverse in a duct gives the following velocity pressure readings for 16 equal areas of duct cross section. The pressure units are in. w.g. Find the average duct velocity in cfm.

   0.26 0.29 0.29 0.25
   0.27 0.32 0.33 0.27
   0.29 0.33 0.34 0.28
   0.27 0.29 0.31 0.25

7. A TAB technician wants to make a four-point Pitot tube traverse of a 4-in. round duct. Show where the Pitot tube should be positioned on a diameter to accomplish the traverse accurately.

8. An 8×14-in. register is designed to deliver 400 cfm. What should be the average velocity over its face? Explain.

9. The following air velocity readings are obtained over the face of a 8×14-in. register that has a K factor of 0.7. What is the average velocity over the face of the register? What is the flow in cfm?

   620 650 660 630
   640 680 700 635
   625 650 630 620

10. Three draft gauges are to be used with a Pitot tube to simultaneously measure total pressure, static pressure and velocity pressure in a duct. Show how the gauges are connected for

    a. A supply air stream.
    b. An exhaust air stream with positive pressure.
    c. An exhaust air stream with negative pressure.

    Explain.

8. Principles of Plumbing

In this chapter, we will introduce you to the field of plumbing with a presentation of the basic knowledge required before actual design can be undertaken. Study of this chapter will enable you to:

1. Understand what plumbing engineers and technologist design.
2. Know the principles on which design is based.
3. Use the applicable codes and other administrative guides.
4. Understand the fundamentals of hydraulics (fluid engineering) as applied to plumbing.
5. Apply units, conversions and dimensional analysis as applied to hydraulics.
6. Have a working knowledge of the piping materials used in plumbing work, including their fittings, joints, supports and installation.
7. Apply control and safety devices as required by the design.
8. Understand the choice of plumbing fixtures of the types commonly found in residential and commercial buildings. This includes obtaining and specifying the required "roughing dimensions" for the guidance of the plumbing contractor.
9. Understand and apply the methods of presenting plumbing information on working drawings.

In Chapters 9 and 10, we will discuss actual design procedures for water supply (including fire standpipes and sprinklers), sanitary and storm drainage and the application of design principles to actual building plans.

8.1 Introduction

The most basic human need is a reliable supply of potable (drinkable) water. As a result, settlements of primitive and ancient man were always located close to such a source. These sources were rivers, streams, springs and wells, both naturally occurring and those dug by men. When settlements grew in size to the point that carrying the water from the source to the dwelling became a major burden, the first man-made aqueducts were constructed. (The word *aqueduct* means in Latin a device to lead or conduct water—that is, a water conduit or duct.) These aqueducts, which started as open trenches more than 5000 years ago, developed into enclosed pressurized pipes used by the Greeks and culminated in the magnificent Roman works of hydraulic engineering. Some of the Roman aqueducts and underground piping is still in use today. Glazed pottery pipe (terra cotta) was in use in ancient Babylon. The Romans introduced the use of lead pipe for water lines, which branched from main aqueducts to public fountains and into the houses of the wealthy. (The fountains were primarily intended as a source of water for the masses in addition to serving a decorative purpose.) Lead was used because it is easily worked and joined, is waterproof and does not corrode. Today, because of our knowledge of the effects of lead poisoning, not only is lead pipe not used, but even lead-wiped joints on other metal pipes are severely restricted. The Latin word for lead is *plumbum* from which we have the English words plumber and plumbing.

The second basic human need after a reliable water supply, is some means for getting rid of human waste products. Liquid waste (urine) was originally simply allowed to sink into the ground. Solid waste (fecal matter) is not so easily disposed of and, if left exposed, putrefies, causing foul odors and attracting insects. The Bible (Deut. 23:13,14) required the Israelites in the desert to bury such material in the earth outside the inhabited area. Settlements near rivers and streams originally used those water channels to carry away these wastes. However, as with domestic water, when towns and cities grew to the point that a trip to the river or stream was not practical, a means of sewage disposal close to the dwelling was developed. The Greeks and other people of the Mediterranean basin developed the first sanitary drainage systems. The original system consisted of water channels that ran open down the center of the street. Human waste was brought in buckets from houses and dumped directly into these open sewers. This system persisted in parts of Europe and the East until modern times and still exists in parts of the world. However, because such an open sewer is obviously unsanitary and odor-producing, particularly when the water supply in the channel runs low, the Greeks and Romans introduced two important improvements. The first was to cover the open channel with paving blocks and the second was to direct part of the water flow to run directly under the local privy, thus eliminating the need to carry the waste manually to the sewer.

With the fall of Rome and the onset of the Dark Ages of Europe, these sanitary refinements fell into disuse. Cities were built and expanded with no sanitary facilities whatsoever. Householders dumped chamber pots of human waste into the street where it putrefied, causing not only foul odors but most of the diseases that wracked Europe, including typhus, typhoid, dysentery and plague. It was not until the 19th century in the United States that human waste began to be collected from private privies by "honey wagons" and that reliable centralized water supply systems were installed. Only in this century has sanitary sewage disposal in major U.S. cities reached the levels achieved at the height of the Roman empire 2000 years ago! The sophisticated interior plumbing systems that we take so much for granted are a relatively modern convenience. It is the design of these systems, along with storm drainage, that will constitute the essentials of the plumbing section of this book. A short section on water piping for fire fighting will also be included.

Modern plumbing engineering and design covers not only these areas, that is, water supply, sanitary and storm drainage and fire fighting, but also a host of other disciplines, which have piping as their common denominator. These include:

- Specialized water systems (chilled, distilled, deionized)
- Gas systems (oxygen, nitrogen, carbon dioxide, cooking gas, nitrogen, nitrous oxide, helium, etc.)
- Extended fire protection (standpipes, halon, etc.)
- Compressed air
- Vacuum systems (clinical, oral, laboratory and cleaning)
- Soap and disinfectant dispensing
- Decorative fountains and swimming pools
- Irrigation systems
- Water treatment and purification systems

The plumbing technologist who acquires a solid grounding in water supply and drainage systems will be in an excellent position to transfer these skills to any of the preceding specialties and thus widen his or her professional and employment horizons. (Industrial and high-pressure piping is a highly technical specialty that is not in the area normally handled by plumbing engineers, designers and technologists. On the other hand, pneumatic tube systems that are actually in the field of material handling are frequently designed by plumbing engineering personnel.)

A competent plumbing technologist-designer has a working knowledge of the following:

- Hydraulic principles as applied to plumbing systems
- Materials used in plumbing systems for conveyance (piping and fittings), control (valves and flow controls), measurement (meters) and usage (plumbing fixtures)
- Design techniques for all the systems with which he or she will be concerned
- Installation procedures and techniques
- Field inspection procedures

The material in the plumbing section of this book is intended to give you this working knowledge.

8.2 Basic Principles of Plumbing

The goal of modern plumbing design for buildings, as it will be discussed in the book, is to safely and reliably provide domestic water, cooking gas and water for fire fighting and to remove sanitary wastes. The word *safely* is emphasized because, although it would not appear so at first glance, plumbing systems can be very dangerous if improperly designed. Dangers from cooking/heating gas are obvious. Less obvious is the explosive potential of hot water systems and pressurized cold water systems, the nauseating effects of improperly vented sanitary drainage systems and the disease-causing potential of inadequate sanitary drainage. System reliability is of primary importance to the beneficial occupancy of a building. Think for a minute about the disruption of normal building use that can be caused by loss of water supply or stoppage of the drainage system. The image is sufficient to confirm the importance of plumbing system reliability. Moreover, reliability means not only long periods of trouble-free service but also a design that permits easy, rapid, economical and effective repairs to be made.

Another important aspect of the plumbing design, as also of the HVAC and electrical design, is flexibility. It is rare that a building's usage remains unchanged throughout the life of the structure. It is, therefore, important that the plumbing system lend itself to alteration. Furthermore, modernization is a continuing process. It is less rapid in plumbing systems than in electrical or HVAC, but it exists. It is, therefore, important that the system design and system materials be such that new developments in fixtures, valving, piping materials and the like can be accommodated with minimum disruption to the building occupants.

All modern plumbing design is founded on basic design principles intended to ensure the previously referenced safe, reliable, effective plumbing systems. A detailed list of these principles can be found in the National Standard Plumbing Code and in other administrative codes. These principles are summed up in the following discussion.

a. Potable Water

All premises intended for extended, continuous human occupancy should be provided with an adequate supply of potable water. Design of the supply shall be such that the purity of the water is always maintained and that contamination of the potable water system from backflow or reverse flow of any sort is prevented.

b. Plumbing Fixtures

Every dwelling unit should have at least one water closet, one lavatory, one kitchen-type sink and one shower or bathtub. Every plumbing fixture in any structure must be supplied with water at the flow rate and pressure required for proper operation. Where hot water is required, it should be furnished at a temperature of not less than 95°F (35°C) and not more than 140°F (60°C) except for commercial fixtures that specifically require higher temperature water. Each fixture directly connected to the drainage system must be equipped with a water seal trap. The traps may be integral, as with water closets, or separate, as with sinks, lavatories and other fixtures. All plumbing fixtures must be made of smooth, nonabsorbent, corrosion-resistant material and shall be installed so that maintenance and cleaning are readily accomplished.

c. Sanitary Drainage System

The sanitary drainage system shall be so designed that clogging and fouling is avoided to the maximum extent possible and so that, when they do occur, they can easily and readily be cleared. Additionally, the system must be designed with proper venting to protect all fixture water seals from siphonage and blowout under ordinary conditions of use. All fixture vents must be pipe connected to a vent stack terminating in fresh air outside the building. The vent system must be designed to maximize fresh air intake and minimize the possibility of clogging and the trapping of fouled air inside the building.

The sanitary drainage system should connect to a public sewer, if such exists, within a reasonable distance. If one does not exist, then an accepted method of sewage treatment and disposal that will accept the effluent from the sanitary drainage system must be designed and constructed. All connections to public sewers or private disposal systems shall be designed so that backflow (reverse flow from the sewer into the building) is prevented.

d. Storm Drainage

Every structure shall be provided with a storm drainage system that will conduct storm water from roofs and all paved areas into an approved storm sewer system. In no case, except where specifically so instructed by local authorities, should storm water be connected into a sanitary sewage system. The storm water drainage system within or on a building must be completely separate from the sanitary drainage system.

e. General Considerations

Building plumbing systems shall be designed using materials that are durable and maintenance-free to the extent possible. Installation shall be such that accessibility for maintenance and, in particular, for clearing of clogged pipes is provided. Shut-off valving should be installed to simplify repair and replacement of parts. All required and recommended safety devices including pressure and temperature relief valves and energy cutoff devices must be provided. Finally, the entire system must be tested in accordance with approved and accepted procedures before being put into service.

8.3 Plumbing System Design Constraints

Plumbing system design is carefully controlled and tightly regulated by the local ordinances of the city or town in which the construction is intended. Some of the large cities have their own plumbing codes that are in general stricter than national codes. Most areas in the United States, however, rely upon one of the four major plumbing codes in wide use today. Of these four, the National Standard Plumbing Code is the most widely used, and it is this code that we mean when we simply use the word *Code* in this book. As a convenience to users of this Code who are working on jobs that require use of one of the other major codes, a cross-reference index of code sections is provided in Appendix G of the National Standard Plumbing Code. The four major codes and their publishers follow:

1. *National Standard Plumbing Code*, published by National Association of Plumbing-Heating-Cooling Contractors, P.O. Box 6080, Falls Church, Va. 22040 (1-800-533-7694)
2. *Uniform Plumbing Code*, published by International Association of Plumbing and Mechanical Officials (IAPMO), 20001 South Walnut Drive, Walnut, Ca. 91789
3. BOCA Basic Plumbing Code, published by Building Officials and Code Administrators International, Inc., 4051 West Flosmoor Road, County Club Hill, Ill. 60477
4. Southern Standard Plumbing Code, published by Southern Building Code Congress International, Inc., 900 Montclair Road, Birmingham, Al. 35213-1206

In addition to these administrative constraints, there are other "external" constraints placed by insurance companies, environmental regulations, regulations governing facilities for the handicapped, and others. These constraints are, in general, not the concern of the plumbing technologist and will, therefore, not be discussed here. What does concern the technologist are physical constraints imposed by the building structure and by the other building systems. These require careful coordination so that the plumbing work will not conflict, spacewise, with other construction. In general, the order of precedence for space allocation is structure first, followed by HVAC duct work because of bulkiness, then plumbing drainage piping because of size and required pitch, followed by smaller HVAC and plumbing piping, and finally electrical work.

8.4 Minimum Plumbing Facilities

All codes state the minimum plumbing facilities that are required in each building type. Since this is normally the responsibility of the project's architect, the complete table will not be reproduced here. Instead, we reproduce in Table 8.1 a portion of a table from the (National Standard Plumbing

Table 8.1 Minimum Number of Plumbing Fixtures[a]

Use Group or Type of Building	Water Closets (Urinals[d] See Note 4)		Lavatories	Drinking Water Facilities[b]	Bathtubs or Showers	Other
	No. of Persons of Each Sex	No. of Fixtures				
Schools[c,e,h,j,k,l]						
Preschool/ Day Care	1–15 each add'n 15	1 add 1	½ no. of water closets	1/30 people	1 service sink/floor	
Mercantile/business[b,c,e,f,h,k,l]						
Customers in stores and carry-out food establishments where seating is not provided	1–50	1	1	1/1000 people		
Dwelling units						
Single	—	1	1	—	1	1 kitchen sink
Multiple	—	1/unit	1/unit	—	1/unit	1 kitchen sink/unit 1 laundry tray/100 units
Workplaces[c,e,g,h,j,k,l]						
Employees—most occupancies, such as stores and light industrial service	1–15 16–40	1 2	1 1	1/100 people	—	1 service sink/floor

[a] This table shall be used unless superseded by building code requirements. Consult fire codes for limitations of occupancy. For handicap requirements, see local, state and national ordinances. Additional fixtures may be required where environment conditions or special activities may be encountered.

[b] Drinking fountains are not required in restaurants or other food service establishments if drinking water service is available. Drinking water is not required for customers where normal occupancy is short term. A kitchen or bar sink may be considered the equivalent of a drinking fountain for employees.

[c] In food preparation areas, fixture requirements may be dictated by local health codes.

[d] Whenever both sexes are present in approximately equal numbers, multiply the total census by 50% to determine the number of persons for each sex to be provided for. This regulation applies only when specific information, which would otherwise affect the fixture count, is not provided.

[e] Not more than 50% of the required number of water closets may be urinals.

[f] In buildings constructed with multiple floors, accessibility to the fixtures shall not exceed one vertical story.

[g] Fixtures required for public use may be met by providing a centrally located facility accessible to several stores. The maximum distance from entry to any store to this facility shall not exceed 500 ft.

[h] In stores with floor area of 150 ft² or less, the requirements to provide facilities for use by employees may be met by providing a centrally located facility accessible to several stores. The maximum distance from entry to any store to this facility shall not exceed 300 ft.

[i] Fixtures accessible only to private offices shall not be counted to determine compliance with these requirements.

[j] Multiple dwelling units or boarding houses without public laundry rooms shall not require laundry trays.

[k] For up to ten persons, one toilet facility with one water closet and with a lockable door is permitted.

[l] Requirements for employees and customers may be met with a single set of rest rooms. The required number of fixtures shall be the greater of the required number for employees, or the required number for customers.

Source. Extracted with permission from the National Standard Plumbing Code, published by The National Association of Plumbing Heating Cooling Contractors.

Code), which shows requirements for various occupancies. This table can be used in the absence of complete or adequate architectural plans.

8.5 Hydraulics

Hydraulics is the study of the physical principles that govern the behavior of liquids at rest and in motion. There are two separate and distinct types of liquid flow with which plumbing designers and technologists are concerned and for which the relevant hydraulic principles will be discussed. The first is flow in a closed pressurized system; a system that is nowhere open to the atmosphere and operates above atmospheric pressure. This is the type of flow that occurs in domestic water systems, both hot and cold, in any of the water and other liquid supply systems listed in Section 8.1 and in water systems for fire fighting. (Flow of gases, which are fluids and not liquids, is governed by other physical principles and will not be discussed here because it is not of general concern to plumbing technologists.) *Pressurized system flow* is the type of flow that will be considered in this section and those immediately following.

The second major type of flow with which plumbing designers are concerned is *gravity flow*. This is the type of flow that occurs in all drainage systems (both sanitary drainage and storm drainage) and is caused simply by the slope of the pipe containing the liquid. These systems are open to the atmosphere. The pipes containing the liquids almost always run only partially full (as compared to completely full in pressurized systems). The physical principles describing this open, unpressurized gravity flow are quite different than those of pressurized systems. The hydraulics of gravity flow systems will be described in Chapter 10 where drainage is studied.

8.6 Static Pressure

Static pressure is caused by the weight of water above any point in the system. Refer to Figure 8.1(*a*). Let us calculate the pressure existing at the bottom of the 10-ft high column of water that is 1 ft square in cross-sectional area. (We will use "English" units throughout since they are the units generally used in plumbing work in the United States. You can convert all calculations to SI units with the help of the conversion tables in Appendix A.)

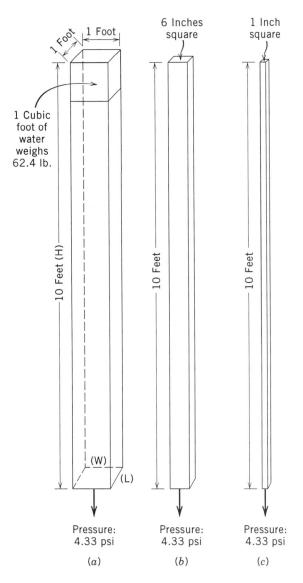

Figure 8.1 The pressure exerted by a 10 ft high column of water is the same (4.33 psi) regardless of the cross-sectional area of the column. This hydrostatic pressure of 4.33 psi is also known as a static (pressure) head of 10 ft of water.

The total volume V of water in the column is

$$V = L \times W \times H = 1 \text{ ft} \times 1 \text{ ft} \times 10 \text{ ft} = 10 \text{ ft}^3$$

Since water weighs 62.4 lb/ft^3, the total weight of this column of water is

Weight = Volume × Density
Weight = 10 ft^3 × 62.4 lb/ft^3 = 624 lb

Since this weight is being exerted over an area of 1 ft^2 ($A = L \times W = 1 \text{ ft} \times 1 \text{ ft} = 1 \text{ ft}^2$), the pressure at the bottom of this column is

$$\text{Pressure} = \frac{\text{Weight}}{\text{Area}}$$

$$\text{Pressure} = \frac{624 \text{ lb}}{1 \text{ ft}^2} = 624 \text{ lb/ft}^2 \text{ (psf)}$$

However, since pressure is normally expressed in pounds per square inch (psi), we can convert using 144 in.^{2}

$$\text{Pressure} = \frac{624 \text{ lb}}{\text{ft}^2} \times \frac{1 \text{ ft}^2}{144 \text{ in.}^2} = 4.33 \frac{\text{lb}}{\text{in.}^2} = 4.33 \text{ psi}$$

We could also have converted the previous answer using simple dimensional conversion. (We strongly recommend that you always perform unit conversions by writing out the units and cancelling, until the desired units are obtained.)

Now, referring to Figure 8.1 *(b)*, we will follow the same steps in order to calculate the pressure at the bottom of this 6-in. square, 10 ft high column of water.

$$\text{Volume} = L \times W \times H = (0.5 \text{ ft})(0.5 \text{ ft})(10 \text{ ft}) = 2.5 \text{ ft}^3$$
$$\text{Weight} = \text{Volume} \times \text{Density} =$$
$$2.5 \text{ ft}^3 \times 62.4 \frac{\text{lb}}{\text{ft}^3} = 156 \text{ lb}$$

$$\text{Pressure} = \frac{\text{Weight}}{\text{Area}} = \frac{156 \text{ lb}}{(6 \text{ in.})(6 \text{ in.})} = \frac{156 \text{ lb}}{36 \text{ in.}^2} = 4.33 \text{ psi}$$

The result is identical to the previous result. To demonstrate that this is not simply a coincidence, let us perform the same calculation for the 1-inch.^{2} column in Figure 8.1*(c)*.

$$V = L \times W \times H = (\tfrac{1}{12} \text{ ft})(\tfrac{1}{12} \text{ ft})(10 \text{ ft}) = \frac{10}{144} \text{ ft}^3$$
$$\text{Weight} = \text{Volume} \times \text{Density} =$$
$$\frac{10}{144} \text{ ft}^3 \times \frac{62.4 \text{ lb}}{\text{ft}^3} = \frac{624}{144} \text{ lb}$$

$$\text{Pressure} = \frac{\text{Weight}}{\text{Area}} = \frac{624/144 \text{ pounds}}{(1 \text{ in.})(1 \text{ in.})} = 4.33 \text{ psi}$$

It should be perfectly clear at this point that static pressure at a point below the surface depends only on the height and, therefore, weight of the water column above that point and is completely independent of area. That means that 10 ft below the surface of a large lake and 10 ft below the surface of a 1-in.^{2} column of water, the pressure is the same 4.33 psi. Since this is so, it follows that static pressure is expressible in height of a water column, that is, in feet of water. If a 10-ft column of water produces a pressure of 4.33 psi, then obviously a 1-ft column of water will produce a pressure of one-tenth this amount, or 0.433 psi. In other words, we could say that the pressure in a system is 0.433 psi

or 1 ft of water. As a double check, refer to Figure 8.2 *(a–c)* and follow the calculations:

1. Figure 8.2*(a)*
$$V = L \times W \times H = (1 \text{ ft})(1 \text{ ft})(1 \text{ ft}) = 1 \text{ ft}^3$$

$$\text{Weight} = 1 \text{ ft}^3 \times 62.4 \frac{\text{lb}}{\text{ft}^3} = 62.4 \text{ lb}$$

$$\text{Area} = (12 \text{ in.})(12 \text{ in.}) = 144 \text{ in.}^2$$
$$\text{Pressure} = \frac{\text{Weight}}{\text{Area}} = \frac{62.4 \text{ lb}}{144 \text{ in.}^2} = 0.433 \frac{\text{lb}}{\text{in}^2} = 0.433 \text{ psi}$$

2. Figure 8.2*(b)*
$$V = L \times W \times H = (0.5 \text{ ft})(0.5 \text{ ft})(1 \text{ ft}) = 0.25 \text{ ft}^3$$
$$\text{Weight} = \text{Volume} \times \text{Density} =$$
$$0.25 \text{ ft}^3 \times 62.4 \frac{\text{lb}}{\text{ft}^3} = 15.6 \text{ lb}$$

$$\text{Pressure} = \frac{\text{Weight}}{\text{Area}} = \frac{15.6 \text{ lb}}{(6 \text{ in.})(6 \text{ in.})} = \frac{15.6 \text{ lb}}{36 \text{ in.}^2} = 0.433 \text{ psi}$$

3. Figure 8.2*(c)*
$$V = L \times W \times H = \left(\tfrac{1}{12} \text{ ft}\right)\left(\tfrac{1}{12} \text{ ft}\right)(1 \text{ ft}) = \frac{1}{144} \text{ ft}^3$$
$$\text{Weight} = \text{Volume} \times \text{Density} =$$
$$\frac{1 \text{ ft}^3}{144} \times 62.4 \frac{\text{lb}}{\text{ft}^3} = \frac{62.4}{144} \text{ lb}$$

$$\text{Pressure} = \frac{\text{Weight}}{\text{Area}} = \frac{62.4/144 \text{ lb}}{(1 \text{ in.})(1 \text{ in.})} = \frac{62.4 \text{ lb}}{144 \text{ in.}^2} = 0.433 \text{ psi}$$

We have now adequately demonstrated that:

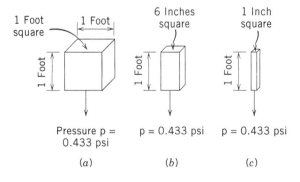

Figure 8.2 As shown in Figure 8.1, pressure is independent of the column area and depends only on the column height. Since all the columns *(a–c)* are 1 ft high, the figure demonstrates that 1 ft of hydrostatic head is equal to 0.433 psi.

(a) Static pressure is independent of surface area or total volume of liquid and depends only on depth, the height of liquid (water) "column" above the point in question.

(b) Static pressure can be expressed in feet of water at the conversion of 1 ft of water equals 0.433 psi. Conversely,

$$1 \text{ psi} = \frac{1}{0.433} = 2.31 \text{ ft of water.}$$

Use the dimensional type of analysis that we recommend, and try not to depend on conversion factors that can be misused. To convert pounds per square inch of pressure to feet of water, multiply by 2.31 or divide by 0.433. Conversely, to convert feet of water to pounds per square inch of pressure, multiply by 0.433 or divide by 2.31.

Since static pressure in a water system is caused by the weight of water, it is also referred to as hydrostatic pressure. Also, since it is expressed in feet of height of a column of water, it is also referred to as *static head* or *hydrostatic head*, where head is a synonym for pressure. The term *pressure head* is also used despite the fact that it is basically repetitive.

As we stated in our discussion and demonstrated by calculation, static pressure in a hydraulic system depends only on the depth at which the measurement is taken and not on the area of water above. This is graphically illustrated in Figure 8.3. As noted on that figure, static pressure at the surface is zero since depth is zero. However, as we well know, the pressure at the surface is not zero; it is atmospheric pressure, which at sea level amounts to 14.7 psi. We must, therefore, differentiate between *absolute pressure*, which includes atmospheric pressure, and *gauge pressure*, which does not. Static pressure in hydraulic systems, unless specifically noted otherwise, is always gauge pressure, that is, zero at the liquid surface and increasing with depth.

We stated that atmospheric pressure is 14.7 psi at sea level. This pressure is caused by the weight of the atmosphere (air) above exactly as hydrostatic pressure is caused by the weight of water above. For this reason, we stated the pressure at sea level. Below sea level—at the Dead Sea, for instance, which is 1300 ft below sea level—atmospheric pressure is about 15.1 psi, whereas at the top of Mt. Everest, it is only about 11 psi. Using the conversion factors we developed, we can express atmospheric pressure in terms of feet of water:

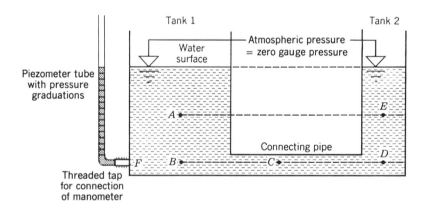

Figure 8.3 The pressure at point *A* equals the pressure at point *E* because both are at the same depth. Similarly, the hydrostatic pressures at points *B*, *C* and *D* are equal. This demonstrates that, although "depth" of point *C* is not obvious, its pressure must be the same as that at points *B* and *D* because all three points are at the same elevation. Note also that hydrostatic pressure is gauge pressure, that is, pressure above atmospheric pressure, which is set arbitrarily at 0 gauge. Absolute pressure equals gauge pressure plus atmospheric pressure (14.7 psi). See text for application of the manometer connection at point *F*. (From Nathanson, *Basic Environmental Technology*, 1986, © John Wiley & Sons, reprinted by permission of John Wiley & Sons.)

$$14.7 \text{ psi} \times \frac{\text{ft of water}}{0.433 \text{ psi}} = 33.93 \text{ ft of water}$$

In physical terms, this means that, in a closed system, atmospheric pressure will support a column of water 33.9 ft high.

You will occasionally encounter pressure expressed in units other than pounds per square inch and feet of water. The most important of these historically is measurement in millimeters of mercury. Since mercury is much denser (heavier per unit volume) than water, pressure expressed in the height of a mercury column will be a smaller number than that of the equivalent column of water, in exactly the density ratio between mercury and water. Thus, atmospheric pressure, which we have noted to be 33.9 ft of water, when expressed in terms of a column of mercury is exactly 760 mm of mercury at sea level. The usefulness of a mercury column as a pressure-measuring device will become clear in Section 8.7. A list of the various units in which atmospheric pressure is expressed and their equivalency is given.

One atmosphere = 0 psi gauge pressure
= 14.7 psi absolute pressure
= 33.93 ft of water
= 760 mm of mercury
= 1.01 bar
= 101.3 kPa (kiloPascal)

8.7 Pressure Measurement

As with all physical properties, it is frequently necessary to measure liquid pressure. The simplest liquid pressure measurement device is a piezometer tube, which is no more than a clear glass or plastic tube that is connected into the system at the point at which pressure measurement is required. If, for instance, we were to connect a piezometer tube into the tap in the tank and pipe arrangement of Figure 8.3 at point F, the water in the tube would rise to the same level as that in both tanks. If the tube were marked with graduations (equally spaced markings) and calibrated in pounds per square inch, we would be able to read the pressure at point F in the tank directly. Let us assume now that, instead of a tank, we wished to measure the pressure in a pipe carrying water at a pressure of between 45 and 80 psi. (This is the range of pressures normally found in public water mains.) A piezometer tube capable of reading this pressure would have to be at least 185 ft high!

$$80 \text{ psi} \times \frac{\text{feet of water}}{0.433 \text{ psi}} = 185 \text{ ft of water}$$

Obviously, this is impractical. This demonstrates that a simple water tube can be used to measure only low pressures. For somewhat higher pressures, a mercury *manometer* can be used. A manometer is basically a column of liquid in a glass tube, where the weight of the liquid is used to balance the pressure being measured and the height of the column indicates the pressure—as with the piezometer tube. A physician's blood pressure machine (sphygmomanometer) is a manometer as the name indicates. It balances the pumped-up air pressure inside the device against the weight of a column of mercury, which is graduated in millimeters. A blood pressure reading of 120 simply means that the column of mercury rose to a height of 120 mm. If we were interested in knowing the actual pressure, we could convert, by remembering that atmospheric pressure of 14.7 psi corresponds to 760 mm of mercury column height. Therefore,

$$120 \text{ mm} \times \frac{14.7 \text{ psi}}{760 \text{ mm}} = 2.32 \text{ psi}$$

If a water column were used instead of mercury, it would have to be 13.6 times as long (since mercury is 13.6 times as heavy as water). This would obviously be impractical, which is the reason that mercury is used.

Returning to the pressure measurement in a 80-psi water main, even a mercury manometer of the simple open type would have to be 14 ft long to be adequate:

$$80 \text{ psi} \times \frac{760 \text{ mm}}{14.7 \text{ psi}} = 4136 \text{ mm} = 4.136 \text{ m} = 13.6 \text{ ft}$$

For this reason, simple open-end manometers are not used except for low pressures. For higher pressures, differential closed-end manometers are used, since they are much smaller and can measure higher pressures.

Most often, however, a Bourdon gauge is used. The design of this gauge is based on an observation made in the mid-1800s by a scientist named Bourdon. He noted that fluid pressure inside a bent tube acts to straighten the tube and that, within limits, the action is linear. That means that the amount of "uncurling" of the bend is proportional to the pressure in the tube. This principle can, therefore, be used to measure pressure, as shown in Figure

8.4(a). Because the tube movement is small, commercial Bourdon gauges use designs other than a simple tube to increase sensitivity and magnify the tube movement, but the action is essentially as described. A commercial Bourdon gauge is shown in Figure 8.4(b). Other pressure-measuring devices include those with diaphragms and bellows whose motion is proportional to the applied pressure and those incorporating strain gauges. These are used principally in control and are not generally of interest to the plumbing technologist. Refer to the bibliography at the end of Chapter 10 for further information.

8.8 Liquid Flow

Water in a closed system under hydrostatic pressure does not move until a valve (a faucet, for instance) is opened, causing water to discharge through that opening. As soon as that occurs, the static pressure in the system, which is in reality a type of stored, or potential energy, causes the water to move toward the opening in the system, that is, the point at which water is being discharged. The stored or potential energy of the system that results from the static pressure is converted into kinetic energy of moving water. In addition, it supplies the energy necessary to overcome the friction in the system. In order to understand the action of moving (flowing) water, we must first express numerically the relationships involved in flow. The first relation is between discharge rate Q (also called *flow rate* or rate of flow) and flow velocity in a closed pipe. It is

$$Q = A V \qquad (8.1)$$

Using conventional units,

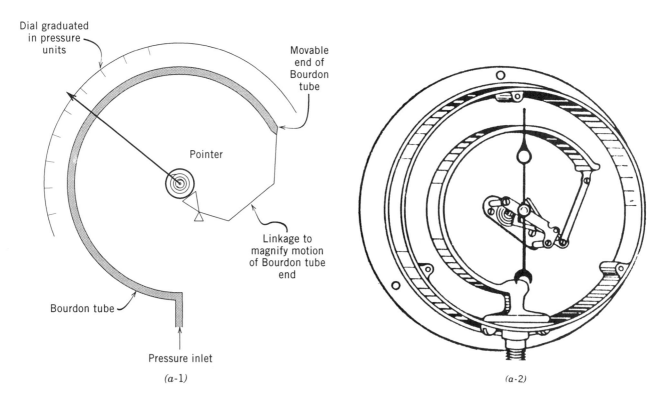

Figure 8.4 *(a-1)* Principles of operation of a Bourdon-type pressure gauge. Liquid under pressure is introduced into the gauge and causes the end of the bent Bourdon tube to move in an "uncurling" motion. The pointer is attached to the tube end via a linkage that magnifies the tube movement. This causes the pointer to move linearly over the entire face of the meter in response to small tube-end movement. *(a-2)* Pictorial representation of the Bourdon gauge internal construction. [Diagram *(a-2)* from Severns and Fellows, *Air Conditioning and Refrigeration*, 1958, © John Wiley & Sons, reprinted by permission of John Wiley & Sons.]

LIQUID FLOW / 411

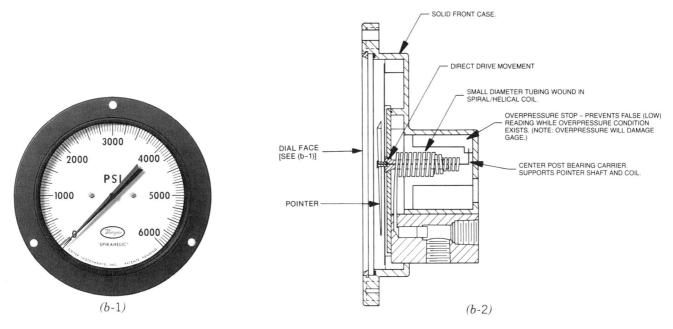

(b-1)   (b-2)

Figure 8.4 *(b)* In commercial Bourdon gauges, the tube is frequently wound as a spiral as in this unit. This type of construction amplifies the tube movement and increases the meter's sensitivity. *(b-1)* The face of a typical high pressure Bourdon gauge using a spiral-helical coil. *(b-2)* Section through a coil-type Bourdon pressure gauge showing internal construction. (Courtesy of Dwyer Instrument, Inc.)

Q is flow rate in cubic feet per second (cfs or ft³/sec), A is cross-sectional pipe area in square feet (ft²) and V is fluid velocity in feet per second (fps) or ft/sec.

Alternately, if fluid velocity were in feet per minute, then obviously flow would be in cubic feet per minute. Obviously, if area A is expressed in square inches, a conversion to square feet would be necessary.

In order to understand the simplicity and derivation of this expression, refer to Figure 8.5. The volume of fluid Q flowing past point P in 1 sec, when the fluid velocity is 1 fps, would be a cylinder of water 1 ft long of cross-sectional area A. Its volume is the cross-sectional area of the cylinder times its length, that is, $A \times 1$ ft³. If the velocity of fluid were 2 fps, then a cylinder of fluid 2 ft long of area A would flow past point P in 1 sec, whose volume would be $A \times 2$, and so on. Therefore, the flow rate Q expressed in volume of fluid flowing past a point P per second is obviously the pipe area times the velocity, that is $A \times V$. Since the volume of liquid flowing past point P is obviously the same

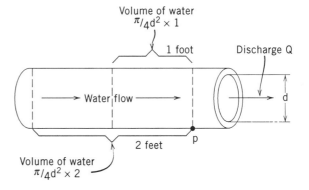

Figure 8.5 The discharge rate Q of a full flowing pipe is expressed as $Q = AV$, where A and V are the pipe internal area and the fluid velocity, respectively. This relation is demonstrated in the drawing. The quantity of water flowing past point P in 1 sec is the pipe area times the length of a water cylinder that is equal to the water velocity, that is, 1 ft long at a velocity of 1 ft/sec and so on. Therefore, the discharge rate or flow Q equals the product of area and velocity or $Q = AV$. If pipe area is expressed in square feet and flow velocity is expressed in feet per second, Q will be expressed in cubic feet per second (cfs). See text for conversion to other units.

as the discharge, then the discharge Q is also equal to $A = V$, that is, $Q = AV$. This expression makes two important assumptions. They are that

1. The pipe is flowing full.
2. The liquid velocity is constant over the cross-sectional area of the pipe.

The first of these assumptions is quite accurate for pressurized plumbing-type water systems such as domestic water and fire sprinklers. It is not correct for gravity flow systems such as drainage. Flow in these systems will be analyzed in our study of drainage systems. The second assumption is accurate if we understand fluid velocity to be average over the entire cross section of the pipe.

An example should clarify the use of the equation just developed.

Example 8.1 What is the discharge rate of a ¾-in. copper pipe type L if the water velocity in the pipe is 8 fps. (These numbers were chosen deliberately to represent typical values in a domestic water system. Maximum water velocity is held to 8 fps to limit noise in pipes.) The inside diameter of this pipe is 0.785 in. (see Table 8.3, page 431).

Solution:

$$Q = AV = \frac{\pi}{4} d^2 (V)$$

$$= \frac{\pi}{4} (0.785 \text{ in})^2 \times \left(\frac{1 \text{ ft}^2}{144 \text{ in}^2}\right) \times \frac{8 \text{ ft}}{\text{sec}}$$

$$= 0.0269 \frac{\text{ft}^3}{\text{sec}}$$

To convert this to the more useful quantity of gallons per minute, we simply use known conversion relations:

$$Q = 0.0269 \frac{\text{ft}^3}{\text{sec}} \times \frac{60 \text{ sec}}{\text{min}} \times \frac{7.481 \text{ gal}}{\text{ft}^3} = 12.07 \text{ gpm}$$

If this pipe size were ½ in. (I.D. = 0.545 in., see Table 8.3), then Q in gpm would be

$$Q = AV = \frac{\pi}{4} (0.545 \text{.in.})^2 \left(\frac{1 \text{ ft}^2}{144 \text{ in.}^2}\right) (8) \frac{\text{ft}}{\text{sec}} = 0.01296 \frac{\text{ft}^3}{\text{sec}}$$

Converting this to gallons per minute, we have

$$Q = \frac{0.01296 \text{ ft}^3}{\text{sec}} \times \frac{60 \text{ sec}}{\text{min}} \times \frac{7.481 \text{ gal}}{\text{ft}^3} = 5.82 \text{ gpm}$$

(This is somewhat higher than the recommended maximum flow of 4 gpm for a lavatory.)

Since the units most commonly used in plumbing work are discharge rate in gallons per minute, pipe diameter in inches and water velocity in feet per second, we can rewrite the discharge rate equation with the conversion factors built into the equations.

$$Q = AV$$

$$= \frac{\pi (d^2 \text{ in}^2)}{4} \times \frac{\text{ft}^2}{144 \text{ in}^2} \times \frac{V \text{ ft}}{\text{sec}} \times \frac{60 \text{ sec}}{\text{min}} \times \frac{7.481 \text{ gal}}{\text{ft}^3}$$

$$= 2.45 \, d^2 V \qquad (8.2)$$

where

Q = gallons per minute (gpm)
d = pipe diameter in inches
V = water velocity in feet per second (fps)

Reworking the last example (½-in. type L copper pipe) using the equation, we obtain

$$Q = 2.45 \, (0.545)^2 \, (8) = 5.82 \text{ gpm}$$

which is the same as the result obtained previously. Using an approximation of

$$Q = 2.5 \, d^2 V$$

introduces only a 2% error and is, therefore, acceptable for most plumbing work.

One further aspect of the flow equation that we should understand is the effect on velocity of changing pipe area. Refer to Figure 8.6, which is representative of a typical domestic water piping arrangement. Since the discharge at the end of the line is 4 gpm, it is also the flow throughout the system. However, since $Q = AV$ throughout and Q is constant at 4 gpm, velocity must vary inversely with pipe cross-sectional area, that is, inversely with the square of the pipe diameter. Thus, referring to Figure 8.6 and using the approximate equation developed previously,

$$Q = 2.5 \, d^2 V$$

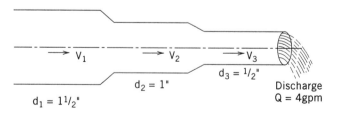

Figure 8.6 Since the flow throughout this piping system must equal the discharge at the end, the velocity in each section must vary inversely with the pipe area in order to maintain the relation $Q = AV$.

we can easily calculate the flow velocity in all the sections of the system:

$$V_1 = \frac{Q}{2.5\, d_1^2} = \frac{4 \text{ gpm}}{2.5\, (1.5 \text{ in.})^2} = 0.7 \text{ fps}$$
$$V_2 = \frac{Q}{2.5\, d_2^2} = \frac{4 \text{ gpm}}{2.5\, (1.0 \text{ in.})^2} = 1.6 \text{ fps}$$
$$V_3 = \frac{Q}{2.5\, d_3^2} = \frac{4 \text{ gpm}}{2.5\, (0.5 \text{ in.})^2} = 6.4 \text{ fps}$$

8.9 Flow Measurement

Measurement of flow in domestic water systems is usually done only at the building service point, by the water utility company, for the purpose of billing. Occasionally, it is also performed by a user in order to determine water usage on a particular line. The water meters most frequently used for this purpose are of the positive displacement type. An example is shown in Figure 8.7. The water passing through the meter causes a disk mounted in the meter to rotate in proportion to the quantity of water passing through the meter. These rotations are recorded in a numerical register on the meter face, as seen in Figure 8.7. Other flow meters of the Venturi and magnetic flow type are available for continuous monitoring of variable flow and for other precise measurement tasks. Some can be equipped with electrical transducers for remote readout.

8.10 Dimensions and Conversions

The plumbing technologist will find it necessary frequently to convert from one unit to another. This we have already seen in preceding sections and will continue to meet in the material that follows. Because this occurs so frequently, and between the same units, it is an excellent idea to memorize a few basic conversions, as was pointed out in Section 1.11. In their absence, the conversion technique described in detail Section 1.11 can be used. Finally, or more likely initially, conversion factors found in Appendix A can be used, with care. It is very easy to use such a factor in reverse and to obtain a completely incorrect result, if you do not know, in advance, approximately what the answer should be. This is the reason that it is so helpful and important to know the important conversions by heart.

Figure 8.7 Typical positive displacement-type water meter. These units are available in ⅝- to 2-in. pipe sizes, with a range of capacities up to 160 gpm. The dial can be calibrated in gallons, cubic feet, imperial gallons, barrels, pounds and metric—liters and cubic meters. (Courtesy of Carlon Meter Company, Inc.)

8.11 Plumbing Materials

As stated at the end of Section 8.1, a competent plumbing technologist/designer has a good working knowledge of plumbing materials. This knowledge is absolutely necessary for the proper design of efficient, safe, reliable and economical plumbing systems. Chapter 3 of the National Standard Plumbing Code is devoted to plumbing system materials. Most of that chapter is devoted to an extensive table that lists the specifications and standards that are accepted in the plumbing industry for approved materials. All materials used in plumbing systems should meet the requirements of at least one of the standards listed there. The principal standards organizations, and the materials for which they issue standards, follow.

a. Principal Standards Organizations

ANSI   American National Standards Institute
      1430 Broadway
      New York, NY 10018

ASSE American Society of Sanitary Engineering
P.O. Box 40362
Bay Village, OH 44140

ASTM American Society for Testing and Materials
1916 Race Street
Philadelphia, PA 19103-1187

AWWA American Water Works Association
6666 W. Quincy Ave.
Denver, CO 80235

CISPI Cast Iron Soil Pipe Institute
5959 Shallowford Road, Suite 419
Chattanooga, TN 37421

FS Federal Specification
General Service Administration
Specification Section, Room 6039
7th & D Streets
Washington, DC 20407

IAPMO International Association of Plumbing and Mechanical Officials
20001 S. Walnut Drive
Walnut, CA 91789

MSS Manufacturing Standardization Society
5203 Leesburg Pike, Suite 502
Falls Church, VA 22041

UL Underwriters Laboratories
333 Pfingsten Road
Northbrook, IL 60062

NSF National Sanitation Foundation
3475 Plymouth Rd
Ann Arbor, MI 48106

b. Plumbing Equipment Fields

These standards organizations publish standards in the following plumbing equipment fields:

1. Ferrous pipe and fittings: ANSI, ASTM, FS, IAPMO, AWWA, CISPI, MSS
2. Nonferrous metal pipe and fittings: ANSI, ASTM, FS, ASSE, MSS
3. Nonmetallic pipe and fittings: ANSI, ASTM, FS, AWWA, NSF
4. Plumbing appliances: ANSI, FS, ASSE, UL
5. Plumbing fixtures: ANSI, FS, ASSE

Since each organization publishes standards only for selected items in each category, the Code and/or the organization's catalog should be consulted for specific items. Every active engineering office keeps an up-to-date file of these standards. A second source is a major standard library or preferably a technical one such as at a college or technical institute. Finally, all standards are available for inspection and purchase at the publisher.

8.12 Piping Materials and Standard Fittings

a. Ferrous Metal Pipe

Iron and steel are available in two principal wall thicknesses: standard weight, also known as Schedule 40 pipe, and heavy wall pipe, also known as Schedule 80 pipe. Basic dimensional data for both types are given in Table 8.2. All piping application must be in accordance with the applicable local plumbing code. In general, galvanized schedule 40 steel pipe with threaded galvanized fittings is used for large (5 in. and above) cold water piping inside buildings and in smaller sizes when the water is very hard. It is also used for gas piping (with threaded malleable iron fittings), fire sprinkler piping (with threaded cast-iron fittings) and as an alternative for cast-iron drainage piping, both sanitary and storm, within buildings, generally with threaded cast-iron fittings. Frequently, however, drainage piping, both storm and sanitary, and vent piping use cast-iron pipe or plastic pipe both underground and inside buildings.

Steel pipe, ungalvanized (also referred to as black iron pipe), is not frequently used even when permitted by Code because of corrosion problems. Galvanized steel and wrought iron are much more corrosion-resistant and are, therefore, preferred, despite their higher cost. Threaded joints in steel pipe have the advantage of strength but the disadvantage that they are difficult and time-consuming to make and require a union fitting for connection to a fixed threaded pipe, such as at a piece of existing equipment or piping. Unions also make possible dismantling and reassembly of piping without the necessity of dismantling the entire pipe run. Typical joints and fittings for threaded steel pipe are shown in Figure 8.8.

Cast-iron (CI) soil pipe is available in diameters from 2 to 15 in. and in standard or heavy weight. Standard weight is used except in applications where the extra strength of heavy-weight pipe is required, such as runs under paved areas. Cast-iron pipe used for (storm) drainage is threaded and uses threaded fittings. The fittings are tapped in such a manner that horizontal branches slope downward at the required ¼ in./ft slope, to allow for gravity drainage of storm water. Vertical storm drains are

Table 8.2 Physical Properties of Ferrous Pipe

Nominal Pipe Size, in.	Schedule	O.D., in.	Wall Thickness, in.	Weight Per Foot, lb	Weight of Water Per Foot of Pipe, lb
3/8	40	0.675	0.091	0.567	0.083
	80	0.675	0.126	0.738	0.061
1/2	40	0.840	0.109	0.850	0.132
	80	0.840	0.147	1.087	0.101
3/4	40	1.05	0.113	1.130	0.230
	80	1.05	0.154	1.473	0.188
1	40	1.315	0.133	1.678	0.374
	80	1.315	0.179	2.171	0.311
1 1/4	40	1.660	0.140	2.272	0.647
	80	1.660	0.191	2.996	0.555
1 1/2	40	1.900	0.145	2.717	0.882
	80	1.900	0.200	3.631	0.765
2	40	2.375	0.154	3.652	1.452
	80	2.375	0.218	5.022	1.279
2 1/2	40	2.875	0.203	5.79	2.072
	80	2.875	0.276	7.66	1.834
3	40	3.500	0.216	7.57	3.20
	80	3.500	0.300	10.25	2.86
3 1/2	40	4.000	0.226	9.11	4.28
	80	4.000	0.318	12.51	3.85
4	40	4.500	0.237	10.79	5.51
	80	4.500	0.337	14.98	4.98
5	40	5.563	0.258	14.62	8.66
	80	5.563	0.375	20.78	7.87
6	40	6.625	0.280	18.97	12.51
	80	6.625	0.432	28.57	11.29
8	40	8.625	0.322	28.55	21.6
	80	8.625	0.500	43.39	19.8
10	40	10.75	0.365	40.48	34.1
	80	10.75	0.593	64.40	31.1
12	40	12.75	0.406	53.6	48.5
	80	12.75	0.687	88.6	44.0

416 / PRINCIPLES OF PLUMBING

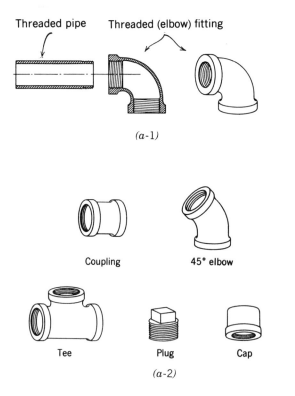

Figure 8.8 *(a-1)* Method of connecting pipes and fittings by threaded joints. *(a-2)* Examples of threaded pipe fittings for ferrous or "iron pipe size" brass pipe.

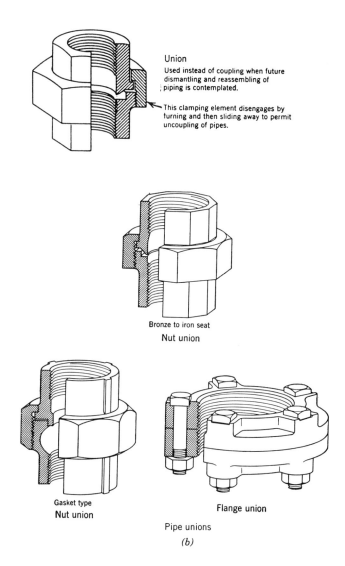

Figure 8.8 *(b)* Union connectors are used when a joint must be dismantled without dismantling the entire pipe run. Nut unions consist of three parts: two sections that attach to the pipe ends and a third part that draws the ends together. Nut unions are made in all sizes up to and including 4 in. Flange unions have two parts that are threaded onto the two pipe ends. The two sections are drawn together with bolts. Flange unions are drawn together with bolts. Flange unions are preferred over nut unions for pipe sizes above 2 in. (From Severns and Fellows, *Air Conditioning and Refrigeration*, 1958, © John Wiley & Sons, reprinted by permission of John Wiley & Sons.)

completely plumb. Underground water lines 3 in. and larger may also use ductile iron bell and spigot piping.

Cast-iron soil pipe is available in two configurations: in the standard hub and spigot design with a hub (bell) at one end and a spigot at the other, and the more recent hubless type with a spigot at one end only. Both types have a complete line of fittings available for joints, couplings and changes of direction. Joints in the hub (bell) and spigot type are traditionally of the oakum and lead type. In this joint [See Figure 8.9(c)], a bead of oakum (tarred jute or hemp rope) is forced into the void between the spigot and the hub, and the joint is then sealed with poured lead. A properly made joint of this type is water and gas tight, permanent and somewhat flexible and has the added advantage that it attenuates sound because there is no iron-to-iron pipe connection. This joining procedure is labor-intensive, time-consuming and expensive and requires a high degree of skill. It is, unfortunately, rapidly becoming obsolete. Other joint types are shown in Figure 8.9(a) and (b).

A few of the fittings that are used with hubless cast-iron pipe are shown in Figure 8.10. Similar data for the standard hub and spigot design are shown in Figure 8.11. Because cast-iron pipe used for sanitary drainage (soil pipe) requires long radius elbows and connections to avoid fouling, a full line of these fittings is available. Elbows are referred to by the extent of their bend: quarter (90°), sixth (60°), eighth (45°) and sixteenth (22.5°). Also, the length of a CI pipe elbow varies and is referred to as either short or long sweep.

b. Nonferrous Metallic Pipe

As already mentioned, black iron (ungalvanized steel) pipe is not commonly used because of severe corrosion problems. Even galvanized steel will eventually corrode. Cast-iron pipe is almost corrosion-free because an initial thin layer of corrosion forms a tight surface bond that tends to prevent any further corrosive action. However, in the small sizes required for water systems, cast-iron piping is not practical. As a result, the most commonly used material for water systems is copper pipe and tubing. The thicker walled, hard temper material is called *pipe;* the thinner, flexible, soft temper material is called *tubing*. Red brass is also suitable for water systems, but because it is more expensive than copper, requires time-consuming threaded connections and is subject to attack by acids in "aggressive" waters that leach out the zinc, it is much less commonly used than copper. An additional advantage of copper (and brass) pipe over steel is lower internal friction. In some cases, this permits the use of a smaller pipe, which helps offset the higher cost of nonferrous pipe. Threaded brass and copper pipe are also used for cooking and heating gas systems, as permitted by the Codes. Figure 8.8 shows the type of threaded fittings used with brass piping.

Copper piping is available in four weights; in descending order of wall thickness, they are types K, L, M, and DWV. Type K, with the heaviest wall thickness, is used for water supply, heating systems and chilled water systems. It is available in both hard temper pipe and soft temper (annealed) tubing. The hard temper pipe is used in exposed locations primarily because of its attractive appearance. Soft temper type K tubing is particularly useful in corrosive situations such as underground, and for runs with many bends. It is also used where it is necessary to feed pipe into inaccessible areas, because no elbows or other fittings are required. Type K hard pipe is made in 20-ft sections. The soft annealed pipe is made in 60 ft coils up to 2 in. diameter. The smaller sizes are made in 100-ft coils.

Type L has a slightly thinner wall than type K (see Table 8.3). It is used above grade for most heating and plumbing applications and below grade for water distribution and storm drainage. Type L pipe, with soldered joints, is the type most commonly used for hot and cold water piping inside buildings. It is available in hard temper sections of 20 ft length and soft temper coils of 60 and 100 ft.

Type M, which is available only in hard temper, is used for light-duty applications such as low pressure on nonpressurized lines, above grade, where corrosion is not a problem.

Type DWV, is used above grade, primarily for drainage, waste and vent piping, as the name implies. It is available only in hard temper and is the nonferrous equivalent of cast-iron piping in drainage application.

Because of its thin wall, copper pipe cannot be threaded easily. As a result, most joints are either soldered or, where permitted by Code, made with flared pressure joints. The soldered connection is made by first cleaning and fluxing the pipe and fitting and then applying solder to the heated joint. The melted solder is drawn into the small space between the tube end and the slightly larger fitting by capillary action, making a perfect cylindrical joint that is gas and liquid tight. This type of

(text continues on p. 431)

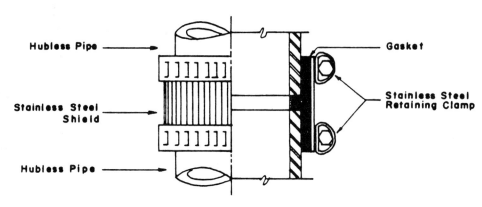

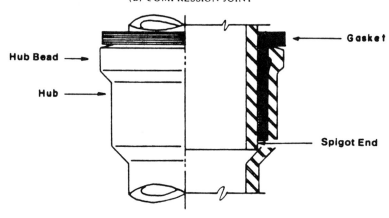

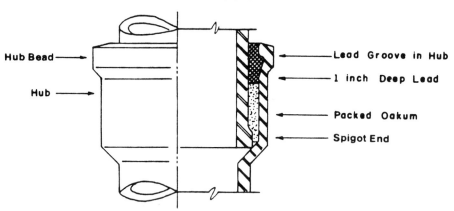

Figure 8.9 Various joint types used with cast-iron soil pipe. *(a)* Joints in plain end (butt-cut) CI pipe are made with a compression-type stainless steel clamp that forces a gasket against the two butt ends of the adjoining pipes. Compression joints *(b)* are simpler to make as they rely only on a preformed gasket for sealing. The most common joint *(c)* is made by forcing oakum (tarred jute or hemp) into the bell around the spigot and sealing the joint with poured molten lead. The joint is liquid and gas tight but requires a high degree of workmanship to make properly. (Copyright © 1989 by The Cast Iron Soil Pipe Institute, reprinted with permission.)

PIPING MATERIALS AND STANDARD FITTINGS / 419

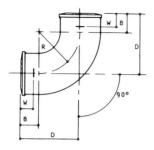

¼ Bend

Size (Inches)	Dimensions in Inches			
	B	D (±⅛)	R	W
1½	1½	4¼	2¾	1⅛
2	1½	4½	3	1⅛
3	1½	5	3½	1⅛
4	1½	5½	4	1⅛

Dimension D is laying length.

(a)

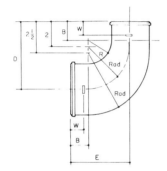

Reducing ¼ Bend

Size (Inches)	Dimensions in Inches				
	B	R	D (±⅛)	E (±⅛)	W
4 x 3	1½	3½	5½	5	1⅛

Dimensions D and E are laying lengths.

(b)

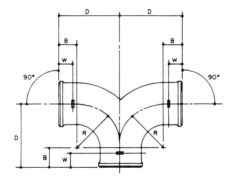

Double ¼ Bend

Size (Inches)	Dimensions in Inches			
	D (±⅛)	R	B	W
3	5	3½	1½	1⅛
4	5½	4	1½	1⅛

Dimension D is laying length.

(c)

Figure 8.10 Typical dimensional data for hubless cast-iron soil pipe fittings for sanitary, storm drain, waste and vent piping applications. (Copyright © 1989 by The Cast Iron Soil Pipe Institute, reprinted with permission.)

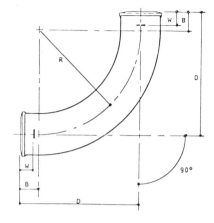

Long Sweep

Size (Inches)	Dimensions in Inches			
	B	D (±1/8)	R	W
2	1 1/2	9 1/2	8	1 1/8
3	1 1/2	10	8 1/2	1 1/8
4	1 1/2	10 1/2	9	1 1/8

Dimension D is laying length.

(d)

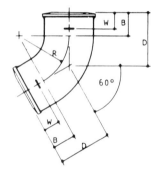

1/6 Bend

Size (Inches)	Dimensions in Inches			
	B	D (±1/8)	R	W
2	1 1/2	3 1/4	3	1 1/8
3	1 1/2	3 1/2	3 1/2	1 1/8
4	1 1/2	3 13/16	4	1 1/8

Dimensions D is laying length.

(e)

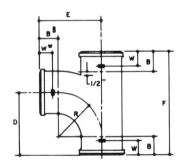

Sanitary Tee Branch

Size (Inches)	Dimensions in Inches							
	B	E (±1/8)	F (±1/8)	D	R	W	W^W	B^B
1 1/2	1 1/2	4 1/4	6 1/2	4 1/4	2 3/4	1 1/8	1 1/8	1 1/2
2 x 1 1/2	1 1/2	4 1/2	6 5/8	4 1/4	2 3/4	1 1/8	1 1/8	1 1/2
2	1 1/2	4 1/2	6 7/8	4 1/2	3	1 1/8	1 1/8	1 1/2
3 x 1 1/2	1 1/2	5	6 1/2	4 1/4	2 3/4	1 1/8	1 1/8	1 1/2
3 x 2	1 1/2	5	6 7/8	4 1/2	3	1 1/8	1 1/8	1 1/2
3	1 1/2	5	8	5	3 1/2	1 1/8	1 1/8	1 1/2
4 x 2	1 1/2	5 1/2	6 7/8	4 1/2	3	1 1/8	1 1/8	1 1/2
4 x 3	1 1/2	5 1/2	8	5	3 1/2	1 1/8	1 1/8	1 1/2
4	1 1/2	5 1/2	9 1/8	5 1/2	4	1 1/8	1 1/8	1 1/2

Dimensions E and F are laying lengths.

(f)

Figure 8.10 *(Continued)* Hubless CI pipe fittings.

Wye

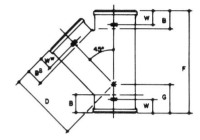

Size (Inches)	B	B^B	D ($\pm 1/8$)	F ($\pm 1/8$)	G	W	W^W
1½	1½	1½	4	6	2	1⅛	1⅛
2	1½	1½	4⅝	6⅝	2	1⅛	1⅛
3 x 2	1½	1½	5⁵⁄₁₆	6⅝	1½	1⅛	1⅛
3	1½	1½	5¾	8	2¼	1⅛	1⅛
4 x 2	1½	1½	6	6⅝	1	1⅛	1⅛
4 x 3	1½	1½	6½	8	1¹¹⁄₁₆	1⅛	1⅛
4	1½	1½	7¹⁄₁₆	9½	2⁷⁄₁₆	1⅛	1⅛

Dimensions in Inches

Dimensions D and F are laying lengths.

(g)

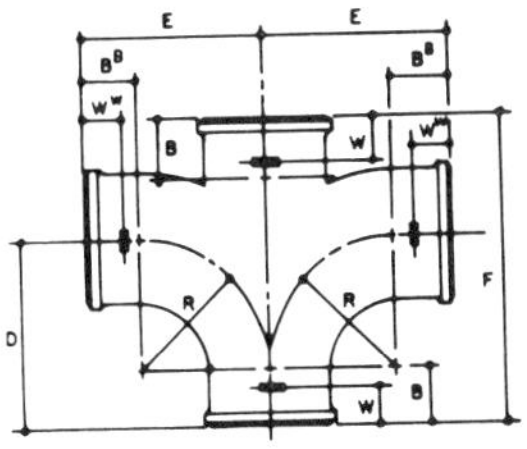

Sanitary Cross

Size (Inches)	B	E ($\pm 1/8$)	F ($\pm 1/8$)	D	R	W	W^W	B^B
1½	1½	4¼	6½	4¼	2¾	1⅛	1⅛	1½
2	1½	4½	6⅞	4½	3	1⅛	1⅛	1½
3 x 2	1½	5	6⅞	4½	3	1⅛	1⅛	1½
3	1½	5	8	5	3½	1⅛	1⅛	1½
4 x 2	1½	5½	6⅞	4½	3	1⅛	1⅛	1½
4 x 3	1½	5½	8	5	3½	1⅛	1⅛	1½
4	1½	5½	9⅛	5½	4	1⅛	1⅛	1½

Dimensions in Inches

Dimensions E and F are laying lengths.

(h)

Increaser–Reducer

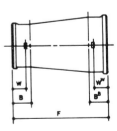

Size (Inches)	B	B^B	F ($\pm 1/8$)	W	W^W
2 x 3	1½	1½	8	1⅛	1⅛
2 x 4	1½	1½	8	1⅛	1⅛
3 x 4	1½	1½	8	1⅛	1⅛

Dimensions in Inches

Dimension F is laying length.

(i)

Figure 8.10 *(Continued)* Hubless CI pipe fittings.

422 / PRINCIPLES OF PLUMBING

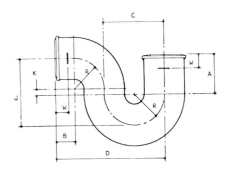

P Trap

Size (Inches)	A	B	C	D (±1/8)	J	K	R	W
1½	2	1½	3½	6¾	3½	—	1¾	1⅛
2	2	1½	4	7½	4	—	2	1⅛
3	3¼	1½	5	9	5½	½	2½	1⅛

Dimensions in Inches

Dimension D is laying length.
Note: A minimum water seal of 2 inches is provided for 2 inch size and smaller; 2½ inches for sizes 3 to 6 inches inclusive.

(j)

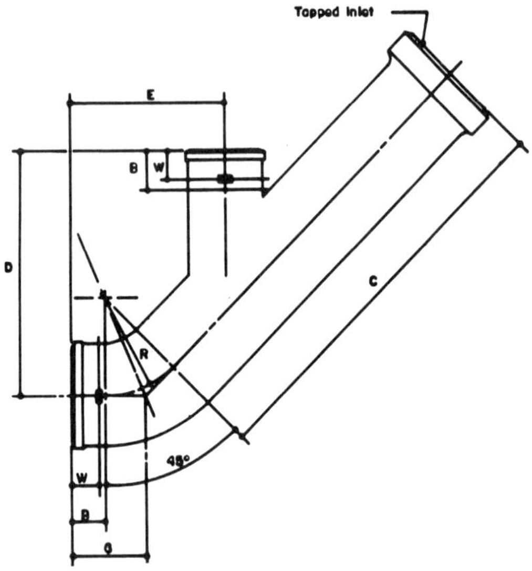

Stack Wye

Size (Inches)	D (±1/8)	B	C	G	E (±1/8)	IPS Tapping	W	R
4 x 3 x 3½	10	1½	16⅞	3⅛	6½	3½	1⅛	4

Dimensions in Inches

Dimensions D and E are laying lengths.

(k)

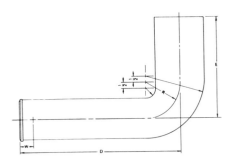

Closet Bend

Size (Inches)	D (±1/8)	E (±1/8)	R	W
3 x 4	Various	Various	3	1⅛
4 x 4	Various	Various	3	1⅛

Dimensions in Inches

Dimensions D and E are laying lengths.
Inclusion of spigot bead and positioning lug optional with manufacturer based on casting method used.

(l)

Figure 8.10 *(Continued)* Hubless CI pipe fittings.

PIPING MATERIALS AND STANDARD FITTINGS / 423

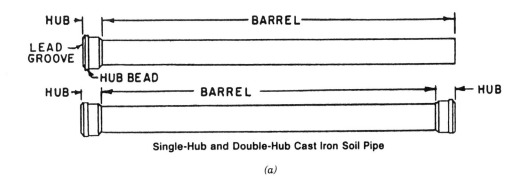

Single-Hub and Double-Hub Cast Iron Soil Pipe

(a)

TABLE 1 Dimensions of Hubs, Spigots, and Barrels for Extra-Heavy and Service Cast Iron Soil Pipe and Fittings, in.

NOTE—1 in. = 25.4 mm; 1 ft = 0.3 m throughout Tables.

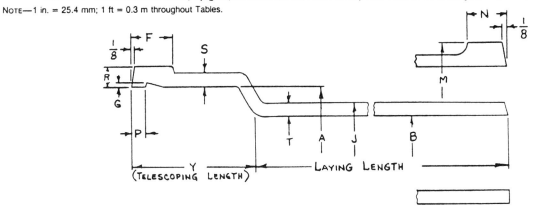

Extra-Heavy Cast Iron Soil Pipe and Fittings:

Size[A] Availability	Inside Diameter of Hub	Outside Diameter of Barrel	Telescoping Length	Inside Diameter of Barrel	Thickness of Barrel	
	A	J	Y	B	T Nom	T Min
2*	3.06	2.38	2.50	2.00	0.19	0.16
3*	4.19	3.50	2.75	3.00	0.25	0.22
4*	5.19	4.50	3.00	4.00	0.25	0.22
5*	6.19	5.50	3.00	5.00	0.25	0.22
6*	7.19	6.50	3.00	6.00	0.25	0.22
8*	9.50	8.62	3.50	8.00	0.31	0.25
10*	11.62	10.75	3.50	10.00	0.37	0.31
12*	13.75	12.75	4.25	12.00	0.37	0.31
15*	16.95	15.88	4.25	15.00	0.44	0.38

Size[A]	Thickness of Hub		Width of Hub Bead[C]	Distance from Lead Groove to End, Pipe and Fittings	Depth of Lead Groove	
	Hub Body S (min)	Over Bead R (min)	F	P	G (min)	G (max)
2	0.18	0.37	0.75	0.22	0.10	0.19
3	0.25	0.43	0.81	0.22	0.10	0.19
4	0.25	0.43	0.88	0.22	0.10	0.19
5	0.25	0.43	0.88	0.22	0.10	0.19
6	0.25	0.43	0.88	0.22	0.10	0.19
8	0.34	0.59	1.19	0.38	0.15	0.22
10	0.40	0.65	1.19	0.38	0.15	0.22
12	0.40	0.65	1.44	0.47	0.15	0.22
15	0.46	0.71	1.44	0.47	0.15	0.22

[A] Nominal inside diameter.
* Indicates this item is made in extra heavy.
[C] Hub ends and spigot ends can be made with or without draft.

(b)

Figure 8.11 Typical dimensional data for cast-iron hub and spigot soil pipe and fittings. (Copyright ASTM, reprinted with permission.)

TABLE 1 *Continued*

Service Cast Iron Soil Pipe:

Size[A] Availability	Inside Diameter of Hub	Outside Diameter of Barrel[C]	Telescoping Length[C]	Inside Diameter of Barrel[C]	Thickness of Barrel[C]	
	A	*J*	*Y*	*B*	*T*	
					Nom	Min
2O	2.94	2.30	2.50	1.96	0.17	0.14
3O	3.94	3.30	2.75	2.96	0.17	0.14
4O	4.94	4.30	3.00	3.94	0.18	0.15
5O	5.94	5.30	3.00	4.94	0.18	0.15
6O	6.94	6.30	3.00	5.94	0.18	0.15
8O	9.25	8.38	3.50	7.94	0.23	0.17
10O	11.38	10.50	3.50	9.94	0.28	0.22
12O	13.50	12.50	4.25	11.94	0.28	0.22
15O	16.95	15.88	4.25	15.16	0.36	0.30

Size[A]	Thickness of Hub		Width of Hub Bead[C]	Distance from Lead Groove to End, Pipe and Fittings	Depth of Lead Groove	
	Hub Body	Over Bead				
	S (min)	*R* (min)	*F*	*P*	*G* (min)	*G* (max)
2	0.13	0.34	0.75	0.22	0.10	0.19
3	0.16	0.37	0.81	0.22	0.10	0.19
4	0.16	0.37	0.88	0.22	0.10	0.19
5	0.16	0.37	0.88	0.22	0.10	0.19
6	0.18	0.37	0.88	0.22	0.10	0.19
8	0.19	0.44	1.19	0.38	0.15	0.22
10	0.27	0.53	1.19	0.38	0.15	0.22
12	0.27	0.53	1.44	0.47	0.15	0.22
15	0.30	0.58	1.44	0.47	0.15	0.22

[A] Nominal inside diameter.
O indicates this item is made in service weight.
[C] Hub ends and spigot ends can be made with or without draft.

(b) (continued)

Dimensions of One-Quarter Bends

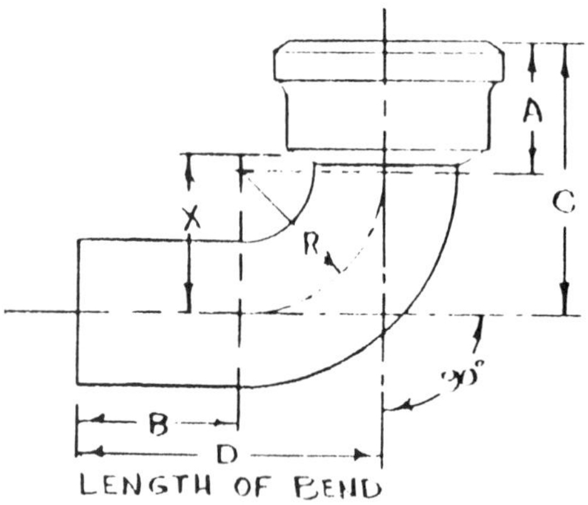

NOTE 1—1 in. = 25.4 mm.
NOTE 2—Dimensions *D* and *X* are laying lengths

Size, in., Availability[A]	Dimensions in in.[B]					
	A	*B*	C	*D*	R	X
2*O	2¾	3	5¾	6	3	3¼
3*O	3¼	3½	6¾	7	3½	4
4*O	3½	4	7½	8	4	4½
5*O	3½	4	8	8½	4½	5
6*O	3½	4	8½	9	5	5½
8*O	4⅛	5½	10⅛	11½	6	6⅝
10*O	4⅛	5½	11⅛	12½	7	7⅝
12*O	5	7	13	15	8	8¾
15*O	5	7	14½	16½	9½	10¼

[A] * indicates this item is made in extra heavy.
O indicates this item is made in service weight.
[B] For details of hubs and spigots, see Table 1.

(c)

Figure 8.11 *(Continued)*

Dimensions of Long One-Quarter Bends

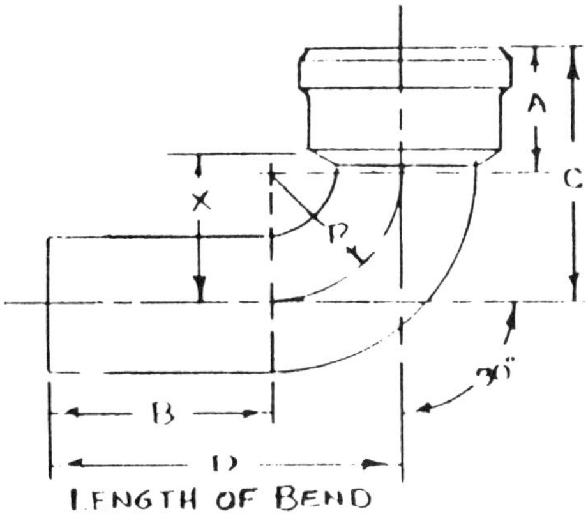

NOTE 1—1 in. = 25.4 mm.
NOTE 2—Dimensions *D* and *X* are laying lengths

Size, in., Availability[A]	Dimensions in in.[B]					
	A	*B*	*C*	*D*	*R*	*X*
2 by 12*O	2¾	9	5¾	12	3	3¼
2 by 18*O	2¾	15	5¾	18	3	3¼
2 by 24*O	2¾	21	5¾	24	3	3¼
3 by 12*O	3¼	8½	6¾	12	3½	4
3 by 18*O	3¼	14½	6¾	18	3½	4
3 by 24*O	3¼	20½	6¾	24	3½	4
4 by 12*O	3½	8	7½	12	4	4½
4 by 18*O	3½	14	7½	18	4	4½
4 by 24*O	3½	20	7½	24	4	4½

[A] * indicates this item is made in extra heavy.
O indicates this item is made in service weight.
[B] For details of hubs and spigots, see Table 1.

(d)

Dimensions of One-Quarter Bends with Low Heel Inlet

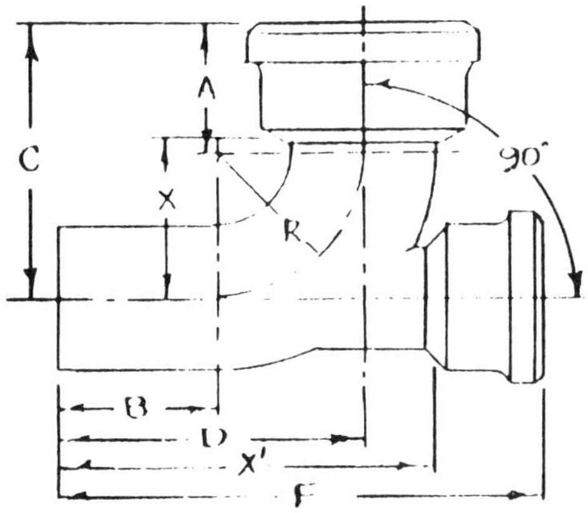

NOTE 1—1 in. = 25.4 mm.
NOTE 2—Dimensions *D*, *X*, and *X'* are laying lengths

Size, in., Availability[A]	Dimensions in in.[B]							
	A	*B*	*C*	*D*	*F*	*R*	*X*	*X'*
3 by 2O	3¼	3½	6¾	7	11½	3½	4	9
4 by 2*O	3½	4	7½	8	13	4	4½	10½
4 by 3*O	3½	4	7½	8	13¼	4	4½	10½

[A] * indicates this item is made in extra heavy.
O indicates this item is made in service weight.
[B] For details of hubs and spigots, see Table 1.

(e)

Dimensions of Long Sweep Bends

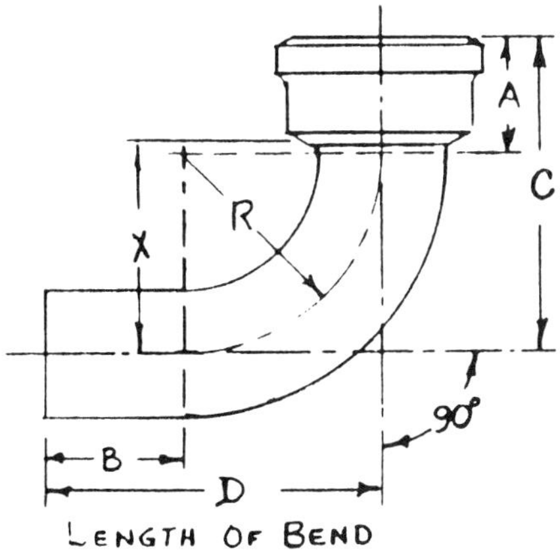

Long Sweep Bends:

NOTE 1—1 in. = 25.4 mm.
NOTE 2—Dimensions *D* and *X* are laying lengths

Size, in., Availability[A]	Dimensions in in.[B]					
	A	*B*	*C*	*D*	*R*	*X*
2*O	2¼	3	10¾	11	8	8¼
3*O	3¼	3½	11¾	12	8½	9
4*O	3½	4	12½	13	9	9½
5*O	3½	4	13	13½	9½	10
6*O	3½	4	13½	14	10	10½
8*O	4⅛	5½	15⅛	16½	11	11⅝
10*O	4⅛	5½	16⅛	17½	12	12⅝
12*O	5	7	18	20	13	13¾
15O	5	7	19½	21½	14½	15¼

[A] * indicates this item is made in extra heavy.
O indicates this item is made in service weight.
[B] For details of hubs and spigots, see Table 1.

(f)

Figure 8.11 *(Continued)*

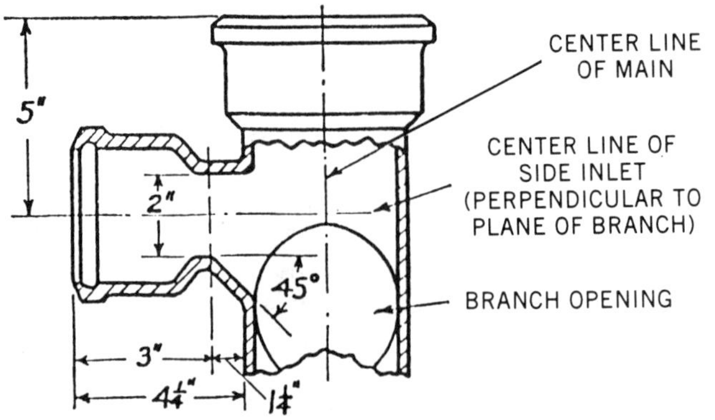

NOTE 1—1 in. = 25.4 mm.

NOTE 2—Dimensions and location of 2-in. side inlet for single or double sanitary T branches and Y branches are shown above. Single and double sanitary T branches and single and double Y branches with 2-in. side inlets are standard in the following sizes only: 4 by 3 by 2-in; 4 by 4 by 2-in.; 5 by 4 by 2-in; 6 by 4 by 2-in.

Dimensions and Locations for 2-in. Side Inlets

(g)

Dimensions of Y Branches, Single and Double

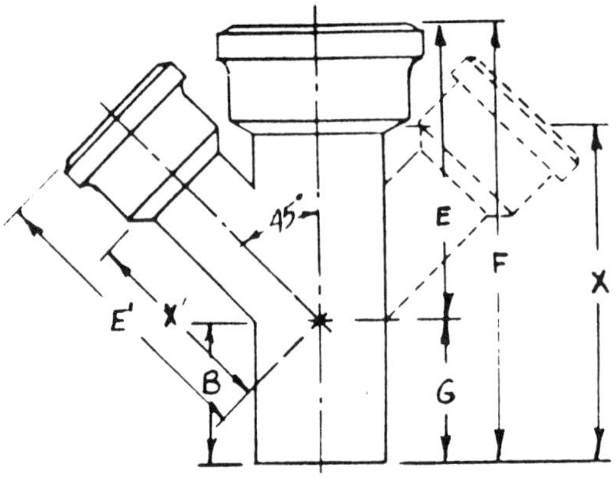

Note 1—1 in. = 25.4 mm.
Note 2—Dimensions *D* and *X* are laying lengths

Size, in., Availability[A]	Single Dimensions in in.[B]						
	B (min)	E	E'	F	G	X	X'
2*○	3 1/2	6 1/2	6 1/2	10 1/2	4	8	4
3*○	4	8 1/4	8 1/4	13 1/4	5	10 1/2	5 1/2
4*○	4	9 3/4	9 3/4	15	5 1/4	12	6 3/4
5*○	4	11	11	16 1/2	5 1/2	13 1/2	8
6*○	4	12 1/4	12 1/4	18	5 3/4	15	9 1/4
8*○	5 1/2	15 5/16	15 5/16	23	7 11/16	19 1/2	11 13/16
3 by 2*○	4	7 9/16	7 1/2	11 3/4	4 3/16	9	5
4 by 2*○	4	8 3/8	8 1/4	12	3 5/8	9	5 3/4
4 by 3*○	4	9 1/16	9	13 1/2	4 7/16	10 1/2	6 1/4
5 by 2*○	4	8 7/8	9	12	3 1/8	9	6 1/2
5 by 3*○	4	9 5/8	9 3/4	13 1/2	3 7/8	10 1/2	7
5 by 4*○	4	10 5/16	10 1/2	15	4 11/16	12	7 1/2
6 by 3*○	4	10 1/8	10 1/2	13 1/2	3 3/8	10 1/2	7 3/4
6 by 4*○	4	10 13/16	11 1/4	15	4 3/16	12	8 1/4
6 by 5*○	4	11 9/16	11 3/4	16 1/2	4 15/16	13 1/2	8 3/4
8 by 4*○	5 1/2	12 1/4	12 1/2	17	4 3/4	13 1/2	9 1/2
8 by 5*○	5 1/2	13	13	18 1/2	5 1/2	15	10
8 by 6*○	5 1/2	13 11/16	13 1/2	20	6 5/16	16 1/2	10 1/2

[A] * indicates this item is made in extra heavy;
○ indicates this item is made in service weight.
[B] For details of hubs and spigots, see Table 1.

Note 1—1 in. = 25.4 mm.

Size, in., Availability[A]	Double Dimensions in in.[B]						
	B (min)	E	E'	F	G	X	X'
2*○	3 1/2	6 1/2	6 1/2	10 1/2	4	8	4
3*○	4	8 1/4	8 1/4	13 1/4	5	10 1/2	5 1/2
4*○	4	9 3/4	9 3/4	15	5 1/4	12	9 1/4
5*○	4	11	11	16 1/2	5 1/2	13 1/2	8
6*○	4	12 1/4	12 1/4	18	5 3/4	15	9 1/4
8*○	5 1/2	15 5/16	15 5/16	23	7 11/16	19 1/2	11 13/16
3 by 2*○	4	7 9/16	7 1/2	11 3/4	4 3/16	9	5
4 by 2*○	4	8 3/8	8 1/4	12	3 3/8	9	5 3/4
4 by 3*○	4	9 1/16	9	13 1/2	4 7/16	10 1/2	6 1/4
5 by 2*○	4	8 7/8	9	12	3 1/8	9	6 1/2
5 by 3*○	4	9 5/8	9 3/4	13 1/2	3 7/8	10 1/2	7
5 by 4*○	4	10 5/16	10 1/2	15	4 11/16	12	7 1/2
6 by 3*○	4	10 1/8	10 1/2	13 1/2	3 3/8	10 1/2	7
6 by 4*○	4	10 13/16	11 1/4	15	4 3/16	12	8 1/4
6 by 5*○	4	11 9/16	11 3/4	16 1/2	4 15/16	13 1/2	8 3/4
8 by 5*○	5 1/2	13	13	18 1/2	5 1/2	15	10
8 by 6*○	5 1/2	13 11/16	13 1/2	20	6 5/16	16 1/2	10 1/2

[A] * indicates this item is made in extra heavy;
○ indicates this item is made in service weight.
[B] For details of hubs and spigots, see Table 1. For details of side inlets see Figure 8.11g.

(h)

Figure 8.11 *(Continued)*

Dimensions of Y Branch Cleanout with Screw Plug on Branch

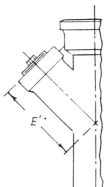

Note 1—1—1 in. = 25.4 mm.
Note 2—Dimensions X and X' are laying lengths.

Size, in., Availability[B]	Dimensions in in.[A]					Minimum I.P.S.[C,D] tapping
	E	E'	F	G	X	
2*O	6⅛	5¼	10½	4	8	1½
3*O	8¼	6⅝	13¼	5	10½	2½
4*O	9¾	7⅞	15	5¼	12	3½

[A] For details of hubs and spigots, see Table 1.
[B] * indicates this item is made in extra heavy.
  O indicates this item is made in service weight.
[C] Iron pipe sizes.
[D] Tappings permit entrance of testing plugs.

(i)

Dimensions of Combination Y and One-Eighth Bend, Single and Double

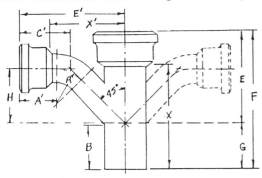

Note 1—1 in. = 25.4 mm.
Note 2—Dimensions X and X' are laying lengths.

Size, in., Availability[B]	Dimensions in in.[A]										
	A'	B (min)	C'	E	E'	F	G	H	R'	X	X'
Single:											
2*O	2¾	3½	4	6½	7⅜	10½	4	3⅜	3	8	4⅞
3*O	3¼	4	4 11/16	8¼	9¾	13¼	5	5 1/16	3½	10½	7
4*O	3½	4	5 3/16	9¾	12	15	5¼	6 13/16	4	12	9
5*O	3½	4	5⅜	11	14	16½	5½	8⅝	4½	13½	11
6*O	3½	4	5 9/16	12¼	15⅞	18	5¾	10 5/16	5	15	12⅞
8*O	4⅛	5½	6⅝	15 5/16	20½	23	7 11/16	13⅞	6	19½	17
3 by 2*O	3	4	4¼	7 9/16	8¼	11¾	4 3/16	4	3	9	5¾
4 by 2*O	3	4	4¼	8 5/16	8¾	12	3 11/16	4½	3	9	6¼
4 by 3*O	3¼	4	4 11/16	9	10¼	13½	4½	5 9/16	3½	10½	7½
5 by 3*O	3¼	4	4 11/16	9½	10¾	13½	4	6 1/16	3½	10½	8
5 by 4*O	3½	4	5 3/16	10¼	12½	15	4¼	7 5/16	4	12	9½
6 by 4*O	3½	4	5 3/16	10¾	13	15	4¼	7 13/16	4	12	10
6 by 5*O	3½	4	5⅜	11 7/16	14½	16½	5 1/16	9⅛	4½	13½	11½
8 by 4*O	3½	5½	5 3/16	12¼	14	17	4¾	8 13/16	4	13½	11
8 by 5*O	3½	5½	5⅜	13	15½	18½	5½	10⅛	4½	15	12½
8 by 6*O	3½	5½	5 9/16	13 11/16	16⅞	20	6 5/16	11 5/16	5	16½	13⅞
Double:											
2*O	2¼	3½	4	6½	7⅜	10½	4	3⅜	3	8	4⅞
3*O	3¼	4	4 11/16	8¼	9¾	13¼	5	5 1/16	3½	10½	7
4*O	3½	4	5 3/16	9¾	12	15	5¼	6 13/16	4	12	9
5*O	3½	4	5⅜	11	14	16½	5½	8⅝	4½	13½	11
6*O	3½	4	5 9/16	12¼	15⅞	18	5¾	10 5/16	5	15	12⅞
3 by 2*O	3	4	4¼	7 9/16	8¼	11¾	4 3/16	4	3	9	5¾
4 by 2*O	3	4	4¼	8 5/16	8¾	12	3 11/16	4½	3	9	6¼
4 by 3*O	3¼	4	4 11/16	9	10¼	13½	4½	5 9/16	3½	10½	7½
5 by 4*O	3½	4	5 3/16	10¼	12½	15	4¾	7 5/16	4	12	9½
6 by 4*O	3½	4	5 3/16	10¾	13	15	4¼	7 13/16	4	12	10

[A] For details of hubs and spigots, see Table 1; for details of side inlets, see Figure 8.11g.
[B] * indicates this item is made in extra heavy;
  O indicates this item is made in service weight.

Figure 8.11 (Continued) (j)

Dimensions of Sanitary T Branches, Single and Double

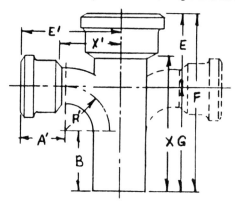

Note 1—1 in. = 25.4 mm.
Note 2—Dimensions X and X' are laying lengths.

Size, in., Availability[B]	Dimensions in in.[A]								
	A'	B	E	E'	F	G	R'	X	X'
Single:									
2*○	2¾	3¾	4¼	5¼	10½	6¼	2½	8	2¾
3*○	3¼	4	5¼	6¾	12¾	7½	3½	10	4¾
4*○	3½	4	6	7½	14	8	4	11	4½
5*○	3½	4	6½	8	15	8½	4½	12	5
6*○	3½	4	7	8½	16	9	5	13	5½
8*○	4⅛	5¾	8¾	10⅛	20½	11¾	6	17	6⅝
3 by 2*○	3	4	4¾	6½	7	3	9	4	
4 by 2*○	3	4	5	7	12	7	3	9	4½
4 by 3*○	3¼	4	5½	7½	13	7½	3½	10	4½
5 by 3*○	3¼	4	5½	7¾	13	7½	3½	10	5
5 by 4*○	3½	4	6	8	14	8	4	11	5
6 by 4*○	3½	4	6	8½	14	8	4	11	5½
6 by 5*○	3½	4	6½	8½	15	8½	4½	12	5½
8 by 5*○	3½	5¾	7¼	9½	17½	10¼	4½	14	6½
8 by 6*○	3½	5¾	7¾	9½	18½	10¾	5	15	6½
Double:									
2*○	2¾	3¾	4¼	5¼	10½	6¼	2½	8	2¾
3*○	3¼	4	5¼	6¾	12¾	7½	3½	10	4
4*○	3½	4	6	7½	14	8	4	11	4½
5*○	3½	4	6½	8	15	8½	4½	12	5
6*○	3½	4	7	8½	16	9	5	13	5½
3 by 2*○	3	4	4¾	6½	11¾	7	3	9	4
4 by 2*○	3	4	5	7	12	7	3	9	4½
5 by 4*○	3½	4	6	8	14	8	4	11	5½
6 by 4*○	3½	4	6	8½	14	8	4	11	5½
8 by 6*○	3½	5¾	7¾	9½	18½	10¾	5	15	6½

[A] For details of hubs and spigots, see Table 1; for details of side inlets, see Figure 8.11g.
[B] * indicates this item is made in extra heavy;
○ indicates this item is made in service weight.

(k)

Figure 8.11 *(Continued)*

PIPING MATERIALS AND STANDARD FITTINGS / 429

Dimensions of Sanitary T Branches, Tapped, Single and Double

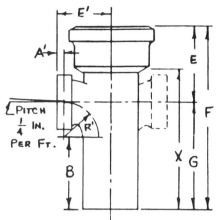

Note 1—1 in.=25.4 mm.
Note 2—Dimension X is the laying length.

Size, in.,[A] Availability[B]	Dimensions in in.[C]							
	A	B	E	E'	F	G	R'	X
2 by 2·O	13/16	4	4 1/2	3 1/16	10 1/2	6 1/4	2 1/4	8
3 by 2·O	13/16	4 3/4	4 3/4	3 9/16	11 3/4	7	2 1/4	9
4 by 2·O	13/16	4 3/4	5	4 1/16	12	7	2 1/4	9
5 by 2·O	13/16	4 3/4	5	4 9/16	12	7	2 1/4	9
6 by 2·O	13/16	4 3/4	5	5 1/16	12	7	2 1/4	9
2·O[D]	...	4 1/2	...	2 13/16	...	...	1 3/4	...
3·O[D]	...	5 1/4	...	3 5/16	...	...	1 3/4	...
4·O[D]	...	5 1/4	...	3 13/16	...	...	1 3/4	...

Note 1—1 in.=25.4 mm.
Note 2—Dimension X is the laying length.

Size, in.,[A] Availability[B]	Dimensions in in.[C]							
	A'	B	E	E'	F	G	R'	X
Double:								
2 by 2·O	13/16	4	4 1/2	3 1/16	10 1/2	6 1/4	2 1/4	8
3 by 2·O	13/16	4 3/4	4 3/4	3 9/16	11 3/4	7	2 1/4	9
4 by 2·O	13/16	4 3/4	5	4 1/16	12	7	2 1/4	9
2·O[D]	...	4 1/2	...	2 13/16	...	...	1 3/4	...
3·O[D]	...	5 1/4	...	3 5/16	...	...	1 3/4	...
4·O[D]	...	5 1/4	...	3 13/16	...	...	1 3/4	...

[A] All sizes of branches are furnished with 1 1/4 and 1 1/2 in. tappings, in addition to the 2 in. tapping.
[B] · Indicates this item is made in extra heavy.
○ indicates this item is made in service weight.
[C] For details of hubs and spigots, see Table 1.
[D] Dimensions for 1 1/4 in. and 1 1/2 in. tapping only.

(l)

Dimensions of Vent Branches, Single

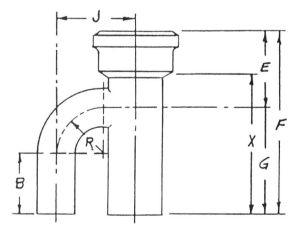

NOTE 1—1 in. = 25.4 mm.
NOTE 2—Dimension X is the laying length.

Size, in. Availa-bility[B]	Dimensions in in.[A]						
	B	E	F	G	J	R'	X
2·O	3 1/4	4 1/4	10 1/2	6 1/4	4 1/2	3	8
3·O	4	5 1/4	12 3/4	7 1/2	5 1/2	3 1/2	10
4·O	4	6	14	8	6 1/2	4	11
3 by 2O	4	4 3/4	11 3/4	7	5	3	9
4 by 2·O	4	5	12	7	5 1/2	3	9
4 by 3·O	4	5 1/2	13	7 1/2	6	3 1/2	10

[A] For details of hubs and spigots, see Table 1.
[B] · indicates this item is made in extra heavy.
○ indicates this item is made in service weight.

(m)

Figure 8.11 (Continued)

430 / PRINCIPLES OF PLUMBING

Dimensions of Double Hub and Long Double Hub

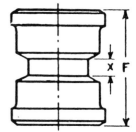

Double-Hub:
NOTE 1—1 in. = 25.4 mm.
NOTE 2—Dimension X is the laying length.

Size, in.[A] Availability[B]	F, in.	X, in.	Size, in.[A]	F, in.	X, in.
2*O	6	1	8	8¼	1¼
3*O	6½	1	10	8¼	1¼
4*O	7	1	12	10	1½
5*O	7	1	15	10	1½
6*O	7	1			

Long Double Hubs:
NOTE—Dimension X is the laying length.

Size, in.[A]	F, in.	X, in.
2 by 30*O	30	25
3 by 30*O	30	24½
4 by 30*O	30	24

[A] For details of hubs and spigots, see Table 1
[B] * indicates this item is made in extra heavy.
O indicates this item is made in service weight.

(n)

Dimensions of Plain P Trap

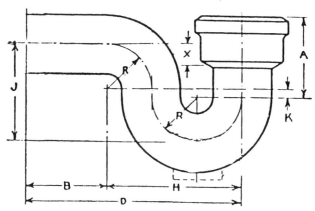

Note 1—1 in. = 25.4 mm.
Note 2—A minimum water seal of 2 in. is provided for the 2-in. size, of 2½ in. for sizes 3 to 6 in., inclusive.
Note 3—Dimensions D and X are laying lengths. Dimension X is measured below the horizontal center line on sizes 5 by 5 in. and smaller.

Size, in Trap by Vents	Availability[A]	Dimensions in in.[B]							
		A	B	D	H	J	K	R	X
2	*O	3	3½	9½	6	4	...	2	1½
3	*O	4½	4½	12	7½	5⅓	½	2½	1¼
4	*O	5½	5	14	9	6½	½	3	1
5	*O	6½	5	15½	10½	7½	½	3½	½
6	*O	7½	5	17	12	8½	½	4	...

[A] * indicates this item is made in extra heavy.
O indicates this item is made in service weight.
[B] For details of hubs and spigots, see Table 1.

(o)

Figure 8.11 (Continued)

Table 8.3 Physical Characteristics of Copper Pipe

Size, in.	O.D., in.	Wall Thickness, in.				Calculated Weight, lb/ft			
		Type K	Type L	Type M	Type DWV	Type K	Type L	Type M	Type DWV
3/8	0.500	0.049	0.035	0.025	—	0.269	0.198	0.145	—
1/2	0.625	0.049	0.040	0.028	—	0.344	0.285	0.204	—
5/8	0.750	0.049	0.042	—	—	0.418	0.362	—	—
3/4	0.875	0.065	0.045	0.032	—	0.641	0.455	0.328	—
1	1.125	0.065	0.050	0.035	—	0.839	0.655	0.465	—
1 1/4	1.375	0.065	0.055	0.042	0.040	1.04	0.884	0.682	0.65
1 1/2	1.625	0.072	0.060	0.049	0.042	1.36	1.14	0.940	0.81
2	2.125	0.083	0.070	0.058	0.042	2.06	1.75	1.46	1.07
2 1/2	2.625	0.095	0.080	0.065	—	2.93	2.48	2.03	—
3	3.125	0.109	0.090	0.072	0.045	4.00	3.33	2.68	1.69
3 1/2	3.625	0.120	0.100	0.083	—	5.12	4.29	3.58	—
4	4.125	0.134	0.110	0.095	0.058	6.51	5.38	4.66	2.87
5	5.125	0.160	0.125	0.109	0.072	9.67	7.61	6.66	4.43
6	6.125	0.192	0.140	0.122	0.083	13.9	10.2	8.92	6.10
8	8.125	0.271	0.200	0.170	0.109	25.9	19.3	16.5	10.6

soldered joint uses "capillary" fittings, named for the capillary action, which they employ. They are also called *sweat joints*. They are quickly and easily made, thus offsetting by low labor cost the higher material cost of the piping. Figure 8.12 shows this type of joint. Flared joints are mechanical and rely entirely on pressure to make them gas and liquid tight. They are used only on soft copper tubing in applications that require disassembly of pipes. The joint is made by flaring the tubing with a special tool and then tightening the threaded fitting collar onto the tube. See Figure 8.12 for details of both

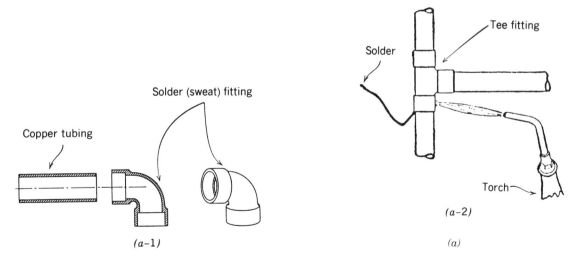

Figure 8.12 (a) A soldered or sweat joint is made by filling the small space between the outside of the pipe and the inside of the fitting with solder after sliding them together. The joint is prepared by cleaning and fluxing the contact surfaces. The joint is then heated with a torch while applying the solder. The solder is drawn into the joint by capillary action, completely sealing the space.

432 / PRINCIPLES OF PLUMBING

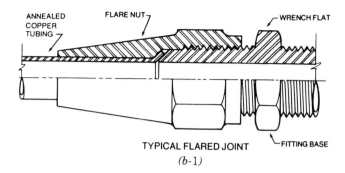

Figure 8.12 *(b-1)* A flared compression joint is made by screwing the compression fitting onto the flared tubing, which, in turn, rests on the fitting sleeve. The flare is made with a special flaring tool. The joint is used only for soft temper (annealed) copper tubing. (Drawing reproduced with permission from National Standard Plumbing Code, published by the National Association of Plumbing Heating Cooling Contractors.)

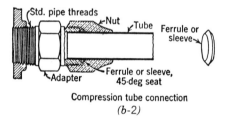

Figure 8.12 *(b-2)* Tubing joints without flaring are made using a ferrule or sleeve, as shown. The nut compresses the ferrule between itself, the tube and the 45° ferrule seat to make a tight joint. (From Severns and Fellows, *Air Conditioning and Refrigeration*, 1958, © John Wiley & Sons, reprinted by permission of John Wiley & Sons.)

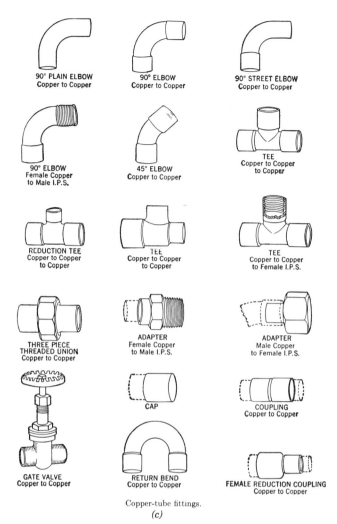

Copper-tube fittings.
(c)

Figure 8.12 *(c)* Typical copper tube fittings (Severns and Fellows, *Air Conditioning and Refrigeration*, © John Wiley & Sons, reprinted by permission of John Wiley & Sons.)

sweat joints and flared compression joints, with typical copper tube fittings.

c. Plastic Pipe

In recent years, plastic pipes of various types have become common in plumbing systems for a number of reasons:

- Plastic pipe is cheaper than metal pipe.
- Plastic pipe is lightweight making handling easier.
- Plastic pipe is simple to join, attach at fittings and, in general, to install. This reduces high labor costs.
- Plastic pipe is very highly resistant to corrosion. As a result, it can be used in corrosive atmospheres and with corrosive liquids that would quickly destroy metallic piping. Some of the plastic pipes available today are so highly corrosion-resistant that they carry a 50-yr guarantee.
- Because so many different types of plastics are available, plastic pipe can be selected to meet specific job requirements and for a large enough project, even specially fabricated for special job conditions.
- Plastic pipe is very smooth. This reduces friction in water systems and may permit the use of smaller pipes and pumps than is possible with steel pipe. This further reduces cost.
- Many of the plastics used for pipes are thermoplastic, that is, they melt when heat is applied. This characteristic is convenient when bends that are beyond the normal flexibility of the pipe have to be made. Like copper tubing, many types of plastic pipe come in rolls that permit snaking the pipe into inaccessible areas.
- Because plastic is an electrical insulator, it can be used where metal would constitute an electrical hazard.

Disadvantages of plastic pipes include:

- They lack physical strength. As a result plastic pipe crushes easily and requires much more support than metal pipe. These supports add to installation costs. The physical weakness also severely limits the system pressure (internal) for which most plastic pipe can be used.
- The thermoplastic properties of some plastics (melt when heated) can be a great disadvantage when pipe is exposed to large temperature variations. This limits the use of most plastics for hot water systems where water temperature exceed 140°F. (Some plastics are available that will carry 180°F water safely. See the listing that follows.)
- Plastic has a much higher coefficient of expansion than metallic pipe [see Section 8.13(c). As a result, long runs of flexible pipe must be snaked, and runs of rigid pipe must have frequent expansion loops.

The principal types of plastic pipe in use in plumbing work today are

ABS (acrylonitrile butadiene styrene),
PE (polyethylene),
PVC (polyvinyl chloride) and
CPVC (chlorinated polyvinyl chloride).

Each type has different characteristics and different applications. We will discuss each briefly. (Type PB, polybutylene, which was once quite common, has been largely replaced by CPVC.)

Type ABS pipe is widely used for DWV (drainage waste and vent) piping. Joints are cemented using a special plastic pipe glue. Once the glue has set, the joint cannot be opened. The plastic will soften at high temperatures so that it is not usable for high-temperature water such as a dishwasher drain. See Figure 8.13 for typical plastic DWV fittings. Type ABS is manufactured only as rigid pipe.

Type PE is a flexible pipe, manufactured in coils of several hundred feet in length. Joints are made with solvent cement. Principal applications of PE pipe are for underground use as in sprinkler and water supply systems. Because of its thermoplasticity, it is not usable with hot water. The pipe is quite strong and can be used in most pressurized water systems. All plastic pipe intended for use with potable water systems must carry the NSF label of approval for this application. PE is not normally used for DWV piping. See Figure 8.14.

PVC and CPVC piping are similar in application except that CPVC is usable with water temperatures of up to 180°F and PVC is not. Both are rigid, can withstand high pressures and are usable in all plumbing applications above and below ground including DWV. Both are made in schedule 40 and schedule 80 sizes with the same dimensions as steel pipe of the same pipe size and schedule. Joints are usually made using a special PVC solvent cement, although schedule 80 pipe can be threaded. Type PVC is highly thermoplastic and must be supported along its entire length when used with hot

434 / PRINCIPLES OF PLUMBING

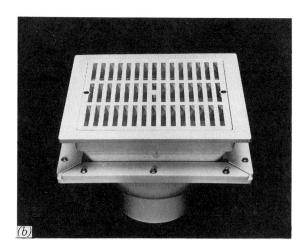

Figure 8.13 *(a)* Assortment of plastic piping and DWV fittings; made of ABS (black) and PVC (white) plastic. *(b)* White PVC floor sink drainage fitting. (Photos courtesy of R&G Sloane.)

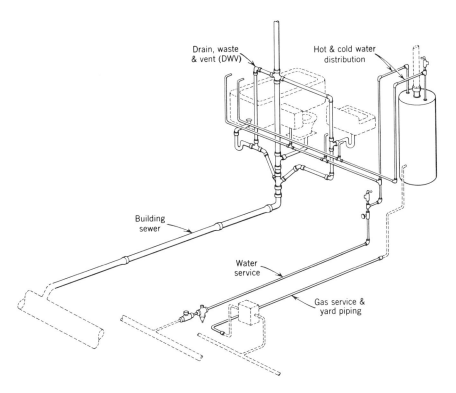

Figure 8.14 Plastic piping (solid lines) for water service, gas service, hot and cold waterlines and DWV. Gas service below grade can be PE, PB or PVC. (Courtesy of Plastics Pipe Institute.)

water or other hot liquids up to 110°F. Above that temperature, CPVC should be used. Like all plastic pipe, both types have a high coefficient of expansion and require frequent expansion loops. A 10-ft section of pipe will expand or contract ¼ in. for a 60 °F temperature change, which is about the temperature difference between warm and cold domestic water.

d. Nonmetallic Pipe Other Than Plastic

This classification includes vitrified clay (terra cotta), asbestos cement and concrete pipe. These materials, as allowed by Code, are generally used in large sizes (8 in. and above) for sewer construction.

e. Joints Between Dissimilar Materials

Where it is necessary to join pipes made of dissimilar metals, such as copper and iron pipe, special connectors should be used that will minimize corrosion. See Figure 8.15. These connectors are called "dielectric" (insulating) connectors because they contain electrical insulation that physically separates the two metals. This reduces the electrolytic action that is always present when dissimilar metals are connected in the presence of an electrolyte (electrically conducting fluid). In this case, the fluid is either domestic water or drainage water. All water, except distilled water, has some degree of electrical conductivity; the harder the water is, the better it conducts, and the better it conducts, the more corrosion there is. (Also, the hotter the water is, the more corrosion there is.) Without such dielectric connectors, corrosion is rapid, and joint clogging and failure follow.

In some cases, it is simply not possible to avoid electrolytic corrosion. To combat this process, a procedure known as *cathodic protection* can be used. The design of these systems is a specialty, not the responsibility of a plumbing technologist. However, as with many such specialties, it is important that a technologist know what the system does. Essentially, a cathodic protection system

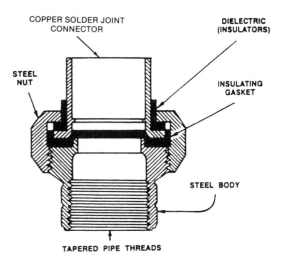

Figure 8.15 Dielectric connectors are used to reduce corrosion that occurs when dissimilar metals are connected. The illustrated section is of a soldered copper pipe connected to a threaded steel pipe. (Reproduced with permission from National Standard Plumbing Code, published by the National Association of Plumbing Heating Cooling Contractors.)

the entire height of the run. These columns are known as *wet columns* because of the piping. See Figure 8.16.

(2) Horizontal runs of water piping are always installed pitched (sloped) to permit draining the pipes for maintenance. This pitch must be constant over the entire length of a run. Low points encourage collection of sediments and eventual blockage. High points cause air pockets that result in noise, reduced flow and accelerated corrosion. Water system piping is usually pitched about ¼ in./ft. (The pitch of drainage piping controls the velocity of flow and is, therefore, a design decision. It is discussed in detail in Chapter 10).

(3) Piping should not be installed over electrical equipment because of the possibility of leaks and condensation dripping. Where such runs are unavoidable, drip pans piped to floor drains should be installed under the piping where it crosses the electrical equipment. This item is particularly important where piping crosses switchgear, panelboards, motor control centers and electrical control panels. These items of

supplies the required electrolytic current from another source, thus preventing corrosion of the protected item. The "other source" may be a sacrificial anode or a electrode connected to a source of voltage. The specific design depends on conditions and is beyond our scope here.

When the different materials are such that no electrolytic cell is created, such as joints between metal and plastic pipe, clay and plastic pipe or even cast iron to threaded steel pipe, special transitional joint materials and joint construction techniques must be used to ensure tight, maintenance-free connections. See the Code, Chapter 4, for details of dielectric and transitional joints.

8.13 Piping Installation

a. General Rules

The general rules governing the installation of plumbing piping follow:

(1) Vertical pipes should be plumb and installed in groups (banks) wherever possible. In high-rise buildings, this grouping is best accomplished by running alongside a column. The space occupied by these pipes is furred out for

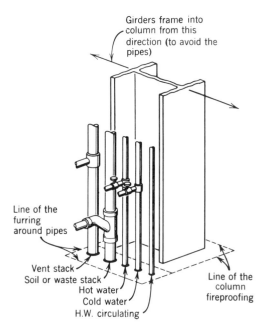

Figure 8.16 Schematic wet column arrangement. Details such as pipe supports at the floor and joints are not shown, for clarity. (From Stein and Reynolds, *Mechanical and Electrical Equipment for Buildings*, 8 ed., 1992, © John Wiley & Sons, reproduced by permission of John Wiley & Sons.

equipment are constructed with sheet metal enclosures and are particularly prone to damage and malfunction from water leakage. In contrast, motors and switches are available in drip-proof enclosures. The project electrical and mechanical designers should be notified if such equipment enclosures are required due to overhead plumbing piping.

(4) Ferrous metal pipe installed underground should be coated with asphaltum compound or an equivalent approved waterproofing material. Trenching, bedding and backfilling for underground piping should be in accordance with Code requirements. See Code Section 2.6 for installation details. Where underground pipes enter a building through masonry walls, the designer must provide a sleeve with sufficient clearance ($\frac{1}{2}$ in. minimum) for caulking around the pipe with approved materials. Asphaltum is the most commonly used of such materials although oakum and lead can still be found. The project structural engineer must be notified so that an appropriate arch or other means can be designed that will remove all structural load from such through-the-wall piping. Where pipes pass through fire-rated walls, the sleeve and packing around the pipe must be fire rated as well, to maintain the wall's fire rating integrity.

(5) Where piping is subject to temperatures below freezing, electrical heating tracer cables wrapped on the piping will probably be required. Consult the project electrical engineer for heating cable details and connections.

(6) Piping that requires frequent maintenance should be designed for easy disassembly. Joints in piping $2\frac{1}{2}$ in. and smaller should be made up with frequent use of unions. Larger pipes should use bolted flanged joints.

(7) Pipes carrying cold liquids will frequently cause the moisture present in the air to condense on the surface of the pipe and then drip off. This can cause mildew, ruin finished surfaces and even cause electrical equipment failure. A glance at the psychrometric chart in Figure 2.17 shows that a pipe with a surface temperature of 60°F (common for cold water piping) will cause the water in 85°F air to condense anytime the relative humidity (RH) exceeds 43%. Since RH in the summer frequently exceeds 43% in many areas of the United States, condensation will form on the cold water piping. To avoid this undesirable effect, pipes and fittings carrying cold liquids should be insulated. Glass fiber of minimum $\frac{1}{2}$ in. thickness is often used for this purpose. A vapor barrier on the outside of the insulation prevents penetration of the wet outside air to the cold pipe. Another benefit of this insulation is that it prevents the cold water from warming up as it travels along the piping. Condensation on cold water pipes, as on other cold surfaces exposed to warm wet air is commonly known as *sweating.* This is because the condensation forms in droplets that look like drops of perspiration.

b. Piping Supports

Horizontal supports spacing for piping depends on the inherent strength of the material being supported. Pipe strength also depends on pipe wall thickness. Steel is stronger than copper, and copper is stronger than plastic pipe. Supports are closer together for smaller and weaker materials. As with all aspects of installation, supports must meet local plumbing Code requirements. Suggested spacing of horizontal supports for piping are:

- Cast-iron soil pipe—every 5 ft
- Threaded steel pipe 1 in. diameter and smaller—every 8 ft; threaded steel pipe $1\frac{1}{4}$ in. and larger—every 12 ft
- Copper tubing $1\frac{1}{4}$ in. or less—every 6 ft; copper tubing $1\frac{1}{2}$ in. and larger—every 10 ft
- Plastic pipe—no more than 4 ft between supports, plus additional supports at changes in direction and elevation, at branch ends and at special fittings (Plastic pipe carrying hot liquids may require continuous support over its entire length.)

Vertical supports also depend on material strength but can be much farther apart than horizontal supports. Vertical piping essentially supports itself. The support fittings at the following intervals provide stiffening so that the pipe "column" does not flex. Support fittings for each type of pipe must adhere to manufacturers' recommendation in addition to meeting Code requirements. Generally, vertical supports are furnished as follows:

- Cast-iron soil pipe—at the riser base and at every floor
- Threaded steel pipe—at every other floor

- Copper pipe—at every floor but not exceeding 10 ft intervals
- Plastic pipe—per manufacturers' recommendations, for each type of plastic

All supports must be securely fastened to the building structure and should be designed to carry the weight of the piping plus the liquid contained in the piping. A few typical pipe support devices are shown in Figure 8.17.

c. Thermal Expansion

The problem of accommodating thermal expansion of piping is particularly important for hot water and steam piping. Although the problem is usually negligible in residential work because of short runs, it is definitely not so in high-rise buildings or in structures with long horizontal runs. The temperature difference to which hot water piping is subjected can easily reach 100°F. If the piping is installed at an air temperature of 60°F and carries hot water for laundry or commercial dishwasher use at not less than 160°F, the differential is 100°F. Table 8.4 lists the expansion coefficients and typical expansions for common piping materials. See also the tabulation in Section 3.10 (page 125) for copper piping.

Example 8.2 In a school building, the distance between the hot water boiler and the cafeteria dishwasher is 120 ft. What is the increase in length of the hot water piping from a "resting" condition (shutdown) of 50°F to an operating condition carrying 140°F water (a) using copper pipe? (b) using CPVC plastic pipe?

Solution:

(a) using the coefficient of expansion from Table 8.4 for copper pipe:

$$\text{Linear expansion} = 11.28 \times 10^{-5} \frac{\text{in.}}{\text{ft} \times °F}$$
$$\times (140-50)°F \times 120 \text{ ft}$$
$$= 11.28 \times 10^{-5} \times 90 \times 120 \text{ in}$$
$$= 1.22 \text{ in.}$$

The same result can be arrived at using the precalculated figures in Table 8.4 for 100 ft and Δt of 100F°:

$$\text{Linear expansion} = 1.128 \text{ in.} \times \frac{120 \text{ ft}}{100 \text{ ft}} \times \frac{90°F}{100°F}$$
$$= 1.22 \text{ inches}$$

(b) For CPVC pipe the expansion is much larger.

$$\text{Linear expansion} = 4.2 \times 10^{-4} \frac{\text{in.}}{\text{ft} \times °F} F \times 120 \text{ ft}$$
$$= 4.54 \text{ inches, or, using the pre-calculated Figure in Table 8.4 for 100 ft. and } \Delta t \text{ of 100F°:}$$

$$\text{Linear expansion} = 4.2 \text{ in.} \times \frac{120 \text{ ft}}{100 \text{ ft}}$$
$$\times \frac{90°F}{100°F} = 4.54 \text{ in.}$$

If this expansion is not provided for in some way, it will cause high physical stress in the

---

Table 8.4 Linear Thermal Expansion of Common Piping Materials

Material	Coefficient of Expansion, in./ft/°F	Linear Length Change, in./ 100 ft Pipe Length			
		±10°F	±20°F	±50°F	±100°F
Cast iron and steel	6.72×10^{-5}	0.067	0.134	0.336	0.672
Wrought iron	7.92×10^{-5}	0.079	0.158	0.396	0.792
Red brass	11.04×10^{-5}	0.11	0.221	0.552	1.104
Copper	11.28×10^{-5}	0.11	0.226	0.564	1.128
Plastic					
PVC	3.6×10^{-4}	0.36	0.72	1.8	3.6
CPVC	4.2×10^{-4}	0.42	0.84	2.1	4.2
ABS	6.6×10^{-4}	0.66	1.32	3.3	6.6
PB	9.0×10^{-4}	0.90	1.80	4.5	9.0

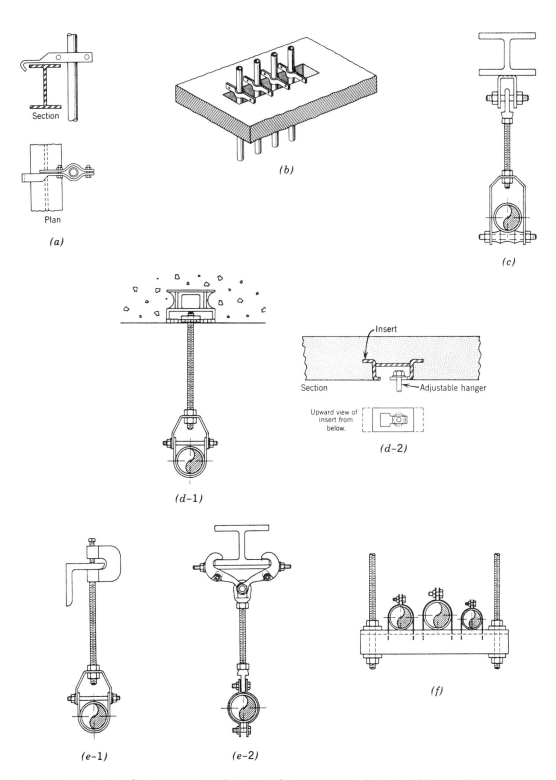

Figure 8.17 Typical pipe supports. *(a)* Vertical riser supported at a steel beam. *(b)* Vertical riser group supported at a slot in the concrete slab with pipe clamps. *(c)* Horizontal pipe support. The pipe is hung from slab above by an adjustable-length clevis hanger. The variable hanger length permits adjustment of the pipe pitch (slope). Note too that the pipe is supported on a roller fitting and is, therefore, free to move horizontally. This allows for thermal expansion movement. *(d-1)* Clevis hanger suspended from a ceiling insert. *(d-2)* Details of the ceiling insert. *(e)* Pipe hangers from steel members. *(e-1)* Clamp and clevis hanger on angle steel. *(e-2)* Beam clamp and rigid clamp hanger. *(f)* Typical trapeze arrangement. Notice that the entire trapeze can be adjusted vertically to achieve proper pipe pitch. [Drawings *(c)*, *(d-1)*, *(e-1)*, *(e-2)*, and *(f)* courtesy of B-Line Systems, Inc.]

pipe, and the pipe will buckle laterally. There are two ways to compensate for this expansion. One is to use linear expansion joints that work either with a type of bellows or by a sliding action within a watertight sleeve. The bellows-type unit is shown in Figure 8.18. The second method is to provide an expansion loop in the piping itself. The loop absorbs the pipe expansion by flexing the pipe and the joints in the loop. Two such arrangements are shown in Figure 8.19. Expansion joints and loops are installed approximately every 50 ft, depending on pipe size, type of material and design of the joint.

d. Test and Inspection

Plumbing technologists may be called upon to inspect plumbing roughing and finished plumbing work and, therefore, should be familiar with the procedures involved. Inspection of the roughing is made before fixture installation and before applying insulation to pipes requiring it. Tests must conform to local code requirements. Water supply systems are tested by sealing all piping, filling the system with potable water and applying a pressure as required by codes. Some codes call for a pressure of 150% of the system design pressure but not to exceed 100 psi. The National Standard Plumbing Code (Section 15) calls for a test at working pressure but not less than 80 psi. Inspectors check for leaks in the system.

Drainage and vent piping roughing (without fixtures) is tested with water or air, for leaks. The water test is done by completely filling the system and checking for leaks. Minimum static pressure should be 10 ft of water (4.3 psi). The air test uses compressed air at 5 psi for the same purpose. After fixture installation, the finished drainage system is tested basically for trap integrity, that is, to ensure that fixture water traps are effective. (Traps and venting are discussed in detail in Chapter 10.) The system is subjected to a pressure of 1 in. of water using smoke or peppermint oil to trace trap failure.

Once all the finished piping is tested, the fixtures are operated to check their adequacy. Finally, the water system is usually drained and refilled with chlorinated water for system disinfection. This is drained, and the system is flushed and then refilled. In some areas, codes require that potable water from the system be tested by a water-testing laboratory for purity before final approval of the potable water system is granted.

8.14 Valves

Flow of fluid in piping is controlled by valves. Valves are available in a great variety of sizes and designs ranging from tiny needle valves that control fluid flow in hydraulic control lines to huge valves that control flow in 7-ft diameter transcontinental pipelines. Fortunately, the plumbing technologist has a much more restricted choice, but even so the variety is huge. Even after selecting the type of valve required (globe, gate, etc.), the plumbing designer must specify material, type of operation (for instance rising stem or fixed stem), connection (threaded, flanged, sweated), packing material, renewability, pressure range, temperature and a half dozen other important characteristics. Here again, standards and experience with similar jobs will assist the technologist in this complex task. The material that follows is intended to provide you with basic knowledge on the operation and application of valves used in standard plumbing systems.

a. General

There are almost as many synonyms for the word *valve* as there are valve types. A valve is also referred to as a bib (a hose bib is a faucet outside a house for connection of a garden hose), a cock (a sill cock is the same as a hose bib), a faucet (which is usually at the end of a line for the purpose of releasing and throttling water flow), a shutoff (normally in a piping run, to permit maintenance work further down the line), a stop-and-waste (a shutoff valve with a drain that is placed upstream from piping to permit draining as, for instance, ahead of fixtures that cannot drain fully or ahead of outside piping subjected to freezing) and so on.

Valves are normally marked by the manufacturer with its name or trademark, size in inches and working pressure. In addition, some manufacturers identify valve types by color coding the plate under the hand wheel nut; red indicates a hot water or steam valve; black indicates cold water, gas or oil. Valve bodies are either cast or forged of brass, bronze, wrought iron or malleable iron. Faucets for bathrooms, washrooms, showers and kitchens are usually nickel-plated brass for appearance purposes. Faucets are also cast directly from white metal alloys for the same reason. Traditionally, the types of valves used in plumbing work were gate, globe, angle and check valves. In recent years, single-handle sink faucets for hot and cold

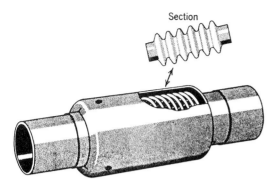

Figure 8.18 The bellows or accordian-type expansion fitting absorbs thermal expansion by linear movement. The rigid pipe is interrupted by a bellows-type fitting, as shown, that absorbs expansion in straight runs, without the necessity for expansion loops. Pipe supports on both sides of the fitting must be of the sliding type. Consult manufacturers' data for sizes and spacing between fittings in straight runs. See also the table of linear distances in Section 3.10 (page 125). (Courtesy of Dunham-Bush, Inc.)

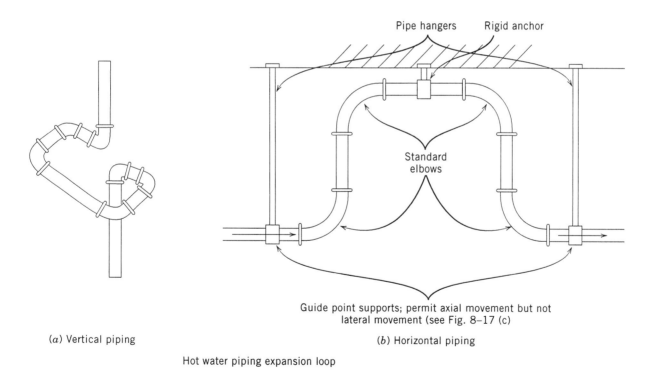

Figure 8.19 Expansion loops in (a) vertical and (b) horizontal piping runs. Thermal expansion is absorbed in the "springiness" of the pipe material.

Table 8.5 Nominal Valve Dimensions[a,b]

Size, in.	Gate[c]				Globe[d]		Angle[e]		Lift Check[f]		Swing Check[f]		Ball[g]		
	A	B[h]	C	E	F	G	H	I	J	K	L	M	S	T	U
Threaded Connections Unless Otherwise Noted															
1/4	1 3/4	4 1/2	1 3/4	3 3/4	1 3/4	2 3/4	7/8	3	1 7/8	1	2 1/8	1 1/2	2 1/2	1 1/4	4
3/8	1 3/4	4 1/2	1 3/4	3 3/4	1 7/8	2 7/8	1	3 1/4	2	1 1/8	2 1/8	1 1/2	2 3/4	1 1/4	4
1/2	2 1/8	5 1/4	2 1/8	3 3/4	2 1/4	3 1/2	1 1/8	3 3/4	2 1/2	1 3/8	2 1/2	1 3/4	3	1 1/4	4
3/4	2 1/4	6 1/2	2 5/8	4 1/2	2 3/4	4	1 3/8	4 1/4	3	1 7/8	3	2 1/8	3	1 1/2	4 1/2
1	2 3/4	7 3/4	2 3/4	5 1/4	3 3/8	4 1/2	1 5/8	5	3 1/2	2	3 3/4	2 1/2	3 1/2	2 1/8	6
1 1/4	3	9 1/4	3 1/8	5 7/8	3 7/8	4 7/8	1 3/4	5 1/2	4 1/8	2 3/8	4 1/4	3	4	2 5/8	7
1 1/2	3 1/4	10 1/2	3 5/8	7 1/4	4 1/2	5 1/2	2 1/4	6 1/4	4 5/8	2 5/8	5	3 1/2	4 1/4	2 3/4	7
2	3 3/4	12 3/4	4 3/4	8 3/8	5 1/4	6	2 5/8	7 1/2	5 3/4	3 1/4	6	4 1/4	4 1/2	3 1/8	8
Flanged Connections Unless Otherwise Noted															
2	8 1/2	18	8	11	8	13 3/4	4	12 1/2	Threaded		8	5	—	—	—
2 1/2	9 1/2	19	9	13 1/4	8 1/2	14 1/2	4 1/4	13	6 7/8	3 7/8	8 1/2	5 1/2	—	—	—
3	11 1/8	19 7/8	10	14 3/4	9 1/2	16 1/2	4 3/4	15	8	4 1/2	9 1/2	6	8	7	12
4	12	23 3/4	12	17 1/2	11 1/2	19 3/4	5 3/4	17 3/4	—	—	11 1/2	7	9	7 1/2	15
6	15 7/8	32 1/2	16	23	16	24 1/2	8	21 3/4	—	—	14	9	10 1/2	9 1/4	30
8	16 1/2	40 3/4	20	30 3/4	19 1/2	26 1/2	9 3/4	24	—	—	19 1/2	10 1/4	11 1/2	10 3/4	37 1/2

[a] Refer to manufacturers' literature for exact dimensions of specific valves.
[b] Sizes are nominal; all dimensions are in inches.
[c] See Figure 8.20.
[d] See Figure 8.21.
[e] See Figure 8.22.
[f] See Figure 8.23.
[g] See Figure 8.24.
[h] Height dimension for rising stem gate valve corresponding to E dimension of Figure 8.20 for nonrising stem.

Source. Data extracted with permission from Ramsey and Sleeper, *Architectural Graphic Standards*, 8th ed., Wiley, New York, 1988.

water have become popular. These faucets use ball-type valves among others. Nominal valve dimensions are given in Table 8.5.

b. Gate Valves

Refer to Figure 8.20. Gate valves are used in the full open or full closed position. They are also known as stop valves when installed upstream and intended to close off all fluid flow. They are unsuitable for throttling applications, because the wedge or disc chatters and vibrates when partially open, causing noise and damage to the seating surfaces. In the fully closed position, the gate wedges firmly against the seat rings, completely closing off all fluid flow. The valve is made in two basic designs; rising and nonrising stem. The latter is illustrated in Figure 8.20. It has a left-hand thread on which only the wedge rises and lowers. In the rising stem design, the wedge is fixed to the stem and the entire stem rises or lowers. The solid wedge design illustrated is usable for steam, water, gas or oil and can be installed in any position.

Gates are also made as split wedges and double discs. These are more prone to catch sediment and are, therefore, used only in the vertical position. Bolted bonnet valves are used on high pressure lines. For low pressure, screwed bonnet valves are used. Since flow is in a straight line through the valve, it offers little more friction, if any, to the liquid flow than an equivalent piece of straight pipe. Gate valves are used where infrequent operation is intended because frequent operation will eventually damage the gate seat and cause leaks. They are usually made of bronze in plumbing work, although large sizes are made of cast iron and steel. Bronze valves are available from 3/8 to 3 in. and in pressures up to 250 psi.

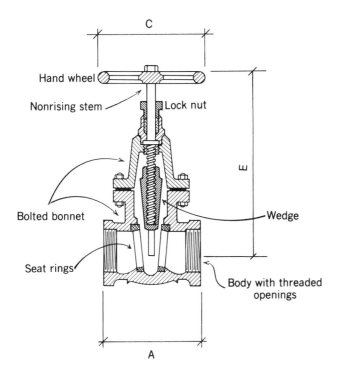

Figure 8.20 Gate valve of the nonrising stem design. The wedge moves down along the stem to seat firmly in the seat rings, completely stopping liquid flow. A gate valve is used only for full-open and full-closed service and is occasionally referred to as a stop valve. In the rising stem design (not illustrated), the wedge (gate) is firmly attached to the stem, and both move vertically when the hand wheel is turned. See Table 8.5 for nominal dimensions.

c. Globe Valves

See Figure 8.21. Globe valves, unlike gate valves, are used to regulate (throttle) flow. They are the type of valve used in conventional faucets, since flow regulation is necessary. In water systems, they are installed with the flow direction as indicated in Figure 8.21. This permits the stem packing to be changed without removing the valve, because the inlet water pressure is under the seat, which remains closed. Turning the stem raises and lowers the disc washer until it seats firmly on the seat ring and closes off the flow. As the disc washer is raised, flow increases until it is maximum with the valve fully open. Because the water must make as many as two 90° turns to pass through the valve (see Figure 8.21), its friction loss (pressure drop) is very high—as much as 50 times that of a gate valve. In lines that must be completely drained, the valve must be installed in a vertical pipe section, with the inlet at the bottom. Otherwise, the valve body will dam a small quantity of liquid. Globe valves placed ahead of a fixture or a group of fixtures, with the sole purpose of shutting off water to permit fixture maintenance, are also referred to as stop valves.

Globe valves are made in straight-through and 90° angle configurations and in three basic disc designs: composition disc (illustrated), conventional disc and plug disc. The *composition disc* design has the advantage of easy disc (washer) replacement and of a choice of washer material for different fluids such as water, oil, gases and steam. The *conventional disc* or washerless valve relies on a metal-to-metal contact to seal off flow. The seat is beveled for this purpose. The thin line contact between the two metal surfaces is useful in breaking down deposits that form on valve seats of valves that are open for extended periods. *Plug-type discs* have a large tapered contact area, which is useful with fluids carrying considerable impurities such as dirt and scale particles. They are also applied in severe service, such as steam throttling.

Globe valves with brass or bronze bodies and threaded connections are made in sizes up to 3 in. In larger sizes, the valves have cast-iron or steel bodies with flanged connections. Pressures in the smaller sizes are available up to 300 psi and in larger sizes up to 125 psi. As with gate valves, globe valves are available with either rising (illustrated) or nonrising stems.

d. Angle Valves

See Figure 8.22. Angle valves are simply globe valves where the inlet and outlet are set at 90° to each other. They are made in the same designs as other globe valves. They can be used in place of a valve and elbow combination to change direction and provide the required throttling action. Since the fluid passing through the valve makes only one 90° turn, the pressure head loss (friction) in the valve is about one-half of that in a straight-through globe valve.

e. Check Valves

See Figure 8.23. Check valves are installed to prevent fluids from flowing in the wrong direction, that is, they prevent reverse flow or backflow. For instance, they prevent polluted water from siphoning back into the potable water system when pressure fails in the water mains. They are also

444 / PRINCIPLES OF PLUMBING

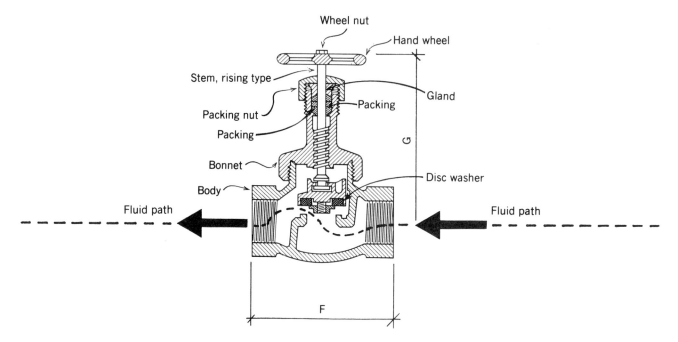

Figure 8.21 Globe valve of the composition disc (washer) rising stem design. These valves are intended for throttling (flow regulation) use. They are the type used in conventional faucets (not the single-handle type). Note that, in water systems, the direction of flow should be as shown; with the line pressure below the disc seal. See Table 8.5 for nominal dimensions.

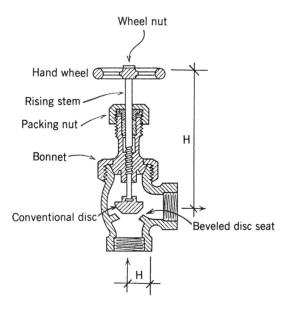

Figure 8.22 Globe valve of the washerless 90° angle type. The valve closure uses a metal-to-metal contact between beveled metal surfaces. This valve has a rising stem. See Table 8.5 for nominal dimensions.

VALVES / 445

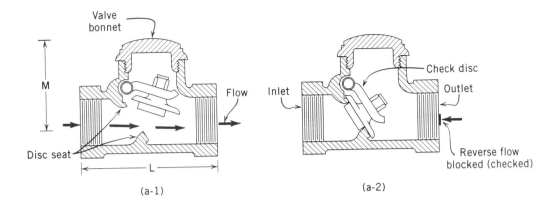

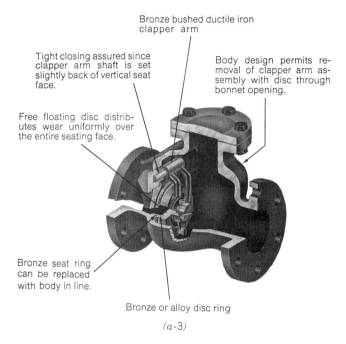

Figure 8.23 *(a)* Operation of a swing check valve. Pressure of flowing water lifts the check disc *(a-1)* permitting flow of fluid (water, oil, etc.). The weight of the disc and the slightly S-shaped path of the water through the valve somewhat increases pressure drop (static pressure head loss). When inlet pressure fails, the check disc drops onto the seat *(a-2)*, preventing reverse liquid flow. A cutaway of a flanged end swing check valve is shown in *(a-3)*. Note that the clapper arm can be removed through the bonnet opening. This simplifies maintenance, since the valve can remain in place during servicing. [Photo *(a-3)* courtesy of American Darling Valve Company.]

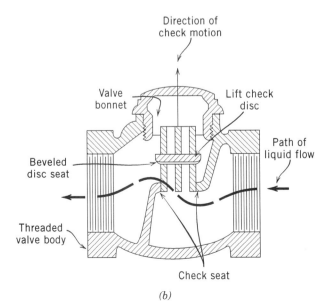

Figure 8.23 *(b)* In the lift check valve, the check disc is lifted vertically by the pressure of the flowing liquid. When pressure falls below a predetermined level, the check disc drops onto its seat and closes off reverse flow. It is held in place by gravity plus the weight of the liquid in the outlet piping, which bears down on the disc. (Drawing courtesy of Crane Valve.)

used in circulating systems, pumped liquid piping and drain lines. When pressure drops as a result of pump failure or a loss of pressure due to a break in a line, the valve automatically closes, thus preventing flow reversal. They are also used to prevent loss of pump priming liquid when placed in the pump's suction line.

Check valves for plumbing use are made in three principle designs: swing check [Figure 8.23(a)], lift check or piston check [Figure 8.23(b)] and ball check. In the *swing check* design, the pressure of the straight-through flow lifts the swinging disc and holds it open [Figure 8.23 (a-1)] as long as flow continues. When flow stops, the tilted disc closes by gravity [Figure 8.23(a-2)], preventing backflow. (Some swing checks are spring loaded for tighter closure.) These valves have relatively low friction loss (pressure head loss). They are used principally in systems operating at low to moderate pressures, that is, up to 125 psi, and are available in sizes up to 12 in., although sizes above 6 in. are rare in plumbing work.

Lift check valves are similar in disc construction to globe valves [Figure 8.23(b)] except that it is the flow of liquid that causes the disc to open rather than turning a stem. Because the fluid must make 2 to 90° turns as in a globe valve, and also hold up the check disc, this type of check valve has a considerably higher pressure drop than the swing check. Its advantage is that it seals more tightly than the swing disc type and, therefore, is usable on higher pressure systems. It is also more versatile than the swing check type, because it can be used on water, steam, gas and air piping.

One common use of a lift check is to prevent hot or cold water from passing to the other line in mixing valves, when pressure is unequal. When the disc check of a lift check valve moves in a vertical sleeve, it is referred to as a piston check valve. This type, with a spring-loaded piston, is commonly used in pump discharge lines. The spring is adjusted to close the valve solidly when fluid velocity is zero. Swing check valves should not be used in pump discharge lines because the backflow resulting from pump stoppage would slam the valve shut, causing a severe shock wave (water hammer) throughout the piping system.

A *ball check* valve is a type of lift check where the liquid flow simply lifts a ball from the valve seat as long as pressure is maintained. Loss of pressure causes the ball to drop back into its seat by gravity, where it is held by gravity and the head of water in the discharge line, thus preventing backflow. Vertical ball check valves are useful in well pump discharge lines.

Check valves in sizes up to and including 2 in. are made with brass or bronze bodies. Larger valves are cast iron or steel. Valves rated up to 125 psi are standard weight; higher pressure units are heavy-duty units. Connections are screwed in smaller sizes and flanged in larger valves. Because the valves operate by gravity, great care must be taken to install them exactly as intended (usually exactly horizontal) and obviously with the proper direction of liquid flow. Some valves have replaceable disc check seats and washers. Most are designed with removable caps that permit access to the valve interior for maintenance without the necessity of removing the valve from the line.

Plug check valves have a circularly tapered plug that seats in a hole aligned with the pipe. They are most commonly used in hot water heating and air conditioning systems and are, therefore, mentioned here only for general knowledge.

f. Ball Valves

Ball valves operate by aligning a hole drilled in a ball in the body of the valve with the valve's inlet and outlet openings. The alignment is accomplished by movement of the valve handle, as seen in the Figure 8.24(a). The openings are sealed by rubber or neoprene O rings.

Standard ball valves have very low friction loss when fully open and operate by one-quarter turn in either direction. Partial opening for flow regulation is generally not used. These valves are made of brass or bronze up to 2 in. and ferrous metal for larger sizes. They are usable with steam, water and gas and are applied where rapid flow control, tight closure and compact design are required.

In some modern single-lever faucets, the ball valve principle is employed in a complex fashion. See Figure 8.24(b). Two holes drilled in the ball align with openings to the hot and cold water inlets so that the relative amounts of each can be adjusted by side-to-side movement of the single-lever handle. Raising the handle lifts the stainless steel ball, permitting a greater flow of both hot and cold water in the same proportions as previously adjusted. Obviously, this faucet must be very carefully engineered and manufactured to close tolerances, to permit satisfactory operation. Valve bodies are nickel-coated brass or white metal alloy, suitable for pressures up to 125 psi and temperatures up to 180°F.

g. Relief Valves

See Figure 8.25. Relief valves are essentially safety devices. In the plumbing field, they are found on

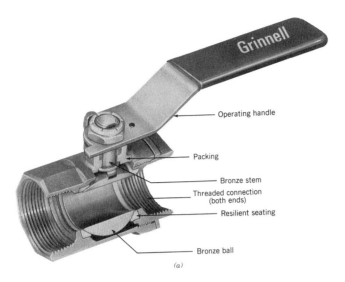

Figure 8.24 *(a)* Cutaway of a typical ball valve. The illustrated valve is suitable for 125-psi steam or 400-psi WOG (water, oil, gas) and is available from ½ to 2 in. in pipe size. Applications include residential, commercial and light industrial use. Full closure is achieved with a 90° turn of the operating handle. (Courtesy of Grinnell Corporation.)

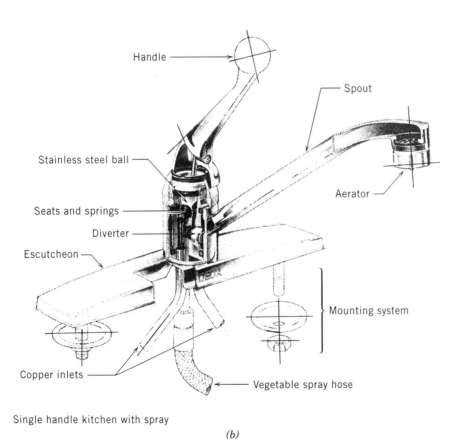

Single handle kitchen with spray

(b)

Figure 8.24 *(b)* Phantom view of a ball valve-type single-handle kitchen faucet. The stainless steel ball has dual-slot drillings, allowing variable amounts of hot and cold water into the spout, as the ball is manipulated by the handle. (Courtesy of Delta Faucet Company.)

448 / PRINCIPLES OF PLUMBING

Figure 8.25 (a) Pressure relief valve (safety valve). The inlet is at the bottom, and the drain connection is at the side. Operation is shown in the cutaway (b). When pressure exceeds the preset limit, the disc held by the spring unseats, permitting pressure to be relieved through the drain, which is open to the atmosphere. (Courtesy of ITT Fluid Handling Sales.)

all hot water heaters and hot water storage tanks. When water is heated, it expands. A 100°F rise causes approximately a 70% volume increase. Since water is incompressible, when contained in a closed vessel such as a hot water heater or tank, the increased water volume increases the pressure inside the vessel. This pressure must be kept within the pressure capacity of the heater or tank, which is normally rated at 125 psi. When operating properly, the system controls will shut down the system when water temperature approaches its preset temperature. Since the controls may malfunction, it is imperative that every water heater (and heating boiler) be provided with a relief valve that will act to relieve the excess pressure and temperature created. Most codes require both pressure and temperature relief. Both features can be, and usually are, built into a single valve, which is referred to as a temperature and pressure (T & P) relief valve. Alternatively, individual valves can be used. It is a false economy to use only a pressure relief valve even when permitted by codes, since studies have shown that excessive water temperature can cause scalding and serious injury.

The pressure relief element is usually a diaphragm or a spring-loaded disc, which is lifted by the pressure inside the tank. The usual setting is 25% above operating pressure or 20–30 psi higher than the system operating pressure, whichever is lower. Maximum setting is 100 psi or 80% of the tank's maximum capacity, whichever is lower. The temperature element is one of many thermostatic element designs that will open the valve at preset temperature.

After operating, the valve should reset tightly automatically when water temperature drops below the usual setting of 180°F. Relief valves must be connected to a discharge pipe that terminates at a point where the scalding discharge will cause no personal injury or property damage (normally at a floor drain). Because relief valves remain in position for long periods without operation, they should be manually operated occasionally to prevent clogging and to clear corrosion that has formed. To this end, most relief valves are equipped with a handle to permit manual operation.

h. Pressure-Reducing Valves

See Figure 8.26. Pressure-reducing valves (PRV) are used in plumbing work to reduce water main pressure to levels usable for building work (60 psi maximum). Although street mains pressure does

(a)

Figure 8.26 (a) Diaphragm-type pressure-reducing valve, set at 45 psi. The threaded connections are marked INLET at the right, BOILER at the left and STRAINER at the bottom. (Domestic hot water boilers require a PRV on the inlet line because they have a low pressure rating.) (Photo courtesy of ITT Fluid Handling Sales.)

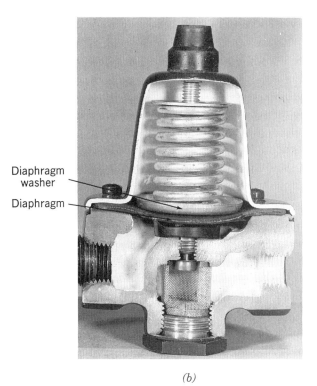

(b)

Figure 8.26 (b) A cutaway of the PRV shows its operation. Excess pressure pushes the diaphragm up, forcing the washer up against its seat and closing the valve. The spring on top of the diaphragm adjusts the pressure, keeping the valve open. Pressure is adjusted by turning the nut on top of the valve. The valve is not intended for field adjustment; it is factory set. (Photo courtesy of ITT Fluid Handling Sales.)

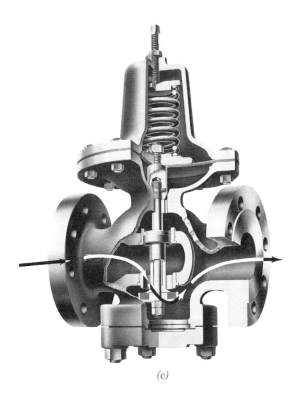

(c)

Figure 8.26 (c) Cutaway of a large-capacity diaphragm-type water pressure-reducing valve with flange connections. Inlet is at the left. The valve is shown in the open position, and arrows indicate water flow. High pressure forces the plunger up, against spring pressure, throttling the bottom opening to preset pressure. Springs are interchangeable for different pressure ranges. The valves are available in sizes up to 6 in. (Photo courtesy of Spence Engineering Company, Inc.)

not usually exceed 80 psi, there are instances of even higher pressures. In addition, the valves act to regulate their output pressure under conditions of rapidly varying water demand such as occurs in facilities using flushometer valves. Pressure-reducing valves are normally built with spring-loaded diaphragms, which rise to allow passage of water at the desired pressure. Valves for normal domestic and commercial service are available in sizes up to 2 in., with bronze bodies and threaded connections. Springs for different output pressure ranges are adjustable within each pressure range. Pressure-reducing valves are frequently installed with a bypass that permits water service to continue while the valve is being changed or repaired. A gauge on the PRV outlet side must be monitored to check that system pressure limits are not exceeded while the PRV is out of service.

i. Controllable Valves

All the previously discussed valves are either manually operated (globe, gate, angle, ball, relief) or passive (check). Passive valves are operated by the flow of fluid. There are, however, many types of valves that are arranged for electrical, hydraulic or pneumatic operation. These are principally diaphragm valves, butterfly valves and gate valves. They were not discussed in detail because they do not normally concern the plumbing technologist.

8.15 Fixtures

a. General

The purpose of the entire plumbing system is to permit the use of the facility's plumbing fixtures. The fixtures required for each facility are normally determined by the architect, owner and plumbing engineer or designer working together. The decision is based first on minimum Code requirements (see Code Chapter 7) and beyond that on the owner's desires, modified by the advice and constraint input of the architect and plumbing designer. One of the design constraints in all public buildings is the requirements of the Americans with Disabilities Act, commonly abbreviated ADA. This act essentially requires, by law, that all buildings financed by public funds be designed to permit full use by handicapped people. With respect to our subject, this requirement means that such buildings must have plumbing fixtures that are accessible and usable by handicapped persons.

This subject is normally the province of the building architect. However, like so many other specialties, a competent technologist must be aware of the subject and its influence on materials, layout and design.

The decision-making process in fixture selection and layout is frequently one of interaction between the interested parties. For instance, an owner might want a certain bath location. The plumbing designer would point out the considerable savings possible with a back-to-back arrangement, and the architect might indicate the effect of the bath window on the building appearance, and so on. These decisions are not usually in the plumbing technologist's area of work although he or she should have the necessary information if called upon to supply it. The specific types of plumbing fixtures frequently found in a building vary somewhat depending on the type of facility being considered. Table 8.6 shows what can be expected, depending on the design of the building. Not every residence has a bidet nor every school a multiple shower. Table 8.6 is for general information only.

In general, a plumbing fixture is a device that accepts clean potable water and, by its use, discharges unclean, contaminated waste water. This waste water is referred to as *grey water* if it contains no human waste products and *black water* if it does. Therefore, the discharge from lavatories, kitchen sinks, laundry tubs, clothes washers, showers and bathtubs is grey water; that from urinals, water closets and bidets is black water. The significance of these terms will be explained more fully in Chapter 10, which deals with drainage. The purpose of the fixture itself is to make possible the desired washing or elimination function. Since drinking fountains do not fulfill either of these functions (washing or waste elimination), some designers do not classify them as plumbing fixtures and place them in a separate category by themselves. We will follow the Code and classify them as a special type of plumbing fixture.

Plumbing fixtures, because of their use, must be constructed of smooth, durable, chip resistant, easily cleaned material. Among the most common are enameled cast iron (sinks and bathtubs), enameled steel (lavatories, laundry sinks and bathtubs), stainless steel (sinks and drinking fountains), vitrified porcelain or china (closets, sinks, lavatories) and acrylic and fiberglass (tubs, showers and shower pans). Less common construction includes all metal vandal-proof fixtures, cast concrete, soapstone and terrazzo fixtures (wash fountains, large sinks and tubs, shower pans) and marble fixtures.

Table 8.6 Plumbing Fixture Usage

Fixture	Residence	Dormitory	Store	School	Industrial
Kitchen sink	*	*		*	
Laundry sink (service sink)	*	*		*	*
Lavatory	*	*	*	*	*
Wash fountains					*
Single shower	*	*			
Multiple shower		*		*	*
Bath tub	*				
Urinal		*		*	*
Water closet (wall hung)		*		*	*
Water closet (floor mounted)	*		*		
Bidet	*	*			
Drinking fountain				*	*

Stainless steel, which had once been almost exclusively restricted to commercial scullery (scrubbing) sinks, is now found in residential application. On such sinks a coating is applied to the sink underside to deaden the sound of water hitting the metal. A satin finish on these sinks reduces the nuisance of spotting. These sinks are easy to clean, which accounts for their commercial kitchen use, but are considerably more expensive than enameled metal or porcelain ware.

All fixtures are designed with smooth curves and no crevices, to facilitate thorough cleaning, both inside and, where applicable, outside. They are also designed for rapid drainage during use and complete drainage after use. Fouled water remaining in a fixture (except by design, as in water closets and urinals) can be a source of odor and will cause spotting. Fixtures are also designed for convenient connection and maintenance of all piping.

Since all piping is completely installed before fixtures are attached, it is extremely important that the roughing-in (piping installed to receive fixtures plus devices intended to support fixtures) be installed by the plumbing contractor at precisely the correct locations and elevations. The tolerances for error are very small. In order to ensure that the roughing is correct, the plumbing contractor must be supplied with exact roughing dimensions for each plumbing fixture. These are supplied by the manufacturer to the plumbing designer/technologist who, after checking the fixture's adequacy against the project working drawings, will furnish the roughing-in dimension drawings to the plumbing contractor. Typical roughing-in dimensioned drawings for standard fixtures are given along with fixture illustrations in the following discussion. The fixtures are discussed in the same order as they appear in Table 8.6.

b. Sinks

See Figure 8.27. The sinks most commonly used are kitchen, bar and laundry sinks in residential-type buildings and service/janitor's sinks in institutional and industrial buildings. Modern kitchen sinks are either satin-finish stainless steel, epoxy-stone chip compounds or china. Enameled steel, although still available, has fallen into disuse. Kitchen sinks are single, double or triple bowl and are generally arranged to be set into a countertop. Faucet hole spacing is standardized at 4 in. Kitchen sinks normally have four openings at the faucet location; two for $3/8$- or $1/2$-in. hot and cold water lines, one for the spout and one for a sprayer. A $3\frac{1}{2}$-in. diameter center drain hole is standard. The same sink is usable with single-lever and two-handle faucet batteries, since the battery covers

452 / PRINCIPLES OF PLUMBING

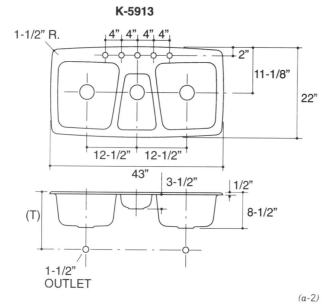

Figure 8.27 *(a-1)* Three-compartment, enameled cast-iron, self-rimming kitchen sink. *(a-2)* Roughing-in data. (All photos and data courtesy of Kohler Company.)

FIXTURES / 453

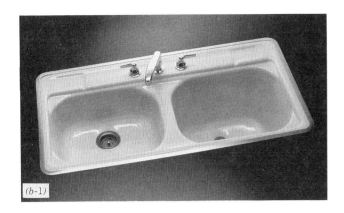

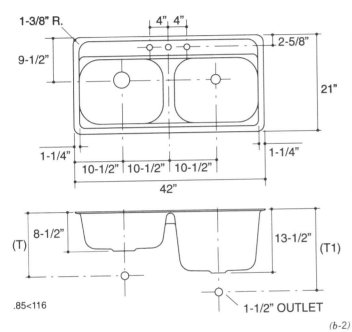

Roughing-In Notes	
Fixture dimensions are nominal and conform to tolerances in ASME/ANSI Standard A112.19.1M.	
(T) Drain typically 13-7/8".	
(T1) Drain typically 17-3/4".	
Cutout	Metal frame required.

Fixture*:		basin area	water depth
Sinks		14" x 19"	6-5/8"
Laundry		14" x 19"	11-5/8"
Sink outlet	3-5/8" D.		
Laundry outlet	2-1/8" D.		
Faucet holes	1-3/8" D.		
* Approximate measurements for comparison only.			

(b-2)

Figure 8.27 *(Continued) (b-1)* Two-compartment, enameled cast-iron laundry/all-purpose sink, for cabinet installation with a metal frame. *(b-2)* Roughing-in data.

454 / PRINCIPLES OF PLUMBING

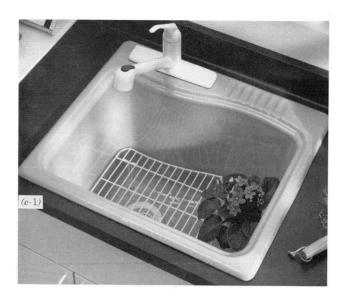

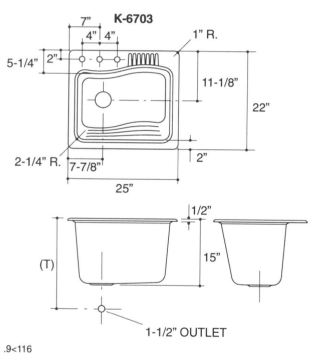

Product Information

Fixture*:		basin area	water depth
Sink		21" x 16"	14-3/8"
Water capacity	19 gals.		
Outlet	3-5/8" D.		
Faucet holes	1-3/8" D.		
* Approximate measurements for comparison only.			

Roughing-In Notes

Fixture dimensions are nominal and conform to tolerances in ASME/ANSI Standard A112.19.1M.	
(T) Drain typically 20-5/16".	
Cutout	23-1/2" x 20-1/2" with 1" radius corners.

Figure 8.27 *(Continued) (c-1)* Enameled cast-iron self-rimming utility sink, with sloping ribbed inside front surface (integral washboard). *(c-2)* Roughing-in data.

FIXTURES / 455

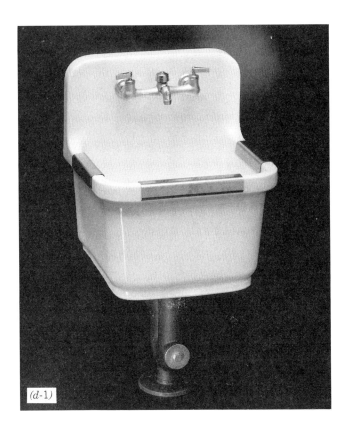

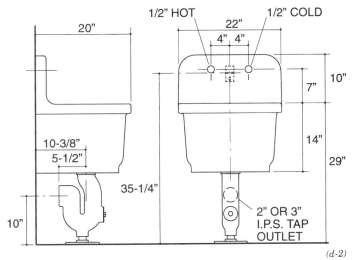

Product Information

Fixture*:		basin area	water depth
Sink		19" x 16"	13-1/8"
Outlet	3" D.		
Faucet holes	1-1/2" D.		

* Approximate measurements for comparison only.

Roughing-In Notes
Fixture dimensions are nominal and conform to tolerances in ASME Standard A112.19.2M.
No change in sink measurements when used with K-6672 or K-6673 trap standard.
Recommended accessories/hardware are shown.

Figure 8.27 *(Continued)* *(d-1)* Vitreous china service sink with stainless steel rim guards and enameled inside trap standard. *(d-2)* Roughing-in data.

the three predrilled holes. (Some sinks come with knockouts rather than predrilled holes.) Despite standardization, the technologist should specify the specific battery to be used with each sink to avoid field problems.

Sinks not intended for cabinet mounting generally have faucet holes in their vertical backsplash. Since such sinks can take single faucets or a hot-cold battery, coordination is important. Janitorial and service sinks almost always have the hot and cold water faucets on the elevated backsplash. Since the sink height varies with different manufacturers, it is important that the contractor receive the roughing-in data for the particular sink involved. Service sinks are usually enameled cast iron or vitreous china. Recently acrylic and fiberglass sinks have appeared as well. Laundry trays (sinks) do not have an elevated backsplash as an integral part of the sink. Faucets are mounted on a horizontal ledge behind the single or double tray (deck mount) or above them (wall mount), with a swivel spout where two compartments must be served. Modern laundry trays are plastic or fiberglass, although enameled iron is still available. All sinks are trapped immediately below the drain. Janitorial (slop) sinks are occasionally supported on a floor-mounted 2- to 3-in. trap containing a cleanout plug.

c. Lavatories

See Figure 8.28. Lavatories (wash sinks) are available in cabinet mount with and without integral faucet batteries, in wall mount with and without integral faucets and in pedestal mount with integral faucets. Modern lavatories are constructed of enameled cast iron and acrylic or glazed vitreous china. Lavatory dimensions are fairly standard, although each manufacturer makes its own specialties. Faucet battery dimensions and spacings are standard. As with sinks, lavatories that are not counter-top mounted frequently have elevated backsplashes with a top ledge. Lavatories intended for use in public areas should have self-closing faucets. These faucets must be adjusted to avoid quick closing, which causes water hammer. Automated faucets with infrared sensors can be used where only cold water is supplied at the lavatory. If both hot and cold are supplied they must be mixed to temper the hot water, and automatic operation is not possible without the addition of an expensive thermostatic mixing valve. This arrangement is not practical for a public lavatory, leading to the wide use of self-closing valves. Foot pedals that close on release are also occasionally used to avoid water waste.

All faucet outlets must be well above the rim of the lav (or sink) to avoid back-siphoning of contaminated water into the potable water system. [For a full discussion of backflow prevention, see Section 9.14, page 534, and Figure 10.1(a), page 559.] Most lav bowls have overflows that drain into the bowl outlet before the fixture trap. Bowls without overflows (drain holes near the top rim of the bowl) should never be used in public areas because of the possibility of flooding due to stoppage of the bowl drain with waste. Similarly, lavatory bowls in unattended public toilets should not have stoppers for the same reason. Lavatories in private installation can have stoppers of either the chain and stopper type or the built-in pop-up lever-controlled type. Neither is entirely satisfactory. The chain and stopper is unsightly and a nuisance when not in use, but it provides a good bowl seal. The pop-up type does not seal tightly, and because it is permanently installed in the drain opening, it constantly accumulates soap scum and traps hair and other small items.

Multiple-position lavatories are normally called wash fountains. They are made of enameled cast iron, precast terrazzo, stainless steel and occasionally acrylic. They are used in public institutions such as schools and in industrial buildings where large numbers of people follow a timed schedule, thus creating short periods of heavy fixture usage. To handle this type of service with individual fixtures would be expensive and waste floor space. The units are available in wall mount and pedestal designs, and in oval, half-oval, quarter round, semicircular and circular shapes, to meet the washrooms' space constraints. Dimensional data for sinks and lavatories is given in Figure 8.29.

d. Showers

Showers are of concern to the plumbing technologist with respect to valves, heads and drains. Individual showers are fed with ½-in. hot and cold water pipes. If shower water temperature is controlled by the individual faucets, a change in demand at some other point in the building will cause a pressure drop in that line. As a consequence, the unaffected line becomes dominant, and the bather is "treated" to a radical change in the

(text continues on p. 463)

FIXTURES / 457

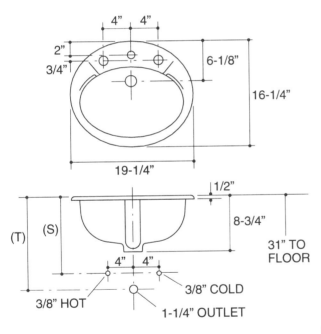

Product Information

Fixture*:		basin area	water depth
Lavatory		11" x 16"	5"
Outlet	1-11/16" D.		
Faucet holes	1-5/16" D.		
Spout hole	1-3/16" D.		
* Approximate measurements for comparison only.			

Roughing-In Notes

Fixture dimensions are nominal and conform to tolerances in ASME/ANSI Standard A112.19.1M.
(T) Pop-up drain typically 14-3/8", drain typically 13-1/2".
(S) 14" (Based on 12" riser which may require cutting).

Figure 8.28 *(a-1)* Enameled cast-iron self-rimming countertop lavatory. *(a-2)* Roughing-in data. (Photos and data courtesy of Kohler Company.)

458 / PRINCIPLES OF PLUMBING

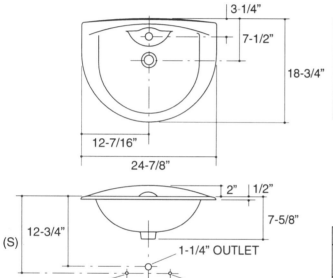

Product Information

Fixture*:		basin area	water depth
Lavatory		19" x 12"	5-7/8"
Outlet	1-3/4" D.		
Faucet holes	1-3/8" D.		
* Approximate measurements for comparison only.			

Roughing-In Notes

Fixture dimensions are nominal and conform to tolerances in ANSI Standard A112.19.2.

(S) 14" (Based on 12" riser which may require cutting).

Figure 8.28 *(Continued) (b-1)* Vitreous china self-rimming countertop lavatory with one-hole drilling for single-handle faucet. *(b-2)* Roughing-in data.

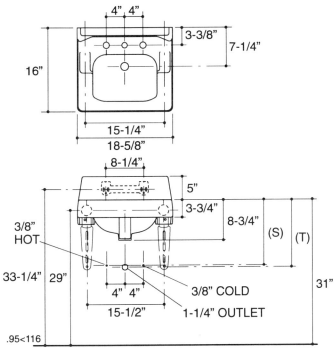

CONCEALED ARM HOLE LOCATIONS

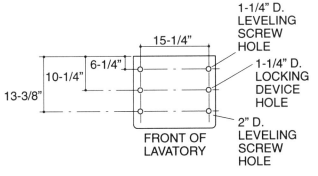

Product Information

Fixture*:		basin area	water depth
Lavatory		13" x 8"	5-1/4"
Outlet	1-3/4" D.		
Faucet holes	1-3/8" D.		
Spout hole	1-3/16" D.		
* Approximate measurements for comparison only.			

Roughing-In Notes
Fixture dimensions are nominal and conform to tolerances in ASME Standard A112.19.2M.
(T) Pop-up drain typically 14-3/8", drain typically 13-1/2".
(S) 14" (Based on 12" riser which may require cutting).

Figure 8.28 *(Continued) (c-1)* Vitreous china wall-mount lavatory. *(c-2)* Roughing-in data.

460 / PRINCIPLES OF PLUMBING

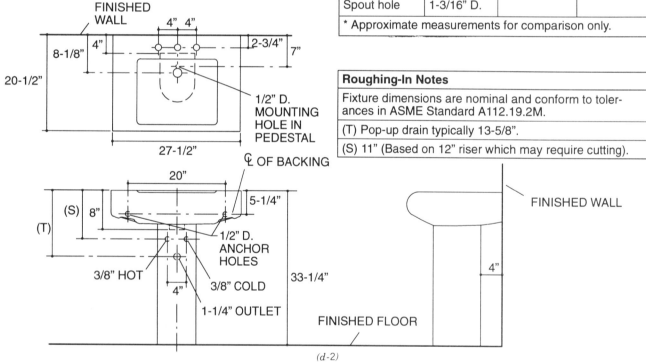

Product Information

Fixture*:		basin area	water depth
Lavatory		18" x 14"	4-5/8"
Outlet	1-3/4" D.		
Faucet holes	1-3/8" D.		
Spout hole	1-3/16" D.		

* Approximate measurements for comparison only.

Roughing-In Notes
Fixture dimensions are nominal and conform to tolerances in ASME Standard A112.19.2M.
(T) Pop-up drain typically 13-5/8".
(S) 11" (Based on 12" riser which may require cutting).

Figure 8.28 *(Continued) (d-1)* Vitreous china pedestal-type lavatory. *(d-2)* Roughing-in data.

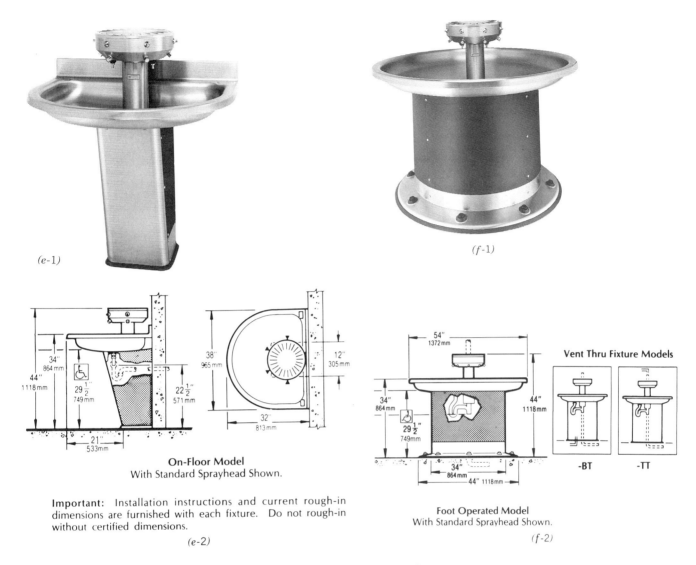

Figure 8.28 *(e)* and *(f)* Multiple-position wash fixtures. Fixture *(e-1)* is a four-position stainless unit. Fixture *(f-1)* is an eight-position unit, normally called a wash fountain because of its shape. Both are built with the barrier-free design that allows use by persons confined to wheelchairs. The four-position unit can be arranged for completely "hands free" sensor operation. Both units can be equipped with push-button-operated time-out valves. Tempered water is supplied via a field adjustable hot/cold mixing valve. Dimensional data is shown in *(e-2)* and *(f-2)*. Roughing-in dimensional drawings are supplied with the units. Multiple lavs are used in schools, factories and other buildings with high usage, short-time washing requirements. (Courtesy of Acorn Engineering Company.)

462 / PRINCIPLES OF PLUMBING

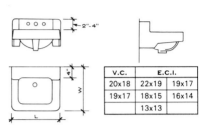

V.C.		E.C.I.
20x18	22x19	19x17
19x17	18x15	16x14
	13x13	

Shelf-back lavatories generally are rectangular with semi-oval basins. Height of the shelf typically is 4 in.; depth is usually 5 in. Support with metal legs and brackets or concealed carrier.

SHELF BACK

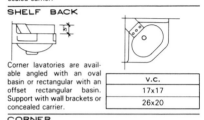

Corner lavatories are available angled with an oval basin or rectangular with an offset rectangular basin. Support with wall brackets or concealed carrier.

V.C.
17x17
26x20

CORNER

Wash sinks supported with concealed wall brackets for E.C.I. or with angle supports for S.S.

E.C.I.		S.S.		STATIONS
18x36	18x48	20x48		2
18x60	18x72	20x60	20x72	3
			20x96	4

WASH SINKS

Wall-mounted service sinks are designed for janitorial requirements of hospitals, plants, institutions, office buildings, and schools. Floor to rim dimension is 2 ft. 3 in. to 2 ft. 5 in. Fittings are mounted either on or above the sink back. "H" designates flushing rim design for hospital use specifically.

V.C.		E.C.I.		S.S.	
28x22	26x20 H	24x20	24x18	25x19	23x18
24x22 H	20x20 H	22x18			
22x20					

SERVICE SINKS

GENERAL NOTES

Lavatories and work sinks are available in vitreous china (V.C.), enameled cast iron (E.C.I.), enameled steel (E.S.), and stainless steel (S.S.). Typically, floor to rim dimension is 2 ft. 7 in., unless otherwise noted. The most commonly used means of support is the chair or wall carrier with concealed arms. Other methods are detailed below. Consult manufacturer's data for specific fixture design and support recommendations.

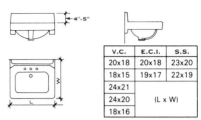

V.C.	E.C.I.	S.S.
20x18	20x18	23x20
18x15	19x17	22x19
24x21		
24x20	(L x W)	
18x16		

Most flat-back lavatories are available with rectangular or semi-oval basins. Typically, floor to rim dimension is 2 ft. 7 in. Support using metal legs with brackets or with concealed carrier.

FLAT BACK

V.C.
20x18
24x20

Slab lavatories generally are rectangular with rectangular basins. A 2 in. escutcheon typically spaces lavatory from finish wall. (4 in. and 6 in. also are available.) Vitreous china leg with brackets can be used as alternate means of support.

SLAB

V.C.	S.S.
20x27	23x19

Wheelchair lavatories must be supported using a concealed arm carrier. Height from floor to rim is 2 ft. 10 in.

WHEELCHAIR LAVATORY

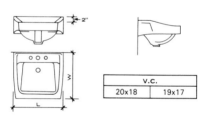

V.C.	
20x18	19x17

Ledge-back lavatories generally are rectangular with rectangular basins. Ledge width usually is 4 in. Typically supported with concealed carrier.

LEDGE BACK

TYPE	DIAM. (IN.)	NO. USERS
Circ.	54	8
	36	5
Semi-circ.	54	4
	36	3
Corner	54	3

In addition to circular designs, semi-circular and corner types are available, most in precast terrazzo, stainless steel, and some in fiberglass. Most have foot controls, and some have hand controls. Supply from above, below, or through the wall. Vents many rise centrally or come off drain through wall or floor.

WASH SINK

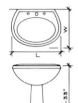

Pedestal lavatories are available in a wide variety of forms, sizes, and basin shapes. See manufacturer for specific designs.

V.C.		
38x22	30x20	28x21
24x19	26x22	25x21
22x21	20x18	

PEDESTAL LAVATORY

V.C.	E.C.I.
14x13	16x14
14x12	

Institutional lavatories have an integral supply channel to spout and drinking nozzle, strainer, and soap dish. Trap is enclosed in wall. Wall thickness must be specified.

INSTITUTIONAL LAVATORY

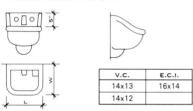

Floor-mounted chair carriers support fixture independent of wall construction. Available with exposed or concealed arms. Wall-mounted carrier with exposed or concealed arms also is available. Additional methods include floor-mounted hanger plate types, floor-mounted bearing plate types, paired metal or single vitreous china leg, in addition to exposed, enameled wall brackets.

METHODS OF LAVATORY SUPPORT

Figure 8.29 Details of typical lavatories and work sinks with dimensions and mounting heights. For exact dimensions of specific lavatories, obtain roughing-in dimensioned drawings from the manufacturers. (Ramsey and Sleeper, *Architectural Graphic Standards*, 8th ed., 1988, © John Wiley & Sons, reprinted by permission of John Wiley & Sons.)

temperature of the shower water. This can be dangerous if the hot water temperature exceeds 110°F. To overcome this shortcoming, automatic pressure balancing valves were developed and are in common use. See Figure 8.30*(a)*. However, pressure control is not sufficient to keep the shower water temperature constant; that requires a thermostatically controlled mixing valve. These valves (automatic pressure balancing and thermostatic water temperature control) have become almost standard in all new construction. See Figure 8.30*(b)*.

Shower head design is normally a matter of owner preference. Multiple (gang) shower heads, as multiple washstations, are found where many people must shower simultaneously as in gyms, factories and penal institutions. Because these stations are free standing as well as wall mounted, the roughing-in piping must be carefully located. See Figure 8.30*(d)*.

Since individual on-off and temperature control in gang showers is neither practical nor generally desirable, modern installations use central temperature and pressure balancing valves with individual electronic, infrared presence-sensing to control the on-off function. See Figure 8.30*(c)*. Sometimes a timed push-button control is added so that both presence-sensing and push-button operation is required to operate the shower head. To avoid water waste that can be caused by accidental or deliberate operation of the sensor, a time-out feature will cut off the water after a predetermined period of time. Reactivation of the shower requires the push-button to be operated again. This feature discourage "playful" operation of unoccupied shower positions. See Figure 8.31.

Individual showers normally are constructed with a precast shower pan in the center of which is placed the shower drain. See Figure 8.32. The drain has a height-adjustable strainer to fit the pan and is connected either to a cleanout basket first or else directly to the shower trap. Floor drains in multiple shower rooms are set at low points in the pitched floor. Coordination with the construction contractor is, therefore, important so that the floor fill pour is properly and adequately pitched to the floor drain.

e. Bathtub

See Figure 8.33. Standard tubs are constructed of enameled sheet steel, enameled cast iron, vitreous china, enameled polymer, acrylic and fiberglass. Material selection is not usually the responsibility of the technologist. Simple standard tubs are in part being replaced with tubs equipped with whirlpools, pressurized jets and other specialty fittings. Each such tub is designed differently with different piping requirements. No general rules can be stated except that careful coordination of the plumbing, electrical and construction work is particularly important with these tubs. Standard tubs take ½-in. hot and cold lines. They terminate in faucets that are tub or wall mounted with either a tub-mounted or over-the-rim spout, respectively. A tub/shower combination is very common, and is supplied by the tub faucets and a transfer valve. Where a tub/shower combination is installed, the pressure/temperature anti-scald valve required for showers (see Section 8.15.d) is required.

Control of tub drainage is generally accomplished with a pop-up drain seal that is controlled by a small lever-handle at the tub overflow. Tubs are designed to self-drain when the rim is installed level. Tubs that are installed against a wall and enclosed in ceramic tile should have an access panel through which access to the water and drain is obtained. Otherwise, almost every maintenance procedure involves removal (breaking) of tiles and their replacement. This is not only expensive, but it is also disruptive to the area and almost invariably results in a patched appearance of the tiling.

f. Urinals

See Figure 8.34. In public buildings, urinals in mens' rooms are desirable for two reasons:

- They reduce the floor and seat soiling that occurs when a water closet is used for urination.
- They increase the toilet capacity of a given space since most mens' toilet usage is for elimination of liquid waste.

(We refer only to men's urinals because women's urinals are rarely used in the United States. They are found in Europe and the East. They look like a narrow elongated bidet with a raised end and are designed to be used without any body contact.)

Urinals are generally constructed of vitreous china. They are available in three designs; individual wall-mounted units, individual full-height stall-type units and a continuous full-height unit with a sloping trough at floor level. Continuous wall-mounted trough-type units are also available but are rarely used. In some areas, trough units are

464 / PRINCIPLES OF PLUMBING

not permitted by local codes. Each type has its advantages and disadvantages.

Individual wall-mounted units have integral traps that connect to a drain line in the wall. Because the floor becomes soiled rapidly, it must be washed (hosed) frequently. This requires a (trapped) floor drain plus a hose bib (also spelled "bibb") for supplying the wash water. The floor drain should be located close to the wall-hung urinals. Another disadvantage of wall-hung units is the problem of mounting height. Wall-hung types have their lower front lip mounted at approximately 24 in. above the floor. See Figure 8.34. This is too high for use by young boys. The solution to this problem is either to provide one or more urinals specially designed for small boys or to mount a standard urinal lower. The latter retains its serviceability for adults, although it has a peculiar appearance in a large bank. In men's rooms with only one or two wall-hung urinals, they are mounted at standard height and young boys are forced to use water closets, with the attendant soiling referred to previously.

Individual full-height stall-type urinals require a separate floor trap for each unit as opposed to the integral trap of the wall hung unit. Floor washing is somewhat simplified since water is drained directly into the stalls, thus eliminating the need for a separate floor drain. Washing between stalls, however, is difficult. A distinct advantage of stall units is that the mounting-height problem of wall-hung units is eliminated. This accounts for the common use of stall types where the number of units is small.

Continuous full-height urinals are found in some public buildings where permitted by Code. They have the distinct advantages of easy floor cleaning, elimination of wall cleaning and height problems and the economy of a single trap for a full wall length of urinals. The disadvantages are the lack of privacy that individual fixtures give, even when mounted close together and the serious problem of unsanitary and highly objectionable splashing and spraying, which is eliminated by the sides of individual fixtures. We thus see that none of the choices are entirely satisfactory. The architect and plumbing designer must decide jointly on the optimum solution.

Urinal fixtures are most often vitreous china although enameled cast iron is also used. The vast majority are flushed using flush valves. Flush tanks, although available, are rarely used today. They were once common when used to provide continuous, low-rate flushing. This, however, is wasteful of water, as is timed periodic flushing. The modern design trend is to use sensor-operated flushing. This is a system whereby a sensor, normally passive infrared, senses the presence of a person in front of the urinal and then his absence, at which point the urinal is flushed. The "presence"

(text continues on p. 475)

Figure 8.30 *(a)* Cutaway of a three-port (shower only) pressure-balancing valve. Action of the valve is illustrated in schematic drawings 1–4, which show the valve action for different hot and cold water pressures. Note that low pressure of either hot (sketch 2) or cold (sketch 4) opens the valve to that side resulting in increased flow. Loss of either pressure (sketch 4) results in a virtual shutoff of water flow. *(b)* Cutaway of thermostatic mixing valve for use in a single shower, bath or combined shower/bath. The valve's thermostatically activated valve mechanism compensates for both pressure and temperature variations in the hot and cold water supply to furnish tempered water at the desired temperature. *(c)* Cutaway of a master mixing valve that is used to supply tempered water to shower rooms, group showers and industrial processes or domestic water to small buildings. The valve compensates for supply pressure and temperature variations by means of its internal thermostatic actuator. Units are available with flow rates up to 200 gpm and pressure differentials up to 100 psi. [Photos *(a–c)* courtesy of Powers Process Controls.]

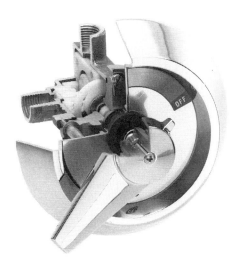

A reliable method of pressure balancing control is the diaphragm and poppet mixing valve. This design uses a balancing diaphragm and poppet type discs and seats (with wide clearances to provide maximum protection against liming and dirt conditions). Should cold water pressure suddenly drop due to demand elsewhere on the supply line (such as flushing toilets), the diaphragm immediately responds to keep the ratio of cold water to hot water constant, thereby maintaining the bather's setting.

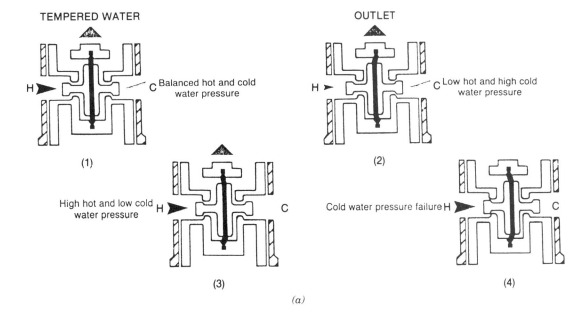

(a)

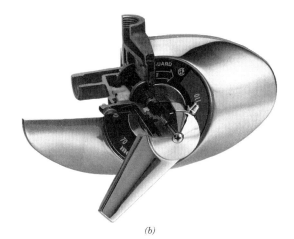

(b)

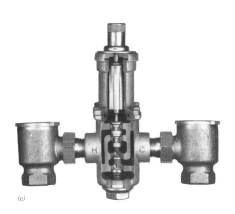

(c)

466 / PRINCIPLES OF PLUMBING

(d-1)

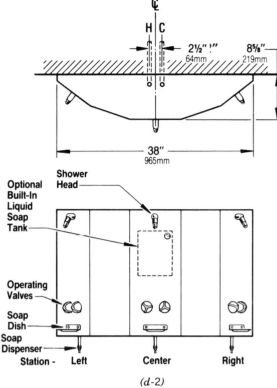

(d-2)

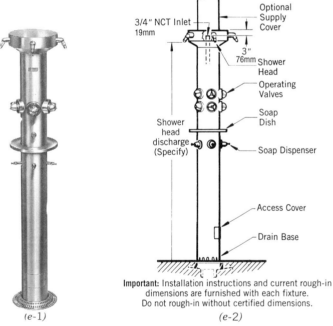

(e-1) (e-2)

Important: Installation instructions and current rough-in dimensions are furnished with each fixture. Do not rough-in without certified dimensions.

Figure 8.30 *(Continued)* Wall-mounted *(d)* and free-standing column-type *(e)* multiple showers have adjustable stream heads and various types of water control schemes. Water temperature is centrally controlled. Multiple showers are used in schools, resident institutions and some types of industrial installations. [Photos *(d,e)* courtesy of Acorn Engineering Company.]

FIXTURES / 467

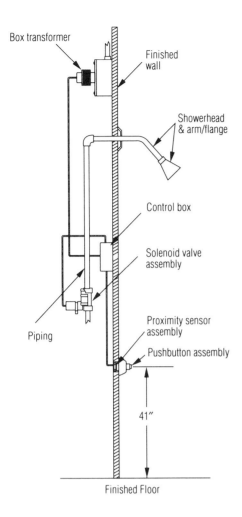

Typical Shower Installation

Figure 8.31 Detail of push-button/proximity sensor-controlled shower. This arrangement requires operation of the push button to start the shower, which then operates for a predetermined maximum length of time unless shut off earlier by the bather. This type of control is common in multiple shower installations where it is desirable to override shower operation due to false sensor signals, when no bather is present. (Courtesy of Power Process Controls.)

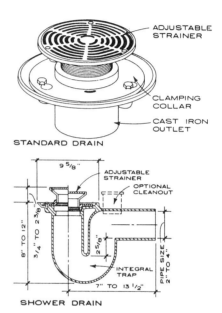

Figure 8.32 Typical shower drain. The collar clamps down onto a waterproof pan or liner (not shown). This completely isolates the shower floor and ensures that the shower water enters the drain. Water that seeps under and around the strainer is channelled into the drain by secondary weep holes in the clamping collar. Dimensions are typical. See also Figure 10.4 *c* (page 564). (Ramsey and Sleeper, *Architectural Graphics Standards*, 8th ed., 1988, © John Wiley & Sons, reprinted by permission of John Wiley & Sons.)

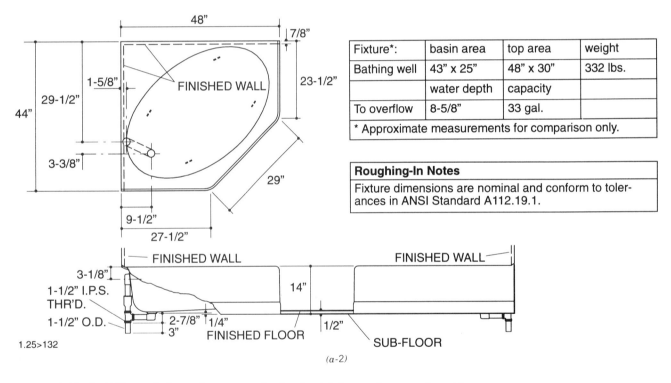

Figure 8.33 *(a-1)* Enameled cast-iron corner bathtub with integral seat and oval basin. Single-handle thermostatic water valve supplies the bath and shower (not shown). Shower diverter valve is shown in the wall-mounted tub faucet. *(a-2)* Roughing-in data. *(b-1)* Enameled cast-iron whirlpool bath with built-in water heater. *(b-2)* Roughing-in data and electrical requirements. [*(a, b)* courtesy of Kohler Company.]

FIXTURES / 469

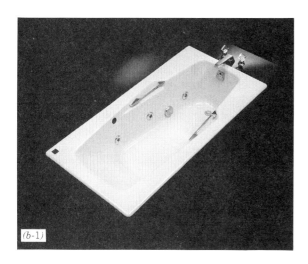

(b-1)

Fixture*:	basin area	top area	weight
Bathing well	47" x 24"	67" x 29"	435 lbs.
	water depth	capacity	
To overflow	12-7/8"	52 gals.	

* Approximate measurements for comparison only.

Pump, 2–speed:	high HP	low HP	V	Hz	A
60 Hz	1	1/8	120	60	12

Required Electrical Service

Dedicated circuits required, protected with Class A Ground-Fault Circuit-Interrupter (GFCI):	
Pump/control	120 V., 15 A, 60 Hz
Heater (60 Hz)	20 A, 1.5 kW at 120 V. or 30 A, 6.0 kW at 240 V.

Roughing-In Notes		
Fixture dimensions are nominal and conform to tolerances in ANSI Standard A112.19.1.		
No change in measurements if connected with drain illustrated.		
Cut-out	70-1/2" x 34-1/2"	
Minimum access:		
Pump/control	30" W x 15" H panel	required
Heater	20" W x 15" H panel	recommended
Heater service	16" clearance	required

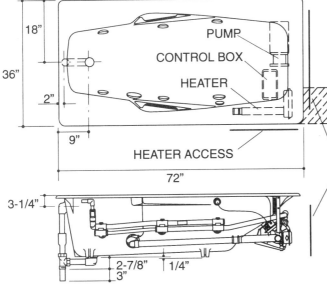
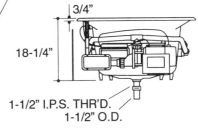

(b-2)

Figure 8.33 *(Continued)*

470 / PRINCIPLES OF PLUMBING

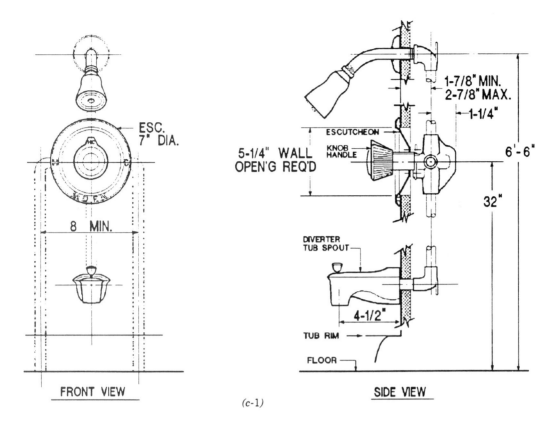

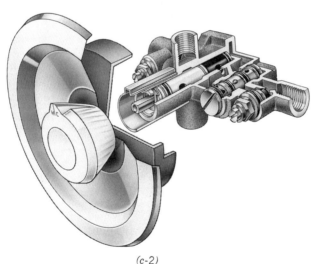

Figure 8.33 *(Continued) (c-1)* Typical dimensional data for a tub/shower faucet arrangement. The single-control handle permits control both of water temperature (by rotation) and volume (push-pull arrangement). The transfer valve on the tub spout is designed to drop back into tub-fill position automatically when the shower valve is shut off. *(c-2)* Cutaway of a single-knob shower valve. It is a four-port pressure-balancing unit with built-in check stops. It will maintain selected temperature to ±2°F and will drastically reduce flow if either hot or cold supply pressure fails. [*(c)* courtesy of Moen.]

FIXTURES / 471

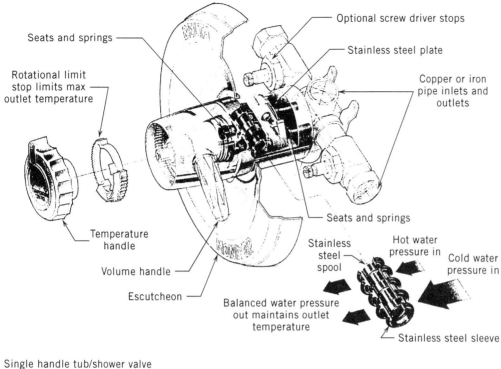

(d)

Figure 8.33 (d) Phantom view of a spool-and-sleeve-type constant temperature pressure-balancing single-handle tub/shower valve. When pressures are balanced, the spool is stationary within the sleeve, and hot and cold water are mixed according to the handle position. When the pressure of either source drops, the spool slides in the sleeve in the opposite direction, reducing the flow of the higher pressure water and increasing the flow of the lower pressure water, thus maintaining the temperature of the mixture. (Courtesy of Delta Faucet Company.)

472 / PRINCIPLES OF PLUMBING

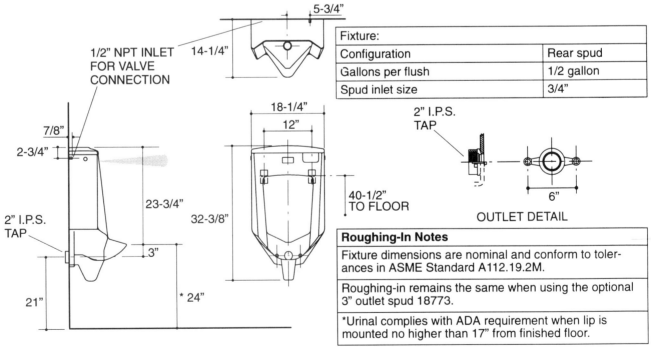

Figure 8.34 *(a-1)* Wall-mounted vitreous china, washout action urinal with integral electronic flush valve that is activated by fixture use. The infrared beam is concentrated to avoid unwanted flushing. The efficient flush uses a ½ gal of water. The unit has an elongated rim and meets the requirements of ADA (Americans with Disabilities Act). *(a-2)* Roughing-in data. *(b-1)* Wall-mounted vitreous china urinal with flushing rim and siphon-jet flush action. The flush valve is furnished separately. This unit meets ADA requirements when installed with the top of the front rim at 17 in. AFF. *(b-2)* Roughing-in data. [Photos and data for *(a, c)* courtesy of Kohler Company. Photo and data for *(b)* courtesy of American Standard. Data for *(d)* from Ramsey and Sleeper, *Architectural Graphics Standards*, 8th ed., 1988, © John Wiley & Sons, reprinted by permission of John Wiley & Sons.]

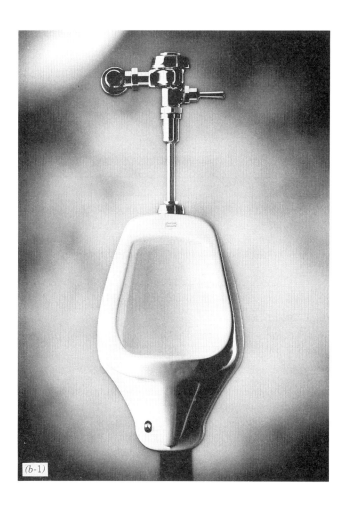
(b-1)

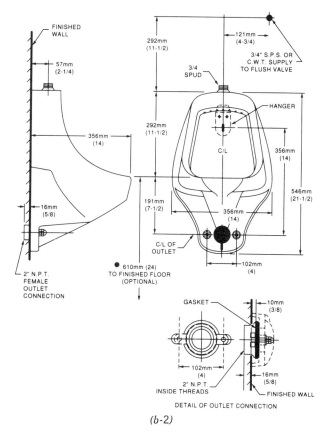
(b-2)

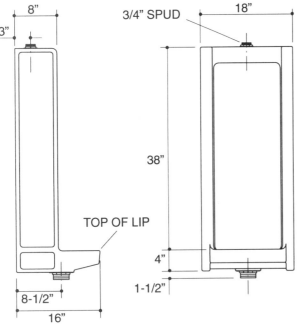

Product Information

Fixture:	
Configuration	Top spud
Gallons per flush	< 1 gallon*
Spud inlet size	3/4"
* Designed to flush with less than one gallon of water when installed with a water saving flush valve.	

Roughing-In Notes
Fixture dimensions are nominal and conform to tolerances in ASME Standard A112.19.2M.
For proper drainage, install lip of urinal below floor level.

(c-2)

Figure 8.34 *(Continued) (c-1)* Stall-type, washout action vitreous china urinal with wall-mounted flush tank. *(c-2)* Roughing-in data.

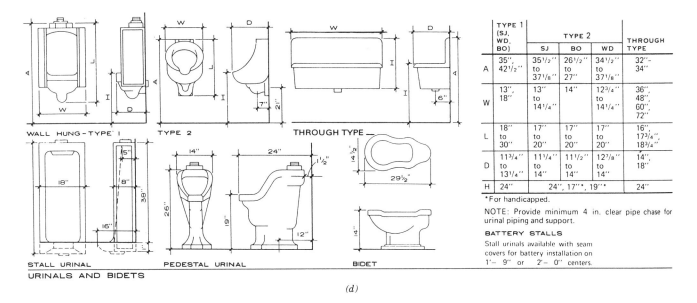

Figure 8.34 *(Continued)* *(d)* Typical dimensional data for urinals and bidets. The abbreviations SJ, BO and WD stand for siphon-jet, blowout and washdown, respectively. For actual dimensions of specific fixtures, obtain roughing-in dimensioned drawings from the manufacturer.

sensor is timed so that a person passing in front of the urinal does not cause it to flush. Where sensing is not used, a standard manually operated disc-handle flush valve is provided, despite its somewhat unsanitary aspects. The preferred flushing action is siphon-jet type if the architect will provide privacy shields between individual wall-hung units. This action uses a moderate water quantity, is quiet and requires little maintenance. If shields are not provided, then either blowout or washout flushing can be used. The former has high water use and a very high noise level but low maintenance. The latter has low water use, acceptable noise level but high maintenance.

g. Water Closets

The common toilet is known in the trade as a water closet or simply as a closet or closet bowl. It is constructed of vitreous china and requires a source of water for flushing. This source is either a tank or a flush valve. Old-style toilets such as the washdown type (which is no longer made in the United States) used a tank mounted high on the wall behind the bowl. The high velocity of flush water that resulted from the large static head, provided adequate flushing with low water consumption but a very high noise level. These bowls were also subject to frequent clogging due to the narrow irregular shape of the built-in siphon. All toilets flush by a siphon action through the integral trap. When (flush) water is added to the bowl, the lower leg of the siphon discharges an equal amount. The moving water carries the waste products in the bowl along with it through the siphon. When design factors dictated lowering the flush tank, the reduced water velocity required redesign of the bowl to increase the siphon action with the now reduced pressure. The solution uses larger quantities of water to achieve the required cleansing. Four types of closet bowls are in use today. All are illustrated in Figure 8.35.

(1) The *siphon-jet* design is the most common type, both in residential and commercial use. It uses a small jet of water injected directly into the siphon to set the siphon water in motion, while the majority of the flush water enters the bowl through the rim and washes down, with the bowl contents, through the siphon. It is an efficient, sanitary and very quiet design. With elevated tanks, it uses less than 4 gal of water. In the close-coupled toilet design (tank rests on the closet bowl), the reduction in pressure (static head) requires a large increase in the water quantity; it uses as much as 7 gal, de-

476 / PRINCIPLES OF PLUMBING

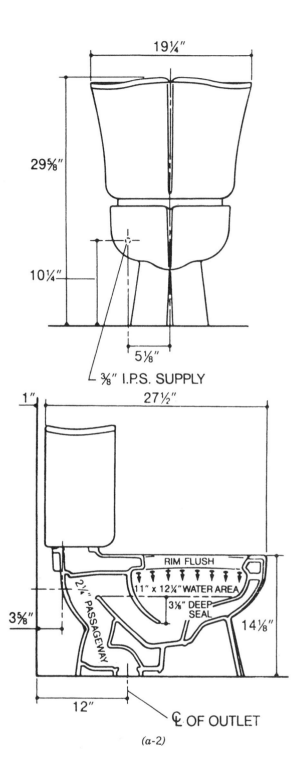

Figure 8.35 *(a-1)* Vitreous china, round front, close-coupled water closet and tank, using a siphon-jet flush of 3.5 gal. The flush is activated by a push button in the tank cover. *(a-2)* Roughing-in dimensions, plus a diagram of the flush action shown on the closet section. [Photos and data for *(a–c)* courtesy of Kohler Company; data for *(d)* from Ramsey and Sleeper, *Architectural Graphics Standards*, 8th ed., 1988, © John Wiley & Sons, reprinted by permission of John Wiley & Sons.]

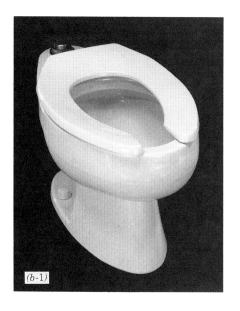

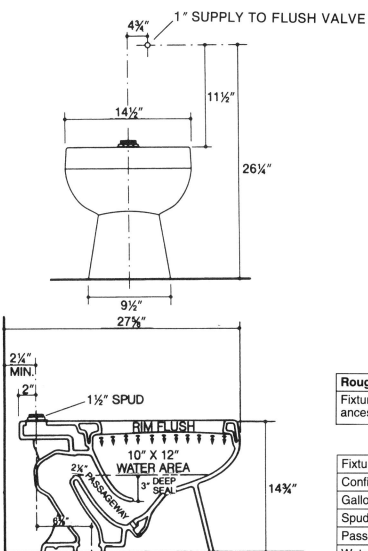

Roughing-In Notes

Fixture dimensions are nominal and conform to tolerances in ASME Standard A112.19.2M.

Fixture:	
Configuration	Top spud
Gallons per flush	1.6 gallons*
Spud size	1-1/2"
Passageway	2-1/4"
Water depth from rim	6"
Seat post hole centers	5-1/2"
* Designed to flush with 1.6 gallons of water when installed with a 1.6 gallon flush valve.	

(b-2)

Figure 8.35 *(Continued) (b-1)* Vitreous china elongated bowl designed for commercial and institutional use. When used with the appropriate flush valve, the siphon-jet design uses only 1.6 gal of water. *(b-2)* Roughing-in dimensions, plus a diagram of the flush action, shown on the bowl section.

478 / PRINCIPLES OF PLUMBING

Product Information

Fixture:	
Configuration	1-piece, elongated
Gallons per flush	1.1/1.6 gallons
Passageway	2-1/4"
Water depth from rim	5-1/2"
Seat post hole centers	12-5/8"
Minimum running pressure required	15 p.s.i.

Pump	HP	V	Hz	A
60 Hz	1/5	120	60	8.5

Roughing-In Notes

Fixture dimensions are nominal and conform to tolerances in ASME Standard A112.19.2M.

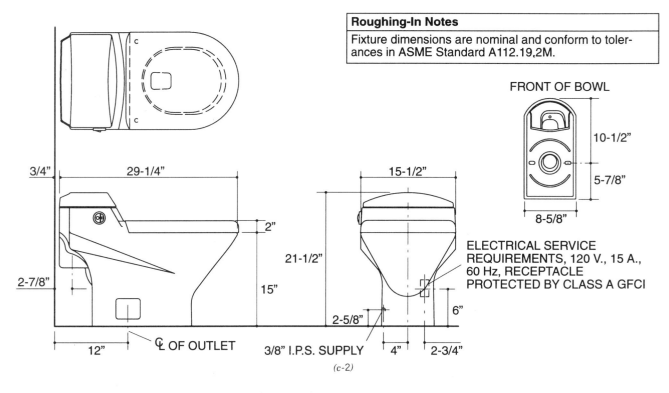

Figure 8.35 *(Continued) (c-1)* One-piece vitreous china bowl and tank with pressure-assisted siphon vortex flushing. A standard 120-v GFCI electrical outlet and a minimum static water pressure of 15 psi are required for proper flush operation. The unit contains an integral 0.2-hp pump. It provides a small flush (1.1 gal) or large flush (1.6 gal), depending on which side of the control button is pushed. *(c-2)* Roughing-in data.

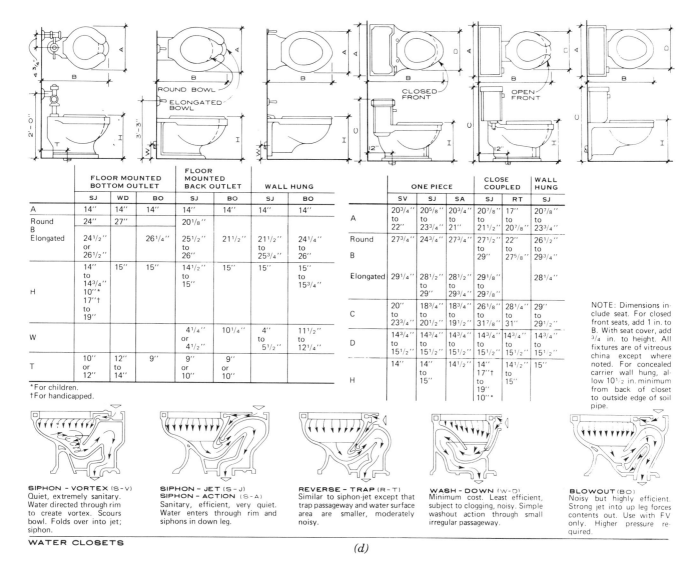

Figure 8.35 *(Continued) (d)* Description of fixture action and typical dimensional data for five types of closet bowls. Washdown (WD) bowls are no longer made in the United States. Reverse trap units are used where space is restricted. For actual dimensional data of a specific fixture, obtain roughing-in dimensioned drawings from the manufacturer.

pending on design. This makes it unsatisfactory from the point of view of water conservation guidelines, which mandate a maximum of 4 gal per flush. Recent design changes in siphon-jet bowl design have reduced their water consumption to well below 4 gal. This is accomplished by using a smaller exposed water surface in the bowl and a smaller siphon cross-sectional area, which is still large enough to prevent clogging. Special low consumption designs using compressed air have also met with considerable success.

When equipped with a flush valve (flushometer), the flush water is delivered at water line pressure, and the quantity of water required is reduced to well below 4 gal. Siphon-jet bowls equipped with flush valves require a minimum pressure of 15 psi for floor-mounted bowls and 20 psi for wall-mounted bowls.

(2) *Siphon vortex* closet bowls are specifically designed for the low pressure and water velocity of close-coupled tanks. The water enters the bowl off-center, creating a vortex action in the siphon. They are extremely quiet, making them ideal for toilets adjoining sleeping areas. They are also highly sanitary due to the waters scouring action. Most designs use up to 8 gal of water.

(3) *Blowout* closet bowls are highly efficient due to a high velocity water design but, as a result, are very noisy. They find wide application, with flush valves, in public and institutional toilets where self-cleansing and low maintenance plus low water usage far outweigh any noise considerations. They are not used with tank flushing. Blowout closets with flushometers require a minimum pressure of 20 psi for floor-mounted units and 25 psi for wall-mounted bowls.

(4) *Reverse trap* bowls are used where front-to-back space is at a premium. Their action is similar to that of a siphon-jet bowl.

Flush tanks are supplied with a ⅜- or ½-in. line. A small line is sufficient because they refill slowly and, therefore, place only a light load on the water line. The disadvantage is that repeat flushing is delayed until the tank refills. Flush valves take a 1-in. line to handle the short burst of high velocity water required. Since the flush valve supplies water directly from the water main, no delay is caused, and immediate repeat flushing is possible.

Closet bowls are made in rounded and elongated front shapes. The latter is considered more sanitary and is used with an open front seat in public toilets.

Bowls are made for floor or wall mounting. The wall-mounted units simplify floor cleaning and are, therefore, used in public toilets whenever the very substantial fixture closet carrier can be accommodated. These carriers must bear the weight of a heavy person and, therefore, masonry construction in the wall behind the closet is preferred. Wherever possible, back-to-back carriers should be used. See Figure 8.36 for details of closet carriers.

h. Bidet

See Figure 8.37. This fixture, which is gaining popularity in the United States (and is widely used throughout Europe), is essentially a modified lavatory intended primarily for female perineal cleaning. It is provided with a hot and cold water battery that feeds a spray outlet, usually vertical. Waste water exits through a drain in the center of the fixture in a manner similar to a lavatory. Bidets are constructed of vitreous china and take ½-in. hot and cold water lines.

See Figure 8.38 for all fixture-mounting clearances as required by Code.

i. Drinking Fountains

See Figure 8.39. These devices, also referred to as water coolers take a ⅜-in. cold water line and a 1¼-in. drain line. The specific design and location is chosen by the project architect. The plumbing designer should remember to notify the electrical designer of the water cooler location, since an electrical outlet is required.

8.16 Drawing Presentation

As with HVAC and electrical construction information, the plumbing design is transmitted to the plumbing contractor on working drawings. The starting point for all types of projects is the architectural floor plan. The plumbing technologist draws the building's floor plan, using the architectural plan as the basis but eliminating all architectural information that is not relevant to the plumbing work. All the plumbing fixtures are then drawn, using standard symbols. If the building is large, and only a small portion of the building contains plumbing fixtures, then he or she would draw only that portion of the building, generally at a larger scale than the architectural plan. This might be the case, for instance, with a large assembly plant such as that shown in Figure 3.48 (page 153), drawn to 1/16-in. scale, which has toilets and a rest area in

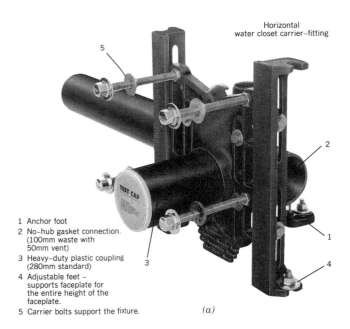

1 Anchor foot
2 No-hub gasket connection. (100mm waste with 50mm vent)
3 Heavy-duty plastic coupling (280mm standard)
4 Adjustable feet – supports faceplate for the entire height of the faceplate.
5 Carrier bolts support the fixture.

(a)

Single Adjustable Carrier Fittings

Double Adjustable Carrier Fittings

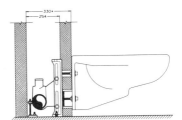
*286mm Minimum Using 38mm Coupling

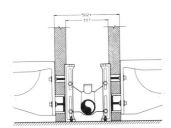
*413mm Minimum Using 38mm Coupling

(b)

8.36 *(a)* Detail of horizontal water closet carrier fitting. *(b)* Carrier fittings are available for either single or back-to-back installations. All dimensions are in millimeters. The entire closet load is carried by the carrier-fitting bolts. (Courtesy of Tyler Pipe, subsidiary of Tyler Corporation.)

482 / PRINCIPLES OF PLUMBING

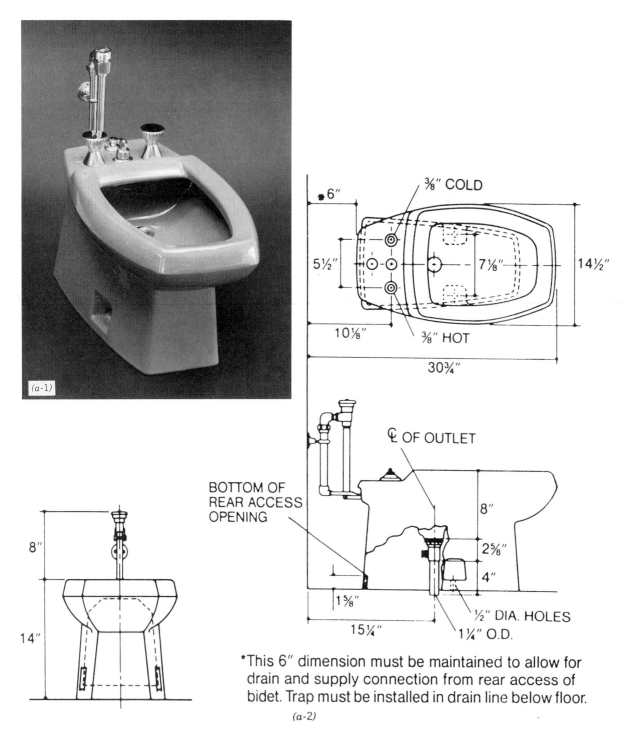

8.37 *(a-1)* Vitreous china bidet with flushing rim, integral overflow and vertical spray. Faucets and pop-up drain control are deck-mounted on the fixture. A vacuum breaker is shown behind the bidet. *(a-2)* Roughing-in dimensions. *(b-1)* Vitreous china bidet single-hole bidet faucet and horizontal spray. Roughing can be installed above or below the floor. *(b-2)* Roughing-in data. (Photos and data courtesy of Kohler Company.)

DRAWING PRESENTATION / 483

(b-1)

Fixture:	
Configuration	Faucet deck
Outlet	1-3/4" D.
Faucet hole	1-3/8" D.
P-trap	1-1/4" O.D.

Roughing-In Notes
Fixture dimensions are nominal and conform to tolerances in ASME Standard A112.19.2M.
P-trap furnished with fixture must be installed for above floor installation. Regular trap can be installed below the floor.

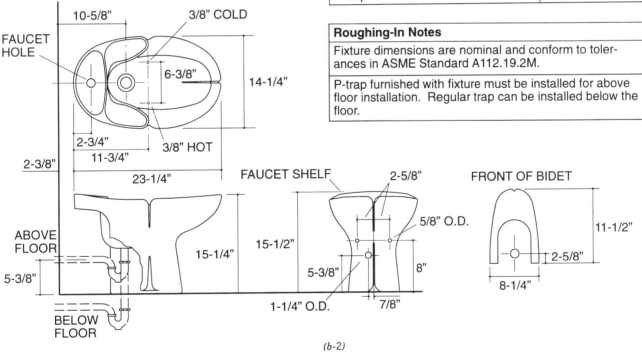
(b-2)

Figure 8.37 *(Continued)*

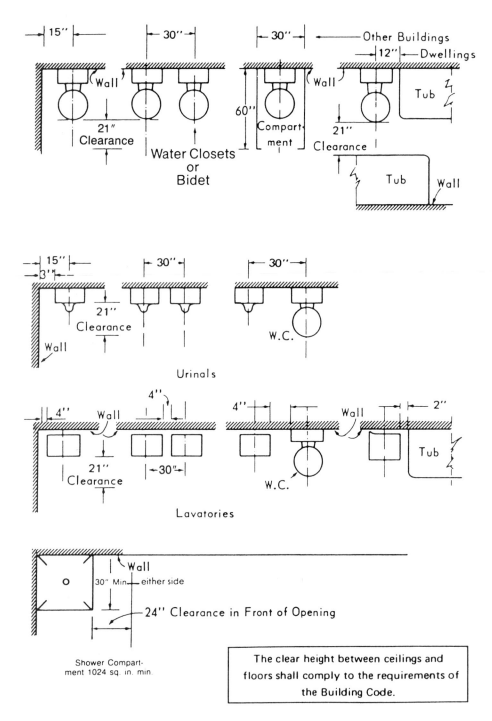

Figure 8.38 Minimum fixture clearances. (Reprinted with permission from The National Standard Plumbing Code, published by the National Association of Plumbing Heating Cooling Contractors.)

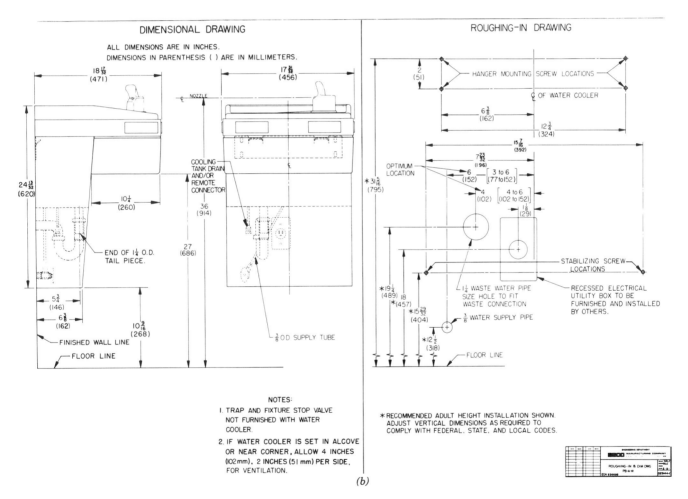

Figure 8.39 (a) Typical wall-mounted water cooler, of barrier-free design. Various designs are available to operate the cooler by sensors or by light pressure on touch pads situated around the cooler. (b) A dimensional drawing (left) is required by the architect to permit adequate space planning. Note too the location of the electrical outlet that is required. A roughing-in drawing (right) permits the plumbing contractor to position the water and drain piping exactly. (Courtesy of EBCO Manufacturing Company.)

486 / PRINCIPLES OF PLUMBING

one corner. The technologist would then draw the toilet/rest area to ⅛-, ¼- or even ½-in. scale, indicating its location in the building by column lines. All the local piping would be shown on this part plan, with connecting water and drainage lines shown there and on a small-scale area plan. This is exactly the approach that was used in preparing the water and drainage plans shown in Figure 9.29 (page 547). [*Note:* We are fully aware that much of today's drawing preparation is done with a drawing or CAD program of some type rather than by manual drafting. We use the word "draw" to include both techniques.] If the building is relatively small, as for instance a residence, then the technologist would probably show the entire floor plan, although he or she might go the part plan route if it were felt that this approach added clarity to the design.

The first step in the design, as already stated, is to show the plumbing fixtures on the architectural background. See Figure 8.40, which is The Basic House plan that we studied in the HVAC and electrical sections. Using Figure 3.32 as the basis, we would eliminate the portions of the main floor and basement that are irrelevant, since they contain no plumbing fixtures, to arrive at the basis for Figure 8.40. To this we add the symbols for the required plumbing connections for each fixture as appear in Figure 8.40. This then constitutes the preliminary-stage planning.

In order to understand the next stage of drawing presentation, you must have some familiarity with plumbing symbols. A fairly complete symbol list is given in Figure 8.41. As with HVAC and electrical work, plumbing work is shown in one line form without any attempt to draw items to scale. See Figure 8.42. Scale drawing would make most equipment, especially piping and valves, so small as to be impossible to draw. Plumbing details are often drawn to scale, and the scale is given, or the detail is dimensioned.

The next stage of plumbing drawing presentation shows the piping on the floor plan that was prepared at the preliminary stage. This is shown in Figure 8.43. In addition, the technologist will prepare plumbing sections for the hot and cold water piping and for drainage. These "sections" show the piping in elevation. They are called sections in small buildings and risers in tall buildings. In small simple jobs, all the piping is shown in a single section. In more complex jobs, the water piping and drainage are shown on separate sections. See Figures 8.44 (page 493) and 10.3 (page 561). Where a two-dimensional representation is not sufficiently clear, as is common with drainage piping, an isometric detail can be drawn showing the various fixtures in both horizontal and vertical relationship. Finally, a plot plan is prepared showing service connections, such as Figure 10.56.

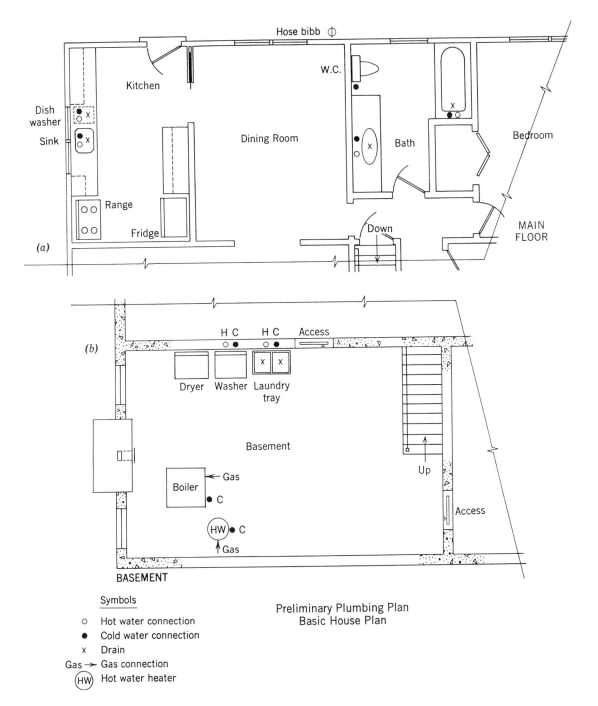

Figure 8.40 Preliminary plumbing plan of The Basic House: *(a)* main floor and *(b)* basement.

488 / PRINCIPLES OF PLUMBING

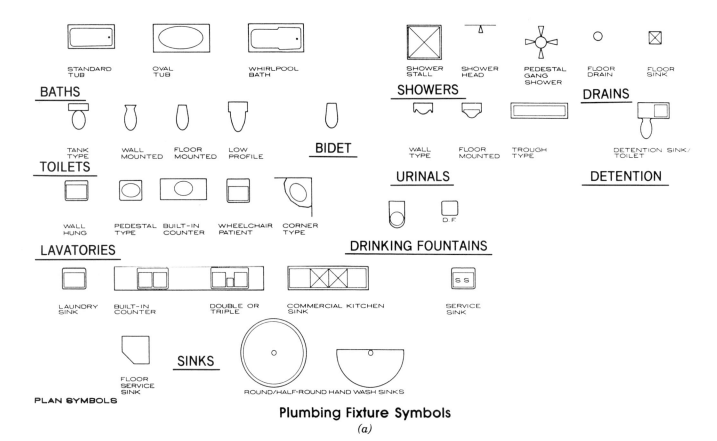

Figure 8.41 *(a)* Plumbing fixture symbols for use on plan drawings. (Ramsey and Sleeper, *Architectural Graphics Standards*, 8th ed., 1988, © John Wiley & Sons, reprinted by permission of John Wiley & Sons.)

PLUMBING PIPING SYMBOLS

COLD WATER	—··—··—··—	SOIL, WASTE OR LEADER (ABOVE GRADE)	————————
HOT WATER	—··—··—··	SOIL, WASTE OR LEADER (BELOW GRADE)	—— —— —— -
HOT WATER RETURN	—···—···—	VENT	- - - - - -
HOT WATER SUPPLY (180°F.)	—··—180°—··—	COMBINATION WASTE AND VENT	——SV——
HOT WATER RETURN (180°F.)	—···—180°—···—	INDIRECT DRAIN	——IW——
FIRE LINE	—F——F—	STORM DRAIN	——S——
WET STANDPIPE	——WSP——	MAIN SUPPLIES SPRINKLER	——S——
DRY STANDPIPE	——DSP——	BRANCH AND HEAD SPRINKLER	——○——○—
		DRAIN - OPEN TILE OR AGRICULTURAL TILE	====

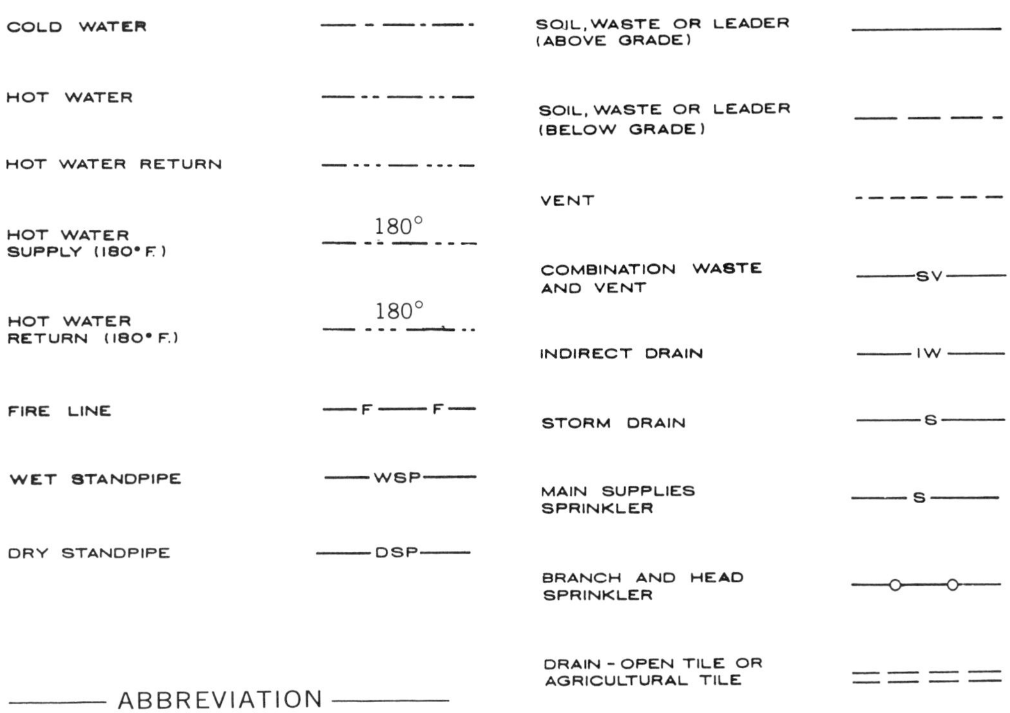

——— ABBREVIATION ———

List of Abbreviations

- A Compressed air
- AW Acid waste
- AV Acid vent
- G Gas, low pressure
- MG Gas, medium pressure
- HG Gas, high pressure
- LPG Liquid petroleum gas
- N Nitrogen
- O Oxygen
- V Vacuum
- VC Vacuum cleaning

Special Abbreviations

- PN Pneumatic tubes
- CI Cast iron pipe
- WI Wrought iron pipe
- CT Clay Tile

(b)

Figure 8.41 (b) Plumbing piping symbols and abbreviations.

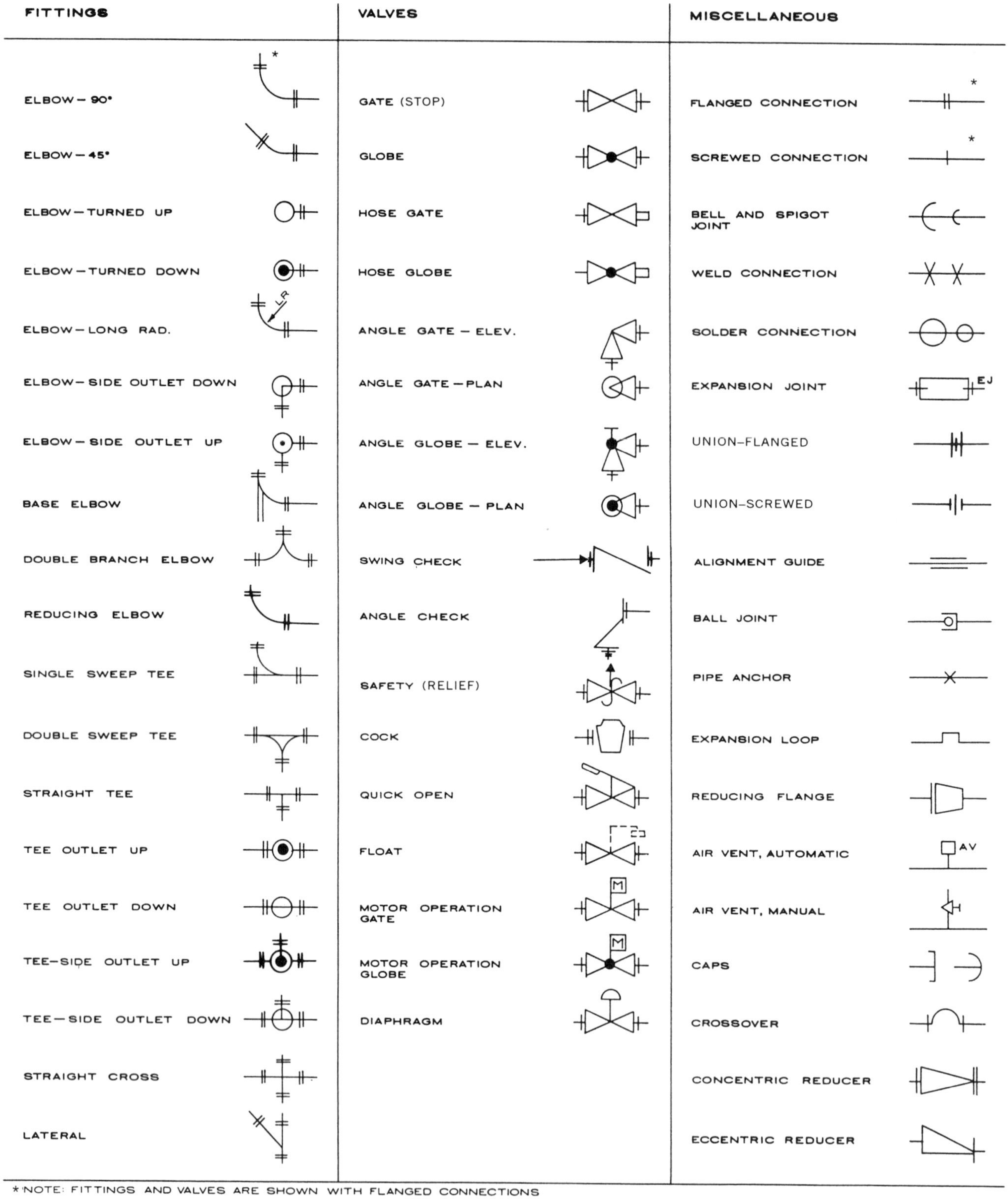

FITTINGS		VALVES		MISCELLANEOUS	
ELBOW – 90°		GATE (STOP)		FLANGED CONNECTION	
ELBOW – 45°		GLOBE		SCREWED CONNECTION	
ELBOW – TURNED UP		HOSE GATE		BELL AND SPIGOT JOINT	
ELBOW – TURNED DOWN		HOSE GLOBE		WELD CONNECTION	
ELBOW – LONG RAD.		ANGLE GATE – ELEV.		SOLDER CONNECTION	
ELBOW – SIDE OUTLET DOWN		ANGLE GATE – PLAN		EXPANSION JOINT	
ELBOW – SIDE OUTLET UP		ANGLE GLOBE – ELEV.		UNION – FLANGED	
BASE ELBOW		ANGLE GLOBE – PLAN		UNION – SCREWED	
DOUBLE BRANCH ELBOW		SWING CHECK		ALIGNMENT GUIDE	
REDUCING ELBOW		ANGLE CHECK		BALL JOINT	
SINGLE SWEEP TEE		SAFETY (RELIEF)		PIPE ANCHOR	
DOUBLE SWEEP TEE		COCK		EXPANSION LOOP	
STRAIGHT TEE		QUICK OPEN		REDUCING FLANGE	
TEE OUTLET UP		FLOAT		AIR VENT, AUTOMATIC	
TEE OUTLET DOWN		MOTOR OPERATION GATE		AIR VENT, MANUAL	
TEE – SIDE OUTLET UP		MOTOR OPERATION GLOBE		CAPS	
TEE – SIDE OUTLET DOWN		DIAPHRAGM		CROSSOVER	
STRAIGHT CROSS				CONCENTRIC REDUCER	
LATERAL				ECCENTRIC REDUCER	

*NOTE: FITTINGS AND VALVES ARE SHOWN WITH FLANGED CONNECTIONS
SCREWED CONNECTION DEVICES USE ONLY A SINGLE CROSSHATCH

Piping Symbols: Pipe Fittings and Valves

(c)

Figure 8.41 (c) Piping symbols; pipe fittings and valves. (Ramsey and Sleeper, *Architectural Graphics Standards*, 8th ed., 1988, © John Wiley & Sons, reprinted by permission of John Wiley & Sons.)

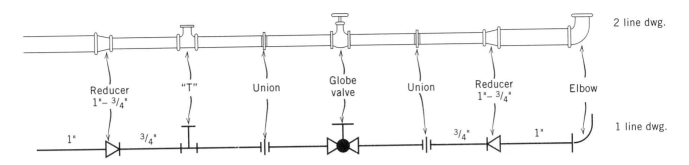

Figure 8.42 Single-line representation of plumbing piping and fittings. See Figure 8.41(c) for symbols.

492 / PRINCIPLES OF PLUMBING

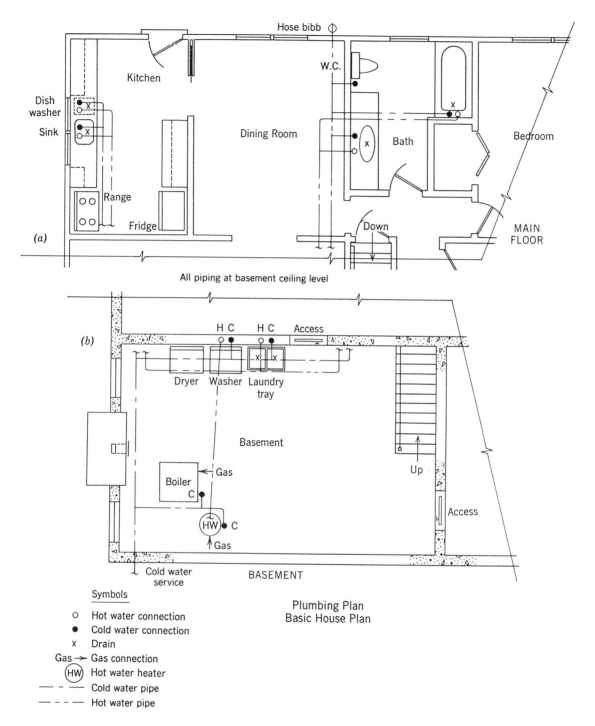

Figure 8.43 Piped plan of The Basic House. All piping is run at the basement ceiling level and turned up for the first floor fixtures. Valving is shown on the riser in Figure 8.44.

DRAWING PRESENTATION / 493

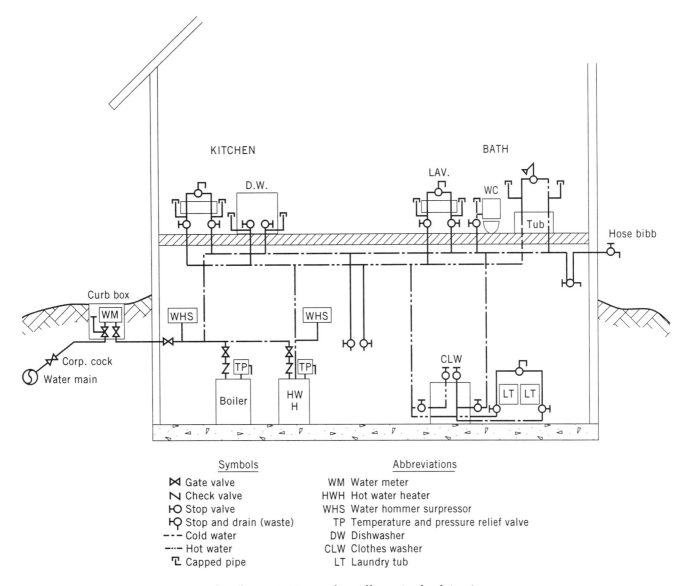

Figure 8.44 Plumbing water riser for The Basic House plan. All required valving is shown.

Key Terms

Having completed the study of this chapter, you should be familiar with the following key terms. If any appear unfamiliar or not entirely clear, you should review the section in which these terms appear. All key terms are listed in the index to assist you in locating the relevant text.

Absolute pressure
American National Standards Institute (ANSI)
American Society for Testing and Materials (ASTM)
Asbestos cement pipe
Asphaltum compound
Atmospheric pressure
Backflow
Ball check valve
Bib (bibb)
Bidet
Black iron pipe
Black water
Blowout closet bowls
BOCA Basic Plumbing Code
Bolted bonnet
Bourdon gauge
Cast-iron (CI) soil pipe
Cast Iron Soil Pipe Institute (CISPI)
Cathodic protection
Check valves
Close-coupled toilet
Closet bowl
Closet carriers
Composition disc
Conventional disc
Dielectric pipe connectors
Differential manometers
Disc check seats
DWV
Effluent
English units
Expansion coefficients
Expansion loop
Faucet
Ferrous metal pipe
Fixture vents
Flared pressure joints
Flow rate
fps
Gauge pressure
Gate valves
Globe valves
gpm
Gravity flow
Grey water
Hard temper pipe

Head (pressure)
Heating tracer cables
Hub and spigot design
Hubless-type soil pipe
Hydraulics
Hydrostatic pressure
Lift check valve
Linear expansion joint
Manometer
Mercury manometer
Millimeters of mercury
National Sanitation Foundation (NSF)
National Standard Plumbing Code
NSF label
Oakum
Piezometer tube
Pipe and tubing
Piston check valve
Pitch
Plastic pipe
Plug disc
Plumbing fixtures
Plumbing sections
Potable water
Pressure-balancing valves
Pressure-reducing valves (PRV)
Pressure relief valve
Pressurized flow
psi
Relief valves
Rising and nonrising stem
Roughing
Sanitary drainage
Schedule 40 pipe
Schedule 80 pipe
Screwed bonnet
Self-closing faucets
Service sinks
SI units
Sill cock
Siphon-jet closet bowl
Siphon vortex closet bowls
Soft temper pipe
Southern Standard Plumbing Code
Static pressure
Stop-and-waste valve

Storm drainage
Sweat joints
Swing check valve
Temperature control
Temperature relief valve
Thermal expansion
Thermoplastic
Thermostatically controlled valve
Type ABS pipe
Type CPVC pipe

Type PE pipe
Type PVC pipe
Types: K, L, M, DWV, copper pipe
Uniform Plumbing Code
Valve seats
Vitrified clay (terra cotta) pipe
Wash fountains
Washerless valve
Water coolers
Water closet

Supplementary Reading

See listing at the end of Chapter 10.

Problems

1. List five building systems for which the plumbing technologist should be able to design the piping.
2. List the basic plumbing fixtures that are required in a
   a. Residence.
   b. School.
   c. Shopping mall.
   d. Dormitory.
3. What is the function of a plumbing code?
4. Define the following terms:
   a. Hydrostatic pressure.
   b. Backflow prevention.
   c. Gravity flow.
   d. Hydraulics.
   e. Manometer.
   f. Gauge pressure.
   g. Absolute pressure.
5. What is the hydrostatic pressure at a point 6½ ft deep in a tank of water open to the atmosphere,
   a. In pounds per square inch.
   b. In feet of water.
   c. In millimeters of mercury.
6. What is the static pressure at a point 10 mm below the surface of a tube of mercury open to the atmosphere in
   a. Millimeters of mercury.
   b. Pounds per square inch.
   c. Feet of water.
   d. Atmospheres.
7. What is the water velocity in a ½-in. copper pipe feeding a cold water faucet that is discharging water at the rate of 4 gpm? Is this velocity within the acceptable limits as far as noise is concerned?
8. Starting with the formula
   $$Q = 2.45\, d^2 V$$
   where

   Q is in gallons per minute,
   d is in inches and
   V is in feet per second,

   develop a formula where

   Q is liters per minute,
   d is millimeters and
   V is meters per second.

   and show all the steps in the derivation. Use the unit cancellation method.
9. Describe the method of connecting:
   a. Iron pipe to fittings.
   b. Copper tube to fittings.
10. Describe briefly three methods of joining lengths of cast-iron pipe.
11. A 150-ft run of DWV PVC plastic drainage line is suddenly subjected to a flow of hot water from an industrial process. The effective tem-

perature of the pipe rises to 120°F from the factory temperature of 60°F. By how many inches will it increase in length?
12. A cast-iron pipe under the same conditions as in Problem 11 will lengthen by how many inches?
13. Describe briefly the tests made on piping roughing and on finished piping.
14. Describe briefly the functioning and use of the following valves:
    a. Gate.
    b. Globe.
    c. Relief.
    d. Angle.
    e. Check.
    f. Pressure reducing.
15. Describe two methods of taking care of the expansion of water in the piping of a domestic hot water system.
16. Name materials suitable for the following services below ground:
    a. Gas piping.
    b. Sanitary building sewer.
    c. Water service.
17. Describe briefly the action of the following closet bowls:

    a. Siphon-jet.
    b. Siphon vortex.
    c. Blowdown.
    d. Washdown.

    Which require flush valves and which require tanks? Which closets and valves would you recommend for installation in public rest rooms in a concert hall? Why?
18. Concerning minimum fixture clearances, what is:
    a. The size of a toilet compartment?
    b. The spacing on centers of water closets?
    c. The clearance in front of a water closet?
    d. The clearance in front of a lavatory?
    e. The clearance in front of a shower compartment opening?
19. Briefly describe the type of plumbing drawings that are prepared at the preliminary and final stages of a small two-story garden apartment.
20. Compare the advantages and disadvantages of copper pipe versus plastic pipe for domestic water systems. Specify the exact types of pipes involved.

9. Water Supply, Distribution and Fire Suppression

One of the two basic functions of a building plumbing system is the supply of water. (The other is drainage.) In the vast majority of buildings, only potable water is supplied, regardless of its eventual use. The source of water is almost always a water utility pipeline in the street. (Artesian wells are used in remote isolated areas.) It is the plumbing technologist's responsibility to design the entire water service and distribution system for all uses, recognizing the pressure and flow limitations of the water supply. The techniques and procedures involved in that design are fully explained in this chapter. Study of this chapter will enable you to:

1. Determine the adequacy of the water supply pressure with respect to any building being designed.
2. Select the water supply system appropriate to any building, considering the flow and pressure characteristics of the available water service.
3. Determine the required water flow and service pipe size for the building being designed.
4. Understand the procedure for calculating friction head in all types of plumbing piping.
5. Calculate water system pipe sizes using the friction head loss method.
6. Calculate water system pipe sizes using the velocity limitation method.
7. Understand the functioning of the various systems of hot water supply.
8. Select a hot water heater for a small building.
9. Understand the different types of hot water circulation systems.
10. Understand and apply the principles underlying proper selection and location of valves in water systems.
11. Be familiar with the means to prevent backflow and water hammer.
12. Design most residential and many straightfor-

ward nonresidential water supply and distribution systems.
13. Understand the layout and application of fire standpipe and hose systems.
14. Be familiar with the various types of sprinkler systems and their usage.

9.1 Design Procedure

The procedure for the design of a water distribution system for a building is straightforward. It is assumed that an adequate reliable supply of clean potable water is available. Whether the source is street mains as is usually the case or a private well is immaterial to the technologist whose work begins at the water service entrance. The design procedure is then as follows:

(a) Determine the pressure of the source. Decide whether to use the source directly, reduce the pressure or increase it.
(b) Determine whether the structure will be treated as a single unit or whether it is necessary to zone it.
(c) Decide whether to use an upfeed or downfeed system.
(d) Determine the pressure and flow requirements of all fixtures and all continuous water uses.
(e) Determine maximum instantaneous water demand. This is a combination of fixture use and other water uses in the building.
(f) Determine the service size on the basis of maximum water requirements.
(g) Determine minimum pipe sizes on the basis of required flow rates and pressure for the water use device farthest from the service. This requires use of pipe friction charts or tables or, alternatively, use of the velocity method.
(h) Decide on the method of supplying hot water. This includes hot water source and type of circulation system, if any.
(i) Determine water pipe sizes for the entire structure. Pressures of hot and cold water should be equal at fixtures using both, to prevent cross-flow during mixing.
(j) Design details of the piping system including water service details, hot water supply details, all valving, location of vacuum breakers, special support details and the like.
(k) Determine location of shock arresters (water hammer eliminators) and any other special devices required.

9.2 Water Pressure

The first major decision to be made in designing the building water supply depends on the supply's water pressure and the type and vertical elevation of the highest fixture outlet in the building. These two factors will generally determine the type of system to be used. In actual practice, this decision is usually made jointly with the architect and the project engineer. They, however, rely heavily on the input of the plumbing design technologist. We are, therefore, presenting the factors involved in this decision in detail, in order to give the technologist the required background.

We begin with the assumption, as already stated, that the water source, whether it is a well or city water mains, has adequate capacity and pressure. City mains normally have a pressure to 30–60 psi. The technologist should verify what minimum pressure is maintained during periods of maximum demand, which is normally during the summer months. A working pressure 10 psi lower than that figure should then be used as the mains pressure in order to ensure adequate flow pressure under all conditions. Line pressure above 80 psi cannot be used directly. It requires installation of a pressure-reducing valve as a Code requirement (with certain exceptions). See Code Section 10.14.6. The reason is that conventional fixtures have a 80-psi pressure limit under no-flow conditions. Inadequate pressure requires installation of a booster pump and pressure tank or use of a downfeed system, as will be discussed later. *Inadequate pressure* is defined as a line pressure that is insufficient to provide the minimum *flow pressures* and *flow rates* at fixtures, as listed in Tables 9.1 and 9.2.

Under no-flow conditions, the street mains pressure is reduced throughout the system vertically, by height. Assume for instance, a five-story building with 10 ft between floors and the highest fixture 3 ft above the fifth floor. The total hydrostatic pressure at the bottom of the water riser due to height to this last fixture would be

$$[4(10 \text{ ft}) + 3 \text{ ft}] \times 0.433 \text{ psi/ft of water}$$
$$= 43 \text{ ft} (0.433) = 18.6 \text{ psi}$$

If the mains pressure were 40 psi, then under no-flow conditions the static pressure at the top fixture would be

$$40 \text{ psi} - 18.6 \text{ psi} = 21.4 \text{ psi}$$

which is well below the permissible maximum fixture pressure of 80 psi. Similarly, the pressure at a similar fixture 2 ft above the first floor, under no-flow conditions, would be

$$40 \text{ psi} - (2 \text{ ft})(0.433 \text{ psi/ft}) = 39.1 \text{ psi}$$

When a fixture operates and water flows, the pressure equation changes completely. Under flow conditions

Table 9.1 Minimum Pressure Required by Typical Plumbing Fixtures

Fixture Type	Minimum Pressure, psi
Sink and tub faucets	8
Shower	8
Water closet—tank flush	8
Flush valve—urinal	15
Flush valve—siphon jet bowl	
floor-mounted	15
wall-mounted	20
Flush valve—blowout bowl	
floor-mounted	20
wall-mounted	25
Garden hose	
5/8-in. sill cock	15
3/4-in. sill cock	30
Drinking fountain	15

Source. EPA Manual of Individual Water Supply System, 1975 and manufacturers' data.

Table 9.2 Recommended Flow Rates for Typical Plumbing Fixtures

Fixture Type	Flow, gpm
Lavatory	3
Sink	4.5
Bathtub	6
Laundry tray	5
Shower	3–10
Water closets	
tank type	3
flush valve[a]	15–40
Urinal flush valve	15
Garden hose	
5/8-in. sill cock	3 1/3
3/4-in. sill cock	5
Drinking fountain	3/4

Source. Data extracted from various sources.
[a] Wide range of flows; depends on flow pressure.

Total (mains) pressure = Static head + Friction head (loss) + Flow pressure

That is, the total pressure available is converted to static head, which is used to overcome height; friction head, which is used to overcome the friction between the moving water and the piping; and flow pressure, which is used to impart kinetic energy (motion) to the water. *Flow pressure* is the pressure that is available at the fixture when the outlet is wide open. It must equal or exceed the minimum pressure listed in Table 9.1 in order that the fixture flow be adequate.

An example should clarify this pressure equation. Referring to the same five-story building as before, let us assume that the highest fixture is a sink faucet and that the total friction head loss from the mains to the fixture, including the water meter, piping and all fittings, is 10 psi. Would the fixture flow pressure be sufficient?

The system pressures follow:

Mains pressure	40 psi
Static head to top fixture	18.6 psi (43 ft high)
Friction head	10 psi (assumed)
Flow pressure	?

In equation form

$$40 \text{ psi} = 18.6 \text{ psi} + 10 \text{ psi} + \text{Flow pressure}$$

Therefore,

$$\text{Flow pressure} = 40 - 18.6 - 10 = 11.4 \text{ psi}$$

Referring to Table 9.1, we see that this is sufficient for a faucet or a water closet tank but insufficient for a flush valve, which requires a minimum pressure of 15 psi.

The actual design procedure reverses the order of the calculation. In design, we begin by using the minimum flow pressure needed, from Table 9.1. We then calculate the maximum permissible system friction, and with that number we then size the piping. For instance, using the same data, and assuming a flush valve as the topmost fixture, we would have the following data:

Mains pressure	40 psi
Static head	18.6 psi
Minimum flow pressure	15 psi
Maximum friction head	?

In equation form, we would have

$$40 \text{ psi} = 18.6 + 15 + \text{Maximum friction head}$$

Therefore,

Maximum friction head = 40 − 18.6 − 15 = 6.4 psi

We would then design the piping (as will be explained later) to give a maximum overall friction loss of 6.4 psi. Then we would design piping so that this fixture, and all others, would have sufficient pressure to deliver the minimum flow as listed in Table 9.2. Before demonstrating this calculation procedure, however, we will see how the system pressure governs the choice of distribution system to be used.

9.3 Water Supply System

When the pressure from the city mains is sufficient to overcome all friction in the system with the calculated flow and still maintain the minimum pressure needed at the highest outlet, the system used is called an *upfeed system*. Refer to Figure 9.1, which represents a typical plumbing section for a two-story residence. To check whether the reported 40-psi minimum maintained city mains pressure is sufficient for this structure, we would make the following very reasonable assumptions in our calculation:

1. Assume a 5-psi friction loss in the water meter.
2. Assume an 8 psi/100 ft pressure loss in piping.
3. Assume that fittings add 50% to effective pipe length.

To perform the calculation, we would measure the distance from the water main to the farthest fixtures, which in this case are the second-floor shower head and the garden hose bibb. Assume that the first is 90 ft and the second is 60 ft. From Table 9.1 we note that the required minimum pres-

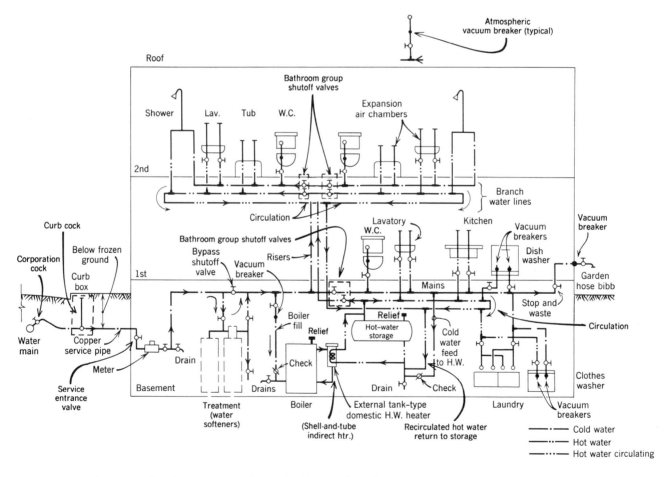

Figure 9.1 Schematic plumbing section of an upfeed water distribution system using street mains pressure. The building is a two-story residence. Not all required valving is shown, for clarity.

sures for these two fixtures are 8 and 15 psi, respectively. (We use a ⅝-in. sill cock, which is appropriate for a residence.) The pressure calculations for the shower and hose bibb follow.

Shower: The shower head is 7 ft above the second floor and 21 ft total above the water main (9 ft floor height plus 5 ft from the mains to the first-floor level). The pressure equation would then be

Fixture pressure = Mains pressure − Static head
  − Total friction

$$= 40 \text{ psi} - 21 \text{ ft} \left(\frac{0.433 \text{ psi}}{\text{ft of water}}\right) - 5 \text{ psi}$$
$$\text{(meter friction loss)}$$

$$- (150\%^* \text{ of } 90 \text{ ft}) \left(\frac{8 \text{ psi}}{100 \text{ ft of pipe}}\right)$$

$$= 40 \text{ psi} - 9.1 \text{ psi} - 5 \text{ psi} - 10.8 \text{ psi}$$
$$= 15 \text{ psi, which is adequate for a shower}$$
  (8 psi required)

Hose bibb: The hose bibb is 8 ft above the water main.

Fixture pressure

$$= 40 \text{ psi} - 8 \text{ ft} \left(\frac{0.433 \text{ psi}}{\text{ft of water}}\right) - 5 \text{ psi (meter friction)}$$

$$- (150\% \text{ of } 60 \text{ ft}) \left(\frac{8 \text{ psi}}{100 \text{ ft of pipe}}\right)$$

$$= 40 - 3.5 - 5 - 7.2 \text{ psi}$$
$$= 24.3 \text{ psi, which is adequate for a hose bibb}$$
  (15 psi required)

Note that we checked not only the highest, most remote fixture but also a lower, closer one that required a higher minimum pressure. If the hose bibb had been ¾ in., requiring 30 psi, the mains pressure would not be adequate. The technologist should, therefore, always spot fixtures requiring high pressure, such as flush valves, and check them. It occasionally occurs that these fixtures, even if closer to the water service entrance, are the critical ones.

As a matter of interest, note on Figure 9.1 that the service entrance water pipe is installed below the frost level. This is done so that there is no danger of ice forming in the pipe, during cold winter nights when water is not flowing, and blocking the water supply. (Flowing water does not freeze until temperatures drop far below freezing.) Note too that the hose bibb is fed through a *stop and waste valve*. This valve's function is to drain the section of pipe leading to the hose bibb. Otherwise, the water in the pipe might freeze in the winter and probably burst the pipe. The "*corporation cock*" at the tap onto the street main is simply a main shutoff valve. A second shutoff valve is installed in a *curb box*. Because the service entrance pipe is quite deep at this point, the shutoff valve is shown with a long extension handle.

Let us now consider a building where the street main pressure is insufficient. Assume a five-story office building with toilets using flush valves on every floor. Assume a 10-ft floor-to-floor height, service piping 5 ft below the first floor level, flush valves 2 ft above floor level, and the same 40-psi minimum street mains pressure. The required street mains pressure for this building would be:

Static head:
[5 ft + 4(10 ft) + 2 ft]0.433 psi/ft   20.4 psi
Minimum friction head   10 psi
Minimum fixture (flush valve) pressure 15 psi
  Total required pressure   45.4 psi

Since this is close to the pressure available, the architect-project manager-technologist decision group would probably decide on a *pumped upfeed system*, rather than a *roof tank* and a *downfeed system*. See Figure 9.2.

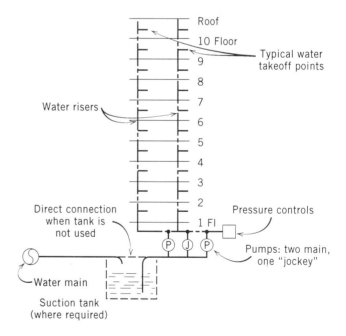

Figure 9.2 Schematic diagram of a pumped upfeed water supply system. Valving and fittings are omitted, for clarity. The suction tank is normally used where the building maximum water demand exceeds about 350 gpm. The pressure controls operate the main pumps alternately to equalize wear. The small jockey pump is used to supply light loads and to assist a large pump.

---

*to account for fittings in the piping

Placing a water tank on the roof has these disadvantages:

- The large structural load (water tank weight) on the roof requires roof supports and additional structural supports in the building.
- The tank requires maintenance such as cleaning, structural maintenance and periodic testing of water purity.
- In cold areas, roof tanks require heating coils to prevent freezing in the tank and in piping.
- A tank is an eyesore and requires screening.
- A *suction tank* in the building basement from which water is pumped to the roof tank may be required by city officials. This avoids a city mains pressure drop that might result from pumping directly from the mains. A suction tank is normally not required with small to medium-size buildings that have limited water demands. It is almost always required when the water demand exceeds 350 gpm.

As a result of these disadvantages, a pumped upfeed system is almost always chosen for medium-size buildings up to about 10–12 floors. In taller or long buildings, the pump installation becomes more expensive to install and operate than the roof tank installation. This type of economic decision, however, is not the responsibility of a plumbing technologist.

Refer again to Figure 9.2. The system operates by increasing city mains line pressure with pumps to the extent necessary. The usual arrangement uses three pumps: two large units that alternate in starting and a small *"jockey" pump* that handles light loads. Pressure sensors control the output pressure and delivery rate of the pumps(s).

When buildings exceed the size that can be economically handled by an upfeed system, a gravity downfeed system is used. See Figures 9.3 and 9.4. This system operates by pumping water from the city mains up to a roof tank, from which the building outlets are fed by gravity. As with the upfeed system, if anticipated demand is large enough to cause a pressure drop in the city mains, a suction tank in the basement will probably be required. Pump action is controlled by float switches in the roof tank and in the suction tank, if used. Pressure considerations for a downfeed system are different than those of the upfeed system, because all pressure results from gravity. It is the top floor outlets that have minimum pressure and that may be a problem. On Figure 9.3, note the dimension labelled "minimum head above top fixture." This is measured at the low water level in the tank, at which level the house pumps begin to fill the roof tank. It is the minimum head available at the top fixture, and it must be sufficient to supply this fixture's requirement.

Suppose that the closest top floor fixture is a lavatory, requiring 8-psi minimum pressure (See Table 9.1). Since all pressure is hydrostatic, we can convert this pressure to feet of water:

$$\frac{8 \text{ psi}}{0.433 \text{ psi/ft}} = 18.5 \text{ ft}$$

To this must be added about 2 psi (4.6 ft) for estimated friction head loss, making a total of 23 feet needed to the tank low water level. Since floor-to-floor height rarely exceeds 12 ft, the tank almost always must be elevated to achieve the required pressure. Assuming a 12-ft floor height, the pressure available at the next floor down would be increased by

$$12 \text{ ft} \times \frac{0.433 \text{ psi}}{\text{ft}} = 5.2 \text{ psi}$$

Note that this would still be insufficient for a 15-psi flush valve, even with an elevated roof tank:

Total pressure one floor down

= 23 ft from low water level in roof tank to the top floor lavatory level of 3 ft above floor

Figure 9.3 Schematic plumbing section of a typical downfeed water system. The basement suction tank is required when maximum water demand is high. The roof tank is partitioned so that it can be serviced during periods of minimum water use. Note that the minimum static head at the top floor fixtures occurs at the tank low water level. (From B. Stein and J. S. Reynolds, *Mechanical and Electrical Equipment for Buildings*, 8th ed. 1992, John Wiley & Sons, New York. Reprinted by permission of John Wiley & Sons.)

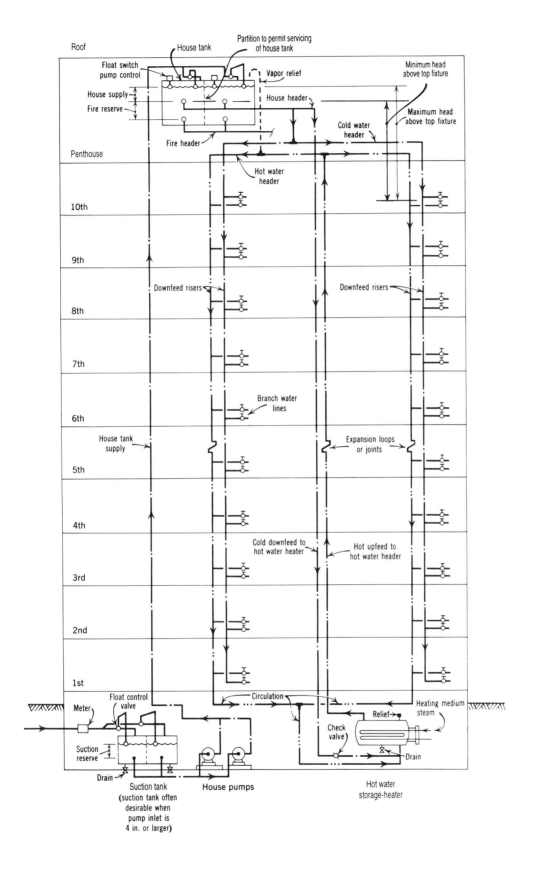

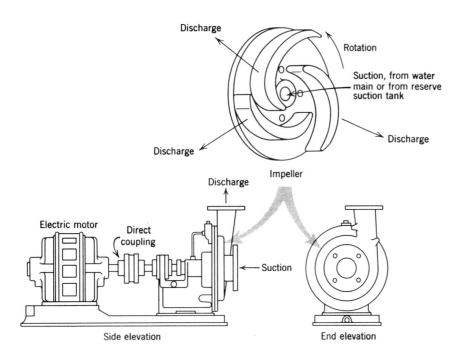

Figure 9.4 Centrifugal house pump. This is the type normally used to draw water from the suction tank and pump it up to the house tank at the roof level. The motor is controlled by float switches in the roof tank. (From B. Stein and J. S. Reynolds, *Mechanical and Electrical Equipment for Buildings*, 8th ed., 1992, John Wiley & Sons, New York. Reprinted by permission of John Wiley & Sons.)

plus
12 ft floor-to-floor height
plus
1 ft height difference to flush valve at 2 ft above floor level
$= (23 \text{ ft} + 12 \text{ ft} + 1 \text{ ft}) (0.433 \text{ psi/ft}) = 15.6 \text{ psi}$

From this must be subtracted about 2 psi for friction head, leaving only 13 psi. (For this reason, such buildings use flush tanks, that require only 8 psi, in their three top floors.)

These calculations were done to demonstrate that the designer must check pressure adequacy not only at outlets close to the roof tank but also at more remote fixtures requiring higher pressure, such as flushometers. The upper limit of building height for a downfeed system is set by the 80-psi no-flow fixture pressure limit set by Code. This pressure, when converted to feet of water, is

$$\frac{80 \text{ psi}}{0.433 \text{ psi/ft}} = 185 \text{ ft}$$

At 10-ft floor height, this means a 15- to 16-story building (because of the tank height above the top floor) and about 12–13 floors for a building with 12-ft floor height. Taller buildings must be zoned. (Pressure-reducing valves can be used to permit a larger number of floors in a building or zone.)

Advantages of a downfeed roof tank system over a pumped upfeed system are:

- A roof tank holds considerable building water reserve so that an electrical outage that would disable a pump does not cut off the water supply.
- The tank can be filled at night when electrical demand charges are low.
- A reserve of water for fire suppression is always available, protecting the entire building including its highest floors.

There are several other systems of water supply in use. One such is a combination system where lower floors are supplied directly from city mains and upper floors are supplied from a (smaller) roof tank. This arrangement is useful for buildings somewhat taller than can be supplied completely from a roof tank but not tall enough for two downfeed zones (which requires a tank on an intermediate floor). Another system is *hydropneumatic*. It uses a compressed air tank to increase available city mains pressure, which thereby increases the height of a building that can be handled without resorting to a roof tank. All these systems have advantages and disadvantages including cost considerations that must be compared to reach a good engineering decision. Although the plumbing technologist will not be expected to make this decision for large buildings, he or she certainly will do so for small buildings. This being so, it is important that the technologist be familiar with the considerations involved, which, when combined with actual office design experience, will fill out his or her required background in the area of system planning.

9.4 Water Service Sizing

Referring to Section 9.1, Design Procedure, we see that we have already considered items a through c. We will now consider items d and e, which together will permit determination of the water service size—item f. The procedure is straightforward. Count the number of each type of fixture in the building. Knowing its average water demand characteristic (flow, frequency of use and duration of each use) and applying a *diversity factor* since not all fixtures are used simultaneously, we can arrive at a maximum probable water demand. To this is added any continuous or extended water demand such as process water, irrigation and makeup water to obtain total demand. This is then used to size the water service.

The procedure is indeed straightforward, but difficulty arises in determining what overall diversity factor to apply to plumbing fixtures that have:

(a) Intermittent use.
(b) Diversity between units.

Put more simply, not every fixture is in continuous use, and not all fixtures are in use at the same time. Therefore, there is a *fixture use demand factor* for each fixture and a *diversity factor* between fixtures. The combination of both factors is the overall diversity factor. This overall factor when applied to the maximum possible demand (the sum of all fixtures uses, operating together) will give the maximum probable flow, also called *peak flow*.

In 1940, the U.S. National Bureau of Standards published a report (BMS65) entitled "Methods of Estimating Loads in Plumbing Systems" by Dr. Roy B. Hunter. This report assigns *water supply fixture units (WSFU)* to each type of plumbing fixture, which are then used instead of gpm. These WSFU values already contain the individual fixture use demand factors. The report also contains probability curves that enable the WSFU totals to be translated into gpm flow, for use in sizing of piping. Before illustrating the use of the fixture unit technique in determining maximum probable water demand, a caution must be stated. Because the system is based on diversity between fixtures in use, it becomes more accurate as the number of fixtures increases. As a result, it should never be applied to installations with only a few fixtures because, in such installations, the additional use of a single fixture can drastically change the total usage pattern. For small installations such as residences, small stores and the like, use the unit of "bathroom groups," as given in Table 9.3 converted to gpm, plus individual fixture flow rates (in gpm) as given in Table 9.4.

Modern plumbing fixtures are much more economical of water use than they were in 1940 when the NBS study was published. As a result, actual measured water demand in buildings is almost always considerably lower (by 30–40%) than that predicted from the NBS data. However, most plumbing codes will not permit piping to be sized for water-saving fixtures. The technologist should check this point with the local code authorities before beginning the design.

Example 9.1 A three story office building has the following plumbing fixtures: Two-risers, each with 6 flush valve urinals; 15 water closet bowls with flush valves; 12 lavatories; and 3 office sinks. During the summer months, the cooling tower requires 2 gpm of make-up water, and the lawn sprinklers use 4 gpm. What flow rate should the service be designed to handle (in gpm).

Solution: Although all the fixtures can be lumped, it is best to calculate the two risers separately since the flow in each will be required in order to size the piping. Refer to Table 9.3 and tabulate the WSFU for each riser.

Note carefully that, when combining loads for various parts of the building, the fixture units, not the gpm, are added together.

Table 9.3 Water Supply Fixture Units and Fixture Branch Sizes

Fixture[a]	Use	Type of Supply Control	Fixture Units[b]	Min. Size of Fixture Branch[d] in.
Bathroom group[c]	Private	Flushometer	8	—
Bathroom group[c]	Private	Flush tank for closet	6	—
Bathtub	Private	Faucet	2	1/2
Bathtub	General	Faucet	4	1/2
Clothes washer	Private	Faucet	2	1/2
Clothes washer	General	Faucet	4	1/2
Combination fixture	Private	Faucet	3	1/2
Dishwasher[f]	Private	Automatic	1	1/2
Drinking fountain	Offices, etc.	Faucet 3/8 in.	0.25	1/2
Kitchen sink	Private	Faucet	2	1/2
Kitchen sink	General	Faucet	4	1/2
Laundry trays (1–3)	Private	Faucet	3	1/2
Lavatory	Private	Faucet	1	3/8
Lavatory	General	Faucet	2	1/2
Separate shower	Private	Mixing valve	2	1/2
Service sink	General	Faucet	3	1/2
Shower head	Private	Mixing valve	2	1/2
Shower head	General	Mixing valve	4	1/2
Urinal	General	Flushometer	5	3/4[e]
Urinal	General	Flush tank	3	1/2
Water closet	Private	Flushometer	6	1
Water closet	Private	Flushometer/tank	3	1/2
Water closet	Private	Flush tank	3	1/2
Water closet	General	Flushometer	10	1
Water closet	General	Flushometer/tank	5	1/2
Water closet	General	Flush tank	5	1/2

Water supply outlets not listed above shall be computed at their maximum demand, but in no case less than the following values:

Fixture Branch[d]	Number of Fixture Units	
	Private Use	General Use
3/8	1	2
1/2	2	4
3/4	3	6
1	6	10

[a] For supply outlets likely to impose continuous demands, estimate continuous supply separately and add to total demand for fixtures.

[b] The given weights are for total demand. For fixtures with both hot and cold water supplies, the weights for maximum separate demands may be taken as three-quarters the listed demand for the supply.

[c] A bathroom group for the purposes of this table consists of not more than one water closet, one lavatory, one bathtub, one shower stall or one water closet, two lavatories, one bathtub or one separate shower stall.

[d] Nominal I.D. pipe size.

[e] Some may require larger sizes—see manufacturer's instructions.

[f] Data extracted from Code Table B.5.2.

Source. Reproduced with permission from The National Standard Plumbing Code, published by The National Association of Plumbing Heating Cooling Contractors.

Table 9.4 Table for Estimating Demand

Supply Systems Predominantly for Flush Tanks		Supply Systems Predominantly for Flushometers	
Load, WSFU[a]	Demand, gpm	Load, WSFU[a]	Demand, gpm
6	5	—	—
10	8	10	27
15	11	15	31
20	14	20	35
25	17	25	38
30	20	30	41
40	25	40	47
50	29	50	51
60	33	60	55
80	39	80	62
100	44	100	68
120	49	120	74
140	53	140	78
160	57	160	83
180	61	180	87
200	65	200	91
225	70	225	95
250	75	250	100
300	85	300	110
400	105	400	125
500	125	500	140
750	170	750	175
1000	210	1000	218
1250	240	1250	240
1500	270	1500	270
1750	300	1750	300
2000	325	2000	325
2500	380	2500	380
3000	435	3000	435
4000	525	4000	525
5000	600	5000	600
6000	650	6000	650
7000	700	7000	700
8000	730	8000	730
9000	760	9000	760
10,000	790	10,000	790

[a] Water Supply Fixture Units
Source. Reproduced with permission from The National Standard Plumbing Code, published by The National Association of Plumbing Heating Cooling Contractors.

Fixture	Riser A or B Load per Fixture, WSFU	No. of Fixtures	Total Load, WSFU
Urinal, general	5	6	30
Flush valve water closet, general	10	15	150
Lavatory, general	2	12	24
Sink, private	3	2	6
Total WSFU per riser			210
Total WSFU risers A and B			420

To the total of 420 WSFU must be added the continuous load of 2-gpm cooling tower water and 4-gpm sprinklers, that is, 6 gpm of continuous load. To do this, we must convert 420 WSFU to gallons per minute. This is done using Table 9.4 (for flushometers) and interpolating.

Load, WSFU	Demand, gpm
400	125
420	x
500	140

The interpolation calculation is

$$\frac{420-400}{500-400}=\frac{x-125}{140-125}$$

$$\frac{20}{100}=\frac{x-125}{15}$$

$$x-125=\frac{20(15)}{100}=\frac{300}{100}=3.0$$

$$x=125+3.0=128 \text{ gpm}$$

To this we add 6 gpm of continuous load to obtain a water service demand of 134 gpm.

Example 9.2 Determine the service capacity of The Basic House plan shown in Figure 8.40.

Solution: The plumbing fixtures in the house and their total water requirements follow.

Fixture	Load, WSFU
Kitchen sink	2
Dishwasher	1
Bathroom group (tank)	6
Clothes washer	2
Laundry tray	3
Total WSFU	14

This translates, from Table 9.4 (flush tank columns), to 11 gpm. To this value must be added the hose bibb flow of 5 gpm, which is obtained from Table 9.5, giving a total of 16 gpm.

Because the number of fixtures in this house is small, and it is entirely possible to have all in use simultaneously, we should redo the calculation using the figures in Tables 9.3, 9.4 and 9.5, as follows:

Bathroom group 6 WSFU =	5 gpm	(Table 9.3)
Kitchen sink	4.5 gpm	(Table 9.5)
Dishwasher	4 gpm	(Table 9.5)
Clothes washer	4 gpm	(Table 9.5)
Hose bibb	5 gpm	(Table 9.5)
Total	22.5 gpm	

Good practice would dictate the use of the higher of the two figures, that is, 22.5 gpm. (The laundry tray would not be in use when the clothes washer is running.)

Table 9.5 Demand at Individual Water Outlets

Type of Outlet	Demand, gpm
Ordinary lavatory faucet	2.0
Self-closing lavatory faucet	2.5
Sink faucet, 3/8 or 1/2 in.	4.5
Sink faucet, 3/4 in.	6.0
Bath faucet, 1/2 in.	5.0
Shower head, 1/2 in.	5.0
Laundry faucet, 1/2 in.	5.0
Ballcock in water closet flush tank	3.0
1-in. flush valve (25-psi flow pressure)	35.0
1-in. flush valve (15-psi flow pressure)	27.0
3/4-in. flush valve (15-psi flow pressure)	15.0
Drinking fountain jet	0.75
Dishwashing machine (domestic)	4.0
Laundry machine (8 or 16 lb)	4.0
Aspirator (operating room or laboratory)	2.5
Hose bibb or sill cock (1/2 in.)	5.0

Source. Data reproduced with permission from National Standard Plumbing Code, published by the National Association of Plumbing, Heating, Cooling Contractors.

9.5 Friction Head

At this point, we have learned about the pressure factors involved in selecting a water distribution system (upfeed, downfeed) and how to calculate water flow (water supply fixture units). Remember that in our pressure calculations (Section 9.3) we assumed a pressure drop in the piping. We will now learn how to calculate this pressure drop accurately. Refer to Figures 9.5, 9.6 and 9.7 and Table 9.6. The charts are applicable to water flow in steel, copper and plastic pipe, respectively. The variables in each chart are pipe size, water flow, water velocity and friction head loss. When two of these factors are known (or assumed), the third and fourth can be found from the chart. Note carefully that friction head loss in these three figures and Table 9.6 is denominated in feet of water. Therefore, when preparing a design where pressures are measured in pounds per square inch, the data from these figures and the table must be converted, with the well-known factor of 1 ft of water = 0.433 psi. For your convenience, we have added a psi scale on Figures 9.4–9.6. We also suggest that you use the chart on page 137 (Figure 3.34) when using copper pipe. It gives friction head in psi directly.

Purely as a matter of interest, the friction head loss in a pipe is directly proportional to the pipe length, roughness and the square of the water velocity and inversely proportional to pipe diameter. Mathematically, the expression is

$$H_{\mathrm{fr}} = K \frac{LV^2}{d} \quad (9.1)$$

where

- H_{fr} is the friction head,
- f is the dimensionless coefficient of friction of the pipe's interior wall,
- K is a constant that depends on the units used,
- L is the total equivalent pipe length, including fittings, and
- d is the pipe diameter.

It is not necessary or even useful to attempt to compute the friction mathematically since charts such as those shown in Figures 9.5–9.7 are readily available.

Example 9.3 To demonstrate the use of these friction head (loss) charts, let us assume that a design requires a friction loss of 6 ft of water per 100 ft of pipe with a flow of 100 gpm. Velocity is not to exceed 8 fps. What size (a) steel, (b) copper or (c) plastic pipe could we use?

Solution:

(a) Refer to Figure 9.5 for iron/steel pipe. A 2½-in. pipe would give slightly excessive friction and

(text continues on p. 514)

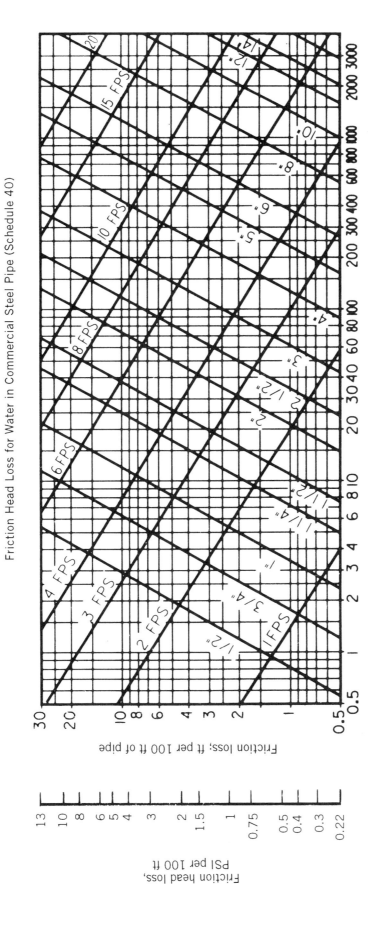

Figure 9.5 Chart of friction head loss in Schedule 40 black iron or steel pipe, for water at 60°F, in feet of water and psi per 100 ft of equivalent pipe length. Pipe sizes are nominal. (Reprinted by permission of the American Society of Heating, Refrigerating and Air-Conditioning Engineers, Atlanta, Georgia, from the 1993 *ASHRAE Handbook—Fundamentals*.)

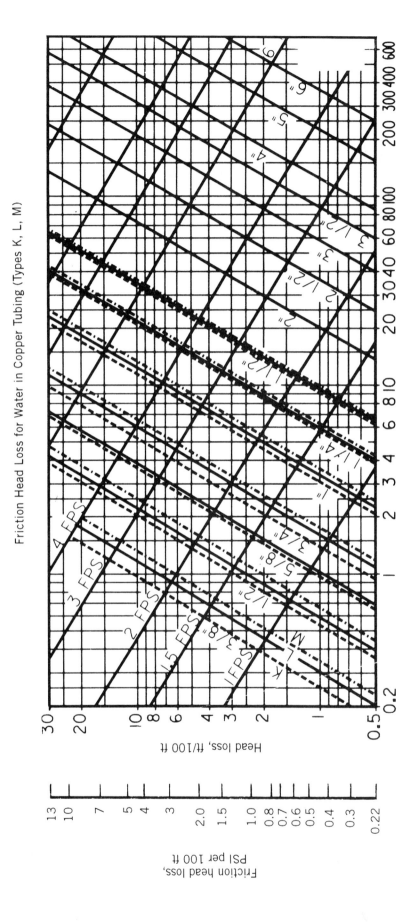

Figure 9.6 Chart of friction head loss in copper pipe and tubing for water at 60°F, in feet of water and psi per 100 ft of equivalent pipe length. Pipe and tubing sizes are nominal. (Reprinted by permission of the American Society of Heating, Refrigerating and Air-Conditioning Engineers, Atlanta, Georgia, from the 1993 *ASHRAE Handbook—Fundamentals*.)

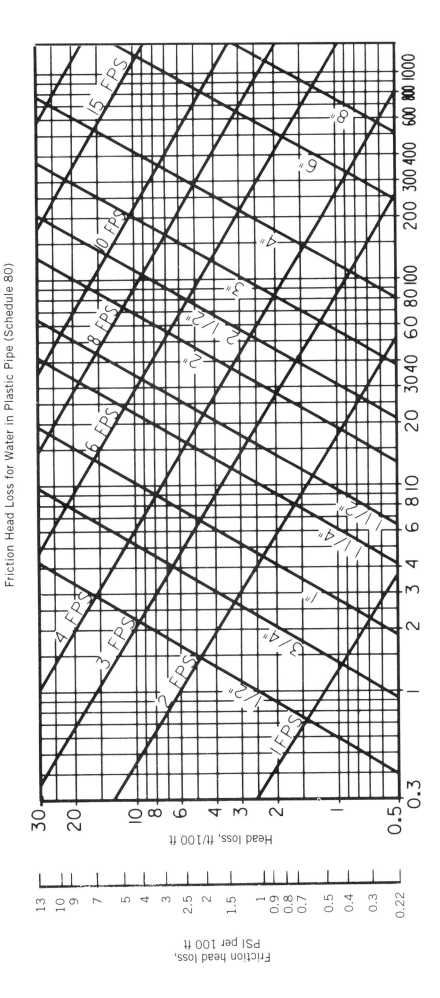

Figure 9.7 Chart of friction head loss in Schedule 80 plastic pipe for water at 60°F, in feet of water and psi per 100 ft of equivalent pipe length. Pipe sizes are nominal. (Reprinted by permission of the American Society of Heating, Refrigerating and Air-Conditioning Engineers, Atlanta, Georgia, from the 1993 *ASHRAE Handbook—Fundamentals*.)

Table 9.6(a) Flow (gpm), Velocity (V, fps) and Friction Head Loss (H, ft of water) for Schedule 40 Thermoplastic Pipe Per 100 ft of Equivalent Length

Flow, gpm	1/2 in.		3/4 in.		1 in.		1 1/4 in.		1 1/2 in.		2 in.		2 1/2 in.		3 in.		4 in.		6 in.		8 in.	
	V	H	V	H	V	H	V	H	V	H	V	H	V	H	V	H	V	H	V	H	V	H
1	1.1	1.2	1.3	1.0																		
2	2.3	4.2	3.2	5.6																		
5	5.6	22.9	4.4	10.4	1.9	1.7	1.1	0.4														
7	7.9	42.7	6.3	20.2	2.7	3.1	1.6	0.8	1.1	0.4												
10	11.3	82.6	9.5	42.8	3.9	6.1	2.2	1.6	1.6	0.7												
15	16.9	175.0	12.6	73.0	5.8	12.9	3.3	3.3	2.4	1.6	1.5	0.5	1.0	0.2								
20			15.8	110.3	7.7	22.0	4.4	5.7	3.2	2.6	2.0	0.8	1.4	0.3								
25					9.6	33.2	5.5	8.6	4.0	4.0	2.4	1.2	1.7	0.5	1.1	0.2						
30					11.6	46.5	6.6	12.0	4.9	5.6	2.9	1.6	2.1	0.7	1.3	0.2						
35					13.5	61.9	7.7	15.9	5.7	7.5	3.4	2.2	2.4	0.9	1.5	0.3						
40					15.4	79.3	8.8	20.4	6.5	9.5	3.9	2.8	2.7	1.2	1.8	0.4	1.0	0.1				
45							9.9	25.4	7.3	11.9	4.4	3.5	3.1	1.5	2.0	0.5	1.2	0.1				
50							11.0	30.9	8.1	14.4	4.9	4.2	3.4	1.8	2.2	0.6	1.3	0.2				
60							13.3	43.3	9.7	20.2	5.9	5.9	4.1	2.5	2.7	0.9	1.5	0.2				
70							15.5	57.5	11.3	26.9	6.8	7.9	4.8	3.3	3.1	1.2	1.8	0.3				
75									12.1	30.6	7.3	8.9	5.1	3.8	3.3	1.3	1.9	0.3				
80									12.9	34.4	7.8	10.1	5.5	4.3	3.5	1.5	2.0	0.4				
90									14.5	42.9	8.8	12.5	6.2	5.3	4.0	1.8	2.3	0.5	1.0	0.1		
100									16.2	52.1	9.8	15.2	6.8	6.4	4.4	2.2	2.6	0.6	1.1	0.1		
125											12.2	23.0	8.5	9.7	5.5	3.4	3.2	0.9	1.4	0.1		
150											14.6	32.3	10.3	13.6	6.6	4.7	3.8	1.2	1.7	0.2		
175											17.1	43.0	12.0	18.1	7.7	6.3	4.5	1.7	2.0	0.2	1.1	0.1
200													13.7	23.2	8.8	8.0	5.1	2.1	2.3	0.3	1.3	0.1
250													17.1	35.0	11.0	12.1	6.4	3.2	2.8	0.4	1.6	0.1
300															13.2	17.0	7.7	4.5	3.4	0.6	1.9	0.2
350															15.5	22.6	8.9	6.0	3.9	0.8	2.3	0.2
400																	10.2	7.6	4.5	1.0	2.6	0.3
450																	11.5	9.5	5.1	1.3	2.9	0.3
500																	12.8	11.5	5.6	1.6	3.2	0.4
750																	19.2	24.5	8.4	3.3	4.9	0.9
1000																			11.2	5.6	6.5	1.5
1250																			14.0	8.5	8.1	2.2
1500																			16.8	12.0	9.7	3.1

Note: The figures have been rounded to one decimal place. For exact data refer to original source.
Source. Data extracted, with permission, from *Chemtrol Thermoplastic Piping Technical Manual*.

Table 9.6(b) Flow (gpm), Velocity (V, fps) and Friction Head Loss (H, ft of water) for Schedule 80 Thermoplastic Pipe Per 100 Ft. of Equivalent Length

Flow, gpm	1/2 in.		3/4 in.		1 in.		1 1/4 in.		1 1/2 in.		2 in.		2 1/2 in.		3 in.		4 in.		6 in.		8 in.	
	V	H	V	H	V	H	V	H	V	H	V	H	V	H	V	H	V	H	V	H	V	H
1/4																						
1/2																						
3/4	1.1	1.3																				
1	1.5	2.2	0.8	0.5																		
3	4.4	17.1	2.4	3.7	1.4	1.0																
5	7.4	44.1	3.9	9.5	2.3	9.5	1.3	0.6														
7	10.3	82.3	5.5	17.6	3.3	5.0	1.8	1.2	1.3	0.5												
10	14.8	159.3	7.8	34.1	4.7	9.7	2.6	2.3	1.9	1.1	1.1	0.3										
15			11.8	72.3	7.0	20.4	3.9	4.9	2.8	2.2	1.7	0.6	1.2	0.3								
20			15.7	123.1	9.3	34.8	5.2	8.4	3.8	3.8	2.2	1.1	1.6	0.5	1.0	0.2						
25					11.7	52.6	6.5	12.6	4.7	5.7	2.8	1.6	1.9	0.7	1.2	0.2						
30					14.0	78.8	7.8	17.7	5.6	8.0	3.4	2.3	2.3	1.0	1.5	0.3						
35					16.3	98.2	9.1	23.6	6.6	10.7	3.9	3.0	2.7	1.3	1.7	0.4	1.0	0.1				
40							10.4	30.2	7.5	13.7	4.5	3.9	3.1	1.6	2.0	0.5	1.1	0.1				
45							11.7	37.5	8.4	17.1	5.0	4.4	3.5	2.0	2.2	0.7	1.3	0.2				
50							13.0	45.6	9.4	20.7	5.6	5.9	3.9	2.5	2.5	0.8	1.4	0.2				
60							15.6	69.9	11.3	29.0	6.1	8.2	4.7	3.4	3.0	1.2	1.7	0.3				
70									13.1	38.6	7.8	10.9	5.5	4.6	3.5	1.5	2.0	0.4				
75									14.1	43.9	8.4	12.4	5.8	5.2	3.7	1.7	2.1	0.5				
80									15.0	49.5	8.9	14.0	6.2	5.8	4.0	2.0	2.3	0.5	1.0	0.1		
90									16.9	61.5	10.1	17.4	7.0	7.3	4.5	2.5	2.6	0.6	1.1	0.6		
100											11.2	21.2	7.8	8.8	5.0	3.0	2.9	0.8	1.3	0.1		
125											14.0	32.0	9.7	13.3	6.2	4.5	3.6	1.2	1.6	0.2		
150											16.7	44.9	11.7	18.7	7.5	6.3	4.3	1.6	1.9	0.2	1.1	0.1
175													13.6	24.9	8.7	8.4	5.0	2.2	2.2	0.3	1.3	0.1
200													15.6	31.9	10.0	10.7	5.7	2.8	2.5	0.4	1.4	0.1
250															12.5	16.2	7.1	4.2	3.1	0.6	1.8	0.1
300															14.9	22.7	8.6	5.9	3.8	0.8	2.1	0.2
350															17.4	30.3	10.0	7.8	4.4	1.1	2.5	0.3
400																	11.4	10.0	5.0	1.4	2.9	0.4
450																	12.9	12.4	5.6	1.7	3.2	0.4
500																	14.3	15.1	6.3	2.0	3.6	0.5
750																	21.4	31.9	9.4	4.3	5.4	1.1
1000																			12.5	7.4	7.1	1.9
1250																			15.7	11.1	8.9	2.8
1500																					10.7	4.0

Note: The figures have been rounded to one decimal place.
Source. Data extracted, with permission, from *Chemtrol Thermoplastic Piping Technical Manual.*

6.4 fps velocity. A 3-in. pipe gives a friction loss of 2½ ft of water and a water velocity of 4.4 fps.

(b) Refer to Figure 9.6 for copper pipe. A 2½-in. pipe gives exactly the friction head desired and a water velocity of 6.6 fps.

(c) Refer to Table 9.6a—Schedule 40 plastic pipe. A 2½-in. pipe gives slightly high friction (6.4 ft) and a velocity of 6.8 fps.

You are encouraged to check these figures for yourself to become familiar with the use of the charts. The data for Schedule 80 plastic pipe is given in both graphic form (Figure 9.7) and tabular form [Table 9.6(b)] for convenience. It is more convenient to use the graph when friction is known and pipe size needed. It is more convenient to use the table when pipe size is known and friction loss is needed.

Note that Figures 9.5–9.7 and Table 9.6 refer to equivalent length of piping. Since fittings such as couplings, elbows, valves and the like have higher friction than the same length of straight pipe, it is customary to convert their resistance to the equivalent length of straight pipe and add these figures to the actual pipe length (measured along the centerline). The total length of pipe thus derived is called the *total equivalent length* of pipe, or simply *TEL*. Tables 9.7 and 9.8 give approximate equivalent pipe lengths for plastic pipe fittings and screwed metal pipe fittings, respectively. In the absence of a detailed fitting count, it is customary to add 50% to the actual pipe length of a run to account for fittings.

The remaining item for which the technologist will require friction head information is the water meter. Refer to Figure 9.8, which shows the pressure drop (friction head loss) in pounds per square inch, in disk-type water meters. Pressure loss is proportional to flow and inversely proportional to the pipe size. Remember that maximum water demand was determined by the total WSFU of the building. Pipe size depends on friction drop requirements and will be explained in the next section.

9.6 Water Pipe Sizing by Friction Head Loss

We are now at the point that we can calculate pipe sizes for water distribution in a building. This corresponds to item g in the overall design procedure outlined in Section 9.1. The procedure follows:

Step 1. Draw a riser (plumbing section). On this riser show floor-to-floor heights, runout distance to farthest fixture on each floor, and lengths of piping from the service point to the floor takeoff points.

Step 2. Show the WSFU for each fixture and fixture unit total on each piping runout. Use separate fixture units for hot and cold water where applicable.

Step 3. Total the fixture units in each branch of the system. Show both cold and hot water fixture units. (It is understood that hot water pipe sizing will require a separate diagram and calculation.) Add the continuous water loads.

Table 9.7 Equivalent Length of Plastic Pipe (ft) for Standard Plastic Fittings

Type Fitting	Size Fitting, in.										
	½	¾	1	1¼	1½	2	2½	3	4	6	8
90° Standard elbow	1.6	2.1	2.6	3.5	4.0	5.5	6.2	7.7	10.1	15.2	20.0
45° Standard elbow	0.8	1.1	1.4	1.8	2.1	2.8	3.3	4.1	5.4	8.1	10.6
90° Long radius elbow	1.0	1.4	1.7	2.3	2.7	4.3	5.1	6.3	8.3	12.5	16.5
90° Street elbow	2.6	3.4	4.4	5.8	6.7	8.6	10.3	12.8	16.8	25.3	33.3
45° Street elbow	1.3	1.8	2.3	3.0	3.5	4.5	5.4	6.6	8.7	13.1	17.3
Square corner elbow	3.0	3.9	5.0	6.5	7.6	9.8	11.7	14.6	19.1	28.8	37.9
Standard tee											
with flow thru run	1.0	1.4	1.7	2.3	2.7	4.3	5.1	6.3	8.3	12.5	16.5
with flow thru branch	4.0	5.1	6.0	6.9	8.1	12.0	14.3	16.3	22.1	32.2	39.9

Source. Data extracted, with permission, from *Chemtrol Thermoplastic Piping Technical Manual.*

Table 9.8 Equivalent Length of Metal Pipe (ft) for Standard Metal Fittings and Valves

Nominal Pipe Dia., in.	90° Ell Reg.	90° Ell Long	45° Ell	Return Bend	Tee Line	Tee Branch	Globe Valve	Gate Valve	Angle Valve	Swing Check Valve
3/8	2.5	—	0.38	2.5	0.90	2.7	20	0.40	—	8.0
1/2	2.1	—	0.37	2.1	0.90	2.4	14	0.33	—	5.5
3/4	1.7	0.92	0.35	1.7	0.90	2.1	10	0.28	6.1	3.7
1	1.5	0.78	0.34	1.5	0.90	1.8	9	0.24	4.6	3.0
1 1/4	1.3	0.65	0.33	1.3	0.90	1.7	8.5	0.22	3.6	2.7
1 1/2	1.2	0.54	0.32	1.2	0.90	1.6	8	0.19	2.9	2.5
2	1.0	0.42	0.31	1.0	0.90	1.4	7	0.17	2.1	2.3
2 1/2	0.85	0.35	0.30	0.85	0.90	1.3	6.5	0.16	1.6	2.2
3	0.80	0.31	0.29	0.80	0.90	1.2	6	0.14	1.3	2.1
4	0.70	0.24	0.28	0.70	0.90	1.1	5.7	0.12	1.0	2.0

Source. Data extracted and reprinted by permission of the American Society of Heating, Refrigerating and Air-Conditioning Engineers, Atlanta, Georgia, from the 1993 *ASHRAE Handbook—Fundamentals*.

Step 4. Show source pressure (minimum) and the minimum flow pressure required at the most remote outlet(s).

Step 5. Determine the pressure available for friction head loss from the service point to the final outlet.

Step 6. Determine the required pipe size in each section, using the friction head loss data calculated in Step 5 and the friction head charts. Selection is normally based on uniform friction head loss per foot throughout and a maximum water velocity—usually 8 fps, except that branches feeding quick

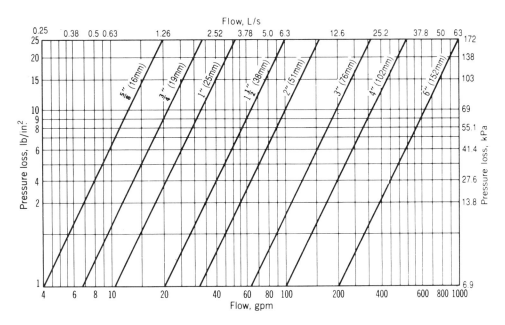

Figure 9.8 Pressure loss (friction head loss) in disk-type water meters. (Reprinted by permission of the American Society of Heating, Refrigerating and Air-Conditioning Engineers, Atlanta, Ga., from the 1993 *ASHRAE Handbook—Fundamentals*.)

closing devices such as flush valves should be limited to about 4 fps to avoid water hammer.

An example should make application of the design procedure clear.

Example 9.4 Design the cold water piping required for the office building described in Example 9.1. For completeness, we will detail the building data here.

(a) Three-story office building.
(b) 12-ft floor-to-floor height.
(c) Two water risers, each feeding two bathroom groups on each floor. One group contains three water closets and two lavatories; the second group contains two urinals, two water closets and two lavatories. In addition, each riser feeds one office-type sink on each floor.
(d) Continuous water demand (summer) consists of 2-gpm cooling tower make-up and 4-gpm sprinklers.
(e) Minimum maintained water pressure from city mains is 50 psi.

Assume that all water closets use flush valves. Determine that sufficient pressure exists to operate these valves with the piping as designed. If pressure is insufficient, indicate the alternatives.

Solution: Follow each step of the solution carefully with Figure 9.9. The steps correspond to the procedure detailed at the beginning of this section.

Step 1. Draw the riser.

It is shown in Figure 9.9. All the pipe distances from the service main to the last fixture are shown. Floor runouts are taken at 20 ft to the last fixture, which is a water closet. Note that the pipe length from the basement tap off to the first floor tapoff is 18 ft, with a 12-ft rise. This indicates 6 ft of horizontal run in the basement, after the tap.

Step 2. Show the water supply fixture units (WSFU) for each fixture and the totals.

The data are taken from Table 9.3. Each fixture requiring both hot and cold water has both WSFU values shown. Each is three-quarters of the total WSFU value (see footnote *b* in Table 9.3). For simplicity, WSFU is noted simply as FU on the drawing. This practice is customary.

Step 3. Total the fixture units in the runouts, risers and basement feeds.

Note that adding the cold and hot WSFU does not give the correct total since each is three-quarters of the total of a fixture, and not half. Therefore, to obtain the total flow (i.e. before the takeoff for the hot water boiler), use the total fixture requirement (e.g., 2 WSFU for a lavatory). After the hot water boiler takeoff, use only cold water fixture units. This is shown Figure 9.9. Before the takeoff, we have 420 WSFU. Refer to the totals in Example 9.1 to see the derivation of this figure. After the hot water takeoff, the totals are split into 405 WSFU cold and 45 WSFU hot. These totals are arrived at by simply adding the floor totals, as shown on the riser:

3 floors (67.5 + 67.5) = 3(135) = 405 WSFU, cold
3 floors (7.5 + 7.5)   = 3(15) = 45 WSFU, hot

All these totals and their gpm equivalents (for cold water) are shown on the riser. Note particularly that the section of pipe between risers carries 97 gpm, consisting of 91 gpm (from 202.5 WSFU) plus 6 gpm of continuous demand.

Step 4. Minimum source pressure has been determined to be 50 psi, maintained, and it is so indicated on the drawing.

Minimum required flow pressure at the most remote outlet on the third floor (floor-mounted siphon-jet water closet with flush valve) is 15 psi. That too is indicated.

Step 5. The pressure available for friction head loss, for the entire length of piping, is equal to the total available mains pressure less static head and minimum flow pressure:

Static head (5 ft + 3 × 12 ft)(0.433 psi/ft)	17.75 psi
Minimum flush valve pressure	15.0 psi
Total	32.75 psi

Total maximum friction head loss = 50 − 32.75 = 17.25 psi. At this point, we must make an assumption about the loss in the water meter. Refer to Figure 9.8. Assuming a 3-in. pipe size for service, the loss in the meter is 5 psi. This leaves 17.25 psi − 5 psi or 12.25 psi for friction head loss in piping. Once we determine actual pipe size, we can revise this assumption, if necessary.

Step 6. The total length of piping from service tap point to farthest fixture is:

Water mains to water meter	50 ft
Water meter to base of second riser	80 ft
Riser length (18 + 12 + 12)	42 ft
Final runout	20 ft
Total	192 ft

Assume 50% additional equivalent length to account for fittings. Therefore, *total developed length*

WATER PIPE SIZING BY FRICTION HEAD LOSS / 517

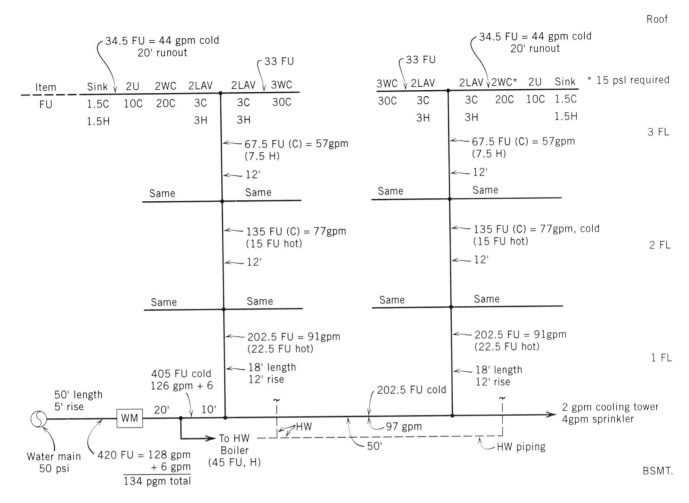

Figure 9.9 Plumbing riser, with all the data shown that is required to calculate friction head loss in the cold water piping system. All fixtures are shown with WSFU data in all branches and continuous flow water demand. Total equivalent pipe lengths (TEL) are indicated for all sections of the system. See text for a detailed discussion of the friction head calculation procedure.

or total equivalent length (TEL) = 192 (1.5) = 288 ft. The usual design procedure aims for uniform friction loss along the entire pipe length. To do this, we establish a friction loss per 100 ft by dividing total friction loss by total length and then size the piping accordingly.

Uniform design friction loss in psi/100 ft is:

$$\text{Friction}/100 \text{ ft} = \frac{12.25 \text{ psi (available head)}}{288 \text{ ft (TEL)}}$$

$$= \frac{12.25 \text{ psi}}{2.88 \times 100 \text{ ft}} = \frac{4.25 \text{ psi}}{100 \text{ ft}}$$

$$= 4.25 \text{ psi}/100 \text{ ft}$$

Assume that all piping is copper. Using Figure 9.6 or Figure 3.34, we find that a flow of 134 gpm in the 70 ft of pipe between the service main and the tapoff to the hot water heater gives 1.9 psi/100 ft drop in a 3-in. pipe, with a water velocity of 5.5 fps. This indicates that 3-in. pipe is too large. We also see on the chart that 2½-in. pipe gives exactly 4.25 psi/100 ft head loss with 8.8 fps water velocity. We would, therefore, select a 2½-in. service pipe. Since a 2-in. water meter would give unacceptably high loss (18 psi), we would use a 3-in. meter with reducing fittings. This confirms the assumption that we made of 5-psi loss in the meter. We can tabulate the results as they are calculated.

As noted previously, branches supplying mainly flush valve loads should be limited to a water velocity of 4 fps to avoid water hammer. Since the runouts here are of this type, carrying about 85% flush valve load (30 FU out of 34.5 FU), velocity must be limited to about 4 fps. For this reason, a 2-in. pipe is chosen, giving a 4.4 fps velocity. The runout carrying 33 FU would also be fed with a 2-in. pipe for consistency.

Note that the total friction drop of 9.66 psi is considerably less than the permissible maximum of 12.25 psi. The pipe sizes could be reduced considerably by using tank flush closet bowls. In practice, this would be brought to the attention of the architect and owner. Alternatively, a small pressure booster pump, which would kick-in whenever line pressure fell below a pressure of about 60 psi, could be used. These additional 10 psi would permit use of smaller piping. These decisions are not made by the technologist, but it is he or she that brings them to the architect or project manager's attention.

Connection between the fixtures and the runout piping is made with a smaller pipe. The minimum pipe size required for a fixture branch pipe is given in Table 9.3.

9.7 Water Pipe Sizing by Velocity Limitation

The friction head loss method of water pipe sizing detailed in Section 9.6 is accurate but time-consuming because of the necessary calculations. For buildings where available water pressure is more than adequate to supply all the fixtures, there exists a simplified pipe sizing method based on water velocity considerations. This method is normally applicable to all private residences, multiple residences, and commercial and industrial buildings

Pipe Section	Equivalent Length, ft[a]	Pipe Size, in.	Friction per 100 ft, psi	Velocity, fps	Section Friction, psi	Cumulative Friction, psi
Service to hot water tap —134 gpm	105	2½	4.25	8.8	4.46	4.46
Hot water tap off to first riser—132 gpm	15	2½	4.2	8.8	0.63	5.09
Between risers—97 gpm	75	2½	2.8	7.0	2.1	7.19
First riser section—91 gpm	27	2½	2.3	6.0	0.62	7.81
Second riser section—77 gpm	18	2	4.7	7.7	0.85	8.66
Third riser section—57 gpm	18	2	2.8	6.0	0.5	9.16
Runout—44 gpm	30	2	1.8	4.4	0.5	9.66

[a]Actual pipe length plus 50% to account for fittings.

up to three stories in height. To determine the method applicability before applying it, a rapid pressure calculation of the type described in Sections 9.2 and 9.3 can be made. If this calculation shows that pressure is adequate, use the following procedure:

Step 1. Prepare a building riser diagram.

Show all fixtures, fixture loads in WSFU, and gpm in each pipe section, exactly as was done for the friction head method. Such a diagram is shown in Figure 9.9. Include all continuous loads in the gpm figures, as in Figure 9.9.

Step 2. Identify all branch piping that feeds quick-closing devices such as flush valves, solenoid valves (as in many clothes washers), and self-closing faucets. The velocity in these branch pipes must be limited to 4 fps to avoid severe water hammer.

Step 3. Size all individual fixture branches according to the Code minimum requirements as given in Table 9.3.

Step 4. Size all other parts of the piping system in accordance with water velocity limitations, for the type of piping selected, using Table 9.9.

Note carefully that the table differentiates between WSFU load that serves flush valves (column B) and loads that do not (column A). The normal velocity limitation for which piping is designed is 8 fps except, as noted in Step 2, where water hammer considerations dictate the use of 4 fps maximum velocity.

These tables have been calculated to include adequate friction head loss for piping runs in relatively small buildings. Obviously, if one is designing a very large low building such as a 500-ft-long, one-story assembly plant, this method should not be used because of the long piping runs and high friction loss involved.

An example should make the use of these tables clear.

Example 9.5 Rework the pipe sizing of the building in Example 9.4, assuming that supply pressure is more than adequate.

Solution: Use Table 9.9(d) for Type K copper pipe. Tabulate the results as was done for Example 9.4, using 8 fps and Column B (flush valves) for mains and 4 fps for runouts.

At the end of Example 9.4, we noted that an increase in line pressure would permit use of smaller piping. This is borne out by the results just

Pipe Section	*Flow, gpm*	*Pipe Size, in.*
Service to hot water	134	2
Hot water tap to first riser	132	2
Pipe between risers	97	2
First riser section	91	2
Second riser section	77	2
Third riser section	57	2
Runout (4 fps)	44	2½

arrived at in Example 9.5. The velocity limitation method used there assumes more than adequate line pressure. This results in a pipe size reduction from 2½ in. to 2 in. for the first four calculations. Runouts were sized as 2 in. in Example 9.5, because we allowed the water velocity to exceed 4 fps slightly. Had we insisted on a 4 fps maximum, 2½ in. pipe would have been required. In actual design, 2 in. probably would be used, since runouts only rarely are larger than the mains from which they are fed.

Domestic Hot Water

9.8 General Considerations

Almost all plumbing fixtures except flush-type units (closet bowls and urinals) require hot water as well as cold. It is, therefore, the designer's responsibility to ensure this supply at the proper temperature, with minimum delay, economically and safely. The usual point of use temperatures are:

Lavatories, showers and tubs	95–105°F
Residential dishwashing and laundry	120–140°F
Commercial and institutional kitchens	140°F
Commercial and institutional laundries	180°F

Note that these are fixture water temperatures. Depending on the design and length of the supply piping from the hot water heater, the water heater outlet temperature will be 5 to 20 F° higher than the fixture temperature, to compensate for temperature loss in the supply piping. Residential systems should be designed for required lavatory temperatures, and the higher water temperature requirement for laundry and dishwasher achieved with

Table 9.9 Sizing Tables Based on Velocity Limitation

Nominal Size,[a] in.	V=4 fps				V=8 fps			
	Column A		Column B		Column A		Column B	
	Flow, gpm	Load,[a] WSFU	Load,[b] WSFU	Friction,[c] psi/100 ft	Flow, gpm	Load,[a] WSFU	Load,[b] WSFU	Friction,[d] psi/100 ft
(a) Copper and Brass Pipe, Standard Pipe Size (Schedule 40) (Smooth)								
1/2	3.8			6.8	7.7	10		24
3/4	6.6	8		5.0	13.2	18		18
1	11.0	15		3.7	22.1	34		13.3
1 1/4	18.3	27		2.7	36.6	70	21	9.7
1 1/2	25.1	40	10	2.3	50.1	125	51	8.2
2	41.6	92	31	1.7	83.2	290	165	6.1
2 1/2	61.2	180	80	1.3	122.4	490	390	4.9
3	91.8	330	210	1.1	183.6	850	810	3.9
4	156.7	680	620	0.8	313.4	1900	1900	2.8
(b) Copper Water Tube, Type M (Smooth)								
1/8	2.0			9.6	3.9			34
1/2	3.2			6.4	6.3	8		26
3/4	6.4	8		5.0	12.9	18		18
1	10.9	15		3.6	21.8	34		13
1 1/4	16.3	24		2.9	32.6	60	15	10
1 1/2	22.8	35		2.4	45.6	110	38	8.5
2	38.6	80	26	1.8	77.2	270	140	6.2
2 1/2	59.5	170	70	1.4	119.0	470	360	5.0
3	84.9	300	170	1.1	169.9	750	730	4.0
4	149.3	625	575	0.8	296.7	1750	1750	2.8
(c) Copper Water Tube, Type L (Smooth)								
3/8	1.8			11.0	3.6			39
1/2	2.9			8.1	5.8	7		29
3/4	6.0	7		5.3	12.1	17		19
1	10.3	14		4.0	20.6	30		14
1 1/4	15.7	23		3.0	31.3	55	15	11
1 1/2	22.2	35		2.5	44.4	100	36	8.7
2	38.8	80	26	1.8	77.2	270	140	6.2
2 1/2	59.5	170	70	1.4	119.0	470	360	5.0
3	84.9	300	170	1.1	169.9	750	730	4.0
4	149.3	625	575	0.8	298.7	1750	1750	2.8
(d) Copper Water Tube, Type K (Smooth)								
3/8	1.6			11.0	3.2			36
1/2	2.7			8.2	5.4	6		30
3/4	5.4	6		5.6	10.9	15		20
1	9.7	13		4.1	19.4	29		14
1 1/4	15.2	24		3.1	30.4	55	15	12
1 1/2	21.5	33		2.6	43.0	97	35	9.0
2	38.8	80	26	1.8	77.2	270	140	6.2
2 1/2	59.5	170	70	1.4	119.0	470	360	5.0
3	84.9	300	170	1.1	169.9	750	730	4:0
4	149.3	625	575	0.8	298.7	1750	1750	2.8

Table 9.9 (Continued)

	V = 4 fps				V = 8 fps			
	Column A		Column B		Column A		Column B	
Nominal Size,[a] in.	Flow, gpm	Load,[a] WSFU	Load,[b] WSFU	Friction,[c] psi/100 ft	Flow, gpm	Load,[a] WSFU	Load,[b] WSFU	Friction,[d] psi/100 ft
	(e) Galvanized Iron and Steel Pipe, Standard Pipe Size (Schedule 40) (Fairly Rough)							
1/2	3.8			8.0	7.6	10		31
3/4	6.6	8		6.0	13.3	18		22
1	10.8	15		4.5	21.5	34		17
1 1/4	18.6	27		3.4	37.3	75	21	13
1 1/2	25.4	41		2.8	50.7	125	51	11
2	41.8	92	32	2.2	83.6	290	165	8.1
2 1/2	59.6	172	72	1.8	119.3	490	390	6.8
3	92.1	330	210	1.4	184.3	850	810	5.4
4	158.7	680	620	1.1	317.5	1900	1900	4.0
	(f) Schedule 40 Plastic Pipe (PE, PVC & ABS) (Smooth)							
1/2	3.8			7.0	7.6	10		24
3/4	6.6	8		5.1	13.3	18		17.5
1	10.8	15		3.7	21.5	34		13.0
1 1/4	18.6	27		2.7	37.3	75	21	9.5
1 1/2	25.4	41		2.3	50.7	125	51	8.0
2	41.8	92	32	1.7	83.6	290	165	6.0
2 1/2	59.6	172	72	1.4	119.3	490	390	4.8
3	92.1	330	210	1.1	184.3	850	810	3.7
4	158.7	680	620	0.8	317.5	1900	1900	2.7
	(g) CPVC Tubing SDR11 (ASTM D2846)							
1/2	2.3	2.7		7.5	4.6	5.5		27
3/4	4.9	5.9		5.9	9.9	13.3		21
1	8.3	10.5		4.4	16.6	24.4	6.2	16
1 1/4	12.4	17.3	4.6	3.5	24.8	40.0	9.2	12
1 1/2	17.3	25.5	6.4	2.8	34.6	68.6	19.4	10
2	29.6	51.0	13.2	2.1	59.2	171.0	72.0	7.4
	(h) Schedule 80 Plastic Pipe (PVC, CPVC) (Smooth)							
1/2	2.9			6.9	5.8	7		22.7
3/4	5.4			4.8	10.8	15		16.0
1	9.4	12		3.4	17.9	23		11.2
1 1/4	16.0	24		2.4	32.0	58	17	11.2
1 1/2	22.0	34		2.0	44.1	100	36	6.5
2	36.87	73	23	1.5	73.65	240	120	5.0

[a] Column A applies to piping that does not supply flush valves.
[b] Column B applies to piping that supplies flush valves.
[c] The friction head loss corresponds to the flow rate shown, for piping having fairly smooth surface condition after extended service.
[d] For pipe dimensions see Tables 8.2 and 8.3.
Source. Data reprinted with permission from National Standard Plumbing Code, published by the National Association of Plumbing Heating Cooling Contractors.

local booster heaters at these appliances. This has the triple advantage of lower heat losses in the piping, slower scale formation in piping and avoidance of scalding water temperatures. Water at temperature above 110°F is uncomfortably hot to the hands and even more so to other parts of the body.

The type of fuel used to heat hot water depends, in large measure, on the type of heater used. When hot water is generated by the building's heating system then, obviously, the fuel is that which the heating system uses and may be steam, coal, oil or gas and, though rarely, electricity. The choice of heater size, type and fuel is not normally the responsibility of the technologist, although he or she must be able to carry out most of the design once the decision has been made. This requires complete familarity with the equipment and piping arrangements normally in use.

There are two basic types of water-heating arrangements: storage units and instantaneous units. Within each classification, there are variations. In storage heaters, there are directly heated automatic storage, tank storage and mass storage. In the instantaneous heaters, there are directly heated, indirectly heated, semi-instantaneous, internal tankless, external tankless and others. Furthermore, with both types of heaters, there are different types of circulation systems. Sections 9.9–9.11 will discuss the most common types of units and their piping arrangements.

9.9 Instantaneous Water Heaters

Instantaneous water heaters are units that heat water only when a hot water faucet opens or another fixture demands hot water. That is, the heat source is triggered by hot water demand, and fuel use is proportional to hot water flow. This makes the unit economical to operate, although it is often slow to respond and, therefore, cannot supply hot water within close temperature limits. The temperature swing for on-off operation can be as much as ±10 F°, depending on the unit design. In addition to economy of operation, the fact that it uses no water storage equipment makes it compact and economical in first cost. In directly heated units of this type, the heat source, whether it is an electrical heating element or a flame, contacts the water coils directly. See Figure 9.10. In indirectly heated units, the heat source is hot water or steam that was produced in a heating or process boiler. This then

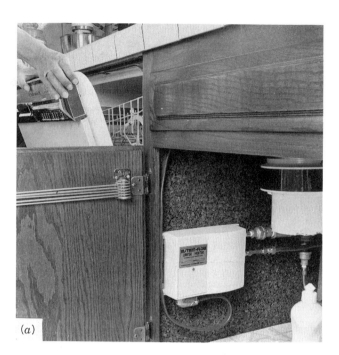

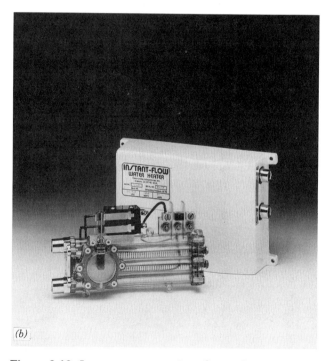

Figure 9.10 Instantaneous, point-of-use, electrical hot water heater. The illustrated unit, which measures $6 \times 9 \times 2\frac{1}{2}$ in. and weighs about 6 lb, is shown (a) in an undercounter installation for use as a dishwasher booster heater. The electrical coils and controls that the unit contain are shown in (b). The illustrated unit is available in flow rates of 1–3 gpm, water temperature rise of 55–95°F (depending on electrical rating and water flow) and electrical capacities of 2.6–9.0 kw. A range of voltage ratings is also obtainable. (Photos courtesy of Chronomite Laboratories, Inc.)

INSTANTANEOUS WATER HEATERS / 523

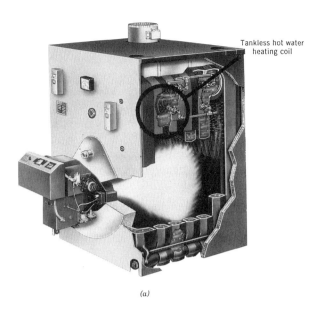

(a)

Figure 9.11 *(a)* Cutaway of hot water (or steam) boiler with tankless, instantaneous, indirectly heated, hot water heating coil. (Illustration courtesy of Burnham Corporation.)

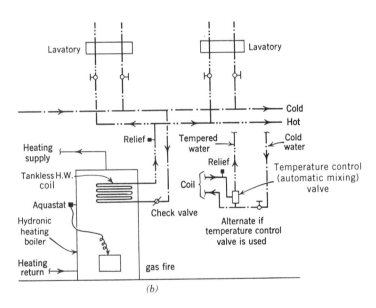

(b)

Figure 9.11 *(b)* Schematic piping diagram for an instantaneous domestic hot water heater. The heating unit is a tankless coil inside the heating boiler, which is indirectly heated by the boiler hot water. Notice the positioning of the relief valve in the hot water outlet pipe(s) and the check valve that prevents reverse flow. Because the hot water is often too hot for use in fixtures such as lavatories and tubs, an alternative arrangement, also shown, tempers the hot water with cold through a thermostatically controlled mixing valve and feeds tempered water at about 110°F to the building's fixtures. Hotter water can be supplied to the dishwasher and laundry outlets. (From B. Stein, J. S. Reynolds, and W. J. McGuinness, *Mechanical and Electrical Equipment for Buildings*, 7th ed., 1986, John Wiley & Sons, New York. Reprinted by permission of John Wiley & Sons.)

Figure 9.12 External tankless hot water heater. Hot water from the heating boiler surrounds the domestic hot water coils, transferring heat to the water in them. Boiler water moves by thermosiphon. (From B. Stein and J. S. Reynolds, *Mechanical and Electrical Equipment for Buildings*, 8th ed., 1992, John Wiley & Sons, New York. Reprinted by permission of John Wiley & Sons.)

heats the domestic hot water coil either within the boiler itself (Figure 9.11) or in an external heat exchanger that takes hot water or steam from the boiler (Figure 9.12). In both cases, no hot water storage facility is used. A distinct disadvantage of the type of system that uses a hydronic heating boiler to supply the domestic hot water is that the boiler must be operated even in the summertime when no heating is required. This causes large heat losses and discomfort in the spaces adjacent to the boiler.

Instantaneous heaters are frequently referred to as tankless heaters because they do not use any sort of hot water storage tank. The water heaters in Figures 9.10 and 9.11 are tankless hot water heaters. Instantaneous heaters must be large enough to supply maximum demand immediately at the required temperature, since no hot water is stored. This requires long coils exposed to the heat source, which causes high friction head loss. Directly heated small units are best applied as point-of-use heaters, to supply hot water at remote or isolated plumbing fixtures. Connecting such fixtures into the building's hot water piping would involve long runs of piping with high losses and high first cost. Point-of-use units are also commonly used as booster heaters at dishwashers and laundry machines that require 140°F water and at commercial/institutional laundry machines that require 180°F water for a sanitizing rinse. Because of the large temperature swings caused by on-off operation, instantaneous heaters are best applied where demand is continuous such as for a heated swimming pool, a commercial laundry or heated process water.

In order to overcome the problem of close control of water temperature, some instantaneous heaters use a small tank (or shell) that stores about half of the maximum gpm flow, in preheated water. Thus, a 50-gpm unit would store about 20–30 gal of hot water that acts as a demand buffer. This permits better temperature control of the hot water and makes such units applicable for use in offices, apartments and small facilities that do not have peaks of large water demand. Of course, this type of heater is no longer strictly instantaneous and tankless—it is in reality a small storage heater. See Figure 9.13.

9.10 Storage-Type Hot Water Heaters

The most common type of water heater in use today is the directly heated automatic storage type. It is used almost universally in private homes, apartments, stores, office and other small to medium-size buildings. Units are available in sizes up to 150 gal. The great advantage of a storage heater is that it makes available a large quantity of heated water on demand, with low fuel demand because the water is preheated and stored in an insulated tank. These units are highly efficient when sized correctly for the hot water demand in the facility. They are simple in construction, relatively cheap, almost completely maintenance-free, long-lived and simple to install, control and use. Fuels are electricity, gas or oil. Tanks are glass-lined or otherwise treated to be corrosion-resistant. The units operate well with almost any type of water. They are, therefore, the unit of choice when scale formation from hard water is a particular problem. Construction and operation are simple and straightforward. See Figure 9.14. The heat source, whether it is a flame or an electric heating element, is controlled by a built-in thermostat, which can be adjusted to the desired outlet water temperature. No additional human intervention is required. Recovery is normally slow, requiring up to 2 hr to heat an entire tank of cold water, depending on the design of the particular unit.

A second type of storage unit is the mass storage type. These units are used in commercial, institutional and industrial buildings that experience large peak loads followed by relatively long periods of low hot water usage. Typical applications would be gymnasium showers, restaurant dishwashers, institutional laundries and the like. The large stor-

STORAGE-TYPE HOT WATER HEATERS / 525

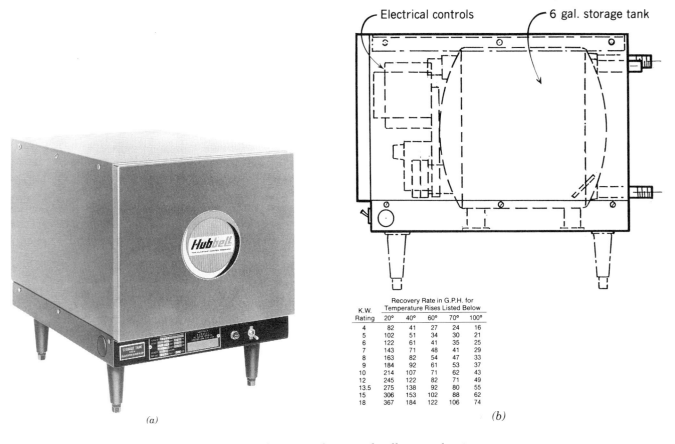

	Recovery Rate in G.P.H. for				
K.W. Rating	Temperature Rises Listed Below				
	20°	40°	60°	70°	100°
4	82	41	27	24	16
5	102	51	34	30	21
6	122	61	41	35	25
7	143	71	48	41	29
8	163	82	54	47	33
9	184	92	61	53	37
10	214	107	71	62	43
12	245	122	82	71	49
13.5	275	138	92	80	55
15	306	153	102	88	62
18	367	184	122	106	74

Figure 9.13 Semi-instantaneous booster-type hot water heater. The illustrated unit, which measures $15 \times 20\frac{1}{2} \times 18\frac{1}{2}$ in. (with legs), contains a 6-gal storage tank plus all the needed electrical heating coils and controls. The exterior (a) shows only an illuminated ON–OFF switch. Interior controls (b) include a T & P safety valve, low water cutoff, pressure regulator and temperature and pressure gauges. The electrical heater rating governs the recovery rate as shown in the accompanying table. (Courtesy of Hubbel, The Electric Heater Company.)

526 / WATER SUPPLY, DISTRIBUTION AND FIRE SUPPRESSION

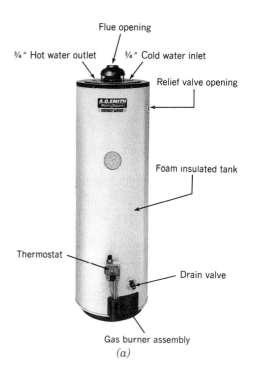

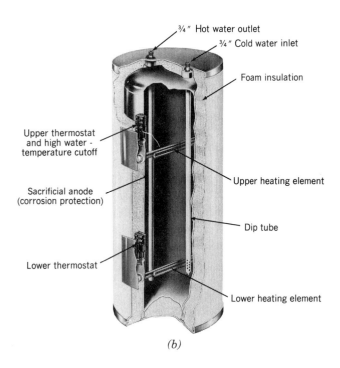

Figure 9.14 (a) Typical residential gas-fired, automatic, storage-type water heater. (b) Cutaway of a typical residential electrical water heater. The sacrificial anode provides corrosion protection. All water heaters are available in a wide range of tank sizes and recovery rates. (Photos courtesy of A. O. Smith Corporation.)

age tank is required to supply the peak demand without the hot water temperature dropping below an acceptable minimum. Water temperatures drop because cold water enters the heater as make-up water to replace the hot water being drawn off. Therefore, the storage tank must contain considerably more water than will be drawn off during peak demand periods. Sizing of storage heaters (as well as other domestic hot water heaters) is quite complex and is not normally done by technologists. Those interested can refer to the references in the bibliography at the end of Chapter 10.

As we stated in Section 8.14g, every hot water system must be provided with relief (safety) valves. These should preferably be of the temperature and pressure (T&P) variety that will relieve excess pressure or temperature in the event of a control malfunction. In addition, instantaneous water heaters should have a hot water tempering valve to prevent the entrance of scalding hot water into the domestic hot water system. Semi-instantaneous heaters frequently have temperature controls that make tempering valves unnecessary.

9.11 Hot Water Circulation Systems

Refer to Figure 9.11(b). This is the piping arrangement that is commonly found in residences and other small facilities. The heat source can be a tankless coil, as shown, or a separate hot water heater of the storage type, as shown in Figure 9.14. Hot water flows only when a faucet is opened. At all other times the hot water in the hot water piping simply stands still and cools off. As a result, when a hot water faucet is opened, the user has to wait until all the cooled water in the piping between the faucet and the hot water heater has run out before the water becomes warm and then hot. This is both frustrating and a waste of water. Indeed, frequently the user utilizes the cold to lukewarm water rather than wait until it becomes hot. This results in again filling the pipes with hot water, which again cools off, wasting energy. One can estimate the waiting time by simply measuring the piping distance between the heater and the faucet. In most small buildings, this pipe length rarely exceeds 50 ft. Assuming a water velocity of 4–6 fps, this means a wait of 8–12 seconds for warm and then hot water. The water is only warm at first because much of its heat is lost to the cold piping. One "cure," therefore, for this problem is to insulate all hot water piping. This will help, but regardless of insulation thickness, the piping will eventually cool off unless hot water is drawn frequently. Hot water pipes should definitely be insulated, but to ensure a prompt supply of hot water, a circulating system is required.

Circulating systems are usually provided when the piping run is about 100 ft long, as would be the case in a large ranch-type residence or a four-story building. With that length of piping between the heater and a hot water faucet, the delay until warm water is received would be on the order of 20–30 seconds, and for hot 30–40 seconds. That long a wait, and the waste of water involved, is considered by most users unacceptable.

Refer to Figure 9.1. The hot water heater shown there is an external (to the boiler) tank-type hot water unit. Notice that in addition to the hot water piping, which is similar to that of Figure 9.11(b), there are two additional items: a hot water storage tank and an additional hot water pipe labelled "circulation." In this arrangement, when no hot water is being drawn, hot water circulates in the piping by thermosiphon action.

When water is heated, it becomes less dense (lighter) and, therefore, rises in the piping. Cold(er) water is heavier and drops down in the piping. If a piping loop is set up with a source of heat at some point in the loop, the water will circulate in this closed loop by the thermal action just described. This action is called *thermosiphon circulation*. It is shown graphically in Figure 9.15.

Now examine Figure 9.1. Notice that a loop exists starting at the hot water storage tank, rising to the hot water pipe at both levels and returning to the tank via the circulation pipe. Hot water constantly circulates in this loop by thermosiphon action. The hottest water in the tank is near the top since it is lightest. It rises into the loop, and, as it circulates, it loses heat to the piping (and eventually to the air), becomes heavier, drops down via the circulation piping, reenters the bottom of the tank, is reheated in the tank, and rises again. This loop makes hot water available at the beginning of the runout piping to each fixture so that only a few seconds of delay is involved in receiving hot water. The disadvantage of this system is, of course, the continuous heat loss from the circulating water. This can be minimized by properly insulating the pipes.

Notice that two more thermosiphon loops exist: one between the hot water storage tank and the external tank-type hot water heater and the second between the boiler and the external hot water heater. As the storage tank water cools below the temperature of the tank heater, it will take hot water from the heater at its top and return cooler water from its bottom. A similar thermosiphon loop exists between the hot water boiler and the external tank heater. As a result, hot water is constantly circulating, drawing heat from the boiler via three loops. Since the movement of water in all the loops depends on the difference in density (weight) of water in the up and down pipes (see Figure 9.15) and since water density depends on temperature, the hotter the water, the faster it moves. Also, since the actual movement depends on the difference of weight of water in the up and down pipes, the taller the building, the better the thermosiphon. This means that a long low building

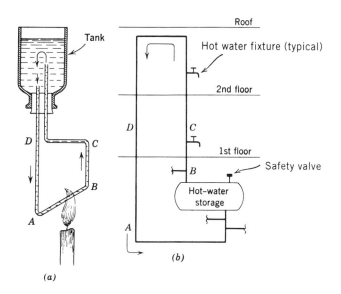

Figure 9.15 (a) Principle of water circulation by thermosiphon. The heated water, being less dense (lighter), rises from B to C; cools off on the way and in the tank; drops to the bottom of the tank as the water becomes cooler and denser (heavier); and returns to the heat source via pipe D-A. (b) The thermosiphon principle can be applied to a building's hot water piping by establishing a closed loop. The speed of water movement is proportional to the temperature difference between outgoing and return water and to the height of the building. Even if no hot water is used, there is sufficient heat loss in the piping loop from C to D to activate the thermosiphon movement, thus making hot water (almost) immediately available at all fixtures. (From B. Stein, J. S. Reynolds, and W. J. McGuinness, *Mechanical and Electrical Equipment for Buildings*, 7th ed., 1986, John Wiley & Sons, New York. Reprinted by permission of John Wiley & Sons.)

will have a very slow thermosiphon not only because of lack of height but also because of the friction in the extensive piping.

To correct this condition for such buildings, a pump is added into the circulation circuit and into the tank circuit. See Figure 9.16. Notice that the hot water source there is an automatic storage-type heater. This type of system is called, logically, a *forced circulation system*. Aquastats control the pump action. (An aquastat is to water what a thermostat is to air.) Aquastat C senses the water temperature at the end of the circulation-return pipe and activates the circulation piping system pump. Aquastat B senses the tank water temperature and activates the tank pump that circulates water between the tank and the heater. Aquastat A is built into the water heater and activates the oil burner when water temperature in the water heater tank drops below its setting. Typical dimensions of hot water tanks are given in Figure 9.17. Obviously, where practical, a thermosiphon system is preferable to a forced circulation system because of energy, equipment and maintenance costs of the latter.

In a recirculating system, the capacity and pressure rating of the piping pump depends on the length and size of piping in the circulation loop, water temperatures and heat losses from the piping. The procedure for these calculations follows.

Step 1. Calculate the heat loss rates of all the piping, after determining the type of pipe and the thickness of insulation to be used. (Because hot water tends to cause rapid corrosion and scale accumulation, copper or CPVC are the materials of choice.)

Step 2. Determine the water velocity rates required to supply water at predetermined temperatures, with the heat losses calculated in Step 1.

Step 3. Calculate the overall friction head loss at the velocities calculated in Step 2 and from this the uniform friction head loss (per 100 ft of TEL).

Step 4. Calculate the individual pipe sizes in the system using the uniform friction head loss determined in Step 3.

Step 5. Size the pump to deliver sufficient pressure and flow.

As you can see, these calculations are technical and complex, and their details are beyond the scope of our discussion here.

As with cold water systems, hot water circulating systems can be arranged as upfeed, downfeed, or a combination system of upfeed and downfeed. See Figure 9.18. A few important items to remember in forced circulation systems are these:

- The water heater can be placed at the top or bottom.
- Since air tends to collect at the top of the system, a means must be provided to release it, because air interferes with circulation. In an upfeed system [Figure 9.18(c)], the circulation riser is connected below the top outlet. That way, when the top outlet is opened, accumulated air is released.
- As mentioned, hot water tends to produce scale. In a downfeed system with the heater at the top, the circulation riser should be connected above the lowest outlet so that loose scale and sediment will be released when the bottom outlet is opened. See Figure 9.18(b).

9.12 Sizing of Hot Water Heaters

Determining the required size of a hot water heater for a specific facility is not a simple task. The calculation involves knowledge of daily consumption, peak load, and the duration of this peak load. With these data, a balance must be made between heating capacity and storage. The larger the burner, the smaller the required storage, and vice versa. For buildings with long periods of fairly uniform demand such as hotels and laundries, a large capacity burner (rapid recovery) with small storage is indicated, because peaks are small. This type of heater is highly efficient and physically small. For buildings with large but infrequent peak loads, such as dormitories and gyms, a small burner (low recovery rate) and a large storage tank are chosen. With that arrangement, the hot water in the tank can supply the large peak load, and the small burner then has a long period in which to heat the (cool) water in the tank.

Use of Table 9.10 will permit calculation of an approximate water heater size for the listed building types. An example will demonstrate use of the table's information.

Example 9.6 Estimate the heating and storage capacity of hot water heater for a small office complex with a normal staff of 150 persons.

Solution: Assume 2½ gal per person per day. Then

Daily hot water use	2.5 (150)	375 gal
Maximum hourly demand	⅕ of 375	75 gal
Storage tank size	⅕ of 375	75 gal

SIZING OF HOT WATER HEATERS / 529

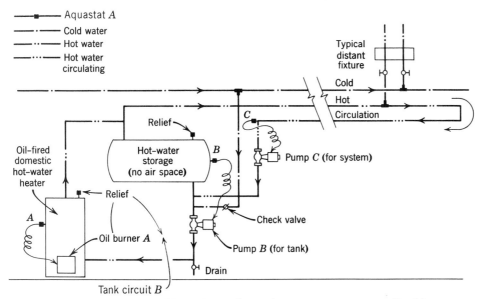

Figure 9.16 Typical piping of forced circulation hot water system, applicable to large low buildings. Aquastats *A*, *B* and *C* control burner *A*, storage tank pump *B* and circulation pump *C*, respectively. Tank circuit *B* circulates water between the storage tank and the heater. Pump *C* circulates the hot water in the piping circuit. (From B. Stein, J. S. Reynolds, and W. J. McGuinness, *Mechanical and Electrical Equipment for Buildings*, 7th ed., 1986, John Wiley & Sons, New York. Reprinted by permission of John Wiley & Sons.)

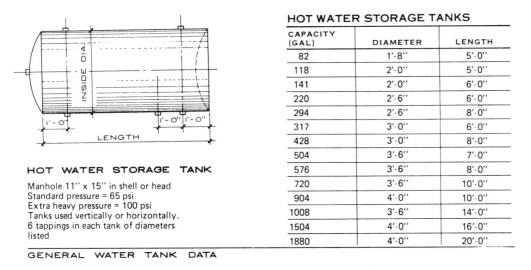

HOT WATER STORAGE TANKS

CAPACITY (GAL)	DIAMETER	LENGTH
82	1'-8"	5'-0"
118	2'-0"	5'-0"
141	2'-0"	6'-0"
220	2'-6"	6'-0"
294	2'-6"	8'-0"
317	3'-0"	6'-0"
428	3'-0"	8'-0"
504	3'-6"	7'-0"
576	3'-6"	8'-0"
720	3'-6"	10'-0"
904	4'-0"	10'-0"
1008	3'-6"	14'-0"
1504	4'-0"	16'-0"
1880	4'-0"	20'-0"

Figure 9.17 Typical dimensional data for hot water storage tanks. (From Ramsey and Sleeper, *Architectural Graphic Standards* 8th ed., 1989 reprinted by permission of John Wiley & Sons.)

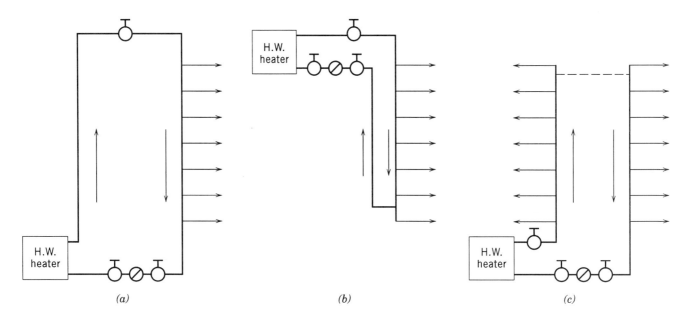

Figure 9.18 Various configurations of hot water circulating systems. *(a)* Downfeed system with heater at the bottom. *(b)* Downfeed system with heater at the top. *(c)* Combination upfeed and downfeed with the heater at the bottom.

Heating capacity   1/6 of 375   63 gph
  (recovery rate)

Since the maximum hourly demand is 75 gal and its duration is 2 hr, we have

Hot water required for peak load = 2 (75) = 150 gal

From this we should deduct three-quarters of the capacity of the tank to determine whether the recovery rate is sufficient. We can effectively use only three-quarters of the tank because, as hot water is drawn off, cold water enters, lowering the water temperature. Therefore,

Peak load   150 gal
Tank supply   3/4 (75 gal) = 56 gal
  2-hr make-up required   94 gal

Since the recovery rate is 63 gph, it is more than adequate for this usage. We would then select a

Table 9.10 Estimated Hot Water Demand

Building Type	Hot Water[a] per Person, gal/day	Maximum Hourly Demand, Portion of Daily Use, gal	Duration of Peak Load, hr	Storage Capacity, Portion of Daily Use, gal	Heating Capacity, Portion of Daily Use, gph
Residences, apartments, hotels[b]	20–40	1/7	4	1/5	1/7
Office buildings	2–3	1/5	2	1/5	1/6
Factory buildings	5	1/3	1	2/5	1/8

[a] at 140°F.
[b] Allow additional 15 gal per dishwasher and 40 gal per domestic clothes washer.
Source. From Ramsey and Sleeper, *Architectural Graphic Standards*, 8th ed., 1989, reprinted by permission of John Wiley & Sons.

heater from the manufacturer's catalog with a 75-gal storage capacity and a minimum recovery rate of 63 gph. The actual recovery rate might have to be larger than this because of the heat losses in the piping. After calculating those losses, the designer can then adjust the storage and recovery rate figures; that is, a higher recovery rate permits use of a smaller tank, and vice versa, as noted before. (For a complete and detailed discussion of this subject, see Stein, B., and Reynolds, J., *Mechanical and Electrical Equipment for Buildings*, 8th ed., Wiley, New York, 1992, or the *ASHRAE System Handbook*.)

An additional illustrative example for a residential occupancy should be helpful.

Example 9.7 Select a water heater for The Basic House plan of Figure 8.40.

Solution: We will present two methods of solution, based on two reference sources, and compare the results. The Basic House has only two bedrooms and one bath. Therefore, we will assume that no more than four people occupy the house.

Method A: Referring to Table 9.10, we obtain the following maximum daily usage:

Occupants (4 × 40)		160 gal
Dishwasher		15 gal
Clothes washer		40 gal
Total		215 gal
Maximum hourly demand	$1/7$ of 215	31 gal
Peak duration		4 hr
Total Peak demand	4 × 31	124 gal
Storage capacity	$1/5$ of 215	44 gal
Heating capacity (recovery)	$1/7$ of 215	31 gph

Again we see that the recovery rate is sufficient:

Required recovery

$$= \frac{\text{Total peak demand} - \text{Storage}}{\text{Peak hours}} = \frac{124 - 3/4(44) \text{ gal}}{4 \text{ hr}}$$

$$= 23 \text{ gph}$$

We would, therefore, probably select a heater with a 40-gal storage and a recovery rate of 30 since piping runs are short and losses are low.

Method B: Refer to Table 9.11. This table shows requirements established by federal housing agencies. Assuming a gas-fired unit, we would select from the table, under one bath and two bedrooms, the following:

Storage	30 gal
Input (1000 Btuh)	36 MBH (36,000 Btuh total)
One-hour draw*	60 gal
Recovery	30 gph

This method yields a smaller tank than method A, which may indicate that a dishwasher is not considered. Since a large percentage of American families living in one-family houses do own a dishwasher, the 40-gal storage capacity seems more appropriate.

Water Supply Design

We have now reached items j and k in the design procedure as outlined in Section 9.1. These items refer to essential details of the design process including all valving, shock arresters, vacuum breakers and the like. After discussing these items, we will proceed to several typical design problems, to demonstrate the application of all that we have discussed relative to water supply in buildings.

9.13 Valving

Correct valving enables a user to properly construct, utilize and especially maintain a water supply and distribution system. Correct placement of valves is an important part of a design technologist's work. In order to accomplish it properly, he or she must know the purpose of each valve in a conventional system. We will, therefore, follow through typical layouts, beginning at the service point.

Refer to Figure 9.1. The first valve encountered is at the tap to the water main. It is usually called a *corporation cock* (company valve) whose purpose is to shut off all water to a building, including the service piping. This demonstrates an important principle; it must be possible to shut off the water supply to any portion of a piping system.

It is also very desirable to be able to isolate different sections of the system, by valving. That way, maintenance can be carried on in one small

---

*The "one hour draw" is the sum of the tank capacity plus 1 hr of recovery.

Table 9.11 Minimum Residential Water Heater Capacities

Number of Baths	1 to 1½			2 to 2½				3 to 3½			
Number of Bedrooms	1	2	3	2	3	4	5	3	4	5	6
Gas[a]											
Storage, gal	20	30	30	30	40	40	50	40	50	50	50
1000 Btu/h input	27	36	36	36	36	38	47	38	38	47	50
1-h draw, gal	43	60	60	60	70	72	90	72	82	90	92
Recovery, gph	23	30	30	30	30	32	40	32	32	40	42
Electric[a]											
Storage, gal	20	30	40	40	50	50	66	50	66	66	80
kW input	2.5	3.5	4.5	4.5	5.5	5.5	5.5	5.5	5.5	5.5	5.5
1-h draw, gal	39	44	58	58	72	72	88	72	88	88	102
Recovery, gph	10	14	18	18	22	22	22	22	22	22	22
Oil[a]											
Storage, gal	30	30	30	30	30	30	30	30	30	30	40
1000 Btu/h input	70	70	70	70	70	70	70	70	70	70	70
1-h draw, gal	89	89	89	89	89	89	89	89	89	89	89
Recovery, gph	59	59	59	59	59	59	59	59	59	59	59
Tank-type indirect[b,c]											
1-W-H rated draw, gal in 3-h, 100F° rise		40	40		66[d]	66	66	66	66	66	66
Manufacturer-rated draw, gal in 3-h, 100F° rise		49	49		75	75[d]	75	75	75	75	75
Tank capacity, gal		66	66		66	66[d]	82	66	82	82	82
Tankless-type indirect[c,e]											
1-W-H rated, gpm, 100F° rise		2.75	2.75		3.25	3.25[d]	3.75	3.25	3.75	3.75	3.75
Manufacturer-rated draw, gal in 5 min, 100F° rise		15	15		25	25[d]	35	25	35	35	35

[a] Storage capacity, input, and recovery requirements indicated in the table are typical and may vary with each individual manufacturer. Any combination of these requirements to produce the stated 1-h draw will be satisfactory.

[b] Boiler-connected water heater capacities (180° F) boiler water, internal or external connection.

[c] Heater capacities and inputs are minimum allowable. Variations in tank size are permitted when recovery is based on 4 gph/kW @ 100°F rise for electrical. A.G.A. recovery ratings for gas heaters, and IBR ratings for steam and hot water heaters.

[d] Also for 1 to 1½ baths and 4 bedrooms for indirect water heaters.

[e] Boiler-connected heater capacities (200°F) boiler water, internal or external connection.

Source. U.S. HUD-FHA, for one- and two-family living units.

section while the remainder of the building operates normally. Obviously, closing the corporation cock shuts down the whole building. However, we do not want to do that in order to repair, say, a lavatory faucet or even a whole water riser. Therefore, the more valving installed, the easier maintenance becomes. The Code specifies minimum valving as we shall see as we go through the system. Additional valving for convenience is optional.

Returning to Figure 9.1, the next valve encountered is the *curb cock*. This is the point at which the building piping begins, and this valve is used to shut down an entire building, including the metering. This valve is normally installed in a concrete box at the curb, called (logically) a curb box. Depth of the box depends on the depth of the piping, which must be below the layer of ground that is subject to winter freezing. Otherwise, the water service line may freeze and burst. Notice that this valve has an extension handle. Both the corporation cock and the curb cock can be *ground key stop cocks*, ball valves or gate valves. Both of these valves are normally specified by the water company and their details, including installation, must be coordinated with the water company.

The next valve encountered is the *service entrance valve*, installed before the water meter and close to the service entry point. This valve is a gate valve with bleed (drain), a stop and waste valve, or any other full-way valve with bleed. (See Code Section 10.12.) The purpose of the bleed (drain) is to permit draining the piping between that (service entrance) valve and the next (set of) shutoff valve(s) down the line. If the water meter is inside the building, it is placed immediately after the service entrance (gate) valve, with another cutoff valve after it. The purpose of the pair of valves is to permit removal of the meter for service or exchange. Alternatively, the water meter can be installed outside, in the curb box. There too it is bracketed by a pair of valves for the same reason. Occasionally, a bypass valve is installed in piping around the water meter so as not to interrupt water service when the meter is serviced. This valving may be prohibited by the local codes or the water utility.

Following along in Figure 9.1, we come to a drain valve. This valve permits draining the system up to the riser (stop and) drain valves (not shown). Drain valves are extremely important not only to allow for maintenance but also to permit draining the piping in an unheated building. The next group of valves controls the flow into and out of the water treatment tanks, if any. Note the bypass shutoff valve between tanks. It is normally closed, but it is opened when maintenance of the water treatment equipment is required. Note also the drain valve that permits draining off the water in the treatment plant piping before service. Without it, a small flood would appear on the basement floor when the water softeners are disconnected for service. The next item of equipment encountered is the boiler. Alternatively, it could be a separate water heater. The piping and valving is similar. The boiler or water heater is supplied cold water through a shutoff valve and a check valve. The shutoff permits servicing of the boiler or water heater. The check valve prevents reverse flow of boiler/water heater, which could occur if the water main pressure dropped below the back pressure level of the piping. For this reason, too, a vacuum breaker is placed in the line. Vacuum breakers are discussed in the next section. Note that a drain valve for the boiler is provided. This valve permits draining the boiler and its feed line. A similar drain valve is found on separate hot water heaters. Following along in Figure 9.1, we have a drain valve for the hot water storage tank and a check valve in the cold water feed line to prevent reverse flow of hot water.

In order to facilitate routine maintenance of residential fixtures, the Code requires that every water closet, lavatory and kitchen sink must have a shutoff valve at the fixture in each hot and cold water line (Code, Section 10.12.4). These are shown on Figure 9.1. Since shutoff valves for showers and tubs are difficult to place in an accessible location, they are not mandated by the Code. However, remember that Code requirements are minimum. In many instances, space for shutoff valves for these fixtures can be found in wall boxes behind access panels. In such case, the slight additional expense is well worth the added convenience. In addition to individual fixture valves, the Code requires that each residential "bathroom group" consisting of two or more adjacent fixtures (normally including a tub and/or shower) shall be valved. Back-to-back bathrooms can be considered as a single group. Again, the purpose is convenience, particularly where the tub and shower are not valved. Without such group valves, a simple tub faucet repair would entail shutting off the water to the entire house! Further, if the tub or shower were on any but the upper floor, the piping would have to be drained as well. These group shutoff valves are frequently placed in a recessed valve box with an accessible, easily removable cover.

534 / WATER SUPPLY, DISTRIBUTION AND FIRE SUPPRESSION

In nonresidential multistory occupancies, each upfeed riser must be valved at the bottom with a shutoff and drain valve, and each downfeed riser, with a stop valve at the top and a stop and drain at the bottom. These valves permit isolation and draining of each hot and cold riser individually without affecting the remainder of the building. Further, the water distribution pipe to each piece of equipment or fixture in a nonresidential occupancy should have a valve or fixture stop that will shut off the water to that fixture or to the room in which the fixture is located. See Code, Section 10.12.6.

Additional valving can be added to separate zones in a building at the discretion of the designer. All valving must be clearly and permanently labelled, easily accessible, and logically located so that occupants or maintenance personnel can quickly find them.

9.14 Backflow Prevention

The Code requires that, in any piping arrangement where it is possible for contaminated water to flow back into the potable water supply, a backflow valve or vacuum breaker be installed to prevent this backflow. (The backflow generally is the result of pressure loss in the water mains.) A vacuum breaker is a device that prevents backflow, due to negative pressure (siphon), from occurring. Negative pressure is, in effect, a vacuum. A vacuum breaker does just what its name indicates; it breaks the vacuum (generally by introducing air at atmospheric pressure), thus preventing contamination and pollution of the potable water system.

Piping connections that can cause contamination exist in boiler feed lines, hot water feed lines, flush valves, cold water supply to flush tanks, and any place else where no air-break exists in the potable water supply pipe. This situation also occurs in any below-the-rim supply (i.e., a water supply line below the rim of the receiving vessel). In such an arrangement, it is possible for the vessel (sink, tub, etc.) to fill and cover the supply outlet. Then, if supply pressure drops, contaminated water can siphon back into the potable water mains and thereby cause extensive pollution to other buildings as well. In all such conditions, a vacuum breaker is required to prevent the (negative pressure) siphonage.

A clothes washer must have vacuum breakers in the machine to prevent back-siphonage from the washer. If the laundry tub has a threaded-end faucet onto which a flexible rubber hose can be attached, then an atmospheric vacuum breaker must be screwed onto the faucet to prevent any possible backflow from a submerged hose in the laundry tub. Note also that the dishwasher, because it is directly connected, without an air break, must have a vacuum breaker on its supply line. This device is usually part of the dishwasher appliance itself.

Additional vacuum breakers are found in every toilet flush tank because the supply line is below the level of water in the tank. It is also a very good idea to place a vacuum breaker on the garden hose bibb to prevent backflow from a garden hose left submerged in a swimming pool or even a child's wading pool. Typical conventional backflow prevention using vacuum breakers is shown in Figure 9.19.

Backflow prevention can also be accomplished using a pressure differential valve. This valve senses a reverse pressure situation that can cause backflow and operates built-in check valves to prevent this backflow. One such unit, with its specifications, is shown in Figure 9.20.

Figure 9.19 *(a-1)* Typical commercial application of an atmospheric vacuum breaker. *(a-2)* Detail of the internal construction of the vacuum breaker and its action. *(b-1)* A (submerged) service sink hose must be provided with a hose-type atmospheric vacuum breaker to prevent possible backflow. *(b-2, b-3)* Action of the valve is shown in the two section illustrations.

(Diagrams *(a-1, b-1, c* and *d* reproduced with permission from the National Standard Plumbing Code, published by The National Association of Plumbing Heating Cooling Contractors. Figures *a-2, b-2* and *b-3* and photos *a-1* and *b-1* courtesy of Watts Industries, Inc.)

BACKFLOW PREVENTION / 535

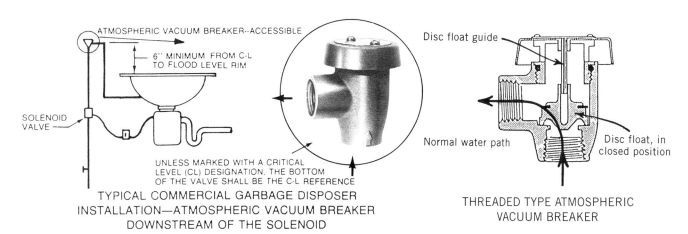

TYPICAL COMMERCIAL GARBAGE DISPOSER
INSTALLATION—ATMOSPHERIC VACUUM BREAKER
DOWNSTREAM OF THE SOLENOID

(a-1)

THREADED TYPE ATMOSPHERIC
VACUUM BREAKER

(a-2)

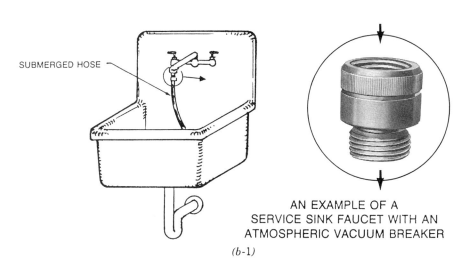

AN EXAMPLE OF A
SERVICE SINK FAUCET WITH AN
ATMOSPHERIC VACUUM BREAKER

(b-1)

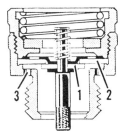

Valve in closed position with supply valve shut off disc (1) seated against diaphragm (2). Atmospheric ports are open (3) during no flow.

(b-2)

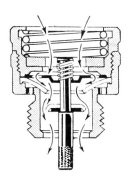

Fully opened valve, illustrating poppet action to provide high capacity with minimum pressure drop through valve.

(b-3)

536 / WATER SUPPLY, DISTRIBUTION AND FIRE SUPPRESSION

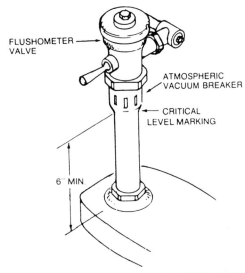

AN EXAMPLE OF A FLUSHOMETER VALVE
WITH ATMOSPHERIC VACUUM BREAKER

(c)

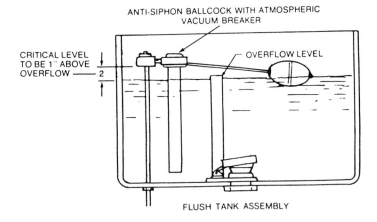

TYPICAL GRAVITY FLUSH TANK BALLCOCK
WITH ATMOSPHERIC VACUUM BREAKER

(d)

Figure 9.19 *(Continued)* An atmospheric vacuum breaker is built into every flush valve *(c)* and flush-tank mechanism *(d)*.

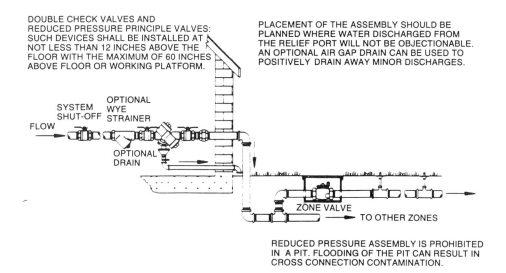

**REDUCED PRESSURE BACKFLOW PREVENTION ASSEMBLY
(INDOOR INSTALLATION)**

Figure 9.20 Reduced pressure backflow assemblies consist of two independent check valves with an intermediate relief valve. They act to prevent back siphonage when loss of line pressure produces a back pressure into the potable water line. (Diagram reproduced with permission from the National Standard Plumbing Code, published by The National Association of Plumbing Heating Cooling Contractors.)

9.15 Water Hammer Shock Suppression

When water flowing in piping is suddenly stopped, all the kinetic energy in the moving water must be absorbed by the rigid piping system. The result is a shock wave that travels at tremendous speed throughout the system (at 3000–3500 mph) rebounding from one end of the piping to the other until all the energy is absorbed in the piping and other equipment. This effect is commonly known as *water hammer* because of the loud hammering noises caused by the shaking of the piping and other connected devices. In addition to noise, however, water hammer can cause serious damage to a system by causing temporary pressures as high as 600 psi. Among the undesirable effects of such high pressures are:

- Loosening of pipe joints and other connections and subsequent leakage.
- Loosening of pipe supports.
- Damage to gauges, meters, regulators, valves, tank outlets, coils and other relatively sensitive and delicate equipment.
- Increased cavitation in system components.
- Loosening of pipe scale with subsequent fouling of faucets, valves and meters

Since the cause of water hammer is usually rapid valve closure, the obvious cure is to avoid such valves. This, however, is not always possible, since valves in such diverse appliances as single-handle faucets, self-closing faucets, domestic dishwashers and domestic clothes washers are, by their nature, rapid-closing. In addition, check valves will snap shut when subjected to reverse flow. What remains, therefore, is to reduce the energy in the shock wave and to absorb it safely. Using low water velocity reduces the system energy. Designers will normally limit water velocity to 4 fps in systems with quick closing valves and, in general, not exceed 8 fps. In addition, water hammer shock arrestors should be installed in every piping system.

The most common type of shock arrestor (also called *shock suppressor*) is simply a piece of sealed pipe installed at fixtures. These can be seen sche-

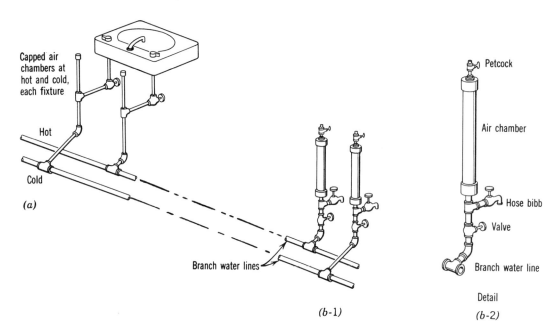

Figure 9.21 (a) Capped sections of pipe at each fixture serve as water hammer shock suppressors. In addition, the capped pipe on the hot water branch line acts as an expansion chamber as well. Rechargeable air chambers (b-1) are superior to sealed pipe air chambers, because they solve the problem of air absorption by the water in sealed pipe extensions. Because rechargeable chambers must be accessible, their use is limited. Air chambers details are shown in (b-2).

matically in Figure 9.1 and in the pictorial drawing of Figure 9.21(a). In addition to absorbing shock by compression of the air in these pipe chambers, the hot water pipe extension acts as an expansion chamber. These simple water hammer shock suppressors gradually lose their effectiveness as the sealed air in the chamber is absorbed by the water. A better arrester, shown in Figure 9.21(b), counteracts this air absorption effect by permitting recharging of the air chamber. These rechargeable units are installed at fixtures instead of the pipe chambers. They must, however, be accessible for periodic recharging. Since neither of these two types is permanent and since the required maintenance is usually not performed, their use is strongly discouraged. Instead, permanent, sealed, maintenance-free arrestors, as described in the following section, should be used.

The best shock suppressors have some sort of permanent, sealed, internal expansion device, which is either a cylinder and piston, a bellows or an expandable flexible wall. See Figure 9.22. (Some of the units, however, are field-adjustable and chargeable. Such units should be installed in an accessible location.) The air in the expansion portion of these devices is isolated from the water in the piping and, therefore, need not be recharged. As a result, these devices can be installed permanently in inaccessible locations (in the wall cavity). They provide the best cure for water hammer problems and should be installed whenever high velocity water (above 4 fps) and quick closing valves are found.

9.16 Residential Water Service Design

At this point, you have the background to perform actual water service and water distribution design if you have carefully followed the discussions in the chapter. We will begin our design example with a residential structure.

Example 9.8 Determine the size required for a water service pipe to serve the Mogensen house shown in Figure 3.38. The actual site plan is given in Figure 9.23.

RESIDENTIAL WATER SERVICE DESIGN / 539

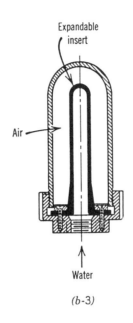

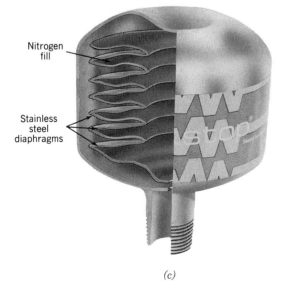

Figure 9.22 Various types of water hammer shock suppressors. *(a-1)* Field chargeable and adjustable piston-type water hammer suppressor operates by compressing the air above the piston to absorb the system shock. Units are applicable to systems with WSFU of 1–430 units and vary in size from 1⅝ × 8 in. to 2⅝ × 18 in. Because the device is field-serviceable, it should be installed exposed. *(a-2)* Nonadjustable piston-type water hammer shock suppressor may be installed concealed (nonaccessible). Dimensions are approximately the same as type *b-2* for the same number of fixture units. *(b-1)* The chamber is filled with air and is separated from the system water by a butyl diaphragm that prevents air loss and water logging. Units are rated by system flow in the range of about 10–100 gpm. For that range, they vary in size from 8½ to 12 in. in height. Diameter is 6⅛ in. *(b-2)* Small diaphragm-type suppressor is field-adjustable and chargeable. Dimensions are 4½– 6 × 3⅜ in. Small air capacity limits application of this unit to small systems. *(b-3)* This diagram shows how a diaphragm-type unit operates. In this unit, the diaphragm, is tubular to conform to the outer body shape. In units *b-1* and *b-2*, the diaphragm is a flat piece of butyl rubber stretched across the diameter of the unit. (Photos *a-1*, *a-2*, *b-1* and *b-2* courtesy of Amtrol, Inc.)

Figure 9.22 *(c)* A bellows-type water hammer shock arrester. The space between the nesting individual stainless steel diaphragms that make up the bellows contains nitrogen. This gas helps to absorb shock and will not be absorbed by the system's liquid. (Photo courtesy of Tyler Pipe, a subsidiary of Tyler Corporation.)

540 / WATER SUPPLY, DISTRIBUTION AND FIRE SUPPRESSION

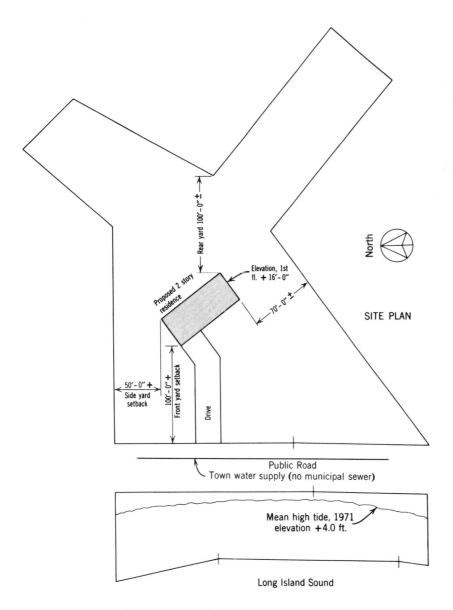

Figure 9.23 Site plan, Mogensen house. This illustration was adapted from the surveyor's plans and the architect's layout. Because the plumbing designer is responsible for water service as well as storm and waste water disposal, an accurate site plan showing existing utilities is important. In this plan, the fact that no municipal sewer is available indicates, as is explained in Chapter 10, that private sewage treatment is required. For this reason, water table data are shown on the site plan. For a completed site plan showing water and drainage service piping, see Figure 10.56.

RESIDENTIAL WATER SERVICE DESIGN / 541

Solution: Assume that the water service pipe will be soft temper type K copper tubing. This material is highly suitable for underground installations. The following additional data are available:

Minimum street mains pressure	45 psi
Total equivalent length (TEL) of pipe from street mains to house valve	175 ft
Depth of street mains below street level	3 ft

We will follow the procedure outlined in Section 9.6, page 514.

Refer to Figures 9.24 through 9.26. Figure 9.24 shows the house plumbing fixtures in plan. This is the drawing that would be prepared at the preliminary stage, as we noted in Chapter 8 (see Figure 8.40). Figures 9.25 and 9.26 correspond to Figures 8.43 and 8.44 and represent the piping plan and the plumbing section (elevation), respectively. Actually it is not necessary to prepare a plumbing plan as shown in Figure 9.25 in order to size the water service. Figure 9.26, the plumbing section (riser), is sufficient. Runout distances can be measured on the architectural plan. We prepared the plumbing plan of Figure 9.25 to show you how such a plan should appear and to assist you in measuring actual runout distances in connection with the problems at the end of this chapter. For residential work, detailed plumbing runout plans are not usually prepared by the designer. It is normally left to the discretion of the plumbing contractor, within the specification limitations.

Referring to Figure 9.26, we see that the two longest TEL runs from the service entrance point are to two hose bibbs: one at the upper level behind the house and one at the lower level in front of the house. The piping is shown according to the actual installation. Since the upper level hose bibb has an additional 9 ft of static head to overcome, we will use it in our pressure calculation. Its runout TEL distance, as shown on Figure 9.26, is 120 ft. The pressure equation is

Total pressure available = Static head + Total friction head loss + Minimum remote fixture pressure

Static head is the difference in elevation between the service water main and the last outlet. It comprises the depth of the main below grade (3 ft) plus the floor-to-floor height (9 ft), plus the elevation of the hose bibb above grade at the rear (3 ft).

Since maximum flow is only 28 gpm even with two hose bibbs fully open (see calculation on figure 9.26) we will assume a maximum pressure loss in the water meter of 4 psi. Refer to Figure 9.8. We will recheck this assumption later in this calculation. The data for our equation are:

Total pressure available	45 psi
Static head (3 ft + 9 ft + 3 ft) (0.433 psi/ft)	6.5 psi
Assumed friction loss in meter	4 psi
Hose bibb minimum pressure	15 psi

Therefore,

Maximum friction head loss = 45 − (6.5 + 4 + 15)
= 19.5 psi

$$\text{Maximum friction per 100 ft} = \frac{19.5 \text{ psi}}{1.75(100) + 1.2(100) \text{ ft}}$$
= 6.6 psi/100 ft

Referring now to Figure 9.6, we find that, at a flow of 28 gpm (maximum), a service pipe of 1¼-in. diameter gives a friction head loss of 6.5 psi/100 ft and a water velocity of 7 fps, both of which are entirely satisfactory. Rechecking the water meter loss in Figure 9.8, we find a loss of 4 psi, exactly as assumed. We, therefore, conclude our service pipe calculation with a decision to use 1¼-in. type K soft temper tubing for the 175-ft water service run between the water main and the building.

This house is unusual in that it has four hose bibbs—two at the front and two at the back. In our calculation of total flow (gpm), we assumed that no more than two, at a flow of 5 gpm each, would be used at once. However, for flow in individual runouts, all the hose bibbs are considered. An actual photograph of the water service tube entering the house at the southeast corner of the garage is seen in Figure 9.27. In plan, it is shown on the completed site drawings of Figure 10.56. Figure 9.28 shows an elevation of the water service and principal water distribution lines as actually installed. This drawing is for educational purposes only. It would not be part of the normal working drawing package.

At this point, given the maximum friction per 100 ft calculated previously, you could calculate the required size of all runouts, using either the exact method as explained in Example 9.4 or the velocity method of Example 9.5. In many instances, however, when sizing is left to the contractors, they will use rules of thumb based on experience, according to the Table 9.12

Since all the outlets in this house are ½ in., we have indicated on Figure 9.26 the runout sizes according to this tabulation. We have left it to you

542 / WATER SUPPLY, DISTRIBUTION AND FIRE SUPPRESSION

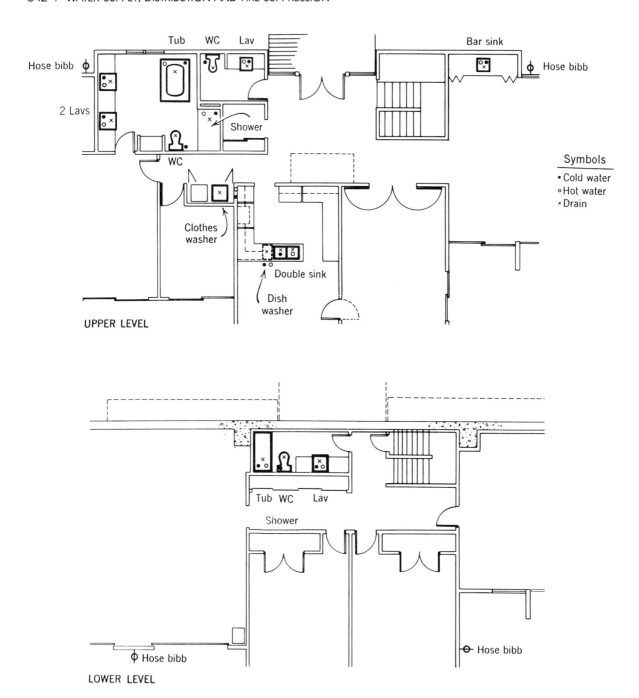

Figure 9.24 Preliminary plumbing plan of the Mogensen house, showing water supply and drainage requirements at fixtures.

RESIDENTIAL WATER SERVICE DESIGN / 543

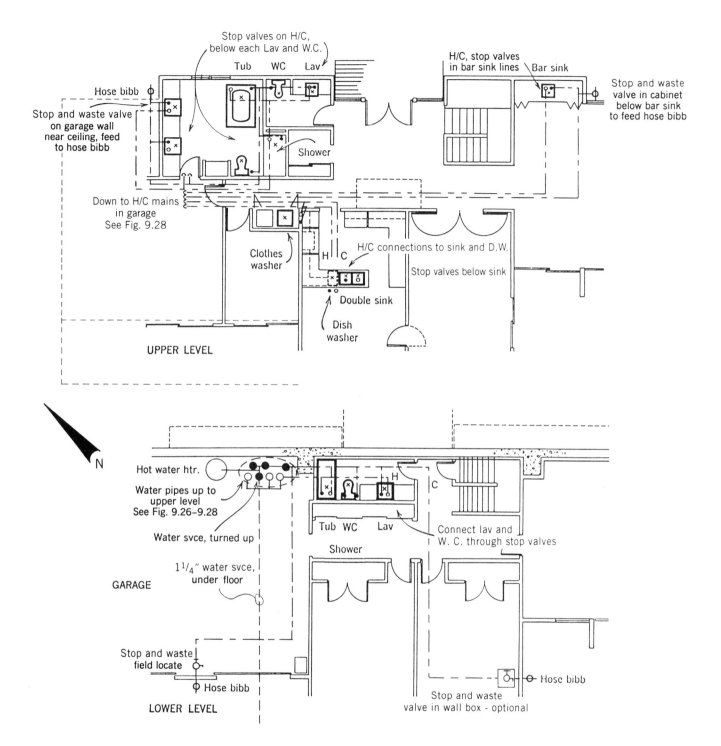

Figure 9.25 Hot and cold water runouts to all fixtures. Hose bibbs should be supplied through stop and waste valves to permit pipe draining before winter. When a lower level (basement) is not available, as in the illustrated house, an (optional) wall box may be used. An elevation of the water service arrangement is shown on Figure 9.28. A plumbing section is shown in Figure 9.26.

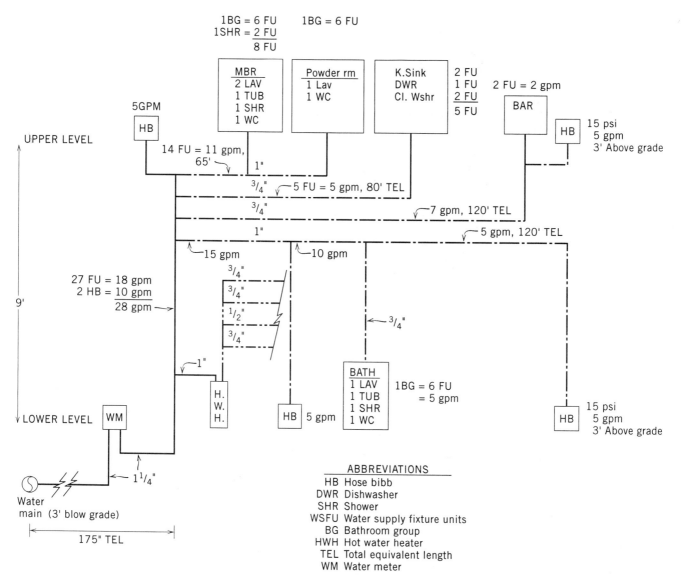

Figure 9.26 Plumbing riser (section) for the Mogensen house. All the data required for calculation of friction loss and pipe sizing are given on this diagram. Runout pipe sizes shown are based on Table 9.12.

Table 9.12 Runout Pipe Size According to Number and Size of Outlets Supplied

Minimum Pipe Size, in.	Maximum Number of 3/8-in. Outlets	Maximum Number of 1/2-in. Outlets	Maximum Number of 3/4-in. Outlets
1/2	3	1	—
3/4	4	3	1
1	15	8	3

to check the sizes by actual calculation and to decide what action to take where the methods yield different answers.

9.17 Nonresidential Water Supply Design

For our illustrative nonresidential building example, we will use the light industry building whose architectural plan is shown in Figure 3.48, page 153. You will also see a building site plan there.

Example 9.9 Design the water service and show the interior water piping for the light industry building shown on page 152. Refer to Figure 3.53, page 160, which shows the location of the boiler room, adjacent to the women's rest room (Room #7). The distance from the corporation cock (water line connection) to the boiler room where the water line enters is 100 ft, with a total rise to the flush valve (or hose bibb) level of 10 feet. Assume all water closets are floor-mounted siphon jet bowls with flushometers. Assume also that minimum maintained street water pressure is 45 psi.

Solution: We will follow the same procedure as for the previous design example. In this example, it is not necessary to prepare a plumbing riser as everything is on one level.

1. Calculate flow Using Table 9.3, the water requirements for this building are:

3 general lavatories @ 2 WSFU	6 WSFU
1 private lavatory @ 1 WSFU	1 WSFU
3 general flush valve closets @ 10 WSFU	30 WSFU
1 private flush valve closet @ 6 WSFU	6 WSFU
1 water cooler @ 0.25 WSFU	0.25 WSFU
1 hose bibb, general use, 1/2-in. piping	4 WSFU
Total	47.25 WSFU

From Table 9.4, under the flushometer section, 47.25 WSFU is equal to 50 gpm (by interpolation).

2. Calculate pressure

Water main pressure = Maximum friction head loss (meter and piping) + Static head + Minimum flow pressure

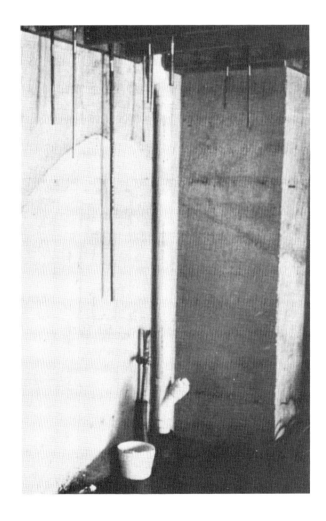

Figure 9.27 Construction stage photograph of the Mogensen house. The view is of the northeast corner of the garage where the water service equipment is installed (under the master bath). The 1 1/4-in. water service is shown at the center of the photo as it rises from below the slab. The domestic hot water heater will be placed where the pail stands in the photo. Branch water lines are shown above. An elevation of this wall is shown in Figure 9.28.

546 / WATER SUPPLY, DISTRIBUTION AND FIRE SUPPRESSION

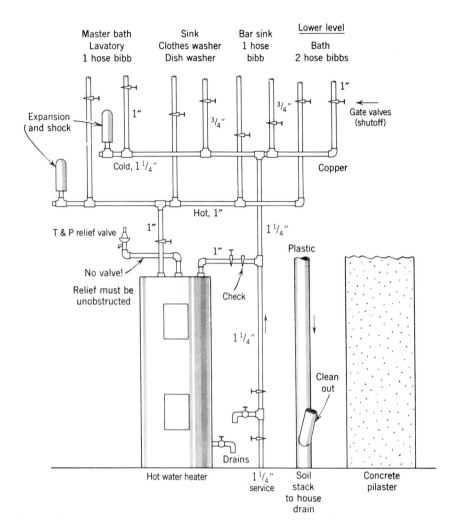

Figure 9.28 Elevation of the back garage wall showing the water service and the principal distribution lines. Since the garage is unheated, all hot water lines will be insulated. See also Figure 9.25–9.27.

From Table 9.1 we see that the minimum flow pressure required for either a floor-mounted siphon jet closet bowl with flushometer or a hose bibb is 15 psi. The TEL distance to the last bowl or the hose bibb, including fittings, is

$$1.5 \times 125 \text{ ft} = 188 \text{ ft}$$

Assuming a 5-psi pressure drop in the water meter, we have

Maximum friction head = 45 psi − [5 psi + 10 ft (0.433 psi/ft) + 15 psi]
= 45 − 24.3 = 20.7 psi

$$\text{Maximum friction head per 100 ft} = \frac{20.7 \text{ psi}}{1.88 \,(100) \text{ ft}}$$
$$= 11.0 \text{ psi}/100 \text{ ft}$$

3. Determine service size Assuming that we are using copper tubing for the service, we refer to Figure 9.6 or Figure 3.34. For a flow of 50 gpm and 11.0 psi maximum friction per 100 ft, we require a 1½-in. service. This gives us an actual friction of 9 psi/100 ft and a water velocity of 8.5 fps. Although velocities in excess of 8 fps are not recommended because of noise, a velocity of 8.5 fps can be considered acceptable for an underground service pipe. A more conservative design would use a 2-in. pipe. Rechecking our assumption of 5-psi pressure drop on the water meter (see Figure 9.8), we find an actual drop of 6 psi for 50-gpm flow and a 1½-in. meter (pipe) size. Recalculating the pressure equation with this increased meter loss, we now have a revised maximum permissible friction head loss of one psi less, or 19.7 psi.

This gives a maximum friction loss per 100 ft of

$$\frac{19.7 \text{ psi}}{1.88 \,(100 \text{ ft})} = 10.47 \text{ psi}/100 \text{ ft}$$

Checking in Figure 9.6, we see that we still require a 1½-in. pipe. This size pipe will carry through the entire run until the taps to the fixtures. See Figure 9.29. Minimum fixture runout sizes are governed by Code; ½ in. to lavatories and 1 in. to flush valves. The header connecting the flush valves can be 1¼ in. and is so indicated on the piping plan. The four ½-in. hot water fixture connections are fed by a 1-in. header, according to the rule stated in Section 9.16. The site plan with all services is shown in Figure 10.48.

Water Supply for Fire Suppression

An integral part of the work of a plumbing technologist is the piping design associated with fire sup-

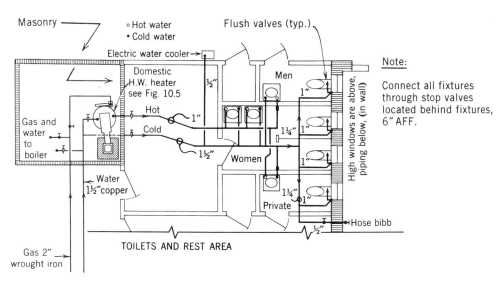

Figure 9.29 Water service and distribution for the toilet area of an industrial building. The gas service line is shown for completeness. See Example 9.9 in the text for details of pipe sizing.

pression. This is a highly specialized field governed by exacting standards issued, among others, by the National Fire Protection Association (NFPA). A complete study of this subject is well beyond the scope of this book. However, its principles are not, and they will be discussed here. Refer to the bibliography at the end of Chapter 10 for this and other topics of special interest. The principal water-based fire suppression systems are wet and dry standpipes and wet and dry sprinkler systems. Other fire suppression systems that are not water-based, such as halon, foam, and carbon dioxide, are also generally "piped" by the plumbing designer, although they are so specialized that they will not be discussed here.

9.18 Standpipes

A schematic diagram of a typical *standpipe* (and hose) system is shown in Figure 9.30. You should follow the discussion with this diagram in front of you. A standpipe is a vertical pipe extending from the lowest to the highest floor of a building. It is normally filled with water under pressure, except in unheated buildings where the pipe is empty *(dry standpipe)*. Connected to each standpipe (more than one in large buildings) on every floor are hose cabinets, which, in addition to a water valve and hose, normally contain a fire extinguisher. See Figure 9.31. The purpose of the standpipe system is obviously to supply water for fire extinguishing at every floor level, at sufficient pressure and quantity to satisfy fire fighting requirements. As mentioned, large buildings have multiple standpipes, so located that every portion of each floor can be reached by a water stream from one of the hoses. Standpipes are required in tall buildings because the upper floors of such buildings cannot be reached by normal fire-fighting apparatus at street level.

The water source for a standpipe system is either the roof tank in downfeed systems or the suction tank in upfeed systems. The amount of water in these tanks is normally sufficient for ½ hr of use. By that time, fire trucks will have arrived, and connection to city fire hydrants or other water source will be made via the standpipe *siamese connection*. See Figure 9.32. All standpipes in a building are interconnected at their lower end, and the connecting pipe is extended to an outside siamese connection. This device will take one or two outside water sources to replace the tank water supply. Each leg of the siamese connection is equipped with a check valve that opens to permit water flow from the outside source. At the same time the check valves in the lines from the roof tank close under reverse pressure to prevent wasteful refilling of the tank by the exterior fire pumps. Standpipes are at least 4 in. in diameter, with 6 in. being used in taller buildings. As stated, exact design of all the system components is governed by applicable fire codes.

9.19 Sprinklers

An automatic sprinkler system consists of a horizontal pattern of pipes, at ceiling height, fitted with sprinkler heads at a fixed spacing along these pipes, plus all the associated piping and control equipment. When abnormally high temperatures in the protected area are detected, the sprinklers release a spray of water that is normally very effective in suppressing a fire. As with standpipe systems, sprinkler design is rigidly controlled by fire codes. Actual design of systems is a highly complex affair, done by specialists in the profession. The function of the plumbing technologist is to lay out the piping on the working drawings, according to the design established by sprinkler experts. Figures 9.33, 9.34 and 9.35 show system piping and some of the details of a typical sprinkler system.

There are four common types of sprinkler systems:

(a) Wet-pipe.
(b) Dry-pipe.
(c) Deluge.
(d) Pre-action.

a. Wet-Pipe Sprinklers

Wet-pipe sprinklers, which are the most common type, have water, under pressure, in the pipes at all times. Pressure is supplied either from a roof tank or from a ground level tank and automatic pump system that maintains pressure in the system. Each sprinkler head acts individually. When unusually high temperature is sensed, the sprinklers in that area only are activated. The usual variety of sprinkler head will continue to spray water until shut off manually. The head that operated must then be replaced. More recently, flow control sprinkler heads have come into use. These automatically close once the ceiling temperature returns to normal (indicating that the fire has been extin-

(text continues on p. 553)

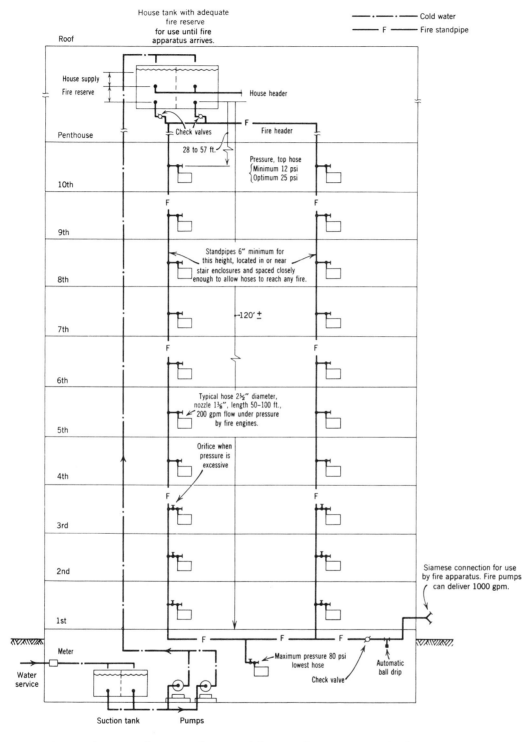

Figure 9.30 Schematic diagram of a typical fire standpipe system. The first half hour of water supply can come from the fire reserve in the roof tank. After that, water must be supplied, upfeed, from the connection to the exterior siamese fitting. (From B. Stein, J. S. Reynolds, *Mechanical and Electrical Equipment for Buildings*, 8th ed., 1992, John Wiley & Sons, New York. Reprinted by permission of John Wiley & Sons.)

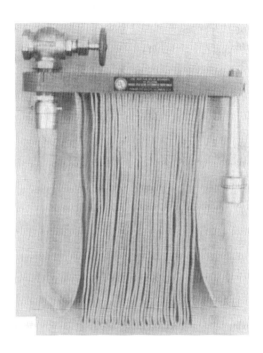

Figure 9.31 Typical hose rack found in a cabinet adjacent to each fire standpipe on each floor. The valve at the upper left opens to permit water to flow into the hose. Some codes require that a pressure-reducing valve be placed at this point to limit hose pressure to about 50 psi. (From B. Stein, J. S. Reynolds, *Mechanical and Electrical Equipment for Buildings*, 8th ed., 1992, John Wiley & Sons, New York. Reprinted by permission of John Wiley & Sons.)

Figure 9.32 Standard siamese (double) connection is mounted between 19 and 36 in. above ground, on the street side of a building. Notice that the standpipe siamese connection is clearly labelled, to differentiate it from the (similar) sprinkler siamese connection. A double connection is provided for fire truck pumps and/or city fire hydrants. (From B. Stein, J. S. Reynolds, *Mechanical and Electrical Equipment for Buildings*, 8th ed., 1992, John Wiley & Sons, New York. Reprinted by permission of John Wiley & Sons.)

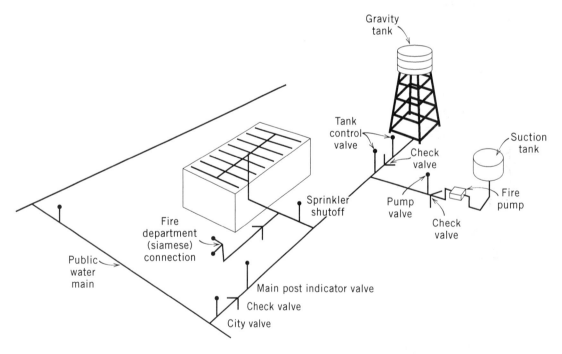

Figure 9.33 A typical sprinkler installation showing all common water supplies, outdoor hydrants and underground piping. (Reprinted with permission from *Fire Protection Handbook*, 17th ed., © 1991, National Fire Protection Association, Quincy, Mass. 02269.)

SPRINKLERS / 551

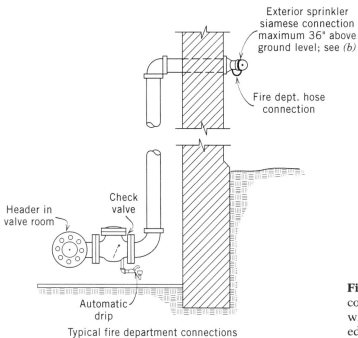

Typical fire department connections
(a)

Figure 9.34 (a) Typical piping from exterior siamese connection to the internal sprinkler header. (Reprinted with permission from *Fire Protection Handbook*, 17th ed., 1991, National Fire Protection Association, Quincy, Mass. 02269.)

Figure 9.34 Combination siamese connections for standpipe and sprinkler systems in front of two major New York City hotels. *(b-1)* Building mounted connections; *(b-2)* free-standing connections. (Photos by Stein.)

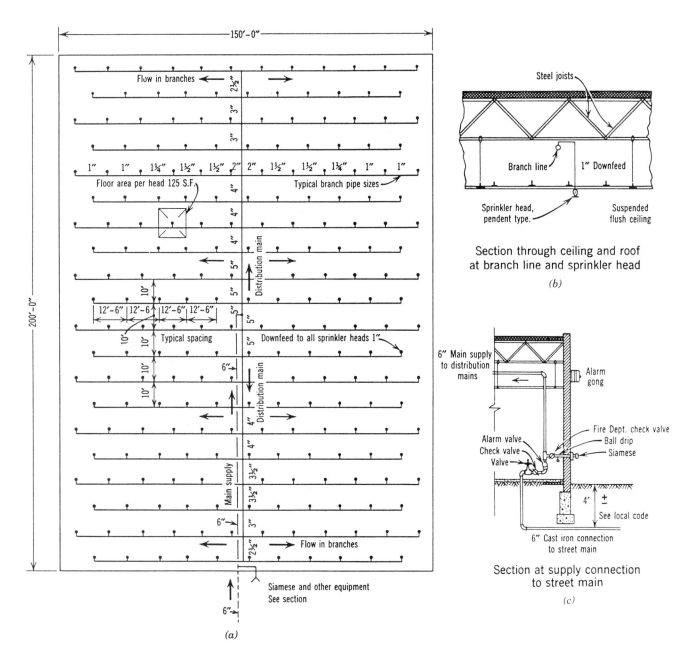

Figure 9.35 *(a–c)* Typical sprinkler plan for an industrial building *(a)* with details of the piping at the ceiling *(b)* and the water supply connection *(c)*. When street water pressure is, or may be, insufficient to meet all requirements, a suction tank and an automated pressurization pump system are provided. Alternatively, an elevated water tank can be used. The outside siamese connection for the sprinkler system is usually separate and distinct from the standpipe siamese connection and is so labelled. See Figure 9.34. (From B. Stein, J. S. Reynolds, and W. J. McGuinness, *Mechanical and Electrical Equipment for Buildings*, 7th ed., 1986, John Wiley & Sons, New York. Reprinted by permission of John Wiley & Sons.)

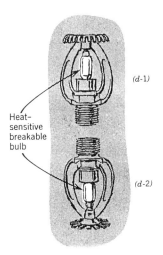

Figure 9.35 *(d)* Typical sprinkler heads. Type *(d-1)* sits up high above exposed piping, where hot gases accumulate. Type *(d-2)* projects through a suspended ceiling. Both types use a bulb that ruptures when heated to a preset temperature, releasing a spray of water. (From B. Stein, J. S. Reynolds, *Mechanical and Electrical Equipment for Buildings*, 8th ed., 1992, John Wiley & Sons, New York. Reprinted by permission of John Wiley & Sons.)

guished). Since the wet-pipe system is always filled with water, it is obviously applicable only to a heated building. Waterflow switches in the supply piping sense the system operation and operate the fire alarm system. See Figure 9.36.

b. Dry-Pipe Sprinklers

Dry-pipe sprinklers are used in buildings where the pipes are subject to freezing such as a dead storage warehouse, a loading area open to the outside, or a cold storage plant. Dry pipes are filled with compressed air. When a head opens, the air is released, allowing water to flow through the piping and out of the operated head. Dry pipe systems must have all piping pitched to permit draining, after a section of piping fills with water. Also, all valves must be placed in heated enclosures to ensure their operation.

c. Deluge Sprinklers

A *deluge sprinkler* system is a dry pipe system. It responds to abnormally high temperature anywhere in the protected area by opening a deluge

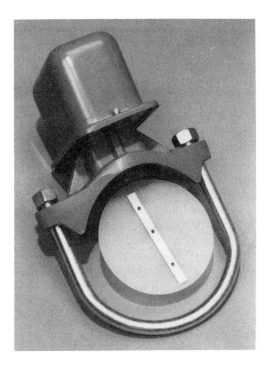

Figure 9.36 Typical water flow indicator. The unit bolts onto a sprinkler pipe with the paddle inside the pipe. Any water motion deflects the paddle, causing a signal to be transmitted to the fire alarm system from the microswitch mounted in the box on top of the pipe. (Courtesy of Notifier Company.)

valve that supplies water to all the heads. Unlike the other sprinkler systems where heads open in response to heat, deluge sprinkler heads are normally open. As a result, all the sprinkler heads operate simultaneously. The purpose of this design is to prevent the spread of fire. This system is used where extremely rapid fire spread can be expected, as in buildings filled with highly flammable materials. It is also used where fire can produce noxious or poisonous fumes, such as in chemical plants.

d. Pre-action Sprinklers

A *pre-action sprinkler* system is a dry-pipe system that operates in conjunction with detectors of a fire alarm system. Operation of any one of these highly sensitive, early warning detectors opens a deluge valve that fills the piping with water without any sprinkler head operating. Afterwards, when a sprinkler head opens due to extreme heat, the spray action begins. The purpose of this type of arrangement is to prevent accidental (water) damage to valuable building contents (such as in a library) that result from a faulty sprinkler head or leaky pipe. The system remains dry until the extremely sensitive fire alarm detector not only causes the system to fill but also turns in an alarm. In the time interval between the alarm and a sprinkler head opening as a result of sensing heat, a fire can frequently be located and extinguished without the extensive water damage that is typical of sprinkler extinguishing. On the other hand, if a sprinkler head does open, water is immediately available, unlike the dry-pipe system, which must first be filled.

Because sprinklers release a huge amount of water directly onto the building floor, means must be provided to drain away the water to avoid extensive flooding. This is usually done with floor drains and wall scuppers. Similarly, all sprinklers must be arranged for controlled flow drainage to an outside drain. This is required for maintenance, construction changes and the like.

Key Terms

Having completed the study of this chapter, you should be familiar with the following key terms. If any appear unfamiliar or not entirely clear, you should review the section in which these terms appear. All key terms are listed in the index to assist you in locating the relevant text.

Air-break
Aquastat
Backflow valve
Back-siphonage
Bathroom group
Below-the-rim supply
Booster pump
Bypass shutoff valve
Corporation cock
Curb box
Curb cock
Deluge sprinkler
Diversity factor
Domestic hot water
Downfeed system
Dry-pipe sprinklers
Dry standpipe
Fixture unit
Fixture use demand factor
Flow pressure
Flow rate
Forced circulation system
Friction head
Gravity downfeed system
Ground key stop cock
Group shutoff valves
Hot water circulating system
Instantaneous water heater
Jockey pump
Mass storage (hot water heater)
Minimum flow pressure

Outlet temperature
Peak flow
Point of use water heater
Plumbing section
Pre-action sprinkler
Pressure differential valve
Pumped upfeed system
Recovery rate
Roof tank
Sanitizing rinse
Service entrance valve
Shock arrestor
Shock suppressor
Siamese connection
Sprinkler head
Standpipes
Stop and drain valves (stop and waste valve)
Suction tank
Tankless heaters
Thermosiphon
Thermosiphon circulation
Total developed length
Total Equivalent Length (TEL)
Upfeed system
Vacuum breakers
Wall scuppers
Waterflow switches
Water hammer
Water supply fixture unit (WSFU)
Wet-pipe sprinklers

Supplementary Reading

See listing at the end of Chapter 10.

Problems

1. a. In a domestic hot water system using a circulation pipe, does the circulation return pipe lead the water to the top or bottom of the domestic hot water storage tank?
   b. Why?
2. Make a water supply plan diagram for:
   a. A specific residence.
   b. A public toilet.
   All diagrams must be based on an actual installation that you have inspected and sketched. Omit pipe sizes.
3. Give two reasons for a valve below the fixture in a water supply line to a lavatory.
4. Name three low points in the water system of a residence where you would require a drain.
5. How many water supply fixture units are assigned to the following fixtures in (a) a private house and (b) a public building?
   Bathtub
   Clothes washer
   Kitchen sink
   Lavatory
   Shower head
   Water closet, flush valve
6. What minimum size water supply pipes would you specify for the following?
   a. A lavatory.
   b. A kitchen sink, commercial.
   c. A water closet, flush-valve type.
7. You are planning a public toilet. It will have a total of 120 WSFU. What would be the demand in gallons per minute if you specified the following?
   a. Flush tank water closets.
   b. Flush valve water closets.
8. A type K copper water service pipe will carry 40 gpm. It must be selected for a pressure drop, due to friction, of 4.4 psi/100 ft of total equivalent length. What size would you select?
9. a. What are the advantages and disadvantages of upfeed water systems? downfeed water systems?
   b. Which would you recommend, and why, for (1) a private residence, (2) a 6-story office building, and (3) a 15-story office building?
10. Why should the pressure of hot and cold water at a fixture be (nearly) equal?
11. Describe a system for hot water supply in a six-story office building and explain why you selected that system.
12. A ranch-type residence contains a master bath, a powder room and a main bath. Show all required fixtures, valves and branch pipe sizes. Assuming that each group is fed from a main pipe, show minimum group feed pipe size. Assume that no problem of low pressure exists. Select the number and type of fixtures that you consider appropriate for each space.
13. To the house of Problem 12, add a kitchen, containing a double sink and dishwasher. Show feeds to these items and size the main feed pipe to the house.
14. Rework Example 9.4 using flush tanks instead of flush valves.
15. Calculate the pipe size of all the runouts in the Mogensen house, using Figure 9.26, by the exact method (Example 9.4) and the velocity method (Example 9.5). Compare the results to the sizing shown on Table 9.12. Prepare a detailed plumbing water riser, showing all lengths and flows (in gpm) and all required and recommended valving.
16. Calculate the size of the final runouts for Example 9.9 by the exact method and the velocity method. Compare the results to the sizes actually used.
17. What is a standpipe? How is it used? What is its water source?
18. Describe the operation of a pre-action sprinkler system. In what type of structure is it best applied?
19. Why is it necessary to drain dry-pipe sprinkler systems. (The answer should include all three types of dry-pipe sprinklers.)
20. What is a siamese connection? Why are the siamese connections of standpipes and sprinklers usually separate?

10. Drainage and Wastewater Disposal

In Chapters 8 and 9, we discussed the principles of plumbing systems, plumbing materials and the design of water supply systems. We also demonstrated the accepted techniques for showing the designs on working drawings. We stated previously that the two principal tasks of the plumbing technologist are the supply of potable water and the provision of drainage systems. We use the plural because there are two separate and distinct drainage systems in a building: sanitary and storm. We will learn about both. In addition, we will also discuss private wastewater disposal systems that are used when an adequate sewer system is not available. Study of this chapter, therefore, will enable you to:

1. Understand the general principles underlying sanitary drainage design.
2. Know when to use direct drainage and when to use indirect waste connections.
3. Understand the hydraulics of sanitary drainage and gravity flow.
4. Size drainage piping as required by the drainage load.
5. Lay out and draw complete drainage piping systems with pipe sizes and slopes, as required.
6. Provide drainage system accessories such as cleanouts and interceptors, as required.
7. Understand fixture trap functioning, design and placement.
8. Design complete vent piping for sanitary drainage systems, including sizes of all pipes and all types of vent connections for residential and nonresidential buildings.
9. Calculate the storm drainage requirements of flat roof and sloped roof buildings.
10. Design complete storm drainage systems, including gutters, leaders, conductors, building drains and building sewers.
11. Understand the functioning of the various

types of private sewage treatment installations.
12. Design private sewage treatment systems for different types of buildings with varying soil conditions and water table levels.

Sanitary Drainage Systems

10.1 Sanitary Drainage—General Principles

It has been said that today's water supply is tomorrow's sewage problem. Although this statement is trite, it is true. It reminds us that there are two distinct systems and that "never the twain shall meet." The water system distributes clean potable water. At each fixture, we destroy this cleanliness. The "sanitary" drainage system carries away the contaminated fluids and solids created at the fixture. The drainage system is not really very sanitary. It does, if effective, ensure sanitation for the occupants.

An important precaution, already mentioned in Section 9.14, is that sewage or polluted "effluent" must never be drawn into the potable water system. To prevent this from occurring, the water is usually delivered above the water level of the fixture. The lavatory (washbasin) is an example of this. See Figure 10.1(a). Because of the high position of the water faucet, an accidental suction in the faucet cannot draw polluted water from the basin into the clean water pipes. If the water must be delivered below the water level of a fixture (flush-valve-type water closet, clothes washer and dishwasher), a vacuum breaker is installed as detailed in Section 9.14. This device, placed in the water supply pipes, breaks the suction that would cause backflow of contaminated water or sewage into the water piping.

In a conventional drainage connection, as shown in Figure 10.1(a), the fixture drain is connected directly to the fixture trap. In some instances, however, this type of direct connection could lead to contamination of the water supply. Such a continuous drain connection, as for instance at a roof tank drain [Figure 10.1(b)] or at a hot water heater drain [Figure 10.1(c)] could pollute the water supply if the trap failed. To avoid this, an arrangement called *indirect waste* is used. This setup introduces an air break in the drain line, similar to the air break in the water supply line seen in Figure 10.1(a). This break prevents sewage backflow and, thereby, the possibility of water supply contamination. The most common example of an indirect waste connection is the common household clothes washer. The outlet of these machines is always a hose that dumps into a laundry tub (tray) or service sink. The vertical distance between the hose outlet and the sink drain provides the required air break. The sink or laundry tray must be deep enough to hold the entire output of the machine without the liquid level reaching within a specified distance of the machine outlet. The distance is specified by local Code authorities. This precaution is necessary in the event of a blockage of the sink drain. A similiar air-break drainage arrangement may be used with a dishwasher discharging into a kitchen sink.

Indirect waste connections are also used where several fixtures drain into a single trap, as for instance in a multiple food preparation sink in a restaurant. There, the air break prevents cross-fixture contamination. A complete discussion and description of indirect waste connections is found in Chapter 9 of the Code.

A normal fixture drainage connection has three parts: the drain (before and after the trap), the *fixture trap* and a vent connection. The function of the drain piping is, of course, to carry away the wastewater. The function of the trap is to provide a seal that prevents sewer gases, foul odors, vermin and other unsanitary substances from entering the building via the drainage pipe. The principal function of the vent pipe is to prevent self-siphoning of the trap (and therefore loss of the trap seal). Functioning of the trap and the vent are shown graphically in Figure 10.2.

Refer now to Figure 10.3, which represents a typical plumbing drainage section for a two-story building. Ignoring for the moment the storm drain, which is not part of the sanitary drainage system, we can clearly see the three essential components of the drainage system: drainage piping, traps and vents. Horizontal piping are called *branches;* vertical sections are called *stacks*. Piping carrying effluent from water closets, urinals and bidets *(black water)* are called *soil pipes*. Piping carrying wastewater from other fixtures *(gray water)* are called *waste pipes*. We, therefore, see on this drawing horizontal piping labelled branch soil, waste and vent, and vertical piping labelled soil, waste and vent stacks. Overall, this piping is known as *DWV piping*, which is an abbreviation for *drainage, waste and vent* piping.

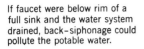
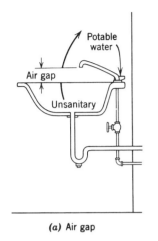
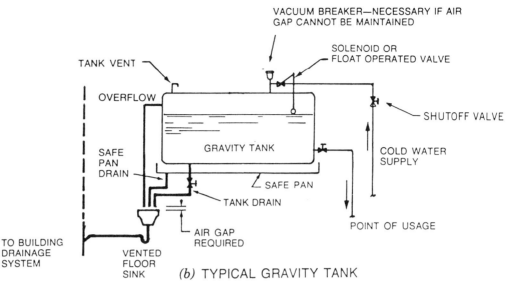
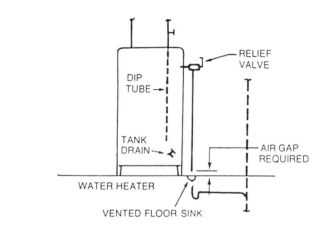

EXAMPLES OF INDIRECT WASTE CONNECTIONS TO
THE POTABLE WATER SUPPLY

NOTE: THE AIR GAP REQUIREMENT ON THESE FIXTURES OR APPLIANCES IS REQUIRED TO
ASSURE PROPER SYSTEM PERFORMANCE WITHOUT THE RISK OF BACKFLOW.

Figure 10.1 *(a)* An air break between the potable water supply and the contaminated water in a fixture is required. This prevents back-siphonage of polluted water into the potable water system. See also Section 9.14. *(b, c)* An air break in the drainage system is required where loss of a trap seal in a direct connected drain could cause sewage backflow into the water system. This type of drain arrangement is called an indirect waste connection. See also Code, Chapter 9, for additional details (Figure 10.1a from B. Stein, J. S. Reynolds, and W. J. McGuinness, *Mechanical and Electrical Equipment for Buildings*, 7th ed., 1986, John Wiley & Sons, New York. Reprinted by permission of John Wiley & Sons; Figures *b* and *c* reprinted with permission from The National Standard Plumbing Code, published by The National Association of Plumbing Heating Cooling Contractors.)

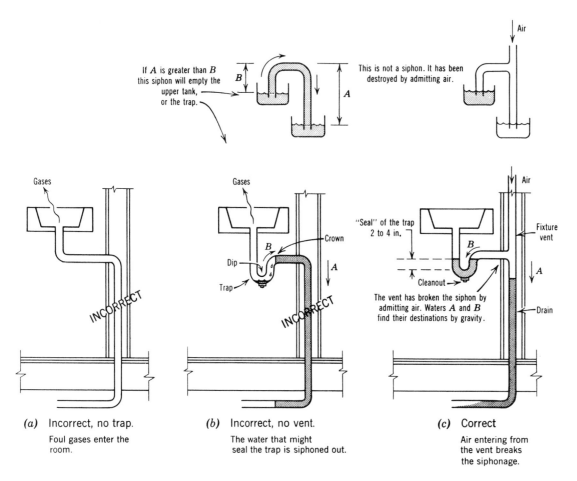

Figure 10.2 (a) An untrapped drainage pipe permits sewer gases and vermin to enter the building via the fixture drain. (b) A trap will self-siphon when the weight of water in the outer leg is greater than that of the inner leg. This would occur, as shown, anytime a slug of water large enough to fill the entire outlet leg is drained from the fixture. The result would be an open drain pipe, as in (a). (c) The vent pipe, which is open to the atmosphere brings air at atmospheric pressure into the drain connection, thus preventing any trap self-siphon. In addition, it introduces fresh air into the drain pipe and permits sewer gases to escape.

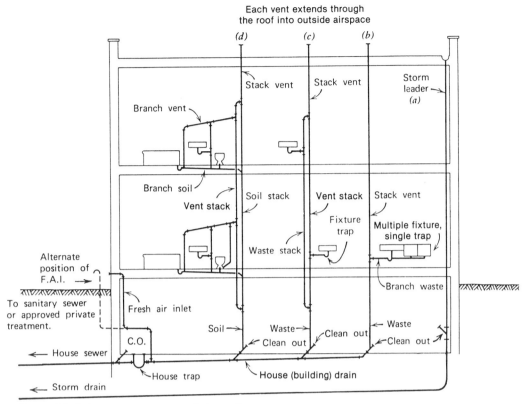

Figure 10.3 Typical sanitary drainage system diagram (plumbing drainage section), showing separation of the sanitary drainage from storm drainage, as is required by most codes. The house trap is generally not used because it is an accumulation point for building effluent solids that can cause stoppages. Also, the modern approach is to use the building vent system to vent the building sewer rather than using a street-level fresh air inlet. Such an inlet could be a source of foul odors at street level. (From B. Stein, J. S. Reynolds, and W. J. McGuinness, *Mechanical and Electrical Equipment for Buildings*, 7th ed., 1986, John Wiley & Sons, New York. Reprinted by permission of John Wiley & Sons.)

10.2 Sanitary Drainage Piping

A sanitary drainage diagram showing the use of cast iron DWV piping is shown in Figure 10.4. Notice that no building trap is used, according to the modern (and Code) approach to design. The basement floor drain is unusual. Its trap is seen just behind the house drain. Floor drains that are not in frequent use can dry out due to evaporation and, thus, open a path for sewer gas and vermin to enter the building. Such traps must either be very deep (4 in. water seal minimum) or else equipped with an automatic priming device that will keep water in the trap constantly. See Code Section 7.16. Diagrams of detailed DWV cast-iron piping are shown in Figure 10.5.

In Figure 10.3 we see that the storm drain is completely separate from the house sanitary drain. This is the preferred arrangement. In many towns that do not have an adequate separate storm sewer system, the drains are combined. This is done only at the specific request of the local plumbing authorities. When it is done, the base of the storm drain must be trapped to prevent sewer gas and other foul odors from entering the storm drain system. Modern design even recommends the separation of waste stack gray water from soil stack black water, where facilities are available for processing and recycling the gray water. When storm drains are combined with sanitary drains, the combined flow during heavy rain storms usually exceeds the capacity of both the sewer and the sewage disposal plant facilities. This leads to overloads, spillage and inadequate processing. As a result, the tendency is to use *dry wells* or adjacent ground absorption for storm water rather than a combined sewer.

10.3 Hydraulics of Gravity Flow

Unlike water piping that flows full in the pipe and under pressure, drainage flows at zero pressure and only partially full. Drainage flow is caused by gravity due to the slope of all drainage piping. As will be explained, drainage piping is deliberately designed to run only partially full; a full pipe, particularly a stack, could blow out or suck out all the trap seals in the system. For a given type of pipe (friction), the variables in drainage flow are pipe slope and depth of liquid. When these two factors are known, the flow velocity V and flow quantity Q can be calculated. Because the calculation is complex, most designers rely on tables and charts for the calculations required in plumbing design. The table normally used by plumbing technologists for horizontal piping is shown in Table 10.1.

The Code requires that horizontal drainage piping be installed with a uniform slope of not less than 1/4 in./ft for pipes 3 in. or less in diameter and not less than 1/8 in./ft for pipes 4 in. in diameter or larger. These minimum slopes result in a liquid velocity of about 2 fps. (See Table 10.1.) This is the minimum velocity recommended to provide the required scouring action of the drainage flow. Below 2 fps, solids such as sand, grit and human wastes will not be carried along with the flow. Instead, they will settle out in the pipe, eventually building up to cause pipe blockages. Greasy effluent requires a minimum velocity of 3 to 4 fps in horizontal piping to provide proper pipe scouring. At lower velocities, the grease tends to stick to the pipe walls, eventually causing blockages. The required higher velocity can be obtained by either increasing pipe slope or pipe size.

To understand this, let us look back at the Code requirement that pipes up to 3 in. be installed at 1/4 in./ft slope minimum and larger pipes at 1/8 in./ft minimum. The reason that large pipes can be installed at a lesser slope than small pipes can be understood if we consider three facts:

- The greater the slope of a pipe, the greater the flow velocity.
- The lower the interior friction of a pipe, the greater the flow velocity.
- Friction in pipe flow is caused by the liquid's drag on the pipe walls.

Large pipes have lower friction than small pipes for the same percentage fill, because the ratio of interior surface to cross-sectional area is lower. This means that for the same percentage fill, flow velocity is higher in a large pipe than in a small one. Therefore, they can be installed at a lesser slope, to achieve the same fluid velocity as a small pipe. This is a particularly useful flow characteristic because it means that the short branch runs are those requiring the 1/4-in./ft slope, whereas the long large building drains require only 1/8-in./ft slope. The difference in elevation between two points on a sloped pipe is called the *"fall"* between those points. Short runs, even at 1/4-in./ft slope, have a small fall and, therefore, usually do not cause

HYDRAULICS OF GRAVITY FLOW / 563

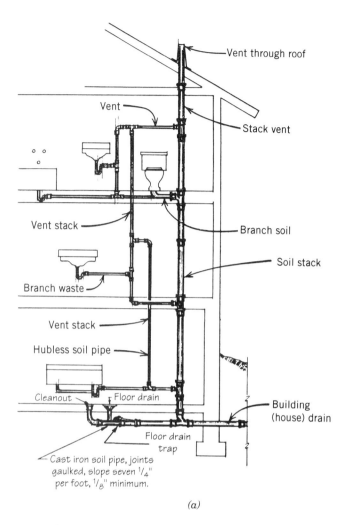

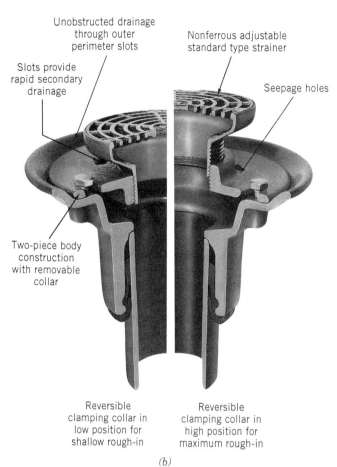

Figure 10.4 *(a)* Typical cast-iron DWV piping for a portion of a drainage system. The piping may be hubless or hub and spigot piping. (Copyright © 1989 by the Cast Iron Soil Pipe Institute, reprinted with permission, labelling added by author.)

Figure 10.4 *(b)* Cast-iron floor drain with flange, integral reversible clamping collar and seepage openings. (Courtesy of Tyler Pipe, subsidiary of Tyler Corporation.)

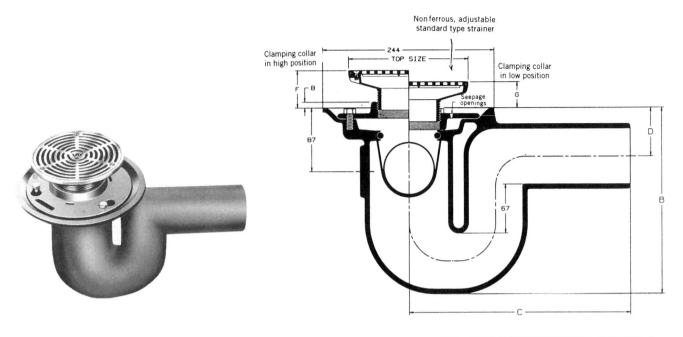

Catalog Number	Pipe Size	Top Size	Wt. Lbs.
W-1120-STD5	50	127	25
W-1120-STD6	**75**	**152**	**30**
W-1120-STD7	100	178	43

(c-1)

Catalog Number	B	C Spigot	D	F Min.	F Max.	G Min.	G Max.
W-1120-STD5	203	279	51	35	54	22	38
W-1120-STD6	254	311	70	41	57	25	41
W-1120-STD7	305	343	76	41	54	25	38

(Dimensions in mm)

OUTLET: Spigot only

(c-2)

Figure 10.4 *(c)* Cast-iron floor drain with integral trap, spigot side outlet, integral clamping collar, seepage openings and nonferrous strainer top. (Courtesy of Tyler Pipe, subsidiary of Tyler Corporation.) See also Figure 8.32, page 467.

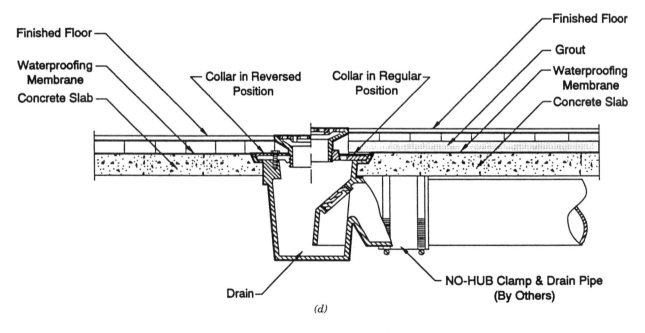

Figure 10.4 *(d)* Installation drawing for a floor drain with integral trap. (Courtesy of Jay R. Smith Manufacturing Company.)

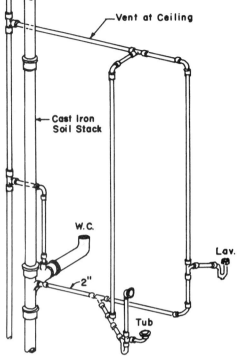
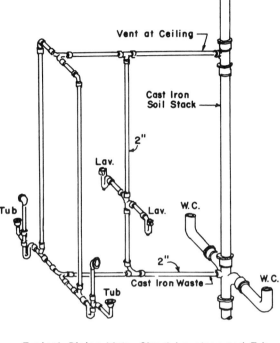

Figure 10.5 Typical piping arrangement for a water closet, lavatory and tub. Piping may be either hubless or hub and spigot piping. (Copyright © 1989 by the Cast Iron Soil Pipe Institute, reprinted with permission.)

installation problems. Long runs, such as building drains, require (fortunately) only 1/8-in./ft slope. If they needed 1/4-in./ft slope as for small pipes, there would not be enough room in many buildings for the required fall! Indeed, very large pipes (8 in. and above) may be installed with only 1/16-in./ft slope, with Code authority permission, for just these reasons. In these pipes, adequate fluid velocity can be maintained even with 1/16-in./ft slope (See Table 10.1), and fall does not become a problem. Summarizing then, we see that, to increase fluid velocity (and flow), we can increase slope or pipe size. For the same flow, a larger pipe requires a smaller slope, and vice versa (a smaller pipe requires a larger slope).

We stated previously that pipes, both horizontal and vertical, must not run full and are sized accordingly. Refer, for instance, to the horizontal drainage shown in Figure 10.3 for the upper floor. As the amount of drain liquid varies in the drain pipe (due to tub drainage as shown there), the amount of air above the liquid in the drain pipe changes. As the liquid increases, the air in the pipe is pushed out, creating a positive pressure. If the pipe were permitted to run full or even three-quarters full, the pressure created could well exceed 1 in. of water, which is the limit permitted by Code. Higher pressure could blow out the sink trap even with the venting installed as required. To avoid this possibility, horizontal branch drains are designed to run at a maximum of 50% fill. Building drains may run at somewhat higher fill.

The preceding discussion has centered on flow in horizontal drainage pipes. The flow in vertical pipes, called *stacks*, is of a much different character. The characteristics of vertical flow in a stack depends on many factors: pipe size, amount of fluid, velocity and direction of the fluid entering the stack, and pipe wall friction (roughness of the pipe wall). All stacks are vented (see Figures 10.3

Table 10.1 Approximate Discharge Rates and Velocities[a] in Sloping Drains Flowing Half Full[b]

Actual Inside Diameter of Pipe, in.	1/16 in./ft Slope		1/8 in./ft Slope		1/4 in./ft Slope		1/2 in./ft Slope	
	Discharge, gpm	Velocity, fps	Discharge, gpm	Velocity, fps	Discharge, gpm	Velocity, fps	Discharge, gpm	Velocity, fps
1 1/4							3.40	1.78
1 3/8					3.13	1.34	4.44	1.90
1 1/2					3.91	1.42	5.53	2.01
1 5/8					4.81	1.50	6.80	2.12
2					8.42	1.72	11.9	2.43
2 1/2			10.8	1.41	15.3	1.99	21.6	2.82
3			17.6	1.59	24.8	2.25	35.1	3.19
4	26.70	1.36	37.8	1.93	53.4	2.73	75.5	3.86
5	48.3	1.58	68.3	2.23	96.6	3.16	137.	4.47
6	78.5	1.78	111.	2.52	157.	3.57	222.	5.04
8	170.	2.17	240.	3.07	340.	4.34	480.	6.13
10	308.	2.52	436.	3.56	616.	5.04	872.	7.12
12	500.	2.83	707.	4.01	999.	5.67	1413	8.02

[a] Computed from the Manning Formula for 1/2-full pipe, $n = 0.015$.
[b] Half full means filled to a depth equal to one-half the inside diameter.

Note: For 1/4 full, multiply discharge by 0.274 and multiply velocity by 0.701. For 1/3 full, multiply discharge by 0.44 and multiply velocity by 0.80. For 3/4 full, multiply discharge by 1.82 and multiply velocity by 1.13. For full, multiply discharge by 2.00 and multiply velocity by 1.00. For smoother pipe, multiply discharge and velocity by 0.015 and divide by n value of smoother pipe.

Source. Reprinted with permission from the National Standard Plumbing Code, Published by The National Association of Plumbing Heating Cooling Contractors.

and 10.4). If this were not so, tremendous turbulence would be created every time a branch drained into a stack, causing violent pressure disturbances throughout the system. These pressure variations would almost certainly empty all the fixture traps.

To understand this action, refer to Figure 10.6. When an unvented bottle (a) is filled, the air in the bottle is compressed until the pressure exceeds the weight of incoming water, at which point it bursts out violently (in bubbles). Placing a straw (vent) in the bottle (b) permits the air to escape without pressure buildup. The same situation, in reverse, occurs when emptying a full bottle. A negative pressure is created inside the bottle, requiring outside entering air to force its way in. This causes turbulence (bubbling) (c). Again a straw (vent) (d) will prevent this turbulence by equalizing pressures. A stack is equivalent to the bottle in Figure 10.6. It must be of sufficient size to run only partially full so that air can escape through the vent or enter via the vent, as required to equalize pressures. If the stack were to run full, then the vent above would be ineffective. As a result, stacks are designed to run between one-quarter and one-third full.

The actual flow in a stack takes various forms. Initially, there are streams down the sides and diaphragms across the pipe for short periods. After only a short drop of 10–20 ft, the wastewater forms a sheet around the inside of the pipe, traveling at a maximum velocity of about 15 fps. See Figure 10.7. The terminal water velocity in a 3-story high stack is the same as that in a 50-story stack. The myth warning of the danger to piping from high velocity falling water is completely false.

When the water reaches the base of the stack, it must turn to again flow horizontally. The flow

HYDRAULICS OF GRAVITY FLOW / 567

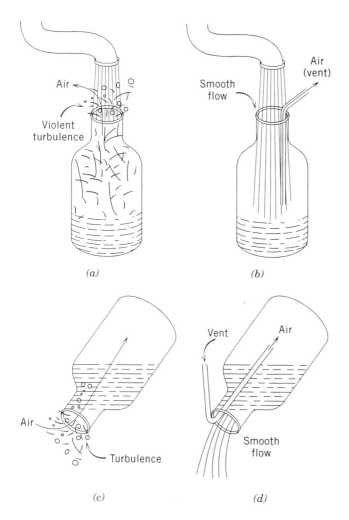

Figure 10.6 Whether a container is being filled *(a)* or emptied *(c)* unless a path for air movement is provided as in *(b)* and *(d)*, the process will be turbulent with sharp pressure variations. With an air vent *(b, d)* pressure inside and outside the container is equalized continuously permitting smooth laminar flow.

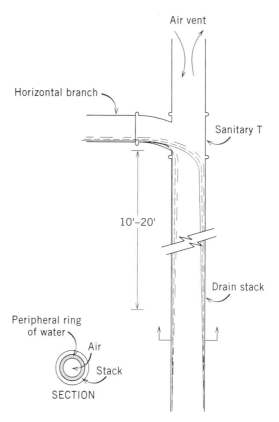

Figure 10.7 Wastewater entering a stack from a horizontal branch will quickly form a peripheral ring around the stack walls and fall at a constant velocity of about 15 fps.

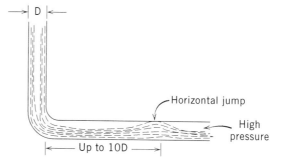

Figure 10.8 At the base of a stack with a sharp 90° bend, wastewater undergoes a rapid change in velocity. Within 10 stack pipe diameters, a horizontal jump occurs in which the water piles up causing large pressure variations. This condition can be avoided by using long radius elbows, larger horizontal drains and additional vents.

velocity in the horizontal pipe, some distance from the stack connection, is controlled only by pipe slope. It is obviously lower than that of the fluid falling in the stack. Therefore, the high velocity water entering the horizontal drain must slow down. For instance, terminal velocity in a 4-in. stack is about 15 fps, while flow velocity in a 4-in. pipe even at a slope of ½ in./ft is only 3.86 fps. In order to adjust to the lower velocity, the pipe fill increases sharply, creating what is known as a *horizontal jump* (see Figure 10.8). The increased fill continues until turbulence and pipe friction smooth out the flow somewhat farther downstream. (Any stack offset in excess of 45° can cause horizontal jump.) To minimize this condition, which causes severe pressure variations, long radius bends are used whenever possible. In addition, the size of the horizontal drain may be increased one diameter, and additional venting may be added.

10.4 Drainage Piping Sizing

Required drainage pipe sizes are calculated by using a concept with which we are already familiar from our study of water supply sizing—the concept of fixture units. The idea is the same. Instead of using gpm of drainage water, we use *drainage fixture units (dfu)*. This unit takes into account not only the fixture's water use but also its frequency of use, that is, the dfu has a built-in diversity factor. This enables us, exactly as for water supply, to add the dfu of various fixtures to obtain the maximum expected drainage flow. Drainage pipes are then sized for a particular number of drainage fixture units, according to tables in the Code. Built into these tables are the fill factors that we discussed previously: one-quarter to one-third fill in stacks and one-half fill in horizontal branches and somewhat more in building drains.

The design procedure follows.

Step 1. Draw an isometric of the entire system showing all fixtures. (Some engineering offices have abandoned the use of isometric drawings except for complex systems where a simple riser would be inadequate.)

Step 2. Assign drainage fixture units to each fixture according to Table 10.2. If a fixture is not listed specifically, base the dfu requirement on its trap size. Minimum fixture trap sizes are listed in Table 10.3. With respect to drainage requirements not due to fixtures, such as nonrecirculated cooling water or process water, use the conversion of 1 gpm = 2 dfu. Thus, for instance, 2 gpm of process water at any drain should be counted as 4 dfu, and so forth.

Step 3. Total the drainage fixture units in each drainage pipe and mark them on the drawing.

Step 4. Determine the required size of horizontal fixture branches and stacks from Table 10.4.

Step 5. Determine the size and slope of the building drain and its branches, and the building sewer using Table 10.5.

Step 6. Determine that the size and slope found in Step 5 meet the requirements of the Code and of Table 10.1.

This procedure will now be applied to a large building. You should follow the discussion with Figure 10.9 in front of you.

Example 10.1 Size all the drainage piping for the building whose drainage piping layout is shown in Figure 10.9.

Solution:
Steps 1 and 2. Prepare the isometric drawing and assign dfu values.

Since a drawing showing all fixtures would be much larger than could possibly be shown on a book page, we have shown only a single horizontal fixture branch with its fixtures. From Table 10.2, we obtain the drainage fixture units for each fixture on this branch *(D)* and these are shown on the drawing: 6 dfu for each general-use water closet and 1 dfu for each lavatory with a 1¼-in. trap (see bottom of Table 10.2). The total for the branch is 14 dfu.

Step 3. Summation of drainage fixture units.

In a similar fashion, fixture units were totalled for all the horizontal fixture branches and are indicated on each horizontal branch. The drainage fixture units were then added as we progressed down each stack into the branch building drains and finally into the building drain.

Step 4. Size the horizontal fixture branches and stacks.

Refer to Table 10.4. In order to apply Table 10.4 properly, we must understand the Code nomenclature for all the different parts of the system. For this purpose, we have added letters to various pipes shown on the drawing and a detail defining branch interval. A *branch interval* by Code definition is "A distance along a soil or waste stack corresponding in general to a story height, but in no case less than 8 feet, within which the horizontal branches of one story or floor of a building are connected to the stack." Applying this definition to the drawing we can proceed with the pipe sizing.

Stack A: Refer to Table 10.4. Since this stack has more than three branch intervals, we will use columns 1, 4 and 5 of Table 10.4. Column 5 applies to horizontal branches, and column 4, to the stack. Beginning at the top, we have, for the first horizontal branch, 6 dfu = 2 in. Proceeding down, the first stack section also has 6 dfu, which from column 4 is only 1½ in. However, the stack size cannot

Table 10.2 Drainage Fixture Unit Values for Various Plumbing Fixtures

Type of Fixture or Group of Fixtures	Drainage Fixture Unit Value, dfu
Automatic clothes washer (2-in. standpipe and trap required, direct connection)	3
Bathtub group consisting of a water closet; lavatory and bathtub or shower stall:	6
Bathtub (with or without overhead shower)[a]	2
Bidet	1
Clinic sink	6
Clothes washer	2
Combination sink-and-tray with food waste grinder	4
Combination sink-and-tray with one 1-in. trap	2
Combination sink-and-tray with separate 1-in. trap	3
Dental unit or cuspidor	1
Dental lavatory	1
Drinking fountain	1/2
Dishwasher, domestic	2
Floor drains with 2-in. waste	3
Kitchen sink, domestic, with one 1-in. trap	2
Kitchen sink, domestic, with food waste grinder	2
Kitchen sink, domestic, with food waste grinder and dishwasher 1-in. trap	3
Kitchen sink, domestic, with dishwasher 1-in trap	3
Lavatory with 1-in. waste	1
Laundry tray (1 or 2 compartments)	2
Shower stall, domestic	2
Showers (group) per head	2
Sinks	
surgeon's	3
flushing rim (with valve)	6
service (trap standard)	3
service (P trap)	2
pot, scullery, etc.	4
Urinal, syphon jet blowout	6
Urinal, wall lip	4
Wash sink (circular or multiple) each set of faucets	2
Water closet, private	4
Water closet, general use	6
Fixtures not already listed	
trap size 1 1/4 in. or less	1
trap size 1 1/2 in.	2
trap size 2 in.	3
trap size 2 1/2 in.	4
trap size 3 in.	5
trap size 4 in.	6

[a] A shower head over a bathtub does not increase the fixture unit value.

Source. Reprinted with permission from The National Standard Plumbing Code, Published by The National Association of Plumbing Heating Cooling Contractors.

Table 10.3 Minimum Size of Nonintegral Traps

Plumbing Fixture	Trap Size, in.
Bathtub (with or without overhead shower)	1 1/2
Bidet	1 1/4
Clothes washing machine standpipe	2
Combination sink and wash (laundry) tray	1 1/2
Combination sink and wash (laundry) tray with food waste grinder unit[a]	1 1/2
Combination kitchen sink, domestic, dishwasher, and food waste grinder	1 1/2
Dental unit or cuspidor	1 1/4
Dental lavatory	1 1/4
Drinking fountain	1 1/4
Dishwasher, commercial	2
Dishwasher, domestic (nonintegral trap)	1 1/2
Floor drain	2
Food waste grinder, commercial	2
Food waster grinder, domestic	1 1/2
Kitchen sink, domestic, with food waste grinder unit	1 1/2
Kitchen sink, domestic	1 1/2
Kitchen sink, domestic, with dishwasher	1 1/2
Lavatory, common	1 1/4
Lavatory (barber shop, beauty parlor or surgeon's)	1 1/2
Lavatory, multiple type (wash fountain or wash sink)	1 1/2
Laundry tray (1 or 2 compartments)	1 1/2
Shower stall or drain	2
Sink (surgeon's)	1 1/2
Sink flushing rim type (flush valve supplied)	3
Sink (service type with floor outlet trap standard)	3
Sink (service trap with P trap)	2
Sink, commercial (pot, scullery, or similar type)	2
Sink, commercial (with food grinder unit)	2

[a] Separate trap required for wash tray and separate trap required for sink compartment with food waste grinder unit.

Source. Reprinted with permission from The National Standard Plumbing Code, published by The National Association of Plumbing Heating Cooling Contractors.

Table 10.4 Horizontal Fixture Branches and Stacks

Diameter of Pipe, in.	Maximum Number of Fixture Units That May Be Connected to			
	Any Horizontal Fixture Branch,[a] dfu	One Stack of Three Branch Intervals or Less, dfu	Stacks with More Than Three Branch Intervals	
			Total for Stack, dfu	Total at One Branch Interval, dfu
1½	3	4	8	2
2	6	10	24	6
2½	12	20	42	9
3	20[b]	48[b]	72[b]	20[b]
4	160	240	500	90
5	360	540	1100	200
6	620	960	1900	350
8	1400	2200	3600	600
10	2500	3800	5600	1000
12	3900	6000	8400	1500
15	7000			

[a] Does not include branches of the building drain.
[b] Not more than two water closets or bathroom groups within each branch interval nor more than six water closets or bathroom groups on the stack.

Note: Stacks shall be sized according to the total accumulated connected load at each story or branch interval and may be reduced in size as this load decreases to a minimum diameter of half of the largest size required.

Source. Reprinted with permission of The National Standard Plumbing Code, published by The National Association of Plumbing Heating Cooling Contractors.

Table 10.5 Building Drains and Sewers[a]

Diameter of Pipe, in.	Maximum Number of Fixture Units That May Be Connected to Any Portion of the Building Drain or the Building Sewer			
	Slope per Foot			
	1/16 in.	1/8 in.	1/4 in.	1/2 in.
2			21	26
2½			24	31
3			42[b]	50[b]
4		180	216	250
5		390	480	575
6		700	840	1000
8	1400	1600	1920	2300
10	2500	2900	3500	4200
12	2900	4600	5600	6700
15	7000	8300	10,000	12,000

[a] On site sewers that serve more than one building may be sized according to the current standards and specifications of the Administrative Authority for public sewers.
[b] Not over two water closets or two bathroom groups, except that in single family dwellings, not over three water closets or three bathroom groups may be installed.

Source. Reprinted with permission from The National Standard Plumbing Code, published by The National Association of Plumbing Heating Cooling Contractors.

DRAINAGE PIPING SIZING / 571

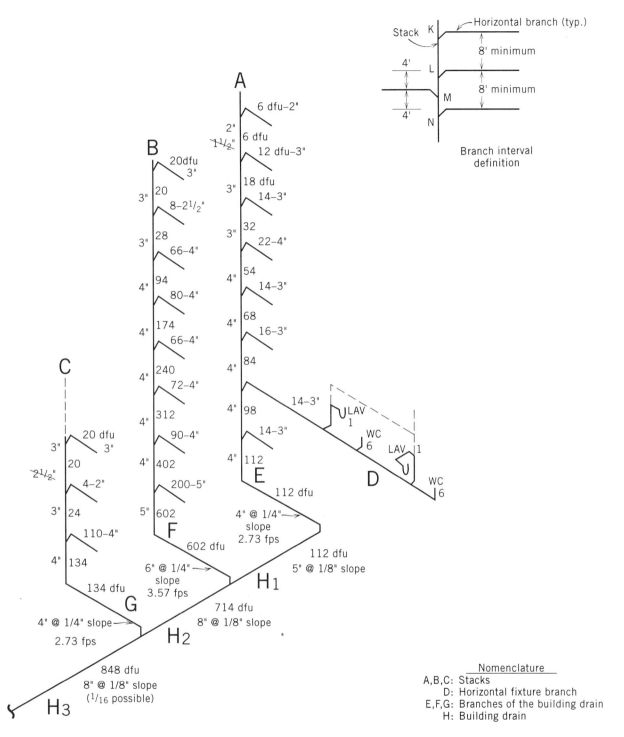

Figure 10.9 Isometric of the drainage piping of a large building showing fixture units and pipe sizing. Horizontal branches and stacks are sized by Table 10.4. Building drains (including branches) are sized by Table 10.5. The inset shows the definition of a branch interval. K-L and L-N are branch intervals; L-M and M-N are not. See text for full discussion of pipe sizing.

be smaller than the horizontal branch size, so we use 2 in. Also, the smallest size permissible is half the largest size in a stack (see Note of Table 10.4). Since the bottom of the stack is 4 in., the top cannot be less than 2 in. The next horizontal branch is 12 dfu = 3 in. from column 5. The stack must also be 3 in., even though, from column 4, 2 in. is sufficient. We proceed down in this fashion. Note that the fifth branch interval down has only 68 dfu. From column 4 this would be 3 in. However, because the preceding section is already 4 in. (because of its horizontal branch), we may not reduce the stack size as we proceed down and, therefore, we use 4 in.

Section E is a branch of the building drain. It carries 112 dfu. From Table 10.5, we see that a 4-in. pipe at a slope of 1/8 in. is satisfactory. It also meets the Code requirement that horizontal pipes larger than 3 in. have a slope of at least 1/8 in./ft. However, Table 10.1 shows that the discharge velocity of a 4-in. pipe at a 1/8-in./ft slope is only 1.93 fps. This is not sufficient. We, therefore, would use a 1/4-in./ft slope, which gives a velocity of 2.73 fps. The rule of thumb to follow with respect to drainage velocities is to try for 2 fps in branch pipes and 3 fps in building drains and sewers. Too low a velocity caused by undersized pipe or insufficient slope does not scour and causes buildup of solids and blockages. Excessive velocity caused by oversized pipe or excessive slope causes noise and fouling.

Stack B: Pipes are sized using the same consideration as explained for Stack A. The principles to remember are:

- The stack cannot be smaller than its horizontal branch.
- The stack size cannot be reduced as it descends from floor to floor.
- The maximum ratio of largest to smallest section of a stack is 2:1.

Stack C: This stack has only three branch intervals. We, therefore, use columns 1–3 of Table 10.4 (column 2 for the horizontal branch and column 3 for the stack section). The three sizing principles already stated continue to apply. Therefore, the top stack section must be 3 in. because the horizontal branch is 3 in. The remainder of the procedure is straightforward. Check out each branch with Table 10.4 to familiarize yourself with the procedure.

The preceding calculation procedure establishes the minimum stack sizes permitted for a given fixture load. In practice, many designers use the maximum size of any portion of the stack for its entire length; that is, they avoid using a tapered stack. The reasoning behind this practice is simply that, once installed, stacks are very difficult to enlarge. Therefore, any appreciable load increase might result in a major construction expense.

Building Drain: This pipe is marked H. The dfu in its three sections are indicated on the drawing. Each section is sized using Table 10.5. Although the first section H_1 could continue the branch drain size and slope (4 in. @ 1/4 in./ft), it might cause a problem due to excessive fall. Also, the remainder of the drain is sloped at 1/8 in./ft. We would, therefore, select a 5-in. pipe with a slope of 1/8 in./ft for section H_1. Section H_2 with 714 dfu requires an 8-in. pipe at 1/8 in./ft slope. Section H_3 also requires an 8-in. pipe. Although a 1/16-in./ft slope is possible, it would create a low point in the building drain that could cause blockages. We would recommend a continuation of the 8-in. pipe at a slope of 1/8 in./ft.

A number of special considerations in sizing of drainage piping exist. When a stack is offset someplace in its rise rather than going straight up, the Code provides special rules. Offsets less than 45° are not considered; that is, the stack is considered to be vertical. For offsets greater than 45°, refer to Section 11.6.5 of the Code.

10.5 Drainage Accessories

In addition to fixture drains, traps and piping, there are a number of devices, connections and accessories to drainage systems with which you should be familiar. They are surveyed briefly next.

a. Cleanouts

See Figure 10.10. No matter how well designed, a drainage system will eventually be blocked. Blockage occurs most frequently at points where drainage pipes change direction and size, and at fittings. Therefore, *cleanout fittings* should be provided:

- At the base of every soil and waste stack.
- At all changes of direction larger than 45°.
- At the point where the building drain exits the building.
- Along all horizontal runs at a frequency of not more than 75 ft apart for drainage pipe 4 in. in

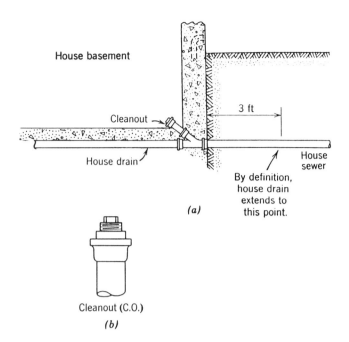

Figure 10.10 *(a)* Location of cleanout at exit of building drain. The cleanout is placed in a wye as shown. *(b)* Detail of cleanout fitting shows threaded removal plug, which allows access to the drain pipe. (From B. Stein, J. S. Reynolds, *Mechanical and Electrical Equipment for Buildings*, 8th ed., 1992, John Wiley & Sons, New York. Reprinted by permission of John Wiley & Sons.)

diameter or less and not more than 100 ft apart for larger pipes.
- Wherever the designer feels that there is a possibility of soil buildup and blockage.

All cleanouts must be accessible, with enough space around the cleanout to manipulate the cleaning equipment.

b. Interceptors

When wastewater from fixtures contains materials that can cause blockages or harm the building drainage system, public sewer or treatment plant, *interceptors* (separators) are used. Harmful materials include sand, grease, oil, hair, gravel and flammable liquids. Because interceptors require periodic cleaning, they must be completely accessible. Grease interceptors are used in commercial kitchens because oil and grease from cooking processes easily solidify inside drain pipes causing blockages. Interceptors are located so that only the wastewater requiring processing discharges through the interceptor. A typical grease interceptor, which is one of the most common types used, is shown in Figure 10.11.

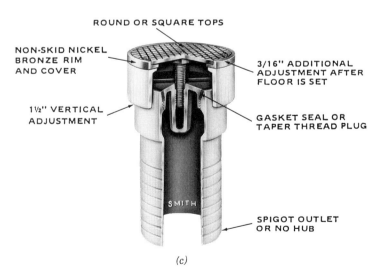

Figure 10.10 *(c)* Cutaway view of a floor-level adjustable cleanout fitting, applicable to finished floors with foot and vehicular traffic. (Courtesy of Jay R. Smith Manufacturing Company.)

574 / DRAINAGE AND WASTEWATER DISPOSAL

(a)

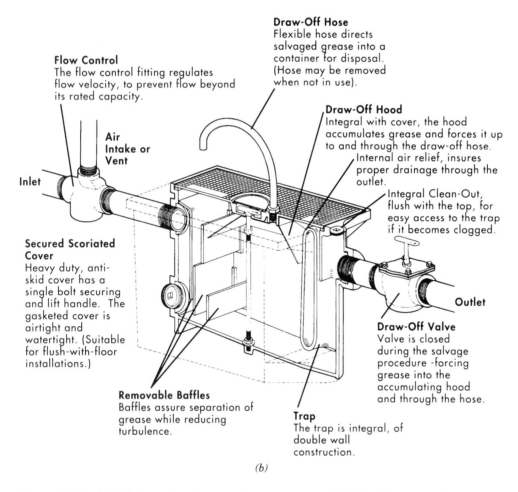
(b)

Figure 10.11 (a) Fabricated steel grease interceptor, available for flow rates of 4–50 gpm. (b) The interior construction of the unit shows the removable baffles, trap and draw-off connections. The inlet and outlet connections are also clearly indicated. (Courtesy of Tyler Pipe, subsidiary of Tyler Corporation.)

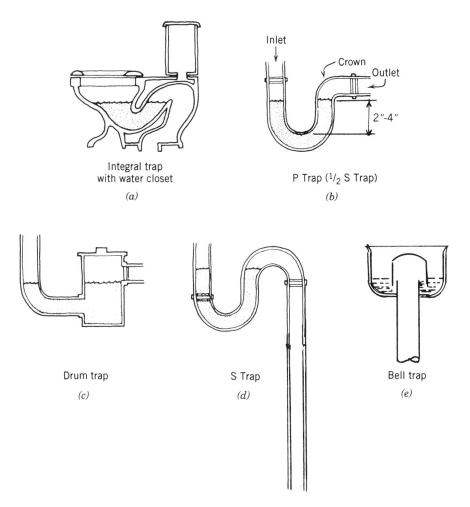

Figure 10.12 Various types of fixture traps. Water closets *(a)* have integral traps. Other fixtures *(b)* normally use the P (½ S) trap. The drum trap *(c)* is prohibited because it is not self-cleaning. The full S trap *(d)* is not permitted because it may self-siphon if the outlet vertical leg fills with waste. The bell trap *(e)* is prohibited because it fouls easily. (Diagrams *a–d* from Bradshaw, *Building Control Systems*, 1985, reprinted by permission of John Wiley & Sons.)

10.6 Traps

In Section 10.1 and in Figure 10.2, we touched on the function of fixture traps. As stated there, the principal function of a trap is to provide a water seal between the drainage piping that connects to the outside sewer and the fixture. This water seal prevents entry into the building of odors, sewer gas and vermin from the sewer, via the fixture. Every plumbing fixture must be trapped, except for a few very special cases specifically referenced in the Code. These include fixtures with indirect (air-gap) waste connections and certain fixtures that discharge through interceptors. The design of the trap, its location and the drainage piping connected to it are each important in its proper functioning.

Refer to Figure 10.12. The only fixture in the illustration that is self-trapped is the water closet, Figure 10.12(a). All traps operate on the principle of siphonage. As water is added to the inlet end, an equal quantity of water leaves the outlet end, provided the pressures at both ends are approximately equal. The Code permits a maximum pressure difference between inlet and outlet of ±1 in. Let us examine this number to understand its derivation. The usual trap depth is from 2 to 4 in. Assuming the minimum depth for our calculation (since this is the easiest trap to blow out or siphon out), we will calculate the pressure required to blow out or siphon out the trap. Refer to Figure 10.13. We see that to destroy the trap seal all we need do is to lift either side of the water seal by 2 in. That means that a pressure that will support 2 in. of water in the trap is the critical pressure. That pressure is obviously 2 in. of water. Therefore, a permissible pressure of ±1 in. gives a safety factor of 2. It follows, of course, that a 3-in. deep trap would require a pressure of 3 in. to blow it out, and so forth.

To appreciate how small a pressure of 2 in. of water is, we can convert it to pounds per square inch:

$$2 \, \cancel{\text{in}} \times \frac{1 \, \cancel{\text{ft}}}{12 \, \cancel{\text{in}}} \times \frac{0.433 \text{ psi}}{\cancel{\text{ft}}} = .0722 \text{ psi}$$

That is less than one-tenth of a psi, or just over one ounce per square inch pressure! These figures hold regardless of the trap pipe size. Only the depth of the trap controls the pressure necessary to destroy the trap seal. This should make very obvious the need for effective venting to equalize pressures on both sides of the trap seal. The question is often asked, If traps are so sensitive to pressure, why not make them deeper? The answer is that a deep trap will not self-scour properly; it will retain foreign bodies and very soon become blocked. That is the reason that the Code limits normal trap depth to 4 in. Most traps are between 2 and 3 in. deep.

Refer again to Figure 10.12. Only the *P trap* is acceptable by Code. Traps must be self-scouring, that is, self-cleansing. That means that all the polluted water that enters the inlet, with all the suspended particles of soap, dirt, waste and the like, must travel completely through the trap, leaving a seal of clean water. The drum trap has a tendency to collect material and will not self-clean. The bell trap and traps with moving parts tend to foul easily. The *S trap* will self-siphon as soon as the outlet leg fills with water. These traps are, therefore, prohibited.

A standard P trap may not be installed more than 24 in. below the fixture drain because the momentum of water falling from a greater height might destroy the trap seal by simply pushing all the water out of the trap. The length of the trap arm may not exceed that shown in Table 10.6. See Figure 10.14. The reason for this limitation is to prevent self-siphoning due to sloping of the trap arm to a point below the weir level of the trap. The trap would then self-siphon exactly like a full S trap. The limited trap arm length also ensures the adequate air movement that is required for proper venting and pressure equalizing.

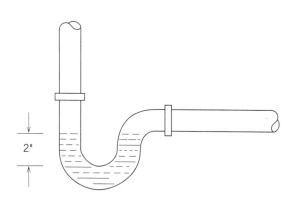

Water in trap at rest; trap seal intact

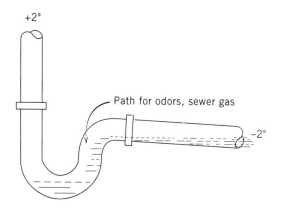

+2" Pressure at inlet or −2" pressure at outlet can break trap seal

Figure 10.13 A negative pressure of 2 in. of water, which is the equivalent of only 1.1 oz/in.² is enough to destroy a trap seal. The Code permits a maximum pressure of ±1 in. of water.

Table 10.6 Maximum Length of Trap Arm

Diameter of Trap Arm, in.	Distance—Trap to Vent, ft
1¼	3.5
1½	5
2	8
3	10
4	12

Note: This table has been expanded in the "Length" requirements to reflect expanded application of the wet venting principles. See Section 10.8 d. Slope shall not exceed ¼ in./ft.

Source. Reprinted with permission from The National Standard Plumbing Code, published by The National Association of Plumbing Heating Cooling Contractors.

A fixture trap should be the same pipe size as the waste pipe to which it is connected. All traps must be accessible for cleaning and must have a cleanout plug, because sooner or later all traps need maintenance. Traps are discussed in Chapter 5 of the Code.

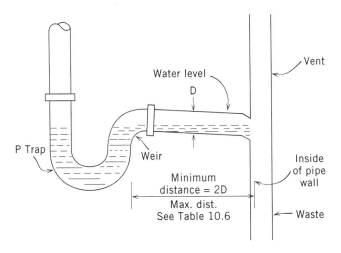

Figure 10.14 The maximum distance between the weir of a trap and the inside wall of the vent pipe to which it connects is specified by Code. See Table 10.6. The trap arm is sloped towards the waste (vent) pipe. It must never be so long that the flow of water (dotted horizontal line) will block the vent pipe. That is, the top of the vent connection must be above the trap weir. The minimum trap arm length, also specified by Code, is two pipe diameters.

Venting

10.7 Principles of Venting

The purpose of venting each fixture trap should be fairly apparent at this point. It is useful, however, to review and summarize the purpose and functioning of vent piping. First, it must be emphasized that, as shown in Figure 10.3, every vent extends through the roof into outside air. This is true for a *vent stack* or a vent extension of a soil/waste stack *(stack vent)*. The stack always extends into fresh air so that it can supply or exhaust air, as required by the flow of waste in the drain piping. See Figure 10.15. Venting performs the following functions:

- It provides an air vent at each fixture trap. This ensures atmospheric pressure on the outlet side of the fixture trap. This, in turn, prevents the trap seal from being blown out or sucked out by pressures generated by drainage flow.
- It provides a safe path to exhaust sewer gases and foul odors that come from the sewer connection via the drainage piping. Building vent piping acts as a sewer vent in the absence (as now recommended) of a building trap and a street level fresh air vent.
- It fills the drainage piping with fresh air, thus reducing odors, corrosion and the formation of slime in the piping.
- It aids in the smooth flow of drainage that occurs when air moves freely in a drain pipe, as was explained in Section 10.3.

In some (rare) instances, a vent to outside air is extremely difficult to install. This might be the case, for instance, at a food concession (containing a sink) located in an open indoor space, such as a sports arena or an airline terminal where a wet vent is not available. See Section 10.8(c). In such cases, the plumbing authorities may permit the use of an indoor vent, terminating in a special vent cap. This cap acts only to equalize pressure, in a manner similar to the action of a vacuum breaker.

The pressure variations in drainage piping are very complicated. A rigid analysis to determine exactly the best place to put vents and how to size them is not practical. As a result, plumbing code officials recommend certain procedures that experience has shown will adequately supply the required venting. Some of these procedures follow.

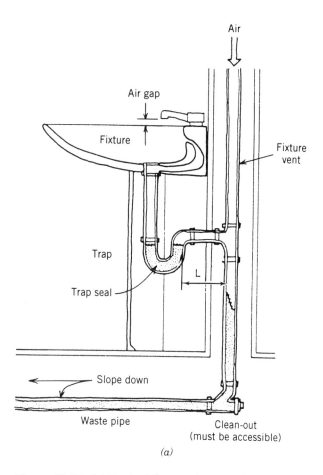

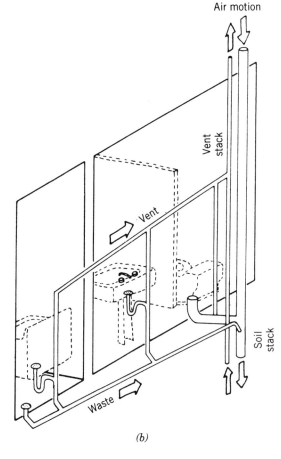

Figure 10.15 (a) Typical fixture drain and vent connection. Trap arm length L may not be less than twice the pipe diameter nor more than the distance shown in Table 10.6. This fixture vent is called a continuous vent, an individual vent or a back vent. See text for an explanation of these terms. (From Bradshaw, *Building Control System*, 1985, reprinted by permission of John Wiley & Sons.)

Figure 10.15 (b) Typical waste and vent connections for a bathroom group. Note the slope of the waste and vent piping. Air is dragged into the waste pipe by the wastewater motion and simultaneously expelled from the vent pipe to equalize pressures. (From Bradshaw, *Building Control System*, 1985, reprinted by permission of John Wiley & Sons.)

(a) Every drainage stack must extend through the roof to fresh air. The size that extends through the roof may not be reduced from that at the top of the stack. The section above the highest horizontal drain connection is called the *vent extension* or more commonly the *stack vent* because it is the vent portion of the soil stack. See Figure 10.16 (a) and (b). The purpose of the extension to open air is to permit free flow of air into and out of the drainage (soil) stack.

(b) Every soil or waste stack more than one story high must have a parallel vent stack. See Figures 10.3, 10.15(b) and 10.16. The purpose of this vent stack is to permit rapid pressure equalization at the lower portions of the drainage stack. This avoids pressures greater than ± 1 in. water column from developing.

(c) The lower end connection of the vent stack to the drain stack must be below the lowest horizontal branch drain connection. The best connection location is just above the point where the drainage stack connects to the horizontal building drain. See Figure 10.16(a). This is the point where maximum pressure is built up due to a rapid change in direction and velocity of the falling water. This is also where horizontal jump occurs if the pressure is not relieved, as explained in Section 10.3. An alter-

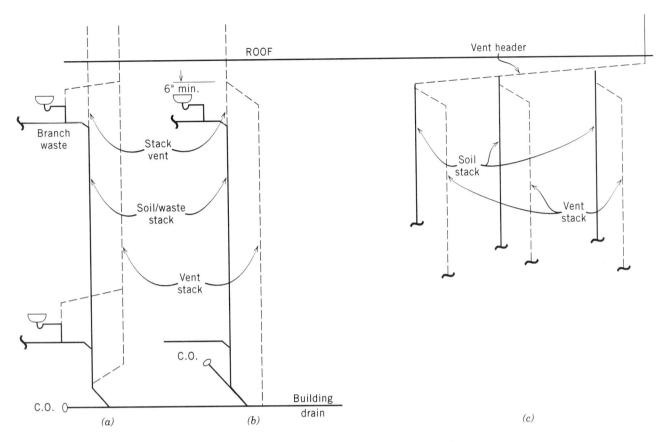

Figure 10.16 All stacks are vented to outside air. The soil, waste and vent stacks can be *(a)* extended into outside air individually, *(b)* connected to each other or *(c)* connected into a vent header. The vent stack connects to the soil waste stack at its base *(a)* or at the joint with the building drain *(b)*.

nate location is at the top of the horizontal building drain, immediately adjacent to the base of the waste stack's base fitting. See Figure 10.16(b).

(d) The vent stack may extend through the roof parallel to the soil stack extension (stack vent) [Figure 10.16(a)]. In practice it is almost always connected to the stack vent just below the roof, at least 6 in. above the flood level of the highest fixture connected to the waste stack [Figure 10.16(b)]. This prevents flooding of vent pipes due to drain stoppage or even normal heavy drainage flow. Another alternative is to connect the top of the stacks into a vent header, in buildings with multiple stacks. See Figure 10.16(c), which shows all the stacks connected to the header.

(e) Because outside vent terminals carry noxious odors and gases, they may not be closer than 10 ft. horizontally from any door, window or air intake unless they are at least 2 ft above the top of the opening. These terminals should terminate 12 in. above a pitched roof and at least 7 ft above a flat roof that can be used for other purposes. Terminals must never exit below building overhangs or any other construction that can block the dispersal of odors or entry of air.

(f) In climates where freezing of the moist warm air exiting a vent terminal is possible, the exposed portion must be kept as short as possible. Also, because freezing tends to block small pipes, such vents should not be smaller than 3 in. In practice, 4 in. is the size commonly used.

(g) All horizontal vent piping must slope to the drainage system. Low points are not permitted, as moisture will condense there to water, and the vent pipe will be partially or fully blocked.

10.8 Types of Vents
a. Individual Vent

The simplest, most direct, most effective (and most expensive) way of venting a fixture trap is to provide an individual vent for every trap. This vent arrangement is also called continuous venting and back venting. It is called *continuous venting* because the vent is a continuation of the drain to which it connects, as seen in Figure 10.17(a). It is called a *back vent* because the vent pipe extends up behind the fixture, and it is called an *individual vent* because there is one for each fixture.

b. Branch Vent

A *branch vent* is a vent connecting one or more individual vents to a vent stack (Figure 10.17a) or a stack vent.

c. Common Vent

A *common vent* is a single vent that connects to a common drain for back-to-back fixtures, as shown in Figure 10.17(a) between the two lavatories. For a common vent for fixture drains at different levels, see Code Section 12.9.2.

d. Circuit and Loop Vents

There is a tendency to call all circuit-type vents by that name. The Code, however, differentiates between them. A *circuit vent* is a branch vent that extends from the downstream side of the last fixture connection on a horizontal fixture branch drain serving a battery of floor outlets. See Figure 10.17(b). It connects to a vent stack. Floor outlets include water closets (except blowout type), tubs, showers and floor drains. A *loop vent* is a circuit vent that loops back and connects to a stack vent [see Figure 10.17(c)] instead of a vent stack. Since the stack vent is the extension of a soil stack above the highest horizontal drain, the loop vent almost always occurs on the top floor of a building. Both the circuit vent and the loop vent were developed to reduce the expense of individual venting. It has been found by experience that both types of circuit vent do an adequate venting job for floor fixture batteries of up to eight fixtures, connected in battery on a single horizontal fixture branch drain. Blowout water closets may not be included in the group. If a wall-type fixture such as a lavatory is included in the battery, it must be individually vented. Circuit vents also require a relief vent taken off in front of the first fixture connection in the battery, as shown in Figure 10.17(b). A *relief vent* is simply an additional vent pipe used where additional air circulation between the drainage and vent systems is required.

e. Wet Vents

See Figure 10.18. A *wet vent* is a vent that not only vents a particular fixture but also serves as a waste line for other fixtures, except water closets and kitchen sinks. Very specific Code rules govern wet vents but vary depending on whether the wet vent is on the top floor or one of a building's lower

TYPES OF VENTS / 581

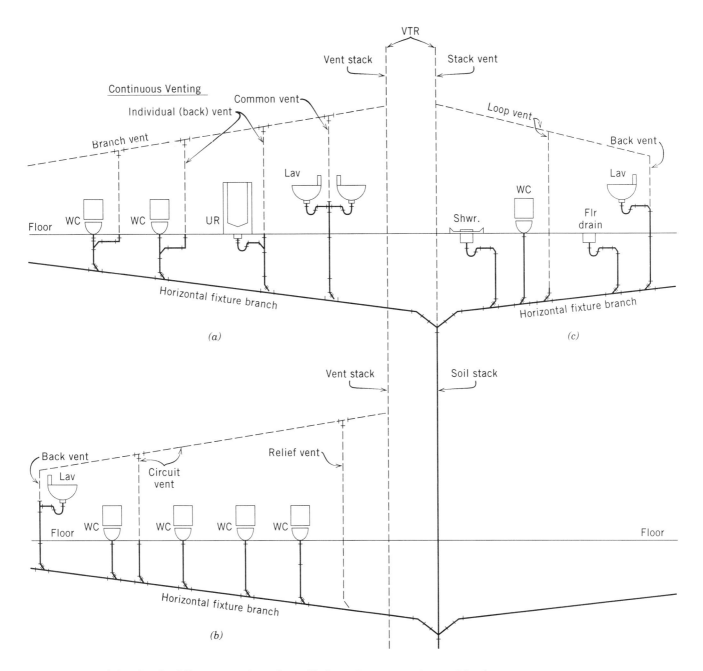

Figure 10.17 *(a)* Individual fixture venting, also called continuous venting and back venting. In certain conditions, it is also called reventing. The drain for each fixture is each connected to a branch vent. Two fixtures that use a common drain, as the two lavatories shown, may also use a common vent. All fixtures drain into a horizontal fixture branch. *(b)* Floor outlets, including tubs, showers, floor drains and most water closets, that connect in battery to a single drain, may be circuit vented. This vent connects to the drain ahead of the last floor fixture and connects to the vent stack. A relief vent, ahead of the entire battery, is also required, to vent the soil stack. All fixtures on the same branch, upstream of the floor battery, must be individually vented, as the lavatory shown. *(c)* A loop vent is the same as a circuit vent except that it connects to a stack vent and not a vent stack. Therefore, no relief vent is required. Back venting of other fixtures is needed, as shown.

582 / DRAINAGE AND WASTEWATER DISPOSAL

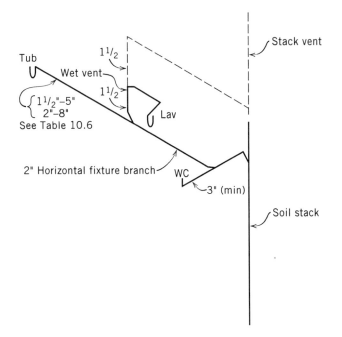

Figure 10.18 A wet vent is a vent for one fixture that is used as a drain for any other fixture except a water closet or kitchen sink. The illustration shows wet venting of a top floor bathroom group. The tub wet vents through the vertical drain sector of the lavatory. It does not require an additional vent if its drain is sized according to Table 10.6. The lavatory drain must be 1½ in. for a single lavatory and 2 in. for draining up to 4 dfu. The horizontal fixture branch must be 2 in. It can connect into the closet bend, as shown, or at the soil stack at the same level.

floors. Wet vents are normally used with bathroom groups in single and multiple residences. They may, however, also be used in commercial buildings, such as venting of an "island" fixture, installed remote from a wall or partition in which a vent pipe could be installed. The purpose of wet venting is usually economy, since a single (larger) pipe is used for two functions—vent and waste. Refer to the Code, Section 12.10, for detailed rules governing wet vents. (Note that a 3-in. soil line is shown for the WC as required by Code. This is minimum; many designers consistently use a 4-in. soil line for water closets, claiming an improved scrubbing action.)

f. Stack Venting

See Figure 10.4(a). As has already been explained, a *stack vent* is an extension of a soil stack to fresh air above the roof. This extension begins above the highest fixture branch connection. Stack venting is used principally in single family homes and on the top floor of multistory buildings. Code rules governing stack vents are given in Code Section 12.11.

The Code rules for stack venting permit certain fixture groups to be installed without individual venting of the fixtures. The Code requires that the fixtures be in a one-story building or at the top of a multistory building—that is, at the level that the stack vent begins. Other conditions require that:

- Each fixture drain must connect independently to the stack.
- The tub/shower and water closet must connect to the stack at the same level. (The tub connection can be a side inlet into a 4-in. closet bend.)

The fixture groups for which this type of venting is permitted include:

- Two bathroom groups, back-to-back. (A bathroom group consists of a closet, a lavatory and a tub/shower.) See Figure 10.19.

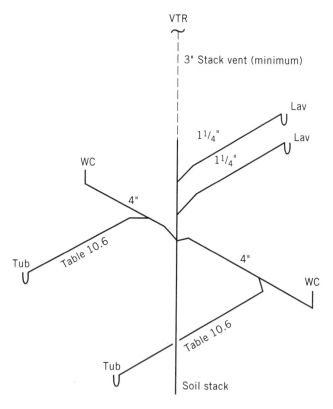

Figure 10.19 Stack venting of back-to-back bathroom groups. Minimum stack vent through roof (VTR) is 3 in.; many designers use a 4 in. minimum VTR.

- A bathroom group back-to-back with a kitchen (as in a single-floor residence). The kitchen may have a sink, food disposal and dishwasher. See Figure 10.20.

A special type of arrangement for stack venting of lower floors is permitted with the use of wye fittings, one-eighths bends and auxiliary 2-in. vents. This procedure is described in Code Section 12.11.2.

10.9 Drainage and Vent Piping Design

The layout of drainage and vent piping is not a simple matter. We have already seen that drainage piping is sized according to the number of drainage fixture units that it will carry. Because the primary purpose of vent piping is equalization of pressures, length is very important. Thus, vent piping is sized according to the size of the drain pipe that it vents (number of dfu) and its own length. See Table 10.7.

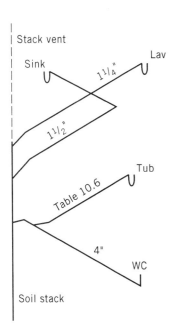

Figure 10.20 Stack venting of bathroom group and kitchen. The kitchen sink may also drain a food disposal unit and a dishwasher. All fixtures must connect directly to the stack, except that tubs may connect into 4-in. closet bends, as shown here and in Figure 10.19. Tub trap arm length must be in accordance with the requirements of Table 10.6. No additional venting is required.

Length is always developed length, that is, the length measured along the pipe center line, including length allowance for fittings. It is measured from its lowest connection at the drainage pipe, to the end of the exterior pipe extension above the roof. As shown in Figure 10.16(c), vent stacks do not always extend above the roof. In such a design, the header is sized in accordance with Table 10.7, as will be explained in detail in Example 10.2.

Additional rules governing vent piping follow.

1. The diameter of a vent pipe may not be less than 1¼ in. or half the size of the drain pipe that it vents, whichever is larger. (Some cities require 1½ in. minimum vent pipe size.)
2. A *relief vent* may not be less than half the size of the drain pipe to which it is connected.
3. When fixtures, other than water closets, discharge downstream from a water closet into a fixture branch, each such fixture shall be individually vented. This procedure is called *reventing*. This requirement does not apply to single dwelling units designed in accordance with Code. See Figure 10.21.

In order to demonstrate the application of the rules and the tables used to design drain and vent piping, we will analyze a few common drainage arrangements in the following examples.

Example 10.2 Develop a drainage/vent plan for a single top floor bathroom group consisting of a tub/shower, two lavatories and a water closet. Draw an isometric showing the piping.

Solution: Refer to Figure 10.22 and Code Section 12.10.1—Single Bathroom Groups. The cheapest piping arrangement uses the common drain for the two lavatories as a wet vent for the tub/shower. Since 2 dfu drain into this (common) wet vent (two lavatories at 1 dfu each), it must be a minimum of 2 in. and is so indicated on the drain and vent. If only a single lavatory were installed, the drain and vent could be 1½ in., which is the minimum wet vent size permitted. The horizontal branch must be 2 in. minimum in all cases. Where the horizontal branch is the topmost load on the stack, it can connect to the stack below the closet bend, as shown in Figure 10.22 (a-1). Otherwise, the connection should be at the same level. Alternatively, the horizontal branch can connect to the closet bend, as seen in Figure 10.22 (a-2). This same arrangement can also be used for a double bath or a double

Table 10.7 Size and Length of Vents

Size of Soil or Waste Stack, in.	Fixture Units Connected	Diameter of Vent Required, in.								
		1¼	1½	2	2½	3	4	5	6	8
		Maximum Length of Vent, ft								
1½	8	50	150							
2	12	30	75	200						
2	20	26	50	150						
2½	42		30	100	300					
3	10		30	100	100	600				
3	30			60	200	500				
3	60			50	80	400				
4	100			35	100	260	1000			
4	200			30	90	250	900			
4	500			20	70	180	700			
5	200				35	80	350	1000		
5	500				30	70	300	900		
5	1100				20	50	200	700		
6	350				25	50	200	400	1300	
6	620				15	30	125	300	1100	
6	960					24	100	250	1000	
6	1900					20	70	200	700	
8	600						50	150	500	1300
8	1400						40	100	400	1200
8	2200						30	80	350	1100
8	3600						25	60	250	800
10	1000							75	125	1000
10	2500							50	100	500
10	3800							30	80	350
10	5600							25	60	250

Source. Reprinted with permission from The National Standard Plumbing Code, published by The National Association of Plumbing Heating Cooling Contractors.

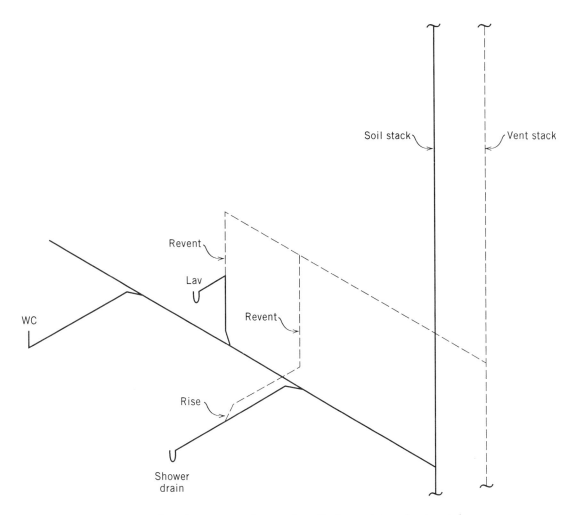

Figure 10.21 Fixtures, other than water closets, that discharge into a horizontal branch downstream from a water closet, must be individually vented. This procedure is termed reventing.

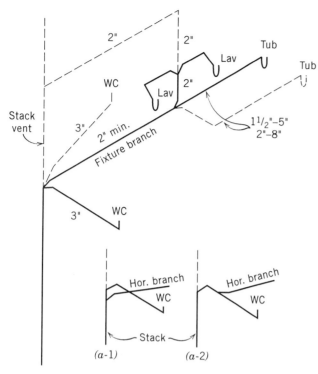

Figure 10.22 Drainage/venting plan for a single (or back-to-back) top floor bathroom group(s) using a stack vent and wet venting through the lavatory back vent. Sizes are shown for back-to-back groups. For a single bathroom group, connection at the stack may be at the same elevation, as shown in the piping diagram, or as shown in the inserts: *(a-1)* when the bathroom group is the highest load on the stack. *(a-2)* when using a 2-in. branch connection to the closet bend.

shower with the pipe sizes shown. The tub drain(s) is sized in accordance with Table 10.6. Wet vents are most commonly used in one- or two-story buildings or on top floors. For this reason, a stack vent is shown.

Example 10.3 Develop a drainage/vent plan for back-to-back bathroom groups on any floor of a building.

Solution: Refer to Figure 10.23 and Code Section 12.10.3. The straightforward approach is to use the 2-in. common vent of the back-to-back lavatories as a wet vent, as seen in Figure 10.23*(a)*. This extends as a 2-in. connection to the vent stack. The two back-to-back water closets must be back vented. As you can see, this arrangement is similar to that of Figure 10.22 except for the individual venting of the closets due to the lower floor location of the fixtures.

The Code, however, provides several alternative acceptable arrangements without the back venting of the closets. This, in a multistory building, can be a source of considerable economy. The possibilities are:

(a) Separate 2 in. wastes from the tub-lav group drain directly into the closet bend with a 45° wye tap. This permits the single large (2 in.) lavatory wet vent to also vent the water closet. See Figure 10.23*(b)*.
(b) Use of a special single closet, double 2-in. pipe fitting on the soil stack that connects one closet and the two 2-in. drains that connect to the tubs and lavatories.
(c) Use of 4-in. closet bends with two 2-in. wye taps each that drain the 2-in. lavatories and tub drains. See Figure 10.23*(c)*. This is very similar to the arrangement of Figure 10.23*(b)*.

Example 10.4 Design the drainage and vent piping for the back-to-back public toilets shown in Figure 10.24.

Solution: A battery of back-to-back toilets on a single horizontal drain can be vented with a circuit vent as shown. The circuit vent must connect downstream of the last fixture. In our example, this point is the common vent of the two lavatories ahead of the last fixture. The total drainage fixture units of all the fixtures connected is

8 public closets at 6 dfu each	48 dfu
6 lavatories at 1 dfu each	6 dfu
Total	54 dfu

From Table 10.4, we see that the fixture branch must be 4 in. From Table 10.7, a 4-in. stack (branch) with 54 units requires a 2-in. vent for 35 ft maximum and 2½ in. for a longer pipe. We select a 2½-in. pipe for the circuit vent. All additional fixtures (in this case, lavatories) discharging above the water closets must be individually vented. We provide them with 1¼ in. back vents. (In practice, 1½ in. pipe would often be used to avoid use of

DRAINAGE AND VENT PIPING DESIGN / 587

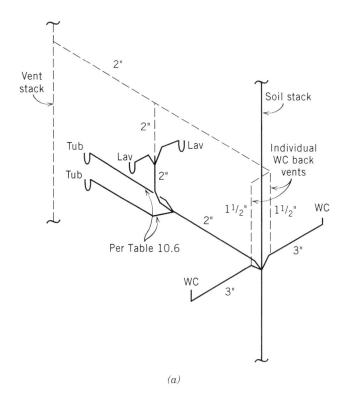

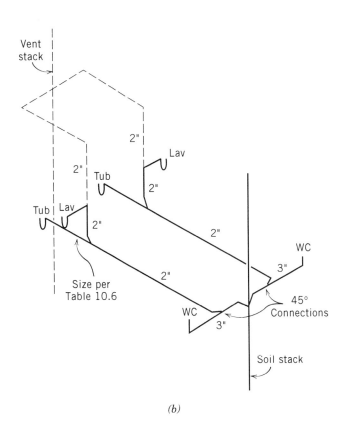

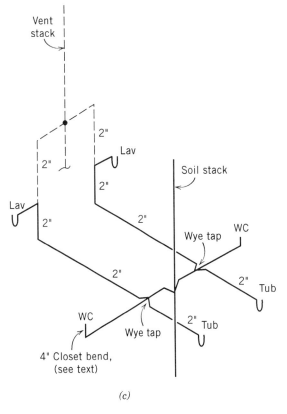

Figure 10.23 Three drainage/vent plans for back-to-back bathroom groups on any floor of a building: *(a)* Using the 2-in. common vent of the back-to-back lavatories as a wet vent and back venting the water closets. *(b)* Using separate 2-in. wastes from the combination tub-lavatory drain into the closet bend, with a 45° wye tap. Additional venting of the closets is not required. *(c)* Using the 4-in. closet bends, each with two 2-in. wye taps. Here, too, back venting of the closet is not required.

588 / DRAINAGE AND WASTEWATER DISPOSAL

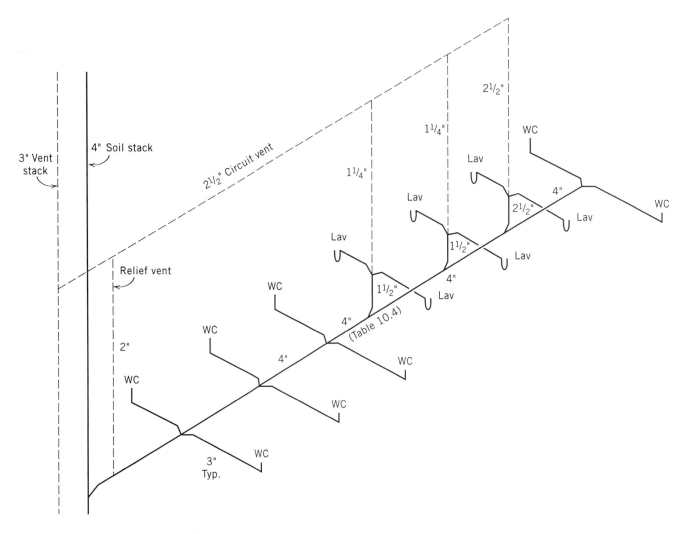

Figure 10.24 Circuit venting of back-to-back public toilets and lavatories. A 2½-in. vent is required (for 54 dfu). The circuit vent extends from the common drain before the two final closets around to the vent stack. In addition, a 2-in. relief vent is required ahead of the fixtures.

reducing fittings, and for standardization.) A relief vent is also required, taken off in front of the first fixture connection in the battery. It is sized at 2 in., that is, one-half the size of the horizontal branch. The drain for the lavatory group is sized at 2 in., per Table 10.4, for 6 dfu. The vent stack would be either 3 or 4 in. depending on the drainage load of the rest of the building. We show it as 3 in. arbitrarily.

Example 10.5 Calculate the vent header size required for the stack layout shown in Figure 10.25.

Solution: You should follow the calculation with Figure 10.25 in front of you. Each section of vent pipe in the vent header B-D-F-G is sized according to the total number of drainage fixture units vented and its total developed length from the lowest point of connection. We, therefore, calculate each vent header section BD, DF and FG separately.

a. Section BD vents stack AB, with a total of 120 dfu. The total developed length for this section extends from its lower connection at A to the vent through roof (VTR) at point G. That is, the entire distance A to G is considered as a single stack with total length of

$$AB + BD + DF + FG = 120 + 40 + 40 + 30 = 230 \text{ ft}$$

Referring to Table 10.7, we find that a vent 230 ft long for a 4-in. drain stack carrying 120 dfu must be at least 3-in. pipe. This is shown on the drawing.

b. Section DF represents the stack from the lower connection at C to the VTR at G. Its total length is

$$CD + DF + FG = 120 + 40 + 30 = 190 \text{ ft}$$

It vents stack AB and CD with a total

$$120 \text{ dfu} + 170 \text{ dfu} = 290 \text{ dfu}$$

A vent with a total length of 190 ft, venting 290 dfu and venting a 4-in. stack, must be 4 in. This is marked on Section DF.

c. Section FG is considered to begin at the lower connection at E and terminates at G. Total length is

$$EF + FG = 120 + 30 = 150 \text{ ft}$$

It vents all three stacks. Therefore, the total drainage fixture units is

$$FG = 120 = 170 = 150 = 440 \text{ dfu}$$

From Table 10.7, a vent of total length 150 ft, with 440 dfu can be 3 in. However, because section DF is already 4 in., we cannot reduce the vent size. Therefore, section FG is also 4 in. and is so marked.

10.10 Residential Drainage Design

The Code permits elimination of individual fixture venting in single dwelling units (apartments, houses) provided a number of conditions are met:

(a) The horizontal branch draining the fixture is uniformly sized.
(b) Trap arm lengths meet the requirements of Table 10.6 and are not shorter than twice the trap pipe diameter (see Figure 10.14).
(c) The fixture vent pipe opening is above the trap wier (see Figure 10.14).
(d) A vent is installed before the first fixture and before or behind the last fixture.
(e) Vents are sized for total drainage fixture units. Where a stack with a vent extension serves an upper floor, it must accommodate the drainage fixture unit load of the upper floor as well. If it vents a closet, it cannot be smaller than 2 in. Figure 10.26 shows some of the possibilities of this venting arrangement.

We are now at a point that we can complete the design of the plumbing system for the Mogensen house, which we began in Example 9.8. The site plan for the building (an actual structure) is shown in Figure 9.23. The architectural plan is shown in Figure 3.38.

Example 10.6 Design a sanitary drainage system for the Mogensen house. The fixture requirements are shown in Figure 9.24, page 542.

Solution: The first step is to lay out a drainage piping plan, Figure 10.27, and a plumbing drainage section, Figure 10.28. We did not prepare the section in isometric form for the sake of clarity and simplicity. The piping is sufficiently clear in two-dimensional drawing form. The plan indicates the actual piping routing, subject to field changes as required and decided upon by the plumbing contractor, preferably after consultation with the architect's plumbing technologist/designer. The riser (section) indicates only pipe sizes and connections, not pipe routing. At this point, you should trace

590 / DRAINAGE AND WASTEWATER DISPOSAL

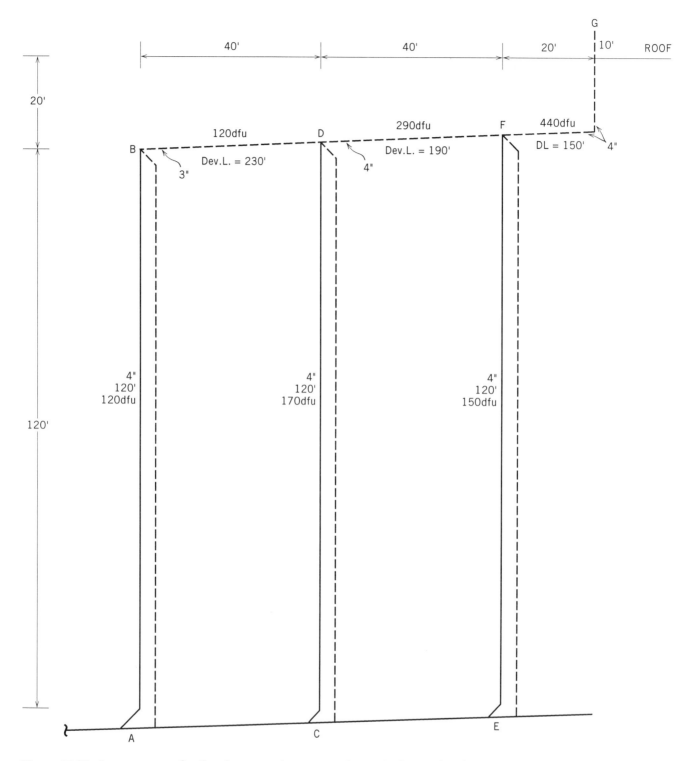

Figure 10.25 Arrangement of soil and vent stacks connected to a single vent header extending through the roof at a single point. Figures for drainage fixture units and vent developed lengths permit calculation of the required vent header sizes. See text.

Single Dwelling Unit

In a single dwelling unit on any uniformly sized horizontal branch or building drain, fixtures may be installed without individual vents provided the requirements of the Code and Table 10.6 are met; and provided that before the first fixture and before or after the last fixture connected, a vent is installed. These vents may be stacks serving upper levels provided the size of the stack can accommodate the total fixture unit load connected, but not less than 2 inches when a water closet is installed in the horizontal branch or building drain.

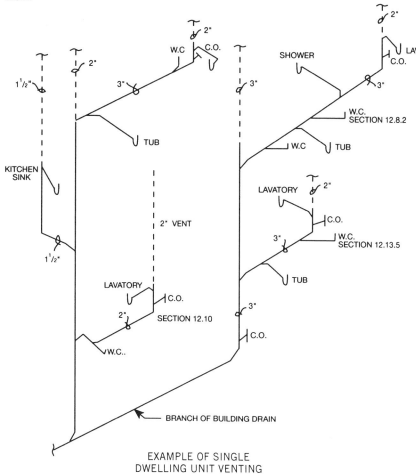

NOTE: Vent sizing as per Table 10.7.

Figure 10.26 Example of single dwelling unit venting. (Diagram and explanation on drawing reproduced with permission from The National Standard Plumbing Code; published by The National Association of Plumbing Heating Cooling Contractors.)

592 / DRAINAGE AND WASTEWATER DISPOSAL

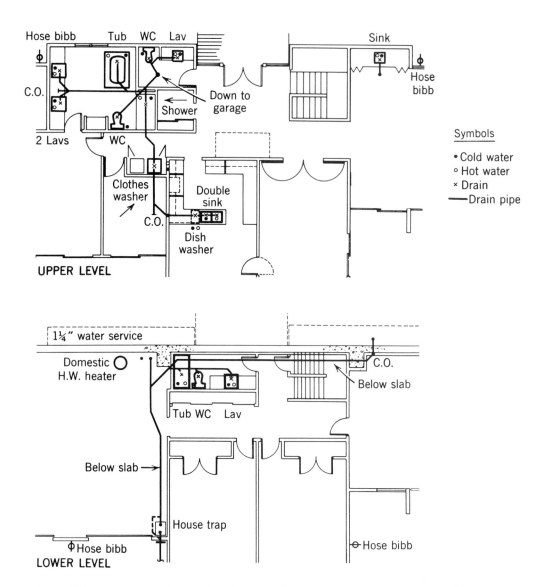

Figure 10.27 Drainage plan for the Mogensen house. Pipe sizes are shown on the plumbing section. Only drainage pipes are shown; vent piping is shown on the section. The plumbing fixture plan is shown in Figure 9.24, page 542.

RESIDENTIAL DRAINAGE DESIGN / 593

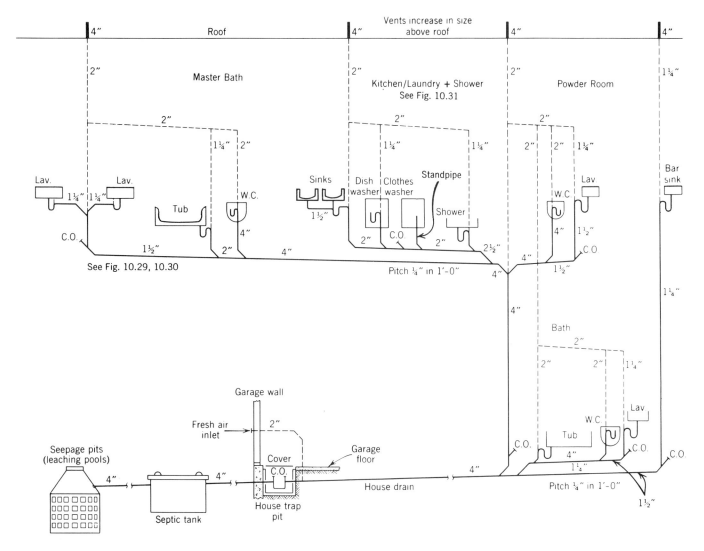

Figure 10.28 Plumbing drainage section (riser) for the Mogensen house. See text for an explanation of pipe sizing and arrangements.

out the drainage on Figure 10.27 and compare it to the elevation in Figure 10.28 to make sure that you clearly understand the layout. Note that the vent piping is shown only on the riser and not on the plan. The reason that this is often done is that only fixture drainage connections show on the plan, not vent connections. The plumbing contractor understands this and routes the vent piping as required by Code.

When the plumbing section (riser) is prepared at first, it does not show pipe sizes. These are added as the design progresses. The first step is to analyze the drainage requirements. Using Table 10.2, we can list the fixtures, in groups, and their drainage fixture units as follows:

Master Bath
    Bathtub group   6 dfu
    Additional lavatory   1 dfu
        Subtotal   7 dfu
Kitchen, Laundry, Shower
    Double sink and dishwasher 3 dfu
    Clothes washer   2 dfu
    Shower   2 dfu
        Subtotal   7 dfu
Powder Room
    Water closet   4 dfu
    Lavatory   1 dfu
        Subtotal   5 dfu
Bar Sink (1¼-in. trap)   1 dfu
Lower Level Bath
    Bathroom group   6 dfu
        Total   26 dfu

Drainage of hose bibbs is into the ground. As we will see in our storm drainage discussion, roof drainage for this design is to dry wells, as is air conditioning condensate.

Referring now to Table 10.5, we see that at a slope of ¼ in./ft, a 3-in. pipe will carry 42 dfu, with a maximum of three water closets, for a single family dwelling. Note, however, from Table 10.1 that the discharge velocity for a 3-in. drain running one-quarter full is only 1.58 fps. We stated in Section 10.4 that the target velocity in a building (house) drain is 3 fps. We, therefore, need to increase the pipe size to 4 in. since the alternative—increasing the slope—is not practical. This increases flow velocity to something over 2 fps, depending on fill. If it were possible, increasing slope to ½ in./ft would be preferable to increasing the pipe size.

At this point, we can size horizontal fixture branches, using columns 1, 2 and 3 of Table 10.4. Beginning at the master bath we use a minimum size drain of 1½ in. This increases to 2 in. after the tub connection. Because a 4-in. closet bend has been used, the horizontal drain, which could otherwise have been 3 in., is increased to 4 in. (A 4-in. closet bend is standard in some areas.) The next group to be considered is the kitchen/laundry/shower group. Note that the double kitchen sink has a single trap. Here too the designer could have used a 1½-in. drain up to the clothes washer but decided, instead, to avoid size change and to use 2 in. up to the shower. Since the entire group is 7 dfu, it requires a 2½-in. drain. Similar considerations govern sizing of the drain piping for the bar sink, the powder room and the first level bath.

Vent sizing is straightforward. A common 2-in. vent is used for the two lavatories that drain together in the master bath. The water closet is back vented with a 2-in. vent, since its closet bend is 4 in. and a vent may not be less than half the size of the drain that it vents. The tub is also individually vented with a minimum size 1¼-in. vent. In the kitchen-laundry combination, each fixture is back vented with a minimum 1¼-in. vent to a 2-in. header. The powder room lavatory vent is 1¼ in. minimum, and the water closet vent is 2 in. as already explained. The header is also 2 in. The lower level bath uses a 1¼ in. minimum size lavatory vent, a 2-in. water closet vent and a 2-in. tub back vent, which also assists in venting the house drain. The tub back vent connects to the powder room header and then to the roof connection. Notice the important fact that all the vents are increased to 4 in. when they pass through the roof to outside air. This is commonly done so that there is no possibility of a vent freezing shut or becoming clogged with snow and ice.

Figures 10.29 and 10.30 show the plumbing work for this house in various stages of construction. Note that in this particular house, local authorities approved the use of a house trap and a fresh air inlet to vent the private sewage disposal arrangement. The design of this private sewage disposal system will be discussed later in this chapter. It is shown on Figure 10.28 for completeness. Figure 10.31 shows a typical residential DWV installation in copper. Most DWV installations today use plastic pipe.

10.11 Nonresidential Drainage Design

Nonresidential drainage/venting design is performed in the same manner as residential design.

Figure 10.29 Horizontal branches of the copper pipe water system and the plastic pipe drainage system are located in the furred space between the garage ceiling and the floor joists of the upper level. Construction stage photograph.

The differences are that the special arrangements permitted for residences are not used. Also, the drainage fixture unit values are the higher "general" units and not the lower "private" units. A fairly simple example will be adequate to demonstrate the methods used.

Example 10.7 Design the drainage system for the industrial building for which we designed the water supply in Example 9.9.

Solution: Refer to Figure 3.48, page 153, which shows the architectural plan, and Figure 9.29, which shows the water supply plan (for information only). The procedure is essentially the same as that used for the residential example in Section 10.10. We prepare a drainage plan (Figure 10.32) and a plumbing section (Figure 10.33). As before, the drainage plan shows the routing of the drainage piping, and the plumbing section shows the pipe sizes. To size the drainage piping, we first total the drainage fixture units in the three toilets, using Table 10.2:

4 lavatories at 1 dfu each	4 dfu
4 flush valve public closets at 6 dfu each	24 dfu
1 drinking fountain	.5 dfu
Total	28.5 dfu

Referring to Table 10.5, we see that a 3-in. building drain at a slope of ¼ in./ft is satisfactory. However, here again, as for the residence analyzed previously, the designer used 4-in. closet bends. This forces us to use a 4-in. drain, since no horizontal branch or drain may be smaller than a connected fixture drain. As with Example 10.6, the 4-in. drain at a slope of ¼ in./ft will give a drain liquid velocity above 2 fps. This should be satisfactory, although a slope of ½ in./ft would be better. Note here that no building trap is used, in accordance with the recommendation of most plumbing codes including The National Standard Plumbing Code.

The three lavatories in the men's and women's rooms total 4 dfu. This requires a 1½-in. drain, according to column 2 of Table 10.4. However, since no 1½ in. pipe is used on the job, a 2-in. fixture branch drain is used for simplicity and it is so shown. Lavatory vents are the minimum size permissible—1¼ in. Water closet vents must be 2 in., because of the 4-in. closet bend. The vent header is, therefore, 2 in. (see Table 10.7), and the roof vent is 4 in. as already explained.

596 / DRAINAGE AND WASTEWATER DISPOSAL

Figure 10.30 *(a)* Water supply and drainage roughing for the two lavatories in the master bedroom. View looking north. Note the copper water tubing and plastic (ABS and PVC) waste and vents. Vertical stack is a vent above and a waste below. The two branch waste pipes lead into the stack through one-eighth bends and wye fittings.

Figure 10.30 *(b)* Plumbing roughing for kitchen, Mogensen house. View looking north. Waste branch at left takes the runoff of sinks and dishwasher. Trap at the right receives the discharge of the clothes washer. Vents are joined and run out through the roof. Hot and cold water tubes are seen in the foreground together with a special high temperature water tube for the preparation of items such as instant coffee.

Note. Concerning the distribution of air for heating and cooling, a vertical return air duct with a grille opening at the bottom is seen in the wall of the study. The back of a similar duct is to the left in the master bedroom. This arrangement offers some advantage over the return air system in Chapter 5. With return grilles at both top and bottom, the lower grille can draw in cool air that could, in winter, collect near the floor, thereby improving thermal comfort.

Figure 10.31 Drainage and vent piping in a typical frame residence. This DWV installation in copper serves two bathrooms at the upper level behind the 6-in. stud partition and a kitchen sink and laundry tray at the lower level, which are on this side of the partition. In the bathrooms, the roughing serves, from left to right, a lavatory, water closet and bathtub and a lavatory, shower and water closet. Bathtub and shower traps can usually be accommodated within the joist depth. The bend below the water closets, however, often leads to a horizontal branch exposed or furred-in below the joists. Some codes permit this branch from a water closet to be 6–10 ft long before joining a vent. The water piping is not yet entirely in place. (Courtesy of Copper Development Association.)

598 / DRAINAGE AND WASTEWATER DISPOSAL

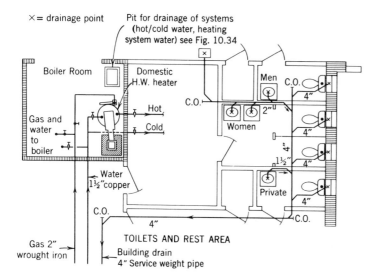

Figure 10.32 Drainage plan for the three toilet rooms of the building. Also shown on this plan are the water and gas service and the building drain and sewer line.

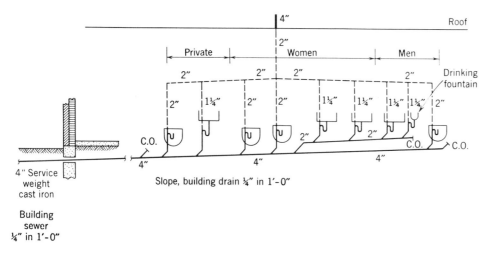

Figure 10.33 Plumbing section for the industrial building toilets of Figure 10.32. See text for an explanation of pipe sizing.

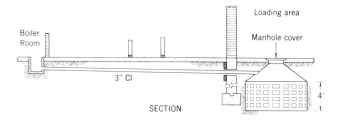

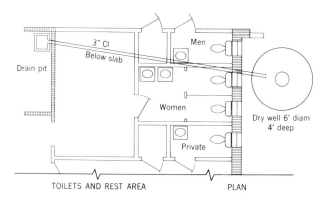

Figure 10.34 Plan and elevation of a dry well that is used to dispose of water in the hot and cold water piping systems. Drainage facilities are necessary when these systems are emptied for maintenance.

Figure 10.34 shows a dry well for drainage of the hot and cold water piping, including the hot water heater. There is no objection to putting this water directly into the ground as it is essentially clean water with no waste or soil in it. This arrangement is necessary because periodic maintenance and repair requires draining of the water piping. Therefore, a drain must be connected to the boiler room sump that receives this wastewater.

A somewhat more complex design example of a small office building illustrates use of the information and tables that we have studied.

Example 10.8 Select sizes for drainage and vent piping for the plumbing in an office building for which the fixtures are shown in the plumbing riser of Figure 10.35.

Solution: Individual fixture branches shall not be less than the sizes indicated in Table 10.3 for the minimum size of trap for each fixture. Drainage fixture units from Table 10.2 are applied to each section of the piping and totalled for each branch and stack and for the building drain and the building sewer. An example of a fixture-unit summary and sample sizes of individual branches that connect into a typical branch of the men's toilet group on any floor are shown in the following table.

Fixtures	Units per Fixture	Total Fixture Units	Diameter, Fixture Branch, in.
1 service sink	3	3	3
3 lavatories	1	3	1½
3 urinals, washout	4	12	2[a]
3 water closets, valve operated	6	18	4[b]
Total fixture units, Men's Toilet Branch		36	

[a] Integral trap; 2-in. drain required.
[b] 3 in. is adequate; 4 in. is used by convention.

Reference to column 2 Table 10.4 indicates that a 3-in. horizontal fixture branch is inadequate for the preceding group, because it will handle only 20 fixture units and not more than two water closets. A 4-in. pipe is selected. Its capacity of 160 fixture units will be more than enough for the 36 needed here. The same table shows that the soil stack can be 4 in. in diameter. It is run this size for its entire height without size reduction, because no section of stack can be smaller than a branch draining into it. Its capacity of 90 fixture units per story is sufficient for the 64 that connect in at each T-Y connection.

According to Table 10.5, the building drain and the building sewer at a pitch of ¼ in./ft should be 5 in. in diameter. Their capacity of 480 fixture units exceeds the 350½ placed upon them. The vent stack, at 70 ft length and 338 fixture units could be 2½ in., but 3 in. is a better choice. This is increased to 4 in. as it passes through the roof.

Figure 10.36 is a photo of plumbing roughing for two lavatory rooms in an office building. Figure 10.37 shows roughing for lavatories in a school. You should carefully study the plumbing riser of Example 10.8 (Figure 10.35) and these two photos until you understand the sizing and function of every component.

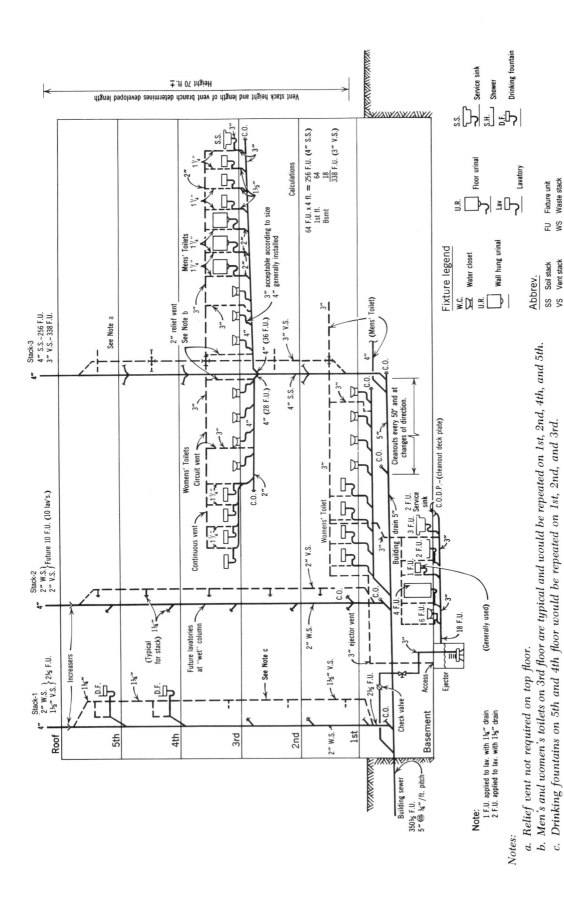

Figure 10.35 (Example 10.8) Office building plumbing section, in general conformity with National Standard Plumbing Code. Circuit vents serve branch soil lines. House trap and fresh-air inlet are not used in the building drain in accordance with modern practice. (From B. Stein, J. S. Reynolds, and W. J. McGuinness, *Mechanical and Electrical Equipment for Buildings*, 7th ed., 1986, John Wiley & Sons, New York. Reprinted by permission of John Wiley & Sons.)

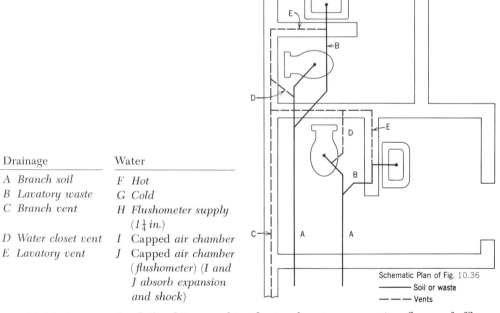

Drainage	Water
A Branch soil	F Hot
B Lavatory waste	G Cold
C Branch vent	H Flushometer supply ($1\frac{1}{4}$ in.)
D Water closet vent	I Capped air chamber
E Lavatory vent	J Capped air chamber (flushometer) (I and J absorb expansion and shock)

Schematic Plan of Fig. 10.36
——— Soil or waste
- - - - Vents

Figure 10.36 An example of plumbing roughing for two lavatory rooms in a fireproof office building. A lavatory and water closet in each room are served by soil and waste branches below and vent branches above. Hot and cold water tubing with air chambers can be seen. The extensions of the water tubing above the two flushometer connections appear to connect into the horizontal vent branches, but they do not. They are capped and merely touch the bottom of the vent branches. Note that soil branches are above the structural slab. A fill of 5 or 6 in. will be necessary to cover the piping. All vertical piping will be within the masonry block used to enclose the cubicles. (Courtesy of Copper Development Association.)

Figure 10.37 Roughing in place for four lavatories in the washroom of a school. At this stage, the waste branches have been capped and the system tested for possible leakage. The waste branches are of cast iron, the vents are of galvanized steel, and the water lines are of copper with soldered fittings. Roughing dimensions have been followed. Vertical capped expansion and shock tubes are seen as extensions of the entering water pipes that serve the hot and cold water branches. The layout course of masonry behind the piping assembly is the position of the partition marking the adjacent room. At the right will be seen a projecting masonry block. The near end of this block is the location of another block wall that will enclose the "furr in" the roughing assembly. (From B. Stein, J. S. Reynolds, and W. J. McGuinness, *Mechanical and Electrical Equipment for Buildings*, 7th ed., 1986, John Wiley & Sons, New York. Reprinted by permission of John Wiley & Sons.)

Storm Drainage

10.12 Storm Drainage— General Principles

As we noted in the beginning of our discussion on drainage, every building has two systems of drainage: sanitary and storm. The first has been analyzed in detail in the preceding sections. It deals with draining the effluent of interior plumbing fixtures and conducting it to a municipal sanitary sewer or to a private sewage treatment installation. The second, which we will discuss in the following material, is concerned with conducting rainwater away from a building. The question that immediately arises is, Why bother? Why not simply allow the rain to run off the roof (and walls) of a building and dispose of itself, just as it did before the building was constructed? The answer has several parts:

(a) Rain frequently does not dispose of itself readily, particularly in built-up areas. It floods, forms small and large puddles, turns earth into mud and, frequently, is not rapidly absorbed into the ground.
(b) Rain that runs down a building wall causes leaks at windows and other wall penetrations and can cause discoloration of stone facings.

(c) Uncontrolled rainwater can cause erosion around foundations and, in extreme cases, building settlement.
(d) Uncontrolled rainwater can, and very frequently does, cause basement flooding.
(e) Buildings frequently have paved areas below grade such as areaways and driveways, that collect rain. This water cannot disperse naturally; it must be collected and conducted to a drain.
(f) Many buildings, particularly in built-up urban areas, have little or no unpaved areas near the building to which water can be conducted for natural absorption in the ground.

For all these reasons, it is just as important to design an adequate and efficient storm drainage system as to design a good sanitary drainage system. Both are the responsibility of the plumbing technologist.

All modern plumbing codes require that the storm drainage system be separate and distinct from the sanitary drainage. Storm drainage piping may not be used for sanitary drainage, and vice versa. Ideally, the two systems will remain separated throughout their course. The sanitary drainage goes to a sanitary sewer and eventually to a sewage treatment plant, while the storm drainage goes to a storm sewer and eventually to a stream, river or open areas for "recharging" of the ground water table. See Figure 10.38. Indeed, this latter is the preferred modern approach because water table levels in many areas have dropped due to overpumping and excessive water use. Many towns and cities do not have adequate storm sewers, if at all. In such areas, storm water is often carried by the sanitary sewers through connection of the rain water conductors into the building sanitary drain and sewer. See Figure 10.39. When these municipalities first constructed sewage treatment facilities, the additional load caused by the rainwater during particularly heavy rainfall frequently overloaded the plants and resulted in bypassing (dumping) of the excess into surrounding land areas, with undesirable results, to say the least. As a result, today all new construction must completely isolate the two drainage systems from each other, unless specifically directed otherwise by the local code authorities.

As shown on Figure 10.38, when a storm drain is

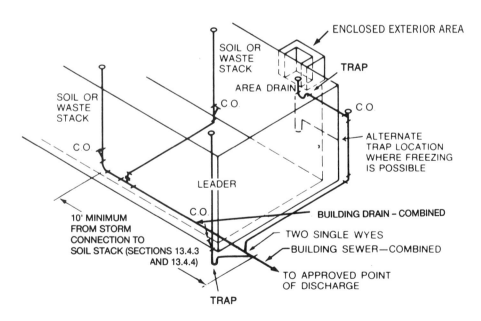

Figure 10.38 An example of a combined sanitary and storm drainage system. Note that the storm drains are trapped and connected at least 10 ft downstream from a soil stack connection. Combined drainage systems must have the approval of local authorities. (Reproduced with permission from the National Standard Plumbing Code, published by The National Association of Plumbing Heating Cooling Contractors.)

604 / DRAINAGE AND WASTEWATER DISPOSAL

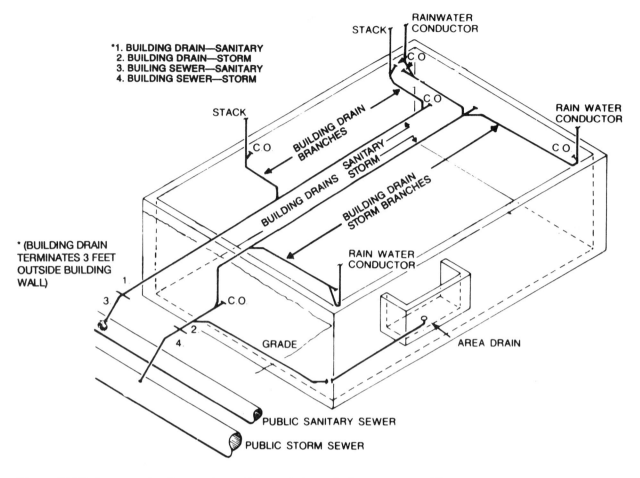

Figure 10.39 An example of separate sanitary and storm drainage systems. The two systems are entirely independent. No trapping of storm drains is required. A cleanout is desirable because of the dirt, sand and the like that is carried with rainwater from a roof through the roof drain strainer. (Reproduced with permission from the National Standard Plumbing Code, published by The National Association of Plumbing Heating Cooling Contractors.)

connected to the building sanitary drain pipe, it must first be trapped and then connected to the drain not less than 10 ft downstream from any soil or waste pipe connection. The purpose of the trap is to prevent sewer gases from entering the storm drain system. The 10-ft distance between connections prevents heavy storm drainage from backing up into the soil lines and also prevents pressure variations that might affect fixture traps. Storm drains, unlike sanitary drains, are designed to run full because there is no problem with pressure variations. In a combined system, however, such pressure changes could affect the sanitary drainage. The 10-ft distance permits pressure variations to equalize. Another reason that storm drainage has minimum pressure problems is that flow is continuous, and not pulsed, (intermittent) as is sanitary drainage flow.

10.13 Roof Drainage

The purpose of the roof drainage design is to arrive at the sizes required for piping, both vertical and horizontal, that will safely and rapidly drain the roof. Because there is some confusion about terms, we will clarify them at the outset. A *downspout* or *leader* is a vertical rainwater pipe on the exterior of a building. A *conductor* is a rainwater pipe on the interior of a building. Because of the different locations, materials are normally different. Leaders and downspouts are usually made of galvanized sheet metal in circular or corrugated cross section. See Figure 10.40. The corrugated shape is preferred because it is stronger. They normally carry rainwater from gutters around the building to ground level, where the water is absorbed by the surrounding ground. See Figure 10.41. Because gutters and downspouts are usually made of sheet metal, they are frequently placed in the HVAC contract along with sheet metal duct work. Their sizing and placement, however, are the responsibility of the plumbing designer. Conductors, being inside the structure, are made of standard pipe materials such as cast iron, galvanized iron or steel, copper or brass.

Conductors connect to roof drains on flat roofs and conduct the rainwater to horizontal drains at the base of the building, where it is usually channelled into a single pipe for disposal. See Figure 10.42. Since the roof drain itself is exposed to outside temperatures and in the winter collects very cold rainwater, an expansion joint of some sort is required at the connection to the conductor piping, which is at inside temperature. An expansion joint in the roof drain or a pipe offset can be used for this purpose. Exterior gutters and leaders are also subject to expansion and contraction due to rainwater and to exposure to sunlight and to large changes in exterior temperature. Expansion joints should be used in all runs, vertical or horizontal, longer than 40 ft. A detail of a standard roof drain for flat roofs is shown in Figure 10.43.

Overflow drains should be specified to prevent the overloading of roofs where the building code has called for a specific maximum water buildup depth. This, of course, would be where parapet scuppers have fallen into disfavor because they create unsightly streaks on the building face. Some codes call for the overflow system to remain independent of the primary leader system to the exterior of the building. The overflow drain remains inactive until the water level reaches its overflow level.

10.14 Storm Drainage Pipe Sizing

The size of the rain collection piping, whether exterior gutters and leaders or interior conductors and drains, depends upon three factors:

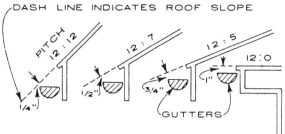

Gutters should be placed below slope line so that snow and ice can slide clear. Steeper pitch requires less clearance.

PLACING OF GUTTERS

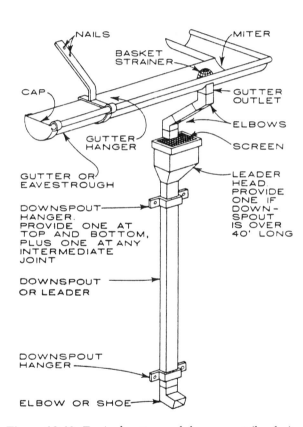

Figure 10.40 Typical gutter and downspout (leader) arrangement. Leaders are normally placed at building corners but not less than 20 ft nor more than 50 ft apart. The illustrated gutter and leader arrangement is used with sloped roofs. From Ramsey and Sleeper, *Architectural Graphic Standards*, 8th ed., 1988, John Wiley & Sons, Reprinted by permission of John Wiley & Sons.)

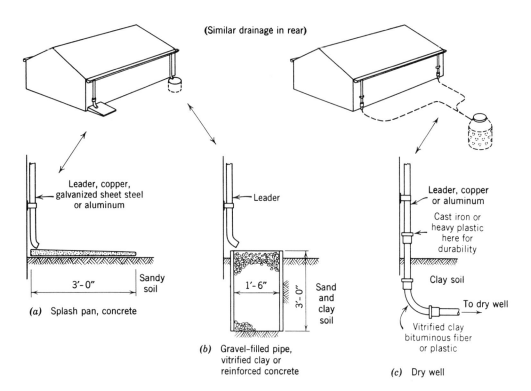

Figure 10.41 Roof drainage for houses. Gutters and leaders are sized with the aid of Tables 10.8 and 10.9. Method (a) is suitable for low rates of flow introduced into very porous soil. When denser soil is encountered, (b) is used to get the water into the ground and thus to avoid surface erosion. For heavy flow or to lead the water farther from the structure, (c) may be used with one or several dry wells.

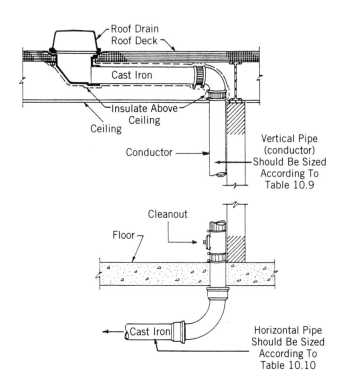

Figure 10.42 Typical cast-iron roof drain and roof leader. Joints may be hubless or hub and spigot when cast iron (as shown) is used for the leader. Insulation on the pipe above the ceiling is required to avoid a wet, discolored ceiling caused by condensate dripping from the leader. Note the use of a cleanout at the base of the leader to clear blockage of dirt, gravel, sand, etc. (Copyright © 1989 by the Cast Iron Soil Pipe Institute, reprinted with permission.)

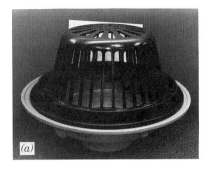
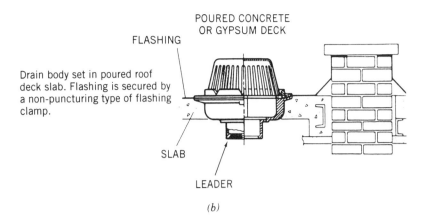
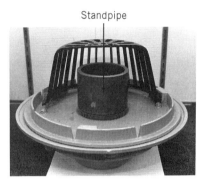

Figure 10.43 Typical roof drain fitting *(a)* for a flat roof. The base of the fitting connects directly to a leader as shown in *(b)*. *(c)* Standpipe-type overflow drains of the type shown are used on flat roofs of any construction where a constant or maximum height of water is desired on the roof. They are also used on conventional roofs as a safety overflow, that drain water only when the water buildup is greater than the setting of the cut standpipe. A standpipe-type water dam can be cut to achieve any desired water depth on the roof. (Courtesy of Jay R. Smith Manufacturing Company.)

- The amount of rainfall in a specified period of time.
- The size of the area being drained.
- The degree of pipe fill, that is, whether a pipe or gutter runs one-third full, one-half full and so on.

a. Rainfall

If longtime weather bureau observations are available for the particular area being designed, those data should be used in preference to general tables or charts. Most codes and many experienced designers recommend that the figures for maximum hourly rainfall over a 10-year period be used. Some designers prefer a 25-year period if reliable data are available. If data are not available, then the chart of the United States shown in Figure 10.44 can be used. Most designers, in the absence of data to the contrary, will use a minimum rainfall of 4 in./hr as the basis of design. Note from Figure 10.44 that even in semiarid areas, periodic heavy rainfall establishes the design criteria. This is because it is maximum short-period rainfall (and not average annual rainfall) that must be accommodated by the piping system.

b. Roof Size

When dealing with a flat roof, there is obviously no problem in determining the rain collection area, as it is simply the area of the roof. Once this area is

**Storm Drainage Sizing Precipitation Rate
Inches/Hour—Based on 15-minute Precipitation**

This table may be used in the absence of pertinent data.

This data in this map is based upon U.S. Weather Bureau Technical Paper No. 40, specifically Chart 4 (10 year with a 30 minute rainfall as recorded in inches).

The rainfall values in Chart 4 were multiplied by 0.72 to obtain the rainfall rates for 15 minutes (multiplier 0.72 presented in Table 3 of same report). Those resultant values were multiplied by 4 to obtain the above designated precipitation rates as listed in inches per hour.

Increase values 20% for a 100 year return period. Decreased values by 20% for a 5 year return period.

In order to ascertain the probable maximum precipitation considerations (where structural failure may result), it is recommended to study Technical Paper No. 40 in greater depth.

*Consult local data.

NOTE: Local weather bureau data should be consulted for specific storm rate patterns, especially when structural failure considerations are being evaluated.

Figure 10.44 U.S. maximum rainfall data map. (Reprinted with permission from the National Standard Plumbing Code, published by The National Association of Plumbing Heating Cooling Contractors.)

established, it is a simple matter to convert rainfall in inches per hour to drainage in gallons per minute. Since most designers and codes use 4 in./hr as the design figure, let us convert it to gallons per minute per square foot of roof. Four inches of rain "piled up" on one square foot of area is $1/3$ cubic foot of water. Therefore,

$$\frac{1/3 \text{ cf}}{\text{hr}} \times \frac{1 \text{ hr}}{60 \text{ min}} \times \frac{7.48 \text{ gal}}{\text{cf}} = 0.04156 \text{ gpm/ft}^2 \text{ of roof}$$

Since this figure is a bit clumsy, we can change it by taking its reciprocal, that is, a rainfall of 4 in./hr will give a drainage flow of

$$\frac{1}{.04156 \text{ gpm/ft}^2} = 24.06 \text{ ft}^2/\text{gpm}$$

Using 24 rather than 24.06 introduces an error of less than $1/4\%$ and is entirely acceptable for all design work. This figure, 24 ft²/gpm, is easy to remember and work with and is the basis of the drainage sizing tables in all Codes. Thus, if rainfall is heavier, say 6 in./hr, we can very simply adjust this figure:

Roof area per gallon per minute for 6 in. rainfall =
$24 \times 4/6 = 16 \text{ ft}^2$

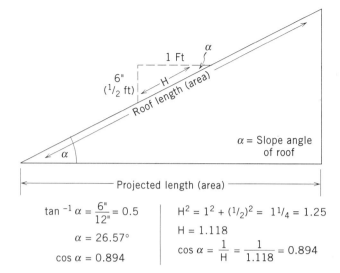

Figure 10.45 Calculation by trigonometry, or by the Pytagorean theorem, of the projected area of a sloping roof.

and so on. Obviously heavier rainfall means a smaller area of roof per gallon per minute, and vice versa.

For pitched roofs, the calculation is not quite so simple. The Code assumes that rain falls vertically, and, therefore, the area of a roof that collects rain is the roof's projected area. The projected area is simply the roof area multiplied by the cosine of the pitch angle:

$$A_{PROJ} = A_{ROOF} \times \cos \alpha$$

where α is the angle of pitch (slope).
See Figure 10.45. If the angle is unknown but the slope is known in terms of inches per foot of distance, the calculation is a bit more complex. See Figure 10.45. The angle of pitch is $\tan^{-1}$ slope, and then the cosine can be taken.

Example 10.9 What is the projected area of a 1000 ft² roof that slopes 6 in./ft.

Solution: Refer to Figure 10.45. The angle of roof slope is

$$\text{arc tan} \frac{6 \text{ in.}}{12 \text{ in.}} = \tan^{-1} 0.5 = 26.570$$

the projected area is, therefore,

$$A_{PROJ} = 1000(\cos 26.57) = 894 \text{ ft}^2$$

If you do not happen to have a calculator with trigonometric functions, the same calculation can be easily accomplished with a simple square root calculation. Since, in a right triangle, the square of the hypotenuse is the sum of the squares of the sides (Pythagorean Theorem), we can easily calculate the length H

$$H = \sqrt{(1)^2 + (0.5)^2} = \sqrt{1.25} = 1.118$$

therefore

$$\cos \alpha = \frac{1}{1.118} = 0.894$$

and

$$A_{PROJ} = 1000 \,(0.894) = 894 \text{ ft}^2$$

As can be seen in Figure 10.46(a) and as previously stated, vertical rainfall falls on the projected roof area of a sloped roof. However, because rain, and in particular the heavy rain for which we are designing, rarely falls vertically, a sloped roof receives more rain than the projected area indicates. The additional area depends on the angle at which the rain falls and is represented in Figure 10.46(b) by the area D. Since the rain angle varies, many designers use a compromise figure that combines slope with about 15°–20° of sloped rain. We would recommend using the following multiplying factors for the projected areas that appear in Tables 10.8–10.10. The following examples will demonstrate the use of these factors.

Projected Area Multipliers for Pitched Roofs

Pitch, in./ft	Factor
0–1	1.0
2–3	1.03
4–5	1.07
6–8	1.10
9–11	1.20
12	1.3

Similarly, the projected area of a flat roof to vertical rain [Figure 10.46(c)] is simply its area. With sloped rain, however, parapets and other vertical surfaces such as elevator penthouses collect rain that add to the total load. See Figure 10.46(d). Again, this depends on the angle of the rain. The proportion of total parapet area that must be added to the roof area follows:

Estimated Angle of Rain, %	Additional Parapet Area, %
10	17
20	35
30	50

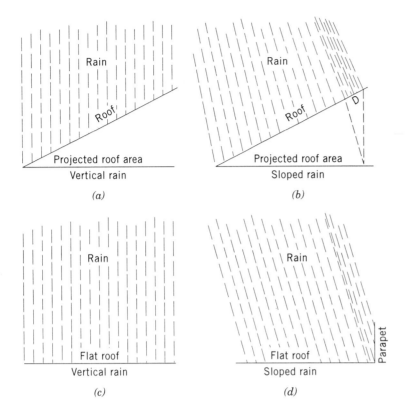

Figure 10.46 Vertical rain (a) falls on projected roof area. With a flat roof (c), the projected and actual area are identical. With sloping rain the projected area of a sloping roof (b) must be increased by area D to obtain the true rain-receiving area. With a flat roof (d), parapets add to the rain load of sloping rain.

c. Pipe Fill

The third, and final, factor in sizing storm drainage vessels is the degree of fill for which piping is designed. Remember that sanitary drainage is designed for approximately 25% fill in vertical pipes and up to 50% in horizontal pipes with 2–3 fps velocity for adequate scouring. Storm drainage is designed for at least 50% fill in vertical piping and 100% fill in horizontal drains and storm sewers. A minimum velocity of 3 fps for adequate scouring in horizontal flow is required.

Table 10.8 gives the required size of semicircular gutters as a function of projected roof area for sloping roofs. Note that the gutter slope is minimal at $1/16$ in./ft, but that all designs must be based on this slope even if the actual slope is greater. Other codes permit larger areas to be used with larger slope, the thought being that at larger slopes water drains out more quickly. We strongly recommend using only the Table 10.8 figures, which give a conservative design. Of course, the projected area figures there must be adjusted for rainfall data, as will be demonstrated.

Table 10.9 gives the size of vertical conductors and leaders required for given projected roof areas. Multiple gutters can feed into a single leader since, as can be seen from Tables 10.8 and 10.9 vertical pipe capacity is much larger than horizontal pipe capacity (because of the 90° "slope").

Table 10.10 gives the projected roof area for horizontal drains and sewers. Note that the areas are much larger than the roof areas for horizontal flow in gutters, as given in Table 10.8 for the same 4 in./hr rain and $1/16$ in. slope. This is because the gutters are semicircular and run about half full, while the horizontal drains are full circles (closed pipes) running full.

A few illustrative examples should make clear the use of the foregoing data and the tables.

Example 10.10 A one-story residence has a 2000 ft² center peaked roof sloped at 6 in./ft in both

Table 10.8 Size of Roof Gutters[a]

Diameter of Gutter,[b] in.	Maximum Projected Roof Area for Gutters	
	1/16-in. Slope[c]	
	ft²	gpm
3	170	7
4	360	15
5	625	26
6	960	40
7	1380	57
8	1990	83
10	3600	150

[a] This table is based upon a maximum rate of rainfall of 4 in./hr. Where maximum rates are more or less than 4 in./hr, the figures for drainage area shall be adjusted by multiplying by 4 and dividing by the local rate in inches per hour. See Figure 10.44.

[b] Gutters other than semicircular may be used provided they have an equivalent cross-sectional area.

[c] Capacities given for slope of 1/16 in./ft shall be used when designing for greater slopes.

Source. Reprinted with the permission of The National Standard Plumbing Code, published by The National Association of Plumbing Heating Cooling Contractors.

Table 10.9 Size of Vertical Conductors and Leaders

Diameter of Conductor or Leader, in.	Design Flow in Conductor, gpm	Allowable Projected Roof Area at Various Rates of Rainfall, ft²				
		1 in./h	2 in./h	4 in./h	5 in./h	6 in./h
2	23	2176	1088	544	435	363
2½	41	3948	1974	987	790	658
3	67	6440	3220	1610	1288	1073
4	144	13,840	6920	3460	2768	2307
5	261	25,120	12,560	6280	5024	4187
6	424	40,800	20,400	10,200	8160	6800
8	913	88,000	44,000	22,000	17,600	14,667

Source. Reprinted with the permission of The National Standard Plumbing Code, published by The National Association of Plumbing Heating Cooling Contractors.

Table 10.10 Size of Horizontal Storm Drains

Diameter of Conductor or Leader, in.	Design Flow in Conductor, gpm	Allowable Projected Roof Area at Various Rates of Rainfall, ft²				
		1 in./h	2 in./h	4 in./h	5 in./h	6 in./h
Slope 1/16 in./ft						
2						
3						
4						
5	100	9600	4800	2400	1920	1600
6	160	15,440	7720	3860	3088	2575
8	340	32,720	16,360	8180	6544	5450
10	620	59,680	29,840	14,920	11,936	9950
12	1000	96,000	48,000	24,000	19,200	16,000
Slope 1/8 in./ft						
2						
3	34	3290	1645	822	658	550
4	78	7520	3760	1880	1504	1250
5	139	13,360	6680	3340	2672	2230
6	222	21,400	10,700	5350	4280	3570
8	478	46,000	23,000	11,500	9200	7670
10	860	82,800	41,400	20,700	16,560	13,800
12	1384	133,200	66,600	33,300	26,640	22,200
15	2473	238,000	119,000	59,500	47,600	39,670
Slope 1/4 in./ft						
2	17	1632	816	408	326	272
3	48	4640	2320	1160	928	775
4	110	10,600	5300	2650	2120	1770
5	196	18,880	9440	4720	3776	3150
6	314	30,200	15,100	7550	6040	5035
8	677	65,200	32,600	16,300	13,040	10,870
10	1214	116,800	58,400	29,200	23,360	19,470
12	1953	188,000	94,000	47,000	37,600	31,335
15	3491	336,000	168,000	84,000	67,200	56,000
Slope 1/2 in./ft						
2	24	2304	1152	576	461	384
3	68	6580	3290	1644	1316	1100
4	156	15,040	7520	3760	3008	2510
5	278	26,720	13,360	6680	5344	4450
6	445	42,800	21,400	10,700	8560	7130
8	956	92,000	46,000	23,000	18,400	15,330
10	1721	165,600	82,800	41,400	33,120	27,600
12	2768	266,400	133,200	66,600	53,280	44,400
15	4946	476,000	238,000	119,000	95,200	79,330

Notes: Tables 10.9 and 10.10 are based on the rainfall rates shown. Local practice of the Administrative Authority should be consulted for the value to use for a particular place. For rainfall rates other than those shown, the allowable roof area is determined by dividing the value given above in the 1-in. column by the specified local rate. For conductors and leaders, the design flow rates are based on the pipes flowing between one-third, and one-half full. For rectangular leaders, the area shall be equal to the area of the circular pipe, provided the ratio of the sides of the leader does not exceed 3 to 1. For horizontal drains, the design flow rates are based on the pipes flowing full. See Figure 10.44.

Source. Reprinted with permission of The National Standard Plumbing Code, published by The National Association of Plumbing Heating Cooling Contractors.

directions. Gutters will be placed along the front and back with a single downspout for each gutter. The building is in the New York City area. Determine the required gutter and leader sizes.

Solution: Since two gutters are used, each will handle half of the roof, or 1000 ft². The slope angle of a roof pitched at 6 in./ft is

$$\text{slope angle} = \tan^{-1}\frac{6\text{ in.}}{12\text{ in.}} = \tan^{-1} 0.5 = 26.570°$$

the projected area of each half of the roof is, therefore,

$$A_{\text{PROJ}} = 1000 \times \cos 26.570 = 1000(0.894) = 894 \text{ ft}^2$$

Because the rain may not be vertical, we would multiply this area by 1.10. This gives a total rain-collecting area for each gutter of

$$A_{\text{TOT}} = 894 \text{ ft}^2 \times 1.1 = 984 \text{ ft}^2$$

Consulting Figure 10.44, we find that the New York City area has a design rainfall figure of 5 in./hr. Consulting Table 10.8, we find that a 7-in. semicircular diameter gutter is adequate for 1380 ft² at 4-in. rainfall. At 5-in. rainfall, this figure would be

$$\text{Roof area for 5-in. rainfall} = 1380 \text{ ft}^2 \times \frac{4 \text{ in.}}{5 \text{ in.}}$$
$$= 1104 \text{ ft}^2$$

Since our total area for each gutter is 984 ft², a 7-in. semicircular gutter, or any other shape with equivalent cross-sectional area, is sufficient. Referring now to Table 10.9 we see that a 3-in. vertical leader is adequate to drain the gutter.

Example 10.11 Design a storm drainage system for the light industry building that we have used as a design example throughout this book. Architectural drawings are found on page 152. A dimensioned plan of the roof is given in Figure 10.47, and a final site plan is shown in Figure 10.48. Assume that the building is to be constructed in a 6 in./hr rainfall area.

Solution: As can be seen on the architectural drawing and the roof plan, four roof drains have been placed in the high roof section, and two, in the low roof section. The roof will be constructed with slopes as shown, to direct water into these roof drains. The roof drains feed into vertical conductors positioned at columns so as not to interfere with manufacturing activities. These conductors, in turn, drain into horizontal drain pipes below the slab and out to a storm sewer.

The four high roof drains take care of an area of
$$A = 2 \times 35 \text{ ft} \times 110 \text{ ft} = 7700 \text{ ft}^2$$
Therefore, each conductor drains one-fourth this area, or

$$A_{\text{COND.}} = \frac{7700 \text{ ft}^2}{4} = 1925 \text{ ft}^2$$

Table 10.9, for 6-in. rainfall shows that a 4-in. conductor will adequately carry this load.

For the low roof area, we have two conductors draining a total area of 97 ft × 26 ft. Thus, each conductor drains

$$\frac{97 \text{ ft} \times 26 \text{ ft}}{2} = 1261 \text{ ft}^2 \text{ per conductor}$$

Again referring to Table 10.9 we see that for this section 4-in. conductors are adequate for the roof and the parapet between the high and low roof sections.

Horizontal drain sizes are selected from Table 10.10, using uniform slope for each continuous run. Although a slope of ¼ in./ft is a very common choice for horizontal sanitary drains, we select ½ in./ft for our storm drains. As you can see in Table 10.10, a larger pipe slope allows a specific size pipe to serve a greater roof area. As branches of the horizontal system join the main flow, pipe sizes increase. Figure 10.47 shows the increasing roof areas served. Pipe sizes are selected from the 6 in./hr rainfall column of Table 10.10 for this building. Starting directly below the slab, the total fall for the longest run (about 130 ft) is

$$130 \text{ ft} \times ½ \text{ in./ft} = 65 \text{ in.} = 5.4 \text{ ft}$$

Therefore, the ½ in./ft slope can be used only if the storm sewer is about 6 ft lower than the finished floor.

You will notice on the site plan of this building on Figure 10.48 and also on the Mogensen house site plan (Figure 10.56) that a number of dry wells are shown. These devices are used to disperse collected rainwater into the soil in locations where:

(a) No storm sewer is available or
(b) It is desired to use the rainwater to recharge the ground water table, and it is not desired to simply disperse the water onto the ground surface and allow it to be absorbed.

Sizing of dry wells is not normally the work of a plumbing technologist, because it requires considerable experience. For this reason, it is not included in our discussion, but it is shown for the sake of completeness.

614 / DRAINAGE AND WASTEWATER DISPOSAL

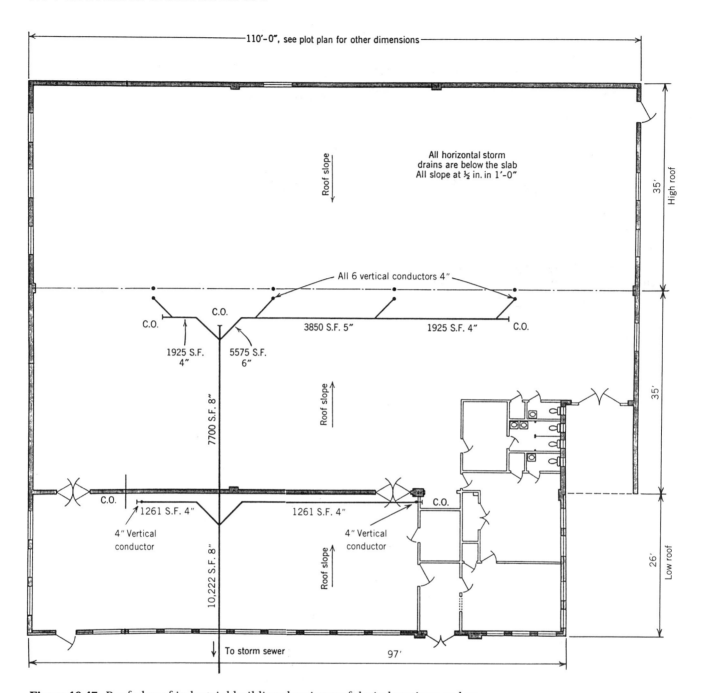

Figure 10.47 Roof plan of industrial building showing roof drain locations and underfloor horizontal storm drainage piping. Since the system is completely separate from the sanitary drainage system, traps are unnecessary, but cleanouts are required and are shown.

CONTROLLED ROOF DRAINAGE / 615

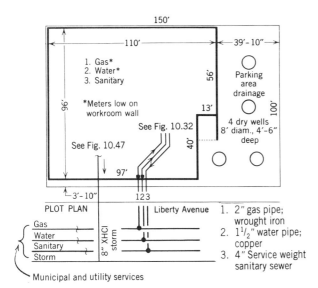

Figure 10.48 Site plan showing all mechanical services, including gas and water connections, plus sanitary storm sewer connections. Dry wells for parking area drainage are shown for completeness.

10.15 Controlled Roof Drainage

We already mentioned that one of the main reasons that storm drainage is separated from sanitary drainage is to avoid overloading the sewage treatment plant. Such overloading leads to bypassing, that is, the extremely undesirable practice of dumping raw sewage into lakes, streams and rivers. To avoid this ecologically disastrous procedure in areas that have combined sewer systems, a method was developed to limit the rate of discharge of storm water drainage from flat roof buildings. Such buildings constitute a major storm drainage load in urban areas. The method is called *controlled roof drainage*. It consists of using special roof drains with built in wiers that permit water to build up to a predetermined depth on the roof, while slowly draining the water away. The usual permitted depth is an average of 3 in. over a sloped roof and 3 in. uniformly on a dead-flat roof. This amount of water weigh 15.6 psf (¼ ft³ of water/ft² of roof surface).

Another advantage of controlled roof drainage is that smaller conductors and building drains can be used. The usual design criteria calls for a maximum drain-down time of 24 hr. With the maximum rainfall data and the capacity of the sewer system known, an appropriate roof drain can be chosen. Typical controlled-flow roof drains are shown in Figure 10.49.

(a)

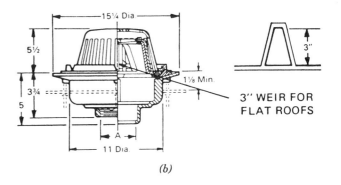

(b)

(c)

Figure 10.49 Controlled roof drainage is accomplished with a drain fitting with a built in weir (a). The standard weir is 3 in. high as seen in dimensioned cutaway drawing (b). This is intended for a flat roof with a 30-psf design load, as a 3-in. height of water will result in approximately a 15-psf load. The drainage flow is shown schematically in drawing (c). (Courtesy of Jay R. Smith Manufacturing Company.)

When performing the required capacity calculations for horizontal drains carrying sanitary and storm drainage together, the Code requires that the figures in Table 10.10 be used. Since this table specifies pipe size in terms of roof area, it is necessary to convert sanitary drainage fixture units to the equivalent square feet of roof area in order to be able to combine the two loads. The Code suggests the following procedure:

- For a sanitary fixture load of 0–256 dfu, add 1000 ft^2 to the projected roof area.
- For every dfu in excess of 256, add 3.9 ft^2 to the roof drainage area.

These figures apply only for 4-in. rainfall areas. For other areas, modify the figures up or down as required.

Private Sewage Treatment Systems

10.16 Private Sewage Treatment

As we stated at the beginning of the sanitary drainage section, we assume that a sanitary sewer is available, since that is the usual situation. However, some suburban and many primarily agricultural areas do not have sewer systems. In these locations, it is necessary to provide private sewage treatment facilities that will handle all the effluent from a building's fixtures, including black and gray water. Since this type of design work is not usually performed by plumbing technologists, we will only review it briefly.

In locations where local treatment is required, it is necessary to provide local digestion (partial purification) of the sewage and disposal of the partially purified liquid effluent to the ground. When there is adequate property area around a house or building and the soil is absorbent, as is sand for instance, the problem is easily solved. On small lots with clay-type soil, the problem is more difficult. In densely populated areas with houses on 40 × 100 ft lots, the situation is very difficult. Country and village health authorities usually consider a private treatment plant as a temporary measure. They sometimes retain the privilege of requiring a complete replacement after a number of years unless a municipal sewer replaces the private disposal system.

The general scheme of a private treatment system is relatively simple. The sewage is retained in a submerged, tightly enclosed septic tank of concrete or steel. Solids sink to the bottom of the tank. Bacterial action breaks up the solids and aids in purifying the fluids. A very small amount of sludge slowly builds up at the bottom of the tank and a scum forms at the top surface of the contents. The outflow pipe that carries the liquid effluent into the surrounding earth is located and protected in a way that prevents its being clogged. The septic tank needs to be pumped out at intervals because of sludge accumulation but usually not oftener than every 5 to 10 years.

The fluid discharges to one of two systems [as shown in Figure 10.50(a) and (b)]. Neither arrangement may be below the groundwater level, since the outflow might pollute it. This requirement often makes system (b) preferable because it is flat and shallow, a quality that is appropriate above a high water table. However, (b) requires a great deal of area. The discharge of raw sewage into a leaching pit or cesspool (c) is fast becoming illegal.

An efficient septic tank is shown in Figure 10.51. Cast-iron pipe fittings at both ends serve important functions. At the inlet, solids are directed to the bottom. The vertical pipes prevent surface scum from fouling either horizontal pipe. Unobstructed flow is ensured by placing the inlet pipe 3 in. above the outlet. Tight-fitting covers provide access for inspection and servicing for tanks that project above the ground. This is a convenience, but a tank may be placed slightly below grade if access is maintained. Several variations in piping arrangements are possible, as shown in Figure 10.52.

In the past, tanks and pits were built in place, but the prefabricated concrete and steel items illustrated in Figures 10.53 and 10.54 have largely superseded this procedure.

In order to demonstrate the techniques involved in designing a private sewage treatment system, we will work out two illustrative examples.

Example 10.12 Design a private sewage treatment facility for the Mogensen house.

System:	Septic tank and seepage pits as shown in Figure 10.49(a)
House category:	Luxury residence
Bedrooms:	Three

(*text continues on p. 620*)

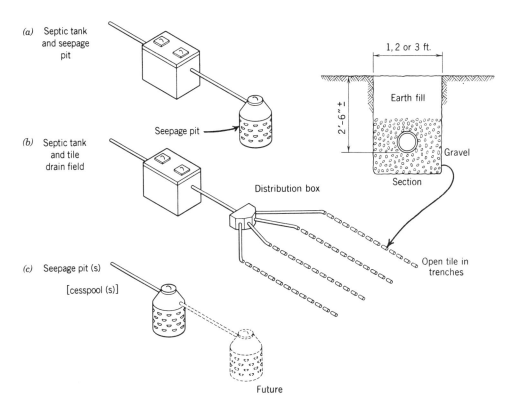

Figure 10.50 Private sewage treatment. *(a)* This method is most suitable in porous soil and where the groundwater level is low. *(b)* This method finds its best use in less porous soils and where the groundwater level is high. *(c)* This cesspool disposal is now discouraged and in some locations is illegal. See also Figures 10.53 and 10.54.

618 / DRAINAGE AND WASTEWATER DISPOSAL

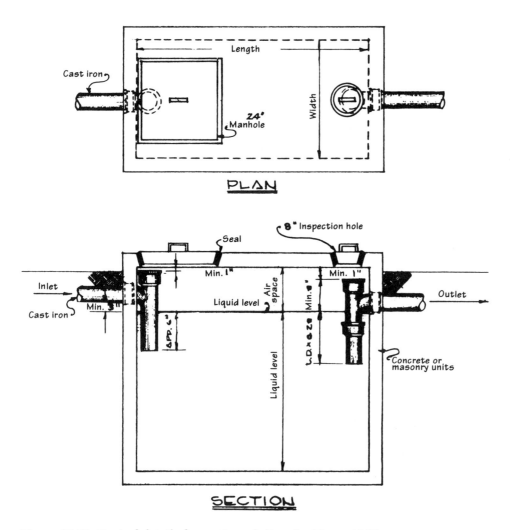

Figure 10.51 Typical detail of a septic tank. See also Figure 10.52.

PRIVATE SEWAGE TREATMENT / 619

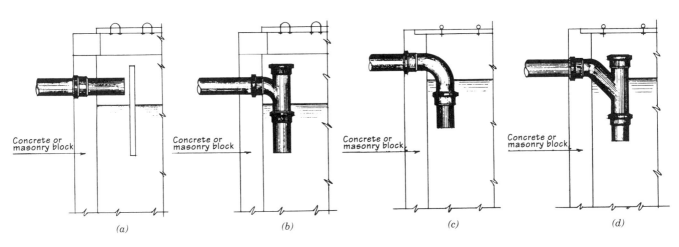

Figure 10.52 Typical piping inlet arrangements for a septic tank.

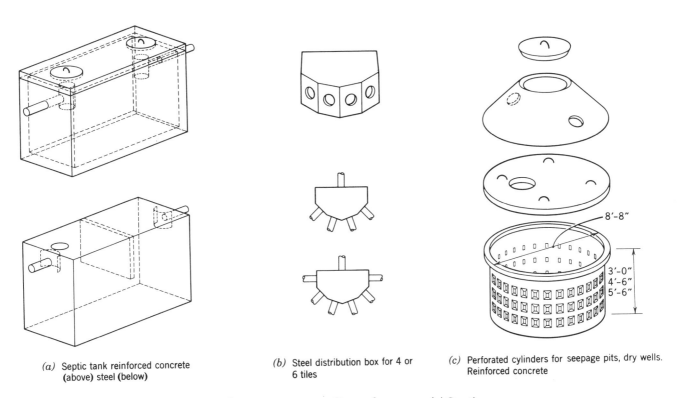

(a) Septic tank reinforced concrete (above) steel (below)

(b) Steel distribution box for 4 or 6 tiles

(c) Perforated cylinders for seepage pits, dry wells. Reinforced concrete

Figure 10.53 Prefabricated elements for private sewage disposal systems. *(a)* Septic tanks; *(b)* distribution box for use with tile fields (see Figure 10.50*(b)*; *(c)* precast seepage pit cylinders (see Figure 10.55).

620 / DRAINAGE AND WASTEWATER DISPOSAL

Figure 10.54 Precast elements used for recharge to the soil of the partially purified effluent of septic tanks. They are known as seepage pits. They are often made of high strength prestressed concrete. Here, shown in the manufacturer's yard, are, right to left, two perforated rings with conical top; alternate perforated cone; and typical extenders for greater depth. Rings are usually available in 8 and 10 ft diameters and in heights of 3, 4, and 5 ft. These same devices are used to disperse collected storm water in locations that lack storm sewers. For that usage they are known as dry wells. (From B. Stein, J. S. Reynolds, *Mechanical and Electrical Equipment for Buildings*, 8th ed., 1992, John Wiley & Sons, New York. Reprinted by permission of John Wiley & Sons.)

Occupancy: Six people
Soil absorption category test result: Two minutes for 1-in. drop

Solution: *Septic Tank.* Based on Table 10.11, we determine the proper size for a septic tank. The house has three bedrooms, but it would be better to select from Table 10.11 a tank that is one size larger than the minimum 1000-gal tank, that is, a 1200-gal tank. Its general arrangement will conform to Figures 10.51 and 10.52. The tank shown in Figure 10.55 was approved by local authorities. Its fluid capacity is

$$4 \text{ ft} \times 8 \text{ ft} \times 5 \text{ ft} \times \frac{7.481 \text{ gal}}{\text{ft}^3} = 1197 \text{ gal}$$

Seepage Pits. The effective absorption area of a seepage pit is the product of the perimeter and the height. The area at the bottom is not included. The effective area that we must provide will depend on the gallons per day of sewage flow and the porosity of the soil. Before calculating these values, we must know how porous the soil is. This is determined by a *percolation test*, which is described in detail in Section 16.5 of the Code. Essentially the test consists of filling with water a hole dug in the ground, and timing how quickly the level of the water drops. This tells us how quickly water will be absorbed (percolate) into the ground. That, in turn, will tell us how large a drain field is required to absorb the effluent flow.

Table 10.12 tells us that, for a luxury residence, the sewage flow is 150 gal/day per person. Since there are six people in residence, the design flow will be 900 gal/day. (gpd).

The data given in the problem state that the percolation test showed a 1-in. drop in water level in 2 minutes. (This indicates a very porous soil.) Referring to Table 10.13, which deals with the required absorption area of seepage pits [Figure 10.50(*a*)], we find that for a 2-minute drop, 40 ft² of absorption area are required for every 100 gal/day of sewage flow. Since we already calculated that this building will produce a maximum flow of 900 gal/day, it follows that the required absorption area is

$$900 \text{ gpd} \times \frac{40 \text{ ft}^2}{100 \text{ gpd}} = 360 \text{ ft}^2$$

Because the groundwater level in the area of the house is relatively high (see Figure 10.55), we

Table 10.11 Capacity of Septic Tanks[a]

Single Family Dwellings, Number of Bedrooms	Multiple Dwelling Units or Apartments—One Bedroom Each, Number of Units	Other Uses, Maximum Fixture Units Served	Minimum Septic Tank Capacity in gal
1–3		20	1000
4	2	25	1200
5 or 6	3	33	1500
7 or 8	4	45	2000
	5	55	2250
	6	60	2500
	7	70	2750
	8	80	3000
	9	90	3250
	10	100	3500

[a] Septic tanks sizes in this table include sludge storage capacity and the connection of domestic food waste disposal units without further volume increase.

Notes: Extra bedroom: 150 gal each. Extra dwelling units over 10: 250 gal each. Extra fixture units over 100: 25 gal/fixture unit.

Source. Reprinted with the permission of The National Standard Plumbing Code, published by The National Association of Plumbing Heating Cooling Contractors.

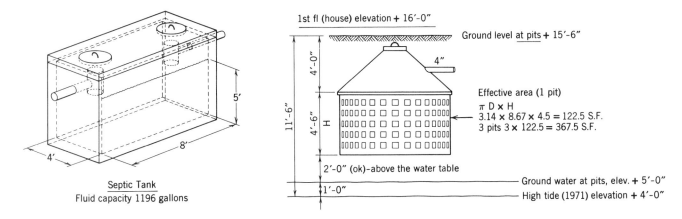

Figure 10.55 Components of the final design of the private sewage treatment system for the Mogensen house.

Table 10.12 Sewage Flows According to Type of Establishment

Type of Establishment	Quantity
Schools (toilets & lavatories only)	15 gal/day per person
Schools (toilets, lavatories, & cafeteria)	25 gal/day per person
Schools (toilets, lavatories, cafeteria & showers)	35 gal/day per person
Day workers at schools & offices	15 gal/day per person
Day camps	25 gal/day per person
Trailer parks or tourist camps (with built-in bath)	50 gal/day per person
Trailer parks or tourist camps (with central bathhouse)	35 gal/day per person
Work or construction camps	50 gal/day per person
Public picnic parks (toilet wastes only)	5 gal/day per person
Public picnic parks (bathhouse, showers & flush toilets)	10 gal/day per person
Swimming pools & beaches	10 gal/day per person
Country clubs	25 gal/locker
Luxury residences & estates	150 gal/day per person
Rooming houses	40 gal/day per person
Boarding houses	50 gal/day per person
Hotels (with private baths—2 persons per room)	100 gal/day per person
Boarding schools	100 gal/day per person
Factories (gallons per person per shift—exclusive of industrial wastes)	25 gal/day per person
Nursing homes	75 gal/day per person
General hospitals	150 gal/day per person
Public institutions (other than hospitals)	100 gal/day per person
Restaurants (toilet & kitchen wastes per unit of serving capacity)	25 gal/day per person
Kitchen wastes from hotels, camps, boarding houses, etc., serving three meals per day	10 gal/day per person
Motels	50 gal/bed space
Motels with bath, toilet, and kitchen wastes	60 gal/bed space
Drive-in theaters	5 gal/car space
Stores	400 gal/toilet room
Service stations	10 gal/vehicle served
Airports	3–5 gal/passenger
Assembly halls	2 gal/seat
Bowling alleys	75 gal/lane
Churches (small)	3–5 gal/sanctuary seat
Churches (large with kitchens)	5–7 gal/sanctuary seat
Dance halls	2 gal/day per person
Laundries (coin-operated)	400 gal/machine
Service stations	1000 gal (first bay)
	500 gal (each additional bay)
Subdivisions or individual homes	75 gal/day per person
Marinas (flush toilets)	36 gal/fixture/hr
Urinals	10 gal/fixture/hr
Wash basins	15 gal/fixture/hr
Showers	150 gal/fixture/hr

Source. Reprinted with the permission of The National Standard Plumbing Code, published by The National Association of Plumbing Heating Cooling Contractors.

Table 10.13 Required Absorption Area in Seepage Pits for Each 100 gal of Sewage per Day

Time for 1-in. Drop, min	Effective Absorption Area, ft^2
1	32
2	40
3	45
5	56
10	75
15	96
20	108
25	139
30	167

Source. Reprinted with permission of The National Standard Plumbing Code, published by The National Association of Plumbing Heating Cooling Contractors.

Table 10.14 Tile Lengths for Each 100 gal of Sewage per Day

Time for 1-in. Drop, min	Tile Length (ft) for Trench Widths of		
	1 ft	2 ft	3 ft
1	25	13	9
2	30	15	10
3	35	18	12
5	42	21	14
10	59	30	20
15	74	37	25
20	91	46	31
25	105	53	35
30	125	63	42

Source. Reprinted with the permission of The National Standard Plumbing Code, published by The National Association of Plumbing Heating Cooling Contractors.

should use a shallow seepage pit ring in order to avoid any possibility of polluting the groundwater. We select a 4 ft 6 in. deep perforated ring, 8 ft 8 in. in diameter (see Figure 10.54). With grade level at 15 ft 6 in. (Figure 10.56) the bottom of the ring is well above groundwater level, and no pollution can occur. Each seepage pit has an inside surface area exposed to earth of 122.5 ft^2. (The calculation is given on Figure 10.55.) Since we require a total of 360 ft^2, we use three such pits, arranged as shown in Figure 10.56.

We selected a septic tank and seepage pits because of the high porosity of the soil. Had we elected to use a tile drain field, the calculation would be quite different. From Table 10.14, we see that for a 2-minute percolation test and, say, a 2-ft wide trench, we would require 15 ft of trench for each 100 gal/day. Therefore, for 900 we would need

$$\frac{900}{100} \times 15 = 135 \text{ ft of trench}$$

If we used four trenches, as shown in Figure 10.50(b), each trench would be

$$\frac{135}{4} = 34 \text{ ft long}$$

This would fit easily on to the lot, as seen on the site plan in Figure 10.56.

Note on Figure 10.56 that dry wells that collect and disperse storm water are shown. As noted previously, their design is somewhat complex and is not covered in our discussion. They are shown for completeness.

624 / DRAINAGE AND WASTEWATER DISPOSAL

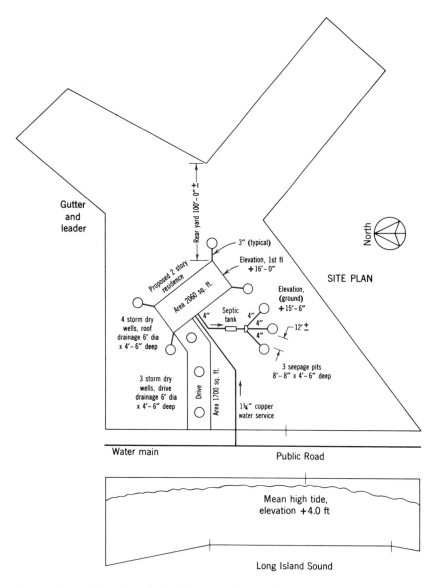

Figure 10.56 Site plan of the Mogensen house showing water service, sewage treatment equipment and dry wells for storm water disposal.

Key Terms

Having completed the study of this chapter, you should be familiar with the following key terms. If any appear unfamiliar or not entirely clear, you should review the section in which these terms appear. All key terms are listed in the index to assist you in locating the relevant text.

Air break
Backflow
Back venting
Black water
Branches
Branch interval
Branch vent
Building drain
Building trap
Circuit vent
Cleanouts
Closet bend
Common vent
Conductor
Continuous venting
Controlled roof drainage
Cross-fixture contamination
Developed length
Downspout
Drain-down time
Drainage fixture units (dfu)
Drainage stack
Dry well
Drainage, Waste and Vent (DWV)
Effluent
Fall
Fixture branch
Fixture trap
Floor drain
Flow velocity
Grease interceptor
Gray water
Gutters
Horizontal fixture branch
Horizontal jump
House drain

Indirect waste
Individual vent
Interceptor
Leader
Loop vent
Offsets
P trap
Private sewage disposal
Potable water
Recharging
Relief vent
Reventing
S trap
Sanitary sewer
Scouring action
Self-scour
Self-siphoning
Sewer gas
Soil pipes
Stack vent
Stack base fitting
Stacks
Storm drainage
Storm sewer
Trap
Trap arm
Trap seal
Vent Through Roof (VTR)
Vacuum breaker
Vent extension
Venting
Vent stack
Waste pipes
Wet vents
Wye fittings

Supplementary Reading

National Association of Plumbing-Heating-Cooling Contractors (NAPHCC); National Standard Plumbing Code—Illustrated, 1993.

Stein, B., and Reynolds, J. R. *Mechanical and Electrical Equipment for Buildings*, 8th ed., Wiley, New York, 1992.

Ramsey, G. G., and Sleeper, H. R. *Architectural Graphic Standards*, 8th ed., AIA, Wiley, New York, 1988.

Nathanson, J. *Basic Environmental Technology; Water Supply, Waste Disposal, Pollution Control*, Wiley, New York, 1986.

Problems

1. What are the scouring velocities of sanitary and storm drainage flow? Why are they different?
2. What are the minimum pipe slopes required for sanitary drainage piping? Why can large diameter pipes be sloped less than small pipes?
3. What are the factors that affect drainage velocity flow?
4. A maximum pressure variation of ±1 in. of water is permitted in drainage piping. Why is this limit so low? Express 1 in. of water pressure in terms of pounds per square inch and millimeters of mercury.
5. What is horizontal jump? How can it be minimized?
6. What is the function of a fixture trap? Why is the vertical distance between a fixture and its trap limited?
7. Name at least three functions of venting and explain each.
8. What is the difference between a vent stack and a stack vent?
9. Define and draw a typical isometric of the following vents:
   a. Individual vent.
   b. Circuit vent.
   c. Loop vent.
   d. Back vent.
   e. Wet vent.
10. Draw a drainage plumbing section for:
    a. A specific residence.
    b. A public toilet.
    All diagrams must be based on an actual installation that you have inspected and sketched. Omit pipe sizes.
11. Concerning minimum fixture clearances, what is:
    a. The size of a toilet compartment?
    b. The spacing on centers of water closets?
    c. The clearance in front of a water closet?
    d. The clearance in front of a lavatory?
    e. The clearance in front of a shower compartment opening?
12. Name materials suitable for the following services below ground.
    a. Gas piping.
    b. Sanitary building sewer.
    c. Water service.
13. A public toilet room is equipped with 20 water closets, flush valve operated; 10 urinals, pedestal, syphon jet blowout; and 10 lavatories, 1¼-in. waste.
    a. Calculate the drainage fixture units.
    b. Select the size for a horizontal fixture branch.
14. What trap size should be used for the following fixtures?
    a. Lavatory, common.
    b. Food waste grinder, commercial use.
    c. Sink, service type with floor outlet trap standard.
15. In disposing of the effluent from a septic tank, what conditions of ground and groundwater would cause you to specify:
    a. Open tile drains in trenches?
    b. Seepage pits?
16. An apartment house in the country will require a septic tank. What tank capacity would you select for this building of eight apartments, each having one bedroom?
17. A building has a daily sewage flow rate of 1200 gal. A soil test records 5 minutes for a 1-in. drop. How many seepage pit rings, 10-ft outside diameter and 5 ft deep, would serve the septic tank?
18. You find that there is a 500-ft run to the nearest storm sewer. The horizontal storm drain for a 40,000-ft² roof at a maximum hourly rainfall of 4 in. is to be selected. You have three choices. Determine the size of drain for:
    a. A ⅛-in./ft slope.
    b. A ¼-in./ft slope.
    c. A ½-in./ft slope.
19. In the 500-ft run of Problem 18, what will be the drops in feet of elevation for cases a, b, and c?
20. The "building for light industry" is being built in Juneau, Alaska. Its roof area is 10,100 ft². Roof drain fixture collects rain that is led away in a single horizontal drain at ½ in./ft slope.
    a. Select the correct pipe size.
    b. Would you enclose the drain in thermal insulation?
21. A 10,000-ft² flat roof has four conductors (vertical pipes) to equally carry away the storm drainage. The location has a maximum hourly rainfall of 3 in. Select a pipe size.

11. Introduction to Electricity

The daily work of the electrical technologist involves the layout, assembly and connection of electrical equipment. This equipment includes common items such as residential lighting and appliances, plus less common items such as commercial heating and refrigeration equipment, meters, conventional and solid-state controls and motor controls. It is not enough to know that a motor runs and a lamp lights when connected. Competent electrical technologists, design draftsmen or entry-level designers must have a sound knowledge of how and why electrical circuitry and equipment work as they do. Only in this way can they increase their usefulness in the technical office in which they are employed. This first chapter is devoted to the basics of electricity and circuitry. The study of the chapter will enable you to:

1. Become familiar with the basic electrical quantities of voltage, current and resistance, plus related terms and concepts.
2. Do circuit calculations in d-c and a-c circuits, in series and parallel arrangements.
3. Calculate power and energy in electric circuits.
4. Understand voltage levels and their uses.
5. Define ampacity and understand its application.
6. Know how electrical quantities are measured and how electrical meters work.
7. Understand the differences between d-c and a-c and apply that knowledge to circuit calculation.
8. Understand the differences in the nature and application of single-phase and three-phase a-c.
9. Acquire a working vocabulary of electrical circuit terms for both d-c and a-c.
10. Draw basic circuit diagrams.

11.1 Electrical Energy

Energy has historically been made available for useful work by burning a fossil fuel such as coal or oil. That is, the energy in the fossil fuel is released in the form of heat. This heat, in turn, is used as we wish—to warm our houses, prepare our food, and generate steam to drive turbo-generators that produce electricity. This last use is relatively recent, since commercial electrical power is barely a century old.

Electricity constitutes a form of energy itself, which occurs naturally only in unusable forms such as lightning and other static discharges, or in the natural galvanic cells that cause corrosion. The primary problem in the utilization of electric energy is that, unlike fuels or even heat, it cannot be stored and, therefore, must be generated and utilized in the same instant. This requires in many respects an entirely different concept of utilization than, say, a heating system with its local fuel tank or piping to a remote fuel storage facility. In the following sections, the concepts of electrical circuiting and application will be illustrated and explained so that you will obtain a thorough understanding of the practical application of electrical power. Our initial discussion will be of *direct current (d-c)*, which was developed earlier than *alternating current (a-c)* and which can be used to illustrate the principles of circuitry and power more easily than a-c. Subsequent sections will introduce a-c, the understanding of which is vital, since the overwhelming proportion of commercial electricity in use today is a-c.

Sources of Electricity

11.2 Batteries

If you are interested, you can follow, in any book on the history of electricity, the development of electric power from its origins in the work of Leyden, Galvani and Volta through Faraday, Darcy and Ampere (note the custom of perpetuating the scientist's name by attaching it to an aspect of his work). It is an interesting and often fascinating tale.

To summarize briefly, however, it was noted that when dissimilar metals were joined by a conducting solution such as salt water or a weak acid, a current would flow between the metals (electrodes). From this simple *galvanic cell*, the battery was developed to its present advanced state. See Figure 11.1. The limitations of batteries as sources of electrical power are fairly obvious. Since the voltage of a cell is determined entirely by its chemical components (1.5 v for a zinc-carbon cell, for instance), to obtain a workable voltage many cells must be placed in additive connection. Similarly, to obtain reasonable amounts of power, multiple cells must act together. Thus, the basic limitations become size and weight. Furthermore, cells discharge and can, therefore, supply power for only a limited time before dropping in voltage and power capacity. They are, therefore, not suitable for continuous, long-term power supply. They are, however, irreplaceable as sources of power for standby electrical service, telephone equipment power, railway signalling, backup power in electronic devices, and many types of portable devices.

Batteries, sometimes also called cells, are generally classified as either primary or secondary (i.e., either nonrechargeable or rechargeable, respectively). Nonrechargeable (primary) cells generate electricity through a nonreversible chemical reaction. Once having used up the chemicals in the unit, the battery is discharged or "dead." Batteries in this category include the common zinc-carbon cell and the mercury, alkaline-manganese, silver-oxide, and lithium cells [see Figure 11.1(a–d)]. Rechargeable (secondary) batteries produce electricity by a reversible chemical reaction. In these cells, the chemical components change as they produce electricity and discharge, but the original chemical composition can be restored by feeding electric current into the battery, that is, by charging the battery. The most familiar battery of this type is the lead-acid automobile battery, which, until recently, was used in every vehicle. Its operation is illustrated in Figure 11.1(e). Recently other rechargeable cell designs have come into use, including nickel-cadmium, lead-calcium and lead-antimony cells. (The latter type is used in the sealed automobile battery.) One such modern battery is shown in Figure 11.1(f).

Each battery type has its unique characteristics of voltage, discharge curve, temperature limitations and, for secondary cells, charging characteristics. Actual battery selection is almost always the responsibility of the manufacturer of the equipment using the battery. The electrical technologist, however, should be aware of the basic characteristics of cells in use for such common applications as emergency lighting, uninterrupted power supplies (UPS), and backup power for electronic controls, so that he or she can properly apply the equipment containing the cells. This can include determining that the ambient conditions of the space containing

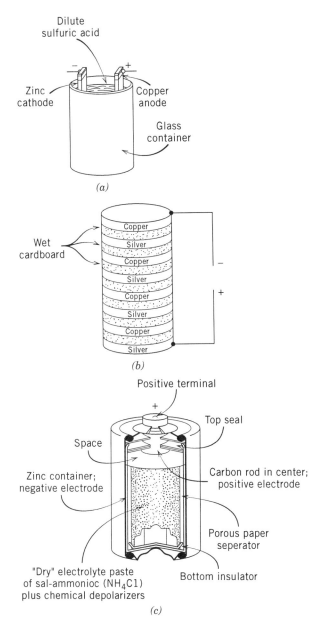

Figure 11.1 (a) Simple galvanic cell type battery, often referred to as a wet cell because of the liquid electrolyte (conducting solution) that connects the two electrodes.

Figure 11.1 (b) Voltaic pile battery, consisting of alternate discs of silver and copper, separated by discs of cardboard saturated in brine. This represents the beginning of the dry cell. The brine is the battery's electrolyte.

Figure 11.1 (c) Construction of common carbon-zinc flashlight battery. In principle, the common alkaline cell is similar, although internal construction is different. It uses zinc powder as its positive electrode (anode), manganese oxide as the negative electrode (cathode) and potassium hydroxide (KOH) as the electrolyte. The KOH electrolyte, which is a strong alkali, gives the cell its name.

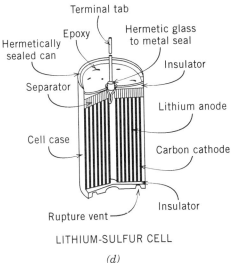

Figure 11.1 (d) Cutaway drawing of a lithium-sulphur dioxide cell. These primary (nonrechargeable) batteries have a number of advantages over conventional cells. Among them are high cell voltage (3.0 v), high power density, good low temperature performance, flat discharge characteristic and long shelf life. (Courtesy of SAFT America, Inc.)

FULLY CHARGED BATTERY

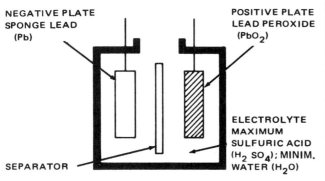

In a charged battery, active material of the negative plate is sponge lead (Pb); active material of the positive plate is lead peroxide (PbO_2); the electrolyte contains sulfuric acid (H_2SO_4) and a minimum of water.

BATTERY DISCHARGING

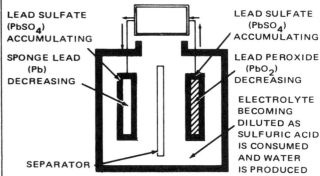

When the battery discharges, the electrolyte reacts with both the positive and negative plates... oxygen from the lead peroxide in the positive plates combines with hydrogen from the sulfuric acid to form water... lead from the lead peroxide combines with the sulfate from the sulfuric acid to form lead sulfate... hydrogen from the sulfuric acid combines with oxygen from the lead peroxide to form more water... lead from the sponge lead in the negative plates combines with the sulfate from the sulfuric acid to form lead sulfate... and electric current flows.

DISCHARGED BATTERY

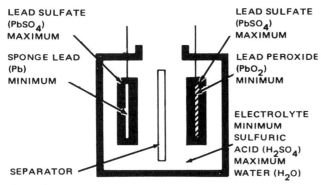

In a discharged battery, most of the active material from negative and positive plates has been converted to lead sulfate ($PbSO_4$), and the electrolyte is greatly diluted with water (H_2O).

BATTERY RECHARGE

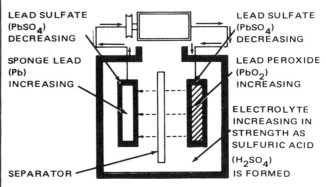

When the battery recharges, the chemical reaction between plates and electrolyte is reversed ...... lead sulfate from positive and negative plates reacts with the electrolyte to form sulfuric acid ... removal of sulfate from the negative plates restores sponge lead as the active material ... oxygen from the water recombines with the lead in the positive plates to form lead peroxide ... and the strength of the battery is restored.

(e)

Figure 11.1 (e) The four stages of one type of rechargeable battery are explained in detail, including the internal chemical changes involved at each stage. The battery illustrated is the common lead-acid automobile battery that was once universally used in motor vehicles. It is being replaced today by sealed lead-antimony batteries, which have the advantage that they do not need maintenance. Although the chemicals involved are different for lead-calcium, lead-antimony and other rechargeable cells, the principle of chemical reversibility by electric charge is the same. (Illustration from GSA publication Automobile Batteries.)

Figure 11.1 *(f)* This battery is a large modern rechargeable lead-calcium battery, designed for standby power use in telecommunications. The unit illustrated has 21 plates (visible through the clear plastic case), measures approximately 11 × 14 × 24 in. and weighs 304 lb, including 84 lb of liquid electrolyte. In this particular battery, all the plates are connected in parallel so that the entire battery is only one (high current) cell. It has an 8-hr capacity of 1680 amp-hr at which point the battery (cell) voltage will be down from a nominal 2.0 v. to 1.75 v. (Courtesy of C&D Charter Power Systems.)

the batteries are appropriate since many batteries are highly temperature-sensitive.

Certain battery designs, such as solar cells and fuel cells, are so highly specialized in design and application that they are rarely encountered by the electrical technologist and are, therefore, not covered here. Refer to the ample literature on the subject if you are interested in a topic not covered here.

11.3 Electrical Power Generation

As we stated earlier, the ability of a battery to produce electricity is limited by the chemical capacity of the cells. Nonrechargeable batteries use up their internal capacity and become discharged; rechargeable cells need another source of power to resupply them. The obvious need for a means of electrical power generation that is independent of chemical action was met with the development of the rotating electrical generator. As a result of the work of Oersted and Ampere, the electromagnet was developed. Soon afterward, Faraday in 1831 performed the crucial experiments that led to his discovery of *electromagnetic induction*. The principle involved is demonstrated in Figure 11.2. A wire moving through a magnetic field has a voltage induced in it. If the wire is formed into a loop, an alternating polarity voltage will be produced at the loop terminals whenever the loop is rotated in a magnetic field. If many of these loops are wound onto a rotor and the assembly rotated, we have the makings of a generator [Figure 11.2*(c)*]. The voltage produced will be alternating (see Section 11.16) but can easily be changed, or rectified, into d-c. With the development of the electrical generator, the search for a continuous source of electrical power ended. The basic generating unit just described has changed only in detail and sophistication to the present day.

Circuit Basics

11.4 Voltage

The electrical "pressure" (potential) produced by a battery is called *voltage* and is measured in units of *volts*. A carbon-zinc dry cell produces an *electrical potential* of approximately 1.5 v. This potential, or voltage, is constant in amplitude (magnitude) because it results from a continuous and unchanging chemical action. Because its direction (polarity) is also unchanging, it is designated d-c voltage. (The term *d-c* originated as an abbreviation for direct-current, but, because of the oddity of an expression such as direct-current voltage or the repetition of direct-current current, the terms universally accepted are *d-c voltage* and *d-c current*.) A d-c voltage of 1.5 units and positive polarity is shown in Figure 11.3*(a)*. A voltage of negative polarity and one unit of magnitude is also shown. In physical terms, the voltage of a battery can be likened to the water pressure in an hydraulic system. The ability to supply current (water in the hydraulic system) exists, but it does not function until a circuit is completed; then current begins to flow. Refer to Figure 11.4. Polarity is similar to flow direction; reversal of polarity causes current (consider the water analogy) to flow in the opposite direction.

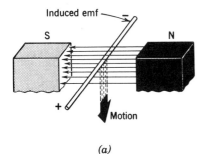

(a)

Figure 11.2 (a) The action fundamental to all generators is illustrated here. When a conductor of electricity moves through a magnetic field, a voltage is induced in the conductor, with polarity shown. The voltage is produced by electromagnetic induction.

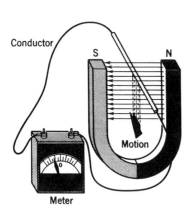

(b)

Figure 11.2 (b) The existence of voltage can be determined by connecting a meter to the conductor and noting current flow.

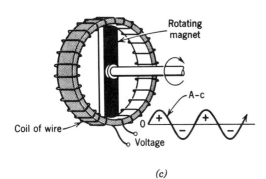

(c)

Figure 11.2 (c) It does not matter whether the conductor moves and the field is stationary, or vice versa, as long as there is relative motion between the two. If the wire or the field (magnet) is rotated, an alternating current is produced. The illustration shows the field rotating.

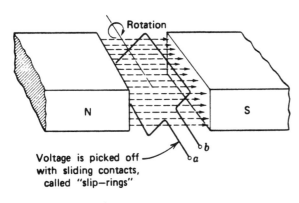

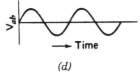

(d)

Figure 11.2 (d) Rotating a coil in a magnetic field produces an alternating sinusoidal voltage at terminals a and b because of the alternating polarity [see Figure 11.2(c)].

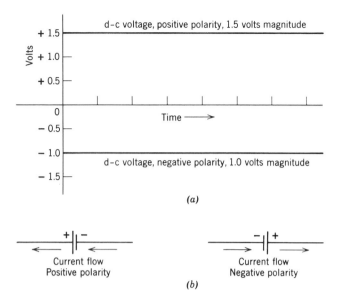

Figure 11.3 (a) Graphical representation of d-c voltage with positive and negative polarity. (b) Circuit symbol representation of a battery source. The longer bar is positive. By convention, current flows from positive to negative around the circuit, and negative to positive within the battery.

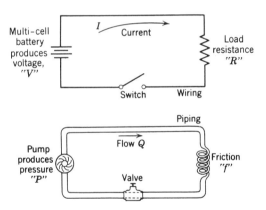

Figure 11.4 Electric-hydraulic analogy. The circuits show that voltage is analogous to pressure, current to flow, friction to resistance, wire to piping, and switches to valves. As a result of Ben Franklin's wrong guess, current is assumed to flow from positive to negative. This convention is still used today.

11.5 Current

When a circuit is formed, comprising a complete loop, and containing all the required components, a *current* will flow. The required components are:

(a) A source of voltage.
(b) A closed loop of components and wiring.
(c) An electric load. (The need for a load will become clearer when Ohm's Law is discussed in Section 11.7.)
(d) A means of opening and closing the circuit.

The electrical circuit shown in Figure 11.4 fulfills all four requirements; therefore, an electrical current will flow. In the hydraulic circuit, the amount of water flowing is proportional to the pressure and inversely proportional to the friction. Similarly, in the electrical circuit, the current in proportional to the voltage and inversely proportional to the circuit resistance (load). The higher the voltage, the larger the current. The higher the resistance, the lower the current. This relationship, known as *Ohm's Law*, is expressed by the equation

$$I = \frac{V}{R} \quad (11.1)$$

where
  I is current,
  V is voltage and
  R is resistance.

We will return to this equation repeatedly, and it is, therefore, very important that it be clearly understood. The letters I, V and R are normally used to represent current, voltage and resistance and should not be used for other electrical factors, to avoid confusion. The unit of current is the *ampere*, abbreviated *amp* or simply a.

11.6 Resistance

The flow of fluid in a hydraulic system is resisted by friction; the flow of current in an electrical circuit is resisted by *resistance*, which is the electrical term for friction. In a d-c circuit, this factor is called simply resistance. The unit of measurement of resistance is the *ohm*. Different materials display different resistance to the flow of electrical current. Metals generally have the least resistance and, therefore, are called *conductors* because they easily conduct electricity through them. The best conductors are the precious metals such as silver, gold and platinum with copper and aluminum being slightly inferior.

634 / INTRODUCTION TO ELECTRICITY

Conversely, materials that tend to prevent (resist) the flow of current, displaying high resistance, are called *insulators*. Glass, mica, rubber, oil, distilled water, porcelain and certain synthetics such as phenolic compounds have this insulating property. Such materials can, therefore, be used to insulate electric conductors. Common examples are the rubber covering on wire, porcelain cable supports, phenolic lamp sockets, glass pole-line insulators and oil-immersed electrical switches.

11.7 Ohm's Law

As mentioned previously, the relationship in a d-c circuit among current in amperes, voltage in volts and resistance in ohms is expressed by Ohm's Law, which is most frequently written in the form

$$V = IR \qquad (11.1)$$

but much more logically is written $I = V/R$, showing the relationship of current to voltage and resistance as already explained. We strongly recommend that this second form be remembered, instead of the mathematical relationship that volts = amperes × ohms, which has no logical basis. Physically, we start with a voltage and resistance and produce current. The equation

$$I = \frac{V}{R}$$

demonstrates this. A few examples will show the logic of Ohm's Law. All the examples refer to Figure 11.4.

Example 11.1 In the circuit of Figure 11.4, assume that the voltage is supplied by a 12-v automobile battery to a headlight R with a resistance of 1.2 ohms. Find the circuit current.

Solution:

$$I = \frac{V}{R} = \frac{12 \text{ v}}{1.2 \text{ ohms}} = 10 \text{ amp (d-c)}$$

Example 11.2 A telephone system load of 10 ohms is fed from a 48-v battery. What is the current drawn? (The current, like the voltage, will be d-c.)

Solution:

$$I = \frac{48 \text{ v}}{10 \text{ ohms}} = 4.8 \text{ amp}$$

Example 11.3 Instead of the switch shown, a 10-amp fuse is used, and the circuit resistance is 12 ohms. What is the maximum voltage that can be applied without blowing the fuse?

Solution:

$$I = \frac{V}{R} \quad \text{or} \quad V = IR$$
$$V = 10 \text{ amp (max)} \times 12 \text{ ohms}$$
$$V = 120 \text{ v}$$

Example 11.4 A 120-v d-c source (either a multicell battery or a d-c generator) feeds a 720-w electric toaster that has a resistance of 20 ohms. Find the current in the circuit.

Solution: For this example, the wattage of the toaster is an unnecessary piece of information. To find the current

$$I = \frac{V}{R} = \frac{120 \text{ v}}{20 \text{ ohms}} = 6 \text{ amp}$$

When we discuss the relation of wattage rating to V, I and R in Section 11.15, it will become clear that, in general, wattage is of more concern than the resistance.

Circuit Arrangements

11.8 Series Circuits

Obviously, circuit components can be arranged in many ways, as for instance in the complicated network shown in Figure 11.5. Still, even that maze can be reduced to two fundamental types of connections—series and parallel. In a series connection, a single path exists for current flow. That is, the elements are arranged in a series, one after the other, with no branches. In such an arrangement, voltage and resistance add. To illustrate, study Figure 11.6. As we explained in Section 11.2, the voltage of a single flashlight battery is about 1.5 v. Since this voltage is too low to use economically, four batteries are arranged in series, with additive polarity, to make a 6-v source. The circuit is then completed with a lamp load and a switch. It is not necessary that a circuit be complete to have a series connection. In Figure 11.7, batteries and *resistors* are shown in series connection, as separate groupings. The only requirement for a series connection is that only a single current path exist.

It is apparent from the previous discussion and from the illustrations shown thus far that component values simply add when connected in series.

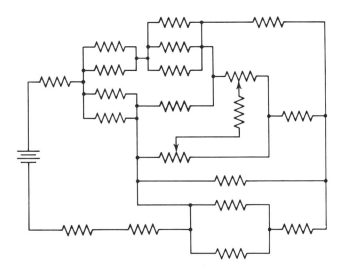

Figure 11.5 A maze of resistances connected in combinations of series and parallel.

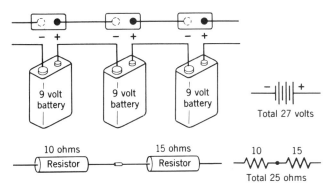

Figure 11.7 In a series connection, voltage and resistance add arithmetically, as shown.

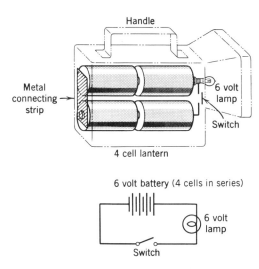

Figure 11.6 Phantom view of a standard four-cell, lantern-type flashlight showing the series (additive) connection of the cells.

Therefore,
$$V_{TOT} = V_1 + V_2 + V_3 + \cdots$$
and
$$R_{TOT} = R_1 + R_2 + R_3 + \cdots$$
where 1, 2, 3, ... represent the items connected in series. Thus, in Figure 11.4, the multicell battery can comprise any number of similar cells in series. Indeed, to supply 120-v emergency power for hospital operating room lights, this is exactly what is done. Some 86 to 95 nickel-cadmium cells, each with a voltage of approximately 1.3 v are connected in series to make the 120 v total. Similarly, a large resistance can be composed of several smaller resistances. An example illustrates this clearly.

Example 11.5 In Figure 11.8, the source is a 12-v automobile battery, and the load comprises two auto headlights connected in series, each with a resistance of 1.2 ohms. What is the current flowing in the circuit?

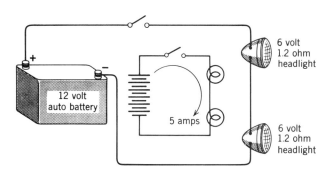

Figure 11.8 Physical and graphic representation of a possible d-c circuit.

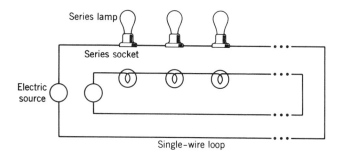

Figure 11.9 Physical and graphic representation of a series lamp circuit. Loss of one lamp disables the entire circuit. Furthermore, the point of fault is not obvious and requires individual testing of lamps.

Solution:

$$V = 12 \text{ v}$$
$$R_{TOT} = R_1 + R_2 = 1.2 + 1.2 = 2.4 \text{ ohms}$$
$$I = \frac{12 \text{ v}}{2.4 \text{ ohms}} = 5 \text{ amp}$$

Note that, although such an arrangement—of two 6-v lights in series on a 12-v service—is possible, it is not used often, since the failure of one unit will cause both to go dark. This is the principal disadvantage of a series connection feeding more than a single load. Since there is only a single current path, a failure in any one unit causes a break in the circuit and thereby kills the entire circuit. Strings of ornamental lights are occasionally made in this fashion to reduce costs, since only a single wire is required around the circuit. See Figure 11.9. Another disadvantage of the series connection is that when a single lamp goes out, not only does the entire string go dark, but the location of the fault is also unknown. This is a great nuisance and accounts for the rarity of this type of light string. To avoid this problem, a parallel connection is used. This connection allows multiple loads to be fed without a failure in one disrupting the others.

11.9 Parallel Circuits

In *parallel (multiple) connection*, the loads being served are all placed across the same voltage and, in effect, constitute separate circuits. In the hydraulic analogy, the connection is equivalent to a branched piping arrangement. See Figure 11.10. Figure 11.11 illustrates this multiple-circuit idea. Obviously, as in Figure 11.12, if the number of devices exceeds the current capacity of the circuit, the fuse will blow (to the surprise of the uninformed homemaker). This aspect of circuitry will be discussed in Chapter 13. For now, observe in Figure 11.12 that, unlike the series circuit, two wires are required throughout. (In proper and safe practice, a separate ground wire is also required. This, however, is an aspect of practical wiring design and is discussed in detail in Chapter 13. For the sake of clarity, at this point in our study, this ground wire is not shown in the illustration.)

The parallel connection is the standard arrange-

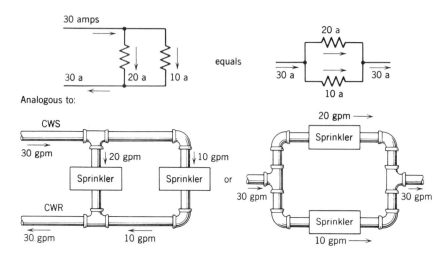

Figure 11.10 In a parallel connection, the flow divides between the branches, but the pressure is the same across each branch.

PARALLEL CIRCUITS / 637

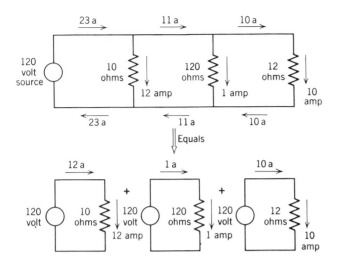

Figure 11.11 Note that loads connected in parallel are equivalent to separate circuits connected as a single circuit. Each load acts as an independent circuit, unrelated to, and unaffected by, the other circuits.

ment in all residential and commercial wiring. A typical lighting and receptacle arrangement for a large room is shown in Figure 11.13. Here the lights constitute one parallel grouping, and the convenience wall outlets constitute a second parallel grouping. The fundamental principle to remember is that loads in parallel are additive for current and that each has the same voltage imposed. This can be seen clearly by a careful study of Figure 11.11.

Notice in Figure 11.11 that the total current flowing in the circuit is the sum of the currents in all the branches, but that the current in each branch is determined by a separate Ohm's Law calculation. Thus, in the 10-ohm load, a 12-amp current flows, and so forth. Study this diagram until the numbers and the related Ohm's Law calculations are perfectly clear.

One additional point is important to appreciate. If we examine Ohm's Law again, we note, as pre-

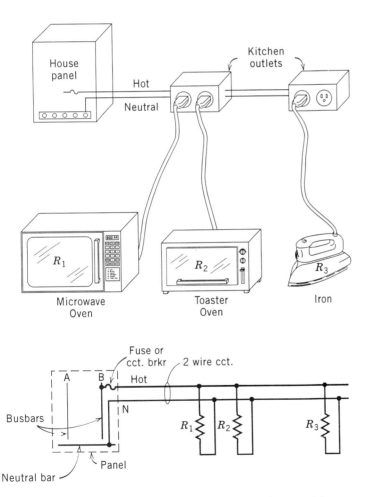

Figure 11.12 A sure way to overload a circuit. (The panel fuse or circuit breaker should open to "clear" the overload).

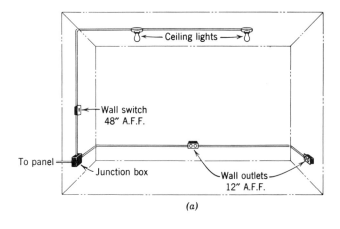

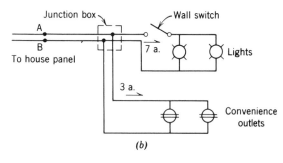

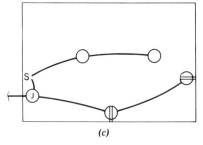

Figure 11.13 Parallel groupings of lights and wall outlets are in turn connected in parallel to each other. Circuit is shown *(a)* pictorially, *(b)* schematically and *(c)* as on an electrical working drawing (architectural-electrical) plan.

viously stated, that current is inversely proportional to resistance. Thus, as resistance drops, current rises. Now look at the circuit of Figure 11.13*(b)*. Under ordinary conditions that circuit will carry 10 amp and will operate normally. But, if by some chance, a connection appears between points *A* and *B*, the circuit is shortened so that there is no resistance in the circuit. The current rises instantly to a very high level, and the condition constitutes a *short circuit*, that is, a circuit with no (appreciable) load. If the circuit is properly protected, the fuse or circuit breaker will open, and the circuit will be disabled. If not, the excessive current will probably generate enough heat to start a fire.

Alternating Current (A-C)

11.10 General

As will be explained in Section 11.16, circuits operating at higher voltages have lower power loss and lower voltage drop and are almost always more economical to construct because of savings in copper. (Power loss is very important in heavy power transmission and distribution but is much less so in small branch circuits.) For these reasons a-c, which allows easy transformation between voltages and much easier generation than d-c, came into favor at the close of the 19th century. A bitter battle ensued between the proponents of d-c electricity such as Edison and the advocates of a-c electricity, including George Westinghouse. Edison opposed a-c on the ground that the high voltages involved in transmission were dangerous. Other opponents derisively labelled a-c as "do-nothing" electricity because the voltage is positive for one-half cycle and negative for one-half cycle, yielding, they claimed, a net of zero. That this is nonsense was not appreciated. The a-c alternation can be likened to the strokes of a saw cutting a piece of wood. Just as a saw cuts on the up stroke and the down stroke, so a-c does work in both the positive and negative halves of its cycle. See Figure 11.14. Fortunately the supporters of a-c won the argument, thus paving the way for the enormous technological advances that we enjoy today, all of which would be impossible without a-c. Reference is made to a-c in Figure 11.2*(c)* and *(d)*. A review of these figures now would be helpful as an introduction to the detailed discussion of a-c that follows.

11.11 A-C Fundamentals

The basic characteristics previously discussed for d-c also apply to a-c, but with some important differences. In addition, certain aspects of a-c, such as frequency, do not apply to d-c. Study Figure 11.14. Note that the a-c current goes through one positive loop and one negative loop to form one complete cycle, which then repeats. The number of

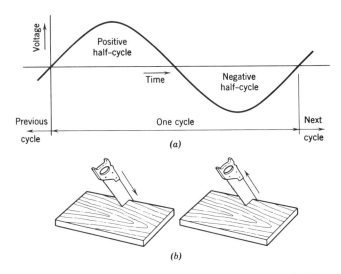

Figure 11.14 Alternating current (a) does work in both halves of its cycle, just as the saw in (b) cuts in both directions of stroke.

and calculation is exactly as for d-c. However, the use of resistance, or impedance, is rare in calculations that are important to the technologist. Of prime importance are power and energy calculations. These will be studied, for both a-c and d-c, in Sections 11.15 and 11.16.

11.12 Voltage Levels and Transformation

We mentioned previously that one of the principal reasons for the victory of a-c over d-c was the ease with which voltages could be transformed with a-c. We also mentioned some of the advantages that come from this ability. As part of the same study it is important to understand the application of the different voltage levels in use today. Voltage level transformation is accomplished with a device appropriately called a *transformer*. A transformer has no moving parts. It consists of a magnetic (iron) core on which are wound primary and sec-

times this cycle of a plus and a minus loop occurs per second is called the *frequency* of a-c, and is expressed, logically, in cycles per second. Because of the tendency of people to say simply *cycles*, which is not the same thing as cycles per second, the electrical profession agreed some years ago to change the expression and at the same time to honor a great physicist who did extensive research in electromagnetism. Therefore, cycles per second are now properly called *hertz* after H. R. Hertz. A correct description of ordinary house current in the United States would be 120 volt, 60 hertz (abbreviated 120 v, 60 hz). In Europe, the normal frequency is 50 hz, and in some parts of eastern Europe and Asia, 25 hz. This latter frequency is so low that flicker is easily noticeable in incandescent lamps. The frequency of d-c is obviously 0 hz, since the voltage is constant and never changes polarity.

In a-c, the quantity corresponding to resistance in a d-c circuit is called *impedance* and is usually represented by the letter Z. It is a compound of resistance plus an a-c concept called *reactance* but, once given, is treated exactly as resistance is in d-c circuits. That is, Ohm's Law for a-c is expressed

$$I = \frac{V}{Z} \quad (11.2)$$

where Z is impedance, expressed in ohms.

For resistive loads, such as incandescent lights and heaters, the impedance is equal to the resistance,

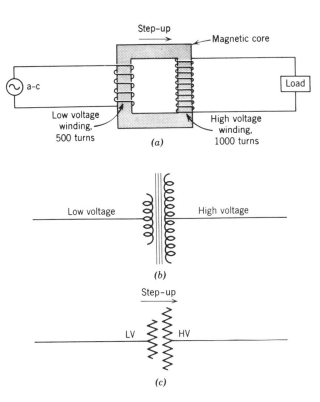

Figure 11.15 Pictorial, diagrammatic and single-line representation of a transformer, in (a), (b) and (c). Representation (c) is most often used in electrical construction drawings.

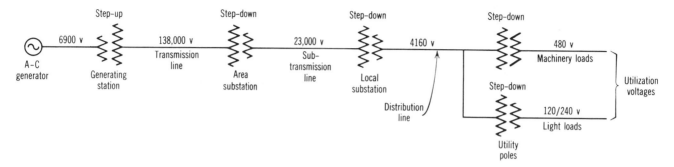

Figure 11.16 One-line diagram of a typical a-c power distribution system, from generation through transmission and distribution, down to utilization level.

ondary windings as shown diagrammatically in Figure 11.15. The voltages appearing at the terminals of these windings are in direct proportion to the number of winding turns. Thus, in Figure 11.15, if 120 v a-c were connected to the left side, which contains 500 turns, 240 v would appear on the right side, which contains 1000 turns. The input side is normally called the transformer's *primary;* the output is normally called the transformer's *secondary.* In this instance, the transformer would be a 120/240 v step-up transformer with a 120-v primary and a 240-v secondary. The very same transformer can usually be used to step-down by reversing the supply and the load. Thus, if a 240-v source were connected to the 1000-turn winding, 120 v would appear on the left, and the unit would be a 240/120 v step-down transformer. The 240 v would then be primary, and the 120 v would be secondary. Simply stated, transformers are, in general, reversible.

Figure 11.16 shows the voltage levels commonly in use and the transformations required to obtain them. Each of the levels was selected after careful engineering and economic studies. The tendency is always to higher levels at all points except at the utilization level. As insulation materials and techniques are improved, voltage levels are raised. Indeed, 345,000 v (345 kv) and 500 kv are becoming common, and voltages up to 750 kv are in use. At the distribution level, 13,800 v is gradually replacing 4160 v, and 46 kv is replacing 23 kv. But at the end of the line, the voltage we find in our house panel will remain 120/240 v for safety reasons, and no increase is contemplated.

Transformers are made in single-phase and three-phase construction. (For an explanation of these terms, see Section 11.14.) They are rated in *volt-amperes (va)* or *kilovolt-amperes (kva).* Like all electrical equipment, transformers have power losses due to the electrical resistance of the windings. These losses show up as heat, which causes the transformer temperature to rise when it carries load. Transformers are rated for full load temperature rise above ambient of 80°C, 115°C and 150°C, depending on the type of insulation on the windings and the unit's cooling medium. The cooling medium is either air (dry-type transformer) or a special liquid with which the transformer is filled.

The cheapest liquid transformer coolant with good electrical characteristics is mineral oil. However, because of its flammability, its application is limited. Nonflammable liquid coolants, called *askarels,* generally containing polychlorinated biphenyl (PCB), were at one time very widely used. However, since 1979, PCB use in new equipment has been banned by federal regulation due to negative ecological impact. Because of this ban, a number of other liquid coolants with low flammability and good electrical and physical characteristics have been developed. As a general rule, transformers used indoors are dry (air-filled) and those used outdoors are liquid-filled. Liquid-filled transformers can also be used indoors, but this generally requires construction of a transformer vault.

An important characteristic of transformers in modern wiring systems is their ability to carry *harmonic currents* without overheating. (Harmonics are multiples of the 60-hz line frequency that are caused by electronic equipment, computers, fluorescent lighting and other equipment found in modern buildings.) This ability is rated by a number called the transformer's *K-factor.* Details of this subject are highly technical and are mentioned here for information only.

11.13 Voltage Systems

In all the preceding material, we have assumed a two-wire circuit. This is generally the case within the building. The usual lighting circuit is 2-wire, 120 v, 60 hz, as is also the normal receptacle circuit that feeds the wall convenience outlets (plugs). See Figure 11.17(a). (In the latter, in modern construction, there is a separate ground wire. The entire subject of grounding is discussed in Chapter 13.) However, the service entrance to the house will most probably be a 3-wire circuit, commonly written: 3-wire, 120/240 v, 60 hz. This system is illustrated in Figure 11.17(b) and has these advantages:

- 120 v is available for lighting and receptacle circuits.
- 240 v is available for heavier loads such as clothes dryers and air conditioning compressors among others.
- The service conductors are sized on the basis of 240 v, rather than 120 v, effecting a large saving in copper, as will be explained in Section 11.18.
- Voltage drop is low.

The National Electrical Code® * (NEC)® (see Section 12.1) requires that all conductors be identified throughout the circuit. Identification can be made by color of insulation, tagging of conductors at each box and outlet or other effective means. The grounded (neutral) conductor (not the equipment ground) must be white or natural gray. The equipment ground conductor (discussed in Chapter 13), when insulated, must be green or green with yellow stripes. All other conductors (phase wires, switching wires, etc.) can have any color or identification means, provided that the identification is used throughout the system. Thus, if phase A is black, it must be black throughout, and so on for the other system conductors.

The illustrations in Figure 11.17 are an introduction to circuit calculation. We shall study these calculations more intensively later in the book. At this point, you are encouraged to study Figure 11.17(c), (d) and (e) until you fully understand the amperage (current) figures shown on the diagrams.

---

*National Electrical Code® and NEC® are registered trademarks of the National Fire Protection Association, Inc., Quincy, Ma. 02269

11.14 Single-Phase and Three-Phase

To make life a bit more complicated, the engineers who championed a-c also developed a system of wiring known as three-phase electricity. The subject is theoretically complex but, in practice, is relatively simple. Instead of single-phase a-c, which can be either 2-wire or 3-wire, as was already explained and illustrated, three-phase a-c uses four wires. It consists of three "hot" legs designated as phases A, B and C, plus a neutral N. This is illustrated in Figure 11.18. (The term *hot* is used in common electrical terminology to mean a conductor or point with voltage.) Three-wire, three-phase a-c, without a neutral wire, was once in common use but is now used only for special applications. For our purposes, three-phase a-c is simply a triple circuit and can be treated as such.

Lighting and most outlet loads are connected between any phase leg and neutral, and heavy machinery loads are connected between phase legs only. This system of wiring is used in all buildings where the building load exceeds approximately 50 kva (kilovolt-amperes) or where it is required by three-phase machinery. Technologists need not be concerned with the complexities of phase relationships; instead, they should understand the application of three-phase and single-phase a-c. Ample opportunity to study practical application will be afforded. Notice in Figure 11.18 that the voltage between phase and neutral is 120 v. The line-to-line voltage (between phases) has intentionally been omitted to avoid confusion. It is 208 v, not 240 v as might be expected. It is important to note here that the neutral conductor, even though it is common to all three of the phase conductors, does not carry triple the phase current. On the contrary, it carries only unbalanced current, and if loads on all three phases are balanced, it carries no current at all. Maximum neutral current is equal to phase current (ignoring harmonics).

Circuit Characteristics

Having learned the basics of a-c and d-c, circuit elements and their arrangements, we are now in a position to begin studying the characteristics of electrical circuits, which are of primary concern to electrical technologists. These are power, energy, current, circuit voltages, voltage drop and current carrying-capacity (ampacity).

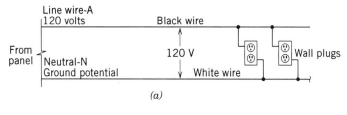

Figure 11.17 *(a)* Typical convenience receptacle (wall plug) circuit. Note designation of lines as A and Neutral, and color coding. Ground connection and separate ground wire are not shown, for clarity.

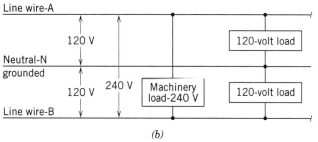

Figure 11.17 *(b)* In a 3-wire, 120/240-v arrangement, which is common for residences and other small buildings, the loads are generally arranged as shown. The 120-v loads are lighting and convenience outlets, plus small appliances. An effort is generally made to balance the loads as is explained in the Chapter 13 discussion on circuitry. Here, also, the separate ground wire is not shown, for clarity.

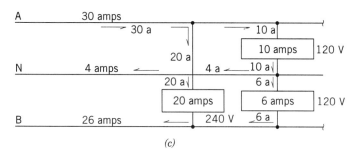

Figure 11.17 *(c)* In a typical 3-wire 120/240-v circuit, the neutral carries only the difference between the 120-v loads on the two line wires. This helps to reduce voltage drop and permits savings in wiring. Note that this connection is single-phase a-c, consisting of two single-phase a-c circuits with a common neutral conductor.

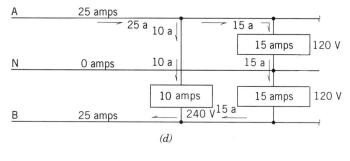

Figure 11.17 *(d)* When the 120-v loads of a 3-wire system are balanced, the neutral carries no current and the line wires A and B carry the entire load. An equivalent 2-wire 120-v system with the same power load would carry 50 amp in each line, or leg, requiring much larger conductors. (Proof of this statement is left to the reader.)

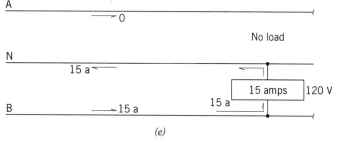

Figure 11.17 *(e)* Maximum neutral current equals the current in one phase and occurs at maximum unbalance, that is, one circuit loaded and the second circuit completely unloaded. Compare to *(c)* and *(d)*.

POWER AND ENERGY / 643

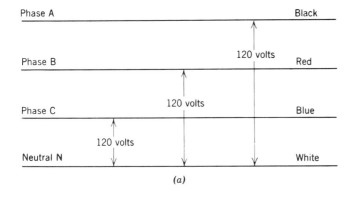

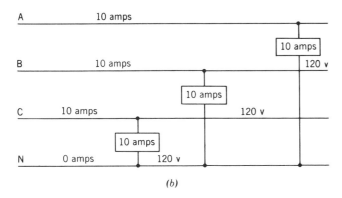

Figure 11.18 *(a)* A typical three-phase wiring system, showing phase to neutral voltages. In effect it constitutes three single-phase circuits with one, common, return wire. *(b)* In a three-phase system, the neutral carries only unbalanced current; when the phases are balanced, the neutral carries no current.

11.15 Power and Energy

We assume that you are familiar with the concepts of power and energy (work) from your physics studies, including scientific terms such as joules, ergs and calories. The world of technology obviously also uses concepts of power and work, but with different units of measurement and a different approach. In electrical work, the units generally used are *horsepower, watt, kilowatt* and *kilowatt-hour*. (When using metric units, horsepower is replaced by kilowatts.) We will now carefully define and explain these terms, along with their physical concepts, and show how they differ from these same concepts as used in physics. This is necessary for a thorough understanding of the practical uses of these quantities.

If we refer to the hydraulic circuit of Figure 11.19, we see that a source of motion energy is supplied to the shaft of the pump. This energy is transferred to the water in the pump and is used in forcing the water through the piping and any hydraulic devices in the circuit. As is shown in the diagram, an energy transfer takes place from rotary motion in the pump to heat (friction), mechanical work in the hydraulic devices and motion (kinetic energy) of the water. Thus, the work done (energy used) in this arrangement consists of transferring energy from one place to another and from one form to another. The longer the system remains in operation, the more energy is used. Physics students would calculate the masses and velocities involved and, then, the work (energy) being expended, by multiplying force and distance. They generally are not interested in the system's *power*, which is the rate of energy utilization.

The technological approach is quite different. Technologists would determine the flow required in gallons per minute and the head (friction) of the system. With these two quantities and a chart for a particular type of pump, they would select a pump and determine the horsepower required to drive it. They are not primarily interested in the system's energy (power × time of operation), since they assume a continuously operating system. They would, however, calculate energy requirements to determine operating costs or to check their energy consumption. To do this, they would convert horsepower to kilowatts (kw), would calculate kilowatt-hours and, with the power company's rate schedule, would then calculate the cost of operating the system. Thus, we see that the technologist's approach is the opposite of the scientists'. Technologists first calculate power and from that energy; scientists do the reverse. Their approaches differ because their purposes differ.

Electrical systems similar to the hydraulic system of Figure 11.19*(a)* are shown in Figure 11.19*(b)*, *(c)* and *(d)*; the difference is the source of power. The energy transfer that would interest physics students is shown on the diagrams. What would interest technologists is quite different. They start with the power rating of each device (its rate of energy use) in watts or kilowatts and work from there. The basic unit of power in electrical terms is the watt, abbreviated w. Since this unit is small for many power applications, we also use a unit 1000 times as large—the kilowatt, abbreviated kw. Obviously, 1 kw equals 1000 w. It is also convenient to remember that 1 horsepower (hp) equals 746 w, or approximately three-fourths of a kilowatt.

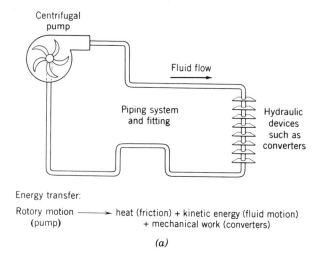

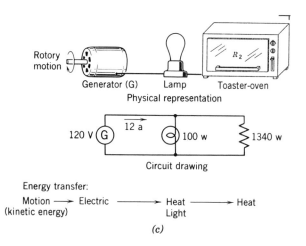

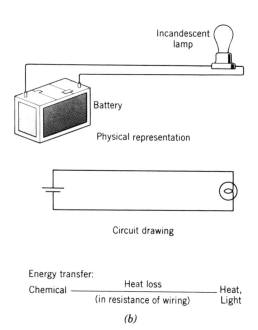

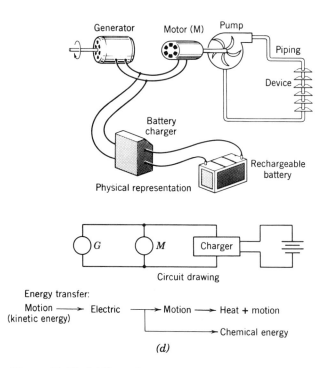

Figure 11.19 (a) Typical hydraulic circuit comprising piping and some mechanical-hydraulic device. Work is done transferring energy from one place to another and from one form to another. (b) In this single battery circuit, chemical energy is converted to electric energy in the battery and then to light and heat in the lamp. The power in the circuit is the rate of energy transfer. The amount of current flowing is proportional to the power in the circuit.

Figure 11.19 (c) Here the generator converts its input into electric power, which lights (and heats) the lamp and heats the toaster oven. (d) Generator supplies power, which transfers energy to heat, motion (kinetic energy) and chemical energy. Only the chemical energy, that is, the recharged battery, is recoverable. The amount of current flowing in each circuit depends on its own electric characteristics.

644

$$1 \text{ hp} = 746 \text{ w} \approx \tfrac{3}{4} \text{ kw}$$

In d-c circuits, power (wattage) is equal to the product of the circuit current and voltage, that is

$$P_{DC} = VI \qquad (11.3)$$

Power calculation in a-c circuits is discussed later in this section.

A few examples of the application of these units and their calculation should help here.

Example 11.6 Referring to the circuit in Figure 11.19(c), assume that the lamp is rated 100 w and that the toaster oven is rated 1340 w. Assume the circuit to be d-c. Find the current flowing in the circuit.

Solution:

$$\text{Generator voltage} = 120 \text{ v d-c}$$
$$\text{Total circuit wattage} = 100 \text{ w} + 1340 \text{ w} = 1440 \text{ w}$$

Since the total power in a d-c circuit is equal to the product of the current and voltage, that is

$$P_{DC} = VI \qquad (11.3)$$

we have

$$1440 \text{ w} = 120 \text{ v} \times I$$

or

$$I = \frac{1440 \text{ w}}{120 \text{ v}} = 12.0 \text{ amp}$$

Example 11.7 Refer again to Figure 11.12 and assume the following data:

Microwave oven   1000 w
Toaster oven   1200 w
Iron   1400 w

Also, assume a conventional house circuit voltage of 120 v. Calculate the circuit current. Assume for this example that the circuit is d-c (although it is actually a-c).

Solution:

$$\text{Total circuit power} =$$
$$1200 \text{ w} + 1000 \text{ w} + 1400 \text{ w} = 3600 \text{ w}$$

Knowing that

$$P = VI \quad \text{or} \quad I = \frac{P}{V} \quad \text{(in a d-c circuit)}$$

We have

$$I = \frac{3600 \text{ w}}{120 \text{ v}} = 30 \text{ amp!!}$$

Since the usual house circuit is sized and fused for 15 or 20 amperes, we have at least a 50% overload. In such a situation, the fuse or circuit breaker must open to prevent overheating and a possible fire.

We stated in Example 11.4 (Section 11.7) that the resistance of the toaster involved did not especially interest us but that the wattage, or power rating, did. Now let us rework that example as it would actually be done.

Example 11.8 Assume a 120-v d-c source feeding a 720-w toaster. Find the current in the circuit. (This is a practical problem, since, as we shall learn, the fuse size is based primarily on the circuit current.)

Solution:

$$\text{Power} = \text{Voltage} \times \text{Current (in d-c circuits)}$$

or

$$\text{Current} = \frac{\text{Power}}{\text{Voltage}}$$

Therefore,

$$I = \frac{720 \text{ w}}{120 \text{ v}} = 6 \text{ amp}$$

Note that this is the same answer we obtained in Example 11.4 but that this time we obtained it from normally available data. Check the nameplate on a few pieces of kitchen equipment. Notice that the data given are wattage and voltage. Resistance is of little concern, especially since, in many instances, hot resistance is quite different from cold resistance; that is, the resistance varies with the temperature of the item.

To review, we have learned that:

- Power is the rate at which energy is used and is expressed in horsepower (hp), watts (w) and kilowatts (kw).
- $1 \text{ hp} = 746 \text{ w} \approx \tfrac{3}{4} \text{ kw}$ (the symbol $\approx$ means approximately equal to).
- For d-c circuits only,

$$P = VI \quad \text{or} \quad I = \frac{P}{V} \qquad (11.3)$$

As an item of interest, 1 hp is approximately the power capability of a horse for a considerable period of time. This was determined by James Watt, whose name is now used as the basic unit of power. A normal man can exert a horsepower for a short period of time by, for instance, running up a flight of ten steps in 2 seconds. (It has also been estimated that a watt equals one rat-power.)

At this point we must learn the extremely important difference between a-c and d-c in calculations of (circuit) power. Indeed, a-c power calculations are much more important because, essentially, all light and power circuitry is a-c. D-C is used only for a very few specialty items in modern building design. As noted previously, in d-c, power is the product of voltage and current. That is, for d-c

$$\text{Volts} \times \text{Amperes} = \text{Watts}$$

In a-c, the product of volts and amperes gives a quantity called volt-amperes (abbreviated volt-amp or va), which is not the same as watts. That is, in a-c,

$$\text{Volts} \times \text{Amperes} = \text{Volt-amperes}$$

To convert volt-amperes to watts, or power, we introduce a dimensionless quantity called *power factor*, which is abbreviated pf.

In a-c circuits, impedance (see Section 11.11) consists of resistance and reactance (a-c resistance of inductance and capacitance) that causes a phase difference between voltage and current. This phase difference is represented by an angle, the cosine of which is called the power factor. This quantity is extremely important in that it enables us to calculate power in an a-c circuit. The a-c power equation is similar to that for d-c with the addition of this special a-c term of power factor.

Power in an a-c circuit is calculated:

$$w = V \times I \times \text{pf} \qquad (11.4)$$
Watts = Volts × Amperes × Power factor

or

$$w = VI \times \text{pf}$$
Watts = Volts-amperes × Power factor

In a purely resistive circuit, such as one with only electric heating elements, impedance equals resistance, power factor equals 1.0, and wattage equals volt-amperage. A few examples here should make applications of these equations clear.

Example 11.9 Recalculate Examples 11.6, 11.7 and 11.8 assuming a-c circuits.

Solution: Since all the appliances listed are basically heating devices for which the impedance equals the resistance, the power factor is 1.0 and the calculations are the same as those given above for d-c.

Example 11.10 Refer to Figure 11.13. Assume the two ceiling lights to be 150 w each, incandescent. [Incandescent lamps are resistive loads and, therefore, have a unity (1.0) power factor.] Also assume the load connected to one convenience outlet to be a 10-amp hair dryer and blower, with a power factor of 0.80. Calculate the current and power in the two branches of the circuit, and the total circuit current, assuming a 120-v a-c source.

Solution: In the circuit branch feeding the lights, we have

$$P = VI \times \text{pf}$$
$$300 \text{ w} = 120 \text{ v} \times I \times 1.0$$
$$I = \frac{300 \text{ w}}{120 \text{ v}} = 2.5 \text{ amp}$$

If we wished to calculate circuit resistance (which is equal to the impedance, since the load is purely resistive),

$$Z \text{ (impedance)} = R \text{ (resistance)} = \frac{V}{I} = \frac{120 \text{ v}}{2.5 \text{ amp}}$$
$$= 48 \text{ ohms}$$

Again, we point out that this latter figure is of little practical use to us and is calculated simply to show technique.

In the second branch we have a 10-amp, 0.8-pf load. Therefore,

Power in watts = Volts × Amperes × Power factor
$$P = 120 \text{ v} \times 10 \text{ a} \times 0.8 = 960 \text{ w}$$

but the circuit volt-amperes are

$$\text{va} = 120 \text{ v} \times 10 \text{ a} = 1200 \text{ va}$$

This latter figure is significant in sizing equipment, as we shall learn later.

To calculate the total current flowing from the panel to both branches of the circuit, we must combine a purely resistive current (lamp circuit) with a reactive one (dryer circuit). The exact value of current can be calculated only by vectorial addition, but that is beyond the scope of this book. In normal practice, the currents are simply added arithmetically. This yields a result that is somewhat higher than actual and is, therefore, on the safe side when we are sizing equipment. Hence,

Approximate total current = 2.5 a + 10 a = 12.5 amp

Actual current is 12.1 amp; our error in approximating is 3.2%, which is acceptable in branch circuit calculation. The preceding calculations and techniques will become routine with practice. One further example at this point will demonstrate the importance of power factor in normal situations.

Example 11.11 The nameplate of a motor shows the following data: 3 hp, 240 v a-c, 17 amp. Assume an efficiency of 90%. Calculate the motor (and, therefore, circuit) power factor.

Solution:

$$1 \text{ hp} = 746 \text{ w}$$

therefore,

$$3 \text{ hp} = 3 \times 746 \text{ w} = 2238 \text{ w (output)}$$

$$\text{Efficiency} = \frac{\text{Power output}}{\text{Power input}}$$

so

$$\text{Power input} = \frac{2238 \text{ w}}{0.9} = 2487 \text{ w}$$

but for a-c,

$$\text{Power} = \text{Volts} \times \text{Amperes} \times \text{Power factor}$$

so

$$2487 \text{ w} = 240 \text{ v} \times 17 \text{ a} \times \text{Power factor}$$

and

$$\text{Power factor} = \frac{2487 \text{ w}}{240 \text{ v} \times 17 \text{ a}} = 0.61$$

Note the large difference between volt-amperes and watts.

$$VI = 240 \text{ v} \times 17 \text{ v} = 4080 \text{ va}$$
$$P = 240 \text{ v} \times 17 \text{ a} \times 0.61 = 2487 \text{ w}$$

Notice here the effect of the power factor. It causes the current drawn to increase with no corresponding increase in power. The motor circuit carries 17 amp at 0.61 pf, for a total power of 2487 w. If pf were unity (1.0), the circuit wattage for this same 17-amp current would be

$$P = VI(\text{pf}) = 240 \text{ v } (17) \text{ a } (1.0) = 4080 \text{ w}$$

This means that, for the same circuit loading of 17 amp, a circuit at 1.0 pf can carry 1593 w (4080−2487) more than a circuit carrying 17 amp at 0.61 pf. Since circuit conductors and circuit loading are both sized on the basis of current, we can readily see the importance of keeping power factor as high as possible. High power factor means lower current. The techniques for raising power factor are called power factor improvement or power factor correction procedures. Details are highly technical and are beyond our scope here. However, when a choice exists between two similar pieces of equipment with different power factors (single-phase motors, for instance), the unit with the higher pf is preferable. Unfortunately, it is almost always more expensive, and the final choice becomes a matter of economics.

11.16 Energy Calculation

Since power is the rate of energy use, it follows, as already stated, that

$$\text{Energy} = \text{Power} \times \text{Time}$$

That means that the amount of energy used is directly proportional to the power of the system and to the length of time it is in operation. Since power is expressed in watts or kilowatts, and time, in hours (seconds and minutes are too small for our use), we have watt-hours (wh) or kilowatt-hours (kwh) for units of energy. Since watt-hours are also too small for normal use, the standard unit of energy is the kilowatt-hour (kwh). One kilowatt-hour equals one kilowatt in use for one hour, or any other two factors that give the same result, such as ½ kw for 2 hr or 2 kw for ½ hr, and the like.

Example 11.12

(a) Find the daily energy consumption of the appliances listed in Example 11.7, if they are used daily as follows:

Toaster oven	15 min
Microwave oven	20 min
Iron	2 hr

Solution:

(b) If the average cost of energy (not power) is $0.08/kwh, find the daily operating cost.

Solution:

Toaster oven (1200 w)	1.2 kw × ¼ hr = 0.30 kwh
Microwave oven (1000 w)	1.0 kw × ⅓ hr = 0.33 kwh
Iron (1400 w)	1.4 kw × 2 hr = 2.80 kwh
Total	3.43 kwh

The cost is

$$3.43 \text{ kwh} \times \$0.08/\text{kwh} = \$0.2744$$

or approximately 27 cents.

The power being used at any specific time during the day by a residential household varies consider-

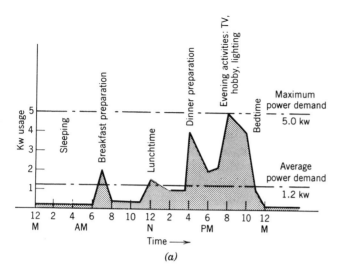

Figure 11.20 *(a)* Hypothetical graph of power usage for a typical household with nonelectric cooking. Total energy use is represented by the area under the curve. Here it totals 28.8 kwh. Therefore, the average daily power demand is

$$\frac{28.8 \text{ kwh}}{24 \text{ hr}} = 1.2 \text{ kw}$$

ably. If we were to graph the power in use for a typical American household during a normal weekday, the plot might look something like that in Figure 11.20*(a)*. Gas cooking is assumed. The average power demand of the household is obviously much lower than the maximum. The ratio between the two is called the *load factor*. This factor runs about 20–25% for a typical household. If the household were to use electric cooking, the meal preparation peaks would be much higher and its power graph might appear as in Figure 11.20*(b)*. Such a household would have an average daily power use of about 1.8 kw (1800 w) and a load factor of 25–30%.

Example 11.13 It has been estimated that the average power demand of an American household with nonelectric cooking is 1.2 kw. Calculate the monthly electricity bill of such a household, assuming a flat rate of $0.085/kwh.

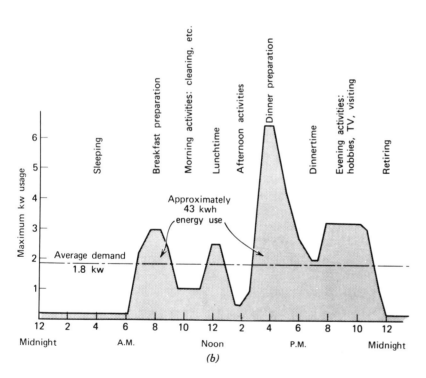

Figure 11.20 *(b)* Hypothetical power graph of a household with electric cooking. This household has a 24-hr energy usage of 43 kwh, giving an average demand of

$$\frac{43 \text{ kwh}}{24 \text{ hr}} = 1.8 \text{ kw}$$

Since maximum demand is 6.5 kw, the load factor is

$$\frac{1.8 \text{ kw}}{6.5 \text{ kw}} \times 100 = 27.5\%$$

(From B. Stein, J. S. Reynolds, and W. J. McGuinness, *Mechanical and Electrical Equipment for Buildings*, 7th ed., 1986, John Wiley & Sons, New York. Reprinted by permission of John Wiley & Sons.)

Solution:

Monthly energy consumption =
$$1.2 \text{ kw} \times \frac{24 \text{ hr}}{\text{day}} \times \frac{30 \text{ days}}{\text{month}} = 864 \text{ kwh/month}$$

Electric power bill =
864 kwh × $0.085/kwh = $73.44/month

11.17 Circuit Voltage and Voltage Drop

To make certain that the ideas connected with basic electrical circuitry are clearly understood, we now analyze several circuits by studying voltage and current division, and voltage drop.

Refer to Figure 11.8. What are the voltages across each component in the circuit? The voltage across each lamp is

$$V = IR = 5 \text{ amp} \times 1.2 \text{ ohms} = 6 \text{ v}$$

This is obviously correct, since the drop across the two lamps must be 12 v to equal the supply voltage. This establishes an important principle. The sum of the voltage drops around a circuit is equal to the supply voltage. This principle is most important in series circuits. In parallel circuits, each item has the same voltage across it and constitutes a circuit by itself, as explained previously. Refer now to Figure 11.11. Note that each resistance has the same 120-v drop across it, equal to the supply voltage. But also notice that in a parallel arrangement all the currents add. This leads us to the second important fact to remember. In parallel circuits, the voltages are the same, and the currents through each branch (may) differ. Look, for example, at Figure 11.12. Note that each appliance has the same voltage imposed, since they are in parallel. Each, however, draws a different current.

Returning to the example of Figure 11.8, we update the information and calculation on the basis of our present knowledge. The lamps would actually be arranged in parallel, as is shown in Figure 11.21(b). We have added a few of the normal automobile accessories. In Figure 11.21(a), we have drawn the same circuit, but it is fed from a 6-v battery (or generator) as was common years ago. Note that the current at 6 v is double that at 12 v because the voltage is half. This is necessarily so, since $P = VI$ and the power rating of the devices is the same in both cases.

At this point you may well ask, What difference does it makes what the voltage is? and Why do most auto manufacturers use the higher voltage?

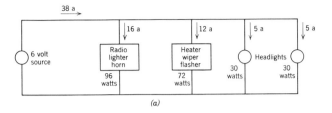

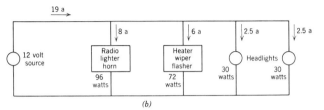

Figure 11.21 Typical auto circuits with identical wattage accessories but at different voltages. Note that the circuit current is halved when voltage is doubled.

The answers to those questions require an understanding of current-carrying capacity of cables. This characteristic, called *ampacity*, will be discussed more fully in the sections on circuitry, but it is introduced here to demonstrate the effect of circuit voltage and the reason for the development of a-c.

11.18 Ampacity

Simply defined, *ampacity* is the ability of a conductor to carry current without overheating. Heat is generated by the resistance of the wire to the current flow. Up to this point, we have ignored the resistance of the wiring, assuming perfect conductors, that is, conductors with no resistance. This is obviously not true. When a wire carries current, the voltage drop in the wire, by Ohm's Law, is

Voltage drop in wire = Current carried × Resistance of wire

The power loss in the wire can be calculated in the same fashion that we have previously calculated it, that is, as the product of voltage and current:

Power loss in wire = Circuit current × Voltage drop

or

$$P = I \times (I \times R) = I^2 R \qquad (11.5)$$

We have here derived the general law of power loss: Power loss is equal to the component's resistance times the current squared. This applies whether the current is a-c or d-c and to all devices and components, not only wire. Thus, a coil of wire can have an impedance Z, which is much higher than its resistance, but the power loss in the coil is still the product of the current squared and the coil resistance. An example should make this clear.

Example 11.14 A coil-type electric room heater is rated 12.4 amp, 120 v. Power factor is 0.90 (90%), and coil resistance is 8.71 ohms. Calculate the heater's wattage.

Solution:

$$\text{Power} = VI \times \text{pf} = 120 \text{ v } (12.4 \text{ a}) (0.90) = 1339 \text{ w}$$

or, alternately

$$\text{Power} = I^2 R = (12.4 \text{ a})^2 \times (8.71 \text{ ohms}) = 1339 \text{ w}$$

The heater's impedance is:

$$Z = \frac{V}{I} = \frac{120}{12.4} = 9.68 \text{ ohms}$$

Note that resistance is the product of impedance and power factor, that is

$$R = Z \times \text{pf} \qquad (11.6)$$

therefore,

$$R = 9.68 \text{ ohms} \times 0.9 \text{ pf} = 8.71 \text{ ohms}.$$

This checks the given resistance value exactly.

Thus, we see that in all circuits, whether d-c or a-c, the power loss in any static circuit component, be it wiring, heater, or lighting, is always the product of the component's resistance and the square of the current in the component. (This is obviously not true for devices such as motors, where a large part of the power taken is converted to kinetic energy.)

Returning now to power loss in wiring, we know that this power loss is converted to heat, which must necessarily raise the temperature of the wire. As we will learn in Section 12.5, the current rating of a wire depends on this temperature rise, which, in turn, as we have just demonstrated, varies with the square of the current being carried by the conductor ($P = I^2 R$). Therefore, reducing the current in a circuit reduces the wiring heat loss by a larger factor, since the heat generated is proportional to the current squared. It, therefore, follows that, for the same temperature rise, a conductor with much higher resistance can be used, that is, a smaller diameter wire.

For instance, if it is desired to carry 1440 w (a typical resistive appliance rating), the current flowing is 12 amp at 120 v, or 6 amp at 240 v. Since basic wire insulation is good for 300 v, the same type of wire insulation can be used for both voltages and the same amount of power can be carried at the higher voltage with a thinner wire, that is, with less than one-half the investment in copper. This accounts for the almost universal use of 220–240 v for basic circuitry in most of the world except for the United States. Referring to Figure 11.21, the circuit in *(b)* is better than the arrangement in *(a)* because

- Smaller wire can be used, giving cost economy.
- Power loss is usually smaller, giving energy economy.
- Voltage drop is lower.

A further consideration that enters the discussion is the fact that a small diameter wire can safely carry more current in proportion to its weight than a large conductor, due to a phenomenon called *skin effect*. This increases the advantage of using small diameter wires still further. Of course, there are limits to raising the voltage of a circuit such as the type of insulation required. However, all other factors being equal, the higher the circuit voltage, the more economical the system.

The inherent advantages of high voltage for transmission and distribution spurred the search for an easy way to change from one voltage to another. This cannot be done with d-c, but it can very easily be accomplished with a-c using a transformer, as we learned in Section 11.12. This one fact was the major cause of the development of a-c and the almost complete abandonment of d-c for general power purposes. Let us examine a practical situation that illustrates these principles.

Example 11.15 A 4.8-kw electrical swimming pool heater is located 300 ft from the residence from which it is to be fed. The residents have a choice of either 120- or 240-v feed from the house panel. Which should they choose? (Assume 100% power factor since the load is purely resistive. This is not exact, but it is close enough for engineering purposes in this example.)

Solution: Let us make parallel solutions for the different voltages.

	120 v	*240 v*
Current drawn	$\dfrac{4800 \text{ w}}{120 \text{ v}} = 40.0$ amp	$\dfrac{4800 \text{ w}}{240 \text{ v}} = 20.0$ amp
Minimum wire size required to carry the current without overheating	No. 8 AWG, copper	No. 12 AWG, copper
Relative cost of wire	2.2	1.0
Voltage drop	17.0 v or 14.2%	8.5 v or 3.5%
Power loss	No. 8 wire $I^2R = 334$ w	No. 12 wire $I^2R = 210$ w

Since a 14% voltage drop for the 120-v feeder is obviously unacceptable, the wire size would have to be increased to at least a No. 2 AWG, making the cost relation about 10 to 1, instead of the 2.2 to 1 shown. This example should make the advantage of higher voltage perfectly obvious.

11.19 Electrical Power Demand and Control

If we refer to Figure 11.20(a), we can see that the maximum electrical power taken (maximum demand) is more than four times as large as the average power demand. The utility company, by terms of its public franchise, must supply the maximum power demand of each customer. This means that it has to build and maintain generating and distribution facilities for maximum coincident (simultaneous) demand. However, it bills users for energy, which is the equivalent of the average demand over the billing period. Therefore, although the customer of Figure 11.20(a) demands 5.0 kw for short periods each day, that same customer pays for only 1.2 kw average. The electric utility must, however, build and maintain the extremely expensive generating and power distribution facilities necessary to supply the maximum, albeit short-time, demand, while billing for the lower average demand. To compensate for this inequity, utility companies have long included in their rate structure a demand charge. This charge is made on measured maximum demand over a short measuring period (15–30 minutes). The charge is, therefore, highest for customers with short time, high power demand or, in other words, users with low load factors. (This charge is usually made only on bills of commercial and industrial users and only rarely for residential customers.)

It should be obvious, therefore, that it is in everyone's best interest to reduce peak loads, that is, to increase load factors:

- It is in the utility's interest, in order to reduce the cost of generating and distributing facilities and to increase overall efficiency of operation.
- It is in the (nonresidential) customers' interest, in order to reduce their electricity bills.
- It is in the national interest, in order to avoid unnecessary power plant and power line construction.

There are various techniques for reduction of high maximum power demand. Utilities offer lower rates for off-peak use and rebates and bonuses for installation of demand control equipment. Users can install the demand control equipment most suitable to their facility, after an

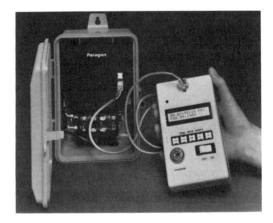

Figure 11.22 Programmable electronic time-control switch designed for application with time-of-use (off-peak) utility price schedules. Programming device (shown hand-held) is separate from the switch and can be used by the utility company to program many customer switches. The illustrated unit is arranged for 365-day scheduling. This permits effective use with utility price schedules that vary with the seasons of the year. Typical controlled loads are water heaters, thermal storage units, water and air accumulators, and any other load that either by its nature or by design can be delayed for several hours. (Courtesy of Paragon Electric Company, Inc.)

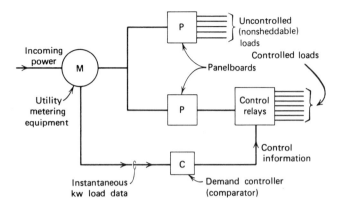

Figure 11.23 Block diagram of a system of automatic electric power demand control. The demand controller receives instantaneous load data from the metering equipment [see Figure 11.30 *(a, c)*,] compares it to preset limits and disconnects and reconnects controllable loads automatically, in order to keep kilowatt demand (load) within these preset limits. (From B. Stein and J. S. Reynolds, *Mechanical and Electric Equipment for Buildings*, 8th ed., 1992, John Wiley & Sons, New York. Reprinted by permission of John Wiley & Sons.)

engineering study of the facility's loads. This equipment, which is known by various names such as automatic load shedding control and peak demand control, basically performs a single function. This function is to control connection and disconnection of electrical loads in accordance with a set scheme. The simplest of these schemes simply controls loads on a fixed time basis. The more sophisticated controllers recognize actual minute-by-minute power use, consider the utility's demand measurements, and then control loads on a priority basis. Although design, selection and application of these units is not usually part of the electrical technologist's responsibilities, an awareness of their functions is certainly required, in view of their increasingly wide use.

Figures 11.22 and 11.23 show a simple time-control unit and the block diagram for a highly sophisticated unit.

11.20 Energy Management

Energy management must not be confused with electric demand control. Energy management systems in buildings most frequently control all the energy used including steam, hot water, electrical heating, gas and oil. Electrical power demand is frequently part of an overall energy management system, and this leads to some confusion. The purpose of an energy management system is primarily energy conservation. In the area of electrical energy, this management takes many forms. One form is the use of energy conservation devices such as occupancy sensors. (These same sensors can be used to control air-conditioning as another aspect of energy conservation.) Another form is control of electric motor loading because, as we shall learn, motors are more efficient when operated at full load than at partial load. Still another aspect of overall energy management is some type of an electrical power demand control system as discussed in Section 11.19. The field of energy management is an engineering specialty by itself. Although it is not the responsibility of electrical technologists to design or apply these systems, it is definitely their responsibility to provide the required power and wiring for the system. As such they must be familiar with system components and their nomenclature, functions and electrical requirements.

Measurement in Electricity

11.21 Ammeters and Voltmeters

In the preceding sections, we explained the fundamental electrical quantities of voltage and current and gave the units involved as volts and amperes, respectively. As is true for other physical quantities such as pressure and temperature, the need existed for a simple means of measuring these electrical quantities. This need was first met by the development of the galvanometer movement illustrated in Figure 11.24. Everyone at one time or another has felt the repulsion between the similar poles of two magnets held close together and, conversely, the attraction between opposite poles. See Figure 11.24a. This principle is used in the galvanometer. It causes a deflection of the pointer as a result of the repulsion between a permanent magnet and an electromagnet. See Figure 11.24*(b)*. The electromagnet is formed when current flows in the coil, and its strength is proportional to the amount of current flowing. Thus, a strong current causes a larger deflection of the needle and, therefore, a higher reading on the dial.

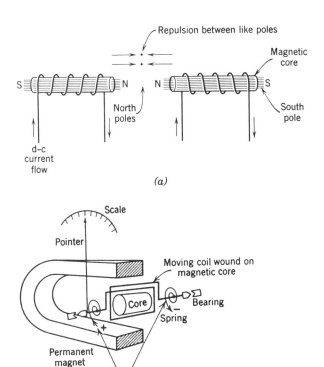

Figure 11.24 (a) Diagram showing basic electromagnetic principal and interaction between electromagnets. Any iron core becomes an electromagnet when current flows in a coil wound on it, as shown. (b) Principle of the electromagnet is used in all galvanometer-type d-c meter movements. Current flowing in the coil forms an electromagnet. It interacts with the permanent magnet to cause a pointer deflection proportional to the current flow. (Courtesy of Wechsler, Division of Hughes Corporation.)

The basic unit [see Figure 11.25(a)] is actually a microammeter, sensitive to millionths of an ampere. To make this highly sensitive device usable for practical currents, we simply divert, or shunt away, most of the current, allowing only a few microamperes to actually flow in the meter coil. The scale is then calibrated in the larger current units (e.g., milliamperes or amperes). This is illustrated in Figure 11.25(b). The size of the *shunt*, that is, its resistance, depends upon the proportion of current that we wish to divert or shunt away. Thus, the higher the ampere scale calibration, the lower the resistance of the shunt, in order that more of the current pass through it.

To use the same microammeter unit as a voltmeter, we put a large resistance *(multiplier)* in series with the meter and, by doing so, again limit the current flowing in the meter coil to a few microamperes. The scale is then calibrated in volts. Note in Figure 11.25(c) that the meter is precisely the same as the ammeter of Figure 11.25(b) except that a different method is employed to limit the actual meter current to the few microamperes permissible. All analog (as opposed to digital) d-c meters are made as shown. Analog a-c meters operate on basically the same principle except that, instead of a permanent magnet, an electromagnet is used. That way, when the polarity reverses, the deflecting force remains in the same direction. A d-c meter connected to an a-c circuit simply will not read, since inertia prevents the needle from bouncing up and down 60 times a second. Typical analog-type switchboard meters are shown in Figure 11.26.

The meters just described are called *analog meters* because they read electrical units in proportion to mechanical forces in the unit, that is, by analogy. Modern electronics has produced solid-state electrical meters (see Figure 11.27) that display the measured electrical values in either digital or analog (dial) mode. They operate in a number of ways, all of which are different from that already described, and all of which are beyond the scope of this book. Remember, however, that just because a meter reads digitally, it is not necessarily highly accurate. Accuracy depends on the quality of the internal circuitry. Digital meters are obviously easier to read since no visual interpretation of the meaning of a dial pointer and a graduated scale is involved. This advantage, plus the constantly dropping cost of sophisticated electronics will undoubtedly give solid-state digital meters an ever larger market share in the future.

11.22 Power and Energy Measurement

The measurement of current, voltage and power factor in practical application is generally not as important to the technologist as the measurement of power and energy, since the power company regulates voltage very accurately, and the other measurements are of more interest to engineers than to technologists. However, power and energy measurements are of great interest in determining loads, costs, energy consumption and proper system operation. To measure power, we take advantage of the fact learned earlier that power in an a-c

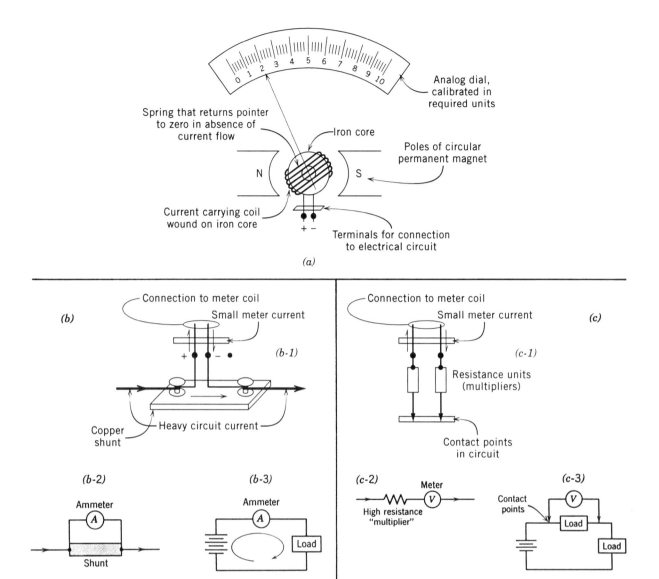

Figure 11.25 *(a)* The basic construction of a d-c analog meter movement is illustrated. [Compare to the schematic of Figure 11.24 *(b)*.] The iron core on which the current coil is wound is fixed, and the spring and pointer pivot on it, with a constant restraining (and damping) force exerted by the spring.
(b) When used as an ammeter, a low resistance shunt is placed in parallel with the meter movement. It carries most of the circuit current. The shunt is sized according to the maximum meter amperage to be measured. The diagrams show the physical *(b-1)*, schematic *(b-2)* and circuit *(b-3)* representations.
(c) When used as a voltmeter, a high resistance multiplier is placed in series with the meter, and it serves to reduce the current to the meter coil to a few microamperes. The multiplier is sized according to the required calibration of the dial, in volts. Diagrams show the physical *(c-1)*, schematic *(c-2)* and circuit *(c-3)* representations.

Figure 11.26 Typical a-c and d-c switchboard electrical meters. (Courtesy of Wechsler Instruments, Division of Hughes Corporation.)

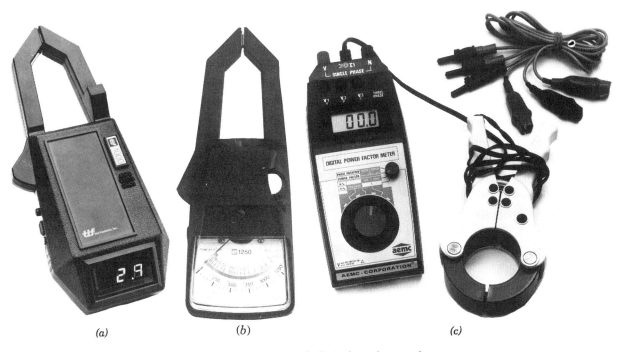

Figure 11.27 *(a)* Solid-state clamp-on-type meter with digital readout and automatic ranging. This latter feature eliminates the necessity to preselect the meter range and is particularly useful where the magnitude of current or voltage is unknown. The clamp-on feature permits use without wiring into, or otherwise disturbing, the circuit being measured. The meter, which is approximately $8 \times 3 \times 1.5$ in., weighs under a pound and is powered by four AA cells. Scales are 0.1–1000 amp a-c, 0.1–1000 v a-c, and 0.1–1000-ohms resistance, with $\pm 2\%$ accuracy. (Courtesy of TIF Instruments, Inc.)
(b) Solid-state auto-ranging clamp-on a-c meter with analog-type readout. Similar in design to the meter in *(a)* except with somewhat larger range scale and additional features such as peak current measurement. (Courtesy of TIF Instruments, Inc.)
(c) Digital power factor meter, measures power factor on single and three-phase circuits. In addition, the meter can measure 0–600 v a-c, 0–2000 amps a-c. Connections to the circuits being measured are made with the clamp-on probes shown. A very similar unit is available that measures power in both sinusoidal and nonsinusoidal wave forms, in a range of 0.1 w through 2000 kw. (Courtesy of AEMC Corporation.)

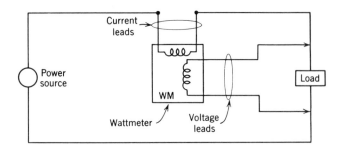

Figure 11.28 Schematic arrangement of wattmeter connections. Note that the current coil is in series with the circuit load while the voltage leads are in parallel. See also Figure 11.29.

circuit is equal to the product of the voltage, current and power factor, that is,

$$P = VI \times \text{pf}$$

Actual physical construction of a conventional (not solid-state) wattmeter is straightforward. We have two coils: a current coil that is similar in connection to an ammeter and a voltage coil that is similar in connection to a voltmeter. By means of the physical coil arrangement, the meter pointer deflection is proportional to the product of the two and, therefore, to the circuit power. The meter can be calibrated as we wish, depending on the size of the shunts and multipliers, from a few milliwatts up to many kilowatts. The schematic arrangement is shown in Figure 11.28.

To measure energy, the element of time must be introduced, since

$$\text{Energy} = \text{Power} \times \text{Time} \quad (11.7)$$

This is done (again referring to conventional, non-electronic meters) by using the voltage and current coils of a wattmeter to drive a small motor and counting its revolutions. Speed of rotation of the motor is proportional to the power being used. In actuality, the motor is a rotating disc, as is shown in the schematic diagram of a kilowatt-hour meter shown in Figure 11.29. These meters are referred

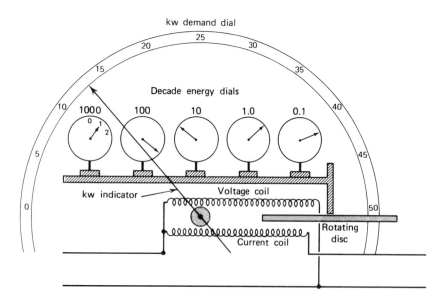

Figure 11.29 Typical induction-type kilowatt-hour meter with kilowatt demand dial. Dials register total disc revolutions that are proportional to energy. Disc speed is proportional to power. Note that the current coil is in series with the load and that the voltage coil is in parallel. (From B. Stein and J. S. Reynolds, *Mechanical and Electric Equipment for Buildings*, 8th ed., 1992, John Wiley & Sons, New York. Reprinted by permission of John Wiley & Sons.)

to as induction-type a-c kilowatt-hour meters and, until the invention of solid-state units, were universally used to measure energy consumption in a-c circuits. (D-C energy meters are also available but are not of general interest due to the rarity of d-c power.)

As can be seen from Figure 11.29 the kilowatt-hour energy consumption can be read directly from the dials. If the numbers involved are too large, or if it is needed for calibration reasons, a multiplying factor is used to arrive at the proper kilowatt-hour consumption. This number is written directly on the meter face or nameplate, and we multiply the meter reading by it to get the actual kilowatt-hours. In the absence of such a number, we can assume that the meter reads directly in kilowatt-hours.

If we want to know the energy consumption of a particular circuit and a kilowatt-hour meter is not available, we may also use a wattmeter and a timer to get the same result by using the relation

$$\text{Energy} = \text{Power} \times \text{Time}$$

However, this is effective only for a constant load such as lighting. Loads that vary in size or turn on and off cannot be measured except with a meter that sums up the instantaneous energy used. This is what a kilowatt-hour meter does. A wattmeter measures instantaneous power, whereas energy, involving time, must be summed. Figure 11.20 illustrates this. Note that the wattmeter measures the amount of power in use at any one time, whereas the kilowatt-hour meter measures the energy used over a period of time. Thus, in Figure 11.20(a), a wattmeter would read 2 kw at 7 A.M.,

Figure 11.30 (a) Several types of modern solid-state electrical meters. (a-1) This modern single-phase electronic meter includes a mechanical register similar to that shown in Figure 11.29, plus a large liquid crystal display. It can be arranged to provide either time-of-use billing or demand/time-of-use billing, in addition to its energy measurement function. It can also be configured to provide load control, demand threshold alert, end-of-interval alert, and load profile recording. (a-2) Field programming of the meter shown in (a-1) is readily accomplished with a hand-held programming device, as shown. (Photos courtesy of Landis and Gyr.)

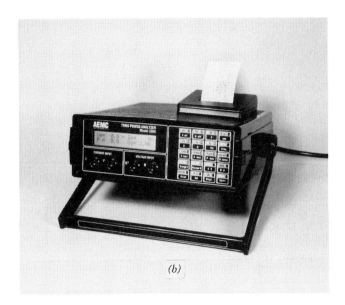

(b)

Figure 11.30 *(b)* This portable programmable power analyzer unit measures approximately 13 × 7 × 12 in. and weighs 6 lb. It is capable of measuring (and computing) 19 electrical parameters among which are current, voltage, energy, power, power factors and frequency and of providing both instantaneous digital readout and printed records (hard copy). Connections permit all functions to be addressed by remote control. (Courtesy of AEMC Corporation.)

(c)

Figure 11.30 *(c)* This device is a multifunction microprocessor-based overcurrent unit. It provides true-root-mean-square (rms) sensing of phase and ground currents, selectable tripping characteristics and trip alarm contacts. In addition, the unit can be used to display and transmit information on phase and ground currents (Photo courtesy of Cutler-Hammer.)

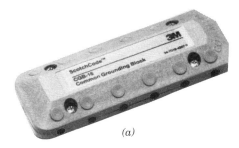

(a)

(c)

Figure 11.31 (c) Infrared heat tracer. This instrument is a highly sensitive and accurate thermometer. It can be used, among other applications, to detect hot spots in electrical machinery, concealed wire and cable, outlet boxes and other electrical equipment. Such hot spots frequently indicate a fault or malfunction, which can then be repaired before they cause a major fault or complete breakdown of the equipment involved. (Photos courtesy of 3M.)

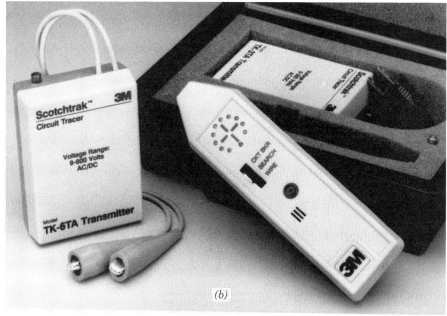

(b)

Figure 11.31 Diagnostic and detection instruments. (a) Multiple wire identifier. This device permits persons concerned with installing multiwire cables to identify the individual wires at both ends of the cable without the time-consuming procedure of ringing-out each wire. (b) Circuit tracer. This electronic instrument enables the technologist using it to identify "hot" and neutral wires in feeder and branch circuit wiring. It is, therefore, extremely useful in renovation projects and troubleshooting work.

0.3 kw at 10 A.M., 4 kw at 4 P.M., and so forth. The energy consumed is represented by the area under the curve and can be measured only by a meter that sums, or integrates, the instantaneous power required over a period of time. The induction-type kilowatt-hour meter just described is one such device. Today, however, a multitude of solid-state meters that can accomplish the same purpose with greater accuracy and dependability are available. They can also perform other functions such as recording and transmitting measured data, limiting power demand and providing alarms, performing load surveys, analyzing power and energy demand trends, and other complex functions. Again, although these usages are more the responsibility of electrical engineers than technologists, it is important for the latter to be aware of the application of modern metering methods. A few of these solid-state power and energy measuring devices, including some that furnish additional functions, are shown in Figure 11.30.

Because manual reading of kwh meters of individual consumers involves so much time and labor cost, even without consideration of routine difficulties such as hard to reach meters and bad weather, electric utility companies have long sought a more cost-effective means of measuring consumer energy usage. Today, thanks to modern microelectronics, a number of interesting, labor-saving kilowatt-hour meters have been developed. One type is equipped with a programmable electro-optical automatic meter-reading system that can be activated either locally or from a remote location. See Figure 11.30(a). The meter data are transmitted electrically to a data-processing center where they may be used by the utility to prepare customers' bills, and to study, in combination with other such data, area load patterns, equipment loading, and so on. Another type is equipped with a miniature radio transmitter that can be remotely activated to transmit the current kilowatt reading. The meter-reader drives along the street and activates the meter from inside his or her vehicle by entering a customer code into a digital pad. A special receiver not only receives the transmitted kilowatt-hour data but also encodes and records it automatically. These and other new kilowatt-hour meters are obviously much more costly than the traditional units but are finding increasing use in areas with high labor costs and in rural areas where the distances between users is relatively large.

11.23 Specialty Meters

In addition to the ammeters, voltmeters, wattmeters and kilowatt-hour meters discussed previously, there are many other types of meters in daily use in the electrical construction industry. These include meters that measure power factor, phase rotation, harmonic content, capacitance, inductance, impedance, resistance, voltage transients and so on. There are also specialty devices that are specifically intended to help technologists, maintenance personnel, electricians and other people engaged in diagnostic and detection work in electrical systems. Three such meters are shown in Figure 11.31.

Key Terms

Having completed the study of this chapter, you should be familiar with the following key terms. If any appear unfamiliar or not entirely clear, you should review the section in which these terms appear. All key terms are listed in the index to assist you in locating the relevant text. Some of the terms also appear in the glossary at the end of the book. You should now be familiar with the following terms:

Alternating current (a-c)
Ampacity
Amperes, amps, a

Analog readout
Balanced, unbalanced loads
Clamp-on meter

Conductor
Current
Demand control
Digital readout
Direct current (d-c)
Electrical potential
Electrode
Electrolyte
Electromagnet
Electromagnetic induction
Energy
Frequency
Galvanic cell
Harmonic currents
Hertz
Impedance
Insulator
K-factor
Kilovolt-ampere (kva)
Kilowatt-hour
Load factor
Multiplier

Ohm
Ohm's Law
Parallel circuit
Parallel (multiple) connection
Phase leg, neutral
Power
Power factor (pf)
Primary, secondary
Reactance
Resistance, ohm
Resistor
Series circuit
Short circuit
Shunt
Single phase
Skin effect
Three-phase
Transformer
Voltage, volts, v
Volt-ampere (va)
Watt, kilowatt

Supplementary Reading

Stein, B., and Reynolds, J. *Mechanical and Electrical Equipment for Buildings*, 8th ed., Wiley, New York, 1992. Material parallel to that in this Chapter is presented in somewhat greater detail and may, therefore, be useful to students seeking a deeper study level.

Problems

1. a. An automobile with a 12-v electrical system is equipped with the devices listed below. Assuming that all devices were operated with the engine off (power drawn from battery), how long will a 40-amp-hr battery last? (For the purpose of this problem, assume that a 40-amp-hr battery will supply any product of amperes and hours equaling 40; for example, 20 amp for 2 hr, 40 amp for 1 hr or 5 amp for 8 hr, even though this is only approximately true.)

    Accessories:

Twin headlights	225 w each
Fan	20 w
Radio	10 w

b. The same automobile has a starting motor that draws 240 amp when cranking. How long will the battery supply the cranking motor?
2. For the circuit shown, calculate the total circuit current and the total circuit power.

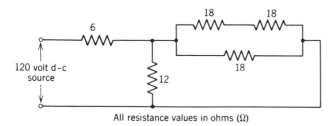

All resistance values in ohms (Ω)

3. The following table shows the relation among current, voltage and resistance for three resistors connected in a battery circuit. Draw a circuit diagram showing the manner in which the resistors are connected and fill in the missing values in the table. What is the battery voltage? (*Hint:* Two of the resistors are in parallel.)

	R_1	R_2	R_3
Voltage	2 v	10 v	?
Current	?	2 amp	4 amp
Resistance	?	5 ohms	?

4. A current of one milliampere (0.001 amp) through the human body can be fatal. Assume that the resistance of the human body is 1000 ohms.

   a. If a person contacts 120 v, what must be the minimum contact resistance of the hands to prevent fatal shock?
   b. Dry skin has a contact resistance of about 100,000 ohms; wet hands, of about 1000 ohms. Calculate the body current with the above voltage, under conditions of wet and dry hands, assuming contact between the hands.

5. An American moved overseas to a country that uses 240-v household circuit voltage. He had with him a 1440-w, 120-v toaster and a 100-w, 120-v immersion heater. He decided that, since each was a 120 v appliance, by placing both in series he could make them operate on a 240 circuit. Was he right? Why? What happened? (Assume that the devices' resistances are the same when hot as when cold, although this is generally not a valid assumption.)

6. A rural farm house generates its own power at 120/240 v, a-c. A 2-wire 240-v copper wire line from the generator shed runs 930 ft to the barn, and there supplies a 4.4-kw, 240-v electric heater. The voltage at the heater is 220 v. Calculate the following:

   a. The line voltage drop.
   b. The line resistance.
   c. The line power loss.
   d. Using Table 12.1 (page 675), determine what size line had been installed.

7. The barn of Problem 6 needed a second heating unit—this one rated 2200 w. Since the load of this unit was one-half that of the first, the farmer ran another circuit but reduced his wire size to No. 10 AWG. Also, he connected this unit to a 120-v circuit, although the unit was dual-rated 120/240 v and could be fed at either voltage. To his surprise, the unit did not perform properly. Why? What would you advise?

8. The farmer of Problems 6 and 7 decided to remove all the previous wiring and to run a single circuit for both heaters. He turned to his electrical-technologist daughter with this requirement:
Design a single 240-v circuit of such size that both heaters will produce at least 95% of their rated nameplate capacity. What size circuit wiring did she recommend? What is the actual power in each heater?

9. A homeowner has recently installed electrical baseboard heaters that are individually controlled. The size of the heaters follows:

Living room   4.0 kw
Kitchen   1.5 kw
Bedroom No. 1   2.0 kw
Bedroom No. 2   1.5 kw
Bedroom No. 3   1.5 kw

The bedroom thermostats have a cycle of 50% on, 50% off. The living room and kitchen thermostats cycle their heater 30% on, 70% off.

Calculate:
a. Maximum electrical heat power demand.
b. Monthly electricity bill, assuming that the thermostat cycle listed here is continuous day and night and assuming a flat rate of 8.5¢/kwh.

10. A single-phase 120-v a-c house circuit supplies the following loads:

Incandescent lighting	300 w
Exhaust fan	2.5 amp, 0.6 pf
Television set	3 amp, 0.85 pf

    Find:
    a. Total circuit volt-amperes.
    b. Total circuit watts.

11. Table 430-148 of the 1996 **The National Electrical Code (NEC)** gives the following motor data for 115-v single-phase a-c motors:

Horsepower	Current, amp
1/6	4.4
1/4	5.8
1/3	7.2
1/2	9.8
1	16
3	34
5	56
10	100

    Assuming an efficiency of 90% for all the motors, calculate the power factor of each.

12. Branch Circuits and Outlets

Now that we have explored the fundamentals of electricity and electrical circuits in theory, we are ready to apply this knowledge to practical electrical circuity as it is found in modern construction. All circuits are constructed basically in the same manner but vary in size, application, control and complexity. We begin our study of applied circuitry with the basic building block of electrical construction—the branch circuit—and proceed from there to circuits of other types. You will soon learn to recognize the common properties of all circuits, be they lighting, receptacle, power, appliance, special purpose, motor or the like. In this chapter, we discuss the characteristics of circuits and demonstrate how they are graphically shown, drawn and wired. We also analyze in detail the materials involved and describe their general and special application. After studying this chapter you will be able to:

1. Distinguish among the three basic types of electrical drawings.
2. Select the branch circuit wiring method best suited to the type of construction being used.
3. Know the construction, application and properties of branch circuit materials including wire and cable, conduit, surface raceways, connectors, outlet boxes and wiring devices.
4. Understand what an electrical outlet consists of, how to draw and count outlets and how to detail them and specify their materials.
5. Describe wiring devices correctly according to their electrical and mechanical characteristics.
6. Draw basic branch circuits that show all the required electrical information.
7. Prepare a symbol list for equipment in branch circuits, including special wiring devices.
8. Draw details of conduits, fittings, supports, raceways and outlet boxes.
9. Understand commonly used symbols and abbreviations.

10. Gain familiarity with the National Electrical Code.

12.1 National Electrical Code

The National Electrical Code (NEC) is the only nationally accepted code of the electrical construction industry. As such, it has become the standard or "bible" of that industry. The NEC, as we will refer to it in this book, is Section 70 of the National Fire Codes published by the National Fire Protection Association. The NEC defines the minimum acceptable quality of electrical design and construction practice necessary to produce a safe installation. It is to the NEC that the electrical inspector refers when inspecting a job. It is, therefore, obvious that the NEC is among the most important reference books in the library of people in the electrical construction industry, and an up-to-date copy should always be at hand. (It is a good idea to save old issues for reference when you examine plans of buildings built under earlier codes.) Some government agencies and some large cities have codes of their own. These generally supplement the NEC but do not replace it. (An exception is the New York City electric code, which is, for the most part, more strict than the NEC.) Since the NEC is revised approximately every 3 years, the current edition must always be on hand, to be consulted for new work. Throughout this book, we make frequent reference to NEC provisions. In particular, definitions will be taken word for word from the NEC. The 1996 edition was used in the preparation of this book.

12.2 Drawing Presentation

Before proceeding any further, we must explain the drawing methods employed in this book, which are the same as those used in construction industry practice. The types of drawing normally encountered are:

- Architectural-electrical plans, also known as working drawings.
- One-line diagrams.
- Wiring diagrams.

To illustrate these methods, let us consider an area in The Basic House, which you have studied in previous chapters. In Figure 12.1(a), we have

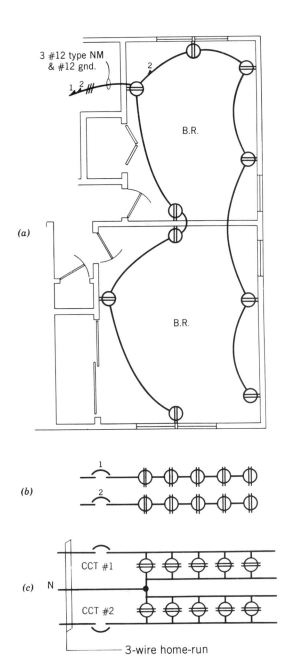

Figure 12.1 (a) Typical circuitry of the Basic House bedroom receptacles. (b) Single-line representation showing circuitry but nothing about arrangement or location. (c) Wiring diagram. (In this case, it adds little information.) Note that only a single neutral wire (labeled N) for both circuits is carried back to the panel. This is typical of a 120/240 v, 3-wire residential wiring system. The separate equipment ground wire that is run in type NM cable is not shown, for the sake of clarity.

spotted wall receptacles, circuited them and shown the circuitry in the manner typically used for this work. The solid lines connecting the receptacles are raceway runs whose purpose is to indicate circuit routing. The number of wires in such a run is generally shown by "tic" or hatch marks on the raceway, although notes may be used instead of hatch marking. Such hatch marks are used when more than two wires are represented. By industry-wide convention the absence of tics indicates two wires, since that is the minimum number of wires that the raceway will contain. Symbol lists frequently state that a line represents (a raceway with) "2 wires unless otherwise noted," the "otherwise noted" generally taking the form of tic marks or notes.

In the home-run of Figure 12.1(a), both tics and a note are shown, although in practice only one or the other would be used. The *home-run* is the connection between the first outlet in a circuit and the panelboard feeding the circuit. It is called a home-run because the wiring from this outlet runs "home," that is, to the panelboard. The panel (board) can easily be thought of as the home (base) of all the circuits, since it is the point from which all circuit wiring originates, as we will see as our study progresses. On a working drawing, a home-run is indicated by a multiple arrowhead. (See the symbol on the symbol list in Figure 12.6.) The number of circuits in a home-run is indicated by the number of arrowheads. The circuit designations (circuit numbers) and wiring is also usually shown. It is important to note that the type of line used—whether solid, dotted, dashed—is entirely up to the design/drawing staff and is chosen on the basis of convenience. No rigid conventions exist as they do in mechanical drafting where solid lines indicate visible edges and dashed lines indicate hidden edges. In circuitry, the solid line is normally used to show the condition most often found, for convenience of drawing, since the solid line is easiest to draw, by hand. (The use of a CAD programs and plotter to prepare drawings eliminates this consideration.) Since in The Basic House most wiring is concealed, the solid line is chosen for this, and a short dash line is used for exposed wiring. The symbols chosen for wiring must be shown on the symbol list for each job.

In Figure 12.1(b), the same electrical work is shown in "one-line" form, so called because multiwire circuits are indicated by a single line. The actual number of wires is understood by the electrical designer or technologist. When it is not obvious, the number is shown. The purpose of these diagrams is to show the circuitry at a glance. Here, for instance, one sees at a quick glance that circuits 1 and 2 each serve five duplex receptacles. Nothing is known about location, but that is not the function of a one-line (single-line) diagram. Such a diagram is purely electrical and has no architectural information.

Finally, a wiring diagram is shown in Figure 12.1(c). This representation shows the number of wires involved and their interconnections. This particular diagram adds nothing to our knowledge in this case and would not be used here. In complex systems, however, a wiring diagram is invaluable. Examples of its usefulness will be shown later. During the course of our discussions, we provide symbol lists relating to the material being discussed. These partial symbol groups combine to make a fairly complete and very useful symbol list that will be immediately applicable to actual jobs. All materials and methods presented are found commonly in actual practice.

Branch Circuits

12.3 Branch Circuits

By **NEC** definition, a branch circuit is "the circuit conductors between the final overcurrent device protecting the circuit, and the outlets." Remember that we stated in Chapter 11 that a circuit consists of a source of voltage, the wiring and the load. After the source of voltage, every practical circuit has some overcurrent protection to protect the circuit components against overloads and short circuits. Overcurrent protection will be considered as a separate subject in Chapter 13. The remaining two portions of every circuit are the wiring and the load. The load is called in **NEC** terminology "the outlets."

The branch circuit itself contains only the wiring, although in everyday trade language a branch circuit is the entire circuit including the outlets, and occasionally even the protective device. It is, therefore, very important to be specific about what is included, when referring to a branch circuit. Since the **NEC** is always used as a reference, it is a good idea to stay with the **NEC** definition and, when outlets are to be included in the work, to state "the branch circuit plus the connected outlets" or words to that effect. We will, however, stay with the official **NEC** definition. See Figure 12.2.

Let us examine the branch circuit, by referring

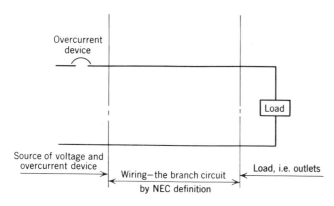

Figure 12.2 Division of an electrical circuit into its components according to the NEC definitions.

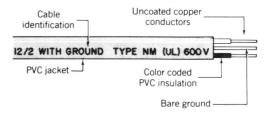

Figure 12.3 Construction of a typical NEC type NM cable. The cable shown is a two-conductor, No. 12 AWG with ground, insulated for 600 v. Also normally shown are the manufacturer, cable trade name, and the letters (UL), which indicate listing of this product by Underwriters' Laboratories, Inc. The ground wire may be bare or covered, and the entire cable can be obtained flat (illustrated), oval or round. (From B. Stein and J. S. Reynolds, *Mechanical and Electrical Equipment for Buildings*, 8th ed., 1992, John Wiley & Sons, New York. Reprinted by permission of John Wiley & Sons.)

to Figure 12.1. In the drawing, the branch circuit consists of the lines connecting the receptacle outlets in the two bedrooms. In actual practice, the wiring method for the branch circuit would consist of one of the following:

- NEC (Code) Type NM nonmetallic sheathed cable. This cable consists of rubber- or plastic-insulated wires in a cloth or plastic jacket and is commonly called by the trade name Romex. See Figure 12.3.
- NEC (Code) Type AC metallic-armored cable, usually called BX. This cable construction comprises two, three or four insulated wires that are covered by an interlocking steel armor jacket. See Figure 12.4.
- Individual insulated wires in aluminum, steel or nonmetallic conduit. Steel conduit is available in three wall thicknesses: thin-wall that is called electric-metallic tubing or simply EMT, intermediate-wall conduit that is called IMC, and heavy-wall conduit that is referred to simply as rigid steel conduit. See Figure 12.5. Nonmetallic conduit is called just what it is made of—plastic conduit. Other types of nonmetallic conduit such as fiber are not normally used in branch circuit wiring.

As will be discussed in detail in the next section, most residential wiring uses nonmetallic sheathed cable (Romex) or flexible metallic-armored cable (BX). This is simply because the common stud wall construction used in residential building requires a wiring system that is flexible and can be threaded through the studs. The obvious disadvantage of this type of wiring is that it cannot be pulled out for replacement as can wiring in conduit. Conduit with pulled-in wires or BX wiring is frequently used in multistory (more than three floors) and multioccupant (more than two families) residential construction and in small residences using masonry (block and concrete) construction.

Nonresidential construction generally uses a conduit wiring system. Some communities have local electrical codes forbidding the use of Romex (type NM) and even BX (type AC) wiring in any occupied structure. It is, therefore, very important to check for applicable codes before beginning any actual design project. NEC restrictions on the use of type AC and type NM wiring are found in Articles 333 and 336, respectively.

The wire material itself is occasionally aluminum, but, because of difficulties sometimes experienced with aluminum connections, copper is almost always specified. Wire size is generally No. 14 or 12 AWG. The wire insulation is usually rated 600 v. The choice of wiring method is left to the designer, but it is coordinated with the architect or owner. Let us return now to the two rooms of the small house of Figure 12.1 and see how the branch circuitry is drawn, circuited and installed, depending on the house construction and the wiring material used.

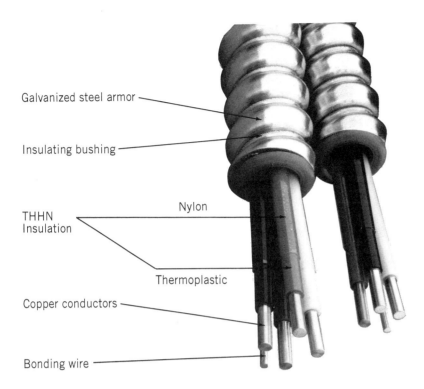

Figure 12.4 Flexible armored cable, NEC type AC; trade name BX. Each cable is provided with a bare copper bonding wire in contact with the cable's steel armor. It acts to reduce the a-c resistance (impedance) of the cable armor, and both together (the armor and the bonding wire) can serve as the circuit's equipment grounding conductor. Note especially the insulating bushing that is always installed on the end of the armor. It protects the wires from damage from the sharp edges of the cut steel armor. The illustrated cable is special because it uses a high temperature wire insulation (THHN), making it suitable for use in areas with high ambient temperatures. Standard construction flexible armored cable differs from that illustrated in that it uses a thermoplastic insulation (type ACT cable) or thermosetting insulation (type AC cable) with a moisture-resistant, fire-retardant fibrous covering. (Photo courtesy of AFC/A Nortek Company.

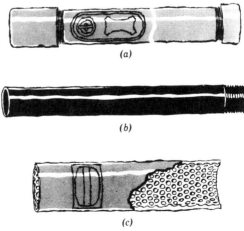

Figure 12.5 Electrical conduits. *(a)* Galvanized, heavy wall, rigid steel conduit and intermediate metal conduit (IMC). *(b)* Black-enameled steel conduit. *(c)* EMT thin-wall steel conduit.

12.4 Branch Circuit Wiring Methods

As was stated in Section 12.2, the symbol used to represent branch circuit wiring depends on the choice made by the electrical technologist or designer. A partial symbol list is given in Figure 12.6, showing symbols generally used for branch circuit wiring. We must again emphasize that there is no standard for these symbols and that they may, and often do, change from job to job. The thoughtful designer will select wiring symbols for clarity of drawing (ease of reading by the contractor) and minimum drawing preparation time. In studying Figure 12.1, let us assume that The Basic House plan is typical American frame construction, wired with BX. Since the house is built with an unexcavated crawl space below the bedrooms, and not on a concrete slab, the actual wiring might be run as is shown in Figure 12.7. Horizontal runs above the floor level require drilling of the wood studs, which are normally on 16-in. centers. Runs under the floor require drilling through the sole plate.

The contractor will compare the two methods and will make the choice that involves the least labor for each run. In this particular case, the contractor would most probably choose a combination of underfloor and through-the-studs wiring, feeding the receptacles on the north, south and west walls with underfloor runs, and the outlets on the east wall with through-the-stud wiring as shown. He would probably not wire in the ceiling space because of the added cost of vertical runs from the receptacle location (12 in. AFF) to the ceiling. The choice of how to wire not only the branch circuit outlets but also the home-run to the panel is almost always economic. That is, when a number of wiring methods are possible, the installing contractor will choose the cheapest one. Since this choice may not correspond to the best choice from an engineering, maintenance or owning cost viewpoint, it is important that, in such cases, the technologist be specific in indicating the wiring method required.

Cable may be fastened to the sides of floor joists and run along them. BX cable may be run exposed in unfinished basements, attached to the underside of floor joists, provided spacing is not excessive. This is not true for small gauge NM cable (Romex). For this type of cable to be run across the joist direction, either a running board must be provided or the joists must be drilled and the cable passed through. The same is true for the through-the-studs wiring as shown in Figure 12.8. BX cable may be installed in notched studs provided metal cover plates are used. Since these requirements change, the electrical technologist should always consult the latest editions of all electrical codes having authority in the locale of the construction, when doing actual design.

The NEC specifies permitted cable installation procedures for type NM cable in Article 336-6 through 336-18, for type AC cable (BX) in Article 333-7 through 333-12, and for all types of cable in Article 300-4. Cables run in attic spaces have somewhat less stringent installation requirements because the space is unused, and, therefore, cables are not exposed to physical damage. As we will discuss later, mixing ceiling and wall outlets on the same circuit is not considered good wiring practice and is best avoided. However, combining separate circuits in the same raceway or multiconductor cable is good practice and is almost always done.

A three-conductor cable in a 120/240 v system can carry two circuits—two circuit conductors, a common neutral and the built-in equipment ground conductor. In BX cable, the metal armor acts as the circuit equipment ground, although not entirely by itself. An internal bare metal strip or wire is carried throughout the length of the cable. It is called the bonding wire or strip, and it acts to bond the spiral metal armor and, therefore, to reduce its impedance (a-c) resistance.

If the building were constructed with metal studs that come with cutouts for horizontal wiring, the walls would probably be used for all the receptacle wiring. It should be noted, however, that frequently, in small residential work, the exact wiring paths are left to the electrician, with only the circuitry indicated. This is because the architect assumes that the electrician will select the most economical wiring method. Therefore, as long as the electrician sticks to the outlet layout, the material and the circuitry, little is to be gained in spending the time to do a detailed layout. This is a valid approach for a single house. For a multihouse development, the technologist would be expected to develop detailed installation drawings from which the electrician would work. These installation drawings show the exact routing of all wiring.

If The Basic House were built on a concrete slab rather than a crawl space, different wiring methods would be available. They are:

SYMBOLS – PART I – WIRING AND RACEWAYS

— ⧸⧸⧸ — NOTE 1 CONDUIT AND WIRING CONCEALED IN CEILING OR WALLS HATCHES INDICATE NO. OF CONDUCTORS EXCLUDING GROUNDS; 2 #12, ¾" CONDUIT UNLESS OTHERWISE NOTED.

— · — CONDUIT AND WIRING CONCEALED IN OR UNDER FLOOR

— — — — CONDUIT AND WIRING EXPOSED

○— · — CONDUIT AND WIRING TURNED UP

●— · — CONDUIT AND WIRING TURNED DOWN

—F–6— FEEDER F–6, SEE SCHEDULE, DWG NO. _____

D———— CONDUIT WITH ADJUSTABLE TOP AND FLUSH PLUG SET LEVEL WITH FINISHED FLOOR

—BX— BX WIRING

—NM— NON-METALLIC CABLE (ROMEX) WIRING

2, 4 / 2PLA / 3# 12, ¾" C HOME RUN TO PANEL 2PLA – NUMERALS INDICATE CIRCUITS, 3 #12 AWG, ¾" RIGID STEEL CONDUIT

← TC2A HOME RUN TO TELEPHONE CABINET TC2A

⟿ FINAL CONNECTION TO EQUIPMENT IN FLEXIBLE CONDUIT

—EC— EMPTY CONDUIT, SUBSCRIPT INDICATES INTENDED USE T – TELEPHINE, IC – INTERCOM, FA – FIRE ALARM ETC. SEE NOTE 2.

◐ ●ₐ SURFACE METAL RACEWAY, SEE NOTE ____ , DWG ____ SIZE AND RECEPTACLES AS SHOWN

⧻ ⧻ ⧻ MULTI–OUTLET ASSEMBLY, SEE NOTE ____ (SEE FIG12-47 FOR ALTERNATE SYMBOL)

NOTE 1. IF THE COMPLETE WIRING SYSTEM IN A BUILDING IS OF A TYPE OTHER THAN CONDUIT AND WIRE, THIS SYMBOL MAY STILL BE USED WITH AN APPROPRIATE NOTE ON THE DRAWINGS OR SPECIFICATIONS, AND ELIMINATION OF THE WORD "CONDUIT" ABOVE.

NOTE 2. TYPE OF LINE (SOLID, DASHED, ETC.) INDICATES METHOD OF RUNNING EMPTY CONDUIT.

Figure 12.6 Architectural-electrical plan (working drawing) symbol list, Part I, Wiring and Raceways. For other portions of the symbol list, see:
Part II, Outlets, Figure 12.50 page 719
Part III, Wiring Devices, Figure 12.51, page 720
Page IV, Abbreviations, Figure 12.58, page 726
Part V, Single Line Diagrams, Figure 13.4, page 733
Part VI, Equipment, Figure 13.15, page 753
Part VII, Signaling Devices, Figure 15.34, page 909
Part VIII, Motors and Motor Control, Figure 16.25, page 944
Part IX, Control and Wiring Diagrams, Figure 16.28, page 947

672 / BRANCH CIRCUITS AND OUTLETS

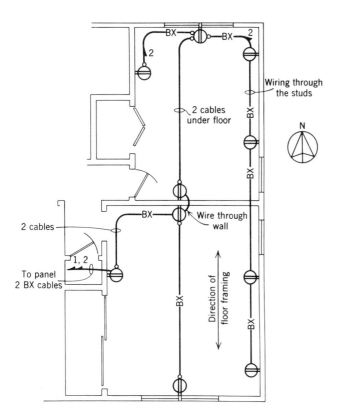

Figure 12.7 One possible wiring solution to layout of Figure 12.1, using BX and underfloor writing. A three-conductor cable can be used instead of 2 two-conductor cables.

- To run BX or Romex in the ceiling spaces and drop down within the walls to wall outlets.
- To bury some form of rigid conduit (plastic or steel) in the concrete floor slab and pull wiring through to wall outlets.
- To install metal-clad cables directly in the concrete slab, turned up at wall outlets. Such cable must be specifically approved for installation in concrete.

The choice of which system to use is again basically one of economics, and as such is generally made by the job engineer. However, since electrical technologists may be called upon to make the decision, they should be aware of both the technical and the cost considerations involved.

Considering each of the three methods individually, we can make the following general remarks.

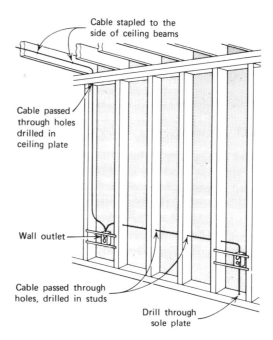

Figure 12.8 Typical method of wiring using type NM nonmetallic sheathed cable in wood frame construction. BX wiring is similar.

Wiring at main level is through-the-studs between devices and through-the-plates to reach basement and ceiling spaces. Where type NM cable passes through the floor, it must be enclosed in metal pipe to provide physical protections (see **NEC** Article 336) and protected from abrasion by special fittings at each end of this metal pipe (conduit) section (**NEC** Article 300). Basement wiring is run along the directions of beams and joists, but through them when run at right angles. For rules governing installation of types AC (BX) and NM (Romex) cable, see **NEC** Articles 333 and 336, respectively. (From B. Stein and J. S. Reynolds, *Mechanical and Electrical Equipment for Buildings*, 8th ed., 1992, John Wiley & Sons, New York. Reprinted by permission of John Wiley & Sons.)

(a) An installation using type AC or NM cable is almost always the most economical except in large jobs, where the cost of field-drilling studs and threading cable more than offsets the cost of wire and conduit or jacketed, metal-clad cable. Also both have the great disadvantage that a faulted cable cannot be replaced. When a cable fault does occur inside a wall, the usual repair involves replacing the faulted section with surface wiring, which is, at best, unattractive. As stated previously, some local codes do not permit this type (type AC and type NM) of wiring and require a conduit system.

(b) When using a conduit system, the material most often chosen for residential work is rigid nonmetallic (plastic) conduit, since it is suitable for use in concrete and is not usually subject either to physical abuse or to high temperatures. (See **NEC** Article 347.) Steel conduit is not normally used because of expense. The use of aluminum conduit in concrete is not advisable. (Some local codes forbid the use of aluminum conduit in concrete.)

(c) Preassembled jacketed metal-clad cables, identified by the manufacturer as suitable for burial in concrete, are expensive, but, because no wire pulling is involved, this option can be very competitive in a large multihouse project (see Figure 12.9).

A layout for the same area as that shown in Figure 12.7, except using plastic conduit in the "mud" (as poured concrete is often called in field work) is shown in Figure 12.10. Note that for a simple case like that of the two bedrooms in The Basic House plan, the branch circuiting can use at least four different materials—Type AC (BX), Type NM (Romex), cable in conduit and jacketed metal-clad cable—and at least four different wiring layouts. You should become familiar with the applications

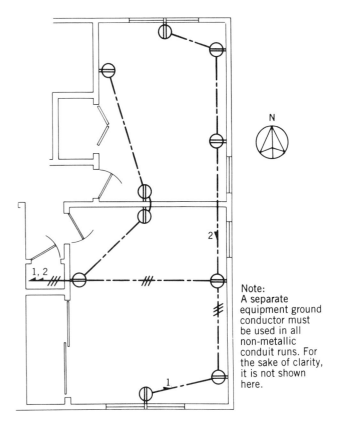

Figure 12.10 Writing of spaces in Figure 12.1, assuming a concrete floor slab and plastic conduit in the slab.

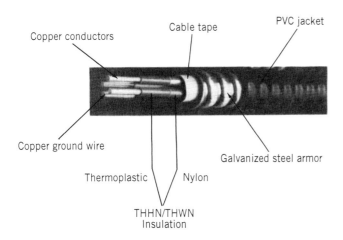

Figure 12.9 Jacketed flexible armored metal-clad cable (NEC type MC, Article 334) is suitable for a wide variety of applications, both exposed and concealed. Where suitability is specifically indicated by the manufacturer, it can be direct-buried in the earth or installed in concrete. (Courtesy of AFC/A Nortek Company).

and limitations of all the common branch circuit materials in order to be able to handle layouts with any of them, with ease and confidence.

In large-scale construction, such as high-rise dwelling units, the floor construction is almost always poured concrete, and the wiring method of choice is wire and conduit. Multistory dwellings, which are constructed with concrete floor but stud partitions, use combinations of wire-in-conduit and preassembled cables (BX, etc.). A typical example of the use of plastic conduit in a concrete floor slab is shown in Figure 12.11.

Conductors

12.5 Wire and Cable

We referred briefly in the preceding section to some of the common wiring materials used in branch

674 / BRANCH CIRCUITS AND OUTLETS

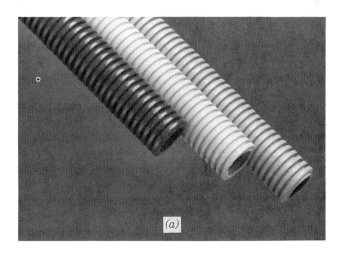

Figure 12.11 *(a)* Flexible plastic conduit (flexible electrical nonmetallic tubing) is available in several colors and in sizes from ½ to 2 in. nominal. It's application includes encasement in concrete floor slabs as shown in photo *(b)*. Color coding aids in electrical system identification. (Photos courtesy of Carlon.)

Table 12.1 Physical Properties of Bare Conductors

Size, AWG/Kcmil	Area, circular mils	Diameter in. Solid	Diameter in. Stranded	D-C Resistance at 25°C (77°F) (Bare copper), ohms/1000 ft
18	1620	0.040	—	7.77
16	2580	0.051	—	4.89
14	4110	0.064	—	3.07
12	6530	0.081	0.092	1.93
10	10,380	0.102	0.116	1.21
8	16,510	0.128	0.146	0.764
6	26,240	—	0.184	0.491
4	41,740	—	0.232	0.308
2	66,360	—	0.292	0.194
1	83,690	—	0.332	0.154
0 (1/0)	105,600	—	0.373	0.122
00 (2/0)	133,100	—	0.418	0.097
000 (3/0)	167,800	—	0.470	0.077
0000 (4/0)	211,600	—	0.528	0.061
250 kcmil (MCM)	250,000	—	0.575	0.052
300 kcmil (MCM)	300,000	—	0.630	0.043
400 kcmil (MCM)	400,000	—	0.728	0.032
500 kcmil (MCM)	500,000	—	0.813	0.026

Source. Reprinted with permission from NFPA 70-1996, the *National Electrical Code®*. Copyright © 1995, National Fire Protection Association, Quincy, MA 02269. This reprinted material is not the complete and official position of the National Fire Protection Association, on the referenced subject which is represented only by the standard in its entirety.

circuitry such as NEC types NM (Romex) and AC (BX) and to some of the types of wire used in wire and conduit work. It is helpful to examine these types more fully here, to better understand their application.

All building circuit conductors consist of an insulated length of wire, usually copper. Aluminum and copper-coated aluminum have made some inroads into the electrical wire field because of lower weight, which leads to lower installation costs. However, to do a proper wiring job with aluminum requires special tools and techniques and men trained in these techniques. Many contractors prefer to leave well enough alone and to stick with copper, particularly for the small sizes encountered in branch circuits where the lower weight advantage of aluminum (cheaper installation labor) is not felt. For this reason, we restrict our discussion here to copper wire.

In the United States at the present time (prior to any change to metric sizing), the universally used gauge for wire is the *American Wire Gauge*, called simply AWG. For some good reason (that we have yet to discover), the AWG wire numbers proceed about one-half way in reverse order to the conductor size and then switch to an order that proceeds in the logical way. Thus, we have the slightly complex situation that, starting with the small size and continuing through the larger sizes, the numbers go from No. 18 AWG to No. 1 AWG, continue from No. 1/0 AWG through No. 4/0 AWG, (also written 0 and 0000) and then switch to a different system entirely. This third grouping used to be called *MCM*, which stands for *thousand circular mils*, where the letter M is the Roman numeral designation for 1000, and circular mils is the square of the copper conductor diameter, where the diameter is expressed in thousandths of an inch (mils). However, because modern terminology always uses the letter k to represent 1000, as in kilowatt (kw) and

Table 12.2 Allowable Ampacities of Insulated Copper Conductors, Rated 0–2000 v, 60–90°C (140–194°F). Not More Than Three Conductors in Raceway or Cable or Direct Buried. Based on Ambient Temperature of 30°C (86°F).

	Temperature Rating of Conductor (see Table 12.3)		
	60°C (140°F)	75°C (167°F)	90°C (194°F)
Size, AWG kcmil (MCM)	Types		
	UF, TW	RHW, THW, THWN, XHHW	THHN, XHHW
14	20[a]	20[a]	25[a]
12	25[a]	25[a]	30[a]
10	30	35[a]	40[a]
8	40	50	55
6	55	65	75
4	70	85	95
2	95	115	130
1	110	130	150
0	125	150	170
00	145	175	195
000	165	200	225
0000	195	230	260
250	215	255	290
300	240	285	320
350	260	310	350
400	280	335	380
500	320	380	430

[a] Unless otherwise specifically permitted by the Code, the overcurrent protection for these conductors shall not exceed 15 amp for 14 AWG, 20 amp for 12 AWG and 30 amp for 10 AWG, after correction factors for ambient temperature and number of conductors have been applied.

Source. Reprinted with permission from NFPA 70-1996, the *National Electrical Code*®, Copyright © 1995, National Fire Protection Association, Quincy, MA 02269. This reprinted material is not the complete and official position of the National Fire Protection Association, on the referenced subject which is represented only by the standard in its entirety.

kilogram (kg), the preferred term today is kcmils. However, because MCM has been in use for so long, many books, catalogs and electricians will continue to use MCM instead of the more modern *kcmils*.

Returning now to wire gauge, the sizes larger than 4/0 AWG go from 250 kcmils on up. As stated, the actual numerical value of kcmils equals the square of the conductor diameter. Thus, for instance, a conductor ½ in. in diameter is 0.500 in. or 500 mils (millinches) in diameter. Then $500^2 = 250,000$ *circular mils* or 250 kcmils. To summarize this unwieldy and complicated system, wire sizes in ascending order are No. 18 AWG through No. 1 AWG, then 1/0, 2/0, 3/0, 4/0 AWG (also written 0,00,000,0000), followed by 250, 300, 350, 400, 500 kcmils or MCM. Of course, there are sizes smaller than No. 18 AWG and larger than 500 kcmils, but they are not frequently found in electrical power systems for buildings. Small sizes such as Nos. 24, 22 and 20 are commonly used in signal and communication wiring.

Table 12.1 shows the important physical characteristics of the conductors we have just discussed. This table is by no means complete, but it is sufficient for our needs. See the **NEC** for more complete tables. Note that normal building wire is solid up to No. 8 AWG and is stranded when larger.

Table 12.3 Conductor Application and Insulation

Trade Name	Type Letter	Maximum Operating Temperature	Application Provisions
Moisture and heat-resistant rubber	RHW	75°C (167°F)	Dry and wet locations
Single conductor, underground feeder and branch-circuit cable	UF	60°C (140°F) 75°C (167°F)[a]	Refer to NEC Article 339
Moisture-resistant thermoplastic	TW	60°C (140°F)	Dry and wet locations
Heat-resistant thermoplastic	THHN	90°C (194°F)	Dry and damp locations
Moisture and heat-resistant thermoplastic	THW	75°C (167°F) 90°C (194°F)	Dry and wet locations Special applications
Moisture and heat-resistant thermoplastic	THWN	75°C (167°F)	Dry and wet locations
Moisture and heat-resistant cross-linked synthetic polyethylene	XHHW	90°C (194°F) 75°C (167°F)	Dry and damp locations Wet locations

[a] For ampacity limitation, see NEC Article 339-5.

Source. Reprinted with permission from NFPA 70-1996, the *National Electrical Code®*, Copyright © 1995, National Fire Protection Association, Quincy, MA 02269. This reprinted material is not the complete and official position of the National Fire Protection Association, on the referenced subject which is represented only by the standard in its entirety.

Conductors larger than No. 8 AWG are often called cables; smaller are called wires. Also, an assembly of two or more conductors in a single jacket is also called a cable. For this reason, BX and Romex assemblies, being multiconductor, are called cable and not wire, but the individual conductors, being smaller than No. 6, are called wires.

All conductors are insulated to prevent contacting each other and short-circuiting. Normal building wire insulation is rated for 600 v, although 300 v BX and Romex is available and in use. The insulation must be able to withstand the heat generated by the current flow through the wire (I^2R loss). Obviously, the larger the current, the more heat is generated. Therefore, a wire with a heat-resistant insulation such as type XHHW (cross-linked polymer), which is rated to withstand a temperature of 90°C (167°F), can obviously carry more current than the same size wire that is insulated with a thermoplastic insulation such as type TW, which is rated to withstand only 60°C (140°F). When you look at Table 12.2 that is exactly what you will find. This safe limit, that is, the amount of current a wire will carry safely with a specific insulation, is called its *current-carrying capacity* or, more simply, its *ampacity*. Table 12.2 is an abbreviated ampacity table. For more complete tables refer to the **NEC**, Section 310-15, and Tables 310-16 through 310-19 with accompanying notes.

The electrical designer/technologist must be very careful to determine the proper ampacity for a conductor according to **NEC** requirements, because higher temperature rating insulation does not always mean higher ampacity. For instance, we have stated several times that type NM nonmetallic sheathed cable is very commonly used in residential wiring. **NEC** Section 336-30 requires that the wire in this cable be rated at 90°C (194°F) but that ampacity shall be that of 60°C (140°F) wires, that is, a reduced current rating. (The reason for this derating is probably because the cable is run in high temperature areas such as attics and in insulation.) The same derating applies to type AC (BX) cable when the cable is run in a building's thermal

Table 12.4 Dimensions of Rubber-Covered and Thermoplastic-Covered Conductors

Size, AWG/ kcmil (MCM)	Type RHW[a]		Types THW[b] TW		Types THHN, THWN		Type XHHW	
	Approximate Diameter, in.	Area, in.^{2}	Approximate Diameter, in.	Area, in.^{2}	Approximate Diameter, in.	Area, in.^{2}	Approximate Diameter, in.	Area, in.^{2}
14	0.204	0.0327	0.162[b]	0.0206[b]	0.105	0.0087	0.129	0.0131
12	0.221	0.0384	0.179[b]	0.0252[b]	0.122	0.0117	0.146	0.0167
10	0.242	0.0460	0.199[b]	0.0311[b]	0.153	0.0184	0.166	0.0216
8	0.328	0.0845	0.276	0.0598	0.218	0.0373	0.241	0.0456
6	0.397	0.1238	0.323	0.0819	0.257	0.0519	0.282	0.0625
4	0.452	0.1605	0.372	0.1087	0.328	0.0845	0.328	0.0845
2	0.513	0.2067	0.433	0.1473	0.388	0.1182	0.388	0.1182
1	0.588	0.2715	0.508	0.2027	0.450	0.1590	0.450	0.1590
1/0	0.629	0.3107	0.549	0.2367	0.491	0.1893	0.491	0.1893
2/0	0.675	0.3578	0.595	0.2781	0.537	0.2265	0.537	0.2265
3/0	0.727	0.4151	0.647	0.3288	0.588	0.2715	0.588	0.2715
4/0	0.785	0.4840	0.705	0.3904	0.646	0.3278	0.646	0.3278
250	0.868	0.5917	0.788	0.4877	0.716	0.4026	0.716	0.4026
300	0.933	0.6837	0.843	0.5581	0.771	0.4669	0.771	0.4669
350	0.985	0.7620	0.895	0.6291	0.822	0.5307	0.822	0.5307
400	1.032	0.8365	0.942	0.6969	0.869	0.5931	0.869	0.5931
500	1.119	0.9834	1.029	0.8316	0.955	0.7163	0.955	0.7163

[a] Dimensions of RHW without outer covering is the same as THW; No. 18 to No. 10 solid; No. 8 and larger, stranded.
[b] Dimensions of THW in sizes Nos. 14 to 8; No. 6 THW and larger is the same dimension as TW.

Source. Reprinted with permission from NFPA 70-1996, the *National Electrical Code®*, Copyright © 1995, National Fire Protection Association, Quincy, MA 02269. This reprinted material is not the complete and official position of the National Fire Protection Association, on the referenced subject which is represented only by the standard in its entirety.

insulation. See **NEC** Section 333-20. Thus, we see that ampacity is determined not only by insulation type but also by the type, location and installation of the wire or cable.

The most common types of wire insulation used in building wiring are types RHW, THW, THWN, THHN and XHHW. See Tables 12.3 and 12.4 for an abbreviated description of these types, their application and their dimensions. Notice from Table 12.4 that type THHN is both thinner than the other types and rated at 90°C (194°F). This dual advantage—easy handling and more conductors in a raceway—plus the ability to be used in high temperature areas, has made this type very popular, despite a material price premium. The choice of which type of wire to use is made by the designer. (Purely as a matter of interest, the cable-type designations are actually abbreviations of the insulation type. Thus, RHW is heat- and water-resistant rubber; THWN is thermoplastic heat- and water-resistant insulation, nylon-jacketed; and XHHW is cross-linked polyethelene, high heat- and water-resistant. See Table 12.3 and fill in the remaining abbreviations on your own.)

It is important for the person preparing the drawings to remember that, when different types of wire are used on a job, as often happens, they must be clearly indicated on the drawings. It is very common to have type TW or type THHN branch circuit wiring and type XHHW heavy feeder wiring. The technologist must remember that these type designations are vitally important and must appear either by general note on the drawings, by description in the specification or along with each wire designation on the drawings. It is poor practice to identify wire type (or, for that matter, anything else) by more than one technique. This is because a change in one place and not the other (as frequently happens in changes) will cause a conflict on the contract documents, and this always results in additional cost. The most common wire sizes for branch circuit work are Nos. 14, 12 and 10 AWG. This will be discussed at length in the sections on circuitry.

As already explained, the purpose of the wire's insulation is to prevent contact between electrically "hot" wires and to withstand the heat generated in the wire by the passage of current. The insulation accomplishes the latter purpose as a function of its nature—that is, cross-linked polymer withstands heat better than rubber or plastic. The ability to insulate electrically (to withstand voltage) depends not only on the type of insulating material but also on the insulation's thickness. Thus, the same material, for example PVC or rubber or even treated paper, can be used to insulate cables at 300 or 3000 v simply by thickening the insulation. There are limitations, of course, both technical and economic, that lead to the use of certain materials and not to others, but, in general, the thicker the insulation, the higher the cable's voltage rating, and vice versa. Table 12.4 gives dimensional data on 600-v insulated cables.

Over the insulated conductor(s), the manufacturer frequently places a covering, or jacket, that protects the cable against all sorts of damage—physical, chemical, heat, water and the like. The interlocked armor on BX cable gives physical protection; the heavy plastic jacket on type UF direct burial underground cables gives water and chemical protection; the neoprene jacket on industrial-use cables gives physical and chemical protection; and so on. Special-use cables are available in literally hundreds of different types, but for our purposes the types shown in Table 12.2 are the most important.

Conductor insulation can be colored during manufacture as desired. This makes possible a standard system of color coding for branch circuits, which allows us to keep wires and phases straight. Without color coding, installation would be much more time-consuming because of having to "ring-out" all the wires to identify them. The standard branch circuit color code as stated in Section 11.13 (See NEC Section 210-5) is

Neutral	white or gray
Phases A, B and C	any consistent color
Ground	green (when insulated)

Phase wire color coding can use the insulation color, a stripe in the insulation or even paint on the

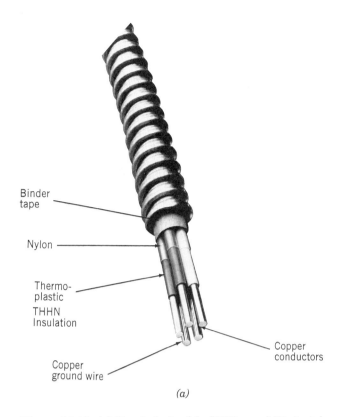

Figure 12.12 (a) Metal-clad cable (NEC type MC, Article 334) with aluminum armor in lieu of the more common galvanized steel armor. Application is similar to that of the steel-armored cable, with the weight advantage of aluminum. Conductors are factory-installed, color-coded and covered with type THHN insulation and nylon jacket. Cables of similar construction, using steel armor, are available for almost all power and control applications. (Courtesy of AFC/A Nortek Company.)

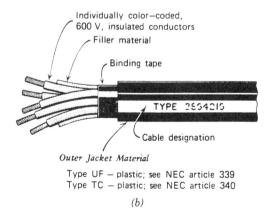

Figure 12.12 (b) Typical construction of a multiconductor jacketed cable. Type of insulation and its voltage rating and the jacket material are selected to suit the application needs.

cable. Neutral and ground, however, must be color coded throughout their lengths. Identification of phase legs at the ends only, is permitted. An uninsulated ground is obviously not color coded. It is quite obviously the ground wire. Whether to use an insulated or uninsulated ground is usually the designer's choice. In type NM cable, the ground wire is generally uninsulated. Two examples of more complex cable assemblies are shown in Figure 12.12.

12.6 Connectors

Connectors for wiring fall into various categories depending on the method used for making the connection between the wires. Similarly, the type

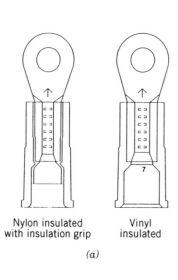

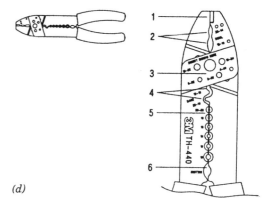

1. Anvil-type wire cutter.
2. Crimp dies for 22–10 AWG insulated terminals, connectors and closed-end connectors.
3. Cut-off dies for mild steel and non-ferrous screws.
4. Crimp dies for 22–10 AWG non-insulated terminals and connectors.
5. Wire stripping stations.
6. Crimp die for 7 and 8 mm ignition parts.

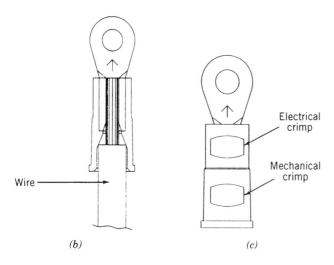

Figure 12.13 Crimp-on lug connectors are frequently used on low voltage signal cables. The connector consists of a nylon- or vinyl-insulated sleeve (a) into which the wire is placed (b). Connection to the metal sleeve is made by crimping (c) with a hand crimping tool (d). A cutaway of such a joint (e) shows its effectiveness. (Figures a–d courtesy of 3M.)

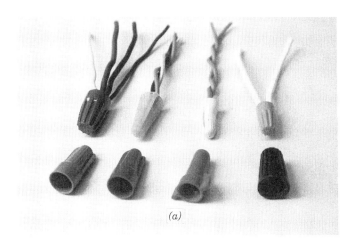
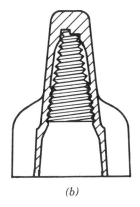

Figure 12.14 (a) A variety of twist-on wire connectors are available for rapid connection of circuit wires varying from No. 22 AWG through No. 6 AWG. Up to six wires can be connected with a single wire connector of this type, although in typical branch circuit wiring no more than four wires are used. (b) Section through a typical screw-on solderless connector, with construction details (c). (d) Typical application of twist-on-type wire connectors. Note the density of wiring in this 4-in. square junction box and the minimal space occupied by five wire connectors. Of interest also is the use of jacketed flexible conduit and special offset fittings in this industrial installation. See Figures 12.20 (b) and 12.25. (Drawings b and c courtesy of Panduit and photo d courtesy of 3M.)

of connector largely determines where and how it is used.

a. Crimp-on Lug Connectors

The first type of connector that we will illustrate is the type that is applied to a single cable, basically as a cable termination. This type of pressure connector, usually called a lug, is applied with a crimping tool and is nonremovable. Its function is to simplify connection of the cable on which it is mounted to a terminal, or to another similarly terminated cable (usually by bolting). These pressure-applied connector-terminators are normally used for signal, communication and other low voltage cables. Some special types of power cables are also crimp-terminated and connected. For these large cables, the crimping tool is power-operated, unlike the hand tool used for signal and other small cables. See Figure 12.13.

b. Twist-on Wire Connectors

Circuit wires, ranging in size from No. 14 AWG to No. 6 AWG, are usually connected to each other with twist-on connectors, frequently called wire nuts after the trade name of one of the major manufacturers. These connectors are also used to make signal and control circuit connections with wires as small as No. 22 AWG. Typical connectors are shown in Figure 12.14.

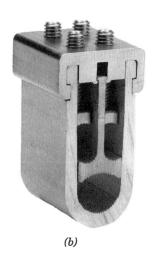

(b)

Figure 12.15 (b) Block-type connector used primarily for tapping a main cable. The straight-through or run cable is laid into the bottom of the U connector, the cover is slid onto the U and then the tap cables are fastened into the cable openings with the set screws at the top of the connector. The illustrated connector can accommodate a run cable of size 3/0 AWG through 500 kcmil (MCM) and up to eight tap cables sizes sized #6 AWG through 1/0 AWG. (Courtesy of Burndy Corporation.)

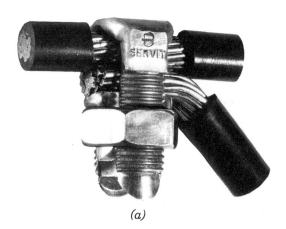

(a)

Figure 12.15 Screw-type pressure connectors for cables are available in various designs for making straight-through or T taps. (a) Classic split bolt design places cables in direct contact with each other. (Courtesy of Burndy Corporation.)

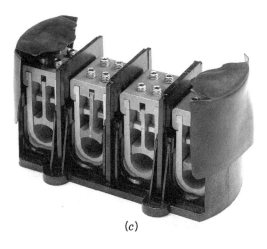

(c)

Figure 12.15 (c) Connectors are available in blocks for compactness, neatness and efficiency when used for tapping multiple cable runs. Block cover, which serves to keep the cable connections clean and dry, is shown cut away. (Courtesy of Burndy Corporation.)

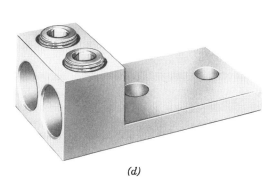

Figure 12.15 *(d)* Solderless lug intended for connection between round cable(s) and flat bus or another lug connector. (Courtesy of ILSCO.)

Figure 12.15 *(f)* Power distribution block connector provides multiple taps from a single or multiple heavy cables. This type of connector is useful in control panels, motor controls, switchboards and similar applications. (Courtesy of ILSCO.)

c. Bolted Pressure Connectors

Connections and taps to heavy cables, in sizes up to 750 kcmils (MCM), are usually made with bolted pressure connectors, which simply clamp the cables either to each other directly as in Figure 12.15*(a)* or to the body of the clamp, thus making connection via the metal in the clamp itself, as in Figure 12.15*(b)* through (f).

All the connectors (and terminals) described so far are known by the general name *solderless connectors*. This is in recognition of the fact that when the electrical industry was young, all connections were soldered. (Busbars were, and to an extent still are, brazed.) With the introduction of the "new" labor-saving connectors, which accomplished the required electrical connections without the use of solder, they become known as solderless connectors, a name that has stuck to this day.

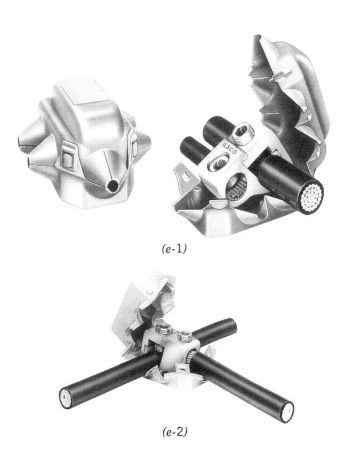

Figure 12.15 *(e)* T-tap connector with insulating cover. As with connector *(b)*, the run-through conductor is laid into the connector after the insulation is stripped from the section inside the connector and the tap cable is introduced from the side *(e-1)* or front *(e-2)*. Premolded cover replaces hand taping of connectors with the additional advantage of labor savings and reusability. Such covers are available for indoor use only. (Courtesy of ILSCO.)

d. Welded Connections

The joints described to this point most often apply to insulated cables. Joints in bare cables such as grounding cables, both exposed and buried, must be made in a manner that will withstand severe corrosion. One of the most reliable of such joints is made by welding the cables together. This is most often accomplished by a process wherein the cables

Figure 12.16 Joints that are subject to corrosion due to exposure to the elements or burial in the earth are effectively made by (exothemic) welding. This process actually fuses the individual conductors into a single metallic mass, as can be seen in the illustrated cutaway. Such a joint is only minimally affected by weather or burial. A typical application is in buried grounding networks (mats) that consist of a grid of interconnected bare cables.

to be joined are encased in a package of exothermic (heat-expelling) materials. When these are activated, they create enough heat to melt the cables together into a single solid metal joint. A section through such a completely corrosion-resistant welded joint is shown in Figure 12.16.

Raceways

12.7 Conduit

By NEC definition, a *raceway* is any channel expressly designed for holding wires. Since round pipe, or conduit, is the most commonly used electrical raceway, we begin our discussion with it. In addition to providing a means for running wires from one point to another, electrical conduit has three additional functions:

- To physically protect the wires.
- To provide a grounded enclosure (in the case of metal conduit).
- To protect the surroundings against the effects of a fault in the wiring.

This last point is often overlooked. We should, however, take note of the number of fires that are caused annually by short circuits and other electrical faults. Many of these fires would probably not have occurred were the faulted and overheated wiring enclosed in steel conduit, since the steel pipe would contain the arcs, help dissipate the heat and tend to snuff out the fire. This is the reasoning behind the NEC tendency to require that the entire electrical system be enclosed in metal. This includes not only the wiring but also the switches, receptacles, panels, switchgear and other components of the wiring system. (Plastic conduit and equipment enclosures are permitted, with restrictions.)

Note in Figure 12.17 how all the components are steel enclosed—by pipe, metal cabinets and boxes. The purpose is, as already stated, protective, and it works two ways. It acts to protect the wiring system from damage by the building and its occupants and, even more so, to protect the building and its occupants from damage by the electrical system. The power available in even the smallest electrical system in a private residence is awesome and must be carefully controlled, limited and isolated. Insulation and conduit do the isolation job. Fuses and other devices do the control and limiting job. These will be discussed in the section on current protection.

In Figure 12.5, we show the most common types of steel conduit. Most branch circuitry that does utilize conduit is run in small-size conduit, namely $1/2$, $3/4$ and 1 in. This is because the small branch circuit wiring, generally Nos. 14, 12 and 10 AWG, fits easily into these size conduits in the quantities normally encountered in branch circuit work. Refer to the NEC Appendix C, which shows the number of wires that can be accommodated by different size conduits. Note, for instance, that a typical run of four No. 12 AWG type TW wires can fit into a $1/2$-in. conduit with room to spare. These small branch circuit conduits are normally installed in walls and floor slabs.

Although the larger-size conduits are not normally encountered in branch circuit work, we discuss their installation problems at this point for convenience. First, however, some idea must be gained of the sizes and weights involved when we speak of electrical conduit. Refer to Table 12.5 for a listing of conduit dimensions and weights for rigid steel, IMC, EMT, and aluminum conduit. Note the very considerable weight of even small conduits and, therefore, the necessity for adequate supports when these conduits are run exposed horizontally or vertically. Minimum spacing of supports is specified in the NEC and we will not duplicate these data, except as required to illustrate use.

CONDUIT / 685

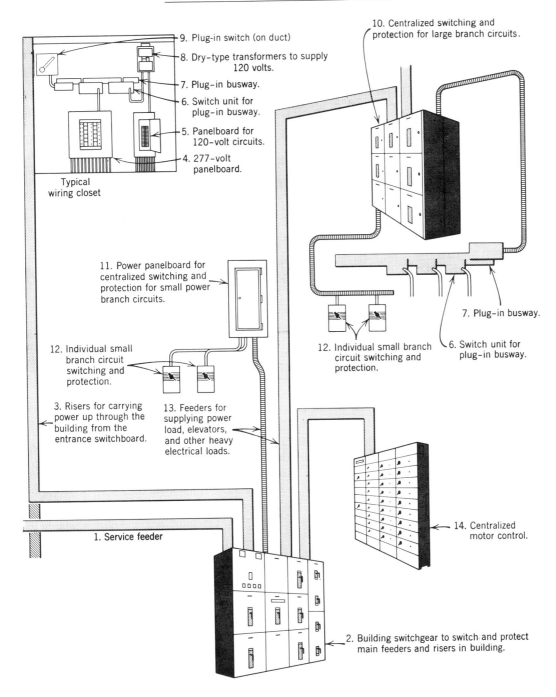

Figure 12.17 Building's basic electrical system shown pictorially, with the power capacity being indicated, approximately, by the size of the conductors. This diagram does not extend beyond the local panelboard and includes only commonly used items. Note that the entire system is, in effect, jacketed in steel. (Courtesy of General Electric Company.)

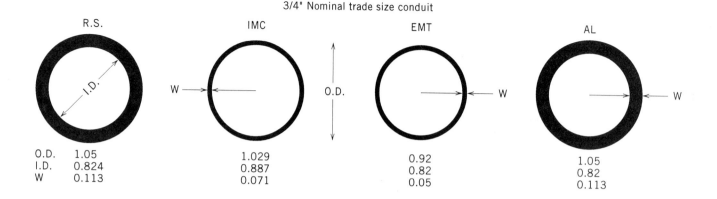

Table 12.5 Comparative Dimensions and Weights of Metallic Conduit[a]

Nominal or Trade Size	Outside Diameter, in.				Inside Diameter, in.				Weight per 10-ft Length, lb[b]			
	RS[c]	IMC[d]	EMT[e]	AL[f]	RS	IMC	EMT	AL	RS	IMC	EMT	AL
½	0.84	0.82	0.71	0.84	0.62	0.69	0.62	0.62	7.9	5.3	2.9	2.7
¾	1.05	1.03	0.92	1.05	0.82	0.89	0.82	0.82	10.9	7.2	4.4	3.6
1	1.32	1.29	1.16	1.32	1.05	1.13	1.05	1.05	16.5	10.6	6.4	5.3
1¼	1.66	1.64	1.51	1.66	1.38	1.47	1.38	1.38	21.5	14.4	9.5	7.0
1½	1.90	1.88	1.74	1.90	1.61	1.70	1.61	1.61	25.8	17.7	11.0	8.6
2	2.38	2.36	2.20	2.38	2.07	2.17	2.07	2.07	35.2	23.6	14.0	11.6
2½	2.88	2.86	2.88	2.88	2.47	2.61	2.73	2.47	56.7	38.2	20.5	18.3
3	3.50	3.48	3.50	3.50	3.07	3.23	3.36	3.07	71.4	46.9	25.0	23.9
3½	4.00	3.97	4.00	4.00	3.55	3.72	3.83	3.55	86.0	54.7	32.5	28.8

[a] Data varies slightly among manufacturers.
[b] Standard length including one coupling.
[c] Standard heavy-wall rigid steel conduit.
[d] Intermediate-weight steel conduit (see NEC Article 345).
[e] Electric metallic tubing.
[f] Aluminum.

With exposed conduit, it is frequently necessary to detail conduit supports and hanging methods. This is particularly true in areas where space is limited, such as in hung ceilings, closets and shafts, where coordination between the requirements of all the different trades is necessary. This type of detailed space coordination work is frequently given to a capable electrical designer or technologist employed by a construction contractor to work out, and, therefore, he or she must be familiar with support methods and hardware. The simplest types of support are the one-hole conduit strap and the "C" clamp [Figure 12.18(a)], which can be used to clamp both vertical and horizontal runs. Where more rigid fixing of the conduit is required, a two-hole pipe strap [Figure 12.18(b)] can be used.

Often, horizontal runs of exposed conduit must be supported by suspension. Individual hangers [Figure 12.18(c)] or trapeze arrangements with individual conduit clamps at each trapeze support [Figure 12.18(d)] are very common. Indeed, the trapeze is frequently assembled into two or more layers of conduit, as is shown. Supports for vertical conduit can be assembled very much like a trapeze support, except, of course, that they are vertical instead of horizontal, or one can use devices manu-

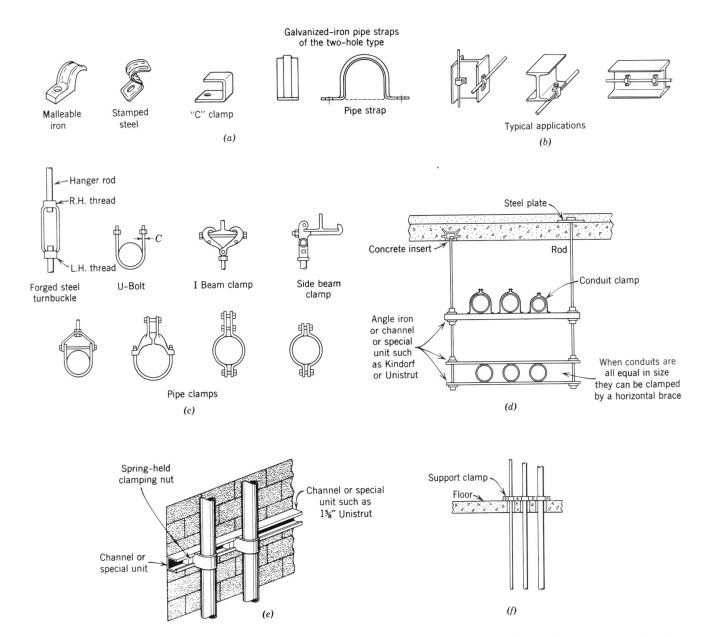

Figure 12.18 Conduit supports. *(a)* and *(b)* One-hole conduit clamps, "C" clamp, two-hole pipe strap and typical applications. *(c)* Single-pipe suspension methods. These fittings are also applicable to pipe hangings. *(d)* Trapeze conduit hangers can be made of standard steel members or assembled of channels specifically intended for the purpose. Each conduit is clamped in place individually. Trapeze hanging can be assembled in one or more layers. *(e)* Vertical conduit support is similar to horizontal. *(f)* When supported at a floor opening, special clamps are used.

688 / BRANCH CIRCUITS AND OUTLETS

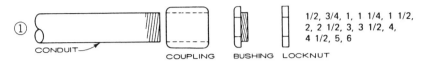

RIGID STEEL CONDUIT (RSC) INTERMEDIATE METALLIC CONDUIT (IMC)
For fireproof construction.
See page on "conduits" for graphic size and weights.

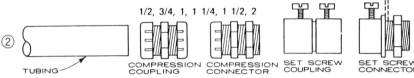

ELECTRICAL METALLIC TUBING
For fireproof construction. Same use as Rigid Conduit above. Walls are thinner, therefore economical.

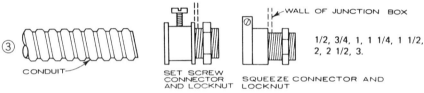

FLEXIBLE METALLIC CONDUIT (FMC)
For fireproof construction.

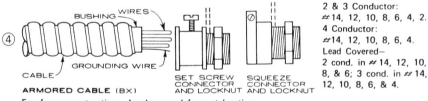

ARMORED CABLE (BX)
For frame construction. Lead covered for wet locations.

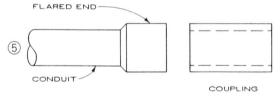

ELECTRIC PLASTIC CONDUIT AND TUBING (EPC AND EPT)

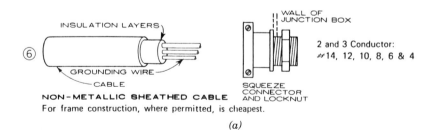

NON-METALLIC SHEATHED CABLE
For frame construction, where permitted, is cheapest.

(a)

Figure 12.19 (a) Schematic details of couplings and termination fittings for:
1. Heavy wall (rigid) and intermediate wall (IMC) metallic conduit
2. Thin wall metallic conduit (EMT)
3. Flexible metal conduit, trade name Greenfield
4. Armored cable, **NEC** type AC, trade name BX
5. Heavy wall and thin wall (tubing) plastic conduit
6. Nonmetallic sheathed cable, **NEC** type NM, trade name Romex
(From Ramsey and Sleeper, *Architectural Graphic Standards*, 8th ed., 1988 reprinted by permission of John Wiley & Sons.)

1. Rigid and IMC fittings

Steel locknut   Steel bushing   Insulated bushing   Compression   Set screw
                                                              Insulated throat connectors

Steel couplings   Rigid-EMT   Rigid-EMT   Rigid-flexible

————————————— C O U P L I N G S —————————————

2. EMT Fittings

Compression   Set-screw   Steel couplings
Insulated throat connectors

3. Flexible metal conduit ("Greenfield") fittings

Straight squeeze – malleable iron   Angle squeeze – die cast zinc

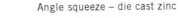

Insulated throat Uninsulated   90°   90°   45°
                              Insulated throat Uninsulated Uninsulated

4. Flexible armored cable fittings

Flexible screw-in   Couplings

Insulated throat   Flex-rigid   Flex-EMT
die cast zinc   Malleable iron

5. Nonmetallic sheathed cable

Die cast zinc   Plastic   All-purpose
                                    die cast zinc

(b)

Figure 12.19 (b) Pictorial details of conduit fittings for:
1. Rigid and intermediate weight conduit (IMC)
2. Electric metallic tubing
3. Flexible metal conduit
4. Flexible armored cable
5. Nonmetallic cable
(Courtesy of Raco, Inc.)

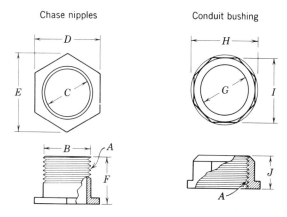

Table 12.6 Dimensions of Conduit Nipples and Bushings[a]

Size, in.	A	B	C	D	E	F	G	H	I	J
½	14	0.82	0.62	1.00	1.15	0.62	0.62	1.00	0.94	0.37
¾	14	1.02	0.82	1.25	1.44	0.81	0.75	1.25	1.12	0.44
1	11.5	1.28	1.04	1.37	1.59	0.94	1.00	1.50	1.37	0.50
1¼	11.5	1.63	1.38	1.75	2.02	1.06	1.25	1.81	1.75	0.56
1½	11.5	2.87	1.61	2.00	2.31	1.12	1.50	2.12	2.00	0.56
2	11.5	2.34	2.06	2.50	2.89	1.31	1.94	2.56	2.37	0.62
2½	8	2.82	2.46	3.00	3.46	1.41	2.37	3.06	2.87	0.75
3	8	3.44	3.06	3.75	4.33	1.50	2.87	3.75	3.50	0.81
3½	8	3.94	3.54	4.25	4.91	1.62	3.25	4.25	4.00	1.00

[a] These dimensions vary slightly from one manufacturer to another.

Note: A = threads per inch.
B = thread diameter.

factured especially for this purpose. See Figure 12.18(e) and (f). These are the basic support elements, which can be used to assemble a good conduit support system.

To lay out a conduit system properly, with the correct dimensioning for tight fits, a knowledge of how conduit joints and terminations are made is required. We restrict ourselves at this point to rigid conduit fittings, which are threaded. Lengths of conduit are joined together by screwed steel couplings. When a conduit that carries wires larger than No. 6 AWG enters a box or cabinet, the **NEC** requires insulated bushings. For smaller wires (and conduits), the joint at the cabinet is usually made up with a locknut and bushing or double locknuts and bushing—insulated or uninsulated.

For many applications, special grounding bushings are used. When attaching a conduit to a ceiling or floor outlet box, a *chase nipple* (see Table 12.6) is generally used. Various fittings for heavy wall and thin wall metallic and nonmetallic conduit, plus fittings for flexible conduit, armored cable (BX) and nonmetallic sheathed cable (type NM) are illustrated in Figure 12.19. Typical dimensional data for rigid conduit couplings, chase nipples and rigid conduit bushings are given in Table 12.6. To lay out rigid steel conduit properly, where close spacing in runs is required, and at cabinet entrances, remember that the conduit terminations are larger than the conduits themselves and that space must be left between locknuts for a wrench. We recommend that this space be ¼ in. minimum. Table 12.7

Table 12.7 Rigid Steel Conduit Spacing at Cabinets (All dimensions in Inches)

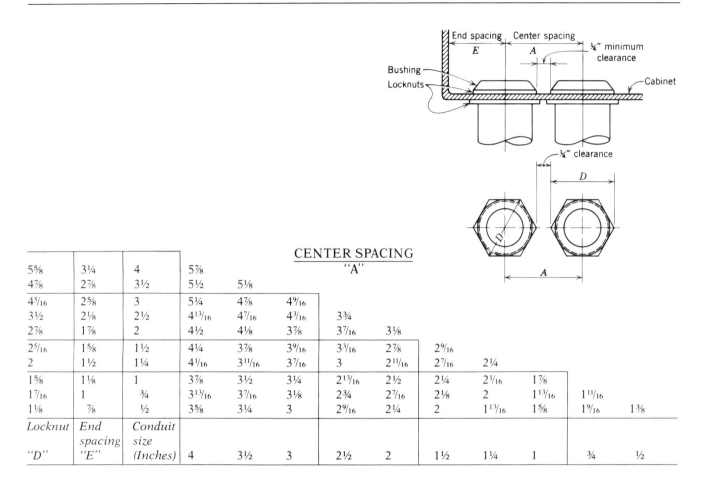

Locknut "D"	End spacing "E"	Conduit size (Inches)	4	3½	3	2½	2	1½	1¼	1	¾	½
5⅝	3¼	4	5⅞									
4⅞	2⅞	3½	5½	5⅛								
4 5/16	2⅝	3	5¼	4⅞	4 9/16							
3½	2⅛	2½	4 13/16	4 7/16	4 3/16	3¾						
2⅞	1⅞	2	4½	4⅛	3⅞	3 7/16	3⅛					
2 5/16	1⅝	1½	4¼	3⅞	3 9/16	3 3/16	2⅞	2 9/16				
2	1½	1¼	4 1/16	3 11/16	3 7/16	3	2 11/16	2 7/16	2¼			
1⅝	1⅛	1	3⅞	3½	3¼	2 13/16	2½	2¼	2 1/16	1⅞		
1 7/16	1	¾	3 13/16	3 7/16	3⅛	2¾	2 7/16	2⅛	2	1 13/16	1 11/16	
1⅛	⅞	½	3⅝	3¼	3	2 9/16	2¼	2	1 13/16	1⅝	1 9/16	1⅜

furnishes spacing data based on this assumption. Table 12.8 presents dimensional data on conduit elbows. Although these data apply to manufactured elbows, they are approximately the same as a good tight field bend and can be used for that, as well as for EMT bends.

12.8 Flexible Metal Conduit

Rigid conduit is used for straight runs and for connection to equipment that is noise and vibration free. However, connection to motors, which vibrate no matter how well balanced, or transformers, which hum, should be made with a loop of flexible conduit that will minimize transmission and amplification of noise and vibration. This type of conduit, which is frequently referred to by the trade name *Greenfield*, is also very useful in getting around obstructions and making multiple tight turns. When it is covered with a plastic jacket, it becomes liquid-tight and suitable for use in wet locations. Flexible nonmetallic conduit is also available. As with rigid nonmetallic conduit, it too requires that a separate grounding conductor be run with the circuit wires, to be connected to all metal boxes in the raceway system.

All types of flexible conduit requires special connectors and fittings specifically designed for this type of raceway. Construction and application of flexible metal conduit and liquid-tight flexible conduit are covered by **NEC** Articles 350 and 351, respectively. Flexible metal tubing, which is covered by **NEC** Article 349, is a liquid-tight, thin-wall flexible metal conduit that is liquid tight without being jacketed. However, unlike the plastic-jacketed flexible conduit, it is limited to use in dry

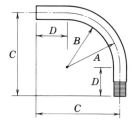

^aVaries for different manufacturer

Table 12.8 Dimensions of Rigid Steel Conduit Elbows

Nominal Trade Size, in.	Actual Inside Diameter, in.	Actual Outside Diameter, in.	Radius, in.				Weight, ea. lb
			A	B	C^a	$D^{a,b}$	
1/2	0.63	0.84	4.00	3.58	6.5	2.5	0.82
3/4	0.83	1.05	4.50	3.98	7.25	2.75	1.09
1	1.05	1.32	5.75	5.09	8.63	2.88	2.01
1 1/4	1.38	1.66	7.25	6.42	10.25	3.0	3.13
1 1/2	1.61	1.90	8.25	7.3	11.25	3.0	4.14
2	2.06	2.38	9.50	8.31	12.5	3.0	7.07
2 1/2	2.47	2.86	10.50	9.06	14.5	4.0	14.11
3	3.06	3.50	13.00	11.25	17.13	4.13	18.50
3 1/2	3.56	4.00	15.00	13.0	19.25	4.25	29.79
4	4.06	4.50	16.00	13.75	20.38	4.38	35.28

^aVaries with different manufacturers.
^bThis dimension represents the straight portion of the elbow.

locations. (Neither type may be used where subject to physical damage.) A frequent application for flexible metal conduit is in hung ceilings, plenums and air-handling spaces. Flexible conduit and some of its applications are shown in Figures 12.20 through 12.25.

12.9 Surface Raceways

In addition to rigid and flexible conduit, which can be run both concealed and exposed (with limitations as listed in the **NEC**), there is an entire line of raceways intended for surface attachment only. These raceways can be either metallic or nonmetallic. Both are covered in **NEC** Article 352, which specifies both permitted applications and use restrictions. Surface raceways are not round in cross section. They have at least one relatively flat side, which is used as the base for attaching the raceway to the building surface or other rigid support.

One-piece raceways are held by special clips attached to the building surface. Two-piece raceways come apart; the base section is attached to the wall or other supporting surface, and the cover is then snapped on or screwed on. Large cross-section raceways can accommodate wiring devices such as switches and receptacles plus the circuit wiring. Metallic surface raceways specifically designed to contain both circuit wires and receptacles are called multioutlet assemblies and are covered by **NEC** Article 353. The receptacles and their associated circuit wiring may be factory- or field-assembled.

Some surface raceways contain one or more internal partitions, which makes them suitable for carrying wires of different systems in the separate

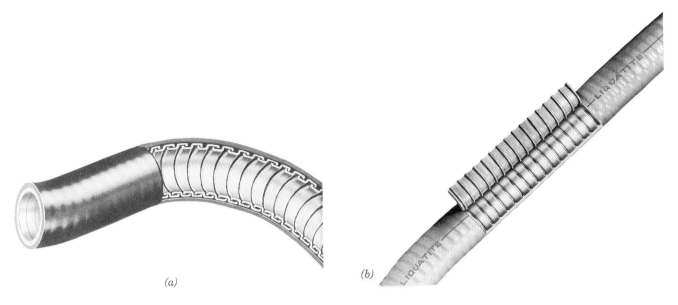

Figure 12.20 Section through liquid-tight conduit shows its construction. Flexibility is obtained from the continuously interlocked spiral metal core *(a)*, and liquid-tightness is ensured by a plastic jacket that is bonded to the metal core as seen in *(b)*. (Courtesy of Electri-Flex Company.)

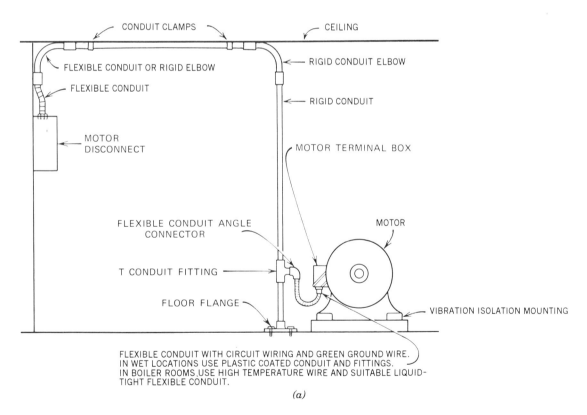

Figure 12.21 *(a)* Installation detail as it would appear on electrical working drawings. Note the flexible conduit connection at the motor, whose purpose it is to absorb vibration. In a wet location, or where subject to splashing, liquid-tight flexible conduit would be used. Note also the ceiling conduit clamps and the method of attachment at the floor. In no case should an unsupported vertical conduit run be installed without some type of rigid floor connection.

694 / BRANCH CIRCUITS AND OUTLETS

Figure 12.21 *(b)* Photo of actual motor installation in a wet location using liquid-tight, plastic-covered flexible conduit. Note the use of a special weatherproof motor switch. In this case, a floor connection for the vertical conduit is unnecessary, since it is rigidly connected to a vertical steel structural member.

Figure 12.21 *(c)* Connection to motor without any slack in the flexible conduit. If the conduit is tight, motor vibration will be transmitted to the adjacent box and from there to the branch circuit conduit and wiring. The result may be loose terminals and objectionable noise.

sections. Such raceways are frequently used where both power and telephone/signal/data cables are required in large numbers and where frequent outlets are needed. See Figure 12.26. Some of the more common types of surface raceways are shown in Table 12.9 along with their wiring capacities, with and without receptacles.

Typical applications of surface raceways are:

- Where the architecture of the building does not permit recessing of a raceway. See Figure 12.27.
- Where recessing of a raceway is possible, but very expensive. This is typically the case in rewiring an existing building. See Figure 12.28(a,b).
- Where expansion is expected, making the large cross-sectional, easily accessible surface raceway an ideal choice. See Figure 12.29(a,b).
- Where many outlets are required close together. See Figure 12.30(a–d).

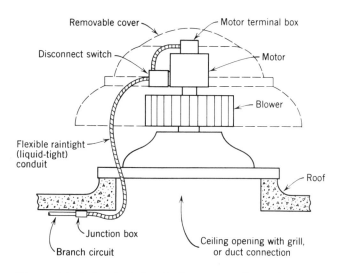

Figure 12.22 Typical detail of power roof ventilator. Note liquid-tight (rainproof) flexible conduit at motor. The disconnect switch is protected from the weather by the ventilator's metal cover.

- Where frequent rewiring is required or anticipated. See Figure 12.31.
- Where it is convenient to mount equipment inside the raceways or where wiring cannot be pulled into a raceway but must be laid in. See Figure 12.32.

Figure 12.23 This is a particularly good application of liquid-tight flexible conduit, since it provides weatherproofing and acoustical isolation of the noise-producing transformer. (Courtesy of Electri-Flex Company.)

Figure 12.24 Because transformers vibrate and hum, all connections should be flexible. In this very poor transformer installation, the conduit connection is rigid. This will transmit the transformer vibration and hum into the conduit system and will cause noise throughout the system. Note also that the transformer is installed without noise-reducing vibration pads. Finally, the case cover is obstructed with a heavy conduit, making maintenance and inspection extremely difficult.

The choice as to whether to use a single two-section raceway as in Figure 12.32 or two separate ones depends on cost, appearance and convenience. On the one hand, a single divided raceway is generally cheaper than two units. On the other hand, separate units are more convenient to wire and maintain. This can be important in installations where frequent access to the signal wiring is required, since good safety procedures normally require that power circuits be deenergized before a raceway is opened. Therefore, using a combination raceway would require a power shutdown even when working only on the signal wiring.

Note, in Figure 12.6, our raceway symbol list, that surface raceways are represented by a double parallel line and identifying letters. The multioutlet assembly is similarly represented. There is no consensus on symbols for this type of raceway, and the electrical designers who prepare the drawings are free to invent any symbol they choose, provided that it is well identified and clear. Multi-outlet assemblies are readily applicable to residential work. Typical applications might be for the sound center in the living room [Figure 12.30(b)]; in the bedroom for the radio, television, telephone answering machine, electric sheet/blanket and other devices commonly used; in the basement workshop; and in any other area where a concentration of electrical devices occurs.

12.10 Floor Raceways

Up to this point, we have discussed basically two types of electrical raceway: conduit and surface-mounted raceways. Both are readily usable to carry wiring to outlets on walls and, with some limitations, to outlets on the ceiling. It is more difficult, however, to supply electrical loads located away from walls. Thus, to supply electrical

696 / BRANCH CIRCUITS AND OUTLETS

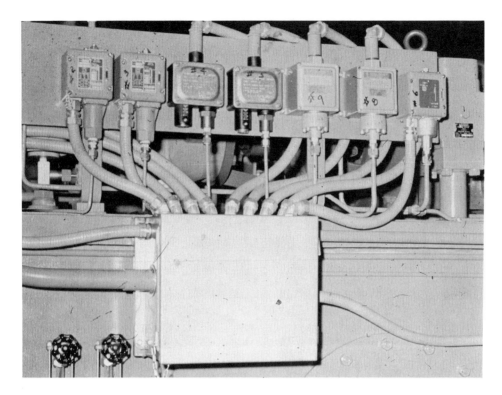

Figure 12.25 Where the density of conduits would make field bending of rigid conduit next to impossible, flexible conduit can be very effective. (Courtesy of Electri-Flex Company.)

and telephone service to a desk or table that is not adjacent to a wall, a floor outlet is usually required. Such a floor outlet can be fed by a conduit in or under the floor or by a surface raceway running across the floor. Any surface-mounted floor raceway (see the last item in Table 12.9), regardless of how well it is installed and how small it is, is unattractive, is a barrier to furniture movement and to movement of handicapped persons, and is a trip hazard to everyone. A floor outlet fed by conduit in or under the floor must be located precisely and it is relatively permanent. As a result, the electrical technologist preparing the electrical plans must know in advance the exact location of furniture and equipment in order to locate floor outlets. Obviously, this is not always, or even frequently, the case. Furthermore, once a floor outlet is installed, it is an expensive and often messy job to move, or even simply to remove.

As a result of these considerations, a number of in-floor raceway systems have been developed that solve the problem of electrical and signal outlet requirements away from walls. They are:

- Underfloor raceways, **NEC** Article 354.
- Cellular metal floor raceways, **NEC** Article 356.
- Cellular concrete floor raceways, **NEC** Article 358.

All three types are used almost exclusively in commercial and industrial construction. Residential construction generally does not require many floor outlets. This is because rooms are rarely so large that requirements cannot be met with an extension cord or wire from a wall outlet. Where a floor outlet is required, it is supplied from a conduit or other approved raceway in, or under, the floor. The basic difference among the three types is that underfloor duct (raceway) is added on to the structure, whereas the two cellular floor raceways systems are a part of the structure itself. All three types are highly specialized, and because they have

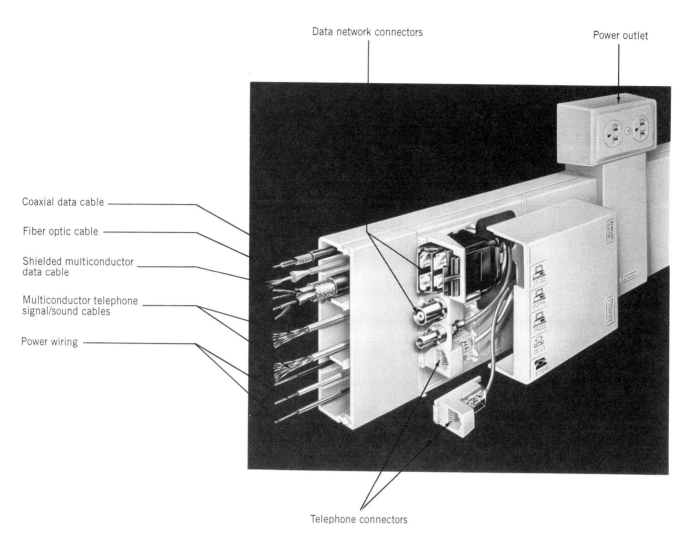

Figure 12.26 Multichannel nonmetallic surface raceway with snap-in connector modules for data network, signal and power wiring systems. (Similar metallic raceways are also available.) The raceway itself measures 4½ in. high by 1 in. deep. The internal dividers are movable or entirely removable, which permits varying the number and size of the wiring channels. Principal application of this type of wireway is in commercial spaces using extensive desktop data processing and communication equipment. (Courtesy of Panduit.)

a major impact on the building's architecture (particularly the cellular floors), their selection and layout are normally done by the job's architect and electrical engineer together. However, even though these raceway systems are not the responsibility of the electrical technologist, he or she should be familiar enough with them to know how they are assembled and how they function. We will, therefore, briefly describe and illustrate each of the three systems.

a. Underfloor Duct

This system is simply a grid of rectangular cross-section ducts laid onto the structural floor and covered with a concrete topping. Most ducts are

Table 12.9 Characteristics of Selected Wiremold® Metallic Surface Raceways

Raceway Type	Section Through Raceway	Wire Type	Number of Single Wires—AWG						Other
			#6	#8	#10	#12	#14	#16[b]	
200[a]		THHN, THWN				3	5	7	
		TW				3	3	5	
		THW				2	2	5	
500[a]		THHN, THWN		2	4	7	9	10	
		TW		2	3	4	6	7	
		THW			2	3	4	7	
700[a]		THHN, THWN	2	3	5	8	11	12	
		TW		2	4	6	7	9	
		THW		2	3	4	5	9	
2000[c,d]			*Without Receptacles*						2–25 pair phone cables
		TW, THW, THWN							
		THHN				7	7		
			With Receptacles						
		THHN, THWN				7	7		
		TW, THW				5	5		
2200[c,d]			*Without Receptacles*						3–25 pair phone cables
		THHN, THWN	11	19	32	51	69		
		TW	7	14	26	34	44		
		THW	7	11	19	23	29		
			With Receptacles						
		THHN, THWN	5	7	13	21	28		
		TW	3	7	10	10	10		
		THW	3	4	8	10	12		
2100[c]			*Without Receptacles*						2–25 pair phone cables
		THHN, THWN	6	10	17	28	37	41	
		TW	4	8	14	19	24	30	
		THW	4	6	10	13	15	30	
G-3000[c]			*Without Receptacles*						9–25 pair; 4–50 pair; phone cables
		THHN, THWN	27	44	76	119	160		
		TW	17	34	62	81	103		
		THW	19	26	45	55	67		
4000[c]			*Without Receptacles; power compartment only*						Low-voltage compartment 7–25 pair; 4–50 pair; phone cables
		THHN, THWN	28	47	81	128			
		TW	18	36	66	86			
		THW	18	28	48	59	72		

Table 12.9 *(Continued)*

Raceway Type	Section Through Raceway	Wire Type	Number of Single Wires—AWG						Other
			#6	#8	#10	#12	#14	#16[b]	
6000[c]			*Without Receptacles*						
		THHN, THWN	122	169	343	540			39–25 pair;
		TW	77	134	282	368			20–50 pair;
		THW	77	106	203	252			10–100 pair; phone cables
			With Receptacles						
		THHN, THWN	66	92	187	295			
		TW	42	73	154	200			
		THW	42	57	111	137			
1500[c,e]		THHN, THWN	2	3	5	7	10		
		TW		2	4	5	6		
		THW			3	3	4		

[a] These are one-piece raceways. Wiring is pulled in, as in conduit.
[b] Not listed by UL Guide Cards. For information when installing low voltage signal wiring.
[c] Two-piece raceway; base is mounted, wiring is installed and cover is snapped on.
[d] This raceway is specifically intended for use with prewired multiple outlets, which are available on 6-, 12- and 18-in. centers.
[e] Designed for installation on hard-surface flooring.

Source. Extracted from published data, and reprinted with permission of the Wiremold Company. For complete current data refer to the current Wiremold® catalog.

Figure 12.27 The exposed wood members of this sloped ceiling construction make the use of concealed wiring raceways impractical. A small cross-section flat surface raceway, as used here for wiring, is almost invisible. The raceway terminates in receptacles (circled) into which the decorative hanging lighting fixtures are plugged. (Courtesy of The Wiremold Company.)

metal, although recently heavy plastic ducts have come into use. Distribution ducts are laid in one direction, and feeder ducts at right angles to them. Distribution ducts are those that carry branch circuit wiring and are tapped to supply floor outlets. Feeder ducts supply wiring to the distribution ducts. Feeder ducts are also known as header ducts. A junction box is placed where the two types of ducts cross. Splices and taps are made in these junction boxes. In a single-level system, all ducts are laid at the same level, whereas in a two-level system, the feeder ducts are installed on top of (and across) the distribution ducts.

The choice of the type of system, the duct spacing, the method of tapping into the distribution ducts to establish a floor outlet (inserts) and other system details are usually beyond the scope of a technologist's work and are, therefore, not included here. If you are interested, you can consult the references listed in the bibliography at the end of this chapter and also manufacturers' catalogs for more information on this system and the cellular floor systems described next. Figure 12.33

700 / BRANCH CIRCUITS AND OUTLETS

Figure 12.29 (a) Three parallel runs of large cross-sectional surface raceway (each approx. 1½ in. deep by 2¾ in. wide) provide ample space for immediate needs plus considerable expansion. Three separate raceways are used to accommodate power wiring, network wiring and telephone/signal/sound wiring. (Compare Figure 12.26.) See also Table 12.9 for wiring capacity of this raceway. (Courtesy of The Wiremold Company.)

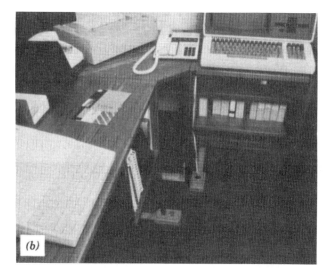

Figure 12.28 Changes in building codes and the huge expansion in the use of signal and data processing equipment are two important factors that lead to add-on wiring facilities. Photo (a) shows surface wiring to an add-on smoke detector. Note how much better the small raceway blends in when run in the same direction as the wood slats of the ceiling. Photo (b) shows that properly placed floor outlets can be added to an existing installation with surface raceway connections, thereby avoiding expensive raceway "burial." (Courtesy of The Wiremold Company.)

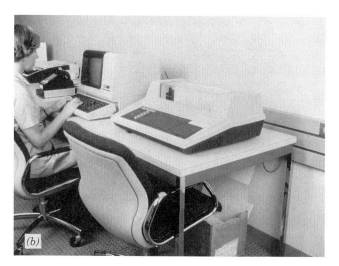

Figure 12.29 (b) A typical application of this type of triple raceway installation. (Courtesy of The Wiremold Company.)

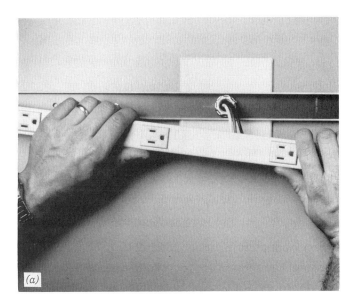

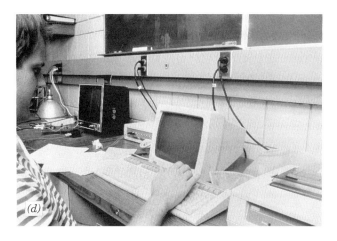

Figure 12.30 Modern occupancies frequently require placing electrical power outlets close together. Using surface raceways, this can easily be done with prewired outlet assemblies *(a, b, c)* or field-wired individual duplex receptacles *(d)*. Typical applications are residences *(b)*, schools *(c)*, and laboratories *(d)*. (Courtesy of The Wiremold Company.)

shows the elements of a modern two-level underfloor duct system. Note in the figures the use of preset inserts. Their use avoids the necessity of drilling into the floor duct at each point where a floor outlet is desired, in order to place an afterset insert. Inserts are of various designs, but their purpose is always to supply the floor outlets.

b. Cellular Floors (Metal and Concrete)

As stated previously, cellular floors are part of the building structure. As with underfloor duct systems, modern cellular floor systems usually use three separate cells to feed each floor outlet loca-

Figure 12.31 The frequent wiring changes required for theatre and exhibition lighting are easily made when the wiring is run in a suspended surface raceway. (Courtesy of The Wiremold Company.)

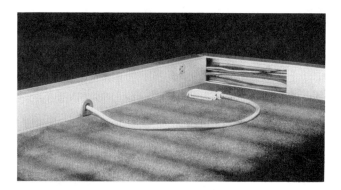

Figure 12.32 The basic raceway illustrated is 1⁷⁄₈ in. deep by 3³⁄₁₆ in. wide. It is shown with a divider installed, permitting use of the top section for power wiring and the bottom section for low voltage wiring. Since data and communication cables are frequently supplied with factory-installed terminations (as in the photo), a raceway, where the cable can be laid in rather than pulled in, is required. Also, terminal strips and other equipment can be installed in the low voltage section of these large raceways, making the use of separate terminal cabinets unnecessary. (Courtesy of The Wiremold Company.)

tion, as can be seen in Figure 12.34. The largest cell is usually reserved for data and network cables; the next smaller, for telephone and signal system wiring; and the smallest cell, for branch circuit power cables. Cells are fed by large cross-section trench ducts that are equivalent to the header or feeder ducts of the underfloor duct system. These ducts are also subdivided into three sections, each of which carries feeder cables for the appropriate system. If cabling demands are not heavy, as in buildings with only a small number of computer terminals, the signal and data cells can be combined into one cell. This results in a simpler and cheaper two-cell system.

As we stated at the beginning of this discussion, the principal reason for the development of floor raceway systems was to be able to feed electrical outlets located away from walls. However, floor raceways are expensive, and, although floor outlets are usually placed under desks so that they are not a tripping hazard, they do place some limitations on floor cleaning and furniture movement. As a result, a much cheaper system of ceiling raceways was developed. These are discussed in the next section.

12.11 Ceiling Raceways

This system of wiring should really be called over-the-ceiling because it is almost always installed inside the hung ceiling that is found in most commercial spaces. Occasionally, it is used in low budget installations that do not have a hung ceiling (see Figure 12.38). In such cases, however, to avoid being very unsightly, the overhead raceways are painted the same dark color as the entire ceiling cavity.

Basically the system consists of a network of rectangular metallic raceways that correspond to those in underfloor systems. Large header ducts carry main cables and feed the smaller distribution ducts. The arrangement can be seen clearly in Figure 12.35. The problem of how to get the wiring down from a duct above the hung ceiling to an outlet at usable height is very neatly solved by the use of a vertical, rectangular cross-section raceway that extends down from the ceiling duct to the floor as in Figure 12.36(a) or to table top level as in Figure 12.36(b). This raceway, which is called by various manufacturers a pole, electric pole, or power pole, is illustrated in Figure 12.37. It not only carries the wiring down but also contains built-in prewired receptacles on the side of the

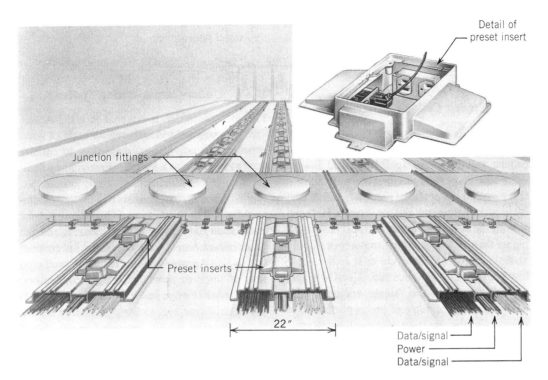

Figure 12.33 Two-level underfloor duct system. In order to reduce the overall height of the system, distribution ducts recess into the feeder (header) ducts at the junction boxes. Note that the preset inserts straddle all three cells of the three-section duct. The preset inserts (shown in the corner illustration) contain two duplex receptacles plus any type of desired connector for the cables in the signal/telephone/data cells. (Courtesy of Walker, a division of Butler Manufacturing Corporation.)

pole. Some poles of this type consist of multiple sections [see Figure 12.37(c)], which are intended to carry telephone and signal wiring in addition to wiring for its power receptacles. A nonhung ceiling application of ceiling raceways and vertical poles is shown in Figure 12.38.

Poles generally come prewired and are connected on the job to the wiring in the distribution ducts at the ceiling level. This connection can be made in the traditional hard-wired fashion, using a junction box and wire connectors, as shown in Figure 12.37. However, this procedure is time-consuming and prevents easy moving of the pole. A more recently developed method of connection for power wiring uses flexible metal-jacketed cables with factory-made terminations at both ends. Such cables are called whips and are shown on Figure 12.35 and labelled as such. A system that uses these precut, preterminated cables is called a *modular wiring system*. The place where such a cable connects to a piece of equipment is called a *modular wiring interface*. One such interface (connection) is shown in Figure 12.37(b). The great advantage of modular wiring systems over conventional hard wiring is that connecting a piece of equipment is simply a matter of plugging it in. Disconnection is simply unplugging. The savings in labor and the ease and speed of making changes in the system are obvious. Of course, the drawback is higher equipment cost. The decision as to the type of wiring system to use (modular or conventional) is made by the project's electrical engineer after making a cost analysis.

In the overhead raceway system shown in Figure 12.35, whips are used to interface between the overhead lateral raceways (distribution ducts) and poles, light switches and lighting fixtures. Whips are also shown connecting telephone lateral raceways to the tops of poles. This indicates that the poles are multisectional and contain telephone as well as power outlets. Overhead raceways systems are also useful in supplying electrical power and

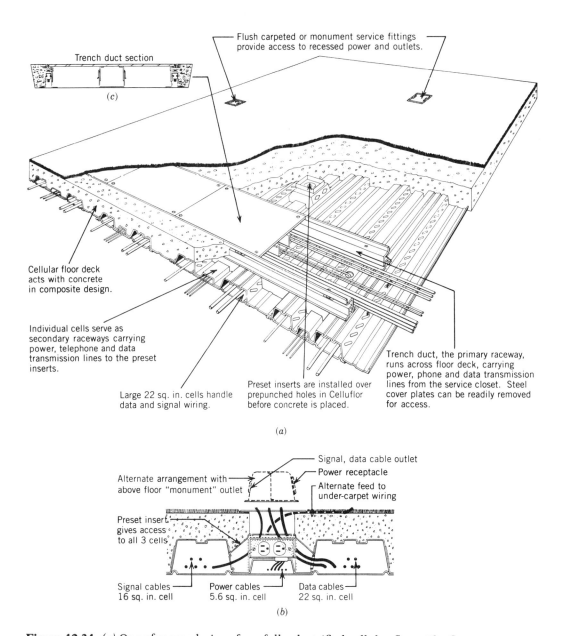

Figure 12.34 (a) One of many designs for a fully electrified cellular floor. The floor cells are available in many designs; the choice depending primarily on the structural requirements. The trench (c) that straddles the cells provides the electrical feeds through precut holes in the cells. The trench itself is completely accessible from the top and, when opened, exposes all the wiring and the cells below. (b) Activated preset insert. Note that the insert straddles the center (power) cell and provides access to the two adjoining low-voltage wiring cells. Power and signal wiring are completely separated at all times by metal barriers. If desired, a standard surface "monument" fitting can be mounted on the floor or connection can be made to undercarpet cables, instead of the flush plate shown. When an insert is to be removed, the flush cover plate is simply replaced with a blank plate. (c) Section through trench duct, which acts as feeder for distribution ducts. Trench is available with or without bottom, in any required height, in widths from 9 to 36 in. and with one, two, or three compartments, depending on floor cell design and cabling requirements. (Courtesy of Walker, a division of Butler Manufacturing Corporation.)

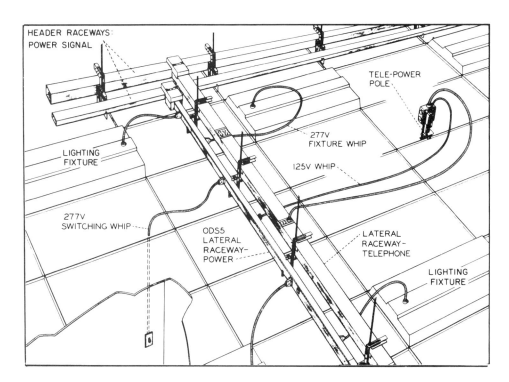

Figure 12.35 Partial view of a two section (power, telephone) overhead raceway system installed above the hung ceiling. Header (feeder) raceways carry the main cables and subfeed into the smaller lateral (distribution) raceways. These smaller distribution raceways feed the vertical floor-to-ceiling raceways known as power/telephone poles. (See Figure 12.37 for details of these vertical raceways.) The wiring system shown is of the modular type, that is, it uses precut, preterminated cables to connect (interface) between raceways and utilization equipment. All parts of the system, except header (feeder) ducts, are factory equipped with modular plug-in-type connectors that match those on the preterminated cables (whips). This permits simple, quick connecting and disconnecting of lighting, switches and power pole connections. Telephone (and data) cables are also of the modular type, with preterminated connectors that snap in to connectors on power/telephone poles. All connectors are polarized so that different systems and different voltages cannot be interconnected by mistake. All equipment names in this figure refer to Wiremold ® equipment. (Courtesy of The Wiremold Company.)

signal connection into movable office partitions, both full-height types and the partial-height type used in office landscaping design. Such partitions must be specifically designed to be electrified. With both types, a *mullion*, which is similar in construction to a power pole, is extended into the ceiling and connected to the power and signal wiring raceways in exactly the same way as a pole raceway. Wiring in the partition itself is usually factory-installed.

An additional use of overhead raceway systems (or hung ceiling conduit systems) is to supply power and signal electrical services to floor outlets on the floor above. This is known in the trade as a *poke-through service fitting*. As seen in Figure 12.39(a), it consists of a fire-rated metal conduit fitting that penetrates (pokes-through) the floor above to provide electrical service to work stations on the floor above. See Figure 12.39(b). Such penetrations must be fire-rated. See **NEC** Section 300-21. Poke-through fittings are particularly useful in feeding outlets in partial-height partitions on the floor above, floor outlets that fall between ducts or cells of underfloor systems, and floor outlets that are not easily accessible by any other wiring method.

Figure 12.36 *(a)* A "landscaped" open office plan uses power poles to bring power to the equipment level from an over-the-ceiling power distribution system. (Courtesy of The Wiremold Company.)

Figure 12.36 *(b)* Power poles extend down from the ceiling to any desired height. In this library, power is required above the base cabinets, and the power pole is easily arranged to supply it. (Courtesy of The Wiremold Company.)

12.12 Other Branch Circuit Wiring Methods

We began our discussion of branch circuit wiring methods with a description of the use of NEC type AC (BX) and NM (Romex) cables. They are self-contained; that is, these cables are installed without a raceway; and their application is very limited. The remainder of our study until this point has covered wire and raceway systems, where the wiring channel, or raceway, is installed first and wiring pulled in or laid in afterwards. This type of installation is permitted in all types of structures. We will now end our discussion of wiring methods with a study of four types of self-contained wiring methods that have very broad application in residential, commercial, institutional, and industrial buildings. They are:

- Lighting track (NEC Article 410 Part R).
- Flat cable assemblies (NEC Article 363).
- Light duty plug-in busway (covered by the general NEC article on busways, Article 364).
- Undercarpet wiring system (NEC Article 328—flat conductor cable).

a. Lighting Track

This item is a factory-assembled metal channel that holds conductors for one to four circuits permanently installed in the track. See Figure 12.40. Strictly speaking, it is not branch circuit wiring, which by definition extends between the circuit protection and the outlet(s). It is really a kind of continuous plug-in device that is specifically designed to support and supply power to track lighting fixtures. By Code restriction, it may be used only for this type of lighting. It is included here because it is similar in construction to the

OTHER BRANCH CIRCUIT WIRING METHODS / 707

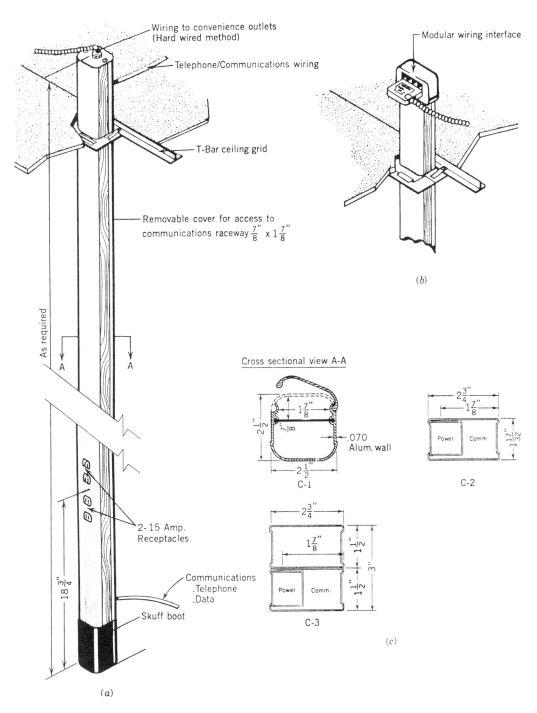

Figure 12.37 *(a)* A typical ceiling-to-floor "pole"-type raceway is fed at the top with both power and communication wiring from raceways in the hung ceiling. These feeds can be made either conventional hard-wired or modular (plug-in). A modular connection (interface) is shown in *(b)*. Modular connectors are also used for communication wiring. Poles are equipped with one or more prewired duplex power receptacles and prewired communication connectors as required. Poles are available in various cross sections to suit the needs of the particular installation. Three typical designs are shown in (c-1) through (c-3). (Illustrations *a,b,c-1* courtesy of Hubbell, Inc., illustrations *c-2,c-3*, courtesy of The Wiremold Company.)

Figure 12.38 In this economical installation, the absence of a hung ceiling is obscured by painting the ceiling, fixture bodies, and raceways black. (Courtesy of The Wiremold Company.)

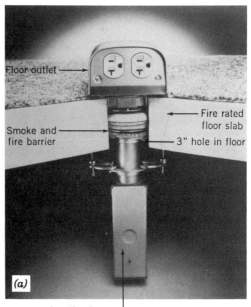

Figure 12.39 (a) Typical poke-through electrical fitting mounts in a 3-in. hole. It is wired from underneath with the required power, telephone, signal and data cables. Power and low voltage cables are separated as required by Code. Units are available prewired, or suitable for field wiring, and adaptable for varying floor thicknesses. The floor fitting is provided with power, telephone and data cable outlets as required for the specific installation. (Courtesy of Hubbell, Inc.)

heavier, multipurpose, self-contained wiring systems discussed next and because, in many installations, it is the only load on a particular lighting circuit. Each separately wired section of lighting track is considered to be one outlet. Also, like fluorescent lighting fixtures, multiple sections of track, when connected as a single continuous run, are also considered to be one outlet.

b. Flat Cable Assemblies

The cable (**NEC** type FC) is a specially designed assembly of two, three or four conductors, size No. 10 AWG, rated 30 amp at 600 v. It is field-installed in a surface metal raceway, specifically identified for this job. The raceway is normally a 1⅝-in. square metal structural channel. The power tap-off devices puncture the insulation when installed, to make contact with the electrical conductors. Wiring from these tap-off devices may feed lighting, motors, switches, or receptacle outlets as desired. The advantage of this wiring system is the same as that of any continuous plug-in arrangement: loads can be added, moved, and removed with a minimum of field labor and materials; and the other loads on the circuit are not disrupted. A similar installation, intended to feed industrial lighting, is shown in Figure 12.41.

c. Light Duty Plug-in Busway

This construction, which is still another form of light duty continuous plug-in electrical feeder, is manufactured in two size ranges: 20–60 amp, 2- or 3-wire, at 300 v and 60–100 amps, 3- or 4-wire, at 600 v. Both groups may be used to feed industrial

NOTES

Poke-through systems are used in conjunction with overhead branch distribution systems run in accessible suspended ceiling cavities to serve outlets in partitions. When services are required at floor locations where adjacent partitions or columns are not available, as in open office planning, they must either be brought down from a wireway assembly (known as a power pole) or up through a floor penetration containing a fire-rated insert fitting and above-floor outlet assembly. To install a poke-through assembly, the floor slab must either be drilled or contain preset sleeves arranged in a modular array. Poke-through assemblies are used in conjunction with cellular deck and underfloor duct systems when precise service location required does not fall directly above its associated system raceway.

With one floor penetration, the single poke-through assembly can serve all the power, communications, and data requirements of a work station. Distribution wiring in the ceiling cavity can be run in raceways. A more cost-effective method is to use armored cable (bx) for power and approved plenum rated cable for communications and data when the ceiling cavity is used for return air. To minimize disturbance to the office space below when a poke-through assembly needs to be relocated or added, a modular system of prewired junction boxes for each service can be provided, although it is more common to elect this option for power only. A different type of wiring system must be selected for a floor slab on grade, above lobby or retail space, above mechanical equipment space, or above space exposed to atmosphere.

Low initial cost of a poke-through system makes it both viable and attractive for investor-owned buildings where tenants are responsible for future changes and for corporate buildings where construction budget is limited. It is effective when office planning includes interconnecting work station panels containing provisions to extend wiring above the floor, reducing the number of floor penetrations needed to services.

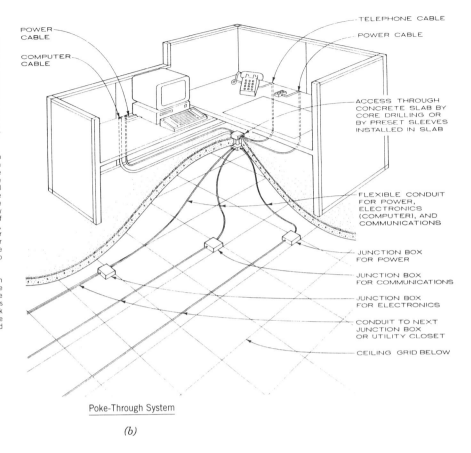

Poke-Through System

(b)

Figure 12.39 *(b)* Typical application of a poke-through fitting to provide power, telephone and data service to a modern work station. This drawing shows the electrical services being tapped at junction boxes in a hung ceiling conduit system on the floor below. The ceiling wiring system can also be a raceway network as in Figure 12.35 and 12.37(b) in lieu of the hard wiring shown here. (From Ramsey and Sleeper, *Architectural Graphic Standards*, 8th. ed., 1988, reprinted by permission of John Wiley & Sons.)

lighting, machine tools, light machinery, and other load within their ratings. Like all busways, their great advantage is the ease of installation and of making power takeoffs. Typical busway designs in both ratings and a typical application are shown in Figure 12.42. Heavy duty busway, commonly known as *busduct*, is discussed in Section 16.5.

d. Undercarpet Wiring

This system was developed as an inexpensive substitute for underfloor ducts or cellular floors. Its basic advantage over ceiling height raceways is that it is installed at floor level and, therefore, eliminates the need for power/telephone poles. The system is based on the use of a factory-assembled flat cable, so thin that it can be installed under carpet squares without creating either a trip hazard or even discomfort while walking. The cable is manufactured in various designs for power and signal wiring. Total cable thickness is approximately $1/32$ in. See Figure 12.43.

The entire cable assembly, which is permitted by the NEC to be installed only under carpet squares, is covered with a metal shield. This shield acts both as physical protection and as a continuous ground path. In addition, a protective bottom shield is required. That shield is either metal or heavy PVC. The cable, which is intended for use in commercial offices, is designed to carry normal

710 / BRANCH CIRCUITS AND OUTLETS

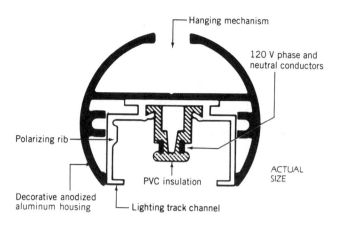

Figure 12.40 Section through a single-circuit lighting track. The decorative circular exterior housing is available in a variety of colors, finishes and hanger arrangements. The actual lighting track, shown full size, is available without the decorative housing, for direct surface mounting. Other track sections are available to accommodate two or three separate circuits. (Courtesy of Swivelier.)

physical loads such as office furniture and personnel traffic, without damage. A complete line of junction, splicing and outlet fittings is readily available. See Figure 12.44.

The great advantage of this system is that it is installed on top of the finished concrete floor, and, except for very complex systems, all wiring including splicing is installed at the floor level, under the carpet. Adding or moving outlets is simply a matter of lifting a few carpet squares and installing the wiring, floor devices and splices. This system of wiring is particularly useful for alteration and retrofit work, although not for frequent alterations because of the high labor costs involved. However, its simplicity and the fact that it is installed at floor level make it a reasonable choice for new construction. See Figure 12.45, which shows the technique, and Figure 12.46, which shows the layout for a typical undercarpet wiring system in a small office.

Outlets

12.13 Outlets

Referring again to the **National Electrical Code** for a definition, we find that an *outlet* is defined as "a point on the wiring system at which current is

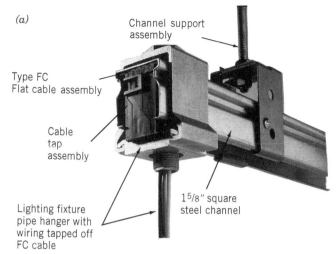

Figure 12.41 A three-phase, 4-wire flat cable assembly installed in a 1⅝ in. square steel channel is shown in (a). Taps into the conductors are made by tightening the tap device shown. Taps can be made between phase wire to ground to give 120 v and phase-to-phase to give 208 v. (If the cable is connected to a 277/480 v three-phase system, then the phase-to-ground and phase-to-phase voltages will be 277 and 480 v, respectively.) (b) Tap feeds lighting fixtures. After a tap device is removed, the puncture made by the tap "heals" itself. (Courtesy of Chan-L-Wire /Wiremold Company.)

taken to supply utilization equipment." In simpler terms, any point that supplies an electrical load is called an outlet. Therefore, a wall receptacle, a motor connection, an electric heater junction box, a ceiling lighting box and the like are all outlets. A wall switch is not an outlet, although in the field

OUTLETS / 711

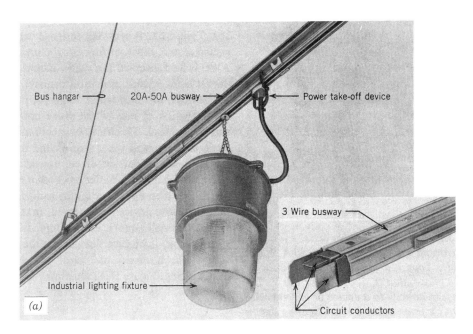

Figure 12.42 *(a)* Light duty busway is rated 20–60 amp, 300 v, and either 2- or 3-wire. Power-takeoff devices twist into the bus to make contact with the circuit conductors. Within NEC restrictions of overcurrent protection, this busway may be used for standard and heavy duty lighting fixtures and for other electrical devices such as electrically powered tools. (Courtesy of Siemens Energy & Automation, Inc.)

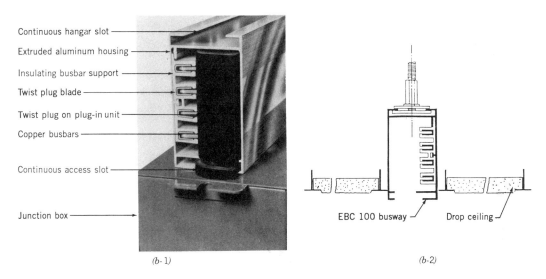

Figure 12.42 *(b-1)* Plug-in busway rated 100 amp, three-phase, 4-wire, 600 v measures approximately 2½ in. wide × 4½ in. high. The twist-in-plug, which is integrally attached to a connection means (junction box in the illustration), is rated 30–100 amp, single- or three-phase, as required. The attached junction box (or receptacle, circuit breaker, or fuse box) then feeds the utilization device (e.g., heavy duty lighting or machinery). This bus can also be installed in a hung ceiling *(b-2)*. (Courtesy of Electric Busway Corporation.)

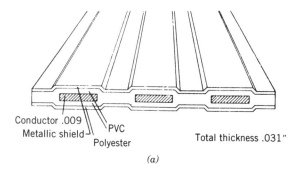

Figure 12.43 (a) Schematic section through one design of NEC type FCC undercarpet cable. The copper conductors illustrated are the equivalent of No. 12 AWG. The PVC acts as insulation, and the polyester, as both insulation and physical protection. All designs require a metallic top shield and a metallic or nonmetallic bottom shield for physical protection. (From B. Stein and J. S. Reynolds, *Mechanical and Electrical Equipment for Buildings*, 8th ed., 1992, John Wiley & Sons, New York. Reprinted by permission of John Wiley & Sons.)

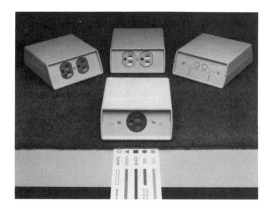

Figure 12.44 Typical components of an undercarpet wiring system. The undercarpet FCC power cable is shown without the metallic top shield that is required in an actual installation. It is a color-coded, five-conductor cable (neutral, equipment ground, and three circuit conductors or two circuit conductors and an isolated ground conductor). The floor outlets shown are: front, single power outlet; rear (left to right), duplex power outlet, one of which has isolated ground, standard duplex power outlet, and combination data cable, communications, and telephone outlet. (Courtesy of Hubbell, Inc.)

Figure 12.43 (b) Preterminated 25-pair undercarpet cable. These cables are commercially available in lengths of from 5 to 50 ft. (Courtesy of AMP, Inc.)

one frequently hears the term *switch outlet*—so much so, that the term is generally accepted and understood. When counting the numbers of outlets on a circuit, switches are not included. They are included when counting outlets for the purpose of making a rough cost estimate. Thus, an electrical contractor will normally quote a figure of X dollars per switch outlet and Y dollars per receptacle out-

let. This same contractor, however, will understand perfectly an instruction to wire no more than six outlets on a circuit and will not include switches in the outlet count.

What then, physically, is an outlet? Generally, it consists of a small metal box into which a raceway and/or cable extends. Outlet boxes vary in size and shape and are specifically designed to suit different kinds of construction. Thus we have the *device box*, also called a jiffy box or gem box, intended for single wall devices; octagonal and round 3½-in. boxes intended for ceiling outlets; the 4-in. square general-purpose box; and the like. The required volume of a box, in cubic inches, depends on the number of wires and devices to be put into the box. It can be calculated from the data in **NEC** Section 370-16. The box required can then be selected from manufacturers' catalog data, which should include box volume. Figure 12.47 illustrates standard boxes that are sufficient for the vast majority of branch circuit outlets. Figure 12.48 illustrates typical outlet box mounting arrangements. Boxes intended for fastening to wall studs are furnished

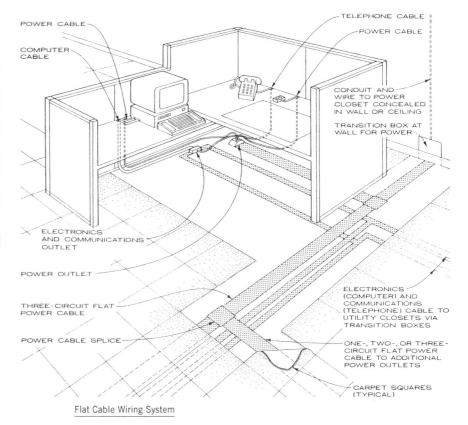

Figure 12.45 Schematic drawing of a section of a flat cable undercarpet wiring system feeding a single work station. The system can be readily adapted to changing requirements, particularly in the area of data and communications. (From Ramsey and Sleeper, *Architectural Graphic Standards*, 8th ed., 1988, reprinted by permission of John Wiley & Sons.)

with mounting holes for nails and screws. Ceiling outlet boxes for construction of this kind are mounted on steel straps that, in turn, fasten to ceiling joists.

Box entrances (knockouts) are arranged to receive cable, BX, or conduit. The important fact to remember is that every outlet has a box. (See NEC Section 300-15.) Boxless wiring devices are actually self-contained box-device combinations. The term *gang*, when referring to a box, indicates the number of wiring device positions supplied. Thus, if three receptacles are to be mounted side by side, a three-gang box would be used. Boxes can normally be fastened to each other, and the dividers removed to form such an assembly.

Although each outlet requires a box, there is only one situation that requires a box that is not an outlet. This is the common *junction box*. When wiring is being run throughout a building, it frequently becomes necessary to make a tap, or junction, in the wiring to take power off to another point. (See for instance Figure 12.49.) These taps, whenever possible, are made at outlet boxes to avoid the expense of an additional junction box. (In effect the outlet box is also serving as a junction box when a tap is made in it.) Sometimes, however, the necessity for a junction box is unavoidable, and one is installed. In it, wires enter, are spliced to each other, and leave to feed outlets. Such a box is not an outlet since, by definition, it does not supply current to a utilization device. Some contractors, incorrectly, count junction boxes as outlets. Care should be used to clarify this point if payment is being made by number of outlets.

On a drawing, junction boxes should be shown by a specific symbol. As we can see in Figure 12.50,

714 / BRANCH CIRCUITS AND OUTLETS

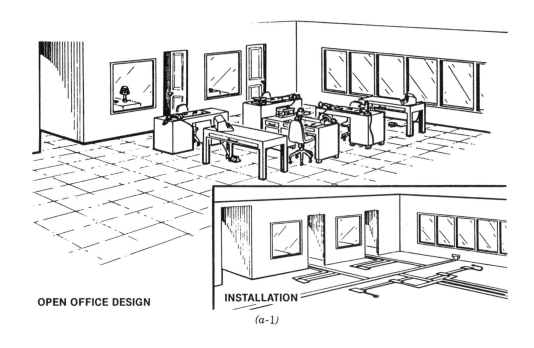

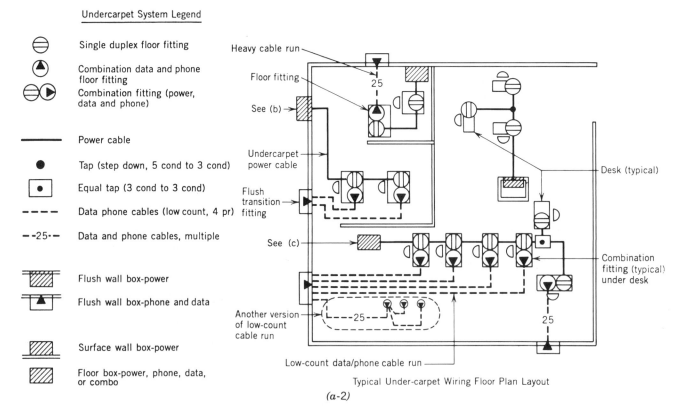

Figure 12.46 *(a-1)* Typical layout of undercarpet cabling for power, telephone and data outlets, in a small office. Each desk is provided with a power outlet plus either a phone connection or a data cable connection. *(a-2)* All outlets are floor mounted adjacent to the desk. (Courtesy of AMP, Inc.)

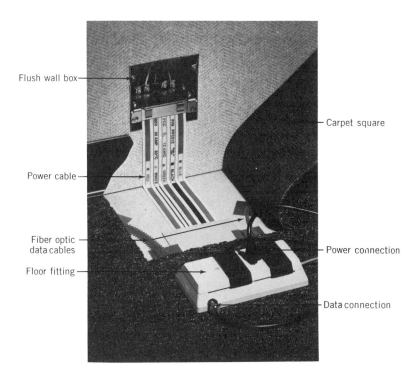

Figure 12.46 *(b)* Photo shows a typical flush wall (transition) box for power cabling. Round cable enters the box and is connected to a color-coded five-conductor flat cable, containing three phase conductors, neutral and equipment ground. Such a cable can be used on a three-phase, 4-wire system. Note that a single floor fitting handles both power and data cabling. In this particular installation, the data cable is fiber optic rather than a low voltage copper cable. (Courtesy of Walker, a division of Butler Manufacturing Corporation.)

the symbol used for an outlet is a circle unless there is a possibility of confusion with round columns or other circular architectural features. This confusion is possible because outlet boxes are obviously not shown to scale. This is true of most electrical work. If we were to attempt to show to scale a 4-in. box on 1/8 in. = 1 ft. scale drawings, the box would be 1/24-in. in diameter—the size of a pencil dot. For this reason, almost all electrical work is shown at a size that makes for easy drawing reading—and that is unrelated to the scale of the drawings. The only possible exception to this (no scale) convention on architectural-electrical plans is fluorescent fixtures. They are normally shown to scale in length and occasionally in width. Of course, construction details are almost always drawn to scale. The convention of not drawing electrical drawings to an architectural scale can occasionally create a problem of space coordination with equipment of other trades, most often with ductwork. Experienced electrical technologists will recognize places where equipment interference may occur and will usually handle the situation by preparing a large scale detail of the area involved. By so doing they will avoid a construction space conflict problem arising in the field. Field problems are always more expensive to solve than similar design problems.

In recent years, nonmetallic boxes and conduits have come into wide use. Their application is somewhat specialized, and you should refer to the NEC for applicability and to manufacturers' literature for equipment details.

One further clarification must be made. The outlet is the point at which current is taken off to feed the "utilization equipment" (**NEC** term for the electrical load apparatus) and does not include the equipment itself. In the case of a receptacle outlet where the receptacle is mounted inside the outlet box, the receptacle device is not an electrical load but rather an extension of the box wiring. Its purpose is to permit easy connection of the load equipment. The outlet is to be considered as separate

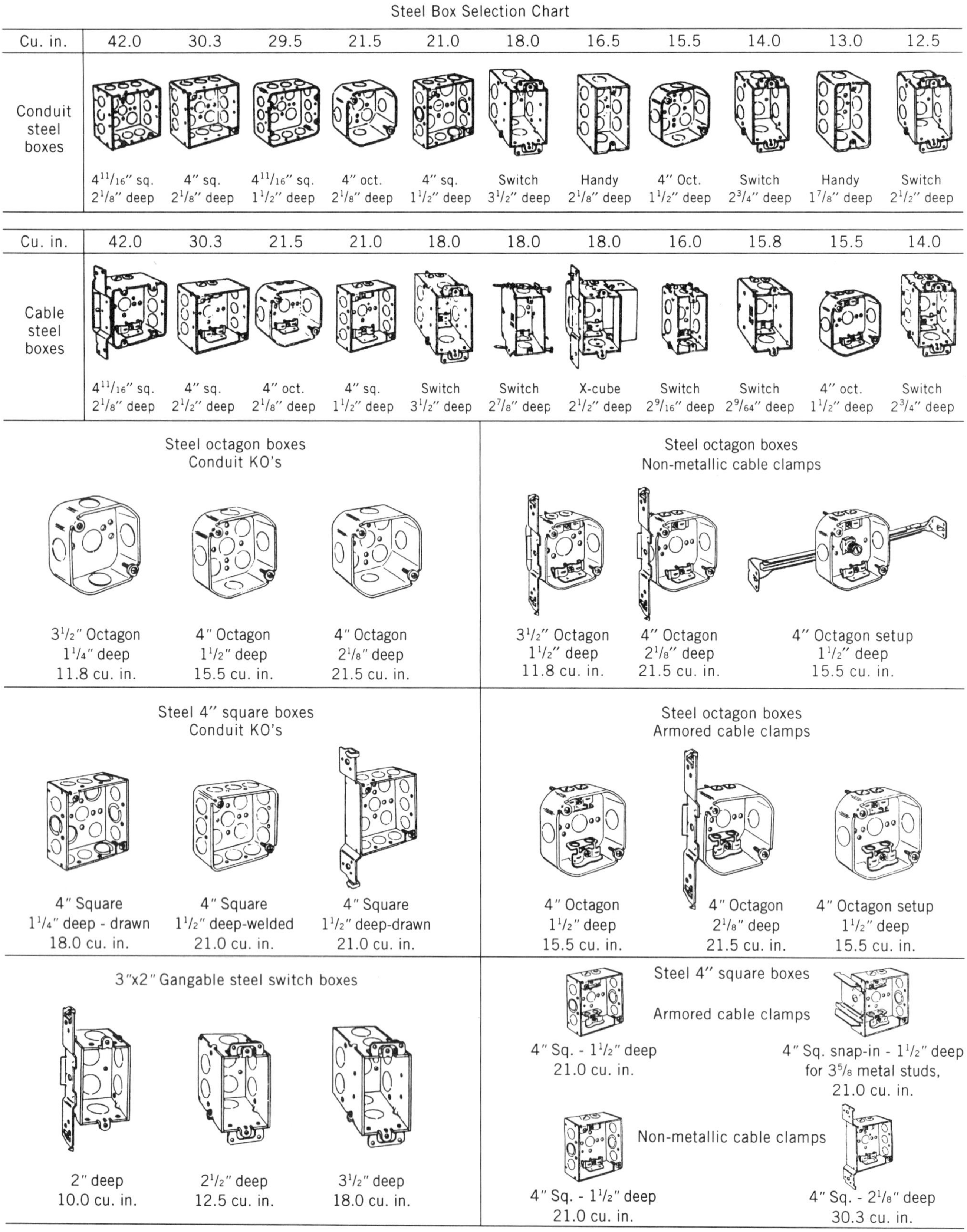

Steel Box Selection Chart

Cu. in.	42.0	30.3	29.5	21.5	21.0	18.0	16.5	15.5	14.0	13.0	12.5
Conduit steel boxes	4¹¹/₁₆" sq. 2⅛" deep	4" sq. 2⅛" deep	4¹¹/₁₆" sq. 1½" deep	4" oct. 2⅛" deep	4" sq. 1½" deep	Switch 3½" deep	Handy 2⅛" deep	4" Oct. 1½" deep	Switch 2¾" deep	Handy 1⅞" deep	Switch 2½" deep

Cu. in.	42.0	30.3	21.5	21.0	18.0	18.0	18.0	16.0	15.8	15.5	14.0
Cable steel boxes	4¹¹/₁₆" sq. 2⅛" deep	4" sq. 2½" deep	4" oct. 2⅛" deep	4" sq. 1½" deep	Switch 3½" deep	Switch 2⅞" deep	X-cube 2½" deep	Switch 2⁹/₁₆" deep	Switch 2⁹/₆₄" deep	4" oct. 1½" deep	Switch 2¾" deep

Steel octagon boxes — Conduit KO's
- 3½" Octagon, 1¼" deep, 11.8 cu. in.
- 4" Octagon, 1½" deep, 15.5 cu. in.
- 4" Octagon, 2⅛" deep, 21.5 cu. in.

Steel octagon boxes — Non-metallic cable clamps
- 3½" Octagon, 1½" deep, 11.8 cu. in.
- 4" Octagon, 2⅛" deep, 21.5 cu. in.
- 4" Octagon setup, 1½" deep, 15.5 cu. in.

Steel 4" square boxes — Conduit KO's
- 4" Square, 1¼" deep - drawn, 18.0 cu. in.
- 4" Square, 1½" deep - welded, 21.0 cu. in.
- 4" Square, 1½" deep - drawn, 21.0 cu. in.

Steel octagon boxes — Armored cable clamps
- 4" Octagon, 1½" deep, 15.5 cu. in.
- 4" Octagon, 2⅛" deep, 21.5 cu. in.
- 4" Octagon setup, 1½" deep, 15.5 cu. in.

3"x2" Gangable steel switch boxes
- 2" deep, 10.0 cu. in.
- 2½" deep, 12.5 cu. in.
- 3½" deep, 18.0 cu. in.

Steel 4" square boxes — Armored cable clamps
- 4" Sq. - 1½" deep, 21.0 cu. in.
- 4" Sq. snap-in - 1½" deep for 3⅝ metal studs, 21.0 cu. in.

Non-metallic cable clamps
- 4" Sq. - 1½" deep, 21.0 cu. in.
- 4" Sq. - 2⅛" deep, 30.3 cu. in.

(a)

Figure 12.47 (a)

OUTLETS / 717

Steel Specialty Boxes

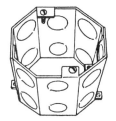
Concrete ring deck box

Concrete deck box

3½" ceiling pans
¾" deep
Nonmetallic cable clamps

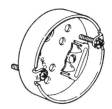
For support of ceiling fan
up to 35 lbs.
Nonmetallic cable clamps

Plenum box and blank covers
4" and 4¹¹⁄₁₆"

Hung ceiling box
3½" deep
⅜" stud, 8–½", 8–¼" KO's
43 cu. in.

Figure 12.47 *(a)* Conventional steel boxes that are used for the vast majority of outlets are shown in a selection chart and in individual detail. Boxes that are intended for use with conduit are provided with appropriately sized knockouts (KOs), that is, prepuched metal discs that can be knocked out by a hammer blow, at each desired conduit entrance. Boxes intended for use with cable, either armored (type AC) or sheathed (type NM), are provided with knockouts for cable entry and with clamps for fastening the cable. The volume of each box in cubic inches (cu in.) is shown, to enable the wiring system designer to select the box with the required dimension to suit the devices, clamps and wires to be placed in it. The data needed to perform the necessary box size calculation are found in the NEC, Article 370-16. *(b)* A few of the vast variety of specialty outlet boxes. As with conventional steel outlet boxes, the volume of these boxes must be sufficient to contain all the intended items as required by the NEC. (Courtesy of Raco, Inc.)

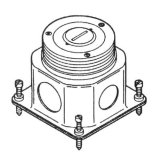
Semi-adjustable steel floor box
26.5 cu. in. capacity

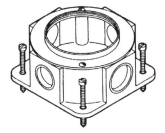
Cast iron floor box with threaded legs
23.8 cu. in capacity

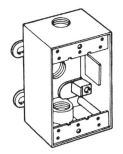
Single-gang weatherproof box
2" deep
17.5 cu. in. capacity

(b)

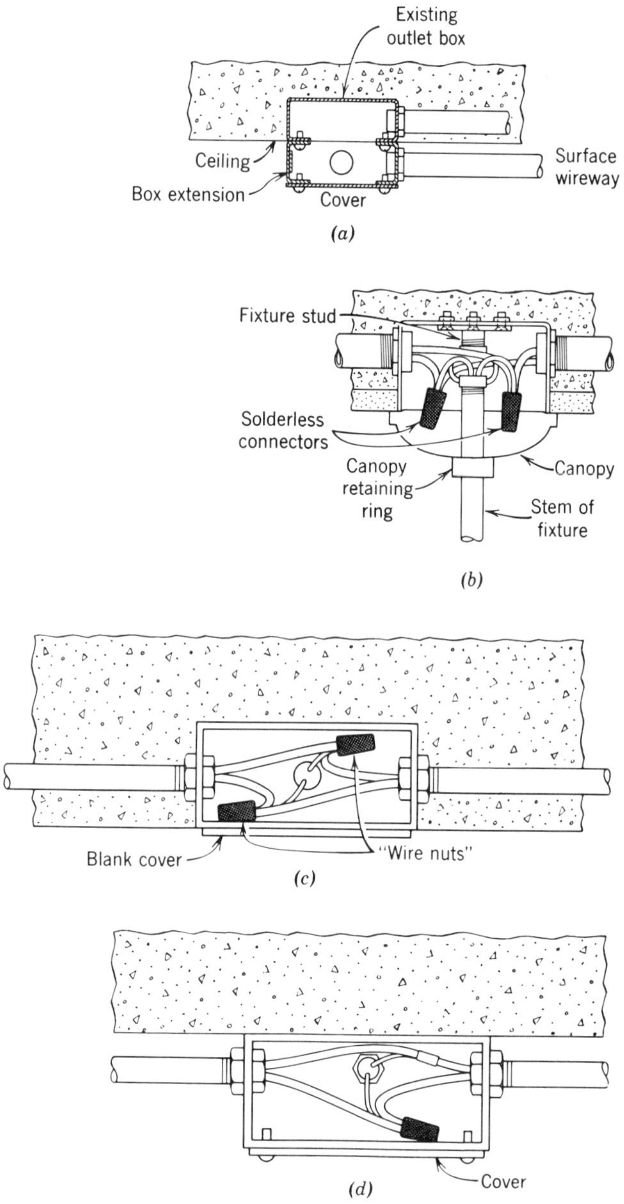

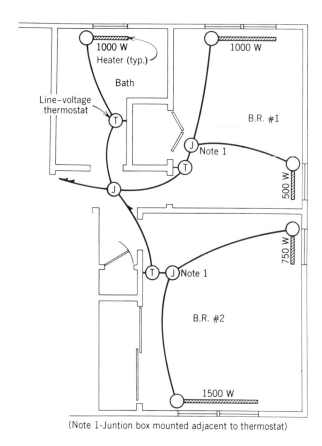

Figure 12.49 Electric heating for The Basic House. Note use of junction and outlet boxes. Absence of pigtail wiring symbol between outlet box and heater indicates internal, factory connection. The heaters in each room are controlled by a single line-voltage thermostat of adequate wattage rating. Due to the wattages involved, the heaters and thermostat would be rated for 240 v. The single arrowhead on the wiring from the thermostat in BR #2 indicates that the two heaters in that room are the total load on that circuit. Since there are only two circuits in the home-run (two arrowheads leaving the junction box in the corridor), the installing electrician knows that the second circuit is split at the junction box to feed the heaters in the bath and in BR #1.

Figure 12.48 Outlet boxes and mounting arrangements. (a) Surface extension to concealed wiring system. (b) One method of mounting a suspended ceiling fixture. (c) Typical recessed ceiling junction box. (d) Typical surface-mounted box, either ceiling or wall. Box is attached to the surface with appropriate mounting device.

from the load device, even if it is included as part of this load device. For instance, a recessed electrical wall heater occasionally comes with an outlet box permanently mounted in or on the unit, but without any wiring between the heating element and the box. In such a case, the drawing must show the final connection to the load device by symbol or note if there is any field work to be done.

This work normally consists of final connections with a short piece of flexible conduit. This final connection is almost always required for motors, transformers and unit heaters (except where connection is made with cord and cap (plug). (See Figure 12.6, the fourth symbol from the end, and Figure 12.50, the last symbol.) If this connection is not shown, a contractor might claim that the circuitry required by the drawings includes only the outlets and not the equipment connections. For these, additional payment would be requested.

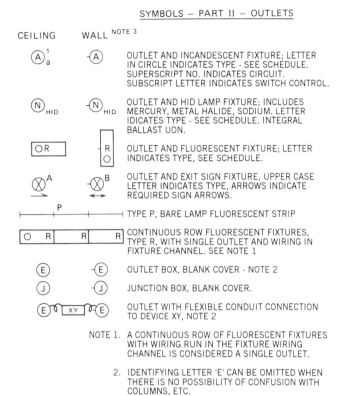

Figure 12.50 Architectural-electrical plan symbol list, Part II, Outlets. For other portions of the symbol list see:
Part I, Wiring and Raceways, Figure 12.6, page 671
Part III, Wiring Devices, Figure 12.51, page 720
Part IV, Abbreviations, Figure 12.58, page 726
Part V, Single Line Diagrams, Figure 13.4, page 733
Part VI, Equipment, Figure 13.15, page 753
Part VII, Signaling Devices, Figure 15.34, page 909
Part VIII, Motors and Motor Control, Figure 16.25, page 944
Part IX, Control and Wiring Diagrams, Figure 16.28, page 947

items of utilization equipment are to be connected on a single branch circuit and where permitted by Code, it is a good idea to place a receptacle in each outlet box feeding an item of equipment and making the final connection to the item with a cord and cap. This practice permits quick disconnection of one unit for service without disturbing the other items on the circuit. The receptacle and cap act as the required load device disconnect means, and the additional cost of the receptacle device in the outlet box is minimal.

In showing ceiling light fixtures, particularly incandescent units (see Figure 12.50), it is customary to show only a circle, symbolizing both the outlet and the fixture. For this reason, ceiling outlets must be clearly shown on the symbol list or otherwise as including the appropriate lighting fixture. The outlet itself, by definition, is only the wired outlet box. Refer to the symbol list, Part II, Figure 12.50. This portion of the list is devoted to outlets. Switches and receptacles are covered in the next section and in Part III of the symbol list.

Your attention is called particularly to Note 1 of Figure 12.50. Only a single outlet is used in a continuous row of fluorescent fixtures. This is always the most economical method of installation. It is, therefore, normally true that an installation with, say, two continuous rows of 6 fixtures each, for a total of 12 fixtures, will be cheaper than an installation of 12 fixtures that are individually installed. This type of economic consideration should be kept in mind when laying out fixtures. As a rule of thumb, a ceiling outlet separately installed costs about the same as a cheap two-lamp fluorescent fixture or a medium-quality incandescent downlight. This rule is obviously very approximate but is close enough so that, in a specific case, the technologist, given the choice of continuous rows or individual fixtures, will have some guidance. He or she can then turn to the senior designer or engineer for a final decision.

Where the wiring between the integral outlet box on a piece of utilization (load) equipment is factory made internally, no connection need be shown on the drawings between the outlet and the device, since no field work is required. See Figure 12.49. However, where the circuit continues beyond the item of load equipment, good installation practice would include an additional outlet (junction) box nearby and a flexible connection to the factory-supplied box on the appliance. This permits the appliance to be removed for servicing without disturbing the circuit. Alternatively, where several

12.14 Receptacles and Other Wiring Devices

The only outlet with a self-contained device is the receptacle. This is, by **NEC** definition, "a contact device installed at the outlet for the connection of single attachment plug." This usually takes the form of the common wall outlet or, as will be illustrated, larger and more complex devices. We must comment here about terms. The common wall outlet is properly called a *convenience recepta-*

(text continues on p. 723)

720 / BRANCH CIRCUITS AND OUTLETS

SYMBOLS – PART III – WIRING DEVICES

⊖	DUPLEX CONVENIENCE RECEPTACLE OUTLET 15 AMP[1] 2P 3W 125 VOLT, GROUNDING, WALL MTD.[2], VERTICAL, ℄ 12" AFF.
●ₐ	SPECIAL RECEPTACLE, LETTER DESIGNATES TYPE, SEE SCHED. DWG. NO._____ WALL MOUNTED.
⊙ᵦ	FLOOR OUTLET TYPE B[4] SEE DWG. NO._____
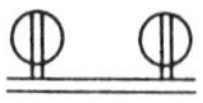 ⌸ ⌸	MULTI-OUTLET ASSEMBLY[4] SEE DWG. NO._____ FOR SCHEDULE AND DETAILS (SEE SPEC.)
	OR
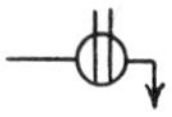 ⌸↓ₓ"	MULTI-OUTLET ASSEMBLY; ARROW INDICATES LENGTH; RECEPTACLE SYMBOL INDICATES TYPE; X" INDICATES OUTLET SPACING
S_a	SINGLE POLE SWITCH, 15 A[1] 125 V, 50" AFF[3] UON. SUBSCRIPT LETTER INDICATES OUTLETS CONTROLLED.
S_L	SWITCH, LOW VOLTAGE SWITCHING SYSTEM.
$_aS_3$	SWITCH, 3 WAY, 15A 125 V, SEE SPEC.[6]; CONTROLLING OUTLETS 'a'[5]
S_{DP}	SWITCH, DOUBLE POLE, 15 A 125 V.
S_4	SWITCH, 4 WAY, 15 A 125 V.
S_K	SWITCH, KEY OPERATED, 15A 125V.
S_D	DOOR SWITCH, SEE SPEC. FOR RATING AND TYPE.
 $S\!\!\!/$	SWITCH/RECEPTACLE COMBINATION IN 2 GANG BOX.
S_P	SWITCH, SP 15A 125V. WITH PILOT LIGHT.
S_{SA}	SWITCH, SPECIAL PURPOSE, TYPE A, SEE SPEC.; SEE DWG. NO._____

ABBREVIATIONS RELEVANT TO SWITCHES:
SP – SINGLE POLE
DP – DOUBLE POLE
SPDT – SINGLE POLE DOUBLE THROW
DPDT – DOUBLE POLE DOUBLE THROW
RC – REMOTE CONTROL

S_{WP}	SWITCH, WEATHER PROOF ENCLOSURE, SEE SPEC.
S_{MC}	SWITCH, MOMENTARY CONTACT.
[R]	OUTLET–BOX–MOUNTED RELAY
[D]	OUTLET–BOX–MOUNTED DIMMER.
[S/D]	OUTLET–BOX–MOUNTED SWITCH AND DIMMER.

NOTE 1. SPECIFY 20 AMP IF DESIRED.
2. ALL RECEPTACLES ARE WALL MOUNTED UON
3. ALL MOUNTING HEIGHTS ARE TO OUTLET ℄; SPECIFY MH. OF EACH OUTLET.
4. ALSO SHOWN IN SYMBOLS, PT. I, FOR COMPLETENESS, USING ALTERNATE SYMBOL.
5. REFER TO SPECIFICATIONS FOR DATA ON SWITCHES.

Figure 12.51 Architectural-electrical plan symbol list, Part III, Wiring Devices. For other portions of the symbol list see:
Part I, Wiring and Raceways, Figure 12.6, page 671
Part II, Outlets, Figure 12.50, page 719
Part IV, Abbreviations, Figure 12.58, page 726
Part V, Single Line Diagrams, Figure 13.4, page 733
Part VI, Equipment, Figure 13.15, page 753
Part VII, Signaling Devices, Figure 15.34, page 909
Part VIII, Motors and Motor Control, Figure 16.25, page 944
Part IX, Control and Wiring Diagrams, Figure 16.28, page 947

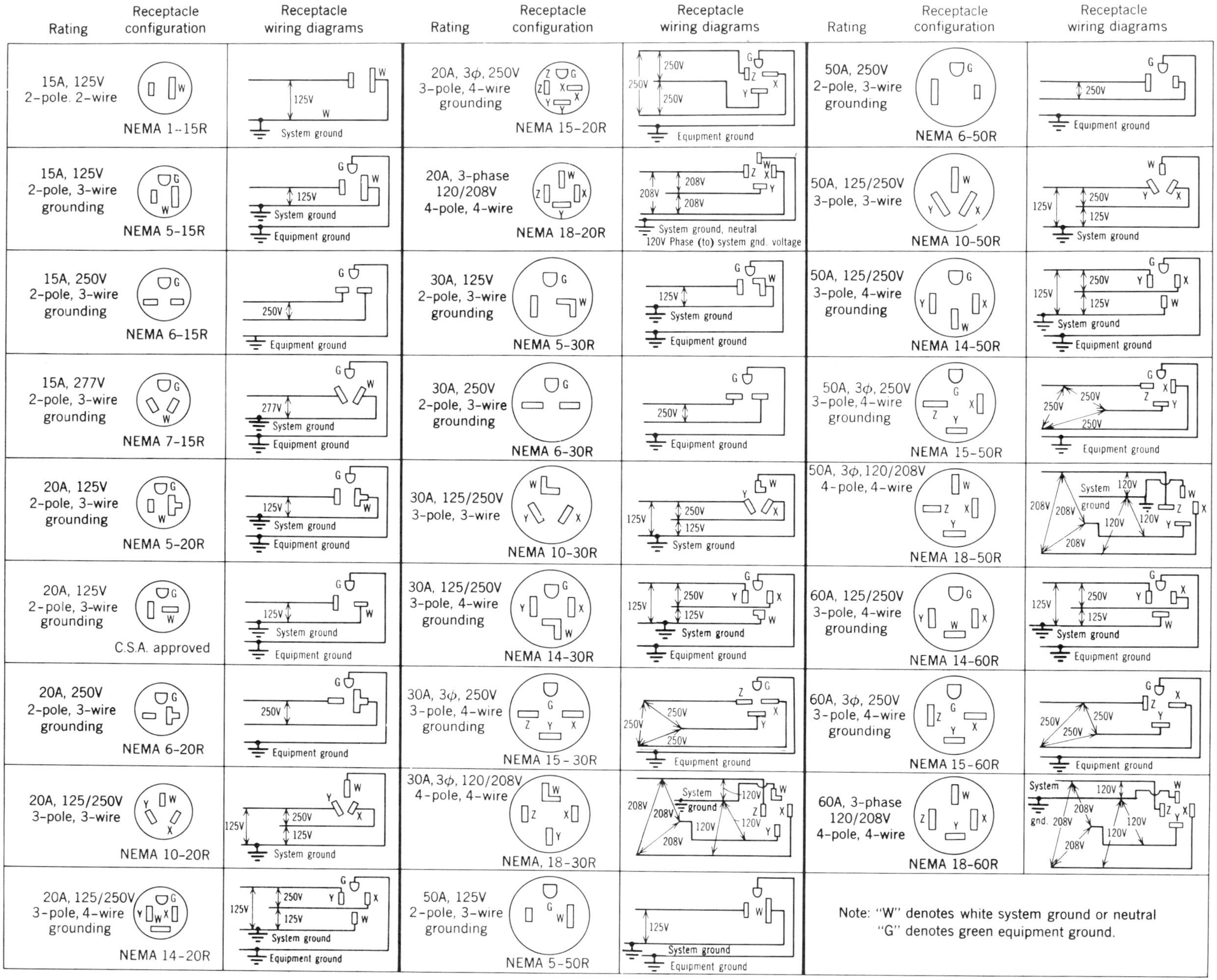

Figure 12.52 Receptacle configuration chart of general-purpose, nonlocking devices, with applicable NEMA configuration numbers and wiring diagrams.

722 / BRANCH CIRCUITS AND OUTLETS

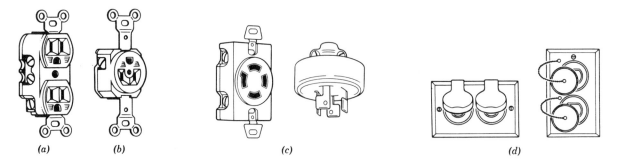

Figure 12.53 Typical receptacle and attachment cap types. *(a)* 2-pole, 3-wire, 15 amp, 125 v duplex grounding type. *(b)* The same as *(a)* except single. *(c)* 3-pole, 4-wire, 20 amp, 125/250 v locking receptacle, with matching cap. *(d)* Outdoor weather-proof receptacles.

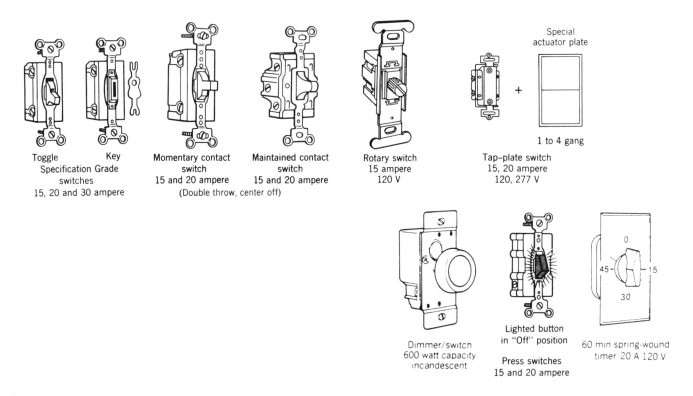

Figure 12.54 Typical branch circuit switches. (From B. Stein and J. S. Reynolds, *Mechanical and Electrical Equipment for Buildings*, 8th ed., 1992, John Wiley & Sons, New York. Reprinted by permission of John Wiley & Sons.)

cle outlet, a *receptacle outlet* or a *convenience outlet*. The term *wall plug*, which is heard so often, is really incorrect. A plug is another name for the attachment plug, or cap, on the wire that carries electricity to a device such as a lamp or appliance. The device at the end of the line cord is plugged into the wall; hence, we have the name plug. However, since the term *wall plug* for convenience outlet is used so often, it has been accepted in trade circles. The technologist should follow proper terminology. See Figure 12.51.

Receptacles belong in the general trade classification of wiring devices, which includes not only all receptacles but also their matching attachment caps (plugs), ordinary wall switches, outlet box mounted controls. Most people restrict the term *wiring device* to any device that will fit into a 4-in. square box, thus excluding devices larger than 30 amp. Switches, too, are limited to 30 amp, single-pole, 220 v in this category; above that size they are separately mounted and are called *disconnect switches*. Essentially, wiring devices are those found in common use in lighting and receptacle circuits, which circuits normally do not exceed 30 amp in rating.

Since by **NEC** definition a receptacle is a contact device for the connection of a single attachment plug and since the normal wall convenience receptacle will take two attachment plugs, it is properly called a duplex convenience receptacle or duplex convenience outlet. Most people shorten this to duplex receptacle or duplex outlet. However, at a single outlet, more than one receptacle can be installed. Just as a row of fluorescent fixtures wired together is considered one outlet (see Note 1, Figure 12.50), so any number of receptacles mounted together in one or more coupled boxes is considered one outlet. Thus, three receptacles so mounted would be a three-gang receptacle outlet. This is extremely important when one is circuiting under conditions limiting either the number of receptacles on a circuit, the number of outlets, or both. Furthermore, as previously stated, the lower the number of outlets, the lower the cost of installation. A circuit with six duplex receptacles individually mounted is normally more than twice the cost of the same six receptacles installed in two outlet groups of three gangs each.

Receptacles are described and identified by poles and wires. The number of poles equals the number of active circuit contacts, thus excluding the grounding pole. The number of wires includes all connections to the receptacle, including the ground wire. This system of description is awkward and

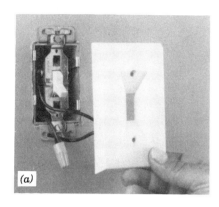

Figure 12.55 *(a)* Time-out switch replaces ordinary wall switch for control of incandescent and fluorescent lighting loads. Its solid-state timing mechanism, which is mounted on the back of the wall faceplate is adjustable to turn off after any time interval of between 10 min and 12 hours. Typical applications are closets, stock rooms, and other short-time occupancy spaces. (Courtesy of Paragon Electric Company, Inc.)

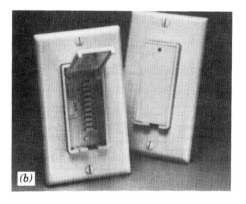

Figure 12.55 *(b)* Programmable lighting switch mounts in a standard wall device box. The unit can operate as a conventional on-off switch, and in automatic programmed mode, with ten 1-hr periods controlled by individual slide switches. (Courtesy of Leviton Manufacturing Company, Inc.)

often confusing, since a normal three-slot (two circuit plus U ground) convenience receptacle is officially a 2-pole, 3-wire device even though it has three slots and is sometimes connected with only two wires, with the ground connection being a bond to the metallic conduit system. For this rea-

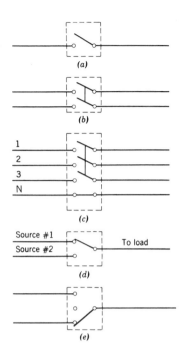

(a) Single-pole single-throw switch.

(b) Two-pole single-throw switch.

(c) Three-pole and solid-neutral (3P and SN) switch.

(d) Single-pole double-throw switch (also called, in small sizes, a 3-way switch).

(e) Single-pole double-throw switch with center "off" position (in control work called a hand-off-automatic switch).

Figure 12.56 Typical switch configurations. Note that switches are always drawn in the open position. (From B. Stein and J. S. Reynolds, *Mechanical and Electrical Equipment for Buildings*, 8th ed., 1992, John Wiley & Sons, New York. Reprinted by permission of John Wiley & Sons.)

son, we strongly suggest that the construction drawing's receptacle symbol list use National Electrical Manufacturer's Association (NEMA) designations and a graphic representation. This is the system we follow to avoid the almost certain confusion of attempting a description by poles and wires.

Receptacles are available in ratings of 10–400 amp, 2 to 4 poles and 125–600 v. The typical wall-mounted duplex convenience receptacle is 2 pole, 3 wire, 125 v, 15 or 20 amp. The quality or grade of the unit is specified in the job specifications and can be economy (cheap), standard (good), specification (excellent), or hospital grade. This latter type is constructed to take heavy abuse without failure and is identified by a green dot on the device face. In preparing electrical construction drawings, many designers use a different symbol for each special type of receptacle other than the standard duplex convenience outlet. This is acceptable practice for a job with up to three or four types, as for example, a small residence. Most nonresidential jobs, however, have five or more receptacle types, and the symbols become confusing. For this reason, we strongly recommend using one standard symbol with an identifying letter for all receptacles other than the common duplex convenience outlet and then including a schedule of these special items on the drawings. This will avoid confusion, save drawing time, allow standardization from job to job and show a systematic professional approach. See the symbol list, Part III, Figure 12.51.

Figure 12.52 gives the physical configurations and NEMA designations of the receptacles most commonly used in electrical design. These designations, that is, the NEMA number, the configuration chart and the wiring diagram should appear in the table of special receptacles on the drawings. The receptacle wiring diagram that appears in Figure 12.52 is for the designer's information and need not appear on the drawings. Other NEMA standard receptacle configurations such as locking types are used only in specialized design. Their configurations and wiring diagrams can be found in catalogs of the major wiring device manufacturers.

Figure 12.53 shows pictorially some of the most common types of receptacles. Mounting height must always be specified with all wall-mounted

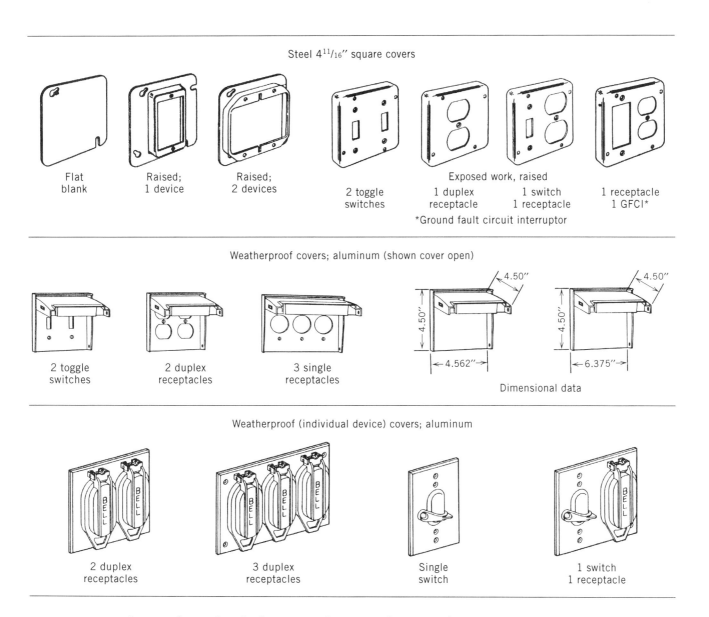

Figure 12.57 A selection of special outlet box covers. (Courtesy of Raco, Inc.)

devices. Wall convenience receptacles are normally mounted vertically, between 12 and 18 in. above finished floor (AFF) to the device centerline. Some architects prefer horizontal mounting. If this is desired, it must be so specified, preferably on the drawings, along with the mounting height. In industrial areas, shops, workrooms and kitchens, where the receptacle outlet must be accessible above counters and tables, mounting height is 42–48 in., and horizontal mounting is preferable so that cords do not hang on top of each other. This, too, must clearly be specified on the drawings.

Three very important special receptacle types have come into increasing use recently. They are

- Receptacles with built-in ground fault protection.
- Receptacles with built-in surge suppressors.
- Receptacles with isolated grounds.

Receptacles with built-in ground fault protection are used in locations where sensitivity to electrical shock is high. These locations include areas near sinks, swimming pools and other wet locations.

Symbols—Part IV—Abbreviations

A,a	Amperes	MH	Mounting height, Manhole
AFF	Above finished floor	N	Neutral
C	Conduit	NC	Normally closed
C/B	Circuit breaker	NO	Normally open
CCT	Circuit	NL	Night light
₵	Centerline	NIC	Not in contract
Dn	Down	OH	Overhead
EWC	Electric water cooler	OL	Overload relay
EM	Emergency	OC	On center
El	Elevation	PB	Push-button, Pull-box
EC	Empty conduit	PC	Pull chain
F	Fuse	S, SW	Switch
FA	Fire alarm	TC	Telephone cabinet
F-3	Fan No. 3	T	Thermostat, Transformer
GND	Ground	TEL	Telephone
GFCI	Ground fault cct-interrupter	TV	Television
GFI	Ground fault interrupter	TYP	Typical
HOA	Hand-off-automatic selector switch	UON	Unless otherwise noted
HP	Horsepower	UF	Unfused
IG	Isolated ground	UG	Underground
L	Line	WP	Weatherproof
LTG	Lighting	XP	Explosion proof
MCC	Motor control center	XFMR	Transformer

Figure 12.58 Architectural-electrical plan symbol list, Part IV, List of Common Abbreviations. For other portions of the symbol list see:
Part I, Wiring and Raceways, Figure 12.6, page 671
Part II, Outlets, Figure 12.50, page 719
Part III, Wiring Devices, Figure 12.51, page 720
Part V, Single Line Diagrams, Figure 13.4, page 733
Part VI, Equipment, Figure 13.15, page 753
Part VII, Signaling Devices, Figure 15.34, page 909
Part VIII, Motors and Motor Control, Figure 16.25, page 944
Part IX, Control and Wiring Diagrams, Figure 16.28, page 947

These devices are normally known as Ground Fault Circuit Interrupter (GFI or GFCI) receptacles. Their construction and application are discussed in Section 13.3.

The great sensitivity of modern electronic equipment to voltage surges and to electrical *noise* (random, spurious electrical voltages) has led to the development of the remaining two special receptacles. Those with built-in surge suppression protect the connected equipment from overvoltage spikes.

Receptacles with isolated grounds separate the device ground terminal from the system (raceway) ground because it had been determined that much of the unwanted electrical noise can be eliminated by such disconnection. Isolated grounds are discussed in Section 13.3. Receptacles intended for use with isolated grounds are identified by their orange color.

The second major type of wiring device is the *wall switch*. See Figure 12.51 for drawing symbols, and Figure 12.54 for a pictorial drawing of a few common types. Figure 12.55 shows two special types that use miniature electronic components to control their switching operations. Switches serve to open and close electrical circuits. Those classified as wiring devices are rated from 15 to 30 amp, single and two pole, single or double throw. Common switch configurations of all types of switches, including wiring device switches, are shown in Figure 12.56. Wiring device switches generally fit into a single gang or at most a two-gang device box. Other types of switches, including service switches, general-use switches, contactors, low voltage switches, remote control switches and time-controlled switches of both the mechanical and solid-state electronic types, will be discussed in following chapters, where their use is explained.

The final feature of outlets that needs description is the device cover. On a blank outlet or a junction box, we use a plain metal cover. With switches and receptacles, the cover is made to match the wiring device, to close the outlet box, and to suit the installation location. In this connection your attention is directed to the Code requirements for receptacle enclosures in damp and wet locations; NEC Article 410-57. Covers are available in one or more gangs and in various materials. As with all other electrical equipment these should be listed on drawings and adequately described in the specifications as to type, quality, material and so on. Although preparation of specifications is normally not the responsibility of technologists, they should refer to the specs during design to determine that no conflict between them and the drawings exists. See Figure 12.57 for an illustration of a common device covers. Floor outlets are mounted in special floor boxes and must be shown as a specific receptacle in a specific type of floor box. Both must be specified. Figure 12.58 is a continuation of the symbol list, that is, Part IV, and lists abbreviations commonly used in architectural-electrical work. Electrical technologists should use only widely accepted abbreviations to avoid possible misunderstanding.

Key Terms

Having completed the study of this chapter, you should be familiar with the following key terms. If any appear unfamiliar or not entirely clear, you should review the section in which these terms appear. All key terms are listed in the index to assist you in locating the relevant text. Some of the terms also appear in the glossary at the end of the book.

American Wire Gauge, AWG
Ampacity
BX
Bolted pressure connector
Bushing, nipple
Ceiling duct
Cellular floor
Convenience receptacle outlet
Current carrying capacity
Electric pole
EMT
Flat cable assembly
Greenfield
IMC
Interface
Junction box
Kcmil, thousand circular mils
Lighting track

MCM, thousand circular mils
Modular wiring
National Electrical Manufacturer's Association (NEMA)
One-line diagram
Outlet box
Poke-through fitting
RHW, THW, TW
Rigid conduit, IMC
Romex
Solderless connectors
Surge surpression
THWN, XHHN, UF
Type AC cable
Type NM cable
Underfloor duct
Undercarpet cable
Wiring device

Supplementary Reading and Bibliography

Earley, M. W., R. H. Murray, R. H., and Caloggars, J. M. *National Electrical Code Handbook*, National Fire Protection Association (NFPA, MA, 1995). Contains the complete text of the National Electrical Code, which is the source and authority on rules of safe practice in electrical design and installation. In addition, the handbook contains extensive explanatory material and diagrams that are very helpful in understanding the NEC.

Stein, B., and Reynolds, J. S. *Mechanical and Electrical Equipment for Buildings*, 8th ed., Wiley, New York. 1992. This book covers the same areas of study as the present book, but in greater detail and scope. Very useful for further study.

American Institute of Architects, Ramsey, G. G., and Sleeper, H. R. *Architectural Graphic Standards*. 9th ed., Wiley, New York. This architect's "bible" provides, in its electrical section, handy physical dimensional data and detailed drawings of many electrical items.

Traister, J. E., and Rosenberg, P. *Construction Electrical Contracting*, Wiley, New York. 2nd ed 1989. An excellent book on electrical construction from the electrical contractor's point of view. Practical data and information on the economics of electrical contracting.

Starr, W. *Electrical Wiring and Design: A Practical Approach*, Wiley, New York. 1983. Covers approximately the same material as the present book, except with less emphasis on contract drawing preparation and more emphasis on calculations. Many illustrative examples of electrical calculations.

Problems

1. For the sleeping area of the house in which you live, lay out the rooms showing all outlets and then lay out the interconnecting raceways as if the wiring method were:
   a. BX.
   b. Romex.
   c. Plastic conduit.
2. Do a material takeoff of the raceway systems of Problem 1, counting outlets, conduit, feet of cable and the like (refer to electrical supply catalogs for fittings).
3. Draw the living room and the kitchen areas of your home showing the outlets, but use surface raceways throughout. Refer to a manufacturer's catalog (we suggest Wiremold Co.). Show any installation details that you think the contractor will need.
4. Repeat Problem 3 for the classroom or lecture room you are now using.
5. Assume that each student in your class requires a source of power and communication at his or her seat. (This situation is rapidly becoming a real-life problem.) How would you wire up the room? Draw a layout for:
   a. New construction.
   b. Retrofit.
6. List four actual places where a surface raceway would be preferable to a concealed raceway and four more where the reverse is true.
7. Using Figure 12.48, make up a table of special receptacles for your school, as you would place it on the drawings.
8. Using Table 12.1, show how the area in circular mils is derived from the diameter of these wire sizes: No. 10, No. 1/0, 500 MCM.
9. a. Using Table 12.2, list the ampacity of each of the following wires: No. 12 TW, No. 12 THW, No. 12 THHN, No. 10 RHW, No. 2 THWN, No. 4/0 XHHW, 250 kcmil UF, 250 MCM THHN, 500 MCM TW, 500 kcmil TW, 500 MCM XHHW. (Remember that Table 12.2 is for up to three wires, in a raceway.) Check these values in the corresponding table in the NEC (1996 edition, Table 310-16).
   b. Adjacent to each ampacity in Part a, place the ampacity for the same wire in free air. Use the appropriate NEC table (1996 edition, Table 310-17).
10. The NEC provides (Chapter 9, Table 1) that four or more conductors in a conduit may occupy 40% of the conduit's cross-sectional area. Using Tables 12.4 and NEC Table 4, Chapter 9, find the conduit size required for the following two groups of single conductors:
    a. 6-No. 2 AWG Type THW plus
       4-No. 1/0 AWG Type TW
    b. 3-No. 4/0 AWG Type THWN
       3-250 MCM Type XHHW
       1-No. 4 equipment ground wire with 600 v TW insulation.
11. Using Table 12.7, lay out, with all dimensions, the end of a pull box that is receiving two layers of conduit as follows:
    Top layer 4–3 in., 2–2 in., 2–1½ in.
    Bottom layer 10–1½ in. conduits
    Use ⅜-in. minimum locknut clearance instead of the ¼ in. shown in the table for conduits in the same layer, and 1 in. minimum locknut clearance between layers.

13. Building Electrical Circuits

In Chapter 12, we discussed two of the components of the typical building electrical circuit—the wiring and the outlets, including the wiring devices connected at the outlets. In this chapter, the third basic circuit element—the circuit protective device—is considered in detail. After discussing this third, and last, basic circuit element, our study will expand in later chapters to cover related areas. These areas are branch circuit criteria, types of branch circuits and basic motor circuits as used in building design. Working drawing presentation of these concepts and techniques is developed. Finally, we study the next stage of electrical design, which is the assembly of branch circuits into a complete wiring system. This stage includes the layout of actual buildings complete with circuitry, panel scheduling, switching and load study. Applications in this chapter will be to The Basic House plan, which you studied previously in the HVAC sections of this book. After studying this chapter, you will be able to:

1. Understand the functioning of overcurrent devices.
2. Determine where in a circuit to place the overcurrent devices and what type to use.
3. Recommend the type of overcurrent devices required for the circuit being considered.
4. Specify fuses and circuit breakers according to size and type.
5. Understand the use of grounding electrodes, grounding conductors and ground fault circuit interrupter (GFCI) devices.
6. Prepare single-phase and three-phase panelboard schedules.
7. Lay out and draw a set of residential electrical working drawings, starting from the architect's plan.
8. Circuit a complete electrical plan (working drawings), including all wiring, devices and loads, and prepare a complete panel layout.

9. Thoroughly understand the principles and guidelines for residential layout and wiring.
10. Apply these layout and circuitry skills to almost any type of building.

Overcurrent Protection

13.1 Circuit Protection

Referring to Figure 12.2, page 668, we see that if we consider the circuit protective device from the point of view of the branch circuit, it represents the source of voltage. In reality, the overcurrent device follows the source of voltage in the circuit. However, since the overcurrent (o/c) device is always connected at its line end to the voltage source and at its load end to the circuit wiring, it becomes the apparent source of voltage. When we study panelboards, of which overcurrent devices are a part, we will learn that the panel's busbars become the source of voltage as we look upstream from the overcurrent devices. The apparent voltage source depends on where in the electrical network we happen to be looking at the moment.

Glance back now at Figure 11.16, page 640. Note that, at each stage of the system, the preceding stage (upstream) is the source of voltage or power, whereas, in reality, the only actual source of power is the generator at the beginning of the line. This same relationship in branch circuit form is shown in Figure 13.1. As you will remember, the **NEC** defines the branch circuit as that portion of the circuit between the overcurrent device and the outlet(s). Therefore, in examining overcurrent devices, we are leaving the branch circuit and moving upstream one step in the electrical network.

What, then, is an overcurrent device? What is its function, appearance and method of operation? How is it shown and how much space does it occupy? These are the questions that we will now answer.

You may have already noticed that there are two different causes of overcurrent in a circuit. One is an overload in the equipment. The second is a fault of some sort, frequently a short circuit or an accidental (unintentional) ground. Both of these conditions result in excessive current flowing in the circuit. The function of an overcurrent device is to protect both the branch circuit and the load device against this excess current.

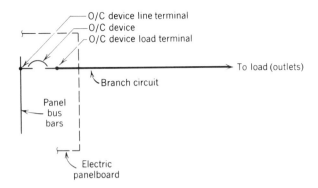

Figure 13.1 Typical electrical circuit. Note that the branch circuit, which extends from the overcurrent device load terminal to the outlets, "sees" the overcurrent device as its power source.

a. Branch Circuit Protection

In our studies in basic electricity and materials in Chapters 11 and 12, we learned that the current in a circuit produces heating in the circuit conductors, usually known as line I^2R loss. We also learned that circuit wiring can safely handle only a specific limited amount of current and that this amount of current, expressed in amperes, is called the ampacity of the conductor. If this ampacity is exceeded, the wire overheats due to the I^2R loss, and a fire may result. The overcurrent device prevents this from happening. It does not matter whether the excess current is being caused by an equipment problem, such as an overload, or by a circuit problem, such as a short circuit or an unintentional ground. The overcurrent device "sees" only excess current—and interrupts it. We will have more to say on this topic later on when we discuss fuses and circuit breakers in detail.

b. Equipment Protection

Consider, for instance, a machine tool that is drawing excess current, due to overloading or some other cause. This excess current, in turn, causes excess internal heating. If this is not relieved, permanent damage to the tool can easily result. Such damage is prevented by the action of the overcurrent device.

As its name suggests, the overcurrent device prevents an overcurrent from continuing by simply opening the circuit when it senses excessive current. It, therefore, acts in the same manner as a

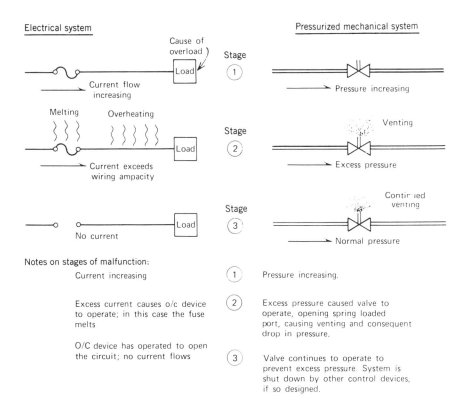

Figure 13.2 Three stages in the action of electrical and mechanical safety devices.

pressure relief safety valve in a mechanical system. There, the valve opens to relieve excess pressure; here, the overcurrent device opens to relieve excess current. In both instances, the action prevents an overload condition from becoming dangerous. But there the similarity ends. The overcurrent device opens the circuits much as a switch would, stopping current flow entirely and deactivating the circuit. With a pressure relief valve, the excess pressure is relieved by venting or bypassing, but the system is normally not shut down by the valve action. (Certain mechanical safety devices, such as low water cutoff in boilers, do cause system shutdown by activating an electrical cutout. They then require manual resetting.) The difference in the two types of action is illustrated in Figure 13.2, where a fused circuit is shown in the three stages of operation: prior to overload, during the protective action, and after clearing the overcurrent. The action of an overcurrent device is called clearing, since it clears the circuit of the overload or the fault.

Notice that the overcurrent device is always upstream of the equipment being protected, that is, it is electrically ahead of the load. This is obviously where it belongs, since current flows downstream. Therefore, to cut off excess current, the overcurrent device must be placed ahead, in an electrical sense, of the protected device. In the case of branch circuits, the overcurrent device is in the electric panel that supplies the branch circuits. The panel (also called panelboard) is the source of current. There, current must be cut off to protect the branch circuits downstream from the panel.

The upstream side of any device is called the *line side*. The downstream side is called the *load side*. In the case of a switch, circuit breaker or any other circuit-interrupting device, the line side remains hot after it is opened, but the load side is dead, or de-energized. Figure 13.3 illustrates the location of overcurrent devices. Observe from this diagram that the line side of a switch remains hot (live, energized), even with the switch open. When we have two disconnect devices in series, the load side of the upstream one is the same as the line side of the downstream one. If this seems confusing, refer again to Figure 13.3 for clarification. The busbars in the main switchboard are hot from connection

732 / BUILDING ELECTRICAL CIRCUITS

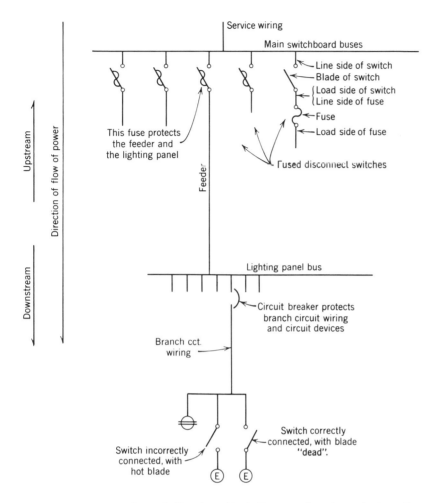

Figure 13.3 Typical single-line (one-line) diagram showing relation of components to each other, proper location of overcurrent devices and terminology in general use.

to the electric service wiring. The line terminals of the fused switches are, therefore, hot. Note that a switch is always connected that way—and never with the blade hot—for safety reasons. Moving downstream, we come to the lighting panel. Here the feeder supplies the panel busbars. The circuit breakers are connected with their line side to the panel and their load side facing downstream, and so on.

It is an accepted custom to show a switch in its open position. (A closed switch would simply look like a straight line.) Circuit breakers and fuses alone are shown closed, since no drawing convention exists to show an open circuit breaker (c/b) or

fuse. The open fuse in stage 3 of Figure 13.2 is drawn that way to illustrate a point. In actual practice, a blown fuse is never shown as a break in the circuit or in any other way. This is because circuits are always shown in normal condition, and a blown fuse or open circuit breaker represents an abnormal condition. When a fuse is shown together with a switch, it is usually shown as in Figure 13.3. Figure 13.4 shows these symbols along with the symbols most often used in single-line (one-line) diagrams. These symbols are very widely used and are, therefore, universally understood. The electrical technologist would be poorly advised to change them without a very good reason.

CIRCUIT PROTECTION / 733

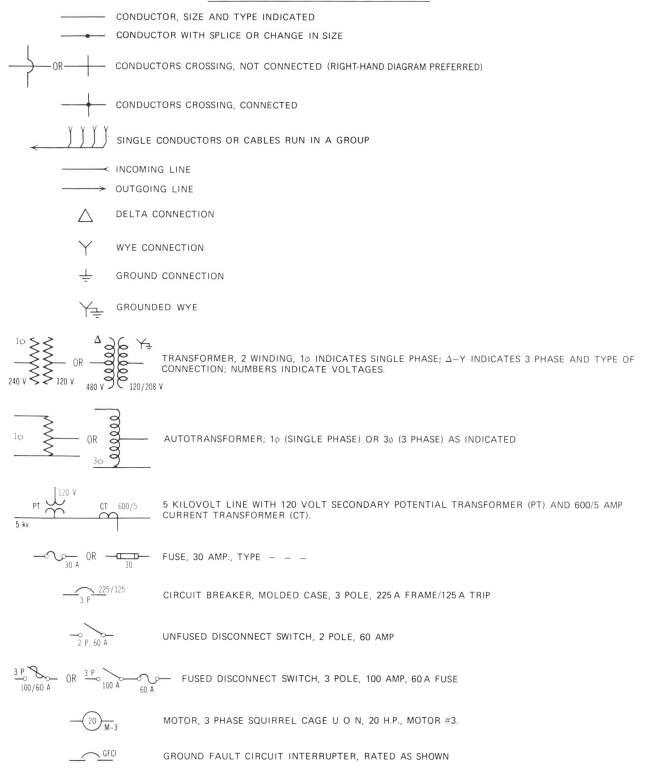

Figure 13.4 Electrical drawing symbol list, Part V, Single-line Diagrams.
  Part I, Raceways, Figure 12.6, page 671.
  Part II, Outlets, Figure 12.50, page 719.
  Part III, Wiring Devices, Figure 12.51, page 720.
  Part IV, Abbreviations, Figure 12.58, page 726.
  Part VI, Equipment, Figure 13.15, page 753.
  Part VII, Signalling Devices, Figure 15.34, page 909.
  Part VIII, Motors and Motor Control, Figure 16.25, page 944.
  Part IX, Control and Wiring Diagrams, Figure 16.28, page 947.

13.2 Fuses and Circuit Breakers

The simplest and most common circuit protective device in use is the fuse. (See **NEC** Article 240, Sections E and F.) Although in modern technology fuses are available in literally hundreds of designs, ratings and shapes, they are all basically the same. A fuse consists of two terminals with a piece of specially designed metal, called the fusible element, in between. Current flows into the line terminal, through the fusible element and out the load terminal. When the current flow is excessive, the fusible element melts due to excessive I^2R loss, and the fuse opens the circuit. It's as simple as that. The technical aspects of fuses—ratings, clearing time and interrupting capacity—are the concern of the design engineers. As far as the electrical technologist is concerned, the fuse is an overcurrent element in a panel, in a switch or even in an individual box or cabinet, to which the circuit to be protected is connected. Several types of fuses are shown in Figure 13.5. Fuse representation in electric plans and diagrams (Section 13.1) is shown in Figures 13.3 and 13.4.

Since the fuse element melts when the fuse operates, it is obviously a one-time device and must be replaced after it clears a fault. This melting, of course, refers only to the fusible element in the fuse, and not to the whole fuse. In one-time or nonrenewable fuses, such as are shown in Figure 13.5, the fusible element in not replaceable. The entire fuse must be discarded after operation, since it has become useless. With renewable fuses, the melted fusible element can be removed and replaced with a new one. This type of fuse is not in common use because it is too easy to replace the element with another of a different rating, thus defeating the purpose and usefulness of the fuse. For this reason, almost all fuses in use are nonrenewable. In part, because of this one-time opera-

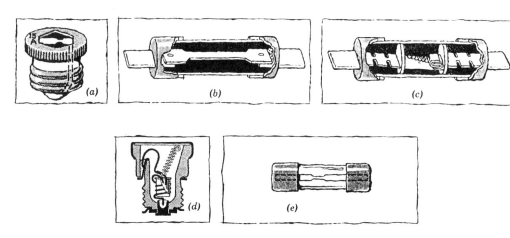

Figure 13.5 Standard types of fuses are *(a)* nonrenewable plug fuse, *(b)* nonrenewable knife-blade fuse, *(c)* nonrenewable dual-element time-delay ferrule cartridge fuse, *(d)* dual-element time-delay **NEC** Type S plug fuse, *(e)* nonrenewable miniature fuse for electronic and instrument applications. Type S fuses are used in new installations. They require a base adapter to fit a standard screw-in (Edison) socket. The adapter is current rated and nonremovable. This prevents deliberate or accidental use of a Type S fuse of incorrect rating.

Since fuses are inherently very fast-acting devices, time delay must be built into a fuse to prevent "blowing" on short-time overloads such as those caused by motor starting. A dual-element fuse such as type *(c)* or *(d)* allows the heat generated by temporary overloads to be dissipated in the large center metal element, preventing fuse blowing. If the overload reaches dangerous proportions, the metal will melt, releasing the spring and opening the circuit. The notched metal portions of the fuse element, at both ends of the dual center element, provide short-circuit protection. The time required to clear (blow) a fuse is generally inversely proportional to the amount of current.

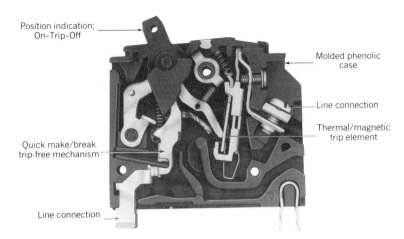

Figure 13.6 Cutaway of a common, thermal/magnetic single-pole, 100-amp frame plug-in-type circuit breaker. This type of circuit breaker is often used in lighting and appliance panelboards in residential and commercial installations. The unit plugs into the panel and contacts the panel's busbars with two metal fingers that are labeled "line connection" in the cutaway photo. See text for a detailed description of the unit's construction and operation. (Courtesy of Square D Company.)

tion characteristic and the replacement and inventory problems it causes, the circuit breaker was developed.

A *circuit breaker* is a device that combines the functions of an overcurrent circuit protective device with that of a switch. Code requirements for circuit breakers are covered in **NEC** Article 240, Section G. Basically, it is a switch equipped with a thermal/magnetic tripping mechanism that is activated by excessive current. Circuit breakers are manufactured in a wide variety of designs including some very complex types with solid-state microprocessor-controlled tripping units. For our purposes, however, we will restrict our discussion to a single type—the molded-case circuit breaker.

Refer to Figure 13.6, which shows a cutaway of a single-pole molded-case plug-in-type circuit breaker. Follow the discussion of circuit breaker construction and operation while referring to the illustration. The unit is called a *molded-case circuit breaker* because the entire operating mechanism is installed inside a molded plastic case. The load connection to the circuit breaker, that is, the connection to the branch circuit wiring that feeds the circuit load, is made with a bolt as shown. The line connection, which is the electrical feed connection to the breaker, is made in this design by pressure.

A pair of spring-type push-on connectors, labelled "line connection" in the figure, grasp one of the panelboard busbars (see Figure 13.12) to make the electrical contact. The breaker is held in place in the panel by the line connection and the mounting spring (see Figure 13.6), which clamps on to a mounting rail in the panelboard. (The terms *breaker* and *circuit breaker* are used interchangeably, as are the terms *panel* and *panelboard*.)

The trip element in this design of breaker is thermal/magnetic. This means that the trip mechanism has two elements. The thermal element, which is similar to the element in a thermostat, reacts to heat; the more current flowing in the circuit, the more heat is generated in the element. When the current reaches overload condition, the bimetallic thermal element will bend sufficiently to trip the mechanism and open the circuit. Because the element is heat activated, most breakers are temperature compensated so that, even in a very hot room, the breaker will not trip out except on true overload. The magnetic element responds to a short-circuit condition, which is like a massive overload, by clearing the fault (opening the breaker and disconnecting the circuit) very rapidly. The calculation of actual time to clear, both on overload and on short circuit, and the selection of the specific circuit breaker with the necessary characteristics is called circuit breaker coordination. This is a highly complex technique that is performed by experienced electrical engineers, either manually or by using computer coordination programs.

Another important characteristic of circuit breakers that is related to their tripping is a unit's *interrupting capacity*. Simply stated, it is the capacity of the unit to break, or interrupt, a short-circuit current. Since short-circuit currents can run into many many thousands of amperes, the breaker must be capable physically of interrupting this huge current. If it fails, because its interrupting capacity is insufficient, it will usually explode, causing great physical damage, and will probably start a fire. For this reason, selection of a breaker with sufficient interrupting capacity is vital. This selection is also done by the project electrical engineer after he or she has done a short-circuit (fault) study and a device coordination study. As we have stated, these two characteristics (clearing time and interrupting capacity), and the studies necessary to set them, are not the technologist's responsibility. However, a competent electrical technologist knows what these terms represent and their importance.

Referring again to Figure 13.6, we note that the operating mechanism is quick make/break and trip free. The *quick make* and *quick break* design means that the mechanism snaps open and snaps closed, so that both operations are full and positive even if the handle is thrown slowly or with hesitation. Thus, unlike a knife switch, slow closing or opening, which causes arcing and pitting of contacts, is impossible. The trip-free mechanism means that if the breaker is closed on to a fault—a short-circuited line, for instance—it will simply trip free and will not close in on the circuit and then trip out. Also, there will be no backlash of the breaker handle, which could injure an operator. The handle of the breaker must indicate the three possible conditions of the breaker by three separate handle positions: ON, TRIP, OFF. The TRIP and OFF positions must be different so that visual inspection will immediately show if a breaker has tripped, as opposed to one that has been manually opened. The circuit breaker shown in Figure 13.6 can be operated manually only. Other designs can be electrically operated (opened and closed) and electrically tripped by a device called a shunt trip.

The most important fact to remember is that circuit breakers, like fuses, react to excess current. The circuit breaker, however, unlike the fuse, is not self-destructive. After opening to clear (disconnect) the fault, it can be simply reclosed and reset. This means that replacement problems are greatly reduced. Furthermore, a circuit breaker can be manually opened so that, if desired, it can also act as a circuit switch. (*Note:* Circuit breakers that have been tripped by a major fault should be checked before reuse. They may require replacement.)

Consider, for instance, a residential circuit that feeds a kitchen electric range, wall oven or other permanently connected appliance. The NEC requires a disconnecting device for each such unit (Article 422 D). This means that, if the house panel is mounted in or near the kitchen and within sight of the appliance, the circuit breaker in the panel may be used as the disconnect and the circuit protective device, saving the cost of a switch or heavy receptacle. With fuse protection of the circuit, this is obviously not possible, and a separate disconnect is required. Such a disconnect can take the form of a switch mounted adjacent to the fuse in what is known as a switch and fuse panel. See Figure 13.13(b). Panels of this kind were once very common; today they are rare. The vast majority of all panels installed today are of the circuit breaker type.

Circuit breakers can also be used as switches for lighting circuits, where the entire circuit is to be switched as a unit. This technique is called panel switching and is frequently used in large, single-purpose spaces such as gyms, auditoriums, large offices and the like. Where used to switch 120- or 277-v fluorescent lighting loads, the breakers must be marked SWD (switching duty). This indicates that the breaker has been especially designed for this switching service.

Several other advantages of the circuit breaker as compared with the fuse are of interest to the technologist. Fuses are single-pole devices; they are put into a single wire and can protect only a single electric line. Circuit breakers, on the other hand, can be multipole. This means that a single circuit breaker can be built with 1, 2 or 3 poles to simultaneously protect and switch one, two or three electric lines. An overcurrent in any line causes the circuit breaker to operate and disconnect all the lines controlled by that circuit breaker. This has an obvious advantage when protecting circuits with two or three hot legs, such as in 208- or 240-v single-phase 2-pole or 208-v three-phase lines. Another advantage of the circuit breaker over the fuse is that it is readily tripped from a remote location—a very useful control function that is impossible with fuses and difficult with a fused switch.

A further advantage of circuit breakers over fuses is that their position—closed, trip, open—shows at the handle. A blown fuse is not easily recognized, since the melted element is inside the fuse casing. As a result of these and other features, the circuit breaker is much more widely used than the fuse for most commercial and all residential use. (It is possible to arrange an auxiliary circuit that will indicate if a blown fuse and to equip some types of fused switches with devices that will open the switch if any of the fuses blow. However, these circuits and devices are add-on arrangements and not part of the fuse/switch itself. They also add considerably to a unit's cost. For these reasons, they are not widely used.)

The single great advantage of the fuse is its stability and reliability. Unlike the circuit breaker, the fuse can stay in position for years, and, when called on to act, it will, just as designed. Circuit breakers, on the other hand, have mechanisms with many moving parts and, therefore, require maintenance, periodic testing and operation to keep them in top shape. A second advantage of fuses is that, for very heavy interrupting duty, fuses are more suitable than circuit breakers. However, for residential and light commercial work circuit breakers will almost always be adequate.

One-line diagrammatic representation of circuit breakers is illustrated in Figure 13.4. Circuit breakers, like fuses, are rated in amperes. However, instead of having a separate design for each ampere rating, manufacturers use frame sizes within each of which a range of ampere trip sizes is available. Unfortunately, there is no real standardization for frame sizes among manufacturers. One major manufacturer makes frame sizes of 100, 225, 400, 600, 1000 and 2000 amp. Another manufacturer covers this range with frame sizes of 150, 250, 400, 600, 1200 and 2000 amp, while a third major manufacturer uses 100-, 250-, 400-, 800-, 1200- and 2000-amp frames. The trip sizes, however, are standard and are listed in **NEC** Article 240-6. There is considerable overlapping between frame sizes, for reasons of design convenience. For exact data, always consult a current manufacturer's catalog.

The **NEC** lists standard ratings of circuit breakers and fuses together, indicating that the same ratings should be commercially available in both. Actually, the **NEC** list represents what is commercially available in fuses. In circuit breakers, depending on the manufacturers, 45, 80 and 110 ampere trip sizes are considered nonstandard. The technologist is cautioned always to consult an up-to-date manufacturer's catalog before calling for a specific item on the working drawings. Where ratings vary from one manufacturer to another, the drawings must not only specify the item but also the manufacturer. A typical drawing note might show: Circuit breaker, 2-pole, 600-v, 100-amp frame, 60-amp trip, in a NEMA type 1 enclosure, ABC Electric, type A-37.2.

System Grounding

13.3 Grounding and Ground Fault Protection

a. Background Material

There are two types of faults that occur in electrical circuits—the line-to-line fault and the ground fault. See Figure 13.7. The **NEC** has given increasing attention over the past decade to ground faults and to the special device that is designed to clear such faults—the *ground fault circuit interrupter (GFCI)*. The reason for this is that line-to-line faults (and line-to-neutral faults) are relatively rare because of the high quality of insulation on cables and equipment. Furthermore, as we have already seen, the entire electrical system is mechanically protected by conduit, cabinets and so on, so that mechanical injury that might cause a line-type electrical fault (short circuit) is rare. Also, a line-to-line short circuit requires a double insulation failure (see Figure 13.7), which also seldom occurs. Finally, if a line (to line) fault does occur, it will cause the branch circuit breaker or fuse to open instantly, and this will clear the fault.

Ground faults on the other hand, are much more common than line faults. This is because the fault requires only a partial insulation breakdown; the fault is frequently between a conductor and the metal enclosure, and because only one item of insulation is involved (not two, as with line faults). Ground faults are particularly dangerous because they often do not cause the branch circuit protection to open. The ground fault circuit interrupter was developed to deal specifically with this type of common, and dangerous, fault. The detailed study of ground faults is a complex subject and not suited to our purposes here. However, an understanding of what a ground fault (circuit) interrupter (GFCI)

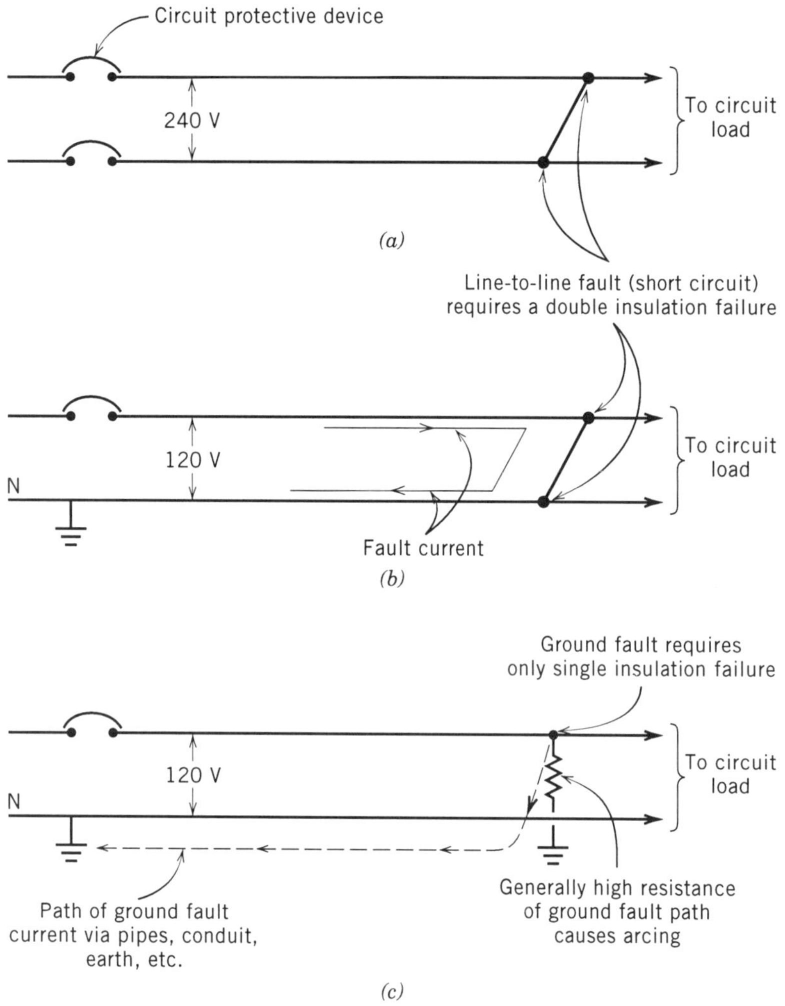

Figure 13.7 Line-to-line faults *(a)* and line-to-neutral faults *(b)* are relatively rare since they require a double insulation failure. When they occur, the fault current is usually high enough to trip the circuit breaker and clear the fault. A ground fault *(c)* results from a single insulation failure. It is usually a poor contact, high resistance path to ground, resulting in a ground current "leak." This leakage current is rarely sufficient to trip the circuit protective device. It is, however, frequently sufficient to cause arcing and extensive physical damage.

is, how it is applied and what it protects against is important. For this reason, we must backtrack a little and first discuss the subject of grounding.

b. System Grounding

Refer again to Figure 11.17, page 642. Notice that the neutral conductor is shown at ground voltage. This is accomplished by physically connecting the system neutral to ground. The purpose of this connection is to establish permanently a zero voltage point in the system. Ground is zero voltage by definition. Without a fixed ground, the system voltages may drift, causing all sorts of control and protection problems. Once a ground is firmly established, it becomes the reference for all voltages in the system. You will hear persons saying that point A is "600 volts above ground" and point B is "120 volts to ground." The physical connection between the system and ground is never broken.

The grounded line of a circuit is never fused, so that a solid, uninterrupted connection to ground is always maintained. The reason for this is essentially for safety, as we explain later. The physical device or object that connects to the ground is called a *grounding electrode*. Acceptable electrodes are listed in the **NEC**. A few of the common ones are listed, with the buried cold water main being the most frequently used:

- Buried cold water main.
- A single or multiple buried ground rods.
- A buried ground plate.

The cable that connects the system neutral to the grounding electrode is called (logically) the grounding electrode conductor. See Figure 13.8, which illustrates a typical electrical service grounding arrangement. The grounding requirements of the **NEC** are covered in **NEC** Article 250.

c. Grounding and Personnel Safety

The importance of electrical grounding to personnel safety is illustrated in the three diagrams of Figure 13.9. In the not too distant past, before grounding outlets were required in all new installations, the situation shown in Figure 13.9(a) existed. This situation still exists in millions of buildings, even though such installations contradict **NEC** requirements. Enforcement of **NEC** requirements is up to local building authorities having jurisdiction. In many areas, these authorities have not required that existing installations be modern-

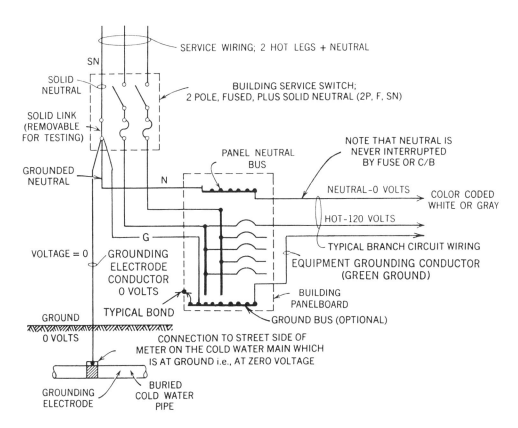

Figure 13.8 Typical service grounding arrangement. Note that the neutral is grounded only at the service point and is then carried unbroken throughout the system. If a wired equipment grounding conductor is used, it is advisable to have an equipment grounding bus in each panel for connection of these green grounds. The equipment ground system and the neutral are completely separate, being connected to each other and to ground only at the service point. For the sake of clarity, bonding corrections, which are required to ground all non-current-carrying metal parts of the electrical system (boxes, cabinets, enclosures, conduits, and the like) are not shown except for a single typical bond at the panelboard.

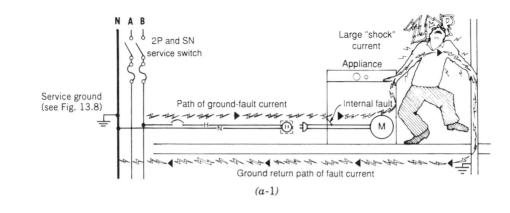

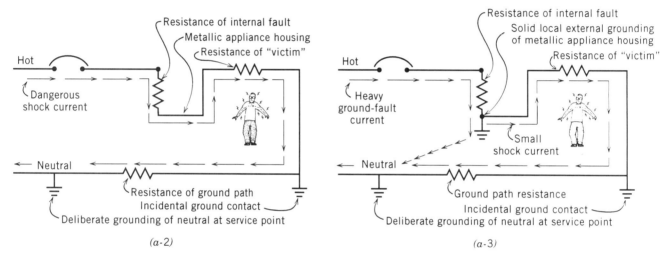

Figure 13.9 Figures *(a-1)* and *(a-2)* show a 2-wire grounded neutral circuit with no means of preventing dangerous shocks from an internal ground fault in a metal-cased appliance. The current passing through the person is limited only by the total resistance of the ground path. This includes the fault's internal resistance, the person's contact resistance and the resistance of the incidental path (earth, conduit, and the like) back to the service ground point. In *(a-3)*, the appliance case has been locally grounded to a cold water pipe. This reduces the resistance of the ground path substantially, and since the person is in parallel with this path, the shock current is lower than that in *(a-1)* or *(a-2)*. It can still, however, be dangerous.

ized to comply with the NEC, with respect to details of grounding systems. As a result, systems using 2-pole 2-wire receptacle outlets remain in service until failure. When repaired or replaced, all new work must comply with the current NEC requirements, and 2-pole 3-wire grounding outlets are installed, along with the necessary wiring changes. These old installations are dangerous because they provide no electric shock hazard protection as will become clear in our discussion and from study of Figure 13.9. Such circuits should use a branch circuit protective device (circuit breaker) equipped with a ground fault circuit interrupter. See Figure 13.11*(a)*.

Refer now to Figure 13.9*(a)*. In this situation, if any electrical contact occurs between the metal enclosure of the appliance (stove, washing machine, dryer, microwave, broiler and the like) and the hot leg of the wiring system, the enclosure also becomes electrically "hot." Such an electrical contact can easily occur if the insulation of the wire is frayed or if the insulation is punctured or cut by a sharp edge of the metal appliance enclosure. Then, anyone touching the appliance and a

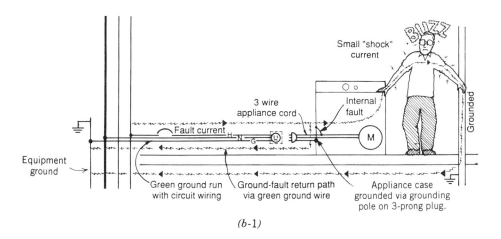

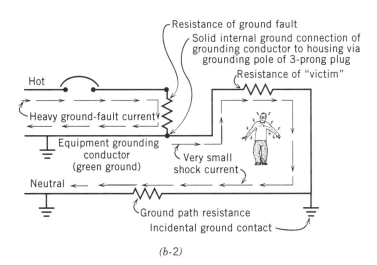

Figure 13.9 The situation represented by the drawing *(b-1)* and the circuit diagram *(b-2)* considerably reduces the shock hazard to the person contacting the metal case of an internally faulted appliance. Here a continuous metallic path back to ground is provided by the equipment grounding conductor. This reduces shock current and increases the ground fault current by reducing the resistance of the ground path. However, as long as this ground current is below the rating of the circuit protective device (usually 15 or 20 amp), it will continue to flow, causing arcing, overheating and possibly a fire. Shock hazard is minimal but remains and can be dangerous under certain conditions.

grounded surface, such as a pipe or a concrete floor (damp concrete is a good conductor), gets a heavy electric shock. [See Figure 13.9*(a-1,2)*.] If this unfortunate person happens to have wet hands, he or she may be electrocuted, as has actually happened many times. To lessen (but not eliminate) this danger, appliance manufacturers have always recommended that metal appliance cases be solidly grounded. With such grounding, the person touching the case is electrically in parallel with the solid case ground. Since body and hand contact resistance is normally much higher than that of a

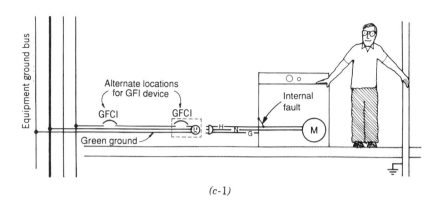

(c-1)

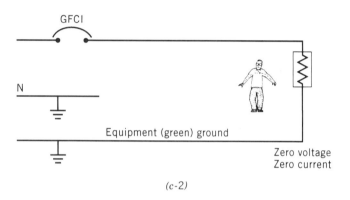

(c-2)

Figure 13.9 *(continued)* Figures *(c-1)* and *(c-2)* show complete elimination of the ground fault current and the shock hazard when the ground fault circuit interrupter opens and disconnects the appliance. The GFCI can be placed either at the receptacle (Figure 13.10) or at the panelboard (Figure 13.11). The GFCI devices are normally set to open when the ground fault reaches 6 milliamperes. Ground fault circuit interrupters are required by the NEC in bathrooms, kitchens, outdoors and other areas where ground faults are both common and dangerous.

solid ground, he or she will receive a lesser shock. See Figure 13.9*(a-3)*. The circuit protective device will generally not trip out even if the appliance case is solidly grounded because the internal ground fault is often one of high resistance. This resistance prevents the ground fault current from reaching a level high enough to trip the branch circuit protective device. This high resistance will also reduce the voltage of the electrical shock since it is in series with the person sustaining the shock. See Figure 13.9*(a-3)* and Section 11.8.

To accomplish the same personnel safety objective as the grounding of the appliance case, but without the necessity of this additional external wiring (that frequently was not installed), the 3-pole grounding-type receptacle and the 3-wire electrical cord were introduced. This arrangement provides a continuous metallic path to ground for any ground fault current. Refer to Figure 13.9*(b)* where this arrangement is shown. The two circuit wires are connected to the electrical device as is *(a)*, but the third wire from the appliance cord is internally and permanently connected to the appliance case. This eliminates the necessity for separately and externally grounding the case, as described previously (although it is still recommended by appliance manufacturers as an additional safety precaution). The grounding pin of the 3-pole cap at the end of the appliance cable connects to the grounding pole of the 3-pole grounding receptacle. This grounding pole of all 3-pole grounding receptacles is connected to the system ground by means of the equipment grounding conductor.

We have shown in Figure 13.9(b) the equipment grounding conductor to be a green ground wire. Actually, the NEC (Article 250-59) permits the use of a bare or insulated conductor run with the circuit wires, metallic conduit (with proper bonding) or a combination of these two methods. (When the grounding wire is insulated, the insulation may be either all green or green with yellow stripes. However, it is always referred to in the profession as either the equipment grounding conductor or the green ground.) Good practice dictates the use of both the green ground conductor and the metallic conduit (where used) together as the equipment grounding conductor. This ensures a continuous low resistance path for ground fault currents. Obviously when nonmetallic conduit is used or when wiring is type NM or type AC, we must rely completely on the internal grounding conductor.

Returning to Figure 13.9(b), note that this arrangement is superior to that of Figure 13.9(a-3) because the ground current has a low resistance metallic path back to the service ground and does not depend on the quality of the locally installed case ground. Therefore, if an accidental contact occurs between the appliance case and the 120-v wire, no severe danger to a person touching the case is present, since the ground current passes harmlessly along the green ground path and back to the panel. As described previously, the person is again in parallel with the ground path. Now, however, the ground path is one of very low resistance. As a result, the grounded person should receive, at most, a very slight shock. Indeed, with a low resistance ground path and a low resistance ground fault, sufficient ground current will flow to cause the branch circuit overcurrent device to trip and clear the fault. Unfortunately, the vast majority of ground faults are high resistance, because they result from such things as weak spots in insulation, imperfect metallic contact and conducting bridges of water, dirt and debris. As a result, not enough current flows to trip the branch circuit overcurrent device. Thus, although shock hazard is minimal, the fault continues to "leak," unnoticed by the system's protective devices. Such leaks continue until they either burn free or become a major ground fault. In both instances, the leak can cause extensive damage and may start a fire. Furthermore, even a slight shock hazard can be dangerous to life where the person has low contact resistance (for instance, a person standing in water) and/or has a weak heart. Thus, even a system with a properly designed and installed equipment ground is not entirely free of personnel hazard and is certainly far from free of equipment damage hazard.

To eliminate this possibly dangerous situation, which occurs anytime there is a leak of current to ground in an electric circuit, the ground fault circuit interrupter (GFCI) was developed. See Figure 13.9(c). This device compares the current flowing in the hot and neutral legs of a circuit. If there is a difference, it indicates a ground fault (dangerous condition), and if the ground fault current is at least 6 ma the device trips out. It should now be clear why we said previously that the GFCI finds a ready application in the old 2-wire circuits. These aging circuit components are prone to ground faults and can best be protected with a GFCI.

Figure 13.10 Duplex receptacle with built-in ground fault protection. This device is useful where a ground fault circuit interrupter is required at one particular outlet on a circuit. It localizes the protection so that a ground fault disconnects only that outlet, leaving the other outlets on the circuit energized. This unit is provided with an indicator light (lower right) to show whether the receptacle is hot. A test button is included to permit periodic testing, and a reset button reconnects the receptacle after it has tripped on ground fault, and the fault has been cleared. (Courtesy of Leviton Manufacturing Company.)

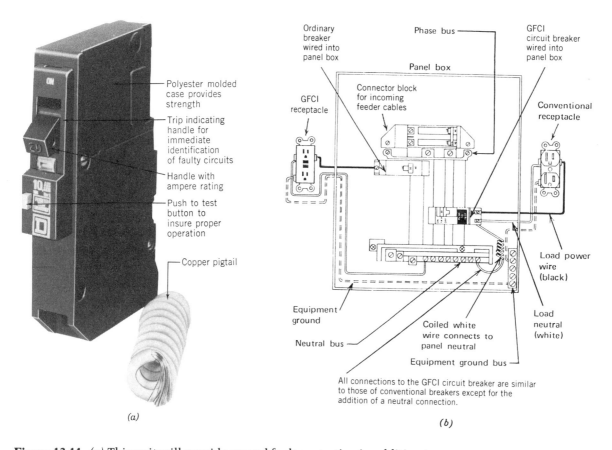

Figure 13.11 (a) This unit will provide ground fault protection in addition to operating as an ordinary circuit breaker. It is designed to fit into the same physical space as a normal molded-case circuit breaker and, therefore, can be used as a replacement in an existing panelboard. The copper pigtail seen at the bottom of the unit is connected to the neutral bus in the panel. This permits the GFCI that is built into the circuit breaker to measure any current difference between the phase (hot) and neutral wires in order to detect a ground current leak (fault). (b) Schematic of a typical small panelboard showing the difference in wiring of a conventional (normal) circuit breaker and a GFCI-equipped circuit breaker. Remember that a GFCI circuit breaker provides ground fault protection for the entire circuit so that ordinary receptacles can be used. Where an ordinary circuit breaker is used (see drawing), a special GFCI receptacle must be provided if ground fault protection is needed at that outlet. [(a) Courtesy of Square D Company.]

In addition, there are many locations in modern installations that are high risk due to the presence of water and ground paths. Among these are bathrooms, kitchens, swimming pools and outdoors. See the NEC (Article 210-8) for a list of locations where GFCI use is mandatory.

The Code permits application of ground fault circuit protection in several ways: at a specific receptacle (Figure 13.10), at the panelboard by use of a GFCI circuit breaker [Figure 13.11(a)] and on a feeder supplying a panelboard or a group of branch circuits. A GFCI receptacle should be used at an outlet where the circuit feeds other outlets that do not require GFCI protection. This will prevent the entire circuit from being tripped out if the GFCI operates at the protected outlet. A GFCI circuit breaker should be used on a circuit where most or all of its outlets require GFCI protection or where for some other reason it is desired to trip the entire circuit. See Figure 13.11(b). A GFCI device on a feeder is unusual because a ground fault downstream, on one outlet in one of the circuits, will trip out all the circuits. A typical application might be on a feeder to a kitchen panel or to a separate swimming pool panelboard.

d. Insulated Ground Receptacles

We stated previously that the grounding pole of all 3-pole grounding-type receptacles is connected to the equipment grounding conductor and that this conductor could include the entire metal conduit system, properly bonded. It was found a few years ago that sensitive electronic equipment fed from receptacles with that type of equipment grounding conductor frequently malfunctioned. These malfunctions were traced to electrical "noise" on the ground connection. (Electrical noise is simple random and unwanted voltages of various frequencies and magnitudes.) To correct this situation, at least partially, the Code permits such sensitive equipment to be fed from receptacles whose grounding terminal is insulated from the wiring system ground. These receptacles are color coded; they are either entirely orange or have an orange triangle on their face. The receptacle grounding terminal is connected to a separate, insulated green ground wire that is carried through the entire system but is only grounded, physically, at the service entrance. See **NEC**, Section 250-74; Exception 4.

13.4 Panelboards

We have referred a number of times in previous sections to electrical panels, also called electrical panelboards or simply panels or panelboards. We have also shown them in Figures 12.17, 13.3, 13.7 and 13.11(b). In this section, we will study in detail this extremely important element of a building electrical system. In residential work, a panel is frequently referred to as a load center. This is particularly true when the panel is a small unit, subfed from a main panel and installed to feed a concentrated group of loads, such as in a kitchen. A kitchen load center is simply an electrical panel that feeds the kitchen electrical loads. To avoid confusion, we recommend not using the term *load center* and instead using a more general term, such as kitchen electrical panel.

An electrical panel is simply a metal box that contains a group of circuit protective devices. If the devices are fuses, we have a fuse panel, and if the devices are circuit breakers, we have a circuit breaker panel. Fuses and breakers are rarely mixed in a panel, except that a circuit breaker panel occasionally has a fused main switch for overall protection of the panel. The **NEC** covers panels in Sections 384-13 through 384-36. It requires that all electrical panelboards be dead front. This means that no portion of a panel that is accessible to a user shall be electrically hot. This is a safety precaution. It eliminates old-style live front panels with open switches, fuses and circuit breakers.

Basically, then, a panel consists of a set of electric busbars to which the circuit protective devices are connected. This assembly is then placed in a metal box called a backbox, which has knockouts to allow for entrance of the branch circuit conductors. These conductors, as we already know, are connected to the load side of the protective devices in the panel. The main feeders that supply power to the panel's busbars enter through larger knockouts in the metal backbox, as can be seen in Figure 13.12(a). Details of panelboard construction can be seen in Figure 13.12(a–c). Figure 13.12(c) shows a panel with a main circuit breaker. Its function is to disconnect the entire panel in the event of a major fault. Figures 13.12(a) and (b) show panelboards with only branch circuit devices and no main breaker or switch. These panels are described as "lugs in mains only". This means that the panel has only connectors (lugs) on the main busbars, for connection of the main feeder cables, and no main protective device.

In most panels the circuit breakers are arranged in two parallel vertical rows. This is true for both single-phase 3-wire (two busbars plus neutral) and three-phase 4-wire (three busbars plus neutral) panelboards. The busbars are equipped with special contact devices that allow for breakers to be arranged in this fashion. These devices can be seen in the bottom of the panel in Figure 13.12(c-1) and in the cutaway section of Figure 13.12(b). Figure 13.13 shows in single-line schematic form the busbars and connections to circuit breakers for a single-phase 3-wire and a three-phase 4-wire panel.

It is very important that the technologist understand the numbering system of poles and circuits in a panel. A pole is a single connection to a bus. Pole numbers, shown in Figure 13.13, do not appear on the physical panel and are used only as a convenience in circuitry, as will become clear later in our study. Refer to Figure 13.13(a). Note that pole connections are made in pairs to the buses: that is, poles 1,2 on phase (bus) A; poles 3,4 on

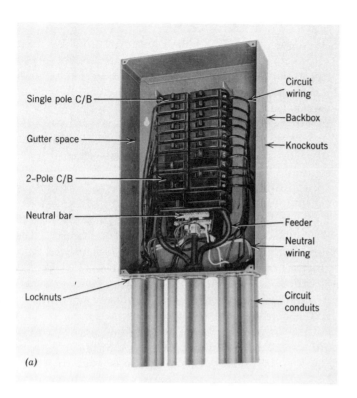

(a)

Figure 13.12 Panelboards may be of the circuit-breaker or fuse type. The panels illustrated in *(a)* and *(b)* are circuit breaker type and contain single- and 2-pole branch circuits. Panels are provided with a minimum 4-in. gutter space to allow routing of circuit wiring and any feed-through conductors. Lighting panels average 4½ in. deep by 16–20 in. wide. Lighting and appliance panels *(c)* by **NEC** definition (Section 384-14) can contain 1-, 2- and 3-pole circuit breakers. Such panels average 6 in. deep, 20–30 in. wide and a maximum of 62 in. high for a 42-pole panel. Panels are mounted with the top circuit device no higher than 78 in. AFF (above finished floor) and the bottom device no lower than 18 in. AFF.

A wired panel and backbox is illustrated in *(a)*; the panel front has been mounted in *(b)*. The 30-pole lighting and appliance panel illustrated in *(c)* has a main circuit breaker and measures 20 in. wide × 60 in. high × 5¾ in. deep. The panels shown in *(a)* and *(b)* have "lugs in mains only," that is, they have connections on the busbars for feeder cables, but they have no main switch or circuit breaker. [Illustration *(c)* Courtesy of Siemens Energy and Automation, Inc.]

phase B; then 5,6 again on phase A. In a three-phase panel [Figure 13.13(b)], poles 5,6 are on phase C, and so on. As a result, since the breakers are arranged in two vertical lines, breakers opposite each other are on the same phase, and breakers next to each other vertically are on different phases. (Check this in Figure 13.13.) Therefore, consecutive phases always skip a pole number. For instance, phases A,B correspond to poles 1 and 3 on the left column of circuit breakers (skipping pole 2) and poles 2 and 4 on the right column (skipping pole 3). This is true whether the panel is single-phase 3-wire [Figure 13.3(a)] or three-phase 4-wire [Figure 13.3(b)]. This numbering system is very important in circuitry as we shall learn.

Returning to numbering, note that a 2-pole circuit breaker is connected to two adjacent busbars but obviously has only one circuit number—for instance, circuits 17 and 18 in Figure 13.13(a) and circuits 11 and 12 in Figure 13.13(b). Since only the

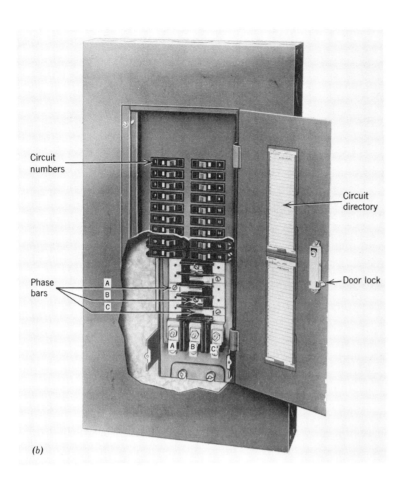

(b)

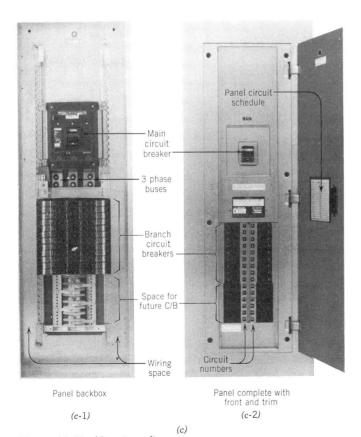

Panel backbox
(c-1)

Panel complete with front and trim
(c-2)

(c)

Figure 13.12 (Continued)

748 / BUILDING ELECTRICAL CIRCUITS

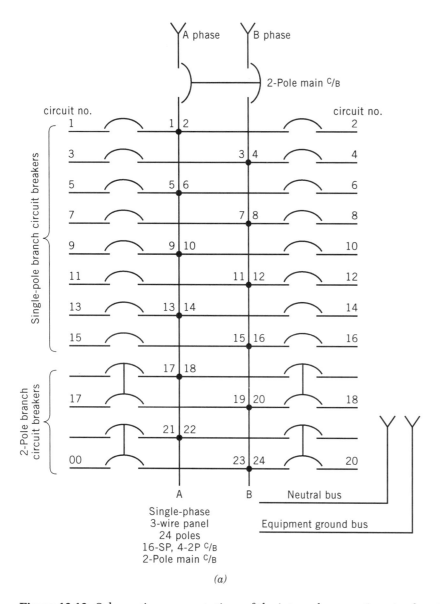

Figure 13.13 Schematic representations of the internal connections in electrical panelboards. Panels are usually either single-phase 2-pole 3-wire as in *(a)* or three-phase 3-pole 4-wire as in *(b)*. The number of poles equals the number of hot buses. The additional wire is the neutral. If an equipment ground bus is provided, it is kept electrically separate from the neutral bus. Both the ground bus and the neutral bus are at ground (zero) potential. The spaces in the three-phase panel are necessary so that the multipole circuit breakers can be connected to phases A, B and C in the proper sequence.

circuit number concerns the user because each circuit number represents a circuit that is protected by a single device in the panel, it should be obvious that pole numbers would only confuse a user and would not supply him or her with any useful information. The same is true for 3-pole circuit breakers, such as those representing circuits 13 and 14 in Figure 13.13*(b)*. Study of Figure 13.13 will also make it obvious why 2-pole and 3-pole breakers are placed opposite each other in the two vertical rows of devices in a panel. It is, of course, possible to place single-pole breakers opposite

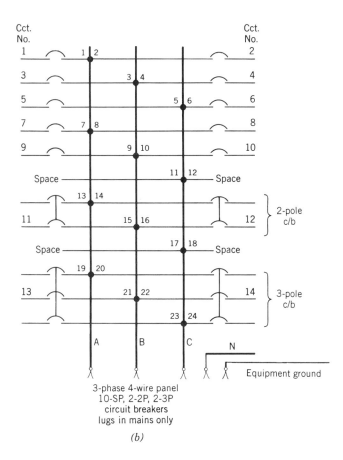

Figure 13.13 *(Continued)*

multipole units, but as a rule it is not done. Single-pole units are grouped together, as are multipole units, as shown. Very small panels, as for instance a two, three or four circuit panel for a very small facility, are built with all the breakers in a single row, connected to a single 120-v bus.

There is no industry-wide standard for panelboard busbar current ratings, although most manufacturers use 100, 225, 400 and 600 amp as their regular product line. To sum up then, panels vary in ampere rating of the buses, type of protective devices installed, whether or not the panel has a main device that completely disconnects and protects the entire panel, and the type of panel mounting, that is, whether the panel is to be flush or surface mounted. A typical drawing note describing a panel might be: Lighting and appliance panel, circuit breaker type, for surface mounting, single-phase, 120/240-v, 225-amp mains, 100/70-amp 2-pole main circuit breaker. Branch breakers—all 100-amp frame; 16–20 amp single-pole, 2–30 amp 2-pole, 2–20 amp single pole with GFCI.

To clarify in our minds the structure of a panelboard and to give us a feeling for what it contains, we will study at this point the preparation of a panel schedule. Almost everyone has seen the schedule of circuits found in a small directory on the inside of a panel door. This schedule, sometimes neatly typed or lettered but more often handwritten, is a listing of the panel circuits and the loads fed by these circuits. This information should properly be available directly from the drawings, and making the door directory should simply be a matter of copying the data off the drawings. Often, however, accurate panel schedule information is not available from the drawings because of circuitry changes during construction. Then, this very useful schedule on the panel door becomes a hit-or-miss affair hastily written by the installing electrician or filled in after being laboriously traced out by the owner or the maintenance personnel. As we will discuss later, the technologist doing the circuiting makes a complete panel schedule. Its failure to appear on the drawings in a complete and easily usable form indicates either that field changes were made or that change orders were issued, both without subsequently updating the drawings. Both represent poor construction practice.

For our analysis, we present the schedule of choice shown in Figure 13.14. We say "schedule of choice" because we do not recommend that a single schedule format be used for all work. The schedule used must meet the needs of the job. Frequently, a single job will use more than one schedule format. Unfortunately, there are almost as many different formats of panel schedules as there are designers and technologists, and it has been our experience that each person is convinced that his or her version is clearly best. The panel schedule formats presented in this book have been arrived at after studying literally dozens of schedules, and they are readily usable and practical.

The format of Figure 13.14 was chosen for illustration here because, in addition to being easy to use and practical, it clearly demonstrates to the novice technologist the "workings" of a panel. Also, it can be transferred as it stands to the physical panel directory. There it exists as a permanent and extremely useful record for the building occupant

750 / BUILDING ELECTRICAL CIRCUITS

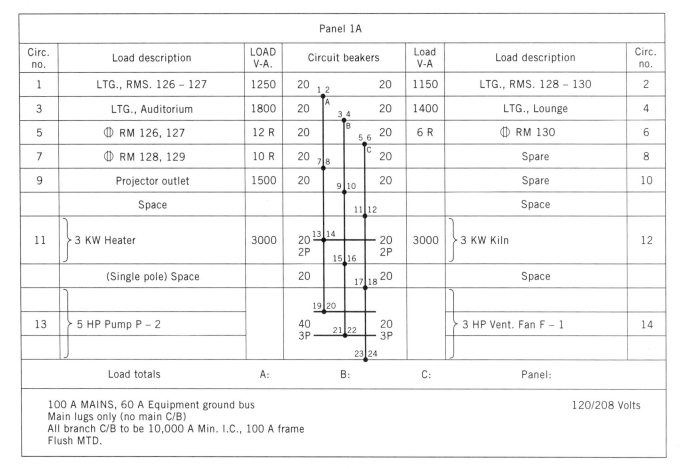

(a)

Figure 13.14 (a) Typical circuit breaker panel schedule, corresponding to schematic diagram shown in Figure 13.13(b). Circuits 13 and 14 feed three-phase motors and are, therefore, 3-pole (see Figure 16.43).

or his or her electrician when it is desired to do any maintenance or repair work. Furthermore, and this is of great importance, it is sufficiently detailed so that changes in the wiring system or loads can be shown on it, and thus it can be kept up to date. The abbreviated panel schedule so frequently found on drawings, which simply consists of a listing of the breakers or fuses, is not useful to the occupant. It does not indicate which outlets are on what circuit or what the fixed loads are. As such, it is useful only at the time of layout and even then causes problems when design changes are made. It should, therefore, be avoided as a false economy.

Refer now to Figure 13.14 and note how the schedule reflects both the physical reality as shown in Figure 13.12(a) and (b) and the schematic drawings of Figure 13.13. One difference is the absence in the panel schedule of the neutral, which is not often shown schematically either. A moment's thought will show that nothing is to be gained by showing the neutral bar. If an equipment ground

PROCEDURE IN WIRING PLANNING / 751

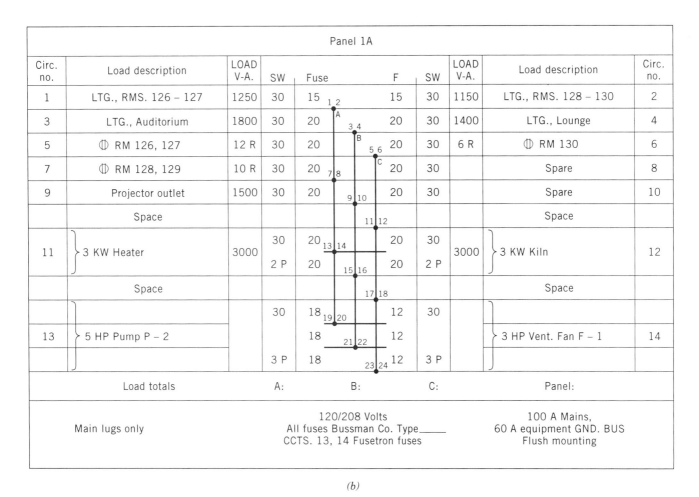

(b)

Figure 13.14 (b) Typical switch/fuse panel, corresponding to circuit breaker panel of Figure 13.14(a) (see Figure 16.44).

bus is called for, it need not be shown but must be specified, and a space at the bottom of the panel schedule exists specifically for that purpose. Using Figure 13.14, note carefully the following:

(a) A pole is a single connection to a bus. Therefore, a multipole circuit breaker such as circuit breaker no. 11 is connected to more than one pole and cannot, without causing considerable confusion, carry the pole numbers. Circuit no. 11, fed through circuit breaker no. 11, is connected to poles 13 and 15. If we wish to know which poles and/or phases are connected to each circuit protective device, we must consult the panel schedule as designed and as shown on the manufacturer's shop drawings.

(b) As already stated, the pole connections are made in pairs from the buses, that is, 1-2 from phase A, 3-4 from phase B and so on. Therefore, to get consecutive phases, we must skip a pole each time. Keep in mind that these are pole numbers and not circuit numbers.

(c) The usual lighting and appliance panel (see the NEC for definition) is limited to 42 poles, not including the main disconnect (see **NEC** Section 384-15). For this reason most panel schedules are made for 42 poles (see Figures 16.41–16.44). If we use less, we simply cut off the panel schedule at the desired point. (Leaving the unused section blank is confusing, since it might be misunderstood as a 42-pole panel with space for future circuit breakers.)

(d) Space should be provided on the panel schedule for busbar and circuit breaker data, as shown. This information is supplied by the job electrical designer.

(e) The load description column should contain a brief but adequate description of the load. We prefer to place the size of the load, in volt-amperes, in a separate column. This makes it very convenient to total the panel loads by phases and to keep the totals at the bottom, as shown. As we shall learn, this is important in order to balance the panel's phase loads and to determine the entire panel load. Were the volt-ampere load entered in the description column, the technologist or designer would have a much more laborious task keeping track of loads, particularly when changes are made.

(f) Panel designations often indicate the panel's location in a large building. For example, panel 2C is in closet C, second floor. Since there are basically only two kinds of panels—lighting and appliance panels and power panels—it is well to avoid misleading titles, such as lighting panel, receptacle panel and distribution panel, and to stick to LP for lighting and appliance panels and PP for power panels. Even simpler is avoiding titles altogether, calling panels simply P_1, P_2 and the like without reference to use.

As mentioned previously, panel schedules change not only during layout as a result of design changes and after panel installation as a result of field changes but also at the manufacturing stage. The manufacturer sends to the design office a panel layout representing the panel as it will actually be built. For one reason or another, the actual arrangement shown in this shop drawing may differ from the panel layout on the drawings. If the designer or electrical technologist who is checking the shop drawing agrees to the manufacturer's version, he or she must make the corresponding changes on the drawings. This will ensure that the drawing used in the field corresponds to the equipment on the job in all respects. Leaving the panel circuit arrangement to the contractor may save office time, but it creates field problems and, what is worse, maintenance problems for the owner. Symbols for panelboards as well as other equipment items that appear on electrical drawings are given in Figure 13.15. This figure is Part VI of the overall symbol list.

Electrical Planning

13.5 Procedure in Wiring Planning

Now that we have studied the basic elements that make up the electrical system of a building, namely, the branch circuit, the panel and protective devices, and the outlets and wiring devices, we can turn our attention to the procedure normally followed in electrically planning a building. For this purpose, we will use The Basic House plan. In Chapter 15, we will generalize the procedure as applied to residential buildings, using a large residence as a study example. We have selected single family residences to begin our study for two reasons:

- This is the type building that most people are familiar with.
- A residence presents many types of wiring and circuitry problems in a single building. Once having gone through residential electrical design, you will have handled lighting, control, kitchen equipment, motors, receptacles, outdoor lighting and various types of electrical signals. No other small building gives such diverse loads. You can then easily go on to the electrical layouts of other types of buildings.

The procedure normally followed in electrical design follows.

Step 1. Identify all spaces on the architectural plan by intended use, both present and future.

Step 2. Assemble all criteria applicable to such spaces.

These criteria will be discussed in part here and more generally in the chapters that follow.

Step 3. Show to scale, and in position where possible, all items of equipment that require electrical connection.

This includes mechanical, heating and plumbing

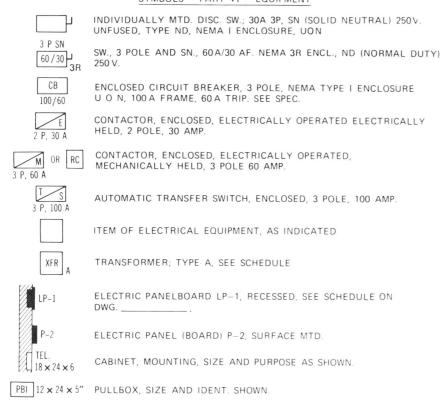

Figure 13.15 Architectural-electrical plan symbols, Part VI, Equipment Items.
Part I, Raceways, Figure 12.6, page 671.
Part II, Outlets, Figure 12.50, page 719.
Part III, Wiring Devices, Figure 12.51, page 720.
Part IV, Abbreviations, Figure 12.58, page 726.
Part V, Single-line Diagrams, Figure 13.4, page 733.
Part VII, Signalling Devices, Figure 15.34, page 909.
Part VIII, Motors and Motor Control, Figure 16.25, page 944.
Part IX, Control and Wiring Devices, Figure 16.28, page 947.

system items. Items that do not require electrical power, but do affect outlet location, must also be shown. Thus furniture placement, and in particular fixed cabinets, should be shown.

Step 4. Locate the remaining electrical devices according to the needs of each space and the applicable criteria.

At some point between steps 1 and 4, a decision has to be made as to the number of drawings needed and the scale of the drawings. This, too, will be discussed as it applies.

Step 5. Circuit all the equipment and also prepare a panel schedule.

This step or the previous one also includes all switching and circuit control elements.

Step 6. Check the work.

We will now apply Steps 1–6 to The Basic House plan while learning the required techniques.

13.6 Computer Use in Electrical Design

Many, and probably most, electrical design offices today use computers to produce their electrical working drawings in a variety of ways. A fully computerized office will receive the architectural drawings in design file form from the architect to be used on their computers. Using one of the many available CAD (computer-aided design) programs, the electrical designer will then reproduce the architectural drawings on his or her system, add the electrical work and produce finished electrical working drawings with a plotter. Changes to either architectural or electrical plans are readily and easily accomplished on the computer using the CAD program, and revised "hard copy" is produced equally as easily by the plotter.

In addition, there exists today a large and ever-growing library of engineering software that will perform a wide variety of electrical design and engineering tasks, including feeder selection, conduit and pull box sizing, lighting calculations and even semigraphic tasks such as preparation of panel schedules and lighting fixture schedules. These software tools are extremely useful and time-saving but should be used only by persons who are completely familiar with the techniques involved and who are capable of doing the work manually. Only such persons can properly input the programs and, more important, judge the output. Furthermore, many offices are only partially computerized. In these offices, engineers, designers and technologists perform part or all the work manually. This is particularly true of the electrical design divisions of multi-outlet store chains, interior design offices, small electrical contractors and contractors specializing in renovation and reconstruction work. For these reasons, we will proceed with our study, assuming that all the design work is to be done manually.

13.7 The Architectural-Electrical Plan

Step 1, as stated previously, is the preparation of the architectural background plan, which is the raw material of our work. This plan is a stripped down version of the architectural plan showing only walls, partitions, windows, door swings and pertinent fixed equipment. Dimensions, room titles and other architectural data are omitted. Such a plan of The Basic House is shown in Figure 3.32, page 130. Since the electrical work must stand out clearly, it is false economy to attempt to use a prepared architectural plan for electrical work. The required background should always be prepared with light line work, showing the elements listed previously.

It was once almost universal practice to trace the architectural background on the back of the sheet. This had two purposes: (1) erasures on the electrical plan would not affect the architectural outline, and (2) the print would show the architectural work as lighter than the electrical plan, which was done on the front side of the tracing. This is still a good idea except when using pencil cloth tracings, which have a glossy back surface. It is a very good idea to obtain a screened print of the stripped architectural plan, which allows the electrical work drawn on it to show as boldly as is desired. Architectural changes, which always occur, can be made on such a photo print with no trouble.

If at all possible, architectural spaces should be identified by room titles placed outside the plan, to avoid conflicting with the electrical work. Where this is not possible or is undesirable, room titles should be penciled in lightly to identify spaces, and only lettered in final form when the electrical work is complete, to avoid space conflicts. This is also true for equipment. The initial drawing must be as open and uncluttered as it is possible to make it. After the electrical work is in, the required space names, descriptions, titles and equipment identification can be added. All these steps can and should be done with a CAD program when using computerized architectural plans in order to produce a perfectly clear electrical working drawing. This is easily accomplished by adjusting the relative line weights of the electrical work and the architectural background. (Adjustable line weights on working drawings are also useful where several trades, such as heating ducts and conduits, are shown overlapping in the same area.)

13.8 Residential Electrical Criteria

By "residential electrical criteria," we simply mean the process by which we decide what is required and where we put it. The answer to those questions is varied. The NEC gives minimum criteria, and the architect and engineer in conjunction with the owner establish additional requirements on the basis of need, convenience and common sense. As a guide to these latter criteria, numerous books and pamphlets have been published by trade and governmental agencies that list room-by-room recommendations for lighting and other electrical outlets. Some of these sources are listed in the Supplementary Reading section at the end of the chapters of this book. You will do well to read these, while keeping in mind that they are recommendations. In contrast to this, the minimum requirements of the NEC are mandatory. Since the technologist must be fully familiar with these NEC requirements, we will review them in this chapter while putting off the optional material until the more detailed discussion of residential buildings in Chapter 15.

Circuitry

13.9 Equipment and Device Layout

This stage of the work includes steps 3 and 4 of Section 13.5. We continue to use The Basic House plan as our reference. Since in Chapters 3 and 4 the heating layout was shown for both hot water and electric heating systems, we show here the electrical layout corresponding to these systems. Refer now to Figure 13.16, which corresponds to the hot water-heated building of Figure 3.33 and to Figure 13.17, which corresponds to the electrically heated house of Figure 4.11.

In accordance with the step-by-step layout procedure described in Section 13.5, we would first show all electrically connected equipment. Referring to Figure 13.16, this would include the electric range, refrigerator, dishwasher and exhaust fan in the kitchen and the laundry equipment and boiler in the basement. In the case of the electrically heated building of Figures 4.11 and 13.17, we have the electric baseboard units and room thermostats in all the spaces, but we eliminate the boiler of Figure 13.16. At this point, we must decide whether showing the electric heating on the drawing, in addition to all the remaining electrical work, will cause the drawings to be cluttered and difficult to read. In this case, we decide that, by use of tables, this clutter can be avoided. The expense of an additional drawing to show the electric heating is thereby avoided.

Since no furniture layout was provided, the remaining devices are located in accordance with the NEC requirements and what the designer or technologist considers to be good practice. The criteria for both are discussed in Chapter 15. The resulting drawings as they stand prior to circuiting are shown in Figures 13.16 and 13.17. A careful comparison between Figures 13.16(a) and 13.17(a) will show that in Figure 13.17(a) we moved a number of convenience outlets when they conflicted with the location of baseboard heating unit. Recep-

tacles should not be installed above baseboard heaters. The reason is that electrical cords hanging down from such receptacles would be subjected to excessive heat that would soon damage them and could easily cause an electrical fault. Factory-installed receptacle outlets in baseboard heating units are permitted. However, it requires less coordination to simply move the receptacles, which is what we have done in Figure 13.17(a). See NEC Section 210-52(a) Exception, and FPN following the Exception.

Note that, at this point, we have already made a symbol list and have shown the switching of lights. The symbol list should be prepared at this stage and not, as is often done, as an afterthought. In this way, all needed symbols are added to the list as they are put onto the drawing, along with required notes, mounting heights and the like, and nothing is forgotten. Similarly, light switching should be indicated as soon as each lighting outlet is located. This avoids any possibility of forgetting a switch. Residences using centralized, remote control switching also use local switching, so that this step is always required. Deciding on the lighting fixture type, location, and control or switching are three parts of a single activity, which are best done together. The type of lighting fixture is selected by the designer or technologist and is shown in a fixture schedule on the drawing.

You may have noticed that no signal devices such as smoke detectors, alarm systems, antenna connections and the like are shown in The Basic House plan of Figures 13.16 and 13.17. This is not an oversight. At this stage of our study, we want to concentrate on the power and lighting aspects of design. When we do an actual full residential design in Chapter 15, these very important systems will be considered.

13.10 Circuitry Guidelines

The preliminary work of layout and switching has been shown; it remains to make the connections to the panelboard, but this is not as simple as it sounds. There are many ways in which the circuiting can be done, and there is often no optimum way. There are, however, certain guidelines to keep in mind, which, if followed, will yield a flexible, economical and convenient layout. Also NEC rules must be followed, and there are a few technical considerations that will help to yield a good layout. The larger and more complex the job, the more

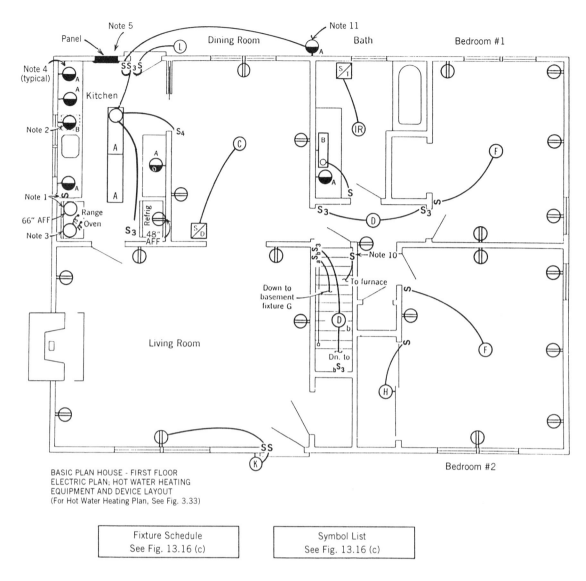

BASIC PLAN HOUSE - FIRST FLOOR
ELECTRIC PLAN; HOT WATER HEATING
EQUIPMENT AND DEVICE LAYOUT
(For Hot Water Heating Plan, See Fig. 3.33)

Fixture Schedule
See Fig. 13.16 (c)

Symbol List
See Fig. 13.16 (c)

Notes:

1. Switch and outlet for exhaust fan. Switch wall mtd. above counter-blacksplash. Outlet with blank cover mounted adjacent to fan wall opening. Separate switch may be omitted if fan is supplied with integral switch.
2. Dishwasher receptacle wall mtd. behind unit, 6" AFF.
3. Range and oven outlet boxes wall mtd., 36" AFF. Flexible connection to units.
4. Receptacles at countertop locations to be wall mounted 2" above backsplash.
5. Max. ht. of top c/b to be 78" AFF.
6. Wiring shown as run exposed indicates absence of finished ceiling in basement level. All BX to be run through framing members. Attachment below ceiling joists not permitted. See Section 12.4.

7. Connect to 2-Type G fixtures ceiling mounted at $1/3$ points of crawl space.
8. Connect to 1-Type G fixture at center of crawl space.
9. Connect to shut-down switch at top of stairs. Boiler control wiring by others. See Note 10.
10. Boiler wiring safety disconnect. Provide RED wall plate, clearly marked "BOILER ON-OFF".
11. Equipped with self-closing gasket WP cover.

(a)

Figure 13.16 The Basic House plan, uncircuited, hot water heat. (a) Street-floor plan. (b) Basement plan. (c) Lighting fixture schedule. (d) Symbol list.

Note: On the circuited plan in Figures 13.20(a) and (b), the wiring from stairwell fixture D to the two three-way switches controlling it does not correspond to that shown here. This is due to wiring economy considerations that are found in the text discussion on Figure 13.26. For the purpose of indicating light fixture switching requirements, the material on this figure is correct. It changes when actual wiring runs are decided upon. The switching functioning remains the same.

CIRCUITRY GUIDELINES / 757

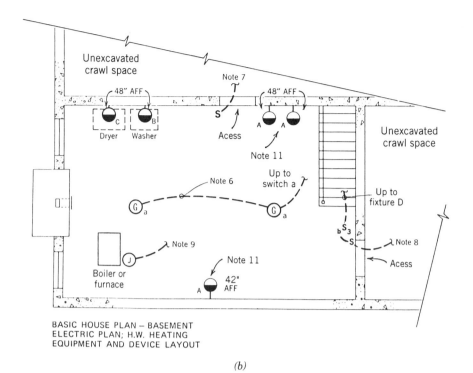

BASIC HOUSE PLAN – BASEMENT
ELECTRIC PLAN; H.W. HEATING
EQUIPMENT AND DEVICE LAYOUT

(b)

| LIGHTING FIXTURE SCHEDULE ||||
TYPE	DESCRIPTION	MANUFACTURER	REMARKS
A	48" L X 12" W X 4" DEEP NOMINAL, 2 LAMP/FLUORES-CENT, WRAP-AROUND ACRYLIC LENS, F 40 WW/LAMPS. SURFACE MTD.	BRITE-LITE CO. CAT. #2/40/KFF OR EQUAL	4" DEPTH MAXIMUM
B	24" L, 1 LAMP 20W FLUOR. FIXTURE, WRAP-AROUND WHITE DIFFUSER, MOUNT ABOVE MEDICINE CABINET.	BRITE-LITE CO. CAT. #1/20/BFF OR EQUAL	MAX. MTG. HT. 78" TO ℄.
C	ADJUSTABLE HEIGHT PENDANT HALOGEN 3-100W MAX.	HOMELAMP CO. CAT. #3/75/DRP OR EQUAL	————
D	10" D. DRUM-TYPE FIXTURE, WHITE GLASS DIFFUSER, CENTER LOCK-UP, 2-60 W INCAND. MAX., SURF. MTD.	BRITELITE CO. CAT. #2/60/HF OR EQUAL	6" MAX. DEPTH.
F	12" D. DRUM FIXTURE, CONCEALED HINGE ON OPAL GLASS DIFFUSER FOR RELAMPING WITHOUT GLASS REMOVAL, 2-75W INCAND. MAX. SURFACE MTD.	DENMARK LIGHTING SPECIAL UNIT #374821	NO SUBSTITUTION WILL BE ACCEPTED.
G	PORCELAIN LAMPHOLDER, PULL CHAIN WITH WIRE GUARD, 100 W. INCAND. SURF. MTD.	————	————
H	SAME AS TYPE G, EXCEPT W/O GUARD.	————	————
K	DECORATIVE OUTDOOR LANTERN, MAX. 150W INCAND., WALL MTD. 84" AFF TO ℄.	TO BE CHOSEN BY OWNER	————
L	UTILITY OUTDOOR LIGHT, ANODIZED ALUMINUM BODY AND CYLINDRICAL OPAL GLASS DIFFUSER. 1-100W INCAND. MAX. 84" AFF TO ℄.	UTIL-LITE CO. CAT. #1/100/BP OR EQUAL	IF VANDALISM IS OF CONCERN, SUBST. PLASTIC DIFFUSER.
IR	RECESSED FIXTURE WITH 150 WATT INFRARED HEAT LAMP	HEAT LIGHT CO.	————

(c)

Figure 13.16 (Continued)

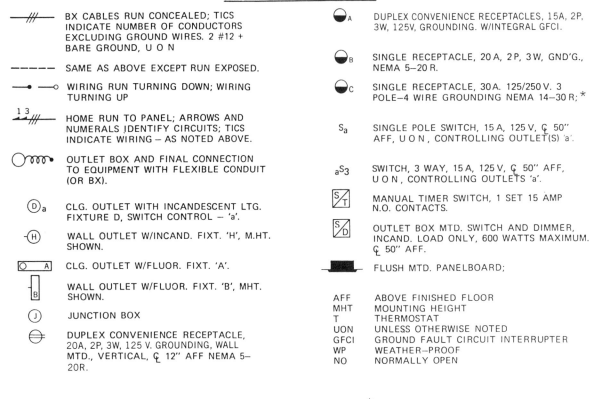

(d)

Figure 13.16 *(Continued)*

ways there are to do it. Yet, if guidelines are followed, all the layouts will be acceptable.

The **NEC** lists minimum requirements for circuits in various occupancies including residences. Since all electrical installations must satisfy the **NEC**, a design in accordance with these requirements will be a minimum design. Good practice almost always calls for more capacity than these minimums, because of today's ever-increasing residential electrical loads. We will, therefore, in the discussion that follows, state both minimum requirements per the **NEC**, and what we consider to be needed for good quality construction. We wish to emphasize that the discussion that follows is based on the 1996 edition of the **NEC**. Since the Code is amended every three years, changes may occur. It is therefore necessary, as stated previously, to use the edition current at the time actual design is performed.

a. Number of General Lighting Circuits

The Code requires a minimum circuit capacity for "general lighting" of 3 v-a/ft^2 of building area, excluding open porches, garages and unused or unoccupied spaces not adaptable for future use [NEC Section 220-3b and Table 220-3*(b)*]. Since so many homeowners finish their basements and convert them to living space, it is good practice to include the area of the basement in the total building area, unless it is certain that the area will remain unoccupied.

The term *general lighting* in the Code is somewhat misleading. This 3 v-a/ft² load includes not only fixed ceiling and wall lighting but also all the convenience receptacles in the building. (Excluded are appliance outlets, which are considered as separate loads.) The idea here is that convenience receptacles are indeed used for general lighting such as floor lamps and table lamps. In reality, of course, these convenience outlets are also used for permanently plugged-in electrical loads in the residence including TV, radio, hi-fi, video and computers, plus incidental loads such as razors, heaters, hair dryers and vacuum cleaners. For this reason the FPN (Fine Print Note) in the Code at that Section says that the 3-v-a/ft² load may not be enough.

In addition, the Code requires that continuous loads be calculated at 125% of their actual amperage. Loads that remain connected more than 3 hours are considered to be continuous (NEC definitions). Good practice dictates that general lighting load, including all the permanently plugged-in devices listed previously, be considered as continuous loads. This reduces the number of square feet per circuit by a factor of 0.8 (reciprocal of 125%).

To summarize this discussion, we will calculate the number of 15- and/or 20-amp circuits required for general lighting in a residence. It is important to understand that a circuit's ampere rating, by definition, is determined by the rating of the circuit's protective device. Therefore, a circuit protected by a 15-amp fuse or circuit breaker is a 15-amp circuit (regardless of the wire size used). Similarly, a circuit protected by a 20-amp fuse or circuit breaker is a 20-amp circuit, by definition. The same is true for all amperage ratings.

The calculation is straightforward. The number of circuits required in a residence for general lighting (including convenience receptacles) follows.

(1) Minimum number of general lighting circuits per NEC:
  (a) Area per 15-amp circuit:
  $$15 \text{ amp} \times 120 \text{ v} = 1800 \text{ v-a}$$
  $$\frac{1800 \text{ v-a}}{3 \text{ v-a/ft}^2} = 600 \text{ ft}^2$$
  (b) Area per 20-amp circuit
  $$20 \text{ amp} \times 120 \text{ v} = 2400 \text{ v-a}$$
  $$\frac{2400 \text{ v-a}}{3 \text{ v-a/ft}^2} = 800 \text{ ft}^2$$

So the minimum number of circuits required for general lighting to meet NEC requirements is one 15-amp circuit per 600-ft² or one 20-amp circuit per 800 ft² of occupied and usable area.

(2) Good practice:
Provide one 15-amp general lighting circuit per 400—480 ft², depending on the anticipated usage. The higher the anticipated load, the lower the square foot area per circuit. When using 20 amp circuits, the areas per circuit are between 530 and 640 ft² again, depending on anticipated usage. These good practice figures will normally provide enough circuits for all but the heaviest loaded residences. However, if an actual design indicates the need for additional circuits, as may be the case in such residences, they obviously must be provided. (See also the illustrative examples in NEC Chapter 9. Those calculations are based on minimum requirements.)

Example 13.1 How many 15-amp general lighting circuits should be provided for a medium quality tract house with 2400 ft² of living space and a 1200 ft² basement?

Solution: Assume that two-thirds of the basement will eventually be occupied. This gives a total living area of

$$2400 + \tfrac{2}{3}(1200) = 3200 \text{ ft}^2$$

Using a medium figure of 440 ft² per 15-amp circuit:

$$\text{Number of circuits} = \frac{3200 \text{ ft}^2}{600 \text{ ft}^2/\text{circuit}} = 7.3 \text{ circuits}$$

Eight circuits should be provided for good practice. The minimum number required by the NEC is 3200 ft²/600 ft² or 6 circuits.

Example 13.2 How many 20-amp general lighting circuits should be provided for a 4000 ft² high quality tract house with a 1200 ft² basement?

Solution: Here again, we assume that the basement will be used for recreational and shop activities and that two-thirds of its area will be in such use. Therefore, using maximum loading (minimum area per circuit):

$$\text{Usable area} = 4000 + \tfrac{2}{3}(1200) = 4800 \text{ ft}^2$$

760 / BUILDING ELECTRICAL CIRCUITS

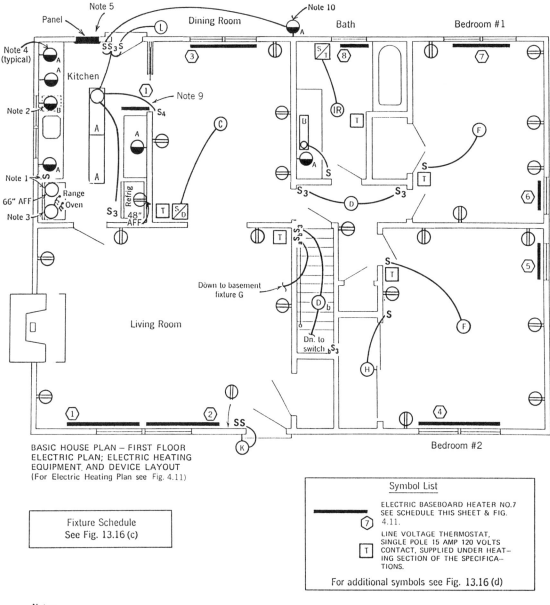

BASIC HOUSE PLAN – FIRST FLOOR
ELECTRIC PLAN; ELECTRIC HEATING
EQUIPMENT, AND DEVICE LAYOUT
(For Electric Heating Plan see Fig. 4.11)

Fixture Schedule
See Fig. 13.16 (c)

Symbol List

ELECTRIC BASEBOARD HEATER NO.7 SEE SCHEDULE THIS SHEET & FIG. 4.11.

LINE VOLTAGE THERMOSTAT, SINGLE POLE 15 AMP 120 VOLTS CONTACT, SUPPLIED UNDER HEATING SECTION OF THE SPECIFICATIONS.

For additional symbols see Fig. 13.16 (d)

Notes:

1. Switch and outlet for exhaust fan. Switch wall mtd. above sink backsplash. Outlet with blank cover mounted adjacent to fan.
2. Dishwasher receptacle wall mtd. behind unit, 6" AFF.
3. Range and oven outlet boxes wall mtd., 36" AFF. Flexible connection to units.
4. Receptacles at countertop locations to be wall mounted 2" above backsplash.
5. Max. ht. of top c/b to be 78" AFF.
6. Wiring shown as run exposed indicates absence of finished ceiling in basement level. All BX to be run through framing members. Attachment below ceiling joists not permitted. See Section 12.4.
7. Connect to 2–Type G fixtures ceiling mounted at ⅓ points.
8. Connect to 1–Type G fixture at center.
9. Mount heater in end of wall cabinet. See detail, on Dwg. 0000. Thermostat integral with heating unit
10. Equipped with self-closing gasketed WP cover.

(a)

Figure 13.17 The Basic House plan, uncircuited, electric heat. See note in caption for Figure 13.16. *(a)* Street floor plan. *(b)* Basement plan.

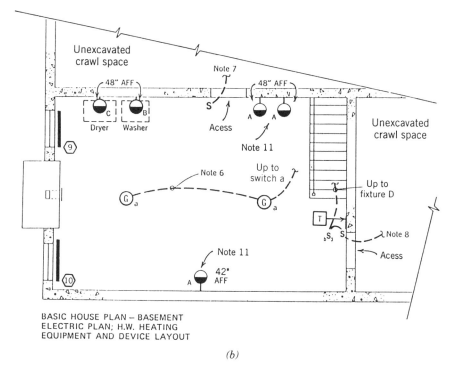

BASIC HOUSE PLAN – BASEMENT
ELECTRIC PLAN; H.W. HEATING
EQUIPMENT AND DEVICE LAYOUT

(b)

Figure 13.17 *(Continued)*

$$\text{No. of 20-amp circuits} = \frac{4800 \text{ ft}^2}{530 \text{ ft}^2/\text{circuit}} = 9 \text{ circuits}$$

The minimum required by the NEC is $4800 \text{ ft}^2/800 \text{ ft}^2 = 6$ circuits.

b. Circuit Loading for General Lighting

Residential general lighting circuits are generally 15 amp, although 20-amp circuits can also be used. Because of the 80% derating that is required for continuous loads such as lighting, a 15-amp circuit may not be loaded to more than 12 amp (80% of 15 amp), and a 20-amp circuit may not be loaded to more than 16 amp. These amperage figures correspond to

15-amp circuit: 12 amp × 120 v = 1440 v-a
20-amp circuit: 16 amp × 120 v = 1920 v-a

These are maximum figures. Since good practice dictates leaving some spare capacity in these circuits, it is wise not to load 15-amp general lighting circuits to more than 1300 v-a and 20-amp circuits to 1800 v-a, depending on anticipated loads.

c. Number of Outlets per General Lighting Circuit

Although the NEC does not limit the number of outlets that may be connected to a general lighting circuit, many local codes do have such restrictions. The technologist who is assigned the residence circuiting task must always check with the project engineer to determine if such a limit exists, before doing the circuiting. A basic guidelines is to figure a normal duplex convenience receptacle at no less than 1.5 amp. This would give eight duplex receptacle outlets to a 15-amp circuit because, as we stated previously, a 15-amp circuit should not be loaded to more than 80% of capacity, that is 12 amp. Therefore,

$$\frac{12 \text{ amp}}{1.5 \text{ amp/duplex receptacle}} = 8 \text{ duplex receptacle outlets}$$

Similarly, a 20-amp circuit could have a maximum of

$$\frac{16 \text{ amp}}{1.5 \text{ amp/receptacle}} = 10 \text{ (or 11) receptacle outlets}$$

Of course, if some of the outlets feed fixed lighting, the actual wattages should be used and not the 1.5-

amp figure. However, as we will note later, mixing fixed lighting and convenience receptacle outlets is not recommended.

We must caution the technologist not to circuit receptacle outlets according to the maximum number permitted without careful attention to their possible use. By this we mean that convenience outlets in the living room, study and family room are likely to have heavy entertainment equipment loads and should be circuited at no more than five or six to a circuit, whereas bedroom, guest room, bathroom and playroom outlets normally have lighter and noncoincident (not at the same time) loads and, therefore, can be circuited as stated previously. The exception to this rule is any receptacle outlet that is likely to be used to supply a room air conditioner. Circuiting of these receptacles is discussed in Section 13.10.e.

d. Small Appliance Branch Circuits

The NEC in Section 220-4(b) requires that, in addition to the circuits required for general lighting in a residence, a minimum of two 20-amp small appliance branch circuits be supplied to feed all the receptacle outlets in the kitchen, pantry, breakfast room, dining room or similar area, and no other outlets except clock outlets and outdoor outlets. By definition, an appliance branch circuit may not supply any fixed lighting. See Figure 13.18. These small appliance receptacle outlets are physically no different than any of the other convenience receptacles in the building except that they will probably feed heavier loads such as kitchen appliances. As a result, good practice requires that these outlets be circuited at no more than four and preferably two or three to a 20-amp circuit. These appliance circuits must be so arranged that all the kitchen outlets are fed from (part of) at least two of these circuits. The requirement for ground fault protection of certain kitchen outlets is discussed in Section 13.10.f.

e. Other Appliance Circuits

Additional 20-amp appliance circuits, similar to the small appliance branch circuits discussed in Section 10.13.d, should be furnished to supply appliance-type outlets as follows.

(1) One outlet in each bedroom of a house that is not centrally air conditioned, where the climate is such that air conditioning is commonly

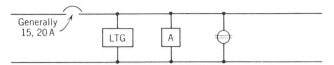

(a) General purpose branch circuit. Supplies outlets for lighting and appliances, including convenience receptacles.

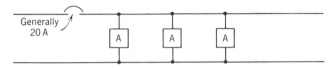

(b) Appliance branch circuit. Supplies outlets intended for feeding appliances. Fixed lighting not supplied.

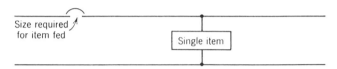

(c) Individual branch circuit, designed to supply a single, specific item.

Figure 13.18 Branch circuit types according to NEC definition.

used. Such outlets are appropriate for small 120-v room air-conditioners. No more than two such outlets should be placed on a single circuit and preferably only one plus up to three other outlets. For large air conditioning units, special outlets and branch circuits are required. All electrical work for air conditioning units must comply with the requirements of NEC Article 440.

(2) At least one receptacle outlet and preferably two or more, in garages and basements, for electrical workbench tools and other appliances. These outlets should be mounted no less than 42 in. above floor level.

(3) The NEC requires at least one receptacle in the laundry area of a dwelling unit, to be fed from a 20-amp appliance circuit that feeds no other outlets. This circuit need not be GFCI protected (because some washing machines have leakage currents high enough to cause the GFCI to trip). See NEC Sections 210-52(f), 220-4(c) and 210-8(a)(5) Exception No. 2. If an electric clothes dryer is anticipated (and it should be unless it is definitely known that a gas dryer will be used), a dedicated individual branch circuit should be supplied to serve this load via a special heavy-duty receptacle. The rating of this dedicated circuit and its receptacle must be in accordance with the rating of the dryer.

f. GFCI-Protected Outlets

Because of the increased shock hazard presented by the presence of water and piping close to an electrical outlet, the NEC requires that such outlets be protected by ground fault circuit interrupters. See NEC Section 210-8. This protection may be either local at the outlet or for the entire circuit by means of a GFCI circuit breaker, as explained in Section 13.3. Included among the required GFCI-protected receptacle outlets are those in kitchens, wet bars, bathrooms, garages, outdoors and basements and crawl spaces. Again we must mention that NEC requirements are minimum. Therefore, if the technologist laying out a residential building knows of an area that has a high shock hazard other than those detailed in the NEC, then he or she should supply a GFCI outlet there as well. Possible applications might be a recreation room with a hobby area supplied with water piping, an outlet intended for a large indoor aquarium, an accessible rooftop outlet intended for antennas and other similar situations.

g. Circuits and Outlets for Electronic Equipment and Computers

The vast increase in the use of sophisticated computers and data processing equipment in residences has resulted in the need for special circuitry and equipment for the electric outlets feeding them. Although this is not generally the responsibility of the electrical technologist but rather that of the engineer and the communication specialist, it is useful to be aware of some of the special requirements. Among them are isolated grounds, electric noise filters and voltage surge suppression. See Sections 12.14 and 13.3.

h. Other Good Practice Circuiting Guidelines Include:

- Avoid combining receptacles and lighting on a single circuit.
- Avoid placing all the lighting in a building on a single circuit.
- Circuit lighting and receptacles so that each space contains parts of at least two circuits. That way, if a single circuit is out, the entire space is not deprived of power.
- Do not use combination switch and receptacle outlets except where convenience of use dictates high mounting, as for example above counters. These combination outlets are often used in economy construction to save a few dollars in wiring. It makes an inconvenient and unsightly receptacle.
- Supply at least one receptacle outside the house. It must be a GFCI type per NEC requirement as noted previously. Switch control of outside receptacles from inside the house is both a safety precaution and a good control function.
- In rooms without fixed overhead or wall lighting, provide switch control of one-half of a strategically located receptacle that is intended to supply a floor or table lamp. See Figure 13.19 for the wiring arrangement in such a case. Kitchens and bathrooms must have fixed dedicated lighting outlets. See NEC Section 210.70 for details of permitted lighting switching arrangements.
- Provide switch control for closet lights. Pull chains are a nuisance.

Keeping in mind these guidelines and the convenience we ourselves would appreciate if we were living in the house, we can proceed to the actual circuitry.

13.11 Drawing Circuitry

At this point, we apply step by step the guidelines listed previously to The Basic House plan in its two versions as shown in Figures 13.16 and 13.17. The difference between them is the heating method: the former is heated by a hot-water system and the latter is heated by electric baseboards. We have deliberately kept all other aspects the same to show the changes in the electrical layout due to electric heating.

a. Preliminary Considerations

The first thing we need is a panel schedule. We use the one shown in Figure 16.41, page 970, as previously discussed in Section 13.4. It is an accepted convention to circuit all single-pole circuits first, followed by 2-pole and then, if any, 3-pole

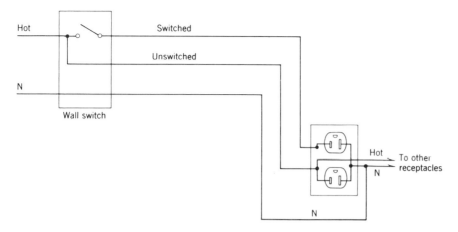

Figure 13.19 Split wiring of a duplex receptacle. Upper half is switch controlled; lower half is hot all the time. This allows wall switch control of a lamp or other device while maintaining part of receptacle for independent use. Note that the receptacle is mounted with its grounding pole up. This is frequently done on the theory that any metallic device falling onto a plug (cap) will contact the grounding pole of the plug and not a hot pole.

circuits. In The Basic House plan, as in most small residences, there are no 3-phase loads and, hence, no 3-pole breakers. A single-phase panel is, therefore, illustrated in Figure 13.20(c). The choice of panel type also depends on power available from the utility, total building load and other factors. This decision is, therefore, generally made by the electrical designer and passed on to the technologist who is doing the job circuiting.

It is also customary that all lighting outlets are circuited first, followed by appliance circuits and individual branch circuits. Although we could use 15-amp branch circuits for lighting, wired with No. 14 AWG wire, we have decided to standardize on No. 12 AWG wire. That then allows us to use 20-amp branch circuits for all lighting and, of course, appliance circuits. The cost savings involved in using No. 14 AWG wire rather than No. 12 AWG is appreciable only in a large house or in a multihouse project. For a single house, the savings are minimal.

Note: An important exception to this combination of No. 12 wire and 20-amp circuit occurs when the permissible current-carrying capacity of the No. 12 wire falls below 20 amp due to derating. Such derating is applied for high ambient temperature (NEC Table 310-16) and/or when a large number of conductors are run in a single raceway (Article 310, note 8 to ampacity tables). Ordinarily, when the ampacity of a conductor does not correspond to the standard rating of a protective device, the next higher setting may be used (Note 9). However, for branch circuit conductors serving receptacles that will feed plug-in devices, the permissible ampacity of the conductors may not be less than the rating of the circuit protective device. Thus, for instance, if No. 12 wire were derated to an ampacity of 18 amp, the circuit breaker protecting the circuit would not exceed 15 amp, and not 20 amp. See NEC Section 210-19(a).

b. Circuiting Guidelines

In accordance with the guideline stated in Section 13.10, we should allow between 530 and 640 ft^2 per 20-amp circuit. The main floor of The Basic House is only 1430 ft^2 gross. We, therefore, need a minimum of three circuits for general lighting and receptacles, assuming that we can reasonably circuit this plan with about 1600 w per circuit. Refer now to Figure 13.20(a). Because of the relatively small size of the house, we find it impossible to follow all the guidelines of Section 13.10. In Section 13.10.h, we say that it is best to avoid combin-

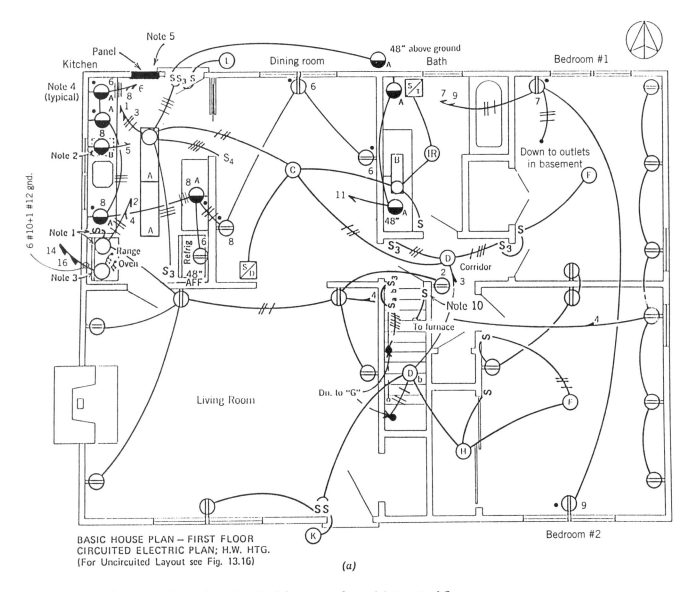

Figure 13.20 The Basic House plan; circuited, hot water heat. *(a)* Circuited floor plan. *(b)* Circuited basement plan. *(c)* Panel schedule.

ing lighting and receptacles on a single circuit. We also say that it is inadvisable to place all the lighting on one circuit. This is so that the failure of one circuit will not black out the entire house. In this building, to place a reasonable load on a circuit, we will have to mix lighting and receptacles, since the main floor lighting totals only about 1000 w.

Note that although convenience outlets (duplex receptacles) do not add to the general lighting load of 3 v-a/ft^2 as noted in Section 13.10.c, in figuring individual circuit capacity they should be counted at no less than 180 w (1.5 amp) each. Thus, six receptacles are 1080 w and eight are 1440 w. These receptacles are the ones not connected to appliance circuits. To show clearly which outlets are intended as appliance outlets, we have placed a dot next to the symbol. This dot should be erased when the circuiting is finished. It simply serves as a reminder during circuiting.

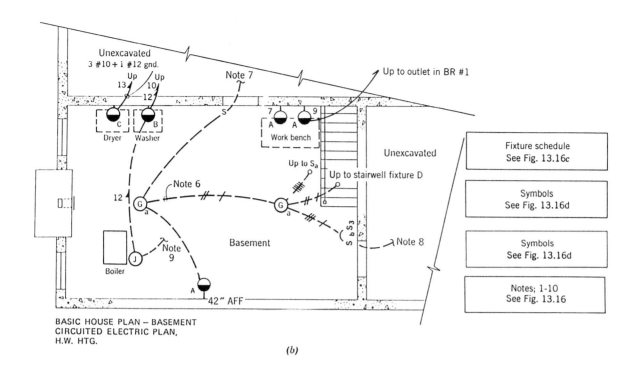

BASIC HOUSE PLAN – BASEMENT
CIRCUITED ELECTRIC PLAN,
H.W. HTG.

(b)

Panel schedule for basic plan of Fig. 13.20						
Circ. no.	Description	LOAD V-A.	Circuit beakers	Load V-A	Description	Circ. no.
1	LTG – { Kit. Dr., Br. #1, Hall, Outside, bath + Φ	1045 1 R	20 1 2 20	30 6 R	Outlets – LR. & corridor + Exh. fan	2
3	LTG – { Outside Lr., Stair, Br. #2, Bsmt. + 4 Φ	935 4 R	20 3 4 20	6 R	Outlets – BR. 1 & 2	4
5	Dishwasher	1500	20 5 6 20	1500	Appliance outlets – Kit., Dr.	6
7	Appliance outlets – BR 1, Bsmt	2 R	20 7 8 20	1500	Appliance outlets – Kit., Dr.	8
9	Appliance outlets – Bsmt., BR 2	2 R	20 9 10 20	1500	Laundry outlet – Bsmt.	10
11	Bathroom outlets	2 R	20 11 12 20	1300	Boiler or furnace	12
13	Spare	—	20 13 14 30 A 2 P	6Kw	Range	14
15	Spare	—	20 15 16			
17	Clothes dryer	5Kw	30 A 2 P 17 18 30 A 2 P	4800w	Oven	16
			19 20			
	Space for 2 – 1P or 1 – 2 pole		21 22		Space for 1 – 2P or 2 – 1 pole	
			23 24			
	Load total – phase A				Load total, phase B	

Panel Data
Mains, GND. BUS: 150 A MNS., 60 A GND. BUS
Main C/B or ~~SW/F~~   150/100
Branch C/B INT. CAP. AMP. 10,000
Mounting – ~~Surf~~/recess
Remarks: Front suitable for painting

Voltage   120/208
1 PH. 3 wire

(c)

Figure 13.20 (Continued)

Since convenience outlets do not count in the building load total but do count in the circuit load, it is necessary to keep a double load record—one for the circuit and one for the building. This is explained in detail in Section 15.8. To avoid confusion with this double load record, it has been our custom to show the panel load as a combination of fixed outlet wattage (actually volt-amperage) plus the number of receptacles. For instance, circuit no. 1, which is detailed later, has 1045 v-a of fixed lighting (including a heat lamp), plus one receptacle. Instead of showing 1045 v-a + 180 v-a or 1225 v-a, we show in the panel schedule 1045 v-a +1R. See Figure 13.20(c). This kind of notation gives more data, because it separates fixed lighting from convenience outlets.

c. Drawing Circuitry—Lighting and Receptacles

Now let us follow the circuiting as it appears in Figure 13.20, keeping in mind that this is only one of many ways that it could be done, and we make no claim to presenting a single best solution. Following the suggestions already listed, we start with lighting and with circuit no. 1, connecting the ceiling lighting outlets in the kitchen, dining room, hall, bedroom and bath. From the kitchen switch, we also feed through to the back patio light fixture L and to the outside receptacle. This gives us a load of

Kitchen ceiling lights	200 v-a
Dining room lights	300 v-a
Bath fixture B	25 v-a
Hall fixture D	120 w max
Bedroom fixture F	150 w max
Back porch	100 w max
Inside total	895 v-a lighting
Outside receptacle	180 v-a
Total	1075 v-a or 895 v-a + 1R

Since this is too light a load for a 20-amp circuit and also since the ceiling heat lamp can be picked up with very little wiring, we add it to the circuit. This gives us a total of 1075 v-a + 150 = 1225 v-a or 1045 v-a + 1R. This information is then placed on the panel schedule in the space reserved for circuit no. 1. The schedule should be ruled up with enough space (3/8 in.) so that two lines of lettering can be used if needed.

Next we consider the remaining lighting in the house, including the basement and unexcavated areas. These are circuited to circuit no. 3, for reasons we will explain later. We start with outlet D in the stairwell and run to fixtures H and F in bedroom no. 2. A second line goes to the switched outlet under the living room window that serves as the lighting outlet in the living room. Then we go to the front lantern and finally down to the basement lights and wall receptacle. This basement wall receptacle was picked up on the lighting circuit so that there is no possibility that a single fault will cause the areas to be without power. This is an important factor in general and all the more so in an isolated area with little or no daylight, such as the basement.

An initial load check for circuit no. 3 shows

Stair fixture D	120 w max
Bedroom fixture F	150 w max
Closet light H (assume 60-w lamp)	60 w
Living room floor lamp (estimated for receptacle instead of normal 180 w)	225 w
Basement ceiling lights	200 w max
Basement unexcavated areas (assume 60-w lamps)	180 w
Maximum total lighting	935 w

Note that we use watts (w) and not volt-amperes (v-a), since all the lighting is incandescent, with a unity power factor. Therefore, watts and volt-amperes are identical. Since it is obvious that the probability of all these lights being on at the same time is just about zero, we can readily add receptacles to this circuit. This we have done by picking up two receptacles in bedroom no. 2 and one in bedroom no. 1. We chose these receptacles because the wiring required to reach them is short and because they present little demand load. By this we mean that it is hard to imagine a situation when the basement, the crawl spaces, the living room and the bedroom light and receptacles are all in use together. Even if they were, the maximum load would be only

Lighting (above)	935 w
Basement receptacle	180 v-a
Bedroom receptacles	540 v-a
Maximum load	1655 v-a

This load (1655 v-a) is still well within the circuit's maximum rating of 1920 v-a. See Section 13.10.b. The load information, that is, 935 v-a lighting and

d. Wiring Paths

At this point, we pause in the actual circuiting to study the reasons for connecting circuits nos. 1 and 3 as we did, in order to explain in detail the technique of circuiting. We began circuit no. 1 at the kitchen outlet, since it is nearest to the panel and thus gives the shortest home-run. A home-run is defined as the wiring between the first outlet on a circuit and the panel. Obviously then, in the interest of economy, the home-run should be taken from an outlet as near as possible to the panel. Since we were picking up lighting on this first circuit, we drew a looped line to the dining room ceiling outlet. From the outlet box at dining room fixture C, several lines emerge. One extends to corridor fixture D and another extends to a wall switch and dimmer. Refer to Figure 13.21 for the basic wiring of a standard single-pole switch. Note carefully that circuiting lines are always drawn curved, to avoid any possibility of confusion with architectural work or equipment. The type of curve used is unimportant.

e. Multiple Point Switching

Corridor fixture D, unlike fixture C, is switched from two points by a method known as three-way switching. The name is somewhat misleading, since the switches are single-pole double throw switches and two of them are used. (See Figure 12.56.) Refer to Figure 13.22, which shows how three-way switching works. Note, however, that the kitchen lighting is switched from three locations—each of the three entrances to the kitchen. This type of switching, which is called four-way switching is more complex than the two-location, three-way switching shown in Figure 13.22. It is explained in Figure 13.23.

Figure 13.24 shows the three alternative methods by which the wiring could be connected between outlets C, D and F in the dining room, corridor and bedroom no. 1, plus their associated switching. We considered all three methods and decided on that shown in Figure 13.24(a), because it is the simplest and cheapest, requiring minimum wiring. Note in Figure 13.24(a) and on the plan, Figure 13.20(a),

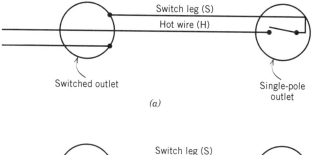

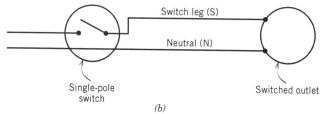

Figure 13.21 Typical standard single-point switching wiring diagrams: (a) Usual arrangement where the hot and neutral wires enter the switched outlet first and (b) where the hot and neutral enter the switch first and feed-through to the outlet. This feed-through arrangement is less desirable because wiring not associated with the switch is present in the switch box. See text for a fuller explanation.

that we feed through from S_3 at the end of the corridor to fixture F, via its wall switch.

f. Feed-Through Wiring

Feed-through between switches is not usually a good idea because it clutters the small switch box with wiring that is not related to the switch. Here, however, it is an excellent arrangement, because the hot leg is already at the switch box and only a run-through neutral leg is required. The same is true for the three switches grouped at the back door in the kitchen. To understand that feed-through is not usually desirable, study the diagram in Figure 13.24(b). There we ran the wiring from dining room outlet C into S_3 at the corridor. Observe how this wiring arrangement requires more wiring than the plan in Figure 13.24(a) and clutters the switch boxes so that a larger than normal box would be required.

The plan in Figure 13.24(c) shows a system whereby all feed-through in switch boxes is eliminated. This is also an acceptable plan, but it is somewhat more expensive than the plan Figure

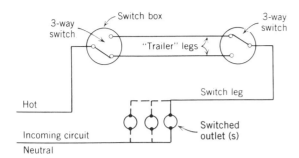

Figure 13.22 Schematic wiring of typical three-way switching arrangement, that is, switching of one (or more) outlets from two points. The switches, which are actually single-pole double-throw, are called three-way because of the three connections to each. Note that two unbroken wires are required between the switches. These wires are variously called trailers, travelers or runners. The circuit is shown with the outlet(s) switched OFF. Changing the position of either switch will energize the outlet(s). The wiring as shown is schematic, since it shows the hot wire entering one outlet and the neutral wire entering another. In actual wiring the hot and neutral always enter one outlet box and are redirected from there. See Figure 13.24.

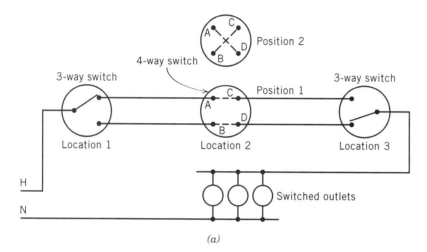

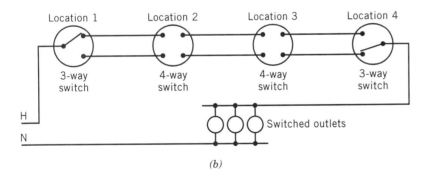

Figure 13.23 *(a)* Schematic wiring diagram of a four-way switching arrangement, that is, switching of one or more outlets from three locations. The system is essentially the same as a three-way (two location) arrangement, with the addition of a four-way (four connections) switch in the two traveler wires that connect the two three-way switches. This four-way switch simply reverses the two traveler (trailer) legs and, by so doing, has the same effect as throwing one of the three-way switches. In position 1, the four-way switch connects the travelers straight through; A to C and B to D. In position 2, it reverses the connections, connecting A to D and B to C. Additional four-way switches connected into the traveler legs in the same fashion can be used to accomplish switching from any number of locations as shown in *(b)*. See Figure 13.24 for a practical application of three-point (location) switching.

770 / BUILDING ELECTRICAL CIRCUITS

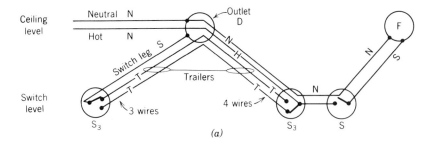

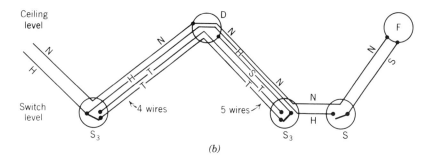

Figure 13.24 *(a)* Wiring between outlets, corresponding to ceiling lights in corridor and bedroom no. 1 of Figure 13.20*(a)* and their switches. A heavy dot indicates an electrical connection or junction. Refer to Figure 13.22 for wiring of three-way (two point) switching. *(b)* Wiring for the same outlets as shown in Figure 13.24*(a)*, except with circuit entering via a switch outlet. The arrangement of *(a)* is clearly preferable. *(c)* Simplified version of wiring between outlets, with wires identified by abbreviations: Hot, Neutral, Switch, Trailer (or traveler). The number of wires is shown by hatch marks (tics). *(d)* Wiring arrangement including fixture B in bath, fed from ceiling junction box J. See discussion in the text. Where the wiring consists of only two wires, no tics are shown. Also, where hatch marks are shown, the neutral is separated from the other marks, as between fixture D and S_3. This is for ease of counting neutrals, which is important, as we will learn. The tics for the two trailer legs can also be separated, as in Figure 13.24*(e)* for further clarity. See the chapter discussion for additional explanation. *(e)* Outlet wiring for kitchen lighting and switching, plus the outside light and receptacle. Note that the feed-through at the triple bank of switches at the back door is the most practical wiring method. For ease of feed-through wiring, these switches are mounted in a three-gang box. The four-way switch is shown schematically between the 2 three-way switches and fed by four trailer wires (two in, two out) as necessary to accomplish the switching. See Figure 13.23 for an explanation of three-point, four-way switching.

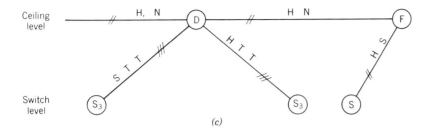

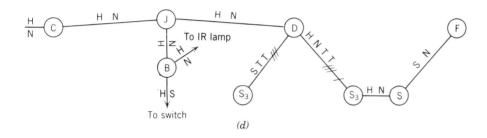

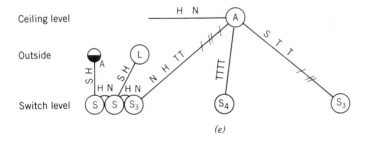

Figure 13.24 *(Continued)*

13.24(a) and was, therefore, not used. It does have the advantage of not running any wiring through switch boxes that is not related to the switches. This is important because switches need periodic replacement, which is made difficult by a cluttered box. Also, when doing such switch replacement in a tight box with old wiring and dried out insulation, a fault may develop. This would not be true if all the wiring not related to the switch were in the ceiling box, where it is normally left untouched for the life of the wiring system.

g. Outlet Box Wiring Capacity

At this point, you may well ask: Why didn't you run the wiring from fixture C to bathroom fixture B and then to D in the corridor? Indeed, that appears to be a good solution at first glance. Look at Figure 13.24(d) to see what it involves. Fixture B is wall mounted, and its box already contains wiring to feed a switch, and an infrared heat lamp. To understand why it is not advisable to add any additional wiring to the box at fixture outlet B, we reproduce in Table 13.1 an excerpt from the NEC that deals with the wiring capacity of outlet boxes. In general, the purpose of restricting the number of wires in these boxes is simply to avoid crowding, which may cause overheating and insulation damage followed by faults. Note, for instance, the capacity allowed for a $3 \times 2 \times 2^{3}/_{4}$-in. device box.

In such a box containing a wiring device and cable clamps for BX, the allowable capacity is reduced to three wires. This is exactly what is required when using such a box to switch a hot leg and also to feed through a neutral. However, if the device were a three-way switch or we were feeding through a hot leg and a neutral (as we did, for instance, at the kitchen door switches), larger single boxes or gang assemblies would be used. In the case at hand, the box behind fixture B, which is normally a $4 \times 1^{1}/_{2}$-in. round, has a nominal capacity of six wires. Deducting one from this for cable clamps leaves a capacity of five wires. But, as you can plainly see from the diagram, we already have six wires in the box, requiring us to use at least a $2^{1}/_{8}$-in. deep box. Any additional wiring would require a collar on the box to deepen it. This is unusual in a wall box, although it is frequently done in ceiling boxes to give them additional capacity. Therefore, we would need to add a ceiling junction box as shown in Figure 13.24(d). The cost of installing this box is greater than the cost of the additional 4 or 5 ft of BX cable in the layout in

Table 13.1 Wiring Capacity of Outlet Boxes

Box Dimensions		Maximum Number of
Inches	Trade Size	No. 12 AWG Conductors
$4 \times 1^{1}/_{4}$	round, octagonal	5
$4 \times 1^{1}/_{2}$	round, octagonal	6
$4 \times 2^{1}/_{8}$	round, octagonal	9
$4 \times 1^{1}/_{4}$	square	8
$4 \times 1^{1}/_{2}$	square	9
$4 \times 2^{1}/_{8}$	square	13
$3 \times 2 \times 1^{1}/_{2}$	device	3
$3 \times 2 \times 2$	device	4
$3 \times 2 \times 2^{1}/_{2}$	device	5
$3 \times 2 \times 2^{3}/_{4}$	device	6

Source. Reprinted with permission from NFPA 70-1996, the *National Electrical Code* ®, Copyright © 1995, National Fire Protection Association, Quincy, MA 02269. This reprinted material is not the complete and official position of the National Fire Protection Association, on the referenced subject which is represented only by the standard in its entirety.

This table is extracted from Table 370-16(a) of the NEC, 1996 edition. In using this table, note that a deduction must be made where the box contains fixture studs, cable clamps, wiring devices, grounding conductors, and other devices. For wiring capacity in these situations, see the NEC Section 370-16. A wire running through the box without a splice is counted as a single wire. All other wires entering the box are counted individually.

Figure 13.24(a). Note also that to use the layout in Figure 13.24(d) without ceiling box J, we would have to go down to outlet B and up again, thus losing some of the advantage we are attempting to gain in shortening the runs.

h. Wire Count

By means of small diagrams of the type shown in Figure 13.23, the technologist can arrive at the number of wires required in each run. Then by using hatch marks, called tics, any run with more than two wires is shown. See Figure 13.20. After some practice, the technologist can use single-line-type sketches as in Figure 13.23(c–e), to show the wiring by abbreviation, without actually drawing the wires as in Figures 13.24*(a)* and *(b)*. Finally, with enough experience, the layout person will be able to work out the choices and wiring mentally, without the necessity of making these little scratch paper diagrams at all, except in the most complex cases.

Summing up, thus far we have learned that:

- Small interconnection wiring sketches help us determine best circuit routing and the number of wires in each run.
- Feeding circuits through switch boxes is generally not advisable but may be used to advantage in specific cases.
- It is good practice to limit the number of cables entering a wall box to three and in ceiling boxes to four. When using more than that number, a count of wires should be made to see whether the additional box capacity that will be needed will fit into the structure without creating space or construction problems.

i. Load Balancing and Neutral Loads

Continuing with the analysis, let us examine the wiring of circuit no. 3. First note that we skipped circuit no. 2 temporarily and passed on to circuit no. 3, in order to reach phase B in the panel, as we explained in Section 13.4. Notice on the panel schedule, Figure 13.20*(c)*, that circuit no. 2 is on the same phase as circuit no. 1, phase A. If we had used circuit no. 2 for the second lighting circuit, we would have ended up with the two lighting circuits on the same phase. This is undesirable for two reasons. First, it is good practice to balance the loads on the panel to the extent possible. This means that not only must the sum of the connected loads on the individual phases be approximately equal, but the actual demand loads, that is, the loads actually being drawn, also must be balanced to the extent possible. Since lighting loads are by nature much more "continuous" than receptacle loads, we attempt to spread the lighting evenly over the phases. (From the NEC point of view, lighting circuits carry continuous load.) Thus we place the second lighting circuit on the other phase. The first available circuit on this second phase, Phase B, is circuit no. 3.

The second reason for choosing circuit no. 3 is a technical one. We are permitted to carry two circuits on a single neutral wire, provided that the circuits are on different phases. If circuit nos. 1 and 3 are connected on phases A and B, respectively, we are permitted to use three wires—two hots and a single, common neutral. (See the hatch marks on the home-run from fixture A.) On the other hand, if we had taken the second lighting circuit from A phase as well, that is, if we had used circuit no. 2, we would need two neutral wires, or four wires in all. This obviously is more expensive and is to be avoided as an unnecessary expense and, therefore, poor practice.

The reason that we are allowed this economy in running neutrals is highly technical and will not be discussed in detail except to say that the single neutral for two circuits on different phases carries no more current than one of the two circuits and, therefore, is not overloaded. As we will discuss in more detail later, this same principal is followed when wiring three-phase panels. Three circuits, all on different phases, can be run with only a single neutral, without any danger of overload. In fact, if the loads are balanced (equal), the neutral current is zero.

We have now learned two more principles of circuiting:

- Circuit loading should be balanced between phases to the extent possible.
- A single neutral will carry up to three circuits provided that these circuits are taken from different phases.

j. Additional Circuitry Details

Now let us return to Figure 13.20*(a)*, circuit no. 3. The loading of this circuit was explained previously. Here again as in circuit no. 1, we use a feed-through technique at the front door switches to feed the outside lantern and the wall outlet. Feed-through of this type to serve an outside light is fairly standard. The wiring coming from stairwell fixture D, to bedroom no. 2 fixtures F and H, plus switches and receptacles, can be arranged in a number of ways. Three of these are illustrated in Figure 13.25, along with the corresponding wiring-routing diagrams with which we have already become familiar. Let us analyze those diagrams and find out which presents the best arrangement and what special considerations are involved.

Wiring as in Figure 13.25*(a)* is a straightforward arrangement using single runs of two-conductor BX. The only disadvantage is the amount of cable used because of the duplication of runs between feeds and switch legs. Look at the runs to fixtures F and H and their switch legs to see what we mean

774 / BUILDING ELECTRICAL CIRCUITS

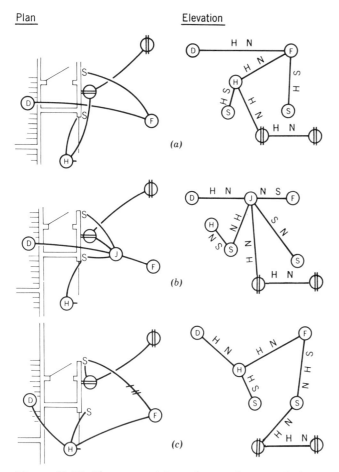

Figure 13.25 Alternate wiring schemes for part of The Basic House plan as shown on Figure 13.20. See discussion in text. Note that the diagrams on the left show the outlets in plan view, whereas the wiring runs on the right are shown in elevation, that is, ceiling level, closet wall level (fixture H), switch level and receptacle level. This enables you to understand where vertical runs are required.

by this comment. This somewhat objectionable condition is corrected in the arrangement of Figure 13.25(b). Note how the addition of a junction box (also called a J box) shortens the runs and avoids the back-and-forth wiring of Figure 13.25(a).

There is a disadvantage here also. The J box has ten wires in it, which if we count the cable clamp as one wire, requires a 4-in. box with extension. All boxes require access. Therefore, unless the house has an unfinished attic, we cannot use this arrangement, unless we are willing to put up with a capped outlet stuck into the ceiling. If we ask why we do not feed across from one switch to another and avoid running back up to the box, the answer lies in the type of construction. We are assuming here a wood stud wall, which requires drilling the studs to make a horizontal run. Vertical runs are much easier and cheaper, and runs in an unfinished ceiling space can be run along boards on top of the ceiling beams. The arrangement of Figure 13.25(c) is the system we have chosen, since in our opinion it offers the shortest runs, does not require junction boxes in the ceiling and has only a short horizontal run, which is more than offset in cost by the cable saved. You are invited to try other arrangements to see if you can come up with a better one than any of those illustrated.

The last area that will be considered here in detail in The Basic House plan is the wiring to stair fixture D. In Figure 13.26, we have drawn three different wiring layouts, each with merits and drawbacks. The layout of Figure 13.26(a) has the advantage that there is no backtracking of wiring. That is, we drop down from the first (main) floor ceiling fixture D to the switches, from there to the ceiling of the basement and, then, to the switches in the basement. We never go back up and down again; therefore, we are being economical with cable. The disadvantage here is the heavy wiring runs between the two banks of switches and fixture G. In one case, we have six wires, and in the other, five. This means two three-conductor cables. Also we have some feed-through at switches, but this is only a neutral wire and is not objectionable.

In the layout in Figure 13.26(b), on the plus side, we have a straight short drop from fixture D to the basement switches. Looking at the plan, we see that this is almost a vertical run, with a turn at the ceiling—an excellent way to wire. This run unloads fixture G, so that we can feed the basement receptacle and the far side switch directly from here without running to the second G fixture as in the layouts in Figures 13.26(a) and (c). On the minus side, we have a double box at fixture D to accommodate the 13 wires in it. (The fixture stud and cable clamps count as two more wires.) The two travelers of the three-way switching arrangement count only once each, since they run through without a splice. Fixture D is in the ceiling of a stairwell—an extremely difficult place to reach if any repairs are required. Therefore, the less wiring in the outlet box at D, the better, unless access from an attic is provided.

Finally in the layout in Figure 13.26(c), on the

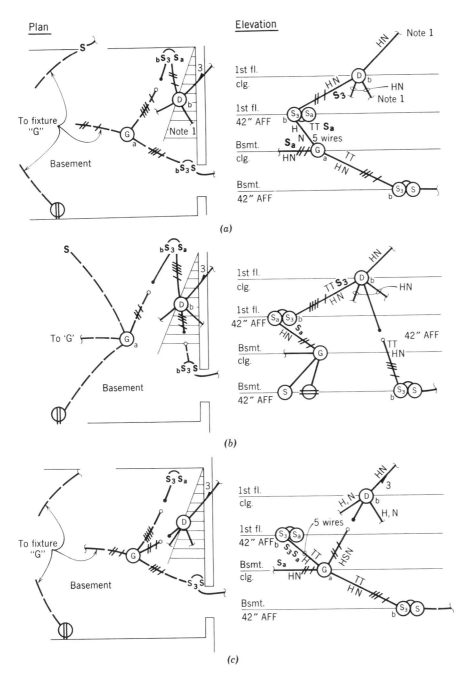

Note 1. Outlet D has 3 cable entries (from corridor fixture D, BR #2 fixture H and living room switches) which are shown for the sake of completeness. Since they do not affect, nor are they affected by, the various arrangements (a), (b) and (c), they would ordinarily be omitted.

Figure 13.26 Three possible wiring arrangements for outlets appearing on The Basic House plan of Figure 13.20. See chapter discussion for a comparison of the three layouts. Observe that both the plan layouts and elevation are necessary to enable you to decide on the best layout. The plan gives distances and physical arrangement, and the wiring sketch shows elevation, connections and number of wires.

plus side, we have no feed-through at switches, minimal wiring at D, and heavy wiring at fixture G, which is highly accessible. The drawback is some duplication of runs, in that we feed up from G to both D and to the switches. It would appear that no layout is obviously superior to the other two. We leave it to you, as an exercise, to evaluate all three layouts in detail and to make a recommendation. We have chosen the layout in Figure 13.26(c) to show on the drawing of The Basic House plan (Figure 13.20).

13.12 Circuitry of the Basic House Plan

We now describe the remainder of the circuitry for The Basic House plan. Make any wiring sketches that you think necessary to follow the description and wiring as shown.

(a) Circuit no. 1 is fully described in the preceding section, covering approximately one-half the lighting in the house.

(b) Circuit no. 3, like circuit no. 1 described in the previous section, is essentially a lighting circuit, covering the remainder of The Basic House plan. It is connected at its closest point to circuit no. 1, that is, from fixture D in the stairwell to fixture D in the corridor. The single arrowhead on this run indicates that circuit no. 3 joins the wiring at this point and is carried with circuit no. 1 through the dining room and kitchen to the combined home-run from the A fixture outlet box. Being on separate phases, the two circuits can be carried with a single neutral as explained previously. Therefore, the home-run is three wires—a hot leg each for circuits 1 and 3 and a neutral.

(c) Circuits nos. 2 and 4 are convenience outlet circuits arranged to cover a number of rooms. Both circuits carry six receptacles, although up to ten can be safely connected to a 20-amp circuit in accordance with the preceding guidelines. Note that the receptacle in the corridor is marked with the figure "2," denoting that it is connected to circuit 2, since otherwise it would not be clear whether it is on circuit 2 or 4. Circuit no. 2 also carries the kitchen exhaust fan, rated 30 w.

Each room of the house has parts of at least two, and normally three, circuits. Thus, no room can be easily blacked out, even if one of the two phases goes out. Although such an occurrence is relatively rare, it happens occasionally as a result of power company outages. Note that the panel schedule shows circuits nos. 2 and 4 as receptacle circuits with a number of receptacles rather than showing a wattage. This is because, as stated previously, convenience outlets are part of the general lighting load of 3 v-a/ft^2. Load calculation is not part of our study here. Consult the NEC for information on this point. Refer to Sections 13.10 and 15.8 for a brief explanation of receptacle loads.

(d) Circuits 6 and 8 are the two small-appliance circuits required by the NEC to feed outlets in the kitchen and dining room. We have arranged the circuits so that the principal counter space on the outside wall has parts of both circuits. Also, the dining room outlets are on both circuits to avoid the possibility of losing all the outlets if one circuit faults. Indeed, in view of the heavy loads taken by kitchen appliances such as microwave ovens, toasters and toaster ovens and the fact that they are frequently used together, a circuiting plan using a third kitchen–dining room appliance circuit would be very reasonable. However, since the dining room outlets are rarely loaded at the same time as the kitchen outlets, two circuits will usually be satisfactory.

The load for these two circuits is recorded as 1500 v-a as required by the NEC, Section 220-16(a). We also indicate on the panel schedule the number of outlets on each circuit, for completeness. The dishwasher, circuit no. 5, is on an individual branch circuit. See Figure 13.18. This completes circuitry of the single-pole kitchen and dining room outlets. The oven and range require multipole circuit breakers and are, therefore, handled after all the single-pole circuiting is complete.

(e) Circuits 7 and 10 pick up the appliance-type outlets in the two bedrooms and the two outlets at the workbench in the basement. The reason for this arrangement is twofold:

- If the climate is such that room air conditioners are required (assuming, of course, that the house is not centrally air-conditioned) then both bedrooms may have such units. A single 20-amp circuit might not be sufficient, particularly in handling coincident starting loads. However, the possibility of heavy loads

at the worktable while the bedroom air conditioning unit is running is remote.
- Since many workbench tools are motor driven and some have heavy electrical loads, it is a good idea to have the two outlets supplied by different circuits.

(f) Circuit 11 feeds the bathroom GFCI receptacles, as required by **NEC** Article 210-52(d). No other outlets are permitted on this 20 amp circuit.

(g) Circuits 10 and 12 are special circuits, as noted on the panel schedule, devoted to the laundry and the boiler (or furnace), respectively. The laundry load is shown as 1500 v-a as required by the **NEC**. The 1300-v-a load for the heating plant is obtained from the project's HVAC technologist.

(h) Looking ahead, we can anticipate that the panel will have at least 20 poles. Most manufacturers make panels in standard sizes of 18, 30 and 42 poles, and a few make a 24-pole panel. In any case, having exceeded 18 poles, it is a good idea to include at least two spare single-pole circuit breakers. This is indicated on the panel as circuits 13 and 15. We can now proceed to circuit the loads that take multipole circuit breakers.

(i) Circuits nos. 14, 16 and 17 are individual branch circuits for single items of electrical equipment. We have assumed here a 6 kw four-burner electric range top plus an 4.8 kw electric oven, since this combination is very frequently found in actual use. Obviously, if gas cooking were used in either of these places, the circuit would be eliminated, and the panel would be changed to suit. The wiring shown on the drawings for these circuits and the size of the required circuit breakers would be arrived at by the technologist in consultation with the project's electrical designer. Refer to **NEC** Articles 210-19(b) and 220-19 for information and circuitry requirements for cooking equipment. Notice that circuit 14 uses poles 14 and 16, circuit 16 uses poles 18 and 20, and circuit 17 uses poles 17 and 19. It should now be perfectly clear that, although only circuit numbers appear on the actual physical panel, pole numbers are required to do the circuitry correctly.

This completes the building's electrical circuitry. However, considering the constantly increasing electrical loads in American homes, it is a good idea to leave at least one 2-pole space. In our panel, we have room for two such spaces, and they are so indicated on the panel schedule. Each such space can accommodate either one 2-pole circuit breaker or two single-pole breakers. This brings the panel up to 24 poles. The remaining panel data relating to buses, circuit breaker interrupting capacity, mounting, and main circuit breaker and panel voltage is filled in by the technologist from data supplied by the engineer.

13.13 The Basic House Plan; Electric Heat

The Basic House plan with electric heat is identical to that of Figure 13.20, with the addition of heaters and thermostats as shown on the uncircuited plan of Figure 13.17(a) and (b). It will, therefore, not be duplicated here. Instead, we will discuss the changes that the technologist and designer/engineer would make to that plan to make it suitable for electric heating. These changes are shown on Figure 13.27.

The designer has decided to add a separate panel to the house to feed the electric heaters—a common practice. He or she, therefore, splits the incoming line at the service entrance point in the kitchen wall and runs a feeder to the electric heating panel (EH), which is centrally located in the house. The single-line diagram shows this arrangement. The switch and feeder sizes have been calculated by the designer and given to the technologist. To avoid cluttering the drawing, the designer and technologist together decide to handle the circuitry of the electric heaters by using the electric heating panel schedule EH and small wiring diagrams on the drawing. See Figure 13.27(b).

The HVAC designer selects the Btu rating (heat output) for each heater (see Figure 4.11, page 180) and tabulates the results on the HVAC drawings. See Table 4.1. In consultation with the electrical designer, the HVAC designer selects the voltage rating of the heaters. Although 120-v heaters are available commercially, it is generally a good idea to use 208- or 240-v units, since they require lighter wiring and, of course, 2-pole circuit breakers. For The Basic House plan, we assumed that the electrical service consists of 2-phase legs of a 3-phase system. This gives a 2-pole line voltage of 208 v, as explained in Section 11.14, and not 240 v as might be expected. The difference is important because the heaters take more current at 208 v than at

778 / BUILDING ELECTRICAL CIRCUITS

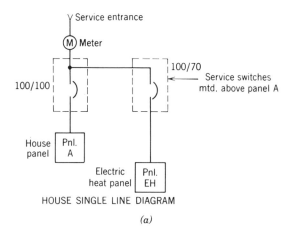

Figure 13.27 (a) Single-line diagram showing service and main feeders to the house panel A and the electrical heat panel EH.

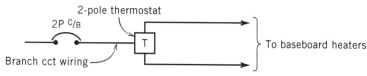

Wiring of multiple heaters controlled by a single thermostat
(living room, bedroom #2, bedroom #1, basement)

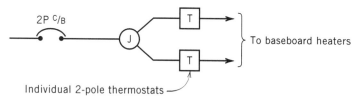

Wiring of multiple thermostats, each of which controls one or more heaters
(kitchen and dining room, bath)

Notes: 1. Bedroom #1 has 2 heaters controlled by a single thermostat, and is fed from the same circuit as the bathroom.
2. Kitchen heater has an integral (built-in) thermostat

(b)

Figure 13.27 (b) Wiring diagrams for baseboard heaters.

	ELECTRIC PANEL EH Electric Heating for Basic Plan						
Circ. no.	Load description	LOAD V-A	\multicolumn{2}{c|}{Circuit beakers}	Load V-A	Load description	Circ. no.
1	Kitchen, dining room	2750	20 A 1|2 2P	15 A 3|4 2P	2500	Bath, bedroom 1	2
3	Living room	3000	20 A 5|6 2P	15 A 7|8 2P	2250	Bedroom 2	4
5	Space only	—	9|10	15 A 11|12 2P	1500	Basement	6
	Load total, phase A	5750V-A	A	B	6250VA	Load total, phase B	
	Panel data: 100 A MNS, 60 A Gnd. bus Lugs in mains only All branch circuit breakers: 100 AF/2 Pole, 10000 A I.C. Flush mount, cover suitable for painting						

(c)

Figure 13.27 (c) Panel schedule for panel EH. Note that the loads on Phases A and B are nearly balanced.

240 v, and that additional current could require a larger circuit breaker and/or larger wiring.

The heater wiring is fairly straightforward. The HVAC designer has elected to use remote line voltage thermostats except for the kitchen unit. These remote thermostats are 2-pole and are rated 240 v. Two small wiring diagrams are shown on the drawings that cover the two heater wiring situations:

- Circuit supplies multiple heaters controlled by a single thermostat, as in the living room, bedrooms and basement.
- Circuit supplies multiple heaters controlled by individual thermostats as in the remainder of the house.

We call your attention to the Code requirement that all permanently connected appliances rated over 300 w must have a disconnect device. (Article 422.21). This device must disconnect all ungrounded conductors feeding the appliance. In the case of the baseboard heaters, a low voltage thermostat controlling a relay would not meet this Code requirement. In such a case, a separate disconnect switch within sight of the heaters would be required. This requirement is obviously a safety precaution. Its purpose is to protect a person working on the heaters from electrical shock. For this reason, the Code permits the disconnect device to be omitted where either the branch circuit switch or circuit breaker is in sight of the heater (or any other appliance covered by this Code section) or the branch circuit breaker or switch can be locked open.

All wiring is No. 12 AWG; therefore, nothing special is shown on the drawings. Showing the actual wiring runs to the heaters would add very little to the information required by the installing electrician and would simply clutter the drawings. The information on panel schedule EH and the two single-line wiring diagrams referred to previously are sufficient.

The house panel remains the same as in Figure 13.20(c) except that circuit 12 becomes a spare, as a result of our eliminating the boiler/furnace. This also means that the home-run from the laundry outlet in the basement contains only circuit 10, since the circuit 12 run to the boiler does not exist. With this we complete our analysis of The Basic House electrical plan, which was presented as a study example and not as a proposed house design. If you have followed the discussion carefully, you should now be in a position to do residential-type circuitry independently from the job engineer. In the chapters that follow, we discuss more specialized wiring and other types of buildings that present particular drawing problems and situations. Also, we consider some of the design factors of residential buildings, to give the technologist a firm background in this subject.

Key Terms

Having completed the study of this chapter you should be familiar with the following key terms. If any appear unfamiliar or not entirely clear, you should review the section in which these terms appear. All key terms are listed in the index to assist you in locating the relevant text. Some of the terms also appear in the glossary at the end of the book.

2-wire, 3-wire, 4-wire panels
3-way, 4-way switching
Appliance (branch) circuits
Circuit breaker
Clearing time
Device layout plan
Dual-element fuse
Equipment ground, equipment ground bus
Frame size
General lighting (branch) circuits
Green ground

Ground fault
Ground fault circuit interrupter (GFCI)
Grounded neutral
Grounding electrode
Home-run
Individual branch circuit
Interrupting capacity
Isolated ground
Lighting fixture schedule
Molded-case circuit breaker
Overcurrent protection

Panel pole number, panel circuit numbers
Panel schedule
Panels, panelboards
Quick-make, quick-break
Renewable and nonrenewable fuses

Switch and fuse panel
Trailer, traveler, runner wires
Trip-free
Trip setting

Supplementary Reading

B. Stein and J. R. Reynolds, *Mechanical and Electrical Equipment for Buildings*, 8th ed., Wiley, New York.

C. G. Ramsey and H. R. Sleeper, 1988. *Architectural Graphic Standards*, 8th ed., Wiley, New York.

Problems

1. Draw a one-line diagram of a branch circuit consisting of a 20-amp circuit breaker, a No. 12 feeder and a 1500-w load. If you feel any additional data are necessary, add them.
2. Draw in single line or block diagram form a circuit, properly showing the following items, with respect to each other: service entrance feeder, wall receptacle, house panel, main service switch, overhead room light, panel branch circuit switch and fuse.
3.
    a. List all the fuse sizes and circuit breaker trip sizes that would protect a wire of 50-amp capacity from overload. Use the NEC or the list in Section 13.2 for sizes.
    b. Which of these sizes would also be useful if this 50-amp feeder were carrying a 40-amp load?
4. Draw schematically a single-phase 120/240-v 16-circuit panelboard. Show 100-amp mains, a 100/70-amp main circuit breaker, and these branch circuit breakers: 8-SP, 20A; 4-SP, 30A; 2-2P, 20A and 2-2P, 30A.
5. On what phase of a single-phase 120/208-v panel are these poles: 1, 4, 8, 10? Answer the same question for a three-phase panel. Use phases A and B for single phase and A-B-C for three phase.
6. In residential wiring:
    a. What is the square foot load allowance for general lighting?
    b. What is the minimum number of appliance circuits?
    c. What is the minimum number of laundry circuits?
    d. Define an appliance circuit, a general-purpose branch circuit and an individual branch circuit.
7. What is the purpose of appliance outlets?
8. Where are GFCI receptacles required by the NEC? Where else would you recommend them?
9. In figuring circuit loads, what v-a load is used for a duplex receptacle? An appliance circuit? (See the NEC.)
10. What is the maximum to which a panel circuit breaker carrying continuous load can be loaded according to the NEC: 50, 60, 80 or 100%? What does this mean in volt-amperes on a 120-v, 20-amp circuit?
11. a. Using the information given in Figure 13.23, show the actual wiring of the four-way switching of the kitchen lighting in Figure 13.20(a).
    b. Show a space with four entrances, a single-ceiling outlet and a switch at each en-

trance. Show complete wiring diagram that will furnish ON-OFF control at each entrance. Include number of wires in each run and indicate switch types. Indicate also the point at which the branch circuit conductors (hot and neutral) are connected.

12. In Figure 13.26, select the wiring arrangement you think is best and explain why. If you can improve on it, do so, also explaining why.

13. A client has examined the electrical plan of Figure 13.16 and requests the following changes be made to the electrical layout, before construction begins.

    a. Kitchen—add a light over the sink and an outlet under the sink for a garbage disposal unit.

    b. Dining room—arrange the ceiling outlet for three-way switching from the two doorways.

    c. Living room—arrange three-way switching for the switch-controlled receptacle, from the two doorways.

    d. Bathroom—add a night-light.

Make all the necessary alterations to Figure 13.20 to accommodate these changes, and add devices, such as switches, made necessary by these changes. Make all appropriate changes in the panel schedule, fixture schedule, notes and the like.

14. How Light Behaves; Lighting Fundamentals

There is obviously a lot more to lighting than simply locating ceiling and wall lighting outlets. So much more, in fact, that lighting design has become a specialty. Much building lighting design work is done at present by the building electrical designer with the assistance of a technologist, and much of this design work is very well done. Once the technologist has mastered the fundamentals of lighting, he or she can pursue its technical and artistic aspects, to the extent of his or her ability. This chapter is devoted to a study of the basics of lighting. It is divided into three parts: how light behaves, how light is produced and how light is used. In the course of this study, we will learn about light sources, illumination levels and lighting fixtures. This information, coupled with the knowledge the technologist has already obtained about building circuits, will give the necessary background to approach an overall building electrical layout. After studying this chapter you will be able to:

1. Understand the fundamentals of the behavior of light, including reflection, transmission and diffusion.
2. Distinguish among the factors that affect the quantity and the quality of light.
3. Calculate illumination in both conventional and SI (metric) units, and convert between the two systems.
4. Understand the effect of luminance ratios, contrast and glare on the quality of a lighting installation.
5. Understand the operating and illumination characteristics of all the major light sources.
6. Select a light source for an installation that will give proper quantity and quality of light, along with operating economy.
7. Design uniform lighting for interior spaces, given illumination requirements.

8. Lay out a uniform lighting system that will produce minimum direct and reflected glare.
9. Draw details of lighting fixtures and architectural lighting elements.
10. Perform illumination and reflectance measurements using conventional meters.

14.1 Reflection of Light

You are able to read this book because light reflected from the page enters your eye. This process is illustrated in Figure 14.1. Reflection is one aspect of the behavior of light that is of particular interest to us. Other factors are absorption, transmission and the particular way in which these processes occur. However, we are not physicists studying light as a form of energy. We are principally interested in how to apply light; in other words, we are interested in illumination. We will, therefore, discuss the five factors that affect our ability to see clearly as well as items related to these factors. The five factors are:

- Luminance.
- Contrast.
- Glare.
- Diffuseness.
- Color.

When light falls on an opaque (non-light-transmitting) object, some of it is reflected, and some of it is absorbed. The ratio between the amount of light reflected and the original amount of light is called the *reflection factor* or, using the more modern term, *reflectance*. The reflection factor of an ordinary mirror is quite high—90% or more. The paper on which this book is printed has a reflectance of about 75%. The light that is not reflected is absorbed by the opaque material and is lost. Therefore, in order for a *lighting fixture (luminaire)* to be efficient, its interior surfaces must be treated to give high reflectance, that is, to have minimum light loss. Actually, the glossy white enamel paint found on the inside of many fluorescent fixtures has a reflectance of about 88%. That means that 12% of the light is lost and 88% of the light from the lamps is reflected and emitted as useful light. Figure 14.2 illustrates this concept.

Although light reflection is obviously necessary to the act of seeing, it can also be disturbing if it is mirrorlike, or what is technically termed *specular reflection*. In specular reflection, the source of light is reflected in the object at which we are looking, causing glare. That is why reading a magazine printed on glossy paper can be very troublesome if the light source is not placed properly. We will have more to say on this subject when we study glare and glare control.

If the surface of the object being viewed is not

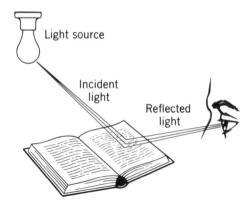

Figure 14.1 We see most objects by reflected light. In the illustration, the ability to see the words on the page is a result of light being reflected from the book onto the eye. Light-emitting objects, such as the light source itself, are seen directly by the light coming from the source, and not by reflection.

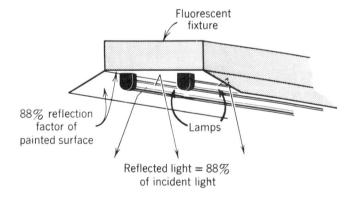

Figure 14.2 The light output of the lamps in the illustrated luminaire is reflected from the inside fixture surfaces as shown. A 12% loss results if the reflectance of the surfaces is 88%. (We are ignoring, for the moment, the losses that result from trapped light and multiple reflections.)

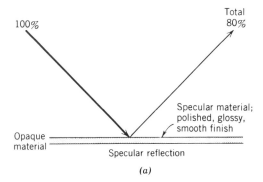
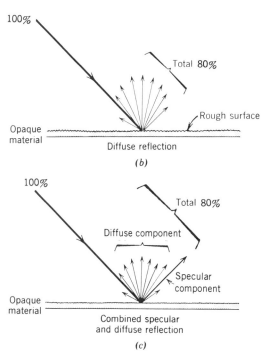

Figure 14.3 The three types of reflection of light from an opaque (non-light-transmitting) surface are illustrated. Notice that in all three cases the total amount of light reflected (80%) is the same.
(a) Mirrorlike or specular reflection from a glossy, polished surface. This type of reflection mirrors the light source and frequently causes glare. Perfect specular reflection does not exist. Optical mirrors such as those used in high quality telescopes approach 99% specular reflectance. Ordinary mirrors have a specular reflectance between 80 and 90%.
(b) In diffuse reflection, light is spread in many directions. This type of reflection, which is found in all flat or matte finishes, is easy on the eye and reduces glare. White plaster and matte-finish white paint have diffuse reflectances of between 75 and 90%.
(c) Most materials show a combination of specular and diffuse reflection. If the specular part of the total reflection is large, as it is with glossy paper, reflected glare can result. If the diffuse portion is large, as it is with the paper on which these words are printed, no glare results, and reading is easy and efficient. See Section 14.8 for a full discussion of glare factors.

glossy (specular), we get a type of reflection that does not interfere with the seeing process. It is called *diffuse reflection*. This is the type of reflection given by a dull, flat finish surface. The difference between diffuse and specular reflection can readily be seen by comparing the appearance of matte finish and glossy finish photographs, particularly when the photos are held in a position that mirrors the source of light. (Actually, most materials give both diffuse and specular reflection, but one kind of reflection is more pronounced than the other.) See Figure 14.3 for a diagrammatic illustration of these two types of reflection.

14.2 Light Transmission

We are probably as familiar with this characteristic of light as we are with reflection. For example, sunlight comes in through the window; light comes from the inside of a frosted incandescent lamp; light comes through the plastic lens of the fluorescent fixture. Just as with reflection, the ratio between the incident light and the transmitted light is called the *transmission factor*, or simply *transmittance*. (Transmittance and reflectance are the preferred terms in the lighting profession.)

As with reflection, we have diffuse and nondif-

786 / HOW LIGHT BEHAVES; LIGHTING FUNDAMENTALS

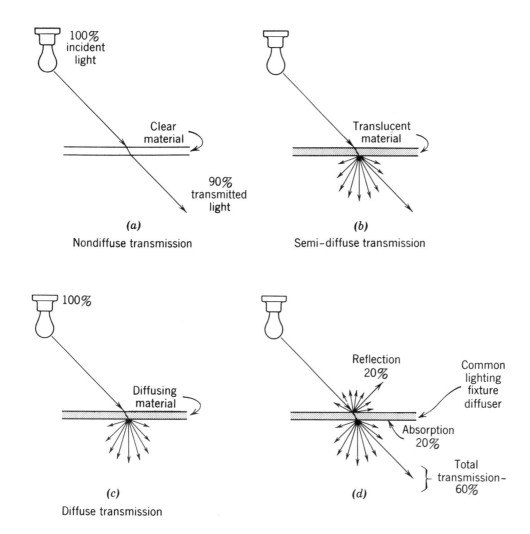

Figure 14.4 (a) Light transmission through different materials. Clear, transparent materials transmit about 90% of incident light.
(b) Instead of the clear glass or plastic of (a), the diffusing medium here is a translucent glass or plastic. It is called semi-diffusing, since the source can be seen somewhat; that is, there is some nondiffuse transmission. The light source can be "read" through the diffuser.
(c) Complete diffusion results in an even brightness on the diffuser and complete hiding of the light source. Milk white glass has this effect.
(d) With most diffusing materials, there is some reflection from the top surface, some absorption in the material, some direct transmission, and some diffuse transmission. For common, commercial light diffusers, a figure of 60% for total transmission of light is reasonable.

fuse transmission of light. See Figure 14.4 for simple ray diagrams that illustrate these two types of transmission. A piece of clear glass or plastic shows almost complete nondiffuse transmission with very little absorption or reflection. A piece of translucent material, however, such as frosted or white glass, milky plexiglass or tissue paper, gives diffuse transmission, low reflection, but relatively high absorption. Frequently, high absorption is the price paid for good diffusion. However, inside-frosted incandescent lamps have only a 2% loss due to the frosting, but give almost perfect diffusion. Unfortunately, the ecological problems caused by the acids used in the frosting process are so great that inside-frosted lamps are disappearing from the market. In their place, coated glass is used, which has slightly higher loss and excellent diffusion.

The light transmission properties of a material are extremely important when that material is to be used to cover the lamps in a lighting fixture. Since the purpose of such a cover is to conceal the lamps (and to keep them clean), we need a material that shows high diffuse transmission. This diffuseness will prevent "reading" (seeing) the lamps through the cover, and the high percentage of transmission will prevent loss of light and maintain high overall luminaire efficiency. These luminaire covers are generally referred to as *diffusers* because of their diffusing action in transmitting the light generated by the lamp inside the unit. The materials generally used for diffusers are various types of glass and plastic that fulfill these requirements fairly well. The topic of diffusers is covered in Section 14.24.

14.3 Light and Vision

In Section 14.1 we mentioned some of the illumination factors that affect how well we see. We say illumination factors because the other principal factor is the eye itself, and that is not our subject. Our concern is to provide the best possible illumination within our budget limits. Also, when we speak of illumination, or simply lighting, we are referring to man-made, electrical lighting. Daylighting is not included in our study here. We must assume a nighttime condition.

When lighting designers talk about lighting, they refer to two things—the quantity and the quality of lighting. The first, quantity, can be calculated and measured and is relatively easy to handle. The second item, quality, is a mixture of all the items related to illumination other than quantity of light. This mixture includes surface luminances, brightness, subjective luminance ratios, contrast, glare, diffuseness and color. In addition, most designers include in a definition of lighting quality, items such as psychological reactions to color, and fixture patterns. All these terms will be defined and discussed in some detail in the sections in this chapter that are devoted to lighting quality.

14.4 Quantity of Light

It is somewhat difficult to speak of the "quantity" of light as if it were an item that can be boxed or bottled. However, we have already overcome this type of difficulty in our study of heat. There, we spoke of the amount of heat generated or lost, as measured in Btu and Btu per hour. Here we are dealing with light, which is simply another form of energy. Quantity of light, assuming continuous light production, is measured by a unit called *lumens*, abbreviated *lm*. This unit is analogous to Btuh in heating and watts in electricity and represents energy per unit time, that is, power.

There is, however, a major difference between this unit of power (lumens) and, say, Btuh. The latter is an objective, physical unit, independent of human reactions to heat. Lumens, on the other hand, are light power as determined by the reaction of a normal human eye, that is, as understood by us. The exact scientific definition of a lumen is beyond our scope here. It is important, however, for the technologist to remember that, because the purpose of lighting design is to enable us to see, we use a lighting quantity unit that is related to the physical act of seeing. That unit is the lumen. In many texts that deal with lighting, the lumen is defined as the unit of *luminous flux*. This is exactly the same definition as previously explained, since quantity of light and luminous flux are one and the same.

To illustrate the use of this unit, the technologist should open any lamp catalog and look for the characteristics of any common lamp. (Specialty lamps are frequently rated by luminous intensity, which is a lighting unit that, for the moment, does not concern us.) You will find that a standard 60 watt inside-frosted lamp produces (initially) 890 lumens of light, continuously. Similarly, a standard 34 w, triphosphor fluorescent lamp produces 3200 lm initially. These light quantity figures can then be used to determine the level of illumination, or illuminance in a space, just as the heat output of a heater is used to determine room temperature.

14.5 Illumination Level; Illuminance

The illumination level or, using the term accepted in the profession, the *illuminance* in a space is a measure of the density of *luminous flux*. Assume that we have a room illuminated with two lighting fixtures, and, for the moment, assume that their light energy is evenly distributed in the room. The result is a certain level of illumination. If we shut off one fixture, the average lighting level is reduced to half of what it was. Similarly, if we double the area of the room and still assume uniform lighting, then the same total light flux is spread over the new doubled area, resulting in an average lighting level, or illuminance, of half what is was. In other words, illuminance in a *uniformly* lighted space is directly proportional to the quantity of light (lumens) and inversely proportional to the area of the space. Expressed mathematically, we say that

$$\text{Illuminance} = \frac{\text{Light flux}}{\text{Area}} = \frac{\text{Lumens}}{\text{Area}} \quad (14.1)$$

Since lumens/area is simply *flux density*, we have shown that illuminance is the same as flux density.

Light flux density or illuminance is measured in units of *footcandles* in the English system of units and in units of *lux* in the metric (SI) system of units; that is:

English units:

$$\text{Footcandles} = \frac{\text{Lumens}}{\text{Square feet of area}} \quad (14.2)$$

or

$$\text{fc} = \frac{\text{lm}}{\text{ft}^2}$$

SI (metric) units:

$$\text{Lux} = \frac{\text{Lumens}}{\text{Square meters of area}} \quad (14.3)$$

or

$$\text{lux} = \frac{\text{lm}}{\text{m}^2}$$

Despite the fact that the United States still uses English units, lux is now used in the lighting profession at least as much as footcandles. It is, therefore, important for the technologist to be able to convert rapidly between the two systems. Since there are 10.76 ft² in one square meter, there are 10.76 lux in one footcandle. Therefore, to convert:

Multiply footcandles by 10.76 to get lux

or

Divide lux by 10.76 to get footcandles

For quick calculation, use 10 rather than 10.76. The error introduced by this approximation is about 7.5%.

Let us illustrate these simple relations. Suppose that we have a light fixture in a room that causes 1000 lm to be distributed evenly on the floor. The room is 10 ft square. What is the illumination on the floor in footcandles? in lux?

$$\text{Footcandles} = \frac{1000 \text{ lm}}{10 \times 10 \text{ ft}^2} = 10 \text{ fc}$$

To calculate lux, let us use the conversion factor;

$$\text{lux} = 10.76 \text{ fc} = 10.76 \times 10 \text{ fc} = 107.6 \text{ lux}$$

(Using the approximate factor instead of the exact one, we would have: lux = 10 × 10 fc = 100 lux.)

We shall learn later how to calculate the room illumination when given the lighting fixture data and the room dimensions and finishes. At this point, we want to emphasize that lux and footcandle are the important units of lighting to the technologist. In practical design and layout work, the technologist will be given the required illumination in footcandles or lux and will be asked to calculate the lighting required and to lay out the fixtures in the room. These footcandle or lux illumination levels are taken from tables of recommended illuminances published by authoritative sources. In the United States, the accepted source of this information is the Illuminating Engineering Society of North America (IESNA), headquartered at 120 Wall Street, 17th floor, New York, NY 10005-4001.

These recommended illuminance tables are fairly complex, since they include considerations of the type of activity for which the lighting is being designed, the reflectance of the visual task and its surroundings, the age of the person involved and speed and accuracy requirements. After taking all these factors into account, the tables then give a recommended illuminance value, which can be adjusted up or down by the designer to compensate for other factors not included in the tables, such as daylight, glare and visual distraction. As should be obvious, the selection of the target illuminance is not a simple matter, but one that requires considerable knowledge and experience in lighting design. For this reason, we are including only a single table of recommended illuminances for generic types of activities. See Table 14.1. Notice in the table that each illuminance entry is composed of 3 numbers: a middle, average figure; a lower figure that is to be used when viewing conditions are excellent; and a higher figure that is used for target illuminance under poor viewing conditions. You can find complete tables in the publications listed in the Supplementary Reading section at the end of this chapter.

In general, the technologist will be given the target illuminance figure by the project electrical designer or lighting consultant. The technologist will then use this figure to calculate fixture require-

Table 14.1 Illuminance Categories and Illuminance Values for Generic Types of Activities in Interiors

Type of Activity	Ranges of Illuminances	
	Lux	Footcandles
General lighting throughout spaces		
Public spaces with dark surroundings	20–30–50	2–3–5
Simple orientation for short temporary visits	50–75–100	5–7.5–10
Working spaces where visual tasks are only occasionally performed	100–150–200	10–15–20
Illuminance on task		
Performance of visual tasks of high contrast or large size	200–300–500	20–30–50
Performance of visual tasks of medium contrast or small size	500–750–1000	50–75–100
Performance of visual tasks of low contrast or very small size	1000–1500–2000	100–150–200
Illuminance on task, obtained by a combination of general and local (supplementary) lighting		
Performance of visual tasks of low contrast and very small size over a prolonged period	2000–3000–5000	200–300–500
Performance of very prolonged and exacting visual tasks	5000–7500–10,000	500–750–1000
Performance of very special visual tasks of extremely low contrast and small size	10,000–15,000–20,000	1000–1500–2000

Source. Courtesy of Illuminating Engineering Society of North America.

ments. A popular rule of thumb for levels is the 10-30-50 rule. This rule states that 10 fc (100 lux) is adequate for halls and corridors; 30 fc (300 lux) is adequate for areas between work stations, and 50 fc (500 lux) is adequate at desks where standard, non-detail office work is done. Notice how closely these figures agree with those in Table 14.1.

14.6 Luminance and Luminance Ratios

As we have stated several times, we see by reflection. If we place a piece of black velvet and a piece of white paper on a table, obviously the paper will look brighter than the velvet even though both are receiving equal illumination. The paper reflects more light and is, therefore, brighter. In lighting terms, the paper has a higher *luminance* than the velvet. It has been found by experience that, when a light colored "seeing task," such as the page you are now reading, is placed on a dark background such as a dark mahogany desk, eye discomfort can result. The cause of this discomfort is the high ratio between the luminance (formerly called brightness) of the paper and the luminance of the scene background. In the case just described, this ratio can be as high as 20 to 1. The Illuminating Engineering Society of North America (IESNA) recommends that the *luminance ratio* between a seeing task and its background not exceed 3 to 1. It is for this reason that modern office furniture is generally light colored—tan or light green being most comfortable to the eye. See Figure 14.5.

14.7 Contrast

In the previous section, we explain that a large difference between the luminance (brightness) of what we are looking at and the background can be annoying. This is true when the background is dark and the object is light as in Figure 14.5. It is also true in reverse. When we pass a person in the street and the light is behind that individual (bright background), we have difficulty seeing the face clearly. All we can really see is the head outline, or silhouette. This effect, that is, emphasis of object outline (silhouette), can also be helpful by provid-

Figure 14.5 *(Continued)*

Figure 14.5 White paper on a desk as in *(a)* gives a luminance ratio of as much as 20 to 1, which causes eye discomfort. A much more desirable condition is shown in *(b)* where the light color of the furniture gives a luminance ratio between work and background of less than the recommended 3 to 1 maximum. In addition to reducing luminance ratios, a light color matte finish on office furniture as on the left of photo *(c)* sharply reduces the extremely disturbing reflection of an overhead luminaire on a dark surface, particularly a polished one, as on the right of photo *(c)*. Notice that the lamps can easily be seen through the diffuser, which in this case is a very high quality prismatic lens. (Photos by Stein.)

Figure 14.5 *(Continued)*

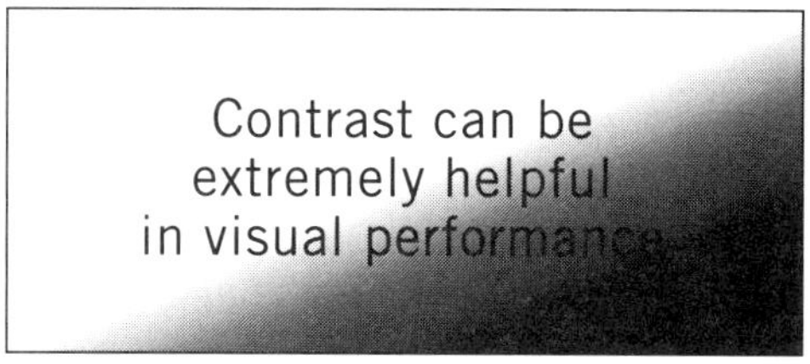

Figure 14.6 The importance of contrast is clearly shown here. Look at this drawing in poor light and notice that the black on white can still be easily read. Compare this with the amount of light you need to read the end of the word *performance* where there is almost no contrast.

ing what is called *contrast*. See Figure 14.6. You are now reading black print on white paper. You see the print clearly because you are reading the shape or outline of dark letters on a bright background. If this book were printed in light gray letters on a white paper, it would be very hard to read. The outline of the letters would fade into the background because of lack of contrast. Similarly, if the letters remained black but the background was also dark, as at the end of the third line in Figure 14.6, we would also have difficulty reading, again because of lack of contrast. We, therefore, see that high contrast is helpful and desirable where outline and shape are important, as in reading. High contrast (luminance ratio) is undesirable between the object being viewed and its surrounding area, as explained in the previous section, when we are trying to see the surface of the object and not (particularly) its outline.

14.8 Glare

More effort and money has been spent in attempting to reduce *glare* than on any other lighting problem. There are two types of glare: direct and indirect. *Direct glare* is the annoyance of bright light sources in a person's normal field of vision. A person sitting at a drawing table in the head-up position can see the ceiling lighting fixtures and the desk lamp directly in front of him or her. See Figure 14.7. To control this direct glare, ceiling fixtures are shielded and designed in such a way that the lamps and their reflections in the fixture are not seen. Also, in a well-designed fixture, luminous areas such as the sides and bottom are not too bright, so that direct glare is not a real problem. The practice of some users in commercial and institutional buildings of removing diffusers from fixtures and also some of the lamps in an effort to reduce energy consumption is extremely bad. The result is a bare bulb installation that brings back the problem of direct glare—a problem that had long since been solved for such fixtures. Worse yet, it aggravates the problem of reflected glare.

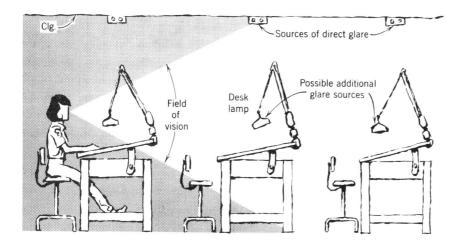

Figure 14.7 The technologist sitting at the table in head-up position can see all the ceiling fixtures in front of him or her and all the desk lamps. Each one is a possible source of direct glare.

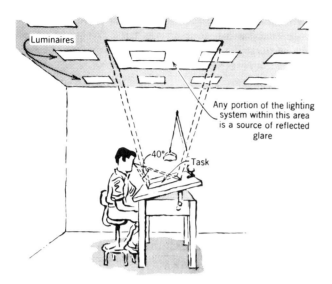

Figure 14.8 The ceiling fixtures in the outlined area reflect onto the table, causing reflected glare from the drawing, triangles, parallel straight edge and instruments.

Reflected glare is much more serious and difficult to control than direct glare. *Reflected glare*, which is called in technical lighting language *veiling reflection*, is just what the term says—glare due to reflection. Look at Figure 14.8. The technologist or draftsperson sitting at the drawing table generally has eyes down, looking at the drawings. Glossy pencil-cloth and plastic triangles reflect the ceiling lights into the eyes, causing reflected glare. You have probably experienced this type of glare and know how bad it can be. It causes the line work and lettering on a drawing to wash out in some places and to shine in others (see Figure 14.9), generally making it impossible to see properly. A draftsperson faced with a serious glare problem will try to do one or more of these things to reduce the glare:

(a) Move the entire table.
(b) Change the angle of the table top.
(c) Reposition the desk lamp (if there is no desk lamp, see about getting one).
(d) Reduce the luminance of the ceiling fixture by removing lamps or changing the diffuser.
(e) Change the type of paper being used for drawing.

Why do these "cures" work? Let's examine each one individually, while we look at Figure 14.8. (The paragraph numbers here refer to the preceding list.)

(a-1) Moving the table will not help in this room, since the entire ceiling is uniformly covered with lighting fixtures. This solution helps only when there is a single fixture or row of fixtures as in Figure 14.10. There, moving the desk so that the fixture is at position *(b)* will eliminate the reflected glare. This is the origin of the rule that for best lighting the light should come from over the left shoulder. (If it comes over the right shoulder, the right hand casts a shadow on the work. Of course, for a left-handed person the situation is exactly reversed.)

(b-1) Since the desk top is acting like a mirror, changing its angle will change what is reflected on it. Notice that as the desk top gets higher and higher, the area of ceiling that can create this problem gets smaller and smaller. See Figure 14.11. (As an exercise, redraw Figure 14.8 with the desk at 60° from the horizontal and note how small the "offending" area of ceiling becomes.) This is why many draftspeople and artists work with tables that are almost vertical. By so doing, reflected glare is almost completely eliminated.

(c-1) Positioning the desk lamp so that the entire work area becomes bright eliminates the glare. This is the same as eliminating the glare of a flashlight in a darkened room by opening the blinds and letting in the sun. The overall light level becomes so high that we no longer see the glare. This method is often the only one a person can use.

(d-1) Referring again to Figure 14.8, notice that, if we shut off the fixture(s) causing the glare, the glare will disappear. Obviously the illumination or footcandle level will drop. Despite this drop, we can frequently see better than with the light on.

(e-1) Reflected glare is caused by a light reflected in a glossy object. In (d-1), we remove one end of the problem by reducing the light. The other end of the problem can also be handled. Remove the glossy objects. Use diffuse white paper, matte-finish triangles, parallel straight edge and the like.

Since drawing is an activity that requires a very high level of illumination—200–500 fc depending on the type of work—the IESNA recommends that this level be achieved by a combination of general and supplementary (local) lighting. This method of lighting is much more energy efficient than lighting an entire space to a high level. In addition, it reduces the problem of reflected glare. For the

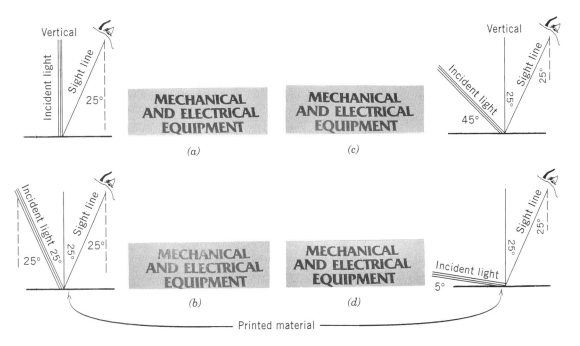

Figure 14.9 Notice the effect of an increasing amount of light in the area of the ceiling that causes reflected glare; contrast is reduced, and the black print washes out. The effect is even worse for a draftsperson working on glossy pencil-cloth with reflective instruments and triangles. The usual viewing angle to a horizontal surface is between 20 and 40° from the vertical; we show 25° because it is the most common viewing angle. With vertical incident light on a diffuse surface (a), such as the pages of a textbook, the print is dark and clear. When the angle of light incidence is equal to the viewing angle (b), we have a mirror reflection situation. Even with diffuse paper, the print is light at best and almost invisible at worst. As the angle of incidence becomes larger (c), reflected glare decreases. When the incident light is at a very low angle (d), there is little reflected glare but all the print appears lighter. (Photos by Stein.)

specific case of a drafting room, the low level overall room lighting would be supplemented by an adjustable desk lamp at each table as in Figures 14.7 and 14.8. Since the overhead lighting is reduced and the desk lamp can be positioned at will, the reflected glare problem in such an office should be minimal.

14.9 Diffuseness

This quality of light is a measure of its directivity. A single lamp produces sharp deep shadows and little diffusion. A luminous ceiling produces a completely *diffuse illumination* and no shadows. Usually, neither extreme is desirable. See Figure 14.12.

14.10 Color

Volumes have been written on color: its definition, effects, characteristics and how to produce it. To the technologist, however, color of lighting is generally a secondary factor. We normally assume all lighting to be white. The fact that incandescent lamps produce a yellowish light while cool white fluorescents produce a blue-white light is normally given little consideration—and with good reason. The eye adapts quickly to the light provided. After a short while in a room, the light produced by all the major lamp sources looks white. We will, however, give some specific recommendations for lamp choice based on color when we discuss the different types of lamps.

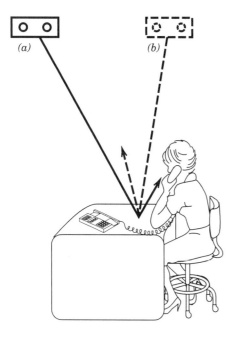

Figure 14.10 If the desk is moved so that the fixture (or row of fixtures) is at position *(b)* with respect to the desk instead of position *(a)*, reflected glare will be almost completely eliminated.

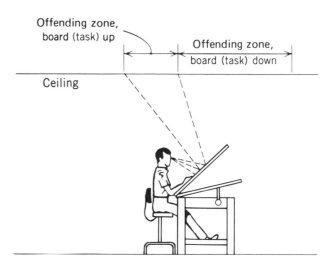

Figure 14.11 If luminaires are kept out of the trapezoidal offending zone, contrast will be excellent. See Figure 14.8. The dependence of the size and location of the glare-producing (offending) zone on table tilt is illustrated. The offending zone becomes smaller as the table is raised, so that with a table near vertical position, glare is all but eliminated. (From Stein, B., and Reynolds, J. *Mechanical and Electrical Equipment for Buildings*, 8th ed., 1992, reproduced by permission of John Wiley & Sons.)

The color of light is important because it affects the way we see colors in objects. There are two criteria of light sources with which the technologist should be familiar: *correlated color temperature* (CCT) and *color rendering index (CRI)*. When a light-absorbing object (called a black body) is heated, it will first glow deep red, then cherry red, then orange and finally blue-white as we continue to heat it. The color of the light given off by the glowing hot body is related to its temperature, in degrees Kelvin, by the term *color temperature* (CT). On this scale, the light from a candle flame is about 1800°K, sunrise is 2000°K, a photoflood lamp is 3000–3500°K, noon sunlight is 5500°K and so on. Of electric light sources, only incandescent lamps have a true color temperature because they produce light by heating an object (the filament). Other sources—fluorescent, halide, mercury and sodium—produce light by using phosphors, as will be explained later when lamp construction is discussed. Such lamps have a correlated color temperature (CCT). This simply means that a glowing black body at this temperature gives light similar in chromaticity to the light from the non-black body source. Therefore, a fluorescent lamp with a CCT of 3000°K gives light that is similar (but not identical) in color to that of a black body (or an incandescent lamp) heated to 3000°K (2540°F, 1343°C).

The second important color criterion of an electric lamp is its color rendering index (CRI). This criterion is just what it says; it is a measure of how well a particular light renders the colors of objects as compared to daylight at the same color temperature as the electric lamp. Daylight, by definition, has a CRI of 100%. Daylight color temperature varies greatly; north light is very blue with a CT of over 10,000°K, while sunrise and sunset are yellow-red at a CT of about 2000°K). Incandescent lamps have a CRI approaching 100%. Fluorescent lamps have CRI values ranging from 40 to 90+; that of halide lamps ranges from 65 to 90+ and so forth. These ratings are discussed further in the material on individual lamp types.

The important facts to remember follow.

- The color of light becomes less yellow and more blue as the source color temperature rises about 3000°K.

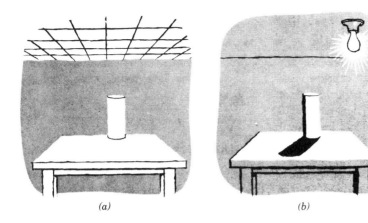

Figure 14.12 A luminous ceiling installation *(a)* provides shadowless, almost perfectly diffuse lighting. By comparison, the single ceiling bulb *(b)* produces sharp shadows and very little light diffusion. (From Stein, B., and Reynolds, J. *Mechanical and Electrical Equipment for Buildings*, 8th ed., 1992, reproduced by permission of John Wiley & Sons.)

- The higher the CRI, the better the source's color rendering. A CRI above 85 is very good, and a CRI above 90 is excellent. A figure between 60 and 80 means some color distortion. A CRI below 60 indicates serious color distortion of certain colors (not all).

14.11 Illuminance Measurement

One of the lighting-related assignments that a technologist is frequently asked to perform is to make a series of illuminance measurements. These can be in connection with either a survey of an existing installation or a field check on the results of a new installation. The instrument used to make these measurements is called (logically) an *illuminance meter*. Several common types are illustrated in Figure 14.13. All good illuminance meters are color and cosine corrected. The color correction ensures that the measurement corresponds to the color response of the human eye regardless of the type of light measured. The cosine correction automatically compensates for light that does not enter the meter cell because it strikes the cell at such a sharp angle that it is reflected from the cell's surface. A meter that is not color and cosine corrected will not give accurate readings and should not be used.

Illuminance meters are calibrated in footcandles, lux or both. When measuring horizontal illuminance levels, the meter should be held with the light cell surface horizontal, and at least 12 in. from the body. If possible, the meter should be placed on a stable surface and read from a distance. Care must be taken that the person doing the survey does not block any of the light. When doing a general illumination check in a room, the meter should be held about 30 in. (75 cm) above the floor, which is desk height. Readings should be taken throughout the room, and the results recorded on a plan of the room. Detailed instructions for conducting field surveys are provided in the IESNA publication, "How to Make a Lighting Survey." This, and other informative publications on lighting, are available from the IESNA at the address given previously. A publication list is also available.

14.12 Reflectance Measurement

It is often important to know the reflectance of a material or of a painted surface, as we will see when we learn how to perform lighting design calculations. Two simple methods of calculating reflectance by use of an illuminance meter are

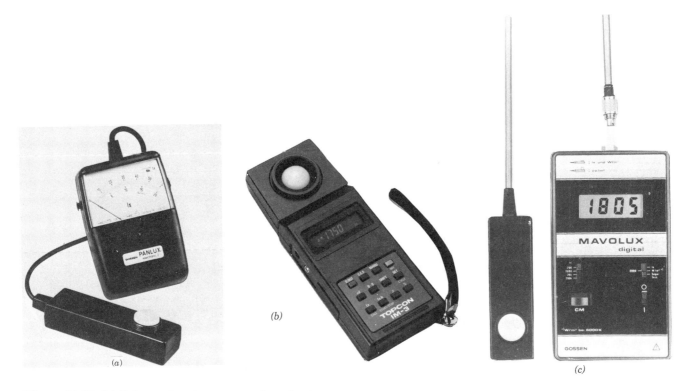

Figure 14.13 (a) Color-and-cosine-corrected analog (not digital) illuminance meter, calibrated in lux and equipped with a remote cable-connected photocell. This unit has nine ranges covering 0–200,000 lux with a maximum ±3.5% error. The unit is battery powered, measures approximately $3 \times 4 \times 1\frac{1}{2}$ in. and weighs 12 oz.
(b) Electronic digital, color-and-cosine-corrected light meter that measures illuminance (0–200,000 lux) in four ranges with a maximum $2\frac{1}{2}\%$ error. It is equipped with a recorder output for use in extended-time monitoring applications. The meter measures $3\frac{1}{2} \times 6 \times 1$ in. thick and weighs under 1 lb.
(c) Portable, highly accurate, autoranging digital illuminance meter has a range of 0.01–200,000 lux (20,000 fc) in five steps with a ±2% accuracy. The unit, which measures approximately $7\frac{1}{4} \times 3 \times 1$ in. and weighs 10 oz, can measure flickering light sources, comparative illuminances, and other convenient photometric measurements by means of a built-in microcomputer and accessories. [(a,b) Courtesy of Gossen GmbH. (c) Courtesy of TOPCON Instrument Corporation of America.]

shown in Figure 14.14. The known sample method is more accurate but requires that the technologist have a material sample that has previously been accurately measured. It is, therefore, a good idea to keep such a sample handy. Its dimensions should be at least 6 in. square. With this method, and assuming an accurate reflectance for the known sample, accuracy is ±3–5%. The reflected/incident light method gives ±5–10% accuracy, depending on how carefully the measurements are made.

How Light Is Produced: Light Sources

There are two major categories of light sources: daylight and man-made light. Daylight, although extremely important, is not normally the concern

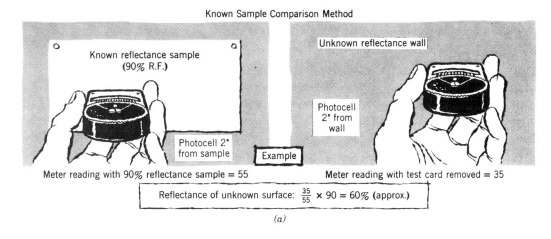
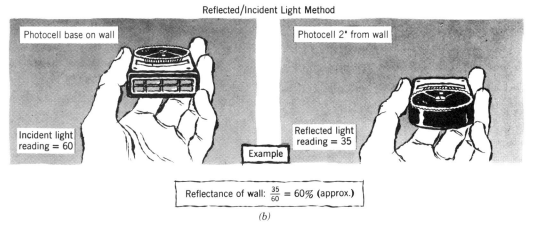

Figure 14.14 Two simple methods of measuring the reflectance of a surface. Method (a) is more accurate. Any of the meters illustrated in Figure 14.13 can be used. When using a meter with a cable-connected photocell such as those shown in Figure 14.13(a) and (b), simply use the photocell component of the meter and not the entire meter.

of a technologist and, therefore, will not be studied here. In the second category, only the widely used electrically energized sources will be studied. These include incandescent, fluorescent and high intensity discharge (HID) lamps. The types of lamp to be used in a particular application are chosen by the engineer or lighting designer or the experienced technologist. The technologist is mainly responsible for proper wiring, switching, circuiting and detailing. He or she, therefore, should know the electrical characteristics, physical shapes and dimensions and something about the application of these lamp types. We use the term *electrical lighting* rather than *artificial lighting* because we think it is more accurate. There are many references in lighting literature to artificial light. This term is meant to distinguish light generated by electrical lamps from natural light (daylight).

14.13 Incandescent Lamps

Construction of a typical general-service *incandescent lamp* is shown in Figure 14.15. Incandescent lamps are made in a wide variety of shapes and sizes with different types of bases. See Figure 14.16 and 14.17 and Table 14.2. The important characteristics of incandescent lamps are briefly discussed next.

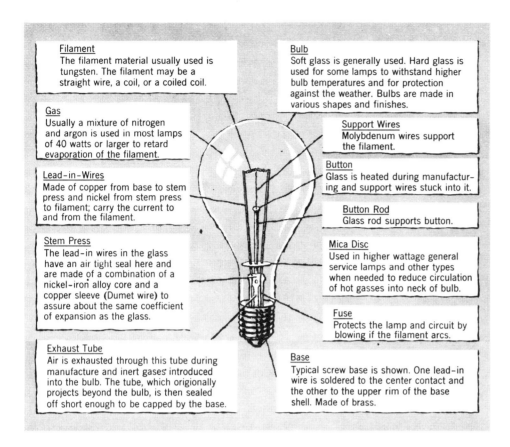

Figure 14.15 Construction details of a typical general-service incandescent lamp. (From Stein, B., and Reynolds, J. *Mechanical and Electrical Equipment for Buildings*, 8th ed., 1992, reproduced by permission of John Wiley & Sons.)

(a) Incandescent lamps are very inefficient producers of light. On the average, less than 10% of the wattage goes to produce light; the remainder is heat. Therefore, incandescent sources are a poor choice from an energy efficiency point of view. Efficiency increases with larger sizes, varying from about 8% for a 25-w lamp to 13% for a 1000-w lamp.

(b) The principal advantages of incandescent lamps are low cost; instant starting; cheap dimming; high power factor; life hours independent of the number of times the lamp is lighted; good warm color which is flattering to the skin; and small size. This last item allows the incandescent lamp to be used as a point source in fixtures that focus the light. This characteristic will be illustrated in our discussion of luminaires. See Figure 14.38.

(c) Incandescent sources have a relatively short useful life, and the life is very voltage sensitive. At 10% undervoltage, life is increased about 250%. It is this effect that is used in "long-life" and "extended-service" lamps. At 10% overvoltage, life is reduced about 75%. This means that for a nominal 1000-hour-life lamp, a swing of 10% in voltage either way can change lamp life from 3500 to 250 hours. Lamps operated at rated voltage give maximum efficiency. Voltage effects are shown in Figure 14.18.

In view of these electrical characteristics, and in particular its very low efficiency, incandescent

Table 14.2 Typical Incandescent Lamp Data (Listing a Few of Many Sizes and Types of 115-, 120-, and 125-v Lamps)

Watts and Life		Lumens		Physical Data		
Lamps Watts[a]	Average Rated Life, h	Initial Lumens	Lumens Per Watt[b]	Shape of Bulb[c]	Base	Description
15	2500	126	8.4	A-15	Med	Long Life[d]
25	2500	232	9.3	A-19	Med	—
40	1500	480	12.0	A-19	Med	—
60	1000	890	14.8	A-19	Med	—
60	2500	750	12.5	A-19	Med	Long Life[d]
75	750	1220	16.3	A-19	Med	—
100	750	1750	17.5	A-19	Med	—
100	2500	1510	15.1	A-19	Med	Long Life[d]
150	750	2850	19.0	A-21	Med	—
150	750	2680	17.9	PS-25	Med	—
200	750	3900	19.5	A-23	Med	—
300	750	6300	21.0	PS-25	Med	—
500	1000	10,850	21.7	PS-35	Mogul	—

[a] Figures in this column are the input watts; thus, 60 means 60 w. All lamps are inside-frosted.
[b] Efficacy (luminous efficiency), in lumens per watt, increases with filament temperature; therefore, it increases with wattage.
[c] Bulb designations consist of a letter to indicate its shape and a figure to indicate the approximate maximum diameter in eights of an inch (see Figures 14.15 and 14.16).
[d] 125-v lamps.

lamps should not be used for general lighting except in residences. They are best used where:

- Lamps are lighted for only short periods.
- Lamps are turned on and off frequently.
- Low purchase cost and/or low cost dimming are important.
- The lamp's color is important; particularly its flattering rendering of skin color.
- Lamps are used as point sources in focusing luminaires.

Reflector (R) and projector (PAR) lamps have built-in beam control and require only a lampholder and not a lighting fixture.

Efficacy is the technical term used in the lighting industry to describe the light producing efficiency of an electric lamp. It is measured in *lumens per watt (lpw)*. A 60 watt incandescent lamp producing 890 lumens (of light flux) has an efficacy of

$$\text{Efficacy} = \frac{890 \text{ lumens}}{60 \text{ watts}} = 14.8 \text{ lpw}$$

Similarly, a 34 watt fluorescent lamp producing 3200 lumens has an efficacy for the lamp alone of

$$\text{Efficacy} = \frac{3400 \text{ lumens}}{34 \text{ watts}} = 100 \text{ lpw}$$

When one includes the fluorescent lamp's ballast loss, the efficacy drops to about 85 lpw.

The Energy Policy Act of 1992 made a number of the most popular incandescent lamps in these designs obsolete, because of the act's minimum efficacy requirements. Among the popular lamps no longer manufactured as of 1995 are the 75-w R30, R40 and PAR38; the 150-w R40 and PAR38; and the 200-w R40. All major manufacturers now produce incandescent reflector lamps that meet the energy act efficacy requirements. These requirements state that R and PAR lamps rated 115–130 v, with medium screw base, bulb diameter greater

BULB SHAPES

Lamps shown at approx. ¼ actual size

A — Standard shape
B, F — Flame shape
C — Cone shape
G — Globe
GA — Combination of G and A
P — Pear shape
K — Arbitrary designation
PS — Pear shape straight neck
PAR — Parabolic aluminized reflector
R — Reflector
S — Straight
T — Tubular

BASE TYPES

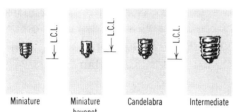

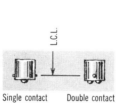

Miniature — Miniature bayonet — Candelabra — Intermediate — Single contact bayonet candelabra — Double contact bayonet candelabra — Disc — Recessed single contact — Medium — Three contact medium

Medium prefocus — Medium side prong — Medium skirted — Mogul — Three contact mogul — Mogul prefocus — Mogul end prong

L.C.L. — Light center length

Bases shown at approx. ⅓ actual size

than 2.75 in. and nominal wattages between 40 and 205, shall have minimum efficacies as follows:

Minimum Efficacy for R and PAR Lamps

Nominal Lamp Wattage, w	Minimum Efficacy (lpw)
40–50	10.5
51–66	11.0
67–85	12.5
86–115	14.0
116–155	14.5
156–205	15.0

Refer to the catalog of any major lamp manufacturer for complete details of sizes and ratings of all incandescent lamps.

14.14 Halogen (Quartz) Lamps

The quartz lamp, or what is technically called a tungsten-halogen lamp, is a special type of incandescent lamp. It produces light by heating a filament just like the common incandescent lamp. However, the filament operates at a much higher temperature than that of the standard incandescent. This is made possible by the addition of some iodine vapor to the gas surrounding the filament. Because of this high temperature, a quartz tube must be used to hold the filament since ordinary glass would melt. This gives the lamp its generic name—quartz lamp. Iodine belongs to a group of elements known as halogens; this gives the lamp its technical description as a tungsten (filament)-halogen lamp. See Figure 14.19. This concentrated high temperature filament makes the lamp essentially a point source, which is ideal for use with a reflector. It has a number of advantages over the common incandescent lamp. Among them are longer life (up to 5000 h, depending on use), slightly higher efficacy and low light depreciation. This last characteristic means that unlike the common lamp, which gradually blackens during its life due to evaporation of the filament, quartz lamps retain almost full output until failure. See the graph in Figure 14.20.

In recent years, a miniature single-ended quartz lamp mounted in a special *multimirrored reflector* has been manufactured. Due to its precise beam pattern, it has found very wide application in all types of accent, display and merchandising lighting. The lamp/fixture is known generically as an MR-16 lamp; named after an early 2-in. diameter model ($^{16}/_8$ diameter, gives 2 in.) Most lamp/fixtures of this design operate at 12 v, although 120-v lamps are also made. Some of the designs for this lamp/fixture use a removable quartz lamp in a reusable multimirrored reflector, while others are made as a single unit with the lamp mounted permanently in the reflector, as a one-piece lamp/fixture. Typical beam and illuminance data are given in Table 14.3 for this lamp design, when applied to illuminate a surface parallel to the lamp face. On angled surfaces, the beam is elliptical, and the illuminance calculations become complex.

The principal disadvantages of tungsten-halogen lamps are their relatively high cost and the fact that they should be operated inside some sort of enclosed fixture. This is due to their tendency to shatter when they burn out, scattering hot quartz fragments.

14.15 Fluorescent Lamps— General Characteristics

The fluorescent lamp is in extremely common use, second only to the incandescent lamp. Like the incandescent lamp, the fluorescent lamp comes in literally hundreds of sizes, types, wattages, shapes, colors, voltages and specific application designs. The original, preheat, fluorescent lamp is a hot cathode type, consisting of a sealed glass tube containing a mixture of inert gas and mercury vapor. See Figure 14.21. The heated cathode causes a

Figure 14.16 Common incandescent lamp bulb and base types. The bulb name indicates type and size; the letter being an abbreviation of the shape and the number equal to the maximum diameter in eighths of an inch. Thus, a PS-52 is a pear-shape bulb, $^{52}/_8$ (6½) in. in diameter, and an R-40 is a reflector lamp $^{40}/_8$ (5) in. in diameter. (From Stein, B., and Reynolds, J. *Mechanical and Electrical Equipment for Buildings*, 8th ed., 1992, reproduced by permission of John Wiley & Sons.)

To find the MR-16 lamp appropriate for straight-on application (face of lamp parallel to face of object being illuminated), measure the distance between the fixture and the object. Select the distance that is nearest this measurement from the following chart (i.e., from 2 to 10 ft), and find the lamp type that offers desired footcandle illumination level and beam size.

The beam pattern from a lamp aimed straight-on is approximately circular. Beam diameter H (ft) is measured at the circle where illuminance B is one-half the beam center illuminance C (fc).

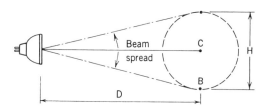

Table 14.3 Minature Mirrored-Reflector Tungsten-Halogen Lamp (MR-16)

Watts	Rated Life, h	Beam Spread, Degrees	Beam Data	D, ft				
				2	4	6	8	10
Very Narrow and Narrow Spot								
20	2000	5.5	C	2725	680	306	172	110
			H	.17	.35	.52	.70	.87
			B	1363	340	152	86	55
42	2500	7	C	3410	852	380	214	138
			H	.24	.49	.73	.98	1.2
			B	1705	426	190	107	69
20	3000	12	C	830	208	94	53	34
			H	0.42	0.84	1.3	1.7	2.1
			B	415	104	47	26	17
50	3000	13	C	—	580	258	146	94
			H	—	0.84	1.3	1.7	2.1
			B	—	290	129	73	47
Spot and Narrow Flood								
42	3000	20	C	706	180	80	44	28
			H	0.72	1.4	2.1	2.9	3.6
			B	352	88	40	22	14
50	3000	26	C	760	190	83	48	30
			H	0.92	1.9	2.8	3.7	4.6
			B	368	92	41	24	15

Table 14.3 *(Continued)*

Watts	Rated Life, h	Beam Spread, Degrees	Beam Data	D, ft 2	4	6	8	10
			Flood					
20	3000	36	C	118	30	13	7.1	4.6
			H	1.3	2.6	3.9	5.2	6.5
			B	59	14	6	3.5	2.2
42	3000	36	C	256	66	30	17	10
			H	1.3	2.6	3.9	5.2	6.5
			B	126	32	14	8.2	4.8
65	3500	38	C	500	130	56	33	20
			H	1.4	2.8	4.1	5.5	6.9
			B	222	62	28	16	10

Source. Data extracted from published material of various manufacturers is for reference only and is subject to change. Photo of Sylvania Tru-Aim Professional lamp, Courtesy of GTE Lighting Europe.

mercury arc to form between the two ends of the tube. This arc produces primarily ultraviolet (UV) light, which is not visible to the naked eye. This UV light strikes the phosphors coating the inside of the tube, which then fluoresce, producing visible light. By changing the type of phosphors, the lamp color and output can be controlled.

Although fluorescent lamps are available today in many shapes, the linear (straight tube) lamp in 2-, 4- and 8-ft lengths is the type most used in commercial work. Compact lamps, which are discussed later, were originally used primarily in residences and institutions. Today, however, they are found in all types of buildings. They are particularly popular in stores. U-shaped lamps and circular lamps are not popular, although they are available. The principal characteristics of standard linear fluorescent lamps are detailed next. Special shapes and types are discussed separately.

(a) Linear tube lamps are large; therefore, a large and relatively expensive luminaire is required to hold them. The lighting fixture also houses the ballast. The fixture must provide the required light control, since the source is a long tube, emitting light along its entire length. Since focusing and accurate light beam control are difficult and expensive for a tubular source, the fluorescent lamp is best applied to general, area lighting.

(b) The efficiency of a fluorescent lamp is much higher than that of an incandescent lamp. Between 16 and 25% of the input energy becomes visible light, with the remainder being converted to heat and a small amount of energy in invisible ultraviolet light. This does not include the energy loss in the ballast, which is all heat energy. Indeed, getting rid of this ballast heat, which amounts to about 10% of the rated lamp wattage, is an important function of the fixture.

As stated previously, the lighting profession does not use the term *efficiency* when referring to lamp output. Instead, the term used is *efficacy*, which is measured in lumens per watt. It, like efficiency, is a measure of how much input energy is converted into visible light, but it is expressed in lighting terms. The technologist would do well to become accustomed to using this term to describe lamp efficiency. Efficacy of a few common types of fluorescent lamps is shown in Table 14.5, including ballast loss. Compare these figures to the efficacy (lumens per watt) figures given in Table 14.2 for incandescent lamps to get an appreciation of how much more efficient fluorescent lamps are than incandescent lamps. When comparing efficacy figures, always include ballast losses. Since the lamp will not operate without a ballast, it is very misleading to use the efficacy of the lamp alone. For fluorescent lamps, lumen output at

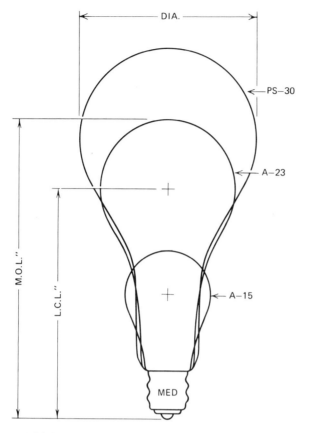

	A – STANDARD SHAPE								PS – PEAR SHAPE							
WATTS	15	25	40	60	75	100	100	150	150	150	200	300	300	500	750	1000	1500
BULB	$A-15$	$A-19_1$	$A-19_2$	$A-19_3$	$A-19_3$	$A-19_3$	$A-21_1$	$A-21_2$	$A-23$	$PS-25$	$PS-30$	$PS-30$	$PS-35$	$PS-40$	$PS-52$	$PS-52$	$PS-52$
DIAMETER"	$1^7/8$	$2^3/8$	$2^3/8$	$2^3/8$	$2^3/8$	$2^3/8$	$2^5/8$	$2^5/8$	$2^7/8$	$3^1/8$	$3^3/4$	$3^3/4$	$4^3/8$	5	$6^1/2$	$6^1/2$	$6^1/2$
M.O.L."	$3^1/2$	$3^7/8$	$4^1/4$	$4^7/16$	$4^7/16$	$4^7/16$	$5^1/4$	$5^1/2$	$6^3/16$	$6^{15}/16$	$8^1/16$	$8^1/16$	$9^3/8$	$9^3/4$	13	13	13
L.C.L."	$2^3/8$	$2^1/2$	$2^{15}/16$	$3^1/8$	$3^1/8$	$3^1/8$	$3^7/8$	4	$4^5/8$	$5^3/16$	6	6	7	7	$9^1/2$	$9^1/2$	$9^1/2$
BASE	MED	MED	MED	MED	MED	MED	MED	MED	MED	MED	MED	MED	MOG	MOG	MOG	MOG	MOG
STANDARD FINISH	IF	IF	IF W	IF	IF	IF	IF	IF	CL IF	CL IF	IF	IF	IF	IF	CL IF	CL IF	CL IF

CL – CLEAR   IF – INSIDE FROSTED

Figure 14.17 Typical dimensional data for general-service incandescent lamps.

100 hr burning is used rather than initial lumens, since at 100 h the lamp output has dropped to its stable operating point.

(c) Fluorescent lamps have outstandingly long life. This life, however, is affected by the number of times the lamp is turned on and off, since switching tends to wear out the cathode. An average fluorescent lamp burned continuously will last more than 30,000 hrs; with 3 burning hours per start, it will last about 18,000 hrs. These figures are constantly being increased by new developments in fluorescent lamps. This

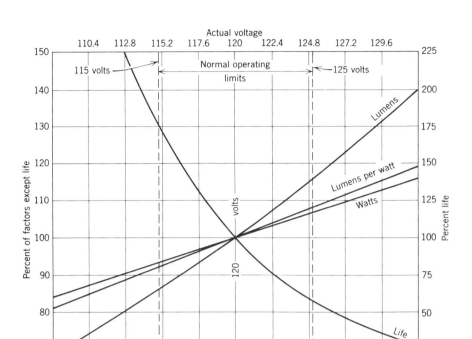

Figure 14.18 Operating characteristics of a standard 120-v incandescent lamp as they vary with voltage.

long lamp life, when the lamp is not switched on and off, is the reason that it used to be general practice to leave fluorescents burning continuously—it turned out to be cheaper. Today, with the high cost of electric power in most areas and, more important, with the need to conserve energy, this is not true. The economic break-even point depends on the type of lamp and the cost of power. It varies from 5 to 15 minutes of off-time, to compensate for lost lamp life. Finally, lamp life figures can be misleading, since they are generally given in catalogs as hours to burnout. Many users replace lamps when they reach about 75% of burnout life because the light output has dropped at that point to about two-thirds. This too is an economic decision that is made by the user.

(d) Fluorescents have the advantage of being cheap, readily available in a very wide range of sizes and colors and relatively insensitive to changes in voltage. This is particularly important in areas where "brownouts" are common.

(e) The color of the light produced by fluorescent lamps was originally very high in blues and greens. As such, it was unkind to human skin color, making people look pale and ill. However, the phosphors have long since been improved so that this problem no longer exists. Lamps are available in a host of other shades and colors including daylight and a color that closely duplicates the color of incandescent lamps. Federal law mandates good color rendering in the most commonly used fluorescent tubes. As a result, the unflattering skin tones that once were typical of fluorescent lighting are rare today.

(g) Full-range dimming of fluorescent fixtures requires the use of a special dimming ballast. These are relatively expensive, and one is required for each lamp. Partial dimming, down to 40–50% output, is possible using conven-

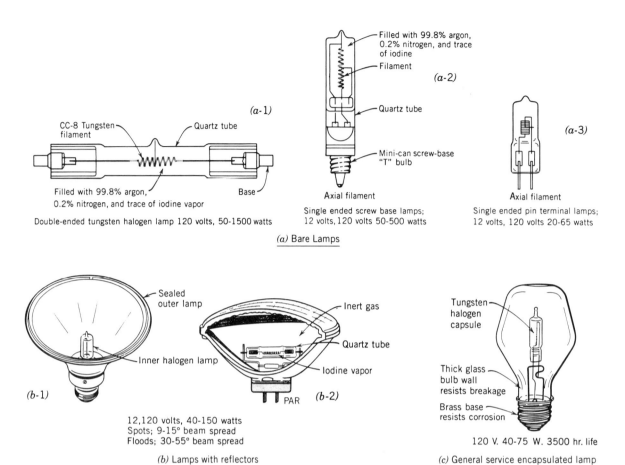

Figure 14.19 Tungsten-halogen lamps are available in a variety of designs. The original design was double ended *(a-1)*, used in reflector fixtures or built into reflector lamps *(b-2)*. Single ended lamps *(a-2, a-3)* are used in the type of reflectors shown in *(b-1)* and in encapsulated general-service lamps *(c)*. (From Stein, B., and Reynolds, J. *Mechanical and Electrical Equipment for Buildings*, 8th ed., 1992, reproduced by permission of John Wiley & Sons.)

tional ballasts and a solid-state dimmer. Full-range dimming is readily accomplished with electronic ballasts.

(h) Other characteristics of fluorescent lamps that should be kept in mind follow:
1. Possible difficulties in starting at low temperatures, which limits outdoor use. In addition, output drops with temperature. Special ballasts are available for satisfactory low temperature operation.
2. Rapid-start lamps require a piece of grounded metal adjacent to the lamp. This is important when using lamps in architectural coves, valances and the like, where the metal of a lighting fixture does not exist. The absence of a starter indicates a rapid-start or other starterless circuit, as will be explained later.

As mentioned previously, there are many types of fluorescent lamps in addition to a huge array of sizes, wattages and special-purpose lamps. The types differ basically in starting techniques and amount of current drawn by the lamp. We will now briefly describe the major types of lamps and their special characteristics. Keep in mind that all the descriptions and tables presented here are brief extracts chosen to illustrate the typical types. Com-

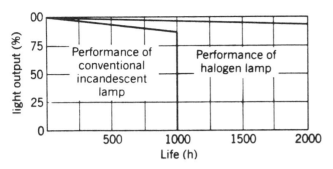

Figure 14.20 The tungsten-halogen lamp retains high output throughout its life, which is at least twice as long as a standard incandescent lamp. (From Stein, B., and Reynolds, J. *Mechanical and Electrical Equipment for Buildings*, 8th ed., 1992, reproduced by permission of John Wiley & Sons.)

plete listings of types and ratings are available in manufacturers' literature, which is up to date and available for the asking.

The three major types of fluorescent lamps and their circuits are preheat, instant-start and rapid-start. Before proceeding with a description of these lamp types, you need to understand the function of the ballast in a fluorescent lamp circuit. The next section is devoted to ballasts, after which the discussion will return to fluorescent lamps.

14.16 Fluorescent Lamp Ballasts

Like all arc discharge lamps, the fluorescent lamp requires a ballast in its circuit. (In the first part of our discussion, when we refer to a ballast, we mean a conventional *core-and-coil ballast*. It consists essentially of an iron core on which is wound a coil. In the latter part of our study, we will discuss *electronic ballasts*. These will always be referred to here as electronic ballasts to avoid confusion.) Refer to Figure 14.22(*a*). As this figure shows, a conventional ballast is basically a coil. Its primary function is to limit the current in the arc circuit. For this reason, simple ballasts are referred to as chokes, since they are no more than a choke coil. Without the ballast in the circuit, the lamp would draw excessive current, and the fuse or circuit breaker would open.

The second purpose of a modern ballast is to improve the power factor of the lamp circuit. See Figure 14.22(*b*). Without this improvement, the circuit operates at a power factor of under 50%. This causes unnecessary power losses and is, therefore, undesirable. Dimensions and weights of ballasts vary from one manufacturer to another. The data in Table 14.4 represent a very small sampling of current manufacture.

a. Conventional Iron Core-and-Coil Ballasts

These are large and heavy, may be noisy and generate a large quantity of heat because of their power loss. An amendment to the National Appliance Energy Conservation Act Amendment of 1988 requires that certain common ballasts have a higher efficiency than the usual core-and-coil ballast. Specifically, as of 1991, the following ballasts must have a minimum *ballast efficacy factor (BEF)* as shown. This factor is the ratio of the *ballast factor (BF)* to the nominal lamp power input. The ballast factor is the ratio of a lamp's output when operated with a test ballast, to the same lamp's output when operated on a standard lab ballast under ANSI test conditions. In other words, a ballast's BEF is a simple measure of the ballasts efficiency as compared to other ballasts operating the same lamp type. What this federal regulation has done is to eliminate the old iron-core aluminum windings ballast that had a 16–20 w heat loss for a two-lamp 40-w unit. To meet the law's requirement, new ballasts of this rating (two-lamp, 40 w) use steel cores and copper windings, with a heat loss of 6–8 w. This is a 60% reduction in heat loss! The affected ballasts are listed next.

Ballast for the Operation of[a]	Nominal Input Voltage	Total Nominal Lamp Watts	Minimum Ballast Efficacy Factor
1—F40T12 lamp	120	40	1.805
	277	40	1.805
2—F40T12 lamps	120	80	1.060
	277	80	1.050
2—F96T12 slimline lamps	120	150	0.570
	277	150	0.570
2—F96T12HO lamps	120	220	0.390
	277	220	0.390

[a] Some 40- and 96-w T-12 lamps ceased to be manufactured after 1994 because of the provisions of the National Energy Policy Act of 1992. These lamps are listed in Section 14.17.

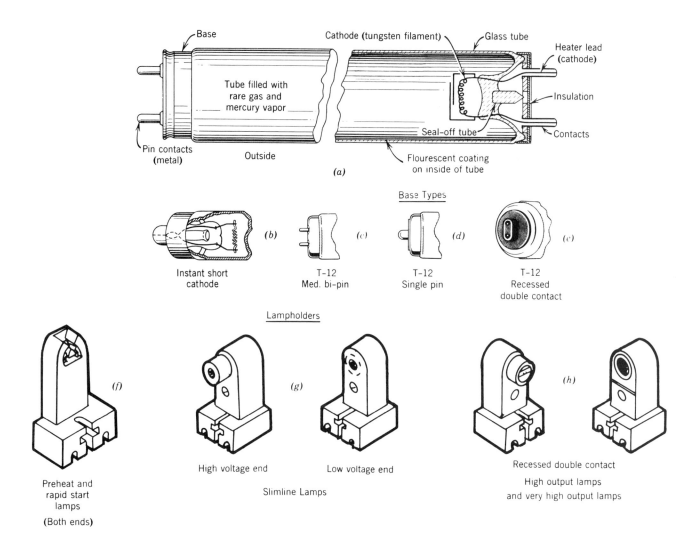

Figure 14.21 Details of typical fluorescent lamps and associated lampholders. (a) Construction of preheat/rapid-start bipin base lamp. This type of lamp has a type (c) base and is held in a type (f) lampholder. Instant-start lamps (b) have a single-pin base (d) and use single-pin lampholders (g), which are different for each end. High output and very high output rapid-start lamps use a recessed dc (double contact) base (e) and type (h) lampholders.

These ballasts also must have a minimum power factor of 90%. All ballasts meeting these requirements will be marked with a capital *E* printed in a circle on the ballast. They are known as E-rated ballasts. Excluded from the requirement of this law are low temperature ballasts, dimming ballasts and low power factor ballasts, manufactured specifically and exclusively for residential use and so marked.

The humming or buzzing noise that is associated with fluorescent lamp installations is created by the conventional core-and-coil ballast. Electronic ballasts are silent. Ballast manufacturers established noise ratings for ballasts ranging from A to

D, with A being the quietest. This allows users to select a ballast that is suitable to the area where the fixture will be used. Class A ballasts are appropriate for residences and other quiet areas. Class D is entirely adequate for a machine shop or foundry. Frequently, the fixture and not the ballast is to blame for a noisy installation. A poorly designed fixture acts as a noise amplifier for the ballast. This should be very carefully checked when examining a fluorescent light fixture for suitability in a particular installation.

There are a number of national organizations that are involved with ballast standards and testing. Among them are Certified Ballast Manufacturers Association (CBM), Electrical Testing Laboratories (ETL) and Underwriters Laboratories (UL). Detailed information on ballast factors, standards and tests can be found in the Supplementary Reading at the end of the chapter and in manufacturer's catalogs.

As a rule of thumb, conventional ballasts lose one-half their life with each 10°C rise in their operating temperature above rated temperature. It is, therefore, important to provide for adequate heat radiation from the fixture. This means, among other things, that standard (not special) fluorescent fixtures recessed into a ceiling should not be directly covered with thermal insulation. Also, because of the heat generated by the ballast and the lamps, the NEC limits how close a fluorescent fixture can be installed to flammable material. See NEC Article 410 and, especially, Sections M, N and P. The NEC also requires that ballasts used in indoor fluorescent fixtures (with a few special exceptions) be protected from overheating by a built-in device that will disconnect the ballast if overheating occurs. These thermally protected ballasts are labelled as Type P by Underwriters Laboratories.

b. Electronic Ballasts

Most electronic ballasts generate a high frequency a-c voltage of 25–30 kHz. At that frequency, fluorescent lamps operate more efficiently. In addition to improving overall lamp-plus-ballast efficacy, electronic ballasts have these additional advantages:

- Almost zero heat loss; therefore, cool operation of the ballast and fixture.
- Sharply reduced lamp flicker.
- Simple and cheap full range lamp dimming.
- Almost completely silent operation.
- High power factor.
- High ballast factor and ballast efficacy factor.
- Low temperature lamp starting.
- Lightweight and small physically.

Principle disadvantages follow:

- High cost.
- High harmonic content in the lamp circuit. (This is an extremely serious problem that can contribute to electrical system overheating and failure.)
- Production of electrical noise that can interfere with the proper operation of sensitive electronic equipment.

The last two disadvantages are highly technical and are not the technologist's responsibility. They are listed here as a matter of interest for further reading and study. These problems are being worked on actively by ballast manufacturers and, no doubt, will be less serious with each new generation of electronic ballast design.

14.17 Fluorescent Lamp Types
a. Preheat Lamps

The preheat lamp is the original (1937) fluorescent lamp. It requires a separate starter, which is a metal cylinder about ¾ in. in diameter and 1¼ in. long, that snaps into the fixture through a hole in the fixture body near the lamp base. The starter allows the cathode to preheat and then opens the circuit, causing an arc to flash across the lamp, starting it. See Figure 14.22. Most of these starters are automatic, although in desk lamps the preheating is done by pressing the start button for a few seconds before releasing it. This closes the circuit and allows the heating current to flow. The absence of a separate starter indicates a rapid-start or other starterless circuit, as will be explained later. All preheat lamps have bipin bases. They range in power from 4 to 90 w and in length from 6 to 96 in. A typical ordering abbreviation for a preheat lamp would be F15T12WW. This means: fluorescent lamps, 15 w, tubular-shape bulb, $12/8$-in. diameter (number represents diameter in one-eighths of an inch), warm white color. See

810 / HOW LIGHT BEHAVES; LIGHTING FUNDAMENTALS

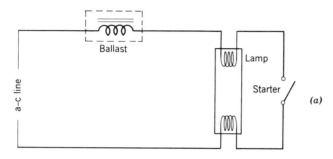

(a) Basic preheat circuit. Starter may be any of several types, manual or automatic.

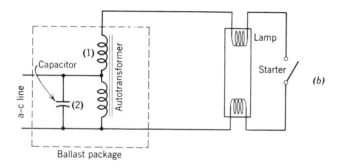

(b) Preheat circuit modified with (1) autotransformer to adjust line voltage to ballast voltage, and (2) capacitor to make the entire device high power factor.

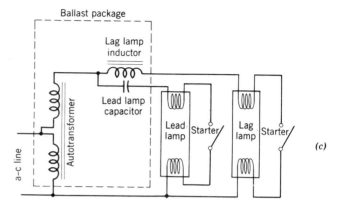

(c) Two lamp preheat circuit—also known as a lead-lag circuit because of the phasing of the lamps. This arrangement gives high power factor and minimizes flicker effects, since the two lamps are exactly out of phase.

Figure 14.22 Simplified preheat lamp circuits for one lamp (a) and (b), and two lamps (c). The circuits do not show compensators and detail elements, for the sake of clarity. Closing the starter circuit causes current to flow in the cathode circuit, to preheat them. Then, opening the starter circuit generates a high voltage across the lamp, causing it to light. Standard output lamps operate at 430 ma.

Table 14.4 Typical Fluorescent Lamp Ballast Data

Lamp		Ballast Characteristics		
Type	Lamp, w (ma)	Dimensions ($L \times W \times H$), in.	Ballast Weight, lb	Sound Rating
Preheat Trigger-Start; Std. Power Factor				
F15 T12	2–15	$6^{1}/_{2} \times 2^{3}/_{8} \times 1^{3}/_{4}$	2.8	A
F20 T12	2–20	$6^{1}/_{2} \times 2^{3}/_{8} \times 1^{3}/_{4}$	2.8	A
Rapid-start; High Power Factor				
F 40 T12	2–40 (430)	$9^{1}/_{2} \times 2^{3}/_{8} \times 1^{3}/_{4}$	3.5	A
F48 H0	2–60 (800)	$9^{1}/_{2} \times 2^{3}/_{8} \times 2^{5}/_{8}$	8	B
F96 H0	2–105 (800)	$11 \times 3^{1}/_{8} \times 2^{1}/_{2}$	12	B
F96 VH0	2–215 (1500)	$15 \times 3^{1}/_{8} \times 2^{3}/_{4}$	14	C
Instant-start (Slimline)				
F48 T12	2–39 (430)	$11^{1}/_{2} \times 3^{1}/_{8} \times 2$	7	B
F72 T12	2–60 (430)	$11^{3}/_{4} \times 3^{1}/_{8} \times 2$	8	C
F96 T12	2–72 (430)	$11^{3}/_{4} \times 3^{1}/_{8} \times 2^{1}/_{2}$	10	C
Electronic Ballast				
F40 T12 RS	2–40	$9^{1}/_{2} \times 2^{3}/_{8} \times 1^{1}/_{2}$	1.5	A

Table 14.5. These lamps operate with a lamp current of 430 ma.

b. Instant-Start (Slimline) Lamps

The instant-start lamp was the second fluorescent lamp developed (1944) and operates without starters. The ballast provides a high enough voltage to strike the arc directly. Since no preheating is required, Slimline, instant-start lamps have only a single pin at each end. A typical catalog description for such a lamp would be F42T6CW Slimline, which means fluorescent, 42-in. length, tubular, 6/8-in. diameter, cool white, instant-start. The T-6 narrow tube indicates a low current, 200-ma lamp, in lieu of the usual 430-ma lamp. Note also that in instant-start lamp designations, the number following F indicates length not wattage. This is true with all lamps that operate at other than 430 ma. To find wattage for these lamps, a catalog must be consulted. See Figure 14.23 for a typical lamp circuit.

c. Rapid-Start Lamps

The third type of lamp, that became available in 1952, is called rapid-start or rapid-start/preheat. The delay in starting the preheat lamp results from

Table 14.5 Typical Fluorescent Lamp Data: Standard Lamps, 60 Hz, Conventional Core-and-Coil Ballasts

Lamp Abbreviations[a]	Lamp Data		Lamp Life, h^b	Initial Output, lm^c	Initial Efficacy, lm/w^d
	Lamp, w	Length, in.			
Preheat lamps,[e] 430 ma					
F-15 T-8 CW	15	18	7500	870	38
F-20 T-12 3000°K	20	24	9000	1300	43
Rapid-start; preheat lamps,[f] 430 ma					
F40 T-12 CW/ES	34	48	20,000+	2950	78
F40 T-12 3000°K	40	48	20,000+	3300	75
F40 T-12 3500°K	40	48	20,000+	3300	75
Rapid-start; high output, 800 ma					
F48 T-12 CW/HO	60	48	12,000	3850	55
F60 T-12 CW/HO	70	60	12,000	5150	64
F72 T-12 CW/HO	85	72	12,000	6350	65
F96 T-12 CW/HO/ES	95	96	12,000	8050	74
Rapid-start; very high output, 1500 ma					
F48 T12 CW/VHO	110	48	10,000	6200	50
F72 T12 CW/VHO	160	72	10,000	10,000	57
F96 T12 CW/VHO/ES	185	96	10,000	12,500	64
Instant-start (Slimline) lamps, 430 ma					
F24 T-12 CW	20	24	7500	1150	40
F48 T-12 CW/ES	32	48	9000	2550	67
F72 T-12 CW	55	72	12,000	4550	69
F96 T-12 4100°K	75	96	12,000	6700	81

[a] Lamp symbol: CW, cool white; 3000° K, color temperature; ES, energy saving. Lamp abbreviations vary among manufacturers.
[b] Life figures are for 3h burning per start.
[c] After 100h burning.
[d] Includes average ballast loss.
[e] Data given for a preheat circuit.
[f] Data given for lamps in a rapid-start circuit.

Basic instant-start lamp circuit. Cathodes are not preheated. Voltage from ballast-transformer causes arc to strike directly. Bases are single pin.

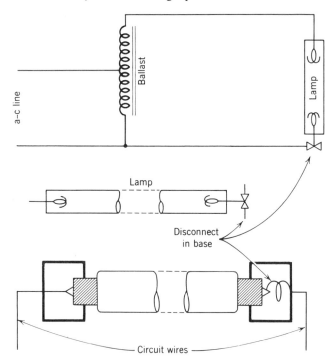

Due to the high voltage involved, the lampholder at one end of the lamp is a disconnecting device that opens the circuit when the lamp is removed.

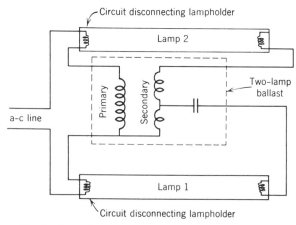

Typical two lamp instant-start circuit. Note disconnecting lampholders and autotransformer in the ballast. The capacitor provides for a phase shift that assists operation. The circuit is called series-sequence, since the lamps start in sequence, in series.

Figure 14.23 Basic instant-start lamp circuits. Notice that lamps are single pin, unlike the rapid-start and preheat types, which have a separate circuit to heat the lamp filament. T-6 and T-8 lamps normally operate at 200 ma; T-12 lamps operate at 430 ma.

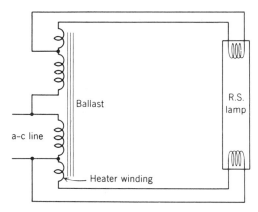

Basic rapid-start circuit. Note the special small end windings used to supply voltage to heat the cathodes. Cathodes (filaments) are heated *constantly*.

Figure 14.24 Typical basic rapid-start circuit. To ensure proper starting, all standard RS lamps must be mounted within ½ in. of a grounded metal strip, extending the full length of the lamp (1 in. for HO and VHO lamps). Normal output lamps are 430-ma T-12. High output lamps operate at 800 ma; very high output lamps operate at 1500 ma.

the time required to heat the cathode. In rapid-start circuits, the cathode is heated continuously by a special winding in the ballast. As a result no heat-up delay is required, and the lamp can be started rapidly. See circuit diagram Figure 14.24. Because of this similarity of operation, rapid-start lamps will operate satisfactorily in a preheat circuit. The reverse is not true because the preheat requires more current to heat the cathode than the rapid-start ballast provides. See Table 14.6 for interchangeability of lamps in the various circuits.

By far the most popular lamp is the nominal 40-w T-12 lamp. As is explained later, certain of these very popular lamps have been made obsolete by the National Energy Policy Act of 1992. A modern energy-efficient lamp in this category might have the ordering abbreviation F40/T12/3000/RS, which translates as fluorescent, T-12 tube (1.5 in. diameter), 3000°K color temperature (close to incandescent), rapid-start. When not specified otherwise, 430-ma lamp current is understood. It must always be kept in mind that the ballast has a wattage (heat) loss. Therefore, a two-lamp fixture using 40-w tubes gives a load of 86–88 w due to the 6 to 8-w ballast loss. Low wattage "energy-saving" lamps reduce this total to about 70 w.

Table 14.6 Fluorescent Lamp Interchangeability: Standard Lamps and Ballasts Only[a]

Lamp Type	Ballast/Circuit Type		
	Preheat	Rapid-start	Instant-start
Preheat	OK	Not good, poor starting	Not good, poor starting, short life[b]
Instant-start (Slimline)	Won't start	Won't start, not good[b]	OK
Rapid-start	OK	OK	Not good, poor starting, short life[b]
Preheat/Rapid-start	OK	OK	Not good, poor starting, short life[b]

[a] Special lamps such as low wattage, high wattage, 265-ma T-8 and so on must be used with matching ballast.

[b] Normally no possibility of interchange. Instant-start lamp is single-pin base; preheat/rapid-start lampholders are for bipin bases.

d. High Output Rapid-Start Lamps

As already mentioned, all preheat, most instant-start and most rapid-start lamps operate at 430 ma. If this current is increased, the output of the lamp also increases. Two special types of higher output rapid-start lamps are available. One operates at 800 ma and is called simply high output (HO). See Table 14.5. The second, which operates at 1500 ma (1.5 amp), is called by different manufacturers very high output (VHO), super-high output or simply 1500-ma rapid-start. There is also a 1500-ma special lamp that uses what looks like a dented or grooved glass tube. This lamp, called Power Groove by General Electric, has somewhat higher output than the standard VHO tube. All these high-output lamps have recessed double contact bases, require special circuits and ballasts and are not interchangeable with any other type of lamp. They are used in applications where high output is required from a limited size source, such as in outdoor sign lighting, street lighting and merchandise displays.

Because of the serious heat problems involved, VHO lamps are frequently operated without enclosing fixtures. This, however, creates a glare problem that limits the application of these very bright sources. Also, the HO and VHO lamps are slightly less efficient than the standard 430-ma rapid-start lamp in the shorter lamps and have considerably shorter life. Typical ordering abbreviations for these high-output lamps are similar to the standard rapid-start lamps, except that the number indicates length, not wattage. For instance, F72T12/CW/HO is fluorescent, 72-in. long T-12 bulb, cool white, high output (800 ma). By consulting the catalog, we find that this lamp is rated 85 w without ballast loss.

e. Minimum Efficacy and Color Rendering Index

The National Energy Policy Act of 1992 mandates minimum efficacy in lumens per watt (lpw) and minimum color rendering index (CRI) for four basic fluorescent lamp types as listed here. These standards were applied at different dates. The last took effect in November 1995. These minimum standards made many very popular types of lamps obsolete. They included the F-40 T-12 lamp in cool white, warm white, white, and daylight; the F-40 T-12 cool white U-shaped tube with 6-in. leg spacing; the 96-in. T-12 cool white Slimline and high output lamps, among others. More efficient lamps in both T-12 and T-8 tubes, with much improved CRI are available to replace these obsolete types.

The minimum standards mandated by the National Energy Policy Act follow.

Minimum Standards for Fluorescent Lamps

Lamp Type	Wattage, w	Minimum CRI	Minimum Efficacy, lpw
F40	>35	69	75
	28–35	45	75
F40/U	>35	69	68
	28–35	45	64
F96T12	>65	69	80
Slimline	52–65	45	80
F96T12/HO	>100	69	80
	80–100	45	80

Specifically exempted from these requirements are specialty lamps including plant growth lamps, low temperature lamps, impact-resistant lamps, reflector and aperture lamps and all lamps with a CRI of 82 or greater.

14.18 Special Fluorescent Lamp Types

a. Low Energy Lamps

Every major lamp manufacturer produces a line of low wattage fluorescent lamps. These lamps generally operate at slightly lower current than standard lamps. As a result, they require special ballasts to operate at maximum efficiency, although many will operate satisfactorily on standard ballasts. Their best application is to reduce lighting and wattage in existing overlighted spaces without the use of dimmers or other special auxiliary circuit devices. Their efficacy when used with matching ballast is equal to, or a little higher than, standard lamps. Some types have shorter life than standard lamps, and most cost more. Here, too, the technologist is strongly advised to consult an up-to-date manufacturer's catalog for current information.

b. Triphosphor (Octic) Lamps

These lamps, which operate at 265 ma are easily recognized because they use a T-8 (1 in. diameter) tube and are clearly labelled. They are called triphosphor, because they use phosphors that produce light in three basic colors that combine to give very good color rendering. As a result, they are widely used in stores, beauty salons, health clubs and other locations where excellent color rendering is important. They operate best with electronic ballasts, have considerably higher efficacy than standard lamps and are available in all standard wattages and shapes. Their principle disadvantage is sensitivity to temperature. This makes them unsuitable for most dimming applications and for outdoor use. See Table 14.7.

c. U-Shaped Lamps

The principal disadvantage of the standard 4-ft fluorescent lamp is its shape. A 4-ft tube requires a long narrow fixture. The U-shaped lamp, which is simply a standard lamp bent into a U shape, answers the need for a fluorescent that would fit into a square fixture. The U-shaped lamp is made with several spacings. Two, three or four lamps will fit into a 2-ft square fixture, depending on the leg spacing. Efficacy is about the same as a standard lamp, and life is slightly lower. As with the standard tubular lamps, the U-shaped lamp is avail-

Table 14.7 Typical Characteristics of Linear Triphosphor Lamps[a]

Watts	Correlated Color[b] Temperature, °K	Initial[c] Lumens	Length, in.
17	3000	1300–1400	24
	3500	1300–1400	24
	4100	1300–1400	24
25	3000	2000–2200	36
	3500	2000–2200	36
	4100	2000–2200	36
30–36[c]	3000	2800–3500	48
	3500	2800–3500	48
	4100	2800–3500	48
40	3000	3600–3800	60
	3500	3600–3800	60
	4100	3600–3800	60
58–62[c]	3000	5500–6200	96
	3500	5500–6200	96
	4100	5500–6200	96

[a] All lamps are linear, T-8 bulb, with medium bi-pin base. Life exceeds 20,000 hours at 3 burning hours per start, except for the 60-w nominal lamp that has a 15,000-hour life.
[b] Color rendering index of all lamps exceeds 75.
[c] Exact figure depends on selection of manufacturer.

Source. Data extracted from current manufacturers' catalogs.

able in a variety of colors and designs, including the triphosphor type described previously.

d. Compact Fluorescent Lamps

These lamps were developed to answer the need for an efficient, long-life lamp that could fit into small fixtures similar to those used for incandescent lamps. Their rapid and wide acceptance led to the development and manufacture of a very wide range of designs and wattages. Some lamps are made with an integral electronic ballast and an Edison screw base. Others are made with pin or pressure connectors that plug into a separate electronic ballast that has an Edison screw base. Still others are made with an outer diffuser so that they can be used as a complete fixture. Another design comes complete with an enclosing reflector. A few of the designs are shown in Figure 14.25. The variety of shapes and sizes is so great that no short tabulation could give a representative sampling. For this reason no tabulation is presented here. Refer to any current manufacturer's catalog for a description of the lamps they make. However, since designs vary among manufacturers, it is advisable to consult the catalogs of several major manufacturers to obtain an overall picture of lamp availability in this general class.

All compact fluorescent lamps have shorter life and lower efficacy than the tubular fluorescent lamps, although they are still much better than incandescent lamps. Like all fluorescent lamps, compact types are life sensitive to burning hours per start and, therefore, should be used only where they are not turned on and off frequently. Typical applications for these lamps are corridor lights, lanterns, desk and table lamps, decorative fixtures and outdoor lights in suitable fixtures. Despite their small size, the entire lamp is luminous, making them unsuitable for use with reflectors that require a small source for focusing.

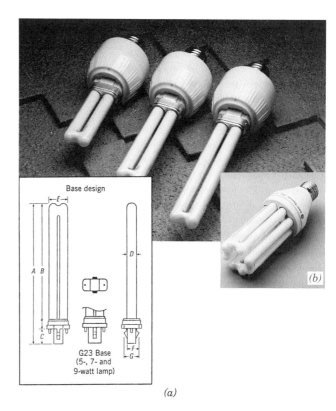

Figure 14.25 (a) Single-folded fluorescent lamps plugged into an electronic ballast that is equipped with a medium screw base. The illustrated lamps rated 5, 7 and 9 w are 4.2, 5.3 and 6.6 in. long, respectively. All use T-4 tubes (½ in. diameter) and have a single-ended two-pin plug-in base. (see inset). All are rated at 10,000h life and have a CRI of 80+. Initial output for the lamps are:
5 watts, 250 lumens
7 watts, 400 lumens
9 watts, 600 lumens

The lamps are available in a range of correlated color temperatures varying from 2700 to 5000°K. (Courtesy of GE Lighting.)
(b) Quadruple-folded lamp with integral electronic ballast and medium screw base. This lamp which is 6.6 in. long and 2.3 in. in diameter is rated 28 watts, 1750 initial lumens, low power factor, 10,000-h life, with a CCT of 2700°K and a CRI of 80+. (Courtesy of GE Lighting.)

14.19 HID (High Intensity Discharge) Lamps

In this category the most common types are mercury, high pressure sodium and metal halide. HID lamps which were once used only outdoors because of poor color and high wattages, are now used extensively indoors. This is the result of great improvements in color, especially in the metal halide family, plus production of low wattage lamps. The high efficiency, small size and long life of these sources make them suitable for many applications that previously used only fluorescent lamps. In applications where color is important but not critical, combinations of high pressure sodium with halide or other sources can be used. The second source improves the yellow sodium color, and the

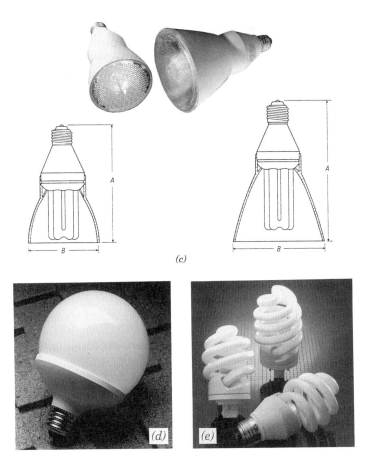

Figure 14.25 *(c)* Two one-piece reflector units. Each comprises a triple-folded lamp, an integral electronic ballast and a built-in reflector. Each entire assembly, with medium screw base is intended as a replacement for an incandescent reflector lamp. The units are rated 15 w (left) and 20 w (right), 10,000-h life, 2700°K CCT and 82 CRI. The 15-w lamp is low power factor (<0.6) and has a total harmonic distortion of 170%. The 20-w lamp has a 800 lumen initial output and is available in either low or high power factor designs. (Courtesy of GE Lighting.)
(d) One-piece, screw-in, globe-shaped, low power factor compact fluorescent is rated 16 w, 10,000-h life, 2800°K CCT, 82 CRI and 750 lm initial output. The lamp is 5.1 in. high and 3.7 in. in diameter. (Courtesy of GE Lighting.)
(e) Helical compact fluorescent is designed to maximize output by reducing the light trapping of multisection-folded fluorescent lamps. This unit operates with an electronic ballast. The lamp is available with plug-in pin base in ratings of 32 and 42 w, and in a screw-in design with integral ballast rated at 20 w. Initial lumen output for the three designs is 2400, 3200 and 1200, respectively. All lamps have a rated 10,000-h life. (Courtesy of GE Lighting.)

combination gives very high efficacy and long lamp life. We will study these three important light sources, discuss their important characteristics and describe typical applications. The characteristics, ratings and sizes of these lamps constantly change because of continuing development work in this field. We, therefore, strongly recommend that the technologist/designer always consult current manufacturer's published material when doing actual design.

14.20 Mercury Vapor Lamps

The mercury vapor lamp was the first HID lamp to be developed. Basic construction details are shown in Figure 14.26. The light is produced in an internal arc tube, which is enclosed by an outer glass bulb. The light produced in the arc tube is in large part visible light, but with a considerable percentage of invisible ultraviolet (UV) light. The visible light has the blue-green color that is typical of clear bulb mercury lamps. This color light produces severe color distortion in most objects (particularly reddish items). This makes the clear mercury lamp unsuitable for any application where color rendering is of any importance at all. Efficacy of clear mercury lamps is about 20% lower than that of standard tubular fluorescent lamps. Life of a mercury lamp is extremely long—in excess of 24,000 hours—provided that it is burned for at least 10 hours per start. Mercury lamps are more sensitive than fluorescent to switching and are, therefore, best applied when they can be burned for long periods without shutoff. (For safety's sake, they should be shut off at least once a week for a minimum of 30 minutes.)

To improve the color of mercury lamps, manufacturers add phosphors inside the bulb and filters on the glass itself. These improve the color to the extent that some of the best color lamps are suitable for indoor use. Lamps are available in clear, white, color corrected and white-deluxe designs, in order of increasingly better color. That is, the poorest color comes from the clear lamp, and the best color comes from the white-deluxe design. Color-improved lamps have lower output than clear lamps, but the same long life. The great advantage of mercury lamps, and indeed all HID lamps, over linear tube fluorescent lamps, is their shape. The compact shape and concentrated arc tube light source make them suitable for use in precision reflectors. Typical data on mercury vapor lamps and sketches of bulb shapes that are available are shown in Table 14.8.

Some important facts to remember about mercury vapor lamps follow.

(a) In common with all discharge lamps, mercury lamps require a ballast to operate. Because conventional magnetic ballasts are noisy, they are frequently mounted remotely from the lamp, particularly when used indoors. Remote ballast mounting (within certain limits) does not affect lamp operation. Magnetic ballasts are large, heavy and cause considerable heat loss. Newer electronic ballasts are much lighter, quieter and more efficient.

(b) Mercury lamps require a warm-up period of up to 6 minutes before giving full output, depending on type of lamp, ambient and lamp temperatures and type of ballast. In the event of a power failure of even only a few seconds, the lamps will not restrike their arcs. They must cool somewhat before they will relight, and then it takes 3–8 minutes to reach full output. It is, therefore, important to include some instant-start sources to provide emergency light after a power outage.

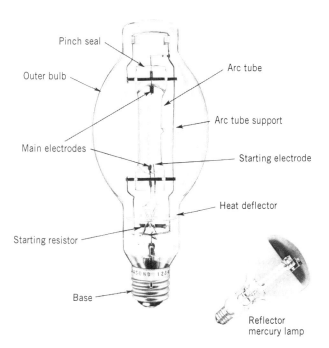

Figure 14.26 Construction details of a typical clear mercury vapor lamp. Color-corrected lamps have phosphors inside the outer bulb, and some have a stain filter on the outside. Note that the arc tube is self contained so that outer glass breakage does not extinguish the lamp.

(c) In the wiring of mercury lamps, care must be taken to keep voltage within +5% unless voltage-compensating ballasts are used.

(d) Since most magnetic mercury lamp ballasts have a high inrush current, the technologist should check with the engineer or specifier before circuiting. The inrush current, not the wattage, may limit the number of fixtures on a circuit.

(e) The lamp designation system for mercury lamps is not as simple as the one for fluorescents. Using the ANSI (American National Standards Institute) system, an H-33 GL-400/DX/BT lamp is a mercury, 400-w, BT-37 bulb, mogul base phosphor-coated, deluxe white lamp. Some manufacturers use their own abbreviation system in addition, which is sometimes clearer, depending on the manufacturer. The only sure way to know is to consult a catalog.

(f) In response to user demand for replacement of low efficiency incandescents with high efficiency lamps, manufacturers some years ago made available direct replacement screw-in low wattage mercury lamps and ballast. These good color, phosphor-coated small lamps are only somewhat more efficient than the incandescent lamp when ballast loss is included. They have the advantage of long life but the disadvantage of not being instant-start.

(g) Self-ballasted mercury lamps are available with a ballast built into the lamp. These lamps have lowered output and are relatively expensive. Their efficacy is only marginally better than incandescents, and their use is only advisable where long life is the deciding factor.

(h) Mercury vapor lamps can be dimmed with the use of an appropriate ballast and solid-state dimming control. These are available for lamps 175 w and larger.

(i) Because the arc tube that produces the light in a mercury lamp is a separate, sealed unit, the lamp will not extinguish if the outer bulb is broken. (See Figure 14.26.) Such a break, however, will release UV radiation that is normally stopped by the glass envelope. This radiation can endanger people exposed to it. As a result, it is recommended that where people can be exposed to UV radiation resulting from lamp breakage, a fixture be used that will provide

Table 14.8 Typical Data for Mercury Vapor Lamps[a,b,c]

Lamp, w	Bulb	ANSI Ordering Description			Approximate Lumens	
		Base	Abbreviation[d]	Type[e]	Initial	Mean
50	ED-17	Med.	H46DL-40-50/DX	G	1550	1250
75	ED-17	Med.	H43AV-75/DX	G, S	2800	2250
100	A-23	Med.	H38MP-100/DX	G	4300	3400
	ED-23 1/2	Mog.	H38HT-100	G, S, B	4100	3450
	R-40	Med.	H38BP-100/DX	RF, FF, VW	2850	2300
175	ED-28	Mog.	H39KC-175/DX	G, S	8500	7600
			H39KC-175/N	G	7000	6000
	R-40	Med.	H39BM-175	RF, FF, W	6100	5150
			H39BP-175/DX	RF, FF, VW	5750	4800
250	ED-28	Mog.	H37KB-250	G, S, B	12,100	10,500
			H37KC-250N	G, S	11,000	8400
400	ED-37	Mog.	H33CD-400	G, S, B	21,000	18,900
			H33GL-400/DX	G, S	23,000	19,100
	R-57	Mog.	H33DN-400/DX	G, SR	23,000	19,100
700	BT-46	Mog.	H35ND-700/DX	G, S	43,000	33,600
1000	BT-56	Mog.	H34GW-1000/DX	G	60,000	45,000
			H36GV-1000	G, S, B	57,500	48,400
			H36GW-1000/DX	G, S	63,000	47,500

Table 14.8 *(Continued)*

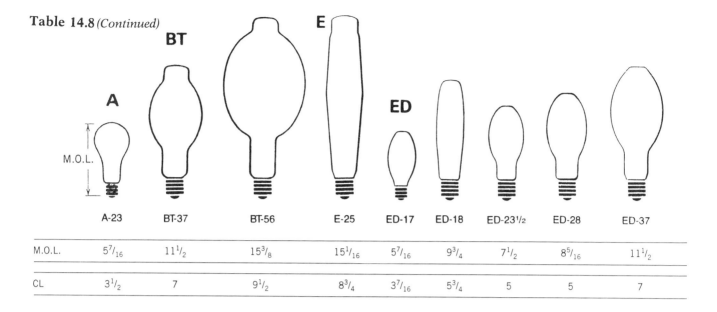

	A-23	BT-37	BT-56	E-25	ED-17	ED-18	ED-23½	ED-28	ED-37
M.O.L.	5⁷⁄₁₆	11½	15³⁄₈	15¹⁄₁₆	5⁷⁄₁₆	9¾	7½	8⁵⁄₁₆	11½
CL	3½	7	9½	8¾	3⁷⁄₁₆	5¾	5	5	7

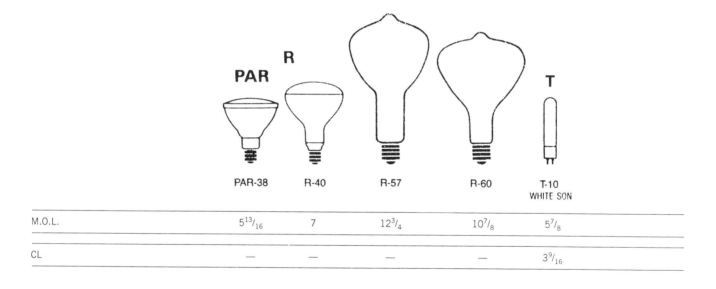

	PAR-38	R-40	R-57	R-60	T-10 WHITE SON
M.O.L.	5¹³⁄₁₆	7	12¾	10⁷⁄₈	5⁷⁄₈
CL	—	—	—	—	3⁹⁄₁₆

[a] For accurate current data, consult the manufacturers' catalogs.

[b] Lamps that self extinguish after breakage of the outer bulb are available in 100-, 175-, 250-, 400- and 1000-w sizes in various types.

[c] Rated average life for all listed lamps is 24,000+ hours except H34GW-1000/DX which is 16,000+ hours.

[d] Explanation of color suffix abbreviations: /DX, deluxe white; /N, style-tone; /C, standard white; no suffix, clear (non-phosphor-coated).

[e] Explanations of Descriptive Symbols: B, black light; FF, frosted face; G, general lighting; S, street lighting; W, wide beam; VW, very wide beam.

Source. Data extracted from published materials of various manufacturers.

adequate shielding. Alternately, manufacturers make a lamp that will automatically extinguish within 15 minutes of glass bulb puncture or breakage.

Application of mercury vapor lamps is generally in industrial areas, both indoors and outdoors, although white deluxe and color-corrected units can be used in commercial spaces indoors. However, because of relatively low efficacy, particularly of the color-improved lamps, many mercury vapor lamp installations are being replaced with more efficient, better color, metal halide lamps. New installations also favor the use of metal halide lamps for the same reasons.

14.21 The Metal Halide (MH) Lamp

This lamp began its life in the early 1960s as a modified mercury vapor lamp. Addition of elements, called halides, to the arc tube results in changes in the output, efficacy, color and life of the lamp. Because of the high efficacy and good color of this lamp, it has undergone intensive development, probably more than any other lamp type. As a result, the metal halide lamp today is available in an extremely varied range of designs and sizes, a small sampling of which appear in Table 14.9. As with all lamps, but more so with lamps under intensive development, a current manufacturer's catalog should be consulted for accurate lamp data. Typical metal halide lamp characteristics are discussed next.

a. Color

Depending on the specific lamp, the color of the light varies from a CCT of 3200 to 5000°K. This corresponds roughly to a range from the yellow-white of incandescent lamps to the blue-white of daylight lamps. CRI varies from 65 to 80, that is from good to excellent. Because the color of a halide lamp changes drastically when it is dimmed, dimming is not recommended.

b. Size, Shape and Efficacy

Metal halide lamps are available in BT, E, ED and BT shapes (which are the standard mercury lamp shapes) and PAR bulbs. (See Table 14.8.) Wattages vary from as low as 35 to 1500 w and larger. Efficacy varies from 70 to 95 LPW not including ballasts losses. This makes the metal halide lamp more efficient than most fluorescent lamps and much more efficient than all but clear mercury vapor lamps.

c. Life

Life of metal halide lamps is lower then that of mercury vapor or fluorescent lamps, ranging from 5000 hours for low wattage lamps to an average of 15,000 hours for larger lamps. The 400-w halide lamp has an exceptionally long life of 20,000 hours. As with mercury vapor lamps, life of metal halide lamps is based on a minimum of 10 burning hours per start. More frequent switching reduces the life of the lamp, and less frequent switching lengthens life. Therefore, like mercury vapor lamps, metal halide lamps are best applied where they can burn for an extended period. Also, as with mercury vapor lamps, metal halide lamps must be shut off at least once a week, for at least ½ hour, for safety reasons.

d. Burning Position

An unusual aspect of halide lamps is their extreme sensitivity to burning position. Lamps are made specifically for base up, base down, horizontal and universal (any) burning position, and they are clearly so marked. Burning a lamp in a position different from its design position severely reduces output and life.

e. Warm-up Time

As with mercury vapor lamps, halide lamps require warm-up time to achieve full output (2–3 minutes), and if extinguished most types require 5–15 minutes to restrike. As a result, spaces lighted with metal halide lamps should also have some instant-on sources, so that a short duration power outage will not cause a long blackout. Special ballasts that will considerably shorten lamp restrike time are available. At least one manufacturer produces a line of lamps with special accessories that will relight instantly after being extinguished. However, even these lamps will give only partial output when re-ignited unless the outage is less than 15 seconds.

f. Physical Shielding of Lamps

Because metal halide lamps have a tendency to shatter when they fail, particularly when operated

Table 14.9 Typical Metal Halide Lamp Data[a]

Watts	Bulb[b]	Base	Description	Average Rated Hours Life, h	Approx. Lumens Initial	Approx. Lumens Mean[c]	Efficacy,[d]
70	ED-17	Med.	Clear	15,000 V 10,000 H	5200	4200	66
			Phosphor-coated	15,000 V 10,000 H	4800	3650	59
100	ED-17	Med.	Clear	15,000 V 10,000 H	8500	6750	75
			Phosphor-coated	15,000 V 10,000 H	8000	5800	71
175	BT-28	Mog.	Clear	10,000 V 7500 H	15,000 V 13,400 H	11,500 V 10,000 H	74 65
250	BT-28	Mog.	Phosphor-coated	10,000	22,000 V 20,000 H	17,000 V 14,000 H	80 69
400	BT-37	Mog.	Clear	20,000 V 15,000 H	36,000 V 32,000 H	29,000 V 24,000 H	80 72
			Phosphor-coated	20,000 V 15,000 H	36,000 V 32,000 H	28,000 V 23,000 H	80 72

[a] Standard Design Lamps: Rated life and mean lumens are based on a minimum of 10 burning hours per start. Rated life and mean lumens are reduced for shorter burning cycles. V, Vertical burning position; H, Horizontal burning position.
[b] For diagram of bulb shapes, see Table 14.8.
[c] Taken at 40% of rated life.
[d] Calculated using average ballast loss figures.

Source. Data extracted from various manufacturers' catalogs.

continuously, manufacturers used to recommend that all metal halide lamps be used only in fully enclosed fixtures, designed to contain a shattering lamp. In recent years, however, metal halide lamp manufacturers have produced a line of lamps suitable for use in open fixtures. As a result, all metal halide lamps are marked for their recommended use—enclosed fixtures only or open/closed fixtures. Since this is a personnel safety factor, all lighting designers must be sure to use the proper lamp/fixture combination. Also, these lamps have the same UV problem as mercury vapor lamps, since they do not extinguish if the outer glass bulb is broken. A special line of lamps that do go out when the glass breaks is also being marketed.

g. Application

The excellent color and color rendering properties of the metal halide lamp combined with high efficacy and long life make it suitable for almost every application. The restrike, burning position, non-dimmability and open/closed fixture limitations must, of course, be kept in mind when selecting an application. Typical metal halide lamp data are shown in Table 14.9. A comparison of mercury vapor (MV) lamps, metal halide (MH) lamps and high pressure sodium (HPS) lamps is given in Table 14.10.

14.22 High Pressure Sodium (HPS) Lamps

The HPS lamp is the third and last HID source that we will study. It was also the last one developed. This lamp is also known as a SON lamp. In basic construction, it is similar to the other HID lamps in that it has a light-producing arc tube inside an

Table 14.10 Comparative Characteristics; Standard Mercury Vapor, Metal Halide and High Pressure Sodium Lamps, with Magnetic Ballasts

Item	Mercury Vapor	Metal Halide	High Pressure Sodium
Color of light	Poor to fair	Good to excellent	Fair to good
Average life	24,000 h	7500–20,000 h	10,000–24,000 h
Efficacy, including ballast loss	20–50 lpw	70–100 lpw	65–130 lpw
Start-up time	3–6 min	2–3 min	3–4 min
Restrike time	3–8 min	8–10 min	½–1½ min
Burning position	Any	As designed only	Any
Safety requirements	Shield against UV exposure	Shield against UV exposure; use in enclosed fixtures unless noted otherwise	None

Table 14.11 Typical Data for High Pressure Sodium Lamps

Watts	Bulb[a]	Base[b]	Approx. Lumens Initial	Approx. Lumens Mean	Lamp and Ballast[c] Description	Efficacy[c]
			Standard Lamps[d]			
50	ED-17	Med.	4000	3600	Clear	60
			3800	3420	Coated	57
70	ED-17	Med.	6300	5700	Clear	68
			6000	5400	Coated	65
150	ED-17	Med.	16,000	14,000	Clear	85
			15,000	13,500	Coated	80
250	ET-18	Mogul	29,000	26,000	Clear	95
	BT-28		26,000	23,400	Coated	85
400	ED-18	Mogul	50,000	45,000	Clear	105
	ED-37		47,500	42,750	Coated	100
			White Lamps[d]			
35	T-10	—	1300	1050	Clear	30
50	ED-17	—	2350	1850	Coated	38
100	ED-17	—	4800	3800	Coated	42

[a] For bulb shapes and dimensions see Table 14.8.
[b] All lamp bases are screw type except T-10 lamps that have prefocus pin-type base.
[c] Efficacy figures include ballast losses for core-and-coil ballasts for standard lamps, and electronic ballast for white lamps.
[d] Lamp life is 24,000+ hours for standard lamps and 10,000 hours for white lamps (based on operation of at least 10 burning hours per start).

Source. Data extracted from various manufacturers' catalogs.

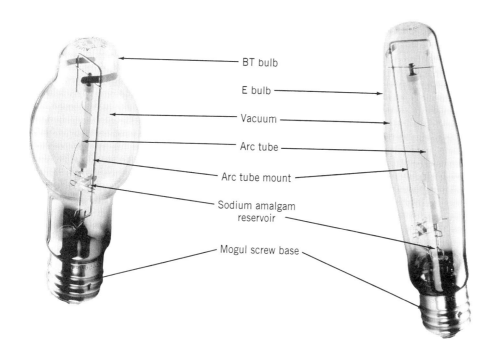

Figure 14.27 The main features of two standard HPS lamps. (Photo courtesy of OSRAM Sylvania, Inc.)

outer glass envelope. There the similarity ends. The sodium, from which the lamp takes its name, is contained in a reservoir adjacent to a ceramic arc tube, as can be seen in Figure 14.27. The lamp has extremely high efficacy and a very long life, is burnable in any position, has shorter warm-up and restrike time than either mercury vapor or metal halide lamps and has excellent lumen maintenance throughout its long life. Also, because it produces very little UV and has no tendency to explode, it can be used in open unguarded fixtures. Its one major disadvantage is its color. The light it produces has a distinct yellow tinge, which makes it unsuitable for indoor use unless it is used along with another whiter light source. The HPS lamp color can be improved by using a phosphor-coated bulb.

See Table 14.10 for a comparison of the principal characteristics of standard HID sources.

A white HPS lamp was developed some years ago. Its light output has essentially the same color as an incandescent lamp. These low wattage lamps (35, 50 and 100 w) are intended specifically for accent and display lighting. They have lower efficacy and shorter life than standard SON lamps. Typical data for both standard and white HPS lamps are given in Table 14.11. The high pressure sodium lamp (SON) must not be confused with the low pressure sodium lamp (SOX). The latter has the highest efficacy of any commercial lamp, but its deep yellow light color makes it suitable only for road lighting.

14.23 Induction Lamps

Several manufacturers have recently introduced similar designs of what is essentially a fluorescent lamp, except that it does not use electrodes. Two such designs are shown in Figures 14.28 and 14.29. The lamp is filled with low pressure mercury vapor. When ionized by the high frequency induction

824 / HOW LIGHT BEHAVES; LIGHTING FUNDAMENTALS

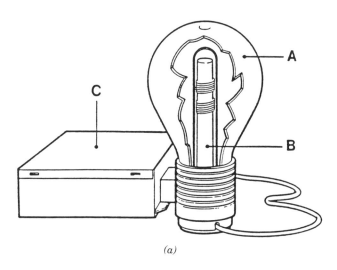

(a)

(b)

Figure 14.28 (a) A schematic cutaway of the induction lamp shows its operating principles. The high frequency generator (C) produces a high frequency current, which circulates in the coil on the power coupler (B). This ionizes the mercury vapor inside the lamp (A) producing UV light. The UV strikes the fluorescent coating inside the lamp, producing visible light.
(b) A photo of the lamp and its high frequency generator. The lamp, which is rated 85 w including all losses, is 4.33 in. in diameter and 7.5 in. high overall. The generator measures 5.5 in. × 5.1 in. × 1.77 in. high. The lamp and generator together weigh 2.2 lb. (Illustration and photo courtesy of Philips Lighting Company.)

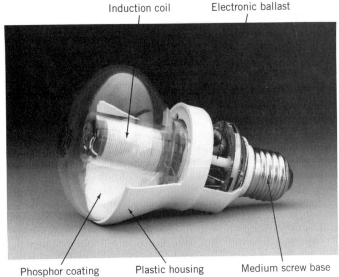

Figure 14.29 Cutaway of the GE induction lamp showing the essential elements: induction coil, phosphor-coated bulb with mercury vapor fill and electronic ballast. This 23-w lamp in a modified R (reflector)-shaped bulb has a height just under 5 in. overall and a 3 in. maximum bulb diameter. (Photo courtesy of GE Lighting.)

coil inside the lamp, the mercury vapor produces ultraviolet light. This, in turn, strikes the fluorescent coating on the inside of the lamp, producing visible light. This is exactly the light-producing process used by standard fluorescent lamps. The difference is that the gas is ionized by an induction coil. As a result, the lamp is known as an induction lamp.

Characteristics of the design shown in Figure 14.28 follow:

Wattage	85 w total
Lumen output	5500 lm ± 10%
System efficacy	5500/85 = 65 lpw
Color of light (2 lamp types)	3000 or 4000°K
Color rendering index	better than 85
Ignition time	under ½ s
Time to 75% output	up to 1 min
Hot restrike time after outage	less than ½ s
Lamp life to 70% output	60,000 h
Burning position	any

Another model is available with a system wattage of 55 w and a light output of 3500 lm.

Table 14.12 Efficacy of Various Light Sources

Source	Efficacy, lpw
Candle	0.1
Oil lamp	0.3
Original Edison lamp	1.4
1910 Edison lamp	4.5
Modern incandescent lamp	8–22
Tungsten halogen lamp	16–20
Standard fluorescent lamp[a,b,d]	35–80
Compact fluorescent lamp[c]	40–75
Mercury lamp[a,d]	30–60
Metal halide lamp[a,d]	70–115
High pressure sodium[a,d]	45–130
Induction lamp[c]	48–85
Maximum predicted by year 2010	150
Maximum theoretical limit	approx. 250

[a] Includes ballast losses.
[b] With electronic ballasts these figures become 40–100 lpw.
[c] With electronic ballast.
[d] With standard core-and-coil ballast.

Characteristics of the design shown in Figure 14.29 follow:

Wattage	23 w
Lumen output	1100 lm
Efficacy	1100/23 = 48 lpw
Color	3000°K
Color rendering index	82
Lamp life to 70% output	10,000 h

The preceding characteristics will undoubtedly be improved and expanded with constant development.

The principal advantages of this lamp over linear fluorescents, compact fluorescents and the various HID sources of comparable output are its small size, which permits accurate focusing with reflectors; its extremely long life; and its excellent color spectrum. One of the special problems involved in fixture design for this lamp is concerned with shielding the lamp's electromagnetic radiation and minimizing harmonic conduction. Both of these items are highly technical and cannot be discussed here. It is sufficient, however, to say that careful fixture design can minimize these effects to the point that they meet all required standards, making the induction lamp a commercially acceptable, and highly desirable, addition to the lamp market.

A summary of the luminous efficiency of the light sources discussed previously, plus other relevant efficacy data, is given in Table 14.12. The color of the light produced by these sources affects, sometimes dramatically, the appearance of colored objects being viewed. A summary of these effects is given in Table 14.13 to assist the technologist/designer in selecting sources for particular application.

The lamp types already discussed—incandescent, halogen, fluorescent, mercury vapor, metal halide, high pressure sodium and induction—are the principal types used in conventional lighting design. In addition, however, there are many other types such as low pressure sodium, xenon, krypton, high output sodium lamp with light pipe, cold cathode fluorescent, neon tubing, miniature lamps, photo-optic lamps and others that are used for special design applications. Technologists who specialize in lighting design soon become familiar with these special lamps and their use.

How Light Is Used: Lighting Fixtures

In the preceding sections we studied some of the important facts about how light is produced and how it acts. We also discussed in some detail how light and vision act together. In other words, we now know how electric light is produced and how we see. The next subject to be studied is how this light is controlled. The lamp sources produce the light, generally radiating in all directions. What do we do with this light to make it useful? How do we redirect the light energy produced so that it provides room illumination? Like most technical questions, the answer is simple in principle but more complex in practice. What we do is to build enclosures for the lamp sources. These enclosures, which are generally called lighting fixtures, or luminaires (the terms are interchangeable) are designed to

- Hold and energize the lamps(s).
- Direct the light.
- Change the quality of the light produced.
- Provide shielding (cutoff) to prevent direct glare.
- Where necessary, protect users from dangerous radiation and/or from flying pieces of lamps that rupture violently.

Table 14.13 Effect of Illuminant on Object Colors

Lamp	Whiteness	Colors Enhanced	Colors Grayed	Notes
Incandescent	Yellowish	Red, orange, yellow	Blue, green	Good overall color rendering
Halogen	White	Red, yellow, orange	Blue, green	Same as incandescent
Fluorescent				
Cool white	White	Yellow, orange, blue	Red	Blends with daylight
Daylight	Bluish	Green, blue	Red, orange	
Triphosphor				
3500°	Pale yellowish	Red, orange, green	Deep red	T8 bulb
4100°	Pale greenish	Red, blue, orange, green	Deep red	T8 bulb
Mercury				
Clear	Blue green	Blue, green	Red, orange	Poor overall color rendering
Deluxe	Pale purplish	Blue, red	Green	Shift over life to greenish
Warm deluxe	Pinkish	Blue, red	Green	Can be blended with fluorescent
Metal halide				
Clear	White	Blue, green, yellow	Red	Shifts to pinkish over life
Phosphor-coated	White	Blue, green, yellow	Red	Can be blended with most sources
HPS				
Standard	Yellowish	Yellow	Red, blue	Blend with a white source for indoor use
White	Yellowish	Red, green, yellow	Deep red, deep blue	Simulates incandescent

Not all fixtures do all of these things. For instance, a simple incandescent lampholder only holds and energizes the lamp, yet most people would call it a lighting fixture, although a very simple one. In the trade, a lampholder is listed as a lighting fixture.

Sometimes, not all five characteristics are required. Reflector-type lamps (see Section 14.13 and Figure 14.16) have their light-directing mechanism built into the lamp and produce a spotlight or floodlight beam when they are installed in a simple lampholder. These lamps have their own optical control. Therefore, the second function on the list is not necessary. Also, we sometimes combine the fixture with the building structure to get the desired lighting. We build coves, coffers, luminous ceilings and other arrangements where the structure acts as a part of the lighting control system. This type of lighting, referred to as architectural lighting elements, will be discussed separately from lighting fixtures. These elements are of great importance to electrical technologists, because they are always detailed on the drawings and that task is generally assigned to the technologist.

14.24 Lampholders

As already mentioned, the most elementary lighting fixture is a simple lampholder. Lampholders can be cord or box-mounted sockets for incandescent lamps as in Figure 14.30, or wiring "strips" for fluorescent lamps as in Figure 14.31. The fluorescent lamp wiring channel also provides mounting for the lamp ballast. This ballast space requirement applies to all fixtures handling discharge-type lamps. HID lamps are almost never used in a simple lampholder. In addition to holding its ballast, the HID lighting fixture should provide some

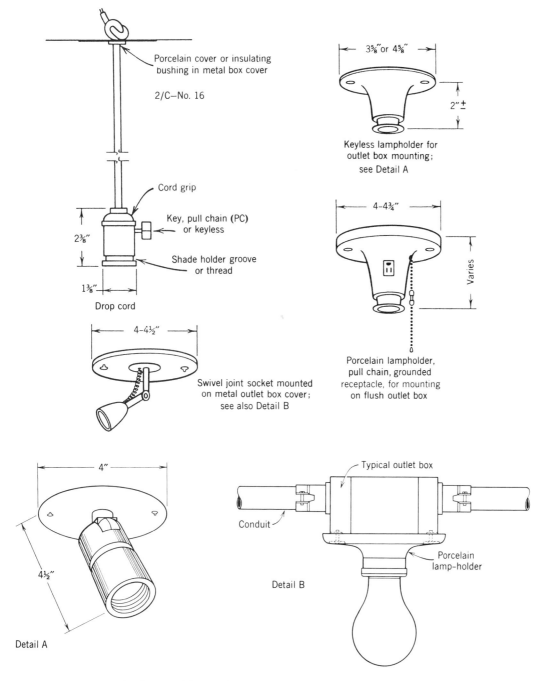

Figure 14.30 Details of typical basic, bare lamp incandescent lampholder. The terms *keyed* and *keyless* refer to the presence or absence, respectively, of a switching mechanism in the lampholder.

828 / HOW LIGHT BEHAVES; LIGHTING FUNDAMENTALS

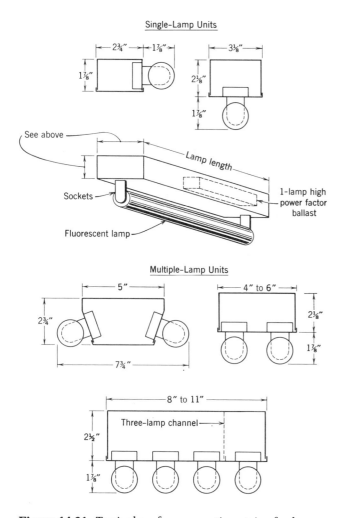

Figure 14.31 Typical surface-mounting strips for bare fluorescent lamps. Dimensions shown are typical and vary among different manufacturers. Lamps shown are T-12 (1½ in. diameter). Dimensions will change when using T-8 lamps. Ballast location within the fixture varies with the strip design. Fixture design should be such that it is not necessary to demount (take down) the fixture in order to replace the ballast.

shielding because the lamps are normally very bright and can cause direct glare.

14.25 Reflectors and Shields

The lampholders described in Section 14.24 can be readily modified by the addition of various types of reflectors. These reflectors are finished with a high reflectance paint or metallizing process, so as to reflect most of the lamp's light. The reflectors provide two of the functions listed previously—direction and shielding or cutoff. Their primary purpose is reflection, but cutoff and some shielding are automatically provided as well. The reflectors can be simple units that fasten onto lampholders (see Figure 14.32) or an integral part of the fixture.

Fixtures consisting of a lampholder channel and a reflector, with no diffuser, are generally called industrial fixtures. This is because that type of unit was originally the lighting unit commonly used in industrial areas. The best quality units of this type carry an *RLM (Reflector Luminaire Manufacturer)* label to indicate that the fixture quality meets the industry standard. Today these RLM and RLM-type open reflector units are used in nonindustrial installations as well. Figure 14.33 shows a number of open reflector fixtures.

Simple reflectors like those in Figures 14.32 and 14.33 provide some shielding, although their primary purpose is to redirect the light by reflection. In the case of fluorescent units, the shielding provided is lengthwise on the fixture, that is, it shields the length of the lamp. This shielding is the important one for fluorescent lamps. If endwise shielding is also desired, *baffles* (also called shields or *louvers*) must be added to the unit as in Figure 14.34. This illustrated fixture provides shielding in both directions. The *shielding angle* is the angle between the horizontal and the line of sight below which the lamp itself cannot be seen—that is, it is shielded. See Figure 14.35, which displays graphically the principle of shielding in both the crosswise (transverse) and endwise (longitudinal) directions for fluorescent fixtures.

Because there is a possibility of confusion when speaking of the direction of shielding, the terms require explanation. Refer to Figures 14.34 and 14.35. The length of the fluorescent tubes is seen when we look at them from the side—that is, crosswise. Therefore, shielding in this viewing direction, which shields the length of the lamps, is called crosswise or *transverse shielding*. See Figure 14.35(a). On the other hand, the shielding provided by the baffles of Figure 14.34 and the louvre of Figure 14.35(b) shields the tubes as they are viewed lengthwise or endwise. In the lengthwise view, we do not see the length of the tube; we see it end on. This shielding is, therefore, correctly called endwise or *longitudinal shielding* (longitudinal means the long way). Do not confuse it with lengthwise shielding, which means, as already explained, shielding of the length of the tube.

The same principle applies to the various types

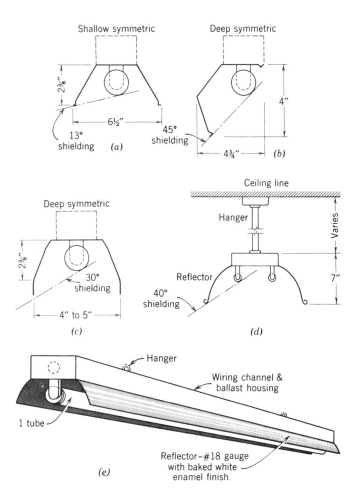

Figure 14.32 Accessory reflectors (a), (b) and (c) provide varying degrees of shielding. Commercial-grade industrial fixture (d) normally provides a minimum of 35° shielding. A typical pendant single-lamp fluorescent strip with type (a) reflectors, is shown in detail (e).

of louvers and baffles that are used to provide lamp cutoff in incandescent and HID fixtures. These are illustrated in Figure 14.36. Shielding in these fixtures is extremely important, because of the very high source brightness. As should be obvious from Figures 14.36 and 14.37, the better the shielding, the poorer the spread of light. Therefore, the cutoff angle is a compromise between the need to reduce direct glare and the need to spread the fixture's light output. Fixtures are available in a wide range of designs with varying *cutoff angles*.

In fluorescent fixtures, good shielding in both directions is provided by a two-way louver. This consists of slats of metal or plastic in both directions, with height and spacing designed to give the shielding angle wanted. Such a louver is called an egg-crate louver, because it looks like an egg crate. See Figure 14.37. This type of two-way louver provides a relatively low brightness surface over the bottom of the fixture instead of the very bright bare fluorescent lamps. The same louvering technique can be used with an incandescent fixture, by making the louvers round. Usually these louvers are concentric circles mounted on the fixture bottom. See Figure 14.38 for an interesting example of this design.

One other aspect of reflectors should be mentioned here, but since it is very complex, it will only be touched on briefly. This is the use of reflectors to focus the light of a point source. You will

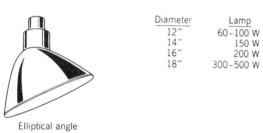

Figure 14.33 Standard shapes for open industrial reflectors for general-service incandescent lamps. The table refers to incandescent lamps only. Reflectors for HID lamps are generally of the deep bowl design because of the need for a large shielding angle, due to the high source brightness.

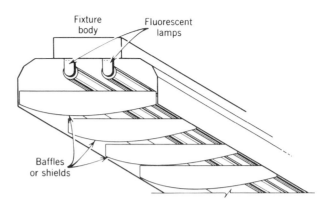

Figure 14.34 Basic fluorescent fixture with lengthwise (endwise) shielding provided by baffles and transverse (crosswise) shielding provided by the reflector.

remember that we stated that one excellent characteristic of incandescent (and HID) lamps was their ability to be focused. This results from the lamp being almost a point source, which can be focused by using a proper reflector. Reflector design is complex and far beyond the scope of this book. However, since the electrical technologist will have occasion to draw and examine fixture details, he or she should be able to recognize reflector types. Figures 14.39 and 14.40 shows two of the principal types in common use. The more important, by far, is the *parabolic reflector*. The parabolic reflector's ability to direct light as shown in Figure 14.40 is used to produce very low brightness fluorescent fixtures by using either a miniature parabolic wedge egg-crate louver as shown in Figure 14.41 or a large parabolic baffle as shown in Figure 14.42. Both types are in very common use in modern installations. The parabolic baffle type of fixture is more efficient and is, therefore, more in favor due to energy savings considerations.

The use of reflectors in fixtures with point sources such as incandescent and tungsten-halogen is extremely common. As a rule, a fixture without a well-designed interior reflector is very inefficient because much of the source output light is trapped inside the fixture. An exception to this rule is fixtures using lamps with built-in reflectors, such as R and PAR lamps. A typical fixture of this type is shown in Figure 14.43. Finally, the characteristics of a lamp with a built-in reflector can be combined with an auxiliary reflector in the fixture to produce special-purpose fixtures, such as wall-lighting (wall-wash) units. One such typical design is shown in Figure 14.44. Although HID lamps are not really point sources (particularly the phosphor-coated variety), they can be treated as such if the fixture is large enough. A typical HID unit is shown in Figure 14.45.

14.26 Diffusers

The next item in the list of functions of a fixture that we will discuss in detail is the item concerned with change of light quality. You will remember that in quality we included contrast, glare, dif-

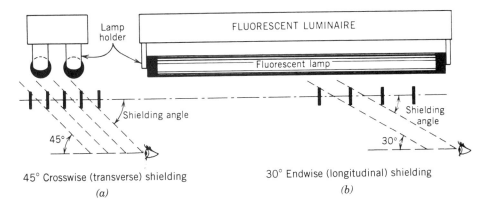

Figure 14.35 Shielding of fluorescent lamps is not as critical as that of HID or incandescent lamps because of their lower luminance (brightness). The 45° × 30° crosswise/endwise shielding shown is considered excellent and 35° × 30° is generally satisfactory. The crosswise shielding is usually higher than the endwise shielding because the lamps are brighter and cause more glare when viewed crosswise than when viewed endwise. Opaque shielding elements reduce direct glare more than translucent plastic units. (From Stein, B., and Reynolds, J. *Mechanical and Electrical Equipment for Buildings*, 8th ed., 1992, reproduced by permission of John Wiley & Sons.)

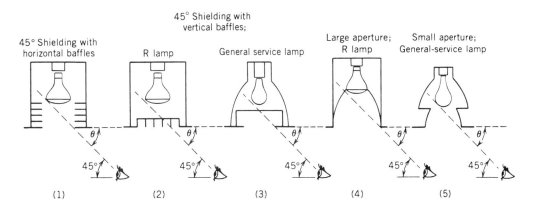

Baffled downlights (1), (2), (3) control unwanted high-angle light by cutoff as illustrated. Black baffles aid by absorbing and appearing dark.

Cones (4), (5) control brightness by cutoff and by redirection of light due to shape. They are either parabolic or elliptical. A light polished finish such as aluminum appears dull; a black polished finish appears unlighted.

Figure 14.36 Methods for shielding light source. Incandescent lamps use circular shields and have the same cutoff angle in all directions. A shielding angle of 45° minimum is recommended because of the extremely high luminance (brightness) of incandescent lamps. Black finishes require high quality maintenance, since dust shows as a bright reflection. (From Stein, B., and Reynolds, J. *Mechanical and Electrical Equipment for Buildings*, 8th ed., 1992, reproduced by permission of John Wiley & Sons.)

832 / HOW LIGHT BEHAVES; LIGHTING FUNDAMENTALS

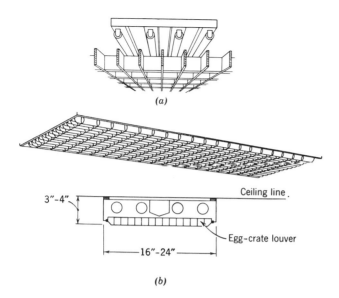

Figure 14.37 Two-way shielding is provided by louvering in both directions. The principle of the design is shown in *(a)*, and an actual fixture with an egg-crate louver is shown in *(b)*. Louver material can be metal or plastic.

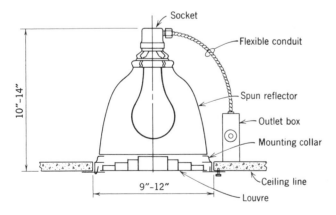

Figure 14.38 Recessed incandescent downlight for general-service lamp. Note that the circular louvers increase in depth as they near the center. This increases the shielding angle so that the source remains shielded as the viewer approaches the fixture. Dimensions vary with the wattage of the lamp.

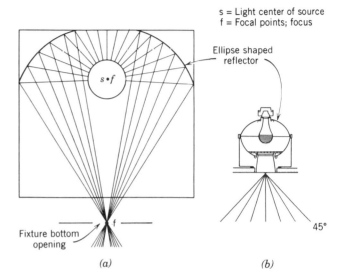

Figure 14.39 The action of an ellipse-shaped reflector in a lighting fixture is shown graphically in diagram *(a)*. An ellipse has two focal points, labelled *f* on the diagram. When a light source *s* is placed at one focal point, as shown, all the light that strikes the ellipse-shaped reflector is reflected back and concentrates at the second focal point. This principle is used in the design of the lighting fixture shown in *(b)*. The light from the source is projected up toward the reflector by using a silvered bowl lamp. It is reflected down and exits the fixture at the bottom, where it spreads out at about a 45° angle. The fixture opening can, therefore, be very small (a "pinhole" opening) without large losses due to trapping of light. The cone at the bottom of the fixture provides high angle cutoff.

fuseness and color. Adding a diffuser to a fixture affects the quality of the light produced by the fixture. The *diffuser* may be a piece of glass or plastic or a complex lens. It is called a diffuser, because the original diffuser was a white or frosted glass bowl surrounding an incandescent lamp. See Figures 14.46 and 14.47. Its function was to diffuse the light. In so doing, it also decreased glare, in-

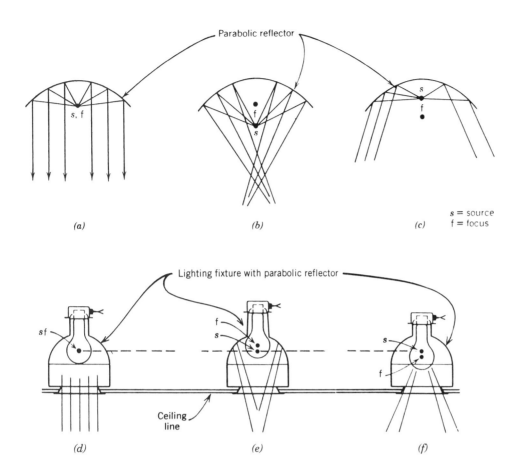

Figure 14.40 The parabolic reflector has one focal point f. *(a)* When the light source s is at the focal point, the light is reflected straight out. *(b)* When the source is below the focal point, the emitted light is concentrated, and *(c)* when the source is above the focal point, the light is spread out. This effect is used in the lighting fixtures shown in *(d), (e)* and *(f)*. All three have the same reflector shape; the difference is the position of the bulb's light source center s. In *(d)* the light exits directly. Moving the bulb down *(e)* concentrates the light, and moving it up *(f)* spreads the light. Therefore, the ceiling opening can be small with *(e)* but must be large with *(f)* to avoid light losses that will reduce fixture efficiency. The location of point s on the bulb is given by the LCL (light center length) listed in lamp catalogs. (From Stein, B., and Reynolds, J. *Mechanical and Electrical Equipment for Buildings*, 8th ed., 1992, reprinted by permission of John Wiley & Sons.)

creased diffuseness, decreased sharp shadows and contrast and also, incidentally, reduced the quantity of light. Such diffusers are still very much in use, as for instance the currently used fixture shown in Figure 14.48. In industrial use, a white glass diffusing element is frequently added to a simple open RLM dome, to reduce direct and reflected glare.

The use of lenses, either glass or plastic, as fixture diffusers is also widespread. These lenses, which are of many different designs, redirect the light, reduce the fixture brightness and glare and, to an extent, diffuse the light. A lens acts much more efficiently than white glass or plastic and is generally more costly. A few examples of fixtures and diffusers are shown in Figures 14.49 and 14.50

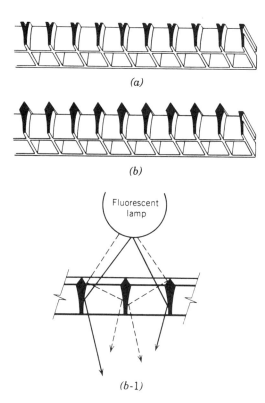

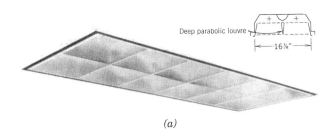

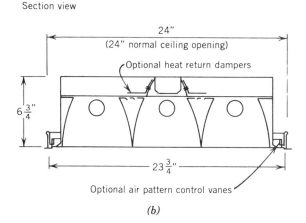

Figure 14.41 *(a)* Section through a common miniature parabolic wedge egg-crate louver. These units give exceptionally low brightness when seen at normal viewing angle (see Figure 14.52) and are, therefore, useful in video display terminal (VDT) areas. Most such units are made of aluminized plastic. Fixtures equipped with these units have low overall efficiency due to the large amount of light trapped by the broad top of each parabolic wedge.
(b) A modified wedge design uses a curved top on each wedge to redirect and utilize light striking the top. *(b-1)* Solid lines represent light rays redirected by the bottom curve. Dotted lines show light redirected by the top curve of the louver *(b)*, which was lost in the design of *(a)*. Typical louver cell dimensions are ½-in. cube for a design *(a)*, which gives a 45° shielding angle and ⅝ to ¾ in. square by ½ in. high for design *(b)* giving a 35–45° shielding angle. (From Stein, B., and Reynolds, J. *Mechanical and Electrical Equipment for Buildings*, 8th ed., 1992, reprinted by permission of John Wiley & Sons.)

Figure 14.42 *(a)* Typical two-lamp parabolic louver fixture. All surfaces are semi-specular aluminum. The entire fixture has a pleasant low brightness appearance as shown in the photograph. *(b)* Section through a three-lamp deep cell parabolic baffle fluorescent fixture. The fixture exhibits the low brightness that is typical of parabolic louver/baffle fixtures. The unit is also extremely efficient. In addition, it has optional heat return dampers that are installed when the fixture is equipped with a heat removal air duct connection.

for incandescents and in Figure 14.51 for fluorescents. An interesting comparison between the surface brightness of a miniature parabolic-cell egg-crate louver and a high quality plastic lens diffuser, is shown in Figure 14.52.

You have undoubtedly noticed that the luminaires illustrated up to this point are ceiling mounted and use general-service incandescent lamps. (Similar designs are available for low wattage HID lamps and compact fluorescents.) However, in order to take advantage of the energy-conserving features of larger wattage HID lamps

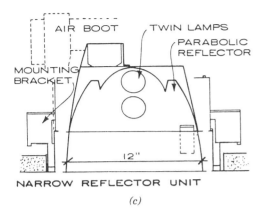

Figure 14.42 *(Continued) (c)* This unusual two-lamp narrow fixture design uses a double parabolic reflector. This permits switching off one of the lamps without changing the light distribution of the fixture. Because of the lamp arrangement, the unit is less efficient than fixtures *(a)* and *(b)*. [*(a)* Courtesy of Columbia Lighting, Inc. *(b)* Courtesy of Lighting Products, Inc. *(c)* Reproduced with permission from Ramsey and Sleeper, *Architectural Graphic Standards*, 8th ed., 1988, © Wiley, reproduced by permission of John Wiley & Sons.]

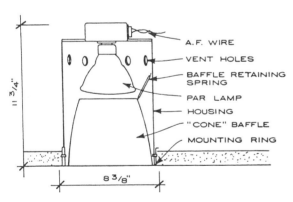

Downlights without reflectors or lenses are commonly called "cans." They have cylindrical housings and rely on a PAR or R lamp for optical control.

Figure 14.43 Fixture with PAR (parabolic aluminized reflector) lamp that contains its own reflector. The fixture does not require any reflector since the required optical control is built into the lamp. The very deep curved cone baffle provides the high angle shielding needed for this very bright source. (From Ramsey/Sleeper, *Architectural Graphic Standards*, 8th ed., 1988, © Wiley, reproduced by permission of John Wiley & Sons.)

(100 w and above) and of the long life, good color, focusability and higher efficiency of tungsten halogen lamps, another approach was needed. This is because these sources are too bright to place in a fixture where the lamp can be seen, even behind a good diffuser. The answer to this problem is to use indirect lighting fixtures, that is, fixtures that throw their light onto the wall and ceiling. In such units, the lamp is completely shielded so that the problem of direct glare from an extremely bright source is solved. Typical examples of this type of fixture design are shown in Section 14.28 that discusses lighting methods, including indirect lighting.

The fifth and last function of a lighting fixture as listed previously is concerned with safety and protection for the lamp users. Protection from the effects of a lamp rupture is a mechanical design function, accomplished by physical enclosure of the lamp. Protection from UV radiation that may result if the outer bulb of an HID source is broken is not quite so simple. It is, however, an aspect of fixture design that, although of interest, does not concern a lighting technologist except for the knowledge that such shielding is required for certain lamps.

14.27 Lighting Fixture Construction, Installation and Appraisal

A number of tasks that are frequently assigned to the electrical technologist relate to lighting fixtures:

- Checking lighting fixture shop drawings to make sure that specifications are met.
- Inspecting fixtures to determine whether materials and workmanship are satisfactory.
- Field checking to determine that installation is satisfactory.

The following list of items will help the technologist understand these tasks and perform them properly.

a. Construction

(1) All fixtures should be wired and constructed to comply with local codes, **NEC** (Article 410)

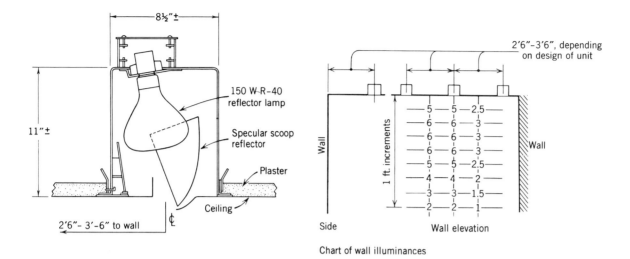

Figure 14.44 A wall-wash lighting fixture, used to produce a lighted wall with smooth changes in the wall illuminance. The optical light control is a combination of beam control from the built-in lamp reflector and beam projection from the fixture to the wall by the fixture's scoop-type reflector. The chart in *(b)* shows the wall illuminance for a particular fixture spacing. Some fixtures of this type have adjustable position lamps that will change the beam pattern. This enables the fixture to be used for varying applications.

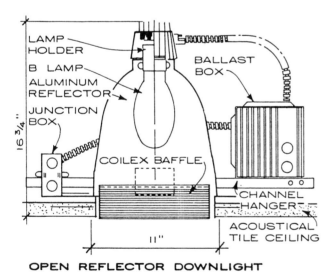

Figure 14.45 Open reflector downlight for an HID source. The coil-type baffle shown helps reduce the fixture brightness. Due to the source size, the fixture is large and requires a deep hung ceiling to accommodate it. (From Ramsey and Sleeper, *Architectural Graphic Standards*, 8th ed., 1988, © John Wiley & Sons, reproduced by permission of John Wiley & Sons.)

and Underwriters Laboratories Standard for Lighting Fixtures and should carry the UL label, where label service is available. The RLM standards should be followed for all porcelain-enameled fixtures.

(2) Fixtures should generally be constructed of 20 gauge (0.0359 in. thick) steel minimum. Cast portions of fixtures should be no less than 1/16 in. thick.

(3) All metals should be coated. The final coat should be a baked-enamel white paint of minimum 85% reflectance, except for anodized or Alzac surfaces. All hardware should be cadmium-plated or otherwise rustproofed.

(4) No point on the outside surface of any fixture should exceed 90°C after installation when operated continuously. For an exception, see NEC Article 410 M.

(5) Each fixture should be identified by a label carrying the manufacturer's name and address and the fixture catalog number.

(6) Glass diffuser panels in fluorescent fixtures should be mounted in a metal frame. Plastic

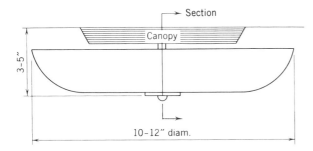
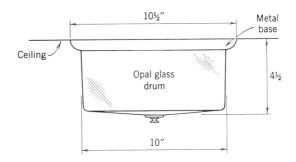

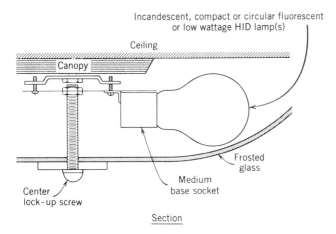

Figure 14.46 Simple ceiling-type diffuser for circular fluorescent or incandescent lamps. Frosted or white glass diffuses light and reduces glare. An open glass dish of this type quickly fills with dust, bugs and the like, sharply reducing light output and increasing the required maintenance. In recent years, fixtures of this basic design have been made for wall mounting. This reduces dirt accumulation, permits air circulation and considerably reduces the required maintenance. Such wall mount luminaires are available for use with compact fluorescent and low wattage HID lamps.

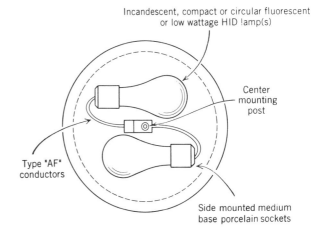

Figure 14.47 Common, enclosed center lockup drum-type fixture. This unit is essentially the same as that of Figure 14.46 except that the diffuser entirely surrounds the lamp. As a result, the fixture is cleaner but runs hotter. Like the open version of that figure, this fixture too is available for fluorescent and HID sources. This type of fixture is also made with spring attached drum, to make relamping easier. In such fixtures, the drum pulls down to expose lamps but does not come off.

diffusers should be suitably hinged. Lay-in plastic diffusers should not be used.
(7) Plastic diffusers should be of the slow-burning or self-extinguishing type with low smoke-density rating and low heat-distortion temperatures. This temperature should be low enough so that the plastic diffuser will bend and drop out of the fixture before reaching ignition temperature. All plastic diffusers and other plastic elements must meet the requirements of local fire codes.
(8) It is imperative that plastics used in air-handling fixtures be of the noncombustible, low-smoke-density type. These requirements also apply to other nonmetallic components of such fixtures.
(9) All plastic diffusers should be clearly marked with their composition material, trade name, and manufacturer's name and identification number. Results of ASTM combustion tests should be submitted with fixture shop drawings.

The characteristics of many plastic diffusers change radically with age and exposure to ultraviolet light. Glass and virgin acrylic plastic are stable in color and strength. Other plastics may yellow and even turn brown,

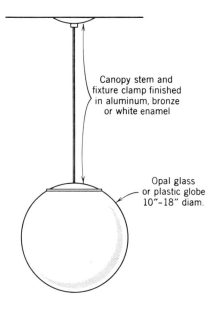

Figure 14.48 Globe-type diffuser for incandescent fixture is identical in principle to the fixtures in Figures 14.46 and 14.47 except for shape and size. To achieve good diffusion and lamp hiding, these fixtures sacrifice efficiency.

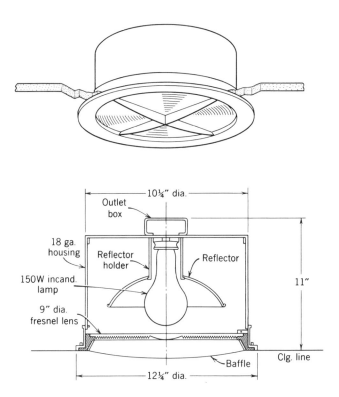

Figure 14.50 An unusual downlight that combines the light-directing effect of a Fresnel lens with the shielding of a decorative baffle. A Fresnel lens is a thin flat lens that performs the same function as a thick curved lens.

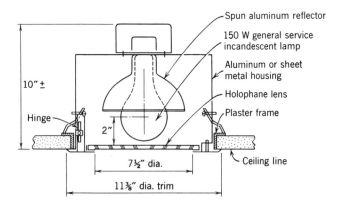

Figure 14.49 Typical recessed downlight with a glass lens-type diffuser. The Holophane Company developed these prismatic lenses, which are known in the fixture trade simply as Holophane lenses. There are many types of such lenses, each having its own particular optical characteristics. In recent years, plastic prismatic lenses have, for the most part, replaced glass lenses. Fixtures with essentially similar design are available for use with low wattage HID and compact fluorescent lamps.

thus reducing light transmission as well as changing the fixture appearance. Some plastics that are initially very tough and vandalproof become brittle with age and exposure to weather or to the ultraviolet light of a mercury, HID or fluorescent source. Therefore, the technologist must investigate the long-range as well as initial characteristics of all diffuser elements before (specification and) approval, particularly when a substitute material is being proposed.

(10) Ballasts should be mounted in fixtures with captive screws on the fixture body, to allow ballast replacement without fixture removal.

(11) All fixtures mounted outdoors, whether under canopies or directly exposed to the weather, should be constructed of appropriate weather-resistant materials and finishes, including gasketing to prevent entrance of water into wiring, and should be marked by the manufacturers, "Suitable for Outdoor Use."

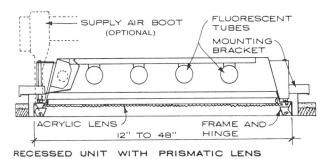

Figure 14.51 Typical recessed fluorescent fixture. These units are designed to fit into a standard hung-ceiling grid. Typical approximate dimensions in feet with the commonly-used associated lamps are: 1×4 ft, one or two lamps; 2×4 ft, three or four lamps; 2×2 ft, two or three U lamps or two to four straight lamps; 3×3 ft with three to six straight lamps; 4×4 ft with four to eight straight lamps. (From Ramsey and Sleeper, *Architectural Graphic Standards*, 8th ed., 1988, reproduced by permission of John Wiley & Sons.)

Figure 14.52 In the illustration, the author is checking the luminance (brightness) of a fluorescent fixture equipped with a miniature parabolic wedge louver, which has very low luminance above 45° (see Figure 14.41). This characteristic is easily visible in comparison to the adjacent fixture, which uses a plastic prismatic lens diffuser.

b. Installation

(12) Although some codes allow fluorescent fixtures weighing less than 40 lb to be mounted directly on the horizontal metal framework of hung-ceiling systems, experience has shown that vibration, metal bending, routine maintenance operation on equipment in hung ceilings and poor workmanship can cause such fixtures to fall, endangering life. It is, therefore, strongly recommended that all fixtures—surface, pendant, or recessed—whether mounted individually or in rows, be supported from the building structure. In no case should the ceiling system itself be used for support. This is particularly important in the case of an exposed Z spline ceiling system. This item should be checked and approved by the project structural engineer and/or architect before installation.

(13) Fixtures installed in bathrooms should not have an integral receptacle and when installed on walls, should have nonmetallic bodies. These are safety precautions.

(14) Recessed fluorescent fixtures, usually called *troffers*, are so common that a system of standards for mounting in hung ceilings was developed by the National Electrical Manufacturer's Association (NEMA). The primary purpose of this standard was to avoid problems of installation in the field, that is, to ensure compatibility between the fixture and the hung ceiling in which it is to be mounted. Although the standard has been rescinded by NEMA, the information in it is still useful in our opinion. The standard, NEMA LE-1 (1963), rescinded, defines five types of recessed troffer. Details of each are shown in Figure 14.53. The technologist must check that the ceiling mounting arrangement of the fixture appears on the shop drawing and that it corresponds to the type of ceiling being installed.

c. Appraisal

The intense competition in the lighting field is a strong temptation to manufacturers to take shortcuts. This necessitates close scrutiny of fixtures. To compare similar lighting fixtures as manufactured by different companies, complete test data plus a sample in a regular shipping carton from a normal manufacturing run are necessary. Checking the photometric data is normally a task performed by

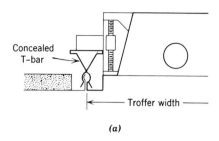

(a)

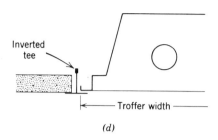

(d)

(a) A Type M luminaire is one having vertical turned-up edges which are parallel to the lamp direction and intended to "snap-in" or otherwise align the luminaire with a concealed T-bar suspension system, the center openings of the Tees being located on modular or other symmetrical dimensional lines.

(d) A Type G luminaire is one having edges which are designed to rest on or "lay-in" the exposed inverted T of a suspension system (customarily described as a grid ceiling system) with the webs of the tees being located on modular or other symmetrical dimensional lines.

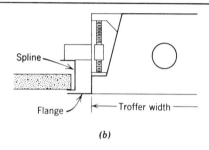

(b)

(b) A Type F luminaire is one having horizontal flanges which are parallel to the lamp direction and designed to conceal the edges of the ceiling opening above which the luminaire is supported by concealed mechanical suspension.

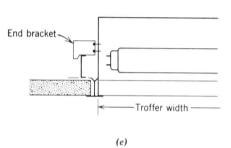

(e)

(e) A Type H luminaire is one having end brackets, hooks or other attachments and designed to be supported at the ends by "hooking-on" to some member of the ceiling suspension system. A *Type HS luminaire* is a Type H luminaire having edges parallel to the lamp direction and dependent on splines of the ceiling suspensions system for concealment of the edges of the luminaire. A *Type HF luminaire* is a Type H luminaire having edges parallel to the lamp direction and designed to conceal the edges of the ceiling opening in which the luminaire is recessed.

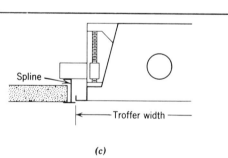

(c)

(c) A Type S luminaire is one which is designed for mechanical suspension from exposed splines and dependent on splines parallel to the lamp direction for concealment of the edges of the luminaire.

Figure 14.53 Standard fluorescent troffer details. [Adapted from NEMA Standard Publication LE-1 (1963) and reprinted by permission of the National Electrical Manufacturers Association. LE 1 is a RESCINDED Standard.]

the job's lighting designer or engineer. Comparison of the physical and other aspects of the fixtures can be done readily by a competent technologist. The following items will be helpful.

(15) Concerning construction and installation, check the sample for workmanship, rigidity, quality of materials and finish, and ease of installation, wiring, and leveling. Installation instruction sheets should be sufficiently detailed. Results of actual operating temperature tests in various installation modes should be included. Air-handling fixtures should be furnished with heat removal data, pressure-drop curves, air diffusion data, and noise criteria (NC) data for different air flow rates. These should be given to the HVAC designers for their approval.

(16) Concerning maintenance, luminaires should be simply and quickly relampable, resistant to dirt collection, and simple to clean. Replacement parts must be readily available.

How Light Is Used: Lighting Design

14.28 Lighting Systems

Most of the lighting fixtures illustrated up to this point radiate light primarily downwards. Some, such as those in Figures 14.47 and 14.48, also radiate to the sides and to some extent upwards. These different light-radiating patterns are recognized by a system of classification by which fixtures are grouped according to their light-radiating patterns; also known as their distribution patterns. There are seven types of distribution in this system, with some overlap between the types. In the following subsections, we will discuss each type briefly, show its characteristics, illustrate typical fixtures in each class and suggest where each type is best applied.

This material is intended to familiarize the technologist with the field of lighting fixtures, in order to help him or her make intelligent use of the literally thousands of fixture types and designs commercially available. The seven types of systems, or more accurately, the seven types of *lighting distribution patterns* are direct concentrated, direct spread, semi-direct, general diffuse, direct-indirect, semi-indirect and indirect.

a. Direct Lighting—Concentrating Light Pattern

Direct lighting means that all the light exiting from the source goes directly downward (or outward in the case of an angled unit); nothing goes to the sides or up. Concentrating patterns mean that the light does not spread out but is focused directly ahead—generally straight down. See Figure 14.54. This type of light is used for highly localized lighting, as for instance on a restaurant table or a countertop, or for merchandise highlighting and accent lighting. Fixtures that give this pattern concentrate their output in a narrow beam, as for instance a parabolic reflector [Figures 14.40(d), 14.45 and 14.54], any miniature tungsten halogen fixture of the MR-16 type with a narrow beam [Table 14.3 and Figure 14.19(b-1)], any fixture with a narrow beam reflector-type lamps (Figure 14.43), and other similar designs. Track lights that use concentrating beam lamps, as shown in Figure 15.11, are also of this type.

b. Direct Lighting—Spread Light Pattern

Direct lighting distribution with a spread light pattern is most common in hung ceiling commercial spaces using recessed fluorescent and spread-type incandescent and HID sources. See Figure 14.55; the fixtures in Figures 14.37, 14.42, 14.49, 14.50, 14.51, and 14.52; and the fixture design of Figure 14.40(f). Surface-mounted ceiling fixtures with metal sides also have this distribution pattern. This type of light distribution results in lighted walls and floor, general uniform diffuse lighting and a relatively dark ceiling. Ceilings in such spaces are usually light colored, to avoid appearing unpleasantly dark.

c. Semi-Direct Lighting Distribution Pattern

The semi-direct lighting pattern is one that has a small amount of uplight that is used to illuminate the ceiling. Otherwise, it is similar to direct distribution. Fixtures that produce this distribution include surface units that emit light from their sides and pendant (suspended) units with bright sides and/or slots on top to emit light. See Figure 14.56.

842 / HOW LIGHT BEHAVES; LIGHTING FUNDAMENTALS

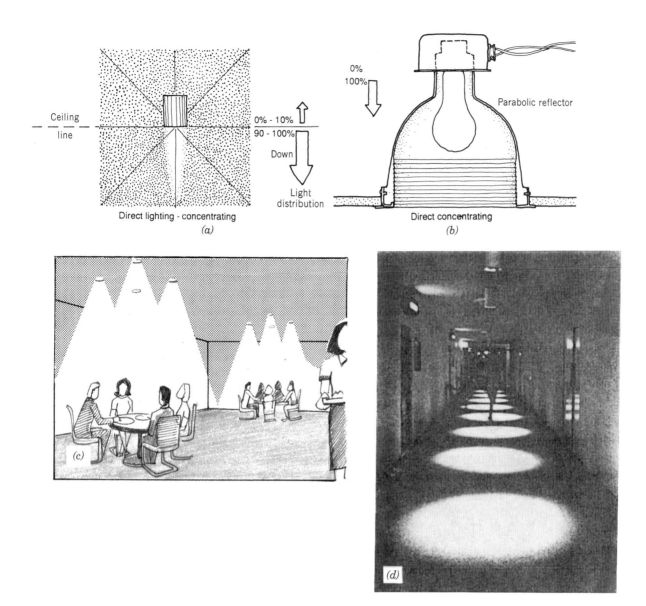

Figure 14.54 Concentrating direct distribution *(a)* gives negligible up light. Typical fixtures of this type are parabolic reflector (recessed) incandescent downlights *(b)* and fixtures with narrow beam-type sources. The result is a space with isolated pools of light in an overall dimly lighted space *(c)* and *(d)*. The walls are all generally dark. The application in *(c)* highlights the tables, as desired. The use in *(d)* has very little to recommend it. [Illustrations [*(a)*,*(b)*] from Stein, B., and Reynolds, J. *Mechanical and Electrical Equipment for Buildings*, 8th ed., 1992, reprinted by permission of John Wiley & Sons; (*d*) reprinted from MEEB, 6 ed.] *(c)* from Sorcar, *Architectural Lighting*, 1987, reprinted by permission of John Wiley & Sons.]

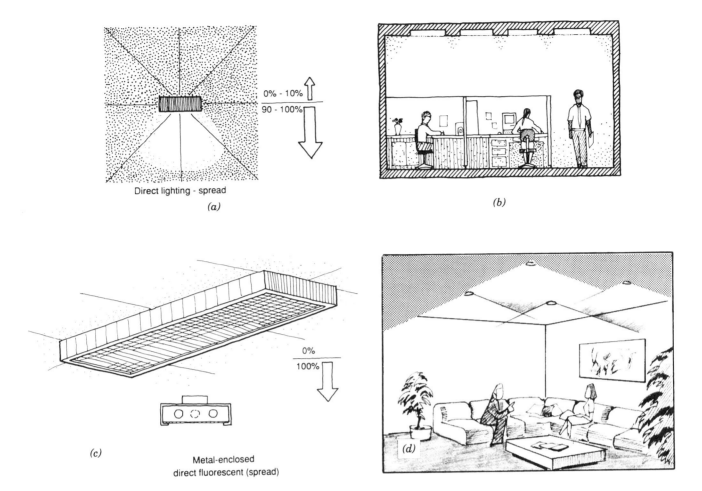

Figure 14.55 Spread-type direct lighting *(a)* illuminates the floors and walls *(b)* and *(d)*. The ceilings are relatively dark. The most common types of fixture with this lighting distribution pattern are recessed fluorescent *(b)*, surface fluorescent with opaque sides *(c)* and spread-type incandescent, compact fluorescent and HID *(d)*. [Illustration *(d)* from Sorcar, *Architectural Lighting*, 1987, reprinted by permission of John Wiley & Sons.] *(a-c)* (From Stein, B., and Reynolds, J. *Mechanical and Electrical Equipment for Buildings*, 8th ed., 1992, reprinted by permission of John Wiley & Sons.)

The illuminated ceiling gives the room a brighter larger look.

d. General Diffuse Lighting Pattern

In a general diffuse lighting pattern, light is emitted in all directions. The type of fixture that produces this pattern is a pendant unit, where the light-diffusing material completely surrounds the light source. A typical unit is shown in Figure 14.48. The result of this distribution is a space where all room surfaces are lighted, giving an overall shadowless (very diffuse) environment. See Figure 14.57.

e. Direct-Indirect Lighting Pattern

The direct-indirect distribution pattern is similar to general diffuse except that walls get little light. This pattern is very common when using sus-

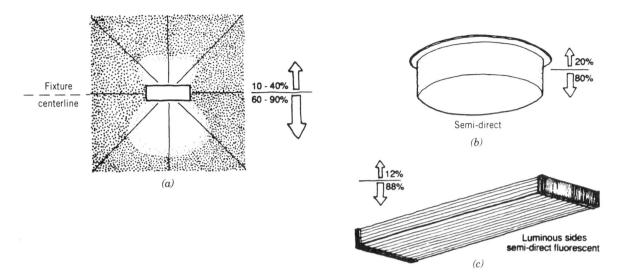

Figure 14.56 Semi-direct lighting provides its own ceiling brightness *(a)*, with surface-mounted fixtures *(b)*, or pendant/surface units *(c)*. The amount of uplight varies from 10 to 40% of the total fixture light output. (From Stein, B., and Reynolds, J. *Mechanical and Electrical Equipment for Buildings*, 8th ed., 1992, reprinted by permission of John Wiley & Sons.)

pended fluorescent fixtures. To supply wall light, wall-washes or sconces are often used. See Figure 14.58.

f. Semi-Indirect Lighting Patterns

The semi-indirect distribution occurs when most of the light from the fixture(s) strikes the ceiling or wall (or both) first and then is reflected into the room. If properly designed, this system of lighting results in a space with no direct glare and, more importantly, no reflected glare. The most common fixture that has this distribution is a tubular fluorescent with open or transparent top and a louver or diffuser bottom. See Figure 14.59.

g. Indirect Lighting Pattern

The indirect distribution pattern is similar to the semi-indirect except that all the light is projected onto the ceiling and/or walls. Absence of glare is characteristic of this lighting system. The traditional way of achieving this distribution is to use suspended fluorescents with opaque sides and bottom, similar to the semi-indirect unit. See Figure 14.60. Another traditional technique is to use a cove around the room, as seen in Figure 14.60*(c)*. However, with the development of high intensity concentrated sources such as compact fluorescent, tungsten halogen and HID (metal halide and low pressure sodium), the modern trend is to use asymmetric reflector fixtures with these sources, as in Figure 14.61.

14.29 Lighting Methods

All these lighting distribution patterns, except direct concentrating distribution, are intended to give overall general room illumination. For many types of close accurate work such as drafting, jewelry making and fine machining, the level of required illumination is so high that providing it for the entire room is wasteful of money and energy. For these special localized areas, the proper design approach is to add supplementary, generally local, lighting. A typical design engineering office, for instance, might be lighted by spread recessed fluorescent troffers for general lighting plus drafting

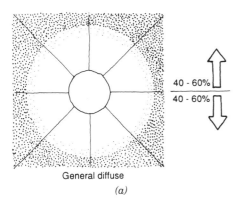
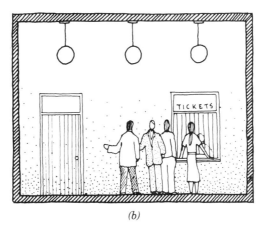

Figure 14.57 General diffuse lighting. All room surfaces are illuminated, and shadows are minimized. Because the source is completely surrounded by a diffuser *(b)*, the light emitted is approximately the same in all directions *(a)*. (From Stein, B., and Reynolds, J. *Mechanical and Electrical Equipment for Buildings*, 8th ed., 1992, reprinted by permission of John Wiley & Sons.)

tables with drawing lamps (see Figures 14.7 and 14.8) and desks with desk lamps. A few of the many types of supplementary lighting fixtures are shown in Figure 14.62.

14.30 Fixture Efficiency and Coefficient of Utilization

A lighting fixture emits less light than is generated by its lamp sources. Light is lost inside by internal reflection and absorption. The ratio of fixture output lumens to lamp lumens is the efficiency of the fixture. Although this figure is important, it is not useful in lighting calculations. What is needed is a number that expresses how well a particular fixture lights a particular room. In other words, what we really need is a number indicating the efficiency of the fixture-room combination. This figure, normally expressed as a decimal, is called the *coefficient of utilization*, or *CU*.

$$CU = \frac{\text{Usable lumens in a particular space}}{\text{Lamp lumens}} \quad (14.4)$$

This means that the same luminaire has a different CU for every different room in which it is used. Tables of coefficients of utilization, according to room characteristics, are published by manufacturers. One such table is shown in Section 14.33.

We cannot go into the derivation of this figure here. What is important to remember is that manufacturers publish CU tables for each fixture they make. With the help of these tables, the technologist, lighting designer or engineer can find the CU for each lighting fixture in each space. Most designers use the zonal cavity method to arrive at the CU. Consult the sources listed in the Supplementary Reading section at the end of the chapter for details of this method.

Because the zonal cavity method is long and time consuming when done by hand, most offices today use one of the many readily available computer programs to do two things: first, to calculate the CU of the fixture selected using the *zonal cavity method* and, second, to calculate the luminaire requirements by the *lumen method* using the CU just calculated. (Actually, most programs will do a great deal more, often including a drawing layout of the luminaires.) These programs use the manufacturer's tables in computer file form, supplied by the manufacturer to the design office on disks. However, as with all computer output, it is absolutely vital that the technologist/designer be able to check the computer results because even a small mistake in input (a misplaced decimal point, for instance) can produce ridiculous results. An experienced designer will immediately recognize that there is an error somewhere and will backtrack to

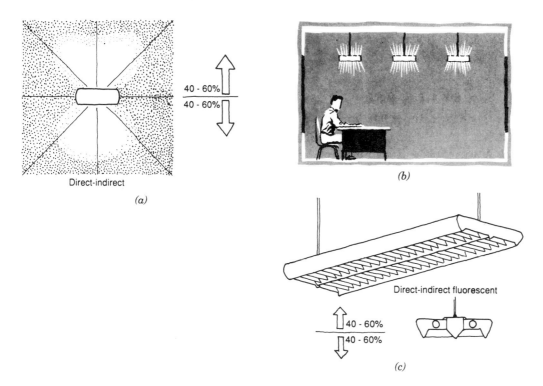

Figure 14.58 (a) Direct-indirect lighting. Upper and lower room surfaces are luminous (b), but the center of walls is not because of the lack of horizontal light from fixtures (c). The principal amount of light on working planes comes directly from the luminaire. [(a,c) (From Stein, B., and Reynolds, J. *Mechanical and Electrical Equipment for Buildings*, 8th ed., 1992, reprinted by permission of John Wiley & Sons.)

find it. An inexperienced technologist will tend to accept the computer output as correct and accurate, since he or she has no basis for doing otherwise. To have such a basis, the novice technologist or designer must make a rapid manual lumen method calculation and compare it to the computer output.

The lumen method calculation itself is simple and straightforward as will be shown in Section 14.31. The difficult portion of the calculation is to determine the CU relatively accurately, without going through the long, involved, tedious full zonal cavity calculation. To do this, a number of shortcut procedures based on the detailed zonal cavity method have been developed. They will give a CU that is within ±10% of the accurate CU figure. This number, when used in a lumen method calculation, can be used for preliminary design, in many cases for final design, and always to check the output of a CU zonal cavity computer program. The author has used one such method for years with excellent results. It is presented in Section 14.33.

14.31 Illuminance Calculations by the Lumen Method

The average illuminance in a space, in footcandles, can be simply calculated if we remember that, by definition, one footcandle equals one lumen per square foot; that is

$$\text{Footcandles (fc)} = \frac{\text{Lumens (lm)}}{\text{Area in square feet (ft}^2)} \quad (14.5)$$

Therefore, all we need do to calculate the average footcandle level in a room is to divide usable lumens by the room area. However, because the usable lumens produced by a fixture in a specific

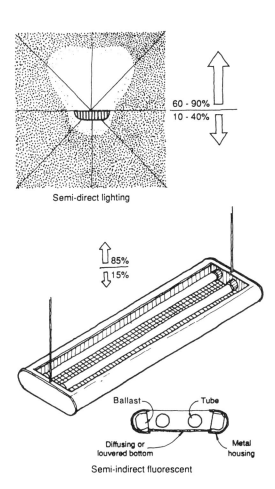

Figure 14.59 Semi-indirect lighting pattern throws most of the light onto the ceiling. From there it is reflected into the room. Finish of the ceiling should be a high reflectance matte white. The suspended fluorescent fixture usually used for this application has one, two or four high output or very high output tubular lamps. (From Stein, B., and Reynolds, J. *Mechanical and Electrical Equipment for Buildings*, 8th ed., 1992, reprinted by permission of John Wiley & Sons.)

room is equal to the fixture lamp lumens times the CU (as we explained in Section 14.30), we must rewrite Equation (14.5) as follows:

$$\text{Initial illuminance (fc)} = \frac{\text{Usable lumens}}{\text{Area (ft}^2\text{)}} \quad (14.6)$$

$$= \frac{\text{Lamp lumens} \times \text{CU}}{\text{Area (ft}^2\text{)}}$$

The quantity thus calculated is the initial room illuminance. We emphasize initial, because the room illuminance drops as time progresses due to decrease in lamp output, dirt in the luminaire, dirt on the walls and the like. The sum total of these loss factors is called the *light loss factor*, abbreviated *LLF*. Therefore, to find the maintained illuminance, that is, the average footcandles in a room after a considerable period of time, we must reduce the initial illuminance by the LLF. The formula for illuminance now becomes

$$\frac{\text{Maintained}}{\text{illuminance}} = \frac{\text{Lamp lumens} \times \text{CU} \times \text{LLF}}{\text{Area}} \quad (14.7)$$

where lamp lumens is simply the number of lamps in the fixture times the initial output of each lamp. This formula is called the lumen method of average illuminance calculation. To show how easy these calculations are, let us try a few examples.

Example 14.1 A classroom 22 ft by 25 ft is lighted with ten fluorescent fixtures, each containing three F40 T12 3500K lamps (40w, 3500°K, rapid-start). Calculate the initial and maintained illuminance in footcandles using the lumen method. Assume a CU of 0.45 and an LLF of 0.65. Use Table 14.5 for lamp data.

Solution:

Total lamp lumens = 10 fixtures × 3 lamps/fixture × 3300 lm/lamp

Lamp lumens = 10 × 3 × 3300 = 99,000 lm

$$\text{Initial fc} = \frac{99,000 \text{ lm} \times 0.45}{25 \times 22 \text{ ft}^2} = 81 \text{ fc}$$

Maintained fc = 81 fc × 0.65 = 52.6 fc, say 53 fc

Or, the entire calculation can be done in one step:

$$\frac{\text{Initial}}{\text{footcandles}} = \frac{10 \times 3 \times 3300 \text{ lm} \times 0.45}{25 \text{ ft} \times 22 \text{ ft}} = 81 \text{ fc}$$

$$\frac{\text{Maintained}}{\text{footcandles}} = \frac{10 \times 3 \times 3300 \text{ lm} \times 0.45 \times 0.65}{25 \text{ ft} \times 22 \text{ ft}}$$
$$= 52.6 \text{ fc, say 53 fc}$$

As a practice exercise, we will do this problem in SI units also. Remembering that maintained illumination (illuminance) equals lamp lumens × CU × LLF, divided by area, we have

$$\text{lux} = \frac{\text{Lumens} \times \text{CU} \times \text{LLF}}{\text{Area (m}^2\text{)}}$$

$$\text{Lux} = \frac{(10 \times 3 \times 3300) \text{ lm} \times 0.45 \times 0.65}{\frac{22}{3.28} \text{ m} \times \frac{25}{3.28} \text{ m}}$$

Therefore,

848 / HOW LIGHT BEHAVES; LIGHTING FUNDAMENTALS

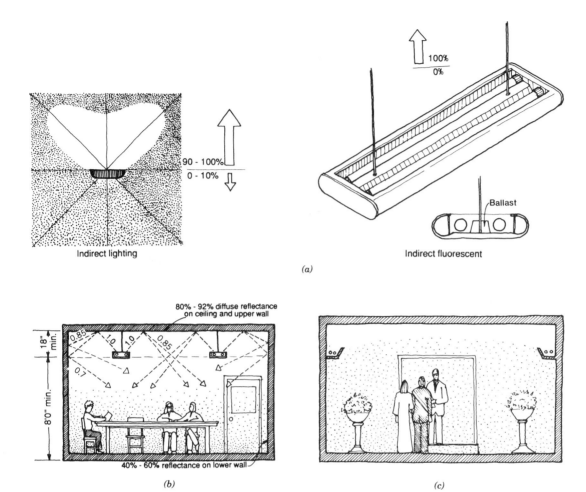

Figure 14.60 Indirect lighting. *(a)* The fixtures deliver 90–100% of their output upwards. *(b)* The ceiling and upper wall surfaces of the room are directly illuminated by the indirect fixtures; they then become secondary sources that, in turn, illuminate the room. The ceiling and upper walls should have a minimum diffuse reflectance of 80%. With 85% reflectance as shown, 77% of the incident light is available even after two reflections, as shown. *(c)* Use of architectural coves, properly designed, gives nearly uniform, glareless illumination in the room. Indirect lighting is particularly useful in spaces with visual display terminals (computer screens). (From Stein, B., and Reynolds, J. *Mechanical and Electrical Equipment for Buildings*, 8th ed., 1992, reprinted by permission of John Wiley & Sons.)

Figure 14.61 Surface-mounted indirect lighting fixtures can be ceiling mounted *(a)*, to project a beam of light onto the wall *(b)* and *(c)* from where it is reflected into the room, or wall mounted *(d)*, to project light onto the ceiling and the wall *(e)*. The projector can be designed enclosed *(f-1)* and *(f-2)* to make its appearance suitable for any use. Source for these asymmetrical projectors include tungsten-halogen, compact fluorescent and HID. (Courtesy of elliptipar, inc.)

ILLUMINANCE CALCULATIONS BY THE LUMEN METHOD / 849

CEILING MOUNTED

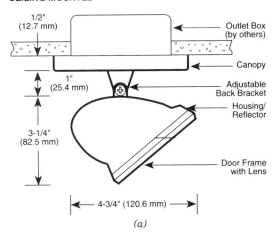
(a)

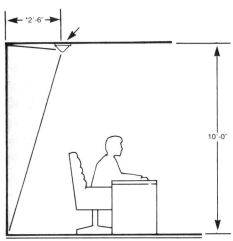
*Typical setback = ¼ times wall height, but not less than 2'-6".
(b)

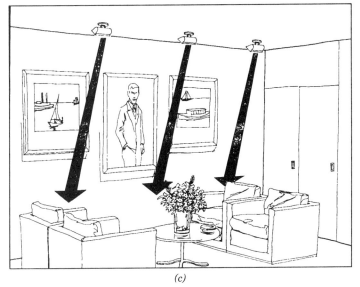
(c)

WALL MOUNTED

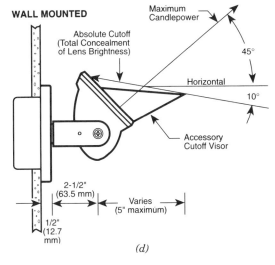
(d)

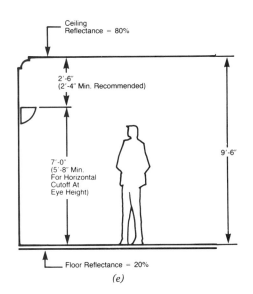
(e)

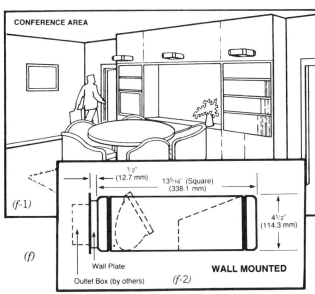
(f)
(f-1)
(f-2) **WALL MOUNTED**

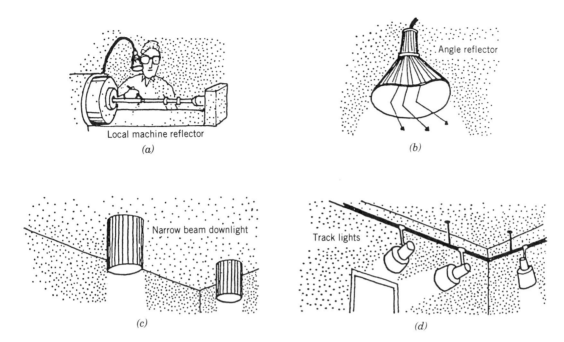

Figure 14.62 Typical supplementary lighting fixtures can be local (at the work) as (a) and (b), ceiling mounted as (c) or ceiling/wall mounted as (d). The purpose of all these types is to supplement the general lighting in order to provide the high level illuminance required for close work. (From Stein, B., and Reynolds, J. *Mechanical and Electrical Equipment for Buildings*, 8th ed., 1992, reprinted by permission of John Wiley & Sons.)

$$\text{Maintained lux} = \frac{99{,}000 \text{ lm} \times 0.45 \times 0.65}{6.71 \text{ m} \times 7.62 \text{ m}} = 566 \text{ lux}$$

To check the accuracy of this answer, we use the conversion factor:

Divide lux by 10.76 to get footcandles

or

$$\frac{532 \text{ lux}}{10.76 \text{ lux/fc}} = 52.6 \text{ fc, say 53 fc}$$

Which checks the previous result.

Most of the time, the problem is presented the other way around. That is, given a room and the desired illuminance, find the number of required fixtures. This is readily done by simply rearranging the equation that we developed previously. We had

$$\frac{\text{Maintained}}{\text{footcandles}} = \frac{\text{Lamp lumens} \times \text{CU} \times \text{LLF}}{\text{Area}} \qquad (14.7)$$

therefore,

$$\text{Lamp lumens} = \frac{\text{Maintained illuminance} \times \text{area}}{\text{CU} \times \text{LLF}}$$

And, once we have lamp lumens, we can easily determine the number of luminaires required. Applying these equations to the same problem, we have the following example.

Example 14.2 A classroom 22 ft by 25 ft is to be lighted to an average maintained footcandle level of 50 fc. Find the number of three-lamp 40-w RS luminaires required. Assume CU = 0.45 and LLF = 0.65.

Solution:

$$\text{Total lamp lumens required} = \frac{50 \text{ fc} \times 22 \text{ ft} \times 25 \text{ ft}}{0.45 \times 0.65}$$
$$= 94{,}017 \text{ lm}$$

Since each F40T12 3500K lamp has an output of 3300 lm,

$$\text{Number of lamps} = \frac{94{,}017}{3300} = 28.49$$

Since we have three lamps per fixture, we have

$$\text{Number of luminaires} = \frac{28.49}{3} = 9.5, \text{ say } 10$$

which corresponds to the data of Example 14.1.

One further technique in illumination calculation should be mastered. Frequently, we must calculate illumination for a very large space such as an entire floor in a building. Instead of doing this in one block, it is easier and more meaningful to calculate the number of fixtures required per bay. This can be done directly or by calculating the area covered by a single fixture as follows. Since

Number of luminaires =

$$\frac{\text{Illuminance} \times \text{area}}{\text{Lamps per fixture} \times \text{Lumens per lamp} \times \text{CU} \times \text{LLF}}$$

it follows that the area lighted by a single fixture is

Area per fixture =

$$\frac{\text{Lamps per fixture} \times \text{Lumens per lamps} \times \text{CU} \times \text{LLF}}{\text{Illuminance requirement}}$$

Expressed in formula form, we have:

$$\text{Number of luminaires} = \frac{\text{fc} \times A}{N \times \text{lm} \times \text{CU} \times \text{LLF}} \quad (14.8)$$

and

$$\text{Area per luminaire} = \frac{N \times \text{lm} \times \text{CU} \times \text{LLF}}{\text{fc}} \quad (14.9)$$

where

fc = required illuminance in footcandles
A = area in square feet
N = number of lamps per luminaire
lm = initial lumens per lamp
CU = luminaire coefficient of utilization
LLF = light loss factor

An example would help here.

Example 14.3 An entire building floor is to be lighted to an average maintained illuminance of 50 fc. The floor measures 320 ft × 150 ft and is divided into 1000-ft² bays, each measuring 40 ft × 25 ft. The space is to be used as an economy clothing store, and the lighting designer has selected a single-lamp, pendant, parabolic reflector fixture. The space is air conditioned. Assume a CU of 0.85 and an LLF of 0.6. (The very high CU is reasonable for such a large open area and a highly efficient luminaire.) Calculate the number of fixtures required per bay and suggest an arrangement. Use the same 48-in. fluorescent lamps as in Example 14.4 (96-in. tubes would also be seriously considered).

Solution A: Let us calculate the number of fixtures per bay using Equation (14.8):

$$\text{Fixtures per bay} = \frac{50 \text{ fc} \times 40 \text{ ft} \times 25 \text{ ft}}{1 \text{ lamp} \times 3300 \text{ lm} \times 0.85 \times 0.6}$$
$$= 29.7, \text{ say } 30$$

This would be either 30 units arranged in three continuous rows of 10 units the long way or five rows of 6 units the short way. In either case, center-to-center spacing is about 8 ft.

Solution B: The same result can be obtained by calculating the required area per fixture using Equation (14.9):

Square feet per fixture =

$$\frac{1 \text{ lamp} \times 3300 \text{ lm} \times 0.85 \times 0.6}{50 \text{ fc}} = 33.66 \text{ ft}^2$$

Since the fixture is 4 ft long, centerline spacing of continuous rows of fixtures is

$$\text{Spacing} = \frac{33.66 \text{ ft}^2}{4 \text{ ft}} = 8+ \text{ ft}$$

$$\text{Fixtures per bay} = \frac{40 \text{ ft} \times 25 \text{ ft}}{33.66 \text{ ft}^2 \text{ per fixture}}$$
$$= 29.7 \text{ fixtures as in Solution A}$$

$$\text{Actual footcandles} = \frac{30 \text{ fixtures}}{29.7 \text{ fixtures calculated}} \times 50 \text{ fc} = 50.5 \text{ fc}$$

This is very close indeed.

See Figure 14.63 for a layout of the luminaires for Example 14.3.

14.32 Lighting Uniformity

In the preceding section, we learned how to calculate the average illumination in a space. Generally, we also want this illumination to be uniform, without bright areas and darker areas. This normally

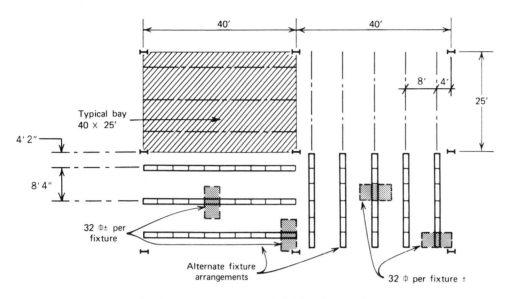

Figure 14.63 Layout of solutions to Example 14.3. The choice of arrangement is up to the designer and depends on space use, furniture layout and other factors.

requires a regular, equally spaced layout of lighting fixtures. This is not enough, however, to ensure uniform lighting. A 30 ft × 30 ft room could have 4 fixtures in a regular pattern or 16 fixtures in a regular pattern. Obviously, the 16 fixture pattern will give greater uniformity, but at a much higher cost.

To help us with this problem, manufacturers publish a recommended spacing-to-mounting height ration (S/MH) for many luminaire types. This figure indicates the maximum fixture spacing permissible in relation to the fixture mounting height above the working plane, which will give acceptable uniformity of illumination. Several examples will illustrate the application of this very useful piece of information. (Remember that mounting height is above the working plane.)

Example 14.4 A room is to be lighted with direct lighting fluorescent troffers that have a S/MH of 1.4. Ceiling height is 8 ft. What is the maximum fixture spacing that will ensure uniform lighting? Assume a working plane of 30 in. AFF, that is, 2.5 ft.

Solution:
$$\frac{S}{MH} = 1.4 = \frac{S}{8 \text{ ft} - 2.5 \text{ ft}} = \frac{S}{5.5 \text{ ft}}$$
therefore,
$$S = 5.5 \text{ ft} \times 1.4 = 7.7 \text{ ft}$$

Maximum side-to-side spacing of the luminaires center-to-center, for uniform lighting, would be 7 ft 8½ in.

Remember that spacing is measured from the fixture centerline. Therefore, if the fixtures in Example 14.4 are 1 ft wide, then the maximum spacing between fixtures, to maintain even lighting, would be 6 ft 8½ in., because 6 in. is deducted on each side to reach the fixture centerline. Notice that we emphasized that the S/MH figures normally given in the catalog for fluorescent fixtures refer to side-to-side spacing. End-to-end spacing is always lower for fluorescents, and if the end-to-end S/MH is not given, an average figure of 1.0 can be assumed. Therefore, assuming the fixtures of this example are 4 ft long, we can calculate the required end-to-end spacing as follows:

Assume that the maximum S/MH for end-to-end spacing = 1.0; therefore,

$$\frac{S}{5.5 \text{ ft}} = 1.0$$

or

$$S = 5.5 \text{ ft, center-to-center}$$

Since the fixtures are each 4 ft long, the end-to-end spacing between fixtures is 5.5 ft minus 2 ft at each end to reach the fixture center, or a net of 1.5 ft (18 in.) between fixture ends.

The light distribution pattern of a (round) HID fixture is symmetrical, and therefore only one S/MH figure is required to calculate maximum fixture spacing.

Example 14.5 A carpet warehouse is to be lighted with pendant dome-type reflectors with metal halide lamps. These domes have an S/MH of 1.3. However, actual spacing, because of architectural considerations, must be on a grid of 16 ft by 16 ft. What is the minimum mounting height? (We ask for minimum mounting height because this corresponds to maximum spacing. Any tighter spacing would allow lower mounting.)

Solution:

$$MH_{min} = \frac{\text{Maximum spacing}}{\text{S/MH}} = \frac{16 \text{ ft}}{1.3}$$
$$= 12.3 \text{ ft} = 12 \text{ ft } 4 \text{ in.}$$

Note that mounting height above the working plane and actual mounting height are the same in this example, since the working plane is at floor level for displaying carpets.

Example 14.6 Check the fixture layout of Figure 14.63, assuming a manufacturers recommended S/MH ratio of 1.5. Assume a 9 ft mounting height.

Solution: Assume the working plane to be desk height of 3 ft. Mounting height above the working plane is therefore

$$MH = 9 \text{ ft} - 3 \text{ ft} = 6 \text{ ft}$$

Since maximum spacing/MH = 1.5 we have maximum spacing = 1.5(6 ft) = 9 ft side-to-side spacing.

Referring to Figure 14.63, we see that in the longitudinal layout, the side-to-side fixture spacing is 8 ft 4 in. and in the transverse layout it is 8 ft. Therefore, either layout is acceptable. Obviously, there is no problem with end-to-end spacing, since we are running continuous rows of luminaires (0 ft end-to-end spacing).

14.33 Determining Coefficient of Utilization by Approximation, Using the Zonal Cavity Method*

In order to understand the approximations, we must first explain the principles of the method. As we stated previously, the CU of a lighting fixture relates the fixture's lighting pattern to the room in which it is installed, that is, to the room's size and surface reflectances (walls, floor, ceiling). Also worked into this CU figure is the mounting height of the luminaire and the level above the floor at which we want to calculate the illuminance, called the working plane elevation. (This latter is important; in an office it is desk height of 30 in., in a drafting room it is table height of 42 in. and in a carpet salesroom it is at the floor level).

Refer now to the sketch at the top of Figure 14.64. Notice that all these factors appear in the sketch. The room is divided into three vertical sections, called cavities—the ceiling cavity above the fixtures, the room cavity between the lighting fixtures and the working plane, and the floor cavity below the working plane. Obviously if the luminaires are ceiling mounted, the height of the ceiling cavity h_{cc} is zero. Similarly if the working plane is at the floor level (carpet showroom for instance), the floor cavity height h_{fc} is zero. The symbols ρ_{cc}, ρ_w and ρ_{fc} refer to the reflectances of the ceiling cavity (cc), the wall cavity (w), and the floor cavity (fc). The CU calculation procedure is outlined on Figure 14.64. Follow the discussion with Figure 14.64 before you.

Step 1. Fill in the first seven general information items at the top of the sheet.

Item (g), total lumens per luminaire, is simply the number of lamps times the initial output of each lamp. In the lamp catalogs, figures are given for initial lumens and maintained, or design lumens. These latter figures represent output after 40% of life, or average over the entire lamp life. For our purpose here, use initial lumens. We will

---

*Based on a method developed by B. F. Jones.

854 / HOW LIGHT BEHAVES; LIGHTING FUNDAMENTALS

ILLUMINATION CALCULATION

Step 1:   **GENERAL INFORMATION**

(a) Project identification: _____
                                     (Give name of area and/or building and room number)

(b) Average maintained illumination for design: _____ lux [footcandles]

Luminaire data: _____   Lamp data: _____
(c) Manufacturer: _____   (e) Type and color: _____
(d) Catalog number: _____   (f) Number per luminaire: _____
                                          (g) Total lumens per luminaire: _____

SELECTION OF COEFFICIENT OF UTILIZATION

Step 2: Determine the equivalent square-room RCR (room cavity ratio):

$$W_{sq} = W + \frac{L-W}{3}$$

$$RCR = \frac{10\ hrc}{W_{sq}}$$

Step 3: Determine ceiling cavity equivalent reflectance by equivalent square room size.

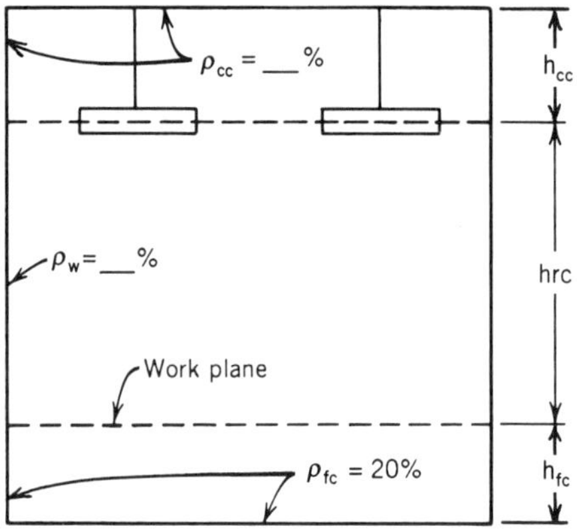

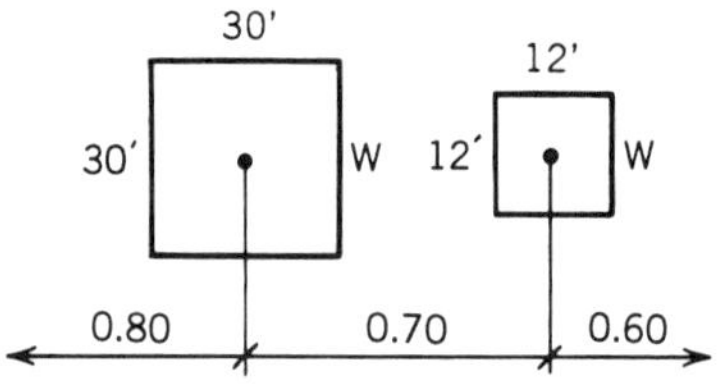

$\rho_{cc} = \_\_\%$ h_{cc}

$\rho_w = \_\_\%$ hrc   L = _____   W = _____

Work plane

$\rho_{fc} = 20\%$ h_{fc}

30'   12'

30' • W   12' • W

$\rho_{cc} = $ 0.80   0.70   0.60

Step 4: Fill in sketch of room above.
Step 5: Assume floor cavity reflectance $\rho_{fc} = 20\%$.
Step 6: Obtain coefficient of utilization CU from manufacturers' data: CU = _____

Step 7: Light loss factor LLF   Good conditions: 0.65
                                     Average conditions: 0.55
                                     Poor conditions: 0.45

Step 8:   **CALCULATIONS**

$$\text{Number of luminaires} = \frac{\text{Recommended illuminance} \times \text{area}}{(\text{Lumens per luminaire}) \times (CU) \times (LLF)}$$

$$\text{OR} \quad \text{Lux [footcandles]} = \frac{(\text{Number of luminaires}) \times (\text{lumens per luminaire}) \times (CU) \times (LLF)}{\text{area m}^2\ (\text{ft}^2)}$$

Figure 14.64 Calculation sheet for illuminance calculation using an approximate zonal cavity method (based on a method developed by B. F. Jones). (From Stein, B., and Reynolds, J. *Mechanical and Electrical Equipment for Buildings*, 8th ed., 1992, reprinted by permission of John Wiley & Sons.)

compensate for lumen reduction in another part of the calculation.

Step 2. Determine the room cavity ratio (RCR).

In order to do this by approximation, we need to "square" the room; that is, we need to calculate an equivalent square room to the room we are dealing with. To do this for a rectangular room, take one-third of the difference between length and width and add it to the smaller dimension to obtain the equivalent square room dimension, W_{sq}. That is, for a room of length L and width W (assuming L is larger than W),

$$W_{sq} = W + \frac{L-W}{3} \tag{14.8}$$

Then calculate RCR as

$$RCR = \frac{10\, h_{rc}}{W_{sq}}$$

An example should make this clear.

Example 14.7 Assume a room 30 ft long by 21 feet wide and 9 ft high, with ceiling-mounted fixtures. Assume the working plane to be 42 in. high (3.5 ft). Find RCR by the approximation method for square rooms.

Solution:

$$L = 30 \text{ ft} \quad W = 21 \text{ ft} \quad h_{rc} = 9 \text{ ft} - 3.5 \text{ ft} = 5.5 \text{ ft}$$

$$W_{sq} = 21 + \frac{30-21}{3} = 21 + \frac{9}{3} = 24 \text{ ft}$$

$$RCR_{sq} = \frac{10 \times h_{rc}}{W_{sq}} = \frac{10(5.5 \text{ ft})}{24 \text{ ft}} = 2.3$$

Step 3. Determine the ceiling cavity equivalent reflectance, using the equivalent square room size.

Where fixtures are ceiling mounted, there is no ceiling cavity, and the ceiling cavity reflectance is simply the ceiling reflectance. That is, when $h_{cc} = 0$, $\rho_{cc} = \rho_c$. You may well ask why the ceiling cavity reflectance depends upon the size of the room. A little thought will answer that question. Since the ceiling cavity is made up partly of ceiling and partly of upper wall, the net reflectance of the combination varies, depending on the ratio of ceiling area to wall area. Another way of saying this is that the larger the room, the higher the combination reflectance, because ceiling reflectance is almost always larger than wall reflectances.

Ceilings are usually white with a reflectance of 0.75–0.85. Walls are usually a light color with a reflectance of 0.4–0.6. The sketch in Figure 14.64 shows the equivalent ceiling cavity ratio in terms of the equivalent square room dimension W_{sq}, as follows:

$\rho_{cc} = 0.80$ for a large room (larger than 30 ft × 30 ft)
$\rho_{cc} = 0.70$ for a medium room (between 12 ft × 12 ft and 30 ft × 30 ft)
$\rho_{cc} = 0.60$ for a small room (less than 12 ft × 12 ft)

Steps 4–6. These steps are self-explanatory and will become very clear when we work out an illustrative example.

Step 7. Determine the light loss factor (LLF).

The LLF relates initial illuminance to maintained illuminance and includes such items as lamp aging, dirt accumulation on the room surfaces as well as in the fixture, component aging and a few other factors, all of which tend to reduce the amount of light in the room. Select the numbers as follows:

(a) Good conditions. This means regular luminaire cleaning, group lamp replacement and clean room conditions (air conditioning). For these conditions LLF = 0.65.
(b) Average conditions. This means occasional fixture cleaning, lamp replacement on burnout, and open windows in an average environment. For these conditions LLF = 0.55.
(c) Poor conditions. This means that the luminaires are rarely cleaned, lamps are replaced only on burnout, and windows are open in a dirty atmosphere (inner city, industrial) environment. For these or similar conditions, LLF = 0.45.

Step 8. This step is a straightforward lumen method calculation step, as it appears on the sheet.

To demonstrate the application of this method, let us rework Example 14.3. (The figures assumed there—CU = 0.85 and LLF = 0.6—were actually calculated using the detailed zonal cavity method for CU and a detailed list of factors that comprise LLF.) Referring to Figure 14.64, fill in the sheet as we work through the steps.

Step 1. Fill in the blanks.

Step 2. Given $h_{rc} = 9$ ft $- 3$ ft $= 6$ ft, $L = 320$ ft, $W = 150$ ft,

Table 14.14. Typical Coefficients of Utilization for a Single-Lamp, Open Reflector Fluorescent Luminaire

ρ_{cc}	80			70			50		
ρ_w	50	30	10	50	30	10	50	30	10
RCR	Coefficients of Utilization for 20% Effective Floor Reflectance ($\rho_{FC}=20$)								
0	0.95	0.95	0.95	0.91	0.91	0.91	0.83	0.83	0.83
1	0.85	0.82	0.80	0.82	0.79	0.77	0.75	0.73	0.72
2	0.76	0.72	0.68	0.74	0.70	0.66	0.68	0.65	0.62
3	0.69	0.63	0.59	0.66	0.61	0.57	0.62	0.58	0.54
4	0.62	0.56	0.51	0.60	0.54	0.50	0.56	0.51	0.47
5	0.55	0.49	0.44	0.53	0.48	0.43	0.50	0.45	0.41
6	0.50	0.43	0.39	0.48	0.42	0.38	0.45	0.40	0.36
7	0.45	0.38	0.34	0.43	0.37	0.33	0.41	0.36	0.32
8	0.40	0.34	0.29	0.39	0.33	0.29	0.37	0.31	0.28
9	0.36	0.30	0.25	0.35	0.29	0.25	0.33	0.28	0.24

$$W_{sq} = 150 + \frac{320-150}{3} = 150 + 56.7 = 206.7 \text{ ft}$$

$$\text{RCR} = \frac{10 \, h_{rc}}{W_{sq}} = \frac{10 \times 6 \text{ ft}}{206.7 \text{ ft}} = 0.29$$

Step 3. W_{sq} is larger than 80; therefore, $\rho_{cc}=0.80$
Step 4. Assume $\rho_w=0.5$.
Step 5. Assume $\rho_{fc}=0.2$.
Step 6. Refer to Table 14.14, which is an actual table for a single lamp open reflector fixture. Find CU by visual interpolation to be 0.92.
Step 7. The space is air-conditioned and a store has average maintenance. However, open fixtures accumulate dust and dirt. Select LLF=0.55.
Step 8. Calculate the number of luminaires per 1000 ft² bay (40 ft × 25 ft) using Equation (14.8):

$$\frac{\text{Number of}}{\text{luminaires}} = \frac{50 \text{ fc} \times 1000 \text{ ft}^2}{3300 \text{ lm.} \times 0.92 \times 0.55} = 29.9, \text{ say } 30$$

This answer is exactly that obtained by the long method. It is obviously satisfactory for preliminary layout, for checking computer output and indeed for final design. As the technologist will discover in actual practice, it is almost always necessary to make a ±10% change in the calculated figures to suit layout requirements. In this case, change is unnecessary.

14.34 Lighting Calculation Estimates

It is often very helpful to be able to take an educated guess at the luminaire requirements of a space. Figure 14.65 will be very helpful in this respect. To use it, simply remember that

$$\frac{\text{Square feet}}{\text{per luminaire}} = \frac{\text{Lighting fixture wattage}}{\text{Chart figure (w/ft}^2)}$$

Since the chart is for an average to large room, increase the wattage up to 10% for small rooms and decrease it up to 10% for very large rooms. Applying it to an example we considered above, we find that the results are quite good for a first estimate.

Example 14.8 Using Figure 14.65, estimate the fixture requirements of Example 14.2.

Solution: For a pendant fluorescent fixture, type G the graph gives

$$50 \text{ fc} = 2.2 \text{ w/ft}^2$$

For the 22 ft × 25 ft classroom at 50 fc, using three-lamp fixtures, and assuming 42 watts per lamp, including ballast loss, we have

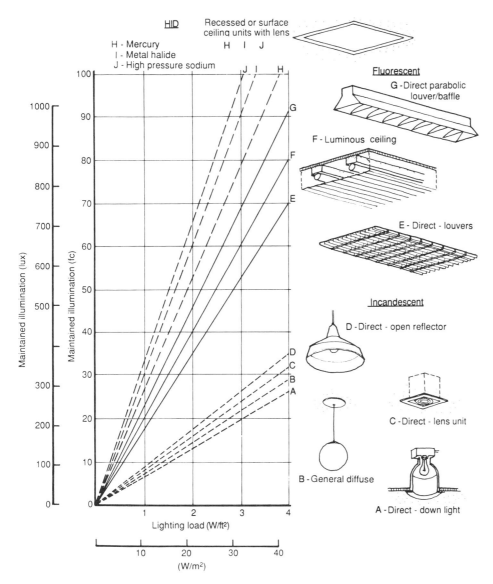

Figure 14.65 Estimating chart for lighting load and related illumination levels for different light sources. The chart was calculated for a fairly large room, approximately classroom size. Although all the figures are approximate because of the variables involved, the chart gives figures close enough for a first approximation. Notice the increase in output as the sources change from incandescent to fluorescent to HID. (From Stein, B., and Reynolds, J. *Mechanical and Electrical Equipment for Buildings*, 8th ed., 1992, reprinted by permission of John Wiley & Sons.)

$$\text{Square feet per fixture} = \frac{126 \text{ w/fixture}}{2.2 \text{ w/ft}^2}$$
$$= 57.3 \text{ ft}^2/\text{fixture}$$

This gives

$$\frac{25 \times 22 \text{ ft}^2}{57.3 \text{ ft}^2/\text{fixture}} = 9.6, \text{ say 10 luminaires per bay}$$

This is exactly the calculated result.

14.35 Conclusion

In this chapter we examined the action of light, the production of electric light and its control and, finally, lighting calculation. The ideas presented here will be applied in Chapters 15 and 16 to specific residential and industrial occupancies to demonstrate practical lighting layout technique.

Key Terms

Having completed the study of this chapter, you should be familiar with the following key terms. If any appear unfamiliar or not entirely clear, you should review the section in which these terms appear. All key terms are listed in the index to assist you in locating the relevant text. Some of the terms also appear in the glossary at the end of the book.

Arc tube
Asymmetric reflector
Baffle
Ballast factor (BF)
Ballast efficacy factor (BEF)
Beam spread
Burning hours per start
Cavity reflectance
Coefficient of utilization (CU)
Color rendering index (CRI)
Color temperature (CT)
Compact fluorescent
Contrast
Core-and-coil ballast
Correlated color temperature (CCT)
Cutoff angle
Diffuse, nondiffuse transmission
Diffuse reflection
Diffuseness of light
Diffuser
Direct glare
Direct-indirect lighting
Direct lighting
Efficacy, lumens per watt
Electronic ballast
Flux density
Footcandle
General diffuse lighting
Glare
HID (High intensity discharge)
High output (HO) lamp
High pressure sodium lamp
Illuminance
Incandescent, reflector, projector lamps
Indirect glare
Indirect lighting
Induction lamp
Instant-start (Slimline)
Light flux
Light loss factor (LLF)
Lighting distribution patterns
Lighting methods
Lighting system

Longitudinal shielding
Louver
Low energy lamp
Lumen (lm)
Lumen method
Luminance, luminance ratio
Luminaire, lighting fixture
Luminous flux
Lux
MR-16
Mercury vapor (MV) lamp
Metal halide (MH) lamp
Multimirrored reflector
PAR (parabolic aluminized reflector)
Parabolic reflector
Preheat start
Prismatic lens
Quartz lamp
Rapid-start
Reflectance
Reflected glare
Reflection factor
Reflector, parabolic reflector
RLM (Reflector Luminaire Manufacturer)
Room cavity ratio
Semi-direct lighting
Semi-indirect lighting
Shield, shielding angle
Spacing-to-mounting-height ratio (S/MH)
Specular reflection
Strike, restrike time
Transmission factor
Transmittance
Transverse shielding
Trigger start
Triphosphor (Octic) lamp
Troffer
Tungsten halogen lamp
Veiling reflection
Very high output (VHO) lamp
Video display terminal (VDT)
Zonal cavity method

Supplementary Reading

B. Stein and J. Reynolds, *Mechanical and Electrical Equipment for Buildings*, 8th ed., Wiley, New York, 1992. This book is discussed in preceding chapters. It covers all the material in this chapter, in greater detail.

C. G. Ramsey and H. R. Sleeper, *Architectural Graphic Standards*, 8th ed., 1988. Wiley, New York. This book, also discussed previously, contains lamp data and fixture details.

ISE Lighting Handbook, Illuminating Engineering Society of North America (IESNA), 1993, New York. This book is the standard references of the lighting industry and contains a wealth of technical material on all facets of lighting design and lighting equipment.

IES Introductory Lighting Education material ED 100.1–100.5, published by the IESNA. These educational materials cover basic theory, light sources, luminaires and calculations. You can also refer to sections ED 100.6–100.9, which cover additional topics in lighting design, exterior lighting and economics.

IES Lighting Ready Reference, IESNA, 1980, New York. This is a compendium of definitions, lamp tables, illuminance tables, calculation data and other useful data.

Problems

1.
   a. What is the reflection factor (reflectance) of an aluminum fixture reflector that reflects 87 lm of every 100 lm falling on it?
   b. What percent is absorbed?
   c. This same fixture radiates 70 lm through its plastic diffuser for every 100 lumens striking the diffuser. What is the transmission factor (transmittance) of the diffuser? What percentage does it absorb?

2. The luminaire of Problem 1 produces 12,800 lamp lumens. Assuming all the lumens radiating from the fixture strike the floor, calculate the average footcandle illuminance level on the floor of a 10 ft × 10 ft room.

3. A classroom is to be lighted with two continuous rows of plastic lens fluorescent fixtures. The manufacturer gives a maximum spacing-to-mounting-height ratio of 1.1. The room is 22 ft wide by 30 ft long. Pendant rows of fixtures are to be run the length of the room. What is the minimum mounting height? Draw a sketch of the room showing the fixtures. (*Hint:* Spacing between fixture and wall is one-half of the fixture-to-fixture spacing.)

4. A supermarket is lighted with continuous rows of two-lamp 96-in. HO rapid-start lamp fixtures mounted on 10-ft centers at 10 ft above the floor. Assuming a CU of 0.5 and an LLF of 0.6, find the average maintained illuminance in the store in footcandles and in lux. Make any necessary assumptions.

5. Assume that the supermarket of Problem 4 is 600 ft long and 300 ft deep and has a ceiling height of 13 ft, with the ceiling above the fixtures painted black (to hide piping and ductwork). The space is air-conditioned. The floor is concrete. Using the approximate zonal cavity method, Figure 14.64 and Table 14.14, make the following calculations.
   a. Determine the maintained luminance at floor level, assuming single-lamp standard output 96-in. lamps (430-ma lamp current), in both footcandles and lux.
   b. Repeat this calculation for a counter height of 36 in.
   c. Repeat a and b for high output (800-ma) lamps.

6. For the same supermarket in Problem 5, using only the data given in Figure 14.65, calculate the lighting fixture requirement for the following conditions:
   a. Fixture type G with a single 96-in. high output (800-ma) lamp.
   b. Fixture type I (metal halide). Calculate the lamp wattage(s) that would be suitable, if the recommended maximum S/MH is 1.6.

Assume that the working plane is at floor level. Justify all assumptions.
  c. Show suggested luminaire layouts for a and b.
7. A 10 ft × 12 ft private office is lighted with 2 four-lamp recessed troffers using F40 RS lamps.
  a. With CU = 0.6 and LLF = 0.7, find the maintained footcandles at desk level.
  b. After redecorating the room with dark wood paneling, the CU dropped to 0.40. What is the new illuminance level in the room? How many of the same type fixtures must be added to restore the former footcandles?
  c. Show a plan of the room with the original two fixtures and with the new fixture layout required to restore the lighting level.
  d. Assuming that the office has a hung ceiling using 1-ft^2 tiles, would the layout of c change? How? Show a dimensioned ceiling plan, with tiles.
8. a. Redraw Figure 14.65 on a separate sheet of paper, using SI units.
  b. Using this new set of curves, recalculate Example 14.7. Do the results check out?
9. An office building has a two-way, coffer-type ceiling with module dimensions of 5 ft 6 in. by 4 ft. Coffer center height is 10 ft 6 in. Fixtures are to be surface mounted in coffers. It is desired to have an average overall maintained illumination of 30 fc. Make a recommendation for lighting the area using fluorescent fixtures. Show the calculations that led to this recommendation. Make any necessary assumptions, but state and justify them. Draw your proposed layout for a bay that measures 28 ft × 22 ft.
10. A classroom is 8 m wide and 10 m long and 3 m high. Make at least two lighting layouts of the room that will give an average maintained illumination of 350 lux. Use two or three lamp fixtures, with 40-w fluorescent lamps. Repeat this problem using metal halide fixtures in a 2.7-m high hung ceiling. Justify all assumptions. Use data from current manufacturers' catalogs.

15. Residential Electrical Work

We will now apply the techniques that we have learned to a large and elegant private residence. In the process, we will carefully examine each space in the house both as a particular case and as an example of the general case. For instance, the living and family rooms will be studied as rooms and also as an example of living spaces. Our purpose is to develop a basic method for handling different kinds of spaces. This will enable the technologist to do the electrical layout for almost any type of residential space and for similar or related spaces in nonresidential buildings. In this chapter, we will also extend our study of the electrical system beyond branch circuitry to load studies, riser diagrams and other electrical service considerations. Finally, we will study the increasingly important fields of automation ("smart" residences) and communications as applicable to residences. Study of this chapter will enable you to:

1. Analyze spaces by usage—present and potential.
2. Understand the electrical design guidelines applicable to residential spaces and apply them to drawing layout.
3. Apply the design guidelines learned here to nonresidential spaces with similar requirements and uses. These include utility, circulation, food preparation, dining, storage and outside areas.
4. Select, lay out and circuit the lighting for residential and nonresidential spaces, using the applicable provisions of the NEC.
5. Draw plans showing some of the important signal and communications devices found in residential buildings, including riser diagrams.
6. Make up panel schedules, complete with loads, wire sizes and other branch circuit data.
7. Do basic load calculations, including electric heat.

8. Be familiar with electrical service equipment for single and multiple residences and other buildings, including metering provisions.
9. Draw an electric plot plan, riser diagram and one-line diagram for basic electrical service.
10. Know the characteristics of underground and overhead service and be familiar with service entrance details.

15.1 General

As stated in Chapter 3, we use for our study, with the architect's permission, the architectural plans of an individually designed residence. We chose this technique rather than using a tract-house plan so that the design would be unaffected by mass-construction considerations. These were considered, in part, in our previous study of The Basic House plan.

The electrical layout that we will develop is not the one that was actually installed in the house. Every designer produces a different layout. Our layout expresses our approach. It is not the best possible layout, since there is no such thing, but it is a good arrangement. It meets not only the minimum requirements but also what we consider to be the special needs of the space. In addition, the layout attempts to anticipate problems and to make the electrical aspects of living in the house easy and comfortable. To accomplish this requires guidelines, experience and common sense. The guidelines are given here; the necessary experience must be gained in the field. We also refer back to The Basic House plan, so that you can see the difference between treatment of large and small rooms that have the same use.

In doing a layout of any space, a furniture arrangement is a great help. This is particularly true in residences where the location of a bed, for instance, makes all the difference in the location of outlets. If a furniture layout is not available (as is the case in mass-produced tract houses), then the designers must do their layout based on an assumed reasonable furniture arrangement. If that, too, is impossible, because the room can have more than one good furniture arrangement, the layout should be made to accommodate several arrangements, even at slight additional cost. Otherwise, the floor will soon be covered with long extension cords, and the accessible wall outlets will be overused. This is not only unsightly but dangerous. A large percentage of fires in homes are caused by worn extension cords and overloaded outlets. Furthermore, living areas are multipurpose and should be treated as such. Houses with family rooms and living rooms are different from those with living rooms only. In the latter, all social functions take place in the living room, and the electrical layout should be flexible enough to meet this need.

In addition to coordinating the electrical layout with the furniture, it is necessary also to coordinate it with the heating/cooling equipment. This work is much more extensive in industrial and commercial buildings than in residences and is discussed in detail in Chapter 16. In houses, it is relatively simple to avoid conflicts in equipment location. We will deal with this in the layout description. We must also take into consideration the changes that computers and advanced telecommunications techniques have caused. The time is not far in the future when many professionals will conduct their work from their homes. The techniques for doing so already exist. Computer modems, telephone conferencing, computer networks and reliable error-free connections make it possible for engineers, computer programmers, analysts of all types, designers, researchers, accountants and a multitude of other professionals to work at home. A good electrical house design will take this into account so that extensive renovations will not be necessary if the owner decides to work at home. Finally, it would be an error not to consider the enormous impact of automation and programmable control (computer and/or microprocessor) on the everyday operation of a home. Here, too, the techniques exist, but the cost factor is high. Since these costs (unlike most others) have shown a continuous downward trend, the automation of residential functions is increasing practically daily. These functions are not only for convenience, as we will discuss, but also for safety.

We have divided the spaces in a residence into six types by use and location, as follows:

(a) Social, meeting and family-function areas, including living rooms and family rooms plus adjoining balconies, play and recreation rooms, and media rooms.
(b) Studies and work rooms.
(c) Kitchen and dining areas, including food preparation areas, breakfast rooms, and dining rooms.
(d) Sleeping areas, including bedrooms, dressing rooms, closets, and adjoining balconies.
(e) Circulation, storage and utility and wash areas. These include the halls, corridors and stairs,

garage, closets, basement, attic, laundry and other work areas, and bath- and washrooms.
(f) Outside areas attached or adjacent to the house.

15.2 Social, Meeting and Family-Function Areas

a. Living Room

As stated in Chapter 13, the wall convenience receptacles in a residence are considered to be part of the general lighting load. This is because these receptacle outlets are frequently used to feed the lighting sources in the room. Refer to Figure 13.15(a). That living room, like most small to medium-size modern living rooms, has no ceiling lighting outlet. Instead, a switched receptacle is used to provide the lighting. The resident will plug a table or floor lamp into this outlet. The reason that a ceiling outlet is not used is that a living room normally requires only a low to medium lighting level, which is easily provided by lamps. Furthermore, in small houses, particularly of the mass-built type, the ceiling is low (7 ft 6 in. to 8 ft). This does not permit use of any type of decorative pendant fixture. Surface-mounted units are generally not attractive, and recessed units are expensive to install and require a deep ceiling space. For all these reasons, living room ceiling lighting outlets gradually disappeared. Furthermore, since lamps are always a part of the furnishings, a ceiling outlet would be used only rarely, even if it were provided. Also a relatively low ceiling outlet tends to attract the eye and cause glare.

On the other hand, multiple ceiling outlets, wallwash units, or lighted coves, valances or cornices are not only acceptable but very desirable. The complete electrical plan for the Mogensen house is shown in Figure 15.1. Refer to the living room in Figure 15.1(b). This room is quite large, and its unusual architecture calls for special lighting solutions, including the type of drawing detailing that a technologist is frequently called on to do. The important architectural features of this space are the sloping ceiling with its breakdown to the low ceiling level, the all-glass front wall leading out to the deck and the brick wall and fireplace. Let us consider these features one by one.

Taking the sloping ceiling first, the break line (in plan) is an ideal location for applying a lighted cove. See the building section in Figure 15.2 and in the architectural plans of Figure 3.37. Now refer to Figure 15.2. Notice that the cove lights up the ceiling, which is visible from outside through the clerestory windows. Since these high windows are either not curtained at all or only lightly curtained, the nighttime effect of the cove illuminated ceiling is particularly attractive from outside. The cove detail is shown in Figure 15.3, and additional cove details are given in Figure 15.4. This type of detailing, whether it is done by hand on a drafting board or with the aid of a CAD program, is an essential part of an electrical technologist's work. Since the installation is a "special," detailing is necessary, so that both the construction contractor and the electrical contractor can build the unit as desired by the designer. The detail shown in Figure 15.3 is schematic only.

An actual working drawing detail includes all dimensions, material details and electrical data. Information that is left out will usually result in an "extra," an additional cost item over and above the contract cost. For this reason, details must carry all the information necessary for construction. Furthermore, the technologist must determine from the appropriate source (architect, engineer) exactly who is responsible for what work and show it on the drawings. (Reliance on the specifications to divide work responsibility is generally not a good idea.) Otherwise, as so very often happens, subcontractors can claim (in some measure correctly) that they were not aware that a specific item of work was their responsibility, and they will demand additional payment.

Let us return now to the technical aspects of the lighting design. The cove lighting is controlled by a switch and appropriate dimmer, which permits level adjustment. In addition to this soft and pleasant cove lighting, the lighting designer has decided on a 3 × 3 pattern of ceiling downlights, which are labelled type Z on the working drawings. These units are also dimmer controlled, by a standard incandescent lamp dimmer, rated 1500 w minimum. The type Z fixture has a black alzac cylinder so that no fixture brightness is reflected in the upper-level glass. The fixture must be obtained with a bottom slope to match the ceiling slope within 10°. Any mismatch in angle larger than 10° will tip the fixture and throw its lighting pattern off center. See Figure 15.5.

Here again, we must remind you that the details shown are schematic. Note that no thermal insulation is shown, although the space between the roof and the sloping ceiling is heavily insulated. As we have already noted in Chapter 13, the installation of lighting fixtures must meet **NEC** (Article 410)

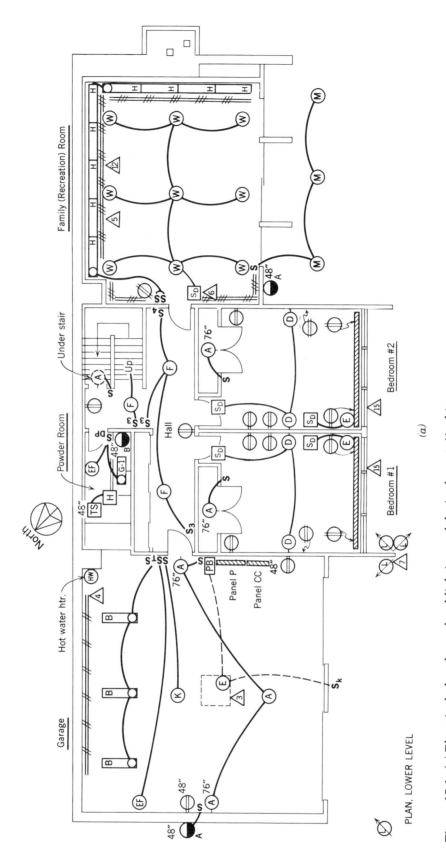

PLAN, LOWER LEVEL

Figure 15.1 (*a*) Electrical plan; lower level lighting and device layout. (Architectural plans courtesy of B. Mogensen A.I.A.)

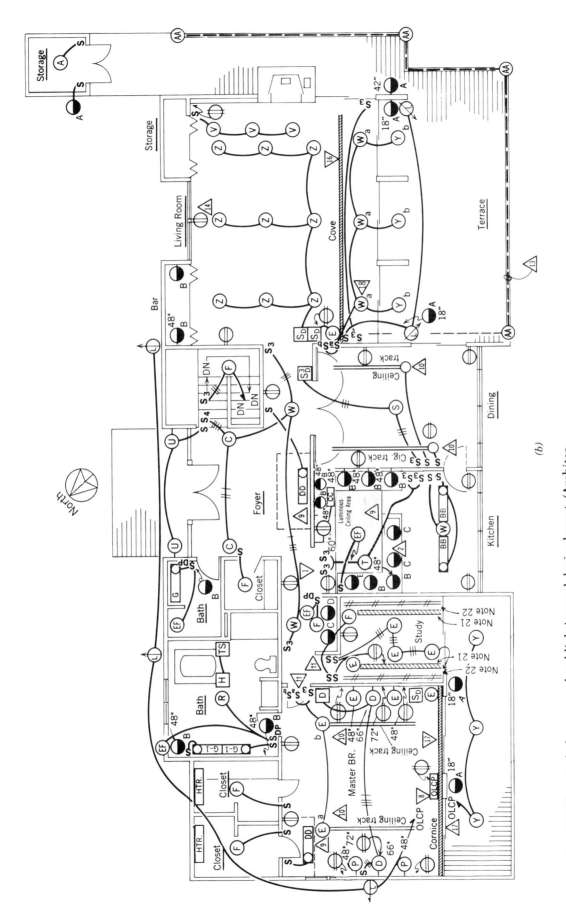

Figure 15.1 (b) Electrical plan; upper level lighting and device layout. (Architectural plans courtesy of B. Mogensen A.I.A.)

866 / RESIDENTIAL ELECTRICAL WORK

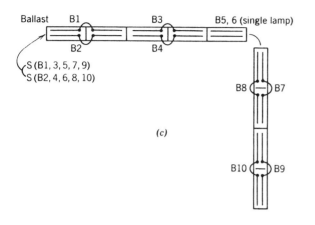

Figure 15.1 *(c)* Wiring arrangement for type H fixtures (wall brackets) in family room. Note that one fixture is supplied with single-lamp ballasts.

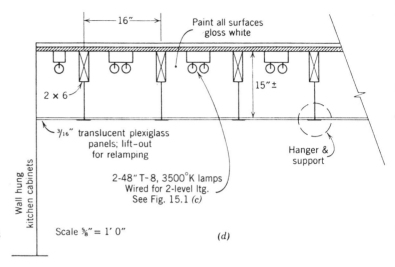

Figure 15.1 *(d)* Kitchen luminous ceiling lighting details.

Notes:
1. Panel face to be suitable for same finish as wall.
2. Outlets under counter.
3. Ceiling-mounted electric garage door opener.
4. Two-circuit wiring channel with 10 single 20-amp, 2-pole, 3-wire grounded receptacles on 24-in. centers, 48-in. AFF. Total 10 outlets; 5 per circuit. Start first outlet 12-in. from one end.
5. Wire fixtures per Figure 15.1(c), so that two-lamp ballast in each fixture controls one lamp in its fixture and one lamp in the adjacent fixture.
6. Single 1500-w switch and dimmer. Install flush in closet wall and enclose rear protruding into closet.
7. Extend floodlight wiring to outside lighting control panel in master bedroom.
8. The contractor is invited to submit a proposal for performing all the lighting switching with low voltage control. The proposal must contain a wiring diagram, complete equipment specifications and a price differential. The control panel would be mounted in the master bedroom, as shown.
9. For detail of recessed daylight lamp fixture see architectural drawings. Note on Drawing 6.4(b) that there is an access to the attic in this area.
10. Ceiling lighting track, field located, 15-amp capacity, with fixtures as selected by architect.
11. Provide a spare switch position.
12. Mount a 3 circuit multi-outlet assembly around the entire perimeter of the family room (excepting doorways and glass areas) Wiremold® 2000 or equal, at 18-in. AFF. Provide 15-amp receptacles, NEMA 5-15R every 24-in. wired in succession on the three circuits indicated. Provide an identical empty raceway immediately below the receptacle raceway. See specs. for fittings to be furnished with both raceways.
13. Run low voltage wiring exposed, under deck. Mount transformer at switch location. Low voltage lighting equipment and wiring shall be in accordance with the requirements of NEC Article 411.
14. At all locations where receptacles conflict with a heating element register or grille, the receptacle can be moved to either side, as directed in the field.
15. Lighting valance above the window; with 48-in. single lamp strips and dimming ballasts. See detail Figure 15.13.
16. Lighted cove approximately 9 ft. AFF. See detail Figure 15.4. 48-in. single lamp strips, with dimming ballasts. Approx. 15 w/ft.
17. Lighted cornice above the drapes. See Figure 15.14. 48-in. single lamp strips with dimming ballasts. Approx. 15 w/ft.
18. Coordinate the location of the 3-type W fixtures with ceiling registers. See Figure 6.4(a).
19. Hood switch supplied integral with hood.
20. Each unit 750 w, 120 v, with integral thermostat and ON/OFF switch.
21. Mount an 8-ft long cove, 12 in. below ceiling, containing two 48-in. T-8, 4500°K lamps in tandem, on each side of the room. See detail, Figure 15.4.
22. Mount a two-circuit multi-outlet assembly along both long walls of the study at 42-in. AFF. Provide 15-amp receptacles, NEMA 5-15R every 24-in. wired in succession on two circuits in both raceways. Provide surge suppression for these circuits as detailed in specifications. Provide an empty raceway, Wiremold® 3000 or equal directly below the multi-outlet raceway. Connect the empty raceway with two 1-in. empty conduits, concealed in wall and under floor; one at each end of the raceway.

(e)

Figure 15.1 *(e)* Notes for electrical device layout.

Type	Mtg.	Maximum Watts	Description
A	—	100	Porcelain lampholder, pull chain, wire guard, mtg as shown
B	Clg	100	Recessed fluorescent, two-lamp 40 w RS industrial type, steel louver, baked white enamel
C	7 ft	75	Decorative incandescent wall bracket
D	Wall	150	Halogen lamp wall sconce, mtd. 66″ AFF to centerline.
F	Clg	100	Recessed square incandescent unit, dropped–dish opal glass diffuser, white enamel frame
G	Wall	50	Fluorescent wall bracket, wrap-around white plexiglass diffuser, one 48-in. T8, 3500°K lamp, mount above wall mirror
G-1	Wall	38	Same as G except one 36-in. T8, 3500°K lamp
H	7 ft	100	Fluorescent, wall bracket with decorative plastic front diffuser; See detail, Figure 15.10(c). Each unit with two 48-in. T12 RS lamps
K	Clg	100	Reel-light, 100 w portable, 25 ft retractable cable, 360 deg swivel mount.
L	—	150	WP flood light for R-40, 150 w bracket mounted, as field directed
M*	Clg	150	WP, surface-mounted cylindrical downlights 150 w, R-40 lamp
P	4 ft	100	Swivel mount, spherical chrome wall bracket fixtures, mounted in pairs, for single 100 w halogen lamp with built-in On/Off switch.
R*	Clg	150	Recessed, gasketed, square, incandescent unit, dropped dish frosted glass diffuser, cast aluminum frame, for single 150 w A lamp
S	Clg	1000	Decorative chandelier, selected by owner
T	Clg	600	Kitchen illuminated ceiling; see Figure 15.1(d)
U	—	100	Decorative wall bracket, WP, 100 w A lamp
V	Clg	150	Special unit; see Figure 15.8
W*	Clg	150	Recessed downlight; see Figure 15.7
Y*	Clg	150	Same as W except WP; see Figure 15.7
Z*	Clg	150	Downlight, one 150 w R-40 lamp; see Figure 15.5
AA	—	25/50	Decorative, WP, post-top lantern, 6 v A-21 lamp, 25 or 50 w, 10-in. white plexiglass globe, with 36-in. redwood post; see Figure 15.9
BB	Clg	100	Fluorescent fixture, surface, wrap-around plexiglass lens. Two-lamp F40 RS WW. Shallow construction. Maximum X-section; 12-in. wide × 3½ in. deep. Diecast end with white baked enamel finish
CC	—	30	Incandescent under-counter fixture. One 17-in. T8 lumiline lamp, disc base. Maximum fixture depth 1¾ in. White plastic diffuser
DD	—	100	Fluorescent strip, 2-lamp 40 W RS, daylight lamp, built into skylight. See detail on architectural drawing, A.00

*May be replaced with similar lighting fixture designed to use compact fluorescent. Dimming devices must be adjusted accordingly.

Figure 15.1 *(f)* Lighting fixture schedule.

SYMBOL LIST

―///―	WIRING RUN CONCEALED. HATCH MARKS INDICATE NUMBER OF #12 WIRES U O N GROUND CONDUCTORS NOT SHOWN.
― ― ―	CONDUCTORS RUN EXPOSED, NOTES AS ABOVE.
—•²‾⁴⟶P-2	HOME RUN OF CCTS 2 & 4 TO PANEL P-2 DOT INDICATES TURNING DOWN
(E)〰•	FINAL CONNECTION FROM OUTLET TO EQUIPMENT.
#-#-#	MULTI-OUTLET ASSEMBLY, HATCH-MARKS INDICATES NUMBER OF CIRCUITS
(A)a -(A)	OUTLET AND INCANDESCENT FIXTURE CLG./WALL MOUNTED. INSCRIBED LETTER INDICATES TYPE. SUBSCRIPT LETTER INDICATES SWITCH CONTROL.
[○ B]a / [○ B]	OUTLET AND FLUORESCENT FIXTURE CLG./WALL MOUNTED. SAME NOTES AS ABOVE.
(E)	OUTLET BOX WITH BLANK COVER.
(J)	JUNCTION BOX WITH BLANK COVER.
⊖•	DUPLEX CONVENIENCE RECEPTACLE OUTLET. 15 AMP 2P 3W 125 VOLT GROUNDING. WALL MT. VERTICAL, ℄ 12" AFF, NEMA 5-15R. DOT INDICATES APPLIANCE OUTLET – SEE TEXT.
⬤ₐ	15A 2P 3W GFCI DUPLEX OUTLET, WP.
⬤ᵦ	15A 2P 3W GFCI DUPLEX OUTLET.
⬤ᴄ	20A 2P 3W SINGLE OUTLET, NEMA 5-20R.
⬤ᴅ	30A 125/250V 3P 4W GND. OUTLET, NEMA 14-30R.
⬤ᴇ	60A 125/250V 3P 4W GND. OUTLET, NEMA 14-60R.
⬤F	20A 2P 3W GFCI SINGLE RECEPTACLE OUTLET WITH WEATHERPROOF COVER
⊕	CLOCK HANGER OUTLET, SEE SPEC. 7'6' AFF TO ℄.
S_a	SINGLE POLE SWITCH, 15A 125V, 50" AFF U O N, LETTER SHOWS OUTLETS CONTROLLED.
S_3	THREE WAY SWITCH 15A, 125V, 50" AFF U O N.
S_4	FOUR WAY SWITCH, AS ABOVE.
S_{DP}	DOUBLE POLE SWITCH, AS ABOVE.
S_K	KEY OPERATED SWITCH, AS ABOVE.
S_T	SWITCH 15A 125V WITH THERMAL ELEMENT SUITED TO MOTOR.
⌀	COMBINATION SWITCH AND RECEPTACLE IN A 2 GANG BOX.
[S_D]	COMBINED SWITCH AND DIMMER.
[D]	DIMMER, RATING AS NOTED, 600W U O N
[TS]	TIMER CONTACTS, MANUAL SET, 15 AMP, 125 VOLTS, 0-20 MINUTES.
[PB]	PUSHBUTTON, 10 AMP MOM. CONTACT U O N
[H]	HEATER, RATING AND DETAILS ON DWG., WITH OUTLET.
(EF)	EXHAUST FAN, 1/12 HP U O N
▬▬	RECESSED PANELBOARD, SEE SCHEDULE.
▬▬ OR ▨▨	SURFACE MTD. PANEL
⊸[M]	INCOMING ELECTRIC SERVICE, METER CABINET AND METER.
▨▨▨▨	ARCHITECTURAL LIGHTING ELEMENT; SEE PLANS.
△	REFERENCE TO NOTE 1. SEE NOTES.

ABBREVIATIONS:

A	AMPERES
AFF	ABOVE FINISHED FLOOR
C/B	CIRCUIT BREAKER
CCT	CIRCUIT
F	FUSE
GFCI	GROUND FAULT CCT INTERRUPTER
GND	GROUND
HP	HOURSEPOWER
LTG	LIGHTING
MH	MTG. HEIGHT
N	NEUTRAL
PC	PULL CHAIN
T	THERMOSTAT
TYP	TYPICAL
UON	UNLESS OTHERWISE NOTED
UF	UNFUSED
WP	WEATHERPROOF

Figure 15.1 *(g)* Symbol list.

requirements, and the technologist, with the help of an experienced designer, must make sure that the data on the details meet these requirements. This is particularly true in heat build-up situations such as insulated and fire-resistant ceilings. (See, for instance, **NEC** Article 410-65.)

The nine type Z fixtures are arranged in a symmetrical pattern in the ceiling, with fixture-to-wall spacing one half of the fixture-to-fixture spacing. Figure 15.6 demonstrates a simple and reliable way of laying out any desired fixture pattern without detailed measurement. This method is based on the preceding universally accepted arrangement: fixture-to-wall spacing is equal to one-half of fixture-to-fixture spacing.

Fixtures W and Y are similar to each other (see Figure 15.7), except that type Y is gasketed since it is installed outdoors. The designer has deliberately provided lighting on both sides of the sliding glass doors. This prevents the glass from acting like a mirror, which happens when the space on the far side of the glass is much darker than on the near side. This can be very annoying when you are trying to see out at night. The solution is, as already explained, to provide lighting on the far side of the glass as well. The fixtures have been selected so as not to show any brightness that will reflect in the glass and be a source of annoyance. This is accomplished by using a regressed (setback) lens and a matte black ring at the bottom of the luminaire, as is shown in the fixture detail in Figure 15.7. The purpose of the lens is to spread the light so as to have even coverage and avoid pools of light on the floor. The fixture detail shows use of an incandescent lamp, as in the basic Z fixture. Of course, if a compact fluorescent lamp is used there, then the same type of lamp should be used in fixtures W and Y, to avoid a change of light color.

Within the living room, the focus of interest is the fireplace and the adjoining brick wall. To accent this, three type V wall-wash lighting units have been supplied. These units are special because they also must be made to fit into the sloping ceiling. See Figure 15.8. Control of these units is local, since they are accent lighting and will not be used all the time. Similarly, control of fixtures W and Y is separate, since they also are special purpose. Control of the general illumination furnished by the cove and the Z fixtures is provided at the entrance to the living room. It is always advisable to use a switch together with a dimmer, so that a preset dimming level can be obtained by simply flicking the switch. Otherwise, one must experiment each time, and the dimmer becomes a nuisance instead of a convenience. Most modern dimmers have a built-in switch for just this purpose. Some electronic dimmers permit presetting of light levels so that particular settings can be duplicated without manual readjustment.

The deck outside the living room is really the outdoor part of the living room. Lighting for this area is, of course, daylighting most of the time. Night lighting should provide enough light for activities on the deck such as eating and recreation. This is easily furnished by the type L adjustable floodlights and the type AA low voltage post lights. Low voltage units were chosen to permit the use of exposed wiring with minimum electrical maintenance problems. These low voltage fixtures are popular for landscape, security and general outdoor use because of the ease of wiring and minimal hazard. Control of the floodlights and the post lights is conveniently placed at the exit to the deck. At the exit to the side deck, a three-way switch is furnished for control of the post lights only. See Figure 15.9 for details of the low voltage units and their wiring.

Receptacles are placed about the room in accordance with the **NEC** rule that no point along a wall shall be farther than 6 ft from a convenience outlet. This rule (**NEC** Article 210-52) specifically excludes sliding panels in exterior walls. Since the entire front wall of the living room consists of sliding panels, no receptacles are shown. However, three weatherproof GFCI-type convenience receptacles are provided on the deck to power electric grills and other devices used outdoors.

All wall outlets will be mounted vertically and at 12 in. AFF to the centerline unless specifically shown otherwise. The outlet in the bar area is shown at 48 in. for convenient use of electric mixers, ice crushers and the like on the bar. This completes the description of the electrical layout of the living room and connected areas. After a similar description of the family room, we will review the layout principles that have been followed.

b. Family Room

This room in many ways is the most important room in the house. Aside from eating, sleeping and formal entertaining, this is where the family spends its time. Here the family relaxes, often using extensive electronic equipment in the process; entertains informally; plays games; and so on. The room, therefore, has many functions. The electrical

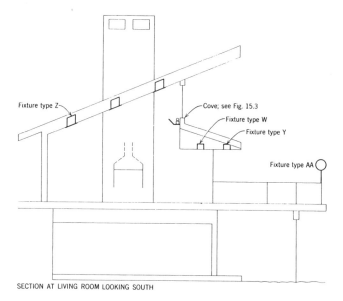

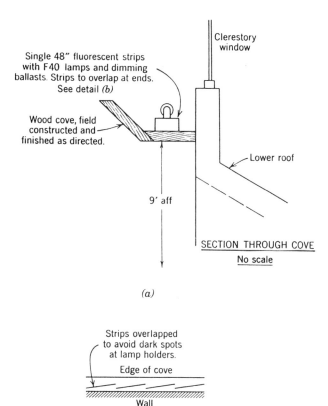

Figure 15.2 Section through living room of Mogensen house showing cove lighting element (Figure 15.3), sloping ceiling downlights (fixture type Z, Figure 15.5) and recessed downlights (types W and Y) on both sides of the sliding glass doors. In addition, low voltage outdoor post light (type AA) is shown. See Figure 3.38 page 141 for architectural plan sections and elevations.

Figure 15.3 (a) Detail of cove construction in the living room at the ceiling break line. Note that the sloped front lip of the cove traps less light than the usual vertical member (see Figure 15.4). Sloped members are particularly appropriate to sloped ceilings, as in this house design. The fluorescent lamp strip can also be placed on its side as in Figure 15.4 for higher efficiency and lower ceiling brightness. (b) Lamp strips should be overlapped as shown to avoid dark spots in the cove.

layout must be flexible enough to serve all these functions.

First, let us consider the lighting. Obviously, many different levels and types of lighting are required to fit all this room's activities. Recreation activities assume a range of levels varying between high for a children's party to soft and subdued for an evening get-together. Similarly, television watching and game and billiards playing all have completely different lighting requirements. We, therefore, decide to supply adjustable level lighting and selectable quality. The type H decorative wall bracket supplies both general room lighting and wall lighting, which adds interest to the room. This is particularly important if the walls are wood paneled. The wall brackets can be mounted at the height desired. This depends on three things: their primary purpose, the ceiling height and the viewing positions as explained in Figure 15.10.

These bracket units can be purchased commercially or constructed on the job, depending on the design. Suggestions for dimensions and components are shown in Figure 15.10. For our application, a translucent plastic front is recommended for this fixture rather than a solid opaque one. This will keep the luminance ratios (Section 14.6) low for eye comfort. Also important to keep in mind are the reflections of the room lighting in the extensive glass walls. As we stated previously, the glass wall acts as a mirror, and all the bright areas in the room will reflect in the glass. Thus, the wall bracket will show up, but the sharp contrast will

LIGHTED COVES

Coves direct all light to the ceiling. Should be used only with white or near-white ceilings. Cove lighting is soft and uniform but lacks punch or emphasis. Suitable for high-ceilinged rooms and for places where ceiling heights abruptly change.

(a)

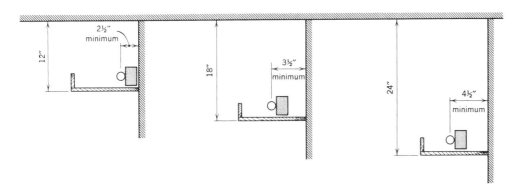

COVE INSTALLATIONS

Good cove proportions: Height of front lip of cove should shield cove from the eye yet expose entire ceiling to the lamp. Orientation of fluorescent strip as shown is an alternate to upright arrangement.

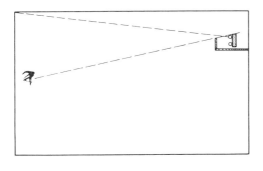

(b)

Figure 15.4 Additional cove details. The lamp strip position shown (strip on its side) reduces ceiling brightness and improves uniformity of lighting on the ceiling. It is most appropriate for shallow coves—12–18 in. from ceiling. [(a) Courtesy of IESNA. (b) From *Westinghouse Lighting Handbook* (out of print).]

be reduced if the front of the fixture is lighted rather than appearing as a dark band caused by an opaque front against a lighted wall background.

A simple and inexpensive switching arrangement that provides two levels of lighting is shown in Figure 15.1(c). This plan lights either one or both lamps in each fixture, giving uniform levels of half or full lighting. Much finer level control with switching can be had by using multilevel ballasts or special impedances that are switched into the circuit. More gradual control requires dimming, which is also available in a variety of designs. Since the technologist is not normally responsible for designing and selecting any of these sophisti-

872 / RESIDENTIAL ELECTRICAL WORK

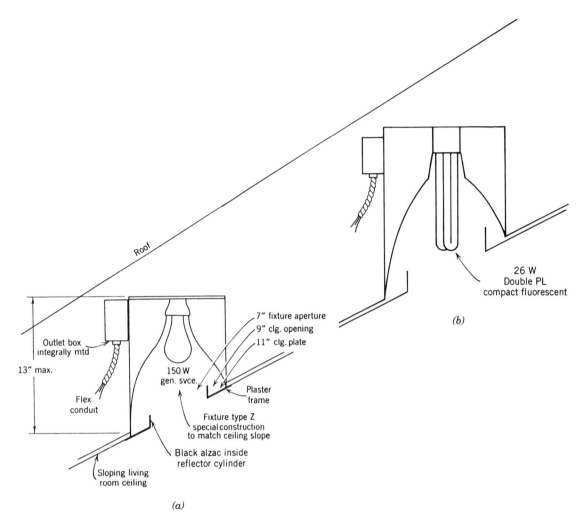

Figure 15.5 Fixture type Z for use in sloped ceiling. Fixtures using incandescent lamps (a) should be equipped with long-life (130-v) lamps because of the inconvenience of relamping in a high ceiling. Standard lamps have only a 750-h life compared to 2500 h for the long-life lamp. The output loss (2300 lm for the long-life lamp versus 2780 lm for the standard lamps) is not critical. Alternatively, a 26-w double PL compact fluorescent lamp fixture (b) can be used. These lamps have an 1800-lm output and 10,000-h life. See Table 14.7, page 814. If PL lamps are used, dimmers suitable to this lamp must be provided.

Figure 15.6 Graphic technique for locating lighting fixtures on a spacing grid that uses a luminaire-to-wall distance of one-half the luminaire-to-luminaire distance. This graphical method eliminates the need for accurate fractional measurements.

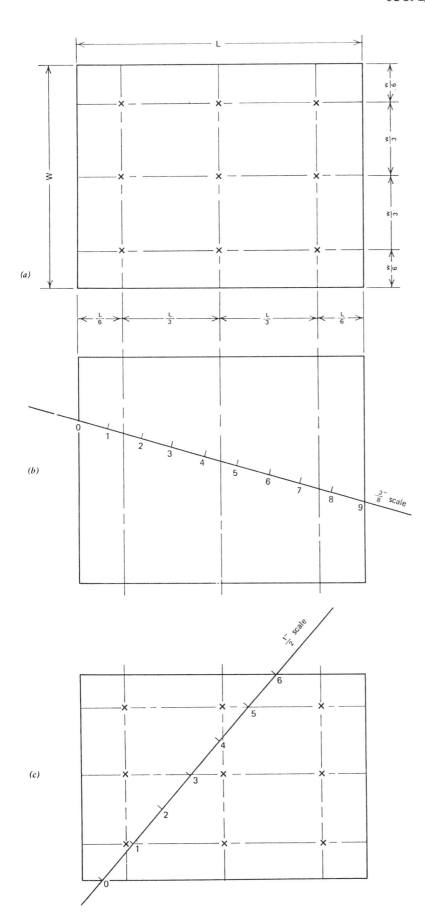

REQUIRED: TO LOCATE 9 LIGHTING FIXTURES IN A SPACE, WITH FIXTURE TO WALL SPACING $\frac{W}{6}$ AND $\frac{L}{6}$, EQUAL TO HALF THE FIXTURE-TO-FIXTURE SPACING $\frac{W}{3}$ AND $\frac{L}{3}$.

TECHNIQUE:

IN LENGTH; LAY AN ARCH. SCALE OVER THE AREA, SELECTING A SCALE DIVISIBLE BY 3. HERE (b), 3/8" SCALE WORKS WELL. NINE DIVIDED BY 3 GIVES 3; HALF OF THIS IS 1 1/2. THEREFORE, HAVING DRAWN THE SKEW LINE, MARK 1 1/2, 4 1/2, 7 1/2 AS SHOWN. EXTEND LINES VERTICALLY THRU THESE POINTS.

IN WIDTH

REPEAT THE ABOVE PROCEDURE IN THE OTHER DIMENSION. HERE (c) 1/2 SCALE FITS WELL, FROM 0 TO 6". DRAW THE SKEW LINE AND MARK 1, 3, 5. EXTEND THESE POINTS HORIZONTALLY TO COMPLETE THE GRID.

THE INTERSECTION OF THE VERTICAL AND HORIZONTAL LINES FORM THE DESIRED CENTERPOINTS OF THE FIXTURES.

874 / RESIDENTIAL ELECTRICAL WORK

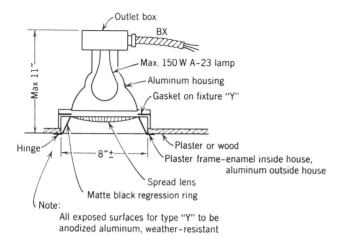

Figure 15.7 Detail of fixtures W and Y. Lamp type should be the same as that chosen for fixture type Z. Since it is exposed to the weather under the overhang, type Y must be marked as suitable for use in damp locations.

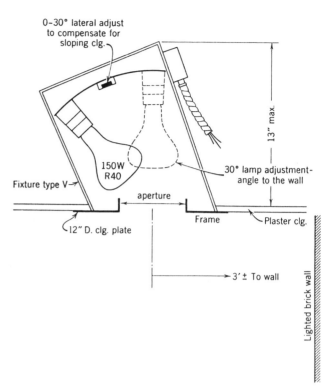

Figure 15.8 Fixture type V. Recessed adjustable wall-wash luminaire, with sloped-ceiling compensation. Details of the fixture construction, including internal reflectors, are not shown, for clarity. The unit has two planes of lamp position adjustability: one lateral (parallel to the brick wall) to compensate for the ceiling slope and one perpendicular to the wall to adjust the position of the light on the wall.

cated control schemes, they are not discussed here in detail. As with other highly technical items, you can consult the books listed in the Supplementary Reading at the end of Chapter 16 for further reading and/or consult with the design engineer in the office, when working on an actual design.

Due to the problem of light reflection in the glass walls at night, the ceiling units chosen have a regressed spread lens and a black, nonreflecting bottom cone. Of course, all the room's lighting could have been supplied by using more of these downlights. This would reduce the mirroring problem. The result, however, would be a ceiling covered with black holes and no wall lighting. The layout presented, with two-level fluorescent wall brackets, is a good compromise. An item to keep in mind when using fluorescent lamps is the type of lamp used. Areas where skin color or color rendering in general is important should use lamps with a high CRI such as the T-8 triphosphor lamp. See Section 14.17.b.

Refer now to Figure 15.7. These lamps are fed through a standard incandescent lamp dimmer for full-range dimming. Keep in mind that compact, inexpensive electronic dimmers can cause interference to sensitive electronic equipment. This is not so with the older, bulkier nonelectronic incandescent-lamp dimmer. In general, downlights will not provide good general lighting unless a wide-spread distribution is used. This can be obtained with lens units. These also have the advantage of low brightness. We are assuming that no curtains are placed on the front wall. We furnish three cylindrical decorative downlights (type M). They will act to cut down the inside mirror effect, as explained previously. The outside lights are controlled by a local switch. Note that, wherever possible, the wiring is run along the ceiling members rather than across them to avoid the labor and expense of drilling.

Note, also, that switches are placed as close as possible to the controlled lights. This avoids large banks of switches at the room entrances. Such banks are unsightly, require nameplates to avoid

SOCIAL, MEETING AND FAMILY-FUNCTION AREAS / 875

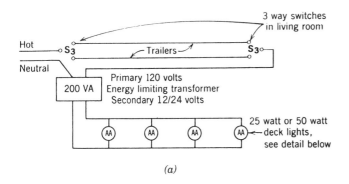

(a)

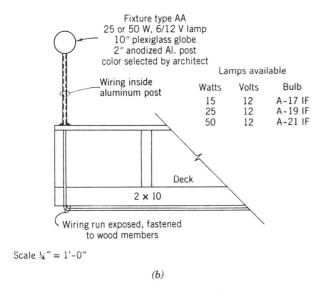

(b)

Figure 15.9 (a) Schematic wiring diagram of low voltage outdoor post lights. Details of wiring and exact equipment location are not shown, for clarity. (They would be shown on contract drawings). (b) Construction details for exterior post lights, type AA.

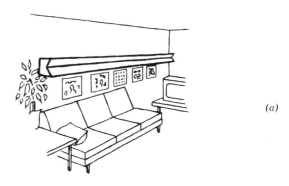

LIGHTED LOW WALL BRACKETS

Low brackets are used for special wall emphasis or for lighting specific tasks such as sink, range, reading in bed, etc. Mounting height is determined by eye height of users, from both seated and standing positions. Length should relate to nearby furniture groupings and room scale.

LIGHTED HIGH WALL BRACKETS

High wall brackets provide both up and down light for general room lighting. Used on interior walls to balance window valance both architecturally and in lighting distribution. Mounting height determined by window or door height.

Figure 15.10 Architectural element wall lighting. (a) Lighted low bracket. (b) Lighted high bracket. [(a,b) Courtesy of Illuminating Engineering Society of North America.]

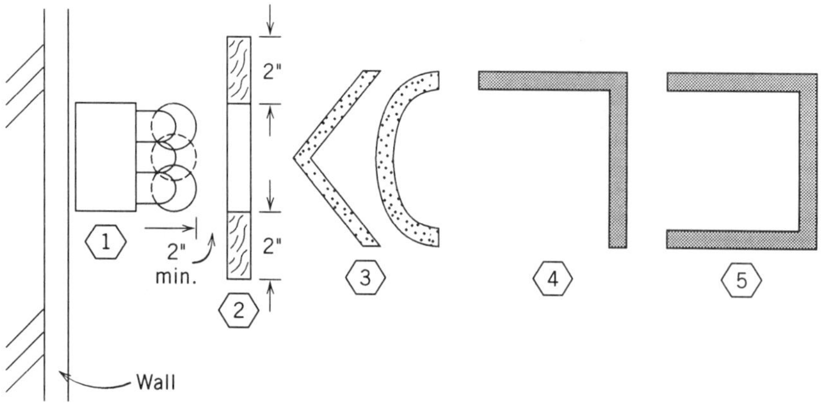

Design Notes for Custom-made High and Low Wall Brackets

⟨1⟩ Fluorescent lamp strip; one or two lamps. High wall bracket mounted 5'0" minimum above floor; low wall bracket mounted 5'0" maximum above finished floor (AFF). Dimension to unit centerline.

⟨2⟩ Simple front shield, translucent or opaque. Extends 2" above strip for low unit or 2" below strip for high bracket.

⟨3⟩ Front sheilds of various shades; preferably translucent.

⟨4⟩ Low bracket shield; front element opaque or translucent; top element attaches to wall, translucent or transparent (usable for display of glass objects).

⟨5⟩ High bracket shield; opaque or translucent front element; translucent top and louvered, baffled or translucent bottom.

(c)

Figure 15.10 *(Continued)* Design details for custom brackets.

confusion and are inconvenient to use. Local switch placement is especially important in controlling special-purpose or accent lighting that is not constantly in use. It is a nuisance to have to go to the opposite end of the room when you wish to turn on the wall-wash lights at this end of the room. The specification should call for identification on all switch banks of three or more switches. This labelling can vary in quality from simple tapes to etched wall plates, depending on the quality of the installation.

Modern consumer electronics is so widely used, that a family (recreation) room in a house of this quality could readily contain a standard TV set, a projection TV, a high resolution TV, one or two video cassette recorders, a satellite dish controller and terminal, a personal computer, extensive sound equipment and speciality devices such as disco-type lighting controllers. All these devices are relatively small and, therefore, easily moveable within a space. For this reason we have provided a multi-outlet assembly (plug-in strip) with receptacles on 24-in. centers around the entire perimeter of the room. This permits placing equipment just about anywhere in the room, without the nuisance (and danger) of long extension cords. The assembly is similar to that of Figure 12.30(b) except with single convenience receptacles on 24-in. centers as stated.

Immediately above the multi-outlet assembly, we provide an identical empty raceway, intended to carry low voltage wiring such as controls, speaker wiring and the like. Although a single dual-section raceway such as that in Figure 12.32 is more attractive than two single units, it requires exposing the power wiring every time the signal wiring cover is removed. Since the low voltage wiring is frequently done by people (including the owner) who are not licensed electricians, it is definitely not a good idea to expose the power wiring every time access to the low voltage wiring raceway is needed. See Section 12.9. The two raceways should be wall mounted 12–18 in. above the floor. We have chosen a surface-mounted raceway because recessing continuous horizontal raceways in wood-frame construction is expensive. Circuiting of the multi-outlet assemblies is shown in Figure 15.16.

c. Guidelines

Layout for living areas should consider the following:

- Switched ceiling outlets in large rooms and switched receptacles in small rooms or rooms known to use only floor or table lamps.
- Capability to change lighting levels and quality. This can be done with dimming or switching and with different types of sources.
- Built-in accent lighting for walls, drapes and paintings, locally switched.
- Architectural elements in lighting such as valances and cornices.
- Avoidance of downlights for general lighting except in rooms with high ceilings.
- Avoidance of any glare sources, either direct or reflected.
- Location of sufficient convenience receptacles to meet **NEC** minimum plus additional ones where they may be useful.
- Avoidance of large banks of switches, unless grouped into a centralized house switch center.
- Detailing of all special fixtures, architectural elements or unusual construction or wiring.
- Provision for the future. It can take the form of capped outlets, empty switch positions, empty conduits in the ceiling or wall, lightly loaded circuits and the like.

15.3 Study; Work Room; Home Office

This category is intended to cover spaces in homes that are used for serious work, whether or not that work is connected with a business or other income-producing activity. This means that dens, TV rooms, and the like are not included here. They are covered in the family-type rooms. The clear division that once existed between place of business and place of work no longer necessarily applies. As mentioned previously, many professionals, academics and business people conduct part of their work activities at home. Furthermore, the numbers of such people is constantly increasing because of the improved communication facilities available. Most such work involves the use of a computer with modem, a fax, and one or more phone lines for telephone communication and data transmission. In addition, a home office uses a variety of electronic and optical devices such as copy machines, scanners and printers. These devices do not require outside connection, but they do require electrical power and cable interconnections.

To supply the required cabling and power for present and reasonable future requirements for a typical home "office," we recommend the following:

(a) A separate 1¼-in. empty conduit (or two ¾-in. conduits) for phone lines and data cables from the work room, terminating at the telephone cable entrance point. (This is shown on Figure 15.33.) The location of the telephone service entrance can be coordinated with the phone company at the design stage of the project.

(b) Power and (empty) signal raceways installed around the room perimeter. (In our case, these raceways are installed on the two long walls of the study.) at 42 in. AFF. These are similar to the raceways installed in the family room, as described in Section 15.2. Here, as there, the raceways can be surface-mounted or, at additional expense, installed flush by notching deep wall studs. The receptacles in the power section of this raceway should be provided with surge suppression protection and should be connected to portions of at least two circuits. If required, such receptacles can be of the insulated ground type to minimize electrical noise interference. (Where these are used, the wiring system must be designed accordingly, as required by the **NEC**. See Article 250-74.) A junction box should be provided at the point at which power is connected to the raceway. This will permit connection of power-conditioning equipment if required. These facilities are shown on Figure 15.1(b). (If one of the walls of the room is to be used in such a manner that a surface raceway interferes, it can readily be moved.)

Lighting for this room is somewhat difficult because of the various seeing tasks that are to be performed here. In order to avoid the extreme annoyance of reflections in a computer monitor screen, we have provided fluorescent coves on both sides of the room, individually switched. A desk will almost certainly be placed on the west (left-hand) wall. A switched receptacle is provided to feed a desk lamp. Finally, in the event that the same highlighting is desired for the east (right-hand) wall or if a lighted cornice is desired above drapes on the front glass wall, prewired blank-covered ceiling outlets are supplied, with appropriate switching. Space is left at the entrance switch bank for additional switching for these outlets. This type of provision for future lighting or a change in intended room use is by far the cheapest and best way of avoiding the problems, expense and disruption that are caused by even a simple wiring alteration.

15.4 Kitchen and Dining Areas

a. Kitchens

Looking at the kitchen of The Basic House plan first, we see that the lighting level provided is quite high for such a small room. This is because the kitchen is basically a work room and usually requires the highest lighting level in the house. A kitchen with a single 100-w incandescent or 32-w circular fluorescent is a depressing and badly lighted place. A minimum illumination level of 50 fc (500 lux) maintained is recommended, plus supplementary lighting on work surfaces in dark corners, under counters or where the occupant shadows the work. This kitchen is not large and is shaped so that no supplementary lighting is needed. However, a light is placed over the sink to provide a low level of lighting for traffic that does not require the main lights. A secondary purpose for this unit is supplementary light.

Switching for the general lighting is placed at the entrances. Since we have three doors, we use four-way switching. The second major design item to attend to is the appliances. The designer will generally furnish the technologist with the necessary location and outlet data. When such data are not furnished, the information given in Table 15.1 should be a considerable help. At the layout stage, only the receptacle type need by chosen. At the circuiting stage (see Section 15.8), the circuit data in Table 15.1 will be very useful.

The portable electrical appliances for a typical kitchen might include a toaster or toaster oven, mixer-grinder-blender, food processor, microwave oven and coffeepot, among others. Many of these devices stand exposed on the counter top, plugged in for ease of use. Since generally not more than two are used simultaneously, the load is low, but the receptacle requirement is large. As a result, a relatively large number of receptacles are required close together, if we wish to avoid the necessity of constant plugging and unplugging appliances. The appliances can then be kept in place, and cords can be kept short for neatness (and safety).

Turning now to the kitchen in the Mogensen house [Figure 15.1(b)], we see that the architect has divided the space into two related, but essentially different, areas. The upper half (northerly) of the room is devoted to food preparation and is defined by the cabinets and the island counter. This area is lighted with a luminous ceiling that supplies the high uniform lighting level required for food preparation. Note in the detail of Figure 15.1(c) that two levels of lighting are possible by means of ballast switching. Dimming is not necessary. Furthermore, special fluorescent lamps with a high CRI are used to provide the good color rendering required in food preparation.

The lower (southerly) portion of the room, adjacent to the windows, will undoubtedly be used as a dinette, or breakfast area. The entrance to this area is through the kitchen-food preparation area. As a result of this traffic pattern, we have provided three-way switching for both light levels at the entrance to the room from the foyer and at the entrance to the dinette area. (Note the 2 three-way switches on the right-hand wall, adjacent to the cabinets.) This location is also close to the entrance to the formal dining room and can be used if the kitchen is entered (or exited) through that door. The switches controlling the dinette lighting are located at the same place. Because large switch banks are confusing, we show the four switches in two banks, separated by at least 6 in. Although this increases costs slightly, the convenience of knowing which switch controls which light just by its location is well worth the expense. Switches must be located according to expected logical traffic patterns. Otherwise, they become a source of annoyance instead of convenience. Lighting of the dinette area is discussed in Section 15.4.b. An undercounter light, type CC, is provided in one location to provide supplemental lighting plus low level midnight-snack lighting.

As with The Basic House plan kitchen, special outlets are supplied to supply fixed and portable appliances. All special receptacle outlets for fixed

Table 15.1 Load, Circuit and Receptacle Chart for Residential Electrical Equipment

Appliance	Typical Connected Volt-Amperes[a]	Volts	Wires[b]	Circuit Breaker or Fuse, amp	Outlets on Circuits	NEMA Device[d] and Configuration (see Figure 12.52)
		Kitchen				
Range[e,c,i]	12,000	115/230	3 #6	60	1	14-60R
Oven (built-in)[c,i]	4500	115/230	3 #10	30	1	14-30R
Range tops[c,i]	6000	115/230	3 #10	30	1	14-30R
Dishwasher[c]	1200	115	2 #12	20	1	5-20R
Waste disposer[c]	300	115	2 #12	20	1	5-20R
Microwave oven	750	115	2 #12	20	1 or more	5-20R
Refrigerator[f]	300	115	2 #12	20	1 or more	5-20R
Freezer[f]	350	115	2 #12	20	1 or more	5-20R
		Laundry				
Washing machine	1200	115	2 #12	20	1	5-20R
Dryer[c,i]	5000	115/230	3 #10	30	1	14-30R
Hand iron[e]	1650	115	2 #12	20	1	5-20R
		Living Areas				
Workshops[e,j]	1500	115	2 #12	20	1 or more	5-20R
Portable heater[e]	1600	115	2 #12	20	1	5-20R
Television	300	115	2 #12	20	1 or more	5-20R
Audio center[g]	350	115	2 #12	20	1 or more	5-20R
VCR[g]	150	115	2 #12	20	1 or more	5-20R
Personal computer and peripherals[g,h]	400	115	2 #12	20	1 or more	5-20R
Fixed lighting	1200	115	2 #12	20	1 or more	—
Air conditioner ¾ hp[i,j]	1200	115	2 #12	20 or 30	1	5-20R 14-30R
Central air conditioner[c,i,j]	5000	115/230	3 #10	40	1	—
Sump pump[j]	300	115	2 #12	20	1 or more	—
Heating plant (i.e., forced-air furnace)[i,k]	600	115	2 #12	20	1	—
Attic fan[j]	300	115	2 #12	20	1 or more	5-20R

[a] Wherever possible, use the actual equipment rating.

[b] Number of wires does not include equipment grounding wires. Ground wire is No. 12 AWG for 20-amp circuit and No. 10 AWG for 30- and 50-amp circuits.

[c] For a discussion of disconnect requirements, see **NEC** Article 422.

[d] Equipment ground is provided in each receptacle.

[e] Heavy-duty appliances regularly used at one location should have a separate circuit. Only one such unit should be attached to a single circuit at the same time.

[f] Separate circuit serving only one other outlet is recommended.

[g] Surge protection recommended.

[h] Isolated ground may be required.

[i] Separate circuit recommended.

[j] Recommended that all motor-driven devices be protected by a local motor-protection element unless motor protection is built into the device.

[k] Connect through disconnect switch equipped with motor-protection element.

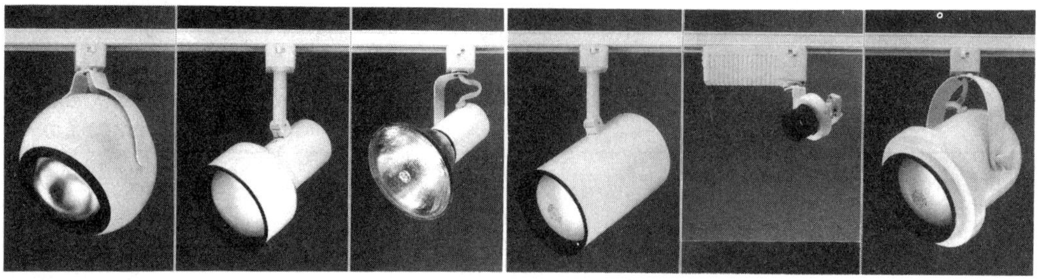

Figure 15.11 Track lighting fixtures are available in a very wide variety of designs. Those illustrated include (left to right) sphere, step cylinder, swivel reflector-lamp holders, flat-back cylinder, external transformer low-voltage spot and gimbal-ring cylinder. Not illustrated are wall washers, adjustable and filtered spots, framing projectors, barndoor shutter units and so on. In addition, a variety of single-and multi-circuit decorative track designs are readily available. (see Figure 12.40). (Photos courtesy of Rudd Lighting.)

appliances will be mounted so that they are accessible. This will satisfy the NEC requirement for use of a cord and plug as an appliance disconnecting means. See Article 422-22(a). The two special outlets under the sink counter (type C) are intended for a garbage disposal unit (switch on the wall at 48 in. AFF) and a dishwasher. The type E outlet feeds the electric range. Because of the NEC requirement that any electric outlet within 6 ft of a sink be equipped with ground fault protection, type B (GFCI receptacles) are located within this 6-ft radius. Beyond this distance, standard receptacles can be used. We take advantage of this by placing a two-circuit multi-outlet assembly on the wall adjacent to the entrance and around the room corner. The apparent multitude of convenience receptacle are not excessive for a modern appliance-loaded kitchen. Also, as in The Basic House plan, a wall switch is provided for an exhaust fan. See Figure 6.17 for ductwork for this fan.

b. Eating Areas

The dining areas of a house must be examined individually and also with respect to each other. For instance, in The Basic House plan, the kitchen has very little eating space—enough perhaps for a small round table and two chairs. Therefore, the dining room becomes the three-meals-a-day eating area. In addition, it is the formal dining area. Lighting for everyday meals should be fairly high level. On the other hand, formal dining calls for more subdued lighting. In The Basic House plan, in the interest of economy, we use a single dimmable pull-down pendant unit. The light is controlled from two locations but is dimmable only from the formal entrance. Wiring is similar to that of Figure 15.9(a), substituting the dimmer for the transformer shown. Obviously, in such a setup, we cannot use the common combination dimmer and switch. Instead, we mount a three-way switch in one gang of a wall box and a dimmer rated at least 300 w in the adjacent gang position.

In the Mogensen residence, everyday meals are taken in the eating area adjoining the kitchen. That leaves the dining room for meals with guests and holiday-style dining. This is also obvious from the double doors that open into the dining room. The dining room is meant to be a show place and is treated as such. A central ceiling outlet with dimming control is provided for an appropriate chandelier. Since the walls will almost surely be used to display art or other objects, ceiling tracks are provided for illuminating both walls. The number and type of fixtures to be mounted on these tracks can be selected to match the decor and function of the room. See Figure 15.11.

The dining room tracks may be recessed or surface-mounted. It should be mentioned that, if this dining room were very formal, tracks would probably not be used because of conflict in decor. Instead, recessed wall-wash units would be installed. Since we are using this house as a vehicle for our study, and we have already used wall-wash units in the living room, we are using ceiling tracks in this room as an example of what can be done to provide flexible room and wall lighting. The track fixtures can also be turned around to provide room lighting, although this almost always creates a direct glare problem. If desired, dimming can be provided for the tracks.

The kitchen dinette area is brightly fluorescent lighted as is desirable for a day-to-day eating place. Our recommendation for lamp type is to match the cooking area, to make the food look more inviting. For nighttime use of this area and when a view outside is desired, the fluorescents can be switched off, and the downlight provided for this purpose can be used instead.

Receptacles are spotted around the eating areas in accordance with the NEC 6-ft rule, as well as additional outlets that we think are necessary. In circuiting, these outlets will be connected to small appliance branch circuits as explained in Section 13.10.d.

c. Guidelines

To summarize, we have established the following guidelines for kitchen and eating areas.

- Lighting for kitchen work areas should be at least 50 fc (500 lux). A combination of general and supplemental lighting is good practice.
- Lighting for eating areas should be at least 30 fc (preferably more) for daily routine dining. More formal or holiday-style dining requires less light but a more subdued quality. Combinations of fixtures, switching and dimming can be used to achieve these results.
- Stationary appliances are supplied by specific-use outlets. Portable appliances are supplied through individual outlets or multi-outlet strips on appliance circuits.
- Switching should be located in accordance with traffic flow. Most often switches at the door strike side are satisfactory.
- If, in the dining areas, a counter, sideboard or table is placed against a wall, an outlet should be supplied 4 in. above it, at the center. This applies to eat-in kitchens as well.
- If plumbing provision is made for a food disposal unit or dishwasher, a wired and capped outlet should also be provided. The food disposal unit's control switch should be placed at such a location that it is impossible to stand at the sink and turn on the unit. Although some commercial units are operated by turning the drain cover in the sink, it is our definite opinion that, in the interest of safety, a wall switch should always be used.

15.5 Sleeping and Related Areas

a. Bedrooms

A wide range of electrical layouts is possible for bedrooms, depending on the size of the room and the uses intended. Some bedrooms are designed to be small, and only for the purpose of dressing and sleeping. Another approach uses this area for resting, reading, and television watching in addition to sleeping. These rooms are larger, and the electrical layout must satisfy the requirements. In either case, a furniture layout will tell a great deal. In its absence, one must be assumed. Refer, for instance, to Figure 15.12, which is the lighting and receptacle layout for the two bedrooms of The Basic House plan, and compare it with the layout of Figure 13.16. Notice that in bedroom no. 2 the layout is the same. There, the east wall is the only logical place beds can be located, and it makes no difference whether twin beds, a double bed or a king size is used. With a double or king size bed, the outlet intended for a lamp between the twin beds is blocked. However, enough outlets remain at good locations.

In bedroom no. 1, which is smaller, the layout of Figure 13.16 assumes twin beds on the east wall. A different arrangement is shown in Figure 15.12. We believe that this second layout is better because it is more flexible. It allows for use of twin or double-sized beds and leaves room for bedroom furniture. The double outlets on the north wall are needed for the twin bed's layout, and only one outlet is blocked.

In bedroom no. 1 of the Basic House plan, the overhead light provides room and closet lighting. In bedroom no. 2, the sliding doors and the larger closet call for a separate closet light. Notice that the closet light is wall mounted above the door, inside the closet. In that position, it lights the clothes and the shelves above. A ceiling light is less useful because of the shadow cast by the shelves, unless it is placed just inside the doors. In all cases, such lights should have a lamp guard to prevent breakage.

In studying the bedrooms of the Mogensen house, we find that the rooms are generally larger than conventional bedrooms. In the two bedrooms on

882 / RESIDENTIAL ELECTRICAL WORK

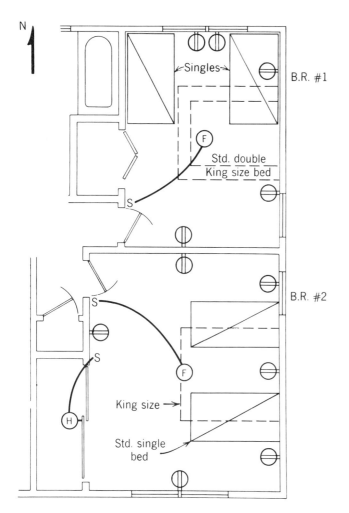

Figure 15.12 Efficient and useful layout of electrical devices depends on furniture arrangement. When furniture arrangement is unknown, devices should be placed to accommodate the most logical layout(s). Note that the device layouts shown will satisfy several furniture arrangements with minimum waste and good accessability.

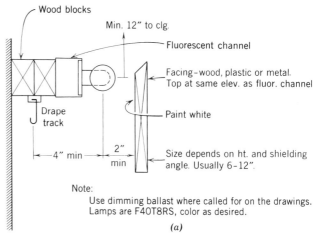

(b) LIGHTED VALANCES

Valances are always used at windows, usually with draperies. They provide up-light which reflects off ceiling for general room lighting and down-light for drapery accent. When closer to ceiling than 10 inches use closed top to eliminate annoying ceiling brightness.

Figure 15.13 Construction detail and application of lighted valances. (Courtesy of Illuminating Engineering Society of North America.)

the lower level, the bed(s) will probably be on the side wall opposite the door, and outlets have been arranged that way. Placement of beds under the high windows is also possible, but less likely. These rooms are large and may also be used as guest rooms. A dimmable lighted valance is installed above the window, to provide a pleasant "sitting room" atmosphere. See Figure 15.13. For general room lighting we have chosen two (dimmed) sconce fixtures with translucent fronts. These fixtures, which are mounted above eye level, throw most their light upwards, giving a lighted wall and

ceiling and an overall soft diffuse atmosphere to the room. Wall fixtures are vastly preferable to ceiling lights that cause extremely annoying glare to a person lying in bed and undesirable shadows throughout the room. Outlets are placed about the room for convenience. The front window wall has high windows and a piece of furniture or chairs will be placed there. Outlets at both ends of this wall will handle any lamps. The outlets on the wall adjacent to the door are intended for entertainment equipment that can be operated remotely while in bed.

Looking at the master bedroom on the upper level, we see that here also, despite the large size of the room, the only two possible locations for bed(s) are on the left and right walls. The other two walls are ruled out because of doors and windows. We have, therefore, set up the lighting for the room symmetrically. Each possible bed wall has a pair of adjustable reading lights (P) with a built-in switch. Between the reading lights, halogen lamp wall brackets (D) similar to those used in the lower level bedrooms are mounted. It is good practice to provide sufficient wiring and outlets for alternative furniture (in this case, bed) locations. Even if the house is being custom-built and the owner selects furniture locations, a layout change or a future owner's desires may very well require different outlet locations. A small investment during construction will avoid expensive and frequently unsightly changes in the future.

A dimmable lighted cornice above the window wall drapes (see Figure 15.14) extends the entire length of the room, giving the room a sitting-room atmosphere. The large wall opposite the beds (either side wall) is illuminated by surface-mounted ceiling track lights. See Figure 15.11. The fixtures on this track can be aimed to highlight paintings, books or other objects. The owner has gone to the expense of providing a skylight, which will give daylight when the blinds or drapes are closed. To take advantage of this desire for daylight, the lighting design has provided an artificial daylight source in this space, which is locally switched.

The deck outside the master bedroom and study is roofed, unlike the deck outside the living room. We have provided overhead lighting for this area. Notice on the front wall an outside lighting control panel (OLCP). This panel controls all the outside lighting, including the deck lighting off the living room. Local control of these lights is also provided. The reasons for centralizing the outside lighting control in the master bedroom are convenience and security. It is convenient because the owners can shut all the outside lights as they retire, without running around this large house. It provides security, so that outside lighting, which acts as security lighting, is in the hands of the owner at his or her nighttime location.

For these same reasons—convenience and security—owners of large houses such as this frequently desire centralized control of all the lighting in the house. To do this with normal, full voltage switching is extremely expensive and clumsy because of the heavy wiring, full size switches and pilot lights required. For this reason, a system of

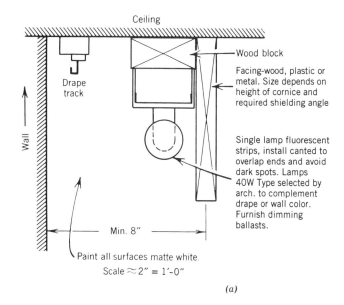

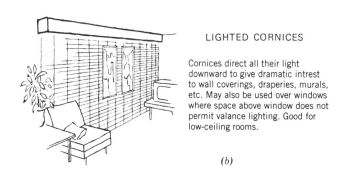

Figure 15.14 Construction detail and application of lighted cornices. (Courtesy of Illuminating Engineering Society of North America.)

switch controls that uses low voltage relays was developed. This system allows the use of very small, low voltage (24-v) control wiring and makes centralized control a relatively simple matter. Of course, local control also remains. A typical wiring diagram and some equipment photos are shown in Figures 15.15, 15.16 and 15.17. Among the advantages of this low voltage switching control system are

- Control can be local or remote.
- Control can be automatic by the use of sensors (heat, light, motion), timers, programmable switches and the like.

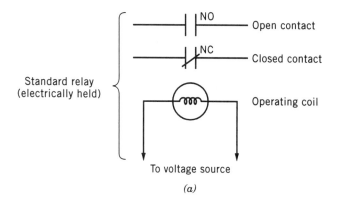

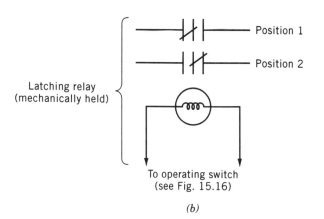

Figure 15.15 (a) In a standard relay, the contacts are "thrown" when the operating coil is energized, and are held in their operating position as long as the coil remains energized. This type relay is called electrically held. (b) In a latching relay, energizing the operating coil throws the contacts, which are then mechanically latched in position. The coil is then energized. This type of relay is called mechanically held and is the type used in low voltage lighting control switching. See Figure 15.16. To change the contacts from position 1 to position 2 or vice versa, the operating coil must be energized momentarily.

- Alterations are easy to make.
- The status of all lights (and other devices) shows at one (or more) centralized locations.

By extending this type of remote relay control to other electrical devices in a house, we can begin to see the basis of the so-called "automated" or "smart" house. In such buildings, all the electrical outlets and fixed equipment are relay controlled.

The control signals come from one or more programmable devices, frequently including a common personal computer equipped with the appropriate software. The principle is simple; the wiring is not. Although reliability of such systems is good, their complexity, at least at this time, requires that a trained technician be available for maintenance and repair.

In Figure 15.1(e) note 8, the contractor is requested to furnish a proposal to perform all the lighting switching using low voltage control. If a more extensive control system is desired, then a complete design should be prepared before requesting a contractor's proposal. In a house of this size, the price of low voltage would be competitive with conventional full voltage lighting switching. If central control of all lighting is desired, it can be accomplished reasonably only with low voltage switching. If, on the other hand, such a centralized control arrangement were desired in The Basic House plan, it could reasonably be done both ways.

Since the preceding discussion deals with new construction, we have illustrated the most common type of wired low voltage switching control. When dealing with existing construction, this type of wired system is often not practical because of the expense and difficulty of running the required wiring. To overcome this problem, a control system that uses the existing power wiring as a carrier for a high-frequency control signal was developed. *This power line carrier (PLC) system does not require rewiring* because the high frequency control signal placed onto the power wires does not affect the power system and is detected and used only at the required control points. You can find details of the system operation and equipment in *Mechanical and Electrical Equipment for Buildings*, 8th edition (Wiley, New York) and in PLC equipment manufacturer's literature.

Receptacles have been spotted around the master bedroom for convenience and in accordance with **NEC** requirements. The receptacles or the exterior balcony are weatherproof, and of the GFCI type. See **NEC** Article 210-8(a)(3).

b. Dressing Rooms and Closets

Lighting of reach-in closets was discussed previously. A guarded, wall-mounted, switch-controlled light is adequate. Walk-in closets and dressing rooms must be treated as small rooms. Switches can be inside or outside, depending on shelf and clothes-pole locations. If the room is to be used for dressing and makeup, appropriate

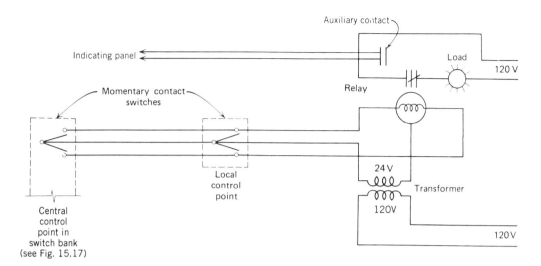

Figure 15.16 Actual wiring diagram of a typical low voltage switching control circuit. The relay is supplied with low voltage (24 v) via a system transformer. The relay is operated by double-throw momentary contacts that change the position of the relay contacts (see Figure 15.15). An extra (auxiliary) contact is shown that is wired to an indicating panel, which then shows the condition of the load (ON or OFF).

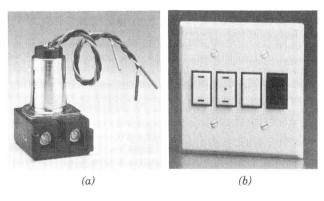

Figure 15.17 Components of a low voltage switching and control system for lighting (or other electrical) loads. *(a)* The basic switching device is a single-pole double-throw latching relay with a 24-v coil and 20 amp, 120- to 277-v contacts. *(b)* Low-voltage wall switches (24 v) can be mounted singly or ganged in groups to two to eight switches. Switches are available in single-button, two-button, rocker, key-operated, and lighted designs. Up to 48 relays can be mounted in a single panel (in this manufacturer's design) with power supply and LED status lights. Control wiring from the panel extends to all the control devices (switches, timing devices, automatic controls) and to the controlled loads (lighting, and if desired, small motors, receptacles). (Courtesy of General Electric Wiring Devices.)

mirror lighting and outlets must be provided. The two walk-in closets in the Mogensen house master bedroom do not function as dressing rooms, and thus only a recessed ceiling light is provided.

c. Guidelines for Sleeping Areas

Reviewing the guideline for sleeping areas, we have learned that we should:

- Furnish indirect, wall-mounted lighting and/or architectural lighting elements such as coves, valances and cornices. Lighting is important near and in closets. Mirrors will probably be placed there.
- Provide overbed lights with control for reading.
- Provide master control of all or parts of the house lighting in the master bedroom, if desirable from a convenience and/or security point of view.
- Provide "atmosphere" lighting if the room is to function as a sitting room.
- Obtain or assume a furniture layout and provide two or more duplex wall-mounted outlets on the bed wall.

- Provide a strategically placed receptacle at 72-in. AFF for a wall-mounted television set.
- In a house without central air conditioning, provide an appropriate outlet near a selected window, for a window air conditioning unit (see The Basic House plan, Figure 15.16).

15.6 Circulation, Storage, Utility and Washing Areas

a. Circulation: Halls, Foyers, Corridors and Stairs

Entrance halls and foyers are generally lighted with decorative fixtures. Inside corridors and stairs use simpler types of fixtures, surface mounted or recessed. Recessed fixtures should have spread distribution—lens or dropped dish-type diffuser—to give diffuse lighting. Concentrating-type downlights are not desirable. Stair lighting must be placed so that the front edge of each step is lighted. Depending on how the stair is constructed, this can be done with a light over the stairs (see The Basic House plan) or lights at the foot and at a center landing (see Mogensen). Lighting at the top is desirable, when there is no center landing or when the landing lighting is blocked.

Switching of circulation lighting is critically important. The technologist must take a mental trip through the building, turning lights on and off as required. Obviously a three-way switch is required at the top and bottom of each stair. Also plain to see is the need for three-way switches at the two ends of any corridor more than 6 ft. long. (See stair lighting in both plans and corridor lighting in the The Basic House plan.) Much less obvious is the switching of lights in the two T-shaped halls in Mogensen. There, because of the choices involved, four-way switching has been provided. Receptacles should be supplied every 10 ft of length for the use of vacuum cleaners, floor polishers, electric brooms and the like. All receptacles should be standard 20-amp, 120 v; 220-v outlets are not required, unless definitely called for by the owner or architect.

b. Storage and Utility Areas

(1) Shallow reach-in closets are sufficiently lighted from the adjoining space. If the closet is deep, has recesses on the sides or top or is walk-in, it should have its own lighting. A switched porcelain lampholder is normally sufficient, if it is provided with physical protection to protect the lamp from breakage. Convenience outlets are generally not required in closets and storage rooms. Dead storage areas such as attics and basement crawl spaces take the same type of basic lighting, switched at the entrance. In these areas a single convenience outlet is a good idea. Crawl space receptacles must have ground fault protection.

(2) Basements serve as storage, utility, recreation and work areas. They start as empty spaces, containing only the heating plant and possibly the laundry, and expand in usage later on. For this reason, basement areas should start with basic utility lighting and a few convenience receptacles. These receptacles must have ground fault circuit protection [**NEC** Article 210-8(a) (5)]. Provision should be made for easy expansion of lighting and easy addition of additional receptacles. Minimum recommended initial requirements are one ceiling lighting outlet for every 200 ft^2 of area or less, receptacles for a spare refrigerator and a freezer, plus one wall-mounted convenience receptacle outlet for every 20 ft of wall space. To allow for future expansion, either spare cables or empty conduits should be run to the basement from the panel.

Normally, an emergency cutoff switch must be placed at the entrance to the space containing the heating plant. This switch must completely shut down the heating system. To indicate the emergency nature of this switch, the plate is usually red in color and is marked with an appropriate legend, such as "Heating System Shutoff." This plate is usually provided by the heating subcontractor. To permit servicing heating and/or air conditioning equipment located in the basement a 120-v convenience outlet should be provided nearby. If the heating system is all-electric, the disconnecting-means requirements are more complex. See **NEC** Article 424.

(3) Laundry areas can be situated in the basement (The Basic House plan), a separate enclosed space (Mogensen), or a porch, kitchen or balcony. In any case, the electrical requirements are the furnishing of appropriate outlets (see Table 15.1) and adequate lighting. Laundry areas must have adequate ventilation, either natural or forced. If space permits, an outlet for a hand iron should be provided. The **NEC** specifically excepts laundry circuit receptacles in a below-grade space from its requirement that receptacles in such areas be GFCI protected. See **NEC** Article 210-8(a) (5) (Ex.2).

(4) Enclosed garages serve many functions in addition to car storage and basic auto maintenance. In houses with no basement or utility room, the garage is also the work and repair area and a storage area for bicycles and garden equipment. Lighting outlets should be placed to illuminate the auto engine compartment with the hood open. Lights over the center of the car are useless. Notice in the Mogensen house garage the wall brackets for general illumination and the reel-light (fixture type K), which greatly assists in repair work. The entire back of the garage is treated as work and utility space and is well lighted with industrial fluorescent lighting fixtures. These heavy-duty steel louver fixtures fit nicely between the 12-in. on-center ceiling beams. Note that recessed fixtures are better than surface units, since the fixtures are subject to physical damage. For this reason also, a steel louver unit has been used. The garage should be amply supplied with receptacles mounted high (48 in.) with a minimum of one on each wall. Most houses use an electronically operated garage door, and interior and exterior switches for control are required (see Figure 15.18). Exterior switches are key operated. Since all the garage receptacles require ground fault protection, we will use ground-fault-protected circuit breakers to supply these receptacles.

c. Bathrooms and Lavatories

(1) Lighting should be over the mirror to illuminate the face head-on. If lights are placed at the sides of the mirror, they should be at both sides so that both sides of the face are illuminated. Overhead lighting fixtures in large bathrooms are acceptable for general lighting, provided that over-the-mirror, full-face lighting is furnished also. Over-the-mirror lighting alone is sufficient in a small room. See the bathroom in The Basic House plan and bath no. 2 and the guest bath in the Mogensen house. In the Mogensen master bath, the overhead fixture provides light for bathing and showering, while the two mirror lights provide the facial illumination required.

Notice that a separate fixture is provided over each basin. Generally, face lighting should be furnished at each mirror location. Normally, although not always, this corresponds to the basin location. Although not absolutely necessary, gasketed fixtures in rooms with baths and showers are advisable. All light should be controlled by wall switches; never local switches or pull chains. Switches should be located conveniently to operate the light. The switch at the door is intended to turn on the general illumination; supplemental lighting is separately switched.

A built-in night-light and switch is a convenience in houses with small children. Devices are available with this combination that fit into a standard outlet box. See The Basic House plan (Figure 13.16). A light inside a stall shower is not a particularly good idea, even though the unit is vaporproof. Enough light is obtained through the shower door to make this an unnecessary expense and a potential safety hazard.

(2) A duplex receptacle at a minimum height of 48 in. should be installed adjacent to each mirror or lavatory basin. Receptacles near washbasins should be equipped with self-closing covers that close while a device is connected. By **NEC** requirement, these must be GFCI types connected on a separate 20A circuit [see Article 210–52(d)]. Receptacles in lighting fixtures are not to be used. A time-switch-controlled heater is a welcome addition to any bathroom, to prevent chills when stepping from a bath or shower. The outlet may be a receptacle (GFCI type), an outlet box intended for a fixed heater (see The Basic House plan) or a fixed heater (the Mogensen house).

An exhaust fan must be provided for all interior bathrooms and are desirable for all bathrooms. Such fans are frequently wired in conjunction with the room light switch. Such a switch is double-pole; one pole for the fan and the second for the light (see the Mogensen guest bath).

15.7 Exterior Electrical Work

This section covers all exterior electrical work not connected with the electrical service equipment. That topic will be discussed in Section 15.11. All exterior lighting fixtures must be weatherproof. This also applies to units installed in exterior soffits. Exterior lighting must be switch controlled from a nearby location inside the house. Adding master control is advisable. When lights are on an automatic time switch, that switch must have an override feature, preferably one that allows override from a remote location. Porches, breezeways, exterior walks, decks, patios and similar areas should be well lighted, for use and security. As in the Mogensen residence, use of low voltage fixtures simplifies wiring and minimizes faults. Low voltage lighting equipment, wiring and installation must conform to **NEC** Article 411.

At least one weatherproof duplex receptacle should be installed on an outside wall of the house. This receptacle has GFCI protection by **NEC** requirement. It should be located for convenience for use with mowers, grills and tools. Switch control from inside will prevent vandalism and will improve security. A covered carport adjacent to the house should have a duplex receptacle within easy reach.

15.8 Circuitry

The next stage in the work is the circuiting of the plans of Figure 15.1(a) and (b). The ground rules for circuiting have already been discussed in detail in Section 13.10. The technique of circuiting has been discussed in Section 13.11. We have followed these guidelines and procedures in the circuitry shown in Figure 15.18(a) and (b), with certain changes. It is very important that you place Figure 15.18 before you and trace out all the circuitry. Because of the density of wiring, this is not an easy task. Particular attention should be given to the wiring of switches.

(a) We recommend in Section 13.10 that lighting and receptacles be separately circuited wherever possible. This is good practice but is less important in residences than in other types of buildings. In residences, convenience receptacles count as part of the lighting load. Refer to Section 13.10.a and **NEC** Table 220-3(b) for this rule. Therefore, fixed lighting outlets and convenience outlets can be combined on a circuit if absolutely necessary. In a room with no fixed lighting outlet, the switched wall receptacle is the lighting outlet and belongs on a lighting circuit. (See the switched outlet in the study.)

(b) Note that, in the interest of economy, we did wire through some switch outlets. Care was taken, however, to avoid, wherever possible, carrying a second circuit through a switch outlet. Also, in the circuited drawings (Figure 15.18), we combined switch leg runs that were shown separated in the layout drawing of Figure 15.1. A good example of this appears in the garage where the switch legs for the exhaust fan and lighting outlets B and K have been combined. This is proper, since they are all on the same circuit. Also notice that we have been careful to provide part of at least two circuits in each space, including bathrooms and halls. This is a rule that should be strictly followed, to avoid blackouts resulting from loss of a single circuit.

(c) Circuiting has taken account of the construction members, and wiring has been run along them whenever possible. Across-the-members wiring has been limited, to avoid the expense of drilling wood members. On the upper level, much of the cross wiring can be done in the attic space, thus avoiding drilling. On the lower level, wiring can be placed in the concrete slab by using conduit. The engineer selects the wiring method required. Home-runs have generally been taken from the outlet closest to the panel. Home-runs from the ceiling outlets are preferable to those from wall outlets. This is done so that where a deep box is required, it can be easily accommodated. The limited depth of walls does not permit this.

(d) In a few places, to avoid confusion, circuit numbers have been placed next to outlets. This is not generally done, since the numbered arrowheads and wiring hatch marks are normally enough to indicate circuit routing. The Mogensen residence, however, is so heavily wired that this type of identification is occasionally desirable. We urge you to follow out each circuit, using the panel schedule, Figure 15.18(c), along with the circuited floor plans of Figure 15.18(a) and (b). While doing so, you should test alternate circuiting routes mentally or by sketching them. By doing this, you will see that there are any number of ways of circuiting, each of which has its good and bad points. Also we emphasize the importance of understanding how the number of wires shown was arrived at. This can best be done with the aid of little sketches of the type shown in Figure 13.24, until you have enough experience to do the counting mentally.

15.9 Load Calculation

a. General Lighting Load

On the basis of square footage (see Section 13.10.a), the house should have a minimum of 10 circuits for general lighting load:

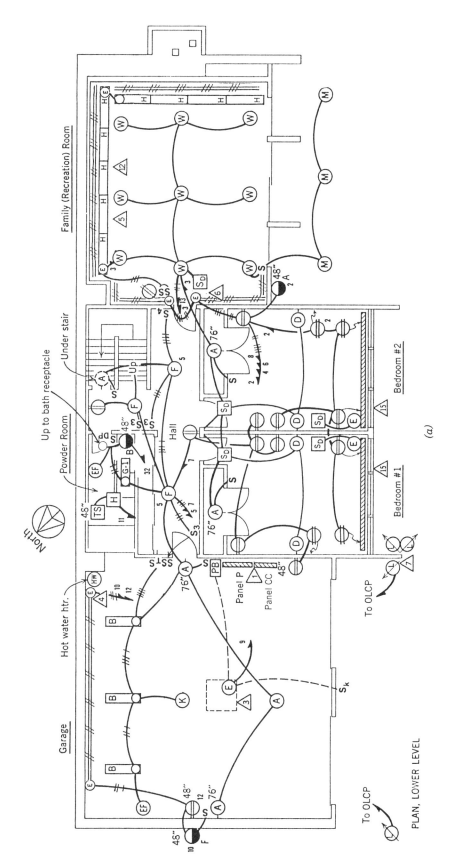

Figure 15.18 (a) Lower level electrical plan circuitry. (Architectural plans courtesy of B. Mogensen A.I.A.)

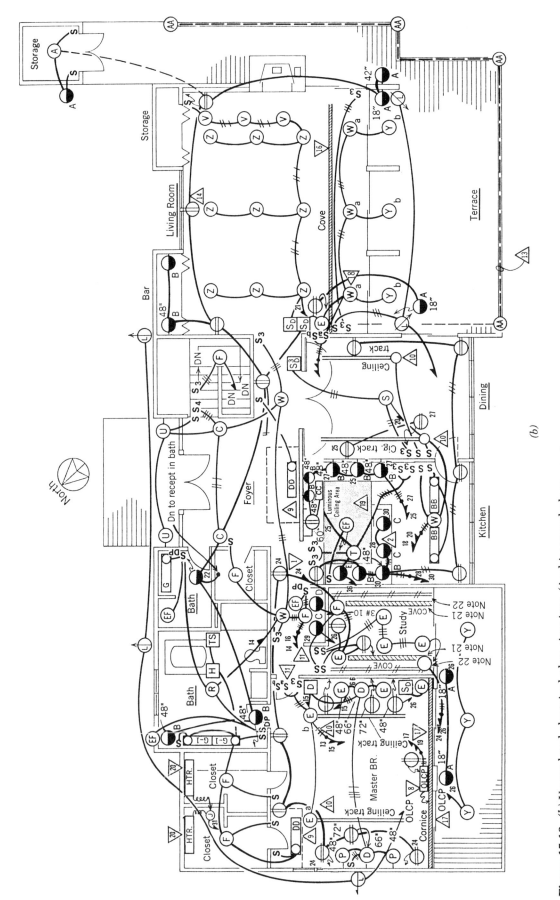

Figure 15.18 (b) Upper level electrical plan circuitry. (Architectural plans courtesy of B. Mogensen A.I.A.)

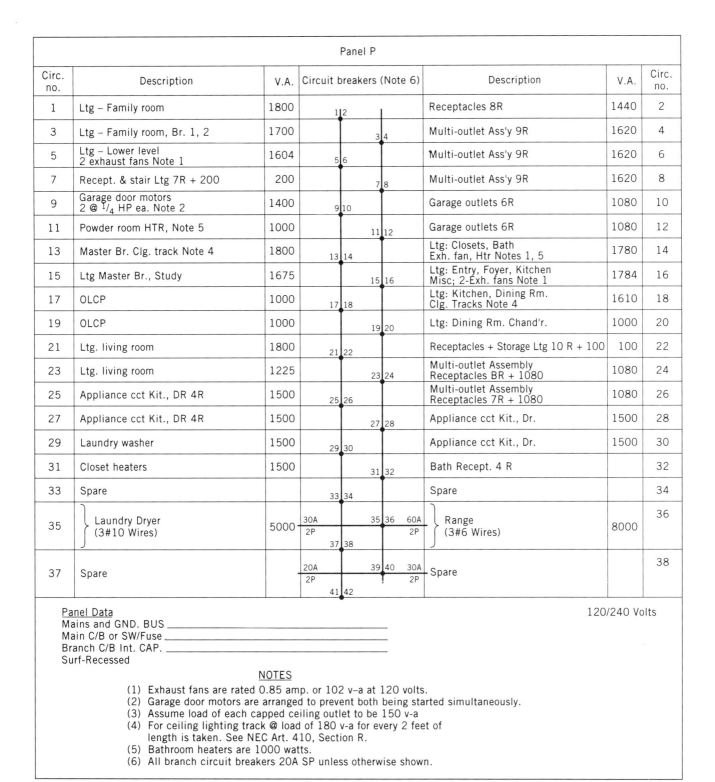

Panel P						
Circ. no.	Description	V.A.	Circuit breakers (Note 6)	Description	V.A.	Circ. no.
1	Ltg – Family room	1800	1|2	Receptacles 8R	1440	2
3	Ltg – Family room, Br. 1, 2	1700	3|4	Multi-outlet Ass'y 9R	1620	4
5	Ltg – Lower level 2 exhaust fans Note 1	1604	5|6	Multi-outlet Ass'y 9R	1620	6
7	Recept. & stair Ltg 7R + 200	200	7|8	Multi-outlet Ass'y 9R	1620	8
9	Garage door motors 2 @ 1/4 HP ea. Note 2	1400	9|10	Garage outlets 6R	1080	10
11	Powder room HTR, Note 5	1000	11|12	Garage outlets 6R	1080	12
13	Master Br. Clg. track Note 4	1800	13|14	Ltg: Closets, Bath Exh. fan, Htr Notes 1, 5	1780	14
15	Ltg Master Br., Study	1675	15|16	Ltg: Entry, Foyer, Kitchen Misc; 2-Exh. fans Note 1	1784	16
17	OLCP	1000	17|18	Ltg: Kitchen, Dining Rm. Clg. Tracks Note 4	1610	18
19	OLCP	1000	19|20	Ltg: Dining Rm. Chand'r.	1000	20
21	Ltg. living room	1800	21|22	Receptacles + Storage Ltg 10 R + 100	100	22
23	Ltg. living room	1225	23|24	Multi-outlet Assembly Receptacles 8R + 1080	1080	24
25	Appliance cct Kit., DR 4R	1500	25|26	Multi-outlet Assembly Receptacles 7R + 1080	1080	26
27	Appliance cct Kit., DR 4R	1500	27|28	Appliance cct Kit., Dr.	1500	28
29	Laundry washer	1500	29|30	Appliance cct Kit., Dr.	1500	30
31	Closet heaters	1500	31|32	Bath Recept. 4 R		32
33	Spare		33|34	Spare		34
35	Laundry Dryer (3#10 Wires)	5000	30A 2P 35|36 37|38 60A 2P	Range (3#6 Wires)	8000	36
37	Spare		20A 2P 39|40 41|42 30A 2P	Spare		38

Panel Data 120/240 Volts
Mains and GND. BUS _____
Main C/B or SW/Fuse _____
Branch C/B Int. CAP. _____
Surf-Recessed

NOTES
(1) Exhaust fans are rated 0.85 amp. or 102 v–a at 120 volts.
(2) Garage door motors are arranged to prevent both being started simultaneously.
(3) Assume load of each capped ceiling outlet to be 150 v-a
(4) For ceiling lighting track @ load of 180 v-a for every 2 feet of length is taken. See NEC Art. 410, Section R.
(5) Bathroom heaters are 1000 watts.
(6) All branch circuit breakers 20A SP unless otherwise shown.

(c)

Figure 15.18 (c) Panel schedule for Panel P.

Minimum number of 20-amp circuits =

$$\frac{\text{Area}}{530 \text{ ft}^2/\text{circuit}} \approx \frac{5000 \text{ ft}^2}{530 \text{ ft}^2/\text{circuit}} \approx 10 \text{ circuits}$$

If we examine the panel schedule [Figure 15.18(c)] carefully and eliminate appliance circuits, special loads circuits and circuits serving multi-outlet assemblies, which in our opinion should not be counted in the general lighting load, we see that the actual number of general lighting circuits considerably exceeds 10. This is entirely reasonable for the electrical system of a house of this complexity and cost.

b. Convenience Outlets

As previously stated, the ordinary duplex convenience receptacle outlets in a residence (excluding multi-outlet assemblies) are counted as part of the general lighting load when figuring the total building load. This creates a situation that requires keeping a double set of loads—one for the circuit and one for the building. For the building and feeder load, the receptacles count as zero load. Look, for instance, at circuit no. 2 of panel schedule P. The panel schedule shows 8R and 1440 v-a in parenthesis. This means that the circuit feeds eight duplex receptacles. These receptacles are classified as general lighting load by Code and, therefore, do not add to the overall building (and feeder) load as given in Code Table 220-3b. We have figured each duplex receptacle at 180 v-a for the purpose of establishing a branch circuit load. This number is arbitrary, but it has proven to be satisfactory in actual design work. Therefore, the branch circuit load required to establish circuit breaker and wire size is shown as 1440 v-a (8 × 180). With respect to building load calculations (feeder calculation), these receptacles do not add to the basic 3-v-a/ft² general lighting load unless it is definitely known that they will serve some special loads. In such a case, these special loads should be included in the total building load. The building load is used in calculating the size of the panel mains and the panel feeder and influences the service size.

These load calculations, that is, the building load, feeder size, panel mains and main protection, are done by the job engineer. We describe them briefly in our discussion of service equipment in Section 15.12. Here, it is important to remember that in residential work the circuit loads should be shown as they appear on panel schedule P. When receptacles are furnished for the specific purpose of supplying plug-in equipment and will most probably not be used for lighting, then they should be calculated at either 180 v-a for each single receptacle or with the actual equipment load, whichever is larger. The multi-outlet assemblies in the family room, study and garage will almost certainly not be used for lighting. Therefore, their load should be included both in branch circuit load and in building (feeder) load.

c. Volt-Amperes and Watts

Note particularly that all loads are figured in volt-amperes and not in watts. The difference was explained in Section 11.15. For purely resistive loads, such as incandescent lighting and electric heating, the volt-amperes and watts are almost identical. However, small motors have a low power factor, and there is a large difference between volt-amperes and watts. An example will help clarify this.

Circuit no. 9 feeds two ¼-hp garage door motors. Referring to NEC Table 430-148, we see that full load current for a 115 v, ¼-hp motor is 5.8 amp. Therefore, the volt-amperes for each motor is

$$120 \text{ v} \times 5.8 \text{ amp} = 696 \text{ v-a}$$

whereas the wattage is only about 250 w. The load shown in the panel for the two garage door motors is, therefore, 1400 v-a, and not 500 w. This very large difference is due partially to the low power factor of standard fractional horsepower motors and partially to the extra-safe figures in the NEC. Most ¼-hp motors actually draw less than 5.8 amp. The NEC, however, takes the worst case to make certain that the electrical circuitry is adequate.

d. Appliance Circuits

Circuits nos. 25, 27, 28 and 30 are appliance circuits feeding the required appliance outlets in the kitchen and dining room. The minimum number of such appliance circuits is two. Here, four are used to supply the large number of outlets. Each appliance circuit is figured at 1500 v-a. The laundry circuit, circuit no. 29, is also figured at 1500 v-a. These load figures are in accordance with NEC requirements [Article 220-16] and are shown on the panel schedule.

e. Large Appliances

The electrical clothes dryer, circuit no. 35, is shown as 5000 v-a. This is the minimum load permitted by the NEC for such an appliance. See Table 15.1 and NEC, 1996 edition, Article 220-18. Of course, if the actual dryer rating is known to be larger than

5000 v-a, it should be used. The electrical range load is shown as 8 kw, even though the unit itself is rated up to 12 kw. This reduction is permitted by the NEC, Table 220-19.

f. Panel Schedule

The panel data at the bottom of the panel schedule [Figure 15.18(c)] has deliberately been left blank. These data are supplied by the engineer to the technologist and are, therefore, omitted here. The remaining panel information, however, is the work of the draftsperson/technologist and is shown filled in.

15.10 Climate Control System

The heating/cooling system for the Mogensen house is described in Chapters 4 and 6. The corresponding electrical work is shown in Figure 15.19(a) and (b). The panel schedule for the climate control panel CC is given in Figure 15.19(c). The system utilizes electrical heating on the lower level [Figure 15.19(a)] and heat pump heating/cooling for the upper level [Figure 15.19(b)].

The NEC in Article 424-3b requires that electrical heating loads be circuited at 125% of their rating. (This is apparently because they are normally "continuous loads," that is, loads that remain on for 3 hours or more). The data shown in Figure 15.19 take this 125% factor into account. The lower level of the house is electrically heated, similar to the heating of The Basic House plan as shown in Figures 13.17 and 13.27. Refer to Figure 15.19(a) and the circuits in the panel schedule of Figure 15.19(c). The upper level heating/cooling system is more complicated. Refer to Figure 15.19(b). This diagram corresponds to Figure 6.9 but in electrical terms. The NEC gives special rules for sizing circuits that feed compressors such as those in the heat pumps. The compressors are shown to be 3 hp each. Figure 6.5(c) gives the electrical data required to prepare the electrical diagram shown in Figure 15.19(b). The 230-v, single-phase blowers (see Figure 6.6) are fed from a 2-pole, 240-v circuit through 2-pole switches equipped with thermal overload elements mounted in the upper level foyer. The electric heating elements in these air-handling units are separately fed from 2-pole power circuits in panel CC. Notice that we specify the use of type THWN conductors. THWN wire (75° rise) permits us to use one size smaller wire than

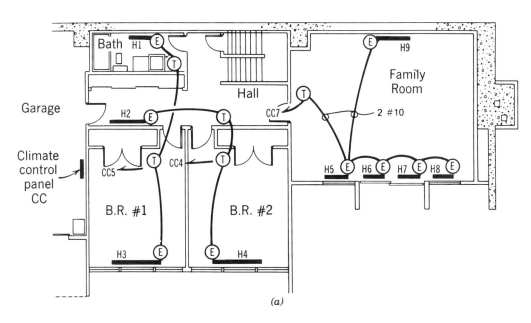

Figure 15.19 (a) Lower level electrical heating plan. See Figure 13.27(b) for a typical wiring diagram and Figure 3.14(b) for the HVAC plan of the heating system. (Architectural plans courtesy of B. Mogensen A.I.A.)

894 / RESIDENTIAL ELECTRICAL WORK

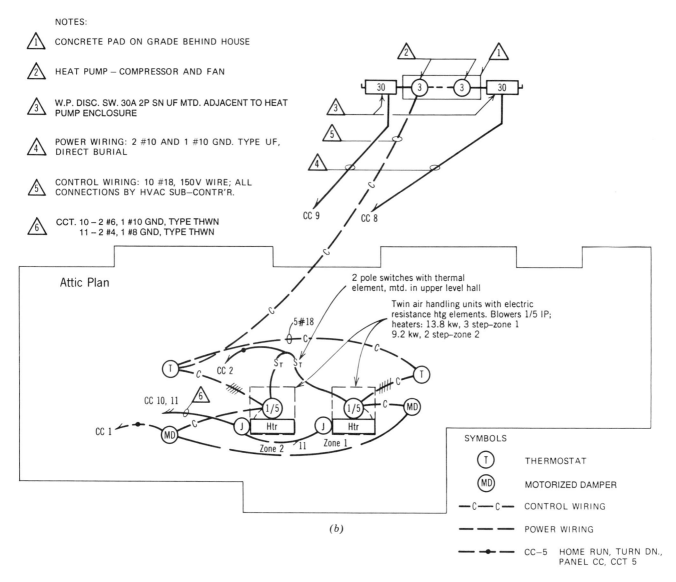

Figure 15.19 (b) Upper level electrical plan of the two-zone climate control system. See Figure 6.9 and the text discussion there for the basis of this plan. (Architectural plans courtesy of B. Mogensen A.I.A.)

the TW (60° rise) that is assumed in the remainder of the job. (See Table 12.2.)

Every motor in a project must be provided with motor overload protection and with a means of disconnection. The rules covering these items are many and varied. Refer to **NEC** Articles 430 and 440 for complete coverage of this complex subject. In this project, an additional factor must be considered. The outside heat pumps are remote from the power source. In the interest of safety, the **NEC** requires that motors of this kind be provided with a local means of disconnection. We have furnished a 2-pole, 30-amp weatherproof switch adjacent to each heat pump. This switch meets the need for a safety disconnect.

Panel CC supplies all the power for the heating and cooling equipment. The panel data at the bottom of the schedule is not usually part of the technologist's layout work. This information is furnished to the technologist by the engineer at a later

ELECTRICAL SERVICE EQUIPMENT / 895

PANEL CC							
CIRC. NO.	DESCRIPTION	VA	CIRCUIT BREAKERS		DESCRIPTION	VA	CIRC. NO.
1	MOTORIZED DAMPERS	NEGLIGIBLE	15 1|2	15	2 – 1/5 HP BLOWERS	1200	2
3	SPARE		20 3|4	20			
5	HTRS, LOWER LEVEL BATH HTR H1; 500 W BR #1 HTR H3; 1500 W	2000	20 5|6 2P 7|8	20 2P	HTRS, LOWER LEVEL, HALL HTR. H2, 1250 W BR #2 HTR, H4 1500 W	2750	4
7	HTRS, FAMILY ROOM H5–9, 750 W. Ea.	3750	30 9|10 2P 11|12	30 2P	SPARE		6
9	HEAT PUMP – 3HP		40 13|14 2P 15|16	40 2P	HEAT PUMP – 3 HP		8
11	13.8 KW HEATER ZONE 1		80 17|18 2P 19|20	50 2P	9.2 KW HEATER ZONE 2		10
13	SPARE		20 21|22 2P 23|24	40 2P	9 KW ELEC HOT WATER HEATER		12
PANEL DATA							
MAINS AND GND BUS.	150A, /SN; 100A GND. BUS				VOLTS	120/240	
MAIN C/B ~~OR SW/FUSE~~	150A 2P, 100 AT						
BRANCH C/B INT. CAP.	10000 AMPS						
SURF /~~RECESSED~~							
REMARKS							

(c)

Figure 15.19 (c) Panel schedule of the garage-mounted climate control electrical panel (CC). This panel supplies all the climate control equipment shown in (a) and (b) plus the electrical hot water heater installed in the garage. See Figure 15.18(a).

stage of the work, to be added to the prepared panel schedule.

15.11 Electrical Service Equipment

The first item to be located on a plan is the electrical panelboard. We are considering it last because it is part of the electrical service equipment. When we look at Figure 15.18(a), we see that the panel is located on the inside wall of the garage. It is placed there for the following reasons.

(a) Convenience to approach. It is located near the door leading inside the house.

(b) Central location. This shortens all home-runs. If the outside wall were used, some of the branch circuit runs would be very long (more than 100 ft), giving excessive voltage drop and requiring increased wire size.

(c) Location near the load center. The heaviest load is the kitchen and the laundry. Both are just above the panel.

Location on the inside wall requires that the service cable be run under the slab, encased in concrete, from the meter location to the service switch at the panel. See Figure 15.20. Such a run is inexpensive when poured together with the garage concrete floor.

The climate control panel is placed alongside

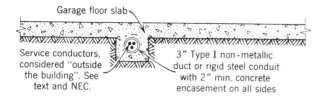

Figure 15.20 Service conductors and conduit encased in 2-in. concrete envelope below garage floor.

the house panel. In an all-electric house, it is a convenience to separate the electric heating and feed it from its own panel. Figure 15.21 is the one-line diagram for the house service equipment. This diagram and all the cable sizes are worked up after the house is laid out and circuited and the panel loads are totaled.

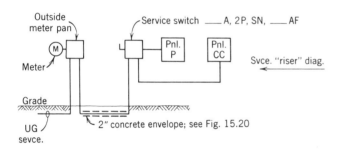

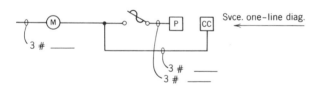

Figure 15.21 The riser diagram and the one-line diagram show essentially the same data but for different purposes. They are, therefore, both usually required. The riser diagram shows the physical arrangement of the main components of the electrical system in block form. It is called a riser, even when, as in this case, everything is at the same elevation. The one-line (single-line) diagram shows the electrical connections of the system, with all cabling shown as single lines and devices shown with their electrical symbols. Equipment and cable sizes have been deliberately left blank; they are supplied to the technologist by the design engineer.

The Mogensen house, like much construction today, uses underground service. This means that the service cable from the utility company line to the building is run underground. Layout of the service run and detailing of service equipment and, sometimes, of the service takeoff is frequently the work of the electrical technologist. For this reason, the paragraphs that follow are devoted to a discussion of low voltage electric service to buildings.

15.12 Electric Service—General

The majority of buildings take service at low voltage, that is, below 600 v. This electric service is provided by the electric utility company at a service entrance point. Service can consist of two, three or four wires, including a grounded neutral wire. The service provided may be 2-wire, 120-v for a very small house, 3-wire, 120/240-v for a house like The Basic House or the Mogensen house, or 4-wire, 120/208 or 277/480 v for larger buildings. In each case, the size of the service (in amperes) varies, depending on the building's load. Typical service capacities are 60, 100, 150 and 200 amp. (Generally, 2-wire, 120-v service does not exceed 60 amp.) Some very large buildings and heavy industrial plants take service at high voltage. Such arrangements are specially designed for each building and cannot be dealt with here. (Some buildings that take low voltage have the utility company's transformers in or adjacent to the building.) The service cables between the building property line and the utility company supply point are generally the property of the utility company in overhead service runs and the property of the owner in underground runs. In the latter case, therefore, the house builder pays for the service run.

15.13 Electric Service—Overhead

The most common form of electrical distribution is overhead lines. At a building requiring overhead electrical service, a service drop is run from the nearest utility pole. This is connected to the building service cables at the service entrance point. Study Figures 15.22 to 15.24 to see how this is done. Figure 15.25 shows the splicing at the service pole. This work is done by the utility company. From here, the overhead service wires are extended

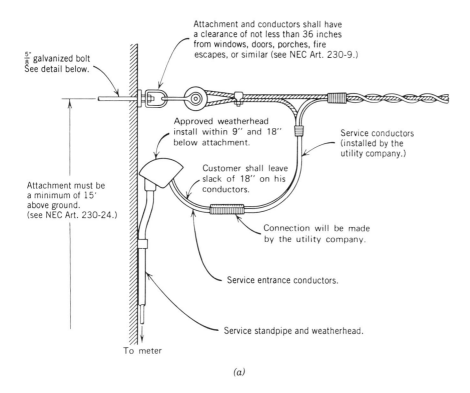

(a)

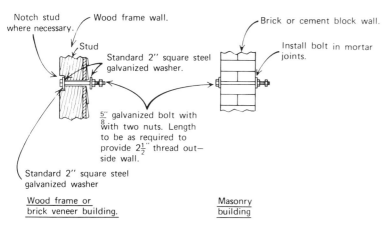

(b)

Figure 15.22 Typical residential overhead electrical service detail. The exact division of responsibility between the utility company and the customer varies from one utility to another. (a) Typical service drop detail. (b) Attachment bolt detail for different wall constructions.

898 / RESIDENTIAL ELECTRICAL WORK

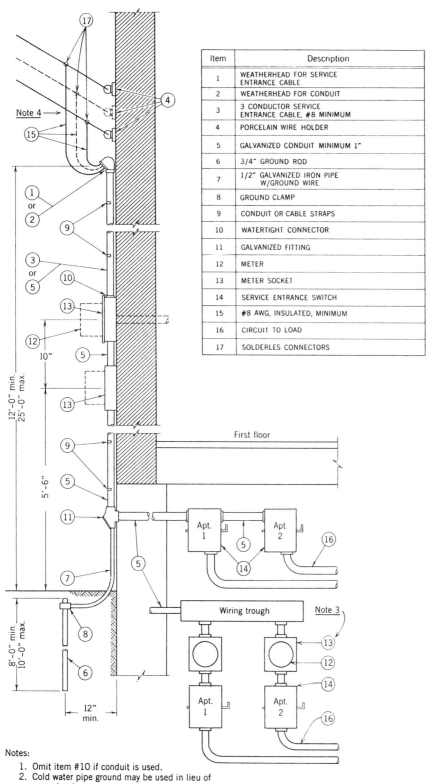

Notes:
1. Omit item #10 if conduit is used.
2. Cold water pipe ground may be used in lieu of ground rod.
3. Meters may alternatively be placed inside the building.
4. See Fig. 15.22 for arrangement with incoming multi-conductor aerial cable instead of individual wires shown here.

Figure 15.23 Typical electrical service detail for a small multiple residential building.

ELECTRIC SERVICE—OVERHEAD / 899

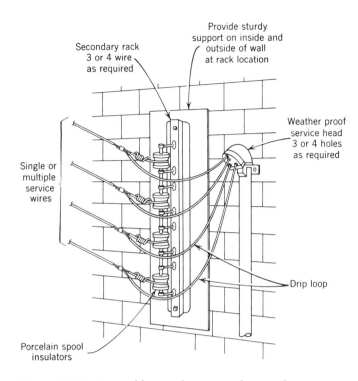

Figure 15.24 Typical heavy-duty secondary-rack-type service entrance, normally used in nonresidential construction.

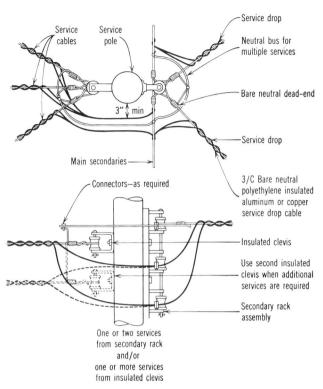

Figure 15.25 Arrangement of secondary cable connections at the service pole. These connections are used with three-conductor bare-neutral insulated service-drop cable. Note the building end of the drop termination in Figures 15.22 and 15.23. The work shown in this detail is generally the responsibility of the utility company, and this drawing is for information only.

to the building. At the building, they are terminated either individually on wire holders (Figure 15.23), on a single-wire holder for an entrance cable assembly (Figure 15.22) or occasionally with heavy drop wires on a building-mounted secondary rack (Figure 15.24). Service entrance wires are brought out of the building from the building panel. They are spliced outside to the service drop cable with solderless connectors, and the joints are taped. (See Figures 15.22 to 15.24.) The service entrance conductors enter the building through the *service weatherhead* and (outside) electric power meter.

The weatherhead is a porcelain and steel device used to bring cables inside without allowing in the rain. The weatherhead varies in number and size of holes and in size of conduit fitting. The detail of Figure 15.23 is for a multiple dwelling. For a single family residence, the service equipment is usually placed on the street level or in the garage or utility room. Note in Figure 15.23 the mounting heights of equipment and the use of a table of materials. In Figure 15.26, the arrangement of Figure 15.23 is shown as it would appear on an electrical plan. Note that all three drawings are necessary: the single line, the plan and the detail. The single line [Figure 15.26(b)] shows the electrical situation at a glance. Frequently, the cable and equipment sizes are shown here. The location sketch [Figure 15.26(a)] shows the physical arrangement and should be to scale. Finally, the detail (Figure 15.23) shows the exact materials and the required construction. Construction details of this type are one of the technologist's most important tasks. Without

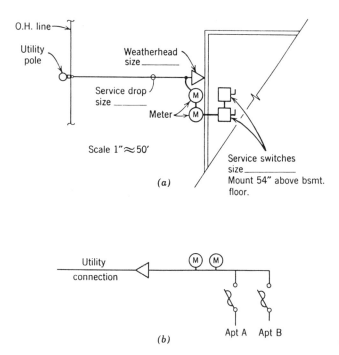
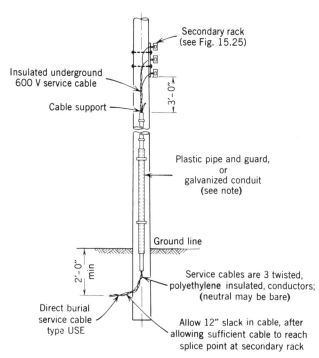

Figure 15.26 Electrical plan (a) and single-line (one-line) representation (b) of the data shown in Figure 15.23.

NOTE: Size of conduit required for service conductors shown, as follows: 1½" conduit up to 3/0; 2" conduit for 4/0. All work must be approved by the Utility Company and must meet the requirements of NEC Article 230.

Figure 15.27 Typical service (riser) pole feeding underground secondary service.

them, the contractor is left to make his or her own choices, and the results may not be what the designer intended.

15.14 Electric Service—Underground

In much new construction the choice is made to bring the service into the building underground. In general, the utility company (electric power company) stops at the building's property line. Therefore, with underground service, the customer must run the service cables from the building to the property line. At that point, the service connection is made. The service work on private property must be done by a private contractor and at the owner's expense. The work is subject to the requirements and inspection of the utility company. This is to ensure a proper grade of work and equipment. The type of equipment used for the underground service cable varies. Some utility companies allow direct burial of cable of acceptable design [type USE *(Underground Service Entrance)* cable] between the service connection and the building. See Figures 15.27 and 15.28. Other companies or terrain situations require greater physical protection for the cable such as heavy wall (Type II) conduit or rigid steel conduit. See Figure 15.29.

Underground nonmetallic conduit is available in two types: Type I, which is intended for concrete encasement (See Figure 16.2), and Type II, which is intended for direct burial in earth, without a concrete envelope. The basic difference between the two types is the conduit's wall thickness. Because Type II is direct buried, it must be stronger than Type I and, therefore, have a thicker wall. Fiber conduit, which was once almost exclusively used, has today been largely replaced by plastic conduit (PVC and styrene). Type I conduit (concrete encased) is normally used for high voltage

ELECTRIC SERVICE—UNDERGROUND / 901

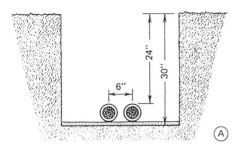

Trench should be deep enough so that cable will be at least 24" below surface. Put a cushion of sand on the bottom of the trench. Lay cable with a slight snaking to allow for earth settling and cable expansion.
When two or more cables are installed in the same trench, space them 6" on centers (no crisscrossing)

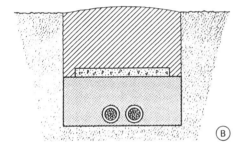

Cover cables to depth of 6" with sand or stone free earth.
Lay a concrete slab or creosote treated plank on refill for protection of cable.
Complete the refill.
Under highways, streets and right-of-way, cable should always be installed in conduit, as in Ⓒ below.

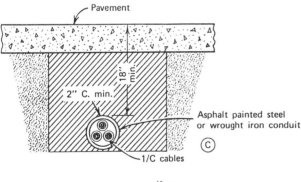

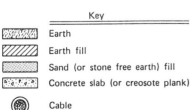

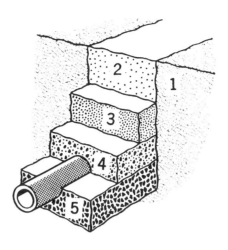

1. Trench wall
2. Ordinary backfill
3. Selected backfill
4. Selected backfill
5. Bedding

Figure 15.29 Underground nonmetallic duct (conduit) installation. The conduit is heavy wall Type II. Depth of burial is 24 in. in ordinary traffic areas and 36 in. in heavy vehicle traffic areas. Each layer of fill is 8–12 in. thick, depending on burial depth.

underground wiring and in installations of all types where the high strength of a concrete duct bank is required. Type II direct burial conduit is normally used for low voltage (under 600 v) power wiring and communication/data cabling.

Low voltage cable is almost never installed in a concrete enveloped raceway. An exception occurs when the service equipment is not at the point at which the underground run meets the building. See **NEC** Article 230-6, "Conductor Considered Outside Building," and Figure 15.18, which shows this situation. In some cases the utility installs cables in empty raceways supplied by the owner. In others, the entire installation is done by the owner. The drawing or specification must clearly state the

Figure 15.28 Installation technique for direct burial cable. Whenever special symbols are used in a detail, a legend or key should be provided, as in this figure.

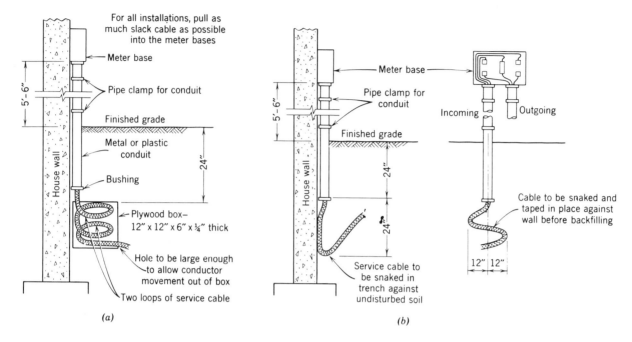

Figure 15.30 Two typical types of termination of service cable at a building. Some utility companies require a similar coil of underground cable at the base of the utility pole. See Figure 15.27.

division of responsibility between the utility and the customer. A typical service detail corresponding to the kind of service taken at the Mogensen house is shown in Figure 15.30.

Because the service wiring, both overhead and underground (including the metering), must meet the requirement of both the electric utility company and the NEC, occasional conflicts arise. These should be referred to the project engineer for clarification.

15.15 Electric Service—Metering

As can be seen in Figures 15.22, 15.26 and 15.30, single meters for residences are normally placed outside. This is helpful to the meter reader, since access to the inside of the house is not needed. Although modern remote-meter-reading techniques are being used increasingly, millions of meters are still read at the meter. In any case, an indoor meter has no real advantage. Even automated meters require periodic adjustment, calibration and occasional servicing, and these are most efficiently done when access to the meter(s) is unrestricted.

For multiple residences and commercial buildings, the metering is normally inside, because (a) the building is open, and (b) the metering installation itself is large. With multiple residences that have individual apartment metering, the practice has been to install the meters in central meter rooms. The advantage of this is that reading the meters is a one-stop affair. To make an installation of this type, multiple meter pans are used. They can be assembled in groups or modules to meet almost any requirement. Examples are shown in Figures 15.31 to 15.33 of typical single- and multiple-meter installations. (Some large rental-type multiple residences and commercial buildings were constructed with master meters only, with the electrical usage cost included in the rental. This practice was discontinued because it led to energy waste. Today, in such buildings, modern electronic sensors that measure the energy usage can be placed on each tenant's service wiring. This can be done without physically cutting into the wiring.

ELECTRIC SERVICE—METERING / 903

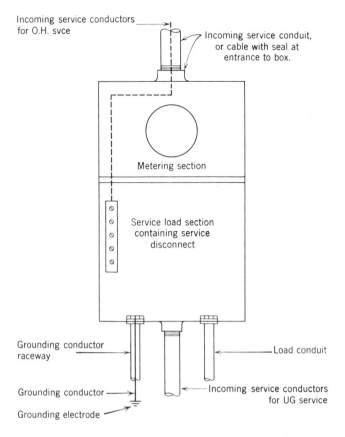

Figure 15.31 Combination meter and service cabinet for overhead or underground service.

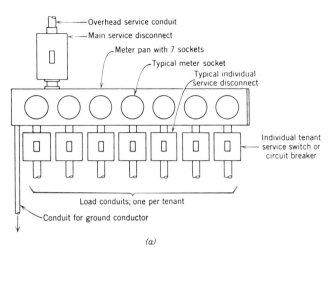

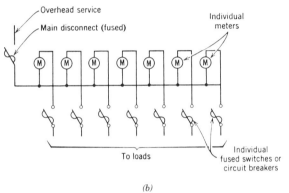

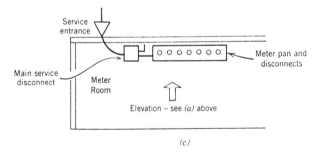

Figure 15.32 Typical arrangement of metering for multiple-occupancy building (seven tenants). *(a)* Physical arrangement in elevation. *(b)* Schematic representation, one-line (single-line) diagram. *(c)* Plan representation.

Figure 15.32 *(d)* Typical modular metering equipment. See also Figure 16.14. The cabinet at the left contains the main service disconnect—in this case a circuit breaker. Adjacent, and fed from this, are sections of meter pan. Below each meter socket is the circuit breaker that is the main protection for the apartment involved. More sections can be added, as needed. (Note the incoming service conduits and the outgoing load conduits.) Complete electrical and physical data are available in manufacturers' catalogs. (Courtesy of Square D Company.)

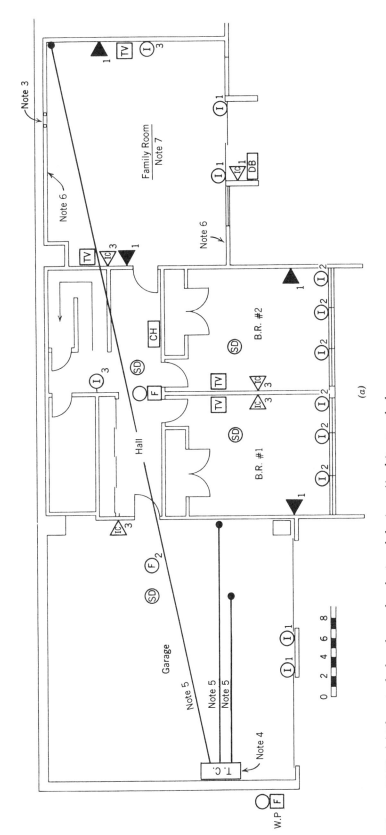

Figure 15.33 (a) Electrical plan, lower level, signal devices. (Architectural plans courtesy of B. Mogensen A.I.A.)

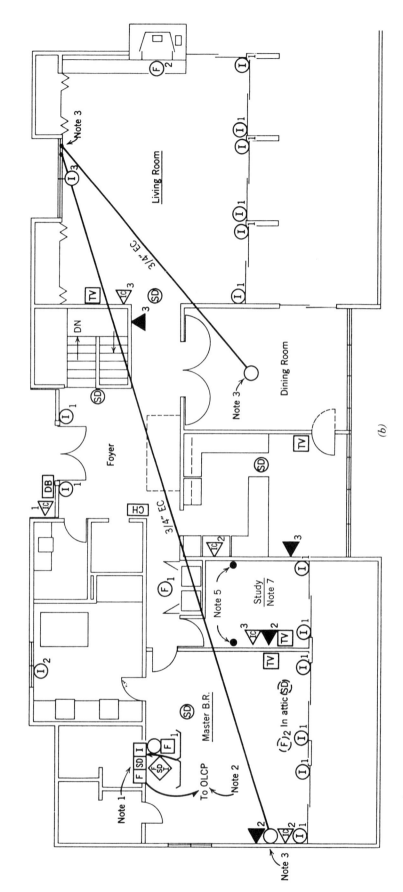

Figure 15.33 (b) Electrical plan, upper level, signal devices. (Architectural plans courtesy of B. Mogensen A.I.A.)

SYMBOLS FOR SIGNAL EQUIPMENT

- ◯ 6" AC VIBRATING BELL, CONCEALED IN RECESSED BOX, WITH GRILL CLOTH COVER, 84" AFF.
- (F) [as above symbol label]
- $(F)_1$ TEMP. DETECTOR; RATE-OF-RISE & FIXED TEMP., RESETTABLE.
- 2 TEMP. DETECTOR; FIXED TEMP., 185°C.
- (SD) SMOKE DETECTOR WITH RESETTABLE FIXED TEMP. DETECTOR.
- $(I)_1$ INTRUSION DETECTOR; MAGNETIC DOOR SWITCH.
- 2 INTRUSION DETECTOR; MAGNETIC WINDOW SWITCH.
- 3 INTRUSION DETECTOR; PASSIVE INFRARED (PIR) MOTION DETECTOR.
- ◇ ANNUCIATOR, CUSTOM DESIGN.
- [CP] CENTRAL PANEL FOR F.A., S.D. & INTRUSION.
- [DB] DOOR BELL.
- [CH] CHIMES SIGNAL.
- ◀$_1$ PREWIRED PHONE OUTLET; JACK 12" AFF.
- 2 PREWIRED PHONE OUTLET; FIXED, 12" AFF.
- 3 PREWIRED PHONE OUTLET; FIXED WALL OUTLET 60" AFF.
- ◁IC$_1$ INTERCOM OUTLET, OUTDOOR, W.P. 60" AFF.
- 2 INTERCOM OUTLET, MASTER STATION 60" AFF.
- 3 INTERCOM OUTLET, REMOTE STATION 60" AFF.
- [TV] (PREWIRED) TV ANTENNA OUTLET, 12" AFF.
- [TC] TELEPHONE CABINET

Figure 15.33 *(c)* Symbols for signal equipment.

Notes:
1. The fire detection, smoke detection and intrusion alarm devices all operate from a single control panel. The alarm bell is common. The annunciator indicates which device operated. See Fig. 15.37.
2. Connection between signal control panel and the Outside Lighting Control Panel (OLCP) activates all outside lights when any signal device trips. Selected lights inside the house can also be connected to go on, as programmed by the owner.
3. Two $3/4$-in. empty plastic conduits extending from two 4-in. boxes in living room wall down to family room and terminating in 4-in. flush boxes. Boxes to be 18 in. AFF and fitted with blank covers. Extend a $3/4$-in. plastic empty conduit (EC) from one 4-in. box in living room to 12-in. speaker backbox recessed in dining room ceiling. Locate in the field. From the second 4-in. box in living room extend a $3/4$-in. empty plastic conduit to an empty 4-in. box in the master bedroom, 18 in. AFF. Finish with blank cover. These raceways are intended to serve audio system wiring and remote loudspeakers.
4. Coordinate the location and size of the telephone entrance service cabinet (or box) with the 'phone company.
5. Extend a $3/4$-in. empty plastic conduit from the telephone entrance cabinet to each of the signal raceways in the study. Extend one $1\,1/4$-in. or two $3/4$-in. empty plastic conduits from the cabinet to the empty signal raceway around the perimeter of the family room.
6. Provide a $3/4$-in. EC through the wall and capped at both ends, for entry of cables from an exterior satellite dish. Coordinate location with TV/CATV/satellite dish contractor.
7. All signal outlets in the family room and study are separate and distinct from the empty signal raceways installed on the walls. At the owners discretion, these raceways can be used for signal system and telephone wiring.

(d)

Figure 15.33 *(d)* Notes for signal equipment drawings.

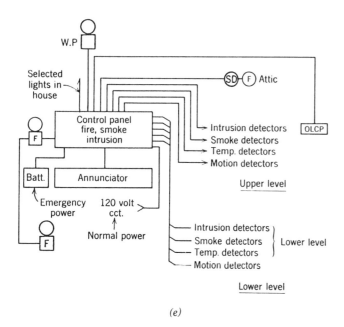

Figure 15.33 *(e)* Signal equipment riser diagram.

15.16 Signal and Communication Equipment

Every residence is equipped with some signal and communication equipment. The simplest house has a doorbell and usually a telephone. An expensive modern house such as the Mogensen plan will most likely have a sophisticated intrusion alarm system, smoke and heat detection equipment combined with an evacuation alarm system, multiple telephone outlets (possibly prewired), multiple television outlets (also possibly prewired) and an intercom system. In addition, as noted in Section 15.3, the study must be arranged to handle data and communication cabling that will serve the needs of a home-based office. Furthermore, the use of closed-circuit television and computer networks in regular teaching programs for students who cannot reach the classroom is increasing. This requires that one or more rooms in the house be equipped to accept the necessary cabling. If you have followed the discussion in this chapter, you have seen that the special requirements of a home office in the study have already been provided. Similarly, the family room is arranged to accept extensive data and communication cabling. That can be part of the teaching programs mentioned previously.

In general, the provision of these facilities involves outlet boxes with the appropriate covers (modular telephone connectors, TV jacks and the like) and an empty raceway system that will permit concealed installation of the required special wiring, usually by specialist contractors. Surface wiring, even in neat rectangular raceways, is normally undesirable. A standard electrical construction contractor is not usually equipped to install these special systems. As a result, specialized contractors, or the owner if he or she is qualified, will do the wiring, adjusting and testing required. These signal alarm and communication outlets are shown on the upper and lower house plans of Figure 15.33(*a*) and (*b*). A symbol list for the systems is shown in Figure 13.33(*c*), and the relevant notes are shown in Figure 15.33(*d*).

When signal and alarm systems were first installed in residences, it was common for each system to be self-contained. The fire alarm system had its control unit and devices, the smoke detection system had its equipment, the intercom had its equipment, and so on. That is still the situation with small "package" systems. However, in large residences, a custom system can be furnished with a single control panel and annunciator. This connects to all the devices and indicates visually and audibly the operation of any device. The control unit is normally placed in the master bedroom. A riser diagram for such a combined, custom-designed system in the Mogensen house is shown in Figure 15.33(*e*). The system can be arranged to do the following when any of the remote units operate:

- Sound audible devices in the master bedroom and on the lower level.
- Show on the annunciator the location of the detector that has operated and what type it is, that is, fire, smoke, or intrusion.
- Turn on all the outside lights, via panel OLCP (Outside Lighting Control Panel).
- Turn on selected inside lights.

Note also that a battery has been furnished to supply emergency power to the unit. The technologist is responsible for the layout of the devices and the riser diagram. To become familiar with the symbols normally used in such diagrams, study the general symbol list of signaling devices furnished in Figure 15.34. This list is Part VII of the overall symbol list. Photographs of some typical residential private signal equipment are shown in Figures 15.35 to 15.38.

ELECTRICAL SYMBOL LIST; PART VII
SIGNALLING DEVICES

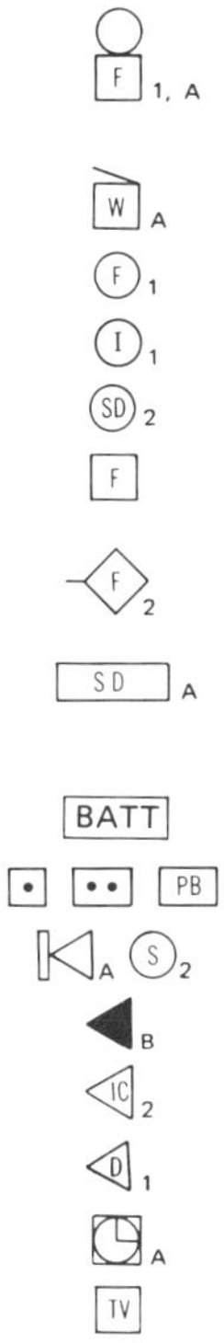

○ / [F]₁,ₐ BELL OR GONG, INSCRIBED LETTER INDICATES SYSTEM (SEE BELOW;) AND SUBSCRIPT LETTER OR NUMBER INDICATES TYPE e.g. A – 8" VIBRATING BELL, 12 V. DC; B – 12" WEATHER PROOF SINGLE STROKE GONG, 120 V. DC etc.

[W]ₐ BUZZER, TYPE A.

(F)₁ FIRE DETECTOR, TYPE 1.

(I)₁ INTRUSION DETECTOR, TYPE 1. } ETC.

(SD)₂ SMOKE DETECTOR, TYPE 2.

[F] MANUAL STATION – WATCHMEN TOUR, FIRE ALARM, ETC. LETTER INDICATES SYSTEM, SEE BELOW.

⟨F⟩₂ ANNUNCIATOR, LETTER AND NUMBER INDICATE SYSTEM AND TYPE.

[SD]ₐ CABINET OR CONTROL PANEL, SMOKE DETECTION, USE IDENTIFYING TYPE LETTER IF MORE THAN ONE TYPE IS USED ON THE PROJECT.

[BATT] AUXILIARY DEVICE.

• •• [PB] PUSH BUTTONS

◁ₐ (S)₂ HORN OR LOUDSPEAKER, TYPE A; TYPE 2

◀ᵦ TELEPHONE OUTLET, TYPE B

◁[IC]₂ INTERCOM OUTLET, TYPE 2

◁[D]₁ DATA CABLE OUTLET, TYPE 1.

⊙ₐ CLOCK SYSTEM OUTLET, TYPE A

[TV] TV ANTENNA OUTLET

SYSTEM TYPES

CATV	CABLE TELEVISION
F, FA	FIRE ALARM
IC	INTERCOM
I, IA	INTRUSION ALARM
NC	NURSE CALL
S, SD	SMOKE DETECTION
S, SP	SPRINKLER
T, TEL	TELEPHONE
TV	TELEVISION
W, WF	WATER, WATERFLOW
WT	WATCHMAN'S TOUR

AUXILIARY DEVICES

BATT	BATTERY
BT	BELL TRANSFORMER
CH	CHIME
CT	CONTROL TRANSFORMER
DB	DOOR BELL
DH	DOOR HOLDER
DO	DOOR OPENER
F.O.	FIBER OPTIC(S)
MOD	MODEM
S, SP	SPEAKER, LOUDSPEAKER
TC	TELEPHONE CABINET

Figure 15.34 Electrical (working) drawing symbol list, Part VII. Signal/communication/low voltage systems symbols in common use. The designer/technologist is free to add to this list as required. Additional symbol lists are found at:

Part I, Raceways, Figure 12.6, page 671.
Part II, Outlets, Figure 12.50, page 719.
Part III, Wiring Devices, Figure 12.51, page 720.
Part IV, Abbreviations, Figure 12.58, page 726.
Part V, Single-line Diagrams, Figure 13.4, page 733.
Part VI, Equipment, Figure 13.15, page 753.
Part VIII, Motors and Motor Control, Figure 16.25, page 944.
Part IX, Control and Wiring Diagrams, Figure 16.28, page 947.

910 / RESIDENTIAL ELECTRICAL WORK

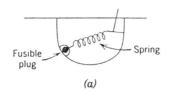
(a)

(b-1)   (c-1)

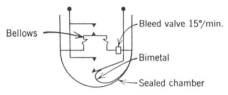

Rate of rise, fixed temp. (auto-reset)

(b-2)

(c-2)

(d)

(e)

Figure 15.39 shows a few of the many types of station termination outlets available for voice/data/communication/network cabling. These devices are equally applicable to residence offices as already described and to business office in dedicated structures.

In multiple residences such as garden apartment blocks and apartment houses, provision for signal equipment is much simpler. Each apartment is, of course, equipped with telephone outlets. It is assumed that the telephone wiring will be adequate for basic computer data transfer needs via telephone modems. When this is not the case, additional cabling can be added in (relatively) unobjectionable surface raceways. These residences are always provided with smoke and fire detectors that contain the required audible evacuation alarm. A few of these devices are shown in Figure 15.40. Some multiple residence intercom equipment is shown in Figure 15.41.

A paging console applicable to an office or institution is shown in Figure 15.42. An administrative intercom console applicable to an office, store or other small facility is shown in Figure 15.43. Provision for these and other signal units by means of raceways cabinets and outlet boxes is an integral part of the basic electrical contract work. As such, it is the concern of the electrical technologist. A typical apartment house telephone conduit riser is shown in Figure 15.44, and a typical apartment conduit layout in Figure 15.45.

Figure 15.35 Spot-type heat detectors. *(a)* Fusible plug melts out at predetermined temperature, opening (or closing) the circuit and causing an alarm. Unit indicates that it has operated and is nonrenewable (one time operation). *(b)* Rate-of-rise unit consists of an air chamber with a bleed valve. Rapid temperature rise causes the bellows to expand before air is lost by bleeding, thereby setting off the alarm. The unit illustrated is combined with a fixed-temperature unit, similar to *(c)*. *(c)* Bimetallic unit action is similar to that of a thermostat and is self-restoring. *(d)* Combination rate-of-rise and fixed-temperature unit installed in a wood-frame house basement adjacent to the furnace. [See also Figure 15.36, which shows a photoelectronic detector with an integral, fixed-temperature detector. Such combined units cover a wider range of hazards than can either single unit.] *(e)* Commercial-grade, surface-mount temperature detector of the self-resetting variety. This unit will alarm both at a preset fixed temperature of 135 or 200°F (depending on type) and, regardless of temperature, when a rapid change in temperature occurs, as at the beginning of the heat state of a fire. The illustrated unit is 6-in. high overall, mounts on a standard 4-in. box and is equipped with an indicating light. [Photo *(e)* Courtesy of Cerberus Pyrotronics.]

Figure 15.36 Typical residential smoke detector of the scattered-light photoelectronic type. This unit can also be equipped with a heat-sensing unit. These detectors are powered by line current and can be wired in tandem on a single circuit for housewide coverage. (Courtesy of Simplex.)

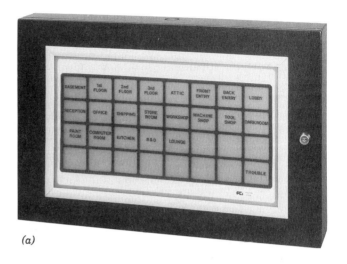

(a)

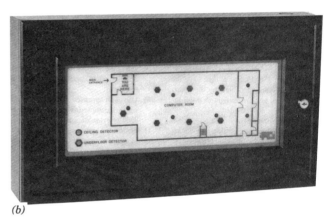

(b)

Figure 15.37 Typical annunciators that can be used in any single or combined system application. *(a)* Tabular back-lighted annunciator can show the location of each sensor, each room or any combination desired. The alarming sensor lights up and can be arranged to flash. *(b)* Graphic annunciator show the location of each sensor on one or more floor plans. Here too the affected sensor lights up and can flash. With both types of annunciator, an audible alarm can sound when a sensor operates. [Photo *(a)* courtesy of Fire Control Instruments, Inc.]

SIGNAL AND COMMUNICATION EQUIPMENT / 913

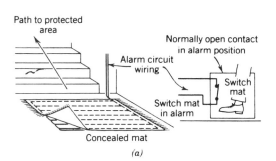
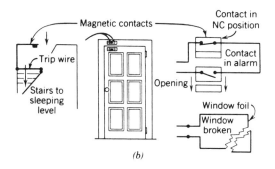
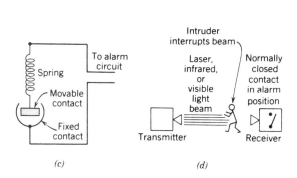
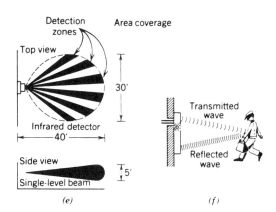

Figure 15.38 Typical intrusion detectors. *(a)* Normally open contact device such as a switch mat is operated by the weight of the intruder. *(b)* Magnetic door contacts are the first line of intrusion alarm at the house. The second may be a low level trip wire at the base of the stairs. *(c)* Vibration detectors are very sensitive to motion and can be used on windows that cannot be conductive-strip-foiled without spoiling their appearance. *(d)* Photoelectric beams form an effective intrusion barrier if placed so that their avoidance is difficult. *(e)* Passive infrared detectors give basically a 30 ft × 40 ft oval protective zone, starting as a narrow beam and widening with distance. Focusability permits exact coverage of any area in a space. Units are also available with multilevel beams that give vertical as well as horizontal coverage. *(f)* Motion detectors detect changes in the frequency of a signal reflected from a moving object. Sensitivity is highest when relative motion is greatest, that is, when the intruder is moving directly toward (or away from) the detector.

914 / RESIDENTIAL ELECTRICAL WORK

Figure 15.39 Station termination modular wiring devices. (a) Single gang wall plates for voice and data cables. (b) Single gang wall plates for video, LAN (local area network) data and voice cabling. (c) Communication and data outlets that match the design of power wiring devices. (d) Surface-mounted housings for communication jacks, usable with surface raceways. (e) Low profile design voice/data/computer/network connection surface module. Faceplates are interchangeable. (Photos courtesy of Hubbell Premise Wiring, Inc.)

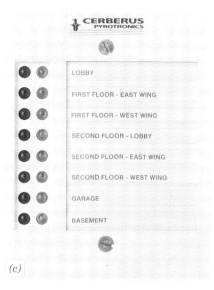

Figure 15.40 *(a)* A typical scattered-light spot smoke detector with integral 135°F (57°C) self-resetting temperature detector. The unit operates on 24 v d-c and mounts on a standard 4-in. outlet box. *(b)* Typical manual station for use in lobbies and corridors. *(c)* Small pilot-light-type annunciator that indicates activated zones in a small to medium size building. [Photo *(a)* courtesy of Protectowire Company; photo *(b)* courtesy of Fire Control Instruments, Inc.; photo *(c)* courtesy of Cerberus Pyrotronics.

916 / RESIDENTIAL ELECTRICAL WORK

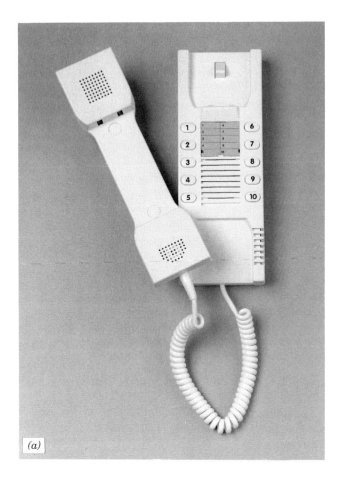

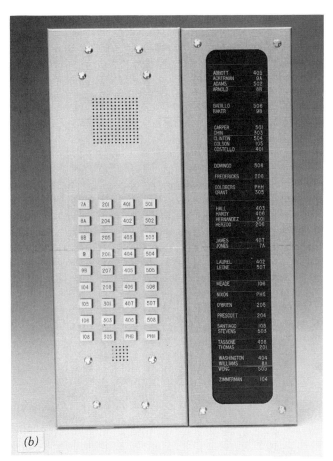

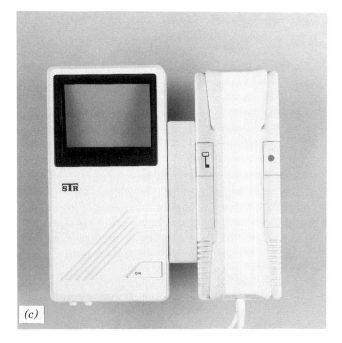

Figure 15.41 Typical multiple residence intercom equipment. *(a)* This unit serves up to ten suites and combines the directory and the phone in a single unit. For larger buildings a directory with a built-in speaker phone *(b)* is common. For a higher degree of security, a TV camera at the entrance provides a video signal in each apartment *(c)* along with the usual audio connection. (Photos courtesy of Alpha Communications, Inc.)

SIGNAL AND COMMUNICATION EQUIPMENT / 917

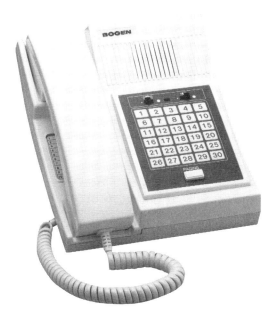

Figure 15.42 A multichannel interphone/paging console is illustrated. The unit serves up to 30 stations and provides paging and intercom facilities. (Photo courtesy of Bogen Communications, Inc.)

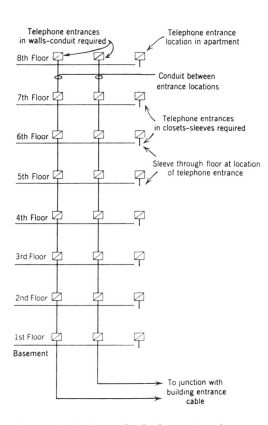

Figure 15.44 Typical telephone riser diagram. Note the need for conduit between apartment when installation is made inaccessible, as in a wall.

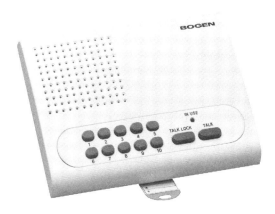

Figure 15.43 A small 11-station intercom console. Such a unit would be used to provide selective intercom and multiple conversation paths. (Photo courtesy of Bogen Communications, Inc.)

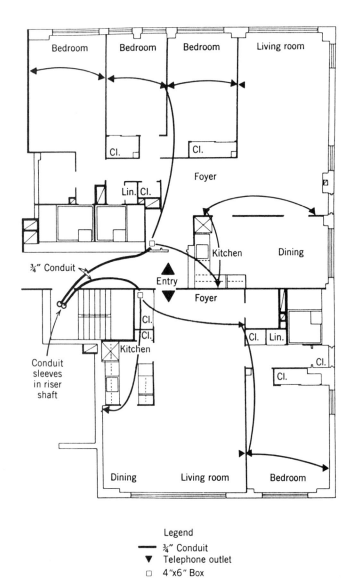

Figure 15.45 Typical telephone floor plan, showing conduit in riser shaft and connections to apartments.

Key Terms

Having completed the study of this chapter, you should be familiar with the following key terms. If any appear unfamiliar or not entirely clear, you should review the section in which these terms appear. All key terms are listed in the index to assist you in locating the relevant text. Some of the terms also appear in the glossary at the end of the book.

Climate control systems
Color rendering index (CRI)
Low voltage switching
Multilevel switching
Overhead service drop
Power line carrier (PLC)
Residential appliance circuits

Service lateral
Service secondary rack
Service weatherhead
Service wire holder
Telephone modem
Type I and Type II conduit and duct
Underground service entrance

Supplementary Reading

Refer to the Supplementary Reading section at the end of Chapter 16.

Problems

1. Convert the basement of The Basic House plan into a family room. Show all partitioning. Provide lighting and outlets for children's and adult's use. Revise the electrical plans and schedules for the basement, which are given in Chapter 13. Draw to scale any necessary details.
2. Draw the detail called for in Note 9, Figure 15.1(e).
3. Based on the data shown on panel schedule H, Figure 15.18(c), compute the difference between the panel load for bus size calculation and the panel load for service calculation. Do not use demand factors.
4. Using the load figures for kitchen equipment given in Table 15.1, calculate the safety factor involved when using the recommended wire size and receptacle size. The safety factor is defined as the ratio of spare capacity to load rating. The ratings for wire and receptacles are found in Tables 12.2 and Figure 12.52, respectively. Tabulate the results.
5. Using the loads shown in Figure 15.18(c) panel P, and Figure 15.19(c) panel CC, calculate:
   (a) Volt-amperes per square foot for the building, not counting receptacles that are considered part of the general lighting load. (Use 180 v-a per duplex receptacle for other receptacles.)
   (b) Volt-amperes per square foot for the building electrical heating load (use 1 hp = 1 kv-a).
   (c) Total volt-amperes per square foot for the building.
6. Using the criteria given in Sections 15.3 to 15.6, analyze (a) a private residence and (b) an apartment of at least 800 ft^2, for which you have complete electrical plans. On a plan, draw the changes you would make to conform the electrical work to these criteria. When completed, these drawings should be suitable to deliver to an electrical contractor for pricing.
7. For the same two residential occupancies, prepare a signal plan providing intercom, security, television outlets and any other devices you think necessary. (Manufacturers' catalogs will be a considerable help.)

16. Nonresidential Electrical Work

In preceding chapters, we discussed the principles of electrical layout work. Branch circuits and circuiting were emphasized. We also gave some general guidelines for electrical layout in residential buildings. We then combined these principles and guidelines and applied them to two residences, complete with circuiting and schedules. We will now turn our attention to the layout of lighting and devices in nonresidential buildings. The guidelines that apply here are very different.

In nonresidential buildings, the service entrance, service equipment and interior electrical distribution become particularly important. These items are called the building's electrical power system. They include service equipment, switchboards, bus and heavy feeders, distribution panels, motors and their control. The technologist is responsible for preparing the drawings for this equipment and showing how it all goes together. To do this, he or she must know what each component looks like, how it is installed and how to wire it. In this chapter, we will explain these functions. A few additional topics of special interest will complete our study. Study of this chapter will enable you to:

1. Assist in the preparation of power riser and one-line diagrams.
2. Draw electrical service details, including emergency power provision.
3. Understand electrical service and metering and arrangements.
4. Calculate required dimensions of pull boxes and draw the related details.
5. Draw a motor wiring diagram.
6. Draw a motor control (ladder) diagram.
7. Show motor wiring on architectural-electrical plans (electrical working drawings).
8. Apply layout and circuitry guidelines to commercial and institutional buildings.
9. Draw stair and exit risers.
10. Prepare the schedules relating to electrical power plans.

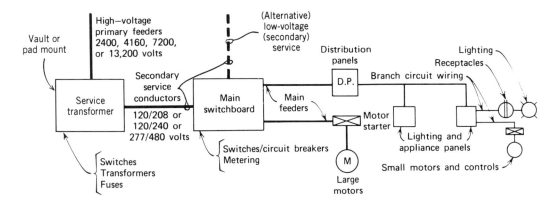

Figure 16.1 Electrical block diagram of a typical nonresidential building's electrical system. These diagrams are called block diagrams because the major items of equipment are shown as boxes, or blocks. Utilization equipment at the branch circuit level is normally not shown. It is added here for completeness. Note that service can be taken either at high voltage (primary) or at low voltage (secondary). We have drawn the interconnecting cables at different line thicknesses, to show relative power-handling capacity. This is not done in actual work. It was done here for information only.

16.1 General

Refer to Figure 16.1. This is a block diagram of the electrical system of a typical commercial building. It corresponds to the pictorial riser shown in Figure 12.17, page 685. The major items are the electrical service and related equipment (not shown in Figure 12.17), the building switchboard, the bus duct and cable distribution system, the motors and their control equipment and finally the branch circuit equipment (that is also not shown in Figure 12.17). Figure 16.1 shows power equipment, that is, panels, switches, controllers, pull boxes, and small transformers. Not shown on either diagram are auxiliary items such as grounding systems, lightning protection equipment, lighting risers and signal equipment. Obviously, not all of these items are found in all nonresidential buildings. Technologists, however, come across these items in their work, and the competent technologist knows how to handle them.

16.2 Electrical Service

The service from the electric utility can be overhead or underground, high voltage (primary) or low voltage (secondary), and any power rating required.

a. Primary Service

Primary service is generally run underground from the utility line to the building. These runs are normally enclosed in rigid steel conduit or in concrete-encased nonmetallic duct. This nonmetallic duct is lighter in weight than the Type II of Figure 15.29 and is called Type I. Such underground power cable runs are often combined with underground (UG) telephone cable runs in a single multiduct bank for economy. See typical detail of Figure 16.2. Although service cables are generally run without splices, occasionally a splice is required. Splicing and pulling of underground cables is done in *manholes* or handholes, depending on the size and voltage of the cable. These manholes and handholes are large concrete boxes set into the earth. Although many are field-poured, precast units are readily available to fill most requirements. Figure 16.3 shows a typical double manhole that could accommodate the ducts of Figure 16.2. Figure 16.4 shows typical duct termination details at the manhole and the building.

A *handhole* is simply a small manhole. The difference is that a person climbs into a manhole but only reaches into a handhole. A typical handhole detail is given in Figure 16.5. A competent electrical technologist can develop details like these from engineer's sketches. For this reason these illustra-

ELECTRICAL SERVICE / 923

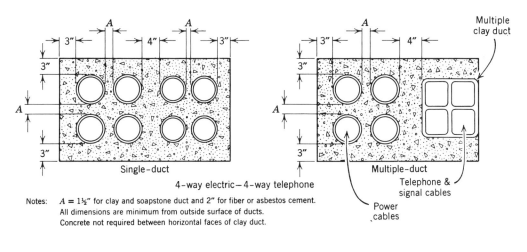

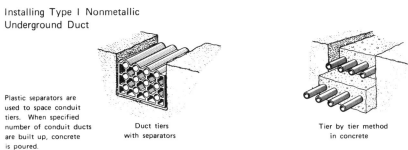

Figure 16.2 Typical underground duct-bank section and details of installation. Although nonmetallic duct is illustrated, steel duct is used where high physical strength is required, such as in filled earth or under paving subjected to very heavy loads. Alternatively, reinforced concrete can be used to provide the required physical duct-bank strength. See also the detail for buried cables in nonmetallic duct without concrete encasement, in Figure 15.29.

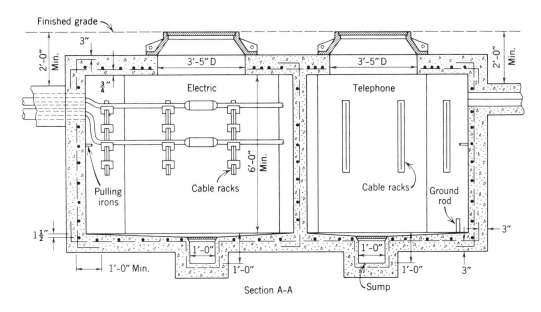

Figure 16.3 Typical details of double power/telephone manhole (a) with required hardware details (b).

924 / NONRESIDENTIAL ELECTRICAL WORK

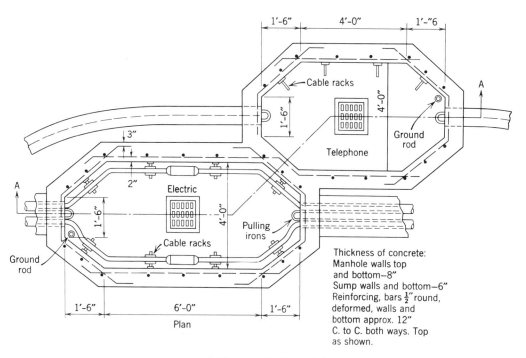

(a) Two-way double manhole

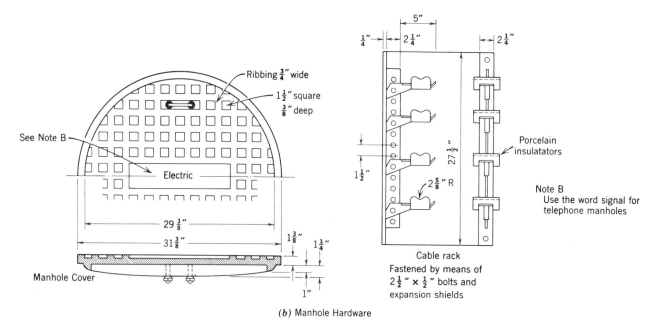

(b) Manhole Hardware

Figure 16.3 *(Continued)*

ELECTRICAL SERVICE / 925

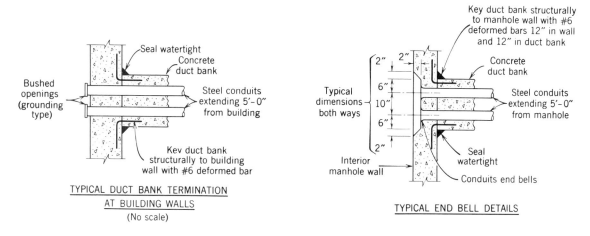

Figure 16.4 Typical underground electrical duct termination details.

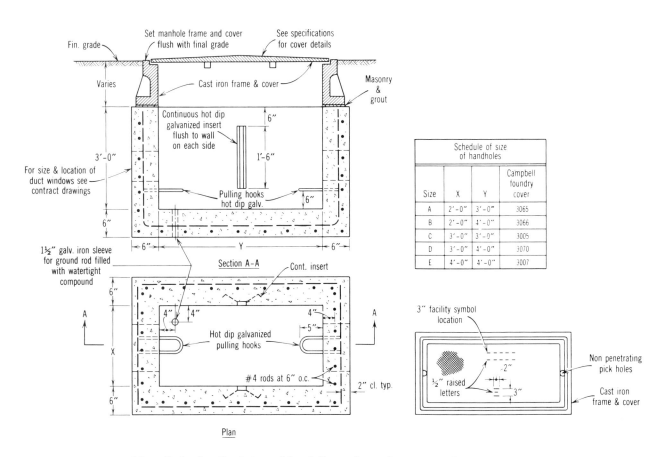

Figure 16.5 Typical handhole detail, giving table of dimensions plus appropriate cover number. Note wall insert for cable support, pulling hooks for cable pulling, and ground rod inside handhole.

926 / NONRESIDENTIAL ELECTRICAL WORK

tions should be carefully studied and fully understood. Where the high voltage (primary) UG cable reaches the building, it is connected to a transformer that changes primary (high) voltage to secondary (low) voltage. See the last step of Figure 11.16, page 640. Since a transformer inside the building can create maintenance problems and usually requires construction of a transformer vault many designers place it outside on a concrete pad. Figure 16.6 shows a typical pad-mount transformer. Dimensions of the concrete pad required and other details including cable entry spaces are available from transformer manufacturers. These data vary among manufacturers and must be coordinated with equipment shop drawings to avoid installation problems in the field.

b. Secondary (Low Voltage) Service

Secondary service is taken either underground or overhead, as explained and illustrated in Sections 15.11 to 15.15. A typical plot plan of a secondary underground service to a nonresidential building is illustrated in Figure 16.7 with related details. Other secondary service details are given in Figures 16.8 and 16.9. Note that, as with the Mogensen

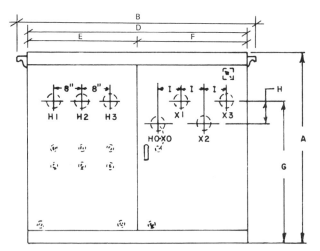

75 through 500 kVA

Figure 16.6 (a) Outside pad-mount liquid-filled transformers such as the one illustrated require a concrete pad sized to accommodate the unit. A slot in the pad permits entry of underground high voltage cables and exit of underground secondary low voltage cables. (b) Dimensions shown are typical. Actual required dimensions vary with each manufacturer. (Courtesy of Uptegraff Manufacturing Company.)

HIGH VOLTAGE 15 KV AND BELOW

KVA	A	B	C*	D	E	F	G	H	I	Approx. Total Net Weight-lbs.
75	48	55	40	47½	30	17½	26	6	3½	2100
112.5	48	55	40	47½	30	17½	26	6	3½	2350
150	52	55	40	47½	30	17½	26	6	4½	3000
225	52	58	52	50½	30	20½	28	8	4½	3850
300	56	58	69½	50½	30	20½	28	8	4½	4000
500	56	58	71½	50½	30	20½	28	8	4½	4850

*Overall depth   (b)

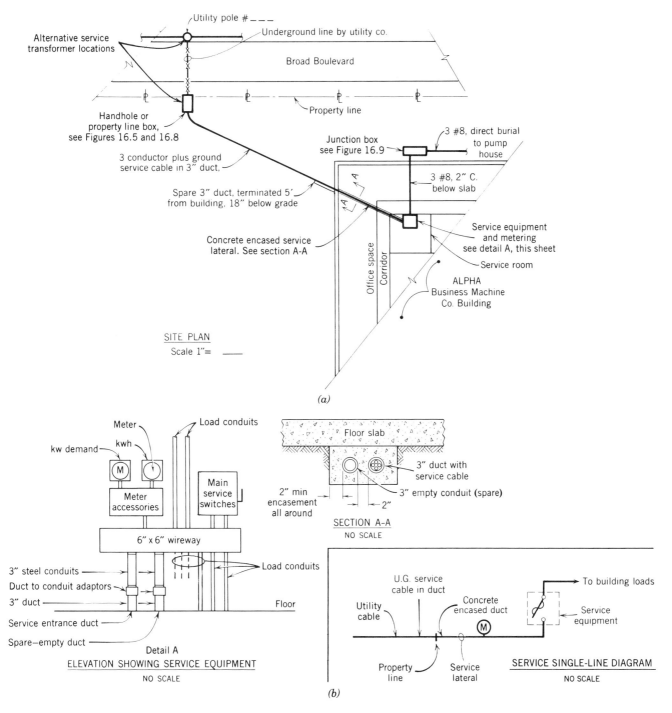

Figure 16.7 Typical electrical site plan for an industrial facility. (a) The site plan to scale, plus related sections and elevations appear on this drawing. Note that if the connection to the utility is at high voltage, then the required transformer can be either pole mounted or pad mounted (see Figure 16.6). In both cases the service lateral is at low voltage. (b) The service single-line diagram shows the equipment schematically. A service transformer would be shown only if it were the property of the customer. Cable and conduit sizes should be shown either on the site plan or the single-line diagram.

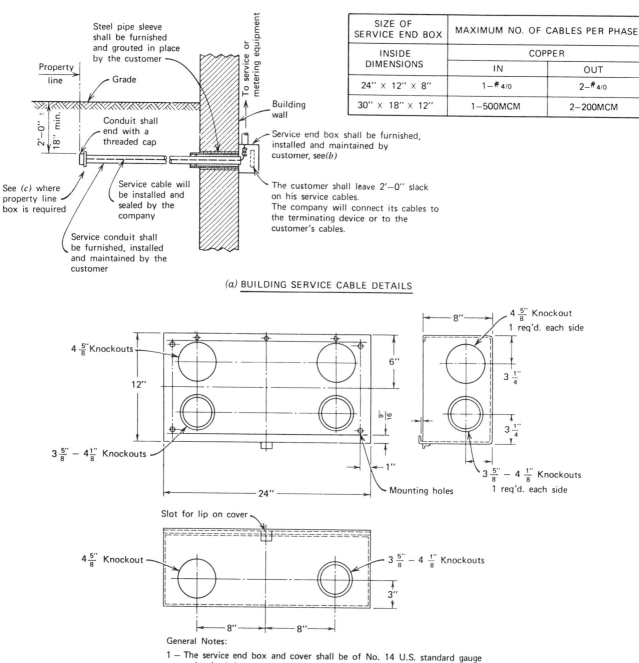

Figure 16.8 (a) Underground low voltage (secondary) service requires a cabinet in which to terminate the service lateral and splice to the service cable. This cabinet is called a *service end box*. Details of this box are shown in (b). These details are furnished by the electrical utility to the building contractor. They vary from one utility to another. Those shown in (b) are typical. If the service cable terminates in a switchboard or similar service equipment, the service end box can be eliminated (if allowed by the utility). If an underground splice is required between the utility's cable and the customer's service cable, a property line box is used, as in (c). Unlike the metal service end box, the property line box is usually precast concrete. The language in the notes on details (a)–(c) is typical for electric service connections from the electric utility's point of view. It defines clearly the responsibility of the customer (contractor) and the electric utility.

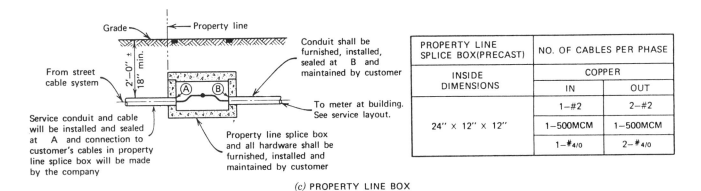

Figure 16.8 *(Continued)*

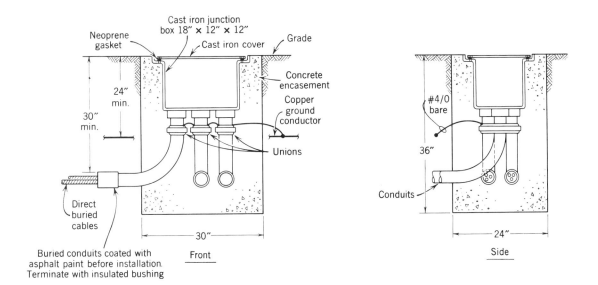

Figure 16.9 Splices in exterior underground cables are made in buried metal or concrete boxes with accessible covers. This detail shows one such design.

house, the service entrance cable is run in a concrete-encased duct. This is to keep it "outside the building" by NEC definition, because the service switch is not at the service entrance point. The technologist should remember that all structural design is the responsibility of the architect and structural engineer. Therefore, all concrete work and reinforcing details must either come from them or be approved by them.

A brief survey of the available secondary service voltages and their typical application should be helpful here. Refer to Figures 11.17 and 11.18, pages 642–643, and to Tables 16.1 and 16.2, while you read the following service descriptions.

(1) 120-v, single-phase, 2-wire, up to 100 amp. This service is used for small residences, farmhouses, outbuilding, barns and the like. Capacity of a 100-amp service of this type is

$$kva = \frac{100 \text{ amp} \times 120 \text{ v}}{1000 \text{ va/kva}} = 12 \text{ kva max}$$

Table 16.1 Nominal Service Size in Amperes[a]

Facility	Area, ft²			
	1000	2000	6000	10000
Single-Phase, 120/240-v, 3-wire				
Residence	100A	100A	200A	—
Store[b]	100A	150A	—	—
School	100A	100A	—	—
Church[b]	100A	150A	200A	—
3-phase, 120/208-v, 4-wire				
Apartment house	—	—	150A	150A
Hospital[b]	—	—	200A	400A
Office[b]	—	100A	400A	600A
Store[b]	—	100A	400A	600A
School	—	100A	150A	200A

[a] Nominal service sizes are 100A, 150A, 200A, 400A, 600A.
[b] Fully air conditioned using electrically driven compressors.

(2) 120/240-v, single-phase, 3-wire, up to 400 amp. This is the usual residential and small commercial service. Maximum power is

$$\text{kva} = \frac{400 \text{ amp} \times 240 \text{ v}}{1000 \text{ v-a/kva}} = 96 \text{ kva}$$

(3) 120/208-v, three-phase, 4-wire, usually not in excess of 2500 amp. This is the normal urban three-phase service taken by most commercial buildings. Maximum power is

$$\text{kva} = \frac{\sqrt{3} \times 208 \text{ v} \times 2500 \text{ amp}}{1000 \text{ v-a/kva}} = 900 \text{ kva}$$

(4) 277/480-v, three-phase, 4-wire, usually not in excess of 2500 amp. This service is taken by commercial and industrial buildings with large loads and heavy motors. Maximum power is

$$\text{kva} = \frac{\sqrt{3} \times 480 \text{ v} \times 2500 \text{ amp}}{1000 \text{ v-a/kva}} = 2078 \text{ kva}$$

To clear up any confusion that may exist between system and utilization voltage, refer to Table 16.3. The system voltage is the nominal voltage the power company agrees to supply. The utilization voltage is the voltage at the power-using devices, after some voltage drop. Motors are rated at utilization voltage. Therefore, a 115-v motor is used on a 120-v line, a 200-v motor is used on a 208-v system, a 230-v motor is used on a 240-v line (system), a 460-v motor is used on a 480-v line, and so on. All drawing notations of voltage should recognize this difference. This means that a transformer should be shown as 240/480 v (system voltage), a motor should be shown at 230 or 460 v, and so on. Showing a motor rated 480 v is incorrect, and this should be avoided.

Table 16.2 Current and Volt-Amperage Relationships

Load, v-a	Load Current					
	120-v, Single-Phase (I = VA/120)	120/240-v, 3-wire (I = VA/240)	120/208-v, Single-Phase (I = VA/208)	120/208-v, 3-Phase (I = VA/360)	277/480-v, 3-Phase (I = VA/830)	277-v, Single-Phase (I = VA/277)
100	0.83	0.41	—	—	—	0.362
200	1.6	0.8	—	—	—	0.72
500	4.2	2.1	—	—	—	1.8
1000	8.3	4.2	4.8	2.77	1.2	3.6
2000	16.6	8.3	9.6	5.5	2.4	7.2
5000	41.7	20.8	24.0	13.9	6.0	18.0
10,000	83.2	41.6	48.0	27.7	12.0	36.0
20,000	—	—	96.0	55.6	24.0	72.0
50,000	—	—	240.0	139.0	60.0	181.0
100,000	—	—	480.0	277.0	120.0	362.0

Table 16.3 System and Utilization Voltages

System Voltage (Transformers)[a]		Utilization Voltage (Motors)[a]	
Nominal	With 4% Drop[b]	New Standard[c]	Old Standard
120	115.2	115	110
208[c]	199.7	200	208
240[c]	230.4	230	220
480	460.8	460	440
600	576.0	575	550

[a] When specifying transformers, use system voltages; for motors, use utilization voltage.

[b] Note that utilization voltage corresponds to a 4% drop from system voltage, which is well within the normal motor tolerance.

[c] Motors for 208-v systems are rated 200 v. Motors for 240-v systems are rated 230 v. They cannot be used interchangeably without seriously affecting motor performance.

16.3 Emergency Electrical Service

In buildings that require electrical service when normal power fails, an *emergency service* is frequently installed. There are many ways to supply emergency power. If all the loads will operate on d-c as well as a-c, and the total load is very light, a battery emergency source can be used. For small a-c only loads, a battery and inverter (d-c to a-c) are used. For large a-c power requirements, a generator is usually furnished. Transfer to the emergency source can be done either manually or automatically by using either a manual or automatic transfer switch, respectively. This latter item senses voltage loss and automatically transfers to the emergency source.

The entire subject of emergency and standby service and equipment is highly complex. It is not normally the responsibility of the electrical technologist, and for that reason is not covered here. You can find a comprehensive discussion in Stein and Reynolds (see Supplementary Reading at the chapter's end). For information only, one of several possible arrangements of normal and emergency electrical service is shown in Figure 16.10.

16.4 Main Electrical Service Equipment

The NEC (Article 230, Part F) requires some means by which the incoming electrical service can be completely disconnected. This disconnecting arrangement can consist of up to six switches or circuit breakers, mounted at the point where the electrical service feeders enter the building. The service disconnect(s) can be combined with the metering in a separate enclosure (Figure 15.31, page 903), mounted in a switchboard or panelboard (Figures 16.11 and 16.12) or mounted entirely separate from all other equipment. When we refer to the service switch, we mean all the service disconnects, whether it is one or six.

The building switchboard, or main panelboard, controls and protects the feeders running through the building. Many types of switchboards are available. Depending on the type of devices in them, their sizes vary greatly. The engineer on the job will normally select this equipment, and the technologist will assist in layout. Some typical building switchboard dimensions are given in Figure 16.13, to give you a "feel" for size. Accurate data are available in manufacturers' catalogs.

16.5 Electrical Power Distribution

The term electrical power distribution (system) generally refers to the network of busduct, conductors, protective devices and enclosures that carries electrical power from the building service to the final distribution point. This final point before the actual power utilization equipment is usually a local electrical panelboard, motor controller or circuit switch. Such a system is shown pictorially in Figure 12.17, page 685. Beyond these last distribution points, the system is normally referred to as "branch circuitry," which we have studied in detail in previous chapters.

A few definitions will be useful here. In a multistory building, the service equipment is normally in the basement (or at grade), in a separate room called the switchboard room. The feeders rise from the equipment in this room to feed the panels on each floor. For this reason, the diagram showing this is called a *riser diagram*. A system on one level, such as in Figure 16.10, is also called a riser diagram for want of a better term. A typical riser diagram is shown in Figure 16.14. Note the panel

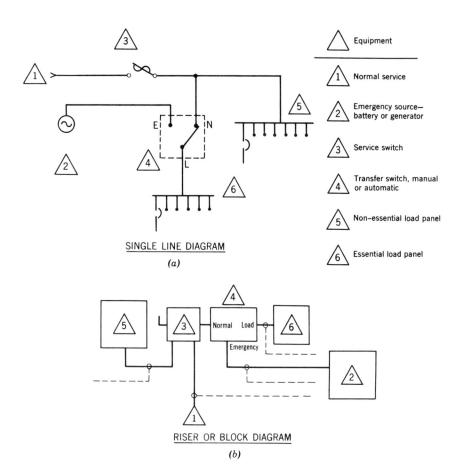

Figure 16.10 Service equipment diagrams showing one possible arrangement of normal and emergency electrical service. (a) Single-line diagram. (b) Riser or block diagram.

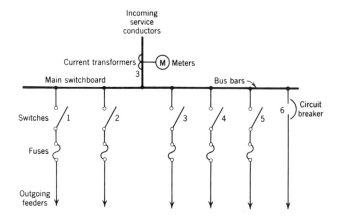

Figure 16.11 Typical switchboard. Switches are normally shown in the open position. The NEC allows up to six switches in parallel as service entrance equipment. Switches must be on the line (supply) side of fuses. Metering is normally placed on the service conductors, and the metering equipment is built into the main switchboard. Each line in a single-line diagram represents a three-phase circuit. If circuit breakers were used, the entire board would consist of units such as illustrated in circuit 6.

Figure 16.12 *(a)* Free-standing low voltage switchboard with individually compartmented main and branch circuit breakers. Identical construction is used for fusible devices. The main device can be a fixed or drawout circuit breaker or fused switch rated up to 4000 amp. Branch devices can be 1- to 3-pole circuit breakers or fusible switches up to 1200 amp. The entire structure has a maximum rating of 6000 amp. See Figure 16.13 for layout and dimensional data.
(b) Circuit breaker type panelboard, using bolt-on and plug-in circuit breakers 1- to 3-pole, 15–100 amp, and main device up to 400 amp. Panel is rated for 240-v a-c maximum. *(c)* Panel similar to that of *(b)* except that it is suitable for 277/480-v a-c service. *(d)* Circuit breaker panel with a maximum rating of 400 amp and 240 v a-c. This panel accommodates bolt-on circuit breakers only. *(e)* Circuit breaker panelboard with maximum rating of 1200 amp and 600 v a-c. Branch circuit breakers can be 15–100 amp, single-pole and 15–1200 amp, 2- to 3-pole. *(f)* Switch and fuse panelboard, similar in ratings and construction to the circuit breaker unit illustrated in *(e)*. (Photos courtesy of Cutler-Hammer.)

934 / NONRESIDENTIAL ELECTRICAL WORK

BE SURE TO INCLUDE
CONDUIT SPACE ON
SECTION DRAWING AND
BILL OF MATERIAL—

1. Customer metering located here for main switches, approximately 15" lower for main breakers.
2. Zero sequence ground fault protection and test panel located here for main FDPW and main breaker, located in switch cover for BPS mains.

Note: 100% rated breakers have the same dimensions as standard rated breakers.

3" TYPICAL → CONDUIT SPACE

2.5" TYPICAL

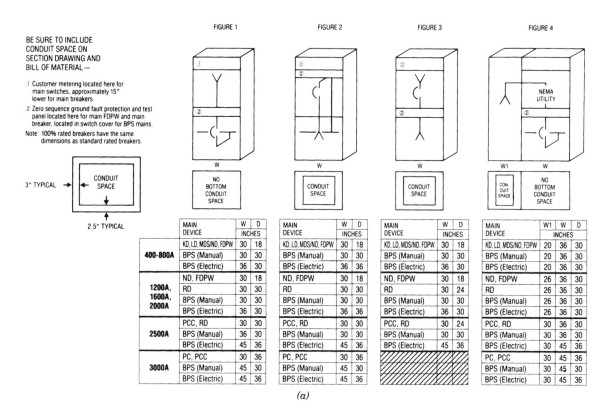

(a)

DEVICE "X" SPACE REQUIREMENT

CIRCUIT BREAKER DISTRIBUTION SIZING

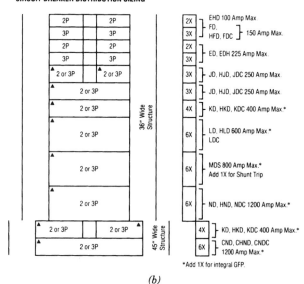

(b)

FUSIBLE SWITCH DISTRIBUTION SIZING

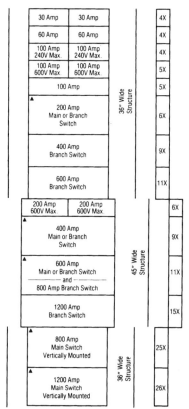

▲ Device may be used as chassis mounted main.
If ground fault is required, use an individually mounted main.
Line side of main device must be cable connected.

(c)

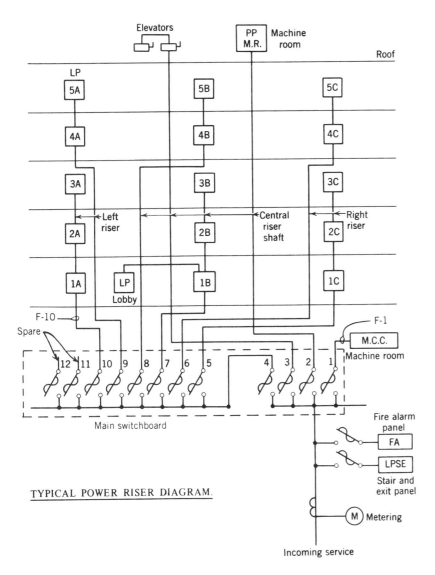

Figure 16.14 Typical electrical riser diagram for a five story building with elevator and machine room penthouse. Ordinarily, the main switchboard (enclosed in dotted lines) would be shown as a large rectangle with feeders F-1 through F-10 coming out of the top and feeding the panels, as shown. The switchboard contents (switch and fuse details) would appear in a schedule, normally on the same sheet as the riser diagram. However, because of this switchboard's unusual split bus arrangement (switches 1–4 are the service switches and switch 4 subfeeds switches 5–12), the switchboard is shown as it would be on a single-line diagram. This switchboard uses fused switches. Note that a fused switch can be shown in either of two ways graphically; as shown here with the fuse drawn on the switchblade or as in Figure 16.11 with the fuse shown separately. Both methods are in common use. Although this switchboard uses fused switches, in actual practice circuit breakers are more frequently used. (The decision is an engineering one, usually based on short circuit capacity considerations.)

Figure 16.13 Typical layout and dimensional data for low voltage distribution switchboards. [See Figure 16.12(a).] (a) Dimensional data for four configurations of service entrance and main device. (b) Dimensional and layout data for a circuit breaker distribution section of switchboard. (c) Dimensional and layout data for a fusible switch section of switchboard. (Extracted and reprinted, with permission, from published Cutler-Hammer data.)

designations. Each riser shaft is given an identifying letter: A, B or C. Panels are then identified by floor and shaft. Panel 4A is fourth floor, shaft A. This particular building has metering that is separate from the switchboard. The panels feeding fire alarm and stair and exit lighting are connected ahead of the four service switches (1–4). This is common practice, so that these panels will remain energized even if power in the building is turned off. (Power is normally turned off in case of a fire, to prevent electrical faults and injury to firefighters.)

We mentioned the term *busduct*. This item of equipment is illustrated in Figure 16.15. It consists of heavy bars of copper or aluminum that are insulated and assembled in a metal enclosure. A busduct is used where it is necessary to carry heavy current. The alternative is to use parallel sets of cables. For instance, to carry 3000 amp requires eight sets of 500 kcmil (MCM) type THW cables in parallel (without consideration of ampacity derating if more than three phase cables are placed in a single conduit). On the other hand, a single copper busduct measuring less than 6 in. × 20 in. (depending on the particular design) can carry the same amount of current. See Figure 16.16.

Certain types of busducts are made with plug-in points, to allow power to be picked off easily. The illustrated unit in Figure 16.17 is of this type. It will carry about 1000 amp. The plug-in points are like giant receptacles at which up to 200 amp can be picked off. Figure 16.18 shows how this type of plug-in busduct is applied. Also shown in the same figure is the feeder duct, which is constructed without plug-in points since it is intended to act as a feeder—that is, to feed power from one point to another without any tap-offs along its length. The figure is not intended to represent an actual installation or current equipment design. Instead, it illustrates typical applications of feeder busduct and plug-in busduct.

Figure 16.16 A sectional view of a two-section compact design busduct. The eight sets of cables shown have the same current-carrying capacity as the busduct. This clearly demonstrates the space savings possible when using busduct. (Reproduced by permission of Square D Company.)

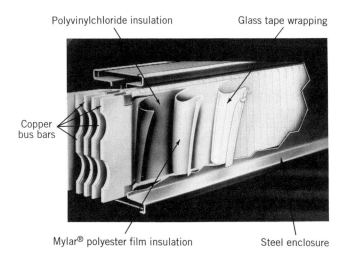

Figure 16.15 Cutaway view showing construction of a typical feeder busduct. This design is highly compact and rigid, which gives desirable electrical characteristics as well as the advantage of small size. (Photo courtesy of Siemens Energy and Automation, Inc.)

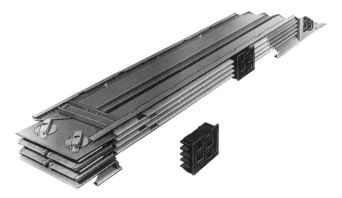

Figure 16.17 Construction of one type of plug-in busduct. Plug-ins are spaced every 12 in. on alternate sides to facilitate connection of plug-in breakers, switches, transformers or cable taps. Notice that bars are insulated over their entire length and are clamped rigidly at plug-ins with spacer blocks of insulating material. Housing is of sheet steel with openings for ventilation. The cover plate is not shown. (Courtesy of Square D Company.)

Now refer again to the riser diagram of Figure 16.14. We stated that the individual risers go up in shafts. These riser shafts are vertically stacked spaces specifically designed for electrical conduits, cables and possibly busducts. On each floor, the shaft usually enters a small closet. See Figure 12.17, page 685. In this closet, a tap is made on the risers, to feed the floor panel. The floor panel, in turn, feeds the branch circuits on each floor. Logically, this closet is called an *electric closet*. Two typical electric closets with their equipment are illustrated in Figure 16.19. Figure 16.20 shows a busduct run through a closet and the metering installation that is tapped off it. See also Figure 15.32.

In a multifloor building, it is very desirable that closets be stacked one above the other. If this is not possible, an offset must be made. Such offsets are made in pull boxes and may require splicing the riser cables. Such splices are best avoided. Offsets in busduct can be made easily with special right-angle fittings. See Figure 16.18. One-story buildings also use electric closets to contain the electric panels. In large buildings or where the local utility companies so require, separate closets are installed for electrical equipment and for telephone equipment. The technologist is responsible for laying out the equipment in the closets, to scale. He or she must make sure that everything fits and that adequate clearances are maintained for safety and for maintenance. The **NEC** specifies minimum clearances and working spaces in Section 110-16. In some buildings, a decision is made to install panels in furred-out spaces in walls rather than in electric closets, or even to have panels surface mounted on walls. The factors leading to these decisions are not usually the responsibility of the technologist.

Manufacturers publish detailed data on the physical dimensions of their equipment. This permits the layout person to size all the electrical spaces accurately. Most manufacturers use a modular sizing code by which each element in an assembly is dimensioned modularly. Refer to Figure 16.21, for instance. This extract from a Cutler–Hammer catalog shows how to arrive at the size of a panelboard once the contents are known. Each item—the branch breakers, the main lugs or main breaker and the neutral space—is dimensioned in modular X units. When these are totaled, the box can be sized. If additional gutter space is required, that too can be provided in the height or width. To help the technologist "get a feel" for the dimensions of the equipment that he or she will deal with day in and day out, we advise that a novice technologist study a current manufacturers' catalog. This will help to understand how the electrical and physical characteristics are presented. Furthermore, because not only dimensions, but also the way in which the data are presented varies among manufacturers, it is a good idea to study the publications of several major manufacturers of electrical equipment.

We mentioned before that riser offsets require the use of a pull box. The purpose of a pull box is, as the name indicates, to provide a pulling point. When a raceway run is long or has many angles, pull boxes enable the pulls to be made easily The exact location and need for a pull box is decided after an examination of the run. The rules covering size, construction and installation of these boxes are found in the **NEC** Section 370-28; Pull and Junction Boxes. Figure 16.22*(a)* shows a typical installation requiring pull boxes. The size of the pull box can be arrived at in the following way.

Draw a sketch of the box as in Figure 16.22*(b)*. The Code gives two methods for calculating the minimum pull box size. Both methods should be tried, and the larger size should be used.

Method 1. For a right-angle-turn box as illustrated, the minimum width of the box is equal to six times the trade diameter of the largest conduit, plus the trade diameters of the other entering conduits. Thus, the minimum box size in Figure 16.22 is

six times the largest conduit	$6 \times 4 =$	24 in.
plus the other conduits	$1 \times 4 =$	4 in.
	$2 \times 3 =$	6 in.
	$2 \times 2 =$	4 in.
width (length) of box		38 in.

Method 2. This method states that the minimum distance between raceways containing the same conductor is six trade diameters. Assuming that the arrangement stays the same entering and leaving, the 2-in. conduits should be 12 in. apart. The triangle in the corner [see Figure 16.22*(b)*], therefore, is an isosceles triangle with a 12-in. hypotenuse. Both sides are, therefore, $8^1/_2$-in. To this we add the minimum conduit spacings obtained from Table 12.7, page 691, as follows:

Corner distance	$8^1/_2$ in.
4 in. to 4 in.	$5^7/_8$ in.
4 in. to 3 in.	$5^1/_4$ in.
3 in. to 3 in.	$4^9/_{16}$ in.
3 in. to 2 in.	$3^3/_4$ in.
2 in. to 2 in.	$3^1/_8$ in.
Total	$31^1/_{16}$ in.

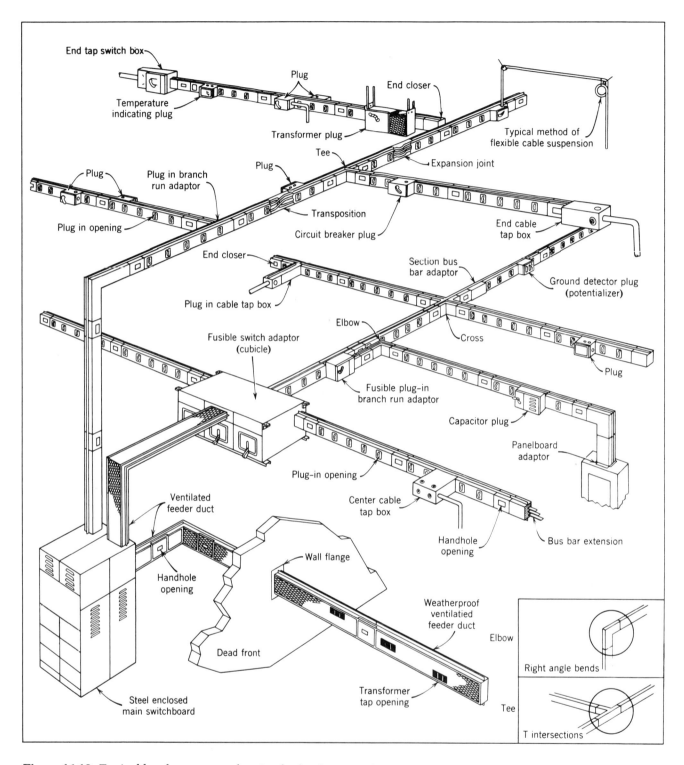

Figure 16.18 Typical busduct system showing feeder duct, weatherproof duct, plug-in duct and various types of plug-in devices. (This drawing is illustrative only, since some of the individual items have been redesigned.) (Drawing reproduced by courtesy of The General Electric Company.)

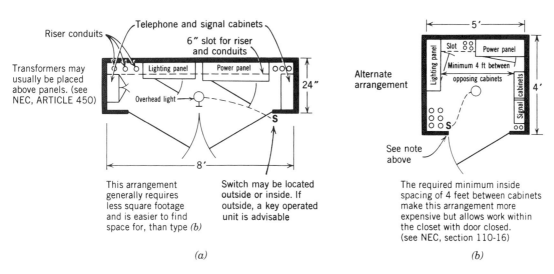

Figure 16.19 Typical electric closets with usual equipment. If required by amount of equipment or by utility companies or by codes, separate closets may be used for low voltage systems conduits and cabinets. These include telephone, data, signal and alarm systems.

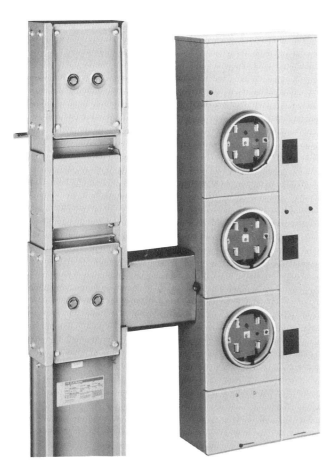

Figure 16.20 Typical submetering installation. A bank of sockets for kilowatt-hour meters is shown tapped directly to a feeder busduct. A fused switch is usually interposed between the meter bank and the busduct tap-off. Such assemblies are placed in electric closets and meter rooms, with access restricted to authorized personnel. See **NEC** Article 230F. (Courtesy of Siemens Energy and Automation, Inc.)

PANELBOARDS
Lighting and Distribution

PANEL LAYOUT AND DIMENSIONS, Continued

Pow-R-Line 4B
Breaker (PRL4B) Type Distribution Panelboards 600V Ac, 250V Dc

Panel Layout and Dimensions
To determine the dimensions of a given panelboard enclosure, make a layout sketch by fitting together the main branch and lug modules according to the appropriate tables in the layout guide. Assign "X" units to each module as shown and obtain a total "X" number.

The height of the enclosure is related to the total "X" units in the layout as shown in Table 4. Three standard box heights are available to accommodate any and all layout arrangements. "X" unit totals that do not exactly match those in Table 4 must be rounded off to the next higher standard (26X, 38X, 50X).

If a calculated "X" total for a main lug panel exceeds 50X, the panel must be split into two or more separate sections with "X" space for through-feed lugs figured in for all but one section. If a neutral is required, a separate neutral bar and appropriate "X" space must be included in each section.

Layout Example
1 – PRL4B panelboard, 480Y/277V, 3Ph, 4W, 65kA, 800 amps, main lug, consisting of:
12 – 20A/1P HFD
2 – 250A/3P HJD
1 – 400A/3P HKD

20A/1P	20A/1P	1X
20A/1P	20A/1P	1X
20A/1P	20A/1P	1X
20A/1P	20A/1P	1X
20A/1P	20A/1P	1X
20A/1P	20A/1P	1X
250A/3P		3X
250A/3P		3X
400A/3P		4X
Main Lugs	800A	10X
Neutral		
		26X

1. From layout guide, total "X" height of panel = 26X, (which is a design standard and no rounding off is necessary).
2. From Table 4, enclosure height for 26X panel = 57 in.
3. Width = 24 in. – directly from layout guide.
4. Total enclosure depth = 11.30 in. – standard for all PRL4 panelboards.

Table 4: Standard Panel and Box Dimensions❶

Panel Height	Box Height, in.	Box Width, in.	Box❷ Depth, in.	Box Catalog Number
26X	57	24❸	10.40	BX2457
38X	73.5	24❸	10.40	BX2473
50X	90	24❸	10.40	BX2490
38X	73.5	36	10.40	BX3673
50X	90	36	10.40	BX3690
38X	73.5	44	10.40	BX4473
50X	90	44	10.40	BX4490

Top and Bottom Gutters (minimum) .... 10.625 in.

Side Gutters (minimum)
24 in. wide box .... 5 in.
36 in. wide box .... 6 in.
44 in. wide box .... 8 in.

Table 5: Layout for Branch and Horizontally Mounted Main Devices. See opposite page for MLO or Neutral and Vertically Mounted Mains Space Requirements

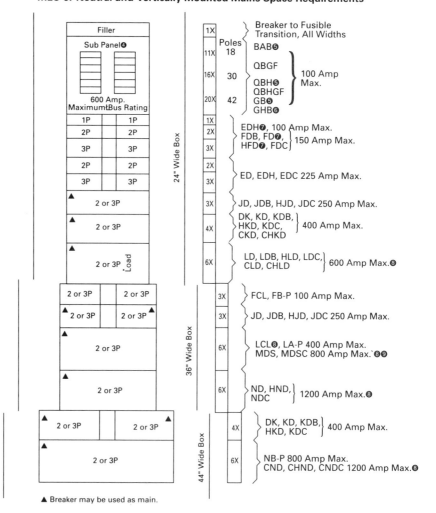

▲ Breaker may be used as main.

❶ Flush trims available on PRL4B panels with Door-in-Door enclosure only.
❷ Box depth is 10.40 in., cover adds .90 in. for overall enclosure depth of 11.30 in.
❸ 800 amperes maximum bus size in 24 in. wide box.
❹ The sum of branch breaker ampere ratings mounted opposite one another cannot exceed 140A.

❺ BAB and QBH breakers with shunt trips require one additional pole space, ie; 1 pole is 2 pole size, 2 pole is 3 pole size and 3 pole size is 4 pole size – see Modification 29.
❻ GB, GHB breakers cannot be mixed on same sub-chassis as BAB, QBH.
❼ When only one single pole breaker of the group is required on either side of chassis, the single pole breaker space required changes from 1X to 2X.

❽ Horizontally mounted breaker with integral ground fault require 1 "X" of additional space.
❾ For vertically mounted M-Frame main breaker in 24 in. wide box, refer to page EE-22.

MOTORS AND MOTOR CONTROL / 941

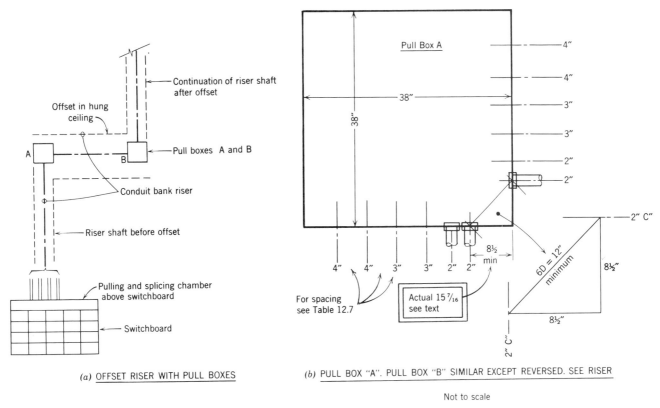

(a) OFFSET RISER WITH PULL BOXES

(b) PULL BOX "A". PULL BOX "B" SIMILAR EXCEPT REVERSED. SEE RISER

Not to scale

Figure 16.22 Pull boxes sizing (see text discussion). Pull boxes must be adequately sized to prevent crowding the cables inside.

Since this is smaller than the figure obtained by Method 1, we use the larger figure of 38 in. The sketch of Figure 16.22 is then revised, increasing the corner distance by the difference between the two methods.

Corner distance = (38 in. − 31 1/16 in.) + 8 1/2 in. = 15 7/16

We now have the actual box layout, which is then shown on the drawings.

16.6 Motors and Motor Control

An extremely important part of electric power work is the layout of electric motors and their control. This subject is complex and is covered in great detail in the NEC. The technologist is responsible for showing the equipment on the working drawings and doing a limited amount of design. He or she must, therefore, know the sizes and functions of the items normally used.

a. Equipment

The chart and list of Figure 16.23 appear in the NEC. Their purpose is to assist the user to find what he or she is after in Article 430 of the NEC, which covers motors and controllers. Using Figure 16.23, we will review each item briefly so that the technologist will be familiar with the terms used in motor circuits.

Figure 16.21 Typical catalog page from an equipment catalog showing the method by which this manufacturer sizes a specific type of electrical panelboard. These data are intended for illustration purposes only. For actual design refer to current manufacturer's data. (Reproduced with permission of Cutler-Hammer.)

Part A. General, Sections 430-1 through 430-18
Part B. Motor Circuit Conductors, Sections 430-21 through 430-29.
Part C. Motor and Branch-Circuit Overload Protection, Sections 430-31 through 430-44.
Part D. Motor Branch-Circuit Short-Circuit and Ground-Fault Protection, Sections 430-51 through 430-58
Part E. Motor Feeder Short-Circuit and Ground-Fault Protection, Sections 430-61 through 430-63
Part F. Motor Control Circuits, Sections 430-71 through 430-74.
Part G. Motor Controllers, Sections 430-81 through 430-91.
Part H. Motor Control Centers, Sections 430-92 Through 430-98.
Part I. Disconnecting Means, Sections 430-101 through 430-113.
Part J. Over 600 Volts, Nominal, Sections 430-121 through 430-127.
Part K. Protection of Live Parts—All Voltages, Sections 430-131 through 430-133.
Part L. Grounding—All Voltages, Sections 430-141 through 430-145.
Part M. Tables, Tables 430-147 through 430-152.

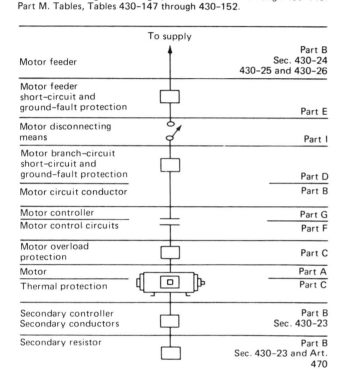

Figure 16.23 Table and chart of the NEC Sections in Article 430 relating to motors, motor circuits and controllers. The purpose of this chart is to assist the designer and technologist in locating the required information in this very large and complex Code article. (Reprinted with permission from NFPA 70-1996, the *National Electrical Code*, Copyright © 1995, National Fire Protection Association, Quincy, MA 02269. This reprinted material is not the complete and official position of the National Fire Protection Association, on the referenced subject which is represented only by the standard in its entirety.)

(1) *Motor Feeder.* These conductors supply the motor(s) with electrical power. The size of these conductors depends on the size, type and number of motors being supplied, plus any additional load on the same feeder. These conductors must carry all the design load without overheating.

(2) *Motor Feeder Short Circuit and Ground Fault Protection.* These devices protect the feeder described in (1) against short circuit and ground faults. The diagram of Figure 16.23 is somewhat misleading. As with all electrical protective equipment, this too is upstream (electrically) from the feeder it protects.

(3) *Motor Disconnect Means.* This device disconnects the motor and its controller. Its purpose is safety. It is frequently part of the controller assembly, although not necessarily. There is usually a disconnect for each motor, although a single disconnect can be used for a group of related motors, as for instance multiple motors in a single machine.

(4) *Motor Branch Circuit-short Circuit and Ground Fault Protection.* This can be either
  (a) the protective device in the panelboard in the case of a separate branch circuit for the motor or
  (b) a separate circuit breaker or fused switch (plus GFCI where required) where the branch circuit also serves other devices. This circuit breaker or fused switch can be part of a combination controller.

(5) *Motor Circuit Conductors.* These are the conductors between the branch circuit protection and the controller and between the controller and the motor. They must carry the motor current without overheating. They are protected by the branch circuit protective device described in item (4).

(6) *Motor Controller.* Also called the motor starter, its purpose is to safely connect the motor to the source of electrical power. It can take many forms: a simple switch, a single or multistep magnetic connector, a device that controls motor speed continuously, and other types.

(7) *Motor Control Circuits.* These auxiliary devices activate the controller, plus the necessary wiring. In this category are push buttons; manual control switches; devices activated by heat, pressure, liquid level, light and elapsed time; programmable devices; and so on. They can be full voltage or low voltage, local or remote.

(8) *Motor Overload and Motor Thermal Protection.* This is running load protection that protects the motor against overheating due to overload. These devices take many forms such as thermal cutouts in the controller and/or in the motor, overload relays or fuses. Motor thermal protection is a type of motor overload protection.

(9) *Secondary Controller, Conductors and Resistor.* These items apply only to wound rotor motors and are not normally the concern of an electrical technologist.

b. Design Procedure

The preceding equipment description followed the NEC chart, but it does not represent the design procedure normally followed. The usual design steps follow.

Step 1. For a given motor, select the required motor circuit conductors (item 5).

Step 2. For these conductors, select the branch circuit protection (item 4).

Step 3. Select the required motor controller, including motor overload protection (items 6 and 8).

Step 4. Design the motor control circuit(s) (item 7).

Step 5. Select and locate the motor disconnect means (item 3).

If the motor is fed from an individual branch circuit, then the design is complete. If, however, the motor is only part of a branch circuit, the remaining design steps are:

Step 6. Repeat steps 1–5 for any additional motors on the same circuit.

Step 7. Select the feeder required for all the circuit loads (step 1).

Step 8. Select the feeder short circuit and ground fault protection for all the combined loads (step 2).

This design work is not normally the responsibility of an electrical technologist. If interested refer to Examples 8 in Chapter 9 of the 1996 NEC for typical calculations. An important item to remember is that if any of the motors involved are parts of air conditioning or refrigeration equipment then the design data are different from ordinary motors, and NEC Article 440 applies.

c. How to Show Motor Electrical Work

When doing motor work, there are four ways of showing the work, each of which has its own special purpose.

(1) The architectural-electrical (working) drawing shows the position of all the equipment on the

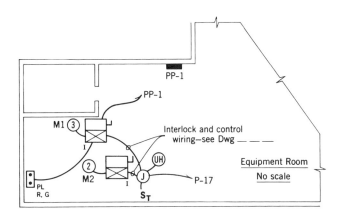

Figure 16.24 Wiring of motors M1 and M2 with interconnections, as drawn on typical electrical working drawings (architectural-electrical drawings). Details of wiring connections are shown in Figure 16.27.

floor plan. See Figure 16.24. The symbols used on these electrical working drawings are shown in Figure 16.25. This list is Part VIII of the master symbol list.

(2) The equipment detail (see Figure 16.26) gives the contractor the construction details he or she needs, to build correctly. It shows actual detailed construction fittings—the nuts and bolts of the job. Details are required when anything special or complex is required. Showing a detail will save much time on the job and will avoid arguments.

(3) A motor wiring diagram as in Figure 16.27 *(a)* shows only the motor and the devices controlling it. Motor feeder and motor feeder overcurrent protection (items 1 and 2 in Section 16.6.a) are not shown. Everything is shown with full wiring, and all terminals are numbered. Power wiring is drawn heavy; control wiring is drawn light. Symbols for this type of diagram are given in Figure 16.28, which also makes up Part IX of the master symbol list. This type of diagram is used in the field to do actual wiring.

(4) The control or *ladder diagram* shows graphically the control wiring only. See Figure 16.27*(b)*. This diagram corresponds to the wiring diagram, except that power wiring is not shown. Terminals are numbered. This diagram, called a ladder because it looks like one, is used by the engineer to develop the control scheme. It is shown on the drawing because it is simple to read. The ladder diagram does not

SYMBOLS – PART VIII – MOTORS AND MOTOR CONTROL

 ⊠ I	MOTOR CONTROLLER, 3 POLE ACROSS–THE–LINE (ATL) UON, NEMA SIZE I, SEE SCHED., DWG. _ _ _
 ⊠ II	COMBINATION TYPE MOTOR CONTROLLER; ATL STARTER PLUS FUSED DISCONNECT SWITCH, NEMA SIZE II, SEE SCHEDULE DWG. _ _ _
 ⊠ I / CB	COMBINATION TYPE MOTOR CONTROLLER; ATL STARTER PLUS CIRCUIT BREAKER, NEMA SIZE I, SEE SCHEDULE DWG. _ _ _
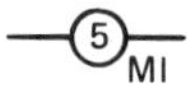 —(5)— M1	MOTOR #1, 5 HP. 3ϕ SQUIRREL CAGE UON
 Ⓣ OR ⊤	DEVICE 'T', SEE LIST OF ABBREVIATIONS, SYMBOLS Part IX
S_T	MANUAL MOTOR CONTROLLER WITH THERMAL ELEMENT.
⊡⊡	PUSH BUTTON STATION – MOMENTARY CONTACT
▪▪	PUSH BUTTON STATION – MAINTAINED CONTACT
 △, ⊠, ①	SYMBOL INDICATING LOCATION OF AN EQUIPMENT ITEM

ABBREVIATIONS

ATL	ACROSS THE LINE STARTER – MAGNETIC
CATL	COMBINATION ACROSS–THE–LINE–MAGNETIC STARTER
FS	FUSED SWITCH
C/B	CIRCUIT BREAKER
RV	REDUCED VOLTAGE
FV	FULL VOLTAGE
SR	STARTER RACK
MCC	MOTOR CONTROL CENTER
S	START BUTTON – MOMENTARY CONTACT
ST	STOP BUTTON – MOMENTARY CONTACT
S/S PB	START–STOP PUSH BUTTON
PL	PILOT LIGHT; COLOR INDICATED BY LETTER; A – AMBER, G – GREEN, B – BLUE R – RED Y – YELLOW
MER	MECHANICAL EQUIP. ROOM
NO	NORMALLY OPEN
NC	NORMALLY CLOSED
LO	LOCKOUT
R	RELAY
UV	UNDERVOLTAGE
OC	OVERCURRENT
REV	REVERSING

Figure 16.25 Architectural-electric plan symbols, Part VIII, Motors and Motor Control. The technologist/designer is free to add other symbols as required, particularly in view of the wide use of logic controls. For other part of the symbol list, see

Part I, Raceways, Figure 12.6, page 671.
Part II, Outlets, Figure 12.50, page 719.
Part III, Wiring Devices, Figure 12.51, page 720.
Part IV, Abbreviations, Figure 12.58, page 726.
Part V, Single-line Diagrams, Figure 13.4, page 733.
Part VI, Equipment, Figure 13.15, page 753.
Part VII, Signalling Devices, Figure 15.34, page 909.
Part IX, Control and Wiring Diagrams, Figure 16.28, page 947.

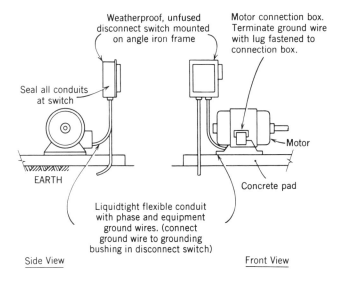

Figure 16.26 Construction details such as the one illustrated give the contractor the information necessary to perform the work in the manner desired. Information not shown on this detail for the sake of clarity, but required on an actual construction detail, includes mechanical details such as method of fastening motor to the pad, including vibration isolations, hardware details, details of the angle iron frame including finish and height of the switch. Also required are electrical details including bonding conductors (see **NEC** Section 250-75), type and size of switch and details of special conduit fittings.

require tracing out of wires as does the wiring diagram. Symbols for it are also shown in Figure 16.28.

On-the-job practice will give the technologist experience with all these forms of information presentation.

d. Motor Controllers

A *motor controller* is a device that, as the name states, controls the action of the motor. This always includes the basic start-stop function but can also include other control functions. Among these, the most common are speed control and reversing. Special functions include jogging, plugging (rapid reversing), sequence speed control (speed control in a series of manually controlled steps), full speed control and acceleration control (manual control of timed acceleration steps between speeds). Although selection of motor controllers is rarely the responsibility of an electrical technologist, he or she must know what the different types are, in order to be able to place and connect them properly. We will, therefore, describe briefly the most important types.

(1) *Full Voltage Manual Controller.* This device is known simply as a starter, since that is its only function. In its simplest form, it is no more than a standard snap switch, which is used for starting small motors such as kitchen and bathroom exhaust fans. For slightly larger fractional horsepower motors, a thermal overload element is added to the switch, and the device is known as a manual motor-starting switch. The smallest of these fits into a standard wall switch box. It is frequently provided with a pilot light that indicates that the motor is running. Manual motor-starting switches can be used with motors as large as 20 HP (horsepower). Two such manual starting switches are illustrated in Figure 16.29.

The important thing to remember is that the switch contacts are operated manually, by pushing a button. That means that remote control of starting is not possible, and, therefore, the starting control is at one location only. Small manual starting switches have built-in overload protection only. Larger units can be equipped with undervoltage protection (the motor trips out when voltage drops below a preset level and must be restarted manually), shunt tripping (stopping, also known as tripping, from a remote location) and auxiliary contacts that can be used for remote running indication or interlocking with another controller.

(2) *Full Voltage Magnetic Controller.* This starter has the same action as the manual starter described previously except that the contacts are closed magnetically. It is, therefore, no different from a magnetic relay or a magnetic contactor, except for design details and the provision of thermal motor overload protection elements. It is commonly called an across-the-line (ATL) starter because it places the motor directly across the line with no intermediate steps. Because the contacts are operated magnetically, a control circuit is required. This permits control to be local or remote, and by push button or by a whole variety of control and pilot devices. See Section 16.6.e.

Further, the magnetic operation simplifies

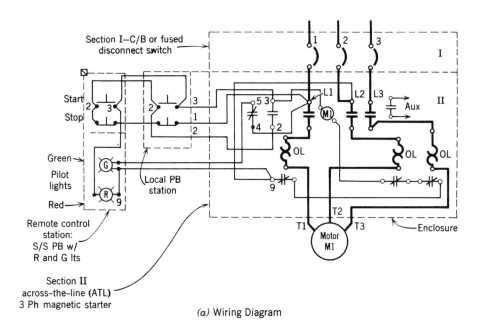

(a) Wiring Diagram

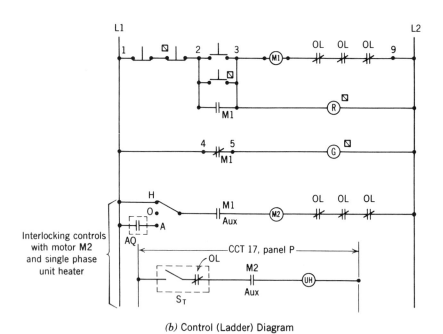

(b) Control (Ladder) Diagram

Figure 16.27 Motor control diagrams. (a) Wiring diagram showing equipment enclosures and actual connections. Shown are a combination circuit-breaker-type across-the-line magnetic starter, three-phase, with integral start-stop momentary contact push-button station. Also shown are a remote start-stop push button with red and green pilot lights. The actual "remote" location of this push-button station is shown on the corresponding floor plan, Figure 16.24.
(b) A control or ladder diagram. The upper section shows the same equipment as in (a). Note that terminals are numbered, for ease in reading the diagram and tracing the circuit. The lower portion of the ladder diagram shows interlocking (interconnection) with motors M2 and unit heater UH. Note that UH is a single-phase motor, connected to circuit 17, panel P, via motor control switch S_T, and is interlocked with M2. Refer to Figure 16.28 for symbols and Figure 16.24 for the same equipment shown on an architectural-electrical plan (electrical working drawing).

SYMBOLS; PART IX
CONTROL DIAGRAMS & WIRING DIAGRAMS

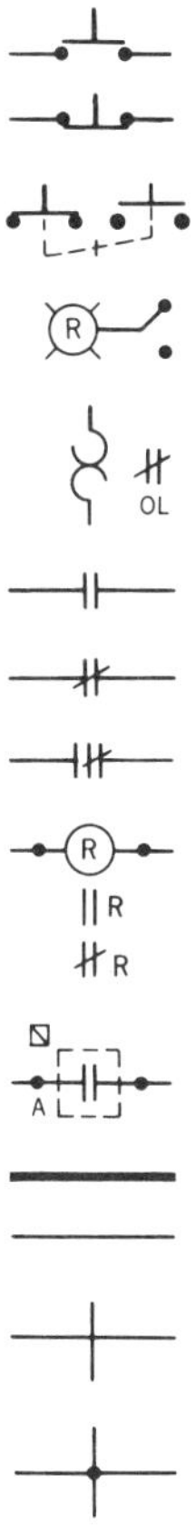

- MOMENTARY CONTACT PUSH BUTTON – NO – (START)
- MOMENTARY CONTACT PUSH BUTTON – NC –(STOP)
- MAINTAINED CONTACT START–STOP PUSH BUTTON ONE N.C. AND ONE N.O. CONTACT.
- PILOT LIGHT, R–RED, G–GREEN, Y–YELLOW (SWITCH INDICATES PUSH–TO–TEST).
- THERMAL OL ELEMENT WITH NC OL CONTACT
- NORMALLY OPEN CONTACT – NO
- NORMALLY CLOSED CONTACT – NC
- DOUBLE ACTION CONTACT; ONE NO AND ONE NC
- OPERATING COIL FOR RELAY OR OTHER MAGNETIC CONTROL DEVICE. WITH ONE NO AND ONE NC CONTACT. LETTERS NORMALLY USED ARE C AND R FOR CONTROL COIL AND RELAY.
- PILOT CONTROL DEVICE TYPE A, SEE LIST OF ABBREVIATIONS. ⊠ INDICATES REMOTE LOCATION.
- POWER WIRING
- CONTROL WIRING
- WIRES CROSSING
- WIRES CONNECTED

LIST OF ABBREVIATIONS

T	THERMOSTAT	MOM	MOMENTARY CONTACT
H	HUMIDISTAT	EP	ELECTRO–PNEUMATIC
SD	SMOKE DETECTOR	PE	PNEUMATIC–ELECTRIC
A,AQ	AQUASTAT	BG	BREAK–GLASS
R	RELAY	F, FL	FLOAT SWITCH
M	MOTOR	PS	PRESSURE SWITCH
MD	MOTORIZED DAMPER	HOA	HAND–OFF–AUTOMATIC SWITCH
PB	PUSH–BUTTON	LS, HS	LOW SPEED, HIGH SPEED
OL	OVERLOAD		

Figure 16.28 Architectural-electrical plan (electrical working drawing) symbols, Part IX. Control Diagram and Wiring Diagrams. For other of the symbol list, see
  Part I, Raceways, Figure 12.6, page 671.
  Part II, Outlets, Figure 12.50, page 719.
  Part III, Wiring Devices, Figure 12.51, page 720.
  Part IV, Abbreviations, Figure 12.58, page 726.
  Part V, Single-line Diagrams, Figure 13.4, page 733.
  Part VI, Equipment, Figure 13.15, page 753.
  Part VII, Signalling Devices, Figure 15.34, page 909.
  Part VIII, Motors and Motor Control, Figure 16.25, page 944.

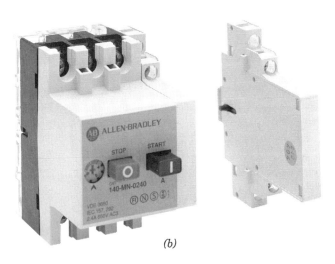

Figure 16.29 Manual motor starters. *(a)* This motor starting switch is provided with thermal overload protection only. It can be used to start and stop single-phase motors up to 5 hp and three-phase motors up to 10 hp. Approximate dimensions are 5 in. W × 7 in. H × 4 in. D. The illustrated unit has a NEMA 12 enclosure. See Table 16.5 for enclosure descriptions. *(b)* This manual controller can be furnished with undervoltage trip, shunt (remote) trip and auxiliary contacts in addition to the standard overload elements. It can control single-phase motors up to 3 hp and three-phase units up to 20 hp. The starter is shown without an enclosure. It measures approximately 3 in. W × 5 in. H × 4 in. D. The device to the right of the starter is an accessory that provides auxiliary contacts. (Photos courtesy of Allen-Bradley Company.)

the function of undervoltage and shunt tripping, interlocking, indication and annunciation. These starters are frequently combined with a switch or circuit breaker in a single enclosure. The switch and circuit breaker provide the required motor disconnecting means (if in sight of the motor) and branch circuit protection (if the switch is fused). See Figure 16.23 and Sections 16.6.a(3) and 16.6.a(4). Such combined units are called combination across-the-line (CATL) starters. See also the first three symbols in the motor control symbol list in Figure 16.25.

The required electrical capacity of a magnetic starter depends on the horsepower of the motor controlled and the line voltage. The National Electrical Manufacturers Association has standardized these sizes. In ascending order the sizes are 00, 0, 1, 2, 3, 4, 5, 6, 7, 8, 9. The most common sizes are 0 through 5. Table 16.4 gives typical physical and electrical data on such starters. A combination fused switch ATL starter is shown in Figure 16.30.

(3) *Special Controllers.* Across-the-line starters are the most common type of motor controller. However, to provide such special functions as speed control and reversing, special controllers are required. You will find technical and physical data on most of these types in manufacturers' literature and in some of the reference books listed in the bibliography at the end of this chapter.

e. Control and Pilot Devices

The simplest control and pilot devices are manual contacts such as push buttons and rotary switches. Automatic contact devices include thermostats, float and pressure switches, timer contacts, interlock contacts and other contact-making devices activated by process equipment. Indicating-type pilot devices include lights, alarms, annunciators and the like. An important schedule that is prepared during the course of electrical design of large facilities with many motors is a motor schedule. A typical schedule is shown in Figure 16.31. This schedule is used to keep track of all the motors on a large job and all the required control elements. It, therefore, serves an important function since, without such a schedule, control devices and interconnections can easily be overlooked.

In recent years electronic logic controls, both fixed and programmable, have become extremely common in the area of motor control. This subject

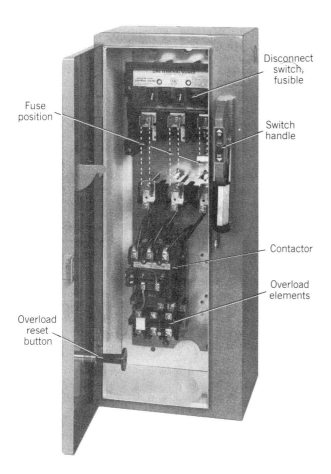

Figure 16.30 Interior of a combination fused-switch type, across-the-line motor controller. Note that the unit is actually a switch and a starter wired together and installed in a single cabinet. (Courtesy of Allen-Bradley Company.)

MOTOR SCHEDULE															
MOTOR								CONTROLLER		CONTROL DEVICES				INDIC. DEVICES	
(1)	(2)	(3)	(4)	(5)	(6)	(7)	(8)	(9)	(10)	(11)	(12)	(13)	(14)	(15)	(16)
										MANUAL		AUTOMATIC			
NO.	USE	LOCATION	MIN. H.P.	P.H.	V.	SPEED	REMARKS	LOCATION	REMARKS	TYPE	LOCATION	TYPE	LOCATION	TYPE	LOCATION
3/4"	2"	1 1/4"	1/2"	1/2"	1/2"	3/4"	3"	1 1/4"	4"	1"	1 1/2"	1"	1 1/2"	1"	1 1/2"

1/2" · 1/2" · 1/2" · 1" · AS REQ'D · 1/4"

NOTES:

COL. 3 LETTERS & NUMBERS INDICATE FLOOR & NEAREST COLUMN LINE.

COL. 6 LETTERS & NUMBERS INDICATE NEAREST COLUMN LINE. M.C.C. #1 DENOTES STARTER IN MOTOR CONTROL CENTER #1.

COL. 11, 13, 15 NOTATIONS ARE AS SHOWN IN SYMBOLS LIST ABBREVATIONS, FIGS. 16.25 AND 16.28

DIMENSIONS SHOWN ARE TYPICAL FOR AN ACTUAL, FULL SIZE FORM.

Figure 16.31 Typical motor and control schedule. A schedule such as this is essential in large design jobs for keeping track of motors, control and indicating devices and their locations.

Table 16.4 Rating and Approximate Dimensions of a-c Full-Voltage Conventional Single-Speed Motor Controllers, Three-Phase Combination Circuit Breaker Type[a]

NEMA Size Designation	Maximum Horsepower[b]	Width, in.	Height, in.	Depth, in.
0	3	10	24	7
1	7½	10	24	7
2	15	10	24	7
3	30	20	24	9
4	50	20	48	9
5	100	20	56	11

[a] All starters are housed in a NEMA 1 ventilated enclosure. See Table 16.5.

[b] Maximum horsepower that can be controlled at 208–230 v. Generally, when operating at 460 v, a starter one size smaller can be used.

is too specialized to be discussed here. As with other specialized and highly technical subjects, you can find additional material in the bibliography at the end of Chapter 12 and in any good electrical library.

f. Equipment Enclosures

In addition to the required electrical characteristics for a motor controller, its enclosure must be suitable for its intended use. NEMA has standardized equipment enclosures by the type of protection provided by each. A brief listing of types and description appears in Table 16.5. A typical NEMA type 3R rainproof (weatherproof) circuit breaker enclosure is shown in Figure 16.32. Details of the enclosures and the tests applied to determine the type of protection provided by each are given in NEMA Standard 250.

Table 16.5 Control Equipment Enclosures

NEMA Designation Type	Description	Application[a]
1	General-purpose	Dry; indoor use
2	Drip-proof	Indoor; subject to dripping
3	Dusttight, raintight, and sleet-resistant	Indoor/outdoor, where subject to windblown dust and water
3R	Rainproof and sleet-resistant	Outdoor; subject to falling rain, snow, and sleet
3S	Dusttight, raintight, and sleet-proof	Outdoor; subject to windblown water, dust, and sleet; most severe exterior duty
4	Watertight and dusttight	Indoor/outdoor; subject to water from all directions; not sleet-proof
4X	Watertight and dusttight	Similar to type 4 except that enclosure is nonmetallic corrosion resistant, applied in chemical plants
6P	Submersible	Indoor/outdoor; submersed at limited depth
7–9	Hazardous	Differing in application by class and group of hazardous use; see NEC
12	Industrial use, dusttight and driptight	Indoor only, general use; industrial and other "dirty" environments
13	Dust, spraytight	Indoor use; subject to dust and sprayed water, oil and noncorrosive liquids

[a] All enclosures provide a degree of protection as defined by NEMA in its publication 250.

Raintight NEMA 3R circuit breaker enclosure.

Figure 16.32 Type 3R outdoor enclosure. This is the type usually intended when weatherproof is specified. (Courtesy of General Electric Company.)

16.7 Motor Control Centers

In spaces using a large number of motors, it is often good design to group the motor controllers in one location, instead of placing one next to each motor. It saves space, money and wiring and gives a central control point. Such an assembly is called a *motor control center*, often abbreviated MCC. Typical units are illustrated in Figure 16.33. Each manufacturer publishes physical data on its units. These data allow the technologist to assist in sizing and arranging the components that will make up a motor control center.

In addition to motor controllers, an MCC can also contain switches, circuit breakers and even panelboards. Manufacturers' catalogs show the space required for each piece of equipment. The electrical designer prepares an MCC schedule during the course of the electrical design. One such typical MCC schedule is shown in Figure 16.34. It lists not only the components of the MCC, but also the load served and other relevant data. This schedule will appear on the electrical working drawings. It provides the electrical construction contractor with the data necessary to perform the wiring between the MCC and its controlled motors. This same schedule is used by the technologist in preparing a preliminary physical layout of the MCC in order to determine space and clearance requirements. It is also used to check manufacturers' shop drawings.

16.8 Guidelines for Layout and Circuitry

Sections 15.2 and 15.6 and Sections 13.9 to 13.13 are devoted to a detailed discussion of the layout of rooms in residences. For nonresidential buildings, the number of different types of spaces is so great that a discussion of that kind is not practical. Instead, we include the following helpful suggestions.

An important decision that is made at the outset of layout work is whether to place all the electrical work on a single drawing, that is, lighting, power (convenience and special receptacles) and signal/data/control equipment. Except for a very simple job, this is usually not advisable due to the resulting clutter on the drawings, making them difficult to read. The usual division is either light and power on one drawing and low voltage on another, or lighting alone on one drawing and power and low voltage on another. Only in very complex jobs are three separate drawings required.

a. Schools

Since schools contain many different types of room use, including those of classroom, lab, shop, office, gym and assembly rooms, it is difficult to discuss each without getting into excessive detail and an extremely lengthy discussion. Furthermore, the traditional classroom, which is no more than a seating area with a blackboard, is rapidly disappearing, even at the grade school level. It is being replaced by the electronic classroom in which each student or group of students works at an electronic console (computer terminal) for at least part of the time. Thus the classroom has become a multiple-use space, and the electrical design must be sufficiently flexible to provide the required services.

The same electronic and computer "revolution" has affected laboratory, assembly and office spaces, but to a lesser degree. Shops and gym space are essentially unaffected. As a result, we will review guidelines that should help an electrical technologist assist in the layout of the lighting and of the power outlets required. We emphasize *assist*, since the electrical design of all but the smallest educational building is usually the responsibility of an electrical design engineer.

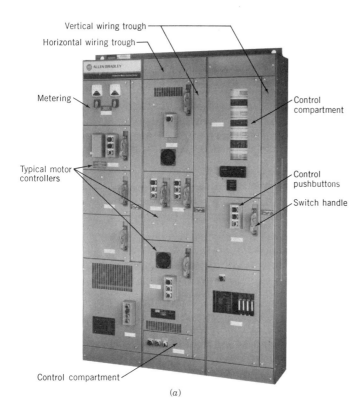

Figure 16.33 *(a)* Modern motor control center containing conventional motor controllers, metering and special control compartments. Standard MCC construction is 90 in. high and 15–16 in. deep, although special controllers may require a 20-in. deep unit. Solid-state controllers can be installed in this MCC along with conventional units.
(b-1) Motor control center suitable for motor starter sizes NEMA 1–6 with heaterless overloads and protection against phase loss, and single phase and ground faults. Also uses solid-state controllers. Dimensions of each section are 90 in. high and 20 in. wide. Depth is 16 or 20 in. for front mounting only (depending on maximum starter size) and 21 in. deep for back-to-back mounting. This unit is also arranged to provide power system communication for data collection, monitoring, remote control and troubleshooting. *(b-2)* Motor control center that can accommodate electromechanical motor starters through NEMA 8, using heater-type overload relays. Similar in dimensions to the MCC of *(b-1)*. *(c)* Back-to-back construction saves space by adding only 5 in. to the front-only-mounting depth. [Photos Courtesy of *(a)* Allen-Bradley Company, *(b)* Cutler-Hammer, *(c)* Furnas.]

(1) *Lighting.* It is not possible to lay out the lighting properly without a detailed architectural plan because of the electronic facilities that require specialized treatment. We can state the following abbreviated guidelines:

- Federal energy regulations plus state and local codes govern the energy efficiency of the lighting and the types of lamp that may (and may not) be used. These considerations are basic to all lighting design.
- Fluorescent lighting in classrooms should be either direct-indirect, semi-indirect or en-

Figure 16.33 *(Continued)*

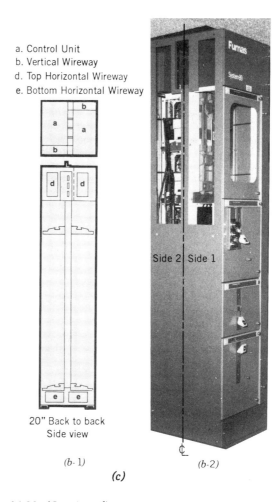

Figure 16.33 *(Continued)*

tirely indirect. HID lighting, because of source luminance (brightness), is usually semi-indirect or indirect.
- Means must be provided to change lighting levels by dimming or switching and to provide different levels in different parts of rooms that are being used for different activities.
- Automatic dimming and/or switching should be provided in areas with sufficient daylight.
- Special lighting must be provided for blackboards, demonstration tables and display areas.
- Where color rendering is important, sources with particularly high color rendering indices must be used. In all cases, minimum CRI is governed by federal standards.

For a detailed analysis of lighting in educational facilities, refer to Sections 21.7 through 21.16 of Stein and Reynolds (see the bibliography at the end of Chapter 12).

(2) *Device Layout.* Here, too, detailed architectural plans are required. As we did for lighting, we can state some useful guidelines:
- Provide sufficient classroom receptacle outlets to handle anticipated electronic teaching equipment, projection television and student-operated electrical equipment. Where such loads are numerous and portable electrical equipment is used, the use of multi-outlet assemblies should be considered.
- Provide special outlets for all fixed electrical equipment in shops, labs, kitchens and the like.
- Use heavy duty devices, key-operated switches where the switches are exposed to students and plastic instead of glass in fixtures. Also use vandalproof equipment wherever possible. All panels must be locked and should be in locked closets.

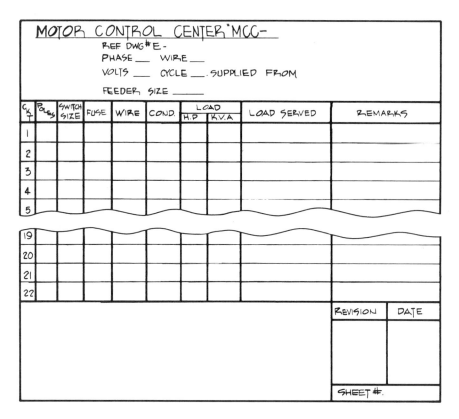

Figure 16.34 Typical design form for a motor control center. This form (or one similar) is filled out as design progresses. It is used in conjunction with a motor schedule of the type shown in Figure 16.31. When complete, this schedule is placed on the electrical working drawings and is also used to determine the physical size of a MCC.

- Keep lighting and receptacles completely separate when circuiting.

b. Office Spaces

By office space we mean generally large open spaces with many desks. The smaller private office is special and must be treated individually. General suggestions for layout and circuitry follow.

(1) *Lighting.* The same energy considerations as stated previously for schools apply to offices as well. Lighting is either fluorescent, HID or a combination of both. Design depends heavily on type of ceiling, type of office work to be performed and budget. For a detailed analysis of office area lighting see Stein and Reynolds, Sections 21.17 to 21.23.

(2) *Device Layout.* Provide at least two duplex convenience outlets at every desk location that will be equipped with electronic data processing equipment. Outlets can be brought to the desk by a service pole (see Figure 12.38), undercarpet wiring (see Figure 12.44) or some type of underfloor raceway system (see Figures 12.33 and 12.34). On-the-floor surface raceways should not be used in new construction. These receptacles do not carry heavy loads (a typical computer terminal in a network will normally take no more than 300 v-a). However, these loads are continuous, since terminals are almost never turned off during working hours. Therefore, they should be wired at no more than six receptacles to a 20-amp circuit, less if auxiliary equipment is standard at each desk.

Fifteen-amp circuits are not advisable in business offices. (A recently developed computer design reduces the power demand of "idling" computers to less than 20% of full load. If computer terminals of this type are anticipated, a diversity factor can be applied to the total computer load, depending on the number of stations involved.) All devices used in a business office must be specification grade, that is, of the highest quality.

c. Industrial Spaces

Industrial spaces are so specialized that no meaningful guidelines can be given. The technologist will receive from the engineers the material he or she needs to do the required layout. Familiarity with the motor, panel, feeder and transformer schedules found in the chapter discussions will be a considerable help.

16.9 Classroom; Electrical Facilities

Our previous discussion and detailed analysis has concentrated on residential buildings. We have done this, as explained, because these buildings contain many different design and layout problems in a single, relatively small area. At this point, using some of the guidelines we have learned, we will lay out the lighting, receptacles and necessary special facilities in part of a school building. We chose a classroom because you are somewhat familiar with them from the preceding discussion and, from the illustrative Examples 14.3 and 14.4. Once the concepts are clear, layout of other types of areas will be easier.

As noted in the previous section, one of the first decisions to be made is whether to use a single drawing for all electrical work or to separate lighting, power and signal work. In addition to lighting and power outlets, we have decided to show only the low voltage work that is associated with teaching. This eliminates all the special devices such as bells, gongs, intercom speakers and closed-circuit TV outlets. As a result, we will place all the electrical work on a single drawing (Figure 16.36). In an actual job, an additional drawing would probably be used to show all the low voltage systems' outlets, devices, raceways and some wiring. Most of the wiring for low voltage systems is done by specialty contractors and not by the electrical construction contractor. The electrical working drawings, therefore, generally show empty raceways—for wiring "by others." For this reason, the conduit between the floor box below the teacher's computer table and the raceway behind the student computers is shown empty. The wiring will be installed by others, that is, by the subcontractor installing the computer equipment.

Refer to Figure 16.35. This is a stripped architectural plan of a single classroom in a school building. (A stripped architectural plan is one from which all architectural material has been removed, leaving only the walls, windows, doors and necessary dimensions. The electrical work can be put directly onto such a plan. Alternatively, a stripped architectural plan can be traced from a full architectural plan. Of course, when using a CAD program to produce the plans, these considerations are not relevant.) Construction is concrete slab and painted masonry block walls. This construction is typical of school buildings.

Room dimensions are 19 ft 8 in. wide by 26 ft 3 in. long and 12 ft ceiling height. (This corresponds to 6 m by 8 m by 3½ high.) The stripped architectural plan shows the location of all items necessary to the electrical designer—blackboard, computer cubicles (partitions between computers are half-height and are, therefore, shown dashed), teacher's desk, files, bookcase and student lockers. The problem given the technologist-designer is to lay out the lighting and electrical outlets to serve the teacher and the students as well as possible. This is typical of the type of work that an electrical technologist is called on to do, and obviously should be capable of doing it well. We will discuss the design procedure in detail as we proceed, as we did with the residential buildings. Where several solutions to the same design problem exist, we will discuss the advantages and disadvantages of each before deciding on one of the alternatives.

The usual procedure in electrical layout work is to design the lighting first. Factors that must be taken into account in the lighting design include:

- The use of computers in the room requires extreme care to avoid reflection of light sources in the visual display terminals (VDT).
- Indirect lighting, which is ideal for spaces with VDT units, will be difficult to design to meet energy guidelines—both federal and local.
- Maintenance in schools is usually poor-to-fair because of budget problems. This means that the annual or biennial painting required to maintain the high reflectances of upper walls and ceilings needed for indirect lighting will not be done.
- Schools are not usually air conditioned, which means a rapid darkening of light color surfaces due to aging and dirt accumulation.
- Adequate lighting must be provided for the

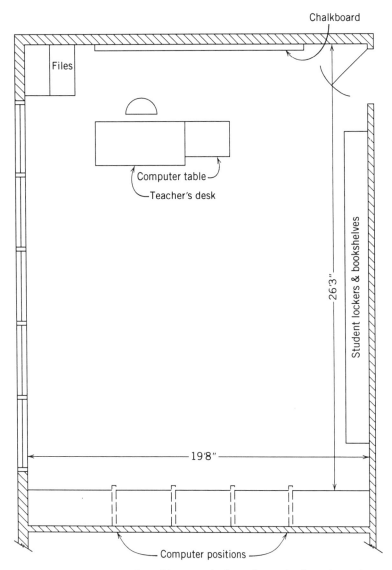

Figure 16.35 Stripped architectural plan of a typical modern classroom. This type of drawing is the starting point for an electrical layout. Dimensions and other items that may interfere with showing the electrical work are moved outside of the plan. See Figure 16.36.

special tasks in the room, including the use of the bookshelves and file cabinets. Care must be taken to position lights to prevent shadowing.
- Since students spend a lot of time in the head-up position facing the teacher, care must be taken to control direct glare from luminaires.
- Reflected glare, which is always a problem, must be minimal and controllable.
- Light sources (lamps) must meet the efficacy and color-rendering requirements of federal and local standards.
- The overall lighting design must also meet the requirements of energy codes.
- Enough flexibility must be built into the lighting design (and controls) to permit several different activities to take place at the same time and to compensate for natural lighting (daylight).
- Costs must be considered in all decisions, since school budgets are almost always tight.

Many of the decisions that are required by these considerations will not be made by the electrical technologist because they require knowledge that he or she usually does not possess. They will be made by the engineer, lighting designer or job

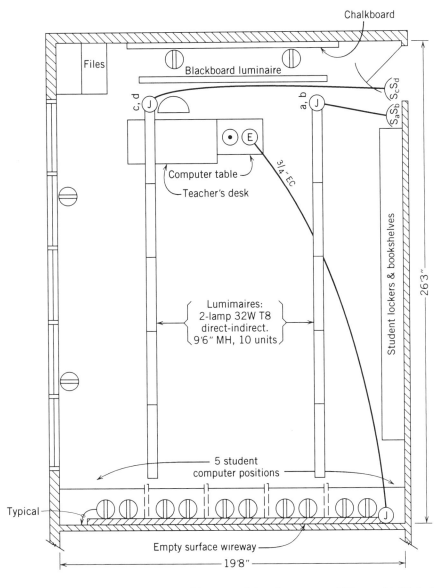

Figure 16.36 Complete electrical layout of the classroom of Figure 16.35, before circuitry. All lighting, receptacle outlets, and switching is shown. Raceway facilities for the wiring of teaching equipment such as students' and teacher's computers are also shown. Signal equipment not directly connected with teaching, such as intercom, bells, fire alarm, closed-circuit TV and the like are normally shown on a separate drawing.

architect, and the decision will be given to the technologist so that the layout can be made. The factors were listed previously to show that what was once a relatively simple, straightforward task is today much more complex. With experience, however, the competent technologist will be able to make many of the needed decisions without assistance.

We will now proceed with the actual lighting design. Refer to Figure 16.36. Two types of lamps will readily satisfy the previously stated requirements: fluorescent and metal halide. (The color-rendering properties of metal halide are satisfactory for ordinary classroom use.) We will work out a design using 32-w, 48 in., T-8 triphosphor fluorescent lamps. [See Table 14.7(a).] The efficacy of these lamps—95 lm/w without consideration of ballast losses—is sufficiently high to meet all modern energy codes. We quote the efficacy of the lamp alone, since ballast loss depends on the type of ballast used, and we will not discuss that item because it is a specialty on its own. There, too, federal and local codes and statutes must be met. The color rendering index of these lamps is 85, which is satisfactory for all but the most specialized type of school activity.

The luminaire chosen is a parabolic baffle unit that has direct-indirect distribution. (See Section 14.26.e and Figure 14.56.) It has very high efficiency and a 60% up–40% down distribution. See Figure 16.37 for its photometric characteristics, an application photo, a table of coefficients of utilization and a rapid design chart. The luminaire will be stem (pendant) mounted at 9 ft 6 in. AFF. The luminaires will be run in continuous rows from front to back of the room at right angles to the blackboard. This orientation reduces reflected glare on the student desk work. It also, very importantly, practically eliminates reflection of the luminaires in the VDT (computer monitors) at the back of the room. Any remaining reflection can be eliminated by adjusting the monitor angle. This luminaire selection and mounting arrangement meets the requirements set forth previously, as follows:

- Luminaire reflection in the VDT unit is minimal.
- The advantage of indirect lighting is obtained by the large (60%) upward light component without sacrificing efficiency.
- The luminaire is open at the top and bottom. It is, therefore, to a large extent, self cleaning.
- Light distribution in the room will be adequate to provide illumination for the files, bookcases and student lockers. (Supplementary blackboard lighting will be provided.)
- Direct glare will be very low because of the high luminaire mounting (9 ft 6 in.), low luminaire luminance, the direction (front-to-back) of luminaire mounting, and the low luminance (brightness) ratio between the luminaire and its background (the ceiling) due to the large upward light component.
- The overall design will meet energy codes.
- Each row of luminaires will be dual-switched to provide half light and full light. This will be accomplished by wiring one lamp in two adjoining fixtures to a two lamp ballast. This will provide manually controlled means for daylight compensation for the area near the windows, and overall illuminance level control.

We can now proceed to the required illuminance calculation. The coefficient of utilization (CU) will be determined by the approximate zonal cavity method, explained in Section 14.31. The actual illuminance calculation will use the lumen method, discussed in Section 14.30. (A rapid review of both of these sections may be helpful to you at this point.) We will follow the step-by-step procedure detailed in Section 14.31 and illustrated clearly in Figure 14.62 to determine the luminaire CU. You should have a copy of that figure in front of you as we proceed.

Step 1. Fill in the data required.

The recommended illuminance level for a classroom is 50 fc (500 lux). Each luminaire contains two lamps. Each lamp produces 3050 initial lumens. Therefore, total lumens per luminaire is 6100.

Step 2. Determine equivalent square-room RCR using Equation (14.8).

$$W_{sq} = W + \frac{L-W}{3}$$

$$W_{sq} = 19 \text{ ft } 8 \text{ in.} + \frac{26 \text{ ft } 3 \text{ in.} - 19 \text{ ft } 3 \text{ in.}}{3} = 21.86 \text{ ft}$$

$$\text{Equivalent square-room RCR} = \frac{10(h_{rc})}{W_{sq}}$$

We need to calculate the room cavity height h_{rc}. Since the fixtures are mounted at 9 ft 6 in. and the working plane is at 2 ft 6 in., the room cavity height h_{rc} is the difference, that is,

$$h_{rc} = 9 \text{ ft } 6 \text{ in.} - 2 \text{ ft } 6 \text{ in.} = 7.0 \text{ ft}$$

Therefore,

$$RCR_{sq} = \frac{10(h_{rc})}{W_{sq}} = \frac{10(7)}{21.86} = 3.2$$

Step 3. Determine the equivalent ceiling cavity reflectance based on the equivalent square room size.

From the sketch in Figure 14.62 and the calculated $W_{sq} = 21.86$, we find that $\rho_{cc} = 0.70$.

Steps 4 and 5. These steps are self-explanatory.

Step 6. Determine CU.

From the CU table in Figure 16.37, using $\rho_{cc} = 0.7$, ρ_w (wall reflectance) = 0.5 (assumed) and $RCR_{sq} = 3.2$ as calculated above, we obtain by visual interpolation that CU = 0.57.

Step 7. Determine the light loss factor.

Assume average conditions. Therefore, LLF = 0.55.

Step 8. Calculate the number of luminaires required. Since we do not need the luminaires to

CLASSROOM; ELECTRICAL FACILITIES / 959

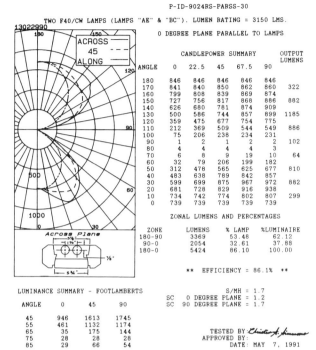

(b)

Figure 16.37 (a) Photo of a schoolroom installation using the luminaire selected. Note that the luminaire brightness is acceptable when viewed either from the side or against the illuminated ceiling. The long stem (2 ft 6 in.) shown, similar to that used in our example, gives a completely uniform ceiling luminance. (b) Typical testing laboratory photometric report for the lighting fixtures selected. (c) Table of luminaire coefficients of utilization. (Photo and data courtesy of LiteControl.)

COEFFICIENTS OF UTILIZATION

EFFECTIVE FLOOR CAVITY REFLECTANCE = .20

CC	80				70				50				30				10				0
WALL	70	50	30	10	70	50	30	10	50	30	10		50	30	10		50	30	10		0
RCR																					
0	.90	.90	.90	.90	.81	.81	.81	.81	.66	.66	.66		.52	.52	.52		.39	.39	.39		.33
1	.83	.80	.77	.74	.75	.72	.70	.67	.59	.58	.56		.47	.46	.45		.36	.35	.35		.29
2	.76	.71	.66	.62	.69	.64	.60	.57	.53	.50	.48		.42	.41	.39		.33	.32	.31		.26
3	.70	.63	.57	.53	.64	.58	.53	.49	.48	.44	.41		.38	.36	.34		.30	.28	.27		.23
4	.64	.56	.50	.45	.59	.51	.46	.42	.43	.39	.36		.35	.32	.30		.27	.25	.24		.21
5	.59	.49	.44	.39	.54	.46	.40	.36	.38	.34	.31		.31	.28	.26		.24	.23	.21		.18
6	.54	.45	.38	.34	.49	.41	.35	.31	.34	.30	.27		.28	.25	.23		.22	.20	.18		.16
7	.50	.40	.34	.29	.46	.37	.31	.27	.31	.27	.23		.25	.22	.20		.20	.18	.16		.14
8	.46	.36	.30	.25	.42	.33	.27	.24	.28	.23	.20		.23	.19	.17		.18	.16	.14		.12
9	.42	.32	.26	.22	.39	.30	.24	.20	.25	.21	.18		.20	.17	.15		.16	.14	.12		.10
10	.39	.29	.23	.19	.36	.27	.22	.18	.23	.18	.15		.18	.15	.13		.15	.12	.11		.09

(c)

extend to the blackboard nor do we want them over the computer area in the back of the room, it is sufficient to use two rows of luminaires, beginning 3 ft from the front of the room and ending before the back computer shelf. This reduces the effective room length to 20 ft. The number of luminaires required for this smaller area is

$$\text{Number of Luminaires} = \frac{50 \text{ fc} \times 19.67 \text{ ft} \times 20 \text{ ft}}{6100 \text{ lm/unit} \times 0.57 \times 0.55}$$
$$= 10.28$$

Using 10 luminaires, the average illuminance in the room is

$$\frac{10}{10.28} \times 50 \text{ fc} = 48.6 \text{ fc} = 525 \text{ lux}$$

Blackboard illumination is assisted by a ceiling-mounted single-lamp asymmetric fluorescent wall-wash-type blackboard fixture at the front of the room. Luminaire side-to-side spacing is based on the principle that the spacing between a row and the wall is half the row-to-row spacing. This gives a row-to-row spacing of almost 10 ft (9 ft 10 in.), and a spacing-to-mounting-height ratio of just over 1.0. This S/MH is well within the limitation of about 1.5 for this type of installation. Room illumination will, therefore, be very uniform.

With respect to convenience receptacles, we have furnished two duplex units at each computer position. This should be adequate for the computers and any accessories. Other receptacles have been spotted around the room, including one floor unit under the teacher's computer table. In addition, we have provided a connection between the teacher's computer and those of the students via an empty conduit connecting a floor box under the teacher's computer table to a wireway on the wall behind the student computer stations. This completes the layout of the classroom. As an exercise, you can work through a lighting solution using metal halide lamps and metric units.

16.10 Typical Commercial Building

To illustrate some of the topics we have already discussed, and to show actual contract drawings, we have chosen a combination office–light industry building. We do not show bare floor plans, since that material should be clear by this point. Instead, we will concentrate on the electrical power aspects of the building.

Refer to the light and power riser diagram in Figure 16.38. Follow the discussion with the drawing in front of you.

(a) The power riser shows the entire electrical system, from the utility through the panels, that is, everything except branch circuits. All elements are shown as boxes, and each is labelled or identified. If a repetitive item occurs such as the 200/90A, 3PSN switch, on the first floor, it is shown once and labelled "Typical." Pull boxes are identified by number such as PB-1, PB-2 and the like. A schedule gives the actual physical sizes. (The schedule is not shown here.)

(b) Notice how panels are identified. There are basically four types—lighting and power on normal service and lighting and power on emergency service. The code is simple:

Normal Service
L1-2   Lighting panel, first floor, panel no. 2
P2-3   Power panel, second floor, panel no. 3

Emergency Service
L3-EM   Lighting panel, third floor
LB-EM   Lighting panel, basement
PB-EM   Power panel, basement

Therefore, at a glance, we see that the third floor has two lighting panels in normal service (L3-1 and L3-2), one lighting panel on emergency power (L3-EM), and nine power panels on normal service (P3-1 to P3-9). Other items easily identified are the power panel feeding the elevator equipment P-EL on the roof level, and the roof level motor control center MCC-R.

In the basement we have:

Normal Service
Lighting panels   LB-1,2,3
Exterior lighting   LB-4,5
Power panels   PB-1,2,3
Kitchen power   PB-K

Emergency Service
Lighting panel   LB-EM
Power panel   PB-EM

In addition, the basement contains two motor control centers (MCC-B1, B2), other special control panels (LC-1, LC-2) and the emergency generator plus its accessories.

(c) The two main switchboards (MS-1 and MS-2) are shown with all their feeders. The switch-

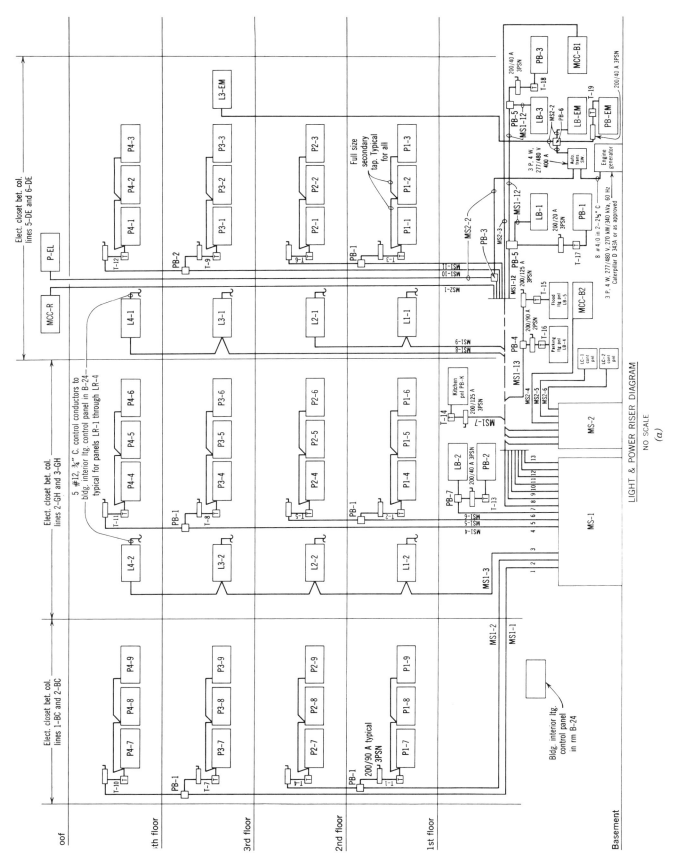

Figure 16.38 (a) Light and power riser diagram. (No scale.)

SWITCHBOARD MS – 1

Main Switchboard Design	Feeder No.	Description of Loads — Item Served	Load (kVA)	Circuit Breaker — Sym. Int. Cap. Min. (kiloamp)	Poles	Frame	Trip	Wiring — Quan	AWG	Ground	Cond. Size	Remarks
MS1 3φ, 4 W. 277/480 v. 4000 amp mains and neutral bus, 1350 amp Gnd. bus 4000/3500 amp Main c/b 100 kA I.C. min.	1	T–7 (RP3–7, 8, 9) T–10 (RP4–7, 8, 9)	150	100	3	600	200	4	3/0	6	2½ in.	
	2	T–1 (RP1–7, 8, 9) T–4 (RP2–7, 8, 9)	150	100	3	600	200	4	3/0	6	2½ in.	
	3	L1–2, L2–2, L3–2 L4–2	280	100	3	600	400	4	500 MCM	—	3½ in.	
	4	T–8 (RP3–4, 5, 6) T–11 (RP4–4, 5, 6)	150	100	3	600	200	4	3/0	6	2½ in.	
	5	T–2 (RP1–4, 5, 6) T–5 (RP2–4, 5, 6)	150	100	3	600	200	4	3/0	6	2½ in.	
	6	LPB–2, T–13 (RPB–B)	100	100	3	600	150	4	1/0	6	2½ in.	
	7	T–14 (P–KP)	112	100	3	600	150	4	1/0	6	2½ in.	
	8	L3–1, L4–1	250	100	3	600	300	4	350 MCM	—	3 in.	
	9	L1–1, L2–1	250	100	3	600	300	4	350 MCM	—	3 in.	
	10	T–9 (RP3–1, 2, 3) T–12 (RP4–1, 2, 3)	225	100	3	600	300	4	350 MCM	2	3 in.	

(b)

Figure 16.38 (b) Schedule of main distribution switchboard MS-1. Repetitive items such as the circuit breaker poles and frame size can be shown by note to save drawing space.

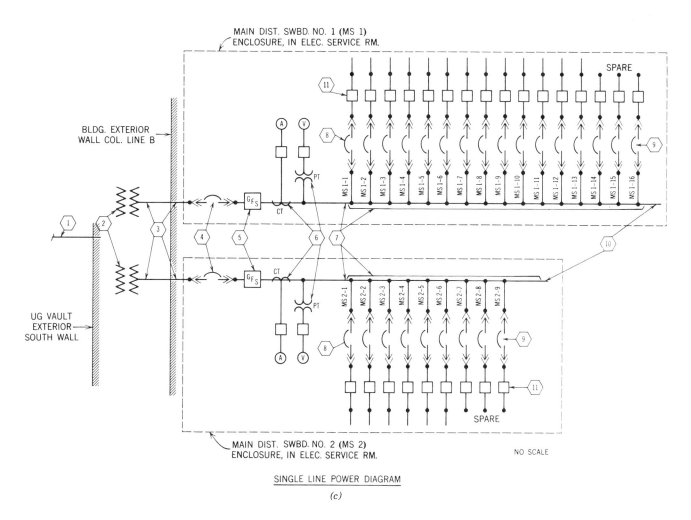

Figure 16.38 *(c)* Single-line power diagram. The hexagonal "bubbles" refer to note numbers in *(d)*. Use of triangles or hexagons to mark note numbers avoids confusion with equipment shown in circles, squares and rectangles.

(PARTIAL) LEGEND OF ELECTRICAL WORK AND EQUIPMENT REQUIREMENTS		
ITEM	DESCRIPTION	REMARKS
1	PRIMARY UG SERVICE; 2–5" (1 ACTIVE, 1 SPARE) FIBER OR ASBESTOS CEMENT CONDUITS IN 3" CONC ENVELOPE, (INCLUDING MANHOLE) PROVIDED UNDER ELEC SECTION OF WORK; HIGH VOLTAGE CONDUCTORS (INCLUDING CONNECTIONS) PROVIDED BY UTILITY CO.	
2	SERVICE TRANSFORMER WITH 3ϕ4W 277/480V WYE GNDED NEUTRAL SECONDARY PROVIDED BY UTILITY CO.	
3	SERVICE FEEDER 3ϕ4W 277/480V. WYE FULL SIZE GNDED NEUTRAL, EACH CONSISTING OF A 4000 AMP FULLY RATED LOW IMPEDANCE COPPER VENTILATED BUS DUCT ASSEMBLY OR 15–3 1/2" STEEL CONDUITS (11 ACTIVE, 4 SPARE) CONTAINING 11 SETS OF 4–500 MCM RHW INSUL. COPPER CONDUCTORS. CONDUCTORS AND CONNECTIONS PROVIDED UNDER ELEC. SECTION OF WORK.	
4	SERVICE MAIN CIRCUIT BREAKER 3P, 4000/3500A WITH SPECIFIED AUXILIARY EQUIPMENT.	SEE SCHEDULES FOR MAIN SWBDS MS1 AND MS2.
5	MAIN BUS GROUND FAULT SENSING EQUIPMENT	
6		

(d)

Figure 16.38 *(d)* Partial legend of the single-line diagram of *(c)*. This legend would be placed adjacent to the single-line diagram on the working drawings.

(PARTIAL) SCHEDULE OF DRY TYPE TRANSFORMERS							
XFMR NO.	KVA	PHASE	INPUT VOLTS	OUTPUT VOLTS	OUTPUT FEEDER	MTG. HT. BOTTOM AFF	REMARKS
T–1	75	3	480V Δ	120/208V Y	4 #4/0 + 1 #2 GND IN 2½"C	7'–6"	TAPS, PRIMARY WINDING, 4 OF 2–2½% ABOVE & 2–2½% BELOW RATED VOLTAGE.
T–2	75	3	480V Δ	120/208V Y	4 #4/0 + 1 #2 GND IN 2½"C	7'–6"	TAPS, PRIMARY WINDING, 4 OF 2–2½% ABOVE & 2–2½% BELOW RATED VOLTAGE.
T–3	112½	3	480V Δ	120/208V Y	4 #350 MCM + 1 #1 GND IN 3"C	7'–6"	TAPS, PRIMARY WINDING, 4 OF 2–2½% ABOVE & 2–2½% BELOW RATED VOLTAGE.
T–4	75	3	480V Δ	120/208V Y	4 #4/0 + 1 #2 GND IN 2½"C	7'–6"	TAPS, PRIMARY WINDING, 4 OF 2–2½% ABOVE & 2–2½% BELOW RATED VOLTAGE.
T–5	75	3	480V Δ	120/208V Y	4 #4/0 + 1 #2 GND IN 2½"C	7'–6"	TAPS, PRIMARY WINDING, 4 OF 2–2½% ABOVE & 2–2½% BELOW RATED VOLTAGE.
T–6	112½	3	480V Δ	120/208V Y	4 #350 MCM + 1 #1 GND IN 3"C	7'–6"	TAPS, PRIMARY WINDING, 4 OF 2–2½% ABOVE & 2–2½% BELOW RATED VOLTAGE.
T–9							

(e)

Figure 16.38 *(e)* Partial schedule of dry-type transformers. This method of data presentation is preferred for clarity, completeness and brevity. It is difficult to forget an item in a schedule. That is not the case for drawing notes or specifications.

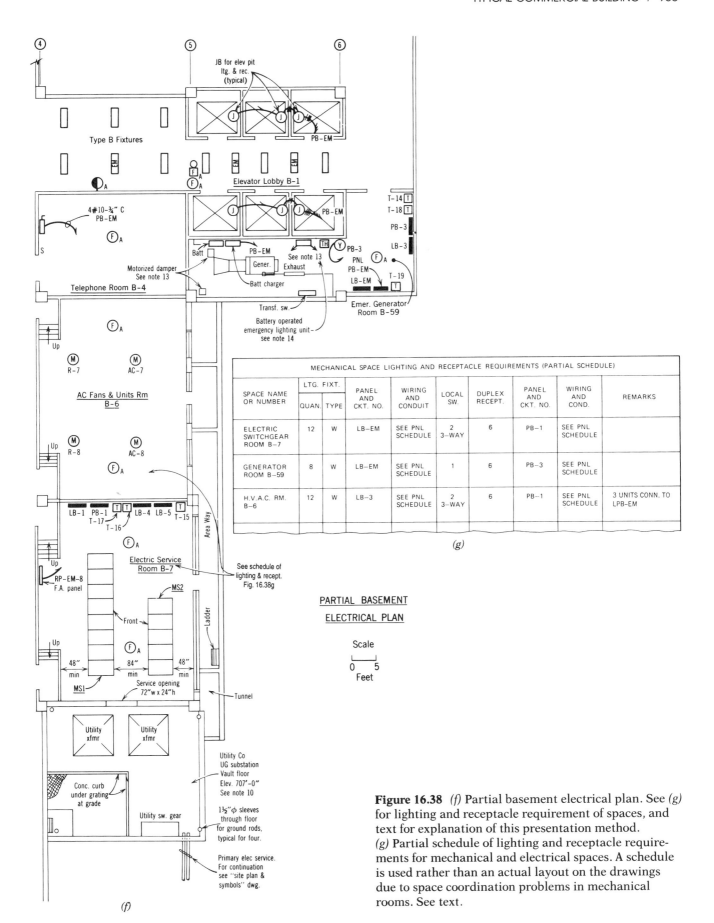

Figure 16.38 (f) Partial basement electrical plan. See (g) for lighting and receptacle requirement of spaces, and text for explanation of this presentation method.
(g) Partial schedule of lighting and receptacle requirements for mechanical and electrical spaces. A schedule is used rather than an actual layout on the drawings due to space coordination problems in mechanical rooms. See text.

board contents are detailed in separate schedules. See Figure 16.38(b), where part of the MS-1 schedule is reproduced. Notice that the schedule contains details on the switchboard mains, each individual circuit, what each circuit feeds and the wiring. This schedule makes it unnecessary to show feeder sizes on the riser. This keeps the riser from becoming cluttered.

(d) Part of the data shown in block form on the riser is also shown in the system single-line diagram and legend [see Figure 16.38(c) and (d)]. The one-line is shown complete to allow comparison with the riser. The notes in the legend are only given in part to show the technique of presentation.

An interesting fact appears in item 3 of the legend in Figure 16.38(d). The designer has given the contractor the choice of using a 4000-amp busduct or 11 sets of parallel 500 MCM cables. Compare this with our remarks about busducts in Section 16.5.

Notice that the single-line diagram goes only as far as the switchboards and does not include the panels. There is no reason to repeat data. On the contrary, it is bad practice to show the same information in two places since, when a change is made, one location can easily be overlooked. The single-line diagram shows the service entrance, the metering (not shown on the riser) and the electrical layout of the switchboards. The amount of information to be shown on the single-line diagram is up to the designer. It might have been helpful to show the emergency generator arrangement here, but the designer chose not to, relying on the riser for this information. The important thing is that the single-line diagram shows the arrangement at a glance. The riser is more difficult to follow and does not show the internal connections in its boxes.

(e) Referring back to the riser diagram, Figure 16.38(a). The utility company supply is 277/480 v. This accounts for the large number of transformers (T-1 to T-18) shown on the riser. They supply 120/208 v. A transformer schedule [Figure 16.38(e)] gives a partial listing of these transformers. The form of this schedule will be useful to the technologist as a guide in work.

(f) Another item of interest on the riser is the building interior lighting control panel shown as located in Room B-24. This panel centrally controls all the building's lighting for convenience and security. Duplicate local control is provided in some areas. The control wiring to permit this central control is shown (by note) running from each lighting panel (LP). The control is accomplished with large contactors inside each lighting panel that energize the entire lighting panel. The action is similar to the low voltage remote control discussed in Section 15.5, only bigger. These contactors are normally called *RC switches,* an abbreviation for remote control switches.

(g) A partial architectural-electrical floor plan of the basement is given in Figure 16.38(f). This now familiar type plan shows the basement equipment in its physical arrangement. Note especially the following:

1. The vault given to the utility company for its equipment is clearly shown. The work to be done by the building contractor is called out. All remaining work in the vault is by the utility.
2. In the electrical service room, the minimum clearances between switchgears are shown. Also, the front of any free-standing item must be specified. Otherwise, it may be put in backwards.
3. The room containing the air conditioning fans and units simply shows the motor locations. All wiring, including control, is shown on the MCC schedule. (Although the room looks empty, a glance at the HVAC drawings would show that the space is filled with ductwork.)
4. Lighting and receptacles for the mechanical and electrical spaces B-7, B-6 and B-59 are not shown on the drawings. This is because the heavy ductwork, piping and conduit work on the ceilings make it extremely difficult to avoid space conflicts. Such conflicts end up as field changes and often mean extra cost. To avoid this, the designer here has prepared a schedule of the lighting and receptacle requirements, to be field located. These are given in Figure 16.38(g).
5. Room B-4 is reserved for telecommunications equipment. The 3-pole, 30-amp unfused switch supplies power to this equipment. Additional power is available from nearby panels, as required.
6. Junction boxes, complete with wiring, connected to an emergency service panel, are provided in each elevator pit. This is standard procedure. In addition, it is normal practice to provide a similar junction box at the midpoint (in elevation) of each elevator shaft.

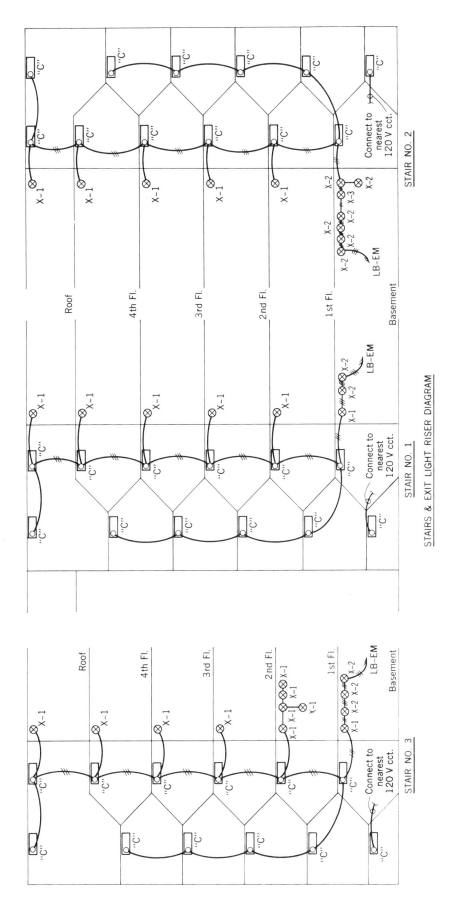

Figure 16.39 Stair and exit light riser diagram. These lights are always connected to an emergency power source.

7. The emergency lighting units throughout the building are easily identified by the letters *EM* on the fixture symbol. Two of these units are shown in the elevator lobby. The lighting in the switchgear and generator rooms is connected to the emergency lighting panels. See Figure 16.38(g). Also, a battery-operated emergency lighting unit is provided in the generator room. This is done so that, if a problem develops with the generator, there will be some lighting in the room to permit servicing of the unit.

8. It is customary to wire the stair lighting and exit lights on vertical risers in a multistory building. This is not done for economy but rather to centralize control of these lights. They can be fed from a separate stair-and-exit panel supplied by the emergency service. Alternatively, the stair-and-exit risers can be fed from a nondedicated emergency lighting panel. This arrangement is shown in Figure 16.39.

16.11 Forms, Schedules and Details

In our discussion throughout the book, we repeatedly refer you to detail drawings. Details of this kind are an essential part of the technologist's work. The engineer relies heavily on the technologist to prepare these details accurately and to scale. Details are as necessary as the floor plans because without them the exact intent of the designer would not be clear. In addition to details, we have also provided throughout our discussions, the forms and schedules needed on most jobs. As we noted previously, schedules are a particularly effective and almost foolproof way of presenting information because they must be filled in. A blank space is immediately noticed. Thus schedules help greatly to prevent oversights and forgotten items. A useful schedule that was not given in the text is given in Figure 16.40. A conduit and cable schedule is very useful in large jobs, where showing the data on the drawings would clutter them, and where frequent changes are expected.

The most commonly used form in electrical work is the panel schedule. A three-phase circuit breaker panel schedule is shown in Figure 13.14(a); a three-phase switch and fuse type is shown in Figure 13.14(b); and a single-phase 3-wire circuit breaker type is shown in Figure 15.18(c). For your convenience, blank forms for circuit breaker and switch and fuse panels of the single- and three-phase types are given in Figures 16.41 through 16.44.

16.12 Conclusion

Our discussion of electrical work is concluded. The material was presented in as direct and useful manner as possible. The technologist who has carefully followed all the preceding discussions and explanations is now in a position to fulfill a major role in the preparation of the HVAC, plumbing and electric drawings.

CONDUIT		WIRE				FROM	TO	REMARKS
SIZE	MAT	NO.	SIZE	TYPE	INS.			
TYPICAL DATA:								
F-16	4"	S	4	500	RHW 600V	SWBD. D-1	MCC-14	
C-12	1"	S	16	22	Tel 300V	MCC #14	Annunc. Pnl A-6	Pilot Lt. Control Cable
		S-Steel		T-Transite				
		E-Emt		F-Fibre				
		A-Aluminum		PVC-Plastic				
		FL-Flex.						

Figure 16.40 Form for a typical conduit and cable schedule. Schedules such as this are used in large electrical projects to keep accurate track of all wiring.

CIRC. NO.	LOAD VA	DESCRIPTION	CIRCUIT BREAKERS	DESCRIPTION	LOAD VA	CIRC. NO.
			PANEL DESIGNATION			
			1 2			
			3 4			
			5 6			
			7 8			
			9 10			
			11 12			
			13 14			
			15 16			
			17 18			
			19 20			
			21 22			
			23 24			
			25 26			
			27 28			
			29 30			
			31 32			
			33 34			
			35 36			
			37 38			
			39 40			
			41 42			

PANEL DATA
MAINS BUS _____
MAIN C/B OR SW/FUSE _____
BRANCH C/B INT. CAP. _____
SURF—RECESSED
REMARKS

EQUIP. GND BUS _____
VOLTAGE _____

LOADS: A: B: PANEL:

Figure 16.41 Recommended panel schedule form for a single-phase 3-wire circuit breaker-type panel.

CIRC. NO.	LOAD VA	DESCRIPTION	SW	FUSE	FUSE	SW	DESCRIPTION	LOAD VA	CIRC. NO.
				1	2				
				3	4				
				5	6				
				7	8				
				9	10				
				11	12				
				13	14				
				15	16				
				17	18				
				19	20				
				21	22				
				23	24				
				25	26				
				27	28				
				29	30				
				31	32				
				33	34				
				35	36				
				37	38				
				39	40				
				41	42				

PANEL DESIGNATION

PANEL DATA
MAINS BUS _____
MAIN C/B OR SW/FUSE _____
BRANCH C/B INT. CAP. _____
SURF—RECESSED
REMARKS

EQUIP. GND BUS _____
VOLTAGE _____

LOADS: A: B: PANEL:

Figure 16.42 Recommended panel schedule form for a single-phase 3-wire switch-and-fuse-type panel.

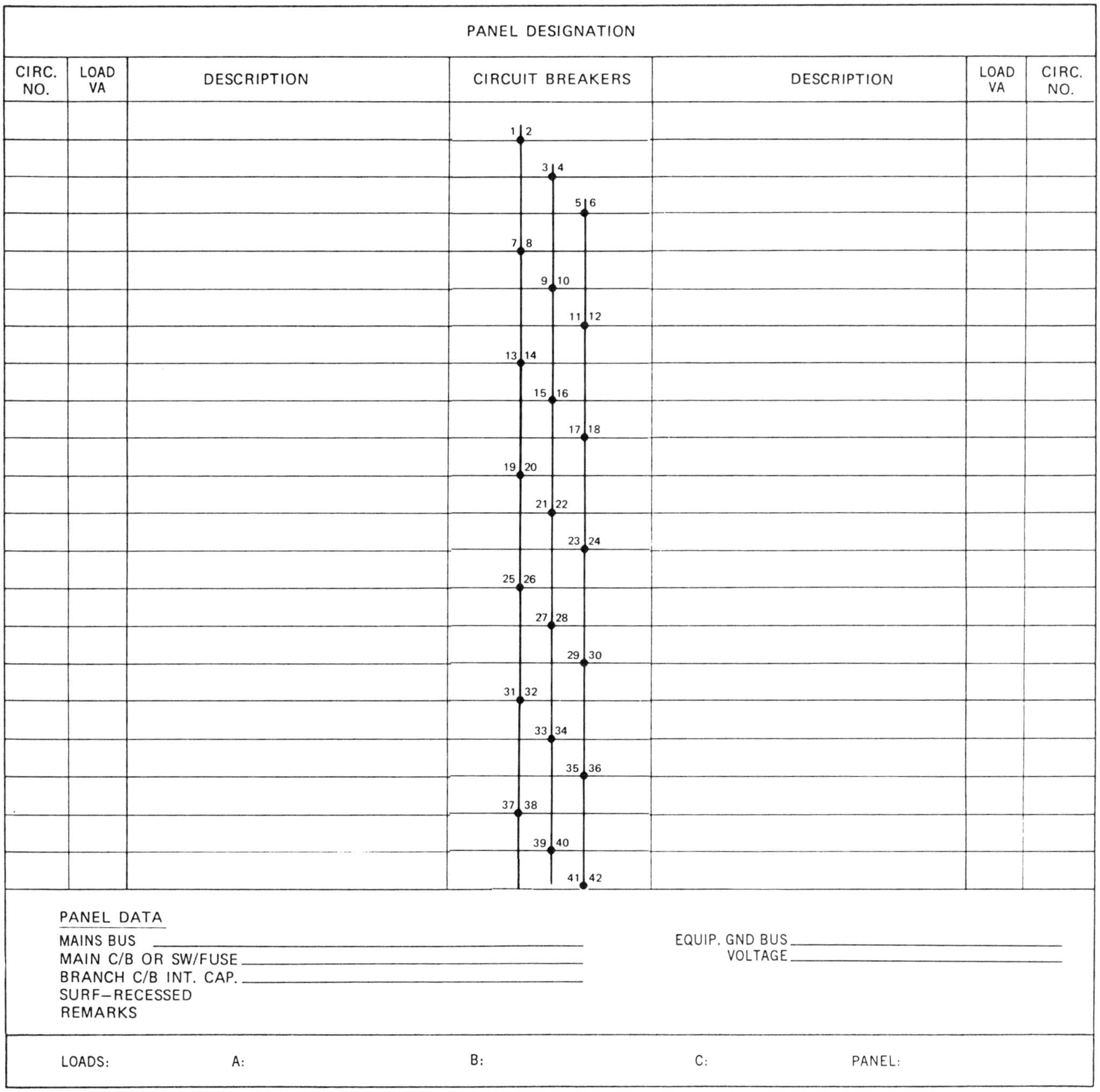

CIRC. NO.	LOAD VA	DESCRIPTION	CIRCUIT BREAKERS	DESCRIPTION	LOAD VA	CIRC. NO.
			PANEL DESIGNATION			
			1 2			
			3 4			
			5 6			
			7 8			
			9 10			
			11 12			
			13 14			
			15 16			
			17 18			
			19 20			
			21 22			
			23 24			
			25 26			
			27 28			
			29 30			
			31 32			
			33 34			
			35 36			
			37 38			
			39 40			
			41 42			

PANEL DATA
MAINS BUS _____
MAIN C/B OR SW/FUSE _____
BRANCH C/B INT. CAP. _____
SURF—RECESSED
REMARKS

EQUIP. GND BUS _____
VOLTAGE _____

LOADS: A: B: C: PANEL:

Figure 16.43 Recommended panel schedule form for a three-phase circuit-breaker-type panel.

PANEL DESIGNATION									
CIRC. NO.	LOAD VA	DESCRIPTION	SW	FUSE	FUSE	SW	DESCRIPTION	LOAD VA	CIRC. NO.
				1	2				
				3	4				
				5	6				
				7	8				
				9	10				
				11	12				
				13	14				
				15	16				
				17	18				
				19	20				
				21	22				
				23	24				
				25	26				
				27	28				
				29	30				
				31	32				
				33	34				
				35	36				
				37	38				
				39	40				
				41	42				

PANEL DATA

MAINS BUS ________________ EQUIP. GND BUS ________________

MAIN C/B OR SW/F ________________ VOLTAGE ________________

BRANCH I.C. ________________

SURF–RECESSED

REMARKS

LOADS: A: B: C: PANEL:

Figure 16.44 Recommended panel schedule form for a three-phase switch-and-fuse-type panel.

Key Terms

Having completed the study of this chapter, you should be familiar with the following key terms. If any appear unfamiliar or not entirely clear, you should review the section in which these terms appear. All key terms are listed in the index to assist you in locating the relevant text. Some of the terms also appear in the glossary at the end of the book.

Across-the-line (ATL); Combination across-the-line (CATL)
Auxiliary contact
Block diagram
Busduct; feeder duct; plug-in duct
Duct bank
Electric closet
Emergency service
Ladder diagram
Manhole, handhole
Manual controller
Motor circuit conductor
Motor control center (MCC)
Motor controller
Motor disconnect means
Motor feeder
Motor feeder protection
Motor overload protection
NEMA enclosures
NEMA starter size
Nominal service size
Normal service
Pad mount
Pilot device
Power riser
Primary service
Pull box
RC switch
Riser diagram
Secondary service
Service disconnect(s)
Service voltage
Shunt trip
Standby service
Starter
Starting switch
Switchboard
System voltage
Thermal overload element
Transformer vault
Types I and Type II ducts
Undervoltage protection
Utilization voltage

Supplementary Reading

See the Supplementary Reading list at the end of Chapter 12.

Problems

1. A residence has a load of 12 kw. What size service is required at

   a. 120-v, single-phase.
   b. 120/240-v, single-phase
   c. 120/208-v, three-phase.

   Which service would you recommend? Why?

2. The residence in Problem 1 has added central air conditioning, including a 5-hp 240-v single-phase compressor. What size service do you now recommend? What voltage?

3. The owner of the residence in Problem 1 has converted the detached garage into a workshop. After renovation, the electrical service includes the following:

   a. UG feeder: 3 No. 1/0 AWG, 120/240 v.
   b. Two service switches in the house: a 150-amp unit for the house and a 100-amp for the garage.
   c. A feeder of 3 No. 4 and 1 No. 8 ground to the garage, run UG in Type II fiber duct. The garage is 50 ft away from the house.
   d. A 100-amp disconnect switch in the garage.

   Draw a plot plan to scale and a riser diagram. Show all these elements and any other data you feel are necessary.

4. Using the data in Figure 16.21, work out the physical size of the panelboards detailed in Figures 13.14(a) and (b) and 15.18(c). Show all data used in tabular form.

5. A conduit run consists of these conduits: 2-2 in., 2-2½ in., 2-3 in., 2-3½ in. The conduits enter a pull box and make a right-angle turn in the same plane but in two directions. Half of the conduits (one of each size) turn left and the other half turn right. Size the pull box and show how you would lay out the conduits to keep the pull box to a minimum size, while avoiding tangling of cables.

6. For the classroom discussed in Section 16.9 and shown in Figure 16.36, repeat the lighting design procedure using metal halide lamps, with metric dimensions. The room dimensions are 6 × 8 × 3.5 m. Explain every decision including type and wattage of lamp selected, type of luminaire, arrangement and mounting heights. Show that the design meets the guidelines and criteria listed in Section 16.9. Select lamp data from the tables in Chapter 14 or from a current manufacturer's catalog. Select a luminaire from a manufacturer's catalog and show the luminaire's construction, dimensions, luminances and CU table.

7. Select a public building in your neighborhood to which you can get access. The building should have at least three floors. Survey the building and from the data obtained prepare a power riser diagram. The riser should show the service equipment, distribution, switchboard (panel), all building panels and all riser feeders. Diagram should be as complete as possible with sizes, designations, and equipment location.

8. For the same building as in Problem 7, prepare a stair riser and an exit light riser.

Appendix A
Metrication; SI Units; Conversions

A.1 General Comments on Metrication

The building profession has been slower than most other professions in adopting the metric [more accurately the Systeme International (SI)] system of units for many reasons, a discussion of which is not relevant here. The change will come. Many of the major professions use a mixture of both systems, because certain units are so common that changing them is almost impossible. However, the advantages of the SI system are well known. In this book, we have followed industry practice, which is to use the traditional English system primarily, with some SI units. For this reason, tables of conversions and approximations are presented here to enable you to work with both systems. Also given later in this appendix are some useful facts that should make using the SI (metric) system a bit easier.

A.2 SI Nomenclature, Symbols

For a full discussion of the SI system, refer to *AIA Metric Building and Construction Guide*, edited by S. Braybrooke (Wiley, New York, 1980). The SI (metric) units in common use include the basic units—meter, kilogram, and second (MKS)—plus many derived supplementary, and non-SI units such as pascal (pressure), watt (power), horsepower (power) and kilowatt-hour (energy). Also, multiple and submultiple units such as liter, metric ton, and millibar are so common that they are established as separate units instead of being expressed as 0.001 m^3, 1000 kg and 0.01 Pa. For this reason, we omit a detailed analysis of the subject and simply supply data that, by experience, we have found useful.

Table A.1 Prefixes That May Be Applied to All SI Units

Multiples and Submultiples	Prefixes	Symbols
$1\,000\,000 = 10^6$	mega	M[a]
$1\,000 = 10^3$	kilo	k[a]
$100 = 10^2$	hecto[b]	h
$10 = 10$	deka	d[a]
$0.1 = 10^{-1}$	deci	d
$0.01 = 10^{-2}$	centi	c[a]
$0.001 = 10^{-3}$	milli	m[a]
$0.000\,001 = 10^{-6}$	micro	μ[a]
$0.000\,000\,001 = 10^{-9}$	nano	n

[a] Most commonly used.
[b] A hectare is a square hectometer (i.e., 100 m × 100 m = 10,000 m²).

Table A.1 lists prefixes with their usual symbols. Symbols do not change in plural. That is, 6 millimeters is written 6 mm and not 6 mms and 20 kilograms is written 20 kg and not 20 kgs. All units and prefixes except Fahrenheit and Celsius are uncapitalized when written out, as in megaton or meter.

A.3 Common Usage

1. *Length:* meter (m), kilometer (km), millimeter (mm), micrometer (μm), nanometer (nm).
2. *Area:* square meter (m²), hectare (ha).
3. *Volume:* cubic meter (m³), liter (L).
4. *Flow:* cubic meters per second (m³/s).
5. *Velocity, air flow:* meters per second (m/s).
6. *Weight:* kilogram (kg), gram (g).

The SI system clearly differentiates between mass (kg) and force (kg·m/s²), the latter being given a separate name and symbol, newton (N). Weight is not used, because it is a force that depends on acceleration and is, therefore, variable. However, the construction industry generally continues to use the terms mass, weight and force interchangeably.

7. *Force:* newton (N), kilonewton (kN). A newton is the force required to accelerate 1 kg at 1.0 m/s².
8. *Pressure:* pascal (Pa), kilopascal (kPa). A pascal is a newton per square meter (N/m²).

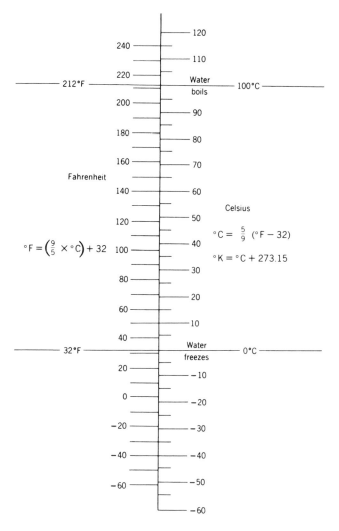

Figure A.1. Conversion chart: degrees Fahrenheit to degrees Celsius (°F to °C).

9. *Energy, work, quantity of heat:* joule (J), kilojoule (kJ), megajoule (MJ). A joule is a watt-second (w-s).
10. *Temperature:* degree Celsius (°C), degree Kelvin (°K).

The SI unit (Kelvins) is not used as commonly as °C. (Celsius is the accepted term; "centigrade" is obsolete.) The Celsius and Kelvin scales are subdivided equally but start at different points, that is, 0°K is −273.15°C. Therefore, to determine Kelvin from Celsius, simply add 273.15. Increments are equal because of equal subdivision; that is, a change of 10 K° is the same as a 10 C° change.

Because of its special importance, a separate Fahrenheit/Celsius conversion chart is given in Figure A.1.

11. *Abbreviations:* See Table A.2.
12. *Illumination:* See Table A.3.
13. *Acoustics:* See Table A.4.
14. *CGS/MKS conversions:* See Table A.5.
15. *Approximations:* See Table A.6.

A.4 Conversion Factors

Table A.6 is an alphabetized list of useful conversion factors. Because normal use involves a hand calculator, we have avoided scientific notation and used decimal notation instead; that is, we write

$$0.00378 \quad \text{not} \quad 3.78 \times 10^{-3} \quad \text{or} \quad 3.78\text{E-}3$$

Table A.2 Typical Abbreviations: All Systems of Units

atmospheres	atm	kilopascals	kPa
British thermal units	Btu	kilowatts	kW
British thermal units per hour	Btu/h, Btuh	kilowatt-hours	kWh
calorie	cal	liters	L
cubic feet	cf, ft³	liters per second	L/s
cubic feet per minute	cfm, ft³/min	megajoules	MJ
cubic feet per second	cfs, ft³/s	meganewtons	MN
cubic meters	m³	megapascals	MPa
feet	ft	meters	m
feet per second	fps, ft/s	meters per second	m/s
gallons	gal	miles per hour	mph
gallons per hour	gph, gal/h	millimeters	mm
gallons per minute	gpm, gal/min	millimeters of mercury	mm Hg
grams	g	newtons	N
hectares	ha	ounces	oz
horsepower	hp	pounds	lb
inches	in.	pounds of force	lbf
inches of mercury	in. Hg	pounds per cubic foot	lb/ft³
joules	J	second	sec, s
kilocalories	kcal	square feet	ft²
kilograms	kg	square inches	in.²
kilograms per second	kg/s	square meters	m²
kilojoules	kJ	watts	W
kilometers	km	watts per square meter	W/m²
kilometers per hour	kph, km/h	yards	yd
kilonewtons	kN		

Table A.3 Lighting Units—Conversion Factors

Unit	Multiply	By	To Obtain
Illuminance *(E)*	Lux (lx)	0.0929	Footcandle (fc)
	Footcandle (fc)	10.764	Lux (lx)
Luminance *(L)*	cd/m²	0.2919	Footlambert (fL)
	cd/cm²	10000	cd/m²
	cd/in.²	1550	cd/m²
	cd/ft²	10.76	cd/m²
	millilambert (mL)	3.183	cd/m²
	Footlambert (fL)	3.4263	cd/m²
Intensity *(I)*	Candela (cd)	1.0	Candlepower (cp)

Table A.4 Acoustic Units and Conversions

Units	MKS	CGS
Force	kilogram-meter/s² = newton	gram-cm/s² = dyne
Intensity	watts/meter²	watts/cm²
Pressure	newtons/meter² = pascals	dynes/cm² = microbars

	In Conversion:		
Quantity	Multiply	By	To Obtain
Force	newtons	10^5	dynes
	dynes	10^{-5}	newtons
Intensity	watts/cm²	10^4	watts/m²
	watts/m²	10^{-4}	watts/cm²
Pressure	pascals	10	microbars
	microbar	10^{-1}	pascals

Note: One atmosphere = 1 bar = 10^6 microbar.

Table A.5 Common Approximations

1 inch	= 25 millimeters
1 foot	= 0.3 meter
1 yard	= 0.9 meter
1 mile	= 1.6 kilometers
1 square inch	= 6.5 square centimeters
1 square foot	= 0.09 square meter
1 square yard	− 0.8 square meter
1 acre	= 0.4 hectare
1 cubic inch	= 16 cubic centimeters
1 cubic foot	= 0.03 cubic meter
1 cubic yard	= 0.8 cubic meter
1 quart	= 1 liter
1 gallon	= 0.004 cubic meter
1 ounce	= 28 grams
1 pound	= 0.45 kilogram
1 horsepower	= 0.75 kilowatt
1 millimeter	= 0.04 inch
1 liter	= 61 cubic inches
1 meter	= 3.3 feet
1 meter	= 1.1 yards
1 kilometer	= 0.6 mile
1 square centimeter	= 0.16 square inch
1 square meter	= 11 square feet
1 square meter	= 1.2 square yards
1 hectare	= 2.5 acres
1 cubic centimeter	= 0.06 cubic inch
1 cubic meter	= 35 cubic feet
1 cubic meter	= 1.3 cubic yards
1 liter	= 1 quart
1 cubic meter	= 250 gallons
1 kilogram	= 2.2 pounds
1 kilowatt	= 1.3 horsepower

Table A.6 Useful Conversion Factors: Alphabetized

Multiply	By	To Get
acres	4047.	square meters
atmospheres	33.93	feet of water
atmospheres	29.92	inches of mercury
atmospheres	760.0	millimeters of mercury
Btu (energy)	0.252	kilocalories
	1.055	kilojoules
Btu/h (power)	0.2928	watts
Btu/h/ft² (energy transfer)	3.152	watts per square meter
BtuF (heat capacity)	1.897	kilojoules per kelvin[a]
Btu/lb/°F (specific heat)	4.182	kilojoules per kilogram per kelvin[a]
Btu/h/°F/ft (thermal conductivity[b])	1.729	watts per kelvin[a] per meter
Btu/h/°F/ft² (conductance[c])	5.673	watts per kelvin[a] per square meter
Btu/Fday (building load coefficient, BLC)	0.022	watts per kelvin[a]
Btu/Fday/ft² (load-collector ratio, LCR)	0.236	watts per kelvin[a] per square meter
cubic feet	0.028	cubic meters
cubic feet	7.481	gallons

Table A.6 *(Continued)*

Multiply	By	To Get
cubic feet	28.32	liters
cubic feet per minute	0.472	liters per second
cubic feet per second	2.832	liters per second
cubic inches	16.39	cubic centimeters
cubic meters	35.32	cubic feet
cubic meters	1.308	cubic yards
cubic meters	264.2	gallons
cubic yards	0.765	cubic meters
feet	0.305	meters
feet	304.8	millimeters
feet per second	0.3048	meters per second
foot-pounds of force per second	1.356	watts
gallons	3.785	liters
gallons per hour	0.00152	liters per second
gallons per minute	0.0022	cubic feet per second
gallons per minute	0.06308	liters per second
grams	0.035	ounces (avoirdupois)
hectares	2.471	acres
horsepower	0.746	kilowatts
horsepower	746.	watts
inches	25.4	millimeters
inches of mercury	0.033	atmospheres
inches of mercury	1.133	feet of water
inches of mercury (60°F)	3377.	newtons per square meter
inches of mercury	0.491	pounds per square inch
inches of water	0.002458	atmospheres
inches of water	0.036	pounds per square inch
inches of water (60°F)	248.8	newtons per square meter
kilocalories	3.968	British thermal units
kilocalories	4190.	joules
kilograms	2.205	pounds
kilograms per cubic meter	1.686	pounds per cubic yard
kilograms per square meter	0.0033	feet of water
kilograms per square meter	0.0029	inches of mercury
kilograms per square meter	0.205	pounds per square feet
kilograms per square meter	0.001422	pounds per square inch
kilojoules	0.948	British thermal units
kilojoules per kilogram	0.430	British thermal units per pound
kilometers	0.621	miles
kilometers	1094.	yards
kilometers per hour	0.621	miles per hour
kilonewtons	0.1004	tons of force
kilonewtons	224.8	pounds of force
kilopascals	20.89	pounds of force per square foot
kilowatts	1.341	horsepower
kilowatt-hours	3.6	megajoules
liters	0.03532	cubic feet
liters	61.02	cubic inches
liters	0.2642	gallons
liters	1.057	quarts
liters per second	2.119	cubic feet per minute
liters per second	951.0	gallons per hour
liters per second	15.85	gallons per minute
megajoules	0.278	kilowatt-hours
meganewtons	100.36	tons of force

Table A.6 *(Continued)*

Multiply	By	To Get
megapascals	145.04	pounds of force per square foot
megapascals	9.324	tons of force per square foot
meters	3.281	feet
meters per second	196.86	feet per minute
meters per second	2.237	miles per hour
miles	1.609	kilometers
miles per hour	1.609	kilometers per hour
miles per hour	0.447	meters per second
milliliters	0.061	cubic inches
millimeters	0.035	fluid ounces
millimeters	0.039	inches
millimeters of mercury	133.3	newtons per square meter
million gallons per day	18.94	cubic meters per hour
newtons	0.225	pounds of force
ounces (avoirdupois)	28.35	grams
ounces (fluid)	28.41	milliliters
pounds	0.454	kilograms
pounds of force	4.448	newtons
pounds of force per square foot	47.88	pascals
pounds of force per square inch	6.895	kilopascals
pounds per cubic foot	16.02	kilograms per cubic meter
pounds per square foot	4.882	kilograms per square meter
pounds per cubic yard	0.593	kilograms per cubic meter
square feet	0.0929	square meters
square inches	645.2	square millimeters
square kilometers	0.386	square miles
square meters	10.76	square feet
square meters	1.196	square yards
square miles	2.590	square kilometers
square yards	0.836	square meters
tons of force	9.964	kilonewtons
tons of force per square foot	107.25	kilopascals
tons of force per square inch	15.44	megapascals
torr (millimeters of mercury at 0°C)	133.3	newtons per square meter
watts	3.412	British thermal units per hour
watts	0.738	foot-pounds of force per second
watts per square meter	0.317	British thermal units per square foot
yards	0.914	meters
Add your own conversion factors here.		
Multiply	By	To Get
Multiply	By	To Get
Multiply	By	To Get

[a] °K or °C.
[b] Thermal conductivity *(K)*.
[c] Thermal conductance *(C)* or transmittance *(U)*.

Appendix B
Equivalent Duct Lengths

Figures B.1–B.9 show the equivalent lengths (of straight duct) in feet to the illustrated duct fittings. These lengths are used in duct friction calculations as explained in Chapter 5 of the text.

The data are reproduced with permission from the following sources as noted:

ACCA   Air Conditioning Contractors of America

SMACNA   Sheet Metal and Air Conditioning Contractors National Association, Inc.

982 / APPENDIX B EQUIVALENT DUCT LENGTHS

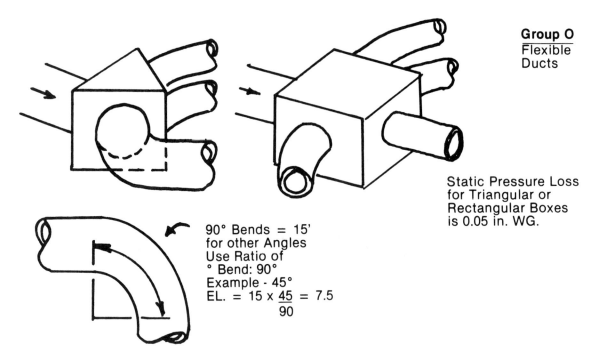

Figure B.0 Equivalent length of flexible duct fittings. (Reproduced with permission from *ACCA Manual D*.)

APPENDIX B EQUIVALENT DUCT LENGTHS / 983

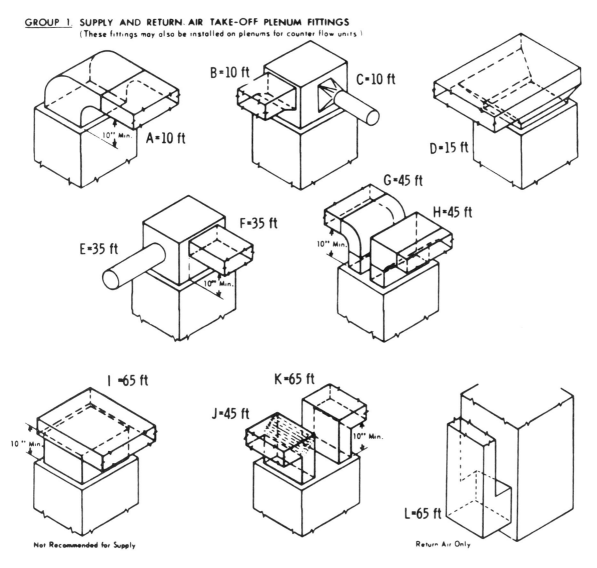

Figure B.1 Equivalent lengths of supply and return air take-off plenum fittings. (Reproduced with permission from *ACCA Manual D*.)

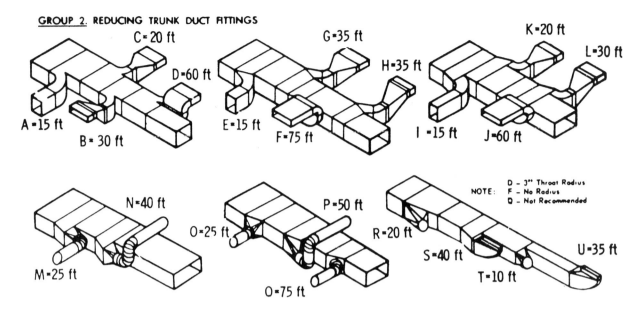

Figure B.2 Equivalent lengths of reducing trunk duct fittings. (Reproduced with permission from *ACCA Manual D.*)

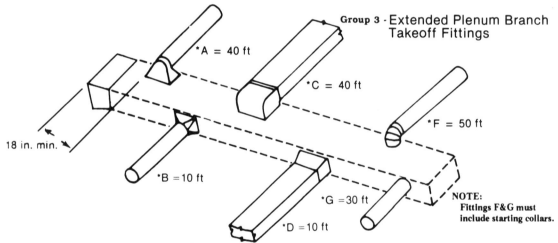

Example: given the system shown above, find the equivalent length for each branch takeoff fitting.
Solution: see schedule below.

Branch	Basic Equiv. Length of Takeoff Fitting	No. of Branches Downstream	Added Equiv. Length	Final Equiv. Length
G	30	0	0	30
F	50	1	10	60
D	10	2	20	30
C	40	3	30	70
B	10	4	40	50
A	40	5	50	90

* Values shown above apply only when the branch is at the end of the truck duct (ex., G above). For all other branches, add to the equivalent length shown ten feet (10 ft) times the number of branches downstream between the takeoff being evaluated and the end of the trunk duct.

Figure B.3 Equivalent lengths of extended plenum branch take-off fittings. (Reproduced with permission from *ACCA Manual D.*)

APPENDIX B EQUIVALENT DUCT LENGTHS / 985

GROUP 4 - Round Duct Fittings

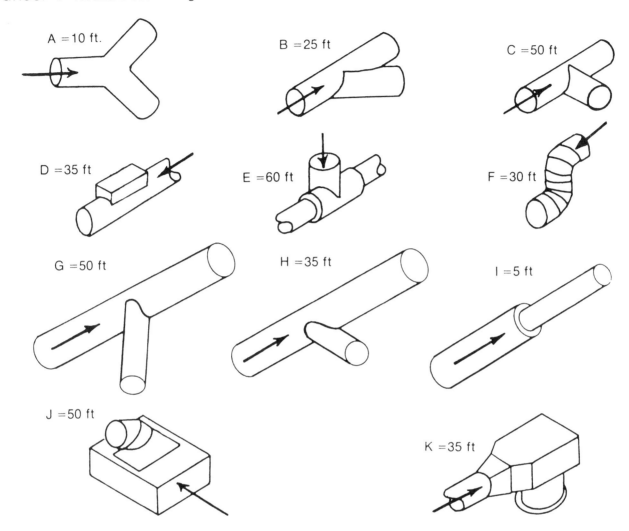

Figure B.4 Equivalent lengths of round duct fittings. (Reproduced with permission from *ACCA Manual D.*)

GROUP 5. ANGLES AND ELBOWS FOR TRUNK DUCTS
(Inside Radius = ½ Width of Duct)

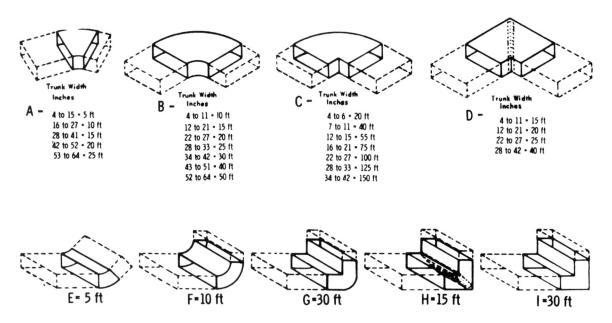

A –
Trunk Width Inches
4 to 15 = 5 ft
16 to 27 = 10 ft
28 to 41 = 15 ft
42 to 52 = 20 ft
53 to 64 = 25 ft

B –
Trunk Width Inches
4 to 11 = 10 ft
12 to 21 = 15 ft
22 to 27 = 20 ft
28 to 33 = 25 ft
34 to 42 = 30 ft
43 to 51 = 40 ft
52 to 64 = 50 ft

C –
Trunk Width Inches
4 to 6 = 20 ft
7 to 11 = 40 ft
12 to 15 = 55 ft
16 to 21 = 75 ft
22 to 27 = 100 ft
28 to 33 = 125 ft
34 to 42 = 150 ft

D –
Trunk Width Inches
4 to 11 = 15 ft
12 to 21 = 20 ft
22 to 27 = 25 ft
28 to 42 = 40 ft

E = 5 ft   F = 10 ft   G = 30 ft   H = 15 ft   I = 30 ft

Figure B.5 Equivalent lengths of angles and elbows for trunk ducts. (Reproduced with permission from *ACCA Manual D*.)

APPENDIX B EQUIVALENT DUCT LENGTHS / 987

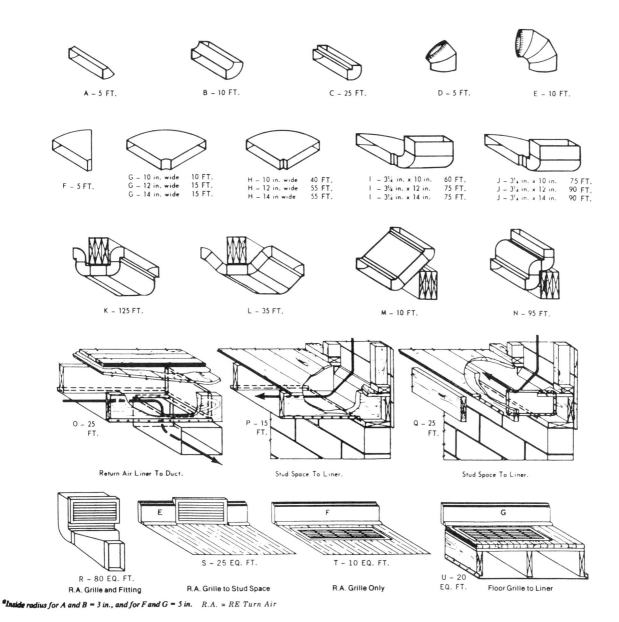

Figure B.6 Equivalent lengths of angles and elbows for individual and branch ducts. (Reproduced with permission from SMACNA—*Installation Standards for Residential Systems*.)

GROUP 7 Boot Fittings

(These values may also be used for floor Diffuser Boxes)

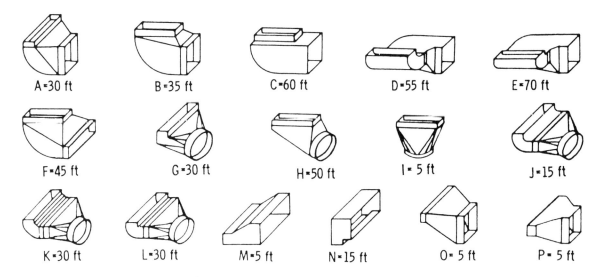

Figure B.7 Equivalent lengths of boot fittings. (Reproduced with permission from *ACCA Manual D.*)

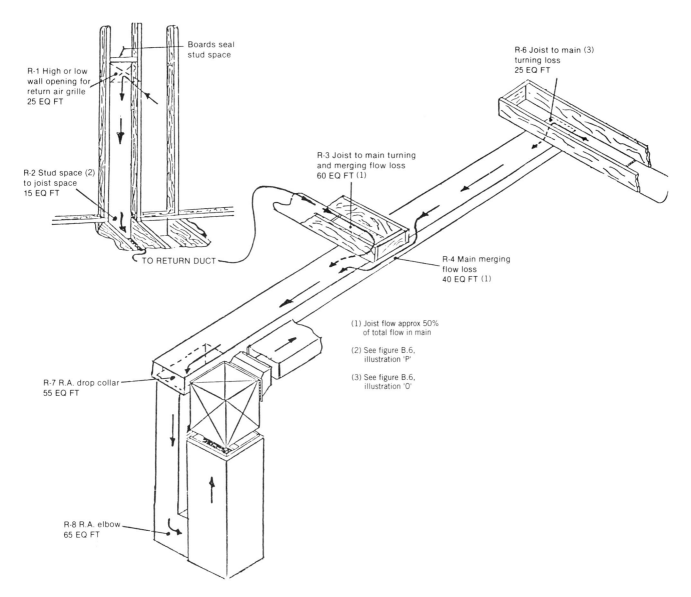

Figure B.8 (part 1) Equivalent lengths of return air fittings and joints. (Reproduced with permission from *ACCA Manual D*.)

990 / APPENDIX B EQUIVALENT DUCT LENGTHS

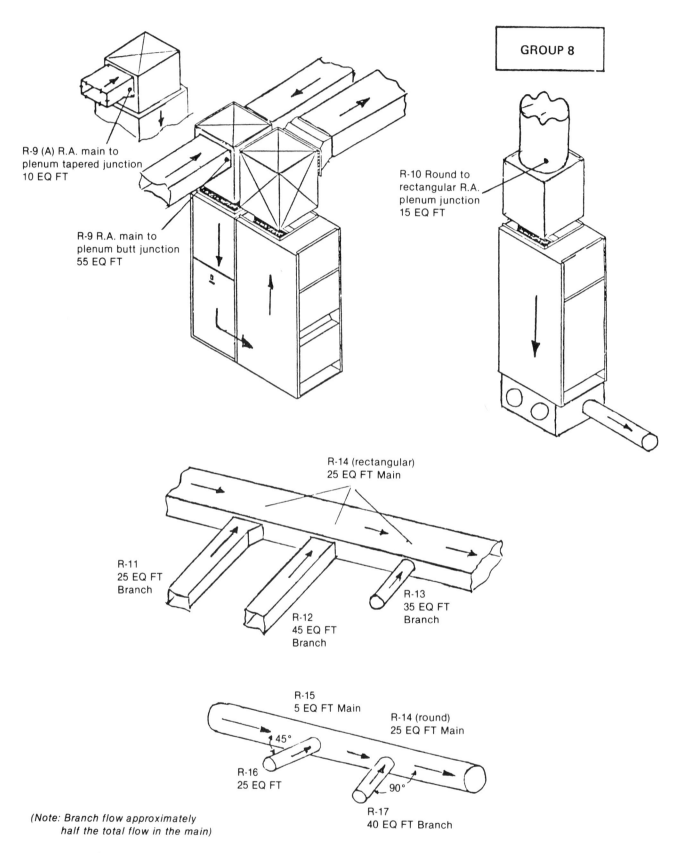

Figure B.8 (part 2) Equivalent lengths of return air fittings and joints. (Reproduced with permission from *ACCA Manual D*.)

Appendix C
Loss Coefficients for Duct Fittings

The figures in this appendix give the loss coefficients for most of the duct fittings used in modern duct systems. All the data are reproduced, with permission from *Manual Q,* Appendix 6, published by the Air Conditioning Contractors of America (ACCA). The use of these data is discussed in Chapter 5.

Section C.1 Loss Coefficients for Elbows
Section C.2 Loss Coefficients for Offsets
Section C.3 Loss Coefficients for Tees and Wyes, Diverging Flow
Section C.4 Loss Coefficients, Diverging Junctions (Tees, Wyes)
Section C.5 Loss Coefficients, Converging Junctions (Tees, Wyes)
Section C.6 Loss Coefficients, Transitions (Diverging Flow)
Section C.7 Loss Coefficients, Transitions (Converging Flow)
Section C.8 Loss Coefficients, Entries
Section C.9 Loss Coefficients, Exits
Section C.10 Loss Coefficients, Screens
Section C.11 Loss Coefficients, Obstructions (Constant Velocities)

Section C.1 Loss Coefficients for Elbows

Elbows may have the same velocity at the entrance and exit or they may produce converging or diverging flow. For all elbows, use the velocity pressure (Pv) that is associated with the entrance (upstream section).

Pt = C x Pv (In.Wg.)

A) Smooth Radius, Round Elbow (Upstream Pv)

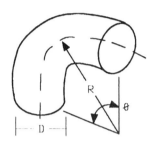

Coefficient C for 90° Elbows

R/D	0.5	0.75	1.0	1.5	2.0	2.5
C	0.71	0.33	0.22	0.15	0.13	0.12

For angles other than 90° multiply by the following factors

ø	0°	20°	30°	45°	60°	75°	90°	110°	130°	150°	180°
K	0	0.31	0.45	0.60	0.78	0.90	1.00	1.13	1.20	1.28	1.40

Adjusted loss coefficient = C x K

B) Round Sectional Elbow, 3 to 5 pieces (Upstream Pv)

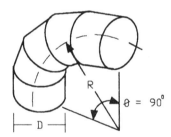

Coefficient C for 90° Elbows

No. of Piece	R/D				
	0.5	0.75	1.0	1.5	2.0
5	---	0.46	0.33	0.24	0.19
4	---	0.50	0.37	0.27	0.24
3	0.98	0.54	0.42	0.34	0.33

Section C.1 Loss Coefficients for Elbows

Elbows may have the same velocity at the entrance and exit or they may produce converging or diverging flow. For all elbows, use the velocity pressure (Pv) that is associated with the entrance (upstream section).

F) Smooth Radius Rectangular Elbow with No Vanes (Upstream Pv)

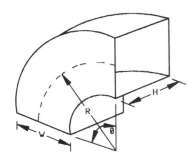

Coefficients for 90° Elbows (C)

R/W	H/W										
	0.25	0.5	0.75	1.0	1.5	2.0	3.0	4.0	5.0	6.0	8.0
0.5	1.5	1.4	1.3	1.2	1.1	1.0	1.0	1.1	1.1	1.2	1.2
0.75	0.57	0.52	0.48	0.44	0.40	0.39	0.39	0.40	0.42	0.43	0.44
1.0	0.27	0.25	0.23	0.21	0.19	0.18	0.18	0.19	0.20	0.27	0.21
1.5	0.22	0.20	0.19	0.17	0.15	0.14	0.14	0.15	0.16	0.17	0.17
2.0	0.20	0.18	0.16	0.15	0.14	0.13	0.13	0.14	0.14	0.15	0.15

Reynolds Number Correction (N)

CFM	R/W < 0.75	R/W > 0.75
50 to 200	1.15	1.60
200 to 400	1.10	1.50
400 to 800	1.05	1.35
800 to 1000	1.03	1.30
1000 to 1500	1.00	1.15
1500 to 2000	1.00	1.05
Above 2000	1.00	1.00

Adjusted loss coefficient = C x N x K

For angles other than 90 multiply C by the following factors:

ø	0	20	30	45	60	75	90	110	130	150	180
K	0.00	0.31	0.45	0.60	0.78	0.90	1.00	1.13	1.20	1.28	1.40

Adjusted loss coefficent = C x N x K

Section C.1 Loss Coefficients for Elbows

Elbows may have the same velocity at the entrance and exit or they may produce converging or diverging flow. For all elbows, use the velocity pressure (Pv) that is associated with the entrance (upstream section).

H) Rectangular Mitered Elbow with Turning Vanes (Upstream Pv)

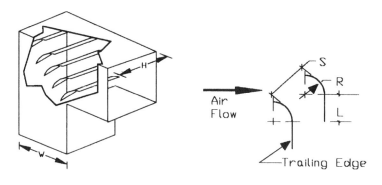

Single Thickness Vanes

	Coefficient C			
No.	Dimensions			
	R	S	L	C
1	2.0	1.5	0.75	0.12
2	4.5	2.25	0	0.15
3	4.5	3.25	1.60	0.18

Double Thickness Vanes

	Coefficent C						
No.	Dimensions		Velocity (V_o), fpm				Remarks
	R	S	1000	2000	3000	4000	
1	2.0	1.5	0.27	0.22	0.19	0.17	Embossed Vane Runner
2	4.5	3.25	0.26	0.21	0.18	0.16	Embossed Vane Runner
3	2.0	1.5	0.33	0.29	0.26	0.23	Push-on Vane Runner
4	2.0	2.13	0.38	0.31	0.27	0.24	Embossed Vane Runner

Section C.2 Loss Coefficients for Offsets

For all offsets use the velocity pressure (Pv) of the upstream section. Fitting loss Pt = C x Pv (In. Wg.)

C) 30° Offsets In Round Duct (Upstream Pv)

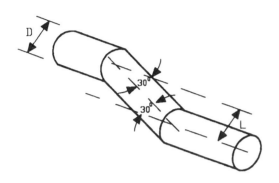

Coefficient C

L/D	0	0.5	1.0	1.5	2.0	2.5	3.0
C	0	0.15	0.15	0.16	0.16	0.16	0.16

Reynolds Number Correction (N)	
CFM	N
50 to 200	1.15
200 to 400	1.10
400 to 800	1.05
800 to 1000	1.03
Above 1000	1.00
Adjusted loss coefficient = C x N	

E) Close Coupled, Rectangular or Round, Radius elbows, S-curve (Upstream Pv)

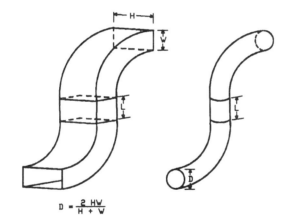

$D = \dfrac{2 HW}{H + W}$

For 0 < L/D < 6
Loss Coefficient C = 1.7 x Loss Coefficient for a single elbow

Section C.2 Loss Coefficients for Offsets

For all offsets use the velocity pressure (Pv) of the upstream section. Fitting loss Pt = C x Pv (In. Wg.)

F) Close Coupled, Rectangular or Round, Radius Elbows, Two Planes (Upstream Pv)

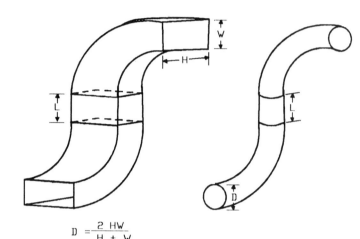

For $0 < L/D < 2$
Loss Coefficient C = 2.0 x Loss Coefficient for single elbow

For $2 < L/D < 6$
Loss Coefficient C = 1.7 x Loss Coefficient for single elbow

$$D = \frac{2\,HW}{H + W}$$

G) Rectangular Duct, Depressed to Avoid an Obstruction (Upstream Pv)

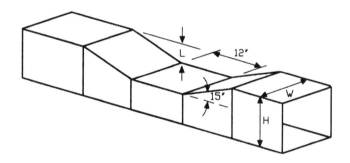

Coefficient C

W/H	L/H			
	0.125	0.15	0.25	0.30
1.0	0.26	0.30	0.33	0.35
4.0	0.10	0.14	0.22	0.30

H) Rectangular Duct with 4-45° Smooth Ells to Avoid an Obstruction (Upstream Pv)

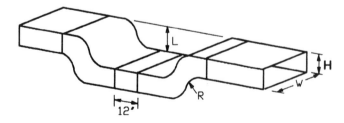

Coefficient C

V (fpm)	800	1200	1600	2000	2400
C	0.18	0.22	0.24	0.25	0.26

Where: W/H = 4
R/H = 1
L/H = 1.5

Section C.3 Loss Coefficients for Tees and Wyes, Diverging Flow

Use the velocity (Vc) in the upstream section to determine the reference velocity pressure (Pv) for either branch.

Pt = C x Pv (In. Wg.)

A) Rectangular Tees, Symmetrical or Unsymmetrical (Upstream Pv)

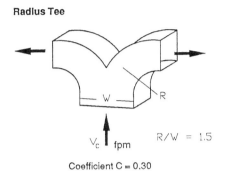

Radius Tee
R/W = 1.5
Coefficient C = 0.30

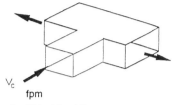

Square Tee, No Vanes
Coefficient C = 1.0
Loss Coefficient for this fitting is not documented. The coefficient listed above is based on the published data for elbows of similar construction with a liberal safety factor applied.

B) Round Tees, Symmetrical (Upstream Pv)

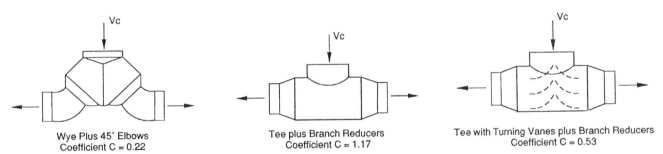

Wye Plus 45° Elbows
Coefficient C = 0.22

Tee plus Branch Reducers
Coefficient C = 1.17

Tee with Turning Vanes plus Branch Reducers
Coefficient C = 0.53

C) Wye, Rectangular and Round (Upstream Pv)

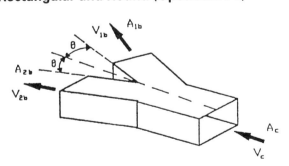

Where: $A_{1b} = A_{2b} = 0.50\, A_c$

Coefficient C

θ, Deg.	V_{1b}/V_c or V_{2b}/V_c						
	0.1	0.2	0.3	0.4	0.5	0.6	0.8
15	0.81	0.65	0.51	0.38	0.28	0.20	0.11
30	0.84	0.69	0.56	0.44	0.34	0.26	0.19
45	0.87	0.74	0.63	0.54	0.45	0.38	0.29
60	0.90	0.82	0.79	0.66	0.59	0.53	0.43
90	1.0	1.0	1.0	1.0	1.0	1.0	1.0

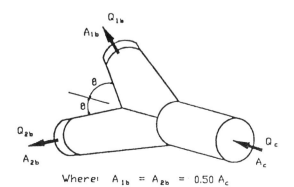

Where: $A_{1b} = A_{2b} = 0.50\, A_c$

θ, Deg.	V_{1b}/V_c or V_{2b}/V_c					
	1.0	1.2	1.4	1.6	1.8	2.0
15	0.06	0.14	0.30	0.51	0.76	1.0
30	0.15	0.15	0.30	0.51	0.76	1.0
45	0.24	0.23	0.30	0.51	0.76	1.0
60	0.36	0.33	0.39	0.51	0.76	1.0
90	1.0	1.0	1.0	1.0	1.0	1.0

Section C.4 Loss Coefficients, Diverging Junctions (Tees, Wyes)

Use the velocity (Vc) in the upstream section to determine the reference velocity pressure (Pv)

Pt = C x Pv (In. Wg.)

A) Tee or Wye, 30° to 90°, Round (Upstream Pv)

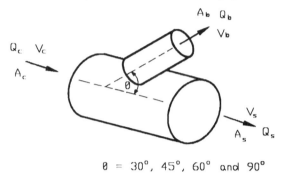

θ = 30°, 45°, 60° and 90°

Main, Coefficient C

Vs/Vc	0	0.1	0.2	0.3	0.4	0.5	0.6	0.8	1.0
C	0.35	0.28	0.22	0.17	0.13	0.09	0.06	0.02	0

Wye ø = 30°: Branch, Coefficient C

Ab/Ac	Qb/Qc								
	0.1	0.2	0.3	0.4	0.5	0.6	0.7	0.8	0.9
0.8	0.75	0.55	0.40	0.28	0.21	0.16	0.15	0.16	0.19
0.7	0.72	0.51	0.36	0.25	0.18	0.15	0.16	0.20	0.26
0.6	0.69	0.46	0.31	0.21	0.17	0.16	0.20	0.28	0.39
0.5	0.65	0.41	0.26	0.19	0.18	0.22	0.32	0.47	0.67
0.4	0.59	0.33	0.21	0.20	0.27	0.40	0.62	0.92	1.3
0.3	0.55	0.28	0.24	0.38	0.76	1.3	2.0	---	---
0.2	0.40	0.26	0.58	1.3	2.5	---	---	---	---
0.1	0.28	1.5	---	---	---	---	---	---	---

Wye ø = 45°: Branch, Coefficient C

Ab/Ac	Qb/Qc								
	0.1	0.2	0.3	0.4	0.5	0.6	0.7	0.8	0.9
0.8	0.78	0.62	0.49	0.40	0.34	0.31	0.32	0.35	0.40
0.7	0.77	0.59	0.47	0.38	0.34	0.32	0.35	0.41	0.50
0.6	0.74	0.56	0.44	0.37	0.35	0.36	0.43	0.54	0.68
0.5	0.71	0.52	0.41	0.38	0.40	0.45	0.59	0.78	1.0
0.4	0.66	0.47	0.40	0.43	0.54	0.69	0.95	1.3	1.7
0.3	0.66	0.48	0.52	0.73	1.2	1.8	2.7	---	---
0.2	0.56	0.56	1.0	1.8	---	---	---	---	---
0.1	0.60	2.1	---	---	---	---	---	---	---

Wye ø = 60°: Branch, Coefficient C

Ab/Ac	Qb/Qc								
	0.1	0.2	0.3	0.4	0.5	0.6	0.7	0.8	0.9
0.8	0.83	0.71	0.62	0.56	0.52	0.50	0.53	0.60	0.68
0.7	0.82	0.69	0.61	0.56	0.54	0.54	0.60	0.70	0.82
0.6	0.81	0.68	0.60	0.58	0.58	0.61	0.72	0.87	1.1
0.5	0.79	0.66	0.61	0.62	0.68	0.76	0.94	1.2	1.5
0.4	0.76	0.65	0.65	0.74	0.89	1.1	1.4	1.8	2.3
0.3	0.80	0.75	0.89	1.2	1.8	2.6	3.5	---	---
0.2	0.77	0.96	1.6	2.5	---	---	---	---	---
0.1	1.0	2.9	---	---	---	---	---	---	---

Tee ø = 90°: Branch, Coefficient C

Ab/Ac	Qb/Qc								
	0.1	0.2	0.3	0.4	0.5	0.6	0.7	0.8	0.9
0.8	0.95	0.92	0.92	0.93	0.94	0.95	1.1	1.2	1.4
0.7	0.95	0.94	0.95	0.98	1.0	1.1	1.2	1.4	1.6
0.6	0.96	0.97	1.0	1.1	1.1	1.2	1.4	1.7	2.0
0.5	0.97	1.0	1.1	1.2	1.4	1.5	1.8	2.1	2.5
0.4	0.99	1.1	1.3	1.5	1.7	2.0	2.4	---	---
0.3	1.1	1.4	1.8	2.3	---	---	---	---	---
0.2	1.3	1.9	2.9	---	---	---	---	---	---
0.1	2.1	---	---	---	---	---	---	---	---

Section C.4 Loss Coefficients, Diverging Junctions (Tees, Wyes)

Use the velocity (Vc) in the upstream section to determine the reference velocity pressure (Pv)

Pt = C x Pv (In. Wg.)

C) 45° Conical Wye, Round (Upstream Pv)

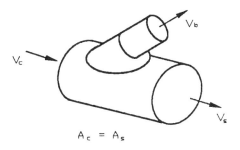

Branch, Coefficient C

V_b/V_c	0	0.2	0.4	0.6	0.8	1.0	1.2	1.4	1.6	1.8	2.0
C	1.0	0.84	0.61	0.41	0.27	0.17	0.12	0.12	0.14	0.18	0.27

Main, Coefficient C

V_s/V_c	0	0.1	0.2	0.3	0.4	0.5	0.6	0.8	1.0
C	0.35	0.28	0.22	0.17	0.13	0.09	0.06	0.02	0

E) 90° Tee, Round, with 90° Elbow, Branch 90° to Main (Upstream Pv)

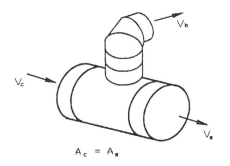

Branch, Coefficient C

V_b/V_c	0	0.2	0.4	0.6	0.8	1.0	1.2	1.4	1.6	1.8	2.0
C	1.0	1.03	1.08	1.18	1.33	1.56	1.86	2.2	2.6	3.0	3.4

Main, Coefficient C

V_s/V_c	0	0.1	0.2	0.3	0.4	0.5	0.6	0.8	1.0
C	0.35	0.28	0.22	0.17	0.13	0.09	0.06	0.02	0

G) 90° Conical Tee, Round, Rolled 45° with 45° Elbow, Branch 90° to Main (Upstream Pv)

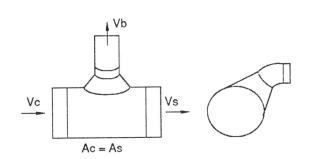

Branch, Coefficient C

V_b/V_c	0	0.2	0.4	0.6	0.8	1.0	1.2	1.4	1.6	1.8	2.0
C	1.0	0.94	0.88	0.84	0.80	0.82	0.84	0.87	0.90	0.95	1.02

Main, Coefficient C

V_s/V_c	0	0.1	0.2	0.3	0.4	0.5	0.6	0.8	1.0
C	0.35	0.28	0.22	0.17	0.13	0.09	0.06	0.02	0

I) 45° Wye, Round, Rolled 45° with 60° Elbow, Branch 90° to Main (Upstream Pv)

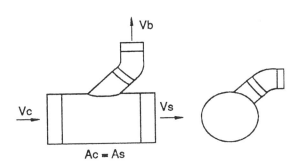

Branch, Coefficient C

V_b/V_c	0	0.2	0.4	0.6	0.8	1.0	1.2	1.4	1.6	1.8	2.0
C	1.0	0.88	0.77	0.68	0.65	0.69	0.73	0.88	1.14	1.54	2.2

Main, Coefficient C

V_s/V_c	0	0.1	0.2	0.3	0.4	0.5	0.6	0.8	1.0
C	0.35	0.28	0.22	0.17	0.13	0.09	0.06	0.02	0

Section C.4 Loss Coefficients, Diverging Junctions (Tees, Wyes)

Use the velocity (Vc) in the upstream section to determine the reference velocity pressure (Pv)

Pt = C x Pv (In. Wg.)

M) 45° Wye, Conical Main and Branch with 45° Elbow, Branch 90° to Main (Upstream Pv)

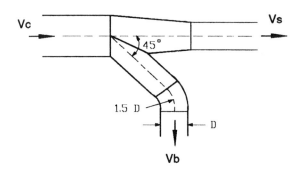

Branch, Coefficient C

Vb/Vc	0.2	0.4	0.6	0.7	0.8	0.9	1.0	1.1	1.2
C	0.76	0.60	0.52	0.50	0.51	0.52	0.56	0.61	0.68
Vb/Vc	1.4	1.6	1.8	2.0	2.2	2.4	2.6	2.8	3.0
C	0.86	1.1	1.4	1.8	2.2	2.6	3.1	3.7	4.2

Main, Coefficient C

Vs/Vc	0.2	0.4	0.6	0.8	1.0	1.2	1.4	1.6	1.8	2.0
C	0.14	0.06	0.05	0.09	0.18	0.30	0.46	0.64	0.84	1.0

N) Tee, 45° Entry, Rectangular Main and Branch (Upstream Pv)

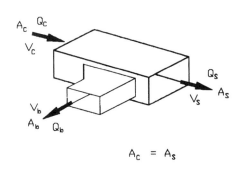

Branch, Coefficient C

Vb/Vc	Qb/Qc								
	0.1	0.2	0.3	0.4	0.5	0.6	0.7	0.8	0.9
0.2	0.91								
0.4	0.81	0.79							
0.6	0.77	0.72	0.70						
0.8	0.78	0.73	0.69	0.66					
1.0	0.78	0.98	0.85	0.79	0.74				
1.2	0.90	1.11	1.16	1.23	1.03	0.86			
1.4	1.19	1.22	1.26	1.29	1.54	1.25	0.92		
1.6	1.35	1.42	1.55	1.59	1.63	1.50	1.31	1.09	
1.8	1.44	1.50	1.75	1.74	1.72	2.24	1.63	1.40	1.17

Main, Coefficient C

Vs/Vc	0	0.1	0.2	0.3	0.4	0.5	0.6	0.8	1.0
C	0.35	0.28	0.22	0.17	0.13	0.09	0.06	0.02	0

P) Tee, 45° Entry, Rectangular Main and Branch with Damper (Upstream Pv)

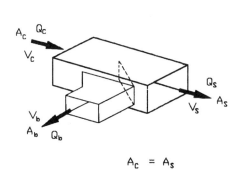

Branch, Coefficient C

Vb/Vc	Qb/Qc								
	0.1	0.2	0.3	0.4	0.5	0.6	0.7	0.8	0.9
0.2	0.61								
0.4	0.46	0.61							
0.6	0.43	0.50	0.54						
0.8	0.39	0.43	0.62	0.53					
1.0	0.34	0.57	0.77	0.73	0.68				
1.2	0.37	0.64	0.85	0.98	1.07	0.83			
1.4	0.57	0.71	1.04	1.16	1.54	1.36	1.18		
1.6	0.89	1.08	1.28	1.30	1.69	2.09	1.81	1.47	
1.8	1.33	1.34	2.04	1.78	1.90	2.40	2.77	2.23	1.92

Main, Coefficient C

Vs/Vc	0.2	0.4	0.6	0.8	1.0	1.2	1.4	1.6	1.8
C	0.03	0.04	0.07	0.12	0.13	0.14	0.27	0.30	0.25

Section C.4 Loss Coefficients, Diverging Junctions (Tees, Wyes)

Use the velocity (Vc) in the upstream section to determine the reference velocity pressure (Pv)

Pt = C x Pv (In. Wg.)

Q) Tee, Rectangular Main and Branch (Upstream Pv)

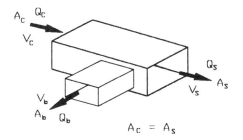

Branch, Coefficient C

Vb/Vc	Qb/Qc								
	0.1	0.2	0.3	0.4	0.5	0.6	0.7	0.8	0.9
0.2	1.03								
0.4	1.04	1.01							
0.6	1.11	1.03	1.05						
0.8	1.16	1.21	1.17	1.12					
1.0	1.38	1.40	1.30	1.36	1.27				
1.2	1.52	1.61	1.68	1.91	1.47	1.66			
1.4	1.79	2.01	1.90	2.31	2.28	2.20	1.95		
1.6	2.07	2.28	2.13	2.71	2.99	2.81	2.09	2.20	
1.8	2.32	2.54	2.64	3.09	3.72	3.48	2.21	2.29	2.57

Main, Coefficient C

Vs/Vc	0	0.1	0.2	0.3	0.4	0.5	0.6	0.8	1.0
C	0.35	0.28	0.22	0.17	0.13	0.09	0.06	0.02	0

R) Tee, Rectangular Main and Branch with Damper (Upstream Pv)

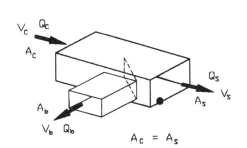

Branch, Coefficient C

Vb/Vc	Qb/Qc								
	0.1	0.2	0.3	0.4	0.5	0.6	0.7	0.8	0.9
0.2	0.58								
0.4	0.67	0.64							
0.6	0.78	0.76	0.75						
0.8	0.88	0.98	0.81	1.01					
1.0	1.12	1.05	1.08	1.18	1.29				
1.2	1.49	1.48	1.40	1.51	1.70	1.91			
1.4	2.10	2.21	2.25	2.29	2.32	2.48	2.53		
1.6	2.72	3.30	2.84	3.09	3.30	3.19	3.29	3.16	
1.8	3.42	4.58	3.65	3.92	4.20	4.15	4.14	4.10	4.05

Main, Coefficient C

Vs/Vc	0.2	0.4	0.6	0.8	1.0	1.2	1.4	1.6	1.8
C	0.03	0.04	0.07	0.12	0.13	0.14	0.27	0.30	0.25

Section C.4 Loss Coefficients, Diverging Junctions (Tees, Wyes)

Use the velocity (Vc) in the upstream section to determine the reference velocity pressure (Pv)

Pt = C x Pv (In. Wg.)

T) Tee, Rectangular Main to Round Branch (Upstream Pv)

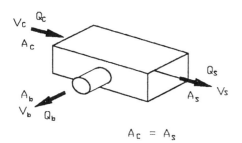

$A_c = A_s$

Branch, Coefficient C

Vb/Vc	Qb/Qc								
	0.1	0.2	0.3	0.4	0.5	0.6	0.7	0.8	0.9
0.2	1.00								
0.4	1.01	1.07							
0.6	1.14	1.10	1.08						
0.8	1.18	1.31	1.12	1.13					
1.0	1.30	1.38	1.20	1.23	1.26				
1.2	1.46	1.58	1.45	1.31	1.39	1.48			
1.4	1.70	1.82	1.65	1.51	1.56	1.64	1.71		
1.6	1.93	2.06	2.00	1.85	1.70	1.76	1.80	1.88	
1.8	2.06	2.17	2.20	2.13	2.06	1.98	1.99	2.00	2.07

Main, Coefficient C

Vs/Vc	0	0.1	0.2	0.3	0.4	0.5	0.6	0.8	1.0
C	0.35	0.28	0.22	0.17	0.13	0.09	0.06	0.02	0

U) Wye, Rectangular (Upstream Pv)

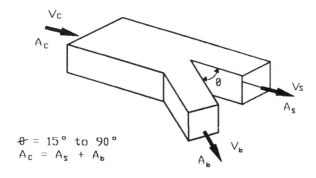

θ = 15° to 90°
$A_c = A_s + A_b$

Branch, Coefficient C

ø, Deg.	Vb/Vc						
	0.1	0.2	0.3	0.4	0.5	0.6	0.8
15	0.81	0.65	0.51	0.38	0.28	0.20	0.11
30	0.84	0.69	0.56	0.44	0.34	0.26	0.19
45	0.87	0.74	0.63	0.54	0.45	0.38	0.29
60	0.90	0.82	0.79	0.66	0.59	0.53	0.43
90	1.0	1.0	1.0	1.0	1.0	1.0	1.0

ø, Deg.	Vb/Vc					
	1.0	1.2	1.4	1.6	1.8	2.0
15	0.06	0.14	0.30	0.51	0.76	1.0
30	0.15	0.15	0.30	0.51	0.76	1.0
45	0.24	0.23	0.30	0.51	0.76	1.0
60	0.36	0.33	0.39	0.51	0.76	1.0
90	1.0	1.0	1.0	1.0	1.0	1.0

Main, Coefficient C

ø	15°-60°	90°				
Vs/Vc		As/Ac				
	0 - 1.0	0 - 0.4	0.5	0.6	0.7	>0.8
0	1.0	1.0	1.0	1.0	1.0	1.0
0.1	0.81	0.81	0.81	0.81	0.81	0.81
0.2	0.64	0.64	0.64	0.64	0.64	0.64
0.3	0.50	0.50	0.52	0.52	0.50	0.50
0.4	0.36	0.36	0.40	0.38	0.37	0.36
0.5	0.25	0.25	0.30	0.28	0.27	0.25
0.6	0.16	0.16	0.23	0.20	0.18	0.16
0.8	0.04	0.04	0.17	0.10	0.07	0.04
1.0	0	0	0.20	0.10	0.05	0
1.2	0.07	0.07	0.36	0.21	0.14	0.07
1.4	0.39	0.39	0.79	0.59	0.39	---
1.6	0.90	0.90	1.4	1.2	---	---
1.8	1.8	1.8	2.4	---	---	---
2.0	3.2	3.2	4.0	---	---	---

Section C.5 Loss Coefficients, Converging Junctions (Tees, Wyes)

Use the velocity (Vc) in the downstream section to determine the reference velocity pressure (Pv)

Pt = C x Pv (in. Wg.)

A) Converging Wye, Round (Downstream Pv)

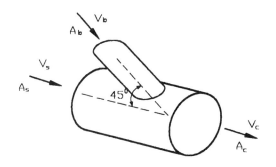

Branch, Coefficient C

Vb/Vc	Ab/Ac						
	0.1	0.2	0.3	0.4	0.6	0.8	1.0
0.4	-.56	-.44	-.35	-.28	-.15	-.04	0.05
0.5	-.48	-.37	-.28	-.21	-.09	0.02	0.11
0.6	-.38	-.27	-.19	-.12	0	0.10	0.18
0.7	-.26	-.16	-.08	-.01	0.10	0.20	0.28
0.8	-.21	-.02	0.05	0.12	0.23	0.32	0.40
0.9	0.04	0.13	0.21	0.27	0.37	0.46	0.53
1.0	0.22	0.31	0.38	0.44	0.53	0.62	0.69
1.5	1.4	1.5	1.5	1.6	1.7	1.7	1.8
2.0	3.1	3.2	3.2	3.2	3.3	3.3	3.3
2.5	5.3	5.3	5.3	5.4	5.4	5.4	5.4
3.0	8.0	8.0	8.0	8.0	8.0	8.0	8.0

Main, Coefficient C

Vs/Vc	Ab/Ac						
	0.1	0.2	0.3	0.4	0.6	0.8	1.0
0.1	-8.6	-4.1	-2.5	-1.7	-.97	-.58	-.34
0.2	-6.7	-3.1	-1.9	-1.3	-.67	-.36	-.18
0.3	-5.0	-2.2	-1.3	-.88	-.42	-.19	-.05
0.4	-3.5	-1.5	-.88	-.55	-.21	-.05	0.05
0.5	-2.3	-.95	-.51	-.28	-.06	0.06	0.13
0.6	-1.3	-.50	-.22	-.09	0.05	0.12	0.17
0.7	-.63	-.18	-.03	0.04	0.12	0.16	0.18
0.8	-.18	0.01	0.07	0.10	0.13	0.15	0.17
0.9	0.03	0.07	0.08	0.09	0.10	0.11	0.13
1.0	-0.01	0	0	0.10	0.02	0.04	0.05

B) Converging Tee, 90°, Round (Downstream Pv)

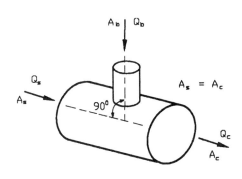

Branch, Coefficient C

Qb/Qc	Ab/Ac						
	0.1	0.2	0.3	0.4	0.6	0.8	1.0
0.1	0.40	-.37	-.51	-.46	-.50	-.51	-.52
0.2	3.8	0.72	0.17	-.02	-.14	-.18	-.24
0.3	9.2	2.3	1.0	0.44	0.21	0.11	-.08
0.4	16	4.3	2.1	0.94	0.54	0.40	0.32
0.5	26	6.8	3.2	1.1	0.66	0.49	0.42
0.6	37	9.7	4.7	1.6	0.92	0.69	0.57
0.7	43	13	6.3	2.1	1.2	0.88	0.72
0.8	65	17	7.9	2.7	1.5	1.1	0.86
0.9	82	21	9.7	3.4	1.8	1.2	0.99
1.0	101	26	12	4.0	2.1	1.4	1.1

Main, Coefficient C

Qb/Qc	0.1	0.2	0.3	0.4	0.5	0.6	0.7	0.8	0.9	1.0
C	0.16	0.27	0.38	0.46	0.53	0.57	0.59	0.60	0.59	0.55

Section C.5 Loss Coefficients, Converging Junctions (Tees, Wyes)

Use the velocity (Vc) in the upstream section to determine the reference velocity pressure (Pv)

Pt = C x Pv (In. Wg.)

C) Converging Tee, Round Branch to Rectangular Main (Downstream Pv)

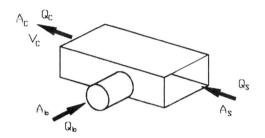

When:

A_b/A_s	A_s/A_c	A_b/A_c
0.5	1.0	0.5

Branch, Coefficient C

Vc	Q_b/Q_c									
	0.1	0.2	0.3	0.4	0.5	0.6	0.7	0.8	0.9	1.0
< 1200 fpm	-.63	-.55	0.13	0.23	0.78	1.30	1.93	3.10	4.88	5.60
> 1200 fpm	-.49	-.21	0.23	0.60	1.27	2.06	2.75	3.70	4.93	5.95

Main, Coefficient C

Q_b/Q_c	0.1	0.2	0.3	0.4	0.5	0.6	0.7	0.8	0.9	1.0
C	0.16	0.27	0.38	0.46	0.53	0.57	0.59	0.60	0.59	0.55

D) Converging Tee, Rectangular Main and Branch (Downstream Pv)

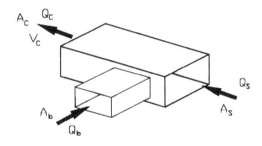

When:

A_b/A_s	A_s/A_c	A_b/A_c
0.5	1.0	0.5

Branch, Coefficient C

Vc	Q_b/Q_c									
	0.1	0.2	0.3	0.4	0.5	0.6	0.7	0.8	0.9	1.0
< 1200 fpm	-.75	-.53	-.03	0.33	1.03	1.10	2.15	2.93	4.18	4.78
> 1200 fpm	-.69	-.21	0.23	0.67	1.17	1.66	2.67	3.36	3.93	5.13

Main, Coefficient C

Q_b/Q_c	0.1	0.2	0.3	0.4	0.5	0.6	0.7	0.8	0.9	1.0
C	0.16	0.27	0.38	0.46	0.53	0.57	0.59	0.60	0.59	0.55

Section C.6 Loss Coefficients, Transitions (Diverging Flow)

Use the velocity (Vc) in the upstream section to determine the reference velocity pressure (Pv)

Pt = C x Pv (In. Wg.)

A) Transition, Round, Conical (Upstream Pv)

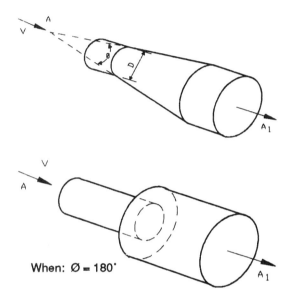

When: Ø = 180°

CFM	A_1/A	Coefficient C ø							
		16°	20°	30°	45°	60°	90°	120°	180°
<2400	2	0.14	0.19	0.32	0.33	0.33	0.32	0.31	0.30
	4	0.23	0.30	0.46	0.61	0.68	0.64	0.63	0.62
	6	0.27	0.33	0.48	0.66	0.77	0.74	0.73	0.72
	10	0.29	0.38	0.59	0.76	0.80	0.83	0.84	0.83
	>16	0.31	0.38	0.60	0.84	0.88	0.88	0.88	0.88
2400 to 12000	2	0.07	0.12	0.23	0.28	0.27	0.27	0.27	0.26
	4	0.15	0.18	0.36	0.55	0.59	0.59	0.58	0.57
	6	0.19	0.28	0.44	0.90	0.70	0.71	0.71	0.69
	10	0.20	0.24	0.43	0.76	0.80	0.81	0.81	0.81
	>16	0.21	0.28	0.52	0.76	0.87	0.87	0.87	0.87
>12000	2	0.05	0.07	0.12	0.27	0.27	0.27	0.27	0.27
	4	0.17	0.24	0.38	0.51	0.56	0.58	0.58	0.57
	6	0.16	0.29	0.46	0.60	0.69	0.71	0.70	0.70
	10	0.21	0.33	0.52	0.60	0.76	0.83	0.84	0.83
	>16	0.21	0.34	0.56	0.72	0.79	0.85	0.87	0.89

Section C.6 Loss Coefficients, Transitions (Diverging Flow)

Use the velocity (Vc) in the upstream section to determine the reference velocity pressure (Pv)

Pt = C x Pv (In. Wg.)

C) Transition, Round to Rectangular or Rectangular to Round (Upstream Pv)

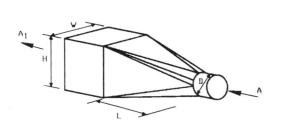
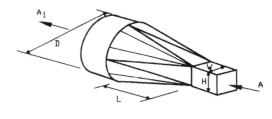

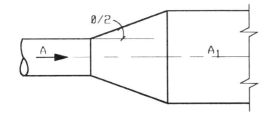
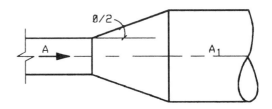

A_1/A	Coefficient C θ								
	8°	10°	14°	20°	30°	45°	60°	90°	180°
2	0.14	0.15	0.20	0.25	0.30	0.33	0.33	0.33	0.30
4	0.20	0.25	0.34	0.45	0.52	0.58	0.62	0.64	0.64
6	0.21	0.30	0.42	0.53	0.63	0.72	0.78	0.79	0.79
>10	0.24	0.30	0.43	0.53	0.64	0.75	0.84	0.89	0.88

E) Transition, Rectangular, Two Sides Straight (Upstream Pv)

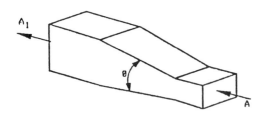

A_1/A	Coefficient C θ							
	15°	30°	45°	60°	90°	120°	150°	180°
2	0.13	0.24	0.35	0.37	0.38	0.37	0.36	0.35
4	0.19	0.43	0.60	0.68	0.70	0.69	0.67	0.66
6	0.22	0.48	0.65	0.76	0.83	0.82	0.81	0.80
10	0.26	0.53	0.69	0.82	0.93	0.93	0.92	0.91

F) Transition, Rectangular, Three Sides Straight (Upstream Pv)

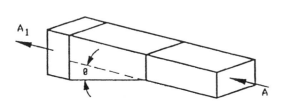

A_1/A	Coefficient C θ						
	10°	15°	20°	30°	45°	60°	90°
2	0.14	0.13	0.15	0.24	0.35	0.37	0.38
4	0.17	0.19	0.22	0.42	0.60	0.68	0.70
10	0.24	0.26	0.36	0.53	0.69	0.82	0.93
17	0.26	0.27	0.40	0.56	0.71	0.86	1.00

Section C.7 Loss Coefficients, Transitions (Converging Flow)

Use the velocity (Vc) in the downstream section to determine the reference velocity pressure (Pv)

Pt = C x Pv (In. Wg.)

A) Contraction, Round and Rectangular, Gradual to Abrupt (Downstream Pv)

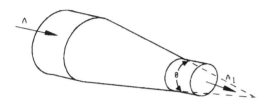

A_1/A	Coefficient C $\emptyset$						
	10°	15°-40°	50°-60°	90°	120°	150°	180°
2	0.05	0.05	0.06	0.12	0.18	0.24	0.26
4	0.05	0.04	0.07	0.17	0.27	0.35	0.41
6	0.05	0.04	0.07	0.18	0.28	0.36	0.42
10	0.05	0.05	0.08	0.19	0.29	0.37	0.43

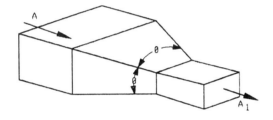

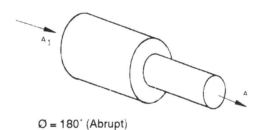

$\emptyset = 180°$ (Abrupt)

C) Contraction, Rectangular, Two sides Parallel (Downstream Pv)

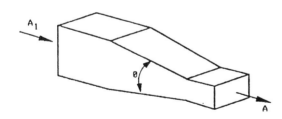

A_1/A	Coefficient C $\emptyset$						
	15°	30°	45°	60°	90°	120°	150°	180°
2	0.05	0.05	0.06	0.07	0.14	0.20	0.25	0.28
4	0.04	0.04	0.06	0.07	0.18	0.28	0.36	0.41
6	0.04	0.04	0.06	0.07	0.18	0.28	0.36	0.42
10	0.05	0.05	0.07	0.08	0.19	0.29	0.37	0.43

Section C.8 Loss Coefficients, Entries

Use the velocity (Vc) in the downstream section to determine the reference velocity pressure (Pv)

Pt = C x Pv (In. Wg.)

A) Duct Mounted in Wall, Round and Rectangular (Exit from Plenum) (Downstream Pv)

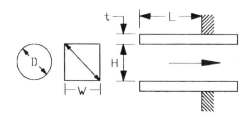

Rectangular: $D = \dfrac{2\,HW}{(H + W)}$

t/D	Coefficient C						
	L/D						
	0	0.002	0.01	0.05	0.2	0.5	>1.0
~0	0.50	0.57	0.68	0.80	0.92	1.0	1.0
0.02	0.50	0.51	0.52	0.55	0.66	0.72	0.72
>0.05	0.50	0.50	0.50	0.50	0.50	0.50	0.50

With Screen or Perforated Plate:
a. Sharp Edge (t/D < 0.05): Cs = 1 + C1
b. Thick Edge (t/D > 0.05) Cs = C + C1

Where:
Cs is coefficient adjusted for screen or perforated plate at entrance.
C is from above table
C1 is from Section 10 (screen or perforated plate).

B) Smooth Converging Bellmouth, Round, without End Wall (Downstream Pv)

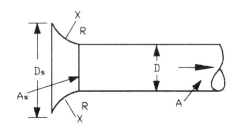

Coefficient C (See Note Below for Screen)

R/D	0	0.01	0.02	0.03	0.04	0.05
C	1.0	0.87	0.74	0.61	0.51	0.40

R/D	0.06	0.08	0.10	0.12	0.16	0.20 & up
C	0.32	0.20	0.15	0.10	0.06	0.03

C) Smooth Converging Bellmouth, Round, with End Wall (Downstream Pv)

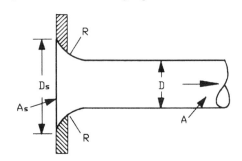

Coefficient C (See Note Below for Screen)

R/D	0	0.01	0.02	0.03	0.04	0.05
C	0.50	0.43	0.36	0.31	0.26	0.22

R/D	0.06	0.08	0.10	0.12	0.16	0.20 & up
C	0.20	0.15	0.12	0.09	0.06	0.03

Note: With screen at As:

$$Cs = C + \dfrac{C1}{(As/A)^2}$$

Where:
Cs is the new coefficient for the fitting with a screen
C is from the table above
C1 is from Section 10

Section C.9 Loss Coefficients, Exits

Use the velocity (Vc) in the upstream section to determine the reference velocity pressure (Pv)

$Pt = C \times Pv$ (In. Wg.)

A) Abrupt Exit from Round or Rectangular Duct (Entrance Into Plenum) (Upstream Pv)

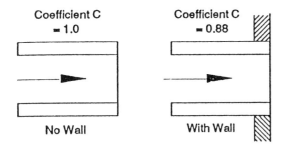

With screen:

$C_S = C + C_1$

Where:
C_S is the new coefficient for the fitting with a screen
C is from the table above
C_1 is from Section 10

B) Exit, Conical, Round, with or without a Wall (Upstream Pv)

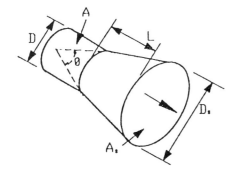

Coefficient C (See Note Below for Screen)

As/A	θ						
	8°	10°	14°	20°	30°	45°	>60°
2	0.36	0.33	0.37	0.51	0.90	1.0	1.0
4	0.24	0.21	0.28	0.40	0.70	0.99	1.0
6	0.20	0.19	0.26	0.37	0.67	0.99	1.0
10	0.18	0.16	0.24	0.36	0.68	0.99	1.0
16	0.16	0.16	0.20	0.36	0.66	0.99	1.0

C) Exit, Plane Diffuser, Rectangular, with or without a Wall (Upstream Pv)

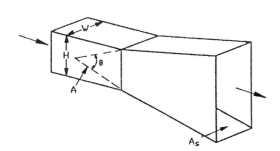

Coefficient C (See Note Below for Screen)

As/A	θ						
	8°	10°	14°	20°	30°	45°	>60°
2	0.50	0.51	0.56	0.63	0.80	0.96	1.0
4	0.34	0.38	0.48	0.63	0.76	0.91	1.0
6	0.32	0.34	0.41	0.56	0.70	0.84	0.96

Multiply C by 0.88 for wall

Note: With screen in opening:

$$C_S = C + \frac{C_1}{(A_S/A)^2}$$

Where:
C_S is the new coefficient for the fitting with a screen
C is from the table above
C_1 is from Section 10

Section C.9 Loss Coefficients, Exits

Use the velocity (Vc) in the upstream section to determine the reference velocity pressure (Pv)

Pt = C x Pv (In. Wg.)

F) Exit, Discharge to Atmosphere from a 90° Elbow, Round and Rectangular (Upstream Pv)

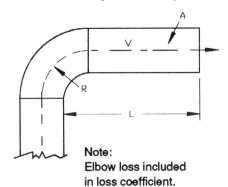

Note:
Elbow loss included in loss coefficient.

Rectangular: Coefficient C (See Note Below for Screen)

R/W	L/W									
	0	0.5	1.0	1.5	2.0	3.0	4.0	6.0	8.0	12.0
0	3.0	3.1	3.2	3.0	2.7	2.4	2.2	2.1	2.1	2.0
0.75	2.2	2.2	2.1	1.8	1.7	1.6	1.6	1.5	1.5	1.5
1.0	1.8	1.5	1.4	1.4	1.3	1.3	1.2	1.2	1.2	1.2
1.5	1.5	1.2	1.1	1.1	1.1	1.1	1.1	1.1	1.1	1.1
2.5	1.2	1.1	1.1	1.0	1.0	1.0	1.0	1.0	1.0	1.0

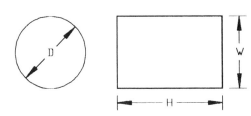

Round: Coefficient C

L/D	0.9	1.3
C	1.5	1.4

When:
R/D = 1.0 (Round)

G) Exit, Duct Flush with Wall, Flow along Wall (Upstream Pv)

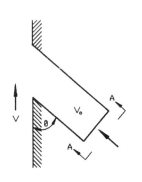

Rectangular:

Coefficient C (See Note Below for Screen)

Aspect Ratio (H/W)	ø	V/V₀				
		0	0.5	1.0	1.5	2.0
0.1 - 0.2	30°-90°	1.0	0.95	1.2	1.5	1.8
	120°	1.0	0.92	1.1	1.4	1.9
	150°	1.0	0.75	0.95	1.4	1.8
0.5 - 2.0	30°-45°	1.0	1.0	1.1	1.3	1.6
	60°	1.0	0.90	1.1	1.4	1.6
	90°	1.0	0.80	0.95	1.4	1.7
	120°	1.0	0.80	0.95	1.3	1.7
	150°	1.0	0.82	0.83	1.0	1.3
5 - 10	45°	1.0	0.92	0.93	1.1	1.3
	60°	1.0	0.87	0.87	1.0	1.3
	90°	1.0	0.82	0.80	0.97	1.2
	120°	1.0	0.80	0.76	0.90	0.98

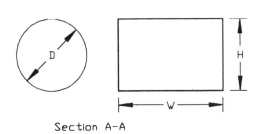

Section A-A

Round:

Coefficient C (See Note Below for Screen)

ø	V/V₀				
	0	0.5	1.0	1.5	2.0
30°-45°	1.0	1.0	1.1	1.3	1.6
60°	1.0	0.90	1.1	1.4	1.6
90°	1.0	0.80	0.95	1.4	1.7
120°	1.0	0.80	0.95	1.3	1.7
150°	1.0	0.82	0.83	1.0	1.3

Note: With screen in opening:
$C_s = C + C_1$

Where:
C_s is the new coefficient for the fitting with a screen
C is from the table above
C_1 is from Section 10

Section C.10 Loss Coefficients, Screens

Use the velocity (Vc) in the downstream section to determine the reference velocity pressure (Pv)

Pt = C x Pv (In. Wg.)

A) Obstruction, Screen, Round and Rectangular (Downstream Pv)

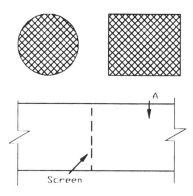

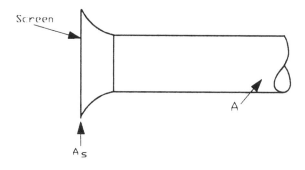

n = free area ratio of screen
A = area of duct
A_s = cross-sectional area of duct or fitting where screen is located

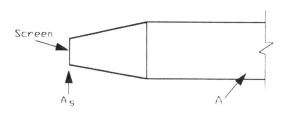

Coefficient C

As/A	n							
	0.3	0.4	0.5	0.6	0.7	0.8	0.9	1.0
0.2	155	75	42	24	15	8.0	3.5	0
0.3	69	33	19	11	6.4	3.6	1.6	0
0.4	39	19	10	6.1	3.6	2.0	0.88	0
0.6	17	8.3	4.7	2.7	1.6	0.89	0.39	0
0.8	9.7	4.7	2.7	1.5	0.91	0.50	0.22	0
1.0	6.2	3.0	1.7	0.97	0.58	0.32	0.14	0
1.2	4.3	2.1	1.2	0.67	0.40	0.22	0.10	0
1.4	3.2	1.5	0.87	0.49	0.30	0.16	0.07	0
1.6	2.4	1.2	0.66	0.38	0.23	0.12	0.05	0
2.0	1.6	0.75	0.43	0.24	0.15	0.08	0.04	0
2.5	0.99	0.48	0.27	0.16	0.09	0.05	0.02	0
3.0	0.69	0.33	0.19	0.11	0.06	0.04	0.02	0
4.0	0.39	0.19	0.11	0.06	0.04	0.02	0.01	0
6.0	0.17	0.08	0.05	0.03	0.02	0.01	0	0

Section C.11 Loss Coefficients, Obstructions (Constant Velocities)

Use the velocity (Vc) in the upstream section to determine the reference velocity pressure (Pv)

Pt = C x Pv (In. Wg.)

A) Damper, Butterfly, Thin Plate, Round (Upstream Pv)

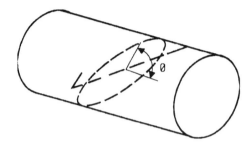

Coefficient C							
θ	0°	10°	20°	30°	40°	50°	60°
C	0.20	0.52	1.5	4.5	11	29	108

0° is full open

B) Damper, Butterfly, Thin Plate, Rectangular (Upstream Pv)

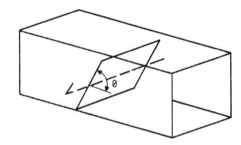

Coefficient C							
θ	0°	10°	20°	30°	40°	50°	60°
C	0.04	0.33	1.2	3.3	9.0	26	70

0° is full open

Section C.11 Loss Coefficients, Obstructions (Constant Velocities)

Use the velocity (Vc) in the upstream section to determine the reference velocity pressure (Pv)

Pt = C x Pv (In. Wg.)

E) Damper, Rectangular, Parallel Blades (Upstream Pv)

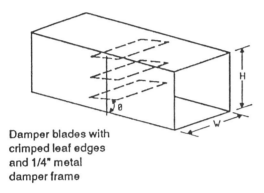

Damper blades with crimped leaf edges and 1/4" metal damper frame

L/R	Coefficient C ∅								
	80°	70°	60°	50°	40°	30°	20°	10°	0° open
0.3	116	32	14	9.0	5.0	2.3	1.4	0.79	0.52
0.4	152	38	16	9.0	6.0	2.4	1.5	0.85	0.52
0.5	188	45	18	9.0	6.0	2.4	1.5	0.92	0.52
0.6	245	45	21	9.0	5.4	2.4	1.5	0.92	0.52
0.8	284	55	22	9.0	5.4	2.5	1.5	0.92	0.52
1.0	361	65	24	10	5.4	2.6	1.6	1.0	0.52
1.5	576	102	28	10	5.4	2.7	1.6	1.0	0.52

$$\frac{L}{R} = \frac{NW}{2(H+W)}$$

Where:
N is number of damper blades
W is duct dimension parallel to blade axis
L is sum of damper blade lengths
R is perimeter of duct
H is duct dimension perpendicular to blade axis

F) Damper, Rectangular, Opposed Blades (Upstream Pv)

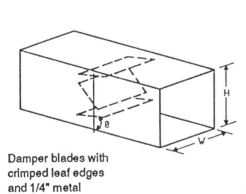

Damper blades with crimped leaf edges and 1/4" metal damper frame

L/R	Coefficient C ∅								
	80°	70°	60°	50°	40°	30°	20°	10°	0° open
0.3	807	284	73	21	9.0	4.1	2.1	0.85	0.52
0.4	915	332	100	28	11	5.0	2.2	0.92	0.52
0.5	1045	337	122	33	13	5.4	2.3	1.0	0.52
0.6	1121	411	148	38	14	6.0	2.3	1.0	0.52
0.8	1299	495	188	54	18	6.6	2.4	1.1	0.52
1.0	1521	547	245	65	21	7.3	2.7	1.2	0.52
1.5	1654	677	361	107	28	9.0	3.2	1.4	0.52

$$\frac{L}{R} = \frac{NW}{2(H+W)}$$

Where:
N is number of damper blades
W is duct dimension parallel to blade axis
L is sum of damper blade lengths
R is perimeter of duct
H is duct dimension perpendicular to blade axis

Appendix D
HVAC Field Test Report Forms

As explained at length in Chapter 7, testing and balancing of systems after construction is an integral part of the construction process. Similarly, existing systems require testing and periodic balancing to keep them operating at maximum efficiency, and as designed. Field test and inspection data are best recorded on prepared forms, using a different form for each type of test and for different types of equipment. Use of prepared forms ensures:

- That all required data are recorded.
- That the necessary tests and procedures have been performed.

The filled in report forms then serve as a permanent record of the last adjustments made to the system. They also serve as an important aid in maintenance and in diagnosing system problems.

The forms and description of use in this appendix are reprinted, with permission, from the Sheet Metal and Air Conditioning Contractors National Association (SMACNA) manual—*HVAC Systems, Testing, Balancing and Adjusting*.

Form D.1. System diagram.
Form D.2. Air apparatus test report.
Form D.3. Apparatus coil test report.
Form D.4. Gas/oil-fired heat apparatus test report.
Form D.5. Electric coil/duct heater test report.
Form D.6. Fan test report.
Form D.7. Rectangular duct traverse report.
Form D.8. Air outlet test report.
Form D.9. Terminal unit coil check report.
Form D.10. Package rooftop/heat pump/air conditioning unit test report.
Form D.11. Compressor and/or condenser test report.
Form D.12. Cooling tower or evaporative condenser test report.
Form D.13. Pump test report.
Form D.14. Instrument calibration report.

Form D.1 System Diagram

This form is to be used primarily for a schematic layout of air distribution systems, but it may be used for hydronic systems as well. A single-line system diagram is highly recommended to ensure systematic and efficient procedures. Be sure to show quantities of outside air, return air and relief air, sizes and cfm for main ducts, sizes and cfm of outlets and inlets, and all dampers, regulating devices and terminal units. All outlets should be numbered before filling out Outlet Test Report form. While diagrams are suggested, the use of this form is not mandatory. If appropriate, a larger (similar) diagram sheet may be used in the report.

SYSTEM DIAGRAM

PROJECT _____ SYSTEM _____

LOCATION _____ DATE _____

Sheet Metal & Air Conditioning Contractors
National Association

Form D.2 Air Apparatus Test Report

The performance of air-handling apparatus with coils is to be reported on this form. In addition, there is space for other information, there is space for other information that will be of benefit to the design engineer, the maintenance engineer and/or the TAB contractor. Motor voltage and amperage for three phase motors should be reported for all three legs (T_1, T_2, T_3). If the design engineer did not specify a design quantity for any item in the test data section, place an X in the slot for the design quantity and record the actual quantity. However, if the equipment manufacturer furnished ratings, enter them in the designs columns. If motor ratings differ from design, provide an explanation at the bottom of the page.

AIR APPARATUS TEST REPORT

PROJECT _____ SYSTEM/UNIT _____

LOCATION _____

UNIT DATA	
Make/Model No.	
Type/Size	
Serial Number	
Arr./Class	
Discharge	
Make Sheave	
Sheave Diam/Bore	
No. Belts/make/size	
No. Filters/type/size	

MOTOR DATA	
Make/Frame	
H.P./RPM	
Volts/Phase/Hertz	
F.L. Amps/S.F.	
Make Sheave	
Sheave Diam/Bore	
Sheave ℄ Distance	

TEST DATA	DESIGN	ACTUAL
Total CFM		
Total S.P.		
Fan RPM		
Motor Volts T_1-T_2 T_2-T_3 T_3-T_1		
Motor Amps T_1 T_2 T_3		
Outside Air CFM		
Return Air CFM		

TEST DATA	DESIGN	ACTUAL
Discharge S.P.		
Suction S.P.		
Reheat Coil △ S.P.		
Cooling Coil △ S.P.		
Preheat Coil △ S.P.		
Filters △ S.P.		
Vortex Damp. Position		
Out. Air Damp. Position		
Ret. Air Damp. Position		

REMARKS:

TEST DATE _____ READINGS BY _____

Sheet Metal & Air Conditioning Contractors
National Association

Form D.3 Apparatus Coil Test Report

This form is to be used for recording performance of chilled water, hot water, steam, or DX coils, and for "run-around" heat recovery systems. The performance of as many as four coils (or two "run-around" systems) can be shown on the same sheet.

APPARATUS COIL TEST REPORT

PROJECT _____

COIL DATA	COIL NO.	COIL NO.	COIL NO.	COIL NO.
System Number				
Location				
Coil Type				
No. Rows-Fins/In.				
Manufacturer				
Model Number				
Face Area, Sq. Ft.				

TEST DATA	DESIGN	ACTUAL	DESIGN	ACTUAL	DESIGN	ACTUAL	DESIGN	ACTUAL
Air Qty., CFM								
Air Vel., FPM								
Press. Drop, In.								
Out. Air DB/WB								
Ret. Air DB/WB								
Ent. Air DB/WB								
Lvg. Air DB/WB								
Air △T								
Water Flow, GPM								
Press. Drop, PSI								
Ent. Water Temp.								
Lvg. Water Temp.								
Water △T								
Exp. Valve/Refrig.								
Refrig. Suction Press.								
Refrig. Suction Temp.								
Inlet Steam Press.								

REMARKS:

TEST DATE _____ READINGS BY _____

Sheet Metal & Air Conditioning Contractors
National Association

Form D.4 Gas/Oil Fired Heat Apparatus Test Report

Data for gas or oil fired devices such as unit heaters and duct furnaces will be recorded on this form. This report is not intended to be used in lieu of a factory start-up equipment report but could be used as a supplement. All available design data should be reported. Motor information can apply to the burner motor, burner fan motor, unit air fan motor, etc., depending on the application or equipment. Therefore, designate the motor of the recorded data.

GAS/OIL FIRED HEAT APPARATUS TEST REPORT

PROJECT_____

UNIT DATA	UNIT NO.	UNIT NO.	UNIT NO.
System			
Location			
Make/Model			
Type/Size			
Serial Number			
Type Fuel/Input			
Output			
Ignition Type			
Burner Control			
Volts/Phase/Hertz			
H.P./RPM			
F.L. Amps/S.F.			
Drive Data			

TEST DATA	DESIGN	ACTUAL	DESIGN	ACTUAL	DESIGN	ACTUAL
CFM						
Ent./Lvg. Air Temp.						
Air Temp. ΔT						
Ent./Lvg. Air Press.						
Air Press. ΔP						
Low Fire Input						
High Fire Input						
Manifold Press./CFH						
High Limit Setting						
Operating Set Point						

REMARKS:

TEST DATE _____ READINGS BY _____

Sheet Metal & Air Conditioning Contractors
National Association

Form D.5 Electric Coil/Duct Heater Test Report

This form is to be used for electric furnaces, or for electric coils installed in built-up units or in branch ducts. "Min.AirVel." is the manufacturers recommended minimum air flow velocity.

PROJECT _____ MFG./MODEL _____

ELECTRIC COIL/DUCT HEATER TEST REPORT

DESCRIPTION			RATED DATA						TEST DATA						
COIL NO.	SYSTEM / LOCATION	KW	STAGES	VOLTS PHASE	AMPS	CFM	FACE AREA, FT.²	MIN. AIR VEL. FPM	KW	VEL. FPM	COIL CFM	EAT	LAT	VOLTAGE T1-T2 / T2-T3 / T3-T1	AMPERAGE T1 / T2 / T3

REMARKS:

TEST DATE _____ READINGS BY _____

Sheet Metal & Air Conditioning Contractors
National Association

Form D.6 Fan Test Report

This form is to be used with supply, return, or exhaust fans. Since housings for various types of fans may have many different shapes and arrangements, not all entry blanks will be needed for testing a particular fan. The performance of up to three fans may be reported on this sheet.

FAN TEST REPORT

PROJECT _____

FAN DATA	FAN NO.	FAN NO.	FAN NO.
Location			
Service			
Manufacturer			
Model Number			
Serial Number			
Type/Class			
Motor Make/Style			
Motor H.P./RPM/Frame			
Volts/Phase/Hertz			
F.L. Amps/S.F.			
Motor Sheave Make/Model			
Motor Sheave Diam./Bore			
Fan Sheave Make			
Fan Sheave Diam./Bore			
No. Belts/Make/Size			
Sheave ₵ Distance			

TEST DATA	DESIGN	ACTUAL	DESIGN	ACTUAL	DESIGN	ACTUAL
CFM						
Fan RPM						
S.P. In/Out						
Total S.P.						
Voltage $T_1\text{-}T_2$ $T_2\text{-}T_3$ $T_3\text{-}T_1$						
Amperage T_1 T_2 T_3						

REMARKS:

TEST DATE _____ READINGS BY _____

Sheet Metal & Air Conditioning Contractors
National Association

Form D.7 Rectangular⁽¹⁾ Duct Traverse Report

This form is to be used as a worksheet for recording the results of a Pitot tube traverse in a rectangular duct. It is recommended that the velocity pressures be recorded in one-half of each of the spaces provided and converted to velocities in the other half of each space at a later time. The velocity shall be averaged (not the velocity pressures).
Instructions for making a traverse are given in the text, Section 7.4a and in Figure 7.7b.

(1) For circular ducts see text, Section 7.4b and Figure 7.10.

RECTANGULAR DUCT TRAVERSE REPORT

PROJECT_____ SYSTEM/UNIT _____

LOCATION/ZONE _____ ACTUAL AIR TEMP. _____ DUCT S.P. _____

DUCT		REQUIRED		ACTUAL	
SIZE _____ SQ. FT. _____		FPM _____ CFM _____		FPM _____ CFM _____	

(SEE REVERSE SIDE FOR INSTRUCTIONS)

POSITION	1	2	3	4	5	6	7	8	9	10	11	12	13	14	15
1															
2															
3															
4															
5															
6															
7															
8															
9															
10															
11															
12															
13															
VELOCITY SUB-TOTALS															

REMARKS:

TEST DATE _____ READINGS BY _____

Sheet Metal & Air Conditioning Contractors
National Association

Form D.8 Air Outlet Test Report

As this form can be used as both a worksheet and a final report form, TAB crews are encouraged to record all readings on this test report form. However, it is not necessary to record preliminary velocity readings on the final sheet unless requested by the design engineer. If more than two sets of preliminary readings are necessary or required, the data can be entered in the two blank columns between "Preliminary" and "Final". The outlet number refers to the number assigned on the schematic layout drawn on form D.1. The column entitled "Type" is to be used for recording the type or model number of the air outlet.
If the final adjusted cfm of any air outlet varies by more than ±10% from the design cfm, a not should be placed in the remarks column indicating the amount of variance. The "remarks" section at the bottom of the sheet should be used to provide known or potential reasons for such deviation.

AIR OUTLET TEST REPORT

PROJECT _____ SYSTEM _____

OUTLET MANUFACTURER _____ TEST APPARATUS _____

AREA SERVED	OUTLET				DESIGN		PRELIMINARY				FINAL		REMARKS
	NO.	TYPE	SIZE	AK	CFM	VEL	VEL. OR CFM	VEL. OR CFM			VEL	CFM	

REMARKS:

TEST DATE _____ READINGS BY _____

Sheet Metal & Air Conditioning Contractors
National Association

Form D.9 Terminal Unit Coil Check Report

This form is used as a worksheet to check the water coil of terminal units. Any of the three alternate methods for determining water flow or heat transfer indicated on the test report form is acceptable.

Abbreviations

ΔP — Pressure difference
ΔT — Temperature difference
EWT — Entering Water Temperature
LWT — Leaving Water Temperature
EAT — Entering Air Temperature
LAT — Leaving Air Temperature

TERMINAL UNIT COIL CHECK REPORT

PROJECT _____ SYSTEM _____ MANUFACTURER _____

ROOM NO.	RISER NO.	UNIT SIZE	DESIGN GPM	ALTERNATE NO. 1			ALTERNATE NO. 2			ALTERNATE NO. 3			ΔP	ΔT
				DESIGN ΔP	ENT. WTR. PR.	LVG. WTR. PR.	DESIGN ΔT	EWT	LWT	DESIGN ΔT	EAT	LAT		

NOTE: USE ONE OF THE ABOVE ALTERNATE METHODS

REMARKS: _____ WATER SUPPLY TEMP. _____ OUT. AIR TEMP. _____

TEST DATE _____ READINGS BY _____

Sheet Metal & Air Conditioning Contractors
National Association

Form D.10 Package Rooftop/Heat Pump/Air Conditioning Unit Test Report

Test data from package units of all types is to be recorded on this form, with most of the data being furnished and verified by the installing contractor. If the unit has components other than the evaporator fan, DX coil, compressor and condenser fan(s), use the appropriate test report form such as: Form D.3 for water or steam coils; Form D.4 for direct fired heaters; Form D.5 for electric coils; Form D.6 for return air fans.

PACKAGE ROOFTOP/HEAT PUMP/AIR CONDITIONING UNIT TEST REPORT

PROJECT _____ SYSTEM/UNIT _____

LOCATION _____

UNIT DATA	
Make/Model Number	
Type/Size	
Serial Number	
Type Filters/Size	
Fan Sheave make	
Fan Sheave Diam./Bore	
No. Belts/make/size	
Type Heating Section*	

MOTOR DATA	
Make/Frame	
H.P./RPM	
Volts/Phase/Hertz	
F.L. Amps/S.F.	
Make Sheave	
Sheave Diam./Bore	
Sheave ₡ Distance	

TEST DATA EVAPORATOR	DESIGN	ACTUAL
Total CFM		
Total S.P.		
Discharge S.P.		
Suction S.P.		
Out. Air CFM		
Out. Air DB/WB		
Ret. Air CFM		
Ret. Air DB/WB		
Ent. Air DB/WB		
Lvg. Air DB/WB		
Fan RPM		
Voltage T_1-T_2, T_2-T_3, T_3-T_1		
Amperage T_1 T_2 T_3		

TEST DATA CONDENSER	DESIGN	ACTUAL
Refrigerant/Lbs		
Compr. Mfr./Number		
Compr. Model/Ser. Number		
Low Amb. Control		
Suction Press./Temp.		
Cond. Press./Temp.		
Crankcase Htr. Amps.		
Compr. Volts T_1-T_2, T_2-T_3, T_3-T_1		
Compr. Amps T_1 T_2 T_3		
L.P./H.P. Cutout Setting		
No. of fans/fan RPM		
Cond. Fan HP/CFM		
Cond. fan Volts/Amps/ϕ		

REMARKS:

TEST DATE _____ READINGS BY _____

Sheet Metal & Air Conditioning Contractors
National Association

Form D.11 Compressor and/or Condenser Test Report

This form may also be used to record data for the refrigerant side of unitary systems, "bare" compressors, separate air-cooled condensers or separate water cooled condensers. This form does not attempt to indicate the performance or efficiency of the machine except as may be determined by the design engineer from the data contained therein.

This form or the manufacturer's form should be substantially completed and verified by the manufacturers' representatives and/or the installing contractor before the HVAC distribution systems are balanced.

COMPRESSOR AND/OR CONDENSER TEST REPORT

PROJECT _____

UNIT DATA	UNIT NO.	UNIT NO.	UNIT NO.
LOCATION			
Unit Manufacturer			
Unit Model/Ser. Number			
Compressor Manufacturer			
Compr. Model/Ser. Number			
Refrigerant/Lbs.			
Low Amb. Control			

TEST DATA	DESIGN	ACTUAL	DESIGN	ACTUAL	DESIGN	ACTUAL
Suction Press./Temp.						
Cond. Press./Temp.						
Oil Press./Temp.						
Voltage T_1-T_2 T_2-T_3 T_3-T_1						
Amps T_1 T_2 T_3						
KW Input						
Crankcase Htr. Amps						
No. of Fans/Fan RPM/CFM						
Fan Motor Make/Frame/H.P.						
Fan Motor Volts/Amps						
Duct Inlet/Outlet S.P.						
Ent./Lvg. Air D.B.						
Cond. Wtr. Temp. In/Out						
Cond. Wtr. Press. In/Out						
Control Setting						
Unloader Set Points						
L.P./H.P. Cutout Setting						

REMARKS:

TEST DATE _____ READINGS BY _____

Sheet Metal & Air Conditioning Contractors
National Association

Form D.12 Cooling Tower or Evaporative Condenser Test Report

This form should be substantially completed and verified by the installing contractor before the system is balanced. The "pump data" section is to be used for the recirculating pump in evaporative condensers, not the system pump used with cooling towers (use Form D.13 Pump Test Report).

COOLING TOWER OR EVAPORATIVE CONDENSER TEST REPORT

PROJECT _____ SYSTEM _____

LOCATION _____

MANUF. _____ MODEL _____ SERIAL NO. _____

NOM. CAPACITY _____ REFRIG. _____ WATER TREAT. _____

FAN DATA	
No. of Fan Motors	
Motor Make/Frame	
Motor H.P./RPM	
Volts/Phase/Hertz	
Motor Sheave Diam./Bore	
Fan Sheave Diam./Bore	
Sheave ℄ Distance	
No. Belts/Make/Size	

PUMP DATA	
Make/Model	
Pump Serial No.	
Motor Make/Frame	
Motor H.P./RPM	
Volts/Phase/Hertz	
GPM	

AIR DATA	DESIGN	ACTUAL
Duct CFM		
Duct Inlet S.P.		
Duct Outlet S.P.		
Avg. Ent. W.B.		
Avg. Lvg. W.B.		
Ambient W.B.		
Fan RPM		
Voltage $T_1\text{-}T_2$ $T_2\text{-}T_3$ $T_3\text{-}T_1$		
Amperage T_1 T_2 T_3		

WATER DATA	DESIGN	ACTUAL
Ent./Lvg. Water Press.		
Water Press. $\triangle P$		
Ent./Lvg. Water Temp.		
Water Temp. $\triangle T$		
GPM		
Bleed GPM		
Voltage $T_1\text{-}T_2$ $T_2\text{-}T_3$ $T_3\text{-}T_1$		
Amperage T_1 T_2 T_3		

REMARKS:

TEST DATE _____ READINGS BY _____

Sheet Metal & Air Conditioning Contractors
National Association

Form D.13 Pump Test Report

This report form may be used as a worksheet. However, the final data on each pump performance must be recorded on this form. The actual impeller diameter entry is that indicated by plotting the head curve or by actual field measurement where possible.

Net positive suction heat (NPSH) is important for pumps in open circuits and for pumps handling fluids at elevated temperatures. NPSH defines the required pressure excess above the fluid flash point at the impeller eye.

PUMP TEST REPORT

PROJECT _____

	DATA	PUMP NO.	PUMP NO.	PUMP NO.	PUMP NO.	PUMP NO.
DESIGN	Location					
	Service					
	Manufacturer					
	Model Number					
	Serial Number					
	GPM/Head					
	Req. NPSH					
	Pump RPM					
	Impeller Diam.					
	Motor Mfr./Frame					
	Motor HP/RPM					
	Volts/Phase/Hertz					
	F.L. Amps/S.F.					
	Seal Type					
ACTUAL	Pump Off-Press.					
	Valve Shut Diff.					
	Act. Impeller Diam.					
	Valve Open Diff.					
	Valve Open GPM					
	Final Dischg. Press.					
	Final Suction Press.					
	Final ΔP					
	Final GPM					
	Voltage T_1-T_2, T_2-T_3, T_3-T_1					
	Amperage T_1 T_2 T_3					

REMARKS:

TEST DATE _____ READINGS BY _____

Sheet Metal & Air Conditioning Contractors
National Association

Form D.14 Instrument Calibration Report

This form is to be used for recording the application and date of the most recent calibration test or calibration for each instrument used in the testing, adjusting, and balancing work covered by the report.

INSTRUMENT CALIBRATION REPORT

PROJECT _____

INSTRUMENT/SERIAL NO.	APPLICATION	DATES OF USE	CALIBRATION TEST DATE

REMARKS:

TEST DATE _____ READINGS BY _____

Appendix E Symbols and Abbreviations

ACCA	Air Conditioning Contractors Association
a-c	Alternating current
ACH	Air changes per hour
ADA	Americans with Disabilities Act
AFF	Above finished floor
AFUE	Annual fuel utilization efficiency
AGA	American Gas Association
AIA	American Institute of Architects
amp	ampere(s)
AMCA	Air Movement and Control Association
ANSI	American National Standards Institute
ASHRAE	American Society of Heating, Refrigerating and Air Conditioning Engineers
ASTM	American Society for Testing and Materials
ATL	Across-the-line (motor starter)
AWG	American Wire Gauge
BEF	Ballast efficacy factor
BF	Ballast factor
Btu	British thermal unit(s)
Btuh	British thermal units per hour
°C	Temperature, degrees Celsius
CAD	Computer aided design
CATL	Combination across-the-line (motor starter)
c/b	Circuit breaker
c, C	Thermal conductance
CCT	Correlated color temperature
cct, ckt	Circuit
cfs	Cubic feet per second
CISPI	Cast Iron Soil Pipe Institute
CLTD	Cooling load temperature difference
cps	Cycles per second
CRI	Color rendering index
Δt	Temperature difference
d-c	Direct current
dfu	Drainage fixture units
DB	Dry bulb (temperature)
DD	Degree days
DHW	Domestic hot water

DL	Developed length	kcmil	Thousand circular mils
DWV	Drainage, waste and vent	kva	Kilovolt-ampere
DX	Dry expansion, direct expansion	kw	Kilowatt(s)
e	Emittance	kwh	Kilowatt-hour
EER	Energy efficiency ratio	lm	Lumen
EMT	Electric metallic tubing	LPW	Lumens per watt
ETL	Electrical Testing Laboratories	LV	Low voltage
F	Fuse	MCC	Motor control center
°F	Temperature, degrees Fahrenheit	MCM	Thousand circular mil
fc	Footcandle	MH	Metal halide (lamp)
fpm	Feet per minute	MV	Mercury vapor (lamp)
fps	Feet per second	NC	Noise criterion
ft	Foot, feet	NEC	National Electrical Code
fu	Fixture units (plumbing)	NEMA	National Electrical Manufacturers' Association
GFCI	Ground fault circuit interrupter		
GFI	Ground fault interrupter	NFPA	National Fire Protection Association
GLF	Glass load factor	OD	Outside diameter
gpd	Gallons per day	PTAC	Package terminal air conditioner
gpm	Gallons per minute	PTHP	Package terminal heat pump
H	Head (pressure), usually in feet of water	PVC	Polyvinyl chloride
HID	High intensity discharge (lamps)	RH	Relative humidity
HO	High output (lamp)	RS	Rapid-start
HPS	High pressure sodium	SC	Shading coefficient
HSPF	Heating season performance factor	SEER	Seasonal energy efficiency ratio
HV	High voltage	SHGF	Solar heat gain factor
HVAC	Heating, ventilating and air-conditioning	SHR	Sensible heat ratio
		SI	Systeme Internationale (metric)
Hz	Hertz	SLF	Shade line factor
I	Symbol for current	SMACNA	Sheet Metal and Air Conditioning Contractors Association, Inc.
I=B=R	Institute of Boiler and Radiator Manufacturers		
		TAB	Testing and balancing
ID	Inside diameter	TEL	Total equivalent length
in.	inch	UL	Underwriters Laboratories
in. w.g.	inches, water gauge	VHO	Very high output (lamp)
IESNA	Illuminating Engineering Society of North America	VTR	Vent through (the) roof
		WB	wet bulb (temperature)
IPS	Iron pipe size	WSFU	Water supply fixture units
k	thermal conductivity		

Glossary

a-c Alternating current, which changes polarity many times per second.

AWG American Wire Gauge, the standard wire size measuring system in the United States.

Access Openings in a building through which equipment can be moved. Also, removable panels in equipment for servicing.

Access box Rust-resistant metal box with hinged cover. Sets flush with floor and allows access to a cleanout or other device.

Acidity An acid condition of water that could cause corrosion.

Air changes In ventilation, the number of times the air in a room is changed per hour.

Air density The weight of air; pounds per cubic foot in English units.

Air flow pattern Methods by which air is introduced to a space, directed through it and removed.

Air foil vanes Flat blades in a register that can be moved so as to direct the air stream.

Air stream Air flow through items such as filters, coils, registers and ducts.

Air vent valve An escape valve for air trapped at high points in a hot water heating system.

Ampacity A wire's ability to carry current safely, without undue heating. The term formerly used was *current-carrying capacity* of the wire.

Appliance branch circuit A branch circuit that supplies outlets specifically intended for appliances. Lighting is not supplied from such circuits, and the number of outlets is generally limited.

Appliance outlet An outlet connected to an appliance circuit. It may be a single or duplex receptacle or an outlet box intended for direct connection of an appliance.

Architectural-electrical plan Architectural plan on which electrical work is shown. Also known as an *electrical working drawing*.

Architectural lighting element A light source built into, or onto, the building structure. Not a commercial lighting fixture.

Auxiliary resistance heating Electrical resistance heaters that supplement heat, as from a heat pump.

Average water temperature Average between temperature of water leaving and returning to a boiler.

BX Trade name of NEC-type AC flexible armored cable.

Baseboard heater Hot water heater or electric heater installed at or near the bottom of a wall.

Blower-coil unit A unit in which a blower moves the air stream across items such as heating coils, cooling coils and filters.

Boiler A device that produces hot water or steam for heating.

Branch circuit Wiring between the last overcurrent device and the branch circuit outlets.

Breathing wall A method such as the incremental system that has an exterior wall opening for heat and moisture rejection and for fresh air supply.

British thermal unit (Btu) Quantity of heat.

Busduct An assembly of heavy bars of copper or aluminum that acts as a conductor of large capacity.

Cable An assembly of two or more wires or a single wire larger than No. 8 AWG.

Centralized system A system with one heating or cooling source and a ducted distribution network.

Check valve A valve that allows fluid to flow in one direction only.

Chilled water The refrigerated water used to cool the air in air systems.

Circuit An electrical arrangement requiring a source of voltage, a closed loop of wiring, an electrical load and some means for opening and closing the loop.

Circuit breaker A switch-type mechanism that opens automatically when it senses an overload (excess current).

Circulating hot water line Piping that permits circulation of domestic hot water for speedy availability at the outlet.

Circulator Centrifugal pump or booster.

Cleanout A removable plug in a drainage system.

Clearing a fault Eliminating a fault condition by some means. Generally taken to mean operation of the over-circuit device that opens the circuit and clears the fault.

Closet carrier An iron and steel frame to support a water closet that hangs from a wall.

Coefficient of utilization The ratio between "usable" lumens and lamp lumens for a particular combination of fixture and space.

Combined sewer One that carries both storm and sanitary drainage.

Common neutral A neutral conductor that is common to, or serves, more than one circuit.

Condensing In a refrigeration cycle, the process of liquifying pressurized refrigerant.

Conduit, electrical A round cross-sectional electrical raceway of metal or plastic.

Connected load The sum of all electrical loads on a circuit.

Continuous circulation An arrangement in which the blower runs continuously, while the evaporator or burner runs intermittently.

Contract documents Legal papers that include the contract, the working drawings and the specifications.

Contrast Difference in apparent or actual brightness between an object and its background.

Control diagram (ladder diagram) A diagram that shows the control scheme only. Power wiring is not shown. The control items are shown between two vertical lines, hence the name ladder diagram.

Convector A heating element that warms the air passing over or through it. The air, in turn, rises to warm the space by convection.

Convenience outlet A duplex receptacle connected to a general-purpose branch circuit and not intended for any specific item of electrical equipment.

Curb box Access to an underground valve at the street curb. It controls water service to a building.

Current (I) The electrical flow in an electrical circuit, which is expressed in amperes (amp).

d-c Direct current, which is unvarying in polarity.

DWV Drainage, waste and vent.

Dead Electrically de-energized, no voltage.

Decentralized system A complete heating and/or cooling system installed locally to serve only the area in which it is placed.

Degree day The number of Fahrenheit degrees that the average outdoor temperature over a 24-hour period is less than 65°F.

Demand (water), the probable maximum rate of water flow as determined by the number of water supply fixture units.

Demand (electric), the actual amount of load on a circuit at any time. The sum of all the loads that are ON. Equal to the connected electrical load minus the loads that are OFF.

Developed pipe length Number of feet of piping in a water circuit, including the equivalent length of all fittings.

Diffuse reflection A type of reflection in which the reflected light is spread out in all directions. The reflection of a matte-finish surface.

Diffuseness A measure of the directiveness of the light. Diffuse light comes from many directions.

Diffuser (light) Material placed between a lamp and a viewer for the purpose of controlling the light flux produced.

Domestic hot water (DHW) Potable hot water as distinguished from hot water used for house heating.

Downstream Electrically speaking, going away from the power source and toward the load.

Drain pit A pit to receive nontoxic water for disposal, often to a dry well.

Dry well Same as a seepage pit, except that it disperses water and not sewage effluent.

Duct liner Acoustic liner to absorb sound.

Duct turns Curved vanes that reduce friction and turbulence when square corners are used in ducts.

Edge loss factor Heat loss, slab to earth.

Efficacy Efficiency of a light source, measured in lumens per watt.

Effluent A fluid flowing away from a process; for example, the effluent of a septic tank.

Electric closet A space containing electric service equipment such as panels and switches.

Electronic air cleaners A filter somewhat more efficient than a bag filter, particularly for the removal of small suspended particles.

Emergency source (electric) Source of electrical power that is used when normal electrical power fails.

Energy (electric) This is expressed in kilowatt-hours (kwh) or watt-hours (wh) and is equal to the product of power and time.

$$\text{Energy} = \text{Power} \times \text{Time}$$
$$\text{Kilowatt-hours} = \text{Kilowatts} \times \text{Hours}$$
$$\text{Watt-hours} = \text{Watts} \times \text{Hours}$$

Engineering layout A drawing of the design, that is the basis for shop drawings.

Equipment ground; equipment ground conductor The conductor that connects non-current-carrying metal ports of a wiring system to the system ground. Bare or covered with green insulation. Green ground.

Evaporation In a refrigeration cycle, the process during which the expanding (evaporating) refrigerant absorbs heat.

Expansion fitting A device to allow for the expansion of tubing or piping due to an increase in temperature.

Facade The face of a building.

Face velocity The speed in feet per minute at which air leaves a register.

Fall per foot The slope of a drainage pipe.

Fault Electrical; a short circuit—either line to line or line to neutral or line to ground.

Feed line A pipe that supplies water to items such as a boiler or a domestic hot water tank.

Feet of head Pressure, measured in feet of water. See conversion tables for other units.

Fill and vent Parts of a liquid storage system.

Finned tube Tube used for heat transfer between water and air.

Fixture clearance Plumbing; distance between fixtures; distance from a fixture to an obstruction; distance between a fixture and a wall.

Fixture unit Plumbing; an index of the rate of flow of water to a fixture (water supply fixture units) or sewage leaving a fixture (drainage fixture units) when the fixture is part of a group. Not used for individual fixtures.

Flow pressure The pressure necessary to supply a plumbing fixture adequately.

Flow rate Cubic feet per minute (cfm) of air circulated in an air system or the number of pounds of water per hour circulated through a hot water heating system.

Flue gas Carbon monoxide, carbon dioxide and other combustion products.

Flush tank A tank that refills automatically and stands ready to flush a water closet.

Flush valve A valve that, when operated manually, delivers a measured amount of water to flush a water closet.

Foot of head The pressure exerted at the bottom of a column of water 1 ft high.

Footcandle (fc) Unit of light flux density, equal to one lumen per square foot.

Fossil fuels Oil, gas and coal.

Four-way switch A four-terminal switch (abcd) that operates ab-cd and ac-bd. Used to control outlets from three or more locations.

Four-way switching Control of (an) outlet(s) from three locations.

Fresh air inlet A vent that admits air, as to the house drain, before it joins the house trap.

Frequency The number of cycles per second for a-c, measured in units of hertz.

Fuse An overcurrent device that opens a circuit by melting out. By nature, it is a single-pole device.

GFI, GFCI [ground fault (circuit) interrupter] A device that senses ground faults and reacts by opening the circuit.

gpm Gallons per minute.

Gang One wiring device position in a box.

Ganged switches A group of switches arranged next to each other in ganged outlet boxes.

General-purpose branch circuit Circuit that supplies a number of outlets for general lighting and convenience receptacles.

Greenfield Trade name for flexible steel conduit.

Grilles Perforated or slotted frames, usually used for air return.

Ground Zero voltage. Also, any point connected to ground.

Ground bus A busbar in a panel or elsewhere, deliberately connected to ground.

Grounding conductor Conductor run in an electrical system, which is deliberately connected to the ground electrode. Its purpose is to provide a ground point throughout the system.

Ground electrode A piece of metal physically connected to ground. Can be a rod, mat, pad, structural member or pipe. See the **NEC**.

Ground fault An unintentional connection to ground.

Groundwater The water below the ground.

Grounded Connected to ground, at zero voltage.

Gutter Rainwater collecting trough at the edge of a roof.

Gutter space The empty spaces around the sides, top and bottom of a panel box, intended as a wiring space.

Handhole Small exterior concrete box, intended as a pulling or splicing point for underground cables.

Hardness Chemical compounds in water, frequently containing calcium. They form a rocklike deposit on the inside of piping.

Head Pressure, generally expressed as feet of water.

Heat pump An electric heating/cooling device that takes energy for heating from outdoor air (or groundwater).

Hertz (Hz) The unit of frequency of a-c. It equals the number of complete cycles per second.

Home-run The wiring run between the panel and the first outlet in the branch circuit. (Looking upstream, it is the wiring run between the last outlet and the panel.)

Horsepower (hp) A unit of power that equals, electrically, 746 w; or 1 hp $\approx$ ¾ kw.

Hose bibb Connection for supply water to a garden hose.

Hot, live Electrically energized.

House trap A trap between the house drain and the house sewer.

Hydronic Heating (or cooling) by water.

Humidifier A device to vaporize water. It is used to increase relative humidity of air.

Impedance (Z) The quantity in an a-c circuit that is equivalent to resistance in a d-c circuit. It relates current and voltage. It is composed of resistance plus a purely a-c concept called reactance and is expressed, like resistance, in ohms.

Incremental HVAC unit Self-contained through-wall unit for heating, cooling and ventilating.

Individual branch circuit Circuit that supplies only a single piece of electrical equipment.

Infiltration Air that leaks into a building.

Junction box Metal box in which tap(s) to circuit conductors are made. Junction box is not an outlet, since no load is fed from it directly.

kcmil Thousand circular mil. Used for large wire sizes; the square of the wire diameter in thousandths of an inch. Replaced the old term MCM.

Latent heat Inherent heat as in water vapor.

Lavatory A wash basin. Also, a room with a lavatory basin and usually a water closet.

Layout Drawings of a system showing the physical relation of the system components, often with dimensions, sizes and notes.

Leader A pipe that carries storm water down from a gutter or a roof drain fixture.

Line side The side of a device electrically closest to the source of current.

Line voltage thermostat A thermostat that is connected directly to the line. Full power is fed through it to the controlled heater or air conditioner.

Light loss factor (LLF) The ratio between maintained and initial footcandles, for a lighting fixture.

Lighting system A method of describing in what directions the light is emitted from a lighting fixture. Systems are direct, semi-direct, direct-indirect, general diffuse, semi-indirect and indirect.

Load side The side of a device electrically farthest from the current source.

Low voltage switching A system of outlet control by low voltage switches and relays.

Lumen Unit of light flux.

Lux Metric unit of lighting flux density, or illuminance. One lux equals one lumen per square meter.

MBH Thousands of British thermal units (Btu) per hour.

MCM This unit has been replaced with kcmil. See *kcmil.*

Manhole Same as handhole except much larger. Intended for underground primary power cables, large secondary cables and telephone cables.

Master (central) control Control of all the outlets from one point.

Mean radiant temperature (MRT) Average interior radiant temperature. Calculated using both temperature and area of interior surfaces.

Meter pan Device intended to hold one or more kilowatt-hour meters. Usually contains buswork and may contain overcurrent devices.

Millinch The pressure exerted at the bottom of a column of water $1/1000$ of 1 in. high.

Millinches per foot The head loss in 1 ft of pipe, caused by friction of water flows expressed in millinches.

Mirroring The effect that causes ordinary glass to reflect like a mirror.

Motor control center A single metal-enclosed assembly containing a number of motor controllers and possibly other devices such as switches and control devices.

Motor controller The device that puts the motor "on the line." Generally, a magnetically operated contactor. Usually called the starter.

Multipole Connects to more than one pole such as a 2-pole circuit breaker.

NEC National Electrical Code, published by the NFPA (National Fire Protection Association) as NFPA 70. Accepted (minimum) standard for safe electrical practice in the United States.

NEMA (National Electrical Manufacturers Association) An American association that establishes standards of manufacture for electrical equipment. NEMA standards are accepted throughout the industry.

Net output As applied to a boiler, the effective value of the MBH delivered by the boiler to heat the building.

Neutral The circuit conductor that is normally grounded. It is insulated white or gray.

Nonintegral trap A trap that is not part of the plumbing fixture as manufactured.

Ohm's Law The relationship between current and voltage in a circuit. It states that current is proportional to voltage and inversely proportional to impedance. Expressed algebraically, in d-c circuits $I = V/R$; in a-c circuits $I = V/Z$.

One-way throw A register that delivers air in only one direction.

Opposed blade dampers Controls for the regulation of the flow rate of air in ducts or through registers.

Outdoor design temperature Outside temperature used in calculating heat loss (and gain).

Outlet A point on a wiring system at which electrical current is taken off to supply an electrical load.

Outlet box A box, usually metal, containing wires from a branch circuit, and connection to wires from an electrical load.

Output Heating; the heat delivered to a room by heating units.

Overcurrent device A device such as a fuse or a circuit breaker designed to protect a circuit against excessive current by opening the circuit.

Overload Electrical; A condition of excess current; more current flowing than the circuit was designed to carry.

Package unit boiler A boiler plus a number of its controls and accessories.

Panel or panelboard A box containing a group of overcurrent devices intended to supply branch circuits.

Panel directory A listing of the panel circuits appearing on the panel door.

Panel schedule A schedule appearing on the electrical drawings detailing the equipment contained in the panel.

Parallel circuit Circuit where all the elements are connected across the voltage source. Therefore, the voltage on each element is the same, but the current through each may be different.

Performance data Ratings for heating, cooling, air-handling capacity (cfm) and the like.

Planting screen Bushes or other plantings that hide an item of equipment.

Plug-in busduct Busduct with built-in power tap-off points. Tap-off is made with a plug-in switch, circuit breaker or other fitting.

Polarity The directions of current flow in a d-c circuit. By convention, current flows from positive to negative. Electron flow is actually in the opposite direction.

Pole An electrical connection point. In a panel, the point of connection. On a device, the terminal that connects to the power.

Potable Water that is safe to drink.

Power *(P)* Expressed in watts (w) or kilowatts (kw) and equal to:

In d-c circuit, $P = VI$ and $P = I^2R$
In a-c circuit, $P = VI \times$ Power factor

Power factor (pf) A quantity that relates the volt-amperes of an a-c circuit to the wattage, or power, that is,

Power = Volt-amperes × Power factor

The power factor cannot be greater than 1.0 and is frequently expressed as a percentage figure. In purely resistive circuits, pf equals 1.0 or 100%, and wattage equals volt-amperes.

Primary air Heated or cooled air directly from the conditioner.

Primary service High voltage service, above 600 v.

Private sewage treatment Sewage treatment other than in central city treatment plants.

Process hot water Hot water needed for manufacturing processes. It is not to be confused with domestic hot water (DHW).

Psi Pounds per square inch pressure.

Public use Fixture use in a public building where toilet room use is greater than in a private residence.

Pull box A metal cabinet inserted into a conduit run for the purpose of providing a cable pulling point. Cable may be spliced in these boxes.

Raceway Any support system, open or closed, for carrying electrical wires.

Radiant cables Electrical cables embedded in the ceiling (or floor) for heating.

Range hood Hood over a stove to collect odor-laden air that is to be exhausted.

Receptacle poles Number of hot contacts.

Receptacle wires Number of connecting wires, including the ground wire.

Recessed convector A convector cabinet that extends partially or fully into a pocket in the wall.

Recharge Putting water back into the ground.

Reflection factor or reflectance Ratio between light reflected from, and light falling on, an object.

Register Slotted frame for control of the direction of air delivered to the space and its flow rate.

Regressed lens An arrangement where the fixture lens is set back into the body of the lighting fixture. A recessed lens.

Remote control (RC) switch A magnetically operated mechanically held switch, normally used for remote switching of blocks of power.

Resistance (R) The unit in an electrical circuit analagous to friction in a hydraulic circuit, expressed in ohms.

Reverse return A return main that does not go directly back to the boiler but connects to convectors in reverse order; farthest first, and so on. Its purpose is to equalize the length of supply plus return piping to each convector.

Riser diagram Electrical block-type diagram showing connection of major items of equipment. It is also applied to signal equipment connections, as a fire-alarm riser diagram. It is generally applied to multistory buildings.

Riser shaft A vertical shaft in a building designed to house the electrical riser cables.

Romex One of several trade names for **NEC** type NM nonmetallic sheathed flexible cable.

Roof drain A metal water collector flashed into a flat roof. It is usually provided with a strainer to exclude debris.

Roof slope Pitch of a flat roof to direct rainwater to a roof drain.

Roughing dimensions Locations of water supply and drainage pipes to ensure proper fit of a plumbing fixture.

Runout With respect to hot water, a branch pipe (supply or return) from a hot water main to a convector cabinet or a convector baseboard.

R-value Resistance rating of thermal insulation.

Sanitary drainage Removal of sewage from a building.

Sealtite Trade name for waterproof flexible steel conduit.

Secondary service Low voltage service, up to 600 v.

Secondary air Air from the space that is drawn along with the primary air, resulting in a tempered mixture.

Seepage pit A chamber that receives the effluent of a septic tank and allows it to seep into the surrounding earth.

Sensible heat Heat that raises the air temperature.

Septic tank A tank in which sewage is held and partially purified.

Series circuit Circuit with all the elements connected end to end. The current is the same throughout, but the voltage can be different across each element.

Service drop The overhead service wires that serve a building.

Service sink A low sink usually used for mopping, sometimes called a slop sink.

Service switch One to six disconnect switches or circuit breakers. The purpose is to completely disconnect the building from the electrical service.

Shielding and cutoff Terms indicating the action of a lighting fixture to shield the lamp source from the viewer. Cutoff is the point in the field of vision where this shielding begins.

Shop drawings Contractor's or manufacturer's drawings giving equipment construction details.

Short circuit An electrical circuit with zero load; an electrical fault.

Shutoff valve A valve for turning off flow, as a water valve near a fixture.

Single pole Connects to a single hot line.

Six-foot rule The NEC rule that no point along a wall area be more than 6 ft from an electrical convenience receptacle.

Soil Major pollutants in plumbing.

Spacing-to-mounting-height ratio (S/MH) Figure provided by lighting fixture manufacturers indicating maximum S/MH for uniform lighting results.

Specific heat of water The heat necessary to raise a given quantity of water by one temperature degree. In the English units system, it is expressed as Btu per pound of water per F°. In the SI system, it is Kilocalories per liter (Kilogram) per C°.

Specular reflection Mirrorlike reflection.

Storm drainage Removal of rainwater from a roof or other area.

Static head Pressure due to the weight of water above a given point.

Sweat fitting A soldered connection of tubing.

System voltage Voltage from the power company; transformer voltage.

Tankless heater A coil in a hot water heating boiler for heating domestic hot water.

Temperature difference Thermal pressure.

Temperature drop (water systems) The difference in temperature of water leaving and returning to the boiler.

Temperature drop In air systems, the difference in temperature of the return air and the heated air delivered.

Thermal transfer Moving heat into or out of a space or between thermal media.

Three-way switching An arrangement for controlling (an) outlet(s) from two locations.

Three-way switch A three-terminal switch that connects c-a or c-b where c is common.

Throw (air) The distance (in feet) from the register that air is "thrown."

Tic mark Hatch mark on drawing raceway symbol, showing number of wires.

U Coefficient Rate of heat transmission.

Upfeed heating system Boiler located below convectors.

Upflow, downflow, horizontal furnace Furnace types classified by direction of air flow.

Upstream Electrically speaking, in the direction toward the power source.

Utilization voltage The voltage that is utilized; motor voltage.

Vacuum breaker A device to prevent suction in a water pipe.

Valve A control device to restrict or shutoff fluid flow.

Vent In plumbing, air-filled piping that prevents siphonage of trap seals or the bubbling of air through trap seals.

Ventilation Controlled use of outdoor air for freshness.

Voltage (V) The electrical pressure in an electric circuit, expressed in volts.

Voltage drop The difference in voltage between two points in a circuit. The voltage drop around a circuit including wiring and loads must equal the supply voltage.

Water cooler An electrical, refrigerated drinking fountain.

Water hammer Banging of pipes caused by the shock of rapid closing of faucets or valves.

Water services Hot and cold water piping and equipment.

Weather barrier A divider between the exterior and interior working parts of an incremental conditioner.

Wire-nut Trade name for small, solderless, twist-on branch circuit conductor connector.

Wireway Term generally used to mean an exposed rectangular raceway.

Wiring diagram Diagram showing actual wiring, with numbered terminals. All wiring is shown.

Wiring device Receptacle, switch, pilot light, small dimmer or any other device that is wired in a branch circuit and fits into a 4-in. outlet box. Normally 30 amp or smaller.

Working plane In terms of lighting, area generally taken to be 30 in. above the floor but that can be set at any desired elevation.

Zone heating, cooling Section of a heating and/or cooling system separately controllable.

Index

a-c, *see* Alternating current
Abbreviations, 977
  Appendix E, 1031–1032
  electrical work, 726
Absolute pressure, 408
ACCA (Air Conditioning Contractors of America), 43
  heat gain (cooling load) calculation forms, 67–69
  heat loss calculation forms, 46, 47
ADA (American with Disabilities Act), 450
AFUE (annual fuel utilization efficiency), 204
Air:
  density, 198
  distribution outlets, 235–264
  equations, heating, 201
  flow measurement, 392–395
  friction, *see* Air friction, ducts and fittings
  heat capacity, 198
  heat capacity of ducted flow, 200
  noise in ducts, 297
  pressure, in ducts, 201, 202
  specific heat, 198
  velocity measurement, 382, 390–392
Air break, *see* Backflow prevention, water systems
Air conditioners:
  A-frame evaporator, 345
  blower evaporator, 345
  DX coil, 328
  rooftop unit, 339–343, 346
  slab-on-grade unit, 353
  split unit, 335, 345, 346–352
  through-the-wall unit, 333, 334, 344
Air cushion tank, *see* Hydronic heating systems
Air distribution outlets, 235–265
  air stream patterns, 254, 256–260, 263–265
  ceiling diffuser, 249–251
  comfort (occupied) zone, 235, 253
  double deflection supply register, 242–244
  face velocity, 262, 392
  four-way register, 245–247
  linear grille, 239–241
  noise ratings, 248, 261
  outlet characteristics, 253, 255, 256, 258
  register area, 255
  return air grille, 238
  selection, location and application, 258–265
  single deflection supply grille, 236, 237
  types, 235, 252
Air equations, 41, 78, 201
Air friction, ducts and fittings:
  altitude and temperature correction factors (chart), 283
  round steel duct (chart), 277
  duct friction calculations, 279, 280
  duct roughness, 276, 280, 281
  equivalent friction, various shapes (table), 286
  equivalent oval ducts (table), 287
  equivalent rectangular ducts (table), 284–285
  fittings, equivalent lengths, 289, 290
  loss coefficients, 291–294
  noncircular ducts, 282
  straight circular duct sections, 275
  TEL (total equivalent length), 295
Air handlers, 351
Air stream, *see* Air distribution outlets
Air systems, heating and cooling:
  all-air systems types, 265–274
  characteristics, 198
  codes and ordinances, 307
  components, 198–200
  cooling air quantity, estimate, 360

Air systems (cont'd)
  cooling design considerations, 357, 359
  cooling load calculation, 357
  design checklist, 309, 310
  distribution outlets, 235–265
  dual duct, 269, 270
  ducts and fittings, 227–234
  duct sizing, 294–307, 308
  friction, see Air friction, ducts and fittings
  furnaces, 199–227
  furnace selection, 308
  heating load calculation, 307
  heating selection, 308
  incremental (distributed) systems, 353
  indoor comfort conditions, 198
  maximum air velocities, 297
  multizone, 266–268
  single duct, reheat, 268, 269
  single duct variable volume (VAV), 268, 269
  single zone, 266, 267, 270–274
  system design, 307–310
  system types, 266
Alternating current (a–c), 638
AMCA (Air Movement and Control Association), 312
Ammeter, 652
Ampacity, 641, 649, 676
Ampere, amp, 633
Analog meters, 653
Anemometer, 15, 379, 386–390
ANSI (American National Standards Institute), 413
Aquastat, 112, 528, 529
Architectural lighting elements:
  brackets, 875–876
  cornices, 883
  coves, 870–871
  valances, 882
ARI (Air Conditioning and Refrigeration Institute), 325
Asbestos-cement pipe, 435
ASHRAE (American Society of Heating, Refrigeration and Air Conditioning Engineers), 15
ASHRAE Standard 90.1, Energy Efficient Design, 91
ASME (American Society of Mechanical Engineers), 87
Asphaltum, 437
ASTM (American Society for Testing and Materials), 414
ATL (across-the-line) motor starter, 945
Atmospheric pressure, 408, 409
AWG (American Wire Gauge), 675

Backflow prevention, water systems, 534–537
Back-siphon, see Vacuum breaker
Balancing damper, 232
Balancing procedures
  air system, 394
  hydronic system, 397
Balancing valve, see Valves
Ball valve, see Valves
Ballast efficiency factor (BEF), 807–809
Ballast factor (BF), 807–809
Ballast, lamp, see type of lamp
Baseboard convector, electric, 170
Baseboard radiation, hot water, 101–102
Basic House, the:
  air system heating and cooling design, 359–362
  architectural plans, 130, 131
  circuitry details, 130, 131, 776
  electric heating design, 141
  floor plans, electrical, circuited, 765–766
  floor plans, electric heating, 142, 760, 777
  floor plans, hot water heating, 134, 135
  floor plans, lighting and devices, 756, 758, 759, 760, 761
  floor plans, plumbing, 487, 492
  hot water (hydronic) heating design, 128–138
  plumbing riser, 493
Bathroom group, 503, 505
Batteries, 628–631
Below-the-rim supply, see Vacuum breaker
Bimetallic element, 380
Black water, 450, 558
BOCA Basic Plumbing Code, 404
Boilers, hot water, 86–92
  boiler code, 87
  cast iron, 86
  commercial, 91
  dip tube, 94
  drain valve, 92
  low water temperature (LWT), 87
  make-up water pipe, 92
  ratings, 87, 91
  residential, 86, 87
  schematic piping diagrams, 89–92, 112
  tankless hot water coil, 86, 87, 91
Bolted pressure connector, 683
Bourdon gauge, tube, 379, 409–411
Branch circuit, 667, 670
  appliance, 762
  general lighting, 758–761

Branch interval (drainage), 568, 571
BTU, definition, 4
Btuh, 14. See also Heat Transfer
Busduct, 931, 936, 938
Busway, plug-in, 708, 710, 711
BX cable, see Wire and Cable

Cabinet heater, electric
  forced air, 171, 173
  gravity flow, 171, 172
Capillary tubing, 381
Cast iron radiation, see Radiators
Cathodic protection, 435
CATL (combination, across-the-line) motor starter, 948
Cavity reflectance, see Zonal cavity method
CBM (Certified Ballast Manufacturer), 809
Ceiling (overhead) electrical raceways, 702–706
Celcius, see Temperature
Cellular floor electrical raceways, 701, 703, 704
Centralized systems, 353
Check valve, see Valves
Circuit breakers:
  clearing time, 735
  frame, trip size, 737
  interrupting capacity, 736
  molded case, 735
  quick make-break, 736
  trip free, 736
Circuiting, 764–776
  load balancing, 773
  multiple point switching, 768–770
  nonresidential, 951
  residential, 888
  wire counting, 772, 774
  wiring paths, 768, 773–775
Circulator (pump), see Hydronic heating systems
Cleanout, 572, 573
Closet, see Plumbing fixtures
Closet bend, 587
Codes and ordinances, 307
Coefficient of utilization, 845–846
Color (lighting), see Lighting
Color rendering index (CRI), 794
Comfort conditions, indoors, 198. See also Thermal comfort
Comfort zone, see Air distribution outlets
Communication wiring, 913–916
Compression tank, see Hydronic heating systems
Condensation, 70, 226
Condenser:
  air cooled, 354

INDEX / 1041

evaporative condenser, 355, 357
  water cooled, 354
Conduction, building materials, 22
Conduction, thermal, 9, 22, 23
Conductor, electric, 633
Conductors, electrical, *see* Wire and cable
Conduit, electrical, 684
  bushings, 690
  dimensional data (table), 686
  EMT (electric metallic tubing), 668, 669
  fittings, 688–690, 692
  flexible, 674, 691, 693–695
  IMC, 668, 669
  rigid steel, 668, 669
  supports, 687
Connectors, electrical, 680–684
Contrast (visual), 790
Controlled roof drainage, 615
Convection, *see* Heat Transfer
Convective air currents, 31
  attics, 31
  windows, 30
Convector, electric:
  baseboard, 170
  floor type, 173, 174
  gravity flow, 171
Convectors (cabinet heaters), hot water, 106, 107
Conversion factors (table), 978–980
Cooling load calculation, *see* Heat gain (cooling load) calculation
  nonresidential, 63–64
  residential, 58–63
Cooling temperature differential, 359
Cooling towers, 354, 356
COP (coefficient of performance), 204, 324
Copper pipe and tubing (K, L, M, DWV), 417, 431
Corporation cock, 501, 531
Correlated color temperature (CCT), 794
Corrosion, 435, 436
Cove lighting, 870–871
CPVC plastic pipe, *see* Plastic pipe
Cross-fixture contamination, 558
Curb box, 501, 533
Current, electric, 633

Dampers, 232
d-c, *see* Direct current
Degree Celcius, *see* Temperature
Degree days (DD), 50
  map, 51
Degree Fahrenheit, *see* Temperature
Degree Kelvin, *see* Temperature
Degree Rankine, *see* Temperature

Design conditions, cooling, 359
Design procedure:
  domestic hot water, 528
  drainage, 568
  electric heating, 179, 180
  electric wiring planning, 752
  hydronic heating, 126, 127
  potable water supply, 531–534
  warm air heating, 498
Dew point, 7, 72
DFU (drainage fixture units) (table), 569
DHW, *see* Domestic hot water
Diaphragm tank, *see* Hydronic heating systems
Differential pressure (meter), 384
Diffuse lighting (system), 843, 845
Diffusers (lighting), 785
Direct current (d-c), 631
Direct-indirect lighting, 843, 846
Direct lighting, 841–843
Disc, faucet, 443
Diversity factor (water service), 505
Domestic hot water (DHW):
  booster heater, 524, 525
  circulation systems, 526–528
  demand, consumption, 528, 530
  design procedure, 528
  forced circulation, 528, 529
  heater sizing, 528–531
  instantaneous heaters, 522, 523
  point-of-use heater, 522, 524
  recirculation, 528, 530
  recovery rate, 528, 530
  residential DHW heaters (table), 532
  storage units, 426, 522, 524, 529
  tankless heaters, 522–524
  thermosphon circulation, 527
  water temperatures, 519
Draft gauge, 203, 382
Drainage, sanitary:
  branch interval, 568, 571
  branch pipe, 558, 570
  building (house) drain, 561, 562, 570, 591, 598, 600
  building (house) trap, 561
  cleanout fittings, 572, 573
  design procedure, 568
  dfu (drainage fixture units) (table), 569
  discharge rate, sloped drain (table), 566
  drainage pipe sizing, 568–572
  drainage section (diagram), 561, 563
  fixture branch, 570, 581
  gray water, 558
  hydraulics, gravity flow, 562

indirect waste, 558, 559
interceptors, 573, 574
light industry building drainage system design, 594
Mogensen house drainage system design, 589–595
office building drainage system design, 599
piping, 562, 563, 568–572
piping design, 583
principles, 558
residential design, 589–594
scouring, 562
size, horizontal fixture branches (table), 570
size, soil stacks (table), 570
soil pipe, 558
soil stack, 562, 565, 570
stack base fitting, 576
traps, 560, 564, 575–577
vent pipe, 560
venting, 577
waste pipe, 578
Drainage, storm:
  conductor, 605, 611
  controlled roof drainage, 615, 616
  design, light industry building, 613, 614
  downspout, 604
  drain size (table), 612
  drain-down time, 615
  gutters, 605
  leader, 604, 611
  pipe sizing, 605
  principles, 602
  rainfall, 607
  roof drainage, 604, 606–611
  roof gutters (table), 611
Drinking fountains, 480, 485
Dry bulb (DB), *see* Temperature
Dry well, 562, 606, 624
Duct design, 227–228
Duct fittings, equivalent lengths, 981–990
Duct fittings, loss coefficients, 991–1013
Duct sizing methods, air velocity (tables), 297
  equal friction method, 295
  extended plenum method, 299–302
  friction rate, 295
  modified equal-friction method, 298
  semi-extended plenum method, 303–306
  TEL (total equivalent length), 295
Ducts:
  air friction, *see* Air friction, ducts and fittings

Ducts: (cont'd)
  air velocity, 288
  aspect ratio, 228
  fittings and air control devices, 229–232
  flexible, 228–229
  insulation, 229, 234
  materials, 227, 228
  noise, 297
  oval, 228
  pressure, in, 201–212
  rectangular, 227
  round, 228
  transitions, 230
DWV (drainage, waste and vent), 558
DX coil, 255, 320

EER (energy efficiency ratio), 204, 325
Efficacy (light sources), 799, 801, 803, 813, 825
Effluent, 558
Electric closet, 937, 939
Electric heating:
  baseboard convectors, 170, 190–193
  centralized system, 168
  decentralized (distributed) system, 168, 172
  duct insert heaters, 169
  embedded ceiling cables, 176, 178
  forced-air convector, 186, 187
  infrared heaters, 169, 173
  radiant heaters, 173–176
  resistance heaters, 168–177
  safety (overheat) cutout, 171
  unit heaters, 177, 178, 188
  unit ventilators, 177, 179
  wall-mounted, 171
Electric potential, 631
Electric service:
  capacity (amps), 929, 930
  diagrams, 932
  emergency service, 931
  handholes, 927
  manholes, 923–924
  metering, 898, 902–904, 908, 937
  normal, 931, 932
  overhead, 896–899
  pad-mount transformer, 926
  primary (high voltage), 922
  secondary (low voltage), 926, 928–929
  secondary rack, 899
  service switch(es), 932
  site plan, typical, 927
  standby, 931, 932
  transformer vault, 966
  underground, 900–902
  underground duct (Type I and II), 922–923
Electrical design, nonresidential:
  circuitry guidelines, by facility, 951–956
  commercial buildings, 960–969
  conduit and cable schedule, 968
  electrical plans, 965
  electrical service, 922, 923, 924, 928, 930, 931
  emergency service, 931–932
  metering, 937
  motors and control, 937–954
  office spaces, 954
  panelboards, 940
  power distribution, 931, 962
  pullbox, 937, 939–940
  schools, 951, 955–960
  stair and exit riser, 967
  switchboards, 932–935, 962–963
Electrical design, residential:
  bedrooms, 881–886
  circuitry, 888
  circulation and utility areas, 887–891
  climate control system, 893
  electrical equipment loads (table), 879
  fire alarm equipment, 910–915
  home office, 877–878
  intrusion alarms, 913
  kitchen and dining areas, 878–881
  living areas, 863–877
  load calculation, 888, 892–893
  multilevel light switching, 869
  plans, Mogensen house, 864–865, 867–868, 889–891, 893–894, 905–906
  service, 895–896
  signals, alarms and communication, 908–917
  telephone and intercom equipment, 916–917
Electrical planning, wiring, 752
  circuitry, 763–776
  circuitry guidelines, 755–763
  equipment and device layout, 755
  nonresidential buildings criteria, 951–956
  residential buildings criteria, 754
Electrical safety, grounding, 739–743
Electrolyte, see Batteries
Electromagnetic induction, 631
Emergency electrical service, 931
Emittance, thermal, 30–32
Energy Policy Act (EPACT), 799, 807, 812, 813
Energy standards:
  ASHRAE (American Society of Heating, Refrigerating and Air Conditioning Engineers): Code 90.1, 91
  NAECA (National Appliance Energy Conservation Act), 91
English units, 406
Enthalpy, 5, 70
ETL (Electrical Testing Laboratories), 809
Equivalent duct lengths, Appendix B, 981–990
Evaporation, 316
Evaporator, 318, 355, 358
  A-frame, 351, 352
  DX coil, 320
  function, 318
Evaporative cooling, 7, 8
Expansion loops and joints, 126, 164
Expansion tank, see Hydronic heating systems
External static pressure, 210, 223

Fahrenheit, see Temperature
Fall (pitched piping), 562
Fan coil units, 109, 110, 345
Faucet, self-closing, 456
Field test report forms, HVAC, 1015–1029
Finned-tube radiation, hot water, 102–106, 158
  commercial, 102
  mounting height, 103, 106
  output ratings, 102, 103, 106
  types, tiers, 104–105, 132
Fire alarm equipment, 910–912, 914
Fire suppression, 547–554
  sprinklers, 548–554
  standpipes, 548–550
Fittings, pipe, see Piping materials
Flared joints, copper pipe, 432
Flat cable assemblies, 708
Floor drain, 563, 564
Floor raceway, electrical, 695–706
Flow control valve, see Valves
Flow rate, water, 410
Flow straightener, 390
Fluorescent lamps:
  ballast data (core-and-coil type), 807–810
  burning hours per start, 804
  characteristics, 803
  characteristics (table), 811
  color, 813
  compact lamps, 825–826
  dimming, 805
  efficacy, 813
  efficacy (table), 811
  electronic ballasts, 809

high-output lamps, 813
instant-start lamps, 811
lampholder designs, 808
life, 804
low energy lamps, 814
preheat lamps, 810
rapid-start lamps, 811
starting circuits, 810
triphosphor (Octic) lamps, 814
U-lamps, 814
Flux density, *see* Lumen
Footcandle, 788
Forms, HVAC field test reports, 1015–1029
Four-way switching, 769
Frequency (of a-c), 639
Friction head loss, water system:
  pipe fittings, 514, 515
  piping, charts, 509–511
  piping (tables), 512–513
  water meters, 515
FU (fixture units), *see* WSFU (water supply fixture units); DFU (drainage fixture units)
Fuels, 52–54
Furnace, warm air, 199–227
  AFUE (annual fuel utilization efficiency), 204
  air delivery data, 222–224
  blower (fan), 224–226
  characteristics, 222–224
  combustion air, 200, 208
  components, 199, 203
  condensing units, 213–221
  dimensional data, 222
  downflow (counterflow) unit, 204, 207
  energy and efficiency, 204, 288
  filters, 227
  GAMA (Gas Appliance Manufacturers Association), 204
  gas units, 208, 217
  horizontal (lateral) unit, 204, 209–212
  humidifiers, 226
  low-boy unit, 204, 208
  noise, 224
  pulse combustion units, 213
  ratings, 222
  upflow (high-boy) unit, 204–206
Fuses:
  clearing time, 735
  types, 734

Galvanic cell, 628
Galvanometer, 652
GAMA (Gas Appliance Manufacturers Association), 204
Gate valve, *see* Valves

Gauge pressure, 408
Generator, electric, 632
GFCI (ground fault circuit interrupter), GFI, 726, 737
  circuit breaker, 744
  locations, 763
  receptacle, 743
  shock hazard elimination, 742
Glare, direct and reflected, 791–793
Glove valve, *see* Valves
Glossary of terms, 1033–1038
Grains of water, 70
Gravity flow, 406, 562
Gray water, 450
Greenfield, *see* Conduit, electrical, flexible
Greenhouse effect, 30
Ground fault, 737, 743
Grounding, electrical, 738
  electrode, 739
  green (equipment) ground, 743
  neutral, 739
  safety, shock hazard, 739

Halogen (quartz) lamps, *see* Light sources
Handhole, electric service, 925
Hard temper copper pipe, 417
Harmonic currents, 640
Heat balance, body, 8, 9, 14
  table, 15
Heat exchanger, 223
Heat gain (cooling load) calculation, 54–70
  building envelope elements, 58–63
  CLTD (cooling load temperature differences), 58–60
  factors, 54, 55, 59
  forms, 65–69
  GLF (glass load factor), 61
  infiltration, 61, 63
  internal latent loads, 58, 63, 64
  internal sensible loads, 56, 63
  latent load factor (LF), 64
  nonresidential, 63–64
  residential, 58–63
  shading, 56
  shading coefficient (SC), 56
  SLF (shade line factor), 63
  temperature range and swing, 58
  thermal lag, 55–57
  windows, 61–62
Heat loss calculation, 41–50
  forms, 44–49
  infiltration and ventilation, 32, 41
  outside design conditions, 50
  through building envelope, 43
  to adjacent unheated spaces, 42
Heat of fusion, 6, 316

Heat of vaporization, 6, 316
Heat pump:
  air-to-air unit, modes, 322, 323, 327, 328
  air-to-water unit, modes, 329
  balance point temperature, 324
  basic theory, 321
  cooling mode, 326
  COP (coefficient of performance), 324
  EER (energy efficiency ratio), 325
  heating mode, performance, 323, 324
  heat sink, 324
  heat source, 324
  HSPF (heating season performance factor), 204, 324
  hybrid systems, 326
  integrated capacity, 325
  SEER (seasonal energy efficiency ratio), 325
  system balance point, 326
  water-to-air unit, modes, 329, 330
  water-to-water unit, modes, 329, 331
Heat source, sink, 324
Heat transfer:
  air spaces, 31–34
  buildings, 22
  built-up sections, 31, 34, 35
  coefficients, 24–28, 37–39
  doors, 34, 41
  perimeter heat loss factor, 40
  principles, 9–14, 22, 23, 28, 29
  slab on (or below) grade, 34, 40
  terms and symbols (table), 29
  windows, 30, 34, 41
Hertz, 639
HID (high intensity discharge) lamps, 815
High sidewall outlet (HSW), 254
Home run (electrical), 667
Horizontal jump, 567
Hose bib (bibb), *see* Valves
Hot water, *see* Domestic hot water
Hot water boiler, *see* Boilers, hot water
Hot water heating, *see* Hydronic heating
HSPF, *see* Heat pump
Humidistat, 226
Humidity, 6
  absolute, 6
  ratio, 6
  relative (RH), 6, 70, 72
  specific, 6, 70, 72
HVAC field test report forms, 1015–1029
Hydraulics, 406

1044 / INDEX

Hydronic heating systems, 83–120
   air cushion tank, 95
   air vents, 93
   average water temperature, 117
   balancing valves, 92, 94
   branch piping, 92
   circulator (pump), 98–101
   components, 84
   controls, 120–125
   definitions, equipment functions, 92
   design procedure, 127–128
   diaphragm tank, 97
   expansion tank, 95, 96, 98
   friction loss in piping, range, 136
   heat, system, 99–101
   inlet water temperature, 132
   I=B=R ratings, 87, 91
   multicircuit piping, 117
   multiple loop piping, 116
   perimeter loop system, 115, 119
   piping arrangements, 114–120
   pump (circulator) performance curves, 98, 99
   radiant floor, 111–113
   radiation, 100, 110, 111
   schematic piping diagrams, 114–120
   temperature drop, system, 117, 128
   terminal units, 100–114
   water temperature control, 124, 125
   zoning, 120
Hydronics Institute, 50
   heat loss calculation forms, 48, 49
Hydrostatic head, see Pressure, water

I=B=R ratings, 87, 91
IESNA (Illuminating Engineering Society of North America), 788
Illuminance (illumination level):
   definition, 787
   levels (table), 789
   measurement, 795
   units, 788
Impedance, electric, 639
Indirect lighting, 844, 848–849
Indirect waste, 558, 559
Induction, see Air distribution outlets, outlet characteristics
Induction lamps, see Light sources
Infiltration. See also Ventilation
   cooling load, 61
   heating load, 41–42
Insulator, electric, 634
Interceptor, 573, 574
Intercom equipment, 916–917

Interface, modular wiring, 703
Internal air quality, see Ventilation
Intrusion alarm equipment, 913
Isothermal jet, see Air distribution outlets, outlet characteristics

Jockey pump, 501, 502

K-factor (transformers), 640
Kcmil (thousand circular mils), 675
Kelvin, see Temperature
Kilocalorie, definition, 4

Ladder diagram, 946
Lamps, see Light sources
Latent heat, 5
   of fusion, 6
   of vaporization, 6
Light, lighting:
   and vision, 787
   calculations, 845–857
   color of, 793–795
   color-rendering index (CRI), 794
   contrast, 790
   correlated color temperature (CCT), 794
   diffuseness, 793
   fixture layout, graphic method, 873
   fixtures (luminaires), 825–841
   glare, 791
   levels, see Illuminance (illumination level)
   methods, 844–845
   quantity, 787
   sources, see Light sources
   systems, 841–844
   uniformity, 851–853
Light industry building:
   air system heating and cooling design, 369–373
   architectural plans, 152–155
   electric heating design, 184–194
   floor plans, electric heating, 185, 189
   hydronic heating design, 151–164
   sanitary drainage design, 594
   site plan, 615
   storm drainage design, 613
   two-pipe reverse return piping, 160, 162–164
Light meters, 796
Light sources:
   base types, 800
   beam spread, MR lamps, 802
   bulb shapes, 800, 804, 819
   color effects (table), 826
   comparative characteristics, HID lamps, 822

   efficacy of sources (table), 825
   fluorescent lamps, 801–816
   halogen (quartz) lamps, 801–802, 806
   HID (high intensity discharge) lamps, 815
   high pressure sodium (HPS) lamps, 822–823
   incandescent lamps, 797–801, 804, 805
   induction lamps, 823–825
   mercury vapor lamps, 817–820
   metal halide lamps, 820–821
   MR lamps, 802
   reflector lamps (R, PAR), 800, 830
Lighting calculations:
   coefficient of utilization, 845, 856
   light loss factor (LLF), 847
   lumen method, 846–847, 850–851
   uniformity, 851–853
   zonal cavity method, 853–856
Lighting fixture layout, graphic, 873
Lighting fixture schedule, 759, 867
Lighting fixtures (luminaires), 825
   appraisal, inspection, 839, 841
   baffles, shields, louvers, 829–832
   coefficient of utilization, 845
   construction, 835–839
   diffusers, 830, 832, 837
   efficiency, 845
   installation, 839
   lampholders, 826–828
   layout, 873
   longitudinal shielding, 828, 831
   parabolic reflector, 830, 833–834
   prismatic lens diffuser, 833
   reflectors and shields, 828–830, 832
   shielding (cutoff) angle, 828–829, 831–832
   supplementary, 850
   transverse shielding, 828, 831
   troffer, 839, 840
   wall-washer, 836
Lighting track, 706–710
Load center, see Panelboard, electric
Load factor (electric), 648
Loss coefficients, duct fittings, 991–1013
Low sidewall outlet (LSW), 254
Low voltage switching, 884–885
Lumen, 787
Luminance, luminance ratios, 789
Luminous ceiling lighting, 866
Luminous flux, see Lumen
Lux, 788

Magnetic pressure gauge, 382, 384
Manhole, electric service, 924

Manometer, 203, 382–386, 408, 409
MBH, 91
MCM (thousand circular mils), 675
Mean radiant temperature (MRT), 11, 12
  calculation, 12
Measurements:
  air pressure in ducts, 203, 382–386
  air velocity, 382
  electrical, 652–661
  temperature, 378–381
  thermal comfort, 15
  water flow, 413
  water pressure, 409–413
  wet bulb temperature, 7
Mercury vapor lamps, see Light sources
Metabolic activity (table), 3
Metabolism, 2, 8, 14
Metal halide lamps, see Light sources
Metrication, 975
Millinch (pressure drop), 137
Mixing box, 270
Modular wiring system, 703
Mogensen house:
  air system heating and cooling design, 362–369
  architectural plans, 140–142
  boiler room details, 150
  electrical design, 863–908
  electric heating design, 180, 182–184
  floor plans, electric heating, 183, 893–894
  floor plans, electrical, 864–865, 889–890, 893–894, 905–906
  floor plans, see Electrical design, residential
  floor plans, hot water heating, 143
  hydronic (hot water) heating design, 142–150
  plumbing plans, 543
  plumbing riser, 544, 546
  preliminary plumbing plan, 542
  sanitary drainage design, 589, 592–594
  sewage treatment system design, 616
  site plan, 540
  water service design, 538–546
Monoflow fitting, 92, 117, 119, 120
Motor control center (MCC), 951–954
Motors, electric:
  control, ladder diagrams, 943–947
  controllers (starters), 945, 949–950
  design procedure, 943
  disconnect, 942
  drawing representation, 943–944
  feeder, 942

motor control center (MCC), 951–954
  NEC requirements, 942
  pilot device, 947
  schedule, 949
  thermal protection, 942, 947
  undervoltage protection, 945
Multi-mirrored reflector halogen lamp, 802
Multiple connection, see Parallel circuit
Multiplier (voltmeter), 653

National Appliance Energy Conservation Act (NAECA), 91, 204, 325
National Electrical Code (NEC), 641, 666
National Standard Plumbing Code, 404
NBS (National Bureau of Standards), 505
NEMA (National Electrical Manufacturer's Association), 724
  equipment enclosure designations, 950
  motor starter sizes, 948
Net static pressure, 223
Noise Criterion (NC), see Air distribution outlets, outlet characteristics
Nonresidential electrical design, see Electrical design, nonresidential
NSF (National Sanitation Foundation), 414

Oakum, 417
Ohm's law, Ohm, 633–634
One-line electrical diagram, 666, 667, 732, 932, 963
  symbols used in, 733
Orifice plate, 117
Outlet box capacity (table), 772
Outlet, electrical, 710
  box capacity, 772
  box covers, 725
  box selection (chart), 716
  ganged boxes, 713
  junction boxes, 713
  mounting, 718
  symbols, 719
Outlets, air distribution, see air distribution outlets
Outside design conditions, 50
Overcurrent protection, 730

Panel schedules, electric, 969–9721
Panelboard, electric, 745

  distribution panel, 940
  lighting and appliance panel, 752
  load balancing, 773
  pole, circuit numbering, 746, 747–752
  ratings, 749
  schedules, 749, 752, 767
  switch and fuse type, 752
  symbols, 753
  2, 3, 4 wire panels, 746
Parallel circuit, 636
PE plastic pipe, see Plastic pipe
Piping:
  branch, 92
  developed length, 133
  expansion loops and joints, 126, 164
  friction pressure loss in copper tubing (chart), 137
  friction pressure loss in copper tubing (table), 135
  friction pressure loss in fittings, 136
  runout, 147
  symbols, 489, 490
  total equivalent length (TEL), 136
Piping (water) installation, 436
  expansion joints and fittings, 441
  supports, 437, 439
  testing and inspection, 440
Piping materials:
  cast-iron soil pipe, 414, 416
  copper pipe and tubing, 417, 431
  dielectric connectors, 436
  ferrous pipe, 414
  ferrous pipe, physical properties (table), 415
  fittings, 416, 418–430, 432, 436
  plastic pipe, 433–436
  thermal expansion (table), 438
Piping, hydronic heating, 114–127
  branch, 92
  design factors, 125–127
  developed length, 133
  expansion loops and joints, 126
  friction head loss, 135, 137
  multicircuit, 117
  multiple loop, 116
  one-pipe system, 117–119
  perimeter loop, 115, 119
  return branch, 147
  series loop, 115, 116
  supply branch, 147
  system friction loss, estimate, 136
  two-pipe systems, 117–118, 120–122
  water velocity, flow rates, 126, 127
  zoned systems, 120, 123, 124
Pitch, gravity flow, 562

Pitot tube, 382, 384, 385, 391, 392, 394
Plastic pipe, 433–436
Plumbing codes and standards, 404, 413, 414
Plumbing fixtures:
 bathtubs, 463, 468, 469, 470
 bidets, 480, 482–483
 characteristics, 450
 flow rates, recommended (table), 499
 lavatories, 456–462
 minimum number, by facility (table), 405
 pressure, minimum required (table), 499
 showers, 456, 464, 466, 467, 470
 sinks, 451–456, 462
 space requirements, clearances, 484
 symbols, drawing, 486
 type, by facility (table), 451
 urinals, 463, 472–475
 water closets, 475–481
Plumbing materials, see specific item
Plumbing riser, water system, 517
Plumbing section, see Plumbing riser
Plumbing sections, 493
Poke-through electrical fittings, 705
Poles, power-telephone, interior wiring, 706–708
Potable water system, see Water system, potable
Power factor (electric), 646
Power line carrier (PLC) system, 884
Power, electric:
 analogies, 644
 calculation, 645–648
 demand control, 651–652
 and energy, 643, 647
 measurement, 652–661
 power factor, 646
 units, 643
Pressure, air
 measurements, 203, 382
 static, 201
 velocity, 201, 202
Pressure (head) water, 98, 99, 406–408
 absolute, 408
 gauge, 408
 measurement, 409
 static head, 408
Pressure differential valve, see Backflow prevention, water systems
Pressure reducing valve, see Valves
Pressurized system flow, 406
Primary air, see Air distribution outlets, outlet characteristics

Psychrometer, sling, 7
Psychrometric chart, 77
 components, 70–71
 dew point, 72
 HVAC processes on, 73–75
Psychrometrics, 6, 70–78
PTAC (package terminal air conditioner), 329, 332
PTHP (packaged terminal heat pump), 329, 332
 rooftop unit, 329, 335–338, 346
 slab-on-grade unit, 353
 split unit, 335, 336, 345, 347–352
 through-the-wall unit, 329, 343
 unitary, 329
Pullbox, electrical, 937, 939–940
Purge valve, see Valves
PVC plastic pipe, see Plastic pipe
Pyrometer, 15, 378, 379, 381

Quartz lamps, see Light sources

Raceways, electrical, 684, 692, 695
 cellular floor, 700–704
 overhead, 702–706
 power-telephone poles, 706–708
 surface, 697–702
 symbols, 671
 type I and II conduit and duct, 901
 underfloor, 697, 703
Radiant floor, hot water, 111–113
Radiant heat, electric, 173–177
Radiant panels, hot water, 110, 113
Radiation, see Heat transfer
Radiation, thermal, 29–32
Radiative heat barrier, 30
Radiator, steel panel, 113, 115
Radiators, hot water, 111, 113, 114, 115
Radius of diffusion, see Air distribution outlets, outlet characteristics
Rankine, see Temperature
RC switches, 966
Reactance, electric, 639
Reflectance, 784
 measurement, 797
Reflectance measurement, 795
Reflection:
 diffuse, 785
 specular, 784
Reflection factor, 784
Refrigeration; process, cycle, equipment:
 compressive cycle, 316–321
 compressor, 357
 condenser, 318, 319, 354
 DX coil, 320, 328

 evaporator, 318, 319, 355
 flow control (valve), 318, 320
 sensible heat ratio (SHR), 328
 thermostatic expansion valve, 318
 ton, definition, 316
Relative humidity, see Humidity
Relief valve, see Valves
Residential electrical design, see Electrical design, residential
Residential water service design, 538–545
Resistance, electric, 633
Resistance, thermal, 23, 24, 35
Riser diagram, electrical, 931, 932, 935, 961, 967
RLM (Reflector Luminaire Manufacturer), 829
Romex, see Wire and cable
Room cavity ratio, see Zonal cavity method
Roughing, plumbing, 600–602
Runout pipe size, 545

Safety valve, see Valves
Sanitary drainage, see Drainage, sanitary
Saturation temperature, 317
Secondary air, see Air distribution outlets, outlet characteristics
SEER (seasonal energy efficiency ratio), 204, 325
Self-siphoning, 558, 560
Semi-direct lighting, 841, 844
Sensible heat, 5
Sensible heat ratio, 328
Series circuit, 634
Service entrance valve, see Water system, potable, valving
Service sink, 455
Sewage treatment, private system:
 design, Mogensen house, 616
 seepage pits (cesspool), 617, 623
 septic tank, 616, 618–620
 sewage flow by facility (table), 623
 tile field, 617, 623
Sewer, sanitary, 604
Sewer, storm, 604
Shading, see Heat gain (cooling load) calculation
Shock arrestor, see Water hammer
Shock suppressor, see Water hammer
Short circuit, 638
Shunt (ammeter), 653
Shunt tripping, 945
Siamese connection, 549–551
Sick Building Syndrome (SBS), 41, 180
Single phase (a-c), 641

INDEX / 1047

Single-line (wiring) diagram, *see* One-line electrical diagram
SI system, 975–976. *See also* Units and conversions
Skin effect (in cable), 650
Sling psychrometer, 7
SMACNA (Sheet Metal and Air Conditioning Contractors' National Association), 312
Sodium (HPS) lamps, *see* Light sources
Soft temper pipe (copper), 417
Sol-air temperature, 56
Solderless connectors, 680–683
Southern Standard Plumbing Code, 404
Spacing-to-mounting-height ratio, *see* Light, lighting, uniformity
Specific heat, 5
   of air, 198
   of water (chart), 5
Specular reflection, *see* Reflection, specular
Spring pressure, 202
Sprinklers, 548–554
   deluge, 553
   dry pipe, 553
   heads, 553
   hose rack, 550
   plan, industrial building, 552
   pre-action, 553
   siamese connection, 549–551
   wet pipe, 548
Stagnant air, *see* Air distribution outlets, outlet characteristics
Static head of air, 201
Static head (pressure), of water, 98, 99, 406–408
Static regain, 202
Stop and waste valve, *see* Valves
Stratification, *see* Air distribution outlets, outlet characteristics
Strike, restrike time, *see* type of lamp
Supply air rise, *see* Air distribution outlets, outlet characteristics
Surface effect, *see* Air distribution outlets, outlet characteristics
Surface raceways, electrical, 692, 697–702
Surge suppression, 725
Sweat joint, 435
Switchboard, electrical, 932–935, 962–963
Symbols:
   ductwork, 233, 234
   electrical control, 947
   electrical equipment, 753
   electrical outlets, 719
   electrical raceways, 671

electrical signal equipment, 908–909
   motors and controls, 944
   one-line (single-line) electrical diagrams, 733
   piping, 490
   plumbing fixtures, 488
   plumbing piping, 489, 491
   wiring and raceways, 671
   wiring devices, 720
System effect, 223

T & P safety valve, *see* Valves, safety
Tab (testing, adjusting and balancing), 377
TEL (total equivalent length), 295, 518
Telephone service manhole, 923–924
Telephone wiring, 917–918
Temperature:
   body, 2
   Celsius scale, 3
   conversion chart, 4, 976
   dry bulb, 6, 70
   Fahrenheit scale, 3
   gradients, 22, 36
   Kelvin scale, 4
   Rankine scale, 3
   wet bulb, 7, 70
Temperature gradient, 22
Temperature measurement, 378
Terra cotta (vitrified clay) pipe, 435
Thermal balance, buildings, 21
Thermal barrier, 30, 31
Thermal comfort, 2, 198
   criteria, 15
   measurements, 15
Thermal conductance (C), 24
Thermal conductivity (k), 23
Thermal emittance, 31–34
Thermal insulation, 23, 32
Thermal lag, 55–57
Thermal resistance (R), 23, 37–39, 41
   rule-of-thumb, 32
Thermal resistivity (r), 24
Thermal stress, 14
Thermal transmission, 29, 35, 37–39, 41
Thermal transmission coefficient (U), 29, 34
Thermodynamics, 8–10
Thermometer, 378–381
Thermosiphon, 527
Three phase (a-c), 641
Three-way switching, 768–769
Tics (hatch marks), 667
Total air pattern, *see* Air distribution outlets, outlet characteristics
Total developed length, 516
Track lighting, 706, 710, 880

Trailer, traveler, runner wires, 768
Transformers, 639, 640, 926, 964
Transmission factor (lighting), 785
Transmission, thermal, 34, 35
Transmittance (lighting), 785, 786
Traps:
   building (house) trap, 561
   fixture trap, 576
   maximum arm length (table), 577
   pressures, 576
   seal, 576
   size, nonintegral (table), 569
   types, 575
Traverse (duct), 391–393
Transmission (lighting), 785

UL (Underwriter's Laboratories), 809
Undercarpet wiring, 709–715
Underfloor electrical raceways, 697, 703
Underground electrical raceways, 901
Uniform Plumbing Code, 404
Unit heaters, hot water, 106, 108, 109
Unit ventilators, 109
Units and conversions, 16–18
   abbreviations, 977
   acoustics, 978
   approximations, 978
   Fahrenheit, Celsius, 976
   lighting, 977
Upfeed water system, *see* Water system, potable

Vacuum breaker, 534–536
Valves:
   angle, 443
   balancing, hydronic heating, 92, 94
   ball, 446, 447
   check, 443, 445
   dimensions (table), 442
   flow control, hydronic heating, 94
   gate, 442, 443
   globe, 443, 444
   hose bib (sill cock), 440
   materials, 440
   mixing, 463, 465, 471
   pressure balancing, 463, 465, 471
   pressure reducing, 94, 448, 449
   purge, hydronic heating, 94
   relief (safety), 446, 448
   safety, pressure relief, 94
   stop-and-waste, 440
   thermostatically controlled, 463, 465
   zone control, hydronic heating, 95

Vane, *see* Air distribution outlets, outlet characteristics
Vapor barrier, 34
VAV, *see* Air systems, heating and cooling
VDT (visual display terminal), lighting, 955
   classroom, 955–960
Veiling reflection, *see* Glare
Velocity pressure, air, 201, 202
Velometer, 389
Ventilation:
   air changes per hour (ACH), 41
   effective leakage area, 41
   internal air quality (IAQ), 54
   Sick Building Syndrome (SBS), 41
Vents and vent piping:
   back vent, 580
   branch vent, 580
   circuit vent, 580, 581
   continuous vent, 580, 581
   individual vent, 580
   loop vent, 580, 581
   piping design, 583–589
   principles, 577
   relief vent, 583
   stack vent, 580
   stack venting, 582
   types of vent, 580
   vent extension, 578
   vent size and length (table), 584
   vent stack, 578
   wet vent, 580, 582
Venturi pipe fitting, 92, 117, 119, 120
Vertical temperature gradient, 253
Viscous impingement filter, 227
Voltage:
   definition, 631
   polarity, d-c, 631, 633
   systems, 641–643
Voltage drop, 649
Voltages, system and utilization (table), 931
Voltmeter, 652

Wall scuppers, 554
Warm air furnace, *see* Furnace, warm air
Warm air heating:
   combustion air, 200
   design procedure, 307
   duct noise, 297
   ducts and fittings, 227–234
   extended plenum, 277
   fresh air supply, 200
   furnaces, 199–227
   heating capacity of ducted air, 200
   maximum air velocities (table), 297
   outlets and their characteristics, 235–265
   perimeter loop, 270–272
   radial ducts, 272, 273
   reducing plenum, 274
   single-zone duct arrangements, 270–274
   system components, 199
   system types, 265–274
Wash fountain, 450
Water column, 203
Water coolers, 480, 485
Water flow, principles, 410–413
Water gauge (w.g.), 203
Water hammer, 537–539
Water meter, 413
Water system, potable:
   backflow prevention, 534–537
   branch pipe sizing (table), 506
   demand, by outlet (table), 508
   demand, gpm, WSFU (table), 507
   design example, office building, 516–518
   design procedure, 498, 531, 533, 534
   downfeed system, 502, 503
   fixture flow pressure (table), 499
   fixture flow rates (table), 499
   friction head calculation, 508–514
   friction head charts, 509–511
   friction head loss, fittings, 514, 515
   friction head loss, water meters, 515
   friction head (tables), 512, 513
   house pump, 504
   house tank, 503
   hydropneumatic system, 505
   light industry building design, 545–547
   loads, WSFU (table), 506
   Mogensen house design, 540–545
   plumbing section, upfeed system, 500
   pressure requirements, 498–499
   residential service, 538
   roof tank, 502, 504
   runout pipe sizing (table), 545
   service size, 505
   suction tank, 503
   system types, 500–505
   upfeed system, 500, 501
   valving, 531, 533, 534
   velocity limitations, pipe sizing, 518–521
Waterflow switch, 553
Welded connections, electrical, 683
Wet bulb (WB), *see* Temperature
Wet bulb, measurement, 7
Wet bulb depression, 7
Wet column, 436
Windows, condensation, 15, 80, 81, 226
Windows, thermal properties, 30
Wire and cable, 673
   ampacity (table), 676
   application and insulation (table), 677
   BX cable (type AC), 668
   color coding, 679
   connectors, 680–684
   current carrying capacity, 677
   dimensions, insulated conductors (table), 678
   flexible armored, 669, 673
   gauge, size, 675
   NM cable, 668
   physical properties, bare conductors (table), 675
   Romex cable, 668
   types RHW, THW, THWN, XHHW, 677–678
Wiring devices, 719
   isolated (insulated) ground receptacle, 726, 745
   receptacle configuration (chart), 721
   receptacles, 719, 722
   switches, 722–724
   symbols, 720
Wiring methods, 670
WSFU (water supply fixture units), 505, 506

Zonal cavity method, *see* Light, lighting, calculations
Zone control valve, *see* Valves